ON4UN's Low-Band DXing

Antennas, Equipment and Techniques for *DXcitement* on 160, 80 and 40 Meters

Author
John Devoldere, ON4UN

Production
R. Dean Straw, N6BV, Editor
Michelle Bloom, WB1ENT
Sue Fagan—Front Cover
Jodi Morin, KA1JPA
Joe Shea
David Pingree, N1NAS
Michael Daniels

Newington, CT 06111-1494
ARRLWeb: www.arrl.org

Copyright © 2005-2007 by The ARRL, Inc.

Copyright secured under the Pan-American Convention

International Copyright secured

This work is publication No. 74 of the Radio Amateur's Library, published by the League. All rights reserved. No part of this work may be reproduced in any form except by written permission of the publisher. All rights of translation are reserved.

Printed in USA

Quedan reservados todos los derechos

4th Edition
Second Printing

ISBN: 0-87259-914-0

CONTENTS

iv	**Foreword**
v	**Preface**
1	**Propagation**
2	**DX-Operating on the Low Bands**
3	**Receiving and Transmitting Equipment**
4	**Antenna-Design Software**
5	**Antennas: General, Terms, Definitions**
6	**The Feed Line and the Antenna**
7	**Receiving Antennas**
8	**The Dipole**
9	**Vertical Antennas**
10	**Large Loop Antennas**
11	**Phased Arrays**
12	**Other Arrays**
13	**Yagis and Quads**
14	**Low-Band DXing from a Small Garden**
15	**From Low-Band DXing to Contesting**
16	**Literature Review**
	Index

FOREWORD

We at ARRL are pleased to publish a major revision of what has become an essential guidebook for low-banders: *ON4UN's Low-Band DXing* by John Devoldere. This book is a must-read for DXers, contesters and even casual operators who venture on occasion onto the 160, 80, and 40 meter bands.

Author John Devoldere, ON4UN, has long been recognized both for his outstanding achievements on the low bands and as a highly successful on-the-air contester and DXer. As in the previous three editions, you will find an amazing wealth of information springing from ON4UN's extensive hands-on experience and that of many, many collaborators and contributors. John must know personally everyone on the low bands! Newcomers and old-timers will both benefit from the vast store of practical information in this major revision.

Those of you familiar with the Third Edition published in 1999 will be pleased to find virtually completely new chapters covering phased arrays and receiving antennas.

Indeed *ON4UN's Low-Band DXing* covers everything you want to know about 160, 80 and 40 meters—with a strong emphasis on antennas, of course. The old adage, "You can't have too many antennas" still rings loud and true, especially on the low bands.

ON4UN also presents solid ideas for the ham with limited space, as well as for those fortunate ones who have the budget and the real estate to erect "monster antennas" where they wish.

As always, we want to hear back from you. Help us make the next edition even better than this new Fourth Edition. We've provided a handy form at the back of this book that you can use to share your ideas with us and with John.

David Sumner, K1ZZ
Executive Vice President
Newington, Connecticut
February 2005

PREFACE

"What can I add to the 4th Edition compared to the 3rd Edition?" I asked myself before starting this writing adventure. I sent out a questionnaire to over 500 low-band DXers and I give many thanks to everyone who took the time to fill out the questionnaire and to share their views and comments with me!

Often I heard the semi-joking comment, *"Tell me how can I do as well as the big guns on 160 meters who have 100-acre antenna ranches!"* Of course, this sounds a bit like, "How can I win a Formula One race with a Volkswagen?" Unfortunately, the truth is that you can't win a Formula One race with a Volkswagen. But maybe there are some things you can do to avoid being the very last car in the race.

Lots of new things have popped up in the past 5 years, especially in the area of small (and not so small!) receiving antennas and high-gain transmit antennas. I deal with these subjects in much greater detail here than in the 3rd Edition of the book. I've completely rewritten the chapters on these subjects and now include two methods for objectively qualifying receiving antennas. No more wild stories; just go by the RDF and DMF numbers, please.

It took me 18 months to write this 4th Edition, clocking 1500 to 2000 hours of work. I personally visited a number of successful low-band DXers and contesters to try to unravel their secrets. I thank all of them for the wonderful hospitality they showed me. In this book they share their techniques with you.

A hobby is something you should enjoy. I enjoy working contests, especially on the low bands. I enjoy CW pile-ups. I enjoy building a competitive low-band and contest station. I get an immense kick out of working N4KG, who uses 5 W on 160 and 80 meters!

And most of all, I enjoy sharing my know-how, my enjoyment and my excitement with other hams. That's why I have written so much over the past 15 years about subjects that are near and dear to my heart. I hope I have been helped bring about some growth of low-band activities in DXing and contesting. My induction into the CQ Contest Hall of Fame in 1997 was a great honor for me and filled me with joy and pride.

Once again, it was a great experience having a number of eminent experts acting as mentors for each chapter in this book. I found friends willing to coach, support, advise and help me with all of the chapters. I am indebted a great deal to these fine gentlemen and true friends. Thank you, Bill (W4ZV), Tom (W8JI), George (W2VJN/7), John (WØUN), John (K9DX), Robye (W1MK), Uli (DJ2YA), Georges (K2UO), Lew (K4VX), Klaus (DJ4AX) and Frank (DL2CC). It was also a great pleasure to work together with Dean, N6BV, who handled the editing of the 4th Edition for the ARRL. Thanks for all your help, Dean. [*You're more than welcome, John!—N6BV, Ed.*]

I am especially grateful to all the authors and ordinary DXers/contesters who let me quote from their work, use figures from their publications or refer to their statements on various E-mail reflectors. Your contributions were essential.

Finally, I would like to thank my many readers. I hope that this new book, which is being made available at the onset of a new low-Sunspot Number era, will help further promote our wonderful hobby. Low-band DXing is something that makes Amateur Radio unique. Unique in a way that satisfaction does not come automatically but as a result of dedication, education, knowledge, perseverance and patience.

I hope that with this book the message from a serious part of the Amateur Radio community gets across to some manufacturers who remained deaf to our wishes and requests all too long. I hope we can all move the frontier of what we can achieve on the low bands one step further ahead.

To you all, enjoy the low bands and have fun! I dedicate this book to my grandsons, Arthur and Louis.

John Devoldere, ON4UN (AA4OI), February 2005

Updates and error corrections will be posted on the ARRLWeb at www.arrl.org/notes/.

About the ARRL

The seed for Amateur Radio was planted in the 1890s, when Guglielmo Marconi began his experiments in wireless telegraphy. Soon he was joined by dozens, then hundreds, of others who were enthusiastic about sending and receiving messages through the air—some with a commercial interest, but others solely out of a love for this new communications medium. The United States government began licensing Amateur Radio operators in 1912.

By 1914, there were thousands of Amateur Radio operators—hams—in the United States. Hiram Percy Maxim, a leading Hartford, Connecticut, inventor and industrialist saw the need for an organization to band together this fledgling group of radio experimenters. In May 1914 he founded the American Radio Relay League (ARRL) to meet that need.

Today ARRL, with approximately 170,000 members, is the largest organization of radio amateurs in the United States. The League is a not-for-profit organization that:

- promotes interest in Amateur Radio communications and experimentation
- represents US radio amateurs in legislative matters, and
- maintains fraternalism and a high standard of conduct among Amateur Radio operators.

At League headquarters in the Hartford suburb of Newington, the staff helps serve the needs of members. ARRL is also International Secretariat for the International Amateur Radio Union, which is made up of similar societies in 150 countries around the world.

ARRL publishes the monthly journal *QST*, as well as newsletters and many publications covering all aspects of Amateur Radio. Its headquarters station, W1AW, transmits bulletins of interest to radio amateurs and Morse code practice sessions. The League also coordinates an extensive field organization, which includes volunteers who provide technical information for radio amateurs and public-service activities. ARRL also represents US amateurs with the Federal Communications Commission and other government agencies in the US and abroad.

Membership in ARRL means much more than receiving *QST* each month. In addition to the services already described, ARRL offers membership services on a personal level, such as the ARRL Volunteer Examiner Coordinator Program and a QSL bureau.

Full ARRL membership (available only to licensed radio amateurs) gives you a voice in how the affairs of the organization are governed. League policy is set by a Board of Directors (one from each of 15 Divisions). Each year, one-third of the ARRL Board of Directors stands for election by the full members they represent. The day-to-day operation of ARRL HQ is managed by an Executive Vice President and a Chief Financial Officer.

No matter what aspect of Amateur Radio attracts you, ARRL membership is relevant and important. There would be no Amateur Radio as we know it today were it not for the ARRL. We would be happy to welcome you as a member! (An Amateur Radio license is not required for Associate Membership.) For more information about ARRL and answers to any questions you may have about Amateur Radio, write or call:

ARRL—The national association for Amateur Radio
225 Main Street
Newington CT 06111-1494
(860) 594-0200
Fax: 860-594-0259
E-mail: **hq@arrl.org**
Internet: **www.arrl.org/**

Prospective new amateurs call (toll free):
800-32-NEW HAM (800-326-3942)
You can contact us also via e-mail at **newham@arrl.org**

CHAPTER 1

Propagation

1. INTRODUCTION

I will try to answer the following questions in this chapter:

- When can I work DX on the low bands?
- Are conditions on 40, 80 and 160 meters the same or similar?
- Are conditions very variable?
- Are conditions predictable?
- Is there any good propagation-prediction software?
- Are there any other tools to help me catch the difficult ones?
- What are crooked paths, when do they happen and what causes them?
- What directions should I aim my antennas?

In the last five years we have seen more publications than ever before covering propagation, antenna modeling and digital communications. It should come as no surprise that the availability of powerful computers is a major reason for this evolution. Speaking of antennas and digital communications, the more that is being published on these subjects, the more we all benefit.

I cannot, however, say the same thing about publications concerning propagation, especially related to the low bands. The number of variables influencing low-band and especially 160-meter propagation appears to be so vast that scientists have only discovered the proverbial tip of the iceberg. So far, no one has come up with a forecasting system that works.

At best, we seem to be able to correlate some of the known parameters with actual observations. But the full "why and how"—the global picture—is still missing. But don't let this scare you off, the elements of mystery and discovery on the low bands makes for half the excitement and fun there!

A number of publications (Ref 101, 103, 104, 105 and 167) cover the basic principles of radio propagation by ionospheric refraction, primarily on HF. Let me recommend in particular Robert Brown's (NM7M) excellent books: *The Little Pistol's Guide to HF propagation* and *The Big Gun's Guide to Low-Band Propagation* (Ref 167 and 179). These books are must-reads for anyone who wants to have a more than casual understanding of propagation.

The existence of books of this caliber makes it easy for me, since I will not have to explain the basics of propagation. This chapter on low-band propagation is thus not meant to be a general-study book on propagation. I have written it for the dedicated low-band DXer, who tries to understand the "how" of propagation (not necessarily the "why") to help him work an elusive new country or maybe to generate a better contest score.

This chapter is mainly based on observations, your observations and mine. You will soon find that the basic rules that govern propagation on the low bands are rather simple. You need a path in darkness, one that exhibits (most often) sunrise and sunset peaks. These are the simple but very important ground rules.

Testimonies of literally hundreds of low band DXers are used to identify what may seem like odd propagation phenomena. I will mention possible or probable mechanisms that govern these phenomena. These can be widely accepted, or in some cases more speculative and yet-to-be-proven mechanisms. What is important is that you be able to recognize "odd" phenomena and circumstances and that you know how to take advantage of them. That's what this chapter is really all about.

I will try to cover propagation on the three low bands (40, 80 and 160 meters), explaining similarities, but also pinpointing important differences. Understanding the basic mechanics is very essential. If you realize that on 160 meters the opening over a 18,000 km path will occur maybe one day a week, and then only during a specific time of a "good" year, and that the opening will last maybe 3 to 5 minutes, you will realize how important it is to know when you have to try to make that contact.

A spectacular example to illustrate this was my QSO on 160 meter with ZL7DK in early Mar 1998. 1998 was a good, low sunspot year. I knew I had a 3 to 5 minute window both in the morning as well as in the evening. After observing the two windows for almost two weeks, one morning the weak signals from Chatham Island finally came through and I was able to work them. That morning "I had the skip." Other mornings their signals made it into England, France or Germany, without any spillover in Belgium. Why? In addition to just noting these facts, I will try to describe and explain a mechanism that seems to fit these facts.

Over the years numerous HF propagation programs have

popped up. While these have proven their use for predicting propagation on the higher bands (the MUF-related bands), I have never had any much use from then on the low bands, certainly not on 160 meters.

In the world of commercial HF-broadcasting and HF-point-to-point communications, the challenge consists in finding the optimum frequency or maybe the best angle of radiation (to select the right transmitting antenna) that will give the most reliable propagation, as a function of the time of day. In low-band DXing, the problem is quite different. The challenge is to determine the best time (month, day and hour) to make a contact on a given (low-band) frequency, with a given antenna setup, between two specific locations.

Cary Oler and Ted Cohen (N4XX) wrote in their excellent article "The 160-Meter Band: An Enigma Shrouded in a Mystery" (Ref 142): "Topband is one of the last frontiers for radio propagation enthusiasts. It involves regions of the Earth's environment that are very difficult to explore and are poorly understood. These factors have led to our failure to predict propagation conditions with any level of accuracy. They also account for our inability to explain some of the puzzling mixtures of conditions that make this one of the most interesting and volatile bands available to the Amateur service."

Bill Tippett, W4ZV, hit the nail on the head when he wrote: "If 160 were perfectly predictable, we would all become bored with it and take up another hobby. Let's just enjoy it as it is because we'll never be able to figure it out!" So, don't expect this chapter to predict all kind of exotic openings on 160 for you!

2. WHEN CAN WE WORK DX ON THE LOW BANDS?

Let's have a look at the following time cycles:
- The 11-Year Sunspot Cycle
- The 27-Day Sun-Rotation Cycle
- The Seasonal Cycle
- The Time of Day

2.1. The Sunspot Cycle

It is well known that radio propagation by ionospheric refraction is greatly influenced by the sunspot cycle. This is simply because ionization is caused mainly by ultraviolet (UV) radiation from the sun, and UV is highly dependent on solar activity.

Solar activity can influence HF propagation in three major areas:
- MUF (maximum usable frequency)
- Absorption from D-layer and E-layer
- The occurrence of magnetic disturbances

The sun's activity is usually expressed in term of the Smoothed Sunspot Number (SSN) or the Solar Flux Index (SFI). You can get the SFI on WWV (eg, on 10 MHz), but if you have a computer with a permanent connection to the Internet, you can use a very nice program called *IonoProbe*, written by VE3NEA. (See **www.dxatlas.com/IonoProbe/**). *IonoProbe* is a 32-bit Windows application that monitors the space weather parameters essential for HF radio, including SSN/SFI, A_p/K_p, X-ray/Proton flux and auroral activity. *IonoProbe* downloads near-real time satellite and ground-station data, stores that information for future use and displays it in a user-friendly way. Time-critical parameters, such as X-ray flux, proton flux and auroral index, are updated every 15 minutes. An alarm can be set up to notify you of a geomagnetic storm within a few minutes after its start (**Fig 1-1**).

You can also view a Solar Terrestrial Activity report in the form of a chart on **www.dxlc.com/solar/**. (See **Fig 1-2**.). **Fig 1-3** shows the progress of the present cycle (daily, monthly and smoothed). The next solar minimum will likely occur sometime in 2006 with Cycle 24 peaking in 2010. (See: **sidc.oma.be/html/sidc_graphics.html** and **www.sec.noaa.gov/SolarCycle/** and **www.wm7d.net/hamradio/solar/index.shtml**.) Note that the smoothed sunspot number in Cycle 19 peaked almost twice as high (200) as

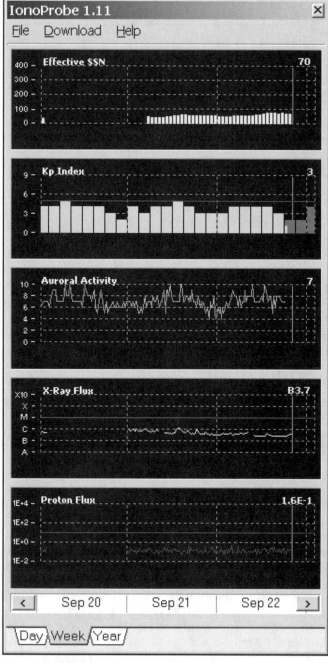

Fig 1-1—A number of ionospheric parameters can be displayed together, for a time frame of 1 day, 1 week or 1 year. For example, the Effective SSN, the K_p Index, Auroral Activity, X-Ray Flux and Proton flux are shown here for the last week. These are updated continuously using the *IonoProbe* program by VE3NEA. (*Courtesy of VE3NEA.*)

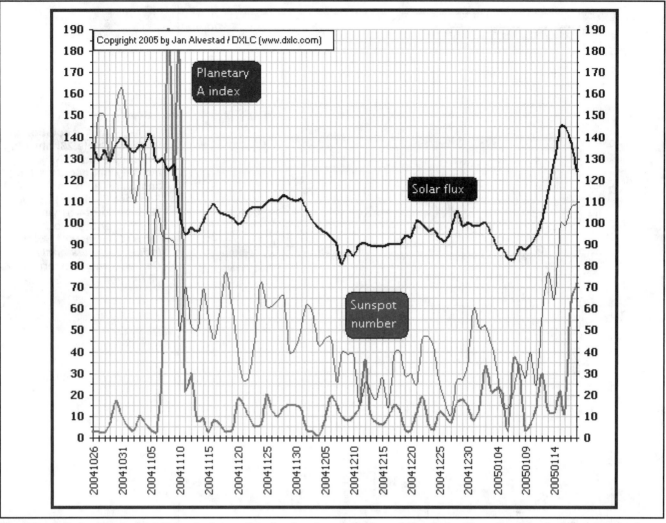

Fig 1-2—A Solar Terrestrial Report shows the day-by-day evolution of the Solar Flux, the Sunspot Number and the Planetary A Index. (*Courtesy of Jan Alvestad.*)

Cycle 23 (just over 100). Old timers will remember the spectacular 10-meter signals we enjoyed in 1957-1958 at the peak of Cycle 19.

2.1.1. The MUF1

The *critical frequency* is the highest frequency at which a signal transmitted straight up at a 90° elevation angle is returned to earth. The critical frequency is continuously measured in several hundred places around the world by devices called *ionosondes*. At frequencies higher than the critical frequency, all energy will travel through the ionosphere and be lost in space (**Fig 1-5**). The critical frequency varies with sunspot cycle, time of year and day, as well as geographical location. Typical values are 9 MHz at noon and 5 MHz at night. During periods with low sunspot activity the critical frequency can be as low as 3 MHz. During those times we can witness dead zones on 80 meters at night.

Fig 1-6A shows a world map with the critical frequencies for midwinter and a low Smoothed Sunspot Number (SSN=65) at 0000 UTC. Fig 1-6B shows the same map for midsummer with a high Smoothed Sunspot Number (SSN=240) at 1200 UTC.

At frequencies slightly higher than the critical frequency, refraction will occur for a relatively high wave angle and all lower angles. As we increase the frequency, the maximum elevation angle at which we have ionospheric refraction will become lower and lower. At 30 MHz during periods of high solar activity, such angles can be of the order of 10° and even as low as 1° (Source: *VOACAP* by NTIA/ITS; see **www.uwasa.fi/~jpe/voacap/**).

The relation between MUF and critical frequency is the *wave elevation angle*, where:

$$MUF = F_{crit} / \sin(\alpha)$$

where α = angle of elevation.

Table 1-1 gives an overview of the multiplication factor ($1/\sin \alpha$, also called secant α) for a number of elevation angles (α). For the situation where the critical frequency is as low as 2 MHz it can be seen that any 3.8-MHz energy radiated at angles higher than 30° will be lost into space. This is one reason for using an antenna with a low radiation angle for the low bands.

The MUF is the highest frequency at which reliable radio communications by ionospheric propagation can be main-

Table 1-1

α (°)	$1/\sin(\alpha)$
10	5.8
20	2.9
30	2.0
40	1.6
50	1.3
60	1.2
70	1.1
80	1.0
90	1.0

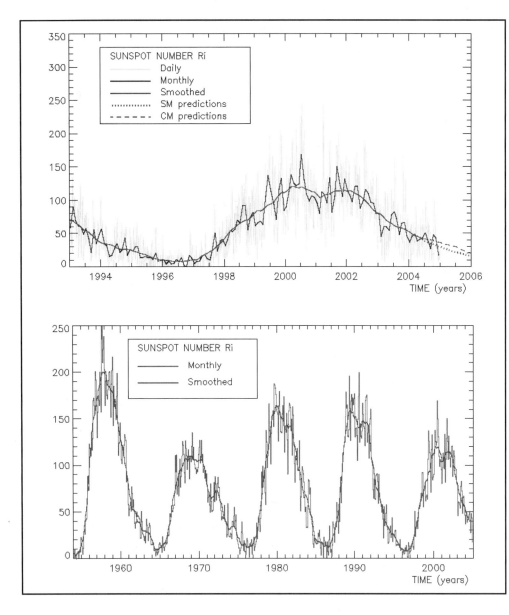

Fig 1-3—At A, evolution of Sunspot Cycle 23, showing smoothed (dark line), monthly (jagged dark line) and daily (light line). At B, Sunspot Cycles 21, 23 and 23 superimposed with monthly and smoothed data. (*Courtesy of Solar Influences Data analysis Center, Royal Observatory of Belgium.*)

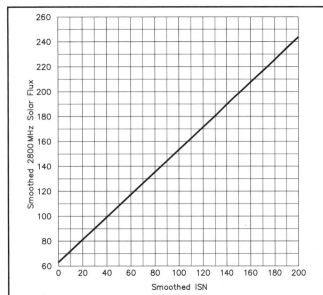

Fig 1-4—Conversion from 2800-MHz solar flux to Smoothed Solar Number (SSN). (*Courtesy* The ARRL Antenna Book, 20th Edition.)

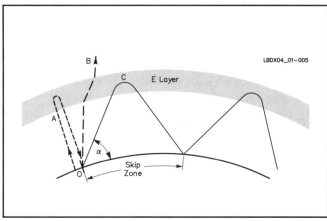

Fig 1-5—Ionospheric propagation: At A, we see refraction of the vertically transmitted wave—this means that the frequency is *below* the critical frequency for that refracting layer. At B, the angle is too high and the refraction is insufficient to return the wave to Earth. At C, we have the highest angle at which the refracted wave will return to Earth. The higher the frequency the lower the angle will become. Note the skip zone, where there is no signal.

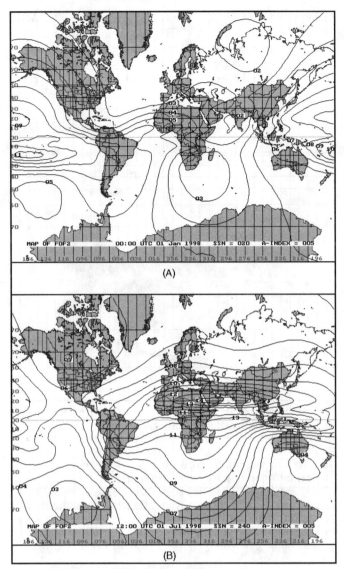

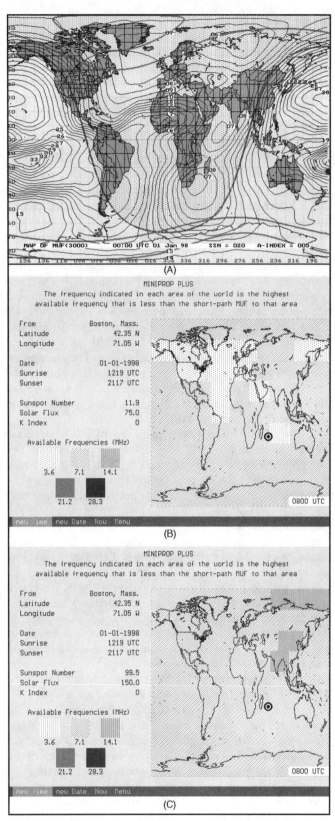

Fig 1-6—At A, Mercator-projection world map with the critical f_oF_2 frequencies for midwinter and a low Smoothed Sunspot Number (SSN = 20) at 0000 UTC. At B, map for midsummer with a high SSN of 240 at 1200 UTC (*Maps generated by* PropLab Pro *software program.*)

Fig 1-7—At A, Mercator projection of the world showing the 3000-km equal-MUF lines for Jan 1 (0000 UTC) for an SSN of 20 (A-index=5). Note that there are areas in the Northern Hemisphere where during the night the MUF is below 7 MHz. Also note the great-circle (short) path between Europe and California. At B, a frequency map for Jan 1, 0800 UTC (SSN=11.9, K=0). The MUF is below 7 MHz on the path between Europe and the USA. At C, a similar map; the only difference is that the SSN = 99.5. This time the path between Europe and the USA is open on 40 meters. (*Map A generated by* PropLab Pro. *Maps B and C were generated by W6EL's* MiniProp Plus *program.*)

tained over a given path—this is called the "classical MUF." Another common use of the term MUF refers to the *median statistical value*. Fifty percent of the time, the actual MUF observed on any given day will be higher than the median (and 50% of the time it will be lower). If you take 85% of the (median) MUF, the path should support signal propagation 90% of the time. This is also called FOT or the *Frequency of Optimum Traffic*. At 115% of the (median) MUF, it should only support signal propagation 10% of the time. This is called the *Highest Possible Frequency* or HPF.

Finally, just because signal propagation is supported does not mean that we will be able to communicate. For that, you have to consider the (S+N)/N, which is a function of the modulation type used and the noise field at the receiver, among other things.

The MUF changes with time and with specific locations on the earth, or to be more exact, with the geographic location of the ionospheric refraction points. The MUF for a given path with multiple refraction points will be equal to the lowest MUF along the path. **Fig 1-7A** shows a typical 3000-km MUF chart on a Mercator projection map produced by *Proplab Pro*. (See Section 3.2.8.1.)

Each point of the MUF map is the midpoint of a 3000-km path. If an MUF of 7 MHz is shown for a given location, then you can expect an MUF of 7 MHz for any 3000-km path for which the mid-point is the given location. In other words, the MUF for a signal, transmitted 1500 km away from the given location and for which the path goes through the ionosphere above that location is the value shown at that location on the map.

It is generally accepted that the optimum communication frequency (FOT) is about 80% to 85% of the MUF. On much lower frequencies, the situation is less than optimum, as the absorption in the ionosphere increases and the atmospheric noise generally increases. We now have computer programs available that will accurately predict MUF and FOT for a given path and a given level of solar activity. These programs are very useful in predicting propagation on the higher bands, as well as for 40 meters. Their usefulness is limited on 80 and they are even less useful on 160 meters, since propagation on that band is not ruled by MUF.

During low sunspot cycle years, the MUF is often below values sustaining 7-MHz long-distance propagation. For example, during low sunspot years the 7-MHz path between Europe and the USA very frequently closes down during the night and contacts are only possible near sunset (eastern end) or near sunrise (eastern end of the path). Even 80 meters can sometimes suffer from this phenomenon. The low sunspot years are therefore not always the best years for low-band propagation—despite the widely held belief of many low banders.

Fig 1-7B shows a so-called *frequency map*, generated using the *MiniProp Plus* program (see Section 3.2.8.1). This frequency map (which really is a MUF map) allows you to quickly assess the frequencies you can use into a given area of the world. Fig 1-7B is for a low sunspot number (SSN=12) on Jan 1 at 0800 UTC (Europe sunrise). Note that the map says there is 3.5-MHz, but no 7.1-MHz, propagation between the USA and Europe, confirming what I showed earlier. Fig 1-7C is for an SSN of 100, which guarantees enough ionization for a 40-meter path between Europe and the USA. In both frequency maps the geomagnetic A-index (see Section 3.2.5.) was entered as 0, which means there would be virtually no solar-induced geomagnetic activity to disturb the path.

Fig 1-8 shows an "Oblique Azimuthal Equidistant" projection (commonly called a *great-circle map*) from *Proplab Pro* that allows us to see the MUF values encountered along a great-circle path between a QTH (the center of the map) and a target QTH. In this example the great-circle map is centered on Western Europe. From these maps we can see that the MUF is lower during local winter and much lower at night than during the daytime.

Another very useful map projection is the "Polar Azimuthal Equidistant" projection in *Proplab Pro*. This map shows either the Northern or the Southern hemisphere and is very suitable for analyzing polar paths. **Fig 1-9** shows an example of such a map centered on Brussels, with equal-MUF lines for midwinter at 0800 UTC (sunrise in Western Europe) for a sunspot number of 20. Note again the low MUF zones between Europe and the USA. **Fig 1-10** shows the same map for a sunspot number of 200, indicating that 40 as well as 30 meters will remain open all night long between Europe and North America at this level of solar activity.

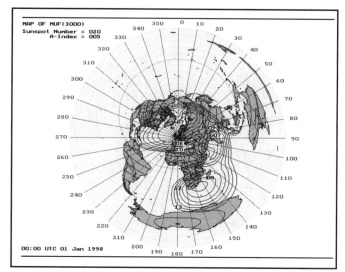

Fig 1-8—Map showing 3000-km, equal-MUF lines, displayed on a great-circle projection (Azimuthal Projection). In this example only the MUF lines up to 10 MHz are shown, in order not to clutter the map. All the radial lines departing from the center QTH (Belgium) are great-circle lines, indicating the straight-line beam headings to all target locations. This same map projection (without the MUF lines) can be used to show beam headings from a particular QTH to DX around the world. (*Map generated by the* PropLab Pro *Program.*)

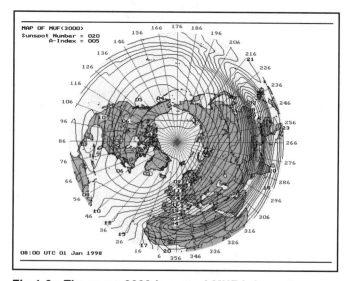

Fig 1-9—The same 3000-km equal-MUF information as shown in Fig 1-8, but now displayed on a Polar Azimuthal Equidistant projection. This example is for 0800 UTC (sunrise in Western Europe). Notice the low MUF over the North Atlantic and North America. This shows why 40 meters can often go dead between Europe and the US during low sunspot years. (*Map created by* PropLab Pro *software.*)

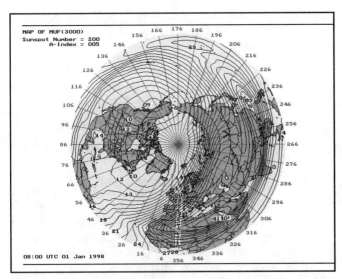

Fig 1-10—The same 3000-km equal-MUF map shown in Fig 1-9, but for a very high sunspot number of 200. Note that even 10 MHz will remain open all night long between the USA and Europe. (*Map created by* PropLab Pro *software.*)

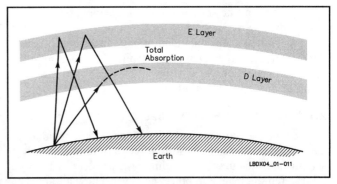

Fig 1-11—D-layer absorption. The high-angle signals pass through the D-layer and are reflected by the E/F layer. Low-angle signals are absorbed. This explains the need for a high-angle antenna to work short distances during daytime.

Finally, the MUF has nothing to do with propagation on 160 meters, since the maximum usable frequencies are always greater than 1.8 MHz, even at solar minimum.

2.1.2. Attenuation through D-layer activity

During the day, the lowest ionospheric layer in existence is the D-layer, at an altitude of 60 to 90 km. **Fig 1-11** shows how low-angle, low-frequency signals are absorbed by the D-layer. The D-layer absorbs signals, rather than noticeably refracting them, because it is much denser than the other higher ionospheric layers. The density of neutral, non-ionized particles, which make up the bulk of the mass in this region, is 1000 times greater in the D-layer than in the E layer. (See Ref 121.) For a low-frequency signal to propagate through any layer without large losses, the number of neutral atoms should be small. Statistically speaking, a free electron in the D layer during the day would collide with nearby neutral atoms about 10 million times per second! Electrons are thus not given much of a chance to refract signals in the D layer and absorption occurs instead. The "collision frequency" is high, resulting in high levels of signal absorption.

During a typical day, the level of ionization of the D layer follows the solar zenith angle, but is greatly influenced by the level of solar x-ray flares. During the night, the ionization level of the D layer drops dramatically but some very small level persists. The remaining ionization in the D layer helps determines the attenuation during the night on 160 meters. Small variation in D-layer ionization can cause large fluctuations in signal absorption on Top Band. This is especially important in multi-hop propagation modes, where the signal has to traverse the D layer twice for each hop.

The absorption level is inversely proportional to the arrival angle of the signal, so high-angle signals pass through the D layer relatively unattenuated. This is one reason our high-angle (low to the ground) dipoles work so well for local traffic on 80 meters.

I think this mechanism may also play a role in the often-reported phenomenon where, for periods shortly after sunrise, high-angle antennas often take over from low-angle antennas for working very long (eg, long-path) distances. (See also Section 2.4.4.4.) How might the sunspot cycle affect this phenomenon? When sunspot activity is low, the formation of the D layer is slower; D layer build-up before noon is less pronounced, while the evening disintegration of the layer occurs faster. This is because there is generally less energy from the sun to create and sustain the high ionization level of the D layer. This means, in turn, that at a sunspot minimum, absorption in the D layer will be less than at a sunspot maximum, especially around dusk and dawn.

The absorption mechanism of the D layer has been studied repeatedly during solar eclipses. Reports (Ref 121, 125 and 180) show that during an eclipse, D-layer attenuation is greatly reduced and propagation similar to nighttime conditions occurs on short-range paths. This means that propagation between two stations that are fully in darkness will only be influenced by the variation in the remaining D-layer density to a very small degree. Most of the attenuation will come from (bottom of) E-layer absorption (see also Section 2.1.3). Things are different when operating in the gray line, where signals have to punch through an already much ionized D-layer. During high sunspot years gray-line signal enhancement will be less pronounced because of greater D-layer absorption, thus producing less chances for signal ducting (propagation without intermediate hops).

2.1.3. The influence of the E-layer on the attenuation mechanism

During the night the D layer that causes high absorption during the day is almost totally absent. Why then is nighttime absorption usually so much higher on 160 compared to 80 meters? The answer is that the absorbing region moves up from the D region in daylight to the lower E-layer region (85-95 km high) in darkness (Ref 169). Even in the dark ionosphere there is still sufficient ionization in the lower E-region (and still a high-enough electron-neutral collision frequency) to cause such absorption, which is much more pronounced on 160 than on 80 meters. The typical absorption (in the ionosphere) for a single hop under these circumstances is 11 dB on 160 meters and 5 to 6 dB less on 80 meters (Ref 169). This limits the multi-hop communication range on 160 meters to approximately 10,000 km, assuming vertical monopoles,

1500 W transmit power and a typical receiving sensitivity at both ends.

We will see later that many longer paths on 160 are by ducting mode, rather than multi-hop modes. Ducting eliminates the losses connected with the E-layer as well as losses incurred through ground reflections (See 2.4.4.4).

2.1.4. Magnetic disturbances

Auroral activity is one of the important low-band propagation disturbances and is still largely a field of research for scientists. Amateurs living within a radius of a few thousand miles from the magnetic poles, however, know all about the consequences of aurora! The most favorable periods during the 11-year sunspot cycle (that is, when there is the least geomagnetic activity) occur during the up-phase of the cycle. The phenomenon of aurora will be covered in more detail in Section 3.2.

2.1.5. Low-band propagation during high sunspot years

For years, the generally accepted notion was that low-band DXing was not favored during high sunspot years, but not everyone shares this opinion now. Since there must be enough ionization to sustain some sort of propagation mechanism (refraction, ducting, etc), higher levels of solar activity should in theory be advantageous for low-band DXing, as well as for DXing in general. On the low bands sunspots affect mainly the 40-meter band.

In the past there were fewer DX signals on the low bands during high sunspot years. To a large degree this was due to the absence of other DXers to work. The relative lack of specific interest in low-band DXing kept run-of-the-mill DXers away from the low bands when 20 meters was open day and night. Multiband and specific low-band awards increased emphasis on low-band operating during contests. Further, the growth of an elite group of Top-Band DXers has been very instrumental in raising the activity on the low bands all through the cycle and all throughout the year.

Some 30 years ago the average DXer "discovered" the low bands and added them to the list of what were considered DX bands. Nowadays, every DXpedition includes 40, 80, and 160-meter work in their operating schedules. Some DXpedition even set out to work only or mainly the low bands! Even in the middle of the summer in the high sunspot years we can often hear several stations calling CQ DX on Top Band.

Still, the sunspot cycle has some impact on low-band propagation, although not as drastic as on, say, 10 meters, where the MUF rules propagation. Low-band propagation, particularly on 160 meters, is rather "digital." Either the waves are reflected in the ionosphere or they are not, and they are lost into space.

On 160 meters, the main limiting mechanism is one of *attenuation*, since 1.8 MHz is always lower than the MUF. There are a number of ways in which Top-Band signals are attenuated during their travels. If enough signal survives being attenuated by all these mechanisms and we can hear the signal above the local noise floor, then we say we have propagation. Distance plays an important role on 160 meters. N6TR recognized this when he introduced the Stew Perry (W1BB) Top Band contest, where scoring goes by distance, just like in VHF/UHF contests. We know that the sun is involved in various mechanisms causing the cumulative path loss.

1. Aurora: It appears that we enjoy the geomagnetically quietest years during the minimum and the rising phase of the sunspot cycle (Ref 142). See Section 3.2. and **www.spacew.com/swim/bigstorm.html**
2. Ionization levels in the E and F layers during the night: On 160 and 80 meters, when the entire path is well into darkness, propagation is primarily by multiple hops in the F layer, with little or no D layer absorption. However, the remaining ionization of the E-layer takes its toll in attenuation (up to 11 dB ionospheric loss per hop on 160 meters). The sun's activity (as witnessed by the sunspot numbers) will influence E-layer absorption. The difference may not be a spectacular number of dB, but we often operate on the verge of what is possible (very low S/N ratio), where a few dB can make the difference. (See Section 2.4.4.)
3. D layer: The remaining ionization of the D region during the night plays a role, especially in the multi-hop propagation. (See Section 2.4.4.) During gray-line twilight periods the absorption in these regions is more substantial than during the night, hence high levels of sunspots do influence gray-line propagation.
4. Ionospheric ducting: Conditions conducive to an ionospheric ducting mechanism are more easily met during low sunspot years. (See Section 2.4.4.5.)

We all know that good propagation, especially over long distances (> 7,000 km) occurs much more frequently during low sunspot years than during high sunspot years. On 160 meters during high sunspot years, I can reach to the US Midwest when conditions are good. But I never work California during the high sunspot years. As a rule, good openings between western Europe and the West Coast of the USA happen *only* during low sunspot years.

2.2. The 27-Day Solar Cycle

The sun rotates around its own axis in approximately 27 Earth days. Sunspots and other phenomena on the sun can last several solar rotations. This means that we can expect similar radiation conditions from the sun to return every 27 days. DXers (both on the low and the higher HF bands) look forward to a repeat of very good conditions 27 days after the last very good ones have occurred—and their expectations are often met. This is probably the only somewhat reliable propagation prediction system for 160 meters that we have at this time! If conditions are good today, the *lack* of bad things is why propagation may possibly be good 27 days from today. If you just have had outstandingly good conditions on 160 meters, always mark your calendar for 27 days later—There is a fair chance you may have the same or similar good conditions again.

However, if conditions today are very bad (maybe due to a solar flare), there's no telling whether conditions will be bad in 27 days, since a solar flare does not repeat every 27 days. A recurring coronal hole, however, would most likely repeat in the next 27-day period, since these tend to hang around for at least several solar rotations. Unfortunately, during the early years of a solar cycle this predicting system may not be very reliable because the sunspots don't hang around for even one full revolution most of the time. As the cycle matures, the 27-day recurrence becomes more important.

2.3. The Seasonal Cycle

We know the mechanism that originates our seasons: the *declination* of the sun relative to the equator. This tilt reaches a maximum of 23.5° around Dec 21 and Jun 21 (See **Fig 1-12**.). This coincides with the middle of the Northern Hemisphere winter propagation season and the middle of the summer propagation season. At those times the days are longest (or shortest) and the sun rises to the highest (or lowest point) at local noon in the non-equatorial zones.

On the equator, the sun will rise to its highest point at local noon twice a year, at the equinoxes around Sep 21 and Mar 21. These are the times of the year when the sun-Earth axis is perpendicular to the Earth's axis (sun declination is zero), and when nights and days are equally long at any place on Earth (*equi* = equal, *nox* = night). On Dec 21 and Jun 21, the sun is still very high at the equator (90° − 23.5° = 66.5°). The maximum height of the sun at any latitude on earth is given by the expression:

Height = 90° − north latitude + 23.5° (with a maximum of 90°).

In other words, the sun never rises higher than 23.5° at the poles, and never higher than 53.5° where the latitude is 60°. The seasonal influence of the sun on low-band propagation will be complementary in the Northern and Southern hemispheres. Any influence will be most prominent near the poles and less pronounced in the equatorial zones (±23.5° of the equator).

But how do the changing seasons influence propagation?

1. The longer the sun's rays can create and activate the D layer, the more absorption there will be during dusk and dawn periods. During local winter, the sun will rise to a much lower apex and the rate of sunrise will be much lower. Accordingly, D-layer ionization will build up much more slowly.
2. If the sun rises quickly (local summer in areas away from the equator) the configuration of the D, E and F layers necessary to set on a wave-ducting mode will last for a much shorter time than in winter. *Gray-line propagation* will thus last longer in the winter than it will during summer.
3. In non-equatorial areas, many thunderstorms are generated in the summer. Electrical noise (QRN) easily masks weaker DX signals and can discourage even the most ardent DX operator. On north-south path (US-to-South America or Europe-to-Africa), the Northern Hemisphere summer is usually the best season, since QRN is likely to be of less intensity than the QRN during the Southern-Hemisphere summer.
4. When the nights are longest during local winter, you will have the greatest possible time for DX openings. Indeed, you must be in darkness or twilight not to suffer from excessive D-layer absorption.
5. The occurrence of aurora is most pronounced around equinox (Mar-Apr and Sep-Oct).

2.3.1. Winter
(15 Oct to 15 Feb in the Northern Hemisphere)

Winter is characterized by lower MUFs, shorter days, lots of darkness, sun rising slowly, longer gray-line duration (see Section 2.4.4.1) and less electrostatic discharges (QRN) from local thunderstorms. This period is best for all stations located in the Northern Hemisphere during the winter. Conversely, this condition will not exist in the Southern Hemisphere. Therefore the winter period in the Northern Hemisphere is ideal for east-to-west and west-to-east propagation between two stations both located in the Northern Hemisphere. Typical paths are US-to-Europe, US-to-Japan, US-to-Asia, etc.

2.3.2. Summer
(15 Apr to 15 Aug in the Northern Hemisphere)

Summer is characterized by higher MUFs, longer days, faster-rising sun, increased D-layer activity at dusk and dawn, and higher probability of QRN due to local thunderstorms. These factors create the worst conditions for east-to-west or west-to-east propagation in the Northern Hemisphere. However, good QRN-free openings to the west can be possible, just

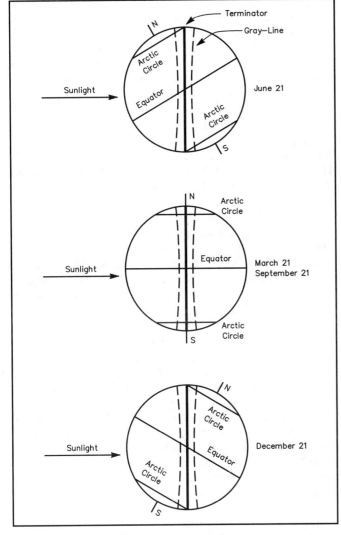

Fig 1-12—These drawings show the declination of the sun at the different positions (solid vertical lines) at different times of the year. The gray line is represented as a zone of variable width (shaded area) to emphasize that its behavior near the poles differs from that near the equator.

around local sunrise at the eastern end of the path. While most amateurs may be fighting the local QRN in the Northern Hemisphere in summertime, our friends down under are enjoying excellent winter conditions. This means that summertime is the best time for transequatorial propagation (for example, from Europe to Southern Africa or North America to the southern part of South America).

2.3.3. Equinox period (15 Aug to 15 Oct and 15 Feb to 15 Apr)

During these periods the ionospheric conditions are fairly similar in both the Northern and the Southern hemispheres: similar MUF values, days and nights approximately 12 hours long on both sides of the equator, reduced QRN, etc. Clearly this is the ideal season for "oblique" transequatorial propagation," on the NE-SW and NW-SE paths. Typical examples are Europe to New Zealand and West Coast US to SE Asia (NA morning) or East Coast NA to Indian Ocean (NA evening).

2.3.4. Propagation into the equatorial zones

In principle, all seasons can produce good conditions for propagation from the Northern or Southern Hemisphere into, but not across, the equatorial zone. On 80 and 40 meters the only real limiting factors can be the MUF distribution along the path and especially the amount of QRN in the equatorial zone itself. Unfortunately, there is no rule of thumb concerning the electrical storm activities in these zones. From Europe, we work African stations and stations in the southern part of South America on 160 meters mainly during the months of June, July and August. A similar situation exists between North America and the southern parts of Africa and South America.

2.3.5. Low bands are open year-around

It is clearly incorrect that DX on 80 and 160 meters can only be worked during the local winter, a popular belief not so very long ago. The equinox period is the best time of year for equatorial and transequatorial propagation, while in the middle of our Northern-Hemisphere summer (if QRN is acceptably low for us), we often work rare DX stations from down under or from the equatorial zones. Even good east-west openings can happen in the middle of the local summer, so long as there is a darkness path and the QRN level makes listening for the signals at all possible. Of course the operators on both sides must be willing to try, and not take for granted that it won't work. Over the years I have worked quite a few rare ones over an east-west path in summertime. Here are a few examples:

XYØRR (Burma) was worked on Sept 3, 1991, on 160 meters, shortly before his sunrise. After a QSO on 80-meter SSB, we moved to 160 meters, where 579 was exchanged. After the QSO XYØRR called CQ a few times, but nobody else came back. Those convinced that summer time is not a good time for 160 meter were wrong once again. The little story behind this contact is that I normally don't have a "long" Beverage up for that direction during the summer. There happened to be 2.5-meter (8-ft) tall corn growing in the field in that direction. That day I spent a memorable couple of hours putting an insulated wire right on top of the cornfield. You must try that sometime for fun!

Another striking example is what happened during the DXpedition of Rudi, DK7PE, to S21ZC in early Aug 1992. The first night he was on 160 meters, I was in the middle of a local thunderstorm (S9 + 40 static crashes) and no chance for a contact. The next day, the QRN was down to S7, and a perfect QSO (579) was made over quite a long path (comparable with a path from ET3PMW to the US East Coast in June, or from the West Coast of Africa to California). See also Section 3.3.

2.4. The Daily Cycle

We know how the Earth's rotation around its axis creates day and night. The transition from day to night is very abrupt in equatorial zones. The sun rises and sets very quickly; the opposite is true in the polar zones. Let us, for convenience, subdivide the day into three periods:

1. Daytime: from after sunrise (dawn) until before sunset (dusk).
2. Nighttime: from after sunset (dusk) until before sunrise (dawn).
3. Dawn/dusk: sunrise and sunset (twilight periods).

2.4.1. Daytime

Around local sunrise, the D layer builds up under the influence of radiation from the sun. Maximum D-layer ionization is reached shortly after local noon. This means that from minimum absorption (due to the D layer) before sunrise, the absorption will gradually increase until a maximum is reached just after local noon. The degree of absorption will depend on the height of the sun at any given time.

For example, near the poles, such as in northern Scandinavia, the sun rises late and sets early in local winter. The consequence will be a late and very gradual buildup of the D layer. In the middle of the winter the sun may be just above the horizon for regions just below the Arctic Circle, situated at $90° - 23.5° = 66.5°$ above the equator. Or the sun may actually be below the horizon all day long for locations above the Arctic Circle. Absorption in the D layer will be minimal or non-existent under these circumstances. This is why stations located in the polar regions can actually work 80-meter DX almost 24 hours a day in winter (provided there are quiet geomagnetic conditions—see Sect 3.2). Contacts between Finland or Sweden and the Pacific or the US West Coast are common around local noon in northern Sweden and northern Finland in winter on 40 and 80 meters (occasionally even on Top Band).

I suppose that this is not a good example of "typical" daytime conditions, since in those polar regions they never actually have typical daytime conditions in midwinter but remain in dusk and dawn periods all day long!

I've mentioned before that during typical daytime conditions, when the D-layer ionization is very intense, low-angle signals will be totally absorbed, while high-angle signals will get through and be refracted in the E layer (160 and 80 meters). Only at peak ionization, just after noon, may the absorption be noticeable on high-angle signals. The signal strength of local stations, received through ionospheric refraction, will dip to a minimum just after local noon. As stated before, to obtain good local coverage on 80 meters during daytime you must have an antenna with a high vertical angle of radiation. This can easily be obtained with a low 80-meter dipole. On 160 meters, middle-of-the-day propagation is essentially limited to ground-wave signals. On the opposite end of the low-band spectrum, 40 meters basically stays open for DX almost 24 hours per day in winter, albeit with much-attenuated signals around local noon.

2.4.2. Nighttime (black-line propagation)

After sunset, the D layer gradually dissipates and almost completely disappears. Consequently, good propagation conditions on the low bands can be expected if both ends of the path, plus the area in between, are in darkness. The greatest distances can be covered if both ends of the path are at the opposite ends of the darkness zone (both located near the *terminator*, which is the dividing line between day and night). During nighttime in a period of low sunspot activity, the critical frequency may descend to values below 3.7 MHz and dead zones (skip zones) will show up regularly. Skip zones are also common on 40 meters during nighttime. Skip zones due to MUF do not occur on 160 meters since the MUF is always higher than 1.8 MHz. In contrast to gray-line propagation, Brown, NM7M, calls propagation with one of the stations at the terminator "dark line propagation." (Ref 140.)

2.4.3. Midway midnight peak

North-south (±30°) paths exhibit a clear propagation peak at local midnight time halfway on the path, both on 80 and 160 meters. When I make schedules on these bands with African or Indian Ocean stations, I will always try to have them at "midway midnight."

Although it has been generally accepted that an east-west path only exhibits a sunrise and a sunset peak, many critical observers have witnessed (at least on 160 meters) a similar halfway midnight peak. I have observed that this is especially true during high sunspot years, when the gray-line (twilight) enhancement seems to be less common than during low sunspot years. This peak certainly does not exist on 40 meters, where the signal peaks are only before and around sunset, and around or after sunrise. Although I have observed this phenomenon several times, it was Peter, DJ8WL, who raised the question on the Internet. The exact mechanism may not be understood, but it probably is connected to the fact that the sun is exactly "on the other side" of the Earth, creating ideal ionization conditions in the E and F regions of the ionosphere on the dark side of the Earth. I will explain how to calculate these peak times, using sunrise-sunset times in Section 5.1.

2.4.4. Dusk and dawn: twilight periods

As mentioned before, the terminator is the dividing line between one half of the Earth in daylight and the other half in darkness. The visual transition from day-to-night and vice versa happens quite abruptly in the equatorial zones and much more slowly in the polar zones. The gray line is a band between day and night, usually referred to as the *twilight* zone. Actually, "gray zone" might have been a more appropriate term than "gray line." Dusk and dawn periods produce very interesting propagation conditions that are not limited to the low bands. However, the mechanisms involved can differ very substantially between high bands (10, 15 and 20 meters) and the low bands (40, 80 and 160 meters).

For low-band operators in particular, it is extremely important that they be able to visualize the situation using maps or globes that show the terminator, the great-circle lines

and the auroral oval. (See Section 4.1.) For a long time many have speculated about what actually produces the enhanced propagation conditions we experience almost daily on the low bands at either dusk or more frequently at dawn, especially during low sunspot years. It has become widely accepted that these twilight effects are due to the onset of a specific propagation mechanism—one that is characterized by lower loss than the standard multi-hop model. The *ducting* mechanisms involved are discussed in detail in Section 2.4.4.

But besides the role of the ducting mechanism at dusk and dawn, there is another reason why we are able to work DX much better during twilight periods. When the sun is rising in the morning, all signals coming from the east (which can often cause a great deal of QRM during the night) are greatly attenuated by the D layer existing in the east. The net result is often a much quieter band from one direction (east in the morning and west in the evening), resulting in much better signal-to-noise ratios on weak signals coming from the opposite direction. This does not necessarily mean that we will be heard better, since the better signal-to-noise ratio obviously only influences receiving and not transmitting.

It is also important to know how long these special propagation conditions exist—in other words, to know how long the effects of the radio-twilight periods last. You should understand that the rate of change from darkness to daylight (and vice versa) depends upon the rate of sunrise (or sunset). There are two factors that determine this rate: the season (the sun rises faster in summer than in winter) and the latitude of your location (the sun rises very high near the equator, and peaks in the sky at low angles near the poles).

It is also clear that propagation where the signals depart or arrive at an angle *perpendicular* to the terminator will enjoy the greatest signal enhancement, since these paths travel the shortest distance through the D layer and the bottom of the E layer. Carl, K9LA, reports that while a multi-hop path is limited to 10,0000 km in total darkness, extensive ray-tracing exercises have proven that propagation *in* the gray zone is limited to half this distance (5000 km), all other parameters being the same. This validates what we already know—The best place for 160-meter propagation is in the dark ionosphere (Ref 169). For 80 meters the maximum distance for a propagation path along the terminator is 8000 km. Here too, the absorption in the lower E-layer region limits the distance a signal can travel and still be heard.

2.4.4.1. The gray line on the low bands

When both ends of a path are in the twilight zone, one side at sunrise and the other at sunset, then we have a so-called *gray-line* situation. However, the term gray-line propagation is also loosely used where only one side of the path is in twilight (usually at sunrise at the eastern end of the path).

The effect of advantageous propagation conditions at sunrise and sunset has been recognized since the early days of low-band DXing. Dale Hoppe, K6UA, and Peter Dalton, W6NLZ, first used the term "gray line" for the zone centered around the geographical terminator (Ref 108). See **Fig 1-12**.

In the past, some authors have shown the gray-line zone as a zone of equal width all along the terminator. This is incorrect so far as radio-propagation phenomenon is concerned. R. Linkous, W7OM, recognized that the zone width varies in his excellent article "Navigating to 80-meter DX" (Ref 109).

The mechanisms that determine the width of the gray lines mean that we have a narrow gray line near the equator and a wider gray line near the poles. The time span during which we will benefit from typical gray-line conditions will accordingly be shorter near the equator and longer in the polar regions. Therefore the gray-line phenomenon is of less importance to the low-band DXer living in equatorial regions than to his colleagues close to the polar circles. This does not mean that there is less enhancement near the equator at sunrise or sunset; it just means that the duration of the enhanced period is shorter.

Some authors (Ref 108 and 118) have mentioned that gray-line propagation always happens along the terminator. On the low bands there have only been occasional instances of such propagation. From the following examples of gray-line propagation, it should be clear that propagation does not happen along the gray line but rather through the dark zone, on a path that is in most cases nearly *perpendicular* to the terminator. In the zone along the gray line there is *more* attenuation due to the absorption in the D layer (and in the lower E-layer region). Gray-line propagation on the low bands is a different affair from what often is called gray-line propagation on the HF bands, where the propagation path does follow the direction of the gray line.

W4ZV states: "*Here is what I have observed many times for what I call 'long-path modes,' SSW before sunrise and SSE after sunset. Signals on these paths typically peak at midway between sunrise/sunset times at each end of the path, and appear to be optimum when there is ~40 minutes of common darkness for 80 meters, and ~80 minutes of common darkness for 160. These paths are doing something different since the arrival and departure azimuths are nearly aligned parallel to the terminator at each end of the path. Here is an example:* **users.vnet.net/btippett/dx_aid_plots.htm**."

Gray-line propagation occurs right at sunrise or sunset. The low bands usually peak from shortly before to shortly after sunrise (sunset). These sunrise/sunset peaks are more pronounced during low sunspot years than during years of high sunspot activity. Usually the sunrise peak is much more pronounced than the sunset peak.

2.4.4.2. Examples of remarkable gray line propagation

Many of us remember the unforgettable DXpedition to Heard Island in Jan 1997 (at the bottom of the sunspot Cycle 22). K9LA (Ref 152) described how US East Coast stations, against all expectations, worked Heard Island on 160 meters day-after-day, on what was considered to be an extremely difficult path. **Fig 1-13A** shows the theoretical great-circle path between Heard Island and New York (mid January at 2300 UTC). Note that the path (heading of 250°) makes an angle of approximately 30° with the terminator on Heard Island. Fig 1-13B shows the same path, with New York as the center of the azimuthal projection map. This path (heading of 130°) makes a similar sharp angle (25°) with the terminator at the US-end of the path. Most, if not all, US East Coast stations who worked VK0IR and who had access to a variety of directive receiving antennas (such as Beverages) noted that the VK0 signals arrived at a heading of approximately 60°, right across Europe. (See also Section 4.3.)

Fig 1-13C shows the path between Heard Island and Spain. The path (beaming 300°) now makes a perfect 90° angle with the terminator on Heard Island. Similarly, the path between the US East Coast and Spain (beaming 65°) also makes a perfect 90° angle with the terminator at the US end of the path (Fig 1-13D). I am convinced that the signals at Heard Island traveled across Europe to the US East Coast. This is supported by testimonies from US stations and it again confirms that enhanced gray-line conditions most often go together with a signal azimuth that is nearly perpendicular to the terminator.

K1GE confirmed that this has happened with other stations from the Indian Ocean as well. I was listening every day during the Heard Island DXpedition and witnessed that the signals faded out completely in Europe at exactly the same time they faded out in North America. This seems to confirm that, indeed, the path to the US was right across Europe. What makes this path skew to more northerly regions is explained in Section 4.3.1.

To be fully correct, I must admit that the QSOs between the US East Coast and Heard Island at 2345 UTC is only half a gray-line QSO. However, stations a little further inland in the USA who worked VK0IR just prior to that did it on a double-sided gray-line path.

The VK0IR expedition to Heard island was a living testimony to how the width of the gray line depends on the latitude of the station involved. Many remember how VK0IR (located at 53° south latitude) was worked almost every day on 160 meters until more than 30 minutes after local sunrise, while 80 meters QSOs were made as late as 0050 UTC, which is 1½ hours after sunrise on Heard island during that DXpedition.

Another striking example of gray-line enhancement involved a QSO I had on 80 meters with Kingman Reef, a particularly difficult path late in the Northern-Hemisphere winter season from Europe. I made QSOs with Kingman Reef and Palmyra around May 1, 1988. If we analyze sunset and sunrise times for that date, we see that sunset on those islands is roughly 40 minutes after sunrise in my location. This means that there is theoretically no opening, but we can force things a little and take advantage of the gray line: Split the 40 minutes in half and try a QSO 20 minutes before sunset in Palmyra (or Kingman Reef) and 20 minutes after sunrise here in Belgium.

Does this sound like a nice 50/50 deal? Certainly not! Those Pacific islands are situated only about 6° north of the equator; Belgium is 51° north. This means that the gray line lasts just seconds out on KH5 and maybe 40 minutes in Belgium. I made the schedules right at Pacific sunset time. On Kingman the QSO was made 5 minutes after Kingman sunset, on Palmyra 4 minutes after sunset, and in both cases 40 to 45 minutes after sunrise in Belgium (where the gray line is fairly "wide"). This is a striking example of how knowledge of the mechanisms involved in propagation can help you make a very difficult QSO. Here's proof of how marginal a situation it really was: From Palmyra only one QSO was made with Europe, and only two from Kingman Reef.

Over shorter paths, stations can often be worked on 80 meters for hours after sunrise (or hours before sunset). Contacts between the US East Coast and Europe are quite common in midwinter as much as 2 hours after sunrise in Europe during low sunspot years.

The width of the gray line on 160 meters is much more restricted than on 80 meters. Even in the middle of the winter,

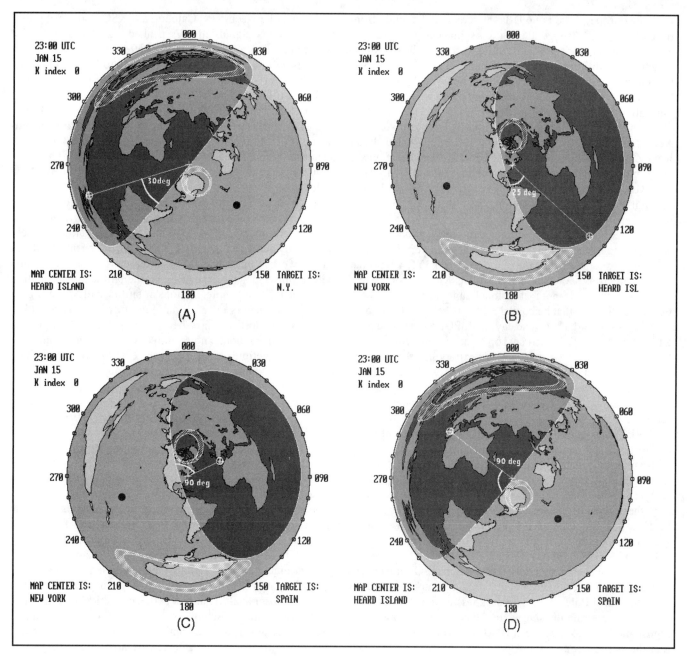

Fig 1-13—At A, the great-circle short path from Heard Island to New York. The angle compared with the terminator at Heard is quite sharp. At B, the other end of the same path, you also see also a sharp angle (25°) compared to the terminator. At C, the path from Heard island to Spain makes for a perfect 90° angle with the terminator, as it does with the bent path from the US East Coast, over Southern Europe, to Heard island. US East Coast stations reported that the Heard Island signals came in a path over Europe, at least for the first days of the DXpedition. The crooked path from Heard to the US East Coast also was launched on a path at right angle to the terminator at Heard Island. (*Figures created using* DX-AID *software, with additions by ON4UN.*)

I have seldom worked real long-haul QSOs more than 15 minutes after sunrise on Top Band. I often copy stations quite late after sunrise, but with weak signals on a very quiet band, which is possible because of the absence of noise from the east. This phenomenon has been confirmed many times by Jack, VE1ZZ, who can copy European stations on 160 meters up to 3 hours before his sunset. This does not mean he was able to contact the many European stations he heard. For that he had to wait almost two hours! Anyhow, at that time of the day the European stations are probably listening to the east. VE1ZZ reported that signals were quite weak and only copiable because there was absolutely no noise whatsoever when this happened.

At this time in the afternoon toward the daylight area, the D layer acts as a shield to ionospheric-propagated noise. This provides a much lower noise floor for the station in sunlight. The stations at the eastern end of the path are subject to high noise levels because they do not have the advantage of the D-layer absorption attenuating the atmospheric noise propagated into their area. Just watch your S meter on 160 meters at 2 PM during winter and then again after sunset, and you will probably see a 20 or 25 dB higher atmospheric noise level, provided you are in a quiet location with little manmade noise. Twenty dB is a difficult spread to overcome. This explains the one-way propagation under such circumstances. At the same time, in Europe, there is no D-layer screen and signals from the east are 50 dB stronger than Jack's signal before his sunset. This one-way propagation is quite common during quiet magnetic conditions (low A and K indices). Similar observations have been made from Europe, where UA9/Ø stations are heard long before European sunset, but again we have to wait until almost sunset to make contact.

On 160 meters there are rare times when I can work DX well after sunrise from my location in Europe—when I can work ZL stations on a genuine long path (21,500 km) about 30 to 45 minutes after our sunrise during low sunspot years. The few ZLs worked on this long path had to be worked right through a wall of English stations, who were enjoying their sunrise peak at exactly that time.

Sometimes, when conditions are really good with no atmospheric noise, signals can also be heard a very long time after sunrise at the eastern end of the path. GW3YDX reported copying many W6/W7 stations as well as KL7 on 160 meters until more than one hour after local sunrise. Really exceptional was the fact that he copied K6SE at 1130 UTC, 3 hours after local sunrise! That same day, GM3POI reported hearing KL7RA at 1230 UTC, also on Top Band (in midwinter)! We have, of course, to be careful and not extrapolate these observations to the whole of Europe—the stations that reported these extraordinary conditions are located at the fringes of Europe, almost at the back door of North America, and quite far north (52 to 53° N).

VE7VV suggests that we should not discount paths (long or short) on 160 meters, where part of the middle of the path is in sunlight. All this means is that the signal must not be making ground reflections in this region because the top of the E layer is where the signal is reflecting back upwards. Actually, these topside E-layer reflections are just what is needed for E-to-F layer ducting propagation with very low loss, since there is no passing through the absorptive D layer. (See Sect 2.4.4.3.)

To me, it is clear that these exceptional QSOs must occur on a *bent path*. (See Section 4.3.) The signals appear to travel near the North Pole, staying in darkness as long as possible. W8LT has reported the same experience on 80 meters, working EI and G stations between 1000 and 1100z. He confirmed that the best antenna for this propagation was a half-square broadside N/S. To him, this indicated that a crooked path was involved. I believe that this kind of propagation can only occur during a short period around winter solstice and when magnetic conditions are exceptionally quiet, with resulting low auroral-zone absorption.

Similar conditions are quite common on 40 meters during the European winter. With a good Yagi antenna, I can work North America 24 hours a day on 40 meters, when there is no geomagnetic disturbance. At local noon, when the sun is highest, I hear W8s and W9s quite commonly, followed somewhat later by W6 and W7/VE7 stations, all on a direct polar path. The West Coast will keep coming through with the beam pointed approximately 350°, until at 1430 UTC the band will also open on the long path. Shortly later the bent short path will close.

These specific propagation paths and times are applicable only to moderate-distance DX when it comes to 160 meters. Really long-haul propagation on 160 seems to follow the rule of enhanced propagation only occurring at dawn/dusk. During a long period of tests (in Nov and Dec) on 160 meters between New Caledonia (FK8CP) and Belgium I found that his signals always peaked right around sunrise (from 3 minutes before, to 3 minutes after sunrise). This "short peak" is valid only for the very long path to Western Europe. FK8CP reports openings into Asia (UA9, UAØ) from much earlier, until a little later after sunrise.

Close-in DX (2000 km or 1500 miles) can be worked as late as 45 minutes after sunrise on 160 meters, again depending on your latitude. FK8CP reports working DX in the Pacific as late as 50 minutes after his sunrise in the middle of his local summer (which is quite late, considering the latitude of New Caledonia).

On 40 meters, the gray line is of course "wider" than on 80 meters (remember the gray line is a zone of variable width, not to be confused with the terminator, which is a line uniquely defined by sun-Earth geometry). In the winter, long-haul DX can be worked until many hours after sunrise (or many hours before sunset), again depending on the latitude of the station concerned. For example, stations at latitudes of 55° or higher will find 40 meters open all day long in winter. Even at my location (51° north), I have been able to work W6 stations at local noontime, about 3 hours into daylight. At the same time, the band sometimes opens up to the east, so we can say that even for my "modest" latitude of 51° N, 40 meters is open for DX 24 hours a day on better days.

2.4.4.3. Propagation without lossy ground reflections?

Multi-hop propagation with intermediate ground reflections has long been the traditional way to explain propagation of radio waves by ionospheric refraction. Not too long ago some scientists stated that these ducts did not exist and that all propagation had to be by means of multiple hops from the Earth to the ionosphere and then back again. In the last 20 years enthusiastic low-band DXers have made literally thousands of observations of propagation "anomalies" that yielded much stronger signals than could be predicted.

This mass of observations spurred propagation scientists to take a closer analytical look at the available data. Theoretical research enables scientists to calculate path losses due to ionospheric absorption (deviative and non-deviative losses), free-space attenuation (path-distance related) and earth (ground or water) reflection losses. The ionospheric reflection loss on 160 meters during the night is approximately 11 dB per hop (about 4 to 5 dB less on 80 meters), excluding ground reflections (Ref 169).

While the theory of propagation with ground reflections and the knowledge of the attenuation involved at each step is satisfactory to explain short- and medium-range contacts (maximum of 10,000 km), the losses through ground reflections and ionospheric (E-layer) losses are no longer accepted by most experts as adequately explaining some of the high signal levels obtained over very long distances, especially when gray-line propagation and genuine long-path situations are involved.

On 160 and 80 meters, when the entire path is well into darkness, propagation is primarily by means of multiple hops in the F layer, with little or no D-layer absorption, but with additional attenuation due to the remains of ionization in the E layer. This seems to be the model that fits observations when neither end of the path is in the twilight zone.

Even this model does not explain why there generally is

a significant signal enhancement where either or both ends of a very long-distance path are in the twilight zone (more specifically, when the eastern end of the path is at sunrise). In the twilight zone there should actually be *more* D- and E-layer absorption, compared to the full-darkness situation. In fact, during low sunspot years, gray-line enhancement is more prominent compared to high sunspot years, because the ionization of these attenuating layers at sunrise and sunset during high sunspot years is more pronounced.

Thus there must be another mechanism involved that compensates for the additional D- and E-layer losses in the twilight zone, a mechanism where we can happily end up with an overall loss that is significantly smaller than the full-darkness, multi-hop model with little or no D- and E-layer absorption. *Signal ducting* could well be that mechanism behind twilight-zone enhancement.

Sometimes, signals are ducted as though they were confined inside a pipe (waveguide) in the ionosphere, without lossy intermediate reflections from the Earth's surface. Such ducting explains strong signals sometimes heard over very long distances. Gray-line enhancement seems to go hand-in-hand with ducting (see Section 2.4.4.3) and this is more pronounced during low sunspot cycle years. Recent propagation-prediction software tools (Ref 153) include models that support three-dimensional ray-tracing and ducting, including geomagnetic effects. *Proplab Pro* is one such program that explicitly computes 80 and 160-meter ducting modes for many long-distance paths.

I first came across a description of the phenomenon of ionospheric ducting in an article by Yuri Blanarovich (ex-VE3BMV, now K3BU) in 1980 (Ref 110). More than 20 years later the phenomenon of signal ducting (ionospheric ducting) on the low bands finally seems to have been accepted also by the scientific community.

I have to admit, however, that the multi-hop-only model without ducting can help to explain why paths that are across saltwater generally produce stronger signals, due to the minimal reflection losses at saltwater reflections. (Of course, paths can include combinations of Earth-ionosphere multiple hops, as well as in-ionsphere ducting mechanisms.)

Dan Robbins, KL7Y, a recent Silent Key who is sorely missed by the amateur fraternity, worked for years in the field of HF radar. He maintained that HF signals really do bounce off the Earth, and that the losses on Earth reflections (especially from saltwater) are mostly insignificant—usually much less than the losses due to an ionospheric reflection, at least for frequencies below the MUF.

Textbooks, including *Ionospheric Radio* by Davies, provide charts that indicate that sea-water reflection loss is a fraction of a decibel on 160 meters for all but the lowest angles (3° or less). A land reflection might typically average several dB, so it is easy to see why long paths over water can produce stronger signals on the low bands compared to paths that require multiple reflections over poor ground. This is one area where I believe many propagation programs fail, since they do not know the geography at the reflection point—They plug in an "average value," something like 2 or 3 dB. On an all-water long path with multiple ground reflections the program could be off by 10 to 20 dB. One of the exceptions is the *Proplab Pro* ray-tracing program mentioned above, which contains an Earth/water/ice geographical database that is used to compute realistic reflection losses.

2.4.4.4. Ionospheric signal ducting

Based on experimental observations (Ref 100) and theoretical studies (Refs 131 and 151), others have come to the conclusion that some very specific ducting modes were allowing exceptionally strong signals to be heard over very long paths.

Due the layered structure of the ionosphere, waveguide-like channels (ducts) appear in which radio waves can propagate over long distances. **Fig 1-14** shows the nighttime electron-density distribution, showing a dip in electron density above the E-layer peak. This valley is responsible for setting up a waveguide-like 160-meter duct, bounded by the F-layer at the top and the top of the E-layer on the bottom (Ref 151). Top Band signals get trapped between the E and F regions rather than propagating between the E-region or the F-region and the ground. The trick is to get the signal to enter this duct, and then later to get it out of the duct at the end of the path.

Cary Oler and Ted Cohen (Ref 142) point out that this kind of E-F ducting is most typical for 160 meters because Top Band signals can be refracted more effectively at higher wave angles than signals at higher frequencies can be. From the launching point of such a duct path, refraction occurs when the signal travels through the E layer, resulting in bending of the waves into a lower angle (in other words, there is not a complete reflection). The wave is propagated further at the required angle to start the ducting. Relatively high launch angles from the Earth are required to punch through the D and E layers, so that the wave can finally be reflected up into the E-F region. This may be another explanation why higher wave-angle transmit antennas sometimes beat out very low-

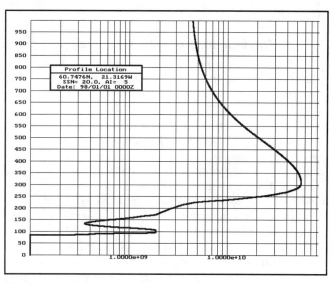

Fig 1-14—Electron-density profile for a point near Iceland, roughly halfway between Europe and North America on the North Atlantic path. Notice the dip in the electron density between the E and the F layer. This dip is responsible for waveguide-like propagation (ducting), with the bottom-side of the E-layer and the topside of the F layer serving as "duct walls."

angle antennas, especially on 160 meters. (See also Section 2.1.2.)

The transition from darkness to daylight causes what is known as *ionospheric tilting*, which can also help bend a signal into (and out of) the duct. In addition, horizontal ionospheric tilting is often required to be able to work into a duct. This condition typically exists at the terminator, provided that the angle between the propagation path and the terminator is close to 90º (Ref 151).

Once inside, how does the signals leave a duct? Towards the sunrise end of the path a change in the slope of the electron-density contours can alter the local refraction angle of the duct, and the signal angle becomes steep enough to break through the E region back down to Earth. This mechanism favors the relatively high elevation angles often observed for such received signals. The exit of the signal from the duct often produces a spotlight-like illumination of the Earth, making signals very strong in a specific location, while being inaudible at another location only a few hundred miles away.

Why is ducting not an everyday occurrence? The ionosphere is particularly turbulent around sunrise, so the well-defined contours of electron density required for signals to leave the duct are not necessarily there every day. Nick Hall-Path, VE7DXR (Ref 177) summarizes the situation as follows: "*I suggest that sunrise enhancement could be caused by ionospheric tilts occurring just before sunrise at the receiver. Those tilts direct signals to the receiver, from a duct between E and F regions. The D-region would tend to absorb such signals rather quickly, however, and no good answer has been offered as to why enhancements occur on some mornings and not on others. Perhaps on mornings when strong signals are heard, there is some retardation of D-layer absorption caused more by terrestrial air movements than by solar and geomagnetic influences.*" This supports the suggestion by Robert R. Brown, NM7N, indicating the importance of the ozone layer in this mechanism (Ref 176).

2.4.4.5. The influence of sunspot cycle on ducting

Higher sunspot numbers means more ionization. How much more? Brown (Ref 170) reports that going from a SSN of 5 to 100, the electron density at the bottom of the E-region increases only modestly, by a factor of merely two. This results, however, in an increase of the critical frequency of the E layer of approximately 30%, which is significant. To penetrate the bottom of the E-layer, the wave angle must now be steeper (again explaining why higher angles seem to do well near sunrise and sunset). This steeper angle results in shorter consecutive hops inside the duct, and consequently more loss on the path.

To sustain a ducting path between E layer peak and the F-region, the valley between these two regions (see Fig 1-14) must be present all along the path. What could cause this valley to disappear? Scientists suggest that even modest increases in electron density in the ducting region will fill up the valley and halt the ducting mechanism. The required levels are levels that will barely increase signal attenuation. Even very modest levels of auroral activity can be disastrous to this mode of propagation, not only by the creation of extra absorption but mainly by stopping the duct itself.

NM7M has stated that sources of ionization for the E valley are starlight, galactic cosmic rays and solar X-rays scattered by the geocorona. Those are listed in order of increasing strength. Starlight and galactic cosmic rays are obviously not directly related to solar activity, although cosmic rays can be affected by a geomagnetic field stirred up after a solar flare or a blitz from a coronal hole. Solar X-rays scattered by the geocorona, however, do increase with increasing solar activity. So the E valley tends to be lowest at times of low SSN and rises with high SSN. This means that there should be more ducting (and better Top-Band DX propagation) in times of low SSN, which confirms our observations on the air.

2.4.4.6 Chordal hops

Others authors have pictured another very specific way of signal ducting called *chordal-hop propagation*. In this ducting mode waves are guided along the concave bottom of the ionospheric layer acting as a "single-walled dunct." The flat angles of incidence necessary for chordal-hop propagation are possible through refraction in the E layer, and because of tilts in the E layer at both ends of the path. Chordal-hop propagation modes over long distances are estimated by some to account for up to 12 dB of gain due to the omission of the ground-reflection losses. Long-delayed echoes, or "around-the-world echoes" witnessed by amateurs on frequencies as low as 80 meters can only be explained by propagation mechanisms excluding intermediate ground reflections.

2.4.4.7. High wave angles at sunrise/sunset

Hams generally accept the notion that low radiation angles are required for DX work on the low bands. Those who can choose between a low-angle antenna and a high-angle radiator (a low dipole, for example) confirm that 95 to 99% of the time, the low-angle antenna is the better one. But there are the occasions where the low dipole will be the winner. This only seems to happen during the gray-line period (dusk or dawn) though, and more specifically, after sunrise. W4ZV wrote: "*I very often saw post-sunrise conditions favor the inverted-V, even though pre-sunrise almost always favored the vertical. Usually the vertical was about 10 dB better before sunrise, then they would both be equal at exact sunrise, then the inverted-V would be 10 dB better. I believe the post-sunrise peak is high-angle for the following reason. Beverages and verticals are both low-angle antennas and are therefore very complementary. After sunrise, in addition to the vertical being down on transmit, the Beverages would become poor for DX stations (but still good for local USA which was probably still low-angle). I remember when I first worked YBØARA (near Jakarta before he moved to /9 in Irian Jaya) well after sunrise. He was perfectly readable on the inverted-V but was inaudible on any Beverage.*"

In the 1960s, Stew Perry, W1BB, speculated that at sunrise and sunset the ionosphere acts like a big wall behind the receiving or transmitting location, and it focuses the weak 160-meter signals like a giant, poorly reflective dish on one area at a time just ahead of the densely ionized region in sunlight. This seems like an acceptable explanation, since it appears that losses at low-incident (grazing) angles are very high near the LUF (Lowest Usable Frequency) of a path. This may also explain why very often at sunrise and at sunset high-angle antennas seem to perform better than low-angle antennas.

Another now generally accepted way of explaining the fact that high-angle antennas often have the edge at sunrise/

sunset is that a high-angle signal on its way to the F-layer can punch right through the absorbing E layer. A lower-angle signal spends too much time passing through the D and the E layer and hence it suffers increased absorption. In other words, over the same distance, a two-hop, high-angle signal can be considerably stronger than a single-hop, low-angle signal due to the effects of the D and E layers.

During the very successful XZØA expedition in 1999 by far the best receiving results at sunrise (working into the USA) were obtained using a horizontal dipole only 6 meters high. Low-angle Beverages were much worse than this low dipole. Over 400 QSOs were made into North America with this receiving antenna, and that was *not* at a sunspot minimum!

Yuri Blanarovich (K3BU) testified that during one of his recent Top Band operations from VE1ZZ's QTH: "*I had an inverted V at 70 ft and 4 square vertical array, and I was able to crack the 'one-way afternoon' Europeans with the inverted V almost two hours before the 4 square was heard. These verticals have ocean of radials under them and are sitting at the ocean shore on a small hill.*" This means the VE1ZZ vertical array can produce low radiation angles, which evidently was not what was needed under those circumstances.

G3PQA notes that his dipole (high-angle antenna) always outperforms his low Beverage (which has very little off-the-side, high-angle radiation) when working ZLs on 80 meters on long path at equinox, when there is a daylight gap and when conditions peak after sunrise. This also seems to confirm that a high angle is required to pierce through the D and E layers to get into some sort of a ducting mechanism.

2.4.4.8. Antipodal focusing

Most low-band DXers know that it is relatively easy to work into regions near the antipodes (points directly opposite your QTH on the globe). This is despite the fact that those are the longest distances you can encounter—You would expect weak signals as a result. The phenomenon of ray focusing in near-antipodal regions explains the high field strengths encountered at those long distances. This is in addition to the gray-line phenomenon. Antipodal focusing is based on the fact that all great circles passing through a given QTH intersect at the antipode of that QTH. Therefore, radio waves radiated by an antenna in a range of azimuthal directions and propagating around the earth along great-circle paths are being focused at the antipodal point. Exact focusing can occur only under ideal conditions—that is, if the refracting properties of the ionosphere are ideal and perfectly homogeneous all over the globe. Since these ideal conditions do not exist (patchy clouds, MUF variation, etc), antipodal focusing will exist only over a limited range of propagation paths (great-circle directions) at a given time.

The smaller the section of the azimuthal shell involved in the focusing (ie, the narrower the beamwidth), the closer the actual properties will approximate ideal conditions. To gain maximum benefit from antipodal focusing, the optimum azimuth (yielding the lowest attenuation and the lowest noise level) has to be known.

Fixed, highly directive antennas (fixed on the geographical great-circle direction) may not be ideal, since the optimum azimuth is changing all the time (winter vs equinox, vs summer). Rotatable or switchable arrays are the ideal answer, but omnidirectional antennas or antennas with a wide forward pattern perform very well for paths near the antipodes (by summing all paths, as with "diversity" antennas). The focusing gain can be as high as 30 dB at the antipodes, and will range up to 15 dB at distances a few thousand kilometers away from the exact antipode.

While the effect on 80 meters seems to spread quite a distance from the theoretical antipode, on 160 meters the focusing appears to be more localized. On 160, the Gs can benefit from this effect into ZL, while on the European continent the effect seems to be all but non-existent. This is very different from 80 meters. Since the G-ZL antipodal-focusing example obviously coincides with gray-line propagation, and since the gray-line period is much more restricted in time on 160 compared to 80 meters, it should be clear that the focusing phenomenon applies to a much narrower region on Top Band. W4ZV, then WØZV, witnessed from firsthand experience this on 160 from Colorado. He was only about 300 km from the antipode to FT5ZB, whose 80 watts to an inverted-L was consistently and amazingly strong.

3. PROPAGATION VS LOCATION

Working the low bands is very different, depending on whether you live near the arctic regions or near the equator. Let's analyze what causes these differences. Previously, I have referred a number of times to the geographical location of the station. There is a close relationship between the time and the location when considering the influence of solar activity. Location is the determining factor in five different aspects of low-band propagation:

1. Latitude of your station vs rate of sunrise/sunset
2. Magnetic disturbances
3. Local atmospheric noise (QRN)
4. Effects caused by the electron gyrofrequency
5. Polarization and power coupling

3.1. Latitude of Your Location vs Solar Activity

This aspect has already been dealt with in detail in Section 2.3. The latitude of the QTH will influence the MUF, the best season for a particular path and the width of the gray-line zone.

3.2. Magnetic Disturbances (Aurora)

In his book *Aurora Australis* F. R. Bond wrote: "*The aurora (Southern and Northern Lights) is mankind's only visible marker of the interactions taking place in the vast and complicated region of the Earth's magnetosphere.*" Brown, NM7M, stated: "*… and Top Band Propagation is another aspect of those interactions.*" (Ref 140).

Auroral absorption, most often evidenced by the aurora at high latitudes, is a very important factor in the long-distance propagation mechanism on the low bands. It is certainly the most important one for those living at geomagnetic latitudes of 60° or more, as well as for all of us living in more southerly regions when we are trying to work stations on paths that cross areas affected by the aurora. We are interested in what effect this phenomenon (which we hams mostly refer to simply as *aurora*) has on low-band radio propagation.

3.2.1. Auroral absorption

Auroral absorption (AA) is very frequent and takes place due to the influx of auroral electrons. The ionization density of the affected areas in the ionosphere is very high and absorption of signals on 1.8 MHz can exceed 35 dB. Auroral absorption is relatively brief in duration, occurring during the times of visible auroral displays. Absorption regions tend to be elongated in longitude and narrow in latitude, just like the aurora display itself.

AA events are always accompanied by geomagnetic activity due to ionospheric current systems. Hence, the interest in the records of auroral-zone magnetometers for predicting times of low magnetic activity (or conversely, periods of high auroral absorption).

3.2.2. Coronal mass ejections (CMEs)

Sporadic outbursts of plasma, called *Coronal Mass Ejections* (CMEs), represent the release of considerable matter/mass from the sun's corona. They are the sources of blasts of solar wind that can disrupt the geomagnetic field, giving rise to auroral ionization and shutting down propagation on the low bands.

Only the plasma from a CME that goes out of the sun in the direction of the Earth may possibly hit the geomagnetic field and cause a magnetic disturbance. CMEs off the backside of the sun do not bother us, since they represent material ejected into space in directions that never can result in an encounter with the Earth.

3.2.3. Aurora

The plasma coming from the solar corona is called *interplanetary* plasma. *Magnetospheric* plasma is plasma that is trapped within the Earth's magnetic field. The solar wind consists mainly of protons and electrons. Magnetic activity here on Earth results from the impact of the solar wind on the magnetosphere.

The solar wind blowing by the Earth's magnetic field acts like a gigantic dynamo, where huge electrical currents are generated. This energy is often pent up in the Earth's magnetosphere. At times the energy is violently released, accelerating electrons in the tail regions of the Earth's magnetosphere. These electrons, since they are charged particles, are constrained to follow the magnetic field lines of the Earth. And since many of these field lines penetrate the Earth in the high-latitude regions, these electrons end up with trajectories that take them into the high-latitude ionosphere, where they collide with constituent particles and ionize the lower regions of the ionosphere. This process also releases photons of light, which we see as auroral activity. The increased electron density and disturbed ionization patterns contribute to increases in auroral absorption and can cause signals to begin experiencing multipathing and fading.

So far as low-band propagation is concerned, the auroral belt at a height of approximately 65 miles (100 km) acts much like the D layer does during the day—it absorbs all low-band signals trying to go through the belt. Sustained periods of low auroral activity appear to be most common during the rising phase of the solar cycle. **Fig 1-15** shows the relation between the solar flux and the A index over a typical solar cycle.

The auroral belt is centered around the magnetic poles. The magnetic North Pole lies about 11° south of the geographic North Pole and 71° west of Greenwich. The magnetic South Pole is situated 12° north of the geographic South Pole and 111° east of Greenwich. The intensity of the aurora determines the diameter, the width and the ionization level of

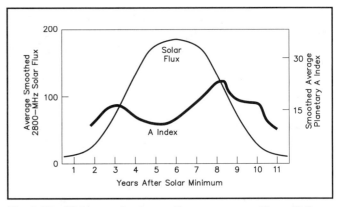

Fig 1-15—This graph shows the geo-magnetic activity (measured A-index) as a function of the solar cycle. It appears that the geomagnetic activity is lowest during the upswing of the solar cycle.

the auroral belt. At very low activity the auroral oval retracts to a major-axis dimension of approximately 3500 km, with a belt width or only a few hundred km. During a very heavy aurora the belt can grow to a major-axis dimension of more than 8000 km, with a belt width of more than 3000 km. Ionization in the auroral oval is usually not constant all the way around. The ionization is—as a rule—minimum at the local noon meridian and maximum at local midnight. Of course, local noon is of very little interest to us low banders since our signals typically propagate only in darkness.

The Earth rotates around the axis going through the geographic poles, while the auroral oval is centered around the magnetic poles. This means that the position of the often irregular-shaped oval changes position continuously with respect to the Earth rotating underneath it.

3.2.4. Effects caused by the auroral oval on low-band propagation

The aurora belts (also called aurora ovals or even auroral donuts) have a profound impact on propagation. If the low-band path over which you are communicating goes along or through the auroral oval, the result is usually degraded propagation caused by strong absorption of the signal. On the higher bands (20 meters and up) fast selective fading (multipathing) is a common sign of aurora. I have very seldom heard this on Top Band, and only infrequently on 80 meters, where these episodes always seem to be of short duration.

During exceptionally quiet geomagnetic conditions (K indices of zero for at least 8 hours), the auroral zone might shrink to a major-axis of about 40% as compared to when geomagnetic conditions are heavily disturbed, while its width might be reduced to a few hundred kilometers. The ionization levels in the shrunken oval can be extremely low during extended fully quiet conditions. Under such circumstances most polar paths will either pass along this small and almost undisturbed area and signals will suffer hardly any degradation.

During disturbed conditions, however, the auroral oval can very rapidly grow to an average size of some 8,000 km. Under such conditions all path that cross or touch this extended oval will be affected by severe absorption in the D and E regions and by other instabilities of the auroral ionosphere.

Cary Oler and Ted Cohen (Ref 142) state that when the auroral zone is contracted, it is possible for Top-Band signals to pass through the auroral zone without suffering heavy absorption by skirting underneath the auroral oval. During periods of very quiet geomagnetic activity, the width (not the diameter!) of the auroral belt is only a few hundred km. On the other hand, radio signals reflected from the E layer can travel over distances of 500 to 2,000 km through the stratosphere and atmosphere, on their way from or to Earth for a propagation hop. This means that with proper geometry low-band signals can literally skip underneath and through the auroral zone into the polar ionosphere inside the auroral belt, where the ionosphere is more stable. They then continue from the polar ionosphere back into the ionosphere at latitudes below the auroral belt, without ever coming in contact with lossy region of the belt itself.

I have often found that propagation into or through polar regions favors the use of low-angle antennas much more than propagation into or across equatorial zones. I suppose this is so because low-angle hopping has more chances to skip underneath the auroral oval, hence suffering less attenuation than would be the case with higher-angle hopping.

Besides "undershooting" the auroral doughnuts, a common way for stations outside the oval to deal with aurora is to launch their signals in a non-great-circle route, called a "bent path" or a "crooked path" away from the auroral oval. A signal launched directly at the auroral oval will bend away from it because the enhanced ionization in the auroral zone creates horizontal ionization gradients. These horizontal gradients refract signals in the horizontal plane. (See Section 4.3.2.)

For stations inside the auroral belt, propagation to the world outside the belt is all but impossible once a geomagnetic disturbance has set in. It has been reported though that stations inside the aurora oval can hear quite well, but they do not seem to get out at all. For example, VY1JA experienced much frustration during the 1998 November Sweepstakes contest hearing strong stations that couldn't hear him. I suspect that the launch angle to skip under the oval was not right at VY1JA's end.

For stations just outside the aurora belt near the North Pole, usually the only (marginal) opening is directly to the south. Frequently, these stations enjoy better propagation towards the equator than stations 1000 or 2000 km further south when the aurora is on.

On at least one occasion on 160 meters, I experienced propagation conditions similar to those on VHF during an extremely heavy aurora. Around 1600 to 1800 UTC on Feb 8, 1986, I heard and worked KL7 and KH6 stations on 80 meters, at the same time that auroral reflection was very predominant on VHF and 28 MHz. From Europe, this was on a path straight across the North Pole and the signals had the buzzy sound typical for auroral reflection. This seemed to indicate to me that under exceptional conditions (the aurora was extremely intense), aurora can be beneficial to low-band DXing. This particular aurora generated an A index of 238. K-index values were reported between 8 and 9. This was one of the largest geomagnetic storms since 1960. A similar situation existed in Jan 1987, when in Europe we could work KL7 stations during several days on 160 meters.

Will, DJ7AA, recently reported a similar happening (Feb 18, 1998): *"Around 0130 I heard K1UO with a very big signal out from nothing working a SP3 station. I went on 1835 and one CQ brought me a huge pile with really big signals banging in here, even from call areas like W5 or W0. I wonder what NA0Y was running, I think he was the loudest ever heard W0 here in my place. Interesting, all signals having a little flutter on it sounds like aurora, and they all coming in over my Beverage to South America, about 3-4 S-units stronger than on my big 500-m Beverage to 320°.... At 0300 the band died. When checking the NOAA home page I saw a very big auroral zone over the Northern Hemisphere at this time, while WWV said: Major Storm..."*

There is almost always a temporary enhancement of conditions right after a sudden rise in the K index. Except for polar paths, however, it seems that a low K index doesn't help for most 160-meter propagation.

Enhanced propagation conditions shortly after a major aurora appear quite regularly. I witnessed a striking example on 80 meters in Nov 12, 1986, only nine hours after a major disturbance. N7AU produced S9 signals via the long path for more than 30 minutes, just before sunset in Belgium. Normally, long-path openings occur to the US West Coast from Belgium only between the middle of December and the middle of January, and even then the openings are extremely rare this far west in Europe. During the November opening, I heard N7UA calling CQ EUROPE with signals between S6 and S9 for almost an hour. The propagation was very selective, since only Belgian stations were returning his calls! A few days earlier DJ4AX was heard working the West Coast and giving 57 reports while the W6/W7 stations were completely inaudible in Belgium, only 200 miles to the Northwest.

I presume that an ionospheric-ducting phenomenon was responsible for such propagation. This means that very specific launching conditions had to be present at both sides of the path. It appears that duct "exit" conditions are very critical and thus area selective—more so for longer path lengths. It also seems that auroral disturbances can occasionally create and enhance such critical conditions.

3.2.5 A and K indexes

The most common way to quantify the level of geomagnetic activity is through the A and K indices. (Ref 158.)

3.2.5.1. The local K-index

The K index indicates the magnitude of irregular variations in the magnetic field over a 3-hour period. This index is calculated from the actual measured value at each observatory station. There are a number of these observatories worldwide. Since magnetic-field measurements vary greatly depending on location, the raw measurements are normalized to produce a K index specific to each observatory.

The K-index scale is quasi-logarithmic, increasing as the geomagnetic field becomes more disturbed. K indices range in value from 0 to 9 (0 = dead quiet, to 9 = extremely disturbed). The K index that we often monitor on radio station WWV is an index derived from magnetometer measurements made at the Table Mountain Observatory located just north of Boulder, Colorado, and hence is referred to as the "Boulder K index." Every 3 hours new K indices are determined and the broadcasts are updated. (See also Section 3.2.7.1.)

3.2.5.2. The local A index

The underlying concept of the A-index is to provide a longer-term picture of geomagnetic activity using measurements averaged over some time frame. The A-index is the

Table 1-2

A_p index	Corresponding K_p
0-2	0
3-5	1
6-10	2
11-20	3
21-35	4
36-61	6
62-102	6
103-166	7
167-268	8
>269	9

Table 1-3

Category	A index range
0-7	Quiet
8-15	Unsettled
16-29	Active
30-49	Minor Storm
50-99	Major Storm
100-400	Severe storm

mathematical average of the *a indices* over the last 24 hours.

The overall A index is an averaged quantitative measure of geomagnetic activity derived from the 3-hour K-index measurements. For each 3-hour K index, a conversion is made to the A index using a conversion table. (See **Table 1-2**.) The A index is the average of the last 8 A indices.

A indices are always linked to a specific day. Therefore, estimated A indices are issued during the day itself. For example, the Boulder A-index (in the WWV announcement) is the 24-hour A index derived from the eight 3-hour K indices recorded at Boulder. The first estimate of the Boulder A index is at 1800 UTC. This estimate is made using the six observed Boulder K indices available at that time (0000 to 1800 UTC) and the best-available prediction for the remaining two K indices. At 2100 UTC, the next observed Boulder K index is measured and the estimated A index is reevaluated and updated if necessary. At 0000 UTC, the eighth and last Boulder K index is measured and the actual Boulder A index is produced. For the 0000 UTC announcement and all subsequent announcements the word "estimated" is dropped and the actual Boulder A index is stated.

A and a indices range in value from 0 to 400 and are derived from K indices based on the table of equivalents. Both A and K indices (for Boulder, CO) are broadcasted by WWV (on 2.5, 5, 10, 15 and 20 MHz) every hour at 18 minutes past the hour or on the web here: **www.sec.noaa.gov/ftpdir/latest/wwv.txt**.

3.2.5.3. Geomagnetic activity terms in English instead of numbers

As an overall assessment of natural variations in the geomagnetic field, six standard English terms are used in reporting geomagnetic activity. The terminology is based on the estimated A index for the 24-hour period directly preceding the time the broadcast was last updated. These are listed in **Table 1-3**.

3.2.5.4. Planetary A and K indices

The Geophysical Institute in Goetteningen, Germany averages the data from 12 observatories (10 in the Northern Hemisphere and 2 in the Southern Hemisphere) to give planetary values, A_p and K_p (the subscript p stands for Planetary).

Table 1-4 shows an example of K, A, K_p and A_p indices from a Boulder report from Jun 24, 1997. It lists the Daily Geomagnetic Data from Fredricksburg, VA; College, Alaska and the Estimated Planetary values from NOAA. You will see differences between the observations (at Fredricksburg and College) and the Estimated A_p and K_p values.

The K values are listed for 3-hour intervals. Both A and K indexes are available from various sources on the Internet (see Section 3.2.7.1).

3.2.5.5. Converting K values to auroral oval average size

For the radio amateur it is important to assess the size and width of the aurora oval to be able to evaluate (on a map or on a globe) whether a given path will touch or pass through the auroral oval. Simplifying somewhat, we can say that the oval is a circular ring, of which the statistical equivalent average radius (at midnight) is given in **Table 1-5**.

If you only have K values, these data allow you to manufacture oval disks of various diameters, which can be used as overlays on maps or on a globe, to help visualize possible crossings of great circle paths with the auroral oval. Of course the oval is a statistical description and does not describe how the ionization is distributed or how energetic it may be. In other words, the local intensity of the aurora is not the same in all points of the oval and at all times of the day.

3.2.6. Viewing the aurora from the satellites

The only source of really reliable information is to use

Table 1-5

K index	Oval Average Radius (km)	Oval Average Width (km)
0	1800	500
1	2050	800
2	2300	1100
3	2550	1400
4	2800	1700
5	3050	2000
6	3300	2300
7	3550	2600
8	3800	2900
9	4050	3200

Table 1-4

Date		Fredricksburg Local		College, AK Local			Estimated Planetary	
June	A	K indices	A	K indices		Ap	Kp indices	
16	8	1-2-1-2-2-3-3-1	3	1-0-0-3-0-1-1-0		5	1-1-0-2-2-2-3-1	
17	6	1-2-2-1-1-2-2-2	1	0-1-1-0-0-0-1-0		5	0-2-2-1-1-2-2-2	
18	3	1-1-1-1-1-1-1-1	0	0-0-0-0-0-0-0-0		4	0-1-1-1-1-2-1-2	
19	11	2-2-4-2-2-2-2-3	6	1-2-3-3-2-0-0-1		10	3-2-4-3-2-2-2-2	
20	6	2-1-2-2-2-2-1-2	2	0-1-2-0-2-0-0-0		5	2-1-1-1-2-2-2-1-2	
21	2	0-0-0-0-0-1-1-2	2	0-0-0-3-0-0-0-0		3	0-0-0-1-1-2-1-1	
22	15	2-4-3-3-3-3-3-2	4	1-2-3-1-1-0-1-0		9	1-3-3-2-3-2-2-2

real-time maps, which now are available from various sources on the Internet. Today we have satellites that produce almost real-time pictures. These can give us much more information than what we've had before. Views of both North and South Poles and the auroral oval are available at: **www.sel.noaa.gov/pmap/**. These are updated when the NOAA Polar-Orbiting Operational Environmental Satellite (POES) satellite passes by about every hour. The satellite maps out the auroral zone for that pass.

The POES images are based on particle-sensor readings the spacecraft makes as it passes over the polar regions. Instruments on board continually monitor the power flux of the protons and electrons that could produce aurora in the atmosphere. These readings are valid only for those longitudes where the spacecraft passes overhead. The readings may be considerably different at other positions along the auroral oval. This is why SEC must examine the results of 100,000 other polar passes in order to form a statistical picture of what is most likely happening elsewhere. This means that what the maps actually shows is based on data obtained from previous passes. It is not a real-time picture, but a combination of real-time data and best-fit extrapolations taken from a huge database.

Fig 1-16A shows a typical POES-generated polar view during a very quiet geomagnetic spell. The black line shows the orbit of the satellite making the measurements, and the dots on either side represent the measurements done in the direction of the stacked black dots. The arrow in the upper-right quadrant shows the local noon meridian, where the width of the oval is usually smallest. Note that the local noon meridian is not related to propagation on the low bands, since propagation at local noon is impossible anyhow due to D-layer absorption. Fig 1-16A shows that the total power in the Northern Hemisphere during quiet geomagnetic conditions is 2.3 GW (that's Gigawatts = billions of watts), and what NOAA calls the auroral "Activity level" is 1, with a good confidence factor n = 0.85. (When the confidence factor approaches n = 2, NOAA is indicating that their statistical model is considered inadequate for this particular time frame because the satellite is not covering an area sufficiently well to make accurate maps.)

Fig 1-16B shows a short history of the Planetary K (K_p)

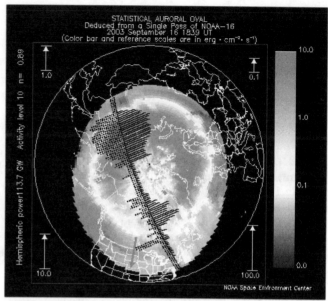

Fig 1-17—North Polar view generated by the POES satellite, during magnetically upset period (K=5-6). Very dark areas inside the auroral doughnut are areas of high ionization, while the lighter tones outside the doughnut show much less ionization. The width of the oval is generally smaller at the local noon meridian. The particular view is for 1839 UTC on Sep 16, the day after the view in Fig 16A. Note that the auroral zone is at its widest around local midnight (across the USA) while the activity is minimal around local noon. The red arrow shows the local noon meridian. (*Source: NOAA Web page.*)

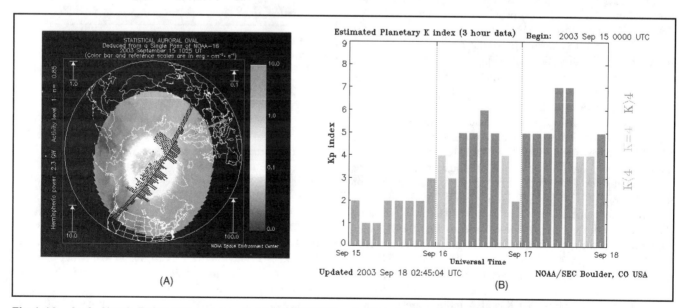

Fig 1-16—At A, North Polar view generated by the POES satellite, during a magnetically quiet period (K_p=1-2). Inside the bright, light-shaded auroral zone "doughnut" dark areas are areas of high ionization. Darker areas outside the doughnut indicate lower levels. The black lines show the orbit of the satellite making the measurements, and the dots on either side represent the measurements done in the directions of the stacked white dots. At B, interplanetary K_p for time period in A. (*Source: NOAA Web pages.*)

Table 1-6

Power (Gigawatt)	K_p index	NOAA Aurora Activity Index
0-2	0	1
2-4	1-	2
4-6	1	3
6-10	2-	4
10-16	2	5
16-24	2+	6
24-39	3	7
39-61	3+	8
61-96	4	9
>96	5	10
>200	8	
>500	9	

indices (at **www.sel.noaa.gov/ftpmenu/plots/2003_plots/kp.html**) over the same period of days shown in Fig 1-16A and **Fig 1-17**. The K_p rose to 6 on Sep 16, 2003, and peaked at 7 on Sep 17, 2003, indicating a magnetic storm was in progress—and indicating that aurora should be possible. Fig 1-17 shows the POES-generated polar view for Sep 17, 2003, during that magnetically upset period. Here, the total Northern Hemisphere power rose to 113.7 GW, with an Activity level of 10. The auroral oval did indeed intensify greatly and did spread to lower latitudes, especially across Northern Europe and Northern Asia.

There are many other pictures available taken from various spacecrafts. **Table 1-6** lists the conversion from Total Hemisphere power, as reported by the NOAA, to the more familiar planetary K_p values and to the latest NOAA Aurora Activity Index.

3.2.7. Putting it into practice
3.2.7.1. Getting geomagnetic data and using them

Both A and K indexes (for Boulder, CO) are broadcast by WWV (on 2.5, 5, 10, 15 and 20 MHz) every hour at 18 minutes past the hour. All DX-clusters provide a command (sh/WWV) that will list the latest WWV numbers. The *IonoProbe* program (See Fig 1-1) also monitors the electromagnetic data relevant to HF radio. The list of parameters monitored includes A_p/K_p indexes and the NOAA POES Aurora Activity parameter on a scale of 1 to 10 (**Fig 1-18**).

You can view a Solar Terrestrial Activity Report, which shows a chart of the solar flux, the sunspot number and the planetary A_p index, on **www.dxlc.com/solar/**. See Fig 1-2.

How do we use this data? Low A and K numbers acquired at stations near the polar regions for a sustained period are prerequisites for good conditions on paths that go near or through these polar regions. It is the K index that is the most important one, since it gives you a more differentiated status than the A index. "Near the poles" means that the Boulder figures are not the most suitable ones! K indices obtained from the observatories in Inuvik, Baker Lake and Cambridge Bay in Canada are ideal because they are located within the aurora belt, when it is active.

We have to realize that K and A indices are measurements derived from what *has already* happened. If these indexes have been zero for 8 hours (or longer) and provided there is no abrupt change, the chances are real that the low bands will be in fair-to-possibly-good shape on polar paths.

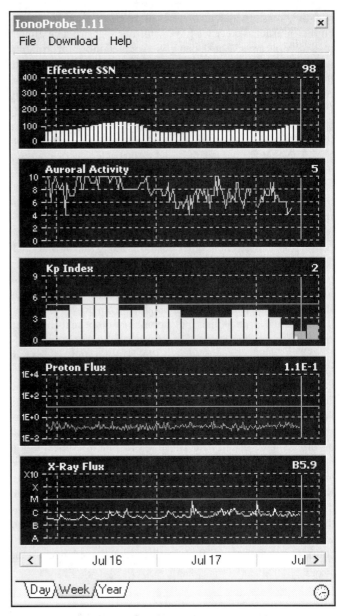

Fig 1-18—*IonoProbe*, which you can have running permanently on your computer, looks for continuous updates on the Internet, and shows you the very latest geomagnetic data.

Because of the sun's 27-day rotation cycle, low geomagnetic activity may be recurrent, especially during the declining and minimum phases of the solar cycle. During the ascending and maximum phases, the recurrent trend often becomes very unreliable (see also Section 2.2). It is a good idea for the serious low-band DXer to make a continuous log of broadcast A and K values. Such logged A indices are particularly interesting to predict the level of magnetic activity in another 27 days. W4ZV keeps a piece of paper marking the distance for 27.5 and 55 days and he uses this "ruler" to quickly calibrate the graph at **www.dxlc.com/solar/**.

I guess most top banders have come to grips with the fact that K and A indices are there to confirm what they have already witnessed—good or bad conditions. N6TR, a well-known Top-Band DXer from the US West Coast, complained: "*I am very skeptical that any of the numbers mean much. I have had good openings with high K numbers, and no openings with longstanding low numbers. About the only thing I can count on is that interesting things seem to happen just as the K starts to rise."*

Since auroral absorption is often initiated by CMEs on the sun, we should be able to predict auroras 2 to 4 days before they hit us (al least for the CME-induced auroras). Before satellite technology was available, we had no detailed information on CMEs and forecasting was based only on the use of recurrence tendencies, extrapolating conditions only from log data of A- and K-values from 27 and 54 days earlier.

You can also subscribe to *Sky & Telescope* magazine's AstroAlert service (actually written by Cary Oler of STD). This gives 24-48 hour notice by e-mail alerts of major CME events. See: **skyandtelescope.com/observing/proamcollab/astroalert/article_332_1.asp**.

3.2.8. More information

Today we have a number of satellites that keep a constant eye on the sun and send a continuous flow data to the Earth, data that is being converted into "readable" reports that are available in abundance on various Web sites on the Internet.

Solar Terrestrial Dispatch has a website (**www.spacew.com/**) where you can find all sorts of information related to amateur radio and radio propagation. Under a special heading "HAM Radio" the following topics are covered on this site.

- MUF map
- Ionospheric X-ray absorption map
- Critical F2-layer frequency map
- Critical E-layer frequency map
- F2-layer max height map
- Daily report on solar activity
- Solar and geophysical indices
- Current 10.7 cm flux
- Latest forecast notes

You can subscribe to a very useful daily summary of auroral activity. This is sent to you by e-mail from **www.spacew.com/www/sublists.html**. These reports forecast magnetic storms based on sun-surface and solar-wind observations, done from satellites. Such reports—together with viewing the NOAA-generated images themselves—are helping low-band DXers to better understand what makes it all tick and to better plan their activities.

Since mid-Mar 1998, Cary Oler from Solar Terrestrial Dispatch has made available a 160-meter Web site, which can be reached at: **www.spacew.com/www/topband.html**. This is called "The Topband radio propagation section" and contains a variety of tools for the 160-meter DXer. In addition to high-latitude K values, which are updated every hour, Oler has created a table where he shows the probability of DX contacts for a large number of polar paths. It also contains the latest auroral-zone pictures and maps (visual and UV) as shown in Section 3.2.6. These predictions are limited in that they only take into account the influence of magnetic disturbances. This is somewhat of a one-way information—When conditions are magnetically disturbed, we know the paths through the polar areas will be dead. But low K-indices, even for a relatively long spell, are no full guarantee that the path will be okay.

There are other mechanisms that enter into the picture on 160 meters and that determine the overall attenuation on a given path. These mechanisms are still largely unknown or, at least, are open to speculation.

Solar Terrestrial Dispatch (STD) also makes available various software packages for those who are vitally interested in the details. *STD Aurora Monitor* is a software package that collects data from a large number of sources and makes those available for the user (real-time and latest data) in a handy format. *STD Aurora Monitor* also includes Ground Based Data (visual and other data from on Earth) as well as a Forecast on geomagnetic activity. You can download a free trial version at: **www.spacew.com/aurora/trial.html**. The program collects:

Fig 1-19—Visible-light image of the Earth. The auroral belt is obviously only visible on the dark side of the Earth. Note the terminator moving across North America. (*Photo courtesy University of Iowa.*)

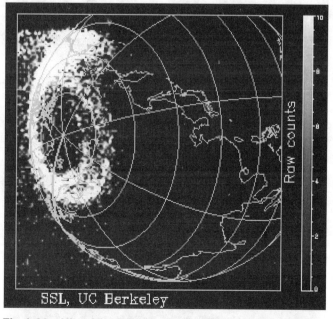

Fig 1-20—Ultra Violet Image (UVI) showing the auroral oval. UV imaging makes it possible to see the aurora in daylight. (*Courtesy STD.*)

- **VIS:** Earth image by the NASA Polar spacecraft's Visible Imaging System. See **Fig 1-19**: **eiger.physics.uiowa.edu/~vis/images/**. Animation images can be seen at: **eiger.physics.uiowa.edu/~vis/images/anims/**.
- **FUV** (Far Ultra Violet) images taken from NASA's Image spacecraft. More information at: **sprg.ssl.berkeley.edu/image/** and **www.sec.noaa.gov/IMAGE/About%20the%20IMAGE%20Spacecraft** and **pluto.space.swri.edu/IMAGE/index.html**.
- **UVI** (Ultra Violet Image): **science.nasa.gov/uvi/default.htm** and **uvisun.msfc.nasa.gov/UVI/LatestImage.html**. See **Fig 1-20**.

If you would like to see a geomagnetic storm from space, have a look at a short movie that you can download at: **pluto.space.swri.edu/IMAGE/wic_197.mpg**.

SWARM (Solar Warning And Real-time Monitor) is another software program developed by Solar Terrestrial Dispatch. Go to **solar.spacew.com/swarm/** for details. *Swarm* monitors everything from geomagnetic and ionospheric conditions to solar activity and solar-wind conditions, all in *real-time*. It is particularly valuable for the prediction of quiet geomagnetic intervals and for the arrival of interplanetary disturbances. *Swarm* also audibly alerts you when geomagnetic activity surpasses certain threshold levels. You will never again be spending valuable time calling CQ on a polar path when the K-index is up to 4 or higher.

Low-band DXers, using *SWARM* can determine precisely when geomagnetic storming is likely to commence. They will also be informed beforehand what is the potential intensity of geomagnetic storming. The software also reports all possible related data, such as solar flux values and sunspot numbers. *SWARM* let you look at real pictures of the aurora (visible and UV). **Figs 1-21** shows a typical *SWARM* scenario.

SWIM (Space Weather Information Monitor) is a state-of-the-art, professional program also created by Solar Terrestrial Dispatch. *SWIM* is a more advanced version of *SWARM*. *SWIM*

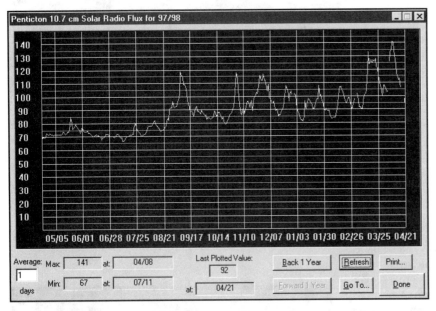

Fig 1-21—One of the many dozens of screens from the *SWARM* program, this one showing the history of the Penticton 10.7 cm solar radio flux. (*Courtesy STD.*)

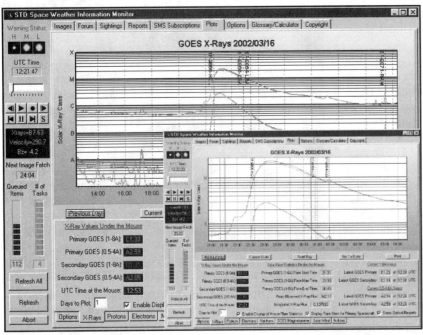

Fig 1-22—One of the many *SWIM* screens, which show all imaginable data related to space weather. Graphs and pictures are automatically updated in "near-real-time," so long as your computer is connected to the Internet. (*Courtesy STD.*)

can monitor, display, animate or print to your printer over 200 space-weather related Internet resources. See **Fig 1-22**. You can expand and manage thousands of additional Internet resources quickly and easily. You simply cut and paste Internet URLs for resources you find interesting and *SWIM* will immediately begin managing those resources for you. It tracks near-real-time geomagnetic A and K indices from as many as 26 global magnetic observatories world-wide. For further info: **solar.spacew.com/swim/**.

SWIM is 100% compatible with the database produced by *SWARM*. The difference is that *SWIM* allows you to manage image resources on the Internet (eg, graphs, maps, anything graphical in GIF, JPEG, or PNG image format). For example, you can use *SWIM* to collect all of the real-time h-alpha solar images, x-ray images, and SOHO images and then review them at your convenience—or animate them like a movie.

SWIM is very sophisticated and requires a Windows NT4, 2000 or XP computer system. Due to limitations in the way Microsoft's Windows 95, 98 and Me operating systems handle memory management, *SWIM* will not function very well, if at all, under these operating systems.

3.2.8.1. Viewing the paths.

If you wish to know whether certain paths will be disturbed, you must visualize the path as well as the auroral oval. These will immediately reveal what's going on for that path for a given auroral intensity. Maps in various projections, as well as globes, can be used. You may have to fabricate your own auroral ovals to use with the map or globe (see Section 3.2.4.).

Nowadays, when every ham has at least one computer in the shack, computer programs do just what we want, with much less hassle. There are numerous propagation-prediction programs around, but only a few address aurora directly. The following four programs all have their own merits and shortcomings, but at least they address geomagnetic conditions: *DX-AID*, *W6ELProp*, *DX-Atlas* and *PROPLAB PRO*. These four programs were extensively used to create figures used in this chapter.

I find *DX-AID* extremely useful for generating great-circle and Mercator-projection maps, including variable-sized auroral ovals based on the K-index. *DX-AID* works well on my computer running *Windows XP Professional*. The program is written by Peter Oldfield. He can be contacted at **poldfield@compuserve.com**. *DX-AID* is not only a mapping program, it also does classic HF-propagation forecasting, which is really of little interest to low banders (except on 40 meters).

W6ELProp, by Sheldon Shallon, W6EL, is another user-friendly program that has some excellent mapping possibilities. (*W6ELProp* is a Windows version of *MiniProp Plus*.) It predicts ionospheric (sky-wave) propagation between any two locations on the earth on frequencies between 3 and 30 MHz. The latest Windows version of the program is freeware and can be downloaded from: **www.qsl.net/w6elprop/**.

PROLAB PRO is a professional ray-tracing program (See **Fig 1-23**). The author, Cary Oler of STD, calls it a "High-Frequency Propagation Laboratory." It is a full-fledged propagation-prediction program that also generates a range of maps. It is probably the most sophisticated and most advanced program available, but it is possibly too professional for the average ham (even a dedicated low bander) because of the very complex user interface. But if you really want to study propagation in fine detail, I can highly recommend this program. It actually does three-dimensional ray tracing that include the effect of the Earth's magnetic field and will predict and plot skewed paths. You can order and download the program from **www.spacew.com/www/proplab.html**.

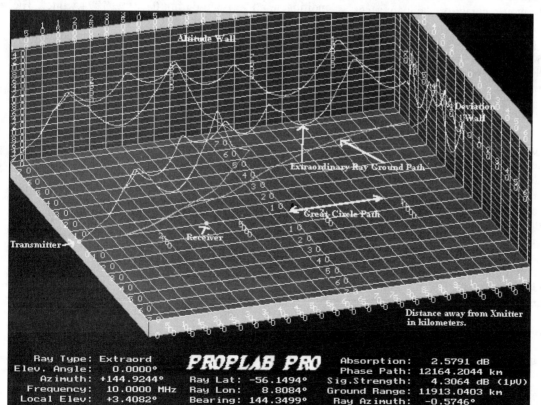

Fig 1-23—3-D ray tracing by *PropLab Pro*. This software even addresses the subject of path skewing, although certainly not all the mechanisms causing this phenomenon are known well enough to be fully described in any present-day modeling program. (*Courtesy STD.*)

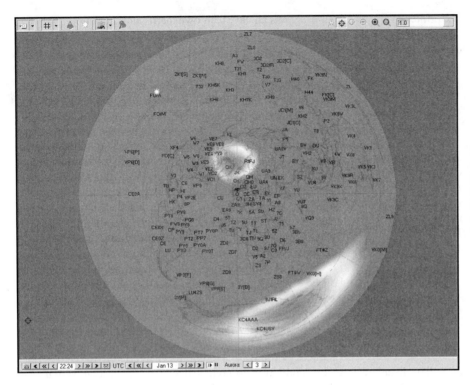

Fig 1-24—Screen shot from the *DX-Atlas* mapping program. Note the aurora ovals on this azimuthal projection. Various other map projections, and a large number of mapping facilities are available. (*Courtesy* DX-Atlas.)

A *PROPLAB PRO Windows* version has been announced for release mid 2003.

DX Atlas is a software mapping program. See **Fig 1-24**. *DX Atlas* displays world maps in rectangular, azimuthal and 3-D globe projections. Overlays are provided for amateur prefixes, CQ and ITU Zones and Grid Squares. For any point selected by the user, latitude, longitude and Grid Square are displayed. The user can select a home location, from which the heading and distance are automatically calculated (both short and long path) to the mouse cursor on the map. Maps can be zoomed in and out, and the gray line can be added to any map display. The gray line automatically reflects the current time and date (as set on the computer), or a fixed time (past or future) can be entered for a specific purpose. Sunrise and sunset times for any point on the map are also shown.

A great variety of ionospheric maps can be called up in *DX Atlas*, such as: MUF (3000) map; F2-layer critical frequency map; F2-layer height; E-layer critical frequency; D-layer peak density; Auroral activity. The following geomagnetic maps can also be displayed: Geomagnetic latitude; Corrected geomagnetic latitude; Magnetic dip; Modified magnetic dip; Magnetic dip latitude. The geomagnetic data input can be obtained automatically from the program *IonoProbe*, written by the same author (see Fig 1-1 and Section 3.2.7.1). You can download a trial version from: **www.dxatlas.com/**.

There are a number of other tools available on the Internet (*GeoClock*, *Geochron*, *DX-Edge*, *DX4WIN*, *VOACAP*), but I have found few that address the issue of the auroral oval, which means they are less-than-ideal for visualizing transpolar propagation paths on the low bands.

3.2.8.2. Correlating geomagnetic data with conditions

Statistical analysis has been done on a representative group of long-haul DX QSOs from the US West Coast on 160 meters for a 2-month period. The occurrences were checked against the K index:

- 62 percent of all QSOs were made on days with a K index of zero
- 30 percent with an index of 1
- Not one QSO with a K index above 3.

In another study, the *Top Band Monitor* did a survey and tried to correlate A-index figures with days of good conditions on Top Band during the 1993/1994 winter. The author tried to link upward swings in A index with good Top Band conditions, and downward swings with bad conditions. My conclusion from studying the data was that only 10% of the good opening on 160 meters were correlated to a downward swing of the A index. (Ref 173.)

KBØMPL, who has a PhD in statistics, did a study on A-indices and 160-meter propagation (Ref 174). She concluded: *"Boulder A-index changes, by themselves, apparently are not related to good or bad propagation days."* She continued adding, *"This does not mean that there is no relationship between good propagation and the A-index."*

Along the same lines Tom Rauch, W8JI, wrote on the Top Band Reflector: *"I've given up totally on watching the A and K indices to estimate how the band is. What I find is generally when ten through twenty meters is good, 160 is poor."* That sounds like a simple and sensible guideline.

But we should not forget that magnetic activity is far from being the only mechanism that rules conditions on the low bands, and more specifically on 160 meters. There are still many unknown mechanisms that make the residual attenuation on 160 meters vary significantly, even when the geomagnetic activity is low.

Sometimes we have weeks of really good conditions (for example at the end of Nov and in Dec 1998) followed by weeks of fairly flat propagation. During both periods there are up and downswings of geomagnetic activity. The low

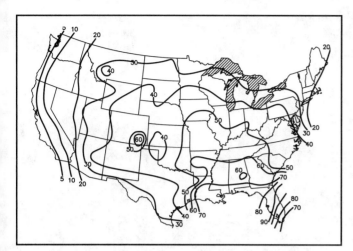

Fig 1-25—This map shows the mean number of thunderstorm days in the US. The figure is related to both mountainous terrain and seasonal weather patterns.

bands—and more particularly Top Band—are still areas where many things still have to be "discovered." That's what makes these bands so interesting and appealing to many!

Trying to assess or predict propagation conditions on the low bands (especially 160 meters) going only by the A and/or the K-index definitely doesn't work. While it is true that high magnetic indices will almost always result in poor propagation for paths near or through the auroral doughnut, paths that don't transit these zones may not suffer at all—and they may even be enhanced. Path attenuation by mechanisms other than aurora is consistently there on Top Band, and scientists have not—so far—been able to unambiguously correlate these mechanisms to any measurable phenomena.

3.3. Local Atmospheric Noise

Most local atmospheric noise (static or QRN) is generated by electrical storms or thunderstorms. We know that during the summer QRN is the major limiting factor in copying weak signals on the low bands, at least for those regions where thunderstorm activities are serious. To give you an idea of the frightening power involved, a thunderstorm has up to 50 times more potential energy than an atomic bomb. There are an estimated 1800 thunderstorms in progress over the Earth's surface at any given time throughout the year. The map in **Fig 1-25** shows the high degree of variation in frequency of thunderstorms in the US. On average there is a lightning strike somewhere on the earth every 10 milliseconds, generating a tremendous amount of radio-frequency energy.

In the Northern Hemisphere above 35° latitude, QRN is almost nonexistent from November until March. In the middle of the summer, when an electrical storm is near, static crashes can produce signals up to 40 dB over S9, and make even local QSOs impossible (and dangerous). In equatorial zones, where electrical storms are very common all year long, QRN is the limiting factor in low-band DXing. This is why we cannot generally speak of an ideal season for DXing into the equatorial zones, since QRN is a good possibility all year long. If you live in the USA check: www.lightningstorm.com/ls2/gpg/lex1/mapdisplay_free.jsp.

The use of highly directive receiving antennas, such as Beverage antennas or small loops, can be helpful to reduce QRN from electrical storms by producing a null in the direction of the storm. Unless directly overhead, electrical storms in general have a fairly sharp directivity pattern.

Rain, hail or snow are often electrically charged and can cause a continuous QRN hash when they come into contact with antennas. Some antennas are more susceptible to this *precipitation noise* than others—Vertical antennas seem to be worst in this respect. Closed-loop antennas generally behave better than open-ended antennas (such as dipoles), while Beverage receiving antennas are almost totally insensitive to precipitation noise.

In very quiet places it is not uncommon for atmospheric noise generated on the other side of the world (often on the other side of the equator) to be propagated just like regular signals and to be heard many thousand miles away. This often shows up as "waves" of noise at the peak time for gray-line propagation between the areas concerned.

3.4. Effects Caused by the Electron Gyrofrequency on Top Band

Modern DXers are aware of some special mechanisms that determine propagation on Top Band. The theory concerning gyrofrequencies on 160 meters is covered in detail in the literature (Ref 142).

The gyrofrequency is a measure of the interaction between an electron in the Earth's atmosphere and the Earth's magnetic field. The closer a transmitted signal is to the gyrofrequency, the more energy is absorbed from the signal by the electron. This is particularly true for radio waves traveling *perpendicular* to the magnetic field. Gyrofrequencies are not influenced by the sun but change with location on the Earth. They vary between 700 and 1600 kHz around the world. A map of the D/E-region electron gyrofrequencies is shown in **Fig 1-26**.

You should remember is that Top-Band signals will be less strongly absorbed and behave more like a conventional signal is expected to behave the farther the frequency is removed from the electron gyrofrequency. Check the map in Fig 1-26 to determine the values of gyrofrequency your signals will encounter for a given path.

Absorption is higher along paths where the signal frequency is closer to the electron gyrofrequency, particularly on paths that are normal to the magnetic field. In other words, north-south paths are less affected than mainly east-west paths, such as from US East Coast to Europe, or the US East Coast to Japan. Similar paths in other parts of the world may not be as sensitive because gyrofrequencies are lower.

If I had to quantify the impact of the gyrofrequency on Top-Band propagation and compare it to the impact of the auroral oval, then I'd say that the auroral oval is the proverbial elephant, while the gyrofrequency is the mere mouse.

3.5. Polarization and Power Coupling on 160 Meters

Power coupling has to do with the way waves, generated by the transmit antenna "couple" into the ionosphere. It appears that the polarization of the antenna plays an important role in achieving optimal coupling (minimum losses). Power coupling is greatest when the E field from an antenna is parallel to the geomagnetic field and the least when the two are perpendicular to each other.

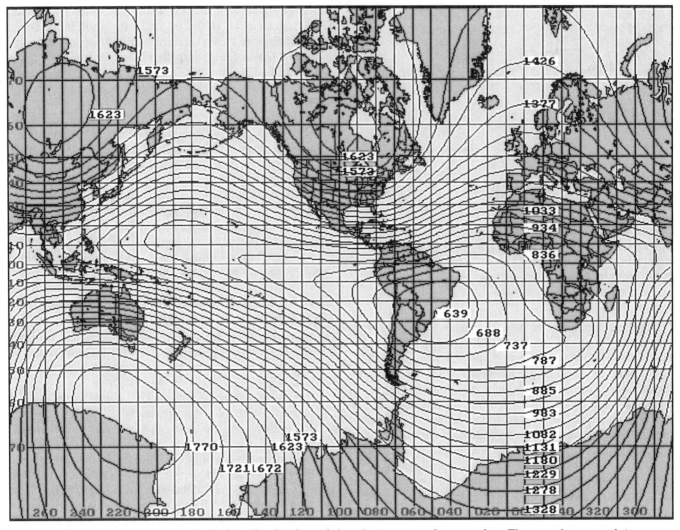

Fig 1-26—This grid shows the worldwide distribution of the electron gyrofrequencies. These values are determined by the Earth's intrinsic magnetic field and are not influenced by the solar cycle. (*Courtesy Cary Oler, STD.*)

In certain areas of the world vertical polarization will produce strongest signals, while in other areas horizontal polarization will. Fortunately, in the US as well as in Europe, vertical is the way to go. This may explain why even 0.5-λ high dipoles do not seem to work well from these regions on 160 meters, while they do fine on 80 meters. There are areas of the world, however, where horizontal polarization on Top Band is the more suitable polarization. This is true for large parts of Asia, Africa and parts of Australia. The geomagnetic latitude of the location is an important factor in this mechanism.

Fig 1-27 shows a Mercator map showing the geomagnetic latitude compared to the geographic latitude and longitude. W8JI, in an e-mail message on the Top-Band reflector, put things in perspective. Losses incurred by TOA (take-off angle) effects (a high-angle antenna vs a low-angle antenna) can be more than 10 dB. Losses incurred if you have very poor ground (in the far field) vs very good ground can be 4 dB. Losses due to improper magnetoionic power coupling can amount to approximately 1 dB.

Most Top Banders use transmitting antennas with vertical polarization, which fortunately seems to be the right

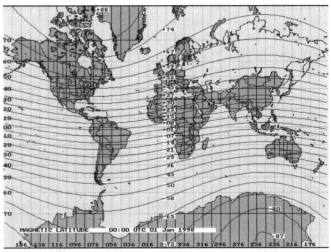

Fig 1-27—Mercator-projection world map showing the geomagnetic latitudes. These are not the same as the geographical latitudes, since the magnetic North and South poles do not coincide with the geographic ones. (*Map generated by* PropLab Pro.)

choice from a power coupling point of view, at least if your QTH is at an average or higher-than-average latitude. It is only stations within 20° of the magnetic equator that may be concerned about power coupling. Even at these low latitudes it is better to have a vertical polarized antenna with a TOA of 25° than a horizontally polarized antenna with a TOA of 90°. This antenna would radiate 10 dB less signal at 25° (typical for a dipole less than 1/2-wave high). With this antenna you may win 1 dB in power coupling but lose 10 dB due to an inappropriate radiation angle!

4. PROPAGATION PATHS

This section discusses the following items to help increase our understanding of low-band propagation paths:

1. Great-circle short path
2. Great-circle long path
3. Particular non-great-circle paths

4.1. Great-Circle Short Path

Great circles are all circles obtained by cutting the globe with any plane going through the center of the Earth. All great circles are 40,000 km long. The equator is a particular great circle, the cutting plane being perpendicular to the Earth's axis. Meridians of longitude are other great circles, passing through both poles.

When we speak about a great-circle map we usually mean an *azimuthal-equidistant projection* map. This map, when covering the entire world, has the unique property of showing the great circles as straight lines, as well as showing distances to any point on the map from the center point. On such a projection, the antipodes of the center location will be represented by the outer circle of the map. Great-circle maps are specific to a particular location. They are most commonly used for determining rotary beam headings for DX work. The advantage of a great-circle map is that headings are straight lines, while the disadvantage is the extreme distortion near the antipodes.

Fig 1-28 shows great-circle maps using *DX-Aid* centered on Boston, Omaha, San Francisco, Brussels, Moscow and Tokyo. There are various sources on the Internet where you can download great-circle maps or programs to make such maps.

4.2. Great-Circle Long Path

A *long-path* condition exists when the station at the eastern side of the path is having *sunset* at approximately the same time as the station at the western end of the path is experiencing *sunrise*. A second, necessary condition is that the propagation occurs on a path that is 180° opposite to the short-path great-circle direction.

We will see further how "crooked-path" propagation can satisfy the first condition, but is not a genuine long-path propagation. One example is the path from Western Europe to Japan at 0745 UTC in midwinter. This involves a path over Northern Siberia and not across South America, as it would be if it were a true long path.

4.2.1. Long path on 40 meters

Long-path QSOs are quite common on 40 meters. From Europe we have a genuine long path to the US West Coast around 1500 to 1600 in midwinter. A very similar long path exists between Japan and Europe around sunrise time in Europe, especially around the equinox. In midwinter, when all the darkness is in the Northern Hemisphere, there still is some long path between Europe and Japan, but there is a generally much-stronger path that is a somewhat-crooked short path across Northern Siberia. In general, the signal direction is about the same as the usual short path direction.

During midwinter, both long and crooked paths exist simultaneously for about 10 or 15 minutes around 0745 UTC (see **Fig 1-29**). This often makes copy very difficult, because of multipath propagation due to the different time delays on each path. During the JA low-band contest in Jan 1998, I had to ask several JA stations to slow down their CW to allow me to copy through the multipath echoes.

4.2.2. Long path on 80 meters

Genuine long paths to areas near the antipodes are very common on 80 meters all through the sunspot cycle, provided there is a full-darkness path and that the long path coincides with areas of lowest attenuation (see also Section 4.3.1.2). Long paths on 80 meters are less common than on 40 meters, except to areas very close to the antipode. Very often paths which we call "long paths" are crooked or bent paths, somewhere between long and short paths (see Section 4.3.).

4.2.3. Long path on 160 meters

Genuine long-path QSOs on 160 meters are almost always to places near the antipodes. Such long-path QSOs are very rare during the sunspot-maxima years. During the low sunspot years I can hear G stations working ZLs long path on 160 meters approximately 30 minutes after my sunrise, but only on very rare occasions have I been able to work ZL on long path myself. Other near-antipode long-path QSOs have been made between VK6HD (Perth) and the US East Coast in midwinter (eg, the QSO between K1ZM and VK6HD at 2115Z on Jan 27, 1985).

Real long-path QSOs (long path that show no path skewing) on 160 meters only seem to occur during a period centered around the one or two years at the minimum of the sunspot cycle. W0ZV remembers a few genuine long-path QSOs made from Colorado; for example, with UA9UCO and JJ1VKL/4S7. Another one that made history was between PY1RO and several JA stations at JA sunrise. Other long-path contacts were made between US East Coast stations and well-known calls, such as 9M2AX, VK6HD and VS6DO.

Many of the often called long-path 160-meter QSOs are really skewed long paths, and they happen at all stages of the sunspot cycle. Examples are the early 2003 QSOs between JT1CO and many US-East Coast stations. See **users.vnet.net/btippett/dx_aid_plots.htm**.

During the 1987-1988 winter, my first winter on 160 meters, I tried for weeks to make a long-path QSO with N7UA, but we never heard signals at either end. During Dec 1992, I ran a daily test with FK8CP on the long path (his sunset is within minutes of my sunrise), but we never made a QSO either (see also Section 4.3.1.4).

During Dec 1997 N7UA, with whom I had many long-path tests back in 1987/1988, made numerous so-called long path QSOs into Eastern and Northern Europe as well as into the UK around 1510 UTC. But were these genuine long-path

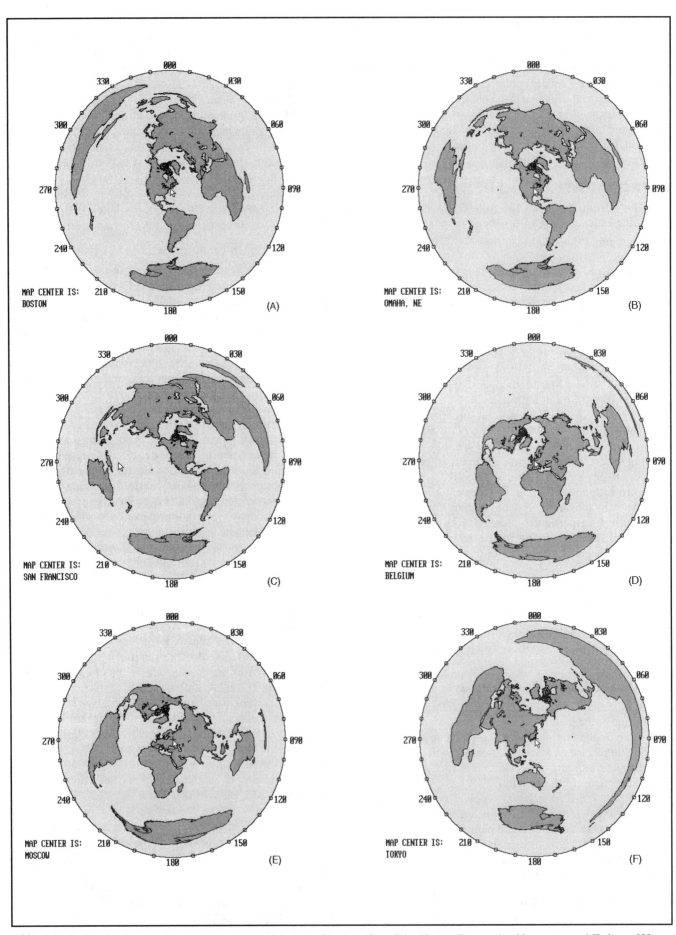

Fig 1-28—Azimuthal projections centered on Boston, Omaha, San Francisco, Brussels, Moscow and Tokyo. (*Maps generated by* DX-AID.)

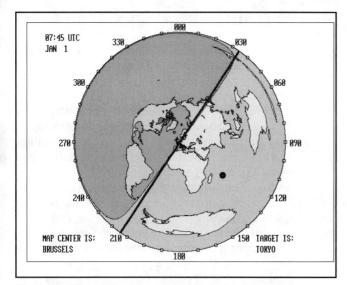

Fig 1-29—The great-circle path for midwinter (0745 UTC) shows the short and the long path that exist simultaneously on 40m from Belgium to Japan. Both run along the terminator, a typical situation for the higher bands, but very uncommon for 80 meters and impossible on 160 meters. (*Maps generated with* DX-AID, *with additions by ON4UN.*)

QSOs? Let's have a look at non-great-circle paths, also called crooked or bent paths.

4.3. Crooked (Skew) Paths

Most propagation paths over relatively short distances on 40, 80 and 160 meters are great-circle paths. We do know, however, that signals often come from anything but great-circle directions. So let's distinguish two categories for the path bending we often observe:

1. **Bending caused by aurora:** This case is the classic one. During periods of high geomagnetic activity (aurora), we identify signals coming from headings off great-circle directions, when the great-circle path would otherwise have to go through the auroral oval. We have all witnessed repeatedly how signals seem to be bending around the aurora belt. In Europe we work US West Coast stations beaming to central or South America under such circumstances.

2. **Bending not caused by aurora:** A second type of bent path on 80 meters (and especially on 160 meters) was witnessed by many operators on the US East Coast in Jan 1997. During the first few days of the VK0IR operation they remember well how the signals peaked right across Europe (60°) instead of on the direct path, which is about 110°. This could not have been a case of seemingly bending *away* from the aurora oval. Rather, it almost was like the signals were being *attracted* to it! During the VK0IR operation geomagnetic conditions were generally very quiet. The reason for this kind of bent path must be different, since there is no aurora involved.

Tom, W8JI, suggested on the Top Band reflector: "*Skew paths are actually fairly common, and don't seem to be tied to anything unusual going on if the path is long. So it seems to me signals simply come from the direction of least absorption. And by the way, there always is the same skew on transmit as there is on receive.*" This lines up perfectly with what Thomas, KN4LF, wrote: "*160 meter propagated signals are always going to travel along the path of least absorptive resistance.*" Sounds very logical, doesn't it? The question here is: "What causes here such path bending?"

4.3.1. The non-heterogeneous ionosphere.

4.3.1.1. The mechanism for deviation from great-circle paths

It is generally accepted that there are only three ways that signals propagate through the ionosphere:

- By *refraction* caused by ionization gradients
- By *reflection* caused by auroral ionization
- By *scattering* of signals by ionospheric or atmospheric irregularities, as well as irregular surfaces (ground or water).

The general mechanism that causes signals to deviate from the great-circle path is the presence of horizontal ionization gradients in the ionosphere. Signals traveling into a layer with a higher degree of ionization will be refracted or reflected away from the gradient. Very steep gradients can be caused by aurora, for example. When there is low geomagnetic activity, however, scattering in the ionosphere itself (or at ground-reflection points) could be another mechanism that can cause path skewing.

The ionosphere is not a perfect mirror, but should rather be thought of as a cloudy and patchy region, with different areas of ionization. Tom, W8JI, stated: "*There is more scattering and skewing going on than most of us ever know about, probably because it isn't a shiny smooth mirror up above.*"

We often visualize a radio wave as a single ray sent in a specific direction, refracted in the ionosphere (which we think of as a perfectly shiny mirror) and reflected from a perfectly flat reflecting surface on the Earth. HF energy, however, in most practical cases is being radiated in a range of azimuths (particularly for a vertical transmitting antenna) and over a range of elevation angles. Some signal is thus taking off in the "wrong" direction (that is, not in the great-circle direction towards our target) and may change course en route by any of the mechanisms described above and yet still arrive at the target!

Part of our signal, of course, actually does take off in the "correct" great-circle direction, but it might encounter the auroral belt and be totally absorbed there. Even if it isn't completely absorbed, it may be reflected or scattered there— And who knows what direction it may end up taking? When and if some signal does reach the destination, there is usually one path (straight, bent or whatever) where the received signal is substantially stronger than those received on other paths. Thus we are mainly aware of the most successful path.

Sometimes we hear signals coming from various directions at the same time. Tom, W8JI, wrote: "*Many times the JAs are SW, and many days the JA signals arrive from multiple directions. When K1ZM and AA1K hear JAs from the NW, I'm hearing them better from the SW.*" This clearly demonstrates that signals from JAs don't travel on just one path. They are propagated in many directions and are received in different places from different directions. The mechanisms behind all of this are very complex ones, and aurora is but just one, but important, cause of path bending.

Reception from multiple propagation paths normally does not cause any problems, since the difference in propagation delays usually is quite small (1 to 10 ms). Sometimes, however, the delays are of an order of magnitude that cannot be explained by a *slightly* bent path. Tom, W8JI, wrote: "*I can hear K9DX, when he is beaming NW, scattering in from the SW with $1/4$ second to $1/2$-second delays on the echo. (Between John's TX antenna and my RX antenna there is probably a 60 dB null on the direct path).*" Such a long-delayed echo must obviously involve some other mechanism than minor path skewing.

The path direction may vary from day to day, and even on a given day may switch continuously and at a very rapid rate. Mike, VK6HD, observed in the Low-Band reflector: "*Last night I had 6 QSOs with NA between 1134 and 1155 Z. With the first one I thought there was very strong QSB, but then I checked my Beverages and I found that when the signal went down on the NE beverage it came up on the SE one, and vice-versa. This switching was happening about every 20 seconds.*"

Clearly, many paths on the low bands are not simple great-circle paths. Testimonies in this respect are overwhelming. W8JI worded it as follows: "*The only nearly 100% agreement you will see is people with directive antennas who do a lot of listening over long periods of time all agree that not much ever comes in through the magnetic pole areas, and that paths on lower bands are not predictable.*"

As to the exact why and how, those questions remain largely unanswered. Cary Oler and Ted Cohen (Ref 142) point out that "*Weak sporadic-E clouds, that might not affect the higher frequencies, can achieve a substantial impact on 160 meter signals by increasing absorption or refracting signals.*" Such sporadic-E clouds can induce a waveguide-like hop between the F-layer and the sporadic-E cloud. They are also considered as a possible cause for skewed paths. There still is a lot of discussion ongoing in scientific circles about the mechanisms that trigger path skewing. Here are some regular, well-documented skew paths.

4.3.1.2. The classic skewed-path example: ZL propagation from Europe

New Zealand is about 19,000 km on the short path from my QTH in Belgium, or about 21,000 km on the long path, very close to being the antipode (see Fig 1-28). From Belgium the short-path heading to New Zealand is 25° to 75° and the long-path heading between 205° and 255°. When I work ZLs on 80 meters on long path during the Northern Hemisphere winter, signals always arrive via North America, at a heading of approximately 300°. This is 90° off the great-circle long-path direction. The path is not a great-circle path, but is inclined as if the signal were trying to leave the Southern Hemisphere as fast as possible (both the ZLs and the Europeans beam across North America in the winter).

As we continue into spring, the optimum path between Western Europe and New Zealand moves from across North America to across Central America (Feb-Mar). Eventually, beaming across South America will yield the best signals later in the year (from Apr onward). Somewhere around the Spring Equinox all three paths produce equally good signals, when the strongest strong signal strengths are heard.

Theory says that there are an indefinite number of great-circle paths to the antipode. Since low-band DX signals travel only over the dark side of the globe, however, the usable number of great-circle headings is limited to 180° (assuming there is no auroral activity screening off part of the aperture). This very seldom means that signals will arrive with equal strength over 180°, not to mention with the proper phase. The relatively short differences in path lengths cause time-delay differences too short to be able to noticed by ear. The strongest signals are received from the direction where the attenuation is least. This confirms W8JI's comment that: "*Signals simply come from the direction of least absorption.*"

These New Zealand-to-Western Europe QSOs are well-documented examples of gray-line propagation, but none of these propagation paths ever coincide with the terminator itself. The actual path *happens* to be more-or-less perpendicular to the terminator at all times of the year (see also Section 2.4.4)!

To summarize, on 80 meters I have observed for over 40 years that long paths and paths to areas near the antipodes are skewed in such a way that the signals will apparently travel the longest possible distance in the hemisphere where it is winter. This is judging from the direction of arrival of these signals.

Similar long-path QSOs between western Europe and ZL are possible on 160 meters during the bottom of the sunspot cycle. But here also, the signals do *not* arrive via the genuine long-path direction (205 to 255°) but from a direction at right angles with this heading (right over North America).

4.3.1.3. South America across North America in Northern-Hemisphere winter

A similar path bending is also quite common on 80 meters over shorter paths. During the European winter, signals from Argentina and Chile regularly arrive in Europe at beam headings pointed directly at North America, up to 90° off the expected great-circle azimuth. The signals from South America appear to travel straight north in order to "escape" the summer conditions in the Southern Hemisphere, and are then propagated toward Europe. One striking example was when I worked 3Y1EE (Peter 1st) on 80 meters (Jan 28, 1987). The signals were totally inaudible from the great-circle direction (190°) but were solid Q5 from 310° (signals coming across North America). Similarly, when I worked CEØY/SMØAGD on 160 meters (Oct 1992), signals were only readable on a Beverage beaming 290°, while the great-circle direction to Easter Island is approximately 250°.

4.3.1.4. The skewed path between Europe and the US West Coast

Early on, some people believed that long path on 80 meters between the US West-Coast and Europe followed the gray-line terminator. We now know that what we commonly call a long-path QSO on the low bands actually involves signals transmitted in a direction different from that directly opposite to the short-path direction. The path cannot actually follow the terminator, since absorption inside the gray line is high. We can safely say that low-band signals never actually propagate far *inside* the gray line. (See also Section 2.4.4.1).

Nowadays, everyone acknowledges the existence of crooked (bent or skewed) paths. This means there is not only a short and a true long path, but any number of *alternative*

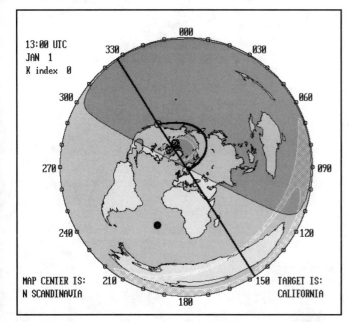

Fig 1-30—The 80-meter long path between Europe and the US West Coast is neither a short path (the line running through the auroral belt), nor a genuine long path (20,000 km in daylight). Instead, it is a crooked path (the curved darker path). In Europe signals generally arrive at headings of 70° to 110° from True North. (*Map generated by* DX-AID, *with additions by ON4UN.*)

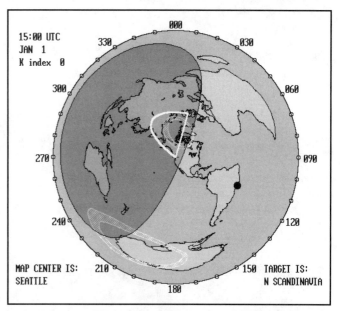

Fig 1-31—Great-circle map centered on Seattle, showing the 160-meter path for contacts into Northern Scandinavia around 1500 UTC in midwinter. The actual path is a crooked one (white curved line). At both ends of the path, the direction is perpendicular to the nearby terminator. Propagation along the terminator in the twilight zones is impossible because of the additional D-layer absorption. (*Map generated by* DX-AID, *with additions by ON4UN.*)

paths that may be available for propagation. The so-called long-path on 80 meters between Scandinavia and Eastern Europe to the US West Coast is an example of such a crooked path. Looking at the darkness distribution on Earth for this in **Fig 1-30** and **Fig 1-31**, it is clear that a genuine (reciprocal) long path is out of the question, since signals would have to travel for nearly 20,000 km in daylight (Africa, South Indian Ocean, Antarctica and South Pacific) around 1300 UTC, or along the entire path in the twilight zone around 1600 UTC.

In Europe the beam headings generally indicate an optimum azimuth angle of approximately 90°, which again is almost perpendicular to the terminator (see Section 2.4.4.). Along their way, the signals will be least attenuated in those areas of the ionosphere where the MUF is lowest, and thus they seem to travel along a crooked path, avoiding areas of higher absorption. OZ8BV reports a 90° to 100° heading when working the West Coast on 80-meter long path from southern Denmark. Ben is using a 3-element Yagi at 54 m (180 feet) and is well placed to confirm this path (the genuine long path would be 150 to 160°).

D. Riggs, N7AM, is using a rotary quad for 80 meters and he wrote: "*We have learned that the 80-meter long path between the Pacific Northwest and Scandinavia is following the LUF (lowest usable frequency). I have always believed that the long path to Europe was not across the equator but leaves us at 240° and since the MUF is highest at the equator it cannot continue at 240° but it bends westerly going under the Hawaiian islands, across the Philippines under Japan and across the Asian continent to Scandinavia. The MUF charts prove this fact. The fact that the long path to Europe lies north of the equator is proven by the northern Europeans and after the West Coast peak.*"

So-called long-path QSOs have been made on 160 meters between the northern part of the West Coast of the USA (N7UA) and Northern Europe (Scandinavia and the UK). Neil, G4DBN, was one of the lucky ones to have done that from the UK, and he wrote: "*Bob N7UA and I had a QSO at around 1505 Z on 29 Dec 1997 and he was only audible on my North-West rx antenna.*" That does not look like a typical azimuthal direction for long path, which should be approximately 150°. This QSO happened on a crooked path, almost but *not* following the gray line. It was outside the gray line at some distance from it because the additional D and E-layer attenuation inside the gray line would make propagation over such distance impossible.

Similar QSOs from Scandinavia are probably easier than from anywhere else in Europe based on much 80-meter long-path experience, where it happens every day, all through the sunspot cycle, for quite a few months in wintertime. On 160 meters, long-path QSOs between the US West Coast and either OH or SM have not been really commonplace, but have occurred a number of times. These occur during both high and low sunspot cycle years.

During the 1997-1998 seasons a number of QSOs were made between the US West Coast and Scandinavian stations on 160 meters. SM4CAN, SM4HCM and SM3CVM all confirmed that the signals were coming from due East (90°), instead of the short path (335°) or long path (145°). N7UA quoted that he was using his JA Beverage, since the SM stations were not audible on the over-the-pole European Beverage. SM4HCM called it "*a skewed path, somewhere between long path and short path.*" (See **Fig 1-32**.) N7UA added that "*...the lower the frequency, the more the long path moves towards North, away from the true reciprocal head-*

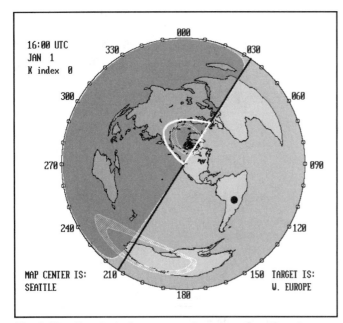

Fig 1-32—Great-circle map centered on Seattle, showing the midwinter so-called long path to Western Europe (light curved line) around 1600 UTC. This is clearly a crooked path that skirts the auroral oval. The real long path, the path directly along the terminator, is open on 40 meters, however. (*Map generated by DX-AID, with additions by ON4UN.*)

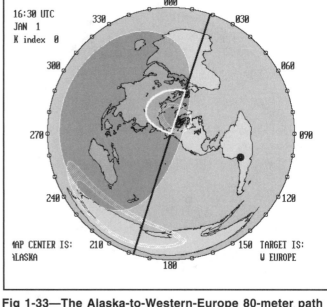

Fig 1-33—The Alaska-to-Western-Europe 80-meter path around 1630 UTC in midwinter. The real long path is totally impossible because it travels for 25,000 km in daylight. The genuine short path travels in the twilight zone; hence there is a high degree of absorption. The most likely path takes off from Alaska in a westerly direction and arrives in Europe East of the short-path bearing. Note again that these two paths are perpendicular to the terminator. (*Map generated by DX-AID, with additions by ON4UN.*)

ing." This observations seem to be related to the principle that the lower the frequency, the more easily a horizontal ionization gradient can cause path skewing (see also Section 4.3.1.1. and Ref 147).

The historic QSO made during the winter of 1999 between N7UA and 5B4ADA is an interesting case. At the time of the QSO, 5B4ADA's sunset was 1436 UTC and N7UA's sunrise 1552 UTC (76 minutes of common darkness). Notice that the QSO (at 1510 UTC) was almost halfway between those times. It seems to be quite common for the path to peak near the mid-point between Europe/Asia sunset and North America sunrise, and it seems to confirm the observation that for best signal-launching conditions the path direction must be at right angles to the terminator. This keeps the signal as far as possible away from the lossy gray-line zone (see also Sections 2.4.4 and 4.3.1.). Similar true long-path QSOs were also made between 4X4NJ and the US West Coast in that same period.

Bill, W4ZV, reports a number of so-called long-path QSOs on 160 (with JT, UA9, S2, XU, XZ, 3W5, 4S7 and 9V1). All occurred with a common darkness path varying between 59 and 109 minutes. Bill also remarks that QSOs over the short path or over near-polar regions seem to be best during low sunspot years, while the so-called long path seems to peak up in higher sunspot years. Scientists owe us an explanation for that remarkable and valuable observation.

While 80-meter long-path QSOs between the US West Coast and Europe occur every winter on an almost-daily basis throughout the sunspot cycle, short-path QSOs (at USA sunrise) only occur when the following conditions are met:

- Near sunspot-minimum years when absorption is minimum.
- When the geomagnetic field is extremely quiet (Ap < 5).
- Centered on the period of maximum Winter Solstice darkness (Dec 21).
- Most common to stations located in the northern part of Europe.

It's clear that the short path suffers from the aurora oval, while the so-called long path stays clear of it!

4.3.1.5. The skewed path between Europe and Alaska

Besides the short path between Europe and KL7, we Europeans can often work Alaska on a non-genuine long path before their sunrise and just after our sunset on 80 meters. At that time (around 1600 UTC in midwinter) signals usually arrive in Europe from the East/North East. This is clearly a bent path across Siberia to Alaska, thus avoiding the auroral belt (**Fig 1-33**), and is certainly *not* a true long path. If the path were a genuine long path, it would go right across the South Pole, which is in continuous daylight in the Northern Hemisphere winter. KL7Y noted the signals come in from the South West, approximately 45° from the genuine long path.

But when geomagnetic condition have been quiet for a long period, it sometimes *is* possible for signals to travel through the heart of the auroral zones. I remember an amazing QSO in the 1970s with KL7U on 80 meters at about 1600 UTC, hearing him only when listening at 350°, which is the direct short path right across the magnetic North Pole (the straight-line short path in Fig 1-33). Going only by the time of the contact, this would usually be called a long-path QSO;

however, it was not, since the signals did not come in from the true long-path direction (approximately 160°) but almost from the regular short-path direction. It is obvious that this can happen only when there is no auroral absorption at all, since this short path goes right across the magnetic North Pole. This path is the equivalent of the JA-Europe path described in Section 4.3.3.1.

4.3.1.6. The Heard Island case

The VKØIR example has been covered in detail previously in Section 2.4.4.2. From the viewpoint of path skewing, the path is very similar to the one between Europe and New Zealand (Section 4.3.1.2.). For several days the signal appeared to be traveling through areas of low MUF (although we know that 160-meter propagation is not directly MUF related). Therefore the path from Heard Island to the US East Coast seems to travel as much as possible through the Northern Hemisphere, avoiding higher MUF areas in the South. The bent path also meets the most advantageous launching conditions where the path direction (at both ends of the path) is perpendicular to the terminator at those points (see Section 2.4.4.1.). This explains why these signals were received on the US East Coast via a bent path across Europe in Jan 1997. (See Fig 1-13C and D.).

4.3.2. Skewed paths avoiding the auroral zones

The second reason for path deviation is to avoid the auroral oval. When there is aurora in the Northern Hemisphere, signals on the low bands—if not totally attenuated—will often appear to arrive from a more Southerly direction than you would expect from great-circle considerations. The path between North America and Europe is greatly affected by this phenomenon, because the magnetic North Pole lies right on that path. Between Japan and Europe there is much less influence.

The aurora was described in detail in Section 3.2. Let us analyze a few paths that suffer frequently from the effects of aurora. To view these paths, get your globe, maps or switch on your mapping program that can show the auroral oval. With *DX-AID* (see also Section 3.2.8.1) you can plot the great-circle paths, enter various values for the K-index and then watch the evil effects of the auroral oval.

The short path between the West Coast of the US and Western Europe has always been a difficult path, because of the interference of the auroral oval with the great-circle direct path. For this reason, the short path is very rare between the West Coast and Northern Scandinavia, which is inside the aurora oval.

Fig 1-34A shows the path between Western Europe and Seattle (in midwinter) at sunrise in Europe, for a K index of 0. The auroral oval has retracted to its minimum size and width. Although the great-circle path goes through the auroral oval, attenuation may be minimal, either because the signal underskirts the ionosphere at the narrow oval or because of slight bending around the oval. Fig 1-34B shows the same for a K-index of 9 (heavily disturbed magnetic conditions). Note the extreme width and major-axis size of the oval. Under such conditions, since Seattle actually lies on the border of the oval, there may be little escape from the aurora. As a rule, stations further south (eg, Southern California stations) may possibly make it into Europe, beaming across South America.

I remember one striking case like this where I made a

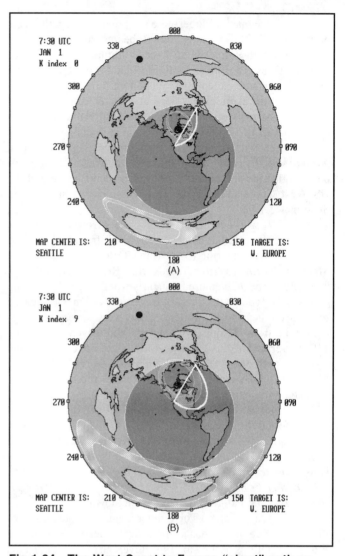

Fig 1-34—The West-Coast-to-Europe "short" path on 80 meters. At A, we have minimum geomagnetic activity (K=0) and the signals either travel the genuine direct path, or maybe very slightly bent southward. At B, the situation for a K index of 9 (major disturbance). Note the extent of the auroral zone. The only way to make a QSO into Europe is for stations at both ends of the path to beam towards South America. (*Map generated by DX-AID, with additions by ON4UN.*)

good QSO with W6RJ on 80 meters where Bob was using a KLM Yagi beaming to South America, while I was using my South-American Beverage to copy him. The direct path, across the aurora oval, was totally dead at that time. Obviously the attenuation on this crooked path is greater than on a straight path when there is no auroral absorption, so only stations with good antennas and some power may regularly experience this path.

Even when the auroral activity is very low, Bob finds that the short path between Northern California and western Europe (a great-circle path of 25°) hardly ever peaks that far North, and he confirms that 75% of the time the path goes right *towards* South America, and the rest of the time peaks at around 40°.

An even more striking example of path skewing due to aurora is the path between Europe and Alaska on 80 meters.

Propagation 1-35

Looking at the map or globe, there is a great-circle path that is only about 7,500-km long, but it beams right across the magnetic North Pole. The distance is similar to the distance between Western Europe and Florida. Straight short-path openings are rare exceptions, happening only a few times a year, when the K index has been at zero for some period of time. The main difference between the Seattle and the Alaska case it that from Seattle it take much more path bending to go around the oval. Because of its more northerly location, Alaska actually lies *inside* the auroral oval. There is, of course, also the so-called long path between Europe and the US West Coast and Alaska. This was covered in detail in Section 4.3.1.5.

W4ZV, who operated for many years from Colorado as WØZV, points out that he never saw skew from Colorado to JA, since the JA bearing (315° from CO) is nowhere close to Magnetic North (13° from True North). He also added that he worked Europe *much* more frequently from Colorado peaking on his 70° Beverage than his 40° Beverage, which was the true great-circle bearing from there. Under very severe geomagnetic conditions (K index = 6), signals would even peak on his 110° Beverage! W4ZV, now in North Carolina, confirms that a similar path exists from the US East Coast to the Far East. Signals from JA quite frequently skew to the south during geomagnetic disturbances (see also Ref 159).

Similar experiences are told by N5JA (Texas), who said that he found the best direction for UU4JMG changing from 40° at 0145 UTC, to 60° at 0215 UTC, to 90° at 0230 UTC and eventually to 120°. He reported shifts going the other way as well: *"I've seen ON4UN in years past become first audible on the 90°, then be best on the 60°, and later best on the 40° or 20° Beverages."* (What goes up must come down!)

KØHA, a top-notch Top Bander from the "black hole" of the USA (Nebraska), has had similar experiences copying European stations best on his 140° Beverage rather than on his 43° or 86° antenna.

Another striking example occurred on 160 meters during the first night of the CQ-160-M contest in Jan 1991: With the exception of VE1ZZ, not one North-American station was heard until 0400 UTC. At that time North-American stations started coming through rather faintly, but they were only audible when beaming to South America (240°). No signals from the *usual* 290° to 320° direction! Between 0400 and 0700 UTC, 80 W/VE stations were worked in 25 states/provinces. All of the signals came through across South America, including K6RK in California. On the North-American Beverage only a few of those stations would have been worked.

The W4s in the Southern part of the East Coast normally have a tough time piercing through the New-England wall of signals trying to get into Europe, except when the aurora is on, says N4UK. That's the only time he can beat the W1s and W2s into Europe on a path skewed to the South. He thinks the skewed path is a great equalizer. It explains that when the aurora is hitting, he can still *bend around* it on his Southerly skewed path, while the W1s and W2s are too close for skewing.

Steve, VE6WZ, who lives right below the auroral belt, says there is not much skewing around the belt on 160 meters from his place. *"On 80 meters, I have noticed that in the last 3 years every 80-meter contact from VE6 into Europe has been skewed, peaking to the E-SE (not the direct path NE)."* When he copied Wolf, DF2PY, on 160 meters he was definitely best on the SE Beverage (South America) and very poor on the N-NE Beverage (Europe).

Bill, W4ZV, remarks: *"The longer the path, and/or the closer to crossing near the magnetic poles, the more often they skew. It's common here to hear JAs on two paths, one west or southwest and the normal northwest path. Often there is multipath as well, a fairly common phenomenon with JAs arriving from both 210 and 330 degrees."*

It is clear that all these skewed paths are caused by magnetic disturbances, as they all happened during periods of fairly high geomagnetic activity. In general, you can say that path skewing on relatively short paths (< 10,000 km) is quite common during disturbed conditions. Skewing can be anything from 30° to almost 90° further south of the normal great-circle direction.

For a station at high or mid-geomagnetic latitudes there are only two ways on 160 meters to make a QSO into an area that is completely hidden behind the auroral doughnut. Either you wait for the day when auroral absorption is very low or non-existent (this happens a couple of days during each 11-year cycle) and work the station right across the magnetic pole. Or you find your way around the problem. That is, you work the station on a crooked path going around the auroral oval. Under such circumstances signals often appear to be coming long path (the direction the signals arrive from are close to opposite from the short-path great-circle direction) and likely travel a long distance on a crooked path. By traveling *around* the auroral doughnut they avoid auroral absorption, but by traveling a very much longer distance (than their direct great-circle path) they suffer considerable additional attenuation.

Along the same lines W4ZV wrote: *"This can be done even in high flux years and even under relatively high geomagnetic activity. Otherwise, we must wait for the very bottom of the cycle and only on days with Kp of 0 or at most 1 to work through the auroral oval to the other side. Taken over an entire solar cycle, these days are rare indeed that contacts may be made by short path through the auroral oval."*

On several occasions these crooked (long) paths have been reported to be quite unstable, and quite abruptly switch directions (over 90°), especially right at or after sunrise. Make sure to use all your Beverages and keep twisting that selector knob.

4.3.3. Crooked polar paths in midwinter, or pseudo long paths

4.3.3.1. Europe to Japan

In Northern Europe we can work Japan in midwinter, just after our sunrise (0745 UTC) on 80 meters. The most common opening to Japan on 80 is at JA sunrise, around 2200 UTC. At first you might be tempted to call the 0745 UTC opening a long-path opening. Careful analysis using directive receiving antennas has shown that the signals come from a direction slightly east of north rather than their true long-path heading of 210°. The opening is rather short at my QTH (typically 15 to a maximum of 30 minutes in midwinter during low sunspot cycle years and almost non existent during high sunspot cycle years). **Fig 1-35** shows the great-circle path *along* the terminator, along the gray-line zone, and is *not* a typical example of gray-line propagation, where low-band signals usually travel more or less perpendicular to the terminator.

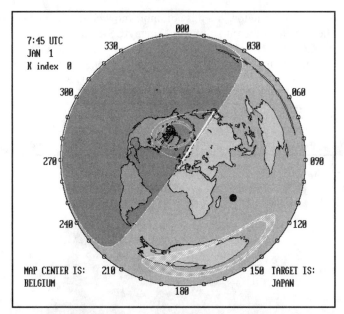

Fig 1-35—The direct short path between Europe and Japan at Europe sunrise in midwinter is an example of quite exceptional propagation. Adding the auroral oval to the great-circle map from Fig 1-29, we see that the alternative path (the light curved line) would skirt around the auroral belt. It would be at right angles to the terminator at both path extremes, and because of this consideration would be the most-favored path. In actual practice, however, I never have seen JA signals come out of the northwest along that path at that time of the day. Instead the signals come on the (heavily attenuated) short path from Japan. (*Map generated by DX-AID, with additions by ON4UN.*)

These are certainly far from ideal conditions for low attenuation and the signals on the JA path are substantially weaker than the signals normally heard from Japan from the same heading, but at JA sunrise. Low-band propagation over long distances along the terminator is clearly not the rule, because of additional D-layer absorption (see Section 2.4.4). If present, signals are weaker than normally expected over similar distances on paths that do not follow the terminator.

If we look at the darkness/daylight distribution across the world at that time (0745 UTC in midwinter), we see that we have indeed more than one path possibility: Paths ranging from true short path (30°) to alternative crooked paths bent slightly east or even west of the magnetic North Pole, all across areas in darkness. These alternative bent paths go right through the North Pole auroral zones, and hence will very seldom produce stronger signals than the path along the terminator.

A number of years ago Hoppe and others (Ref 108 and Ref 118) considered that low band gray line propagation went *along* the terminator, as is the case on the higher bands. It has, however, been proven over and over again that this is *not* the rule on the low bands (see Section 2.4.4.1), where the most spectacular propagation enhancements seem to occur when the propagation path are *perpendicular* to the terminator. The Europe-to-Japan short path at 0745 UTC in midwinter is a remarkable exception to the general rule, as well as some of the SSE – SSW crooked path QSOs mentioned by W4ZV in Section 2.4.4.1.

Often such specific propagation paths (more or less along the terminator) are very area selective, probably because ducting phenomena are involved (see Section 2.4.4.3), occurring only when the necessary launching and exiting conditions exist. In several cases I was able to work JA stations with signals up to S9, while the same stations were reported to be undetectable in Germany only a few hundred kilometers away. In all cases the signals were loudest when they were coming in about 10° east of north. In Japan the openings seem to be very area-selective as well, as can be judged from the call areas worked. Northern Japan (JA7 and JA8) obviously leads the opening. Central Japan follows 30 minutes later, and southern Japan often is too late for this kind of opening at my QTH. In midwinter there is a 1½-hour spread in sunset time between northern and southern Japan.

On 40 meters the situation is somewhat similar, and yet very different at the same time. At 0745 UTC in midwinter, JA 40-meter signals arrive from north or northeast (as on 80 meters), but when that path fades about 15 to 30 minutes later,

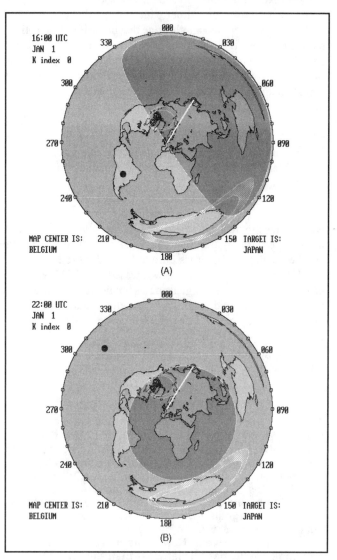

Fig 1-36—The classic path between western Europe and Japan. At A, the situation at sunset in Europe, and at B, for sunrise in Japan. At both times, the path direction is almost perpendicular to the terminator. (*Map generated by DX-AID, with additions by ON4UN.*)

it is immediately replaced by a genuine long path, where the signals now come in across South America, along the terminator as shown in Fig 1-35. I have never observed this genuine long path across South America on 80 meters, let alone 160 meters.

At JA sunrise time at 2200 UTC, the short-path direction is almost at right angles to the terminator (see **Fig 1-36B**). A similar good launching angle (with respect with the terminator) occurs at sunset time in Europe around 1600 UTC (Fig 1-36A). From that point of view the 1600-UTC and the 2200-UTC openings are almost identical. In real life however, we find in Europe the 2200-UTC opening much better. The reason is obvious: At 2200 UTC we have a sunrise peak, while a sunset peak is much less pronounced (if it occurs at all). Further, 1600 UTC is in the middle of the night in Japan.

4.3.3.2. New England to Japan

Brown, NM7M, analyzed a seemingly similar path (Ref 140) on 160 meters: W1 to JA. It is quite different, however, since the auroral zone lays right in the middle of the direct path between New England and Japan. He also found two openings; one he calls a "gray-line" path (the opening around 2140 UTC in midwinter) and another one that he calls a "black-line" path. I consider the former path as a most atypical gray-line path for low frequencies, even though both ends are in twilight (a double gray-line situation). NM7N claims that the few signals heard or worked on such occasions came out of the northeast direction, which is a crooked path across Europe. See **Fig 1-37A**. K1ZM, however, who actually made these QSOs, said the path was from the southeast and not over Europe (both in Dec 1996 and Jan 2000). The morale of this story: Make sure you have receiving antennas covering *all* directions, and keep switching them.

The theoretical great-circle path between W1 and Japan at 2140 UTC (East Coast sunset in midwinter) goes parallel with the terminator, hence suffering severe attenuation due to the D-layer already building up in that zone. The alternative crooked path travels across Europe (according to NM7M) or even further to the southeast, as witnessed by K1ZM. Whether beaming NE or SE, the signals paths are nearly perpendicular to the terminator and hence enjoy excellent launching conditions, since they spend little time in dusk/dawn where D-layer absorption is enhanced.

The great-circle path from New England to Japan in midwinter at 1200 UTC makes an angle of approximately 50° with the terminator, and it is common knowledge that many of these paths are bent southward (many US East coast station claim paths from 270° to 205°), making the angle with the terminator more like 90°. The 1200-UTC opening is a one-sided gray-line signal enhancement (Fig 1-37B), since it occurs at East Coast sunrise while Japan is still fully dark.

QSOs have, of course, been made much earlier (0945

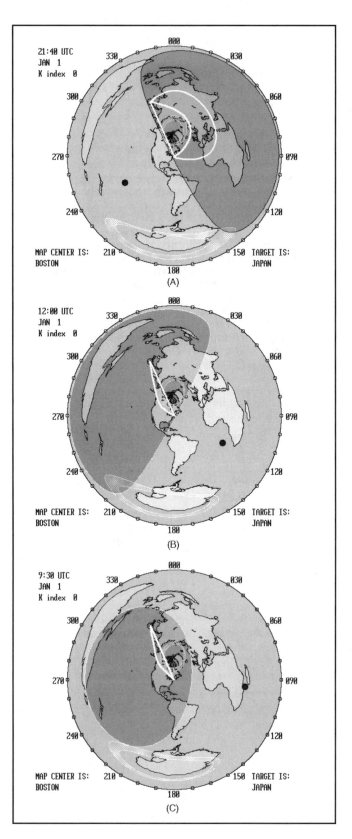

Fig 1-37—The two openings (for both 160 and 80 meters) from New England to Japan. At A, the theoretical 2140-UTC straight-line short path, which goes along the terminator and right through the aurora oval. This is, however, not the way the signals actually travel. N7MN quotes that propagation was across Europe (curved light line), while K1ZM, who made the contacts, received the signals from the southeast (outside curved line). At B, the morning 1200-UTC opening. Actual signals arrive in New England from beam headings somewhat further west than the short path heading in order to skirt the auroral oval. At C, the geometry for an all-night ("black-line") path (shown in white) at 0930 UTC. Even at this time, the beam heading is often bent further south than the direct path. (*Map generated by DX-AID, with additions by ON4UN.*)

UTC). At that time we have a most typical *black-line* path (see Section 2.4.2) occurring at midnight for the point halfway between the path ends (Fig 1-37C). Here too, a slight skewing around the polar regions is very common. At this time of the day there is no signal enhancement by gray-line phenomena, since both ends of the line are well into darkness.

4.3.3.3. Other paths

Similar paths exist on 80 meters in midwinter between California and Central Asia (Mongolia) around 0030 UTC, between Eastern Europe (Moscow) and the northern Pacific (Wake Island) around 0615 UTC, and between the East Coast (and the northern part of the Midwest) of the US and northern Scandinavia around 1230 UTC. All of these polar-region paths are east of the North Pole and should not be influenced by aurora as much as paths going west of it. Use your globe, map or mapping program (*DX-AID* or *DX Atlas*) to visualize the paths that appear to avoid the auroral belt.

4.3.4. Selective paths/areas

I have often wondered how a DX station, or worse yet a DXpedition, can make low-band contacts while hoards of stations keep calling after a QSO has started. We've all seen cases where 80% of the pileup just keep calling (especially in Europe). For an observer, located in the middle of all these callers this is pure chaos.

I am convinced that *selective skip* is the reason why the DX station can work stations despite this seemingly chaos. ZS6EZ, an eminent low bander and DXpeditioner stated: "*On my DXpedition, it was very often like a searchlight, where only one very small area is audible at a time, and that area moves around with time. There is no regular movement either. You might have W1, W8, W9, WØ, W3, W4, W5, WØ, W7, WØ, W9, etc. in a single opening. The pileup is never audible; just a few stations at a time.*" He adds: "*This is why I contend that the argument of the East Coast Wall doesn't hold in most sunrise openings to Africa. If the "spotlight" is on the West Coast, the other coast is not even audible.*"

John, K4TO, commented to me on the same subject: "*I have also observed that given propagation into a particular area, one signal might be ten dB stronger than another, even though both stations were running the same power and comparable antennas. Given a few seconds time lapse, and the situation reverses. The station that was stronger earlier on, is now the weaker of the two. Having observed this first hand, I find that my own anxiety level about working a particular DX station is now reduced to near zero. I feel that sooner or later, the propagation will come to favor me. I wish more DXers would have this same experience. I believe everyone would relax and enjoy the hobby more. Operating manners and techniques would improve greatly.*"

I think that this area-selectivity may also be partly due to polarization rotation of signals on 160 meters. This can also explain the slow QSB that we often witness on Top Band. Sometimes we must wait a few minutes for polarization to "come back" in order to hear a marginal signal again. This is why on Top Band, where weak signals are involved, it is often essential to get a full call at a first try. Often a second try at a partially received call brings dead silence, until a minute or so later the same signal slowly comes out of the noise again.

Signal ducting between the E and F-layers provides another possible explanation for the "searchlight effect." The landfall of the searchlight would relate to finding stations with appropriate launch angles, such that the RF could enter (or exit) the duct.

A most striking example was a QSO I made on Mar 13, 2003, with VP6DIA on 160 meters. During the 15 minutes before my local sunrise, his signals were solid Q5 and I worked him on a single call. I heard nobody else in Europe calling, and I was in touch with a number of East Coast stations on the Internet top-band chat channel. They all confirmed not hearing a beep, while the signals remained solid Q5 all the time here.

The late Dan Robbins, KL7Y, came up with another theory that might help explain selective propagation. He concluded from his work with HF radars that the ionosphere can act as a "filter" for the angle of radiation: "*Near the MUF, high angles are "filtered" out. But below the MUF, low angles may also be filtered out. Once angles are filtered out they are gone, and the same narrow range of angles will propagate over and over again, even if the filtering conditions disappear on later hops. If the range of allowable angles is narrow, the propagation will occur in narrow distance bands from the transmitter, since distance is a function of angle of radiation. This is why one guy works the S9 DX and his buddy 300 miles away hears nothing. This is why the band appears dead, except for that loud 3B8 or whatever.*"

In other words, for any given path there will be an optimum frequency and angle of radiation that produces the maximum signal. If we are restricted to one frequency then there will be one range of optimum radiation angles for that path. According to Dan Robbins the optimum angle may be surprisingly narrow at times—narrow enough to account for almost all instances of selective propagation.

Fortunately, most of the antennas we use on the low bands have rather broad vertical lobes, nothing like the arrays used for OTH radar. This effect of "angle selectivity" is thus largely smoothed by the broad lobe of our transmit antennas.

4.3.5. Path skewing: a summary

On the low bands, and more specifically on 160 meters, we continuously witness bent propagation paths (crooked or skewed paths). At one end of the path (or more likely at both ends of the path), a signal arrives from a direction that is substantially different from the great-circle heading. Documentation of these facts is so overwhelming that there is no doubt about these bent paths.

Robert R Brown, NM7M, wrote on the Top Band reflector: "*In summary, path skewing results from horizontal gradients in electron density. For interpretation, the problem is to locate the gradient and identify its source. Without any gradient or source, reports of skewing are incomplete.*"

I am a radio amateur, not a scientist, and so I take issue with this statement. Most skew-path reports are by hams, not scientists. By looking at the similarity of many reports hams can benefit from them, without necessarily knowing the source of the gradient(s) causing the bending. So, fellow low-band DXers, keep on reporting such odd paths on the Top-Band reflector.

For the low-band operator it is important to know that these bent paths are quite common, both to be able to anticipate them and to make good use of them. Many of the observations are clearly linked with high geomagnetic activity. In these cases signals arrive from directions away from the auroral zones, which seems logical. In this case some sort of horizontal ionization gradients are causing reflection and bending of the path away from the auroral wall.

What is not so clear to me is how waves, bent away from the auroral doughnut (bent southward in the Northern Hemisphere when you are located south of the doughnut), are bent back north in order to *skirt around* the aurora belt. If that were not the case, a target that is *behind* the auroral doughnut would still remain unreachable. In my opinion, such crooked paths must involve two bending mechanisms, one bending away from the aurora zone, and one bending back in northerly direction beyond the auroral doughnut.

In addition to these aurora-related crooked paths there are also many documented cases of signal-path bending on paths that do not go through high-latitude areas where aurora may be present. These are clearly not related to strong geomagnetic activity. A number of witnesses of both cases are listed in Section 2.4.

Many crooked paths cannot be explained by using the present state-of-the-art (ray tracing) computer models. Our present understanding of wave propagation makes it impossible to predict propagation along the terminator for various reasons, such as instability and enhanced D-layer absorption. Since these crooked paths happen in the absence of geomagnetic disturbances, reflection by steep ionization gradients wouldn't seem to be involved.

Wave scattering is another alternative mechanism for skew paths. Signals received from a given direction (determined using a directive Beverage receiving antenna) may not actually travel in that direction over the entire path. The areas of the ionosphere that cause scattering are thought to be near the terminator, where the ionosphere is turbulent around sunrise and sunset. This could be an explanation for some of the crooked paths we enjoy on 160 meters—where electromagnetic disturbances are not involved and where the apparent propagation direction is along the terminator. Many mysteries, however, exist! (Ref 178).

It appears to me that a great number of possible causes and mechanisms are involved on Top Band, making our signals skew and travel along paths that are not always along a great circle. What we all observe—skewed paths—occurring regularly, day after day, is a fact. Explaining the mechanisms behind such skewed paths is the role of the scientists. Once they know all the mechanics behind it, they may even be able to write propagation-prediction software that works, and half of the fun of DXing on 160 meters will be gone! Not having all the answers yet, however, should not stop us from taking advantage of so-called anomalies and enjoying DXing on the low bands.

4.3.6. Path skewing: Concluding remarks

In many of the great-circle maps used in this section, the crooked paths are represented as a nicely curved lines. This is only a symbolic representation of the real paths. What we observe is the azimuth takeoff angles at both ends of the path represented in these maps. The path the signals *actually* travel on their way between the terminal points is probably not along a nicely curved line. If we accept a scatter-reflection mechanism to explain these bent paths, the signal may actually travel in a straight line towards the scattering area (a region with a horizontal ionization gradient of some sort). Often there is a combination of various mechanisms that determine the actual path over its entire path. The real path will likely follow a jagged line rather than a nicely curved one (see Ref 159).

Since DX signals often come in from other directions than the great-circle direction, any serious low-band DXer should have a choice of receiving antennas. You must switch them continuously, searching for the direction that produces the best received signal. Once this direction is established, you should transmit in that same direction as well.

5. TOOLS FOR SUCCESSFUL DXING ON THE LOW BANDS

5.1. Sunrise/Sunset Information

Now that you have read through the foregoing paragraphs, let's get practical. The most important tool is still *sunrise/sunset* information. Today all computer-logging programs include a SR/SS calculator. But I still use the SR/SS tables in a booklet form—It's often faster than a computer program if you want to look up information for a certain date or for a certain period of time.

5.1.1. The ON4UN Sunrise/Sunset tables

The booklet of sunrise/sunset tables I created some years ago shows sunrise and sunset times for over 500 different locations in the world (including 100 different locations in the US) in tabular form. Increments are given per half month. **Fig 1-38** shows an example of a printout for one location.

I have tried all the propagation aids that are described in this book, and many others as well. The only aid I regularly use is the sunrise/sunset tables. Why? You can grab the tables any time and look up the required information in seconds. Just

```
                -W6-CAL-
  USA                   (SAN FRANCISCO)
  ===============================
  37.78 DEG.N.          122.41 DEG.W.

  DATE          SUNRISE        SUNSET
  ----          -------        ------
  JAN  1         15.25         01.01
  JAN 16         15.24         01.15
  FEB  1         15.14         01.33
  FEB 16         14.58         01.49
  MAR  1         14.41         02.03
  MAR 16         14.19         02.17
  APR  1         13.55         02.32
  APR 16         13.33         02.46
  MAY  1         13.14         03.00
  MAY 16         12.59         03.13
  JUN  1         12.49         03.26
  JUN 16         12.47         03.33
  JUL  1         12.51         03.36
  JUL 16         13.00         03.31
  AUG  1         13.13         03.19
  AUG 16         13.25         03.03
  SEP  1         13.39         02.40
  SEP 16         13.52         02.40
  OCT  1         14.05         01.54
  OCT 16         14.19         01.32
  NOV  1         14.35         01.12
  NOV 16         14.51         00.58
  DEC  1         15.06         00.51
  DEC 16         15.19         00.52
```

Fig 1-38—Example printout for one of the more than 500 locations in ON4UN's Sunrise/Sunset tables. Times are given for half-month increments.

```
Location: E003°45', N51°00'                      ON4UN                    Astronomical Applications Dept.
                                        Rise and Set for the Sun for 2003   U. S. Naval Observatory
                                                                             Washington, DC 20392-5420
                                                Universal Time

       Jan.      Feb.      Mar.      Apr.      May       June      July      Aug.      Sept.     Oct.      Nov.      Dec.
Day  Rise Set  Rise Set  Rise Set  Rise Set  Rise Set  Rise Set  Rise Set  Rise Set  Rise Set  Rise Set  Rise Set  Rise Set
      h m h m   h m h m   h m h m   h m h m   h m h m   h m h m   h m h m   h m h m   h m h m   h m h m   h m h m   h m h m
01   0748 1549 0722 1635 0631 1725 0522 1817 0419 1906 0336 1950 0335 2003 0410 1932 0458 1832 0545 1724 0636 1620 0725 1542
02   0748 1550 0721 1637 0629 1727 0520 1819 0418 1908 0336 1951 0335 2002 0411 1930 0459 1829 0546 1722 0638 1618 0727 1541
03   0748 1551 0719 1639 0626 1729 0518 1820 0416 1909 0335 1952 0336 2002 0413 1929 0501 1827 0548 1720 0640 1617 0728 1541
04   0748 1552 0718 1641 0624 1730 0515 1822 0414 1911 0334 1953 0337 2002 0414 1927 0502 1825 0549 1717 0642 1615 0729 1540
05   0748 1553 0716 1642 0622 1732 0513 1824 0412 1912 0334 1954 0337 2001 0416 1925 0504 1823 0551 1715 0643 1613 0731 1540
06   0747 1554 0715 1644 0620 1734 0511 1825 0410 1914 0333 1955 0338 2001 0417 1924 0505 1821 0553 1713 0645 1612 0732 1540
07   0747 1556 0713 1646 0618 1735 0509 1827 0409 1915 0333 1955 0339 2000 0419 1922 0507 1818 0554 1711 0647 1610 0733 1539
08   0747 1557 0711 1648 0616 1737 0507 1828 0407 1917 0332 1956 0340 1959 0420 1920 0508 1816 0556 1709 0648 1608 0734 1539
09   0746 1558 0709 1650 0613 1739 0504 1830 0405 1919 0332 1957 0341 1959 0422 1918 0510 1814 0558 1706 0650 1607 0735 1539
10   0746 1600 0708 1651 0611 1741 0502 1832 0404 1920 0331 1958 0342 1958 0423 1916 0512 1812 0559 1704 0652 1605 0737 1539
11   0745 1601 0706 1653 0609 1742 0500 1833 0402 1922 0331 1959 0343 1957 0425 1915 0513 1809 0601 1702 0654 1604 0738 1538
12   0744 1602 0704 1655 0607 1744 0458 1835 0401 1923 0331 1959 0344 1957 0426 1913 0515 1807 0602 1700 0655 1602 0739 1538
13   0744 1604 0702 1657 0605 1746 0456 1837 0359 1925 0330 2000 0345 1956 0428 1911 0516 1805 0604 1658 0657 1601 0740 1538
14   0743 1605 0701 1659 0602 1747 0454 1838 0357 1926 0330 2000 0346 1955 0430 1909 0518 1803 0606 1656 0659 1559 0740 1538
15   0742 1607 0659 1700 0600 1749 0451 1840 0356 1928 0330 2001 0347 1954 0431 1907 0519 1800 0607 1654 0700 1558 0741 1539
16   0741 1608 0657 1702 0558 1751 0449 1842 0355 1929 0330 2001 0348 1953 0433 1905 0521 1758 0609 1651 0702 1557 0742 1539
17   0740 1610 0655 1704 0556 1752 0447 1843 0353 1931 0330 2002 0350 1952 0434 1903 0522 1756 0611 1649 0704 1556 0743 1539
18   0740 1612 0653 1706 0553 1754 0445 1845 0352 1932 0330 2002 0351 1951 0436 1901 0524 1754 0612 1647 0705 1554 0744 1539
19   0739 1613 0651 1708 0551 1756 0443 1846 0350 1933 0330 2003 0352 1950 0437 1859 0526 1751 0614 1645 0707 1553 0744 1539
20   0738 1615 0649 1709 0549 1757 0441 1848 0349 1935 0330 2003 0353 1949 0439 1857 0527 1749 0616 1643 0709 1552 0745 1540
21   0737 1616 0647 1711 0547 1759 0439 1850 0348 1936 0330 2003 0355 1947 0440 1855 0529 1747 0617 1641 0710 1551 0746 1540
22   0735 1618 0645 1713 0544 1801 0437 1851 0347 1938 0330 2003 0356 1946 0442 1853 0530 1744 0619 1639 0712 1550 0746 1541
23   0734 1620 0643 1715 0542 1802 0435 1853 0345 1939 0331 2003 0357 1945 0444 1851 0532 1742 0621 1637 0713 1549 0747 1541
24   0733 1621 0641 1716 0540 1804 0433 1855 0344 1940 0331 2003 0359 1944 0445 1849 0533 1740 0623 1635 0715 1548 0747 1542
25   0732 1623 0639 1718 0538 1805 0431 1856 0343 1941 0331 2004 0400 1942 0447 1847 0535 1738 0624 1633 0717 1547 0747 1543
26   0731 1625 0637 1720 0535 1807 0429 1858 0342 1943 0332 2004 0401 1941 0448 1845 0537 1735 0626 1631 0718 1546 0748 1543
27   0729 1627 0635 1722 0533 1809 0427 1859 0341 1944 0332 2003 0403 1939 0450 1842 0538 1733 0628 1629 0720 1545 0748 1544
28   0728 1628 0633 1723 0531 1810 0425 1901 0340 1945 0333 2003 0404 1938 0451 1840 0540 1731 0629 1627 0721 1544 0748 1545
29   0727 1630           0529 1812 0423 1903 0339 1946 0333 2003 0406 1936 0453 1838 0541 1729 0631 1626 0723 1543 0748 1546
30   0725 1632           0526 1814 0421 1904 0338 1947 0334 2003 0407 1935 0454 1836 0543 1726 0633 1624 0724 1543 0748 1547
31   0724 1634           0524 1815           0337 1949           0408 1933 0456 1834           0635 1622           0748 1548
```

Fig 1-39—Printout of the Sunrise/Sunset times for a whole year on a day-by-day basis. This is a service provided by the US Naval Observatory.

September 2003
Brussels, Belgium

Sunday	Monday	Tuesday	Wednesday	Thursday	Friday	Saturday
	1 Sun Rise: 6:54am Sun Set: 8:30pm	2 Sun Rise: 6:55am Sun Set: 8:28pm	3 Sun Rise: 6:57am Sun Set: 8:25pm	4 Sun Rise: 6:58am Sun Set: 8:23pm	5 Sun Rise: 7:00am Sun Set: 8:21pm	6 Sun Rise: 7:01am Sun Set: 8:19pm
7 Sun Rise: 7:03am Sun Set: 8:17pm	8 Sun Rise: 7:05am Sun Set: 8:14pm	9 Sun Rise: 7:06am Sun Set: 8:12pm	10 Sun Rise: 7:08am Sun Set: 8:10pm	11 Sun Rise: 7:09am Sun Set: 8:08pm	12 Sun Rise: 7:11am Sun Set: 8:05pm	13 Sun Rise: 7:12am Sun Set: 8:03pm
14 Sun Rise: 7:14am Sun Set: 8:01pm	15 Sun Rise: 7:15am Sun Set: 7:59pm	16 Sun Rise: 7:17am Sun Set: 7:56pm	17 Sun Rise: 7:19am Sun Set: 7:54pm	18 Sun Rise: 7:20am Sun Set: 7:52pm	19 Sun Rise: 7:22am Sun Set: 7:49pm	20 Sun Rise: 7:23am Sun Set: 7:47pm
21 Sun Rise: 7:25am Sun Set: 7:45pm	22 Sun Rise: 7:26am Sun Set: 7:43pm	23 Sun Rise: 7:28am Sun Set: 7:40pm	24 Sun Rise: 7:30am Sun Set: 7:38pm	25 Sun Rise: 7:31am Sun Set: 7:36pm	26 Sun Rise: 7:33am Sun Set: 7:33pm	27 Sun Rise: 7:34am Sun Set: 7:31pm
28 Sun Rise: 7:36am Sun Set: 7:29pm	29 Sun Rise: 7:37am Sun Set: 7:27pm	30 Sun Rise: 7:39am Sun Set: 7:24pm				

DST is in effect for the entire month.
Courtesy of www.sunrisesunset.com

Fig 1-40—Printout of the sunrise/sunset table generated from **www.sunrisesunset.com/**.

keep the little booklet within reach on your operating desk. The tables never get outdated, since sunrise/sunset times hardly change over the years. Most graphical systems (globe, slide rule or computer-screen world maps) are far too inaccurate to be useful for 80 and especially for 160 meters.

There are a few copies of this handy sunrise/sunset booklet (100 pages) still available. Send $10 plus $5 for worldwide airmail postage to John Devoldere, ON4UN, Poelstraat 215, B9820, Merelbeke, Belgium.

You can print you own tables that show the SR/SS times on a day-by-day basis for a whole year. Go to **aa.usno.navy.mil/data/docs/RS_OneYear.html**. **Fig 1-39** shows an example print out for the author's QTH. A similar program listing sunrise and sunset times in a table covering one month can be generated from: **www.sunrisesunset.com/**. (See **Fig 1-40**.) At **aa.usno.navy.mil/data/docs/RS_OneDay.html** you can find the sunrise and sunset times for just one day. If you need the exact coordinates to input a particular city or island, you can find them at **gnpswww.nima.mil/geonames/GNS/index.jsp**.

5.1.2. General rules for using sunrise/sunset times

For all E-W, W-E, NW-SE and NE-SW paths you can expect normally two propagation peaks (for short path):

1. The first peak will occur around sunrise of the station at the eastern end of the path.
2. The second (generally less pronounced) peak occurs around sunset for the station at the western end of the path.

For N-S paths there are no pronounced peaks around either sunset or sunrise. Often the peak seems to occur near midnight (see Section 2.4.3). The use of the tables can best be explained with a few examples.

5.1.3. Example 1

What are the peak propagation times between Belgium and Japan on Feb 15? From the tables:

Belgium: 15 Feb: SRW = 0656 SSW = 1659
Japan: 15 Feb: SRE = 2130 SSE = 0824

where:

SRE = sunrise, eastern end
SRW = sunrise, western end
SSE = sunset, eastern end
SSW = sunset, western end

The first peak is around sunrise in Japan or SRE = 2130 UTC. This is after sunset in Belgium (SSW = 1659), so the path is in darkness. Always check this. The second peak is around sunset in Belgium or SSW = 1659 UTC. This too is after sunset in Japan (0824 UTC) so the path is in darkness.

5.1.4. Example 2

Is there a possibility for a long-path opening on the lower bands? The definition of a long-path opening (see Section 4.2) says we must have sunset at the eastern end before sunrise at the western end of the path. In the example this is not true, because SRW at 0656 UTC is not earlier than SSE at 0824 UTC.

5.1.5. Example 3

Is there a long-path opening from Japan to Belgium on Jan 1?

Belgium: 1 Jan: SRW = 0744 UTC SSW = 1549 UTC
Japan: 1 Jan: SRE = 2152 UTC SSE = 0740 UTC

Here, SRW at 0744 UTC is later than SSE at 0740 UTC. This is indeed a valid condition for a long-path opening. It will be of short duration and will be centered on 0746 UTC. Being a so-called long path does not mean that the direction of signal arrival is the opposite of the short path! In Section 4.3, I explained that this so-called long path to Japan is a typical example of a midwinter crooked path.

5.1.5. Pre-sunset and post-sunrise QSOs

In practice, long-path openings are possible even when the paths are partially in daylight. Near the terminator we are in the so-called *gray-line zone* and can take advantage of the enhanced propagation in these zones. The width of the gray line has been discussed earlier (Section 2.4.4.1.). A striking example of such a genuine long-path QSO was a contact made between Arie, VK2AVA, and me on Mar 19, 1976, at 0700 UTC on 80 meters. The long-path distance is 22,500 km. Note that the QSO was made almost right at equinox (Mar 21), and the path is a textbook example of a NE-SW path. On that day we had the following conditions:

Sunrise west (Belgium) = 0555 UTC
Sunset east (Sydney, Australia) = 0812 UTC.

This means that the long path was in daylight for more than two hours. The QSO was made one hour after sunrise in Belgium and more than one hour before sunset in Australia.

Another similar example was a QSO with VKØGC from Macquarie Island (long-path distance 21,500 km). On Jan 21, 1985, a long-path contact was made on 80 meters that lasted from 0800 until 0830 UTC, with excellent signals. This was more than one hour before sunset on Macquarie (0950) and almost one hour after sunrise in Belgium (0731). Because the locations of these stations (VK2 and VKØ) are fairly close to the antipodes from Belgium, the long paths can safely be considered genuine long paths. Indeed there are no crooked paths that could provide an alternative to the genuine long paths. The gray-line globe is a unique tool to help you visualize a particular path like this. Another example (Palmyra/Kingman Reef) was described in detail in Section 2.4.4.2.

Every now and then we hear about almost magical QSOs where on 160 meters contact was made well after sunrise (UUØJZ to KL7 four hours after sunrise), but these QSOs can certainly never be predicted (see Section 2.4.4.2)!

5.1.6. Calculating the half-way local midnight peak

For east-west paths (± 45°), in addition to the usual sunrise and sunset peaks there is often a so-called *mid-way midnight peak* (see Section 2.4.3.). To calculate the time of this peak use the sunset/sunset tables or a program to determine both sunrise and sunset times for both ends of the path. For example, a path between Denver (CO) and Belgium on Jan 15. The sunrise/ sunset data are:

Colorado: sunset: 0001 UTC, sunrise: 1419 UTC
Belgium: sunset: 1606 UTC, sunrise: 0738 UTC
Midnight in Colorado: 0001 + (1419 − 0001)/2 = 0710 UTC
Midnight in Belgium: 1606 + (2400 + 0738 − 1606)/2 = 2352 UTC

The halfway midnight peak time is calculated as the

mathematical average between the two midnight times:

Local half-way midnight is: (2400 − 2352 + 0710) / 2 = 0339 UTC

For North South paths (± 30°) there is a distinct propagation peak at local midnight at the half-way spot for both 80 and 160 meters. This peak is commonly called the *midnight peak*. How do we calculate the exact time of this midnight peak? Example: Path between New York and Paraguay on Jun 15.

New York sunset: 0028 UTC, sunrise 0925 UTC
Paraguay sunset: 2108 UTC, sunrise: 1034 UTC
Calculate the two local midnight (sun)times as follows:
Midnight-New York: (0925 − 0028) / 2 = 0429 UTC
Midnight Paraguay: (1034 + 2400 − 2108) / 2 = 0643 UTC
Halfway midnight time: (0429 + 0643) / 2 = 0536 UTC.

5.1.7. Calculating Sunrise/Sunset times

On the CD-ROM is the listing for a basic program developed by Van Heddegem (ON4HW) to calculate sunrise and sunset times. It is based on classical astronomy and calculates precise sunrise and sunset times. The program does not use any arcsine or arccosine functions since they are not available in all BASIC dialects.

For low banders, especially on 160 meters, it is important to know precise sunrise and sunset times for both ends of a path. Graphical tools such as slide-rules are generally not accurate enough for this purpose. On some of the long-haul paths, openings may only last a few minutes right at sunrise (sunset), so accurate information produced with computer programs or the ON4UN sunrise-sunset booklet is critical.

5.2. Propagation-Predicting Computer Programs

Earl, K6SE, in a message on the Internet, wrote: "*I gave up long ago on trying to predict DX conditions on 160 meters. One major observation I've made over the years is that on a night conditions are good to EU from here (northwest), conditions to JA (northeast) that same morning are not exceptionally good. And, if conditions are good to JA in the morning, that evening conditions to EU are not good. The only "prediction" I use now is: if conditions are exceptionally good in any particular direction on a given night (ie, to EU 28 Dec 97 in the Stew Perry contest), I hope there will be a repeat one solar rotation later (27 days). Generally, I've come to the conclusion that 160-meter conditions are unpredictable, hi, so I just check the band every night to see what's happening.*"

To be perfectly honest, in my 40+ years of DXing on 80, and in my 15+ years on 160, I have never successfully used a propagation-prediction program. Usually when operating one of the major CQ WW contests, one of my friends will run the propagation predictions on his computer and bring them along. I have never found anything there that I did not know. The easy paths, the well-known conditions were there, clearly on paper, but the marginal ones—the exotic ones—were not. And by the way, these prediction programs usually do not cover 160 meters. Rightfully so, because basically propagation on Top Band uses quite different mechanisms than the higher bands (that is, 160-meter propagation is not MUF related).

Tom, N4KG, who says that he *lives* on the low bands, recently wrote: "*I find very little correlation between solar flux and low band propagation, particularly on 160 and 80 meters. Each of the low bands has a distinct characteristic. Forty meters does follow solar flux in that the MUF can drop below 7 MHz when the flux levels are very low. During these times, 40 meters will close shortly after sunset on Northern paths and there will be no European sunrise opening to the USA. With slightly higher solar activity, 40 meters will stay open all night. On 80 meters, solar flux is not so much of a factor. The BEST times are just before local sunset and after local sunrise. There is a significant enhancement at the terminator, which seems to be a daily event.*

There seems to be some validity to the theory that propagation is enhanced at the very start of a solar disturbance, but then the band goes flat for several days until the ionosphere stabilizes again. 160 meters does not correlate with anything! You may see nice enhancements at sunset and sunrise, or you may see nothing at sunset and sunrise but find good openings between 0200 and 0500 to Europe (from USA). On signals from the east, I have seen peaks at their sunrise and I have seen them peak a full hour before sunrise (before any daylight) and then vanish. Propagation prediction programs seem to be almost useless and often misguiding on the low bands because they fail to model the focusing effects at sunrise and sunset and they do NOT look at non-great circle paths. Most over the pole paths to the opposite side of the world come via skewed LP, at least here in eastern USA...."

I could not have said it better myself! This being said, current propagation prediction programs are useful for predicting 40-meter propagation, and some of them include very useful viewing and mapping facilities (see Section 3.2.7). For me, the final *acid test* that would make me a firm believer in propagation-forecasting programs, is when such a program would successfully forecast the odd 80- and 160-meter openings *and paths* (directions), like the Europe to ZL path described in Section 4.3.1.2. I also would like to see a quantitative confirmation of bent paths that we so often experience. For anyone to accurately model the entire system that determines attenuation on 80 and 160 meters, we will have to know *all* possible mechanisms. Today I have the impression that we see only the tip of the iceberg but the issue is still being addressed.

There may be light at the end of the tunnel. Rod Graves, VE7VV, himself an active low-band DXer, has written a program that attempts some new approaches to predicting the more *odd* openings, the ones that other programs don't seem to recognize. The concept of the program is quite novel and employs a zone method developed by the author. It supports (E-F region) ducted paths and also directly addresses skewed paths. These are not skews due to magnetic disturbances, but skews due to the structure of the ionosphere at a given time. The program seems to predict very long-distance and long-path openings on 80 and 160 meters much better than any other. The limitation of the software is that it does not take into account auroral absorption. The software is still in a development phase, but may become available at a later date. VE7VV also admits that signals from locations near the antipode may travel paths other than the ones his program checks.

VE7VV comments: "*Like most others, I do not use a program to tell me when the bands might be open to unknown locations. I just turn on the receiver to see what, if anything, is coming in from wherever and enjoy the discovery of the unexpected DX. However, when I want to make a sked with someone, or when there is a new DXpedition, I run the program to determine when the band might be open to that specific location and when the predicted optimum times are. I use this as my guide for when to be sure to be listening or for when to make the sked. Sometimes the program shows only what is common lore, or just provides another way of determining what could have been done with sunrise/sunset tables, or grayline devices, but the program does it very conveniently for me. Sometimes, however, the program reveals times of openings, or times of peak signal strength, that are not obvious. This is especially true for paths over 10,000 km, particularly on the low bands (40-160) where refraction and ducting phenomena are more prominent."*

I believe that the sunrise/sunset tables and a good mapping program (eg, *DX-AID* or *DX Atlas*) are the best tools in your search for a new one on 80 or 160 meters. Know the dark path, and you're all set. All you need then is luck and that makes 160 meters an adventure!

5.3. ON4UN Low-Band Software
5.3.1. ON4UN propagation programs

When I was writing the original *Low-Band DXing* book, I developed a number of computer programs as tools for the active DXer. The programs are fully color compatible, and available only for MS-DOS on a 3½-inch diskette or a Zipped-file by E-mail attachment (see order form in the back of this book). The software runs perfectly under Windows (also on XP) in a DOS box. While the majority of my programs are technical programs related to antenna design (and are covered in the antenna chapters of this book), a group of programs deals with the propagation aspects of low-band DXing. The Propagation Software contains the following modules:

1. SUNRISE/SUNSET TIMES

This program lists the sunrise and sunset times in half-month increments for the user's QTH. The user's QTH can be preprogrammed and saved to disk. The QTH can be changed at any time, however. On the screen, the sunset and sunrise times of a "target QTH" are listed side-by-side with your own time. The target QTH can be specified either by coordinates or by name. On screen you also see the great-circle direction as well as distance. The software works in miles as well as kilometers. You can also modify the display increments and can also list the times in single-day increments if you wish, which is very handy for following the gray line on a DXpedition. **Fig 1-41** shows a screen print, which gives all relevant data for the ZL7 (Chatham) expedition from Feb 15 to Mar 10.

2. The DATABASE

When specified by name, the latitude/longitude coordinates are looked up in a database containing over 550 locations worldwide. This database is accessible by the user for updating or adding more locations. The database can contain data for up to 750 locations, and can be sorted in alphabetical order of the country name or the radio prefix. You can also print the data on paper.

3. Listing Sunrise/Sunset Times

The program also allows you to list (scroll on screen) or make a full printout of the sunrise and sunset times (plus directions and distances from your QTH) for a given day of the year. This is a really nice feature for DXpeditioners and for contesting.

4. Gray-Line Program

This section of the program uses a unique algorithm that adapts the effective radio width of the gray-line zone to the location and the time of the year. This width is also different for 80 and 160 meters. In addition, the user can specify a minimum distance under which he is not interested in gray-line information. The printout (on screen or paper) lists the distance to the target QTH, the beginning and ending times of the gray-line window, as well as the effective width of the gray line at the target QTH. **Fig 1-42** shows a screen dump of a gray-line run for Belgium on Feb 27. Notice the short Chatham island opening predicted between 0614 and 0623 UTC. I made the QSO on 160 meters at 0623, and copied ZL7DK until 0634 UTC.

The ON4UN LOW BAND SOFTWARE is on the CD bundled with this book.

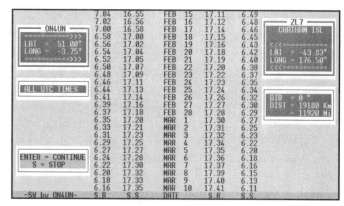

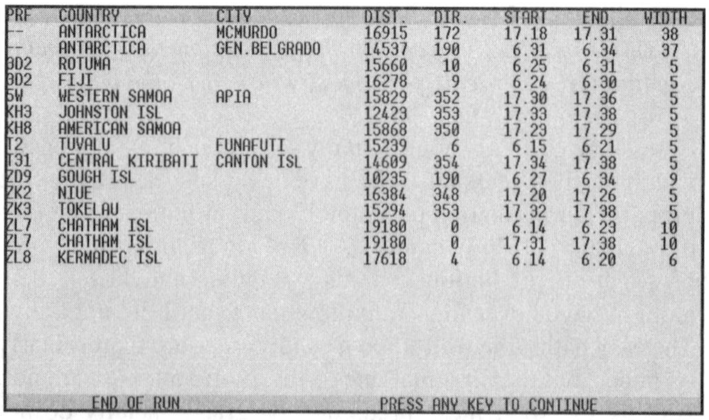

Fig 1-41—Screen dump of the Sunrise/Sunset module of the *ON4UN LOW BAND SOFTWARE*. This example shows, side-by-side, the sunrise/sunset times for Belgium and ZL7, in one-day increments from Feb 15 to Mar 10. These precise data are invaluable when short windows are available, such as in this case.

Fig 1-42—The *ON4UN LOW BAND SOFTWARE* gray-line module calculates all possible gray-line openings for a given day. This example is for Mar 6, 1998, when ON4UN worked ZL7DK on 160 meters. Gray line was predicted between 0626 and 0636 UTC and a QSO was made at 0633 UTC.

6. DIFFFERENCES BETWEEN THE 40, 80 AND 160-METER BANDS

6.1. 40 Meters

- Forty meters is like an HF band that works at night time (it's almost like VHF for a Top Bander!).
- Propagation prediction can be done by classic MUF-based programs.
- Gray-line propagation also happens *along* terminator, as on the higher HF bands.
- Gray-line zones can be very wide (many hours even at medium latitudes).
- Skewed paths are not as common as on 80 and 160 meters.
- 40 meters allows you to work any distance, if properly equipped.

6.2 80 Meters

- With well-equipped stations at both ends, just about any distance and path can be covered at the right time of the year.
- During low sunspot years, 80-meter propagation may be influenced by MUF.
- Gray-line enhancement always occurs on paths perpendicular to the terminator.
- Working DX through the auroral belt is not uncommon, even in high sunspot years.

6.3. 160 Meters

- Propagation is not at all dictated by MUF, and only marginally by the solar cycle.
- Besides the auroral phenomenon we still do *not* know what makes a good DX night or a bad one. Mystery is still a big part of Top Band!
- Auroral absorption is most pronounced on 160 meters.
- Skewed paths most frequently occur on Top Band.
- Gray-line enhancement always appears to occur on paths perpendicular to the terminator.
- 160 meters has a distinct area in which working DX is more or less like a piece of cake— anything in a circle of approximately 5,000 km around your own QTH. For instance, Western Europeans can work the East Coast of the USA almost daily. The "light-gray" zone is W5, W8 and W9 land. WØ land is "dark gray" and for anything beyond that, conditions must be well above normal. This is quite different from 80 meters, where longer distances are possible every day and where the transition between "easy" and "difficult" seems to be much more vague.
- Real long path (that is, without path bending) on 160 meters is rather exceptional, except for stations very near the antipodes, and as a rule only occurs during low sunspot cycle years.
- If 80 meters is swinging, there is no guarantee that 160 meters will be any good. When 80 is bad, though, 160 will likely be bad as well. So don't extrapolate from the higher band to the lower band. This very often does not work.
- Very typical for 160 meters is a slow and deep QSB, especially on very marginal paths. It's advisable to get a call right the first time—There may not be a second time or it may be minutes later. I have seldom seen this on 80 meters. Patience is important also…you may have to wait for propagation to peak to you.
- During low sunspot cycle years, 160 meters usually has very pronounced peaks at sunrise (sometimes also at halfway midnight), especially for really long-haul paths—where the sharp peak is usually within minutes of sunrise. You can almost set your watch by it. The sunset peak on 160 meters is also much less pronounced. There seems to be a broad "peak" within an hour or so after sunset.
- On 160 meters skip is often very selective (for various reasons).
- Working DX through the auroral donut is very difficult during high sunspot years.
- The thrill of working a new one on 160 meters is ten times the thrill of doing it on 80 meters!

7. THE 160-METER MYSTERY

Understanding and predicting propagation on 40 meters is pretty straightforward and 80 meters is well understood as well. With the right equipment and knowledge on both ends, you could probably work 300 countries in a year on 80 meters.

One-sixty is a totally different ball game. The more I have been active on 160 meters, the more I am convinced of how little we know about propagation on that band. True, we know a few of the parameters that influence propagation, but far from all. For a long time I have kept daily records of the K and A indexes, sunspot numbers, etc, together with my own observations of conditions on 160, to find a correlation between the data and actual propagation. But I have found very little or none; only negative correlations. We know more or less when it definitely *will not* work, but not for sure when it *will* work.

Of course, we must realize that on Top Band we are in a gray area, where things are sometimes possible but often not. There are dozens of parameters that make things happen, or not happen. They all seem to influence a delicate mechanism that makes really long-haul propagation on 160 meters work every now and then. Understanding all of the parameters and being able to quantify them and feed them into a computer that will tell exactly when we can work that elusive DX station halfway around the globe will probably be an illusion forever.

There is no interest from the broadcasters in this subject. Broadcasters and utility traffic operators are interested in knowing the frequency that will give them best propagation. They are not interested in studying the subject of "marginal propagation," just on the edge of what is possible. Therefore, long-haul DXing on 160 meters will probably always remain a real hunting game, where limited understanding, feeling, expertise, and luck will be determining factors for success. Don't forget your hunting weapons—your antennas and your equipment.

CHAPTER 2

DX-Operating on the Low Bands

I asked Bill, W4ZV to be my critic, guide, counselor and support for this chapter. As expected, Bill did an excellent job, and fast as well. Bill needs no introduction to the active Low Band DXers, but let me just introduce W4ZV to the newcomers to this playground of our hobby. Bill is one of the very few hams in the world having over 300 countries confirmed on 160 m, and he was the first to reach that number, in 1998. That says enough. You don't achieve this unless you have a profound know-how. And Bill was found willing to use his know-how and expertise to help me with this chapter.

Bill was first licensed at age 12 in 1957 as KN4RID. The DX bug bit hard as soon as he made his first DX contact on 15 meters, and he went on to become the first US Novice Class licensee to achieve DXCC. Bill made the DXCC Honor Roll seven years later in 1964 while he was still a teenager. School, work and marriage mostly curtailed his operating until he got the bug again in 1976 while he was working in Colorado when he became W0ZV. In 1980, he moved to 40 acres in the country and began seriously chasing new ones on 80. In October 1984, he put up a "temporary" 160 antenna for a few multipliers in the CQ WW SSB and became addicted to the band. Bill said, "When I first got on 160, some of the locals said I would do well to make WAC". By April 1985, he was issued the first 160 DXCC for the W0 area, and was

Bill Tippet, W4ZV, the first to work 300 countries on 160 meters.

thoroughly addicted to Topband. In 1993, Bill moved back to his home state of North Carolina and received the call W4ZV in 1996. Now Bill mostly concentrates his low band DXing on 160 since he only needs five more to have them all on 80. Bill is also the skillful moderator of the Topband Reflector on the Internet.

Thank you Bill!

Tree, N6TR, considers 160-meter DXing as a disease. But the symptoms he described apply to the other low bands as well:

- Desire to be on the radio at sunrise.
- Desire to be on the radio at sunset.
- Desire to be on the radio at all times in between Sunset and Sunrise.
- Desire to struggle for months to work a single station in a new country.
- Never being satisfied with the antenna system and constantly trying new ones.
- Only comes down to see the family after working a new country (to gloat). During the rare fantastic opening, will come down after each new country and hold up fingers indicating how many new countries were worked so far. These events are rare and occur about once or twice in a century.
- Drinks lots of water before going to bed with the sole purpose of waking up in the wee hours of the morning to see if a new country can be found.
- Has problems getting to work on time during the winter months.
- Sends equipment and wire to people in unworked countries, hoping that the end result will be their QSL card on the wall.
- Spends thousand of dollars going to rare countries just so other people can work it.

And these are only some of the better-known symptoms. According to Rush Drake, W7RM, it's a painful disease: *"To work DX on 160 you've got to love pain."* Earl, K6SE, changed that to: *"You've got to love torture..."* Who am I to disagree with such eminent low-band DXers?

One-sixty meters is usually referred to as Top Band, the band at the top of the wavelength spectrum, the band with top-notch operators, the band that's a top challenge and that gives you top excitement and satisfaction. Gary, NI6T, says: "One sixty? Not a band, but an obsession."

All kidding aside, low-band DXing is a highly competitive technical hobby. It is certainly not a communications sport for the appliance operators. It is one area of amateur radio where it really helps to be knowledgeable. This is not a "plug and play" hobby!

1. MYTHS

Gerry, VE6LB, who is a successful Low Band DXer from an urban QTH, from the middle of nowhere, right in the auroral doughnut, using simple antennas, summed up a few myths:

- There is no (or little) DX on the low bands!
- You need a big antenna and high power (it's only for the big guns) to work DX on the low bands!
- DX is so scarce that you need to spend many hours (mostly late at night) to find DX on the low bands!
- Any DX to be found on the low bands is on CW.
- There is no low band DX during the summer.
- The low bands are too noisy to work DX.

2. REALITY

Let's look at some facts:

1. All countries have been available on 40 meters, and there are quite a few DXers that have all of them on 40, except for P5. At this time, all countries (with the exception of P5 and BS7H) have been available on 80 meters, and probably not more than a handful countries have not—so far—been available on Top Band. Every year several Top-Band DXers work DXCC in less than a year (as reported in the *Low Band Monitor*).
2. You will probably never win the CQ Worldwide 160-Meter Contest from a suburban lot with a 50-foot antenna-height restriction. But you can work DXCC on the low bands, even with 100 W from a typical suburban lot. KH6DX/W6 worked over 100 countries from a mobile! I have friends who have never run power (more than 100 W) and have over 100 countries on Top Band. It is true—of course—that, the better the means, the more you'll be sitting in the front row when the show is on.
3. Most of the DX on the low bands can be worked around sunset or sunrise. This is a better arrangement than on 10 meters, where the DX shows up in the middle of the day when most of us are at work.
4. Too bad not all the low-band DX is on CW. (That's a personal note. I love CW so much better than phone!) Seriously, there are countries that are only available on Phone and others only on CW. That's the name of the game. When it comes to Top Band though, CW is the name of the game! It's Top Band, and CW, that separate the players.
5. Ever consider that when it's summer here, it's winter on the other side of the equator?
6. Noise, whatever its origin, is one of the main challenges for the low-band DXer, but it certainly does not stop real men from DXing. This is not a broadcast hobby, neither a communicator's hobby. In this low-band hobby we are driven to move the boundaries of what is possible.

Well, all of this does *not* mean that working DX on the low bands is just a piece of cake, a nice pastime for the appliance-type operator. But what makes so many love the low bands for chasing DX?

3. WHAT MAKES PEOPLE CHASE DX ON THE LOW BANDS?

I included this question in my questionnaire that I sent out early 2003 via the Internet. The answers are the same as I the one received 5 years ago when I last ran the survey. For literally everyone who works the low bands what makes them chase DX is the challenge, the sense of fulfillment and having done something difficult. (See Section 17.)

Low-band DXers are always near the edge of what is possible. The most successful low-band DXers are the pioneers who keep moving this edge. Improved understanding of propagation, together with better equipment, and most of all, better antennas, make it possible to dig deeper and deeper into the noise to catch the previously evasive layer of buried signals. Top-Band DXers are those balancing themselves on the cutting edge of the DXing sword!

If you are looking for an easy pastime, stay away from low-band DXing. Maybe one of the many lists that are abundant on the higher bands is something for you. K1ZM, who now is one of the few US stations having worked over 300 countries on 160 meters, wrote on his survey reply: *"160 is truly a MAN's as well as a GENTLEMAN's band. You want a challenge? Get on 160."*

On 80 and 40 meters you do not need to have a genuine "antenna farm" to work DXCC, even within one year's time. Even on 160 meters urban QTHs with small and low antennas regularly produce DXCCs on Top Band. I have included in this book a short chapter on "Working 160-Meter DX From a Small Suburban Lot." There are many examples of rather modest stations on a small suburban lot that have done extremely well. My friend George, K2UO, worked over 200 countries on 160 from a $^1/_2$-acre suburban lot. To be so successful from an average QTH requires a better-than-average knowledge of propagation, as well as a substantial dose of perseverance.

4. THE FREQUENCIES

The frequencies used for DXing on the low bands are not the same in all countries. Therefore it is important that you know where to look for the DX. There are four levels we should look at:

1. What are the allocations in the three ITU regions?
2. What are the frequencies each individual country has allocated on the low bands, and are there mode-related subbands that are enforced?
3. Since subbands do not exist in a most countries, what is the (mode-related) band planning that radio amateurs have agreed upon in a worldwide contest? In others words, what does the IARU band plan say?
4. What is the common-sense band planning that low-band DXers apply, in case the IARU band planning does not meet our goals?

4.1. The ITU Allocations

The ITU has divided the world in three regions:
- Region 1: Africa, Europe, former USSR countries, Middle East (excluding Iran) and Mongolia
- Region 2: North and South America including Hawaii,

Johnston and Midway Island
- Region 3: The rest of Asia and Oceania

The allocations are described in the *Radio Regulations* published by the ITU. Article S5 describes the allocations in detail. The RR publication can be bought from **www.itu.int/ITU-R/publications/rr/index.asp**. The following analysis is based on the 2001 publication of the RR.

4.1.1. 160 meters
Region 1:

1800 to 1810: No Amateur Radio allocation (used for Radiolocation)

1810 to 1850: In principle, primary allocation for Amateur Radio, but....

1810 to 1830: In more than 50 countries, primary allocation is for Fixed and Mobile services. Amateur radio is secondary.

1850 to 2000: Primary allocation to Fixed and Mobile services. In about 30 countries (eg, DL, OZ, OH, ON, HA, EI, 4X, OK, G, U, SM, etc) this section can be allocated to Amateur service but with a power limit of 10 W.

Region 2:

1800 to 1850: Allocated exclusively to Amateur Radio

1850 to 2000: Shared between Amateur, Fixed, Mobile, Radiolocation and Radionavigation. In most South American countries this section is allocated to Fixed and Mobile services on a primary basis (which means Amateur Radio is secondary).

Region 3:

1800 to 2000: Shared between Amateur, Fixed, Mobile, Radionavigation and Radiolocation. There are still special previsions to protect the Loran frequencies, but as the system is no longer used, it is of no impact.

4.1.2. 80 meters
Region 1:

3500 to 3800: Shared between Amateur, Fixed and Mobile services

Region 2:

3500 to 3750: Exclusively Amateur Radio
3750 to 4000: Shared between Amateur Radio, Fixed and Mobile services. In LU, CP, CE, HC, ZP, OA and CX the Amateur Radio allocation is secondary. In VE and OX 3950-5400 can be used for Radiobroadcasting as a primary service.

Region 3:

3500 to 3900: Shared between Amateur Radio, Fixed and Mobile services
3900 to 3950: Aeronautical Mobile and Broadcasting
3900 to 4000: Fixed services and Broadcasting

4.1.3. 40 meters
Region 1:

Before March 29, 2009:

7000 to 7100: Exclusive Radio Amateur (in some African countries, 7000-7050 is also allocated to Fixed service as primary service)

7100 to 7300: Broadcast (as if we didn't know that...)

After March 29, 2009:

7000 to 7200: Exclusive Radio Amateur (in some countries Fixed service can be used as primary service)

7200 to 7400: Broadcast

Region 2:

7100 to 7300: Exclusive Amateur Radio

Region 3:

Before March 29, 2009:

7100 to 7300: Broadcast

After March 29, 2009:

7000 to 7200: Exclusive Radio Amateur (in some countries Fixed service can be used as primary service)

7200 to 7400: Broadcast

4.2. The IARU Bandplan

The IARU (International Amateur Radio Union) groups one National Radio society from each member country (the most representative one) and sets out as one of its goals to establish and maintain a band plan that has been approved by all of the IARU radio societies. The IARU is organized in the same 3 regions as the ITU (see section 4.1).

Table 2-1 (Source: IARU Web page) gives an overview of the band plan regarding 160, 80 and 40 meters in the three regions. This table does not mean that all countries in a given region are permitting operation in all of the segments mentioned in the table! It is obvious that the 7 MHz band plan will substantially change after March 29, 2009, when the new frequency allocations in the 40m band will come into effect.

4.3. Let's be Practical
4.3.1. Let's be practical on 160 meters

For successful DXing on 160 meters, knowing that most of the serious DXing on this band is done on CW, the band plan should reserve a window exclusively for CW. So far the IARU band plans have provisions for CW subbands and Phone + CW subbands. This probably stems from the historic days of Amateur Radio, but I cannot see any reason why the Phone bands should not be as exclusive as the CW subbands!

New to the IARU 160 band plan is the inclusion of "digimode" windows. In creating these new windows, consideration should be given for everyone already on the band

Fig 2-1—W1BB, Mister 160 Meters, in his shack late in his career.

Table 2-1
IARU Band Plan

Region 1	Region 2	Region 3
1810 to 1838: CW	1800 to 1830: CW, Digimode	1800 to 1830: CW
1838 to 1840: Digimode Packet, CW	1830 to 1840: CW, Digimode (DX CW window)	1830 to 1834: RTTY, CW, DX
1840 to 1842: Digimode except Packet, Phone, CW	1840 to 1850: (DX Phone window), CW	1834 to 1840: CW
1842 to 2000: Phone, CW	1850 to 2000: Phone, CW	1840 to 2000: Phone, CW
3500 to 3510: Intercontinental DX CW	3500 to 3510: (DX CW window)	3500 to 3510: DX, CW
3500 to 3560: CW, Contest preferred CW segment	3510 to 3525: CW	3510 to 3535: CW
3560 to 3580: CW	3525 to 3580: CW, (Phone permitted, non-interference basis)	3535 to 3775: Phone, CW
3580 to 3590: Digimode, CW	3580 to 3620: Digimode, (Phone permitted, non-interference basis), CW	3775 to 3800: DX Phone, CW
3590 to 3600: Digimode, Packet Preferred, CW	3620 to 3635: Packet Priority, (Phone permitted, non-interference basis), CW	3800 to 3900: Phone, CW
3600 to 3620: Phone, Digimode, CW	3635 to 3775: Phone, CW	
3600 to 3650: Phone Contest preferred phone segment, CW	3775 to 3800: Phone (DX Phone window), CW	
3650 to 3775: Phone, CW	3800 to 3840: Phone, CW	
3700 to 3800: Phone Contest preferred phone segment, CW	3840 to 3850: SSTV, FAX, Phone, CW	
3730 to 3740: SSTV, FAX, Phone, CW	3850 to 4000: Phone, CW	
3775 to 3800: Intercontinental DX Phone		
7000 to 7035: CW	7000 to 7035: CW	7000 to 7025: CW
7035 to 7040: Digimode (except Packet), SSTV/FAX, CW	7035 to 7040: Digimode with other Regions, CW	7025 to 7030: NB, CW
7040 to 7045: Digimode (except Packet, SSTV/FAX), Phone, CW	7040 to 7050: Packet with other Regions, CW	7030 to 7040: NB/Phone, CW
7045 to 7100: Phone, CW	7050 to 7100: Phone, CW	7040 to 7100: Phone, CW
	7100 to 7120: Digimode, Phone, CW	7100 to 7300: Phone, CW (See footnote 5). This segment is allocated on a secondary basis to amateur service in Australia and New Zealand
	7120 to 7165: Phone, CW	
	7165 to 7175: SSTV, FAX, Phone, CW	
	7175 to 7300: Phone, CW	

before a new exclusive area is created, and this should be done through planning with existing operators (CW and Phone)—rather than dictatorship—or the result will be a real mess and many hard feelings will be created. Before isolating several kHz from a prime area, the rule makers should make sure the operators already there (for many years) will have a suitable place to move and will be found willing to do so. As Tom, W8JI wrote on the Topband Reflector: "*We need a long-term plan that does not displace primary users. The IARU needs to seek input for 160 operators before they mess things up for everyone, and cause a lot of hard feelings that last for many years.*"

Let's look at the ARRL-published 160-meter band plan (Feb 2003). For some unknown reason, this is different from the IARU Region 2 band plan (same publication date). The ARRL band plan certainly makes more sense than the IARU-published one when it comes to digital modes on 160 meters.

ARRL Band Plan 160 Meters

1800 to 1810	Digital Modes
1810	CW QRP
1800 to 2000	CW
1843 to 2000	SSB, SSTV and other wideband modes
1910	SSB QRP
1995 to 2000	Experimental
1999 to 2000	Beacons

There is a major problem area in the choice of the frequencies for digital modes in the present (Feb 2003) IARU band plan. This band plan, where digital modes are squeezed in-between the CW and the Phone (+CW) section, is really unacceptable. Once 160 meters is hot again at the bottom of the solar cycle, this band plan is a guarantee for continuous conflicts and battles. The only reasonable solution is to use 1800 to 1810 kHz in Region 2 and 1810 to 1815 kHz in Region 1 (so long as Region 1 does receive 1800 to 1810) for digital modes. I suggest that all readers contact their national radio societies and ask them to submit a proposal along these lines to the IARU HF committee in your Region.

The clear distinction existing between the CW band and the Phone band is, in my view, not realistic during major contest weekends (CQ 160 contests, CQ WW, ARRL 160, etc). It does not make sense to have a rule that nobody follows. Take for example Europe, where today still many countries only have 1820 to 1850 kHz for both modes (CW and SSB). And if they have an spectrum above 1850, then the power there is (or should be) limited to 10 W (Remark S96 of the ITU frequency-allocation table). The IARU Region 1 band plan calls for no phone signals below 1840 kHz (thus, a carrier frequency of higher than 1843 kHz). This means that all these European stations have five "channels" to use for the entire phone contest. This obviously does not make any sense.

It is my opinion that during the major contest weekends the band plan should be set aside. Compare it to the following situation: In Europe most major roads have bike tracks alongside major highways. A few weekends every year though, when major cycling events take place (Tour de France, for example), bikers can use the entire width of the road. Let it be like that during a few of the major contests. Why does the band plan allow CW fanatics (and I am one of these) to transmit all over the band, while the poor phone guys, who actually need much more room, can only occupy a small portion (in Europe)? Can't CW fans relax during two or three contests every year and let the phone guys enjoy their contest?

And can't the Phone operators relax a few weekends every year when the major CW contests are on? They could take the XYL out for a weekend. Why do some operators have to start QRMing QSOs under those circumstances? It saddens me to see that many people cannot appreciate that other people also want to enjoy the hobby. Here too, we Top Banders should ask our IARU societies to come forward with more realistic band plans.

Let's use the entire width of the road when the Tour de France is on! I love the way Mike, N2MG, put it on the Contest Reflector: *"A band plan, to me, is a lot like handicapped parking. Nothing is more frustrating than driving around a small parking lot over and over trying to find a place because I don't want to offend anyone by using one of several empty handicap spaces... When the lot is fairly empty, the dedicated spaces make sense—as do band plans. When at capacity, they do not. Blindly following band plans during a contest is like telling someone (those supposedly protected by the plan) that their transmissions are more sacred than the contesters."*

It is amazing to see vastly different frequency allocations in an area like Europe, with its many relatively small countries. It looks like politicians and administrators love borders—But they should realize that radio waves ignore borders! If there is one area where legislation should be made at a European level, it is in the area of frequency allocations and power. Let's all press our national radio societies to talk to the bodies governing frequency allocations to better align the allocations and to talk to the politicians to apply European-level rulemaking in this matter. Unless we push, little will happen.

Table 2-2 shows the Top-Band European frequency/power allocations as of February, 2003. Note that a number of countries have no allocation above 1850 KHz and that most of the countries that do have such an allocation impose lower-power limits there. This makes it hard to consider this band section as a DX-hunting ground. Note that according to the ITU *rules* (not merely recommendations) the power is supposed to be limited to 10 W in those Region-1 countries that make frequencies above 1850 kHz available (Remark S5.96 of Article S5 of the RRS5 by the ITU).

In Europe, most countries today do have a high-power limit, at least at the bottom end of the band (1810 to 1850), where previously many were accustomed to only 10 W! So

Table 2-2
European 160-Meter Allocations/Powers

Country	> 1850 kHz	Allocation	Power
Belgium		1810 to 1830	1000 W
		1830 to 1850	1000 W
Bulgaria	x	1810 to 1850	1500 W
		1850 to 1880	1500 W
Croatia		1810 to 1850	600 W output
Cyprus	x	1800 to 2000	26 dBW
Czech Rep.	x	1810 to 1850	750 W
		1850 to 2000	20 W
Denmark	x	1810 to 1850	800 W
		1850 to 1900	10 W
		1930 to 2000	10 W
Estonia	x	1810 to 1850	800 W
		1850 to 1955	100 W
Finland	x	1810 to 1850	1000 W peak
		1850 to 1855	60 W peak
		1861 to 1906	60 W peak
		1911 to 2000	60 W peak
France		1810 to 1850	500 W
Germany	x	1810 to 1850	75 W PEP
		1850 to 1890	75 W PEP
		1890 to 1950	10 W PEP (On special request)
Israel	x	1810 to 1850	1500 W
		1850 to 2000	40 W
Italy		1830 to 1850	500 W
	Expect 1820 to 2000 before year-end 2003		
Lithuania	x	1810 to 2000	1000 W
Monaco		1820 to 1850	100 W input
Montenegro		1810 to 1850	300 W
Netherlands		1810 to 1880	400 W pep
Norway		1810 to 1850	1000 W
		1850 to 2000	10 W (1 KW during selected contests)
Poland		1810 to 1850	500 W
		1850 to 1980	10 W
		1830 to 1850	1500 W
RSA	x	1810 to 1860	400 W PEP
Russia	x	1810 to 2000	10 W
San Marino	x	1810 to 1900	1000 W
Slovenia	x	1810 to 2000	300 W
Spain		1830 to 1850	200 W
Sweden	x	1810 to 1850	1000 W
		1930 to 2000	10 W
UK	x	1810 to 1830	400 W erp
		1830 to 1850	400 W erp
		1850 to 2000	32 W erp

we should not forget that we have seen dramatic improvements in the regulatory scene in the past 5 years. There is reason to hope that further improvements can be realized in the next few years.

In Europe 1840 kHz is the usual bottom end of the phone band. But it appears many operators are not aware that if they operate on a carrier frequency of 1840 kHz on LSB their sidebands spread 3.0 kHz down and that they are therefore taking out 40% of the primary DX CW window in Europe. Fortunately, the IARU band plan now clearly stipulates that the 1840-kHz bottom-end means no one should transmit (on LSB) below a carrier frequency of 1843 kHz.

Note that contrary to what has been done on 80 meters, the IARU never created a DX portion on 160 meters, reserved only for intercontinental work. Common sense, however, has created de-facto DX segments on 160 meters. 1830 to 1840 kHz is generally considered the European CW-transmit segment, while the DX segment of 1820 to 1830 kHz is generally considered the DX window in Europe (that's where the DX is, and where the Europeans—as well as US stations—should stay out of).

DXpeditions seem to use the 1823 to 1828-kHz window most of the time. More recently they have made the wise decision to work on the so-called half-frequencies (eg, 1823.5 kHz). This avoids the spurs and birdies often present in some receivers on even-kHz frequencies. In addition, always avoid 1818 kHz, the W1AW broadcast frequency used for code practice and bulletins. Other frequencies to avoid are exact multiples of 10 kHz (1810, 20, 30, etc) for North American stations and multiples of 9 kHz (1809, 18, 27, etc) for stations in IARU Regions 1 and 3. This is because of BCI images from broadcast stations in the MF band (10-kHz spacing in NA and 9 kHz elsewhere).

While it is true that frequency assignments are not the same in all places, it seems that the minor differences are not a huge problem. Over the years it seems that the different administrations are indeed trying to align themselves.

Band plan for Japan (recently changed):
- 1907.5 to 1912.5 kHz: CW only window (original allocation)
- 1810 to 1825 kHz: CW only (new allocation)

Whether or not this is an improvement is not clear. It is obvious that this new window is now clear from low-power Russian AM-stations, but since most QSOs in the 1810 to 1825-kHz window are no longer made in split-frequency mode, there are now other sources of QRM (the calling stations).

Band plan for Russia, as well as in the CIS (former USSR) countries:
- 1810 to 2000 kHz: CW
- 1840 to 2000 kHz: SSB and CW

Band plan for Australia:
- 1800 to 1810 kHz: Digital modes
- 1810 to 1835 kHz: CW
- 1835 to 1870 kHz: SSB. In international contests SSB may be used down to 1830 kHz.

Dennis, KØCKD has compiled a list of the frequency allocations on 160 meters for all countries. See **www.machlink.com/~k0ckddennis/index1.html**.

4.3.2. Let's be practical on 80 meters
4.3.2.1 The DX windows:

Although the 80-meter band is not allocated uniformly for all continents and countries, this does not really represent a problem for the DXer. On CW all countries have an allocation starting at 3500 kHz. The DX window for CW is the same all over the world: 3500 to 3510 kHz. A secondary de-facto window exists between 3525 and 3530 kHz, which is the lower limit for General and Advanced Class amateurs in the US.

The SSB 80-meter SSB DX window is 3775 to 3800 kHz. While the 3500 to 3510 CW DX window has been internationally recognized by the IARU in both Region 1 and 2 (see Section 4.1), this is not the case for the Phone DX window, which is only recognized by the IARU as a DX window in Region 1. This is not good, and some alignment in these matters in order.

Fortunately, common sense sometimes achieves more than rules, and in Region 2 and 3 these same 25 kHz are also accepted as DX bands by most operators. Anyhow it's common sense that reigns, since IARU band plans are not enforced by law in a great majority of countries. It really is a gentleman's agreement that we should all follow, at least if it makes common sense! If not, we should ask our societies to change their band plans.

In the middle of the day, the DX segments can be used for local work, although you should be aware that local QSOs can cause great QRM to a DXer (at, say, 500 miles) who is already in the grayline zone, and who might just be enjoying peak propagation conditions at his QTH. In Europe situations like this occur almost daily in the winter, when northern Scandinavian stations can work the Pacific and the West Coast of the US at 1300 to 1400 UTC, while Western Europe is in bright daylight and does not hear the DX at all. Western Europeans can hear the Scandinavians quite well, and consequently the Scandinavians can also hear Western Europe well enough to be QRMed. The same is true for NA when Eastern NA local rag chews can interfere with DX for more westerly stations that still have darkness. Hams must be aware of these situations so they don't interfere with DXers in other adjacent areas.

Most countries in Western Europe can operate anywhere between 3.5 and 3.8 MHz, and in most countries there are no mode subbands imposed by the government. The band plan for Russia and CIS countries (former USSR) has changed and is now the same as in all Western European countries.

Band Plan for Western Europe and CIS Countries:
- 3500 to 3580 CW
- 3580 to 3600 RTTY, Packet, CW
- 3600 to 3800 SSB, CW

Band plan for Australia:

Australia has a somewhat peculiar band-plan. The most important change in recent years is the coming expansion of the SSB DX section from 3775 to 3800 (starting in Jan 2004).
- 3500 to 3700 and 3795 to 3800 CW
- 3535 to 3620, 3640 to 3700, 3775 to 3800 SSB
- 3620 to 3640 Digital modes

Band plan for Japan:
- 3500 to 3520 CW only
- 3520 to 3525 Digital modes and CW
- 3525 to 3575 All modes
- 3747 to 3754 kHz All modes
- 3791 to 3805 kHz All modes

Band plan for the USA:

Since the FCC decided to expand SSB privileges in the US, first to 3775 and later to 3750 kHz for Extra-Class amateurs, the DX window has de facto expanded from below 3750 to 3800 kHz during openings to the US, although the top 25 kHz is the focal area.

4.3.2.2. Recommendations for 80 meters

Many amateurs are unaware that 80 meters is a shared band in many parts of the world. In the USA, 80 meters sounds like a quiet VHF band compared to what it sounds like in Europe. Because of the many commercial stations on the band in Europe, the 25-kHz DX window can often hold only five QSOs in-between the extremely strong commercial stations in the local evening hours. If you are fortunate enough to live in a region where 80 meters is either exclusive or not heavily used by commercial stations, please be aware of this and bear with those who must continuously fight the commercial QRM.

Fortunately there is a common (IARU) DX window in Regions 1 and 2, but this is not enforced and is not respected by all. US stations complain bitterly about poor cooperation from rag chewers ("pig farmers" as they're sometimes called in the US), who have another 200 kHz that could be used for their local contacts.

The increased popularity of 80-meter DXing, together with the few DX channels available in the phone DX window, have created a problem where certain individuals would sit on a frequency in the DX window for hours (it seems like days) on end, without giving anyone else a chance. This problem is nonexistent on CW, where you have an abundance of DX channels in the DX window. This situation is also an excuse for creating DX nets, where ethics are not always the highest.

4.3.3. Let's be practical on 40 meters

Forty meters is pretty straightforward. CW DX QSOs all happen between 7000 and 7010 kHz, with rare exceptions around 7025 kHz. Although no formal DX subband has been created by IARU, the 7000 to 7010-kHz window is the defacto DX subband on 40 meters.

In the European or non-US phone band, being as narrow as it is, you can find DX anywhere between 7040 and 7100 kHz, with 7045 to 7080 kHz as the prime focal area.

The big news for radio amateurs coming from the 2003 World Radiocommunication Conference is that there will be a dramatic improvement in the 40-meter band soon. The conference agreed to shift broadcasting stations in Regions 1 and 3 out of the 7100 to 7200-kHz band and to reallocate the band to the Amateur Radio service. The allocation in Region 2 of 7000 to 7300 kHz remains exclusively Amateur. The broadcasting band in Regions 1 and 3 will become 7200 to 7450 kHz and in Region 2, 7300 to 7400 kHz. The changes will take effect on 29 March 2009. How the IARU societies will handle band planning is not certain at the time of writing.

5. SPLIT-FREQUENCY OPERATION

The split-frequency technique is highly recommended for a rare DX station or DXpedition working the low bands. It should logically also be the way we work DX on 160 meters. Signals on Top Band are often so weak that working split should really be the rule rather than the exception!

It is the most effective way of making as many QSOs as possible during the short low-band openings, because the marginal conditions often encountered are conducive to chaos if stations are calling the DX on his frequency. It also gives a fair chance to the stations that have the best propagation to the DX station. With list operations this is not necessarily so, and stations having peak propagation can bite their fingernails off while the MC is passing along stations who barely make contact and have to fight to get a 33 report. With split operation the DXer with a good antenna and with good operating practice is bound to have a lead over the modest station. This is only fair. Why else would we build a station that performs better than the average?

There are two good reasons for the DX station to work split frequency:

1. First he must realize that when he stops his CQ, there are likely to be many stations calling him. Though he might pick out a good strong signal, others may still call him, and his reply to a particular station may be lost in the QRM. This will result in a slow QSO rate, even though the DX station hears the callers well. If he works split, the callers will have more chance to get the DX's reply right the first time. The reason here is obviously that callers cannot cope with the QRM they are creating themselves on the DX's frequency. In this case the DX station should simply specify a single frequency (eg, up 5) where he will be listening.

2. Another reason is that there are such large numbers of stations calling the DX station that the DX cannot discriminate the callers. In this case it is the DX station that will not be able to handle the situation without going split. In this case he will specify a frequency *range* where he will be listening, in order to spread out the callers, and make the layer less thick.

Fig 2-2—W8LRL started DXing on 160 in 1972. In January, 2003, Wally worked VU2PAI for #310 on 160 meters.

A few general rules apply for split-frequency operation:

1. If possible, the DX station should operate in a part of the band where the stations from the area he is working cannot operate, or in a section of the band that is generally considered the DX section.
2. The DX station should indicate his listening frequency at least every minute. It only takes a second to do so, and it goes a long way toward keeping order.
3. The listening frequency should be well *outside* the DX window. Too often I hear a DX station on 3503 listening 5 up, ruining a major part of the DX window. There is no reason why he should not listen 10 or 20 up. The same applies to phone operation, where the DX station transmitting in the window should listen outside the DX window for replies.
4. If the DX station is working by call areas, he should exercise authority to reject those calling from areas other than those he specifies. He should not stay with a particular call area too long. At five stations from each area, at a rate of three QSOs a minute (that's fast!), it takes almost 20 minutes to get through the 10 US call areas!
5. The DX station should check his own part of the world to make sure the frequency is clear. This can be done periodically, especially if there is a sudden drop in QSO rate. Changing the transmit frequency a few kHz may bring relief.
6. If the DX station's listening frequency is being jammed, he should specify a frequency range instead of a single listening frequency.
7. On CW the split should be at least 5 kHz. For splits less than 5kHz the pileup's key clicks are likely to spread onto the DX-station's frequency.
8. Depending on the band plan in the country of the DX station, split-frequency operation may be unavoidable. This is the case when working the US from Europe on 40-meter phone. Under such circumstances, always make it a point to indicate your receiving frequency accurately, and make it a single frequency. If the pileup is too big, make it a reasonable range—10 kHz is usually sufficient. There is really no need to take more of the frequency spectrum than is absolutely necessary. (However, one valid reason to use a wider range than normally necessary is to elude deliberate jammers.)
9. Don't forget that the quality of the operator at the DX end (or DXpedition) is often judged by how wide a listening range he needs to handle the pile-up!

Some time ago I read on the Topband Reflector: "*I've never seen the reason to operate "split frequency" unless the country's band plan does not allow a station to operate on a desired frequency used by another country. I see no reason for a DX station to transmit on one frequency, listen on another and have a bunch of folks that didn't hear the frequency change transmission clutter up the band calling blindly on 2 or 3 different frequencies.*" W4ZV refutes this statement as follows: "*The DXer realizes that his signal is weak compared to the hordes calling. In light of the increased numbers of callers brought by packet spots, this is even more understandable. He understands that HIS signal is apt to be covered up by those stations. Since many callers have adopted the 'call until doomsday' technique, the DX is much less likely to complete a QSO within a reasonable period of time.*

Spreading the pile has only one goal: to make it more likely for the DX-station to pull a call out of that mass of noise... The split frequency method attempts to make the pileup more efficient and to work the maximum number of stations in a given period of time. In some latitudes, that window of opportunity can slam shut very quickly. The DX op may grow weary of getting up well before dawn for mornings on end in order to be able to log just a few QSOs even though he hears a swarm of callers...What we need to do is listen to what the DX op wants. If it appears at odds with your personal operating ethics, don't call him. If you'd like to be in his logs, follow his instructions and observe what he is doing. What we really need is some restraint on the part of callers when the DX station comes back to someone rather than a continuation of this mindless calling, calling and calling."

Tom, W8JI's comments on this same subject are: "*The real problem is many people who rarely work 160, or who are parked on a coastline with all the US behind them, think 160 is like 80 or 40. Factually if you are inland, DX signals are not that strong and local signals are devastating. 160 is a band where DX should always be split, unless it is a ragchew among friends. This is especially true in contests. Simplex DX operation in a 160 contest is just plain silly, unless we want only the big stations to work DX.*"

Who am I to argue with such arguments? The question is to which extent this can really be realized (see also Section 20).

5.1. 160 Meters

Rare DX stations should as a rule operate split frequency. The generally accepted transmit window for the DX stations is 1820 to 1830 kHz, with 1822 to 1828 kHz as the most popular range. It is good practice to use half frequencies; eg 1823.5, 1824.5 kHz. (See also Section 4.1.)

JA stations now can use 1810 to 1825 kHz, in addition to the old 1907.5 to 1912.5-kHz window (both CW only). You should stay out of the 1810 to 1825-kHz window for everything except for JA stations during the opening hours between Japan and the USA or Europe. JA stations should always work split frequency to keep the EU or USA stations out of their window so that their strong local signals don't cause havoc.

Fig 2-3—Greatly missed Peter, DJ8WL (SK), the German "Mister 160 Meters" for many years.

JAs should listen *above* 1830 kHz. For European and US stations: Never call JA stations on their frequency. Force them to go split frequency. They will be happy to listen above 1830 and this will greatly improve the QSO rate.

If you want to call CQ JAPAN, do it above 1830 and indicate your listening frequency either as "QSX 15" or simply as "DWN 15."

I've seen some advocating the use of split frequency during contests. Nowadays our narrow 160-meter band is already fully congested during these contest periods. If we all take two channels for a QSO it will become much worse. Split frequency during contests is not realistic to me, despite what the ARRL band plan advocates.

5.2. 80 Meters

On CW the main reason for the DX station to go split is when the pileup gets too big. Another nice reason for the DX station to listen "up 25" to cover General-class stations in the US.

On SSB I can think of many good reasons to go split: in the first place, not to occupy the DX window more than necessary. Therefore the DX station should always indicate a listening frequency outside the DX window (below 3750 kHz). US stations wanting to work Europe should transmit above 3800 and listen below 3750 to keep the DX window as uncongested as possible.

Middle-East stations should transmit on a frequency below 3750 kHz when working North America to avoid QRMing European stations. Stations in the Pacific working Europe should transmit above 3800 kHz and listen below 3800 kHz to avoid US QRM.

It is not reasonable for a European to transmit inside the US phone band (3780 kHz, for example) and listen on 3805 kHz. If this is done, two windows inside the US subband are occupied for one QSO, and the potential for QRM and confusion is increased. The inverse situation is equally undesirable.

Every year I hear hordes of European stations trying to work USA stations in the ARRL DX phone contest in the shared band 3750 to 3800 kHz, where they must overcome local US and Caribbean-made QRM. I always enjoy doing the contest just below 3750, listening above 3800 kHz, and never have any such problem.

5.3. 40 Meters

The nature of the pre-WRC2003 frequency allocations in different regions made split-frequency operation a very common practice on 40-meter phone. After March 29, 2009, a new internationally agreed band plan will come into effect. It seems that finally Europe and the USA will be able to work each other on SSB without having to go split.

Until March 2009, however, US stations in the 7150 to 7300-kHz window should be aware that they operate in the midst of very strong broadcast stations in Europe. These broadcast stations are not on the same frequency 24 hours a day, and what may be a clear frequency one minute can be totally covered by a 60-over-S9 BC station the next minute. These BC stations usually appear on the hour or on the half-hour. Especially in contests, make sure your supposedly "clear" transmit frequency remains clear! During contests various DX stations may be using the same listening frequency when working split. Therefore it is essential that the caller not only gives his own call, but also the call of the station being called. Going just by timing does not always guarantee a real QSO.

Not only the BC stations cause problems, but also non-Amateur-Radio phone traffic between 7000 and 7100, much of which comes from Mexico and South America. Many pirate stations transmit on channels that are organized in 5-kHz steps. Therefore it is a good idea for the DX station working the USA to try multiples of 2.5 kHz to avoid this kind of QRM.

5.4. DX Subbands in 160-Meter Contests?

Over the past years we've seen rules for DX subbands on 160 meters come and go. This is especially a critical issue on Top Band, since local stations are extremely strong and DX stations are usually very weak, much more than on 80 and 40 meters.

The classic scenario we had on 160 meters was to reserve 1820 to 1830 kHz only for a DX station to call CQ Contest, after which others could reply to his CQ. The problem is "What is DX?" Usually DX means a station outside your continent. This means a P4Ø or a PJ2 in South America could sit in the DX window and works hordes of US stations in North America, while a KP2, FM or FG cannot because they are also in North America.

And when the band is open between US and Europe who's DX? The Ws are DX to me and I am DX to the Ws, so who should be in the DX window? What should I do if I'm in the window and a European comes back to me—He may even be a new multiplier for me. Typically he is just a nice guy, who wants to give me some points. I could ignore him but he will probably keep on calling. So do I have to chase him off with a curt "QSY—I can't talk to you."? That's not a very nice thing to say to someone who's new to the game or who just is trying to do me a favor.

Well, how about considering two windows—one where the USA can call CQ, and one for Europe. But where should the Africans and others go in this scenario? Well, then I guess we need four DX windows, one each for the USA, Europe, Africa, and Asia (the Pacific can use the European window since opening times do not really overlap). But the 160-meter CW band is only 30-kHz wide. Let's see; we can reserve 10 kHz for Europe, another 10 kHz for the USA, and 5 kHz each for Asia and Africa. But that does not solve the problem—Where do the European stations go that want to work Europe? And how about the US stations that want to work US stations? They can do it in the US window, which means they will have 10 kHz, and they will all be sitting one on top of another in this crowded space. Impossible! All of that chaos, while the African and the Asiatic windows will be half empty with the small amount of activity from there.

Well, maybe, we could give zero points for working your own continent. That would be the end of my contesting on Top Band. I don't want to spend 30 hours working only 150 DX stations outside of Europe. That would be really boring. Today we have a vibrant and exciting worldwide contest for everyone to participate in. If we make it a pure DX-to-DX contest it would be a dull, boring and insignificant contest.

So, let's forget about these DX windows and find technical solutions to the problems we face on Top Band. Let's clean up our transmitted signals and use better, more directive and more selective receiving antennas. Let's concentrate our energy on creating solutions rather than workarounds. The more I think about it, the less sense these DX Windows make to me. Considering all of the above it looks like it was a wise decision for CQ 160 to abandon the DX Window.

6. RIT (THE CLARIFIER)

Zero beat is a term indicating that the two stations in contact are transmitting on exactly the same frequency. Unless working split frequency, it is common practice to zero beat on phone. The RIT (Receiver Incremental Tuning, also called the receiver *Clarifier*) on some older transceivers created problems where stations in QSO drifted apart, after which the operator used RIT to compensate. They would be better off making sure that they stayed on the same frequency.

Fortunately, modern equipment is practically immune to frequency drift, so this is not as much of a problem as it used to be, especially with some of the home-built equipment of old. Bill, W4ZV, recommends checking your actual transmit frequency with a separate receiver, especially when you get a new transceiver, to make sure you are placing your TX signal where you think you are! He added "*My 1000MP has some quirk I have never figured out which makes me need to add 70 Hz to the TX frequency to be zero beat.*"

As an example of where this is a problem, let us assume station A does not have a stable VFO. If station A and station B start a QSO at zero beat (both on exactly the same frequency), there are three sequences of events that a monitoring station might observe:

1. Neither station uses RIT: One station always follows the other. The QSO may wander all over, but at least there will be no sudden frequency jumps when passing the microphone, and the QSO will be on one drifting frequency.
2. One station uses RIT, the other does not: If we are still listening on the same frequency that the QSO began on, there will be a frequency jump at the start of transmission of one of the stations, but not for the other station. The QSO will still drift.
3. Both stations use RIT: Again, if we are listening on the original frequency, there will be a frequency jump at the start of transmission for both stations, and the two stations may drift away from one another. The QSO will take up more space on the band, and it will be very annoying to listen to. Should RIT be used in such a case? Decide for yourself.

Some recent transceivers not only have an RIT control but also an XIT control (TX clarifier). This one makes things even more complicated. Be careful when using XIT. There are some instances, however, where RIT can be a welcome feature:

1. Some operators like to listen to SSB signals that sound very high-pitched, like Donald Duck. That means that they tune in too high on LSB. To the other operator, their transmission will sound too low-pitched, because they are no longer zero beat. Tuning in a station using RIT will allow one to listen to the voice pitch he prefers, while staying zero beat with the other station(s) on frequency.
2. When trying to get through a pileup, it can be advantageous to sound a little high in frequency. Adjusting the RIT slightly in such a case will yield that result.

3. Let us assume our transceiver is designed for working CW at an 800-Hz beat note, and it is only when listening at this note that the transmit frequency will be exactly the same as the receiving frequency. You can use RIT to offset the transmit frequency (by, say, 300 Hz) to bring the note down to 500 Hz, and still transmit on the receiving frequency. So you can see that RIT can be a useful feature without creating unnecessary QRM at the same time.

As previously stated, except for intentional split-frequency operation, most DX QSOs on 80 meters are on one frequency. There is no need to waste space on the band by working a station slightly off your frequency.

By the way, I never use the RIT on my transceiver. So far as I am concerned, the modern transceiver with dual VFOs simply has no need for RIT or XIT. My FT1000 second VFO is, under normal circumstances, my TX VFO, and I can do all the tuning around I want with the first VFO without changing my transmit frequency. Why would I want to use that tiny RIT knob when I can use the big main knob?

7. ZERO BEAT

The terminology *zero beating* stems from AM days. On AM one used to really zero-beat. When the transmitter VFO is tuned to the receiving frequency, a beat note is produced. This audio note is the mixing product of the two signals. When the beat tone becomes 0 Hz, the transmitter and receiver are on the same frequency.

On CW, most transceivers are designed so you are transmitting on the same frequency as the station you are working *only if the beat note is some specific frequency*. With older varieties of transceivers this was a fixed beat note, usually 800 Hz. This beat-note frequency is usually specified in the operating manual. Because many hams do not care for the specified 800-Hz beat note, they just listen to what pleases them (450 Hz is my preference). As a result of this, those operators are always off frequency by 350 Hz or so on CW. This is not a problem if the receiving station uses a 2-kHz filter, but it could be a real problem if he uses a 250 or 500-Hz filter. Also, think of all the wasted space on the band. This is the reason that in the past I advocated the use of a separate receiver and transmitter on CW. Then, at least you could really listen to your own frequency!

More modern transceivers on CW provide for operating right on frequency. A good transceiver should at least have an adjustable pitch control. The CW monitor note should also shift accordingly—continuously adjustable down to 200 Hz is the best. Some people like to listen at very low pitches. (W4ZV likes 250 Hz.) The only precaution here is to tune in the station you want to work at exactly the same beat note as your CW monitor note. That's all there is to it. This way, you can get easily within 50 Hz of the other station and still listen to your preferred beat note.

There is a lot of personal preference involved in choosing the beat note itself. It is very tiring to listen to a beat note higher than 700 Hz for extended periods of time. The ability of the ear to discriminate signals very close in frequency is best at lower frequencies. For example, listen to a signal with a beat note of 1000 Hz. Assume a second signal of very similar signal strength and keying characteristics starts transmitting 50-Hz off frequency (at a 950-Hz or 1050-Hz beat note).

Separating these two signals with IF or audio filters would be very difficult. Let us assume we have to rely on the "filters" in our ears to do the discrimination. The relative frequency difference is:

$$\left(\frac{1050-1000}{1000}\right) \times 100 = 5\%$$

If you were using a 400-Hz beat note, the offender would have been at 450 Hz (or 350 Hz), which is a 13% relative frequency difference. This is much more easily discernible to the ear.

It's a good idea to do some checks with a local station (or with a second receiver) to make sure you are truly "zero beat" on CW. This will save you lots of frustration. In contests I have often cleaned up my QRG and worked even the very weak signals, only to find out that there was a guy with an S9 signal calling me 400 Hz up. He was strong but I never heard him, while I easily worked stations that were 40 or 50 dB weaker than he was...

8. THE DXPEDITION—BEING RARE DX

You don't have to be on a DXpedition to be a rare one. There are still dozens of countries where the number of licensed radio amateurs can be counted on one hand. Operating as a resident or temporary resident from such much-wanted country is very similar to working from a DXpedition. The required expertise to make Low Band DXing a success is the same as required from top notch DXpeditioners.

8.1. DXpeditions and the Low Bands

Thirty years ago it was rare to have a DXpedition show up on 80 meters and 160 meters was out of the question. That was just the "Gentlemen's Band" for daytime rag chews. Fortunately there has been positive change over the years, and for most expeditions 80 or 160 meters has become just another band. During the lower parts of the sunspot cycle they are definitely capable of bringing in a lot more DX than 21 or 28 MHz! The 5-Band DXCC, 5-Band WAS, and 5-Band WAZ awards have also greatly promoted low-band DXing. So have the single-band scores

Fig 2-4—Bob Eshleman, W4DR, DX Hall of Fame member, has 310 countries on 160, 346 on 80 , and 358 on 40 meters.

Fig 2-5—Jack, VE1ZZ, always first in and last out when it comes to Top Band openings into Europe. (This reputation is now under attack by VY2ZM—K1ZM's new station on Prince Edward Island!)

and record listings in DX contests.

Until a few years ago some DXpeditioners would only appear on the low bands in the last one or two days of operation. Staying on bands with the best QSO rates will not result in many Top-Band QSOs. But logic tells you to tackle 160 and 80 meters from the first day, as there may not be low-band openings every day. W0CD writes in his survey reply: *"DXpeditions going to new countries should give more time to 160 to be sure there is decent propagation. Not just a few hours the last night."*

A DXpedition should prepare well for the low bands. They should ask an experienced low-band DXer to determine band openings for the low bands. But fortunately, most of the well-organized DXpeditions now include at least one low-band expert. Nothing is more frustrating than to hear a Far-East station working Europe on 80 or 160, during the 10-minute window that this path is open to the US East coast. Joerg, YB1AQS, from the famous ZL7DK team said it so well, *"As we've found all the years—the 160-meter antenna has, if possible, to be the first one up and the last one down."*

8.2. DXpedition Frequencies

On 80 and 40 meters, typical DXpedition frequencies are in the bottom 10 kHz of the bands for CW, usually listening 5 to 10 kHz up, or sometimes operating around the 25-kHz mark. On phone frequencies are usually in the 3795 to 3805 window for 80, or on 40 meters anywhere between 7040 and 7100 kHz. DXpeditions should specify a listening frequency *outside* the DX windows!

On 160 CW the range from 1823 to 1826 kHz (with 1825 as a focal point) is widely used by DXpeditions, with QSX 1830 to 1835 for areas where these frequencies are available.

A DXpedition should announce its frequencies well beforehand. The Internet is the ideal place to do this. It's also a good idea to publish several "escape frequencies" in case of QRM, intentional or not. Stick to the published frequencies, otherwise your credibility may suffer. When leaving one band, always announce where you are going and repeat the information several times (not too fast on CW!).

8.3. Split Frequency

DXpeditions usually operate split frequency, both on CW and SSB. The advantage of split-frequency operation on the low bands is even more outstanding than on the higher bands, because the openings are much shorter and signals can be much weaker than on the higher bands. Working split makes it easier for calling stations to hear the DX. Otherwise, the strong pileup of callers will inevitably cover up the DX station, resulting in a very low QSO rate.

Sometimes we hear DXpeditions spreading the pileups over too wide a portion of the band. This is not generally advantageous for the QSO rate, and most of all, is very inconsiderate to other users of the band. It is also common for two DXpeditions to be on at the same time, both listening in the same part of the band. The net result of this is maximum confusion and frustration for everyone involved. There will inevitably be many "not-in-log" QSOs, where people ended up in the wrong log.

Calling by call areas seems to have become the standard approach to handle a pileup that's become too big to be handled without instructions. This is a fine procedure, provided you don't stay with the same call area for, say, more than five QSOs. Otherwise you might lose propagation to certain areas before going around all call areas. Even at a 2-QSO-per-minute rate (which is high for the low bands), it takes almost half an hour to go through the 10 US call areas! In fact, when working the US on 160 meters it makes no sense to work by call areas, since the propagation usually is very area-selective anyway. Do not call by country. This inevitably leads to frustration. Why did he call for Holland and not for Belgium? Holland is only 20 km from here; why do they get a chance and not me?

On long haul paths on 160 meters skip is very often area selective and moving around. The secret to success on Top Band is to keep things simple. Simple instructions like, USA 5/10 UP or EU 7 UP are okay. More complicated instructions will inevitably lead to chaos on 160. However, do not just send UP. This will result in people calling less than 1 kHz from your frequency. Instead, specify QSX 5, or UP 5/10. If the pile is not too big, specify a single frequency (rather than a range) on which to listen.

On the other hand, if your pileup grows too big you can eliminate those that copy you from those that "pretend" to copy you by suddenly changing your QSX and then quickly working the ones who really are copying you!

During the preparation phase of a DXpedition, it is a good idea to ask DXers in different parts of the world for their best low-band receiving frequencies. This way they can avoid trying to hear on a frequency where there is always a carrier or where every few minutes a commercial station pops up.

8.4. Controlling the Pileup

Sometimes, you hear a beautifully smooth pileup. A dream! A pleasure to listen to! Pure music! Sometimes, it's pure chaos. Let me be blunt: It is the DXpeditioner who's responsible for either situation. Here are a few hints on how to control a pileup:

- Avoid frustrating your public.
- Avoid sounding frustrated; inspire confidence.
- Show authority, but not temper.
- Keep your instructions simple.
- Stick to your instructions yourself. Never make any "out-of-turn" QSOs.
- Change the QSX frequency if the pileup grows too big. Those that copy you well will immediately follow. I saw BQ9P (October, 2003) doing this with Europe on 160 and it was very effective and efficient.
- Avoid copying half calls, this just slows down the QSO rate (especially on 160 where slow and deep fading is commonplace). Working with half-calls *only* works with JA stations, certainly *not* with a European pileup!
- Always repeat the full call to tell the DX station that he's in the log, so he won't call you again for an insurance QSO.
- On 160 and 80 meters when paths are very marginal, send the call of the station you are replying to several times. After sending the report, send his call again and use a standard way of ending each QSO (TU, 73, etc). This is the signal for the crowd to start calling.
- Do not change your way of operating. Have a well thought-out strategy and rhythm, and stick to it. This will inspire confidence rather than frustration in your public.
- From time to time ask if your frequency is still OK (especially if your rate suddenly drops).
- Ask your audience to look for a new transmit frequency for you.

8.5. Calling the DX Station

Before calling a DX station, make sure you hear that station. Often we see guys calling a DX station as soon as it's been announced on the DX cluster, often without even having heard the station. Such a caller is just making a fool of himself.

Another thing you need to do before calling is to listen to his pitch on CW and to his rhythm. Take the time to tune the receiver for best copy and select your best receiving antenna. When it's time to call, never give half of your call. Chances are that when the DX station comes back with "ABC?" you'll find there are a couple of "ABC" stations on the frequency.

If the DX station does not work spit, you're in for a nerve-wracking session. Give your call two or three times and listen. If there are other stations still calling, don't start calling again—Maybe the DX has already called you. Don't call endlessly! Throw in short calls every now and then. Stay relaxed, be patient and pray that eventually the DX station will go split.

If the DX station is working split, first determine where he is listening. Listen in the pileup, and see what his operating strategy is: Does he stay on the same frequency; does he move up or down a small amount; or is he really jumping around? Don't start calling him unless you know what he's doing. In such a pileup it's good to listen more to the calling "mob" than to the DX station! And under no circumstances make any comments about the rude behavior of some other people. Bite off your fingernails instead.

8.6. Calling CQ on a Seemingly Dead Band

We frequently hear that every wise DXer spends all his time listening, and only transmits when he's sure to make a contact. He never calls CQ DX; he just listens all the time and grabs the DX before someone else does. This rule for sure applies to the DXer, to the "hunter."

However, this rule does *not* apply to the DXpedition (the "hunted"). If the golden rule for a DXer is to "LISTEN, LISTEN, LISTEN," then the golden rule for the DXpedition should be "CALL, CALL AND KEEP CALLING CQ!" And please, don't give up after just a few minutes. DXpeditions should call CQ, even on a seemingly dead band, at the times they publish. You can be assured that there are hundreds of faithful low-band DXers digging for your signal.

And don't go away after just one or two contacts, even if there are no replies for a while. You probably will be announced on the DX Cluster, but it takes some time before the news gets out.

The ZL7DK guys said it so well: *"During our stay we got at least one good opening in all possible directions, but on average not more than two per destination. The openings in the critical directions (mostly the polar paths) have to have absolute priority. The paths are open maybe 5 or 10 minutes a day, if they are open. If you are dedicated to work stations on these difficult bands and difficult paths, you must be there every day (to call CQ) in order not to miss any opening."*

8.7. Pilot Stations: Information Support for DXpeditions

After a rather tentative attempt during the AH1A expedition in 1993, the Pilot-Station concept was first introduced on a larger scale during the 3Y0PI expedition in 1994. (Mark, ON4WW, seems to have the honor of being the very first Pilot Station.) Three years later the famous VK0IR expedition in January, 1997, set the standard for how excellent logistics and a smoothly working Pilot Station can help a difficult DXpedition be a huge success. Both these expeditions were led by Bob, KK6EK.

In the past, DXpedition feedback and information had to be forwarded on-the-air during prime operation time. As a consequence information flow was minimal. Well-organized expeditions can use Pacsat (packet radio via satellite) or HF digital communications to establish a solid link between the rare spot and its home base. They can also use Internet e-mail, if available, perhaps with a satellite-telephone system. The only limitation is that this kind of communication link is likely not to be continuously available. Nowadays, however, DXpeditions can use wideband commercial communication networks from even the most remote spots on earth.

The DXpedition pilot takes care of all the information flow to and from the DXpedition via one of these links. He organizes himself to have a maximum of information from the "public" and to feed a maximum of information from the rare spot back to the public. He is the DXpedition's spokesman, the public-relations man, dealing with:

- What does the DXpedition hear during the low-band openings; what are the problems; what are the schedules (times and frequencies)?
- What, and when, is the public hearing the DXpedition, and are there suggestions for improvement?
- Making the log available in (almost) real-time.

The first two items are there to optimize the results and to create confidence that all is being done to "make" it. The real-time logs are important to avoid stations from making a "back-up" contact (I'm not 100% sure my first QSO was a good one.). I once missed a country (Malpelo) on 160 by not making a backup QSO, so I really cannot blame anyone for doing so if not 100% sure about the first try. Having the logs available on the Internet avoids this situation.

We should never forget that a DXpedition must be there for the DXers (the public) in the first place. To successfully add value to the DXpedition the Pilots must have a high esteem from the DXpedition leadership and must be fully integrated with the team. The Pilots should be part of the decision-making process and not just the poor in-between guy, who takes a beating from both sides. Some examples of bad attitude and bad answers from a pilot are:

- *"You are all complaining about the same thing."* (In other words, leave me alone.)
- *"I report what the operators tell me they will do."* (Which means I'm just the in-between guy and I am far from sure that they will do what they say.)
- *"We are making every effort to have a CW operator work from Eastern Europe across the Continent and across the United States on 80 CW at your sunrises."* (This is an empty phrase with no message. A message with real content might be: *"They will be on 3502, from 0300 to 0500Z on Feb 10."*)
- *"I have discussed the problem with the leader and while we are trying we cannot promise anything."* (I other words, don't count on it.)
- *"Alert: they will try 160 tonight. Time unknown."* (There is no useful message here. We expect them on 160 and 80 every day anyhow!)
- *"Golly, it's easier to criticize an operation than it is to put a rare one on the air."* (A pilot must expect to receive criticism. That is part of his role. Criticism is only expressed when one thinks something is wrong. It's the role of the pilot to analyze the criticism, and to do something about it, to provide a solution, or at least an explanation.)

These are not fictitious situations. They were heard during a 2003 DXpedition that lead to great frustration from many low-band DXers.

In the future we will see further improvements in Pilot Stations. Before too long I expect to see real-time logs on the Internet and maybe even the actual logging screens of the DXpedition stations as they work people. A web camera "in situ" could make DXpeditioning a spectator's sport. Almost-real-time spectating was introduced during WRTC 2002, where everybody could follow the scores of all 52 WRTC competitors on the Web in 1-hour increments.

8.8. DXA—Beyond the Pilot Station

Bob Schmieder, KK6EK, of 3Y0PI, VK0HI and XR0Y/Z fame, has recently introduced the concept of *DXA* (DX and Extended Access), which takes DXing a giant step further. The *DXA* Website will go far beyond mere presentation of information. It will be highly interactive, dynamic and fast. *DXA* will use a central computer to maintain a database updated directly from the DXpedition site. This database supports a variety of users, including casual visitors, logged-on visitors, subscribers and other users.

Depending on the user's status, his Web browser will display real-time status of the DXpedition, plus pre-generated streamed content and near-real-time display of his own log status. *DXA* will have the ability to talk to the radio through a data interface, where the radio processes RF signals and the computer processes information, both to/from the radio and to/from the Internet. Merging of the two technologies will provide an exciting, qualitative advance for the radio amateur!

9. CW ON THE NOISY LOW BANDS

There's no doubt about it. CW is superior to Phone when it comes to making a QSO under marginal conditions. CW can use a much narrower bandwidth, which means a better Signal-to-Noise ratio. I typically use a 250-Hz bandwidth on CW, versus 2.1 kHz on SSB, so the advantage is obvious.

What about PSK31? It is a fact that a well-trained CW operator can copy weaker signals in low-band noise much better than PSK31 does. This is because the decoder (the operator's brain) is vastly superior to the PSK demodulator/software. But I must admit, PSK31 comes close.

One of the situations that makes copying signals very difficult is QRN. It appears there are two families of QRN: high-latitude QRN and tropical QRN. The difference is that crashes of tropical QRN generally last much longer than those generated by high-latitude QRN. With higher-latitude QRN the pauses between crashes usually last longer.

If you want your call to make it through high-latitude QRN, high-speed CW can sometimes be a solution. Dan, K8RN, who operated VK9LX on Top Band said: "...*QRN was very bad even with Beverages for receiving. It seemed to me that if the stations calling sent their call fast, they had a better chance of making it through (between) the static crashes. If the speed was too low nothing made it through.*" But high-speed CW is no good at all to pierce through tropical QRN.

Rolf, SM5MX and XV7SW, recently commented on the Top Band reflector: "*In this kind of tropical QRN, each QRN bang often lasts long enough to mask a call sign completely. From the DX end you may just understand that somebody is there and call QRZ?, but the same thing will happen again at the next bang, the next one—and the next and so on, if the speed is too high. So I found it tremendously helpful when people reduced the speed. Once you are able to pick out a letter here and there, you may be able to paste together a full call sign and eventually make it.*"

Referring to another issue regarding high-speed CW on the low bands, Tom, N4KG, commented "*High-speed CW on the low bands by DX stations contributes to confusion and disorderly conduct in the pile-ups. Half of the callers can't copy anything but their own call signs, even with a good signal on a quiet band.*"

The DX station should determine the CW speed. His sending speed should be the speed he expects replaying stations to use. Tom, N4KG, added: "*DX stations sending above 30 WPM on the low bands actually reduce their rate and promote more broken calls. 25 to 28 WPM seems to work well for most cases. On long polar routes, with weak signals, QSB, and QRN, high speed is counterproductive. Sending a call twice at 25 WPM takes less time than three times at 30 WPM and is more readily copied.*

Fig 2-6—Bob Schmieder, KK6EK, who has organized some of the best and most successful DXpeditions in the last decade (3Y0PI, VK0IR, XR0Y).

Joerg YB1AQS formulated it as follows: "*Even if you can hear everybody crystal clear—don't shoot at them in CW with 35 WPM! 22 WPM on 160 m and 28 WPM on 80 meters are enough. Repeat their call sign two times before the report and at least once at the end.*"

Chris, ZS6EZ made an important remark along the same lines: "*Never, never, NEVER screw up the spacing in your call. If you use standard Morse spacing, the receiving station can often recover dits that are inaudible, by listening to the timing of the characters. For some reason, some people think the call is easier to copy if they leave exaggerated spaces between letters.*" Well it simply does not work that way. The rhythm is very important!

One more piece of advice: Send your full call sign, not just part of it, like "XYZ k", expecting the DX to know that it's you. If the DX has to ask "XYZ?"—and several times at that—you are wasting everybody's time, including your own. In addition identifying like this is an illegal practice in many countries. Now, if the DX station comes back with one letter wrong in your call (he sends ON4UM, then I go back to him as ON4UN UN UN UN pse cfm k. But as a rule, let's avoid half calls. They are just a waste of time.

10. NETS AND LIST OPERATIONS

The use of lists, which occur daily on the HF bands, started with net operations on the HF-bands in the 1960s. In most of these nets, a "master of ceremonies" (MC) will check in both the DX and the non-DX stations, usually by area. After completing the check-in procedure, the MC directs the non-DX stations—one at a time, in turn—to call and work the DX station. In most cases the non-DX station has indeed worked the DX station, but there was no competition, no challenge, no know-how involved. Some write the MC a letter, or send him an e-mail message or even call him on the telephone to get on his list!

What satisfaction can you derive from such a QSO? Yes, it gives the QRP operator a better chance to work the DX station, and the only thing you have to do is copy your report—and even that may be relayed to you. When it's your

Fig 2-7— List operations. Sigh...

turn, the MC will call you and invite you to make a call. Doing so, he has used your call so the DX station already knows your call sign. And if the DX station is a DXpedition, there is a good chance that he will give everyone a 59 report, so it becomes even simpler. Just like shooting fish in a barrel, in my opinion. Fortunately, lists have never made it on CW.

A list cannot be used if the DX station refuses to take part. Fortunately, we rarely see a DXpedition worthy of the name doing this. I remember hearing stations asking Carl, WB4ZNH, operating as 3C1BG on 80 meters, if they could run a list for him. Carl was insulted by the proposition.

If a DX station is involved in a list operation, it generally means he cannot cope with the situation. The ability to cope with a pileup is part of the game for rare DX stations. There should definitely be no excuse for such things to happen on DXpeditions. If you are not a good enough operator to handle the situation yourself, you should not go on a DXpedition.

In almost all cases, list operation can be avoided by working split frequency. I think it is always a poor solution. Because there will always be a number of poor operators, as well as newcomers, it is likely that we will have to accept lists every now and then.

11. ARRANGING SKEDS FOR THE LOW BANDS

Once you work your way up the DXCC ladder, you will inevitably come to a point where you will start asking stations on the higher bands for skeds on the low bands. You will often be asked to specify the best time for the schedule. Remember that you asking for a favor, so try a time that is not in the middle of his night. Rather, get yourself up in the middle of the night! Also, don't go by a single schedule. Arrange a minimum of three skeds, or maybe a week's skeds, to hit the day with the right propagation. Tell the other party that the band may be okay only one day out of three or one day out of five. Find out how much power he runs on 160 and what antenna he is using, so that you know what signal to expect. Don't forget to have your sunrise and sunset information ready at all times. Most computer logging programs nowadays include it.

Tell your sked that you will call him. Don't give his full call; just his suffix when you call. Or just call CQ DX at the sked time exactly on the agreed frequency. You don't want to give away your sked to strangers. If you work split, don't give away your listening frequency before you've worked him! Spot him after you make your QSO.

12. GETTING THE RARE ONES

Working the first 100 countries on 80 or 40 meters is fairly easy. Well-equipped stations have done it in one contest weekend. Anyone with a good station should be able to do it easily within a year and a growing number of stations have achieved DXCC on 160 meters. The major DX contests (CQ Worldwide DX, ARRL International DX, WAE, All Asia, CQ Worldwide 160-Meter, ARRL 160 Meter, etc) are excellent opportunities to increase low-band scores. A good time to look for semi-rare ones, by the way, is just before and after a contest, since that may actually be an easier time to work them due to less QRM.

13. GETTING THE LATEST INFORMATION, AND FAST

In the old days we had dozens of DX bulletins, all over the world, to inform us. Then came packet radio, and along with it the DX Clusters. Information was much "fresher," and within a day or so a message sent from the US would arrive in Europe. Local DX-information nets, mainly on 2-meter FM, which were thriving 10 years ago, have all but disappeared.

Today all dedicated DXers use the Internet, e-mail and the Web to get the latest information, quickly. During the VK0IR DXpedition there were up to three bulletins a day sent on the Internet (to various reflectors on e-mail plus a dedicated Web page). News was everywhere within hours of being released on Heard Island. DXers quickly got used to the Internet and its powerful possibilities.

13.1. DX Clusters

DX Clusters have been with us for quite a few years now. The DXer sends to a DX Cluster station his real-time information (the call and frequency of a DX station he has heard) and he receives the same kind of information back, contributed by others connected to the network. Nowadays, most DX Clusters are interconnected through the Internet rather than through RF links on 2 meters and/or 70 cm. It is likely that packet-radio RF networks will be completely replaced by much faster Internet connections in the future (although backup RF links are still very useful when Internet connections crash).

All of this means that you don't need a 2-meter/70-cm station and a TNC to connect to your local DX Cluster. Many hams have continuous Internet access (by CATV cable or wideband twisted pair) and can connect to any DX Cluster anywhere in the world. DX Clusters are thriving better than ever!

DX Clusters and the Internet have changed DXing in general. Some publications have pictured DX Clusters as the greatest evil in Amateur Radio. They are said to undermine the art of listening. Scott, W4PA, has a very strong view about this issue: *"Shut off packet radio, and do it like a man."* The fact is, of course, that DX Clusters and the Internet are here to stay. We will all have to develop different skills that help us keep this technological advantage over competitive fellow DXers in this changing world. I am personally convinced that the DX Clusters and the reflectors and Web pages on the Internet are just a set of superb tools that have evolved in our wonderful hobby.

Today most DX Clusters are interconnected worldwide, and a few operators complain about spots from distant geographic areas. What is the use, they say, for a European DXer of a Japanese station spotting a ZL7 in the middle of the day (in Europe) on 3.5 MHz? This is certainly true to some extent, but such information may help find rare openings on the higher bands, at times nobody would normally expect them. During contests you can also see how propagation moves in a certain direction, and it is always interesting to see how many times you get spotted during a contest!

Here are some addresses of DX Clusters available by telnet:

www.ve9dx.com/telnet/sites.html
www.cpcug.org/user/wfeidt/Misc/cluster.html
www.cestro.com/pcluster

The latest and most advanced of so-called DX Cluster "concentrator programs" is *DX Cluster Concentrator* (*DXC*) by ON5OO, who besides being an avid contester, is a professional programmer. *DXC* any logging and contesting programs having provision for telnet access, such as *DX4WIN* (**www.dx4win.com/**), *DXBase* (**www.dxbase.com/**), *Swisslog*, *LOGic6* (**www.hosenose.com/**), *Eurowinlog* (**www.eurowinlog.de/**), *TRX Manager* (**www.trx-manager.com/**) and *XMLog* (**www.xmlog.com/**).

A concentrator program like *DXC* allows you to be one of the first to receive a spot, even if it's originated half way across the world. The programs includes a lot of additional features, such as selection criteria (eg, which bands), alarm criteria, etc. You can even send an MSM message to your own cellular phone telling you that one of the countries you still need has been spotted!

You can also use *DX Summit*, a popular Web site (**oh2aq.kolumbus.com/dxs/**) built and operated by the members of OH9W/OH2AQ Radio Club. DX Summit collects DX spots from a wide range of DX, making them available to us every 1 to 3 minutes. As such it is *not* a real-time affair, and is *not* faster than your local DX Cluster on packet radio, but it has information from all over. Another advantage is that you can select the spots (eg, only 160 meters). **www.oh2aq.kolumbus.com/dxs/1.html** gives you the last 100 Top-Band spots, updated every 3 minutes.

13.2. Internet Chat Channels

Chat channels are the most real-time gadgets around nowadays. ON4KST set up a dedicated Low-Band Channel (**www.on4kst.com/chat**). The screen has three windows: the main window with chat text, a window showing the users that are logged on, and a third window showing you the latest spots for the low bands (40-80-160) collected from some 20 different DX Clusters worldwide.

This chat channel has become quite busy and many of the big Top Band guys use it to exchange information. In the first days after its creation W8JI, AA1K, K7ZV, JA5AQC and W4ZV were already on it. While such a chat channel is undoubtedly interesting and useful, it could also be a dangerous tool. Even before such Internet chat channels came into existence we have seen some would-be DXers using the "announce" function on a DX Cluster to send messages like: "I am calling you on xyz kHz. Do you hear me?" and "You are 339 did you copy my report?" Such message are ludicrous, unethical and unfair. Let's make sure we all use such chat channels correctly. Then we will have another technological tool with which we can responsibly enjoy our hobby.

13.3. Internet Reflectors, DX Magazines, Etc

While many years ago, printed DX magazines served the noble purpose of informing the DXers of upcoming activities, this role is now taken over by DX Clusters and Internet-reflectors. To my knowledge, the only monthly printed publication that specializes in low-band affairs, is *The Low Band Monitor*, published by Lance Johnson (see **www.qth.com/lowband**). This little magazine has monthly activity reports, stories on recent low-band DXpeditions (and logs for 160 meters), articles on low-band antennas, etc.

Several special-interest groups are interesting to the low-band DXer on the Internet. These interest groups use so-called *reflectors* to exchange information among their sub-scribers. Reflectors are semi-open mailboxes, to which anyone can subscribe, free of charge. Once subscribed, you will get copies of all the mail that is being sent to this reflector. By addressing a mail to the reflector, you reach everyone who is currently subscribed to that particular reflector.

The Topband Reflector (**lists.contesting.com/mailman/listinfo/topband**) is the place to be for all Top-Band related information. Bill, W4ZV, manages this reflector. The archives for this reflector can be found at **lists.contesting.com/pipermail/topband/**. Other popular Internet sources for DX and contest information (related to any HF band) are:

- The DX Reflector (**njdxa.org/dx-news/index.shtml**). The DX Reflector archives can be found at: **www.mail-archive.com/dx-news@pro-usa.net/index.html**.
- If you are into contesting, then the Contest Reflector **www.contesting.com/FAQ/cq-contest** is a very good source of information. The archive containing all the E-mails from the Contest Reflector can be downloaded at **lists.contesting.com/_cq-contest/**.
- Information on planned, current, as well as past DXpeditions, can be found at: **www.cpcug.org/user/wfeidt/Misc/adxo.html**.

13.4. DX Bulletins

DX bulletins on the Internet have replaced paper DX bulletins. Most of these are weekly publications.

- Probably the most popular DX-information sheet is the weekly *DX425News* **www.425dxn.org/**, a no-charge Italian weekly DX-bulletin with almost 10,000 subscribers worldwide.
- Ted, KB8NW, edits the Internet edition of the *OP-DX Bulletin*, which is also a weekly DX bulletin: **www.papays.com/opdx.html**.
- The North Jersey DX Association has an interesting page with good DX tips: **usats.com/ce-dx.html**.
- Bernie, W3UR (**www.dailydx.com/**) publishes *The Daily DX* Monday through Friday. *The Daily DX* is available daily as an e-mail containing a collection of all the latest DX news.

14. THE 8 COMMANDMENTS FOR WORKING THE RARE DXPEDITION

Joerg, YB1AQS / DL8WPX (from ZL7DK, VK9CR, VK9XY, S21XX and P29XX fame), formulated the following rules:

- **Rule #1:** Listen, listen, listen! It's much harder than to transmit.
- **Rule #2:** Don't give up before the DXpedition leaves. If you're serious, you can't miss any possible opening (and your opening may come only on the last day).
- **Rule #3:** Long-haul propagation is always very area selective. Don't forget to monitor closely who has been worked, and in which direction the propagation is moving to determine your skip.
- **Rule #4:** For medium-range distances there are not only the gray-line openings. Don't always wait for the gray line, getting up one hour earlier has often been a winner.
- **Rule #5:** In big pileups try to avoid calling zero beat with anybody else. One hundred Hertz up or down can readily make the difference. On 160 meters I would go even further away. If you ever have tried to work a full-blown pileup covered by two layers of tropical noise, you'll know what I mean.
- **Rule #6:** Tail ending means Tail *Ending*. It's definitely an art and not many DXers can do it right. Don't break-in with your call as long as the previous QSO is not 100% clear. The timing of sending your call is very critical and you have to be synchronized with the behavior of the DX station. But that means clearly you have to hear the DX station well. If not, don't try it.
- **Rule #7:** In case of turmoil on the DX frequency, stay calm and monitor. A good DX operator will soon be aware of the situation and usually try to move just a bit.
- **Rule #8:** If you call, do it with moderate speed and take into account that the DX may have much more difficulties to copy you, especially if he's in the tropics. On Top Band, sending your call only one time is often not enough if the DX operator has to interpolate your call, but more than three times in a row is also not productive.

Dave, NR1DX commented along the same lines on the Topband reflector: *"Listen and understand the 'rhythm' of the station you are trying to work. Does he slide his QSX up/down after every QSO? By how much? As in duck hunting you have to learn to "lead the bird;" i.e. shoot ahead of him so he flies into the shot when it gets there. If he is not taking tail-enders then don't tail end. If he is taking tail-enders, listen and see what the timing of other successful tail-enders is. Sending a tail too early or too late is useless QRM. Adjust your speed to the speed of the stations he is working most. Are the guys that are getting through sending their call, once, twice or three times? In other words you have to spend as much time listening to whom he is working as you do listening to the DX you are trying to work."*

15. ACHIEVEMENT AWARDS

There are a number of low-band-only DX awards. The IARU issues 160 and 80-meter WAC (Worked All Continents) awards. These are available through IARU societies including ARRL (225 Main Street, Newington, CT 06111, USA). ARRL also issues separate DXCC awards for 160, 80 and 40 meters. More information on these awards can be found at the ARRL Web site at: **www.arrl.org/award/**.

CQ magazine issues single-band WAZ awards (for any band). Applications for the WAZ award go to: Floyd Gerald, N5FG, 17Green Hollow Rd, Wiggins, MS 39577-8318.

In addition, there are very challenging 5-band awards: 5-Band WAS (Worked All States), 5-Band DXCC (worked 100 countries on each of 5 bands), both issued by ARRL, and 5-Band WAZ (worked all 40 CQ zones on each of the 5 bands, 10 through 80 meters), issued by CQ (**www.cq-amateur-radio.com/awards.html**).

The *Low Band Monitor* (**www.qth.com/lowband**) sponsors awards for the low-band DXer that begin each season on September 1 and end March 31:

- 160-Meter WAC—The first LBM subscriber to complete a WAC receives a beautiful plaque to commemorate the achievement. Other subscribers who qualify receive individualized, numbered 160-Meter WAC Certificates.
- 80-Meter 100—The first LBM subscriber to work 100 DXCC Countries on 80 meters receives a beautiful plaque. Other subscribers who qualify receive individualized, numbered 80-Meter 100 Certificates.
- 40-Meter 150—The first LBM subscriber to work 150 DXCC Countries on 40 meters receives a beautiful plaque. Other subscribers who qualify receive individualized, numbered 40-Meter 150 Certificates.

The achievement awards issued by the sponsors of the major DX contests that have single-band categories are also highly valued by low-band DX enthusiasts. The major contests of specific interest to low-band DXers are:

- The CQ Worldwide DX Contest (phone, last weekend of October)
- The CQ Worldwide DX Contest (CW, last weekend of November)
- The CQ Worldwide 160-Meter Contest (CW, usually last weekend of January)
- The CQ Worldwide 160-Meter Contest (phone, usually the last weekend of February)
- The ARRL International DX Contest (CW, third weekend of February)
- The ARRL International DX Contest (phone, first weekend of March)
- The ARRL 160-meter contest (first weekend of December)
- The Stew Perry 160-Meter Contest (last weekend of December)

Continental and world records are being broken regularly, depending on sunspots and improvements in antennas, operating techniques, etc.

Collecting awards is not necessarily an essential part of low-band DXing. However, collecting the QSL cards for new countries is essential, at least if you want to claim them. Unfortunately, there are too many bootleggers on the bands, and too many unconfirmed exchanges that optimists would like to count as QSOs. These factors have made written confirmation essential unless, of course, the operator never wishes to claim country or zone totals at all. Many other achievements can be the result of a goal you have set out to reach.

The ultimate low-band DXing achievement would be to

work all countries on the low bands. This goal is quite achievable on 40, possible on 80, but quite impossible on 160 meters, although we see the 160-meter scores slowly climbing steadily above the 310 mark.

16. STANDINGS ON TOP BAND

Six year ago, when I wrote the Third Edition of this book, the big question was: "Who will be the first to get 300 countries on Top Band?" Today five have passed that limit and Wal, W8LRL, worked his 310th country in February, 2003, with VU2PAI. Wal started DXing on Top Band in 1972, and in November, 1976, he sent in 100 QSLs to DXCC and obtained award #3. W1BB and W1HGT beat him to #1, as they hand carried their cards to the ARRL! In 1986 Wal updated his score to 201 countries, and he made a third update in September, 2002, when he hand-carried 108 very valuable cards to ARRL HQ for 309 confirmed countries on 160.

Today, as far as I can tell, fewer than 6 DXCC countries have never been available on 160 meters. These are 7O, BS7H, FR/G, FT/W, FT/X and P5. On 80 meters only P5 and BS7H have never been made available so far.

The ARRL publishes every year a *DXCC Yearbook*, where you can check your ranking in the various DXCC listings. The long-term intention of the ARRL is to make online listings updated monthly and do away with the September-30 logjam for the *DXCC Yearbook*.

Nick, VK1AA/VK9LX, (see **www.qsl.net/160/**) publishes *"Who's Who on the Top Band"* on his very nice Web page. It lists standings for World, Europe, USA East Coast, Mid-West and West Coast, JA and finally the Southern Hemisphere "QRN fighters." Yuri, K3BU makes available a listing of the major 160-meter contest records on: **members.aol.com/k3bu/160Records.htm**.

I analyzed the top 160-meter scores from the US East Coast, the combined countries worked by W4ZV (NC only), K1ZM, W4DR and W8LRL (spread over a 900-km stretch of the US East Coast), from Western Europe (my score) and from Japan (countries worked by all top 160-meter DXers combined—information from JA7AO). According to the early 2002 DXCC list there are 24 missing countries from the US East Coast, 46 countries from Western Europe and 66 from Japan.

I plotted these missing countries on three different

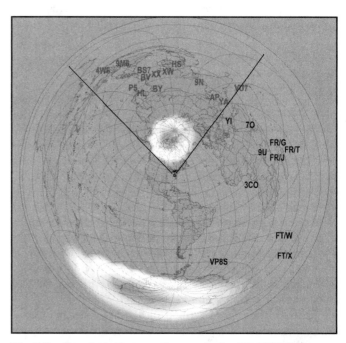

Fig 2-9—Great-circle map (generated with *DXAtlas*) centered on Washington DC, showing the countries needed on 160 meters by K1ZM, W4DR and W8LRL.

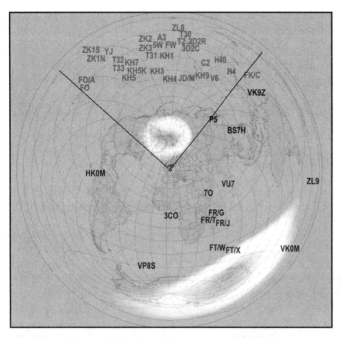

Fig 2-8—Great-circle map (generated with *DXAtlas*) centered on Belgium, showing all the countries needed (January, 2002) by ON4UN on 160 meters. Note that the large majority is in the Pacific, behind the auroral oval.

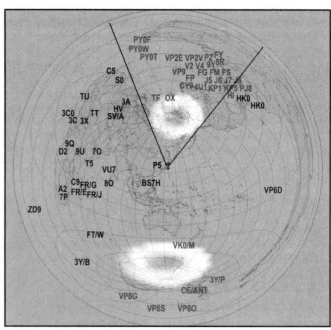

Fig 2-10—Great-circle map (generated with *DXAtlas*) centered on Tokyo, showing the countries that have not been worked by any JA station on 160 meters (source JA7AO, dated January, 2003).

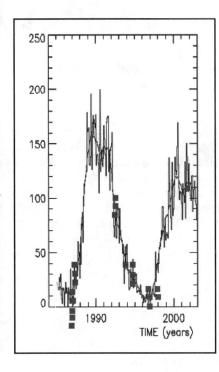

Fig 2-11—The squares on this graph show the dates in which countries in the Pacific hidden behind the auroral donut were worked by the author. The jagged line shows the raw monthly sunspot numbers; the dark line shows the smoothed sunspot numbers (SSN) for the same period.

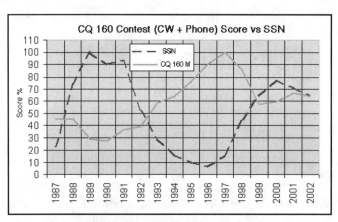

Fig 2-12—The solid line shows the normalized score of the leading six scores in the CQ 160-meter contests (both CW and Phone) for both US and DX. This is compared to the normalized Smoothed Sunspot Number (dashed line), from 1987 until 2002. Note how the high scores are clearly linked to low-SSN years.

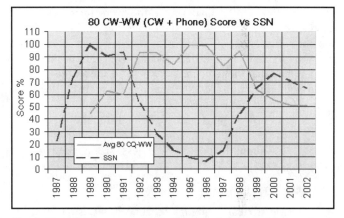

Fig 2-13—The same comparison as in Fig 2-13, but for 80-meter scores from the CQ-WW Phone and CW contests. Again, higher 80-meter scores were made during low SSN years, although the effect is less dramatic on 80 compared to 160 meters.

great-circle maps, centered on Washington, Belgium and Tokyo, and showing the size of the auroral doughnut with low moderate aurora activity. See **Figs 2-9**, **2-10** and **2-11**. For the four US East Coast stations there are approximately 30 countries hidden behind the auroral doughnut, of which they together need only 14 (47%), in addition to 10 "easy ones." In Europe I see 47 countries hidden behind the auroral wall, of which I still need 32 (68%), plus another 14 "easy ones." As for our JA friends, they have 51 hidden countries, of which they need 37 (73%) in addition to 29 "easy ones." Besides shadowing by the auroral region, what can explain these differences in countries actually worked?

- JAs have had a handicap of a small allocation around 1900 kHz for too long.
- The Europe tally (my score) was set over a period of only 15 years.
- The US East Coast tally is from four stations spread over approximately 500 miles of the coast, and they have been active at least one more sunspot cycle than I have.
- Perhaps US Top Banders just spend more time on their radios, rather than writing books?!

It is obvious that the type of countries needed are not the same in all three cases. From Europe the "hidden countries" are all in the Pacific. They are all islands that are only activated now and then by DXpeditions. For the USA East-coast, only a few Pacific islands are on their short list, but the missing countries are not generally populated by active Top Banders either. Our JA friends apparently are in a better position to catch up with the US and Europe on the 160-meter country list. The large number and the nature of the countries in their black hole beyond the auroral doughnut seem to make Western Europe case the toughest one!

Interestingly, 160 WAZ is easier from Europe than USA or JA because of relatively high activity in Europe's toughest Zones: Zone 1 (KL7) and Zone 31 (KH6). Japan's toughest Zone is Zone 2 (VE expeditions only) and the US East Coast's toughest Zone is 23 (JT and UAØY), both of which have relatively low activity and are directly behind the Magnetic North Pole.

We have seen in Chapter 1 that aurora is a major limiting factor for working DX on Top Band. The fact that we do work countries in the black holes beyond the auroral doughnut merely means that sometimes we can get through, but these openings are rare indeed. Luck and patience will help you hit the right opening. This again emphasizes that DXpeditions should be on Top Band from the first day until the last day. Difficult QSOs beyond the auroral doughnut can be made during high-SSN years, but these are much more difficult than during low-SSN years. **Fig 2-12** shows new countries for ON4UN plotted against raw and smoothed sunspot numbers over Solar Cycle 23.

DX-Operating on the Low Bands   2-19

A recent striking example of a crooked-path QSO to an area right behind the aurora doughnut happened in early February, 2003, when a number of US East Coast stations one morning worked JT1CO for their last zone on 160 meters. W4ZV and his friends VE1ZZ, W1JZ, K9HMB, K3UL, K9RJ, K1UO, K1ZM and W1FV had been trying for a long time, and they hit the right day. During high-sunspot years patience is very much a virtue that pays off!

I attempted to correlate contest results (and hence propagation conditions) with Smoothed Sunspot Numbers (SSN). I decided to analyze the 160-meter CQ WW contest scores from 1987 onwards. From **Fig 2-13** we see that the relation is remarkable. The values shown are normalized values. The highest SSN number in this period (approx 155) occurred in 1989. The normalized values shown are for the months of September in each year. The solid line represents the normalized average contest score of the six leading scores in both US/VE and Europe and for both the CW and the Phone contests, making each point the average of 24 scores each year. It is amazing how one curve is the mirror of the other.

I made a similar analysis for the 80-meter single-band scores from the CQ WW CW and Phone contests. I calculated the normalized average score from the 24 highest scores each year (again, six from the USA and six from Europe, for both the CW and the Phone contests). **Fig 2-14** shows the general trend again mirroring the SSN, although the score curve is less peaked than in the case of 160 meters. On 160 meters the scores are about three times higher during the dip in the sunspot cycle as compared to sunspot-peak years, while the ratio is about 2 to 1, a little less pronounced on 80 meters.

17. THE SURVEY

During the early months of 2003 a new survey questionnaire was sent out by e-mail to nearly 500 active Low-Banders. Other volunteers were able to download the questionnaire from Nick's (VK1AA) website, which lists 160-meter DXCC standings. I would like to thank the 270 low-band DXers who contributed to the poll and who made it possible to do some statistical analysis.

17.1 Age and Activity

Instead of concentrating on peoples' ages, I asked how long they'd been DXing on the low bands. Nearly 60% of the active DXers have been active on Top Band for less than 20 years. The average 80-meter DXer has been at this game for somewhat longer, 35% for 20 years or less. It looks like Top Band still must be attractive for newer DXers. The continuous growth in the number of participants in the CQ WW 160-meter contest over the years confirms this trend.

Based on the survey, the average low-band DXer has been on 40 meters for 35 years and on 80 meters for 26 years, but has only been on 160 meters for 22 years. Comparing these figures with those of my last poll (6 years ago), the average age has gone up by approximately 6 years, which would seem to indicate that there are no young newcomers. Going by the average age of 50 determined 6 years ago, it seems to be safe to say that the average age is now approx 56 years. Where is any young blood coming into the hobby itself?

17.2. Time Spent on the Low Bands

What is the split for time spent on the 40, 80 and 160-meter bands by low-band DXers? A large number (almost $1/3$) operates 40 meter only occasionally (less than 10% of the time), while less than 10% work 40 meters more than 50% of their time. I guess this means that 40 is not really a low band: Perhaps there is not enough challenge! While a similar percentage of diehards (< 10%) can be found working more than 50% of their time on 80 meters, the picture is very different for Top Band: Over 30% spend more than 50% of their time on 160 meters! The "average ham" in the poll spends 45% of his "low band time" on 160 meters.

Average Time Spent on Each Band

40 meters	24%
80 meters	31%
160 meters	45%

17.3. Time Spent on Each Mode

Top Band is a CW band: More than 80% operate more than 80% of their Top-Band time on CW and almost $2/3$ say they spend 90 to 100% of their time on CW. Actually all three low bands are CW bands. There is a little more SSB activity on 80 meters, with 33% of the respondents working phone 50% of the time. However, this still means that 67% of the respondents spent 50% of their time working CW.

If we examine the split for stations with a high DXCC score, meaning >275 countries on 40 meters, >250 on 80 meters and >225 countries on 160 meters, the preference for CW becomes even more pronounced. In this category, averaged over the three low bands, 33% work CW 90% of the time or more. On 160 meters 70% of the stations responding say they work CW at least 90% of the time.

Average Time Spent on Each Mode

	Total Group	
Band	CW	SSB
160 meters	90%	10%
80 meters	70%	30%
40 meters	79%	21%

The reason for this CW/SSB split are obvious—CW is by far the more efficient mode when it comes to dealing with weak signals under marginal conditions. On 40 meters the percentage split is higher even than on 80 meters. This probably reflects the presence of strong interfering broadcast stations in the US phone band that drive DXers to the quiet of the CW subband.

17.4. Achievements

The achievement figures are listed in **Table 2-3**. The listing is alphabetically by call. Columns 2,3 and 4 give the year that the station started chasing DX on 40, 80 and 160 meters. The DXCC status shown is the all-time status. WAZ status is shown as well.

Table 2-3
Low-Band DXer's Survey (2003)

Call	% 40	% 80	% 160	On 40 Since	On 80 Since	160 Since	5B DXCC	5B WAZ	160m WAS	40m WAZ Wkd	40m WAZ Cfd	40m DXCC Wkd	40m DXCC Cfd	40m % CW	40m % SSB	80m WAZ Wkd	80m WAZ Cfd	80m DXCC Wkd	80m DXCC Cfd	80m % CW	80m % SSB	160m WAZ Wkd	160m WAZ Cfd	160m DXCC Wkd	160m DXCC Cfd	160m % CW	160m % SSB
3B8CF	40	40	20			1985	No	No	No					80						80	20					100	0
4X4DK	20	30	50	1953	1952	1960	Yes	No	No	40	40	332	329	95	5	40	40	295	289	95	5	37	37	210	209	99	1
4X4NJ	10	5	85	1957	1957	1985	Yes	No	No					100	0					100	0	40	40	255	250	100	0
5B4ADA	20	15	65	1973	1975	1992	No	Yes	No	40	40	263	153	57	25	40	40	188	112	50	50	40	40	159	110	95	5
9M2AX	10	40	50	1966	1969	1972	No	No	No	40	40			99	1					95	5	32	32			0	32
AA4MM	0	50	50	1981	1977	1981	Yes	Yes	Yes		38			0	0	37	37	200	130	10	90	37	37	244	242	20	80
AA4V	20	10	70	1967	1970	1972	Yes	Yes	Yes	40	40	328	326	90	10	40	40	274	268	85	15	40	40	230	229	95	5
AB0X	10	20	60	1975	1975	1980	Yes	Yes	Yes	40	40	279		100	0			286		75	25	33	33	187		90	10
AC0M	10	40	50	1979	1979	1985	Yes	Yes	Yes	40	40	294	290	95	5	40	40	199	190	90	10	21	19	65	59	95	5
AE1Q	18	18	64	1978	1978	1990	No	No	No	32	32	155	155		100	34	34	160	160	0	100	2	2	3	3	0	100
AE9B	33	22	45	1999	1999	2001	Yes	Yes	Yes					60	40					70	30					95	5
AG6X	0	100	0	1981	1981		No	No	No					5	25				*	5	55						
AI9L	18	23	59	1995	1995	1997	Yes	Yes	Yes	36	36	202	175	97	3	29	29	160	139	50	50	27	27	104	97	99	1
AJ1H	0	50	50	1969	1969	1997	No	No	No					0	0	40	40	280	278	100	0	29	29	171	168	100	0
CT1EEB	20	20	60	1990	1997	1999	Yes	No	Yes	40	40	291	284	65	35	40	40	220	211	50	50	30	29	133	121	99	1
DF2PY	0	0	100			1996	No	No	Yes											70	30	40	40	180	127	99	
DJ2YA	10	20	70	1955	1955	1980	Yes	Yes	No	40	40	324	324	90	10	40	40	300	300	60	40	40	40	230	230	95	5
DJ4AX	10	70	20	1957	1957	1977	Yes	Yes	No	40	40	230	205	50	50	40	40	265	251	10	90	25	21	75	59	40	60
DJ6QT	30	20	50	1961	1961	1975	Yes	Yes	No	40	40	290	270	50	50	38	38	220	200	50	50	32	32	160	145	40	60
DK7PE	20	30	50	1973	1973	1973	Yes	Yes	No	40	40			95	5	40	40			95	5					100	0
DL3DXX	20	30	50	1981	1981	1985	No	Yes	No	40	40	337	335	99	1	40	40	307	306	99	1	40	40	249	248	99	1
DL7CX	20	20	30	1999	2000	2001	Yes	Yes	No	39	37	262	204	75	25	35	35	225	182	65	35	32	29	150	106	95	5
EA3VY	1	4	95	1975	1975	1981	No	No	Yes						100					100	0	40	40	240	100	95	5
EA6ACC	25	15	60	1992	1992	1995	Yes	Yes	Yes	38	36	186	141	100	0	32	32	145	120	50	50	33	33	150	147	100	0
EA6NB	20	30	50	1985	1985	1985	Yes	Yes	No	40	40	337	337	30	40	40	40	296	296	100	0	39	39	210	208	20	0
EA7NW	40	40	20	1995	1995	2001	Yes	Yes	No	38	32	150	105	99	1	25	25	125	100	100	0	10	10	50	10	100	0
EY8MM	40	50	10	1985	1985	1993	Yes	Yes	No	40	40	298	289	80	20	40	40	252	240	99	1	30	30	117	94	99	1
F6BKI	10	30	60	1975	1975	1983	Yes	Yes	No	40	40	328	312	50	50	40	40	308	305	80	20	40	40	245	242	80	20
FM5BH	20	20	60	1983	1983	1987	No	Yes	No	40	40			80	20					100	0	26	26	124	124	100	0
G3KMA	20	10	70	1955	1972	1979	Yes	Yes	No	40	40	327	326	85	15	40	40	301	300	80	20	40	40	254	252	95	5
G3XTT	27	27	46	1969	1968	1968	Yes	Yes	No	40	40	326	326	95	5	40	40	285	284	95	5	39	39	237	236	90	10
G4OBK	10	10	80	1982	1982	1982	Yes	Yes	No	40	38	279	261	85	15	38	38	232	214	85	15	35	35	180	174	90	10
G4VGO	0	2	98	1980	1985	1984	No	No	No	40	40	130	125	100	0	34	34	127	116	50	50	40	38	223	205	85	15
GM3PPE	20	30	50	1978	1979	1980	Yes	Yes	Yes	35	30	259	205	98	2	30	30	212	168	80	20	35	35	150	149	90	10
GM3YOR				1975	1975	1970	Yes	Yes	Yes	40	40	251	219	99	1	39	39	193	169	99	1	31	31	134	133	99	1
GM3YTS	40	30	30	1985	1985	1998	Yes	Yes	Yes	40	40	294	290	100	0	40	40	229	228	80	20	30	30	128	115	100	0
GW3YDX	40	30	30	1969	1969	1969	Yes	Yes	Yes	40		343	100	98	2			315	100	100	0	40	?	262	?	99	1
HA0DU	10	20	70	1974	1974	1988	Yes	Yes	No	40	40	328	328	80	20	40	40	315	315	80	20	40	40	261	259	99	1
HB9AMO				1977	1978	1983	Yes	Yes	No	40	40	320	318	80	10	38	38	287	287	60	40	40	40	235	235	98	2
HC8N	30	40	30	1992	1992	1992	Yes	Yes	No	40	40	223		90	10	40	40	194		90	10	31	31	136		90	10
HL3IUA	20	20	60	1988	1992	1999	Yes	Yes	No	40	40	319	316	30	70	40	40	272	271	90	10			83	67	100	
I4EAT	10	30	60	1981	1982	1995	Yes	Yes	No	40	40	333	338	90	10	40	40	323	323	50	50	40	40	261	261	99	1
IC8WIC	40	0	0	1990	1988	1986	No	Yes	Yes	38	38	220	200	0	100	33	33	200	180	0	100						
IN3ASW	40	40	20	1995	1997	2002	Yes	Yes	No	39	39	239	179	30	70	36	33	168	115	50	50	7	7	28	28	80	20
IT9ZGY	10	10	80	1969	1969	1987	Yes	Yes	No	40	40	309	310	90	10	40	40	270	266	90	10	40	40	253	251	95	5
IV3PRK	0	0	100			1986	Yes	Yes	Yes													40	40	264	262	99	1
JA0DAI	30	10	70	1973	1980	1987	Yes	Yes	No	40	40	326	314	96	5	40	40	281	273	95	5	35	35	172	167	100	0
JA1EOD	60	20	20	1970	1976	1995	Yes	Yes	No	40	40	338	338	60	40	40	40	280	280	70	30	33	33	146	144	100	0
JA1UQP	20	70	10	1968	1983	1994	Yes	Yes	No	40	40	345	345	30	70	40	40	332	332	20	80	25	23	78	72	100	0
JA2PJC	0	0	100			1990	No	No	No													37	37	202	200	100	0
JA2VPO	40	20	40	1980	1985	1990	Yes	No	No	40	40	335	335	50	50	40	40	321	321	60	40	35	34	155	151	100	0

Call	% 40	% 80	% 160	On 40 Since	On 80 Since	160 Since	5B DXCC	5B WAZ	160m WAS	40m WAZ Wkd	40m WAZ Cfd	40m DXCC Wkd	40m DXCC Cfd	40m % CW	40m % SSB	80m WAZ WKD	80m WAZ CFD	80m DXCC WKD	80m DXCC CFD	80m % CW	80m % SSB	160m WAZ WKD	160m WAZ CFD	160m DXCC WKD	160m DXCC CFD	160m % CW	160m % SSB
JA3CZY	10	90	0		1980		No	No	No					0	100					0	100						
JA3FYC	5	15	80	1975	1976	1982	Yes	No	No	40	40	331	330	98	2	40	40	294	288	98	2	35	35	190	189	100	0
JA4LKB	15	25	60	1975	1979	1981	Yes	Yes	No	40	39	312	305	70	30	39	39	272	265	50	50	38	38	215	209	100	0
JA5AUC	25	50	25	1970	1978	1985	Yes	No	No			341	339	90	10			322	320	20	80			181	172	100	0
JA5IU	40	30	30	1980	1990	1990	Yes	No	No	40	40	326	327	90	10	39	39	268	269	50	50	19	18	116	117	100	0
JA7KAC	40	50	10	1990	1975	1995	Yes	No	No	40	40	204	281	70	30	39	39	215	198	50	50			85	55	100	0
JE1SPY	30	30	30	1975	1980	1978	No	No	No					30	10					30	30					50	5
JE1SPY	30	30	30	1975	1980	1978	No	No	No					30	10					30	30					50	5
JE1TSD	0	10	90	1983	1983	1998	Yes	No	No	30	38	303	270	70	30	30	30	120	105	95	5	20	16	34	26	100	0
JF6OJX	40	50	0	1981	1983	1993	No	No	No	37	40	325	325	15	85	37	37	240	220	80	20	7	6	7	6	100	0
JG1XLV	100	0	0	1972			No	No	No			270	210	90	10			208	182	95	5						
JH2FXK	20	10	70	1986	1986	1985	Yes	No	No	37	35					35						37	37	200	196	100	0
JH2RMU			90			1995	No	No	No													31	31	132	130	100	0
JH3VNC	30	30	40	1978	1978	1990	Yes	No	No	40	40	330	329	75	25	39	39	284	283	90	10	36	36	178	178	100	0
JH4IFF	20	50	30	1982	1982	1993	Yes	No	No	40	40	336	331	80	20	39	39	276	274	50	50	35	33	162	160	100	0
JH4UYB	40	40	30	1978	1984	1994	Yes	No	No	40	40	290	260	70	30	39	39	250	200	60	40	33	30	160	135	100	0
JK2VOC	20	10	5	1987	1991	1995	No	No	No					40	40					30	10					5	
JL1UXH	60	30	10	1984	1993	1998	Yes	No	No	40	40	297	292	70	30	39	39	192	189	80	20	9	7	15	11	100	0
JX7DFA	50	30	20	1989	1989	1996	No	No	No	37		144		90	10	34		114		75	25	18		75		98	2
K0CS	5	10	85	1971	1976	1980	Yes	No	Yes	40	40	260	260	90	10	40	40	266	266	90	10	35	35	201	201	90	10
K0HA	15	25	60	1970	1970	1972	Yes	No	Yes					90	10					85	15					90	10
K0XM	60	30	30	1993	1993	1993	No	No	Yes	36	29	206	84	90	10	29	20	113	41	85	10	14	10	29	17	95	5
K1FK	40	30	30	1976	1976	1976	Yes	No	Yes	40	39	242	207	100	0	37	37	200	177	100	0	31	31	135	116	100	0
K1GUN	0	0	95	n/a	n/a	1982	Yes	Yes	Yes													36	36	255	255	99	136
K1MY	15	15	0	1990	1990		Yes	No	No	40	40	225	225	50	50	35	35	203	203	25	75						
K1QX	22	22	50	1967	1967	1972	No	No	Yes					80	20					80	20						
K1TTT	40	35	25	1984	1984	1984	Yes	No	Yes			180		50	50			160		50	50						
K1VR	30	35	35	1958	1960	1990	Yes	No	Yes	40				90	10	39	38	242	177	60	40	33	31	220	210	80	20
K1ZM	5	25	70	1959	1960	1973	Yes	No	Yes		40			75	25		40			25	75	40	40	309	309	50	50
K2RD	30	20	50	1965	1980	1981	Yes	No	No	39	35	285	270	99	1	35	33	235	205	95	5	31	30	185	170	90	10
K2TQC	60	20	20	1958	1958	1958	Yes	No	No	40	40	334	334	95	5	40	40	332	332	50	10	39	39	288	288	95	1
K2UO	30	10	60	1975	1975	1981	Yes	No	Yes			316	316	99	1			248	244	90	10			224	222	99	1
K2UOP	40	40	20	1959	1960	1993	No	No	No					60	40					80	20					98	2
K3JGJ	30	40	30	1980	1980	1994	Yes	No	Yes	40	40	314	312	30	0.5	38	38	298	296	100	0	30	30	159	156	70	30
K3NA	35	35	40	1972	1974	1982	Yes	Yes	Yes	40	40	338	338	60	40	40	40	303	303	60	40						25
K3UA	20	30	50	1960	1960	1985	Yes	Yes	Yes	40	40	290	199	95	5	40	40	242	177	95	5	37	37	247	247	90	10
K3WW	60	25	15	1960	1968	1970	Yes	Yes	No	39	37			90	10	37	33			60	40	33	29	195	113	90	10
K4BAI	60	30	30	1955	1955	1960	Yes	No	No			205		90	10			121		95	5			48		98	2
K4CIA	10	10	80	1956	1967	1967	Yes	No	Yes	40	40	342	342	90	10	40	40	314	314	95	5	33	33	206	206	100	0
K4ESE	0	95	5	1948	1964	1976	No	No	No					99	1			319	319	90	60	32	20	172	43	100	0
K4MQG	30	70	0	1980	1986	1975	Yes	No	No	40	40	343	342	95	5	40	40	339	339	40	60						
K4PI	10	60	30	1957	1957	1982	Yes	No	Yes	40	40	344	344	10		40	40	315	315	20	80	30	30	252	254	75	25
K4TEA	5	5	90	1965	1976	1985	Yes	No	Yes	40	40	191	137	98	1	39	39	303	303	60	2	37	37	247	247	90	10
K4TO	40	10	50	1957	1959	1970	Yes	Yes	Yes	27	22			100	2	33	29	177		98	2	29	26	125	113	90	10
K4UEE	15	15	20	1960	1968	1959	Yes	Yes	No	39	30	299	205	95	5	40	40	229		95	5	37	37	231	206	98	2
K4VX	60	30	10	1948	1955	1976	Yes	No	No					90	10				114	100						100	
K4YP	50	50	0	1980	1980		No	No	No	40	40	299	295	99	0	38	38	238	229	15	85						
K4ZW	10	50	40	1979	1979	1979	Yes	Yes	No	40	40	328	328	15	85	40	40	310	306	70	30	37	37	228	228	30	1
K5AQ	50	20	30	1950	1965	1980	Yes	No	No	40	40	317		90	10	40	39	281	280	85	15	33	33	179	179	99	5
K5OVC	40	60	0	1978	1978		No	No	No	30	19	303	300	0	100	30	22	306	306	0	100					100	0
K5PC	20	30	50	1980	1995	1983	Yes	Yes	Yes	40	40	303		90	10	38	38	244	248	75	25	33	33	136	136	90	10
K5RA	40	40	20	1995	1993	1993	Yes	No	No	32	31	149	123	50	50	20	20	54	37	50	50	18	17	37	29	98	2
K5ZD	40	20	40	1993	1993	1978	Yes	No	No	40	36	269	130	90	10	33	20	196	97	60	40	34	18	145	61	95	5
K6ANP	45	15	40	1960	1960	1987	No	No	No			200	125	95	5			167	139	95	5		148	45		99	1
K6EID	20	40	40	1980	1985	1985	Yes	No	No	40	40	319	319	99	1	40	40	294	294	90	10	36	36	187	187	95	5

Call	% 40	% 80	% 160	On 40 Since	On 80 Since	160 Since	5B DXCC	5B WAZ	160m WAS	40m WAZ Wkd	40m WAZ Ctd	40m DXCC Wkd	40m DXCC Ctd	40m % CW	40m % SSB	80m WAZ WKD	80m WAZ CFD	80m DXCC WKD	80m DXCC CFD	80m % CW	80m % SSB	160m WAZ WKD	160m WAZ CFD	160m DXCC WKD	160m DXCC CFD	160m % CW	160m % SSB
K6SE	10	10	80	1955	1956	1966	Yes	No	Yes	40	40	150	150	100	0	?	?	120	120	100	0	35	35	187	187	95	5
K7EM	50	50	0	1982	1985		Yes	Yes	No	40	40	318	267	75	25	40	40	256	222	25	75						
K7FL	10	30	60	1985	1985	2003	Yes	Yes	Yes	40	40	227	225	90	10	40	40	158	158	95	5	14	14	14	5	100	0
K7OX	20	5	75	1960	1970	1961	Yes	No	Yes	40	40	223	223	85	15	No	40	186	Yes	80	20			87		90	10
K7ZV	5	90	5	1976	1994	1998	Yes	No	No	40	40	275	246	60	40	40	40	285	270	20	80	27	27	88	66	95	5
K8BHZ	0	0	100			2002	No	No	No													25	25	84	71	98	2
K8EJ	20	45	35	1959	1975	1987	Yes	Yes	No	40	40	331	331	99	1	40	40	305	305	95	5	33	33	190	187	99.9	0.1
K8GG	20	40	40	1960	1975	1984	Yes	Yes	Yes	35	35	200	200	40	60	35	35	250	250	35	65	37	37	240	239	90	10
K8IP	0	0	100	1947		1979	Yes	Yes	No					100	0					100	0	34	34	204	204	95	5
K8MFO	10	50	40	1971	1971	1983	Yes	Yes	Yes	40	40	340	340	100	0	40	40	301	300	100	0	38	38	282	282	100	0
K8ZR	5	50	45	1981	1990	2002	Yes	No	No	36	36	190	188	100	0	31	30	188	184	95	5	6	6	14	6	95	5
K9EL	40	50	10	1970	1970	1985	Yes	Yes	Yes	40	40	321	321	98	2	40	40	286	285	30	70	31	31	156	155	98	2
K9FD	20	30	50	1980	1980	1985	Yes	Yes	Yes	40	40	331	330	98	2	40	40	313	312	75	25	38	38	222	222	99	1
K9jf	25	40	35	1968	1978	1985	Yes	Yes	Yes	40	40	250	240	95	5	40	40	225	220	75	25	32	32	140	138	99	1
K9KU	40	40	20	1978	1982	1982	Yes	Yes	Yes	40	40	297	295	98	2	40	35	226	219	98	2	27	26	130	122	98	2
K9MA	50	25	25	1966	1989	1992	No	No	No		35			100	0					100	0					100	0
K9RJ	40	30	30	1963	1964	1971	Yes	Yes	Yes	40	40	331	321	90	10	40	40	266	255	70	30	37	36	261	257	95	5
KA7T	25	25	50	1982	1998	1983	Yes	No	Yes	40	40	252	239	100	0	37	37	162	153	100	0	32	32	154	149	100	0
KB0ETC	62	30	8	1995	1995	1995	Yes	Yes	No	38	32	312	304	30	20	37	29	286	226	5	20	19	16	110	94	1	4
KG6I	30	20	50	1964	1964	1980	Yes	No	Yes	40	38	267	253	90	10	35	35	212	202	90	10	33	33	164	162	99	1
KH6DX/M	10	30	60	1989	1989	1989	Yes	Yes	Yes	37	39	195	171	99	1		26	126	118	95	5	27	24	102	102	99	1
KJ9I	60	20	20	1980	1980	1987	Yes	No	No			306	300	99	1	31		267	264	75	25			235	235	99	1
KL7RA	5	25	70	1965	1965	1965	Yes	No	Yes	19	7	57	8	100	0	13	1	35	3	100	0					100	0
LA5HE	30	60	30	1957	1970	1988	Yes	Yes	Yes	39	25	210	175	95	5		17	160	130	90	10	?	?	90	90	99	1
LY3UM	5	25	70	2001	2002	2002	No	No	No	40	24	230	220	98	2	34		143	39	99	1	32	17	46	46	99	1
LZ2JE	25	25	50	1979	1979	1979	Yes	Yes	Yes	39	40	178	61	99	1	40	40	218	204	70	30	35	31	142	142	70	30
N0AT	20	40	40	1972	1985	1988	Yes	Yes	Yes	40	40	268	234	70	30	40	40	281	264	40	30	32	32	189	186	40	1
N0AX	30	30	40	1988	1988	1995	Yes	No	No	36	34	321	314	20	5	33	27	117	93	100	10	20	17	43	33	100	0
N0IJ	20	60	20	1980	1980	1994	Yes	Yes	Yes	40	40	213	188	100	0	38	37	240	210	65	35	29	29	111	109	98	2
N2CG	30	10	60	2001	2001	1997	Yes	No	Yes	19	7	300	265	95	5	13	1		3	100	0	18	18	63	59	100	0
N2NT	40	10	50	1972	1972	1972	Yes	Yes	Yes	39	25	210	175	10	1	35	30	200	125	10	5	15	15	125	100	50	5
N1QT	10	30	60	1991	1991	1995	Yes	Yes	Yes	40	40	327	327	80	20	39	39	290	285	80	20	36	36	216	213	99	1
N4CC	40	40	20	1970	1970	1984	Yes	Yes	Yes	40	40	339	336	95	5	40	40	310	310	75	25	36	36	209	207	99	1
N4JJ	5	15	80	1963	1963	1964	Yes	Yes	Yes	40	40	347	347	99	1	40	40	333	333	95	5	40	40	279	279	95	5
N4JJ	0	25	75	1963	1963	1964	Yes	Yes	Yes	40	40	347	347	99	1	40	40	333	333	95	5	40	40	279	279	95	5
N4MM	40	30	30	1960	1960	1965	Yes	Yes	Yes	40	40	324	324	99	25	38	38	274	274	75	25	29	29	167	167	80	20
N5FG	60	20	20	1975	1975	1991	Yes	Yes	Yes	40	40	326	326	98	2	40	40	238	236	5	95	29	29	110	108	95	5
N5KO	10	30	60	1977	1978	1979	Yes	No	Yes	40	39	258	254	95	5	29	26	103	73	90	10	6	6	13	4	90	10
N5RG	30	60	10	1986	1986	2000	Yes	Yes	Yes				338	90	10				298	50	0				207	20	0
N5UL	50	50	0		?	?	No	Yes	No	40	40	320	319	0	100	40	36	267	81	0	100	29	29	108	68	99	1
N6FF	10	20	70	1987	1997	1990	Yes	Yes	Yes	40	40	259	178	90	20	40	40	211	146	95	5	32	32	140	139	99	1
N6ZZ	50	30	20	1997	1997	1995	Yes	No	Yes			198		100	0					100	10	38	38	217	215	95	5
N7JW	0	30	70	1958	1958	1958	No	No	Yes	40	40	310	310	80	10	40	40	260	259	90	1					100	0
N7RK	30	60	10	1955	1973	1996	No	No	No					70	30					60		37	37	210	202	90	10
N7RT	5	60	10	1976	1965	1968	Yes	Yes	Yes	39	39	257	203	95	5	39	39	270	265	60	40	19	19	72	50	80	20
N9AU	90	5	5	1979	1979	1984	Yes	Yes	Yes	40	40	335	335	95	5	40	40	276	276	100	0	31	31	168	165	100	0
NA0Y	6	6	88	1961	1971	1985	Yes	Yes	Yes	40	40	335	334	98	2	35	35	240	236	98	2	32	31	173	171	98	2
NI0C	30	60	10	1984	1984	2002	No	No	Yes	40	39	200	200	95	5	40	26	103	73	99	1	38	38	243	243	80	20
NR1R	50	25	25	1975	?	?	Yes	Yes	Yes			258	254									6	6	13	4	95	5
NT5C	50	50	0	1989	1989	1979	Yes	Yes	Yes	40	40	320	319	0	100	40	40	267	266	0	100			207	207		
NX4D	5	10	85	1997	1997	2000	Yes	Yes	Yes	40	40	259	178	90	10	40	40	211	152	95	5	28	27	141	132	95	5
OH1MA	10	50	40	1984	1978	1978	Yes	Yes	Yes			330		80	20			316		50	50			250		95	5
OH2BO	5	15	80	1970	1978	1973	Yes	Yes	No					70	30					70		39	39	236	231	85	15
OH2BU	20	30	50	1972	1972	1972	Yes	No	No	40	40	334	334	75	25	40	40	320	320	25	75	38	38	263	255	95	5
OH2KI	30	40	30	1973	1973	1973	No	No	No	35		250	250	80	20	30		150		80	20	30	30	100	207	90	10

DX-Operating on the Low Bands

Call	% 40	% 80	% 160	On 40 Since	On 80 Since	160 Since	5B DXCC	5B WAZ	160 WAS	40m WAZ Wkd	40m WAZ Ctd	40m DXCC Wkd	40m DXCC Ctd	40m % CW	40m % SSB	80m WAZ WKD	80m WAZ CFD	80m DXCC WKD	80m DXCC CFD	80m % CW	80m % SSB	160m WAZ WKD	160m WAZ CFD	160m DXCC WKD	160m DXCC CFD	160m % CW	160m % SSB
OH3SR	1	2	97	1964	1964	1973	Yes	No	No	40	40	333	333	70	30	40	40	321	321	80	20	39	39	239	235	90	10
OH4MFA	10	20	70	1992	1992	1996	No	No	No	36	34	196	127	100	0	30	22	123	69	100	0	28	27	126	98	100	0
OK1DOT	0	50	50		2001	1982	No	No	Yes									205	104	50	50	40	40	266	265	90	10
OK1MP	66	25	9	1957	1969	1982	Yes	Yes	No	40	40	325	314	66	34	40	40	275	269	30	70	25	24	118	115	80	20
OK1RD	12	35	53	1968	1968	1964	Yes	Yes	No	40	40	328	328	85	15	40	40	310	310	60	40	40	40	255	252	95	5
OM2XW	0	0	100			1994	No	No	No													40	40	230	226	80	20
ON4UN	30	30	40	1985	1962	1987	Yes	Yes	Yes	40	40	339	339	95	5	40	40	353	353	95	5	40	40	296	294	99	1
ON5NT	35	35	30	1974	1974	1987	Yes	Yes	No	40	40	334	334	90	10	40	40	310	310	80	20	40	40	260	260	90	10
ON7GB	20	40	40	1992	1992	1994	Yes	Yes	No	40	40	334	334	90	10	40	40	302	301	50	50	38	38	216	205	90	10
OY9JD	20	30	50	1987	1987	1988	No	No	No	40	40	250		45	55	40	40	272	251	25	75			203	171	75	25
OZ1BTE	40	40	5	1984	1984	1993	Yes	Yes	No	40	40	335	339	70	30	40	40	314	314	30	70	40	40	252	251	95	5
OZ1CTK	11	22	67	1977	1977	1993	Yes	Yes	No	40	40	335	335	90	10	40	40	313	311	90	10	40	40	256	243	95	5
OZ1ING	20	35	45	1985	1988	1986	Yes	Yes	Yes	40	40	305	275	85	15	40	40	286	268	60	40	40	37	206	204	98	2
OZ1LO	20	40	40	1962	1961	1962	Yes	Yes	No	40	40	348	347	95	5	40	40	313	312	95	5	40	40	266	266	99	1
OZ3PZ	5	20	75	1970	1970	1990	Yes	Yes	No	40	40	341		95	5	40	40	320	319	95	5	40	40	263	262	99	1
OZ7YY	25	15	60	1960	1960	1980	Yes	Yes	Yes	40	40	340	325	75	25	40	40	313		70	30	40	40	271	270	98	2
OZ8ABE	30	30	40	1991	1992	1992	No	No	No	40	40	326		90	10	40	40	299	299	75	25	40	40	250	248	99	1
PA0LOU	30	30	40	1956	1955	1960	Yes	Yes	Yes	40	40	271		100	0			221	203	100	0			167	161	95	5
PA0ZH	40	60	0	1980	1980		Yes	Yes	No	40	40	286	286	0	100	40	40	270	270	0	100	19	19	145	145	0	100
PA3FQA	16	32	52	1992	1992	1997	No	Yes	No	40	40	296	288	70	30	40	40	240	228	90	10	37	35	147	139	90	10
PA0TAU	22	33	45	1950	1950	1980	Yes	Yes	No	40	40	304	304	15		40	40	267	268	25		37	37	192	187	38	2
PY1BVY	3	0	100	1979	1979	1979	Yes	No	No																		
PY2FUS	5	5	90	1999	1972	1974	No	No	Yes	40	40											40	40	240	235	90	10
RA3DOX	10	0	90	1987			No	No	No			164		95	5			140		95	5	30	30	101		95	5
RV1CC	15	35	50	1994	1994	1994	Yes	Yes	Yes	39	38	218	194	50	50	40	38	214	200	60	40	32	32	143	139	90	10
S50A	30	30	40	1965	1970	1975	Yes	Yes	No	40	40	250	205	80	20	All	All	205	183	80	20	30	30	178	154	90	10
S56A	50	40	30	1962	1962	1972	No	No	No					80	20					75	25					90	10
SM4CTT	30	40	30	1975	1975	1990	Yes	Yes	No	40	40	309	309	75	25	40	40	269	269	50	50	35	35	151	151	90	10
SM6CTQ	20	30	50	1960	1960	1985	No	Yes	No	40	40	318	315	90	10	40	40	292	289	90	10	40	40	257	256	90	10
SM6GZ	2	98	0	1965	1965	1986	No	No	No	40	40			50	50	40	40	313	312	1	99						
SP2FAX	40	40	20	1996	1980	1997	Yes	Yes	Yes	40	40	326	321	50	50	40	40	314	311	30	70	40	40	246	244	95	5
SP5EWY	5	25	70	1972	1973	1988	Yes	Yes	No	40	40	342	342	99	1	40	40	318	318	85	15	40	40	274	273	99	1
SV1AOZ	10	10	80	1994	1994	1998	Yes	No	No	40	40	280	218	80	20	39	39	220	169	60	40	33	33	140	134	90	10
SV8CS	10	30	60	1969	1969	1984	Yes	Yes	No	40	40	255	180	10	90	38	33	210	178	10	90	36	31	201	188	30	70
SV8JE	10	20	70	1970	1970	1981	Yes	Yes	No	40	30	230	210	70	30	40	40	212	205	30	70	40	40	206	203	75	25
T77C	30	20	50	1972	1972	1984	Yes	Yes	No	40	40	326	326	70	30	40	40	270	266	70	30	40	40	193	191	90	10
TF8GX	40	40	20	1997	1997	1997	No	No	No	25	20	110	74	70	30	17	13	71	48	70	30	5	5	36	24	90	10
UA0MF	10	40	50	1969	1969	1979	Yes	Yes	No	40	40	325	325	50	50	40	40	305	305	60	40	40	40	205	205	95	5
UA2FF	20	40	40	1976	1976	1979	Yes	Yes	Yes	40	40	260	184	90	10	39	39	204	152	80	20	40	40	259	254	90	10
UA3AB	20	30	50	1985	1985	1985	No	Yes	No	40	40	311	275	50	50	40	40	243	190	40	60	37	36	172	157	99	1
UA9AT	1	2	97	1988	1988	1988	No	No	No													33	32	149	145	90	10
VA3DX	40	40	20	1976	1976	1984	Yes	Yes	Yes	40	40	332	328	75	25	40	40	300	277	80	20	36	36	216	216	95	5
VA5DX	30	30	40	1973	1973	1975	Yes	Yes	No	40	35	265	258	80	20	32	31	213	215	50	50			133	132	90	10
VE3OSZ	10	30	60	1951	1951	1951	Yes	Yes	Yes	35	35	244	235	100	0	38	??	173	166	100	0	28	28	149	149	100	0
VE6JY	15	75	10	1995	1990	1995	Yes	Yes	No	??	??	236	??	10	90	40	??	161	??	5	95			56		20	80
VE6LB	60	25	15	1956	1956	1990	Yes	Yes	Yes	40	35	224	185	90	10	27	27	130	128	95	5	25	28	108	106	95	5
VE6WZ	50	40	10	1998	1998	1998	No	No	No	40	34	280		60	40	40		152		50	50	25	25	49		95	5
VE7BS	5	10	85	1946	1946	1946	Yes	No	No	30	20			100	0					100	0	17		108	137	95	5
VE7ON	15	40	0	1985	1985		Yes	Yes	No			175	175	1	99			145	145	1	99	30	30	138		90	10
VE7VV	5	5	90	1958	1958	1958	No	No	No	40		240		90	10			150		90	10				143		
VK5GN	40	20	40	1972	1972	1968	No	No	Yes					30	70					75	25	30	30			80	20
VK6HD	10	45	45	1969	1969	1969	Yes	Yes	Yes	40	40	328	328	99	1	40	40	312	312	99	1	39	39	224	224	99	1
VK6VZ	10	2	88	1972	1995	1995	Yes	Yes	Yes					99	1					99	1	35	35	185	185	99	1
W0AIH	10	30	60	1950	1950	1950	Yes	Yes	No					95	5					95	5					90	10
W0BV	30	40	30	1985	1985	1993	Yes	Yes	No	40	40	332	332	95	5	40	40	292	290	90	10	30	30	199	198	95	5

Call	% 40	% 80	% 160	On 40 Since	On 80 Since	160 Since	5B DXCC	5B WAZ	160m WAS	40m WAZ Wkd	40m WAZ Ctd	40m DXCC Wkd	40m DXCC Ctd	40m % CW	40m % SSB	80m WAZ Wkd	80m WAZ CFD	80m DXCC WKD	80m DXCC CFD	80m % CW	80m % SSB	160m WAZ WKD	160m WAZ CFD	160m DXCC WKD	160m DXCC CFD	160m % CW	160m % SSB
W0CD	5	10	85	1970	1970	1970	Yes	No	Yes	?	?	59	59	95	5	?	?	69	69	90	10	38	38	269	269	95	5
W0EJ	40	30	30	1976	1976	1980	Yes	No	Yes	38	31	216	183	85	15	29	29	153	135	85	15	27	26	128	127	90	10
W0GJ	35	40	25	1989	1989	1992	Yes	Yes	Yes	40	39	285	252	99	1	39	39	273	247	90	10	35	33	171	155	100	0
W0SD	20	60	20	1962	1962	1962	Yes	Yes	Yes	40	40	150	110	50	50	40	40	150	130	30	70	20	20	60	55	50	50
W0SF	50	40	10	1997	1998	2000	Yes	Yes	No	40	40	310	305	99	1	38	38	244	240	99	1	24	24	83	60	99	1
W0YG	6	41	53	1962	1962	1985	Yes	No	Yes	39	39	321	311	100	0	40	40	267	264	95	5	40	39	238	236	100	0
W1FV	5	65	30	1969	1969	1971	No	No	No	40	40			99	0	40	40			80	20	36	36			100	0
W1JZ	2	5	93	1955	1955	1957	Yes	No	Yes	40	40			99	1	39	39			99	1	36	36	269	269	99	1
W1NG	9	9	4	1960	1965	1979	Yes	Yes	Yes	40	40	344	344	80	20	40	40	340	340	66	33	40	36	301	301	90	10
W1TE	40	40	20	1995	1995	1995	Yes	Yes	Yes	40	40	250	243	80	20	32	32	154	146	70	30	32	31	122	115	90	10
W1WAI	30	30	40	1953	1953	1983	Yes	No	No	40	40	326	326	98	2	39	39	291	291	95	5	33	33	223	223	99	1
W1ZK	50	10	40	1980	1983	1985	Yes	No	No			239	238	100	0			258	257	85	15			198	197	98	2
W2VJN	30	30	40	1947	1947	1979	No	No	No					100						100						100	
W3GH	10	30	60	1953	1960	1960	Yes	No	Yes	40	40	300	300	95	5	40	38	300	231	50	50	38	38	265	265	95	5
W4DC	15	45	40	1988	1988	1993	Yes	Yes	Yes	40	40	266	233	70	30	38	38	276	231	50	50	?	?	106	98	80	20
W4DR	10	30	60	1950	1967	1970	Yes	Yes	Yes	40	40	358	358	60	40	40	40	346	346	50	50	39	39	304	304	95	5
W4NL	40	32	28	1954	1965	1974	Yes	Yes	No	40	40	320	200	90	10	40	40	292	200	99	1	36	36	218	215	99.9	0.1
W4TO	20	30	50	1991	1991	1991	Yes	Yes	Yes	40	39	323	318	20	5	40	39	297	294	50	50	37	36	211	203	90	10
W4ZV	0	1	99	1980	1980	1984	No	No	No	40	40	300	300	0	0	40	40	338	338	99	1	40	40	310	310	99	1
W5EU	6	80	14	1952	1984	1995	Yes	Yes	No	40	?	?	?	95	5	40	40	312	307	25	75	32	?	105	?	90	10
W5FO	48	48	4	1954	1955	1997	Yes	Yes	No	40	40	240	175	95	5	35	34	175	150	70	30	15	15	39	30	99	1
W5ODD	20	45	35	1979	1979	1997	Yes	Yes	Yes	40	40	287	280	100	0	34	34	191	180	100	0	25	24	75	65	100	0
W5WP	25	30	20	1984	1984	1984	Yes	Yes	No	38	38	169	169	60	40	32	32	147	147	60	40	18	18	54	54	80	20
W5YU	2	96	2	1941	1941	1941	Yes	Yes	No	40	40	116		33	67	40	40	306	300	2	98	40	40	119		50	50
W6DZ	40	50	10	1953	1956	1957	No	No	No					40	0					50	50					10	0
W6SR	30	60	10	1980	1978	1996	Yes	No	Yes	40	40	310	269	90	10	40	39	221	207	50	50	16	14	39	35	90	10
W6YA	95	4	1	1981	1982	1990	No	No	No	40	40	339	339	99	0.5	40	40	278	278	99	1	28	28	121	121	100	0
W7DD	0	100	0		1961		No	No	Yes					2		40	40	268	227	5	95						
W7UT	20	40	40	1975	1975	1984	Yes	Yes	Yes	40	40	341	341	90	10	40	40	327	327	85	15	40	40	269	269	95	5
W8LRL	1	1	98		1972	1972	No	No	Yes													40	40	310	310	99.9	0.1
W8RU	33	33	34	1996	1996	1996	Yes	Yes	Yes	37	35	242	232	90	10	25	24	126	114	98	2	26	25	103	100	99	1
W8UVZ	5	40	55	1978	1978	1982	Yes	Yes	Yes	40	40	337	337	5	5	40	40	315	313	30	10	38	38	277	276	55	0
W8WEJ	25	25	50	1969	1969	1969	Yes	No	No			156	150	90	10	38	38	127	125	95	5			111	110	90	1
W8XD	7	18	75	1962	1973	1992	Yes	Yes	Yes	40	40	200	200	99	1	38	38	200		95	5	36	36	210	207	99	1
W9AJ				1961	1985	1991	Yes	No	No	39	38	253	250	30	70	34	34	197	191	80	50	25	25	112	111	90	10
W9LYN	10	0	90	1956	1957	1998	Yes	No	Yes	27	27	133	132	50	50	25	25	103	103	50	50	15	13	41	37	90	10
W9UCW	5	10	85	2000	1992	1960	Yes	No	Yes					2						50						60	40
WA1RKS	2	2	96	1972	1972	1978	No	No	No												80			24			
WA2UUK	5	10	85	1974	1984	1976	Yes	No	Yes	40	40	320	320	90	5	36	36	284	283	20	5	34	34	218	217	55	5
WB9Z	10	45	45	1974	1983	1985	Yes	No	Yes	40	40			90	25	40	40	315	315	75	25	40	40	282	282	95	5
WD5T	5	5	90	1985	1985	1985	Yes	No	Yes	38	38	175	175	5	0	38	38	155	155	5	0	26	26	110	110	90	1
W0RI	15	10	75	1950	1952	1984	Yes	No	Yes	39	39	195	195	98	2	35	35	150	150	50	50	32	32	159	159	99	1
WX0B	25	50	25	1962	1962	1989	Yes	No	Yes					75	25					75	25					80	20
WZ6Z	75	24	1	1986	1989	NA	Yes	No	No	40	38	276	183	90	10	37	30	182	132	50	50	11	9	24		50	50
XE1KK	60	35	5	1995	2001	1996	No	No	No	38	33	170	117	10	90	27	25	97	77	95	95	20	16	22	15	50	50
XE1RCS	0	0	100			1996	No	No	No					2										75	56	0	100
XE1VIC	10	75	15	1978	1978	1989	Yes	No	Yes	40	40	255	249	0	100	36	40	225	222	0	100	27	26	109	106	0	100
YT6A	40	40	20	1980	1980	1995	Yes	No	No	40	40	254	254	90	10	40	40	295	287	30	70	23	23	156	156	90	10
ZC4DW	60	35	5	2000	2000	2000	No	No	Yes	23	12	116	54	83.1	5.2	18	9	86	45	89.2	4.7	13	8	29	29	99.6	0.4
ZS4TX	60	40	20	1983	1983	1996	Yes	Yes	No	40	40	315	250	90	10	40	40	264	218	90	10	40	40	172	145	98	2
ZS6EZ	70	20	10	1983	1983	1986	Yes	Yes	No	40	40	308	290	70	20	40	40	240	230	60	40	28	27	130	125	99	1

DX-Operating on the Low Bands

17.4.1. Achievement summary

Percentage of Respondents Holding Award

5-Band DXCC	5-Band WAZ	160-m WAZ
68 %	38 %	38 %

Average All-Time DXCC Count:

40 meters	260
80 meters	225
160 meters	164

Average WAZ Count:

40 meters	38.2
80 meters	36.2
160 meters	31.0

The purpose of the listings is not to give an accurate DXCC status report, but to show what some of the leading low-band DXers have achieved and what they are using to do it. A few well-known DXers are missing in the tables. They have chosen not to reply to the questionnaire or could not be reached via e-mail. Rankings of Top-Band early award winners can be found in K1ZM's excellent book *DXing on the Edge* (Ref 511).

17.5. Antennas and Equipment

Table 2-4 at the end of this chapter gives an overview of the antennas used by the participants in the poll. Since a number of respondents mentioned more than one antenna for any particular low band, the sum of the percentage in each group is not necessarily 100%. In all of the data below the term "total group" refers to all respondents to the poll (270).

40-Meter Antennas

Antenna Type	Total Group	Top Group
Yagi/Quad	49%	56%
Dipole	23%	22%
Vertical Antennas	16%	15%
Vertical Array	8%	3%
Delta Loop	5 %	3 %
Horizontal Array	3 %	6 %
Other	3 %	3 %

Note: The category Vertical Antennas includes shunt-fed towers and inverted Ls. Dipoles include inverted Vs; Horizontal Arrays includes Double-Extended Zepps, for example. Top group: Stations with at least 300 countries confirmed (86 stations in our poll). Many stations use 2-element reduced-size 40-meter Yagis (eg, Force 12, CushCraft); one is using Rhombics up 30 meters.

Special Receiving Antennas Used on 40 Meters

Antenna Type	Total Group	Top 100
None	72%	73 %
Beverages	21%	20 %
Flags	4 %	5%
Other	1 %	2 %

Note: The category Flags include EWEs, Flags, Pennants, K9AY arrays, etc. The big guns have rotatable Yagis of Quads, and generally do not use separate receiving antennas, although a few big guns few use Beverages occasionally. One says he's using his gutter as a receive antenna!

80-Meter Antennas

Antenna Type	Total Group	Top Group
Vertical Antennas	33 %	24 %
Vertical Array	20 %	40 %
Dipole/Inv V	30 %	20 %
Yagi/Quad	8 %	8 %
Sloping Dipole	5 %	2 %
Half Sloper	10 %	13 %
Delta Loop	5 %	7 %
Other	5 %	<1 %

Top Group includes stations with at least 300 countries confirmed on 80 meters (45 stations). If we compare the results with those obtained 6 years ago we see a substantial increase in the use of Vertical Arrays. In the Top Group this rose from 23% to 40%.

Special Receiving Antennas Used on 80 Meters

Antenna Type	Total Group	Top Group
None	41%	31%
Beverages	43%	65%
Flags	10%	11%
Magnetic Loops	5%	0%
Low Dipoles	3%	0%
Other	2%	1%

Note: The category Flags include EWEs, Flags, Pennants, K9AY arrays, etc. The number of 80-meter respondents who do not use special receiving antennas has gone up remarkably. The reason is that more are using directive transmit/receive antennas (shorted Yagis, arrays of verticals, etc).

160-Meter Antennas

Antenna Type	Total Group	Top group
Vertical Antenna	26 %	33 %
Inverted-L/T	24 %	18 %
Shunt-Fed Tower	15 %	16 %
Dipole/Inv V	24 %	14 %
Long Wire	2 %	2%
Vertical Array	9 %	16 %
¼-Wave Sloper	9 %	7 %
Delta Loop	2 %	0 %
Other	4 %	6 %

Notes: Top Group means stations who have worked at least 225 DXCC countries. Verticals in all shapes and forms (dedicated verticals, inverted Ls, Ts and shunt-fed towers) make up 66% of Top-Band antennas.

Special Receiving Antennas Used on 160 Meters

Antenna Type	Total Group	Top Group
None	25 %	14 %
Beverages	52 %	68 %
Flags	14 %	16 %
Magnetic loops	9 %	4 %
Low dipoles	6 %	7 %
Array vert. short ele.	2 %	4 %
Other	2 %	2 %

Notes: The category Flags include EWEs, Flags, Pennants, K9AY array, etc. It's obvious that Beverages are the secrets to success for most 80 and 160-meter

Table 2-4
Low-Band DXer's Antenna Survey

CALL	40-m TX Antenna	80-m TX Antenna	160-m TX Antenna	160-m Special RX Antenna
3B8CF	Rot. Dip.	Dip./Inv. V	Crooked Dip.	Flag, Rot. Diamond, Low Dip.
4X4DK	LPDA, Vert. Titanex V160HD	Vert. Titanex V160HD	Vert. Titanex V160HD	5 Beverages
4X4NJ	Rot. Dip. @30m	Inv. L	Inv. L	K6SE Loop
5B4ADA	Rot. Dip. Force 12	Two 1/2-λ Slopers	Inv. V @16m	100
9M2AX	2-Ele. Force 12	GP @45m	Sloping Dip. @45m	9 Beverages
AA4MM	GP @45m	3-Ele. Delta Loop	2-Ele. Vert. (parasitic 1/4-λ spacing)	Pennants, Beverages, Loops
AA4V	None	Force 12 180EV Vert. (Coast)	Inv. L	Small Loops, Pennant
AB0X	Vert. (Seaside)	Vert. + Half Sloper	Shunt-Fed Tower	
AC0M	Phased Verts. + Half Sloper	1/4-λ Slopers	1/4-λ Sloper	
AE1Q	Dip.	Dip.	Inv. L	
AE9B	Dip.	1/2-λ Vert., Dip.	1/4-λ Vert., Dip.	Beverages, K9AY Loop
AG6X	2-Ele. Beam, 4-Square	Butternut HF2V	None	
AJ9L	Butternut HF2V	HF2V Phased Verts.	Vert. (by W9UCW), Inv. V	3 Beverages
AJ1H	Vert. and 2-Ele. Yagi	Vert. Designed by W9UCW	22m Vert.	3 Beverages
CT1EEB	None	Top-Loaded Vert., 22m	Inv. L	
DF2PY	3-Ele. KLM Yagi	Inv. L + Inv. V	4 Ele. In-line Array	4 Beverages
DJ2YA	None	None	Vert., Inv. V @17m	Beverages
DJ4AX	Sloping Dip., Inv. Vee, 1/4 wl. GP	Inv. L	60m wire @50m	
DJ6QT	Vert., Base @50m, Long Wire @50m	2-Ele. Wire Yagi @50m + Sloper @45m	Inv. L, Inv. V	Unterminated Beverages
DK7PE	2-Ele. Full-Size Quad @40m	Sloper	Different Types at Different Locations	Beverages, magnetic Loops
DL3DXX	Different Types at Different Locations	Different Types at Different Locations	20m High Vert. (Top Loaded)	4 Two-Wire Beverages
DL7CX	Full-Size Vert. or 2 Phased Verts	Full-Size Vert.	Inv. L	
EA3V	Triple Leg	Triple Leg	33-m Vert., Dip.	Beverage to USA
EA6ACC	2 element beam	Delta Loop fed in lower corner	25.5-m Vert.	Four EWEs
EA6NB	1/4 λ Dip. @ 14m	1/4 λ Dip. at 14m high.	Shunt-Fed Tower	Small Rot. magnetic Loop
EA7NW	402CD 2-Ele. Yagi	2 Half Slopers	Inv. V Trap Dip.	
EY8MM	Inv. Vee Trap Dip.	Inv. V (with Traps)	Delta Loop, Inv. L	
F6BKI	1/4-λ Vert.	1/4-λ Vert.	27-m Vert.	Two 2-wire Beverages
FM5BH	402CD 2-Ele. Yagi @ 20m	Rot. Dip. @23m	Inv. L (17-m Vert.)	K9AY Loop
G3KMA	402CD CushCraft	Inv. V @23m	Inv. L (24-m Vert.)	
G3XTT	Sloping Dip.	Delta Loop	3/8-λ. Inv. L 50/50 Vert./Hor.	Beverages in Winter
G4OBK	Rot. Dip. @18m	Inv. V @18m	3/8-λ. Balloon Supported Vert., Inv. L @22m high	Six Beverages (60 Deg Spacing)
G4VGO	Butternut HF2V	Butternut HF2V + Inv. V	Inv. L, @ 18-m High	200-m Beverage
GM3PPE	None	20-m Vert.	Sloping Dip. @45m	
GM3YOR	Delta Loop	1/4-λ Vert.	Inv. L , Dip.	
GM3YTS	Dip. and Vert.	Dip. or Vert.	Delta Loop @36m	Beverages
GW3YDX	Rot. Dip.	Dip. or Vert.	Bottom Loaded 25-m Vert.	11 Beverages
HA0DU	402CD 2-Ele. Yagi @33m	Delta Loop @24m	Extended Zepp	
HB9AMO	5/8-λ Vert.	3/8-λ. Vert.	1/4-λ Vert., Sloping Dip.	
HC8N	Extended Zepp	Extended Zepp	4 Square	Beverages
HL3IUA	402CD @42m, 3 Ele. Full-Size on Eu @26m	2 Dips. @42m (right angles)	Dip., Vert. 28m	
I4EAT	4-Ele. Yagi @38m	2-Ele. Phased Array, @40m	None	
IC8WIC	Dip., Vert., 3-Ele. Yagi	Square Array	Inv. V	Pennant with ECP-1
IN3ASW	2-Ele. Yagi	Rot. Dip.	two Inv. V's	
IT9ZGY	Hy-Gain 402BA, Delta Loop	Sloping Dip., Half Sloper	Shunt-Fed Vert. 29-m high	4-Squaremini-Phased Vert. Array, Beverages, 6 Pennants
IV3PRK	Delta Loop	Inv. V	Half Sloper	Small Loop
JA0DAI	None	Half Sloper	Shunt-Fed Tower	
JA1EOD	2-Ele. HB9CV	Sloper	Shunt-Fed Tower	
JA1UQP	AFA-40 (Create) 2-Ele. Phased Array	Rot. Dip.	None	
JA2PJC	2-Ele. Full-Size Phased Array Yagi	None	2-Ele. Vert. (Parasitic 1/4-λ Spacing)	
JA2VPO	None	2-Ele. Beam (CW) + Rot. Dip. (SSB)	Inv. V, Shunt-Fed Tower	
JA3CZY	2-Ele. Beam	2-Ele. Create AFA75	None	
JA3FYC	Rot. Dip.	Short Rot. Dip.	Shunt-Fed 20-m Tower	80-m ant. used as RX ant on 160

CALL	40-m TX Antenna	80-m TX Antenna	160-m TX Antenna	160-m Special RX Antenna
JA4LKB	M2 7&10-30LP8	Dip.	33-m Shunt-Fed Tower	Beverage
JA5AUC	3-Ele. Yagi @28m	Rot. Dip. @30m	Sloper	Mini Loop
JA5IU	3-Ele. Yagi	Dip. @35m	Inv. V @30m	Small Loop
JA7KAC	2-Ele. Phased Array	rot. Dip., 4 sloping Dips, Inv. V	Shunt-Fed Tower, Sloper, Inv. L	Small Loop
JE1SPY	None	Vert., Inv. L, Inv. V	Inv. L, Long-wire, Inv. V, Sloper, Vert.	Small-Loop, Bar-Antenna, Magnetic Loop, Inv. L, Inv. V, Small-Loop
JE1TSD	None	Ground Plane, Half Sloper	Shunt-Fed Tower (20m)	
JF6OJX	2-Ele. Yagi	Rot. Dip.	Dip.	
JG1XLV	AFA-40 (Create) 2-Ele. Phased Array @23m	None	None	
JH2FXK	Dip. @23m	Half Sloper and 1/4-λ Vert.	Shunt-Fed Tower, Vert. Dip.	2 Beverages, Small Loop @15m
JH2RMU	None	None	Inv. V @20m	Inv. V @20m
JH3VNC	4-Ele.multiband Yagi	Dip.	25-m Shunt-Fed Tower	2-1 Square Loop (2-m Sides)
JH4IFF	AFA-40 (Create) 2-Ele. Phased Array	Sloper for 80, AFA75 for 75m	Shunt Feed Tower and Inv. V	
JH4UYB	3-Ele. Full-Size Yagi @32m	2-Ele. Create AFA75 @35m	Inv. V @32m	6 Beverages, Small Loop, Short Rot. Dip.
JL1UXH	2-Ele. Yagi 30/40m @35m	Rot. Dip. @33m	Long Wire	
JX7DFA	Delta Loop	Conical monopole	Inv. L	
K0CS	Cushcraft 40-2CD	1/4-λ Vert.	1/4-λ Vert.	165-m Beverages
K0HA	9-Ele, Parasitic Vert. Array	7-Ele. Parasitic Vert. Array	4-Ele. Parasitic Vert. Array	Beverages
K0XM	2-Ele. Force 12 C4XL @21m	Phased 13-m Top Loaded Verts.	Phased 13-m High Verts.	K9AY, Rotatable Flag
K1FK	4-Wquare	22-m Vert.	23-m Top-Loaded Vert.	Beverages
K1GUN	None	None	1/4-λ Vert.	6 Beverages
K1MY	4 Ele.m2 Yagi (12-m boom)	Sloping Dip.	None	
K1QX	40-2CD CushCraft Yagi	Sloping Dip.	1/4-λ Vert.	Beverages
K1TTT	Stacked 402-CD, 4 Square	2-Ele. Revers. Wire Yagi @45m + 4 Square + Inv. V @21m	Inv. V @45m, Array of 2 Inv. L's (27-m Vert.)	3 Two-Wire Beverages
K1VR	402-CD @27m, 4-Ele. Vert. (Colatchco)	2-Ele. Phased Array (Colatchco)	3 Ele. Vert. Parasitic Array (Spitfire FVR)	Beverages
K1ZM	3 Ele. Yagi @50m	4-Square	4 Square @K1ZM, 4 Rect. @VY2ZM	105-m Beverage, 1.5-m Loop.
K2RD	4 Sloping Dip. Array	Vert., Windom @25m	Inv. L (15-m Vert.), Inv. V @24m	Multiple Beverages
K2TQC	3-Ele. Delta Loops (swingable, switchable)	2-Ele. Delta Loops (switchable, swingable)	Inv. Ls	Beverage
K2UO	Inv. V	Inv. V	Zigzag Dip.	Beverages, Snake
K2UOP	Phased Verts	Butternut Vert.	Inv. L, Sloper	K9AY, Rot. Loaded Loop, Slinky Beverage
K3JGJ	4 Square	4 Square Vert. Array, Sloper	Dip. at 40m	Short 38-m Beverage
K3NA	Dip. @25m	Dips. at 25m	Inv. L (11m Vert.)	50-m Wires to Eu and South
K3UA	Inv. V @12m	Dip. at 10m	Inv. L	
K3WW	402-CD @23m	1/4-λ Vert.	End-Fed Wire	BOG, Flag, K9AY
K4BAI	Dip.	Inv. V	Inv. V, Inv. L	Beverage, Pennant, 80-m 4-Sq
K4CIA	Rot Dip. @34m; Half Square; Vert.	Inv. V + Vert. + Delta Loop	Inv. L	
K4ESE	None	4-Square	Shunt-Fed Tower (30-m high) With Lots of Top Loading	Beverages
K4MQG	40CD @40m	2-Ele. Create @37m, 4 Square, 1/2-λ Sloper, 1/4-λ Vert.	Half Square Vert. Loop	Beverages
K4PI	2-Ele.mosley @30m	K8UR Style 4-Square	None	
K4TEA	2-Ele. Yagi Force 12 @27m	160-m Half Square Vert. Loop With a Tuner	1/4-λ Vert. Base at 8m	Four 150-m Beverages
K4TO	2-Ele. Yagis Stacked @70 and 140 ft	4 Square	1/4-λ Inv. L	Beverages
K4UEE	2-Ele. Force 12	1/4-λ Vert.	1/4-λ Vert.	
K4VX	2-Ele. Yagi (48' Boom) @130'. Ext. Double Zepp	Folded Dips. @48m	None	Beverages
K4YP	Inv. V @20m	Inv. V @20m	1/8-λ Phased Inv. Ls	Beverages
K4ZW	2-Ele. Yagi @32m	4-Square	Shunt-Fed Tower (17m)	Five 100-m Beverages
K5AQ	KLM Rotatable Dip. @21m	Shunt-Fed Tower (16m)	None	
K5OVC	2-Ele.mosley Yagi	4-Ele. Vert. Array (Colatchco)	1/4-λ Vert. Base at 18m	Beverage, EWE, Pennant
K5PC	Collinear Dip. @21m	4-Square, K8UR style	Shunt-Fed Tower	
K5RA	2-Ele. Full-Size Yagi @ 27m	Vert.	1/4-λ Vert., Base at 6m	150-m Beverage
K5ZD	2-Ele. Cushcraft @35m	Inv. V @29m, 4-Square	Delta Loop	
K6ANP	3 Ele. Yagi	Delta Loop	Inv. V	
K6EID	402BA 2-Ele. Yagi	Half Sloper	2 Shunt-Fed Towers spaced 1/4 λ	
K6SE	402BA at 23m	4-Ele. Vert. Array (Shunt-Fed Towers)	None	
K7EM	3/3 Full Size Yagi Stack	4-Square	T Vert. (23-m Vert. Section)	
K7FL	Hy-Tower	Hy-Tower	Shunt-Fed Tower	Beverage
K7OX	Phased Verts.	Sloper		

2-28 Chapter 2

CALL	40-m TX Antenna	80-m TX Antenna	160-m TX Antenna	160-m Special RX Antenna
K7ZV	Rot. Dip. @21m	3-Ele. Yagi with High-Q Coils	Shunt-Fed Tower	6-Ele. Vert. Array in 30-m Dia. Circle (60-Deg Steps)
K8BHZ	None	None	2 Inv. Ls, spaced 90m, Fed In or Out of Phase.	
K8EJ	2-Ele. Force12 EF-240X @26m	Phased Verts.	Vert. T	
K8GG	Short Rot. Dip., Full λ Dip.	Inv. V @21m	Shunt-Fed 21-m Tower	Beverages
K8IP	None	None	Inv. L	
K8MFO	2-Ele.mosley	Inv. L	Inv. L (55% Vert., 45% Hor.)	105-m Beverage
K8ZR	1/4-λ Vert.	Vert.	1/8-λ Vert.	Beverages
K9EL	2-Ele. Yagi	1/4-λ Vert.	N4kg Reverse-Fed Tower	Loop
K9FD	2-Ele. Yagi	1/4-λ Slopers	Shunt-Fed Tower	Beverages
K9JF	CushCraft 40-2CD (modified)	4-Square	Shunt-Fed Tower	House Gutter
K9KU	2-Wire Ground Planes	Ground Plane	Shunt-Fed Tower	Magnetic Loops, Loop Array
K9MA	Rot. Dip.	Shunt-Fed Tower	30m Shunt-Fed Tower	Pennant, Short 2-wire Beverage
K9RJ	2-Ele. Hy-Gain Yagi @39m	1/4-λ Vert.	Phased Verts.	Small Loop
KA7T	2-Ele. Yagi @24m	Phased Verts.	Dip.	
KB0ETC	Double Extended Zepp	Dip.	40-m Double Extended Zepp	
KG6I	160-m Loop on 40	Delta Loop	Vert. Full-λ Loop, Top @36m	3 Unterminated Beverages
KH6DX/M	N9JMX Predator Screwdriver mobile Antenna	N9JMX Predator Screwdriver mobile Antenna	N9JMX Predator Screwdriver + Loading Coil (Height = 5m)	
KJ9I	3-Ele. Shortened Yagi @35m	1/4-λ Vert.	1/2-λ Sloping Dips. @60m	
KL7RA	3-Ele. Telrex Yagi @57m	Dip. @50m	1/4-λ Vert.	Beverage
LA5HE	Titanex V160 HD Vert. + Dip.	Titanex V160 HD Vert. + Dip.	Titanex V160 HD Vert.	
LY3UM	Cushcraft R8	21-m GP (T-mode) + 4 Radials	21-m Vert.	5 Beverages
L22JE	GP on 9-Floor House	Inv. V + Delta Loop	Inv. V	Sloping Long Wires
N0AT	Force 12 EF4020	2 Phased Verts.	Two 1/8-λ Verts., 1/8-λ Spacing	Beverages
N0AX	40-2CD Yagi	2 Half-Slopers @15m	Inv. L	Rot. Coax Loop
N0IJ	Mosley 2-Ele. at 37m	Force 12180C @35m	Alpha Delta DX Sloper	Dip.
N2CG	End-Fed Long Wire	End-Fed Long Wire	1/4-λ Inv. L	2-m Loop
N2NT	2-Ele. Yagi @35m + 3 Ele. Wire Beam @20m	2-Ele. Wire Beam NE/SW, Dip.	4 Square Inv. L, Dip.	Beverages, K9AY Loop
N2QT	40-2CD @18m	Squashed Delta Loop	Shunt-Fed Tower @20m	Ewes
N4CC	2 Ele. Full Quad @42m	1/2-λ Sloper System	1/4-λ Sloper	Beverages
N4JJ	402-CD @32m	8-Ele. Vert. Array	Inv. V (@15m)	2 Pennants
N4MM	Dip. @18m	3-Ele. Vert. Array, Inv. V @32m	10-Ele. Vert. Array 7dB re 1 Vert.	8-Ele. Vert. Array, same as TX
N5FG	2-Ele. Yagi @25m	Phased 1/4-λ Verts.	Vert.	
N5KO	Inv. V in a Tree	Dip. @23m	Shunt-Fed 33-m Tower, Inv. V @30m	
N5RG	402-CD @17m	Inv. V @20m	Dip. @24m	Beverages
N5UL	Delta Loop	1/4-λ GP in a Tree	Delta Loop	Flag
N6FF	HT-18 Hy-Tower	1/4-λ Vert.	1/4-λ Sloper	Short Beverage
N6ZZ	Force 12 C4SXL	Dip.	Sloper at 23m	
N7JW	Log Periodic	HT-18 Hy-Tower	Inv. V	Beverages
N7RK	1/2-λ Vert.	Sloping Dip.	Vert. T (Top @26m)	6 Beverages
N7RT	4 Ele. Yagi @43m	8-Ele. Vert. Array	Shunt-Fed Top-Loaded Tower (24m)	2 Beverages
N9AU	2-Ele. Yagi	Phased 1/4-λ Verts.	Inv. V (@15m)	8-Ele. Vert. Array, same as TX
NA0Y	Inv. V @22m	2-Ele. Yagi	Vert.	
NI0C	HF-2V Vert.	1/4-λ Sloper	Delta Loop	Beverages
NR1R	2-Ele. Yagi @32m	Sloper @22m	1/4-λ Sloper	Flag
NT5C	40-2CD @27m on hilltop	HF-2V Vert.	Sloper at 23m	Short Beverage
NX4D	2-Ele. Linear-Loaded Yagi	2-Ele. Phased Array (Colatchco)	HF-2V Vert. with 160-m Kit	Beverages
OH1MA	3 Ele. Yagi @35m	2 Sloping Dips. @21m	Dip. @33m	6 Beverages
OH2BO	Dip.	Top-Loaded Shunt-Fed 18-m Tower	None	Rot. Phased Broadband Loops
OH2BU	Two 3-Ele. Yagis, both @34m	Force 12 2-Ele. Yagi @40m	Top-Loaded Shunt-Fed 18-m Tower	2 Beverages
OH2KI	2-Ele. Yagi	4-Square, GP	Inv. L, 1/4-λ Slopers	
OH3SR	2-Ele. Yagi @33m	2 Sloping Dips.	Four Square	
OH4MFA	Dip., Inv. V, Delta Loop	4 Half Slopers	3 1/2-λ Slopers, E/W/S	Beverages
OK1DOT	None	Inv. L	Dip. @30m	Sometimes 80 or 40-m TX Ant.
OK1MP	Inv. V	Quad	1/4-λ Slopers	Beverages
OK1RD	3 Over 3 Ele. Stack	2-Ele.m2 Yagi @42m	1/4-λ Vert.	Beverage
OM2XW	None	None	Quad, Inv. V	
			Short End-Loaded Inv. V	Titanex SES160 Rot. Array , 2 Short Beverages
			Shunt-Fed Tower @50m	Unterminated Snakes
			Grounded Half Loop, Top Loaded	

DX-Operating on the Low Bands

CALL	40-m TX Antenna	80-m TX Antenna	160-m TX Antenna	160-m Special RX Antenna
ON4UN	3 Ele. Yagi @30m	4 Square	1/4-λ Vert.	12 Beverages (Some Phased, Staggered)
ON5NT	Sloper	Inv. V	Shunt-Fed Tower	Beverages
ON7GB	Inv. V, 2-Ele. Yagi, Delta Loop	Inv. V, 1/4-λ Slopers, Rot. Dip., GP	Inv. V, Inv. L, Sloping Top-Loaded Vert., 1/4-λ Sloper	
OY9JD	1/4-λ. GP, 1/2-λ. Vert. Dip.	Shunt-Fed Tower	Inv. L	
OZ1BTE	2-Ele. HB9CV	27-m Vert.	27-m Vert. With Linear Loading	
OZ1CTK	3 Ele. Yagi (Mosley PRO96)	1/4-λ. Sloper	1/8-λ. Inv. L (22-m Vert.)	EWE, 2 Beverages
OZ1NG	2-Ele. Force 12 Yagi @40m	Sloping Dip. @30m	Inv. L	Beverages, Rot. Flag
OZ1LO	2-Ele. Yagi	1/4-λ Yagi	1/4-λ Inv. L	Beverages, Inv. V
OZ3PZ	Force 12 240	Delta Loop @23m		
OZ7YY	2-Ele. Yagi @34m, 24-m Vert., Inv. V @15m	24-m Vert, Inv. V @15m	24-m Vert.	4 Beverages, 4 Pennants, Loop 2m Square, Ferrite Stick, Active Whips
OZ8ABE	Short Dip. @23m	Sloper	80-m Sloper	
PA0LOU	Delta Loop	Delta Loop	Inv. L	
PA0ZH	4 Square (Sloping Dips.)	4 Square	Delta Loop, Center @36m	
PA2FQA	Butternut HF2V	Butternut HF2V, Sloper	Inv. L	K9AY Loops
PA0TAU	2-Ele. Yagi	Vert., Delta Loop	Vert.	K9AY
PY1BVY	None	None	1/2-λ. Dips., Inv. V, Vert., 2 Long Wires	
PY2FUS	Inv. V	Sloping Dip.	Inv. V	Dip., Inv. V
RV1CC	Hor. Delta Loop 340-m Perimeter @30m	Hor. Delta Loop 340-m Perimeter @30m	Hor. Delta Loop 340-m Perimeter @30m	Beverages to USA
S50A	3-Ele. Yagi. 2-Ele. (Shortened) Yagi	3 Ele. Short Yagi, Inv. V	Three 1/4-λ Slopers	
S56A	2-Ele. Yagi	Dip., Windom	Windom	
SM4CTT	Sloping Dip. K8UR style	Inv. V	Inv. V, Inv. L	
SM6CTQ	Vert.	Vert., Inv... V	Inv. L, Dip.	Inv. V
SM6GZ	Inv. V	4 Square Array	None	
SP2FAX	3-Ele. Full Size	4 Square Array	44-m High Vert.	11 Beverages
SP5EWY	2-Ele. Delta Loop	Delta Loop, Half Sloper @18m	1/4-λ. Sloper @18m	K9AY, Beverages in Winter
SV1AOZ	40-2CD	Shunt-Fed Tower, Inv. L	Inv. L (20-m Vert., 35-m Hor.)	Loops (EWE-Flag), Shielded Loop
SV8CS	3-Ele. Yagi @30m	Inv. V @30m, Vert.	33-m Vert.	5 Beverages
SV8JE	Delta Loop	Delta Loop	Dip. @20m	Receiving Loops
T77C	Vert. + Half Sloper	Dip., Vert., Half Sloper	Half Sloper, Inv. V	
TF8GX	Gap Titan Vert.	Dip., Gap Titan Vert.	Dip.	
UA0MF	GP, 3 Switched Slopers	Inv. V @40m, 1/2-λ Sloper	Inv. V @40m, 1/4-λ Sloper	Beverage
UA2FF	3-Ele. Yagi	Full Size Vert.	Full-Size Vert.	Beverages
UA3AB	Force12mAG340N	W8AV Triangle Array	26-m Shunt-Fed Tower	K9AY Loop
UA9AT	None	None	Inv. V., 2 Delta Loops	
VA3DX	2-Ele. 4O2BA 25m	Inv. V @21m	Inv. L (21-m Vert.)	
VA5DX	Cushcraft XM240 @30m	4-Square	Two 1/4-λ. Slopers @30m	
VE3OSZ	Dip. @10m	Inv. L @20m	Inv. L (20-m Vert.)	Beverage, Pennant
VE6JY	3-Ele. Yagi @36m, 5-Ele. @45m, 4-30mHz LPDA @29m	4-Ele. Full-Size Yagi @45m, 2 Bobtail Curtains	Inv. V @48m	Rosette of Beverages Covering All Directions
VE6LB	HF-2V, Inv. L, GAP Challenger	HF-2V, Inv. L, GAP Challenger	Modified HF-2V	Shielded Loop
VE6WZ	2-Ele. Yagi Cushcraft XM-240	2-Ele. Loaded Yagi	Shunt-Fed 30-m Tower	3 Beverages.
VE7BS	160-m Hor. Loop	160-m Hor. Loop	Hor. Loop 1-λ, Vert. Half Diamond	Beverages
VE7ON	Cushcraft 2-Ele. Yagi	3-Ele. Vert. Array	None	
VE7VV	Rot. Dip.	1/4-λ Vert.	Triangular Phased Vert. Array	
VK5GN	Delta Loop	Sloper	1/4-λ Sloper	Loop, Short Beverages
VK6HD	Sloping Dip.	1/4-λ. Vert.	Shunt-Fed Tower	Two-Wire Beverages
VK6VZ	Inv. V @27m	Dip.	24-m Top-Loaded Vert.	6 Beverages
W0AIH	3-Ele. Yagi @23m, 3-Ele. 403Bs Stacked @58 and 27m	2-Ele. Phased Array	Inv. L (21-m Vert), Inv. V @28m	Beverages
W0BV	2-Ele. KLM Yagi @20m	Phased Bobtails, Zepps @50m, 5-Ele. Vert. Array, etc	1/4-λ Vert., Base @15m, 3-Ele. Wire Beam @24m, 2 Zepps, Array of 4 Verts etc Beverages	
W0CD	Rot. Dip.	1/4-λ Vert.	1/4-λ Vert., Base @15m, N4KG Reverse-Fed Tower Vert.	Beverage
W0EJ	Inv. V	3-Ele. Vert. Array	Hor. Loop 1-λ, Vert. Half Diamond	2-Ele. (2-m Square) Phased Loop Arrays (4 directions)
W0GJ	4 Ele. @45m (50' Boom), 4-Square Array; Dip. @30m	Inv. V	1/4-λ Sloper	
W0SD	Full-Size Yagi, 2-Ele. Full-Size Quad to EU	4 Square, Rot. Dip. @60m, Dip. @25m	1/4-λ Vert., Inv. V @42m	5 Beverages
W0SF	Hy-Gain HyTower Vert.	1/4-λ Vert., Dip. @54m	1/4-λ Vert. with 125 Rads, Dip. @ 195'	3 Beverages
W0YG	Force 12 340N	Hy-Gain HyTower Vert.	Hy-Gain HyTower Vert.	5 Beverages
W1FV	3-Ele. Triangular Phased Vert. Array	4 Square	2 Phased Verts.	4 Two-Wire Beverages
		3-Ele. Phased Array	Short (18-m) Vert.	

CALL	40-m TX Antenna	80-m TX Antenna	160-m TX Antenna	160-m Special RX Antenna
W1JZ	2-Ele. Yagi @33m	Shunt-Fed Tower	50-m Shunt-Fed Tower	Beverages
W1NG	Mostly Wires	Wires, Inv. L	Dip., Inv. L, Phased Inv. L's	Beverages
W1TE	4 Square	Inv. V @22m	Inv. L	K9AY Loops, Pennants, Beverages
W1WAI	Dip. @15m	Dip. @15m	Inv. V @18m	
W1ZK	Rot. Dip.	1/4-λ Vert.	Sloper	Beverage
W2VJN	2 over 2 @32m and 50m	Inv. Vs @44m	4-Ele. Array Sloping Dips @45m	
W3GH	3-Ele. Yagi @25m	3/8-λ Vert.	Linear Loaded Vert. (24m)	Low 1/2-λ Dip.
W4DC	Dip., Vert.	Windom	Vert. Dip.	7 Beverages
W4DR	3-Ele. Wilson Yagi @36m	4 Square	4 Square	
W4NL	3-Ele. Yagi @27m	1/4-λ Vert.	Inv. L (20-m Vert.)	2 Unterminated Beverages
W4TO	2-Ele. Yagi @27m	4 Square	Inv. V @33m.	Beverages
W4ZV	None	Sloping Dip.	54-m Vert.	7 Beverages
W5EU	4 Square	4 Square	35-m Vert.	Beverages
W5FO	Dip. @30m	1/4-λ Vert., Inv. V @27m	Inv. L	
W5ODD	2-Ele. beam at 125'	Inv. V @38m	Inv. V @50m	
W5WP	3 18HT HyTowers, B&W Dip.	3 18HT HyTowers, Windom, Dip.	Windom, 3 18HT HyTower Verts.	
W5YU	Rhombic @30m	3-Ele. Full-Size, 4 Square, 4 1/2-λ Slopers @30m	Rhombic @30m, Inv. L	
W6DZ	V-beam 85-m Legs @21m	6-m High Vert.	6-m High Vert.	
W6PBI	V-beam 85-m Legs @21m	Inv. V @21m, 21-m Vert.	21-m Vert.	Beverages, Pennant, Rot. Loop
W6SR	3-Ele. Vert. Array	2-Ele. Sloping Wire Beam	Loaded 21-m Tower, Inv. L	4 Beverages
W6YA	2-Ele. Delta Loop	Vert.	Vert.	
W7DD	None	4 Square Using Sloping Dips	None	
W7JT	13-m Vert.	15-m Vert.	14-m T-Top Vert.	
W8LRL	Vert. Dip.	1/4-λ Vert.	60-m Vert.	8 Beverages, 2 End-Fire Phased Bevs, 8-Ele. Short Vert. RX Ant 320' Dia. Circle (8 Directions)
W8RU	Force-12 C4SXL 2-Ele. Yagi @20m	Inv. V @18m	1/4-λ Inv. L (15-m Vert.)	Rot. Coaxial Loop
W8UVZ	2-Ele. Yagi at 27m	5-Ele. Vert. Array	5-Ele. Phased Vert. Array	2 Pairs of Phased Beverages
W8WEJ	Loop	Top-Loaded Wire Vert.	Top-Loaded 17-m wire Vert.	Slinky, K9AY Loops
W8XD	Butternut HF2V	Butternut HF2V Vert.	Inv. L	Long Wire
W9LYN	Hustler 4BTV Trap Vert.	None	Inv. L (14-m Vert.), Dip. @25m	
W9UCW	22-m Vert., 160-m Hor. Square Loop @15m	22-m Vert., 160-m Hor. Square Loop @15m	22-m Vert., 160-m Hor. Square Loop @15m	
WA1RKS	2-Ele. Yagi	Dip.	Dip.	Beverages
WA2UUK	3-Ele. Full-Size Yagi @42m	3-Ele.m2 Yagi @42m	48-m Vert.	Beverages
WB9Z	Dip.	Dip.	Dip.	
WD5T	Inv. Vee @22m	1/4-λ Sloper	24-m Shunt-Fed Tower	
W0RI	Stack Cal-AV beams @24/45m	Stack Cal-AV sloping Dip. Array	1/4-λ Vert.	Beverages
WX0B	3-Ele. Yagi @45m	Rot. Dip. @48m	Sloping Dip.	
WZ6Z	3-Ele. Yagi	Inv. V	Short Inv. V	Coax Loop
XE1KK	None	None	1/2-λ Vert. Tower, Inv. Vs	8 Beverages
XE1RCS	2-Ele. Hy-Gain Yagi	Dip.	Shunt-Fed Tower, Dip., Sloper	Beverages
XE1VIC	5-Ele. 35-m boom to NW, 2-Ele. to JA	3 Ele. Rot.	1/4-λ Sloper	
YT6A	Inv. L	Inv. L	Inv. L	
ZC4DW	3-Ele. Force 12 Yagi @30m	2-Ele. Force 12 Yagi @30m	Shunt-Fed 30-m Tower	Beverage
ZS4TX	3 Ele. @36m, 2-Ele. @22m	2-Ele. @36m	1/4-λ Sloper @34m	
ZS6EZ				

DX-Operating on the Low Bands

DXers. Many of those not using Beverages say: "...*I wished I had enough room...*). Although sophisticated arrays of short verticals can equal the performance of the best Beverages, only few actually use them. As we will see in Chapter 7 these antennas are much more complex to put up and get working than Beverages.

17.6. The Low-Band DXer's Equipment

Transceiver	Total Group	Top Group 160 meters	Top Group 80 meters	Top group 40 meters
Yaesu FT1000 (*)	51%	60%	59%	60%
Yaesu (other types)	5%	4%	1%	2%
Kenwood	20%	12%	21%	20%
Icom	27%	21%	22%	21%
Ten Tec	7%	2%	3%	3%
Other	4%	2%	4%	3%

Notes: The category Total Group 160 includes stations with at least 225 DXCC countries. Total Group 80 includes stations with at least 250 DXCC countries. Total Group 40 includes stations with at least 275 DXCC countries. (*) includes FT-1000(D), FT-1000MP and FT-1000MKV. Other category includes Collins, JRC, and Kachina.

It's interesting to see how these figures change over the years. Yaesu has made remarkable progress in popularity through the different Editions of this book. Ten years ago Yaesu ranked only in third place, with a mere 12% score. Six years ago this had grown to 44% (52% in Top 100) and now Yaesu has reached the 60% Total Group score.

The number of Kenwood users has dwindled from 30% to 20% in the Total Group. The TS-830 is still considered a very good Top-Band transceiver by many (good tuned front-end selectivity, low first IF making for good close-in IMD performance). The TS-850 and TS-930 also remain popular radios, while the TS-2000 was used by only one station.

ICOM has made significant progress, especially in the non-Top Group (17% to 27%), while it remained at a constant level in the Top Group (20% to 21%). The percentage for Ten-Tec remained constant but with the new Orion transceiver, a real breakthrough in many aspects, I expect them to take an important slice of the cake in the near future.

17.7. Why Does the Low-Band DXer Operate the Low Bands?

Since I asked that general question in the last survey as well, I knew the answer: "For the challenge!"

- DXing on the high bands is like shooting fish in a barrel. (AA4MM)
- Low-band DXing is the greatest challenge in amateur radio. (AB0X)
- I love a good "static salad." (K1UO)
- Anyone can do it if it's easy. (K4PI)
- I experience the same thrills as 40 years ago that hooked me on radio, high bands are too easy. (K4TEA)
- Top Band is the only band that still gives me a thrill. (K6ANP)
- 160 is an addictive band, 160 is not easy to be good at. (KO1W)
- Worked a new one on 160 is not so cut-and-dried. (KX4R)
- On 160, CW shines. (VO1NA)
- Fewer lids than on high bands. (W0GJ)
- See how much pain one can endure before taking the headphones off. (W7TVF)
- Fun. (WB9Z)—*These two guys should get together.*
- The low bands are where you can test the station and the operator's skills. (UA3AB)
- Challenge of hearing, silence the utility poles, be ready all the time. (N7RT)
- Best demonstration of operating skill, station design and knowledge of propagation (like 6 meters). (W4DR)
- Pushing the operator and the station to the limits. (N4KG)
- 160: No nets, no lists, no deliberate QRM, moving on the edge, alone with QRN... (IV3PRK)
- 160: This is a new mountain to climb (the tallest one). (N6RK)
- 160 requires more technical skills and operating skills: The ultimate DXing challenge. (K9RJ)
- On the low bands success comes through knowledge (antennas), not money. Few do it well. (K1VR)
- It's not that easy but I like difficulties (easy things are for everyone). (RA3AUU)
- DX nets on high bands make many contacts phony; playing field on low bands is more level. (ZS6EZ)
- Why do you climb mountains? ...because they're there. 160 is the highest mountain with no worn path. (K0HA)
- Try to get the impossible, work all countries on all bands. (HB9AMO)
- I like difficult things, and... if you can't hear them you can't work them. (ON7TK)
- 160 is like the BC band, I was a BC SWL as a child. (N5SV)
- 160-meter DX requires the best of everything: antennas, equipment, QTH, operator skills. (4X4NJ)
- I think I was dropped on my head when I was a baby. (K4SB)
- On 160 you can be competitive using your hands, not your checkbook. (NW6N)
- Doing the impossible from a city, camaraderie on West Coast. (K6SSS)
- Satisfaction of achieving the seemingly impossible. (PA3DZN)
- The intellectual challenge of dealing with all the odd variables of propagation makes it a thrilling activity. (N0AX)
- I feel more at ease with my fellow low-band DXers than some of the "stuffed shirts" that hang out on 20 meters. (W0FS)
- Ties with early pioneers who did so much with so little. (K8MN)
- 160 is the absolute end in DXing, the last frontier. (K9UWA)
- To make the impossible possible: 160 DXCC from the worst place on earth. (YB1AQS)
- K6SE got infected at an early age: "As an 8-year old in Detroit I would stay up late at night do DX on the AM broadcast band."
- Chance to do something everybody thinks is impossible. (G4DBN)
- 160 is more a gentleman's band: Lids are too lazy to fight QRN. (W9WI)
- No pain, no gain, and no nets on 160 yet. (GW3YDX)

- Creates great friendships. (W6KW, ex-W6NLZ, ex-K2RBT)
- It helps to be insomniac. (W8RU)
- I am a man whose life begins after sundown. (AA4V)
- The challenge both on the technical side (antenna design and propagation) and DX techniques. (CT1EEB)
- More difficult and more value for each QSO. (EY8MM)
- You tend to find better operators on LF. (GM3YTS)
- Because it is FUN… (HAØDU)
- For contest multipliers that are harder for others to get. (K1TTT)
- Operating skill as important as hardware. (K2RD)
- The challenges make it fun. (KØXM)
- Unusual and unexpected propagation and openings. (K3NA)
- Unpredictable! Fun! (K4CIA)
- Success is not automatic… (K4TEA)
- The challenge of the fight. (K6EID)
- It's a challenge! Plus it gets back to the roots of ham radio DXing. (K8BHZ)
- On 160 most operators are DX oriented gentleman, good camaraderie. (K9FD)
- The engineering needed to be competitive on these bands. (K9JF)
- The sense of accomplishment, especially for my limited antennas and real estate. (K9KU)
- It's not something the average ham can do well, with the high noise, strange DX hours required, the skill and dedication needed to be successful. Nothing like working a new one" on Top Band (except for receiving the QSL!). (KG6I)
- The challenge doing it from my mobile. (W6/KH6DX/M)
- Results indicate antenna competency and operator savvy. (N4JJ)
- The lower the frequency, the higher the challenge. (NX4D)
- The challenge of propagation, the valuable awards. (S5ØA)
- Frees up some daylight time! (VE7BS)
- ANYONE can work DX on the high bands. (WØGJ, W1JZ, VE7ON, etc)
- 160 is the most challenging in terms of propagation and technical. (W4ZV)
- Working the seemingly unworkable. (W9AJ)
- Testing antenna systems. (WXØB)
- Is there any challenge left in high band operating? (ZS6EZ)
- The challenge on the low bands reminds of my early days as a new ham. (K4UEE)
- It has a charm of its own, and reminds me of early days with W1BB and W2EQS. (WØAIH)
- The lower the frequency, the higher the challenge. (NX4D)
- It's the only challenge left. (K2UO)
- I was inspired by W1BB; Stu gave me my Novice license test when I was ~12 yrs old. (AJ1H)

Trying to break up the answers of the 2003 poll into categories I came to the following overview:

- 68% mentioned the challenge.
- 17% mentioned the fact that you had to home build and design antennas.
- 15% mentioned the unpredictable propagation on 160.
- 13% mentioned the competition aspect (including getting multipliers in contests).
- 11% mentioned thrill and excitement.

Fig 2-14—Contests offer good opportunities for newcomers on the low bands to work new countries. XE1RCS (managed by XE1KK), a top-notch phone contest station from Mexico, has given a new country to many 160-meter DXers worldwide.

- 7% mentioned they like the company of the low-band operators better (160 is a gentleman's band).
- 6% mentioned that better operator are required to work the low bands.
- 3% mentioned they operated the low bands because they are night bands.
- Only 3% mentioned it was MORE FUN.

Isn't fun essential to any hobby?

17.8. QSL Cards

About 87% of the Low band DXers in the poll said they collect QSL cards vs 96% six years ago). This time the same 87% said they answer all cards received. Only 15% uses E-QSLs. To me an E-QSL is to QSLing what lists are to amateur radio. I want to be able to hold the card in my hands. In my questionnaire I asked if the addressee was using electronic QSLs (E-QSL). No one did. Maybe the low-banders are a little old fashioned, or are most of them just very straight?

17.8.1. Logbook of the World (LoTW)

On September 15, 2003, ARRL announced the on-line availability of the "Logbook of the World" (LoTW). See **www.arrl.org/lotw/intro** and **www.arrl.org/lotw/faq**. ARRL states: "ARRL's *Logbook of the World* (LoTW) system is a repository of log records submitted by users from around the world. When both participants in a QSO submit matching QSO records to LoTW, the result is a QSL that can be used for ARRL award credit." LoTW incorporates an elaborate set of safeguards to ensure that QSOs are secure. Each and every QSO is automatically 'signed' electronically to prevent fraud or manipulation.

Answering a frequently asked question concerning printed QSLs: "Logbook of the World is initially designed to create awards credit, that is to say, that if your QSO matches that of another station, either you or the other operator may be able to apply that confirmed QSO to various awards. Creating an image based in part on the

QSO information for the purpose of making a file that can be printed, or creating a QSL card, is not presently part of LoTW. There are other services available that can do that. LoTW goes a step or two beyond the conception of a QSL card (which is essentially a one-sided request for a confirmation from the other side of the QSO) by verifying that a QSO occurred between two stations, based on the 'signed' data submitted by each."

In the first month of operation approximately 15 million QSOs were uploaded to the LoTW database. For active contest stations, who often make tens of thousands of QSOs each year, Logbook of the World promises to relieve much of the burden of sending physical QSLs to bureaus around the world.

18. THE SUCCESSFUL LOW-BAND DXER

If we want to analyze what's required to become a successful low-band DXer, we must first agree on what is success. Success can be very relative. If you have only a $^1/_8$-acre city lot and you want to work the low bands, your goals will have to be different from the guy who's got 10 acres and a well-filled bank account. But you can be successful just as well, in your own way, relative to your own goals.

There are a few essential qualities that make good low-band DXers, I think. They apply even to the low-band DXer with a modest setup.

Knowledge of antennas: For the low bands, it is not like opening a catalog and ordering an antenna. You have to understand antennas—the Whys and the Why Nots. You will have to become an antenna experimenter to be successful, even more so if you'll have to do it from a tiny city lot!

Knowledge and experience in propagation: Don't expect to turn on the radio any time of the day on 80 or 160 meters and work across the globe. You must understand that you are trying to do something that is very difficult, something that requires a lot of experience to be successful. You'll have to be able to predict openings, sometimes with an accuracy of minutes. The successful low-band DXer must build up his propagation expertise over a long period of time.

Willingness to learn: Isn't improving our technical knowledge and ability what our hobby is all about? Working DX on 160 meters makes you feel like you are doing it like Marconi!

Equipment and technologies: Receivers are getting better at every vintage, even if the evolution isn't moving as fast as we might like. The successful low-band DXer uses the best equipment available and he uses it in a professional way. He gets involved with the latest technologies in radio communication, such as packet radio and DX Clusters. These provide real-time information about activity on the different low bands.

Good QTH: Successful operators work DX from excellent QTHs. They are not all mountaintop QTHs, but each success story has been written from an "above-average" QTH. This does not mean that a successful low-band DXer has to be a rich land owner. I, for one, have just over half an acre, but my location is excellent. The neighbors are nice and I can use their fields in the winter for my Beverage antennas.

Perseverance, persistence, dedication: If you are not prepared to get up in the middle of the night five days in a row to try to work your umpteenth country on 80 or 160, you will not be successful. If you think it's too hard to go out at night, in the fields or through the woods, in the dark and roll out a special one-time Beverage for that new country you have a sked with in a few hours, then you better forget about becoming successful in the game, or rather the art, of low-band DXing!

Operating proficiency: Your "know-how-to-do-it" is probably the best weapon that can make or break a low-band DXer with a modest station.

Willingness to become a good CW operator. I don't think this needs to be explained!

19. THE 10 LOW-BAND COMMANDMENTS

Mark Twain once said: "If we were supposed to talk more than we listen, we would have two mouths and one ear." How true this is for low-band DXing—And for most other human endeavors.

Jeff, K1ZM, published in his excellent book *DXing on the Edge* (Ref 511) a set of rules, from the hand of Bill, W4ZV, and which had been published earlier on, on the Top Band Reflector. It goes without saying that these rules equally as well apply to the other low bands. A chapter on operating would not be complete without these rules, which I like to call the 10 Low-Band Commandments:

Rule #1: When the DX station answers someone else, listen; do not call. Instead try to find where *he* is listening. Most good operators spread the pileup over at least 1 to 2 kHz. If you listen for the station he is working, you will maximize your probability of being heard since you will know where he is listening. You may also recognize the pattern the operator uses. That is, is he slowly moving up in frequency, down in frequency or alternating picks on either side of the pileup? You will also know when to transmit (ie, when *he* is listening). It's very hard for him to hear you calling while he is transmitting!

Rule #2: Listen carefully! He may change his QSX frequency or QSY. If you're calling continuously, you will never know it. I can't tell you all the good stuff I've worked easily because I was one of the first on a new QSX frequency. If you're transmitting continuously, you'll be one of the last to know. For those of you with QSK, you have an advantage here. If you don't, use a foot switch so that you can listen between calls and stop sending when he starts.

Rule #3: Do not transmit on the station answering. Why? Because a good operator will stay with that station until he finishes the QSO. Repeats necessitated by your QRM just reduces the amount of time *you* will have to work him before propagation goes out. The name of the game is for the DX to work as many stations as quickly as possible. Continuously calling only slows down the whole process and reduces *your* probability of a QSO. It might also encourage some DX operators to make a mental note in their head to never "hear" you again!

Rule #4: Learn your equipment so you know how *exactly* to place your transmit signal properly on frequency. No, this does not mean exactly zero beat on the last listening frequency where all the other guys are. It's far better to offset by a few hundred Hz based upon which way you think the DX is tuning (see Rule #1). Also *please* learn to use your equipment so you don't transmit on the DX frequency inadvertently. This only slows things down for everyone and wastes precious opening time on 160 meters.

Rule #5: If you have limited resources on 160, focus on your receive-antenna capability. You will work far more 160 DX with good ears than with a big mouth. Being an "alligator" that cannot hear anything is not productive on Top Band.

Rule #6: Send your full call. Partial calls only slow things down on Top Band. (From Rolf, SM5MX, XV7SW)

Rule #7: Use proper and consistent spacing when sending your call on CW. There are some very well known DXers who don't understand this. They will break the cadence of their calls with pregnant pauses—this can confuse the DX station trying to decipher your call through 160-meter QSB and QRN.

Rule #8: Send the DX station's call if you are in doubt whom you are working. You will not be happy if you log a DX station while you actually worked another station! This is especially important if more than one DX station is listening QSX in the same general area of the band. (From 4S7RPG)

Rule #9: Listen to the DX station's reports and match his sending speed. If he is giving 459 at 18 WPM, don't reply at 35 WPM! If the DX station is missing part of your call, or if he has incorrectly copied part of your call, repeat only that part of the call several times, at a constant pace. (From 4S7RPG)

Rule #10: Listen, ... listen, ... listen!

CHAPTER 3
Receiving and Transmitting Equipment

I would like to thank George Cutsogeorge, W2VJN, for once again being my critic and proofreader for this chapter. George has held the same call since 1947. He is an electronic engineer, retired after a 40-year career, including 17 years with RCA designing spacecraft and ground-station equipment and another 17 years with Princeton University, where he was involved with fusion energy research. George now owns half of International Radio and Top Ten Devices, which keep him heavily involved with electronics and radio amateurs. He has been on the top of the DXCC Honor roll since 1986 and is waiting for a P5 to show up on CW. Thank you also, George, for letting me quote from your publications.

The performance of communication equipment has progressed by leaps and bounds over the years. However significant improvement can still be made to present-day radios. Low-band DXing and contesting are two areas that demand the utmost from our equipment, due to the almost continuous presence of all kind of noise, as well as strong nearby signals.

1. THE RECEIVER
1.1. Receiver Specifications

Until about 30 years ago, receiver performance was almost exclusively defined by sensitivity and selectivity. In the 1950s and early 1960s a triple-conversion superheterodyne receiver was a status symbol, more or less like a transceiver with "IF DSP" nowadays. It was not until the mid-1960s that strong-signal handling became an important parameter (Ref 250).

In the following, I review in detail performance parameters that characterize a modern communication receiver, highlighting their impacts on successful low-band DXing.

1.2. Sensitivity

The nomograph in **Fig 3-1** shows the voltage and power relationships of the RF signals found at a receiver's input. The table can be used to convert between the many different units used to express signal strength. (See also **www.qsl.net/ve7ca/DesLev.htm** or the spreadsheet

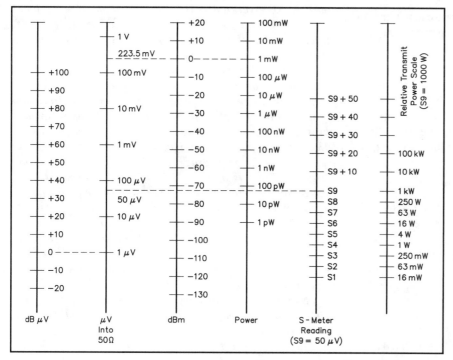

Fig 3-1—This nomograph shows the relationship between receiver input voltages, standard S-meter readings and transmitter output power.

Receiver_Levels.xls on the CD bundled with this book.) In the spreadsheet you can change the system impedance (for example, to 75 Ω instead of 50 Ω).

Sensitivity is the ability of a receiver to detect weak signals. The most important concept related to sensitivity is the concept of *signal-to-noise ratio*. Good reception of a weak signal implies that the signal is substantially stronger than the noise. It is accepted as a standard that comfortable SSB reception requires a 10-dB signal-to-noise ratio. CW reception requires a lower S/N, and any moderately experienced CW operator can rather easily deal with a 0-dB S/N. A really good operator can dig CW signals out of the noise at –10 dB S/N in a 500-Hz bandwidth, mainly because his built-in "brain filter" narrows the noise bandwidth much further. This shows the inherent advantage of CW over SSB for weak-signal communications.

1.2.1. Thermal noise

The noise present at the receiver audio output terminals is generated in different ways. Inherent *internal receiver noise* is produced by the movement of electrons in any substance (such as resistors, transistors and FETs that are part of the receiver circuit) that has a temperature above absolute 0 kelvin (0 K or –273° Celsius). Absolute zero is where all electrons have stopped moving. Above 0 K electrons move in a random fashion, colliding with relatively immobile ions that make up the bulk of the material. The final result is that in most substances there is no net current in any particular direction on a long-term average, but rather a series of random pulses. These pulses produce what is called thermal-agitation noise, or simply thermal noise.

The Boltzmann equation expresses the noise power in a system. The equation is written as:

$$p = kTB \quad \text{(Eq 1)}$$

where
- p = thermal noise power, watts
- k = Boltzmann's constant (1.38×10^{-23} joules/kelvin)
- T = absolute temperature in kelvin
- B = bandwidth, Hz

Notice that the power is directly proportional to temperature, and that at 0 kelvin the thermal noise power is zero.

Expressing equivalent noise voltage, the equation is rewritten as:

$$E = \sqrt{ktBR} \quad \text{(Eq 2)}$$

where R is the system impedance (usually 50 Ω).

For example, at an ambient temperature of 27°C (300 K), in a 50–Ω system with a receiver bandwidth of 3 kHz, the thermal noise power is:

$$p = 1.38 \times 10^{-23} \times 300 \times 3000 = 1.24 \times 10^{-17} \text{ W}$$

This is equivalent to $10 \log (1.24 \times 10^{-17}) = -169$ dBW or –139 dBm (139 dB below 1 milliwatt), and is equivalent to 32 dB below 1 μV or –32 dBμV (Ref 223). This is the theoretical maximum sensitivity of the receiver under given bandwidth and temperature conditions. If you want more sensitivity this can be achieved by reducing the bandwidth or by cooling your equipment.

1.2.2. Receiver noise

No receiver is noiseless. The internally generated noise is often evaluated by two measurements, called *Noise Figure* and *Noise Factor*. Noise Factor is by definition the ratio of the total output noise power to the input noise power when the receiver's input is at the standard temperature of 290 K (17° C). Being a ratio, it is independent of bandwidth, temperature and impedance. The Noise Figure is the logarithmic expression of the Noise Factor, in dB:

$$NF = 10 \log F \quad \text{(Eq 3)}$$

where F is the Noise Factor.

1.2.3. Minimum Discernable Signal (MDS)

The Minimum Discernable Signal (MDS) produces an output that is the same as the internal noise level of the receiver. MDS can be expressed as:

$$\text{MDS (dBm)} = -174 \text{ dBm} + 10 \log (BW) + NF \quad \text{(Eq 4)}$$

where BW is expressed in Hz.

A conversion tool (*RX_noise_figure_andMDS_calculator.xls*) that calculates the level of the receiver internal noise as well as the receiver's Noise Figure (given the receiver's bandwidth, temperature and MDS) is available on the CD in this book. A conventional signal generator can be used to measure the MDS. This is where the signal plus noise is 3 dB higher than the noise floor. This is measured with an audio RMS meter on the receiver output.

1.2.4. External Noise and Receiver Sensitivity Margin

Besides the noise generated inside a receiver, the factor that limits the sensitivity of the overall receiver system is the noise coming from the antenna. This noise is mainly atmospheric and/or manmade noise.

From W8JI's Web site:

"The noise that limits our ability to hear a weak signal on the lower bands is almost always an accumulation of many signal sources. Below 18 MHz, the noise we hear on our receivers (even at the quietest sites) comes from terrestrial sources. Receiver noise is generally a mixture of local groundwave and ionosphere propagated noise sources, although some of us suffer with dominant noise sources located very close to our antennas.

Urban: *In urban-type noise situations, noise arrives from multiple random sources through direct and groundwave propagation from local sources. One or more sources can actually be the induction-field zone of our antennas (in most cases the induction field dominates at distances less than $^1/_2\lambda$). Urban locations are the least desirable locations because typical noise floors average 16 dB higher than suburban locations. There is often no evidence of winter night noise increase on 160 meters, since ionosphere-propagated noises are swamped out by the combined noise power of multiple local noise sources. Much of the noise sources are utility distribution lines, because of the large amount of hardware required to serve multiple users. Other noise sources are switching power supplies, arcing signs, and other unintentional man-made noise transmitters.*

Suburban *locations average about 16 dB quieter than urban locations, and are about 20 dB noisier than rural locations. Noise generally is directional, arriving mostly from areas of densest population or the most noise-offensive* power lines. Utility high-voltage transmission lines are often problematic at distances greater than a mile, and occasionally distribution lines can be problems. The recent influx of computers and switching power supplies has added a new dimension to suburban noise.

There is often a small increase in nighttime winter noise at exceptionally quiet suburban locations. This increase occurs when propagated terrestrial noise equals or exceeds local noise sources.

Rural locations, especially those miles from any population center, offer the quietest environment for low-band receiving. Daytime 160-meter noise levels are typically around 35-50 dB quieter than urban, more than 20 dB quieter than suburban locations. Nighttime brings a dramatic increase in low-band noise, as noise propagates in via the ionosphere from multiple distant sources.

Primary local noise sources are electric fences, switching power supplies, and utility lines. I can measure a 3 to 5 dB daytime noise increase in the direction of two population centers, Barnesville (population 7500) and Forsyth (population 10,000) both 10 km from my QTH.

Typical daytime noise levels, measured on a 200-foot omnidirectional vertical, are around –113 dBm with a 350-Hz bandwidth (noise power is directly proportional to receiver bandwidth). Noise power increases about 5 to 15 dB at night, when the band "opens." As in the case of suburban systems, directional antennas reduce noise power. Nighttime is the "big equalizer," reducing the advantage of location as distant noises increase with improved propagation."

Recently the ARRL Laboratory asked Tom, W8JI, to make some manmade noise-level measurements at his super-quiet QTH, in preparation for ARRL's filing comments about BPL (Broadband Over Power Lines). I visited W8JI to find out firsthand just how quiet a rural environment can be on 160 meters. During the daytime I could switch his Beverage antennas and tell from the rise in noise level the directions to the nearest towns about 10 km away.

Tom measured a typical daytime manmade noise level on 160 meters of –113 dBm in a 350-Hz bandwidth. Using a large omnidirectional vertical antenna, Tom has a noise level between S2 and S3. These readings are in the absence of local QRN (static and thunderstorms). W8JI's margin over the receiver MDS (–141 dBm) is thus 28 dB during the day. This assumes that Tom would be using a large vertical for receiving, although this is usually not the case since he usually uses Beverage antennas, which can discriminate against noise coming from other directions.

On 80 and 40 meters typical manmade noise levels are lower by about 8 dB on 80 meters and 18 dB on 40 meters. **Table 3-1** summarizes W8JI's noise numbers for 160 meters.

The manmade plus atmospheric noise noise-level data in Table 3-1 are referenced to a level of –113 dBm (at 0 dB) for the best-case (rural) daytime local manmade noise propagated by ground wave. For this level of noise, a receiver noise figure of 35 dB would be adequate to maintain a S/N of 0 dB in a 350-Hz bandwidth. This means that a 25-dB attenuator could be placed at the input of a typical receiver with a 10-db-noise figure and the desired signal would drop down to the thermal-noise level of the receiver itself. The output S/N would thus still remain adequate for copy by a good CW operator.

Table 3-1
160-meter manmade and atmospheric noise data, from W8JI.
Total noise reference level 0 dB = –113 dBm in 350-Hz bandwidth.

	Daytime Noise Level dB	Nighttime Noise Level dB
Rural	0	5 to 15
Suburban	20	20
Urban	35 to 50	35 to 50

In quiet areas the total noise will be higher during the night, since additional noise arrives by atmospheric propagation, adding to the local manmade noise found during the daytime. In urban residential areas the local manmade noise is so high that you never can hear such propagated noise. This means that the receiver sensitivity is essentially a moot point. Almost any receiver is sensitive enough in such a hostile environment!

Noise levels can, and do, vary tremendously from one location to another. We have many bad 160-meter locations, and only a few very good ones. Indeed, the difference in manmade noise levels can be up to 50 dB! The figures published by Tom Rauch are the first ones I have seen from an active and knowledgeable radio amateur. Therefore they should be considered very important.

Assuming all noise is evenly distributed in all directions, well-engineered special receiving antennas (see Chapter 7) will receive up to 12-15 dB less noise than an omnidirectional antenna. To benefit from this directivity advantage, the receiver will require 12-15 dB better noise figures than the levels shown in Table 3-1. In addition, such directive noise-reducing antennas are usually low output antennas (such as Beverages with typically –10 dBi gain) and often used with long "lossy" feed lines (eg, 1 dB loss). Add all of this together, and you need 12 + 10 + 1 = 23 dB better receiver noise figures and a lot of your surplus sensitivity for 160 meters goes away. Therefore we often use a preamp (10-20 dB gain). The use of preamplifiers for low-noise receiving antennas is covered in more detail in the Chapter 7 on Special Receiving Antennas.

1.3. Intermodulation Distortion

Intermodulation distortion (IMD) is an effect caused by two (or more) strong signals that drives one (or more) of the stages in the receiver beyond its linear range, so that spurious signals called *intermodulation-distortion* (IMD) products are produced. Third-order IMD is the most common and annoying front-end overload effect. **Fig 3-2** shows the IMD spectrum for an example where the two parent signals are spaced 1-kHz apart. (The third-order products are: $2F_1-F_2$, and $2F_2-F_1$.)

Third-order IMD products increase in amplitude three times as fast as the pair of equal parent signals (Ref 210, 211, 213, 226, 239, 247, 255, 274, 281). **Fig 3-3** shows three examples of third-order intercept points. The vertical scale is

Receiving and Transmitting Equipment 3-3

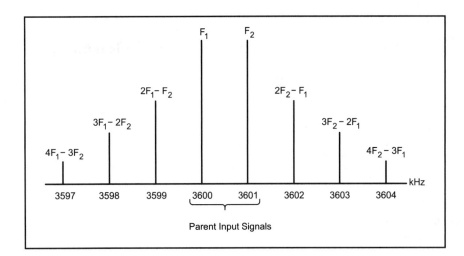

Fig 3-2—Third, fifth and 7th order intermodulation products generated by parent input signals on 3600 and 3601 kHz. Note that the "order" of a set of IMD products is determined by adding the multipliers for each frequency. For example, $3F_1–2F_2$ is 3 + 2 = 5th order.

the relative output of the receiver front end in dB, referenced to an arbitrary zero level. The horizontal axis shows the input level of the two equal-amplitude parent signals, expressed in dBm. Point A sits right on the receiver noise floor. Point A′ is the floor with a 20-dB attenuator at the receiver input. Increasing the power of the parent signals results in an increase of the fundamental output signal at a one-to-one ratio. Between –129 dBm and –44 dBm, no IMD products are generated that are equal to or stronger than the receiver noise floor for the no-attenuator Case 1. At –44 dBm (point B), the third-order IMD products have risen to exactly the receiver noise floor level.

Point B is called a two-tone IMD point, expressed in dBm. Further increasing the power of the parent input signals will continue to raise the power of the third-order IMD products *three times faster* than that of the parent signals. At some point, the fundamental and third-order response lines will flatten because of *gain compression*. Extensions of both response lines cross at a point called the *third-order intercept point*. The level can be read from the input scale in dBm. The intercept point (I_P) can be calculated from the IMD point as follows.

$$I_P = \frac{2 \times \text{MDS (noise floor)} + 3 \times \text{IMD}_{DR}}{2} \quad \text{(Eq 5)}$$

where
  MDS = minimum discernible signal
  IMD_{DR} = IMD dynamic range

Applying the three examples from Fig 3-3, we find:
Case (1): $I_P = (2 \times –129 + 3 \times 84)/2 = –3$ dBm. This is for a receiver with an 84 dB IMD_{DR} and no front-end attenuator.
Case (2): $I_P = (2 \times –109 + 3 \times 84)/2 = +17$ dBm. This is for the same receiver as in Case (1) but with a 20-dB attenuator.
Case (3): $I_P = (2 \times –129 + 3 \times 104)/2 = +27$ dBm. This is for a receiver with a 104 dB IMD_{DR} and no attenuator.

Conversely, the two-tone IMD point can be derived mathematically from the intercept point.

$$P_{IMD} = \frac{2 \times I_P + Nf}{3} \quad \text{(Eq 6)}$$

where Nf = noise floor.

The three examples from Fig 3-3:

Case (1): $P_{IMD} = –(2 \times –3 + –129)/3 = –45$ dBm

Case (2): $P_{IMD} = –(2 \times +17 + –109)/3 = –25$ dBm

Case (3): $P_{IMD} = –(2 \times +27 + –129)/3 = –25$ dBm

What does this mean? For Case (1) this means that two signals below –45 dBm will not create audible IMD products.

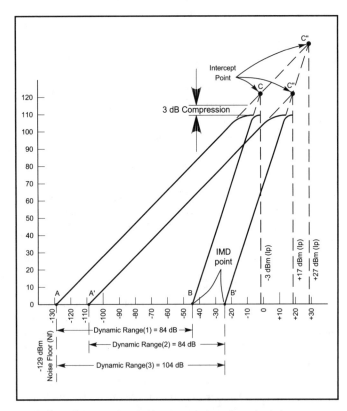

Fig 3-3—Third-order intercept point showing three examples, with and without 20-dB front-end attenuation. The intercept point increases by the same amount as the attenuation is increased. This is for an average receiver with an 84-dB intermodulation distortion dynamic range (IM3 DR) and with a 104-dB dynamic range.

Two signals at around S9 + 30 dB will start generating audible IMD products for Case (1). In Europe this is an everyday situation on the 7-MHz band, where 30 to 50-mV signals (–17 to –13 dBm) are common.

When evaluating third-order intercept points, we must always look at receiver noise floor levels at the same time. When we raise the noise floor from –129 dBm to –109 dBm—for example by inserting 20 dB of attenuation in the receiver input line as in Case (2)—both response lines and the intercept point will shift 20 dB to the right. This means that the intercept point has been improved by 20 dB, while the dynamic range remains the same. An average receiver with a +5-dBm intercept point can be raised to +25 dBm merely by inserting 20 dB of attenuation into the input. Remember that this can frequently be done with present-day receivers, since they often have a large surplus sensitivity. Case (3) shown in Fig 3-3 represents a better solution, where the improvement is obtained by designing the receiver to handle stronger signals before becoming non-linear.

The frequency separation of the two parent input signals can greatly influence the intermodulation results. The worst case applies when there is no selectivity in the front end to attenuate one of the signals. This happens when both input signals are within the passband of the first-IF filter and the intermodulation products are produced in the second mixer. Most present-day receivers have a rather wide first IF so they can accommodate narrow-band FM and AM signals without requiring filter changes.

This is one of the greatest problems with modern "all bells and whistles" general-coverage transceivers, using fixed-tuned RF input circuits. To obtain sufficient image rejection a very high IF outside the operating range of the equipment is required. So-called *roofing filters* at IFs between 50 and 100 MHz are very inferior in shape factor to what can be obtained on much lower frequencies.

Measurements at 2-kHz spacing using a 500-Hz second-IF filter (CW filter) are often used to find the worst-case IMD performance. In this case there will be *no* selectivity from the first-IF roofing filter. Measurements at 20 and 100-kHz spacing are also used for assessing receiver intermodulation performance, in which case the roofing filter does play a role. With this much spacing, the first-IF filter generally improves the IMD picture considerably. Close-spaced measurements show the real picture for low banders, however, because that is the situation we encounter when operating on crowded bands—especially during contests. Therefore a top-notch low-band receiver should not be a general-coverage receiver, and it should have a "low" first IF (perhaps, 9 MHz), where narrow first-IF filters (roofing filters) can be used.

Often measurements using 100-kHz tone spacings are made to evaluate the strong signal handling performance of receivers where the receiver's local oscillator (LO) noise limits measurement accuracy at 20 kHz and closer spacings (see Section 1.7).

1.4. Gain Compression or Receiver Blocking

Gain compression occurs when a strong signal drives an amplifier stage (for example, a receiver front end) so hard that it cannot produce any more output. The stage is driven beyond its linear operating region and is saturated. Gain compression can be recognized by a decrease in the background noise level when saturation occurs (Ref 223, 239, 281). Gain compression can be caused by other amateur stations nearby; such as in a multi-operator contest environment. Outboard front-end filters are the answer to this problem. (See Section 1.11.)

1.5. Dynamic Range

The lower limit of the dynamic range of a receiver is the power level of the weakest detectable signal (limited by the receiver's internal noise floor). The upper limit is the power level of the signals at which IMD becomes noticeable (where intermodulation products are equal to the receiver's internal noise floor). In other words it's the ratio between the weakest signal that can be heard to the level where problems start. Refer back to Fig 3-3 for a graphical representation of dynamic range. Dynamic range can be calculated as follows:

$$DR = P_{IMD} - Nf \qquad (Eq\ 7)$$

where:

DR = dynamic range, dB
P_{IMD} = two-tone IMD point, dBm
Nf = receiver noise floor, dBm

If the intercept point is known instead of the two-tone IMD point we can use the following equation:

$$DR = \frac{2(I_P - Nf)}{3} \qquad (Eq\ 8)$$

where Ip is the intercept point in dBm.

The dynamic range of a receiver is important because it allows us to directly compare the strong-signal-handling performance of receivers (Ref 234, 239, 255), since it takes into account the sensitivity as well.

1.5.1. Intermodulation Dynamic Range

On his Web site, W8JI distinguishes between two types of dynamic ranges: Intermodulation dynamic range and blocking dynamic range. Intermodulation dynamic range (IMD_{DR} or $IM3_{DR}$) is also called the *two-tone dynamic range*. This is measured using two equal-strength signals (from low-noise oscillators) into the receiver with a specified tone spacing. This test is equivalent to having two strong signals very near each other, with just the right spacing to cause an intermodulation mixing product to fall on top of a noise-floor level signal you are trying to copy. When the signal level of the intermodulation product is just audible above the noise floor, the ratio of the strong interfering signals to the MDS (minimum discernable signal) is the IM dynamic range.

What does this mean in plain language? Imagine we have two equally strong CW signals spaced 1 kHz apart, one at 3600 kHz and another at 3601 kHz. See Fig 3-2. As the level is increased, the receiving system shows an increasingly non-linear response. The second harmonic of 3600 mixes with the fundamental at 3601, and the result is a new signal at 2 × 3600 − 3601 = 3599 kHz. Another signal appears at 2 × 3601 −3600 = 3602 kHz. The level of the mixing products increase faster than the level of either individual interfering signal. When we can hear the "phantom signal" above the noise floor of the receiver, it adds interference (showing up as "*bloops, bleeps, and random musical thumps or phantom signals on CW*" as W8JI puts it) to the weak signals we are trying to hear. We reference the main signal level at which this occurs to the receiver's internal noise floor, because that is the level where it would just start to be noticeable. For example, a receiver with an MDS of –135 dBm (in a 500-Hz bandwidth), and which has a measured IMD_{DR} of 100 dB (at a given spacing between the two offending strong signals) will produce the unwanted signals for levels greater than (–135dBm) – (–100dBm) = 35 dBm, which is approx S9 + 35 dB (see Fig 3-1).

1.5.2. Blocking Dynamic Range

Blocking dynamic range (BDR) is measured by setting a signal generator to a frequency either 2 or 10 kHz above the interfering signal. This is equivalent to having a single strong station near a very weak station you are trying to copy. The interfering signal may drop your receiver volume or it may generate a hiss (noise translated either from the interfering signal itself or from the receiver's noisy local oscillator). In either case it deteriorates the signal-to-noise ratio of a weak signal. The level of a strong, clean interfering signal is adjusted until the slightest detectable change in S/N of the MDS-level desired signal occurs. The difference between the MDS and the level causing the blocking is the blocking dynamic range.

While most published figures for BDR use a wide spacing for the test signals (*wide* means wider than the –60 dB passband of the roofing first-IF filter of the receiver) this test is not very meaningful because when we have on-the-air interference problems, it is almost always with a station a few kHz or less away. It is clear that even with excellent wide-spaced performance, close-spaced performance can be horrible, because of the wide passband of the roofing filter in general-coverage radios with a very high first IF (typically 40 to 70 MHz). It is also obvious that, when close-spaced performance is good, wide-spaced performance is just as good or better. Performance test at 20 kHz spacing or wider are not very meaningful for the low-band DXer.

Dynamic range is the most important feature of a good low-band receiver, because low banders are chasing very weak signals in the presence of very strong signals, be it during contests or in a pileup on a new country. For more details on this extremely important issue, visit **www.w8ji.com/receiver_tests.htm**, where W8JI shows some measurement results with comments. Tadeusz, SP7HT, wrote an excellent article in *QEX* (Ref 446) reviewing the results on BDR and IMD_{DR} testing done by G3SJX and W8JI on several radios.

1.6. Cross Modulation

Cross modulation occurs when modulation from an undesired signal is partially transferred to a desired signal in the passband of the receiver. Cross modulation starts at the 3-dB compression point on the fundamental response curve as shown in Fig 3-3. Cross modulation is independent of the strength of the desired signal and proportional to the square of the undesired signal amplitude, so a front-end attenuator can be very helpful to reduce cross modulation. Introducing 10 dB of attenuation will reduce cross modulation by 20 dB. This exclusive relationship can also help to distinguish cross modulation from other IMD phenomena (Ref 223, 247).

1.7. Reciprocal Mixing (Local Oscillator Noise)

Reciprocal mixing (oftentimes referred to as *phase noise*) is a large-signal effect caused by noise sidebands of the local oscillator feeding the input mixer. Oscillators are mostly thought of as single-signal sources, but this is never so in reality. All oscillators have noise sidebands to some extent. One example of the sidebands produced by an oscillator is shown in **Fig 3-4**.

The detrimental effect of these noise sidebands remained largely unnoticed until voltage-controlled oscillators (VCOs) were introduced in state-of-the art synthesized receivers (Ref 286 and 289). Fig 3-4 shows the levels of the interfering signals produced versus frequency spacing, the standard method of evaluating the effects of the LO noise in a receiver.

VCOs (voltage-controlled oscillators) are much more prone to creating noise sidebands than crystal-controlled or standard LC oscillators. Wide-range phase-locked loops in VCOs are responsible for the poor noise spectrum (Ref 209). Reciprocal mixing introduces off-channel signals into the IF at levels proportional to the frequency separation between the desired signal and the unwanted signal. This effectively reduces the selectivity of the receiver. In other words, if the static

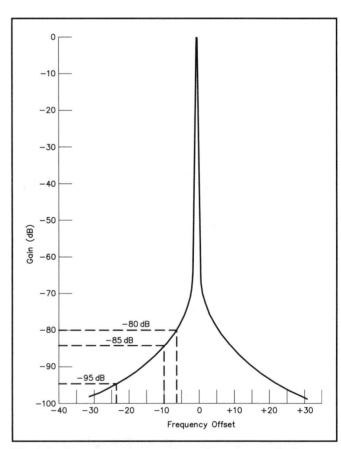

Fig 3-4—Output spectrum of a voltage-controlled oscillator. If the measurement is made in a 3-kHz bandwidth, the oscillator sideband performance referred to a 1-Hz bandwidth is –85 + –34 = –119 dBc/Hz (dB referenced to the carrier per Hz).

response of the IF filters is specified down to –80 dB, the noise in the LO must be down at least the same amount in the same bandwidth in order not to degrade the effective selectivity of the filter.

According to the thermal-noise Eq 1, the noise power is –174 dBm at room temperature for a bandwidth of 1 Hz. The noise in an SSB bandwidth of X Hz can be scaled to a 1-Hz bandwidth using the factor (10 log X). This yields a factor of 34.8 dB for a 3-kHz bandwidth, 34.3 dB for 2.7 kHz and 33.2 dB for 2.1 kHz. Continuing with the example where the static response of the IF filter is –80 dB and the filter has a 3-kHz bandwidth, the noise of the LO should be no more than –80 – 34 = –114 dBm in a 1-Hz bandwidth. The carrier noise in a 1-Hz bandwidth is usually stated in dBc/Hz—in this example, –114 dBc/Hz.

1.7.1. Measuring reciprocal mixing noise

Reciprocal mixing noise is most easily measured with a single tone. A signal source, such as a low-noise crystal oscillator, is connected through an attenuator to the receiver input. The receiver frequency is offset from the crystal frequency by various values and the level required to reach the receiver noise floor (MDS) is recorded with an audio rms voltmeter. (Ref 281, 274, 247). I encourage all equipment manufacturers to specify their reciprocal-mixing noise specifications at 1 kHz and 10 kHz spacings.

1.8. Selectivity

Selectivity is the ability of a receiver to separate (select) a desired signal from unwanted signals.

1.8.1. SSB bandwidth

On a quiet band with a reasonably strong desired signal, the best sounding audio and signal-to-noise ratio can be obtained with selectivities on the order of 2.7 kHz at –6 dB. Under adverse conditions, selectivities as narrow as 1 kHz can be used for SSB, but the carrier positioning on the filter slope becomes very critical for optimum readability. The ideal selectivity for SSB reception will of course vary, depending on the degree of interference on adjacent frequencies.

1.8.2. CW bandwidth

There appear to be two schools in this area: those that swear by the narrowest possible bandwidth and those that like to keep it wide (500 Hz or even more). We use filters not only to discriminate against other nearby signals, but also to reduce the noise level. On the low bands we are confronted mainly with propagated noise (eg, thunderstorm QRN, clicks, etc). This is very different from the EME guys, who use really narrow filters to dig for weak signals in a different type of noise: white galactic noise.

I have talked to many low banders, and certainly a large majority prefers a relatively wide filter (typically 500 Hz). They let their brains do the required signal processing. I belong to that school, and using my FT-1000MP or MP-MK5 (equipped with the Inrad 400 Hz and 250 Hz filters) I use the 400-Hz filters 99% of the time. I would typically switch to 250 Hz only when someone would be almost stepping on my toes.

Tom, W8JI, wrote: "*I use 250-Hz filters when the band is quiet with only white noise, and 600-Hz filters when there is QRN or "rough" noise. A wider filter always works better*

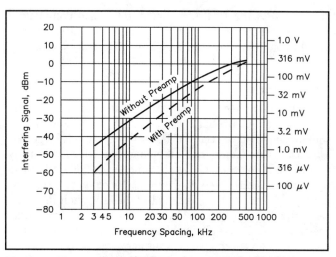

Fig 3-5—The levels of interfering signal (vertical axis) at a given signal spacing (horizontal axis) that causes the AF noise to increase by 3 dB. A 2.7-kHz IF bandwidth is assumed. This is the standard method of evaluating the effects of the LO noise in a receiver.

when there are static crashes. I do have receivers with very fast very good AGC systems, and they work very well during static crashes with AGC on, but I still find that wider selectivity helps. Wider selectivity helps because the sharp waveform of the static crash is not lengthened and blurred, and so my ears can do a better job of filtering the noise from the signal."

1.8.3. Passband tuning

Passband tuning (IF shift) allows the position of the passband on the slope to be altered without requiring that the receiver be retuned. The bandwidth of the passband filter remains constant, however. In some cases interfering signals can be moved outside the passband of the receiver by adjusting the passband tuning. In better receivers a combination of passband tuning and continuously variable bandwidth is available.

1.8.4 Continuously variable IF bandwidth

Until recently continuously variable bandwidth was achieved by moving the passbands of two filters (at different IFs) one across the other. This system was and still is available in two configurations. In one the filter can be independently narrowed down from both sides (low pass and high pass). The other approach is using a WIDTH plus a PASSBAND tuning control. This feature is available on all state-of-the-art receivers.

Producing a continuously variable bandwidth involves passing the signal through two separate filters at two different IFs (such as, 9 MHz and 455 kHz). The mixing frequency is slightly altered so the two filters do not superimpose 100%, but have their passbands sliding across one another. You must understand, however, that a variable bandwidth system such as this can never have as good a shape factor as individual well-shaped crystal filters, since the shape factor always worsens when you narrow the bandwidth in this fashion.

More recently, receivers have coming to market where the final selectivity is obtained at a very low (typically

30 kHz) IF using DSP techniques. If used in conjunction with narrowband roofing filters after the first mixer, this is the ultimate solution. The Ten-Tec ORION is a state-of-the-art transceiver using this technique.

1.8.5. Filter shape factor

The filter shape factor is expressed as the ratio of the bandwidth at –60 dB to the bandwidth at –6 dB. Good filters should have a shape factor of 1.5 or better. This 1.5 figure is a typical shape factor for an 8-pole crystal filter. Too many transceivers are equipped with rather wide SSB IF filters (typically 2.7 kHz at –6 dB) with mediocre skirt selectivity. On a quiet band these give nearly hi-fi quality, but is this what we are really after? For the average operator this may be an acceptable situation, although the serious DXer and contest operator will want to go a step further.

International Radio (formerly RCI/Fox Tango), offers modification kits for modern transceivers. See **www.qth.com/inrad/index.htm**. See **Fig 3-6** for a comparison of a stock mechanical filter and an International Radio replacement.

DSP filters can be designed with excellent shape factors. The DSP filters in the Ten-Tec ORION, which allow you to change bandwidth in 10-Hz increments, are *finite-impulse response* (FIR) filters. The skirt (transition band) of these filters has roughly the same shape no matter the bandwidth selected. As a result, the shape factor changes with bandwidth. There is a way to avoid that using higher decimation ratios, but that involves increasing the delay through the receiver, which is unacceptable for AMTOR and other modes. Ten-Tec decided to maintain a low delay time through the receiver. The measured bandwidth and shape factors are:

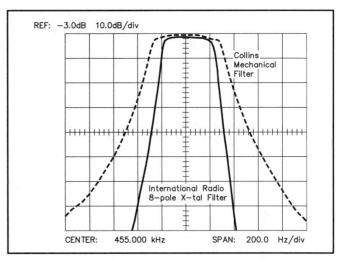

Fig 3-6—Selectivity curves for typical 455-kHz IF, 500-Hz passband Collins mechanical filter and the 400-Hz crystal filter offered by International radio. Notice that at –60 dB the replacement crystal filter has only half the bandwidth of the stock mechanical filter.

Table 3-2
Measured Bandwidth and Shape Factors

Nom. BW	–6 dB BW	–60 dB BW	Shape Factor
100	150	440	2.9:1
250	240	470	1.95:1
500	510	820	1.6:1

above 1000 Hz: Shape factor < 1.2:1

Although the shape factor at narrow bandwidths may not look spectacular, I have found this setup—where ringing is totally absent—to be the smoothest and most efficient way of obtaining the most suitable bandwidth for each individual situation.

1.8.6. Filter group delay

Generally speaking, very selective filters have group-delay problems, where the time for a signal to pass through the filter is different for different frequencies in the passband. The result is *ringing* or stretching of the signal you hear, because you might hear, for example, the upper passband noise before the lower passband noise when the same broadband noise pulse hits the input of the filter. The noise peak amplitude is reduced but the duration of a noise pulse is stretched out.

That's why mechanical filters, which have a very flat delay curve although they have poorer skirts, are often better than crystal filters when digging weak signals out of *rough* noise and static crashes. The Inrad CW crystal filters, however, have excellent group-delay characteristics that are very similar to those of the amateur-grade Collins mechanical filters.

When there is only white noise (quiet band with no atmospherics) you can use a 250-Hz bandwidth in a crystal filter, but when there is QRN, it is better to use a wider bandwidth. When using DSP for obtaining final selectivity, there is no group-delay issue. Ten-Tec advises its ORION users *not* to use the 250 or 500-Hz roofing filters when digging for weak signals in static crashes.

Tom, W8JI, reported on a test he did with a particular 250-Hz filter: "*I just took a moment to measure group delay in a 250-Hz wide crystal filter. The swept pattern looks like a 45-degree tilted Z. Group delay error delta totaled 18 mS within the –6 dB attenuation points of the filter. There is one point in the passband, near the flat top of the passband, where a sudden spike in delay time occurs. This spike results in a 10-mS time delta in a frequency span of only a few Hz! When a signal is near "rough" noise, such changes are absolutely devastating. Sharp noise pulses smear out to cover weak signals, and the rise and fall of the signal are distorted and blended with the noise, so I never used this filter. Now I know why!! After looking at this filter, from a receiver that really stinks on high selectivity with weak signals when noise is present, I'm convinced filter design is a major player in weak signal work in the presence of noise.*"

1.8.7. Static and dynamic selectivity

Fig 3-7 shows the typical *static selectivity* curve of a filter system with independent slope tuning. The static selectivity curve is the transfer curve of the filter with no reciprocal mixing with noise in the local oscillator. The *dynamic selectivity* of a receiver is the combination of the static selectivity and the effects of reciprocal mixing. Note that the static selectivity can be deteriorated significantly by the effect of reciprocal mixing.

If the amplitude of the reciprocal mixing products are greater than the stop-band attenuation of the filter, the ultimate stop-band characteristics of the filter will deteriorate. Good frequency-synthesizer designs can yield 95 dB (−129 dBc), while good crystal oscillators can achieve over 110 dB (−144 dBc) at a 10-kHz offset. This means it is pointless to use an excellent filter with a 100-dB stop-band characteristic if the reciprocal mixing figure is only 75 dB.

Hart, G3SJX, uses an interesting graphical representation of the main receiver parameters. **Fig 3-8** shows the characteristics of a "dream receiver" using Hart's graph technique. Fig 3-8 was first published in 1987. The specifications for such a high-performance receiver would read:

- Spurious-free dynamic range: 100 dB minimum.
- Noise floor: MDS = −130 dBm (500 Hz bandwidth).
- IMD3: > 100 dB.
- Third-order intercept point: +40 dBm at full sensitivity (with preamp) as derived from MDS and IMD3 above.
- Blocking dynamic range: > 120 dB.
- LO sideband noise performance: Better than −135 dBc at close spacing (2 kHz).

Only recently, with the advent of the Ten-Tec ORION, have such specifications actually been met in a production transceiver.

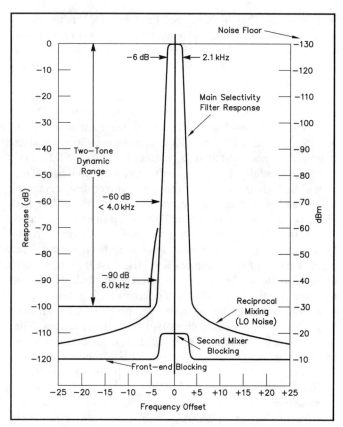

Fig 3-8—Merit graph for a "dream receiver," which was first published in 1987 in the First Edition of this book. Only the Ten-Tec ORION has met this merit graph with its two-tone 3rd-order dynamic range of 100 dB and sharp selectivity skirts.

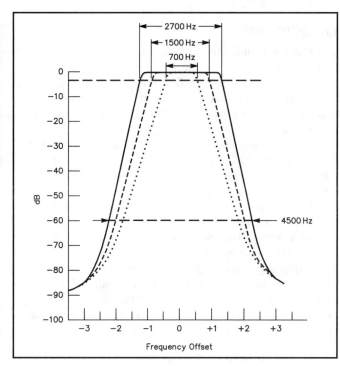

Fig 3-7—Static selectivity curve of a receiver using continuously variable bandwidth. This result is obtained by using selective filters in the first and in the second (or second and third) IFs, and shifting the two superimposed windows slightly through a change in the mixing frequency. Note that the filter shape factor worsens as the bandwidth is reduced. It is not ideal to use this method, since the shape factor may deteriorate to 4 and more, while a good stand-alone CW filter can yield a shape factor of less than 2.

1.8.8. IF filter position

The filter providing the bulk of the operational selectivity can theoretically be inserted anywhere in a receiver between the RF input and audio output. When considering parameters other than selectivity, however, it is clear that the filter should be as close as possible to the antenna terminals of the receiver. In Section 1.3 we saw that front-end selectivity will help reduce IMD products.

Most modern receivers use triple or even quadruple conversion. In order to be most effective, the selectivity (filter) should be as far ahead in the receiver as possible. The logical choice is the first IF. Most modern designs use a first IF in the 40- to 100-MHz range, for image rejection reasons. This is not the most ideal frequency for building crystal filters with the best possible shape factor. In general, we find rather simple 2-pole crystal filters with a nominal selectivity of 15 to 20 kHz (at −6 dB) in the first-IF chain. The reasons for this very wide bandwidth are threefold:

- To retain the original impulse noise shape (short rise time) to be able to incorporate a noise blanker.
- Most of the modern (all bells and whistles) transceivers must operate on FM as well, where a selectivity of less than 15-20 kHz cannot be tolerated.
- It is difficult and expensive to make low-loss, narrowband crystal filters at VHF.

I am convinced that most successful low-band DXers live

in quiet areas. They have no need for noise blankers to reduce manmade noise. If you are plagued with this kind of noise, you must cure the problem at the source. I also am not at all interested in being able to receive FM on my transceiver. This means we really could use better (narrower) first-IF filters.

This whole problem of poor roofing filters in modern transceivers is the reason why the Sherwood-modified Drake C-Line receiver, with a 600-Hz first IF filter and a 500, 250, or 125 second IF filter, is still today, more than 30 years after it was designed, found in a number of low-band DXers' shacks, especially those regularly operating contests. This should be a clear message to modern receiver designers.

Finally one manufacturer has heard our voices. Ten-Tec recently came out with the ORION, a top-of-the-line transceiver that incorporates many features for which serious DXers and low-band operators have long been asking. The main receiver of the Orion has a 9-MHz first IF, where very narrow (switcheable) roofing filters are located. The main receiver is not a general-coverage receiver, but who cares? The front-panel layout is much like the FT-1000 with two large tuning knobs controlling two separate receivers. The second receiver has a high IF and is a general-coverage receiver, so that you have access to all HF frequencies.

Bill, WØZV, worded it very well on the Internet: *"KUDOS to Ten-Tec for LISTENING to actual users! Japanese manufacturers must surely be watching the success Elecraft and Ten-Tec are having by incorporating real-time user feedback into their products. If they don't soon start doing the same, I believe they will all be history in a few years."*

In other modern-day transceivers with a high first IF, the second IF is often in the 9-MHz region and the third IF usually at 455 kHz. Both these lower frequencies are well suited for high-quality crystal filters with excellent shape factors. In some receivers, however, ceramic or mechanical filters are used at 455 kHz, but they have an inferior shape factor compared to a good crystal filter (see also Section 1.8.5.) I strongly suggest that you not compromise in this area. Install optional filters with excellent shape factors at the lowest IF. Otherwise, the performance of the variable-bandwidth control will be very mediocre.

Ideally, the last-IF filters should be placed just ahead of the product detector, to reduce wideband noise generated in the IF amplifier stage. Many modern receivers show an annoying wideband hiss, which is especially noticeable on narrow CW when the band is very quiet.

For several year now DSP signal processing has been introduced at the lowest IF, usually in the 10 to 50-kHz range. DSP allows unlimited flexibility so far as varying bandwidths, but it can never replace the very essential filters in the earliest stages of the transceivers. Transceivers that have tried to do it this way have quickly becomes famous for their very bad behavior with strong signals.

I am convinced that somehow the designers have to move away from these very high first IFs, and compensate for the loss of image rejection by going back to tuned input circuits, or narrowband filters (not octave filter) covering just the amateur bands (as was done in the Ten-Tec ORION).

1.8.9. DSP filtering

Digital signal processing digitizes (ADC) the analog signal (eg, an audio signal or a low IF signal) so that a digital processor can handle the signal and do whatever is needed before converting it back to an analog signal (DAC). (Ref 290 and 291). To be able to handle the digitized signals, a CPU with a very high clock frequency is used. The heart of a DSP system is the software.

1.8.9.1. Audio DSP

Outboard DSP signal processors for use after the receiver audio chain were the first ones available on the commercial market. There are still a great number of excellent transceivers without DSP on the market that score very well on the low bands; eg, the Drake R4C, the Yaesu FT-1000D, the Kenwood TS-830 and TS-930. The basic performance of these receivers can be further enhanced with the use of external AF DSP systems. DSP filters as a rule perform three different sorts of tasks:

1. They add variable high-skirt selectivity.
2. They perform as automatic (multi-frequency) notch filters: This is an area where DSP can excel. DSP units are available that can handle multiple carriers in the audio spectrum and carriers are notched out before the user even notices that one came on. The great disadvantage of any notch filter at AF is that the offending carrier is still present all through the receiver IF chain and will desensitize the receiver through AGC action. Ideally these notch filters would be inserted ahead of the AGC detector, and as far forward in the receiver chain as possible.
3. Noise reduction: Noise reduction DSP works because information-carrying signals have some patterns, while noise is totally random.

1.8.9.2. IF DSP

Modern manufacturers have started putting their DSP circuits in the last IF of the receiver, mostly in the 10 to 50 KHz range. This very low IF is today still necessary, since CPUs operating at frequencies high enough to allow operation at much higher IFs are either still too expensive or still under development.

The newest transceivers do many functions in DSP: operational bandwidth filtering, noise reduction, automatic notch, AGC, detection, etc. Using *only* the last IF for achieving the operational selectivity, without having sharp filters closer to the front end, proves to be very disastrous especially when strong signals are involved on nearby frequencies. A particular transceiver using this approach turned out to be totally useless in a contest environment.

Using narrow roofing filters right after the first mixer (for example at 9 MHz) in combination with IF-DSP (as is done in the Ten-Tec ORION) can give the best of both worlds: Very high dynamic range at close signal spacings and utmost flexibility regarding operational bandwidth.

1.8.10. Audio filters

Audio filters, just like AF DSP circuits, can never replace IF filters. They can be welcome additions, however, especially with older receivers/transceivers that lack good built-in CW filtering. Introducing some AF filtering reduces any remaining wideband IF noise, and can improve the S/N ratio. Removing some of the higher-pitched hiss can also be quite advantageous, especially when long operating times are involved, such as in a contest (Ref 237).

1.9 Stability and Frequency Readout

State-of-the-art fully synthesized receivers have the stability of the reference source. All present-day receivers have achieved a level of stability that is adequate for all types of amateur work and most modern transceivers have a frequency readout to at least the nearest 10 Hz, often down to 1 Hz.

1.10. Switchable Sideband on CW

Switchable CW sidebands is a very useful feature that was introduced in the Kenwood TS-850. The user can switch CW reception from lower sideband to upper sideband, just like in SSB. Although the terminology of lower and upper sideband is not so common on CW, CW signals are indeed received with the beat oscillator frequency either above (as an LSB signal) or below (as a USB signal). This feature can be quite handy in the daily fight against QRM. Together with bandpass tuning, sideband switching can often move an offending signal down the skirts of your filter to a point where no harm is done. The default on commercial receivers CW reception should always be LCW (lower sideband).

1.11. Outboard Front-End Filters

Most present-day amateur receivers and receiver sections in transceivers are general-coverage (100 kHz to 30 MHz). They make wide use of half-octave front-end filters, which do not provide narrowband front-end selectivity. Older amateur-band-only receivers used either tracked-tuned filters or narrow band-pass filters, which provide a much higher degree of front-end protection, especially in highly RF-polluted areas. Instead of providing automatic antenna tuners in modern transceivers, I believe that same space could more advantageously be taken up by some sharply tuned input filters that could be switched into the receiver when needed.

Some excellent articles describe selective front-end receiving filters (Ref 219, 221, 251 and 266). Martin (Ref 219) and Hayward (Ref 221) describe tunable preselector filters that are very suitable for low-band applications in highly polluted areas (Ref 294).

Whether or not such a front-end filter will improve reception depends on the presence of very strong signal within the passband of the half-octave filters. Several outboard narrow-tuned-filter designs have been published over the years by K4VX (Ref 295), W3LPL (Ref 2953), K1KP (Ref 2954), and N6AW (Ref 2952). Another popular and rather simple filter was designed by the members of the Bavarian Contest Club (Ref 2951).

An excellent (but expensive) commercially made bandpass filter is available from the German manufacturer, Braun (Karl Braun, Deichlerstrasse 13, D-90489, Nuerenberg, Germany). The Braun preselector SWF1-40 covers all bands (including WARC bands) from 10 through 160 meters, and includes an excellent preamp. The attenuation of the filters is 8 dB. The preamp can compensate for this, or add an extra 8 dB, which may come in handy when using low gain receiving antennas (see **Fig 3-9**).

International Radio (formerly RCI/Fox Tango), also offers front-end crystal filters, which are the ultimate solution for multi-multi contest stations for protection against interference from a multiplier station operating at the same location on the same band. Information can be obtained at **www.qth.com/inrad/index.htm**.

Many Top Banders experience problems with overload from local BC stations on 160 meters. Each situation requires a different approach to solve the problem. If the problem occurs with a special receiving antenna, such as a Beverage, then a tuned preselector (as described above) may help. The

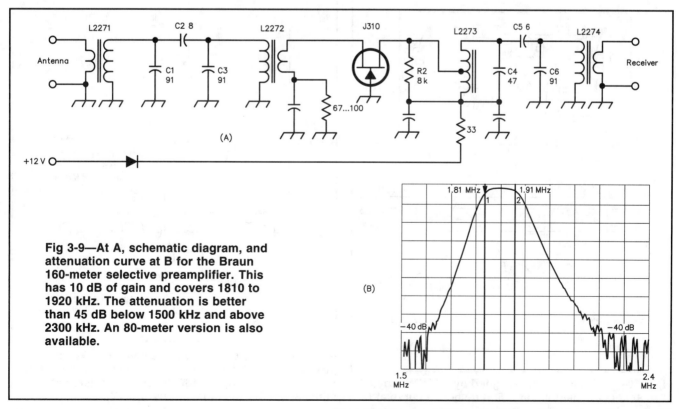

Fig 3-9—At A, schematic diagram, and attenuation curve at B for the Braun 160-meter selective preamplifier. This has 10 dB of gain and covers 1810 to 1920 kHz. The attenuation is better than 45 dB below 1500 kHz and above 2300 kHz. An 80-meter version is also available.

Braun selective preamp uses a double-tuned input circuit plus a double-tuned output circuit, with a transistor preamp that is adjustable for a maximum gain of 10 dB. This little preselector attenuates signals in the BC band by at least 45 dB. Similar circuits are available from other sources. **Fig 3-10** shows the schematic diagram, the layout and the bandpass curve of a highly effective and popular passive BCI filter designed by W3NQN (Ref 298, 299).

If you use no separate receiving antenna, and the problem exists when listening on your transmit antenna, you will need to install a similar filter, which you must bypass while transmitting.

1.12. Band Splitter for Beverages

All modern receivers have a separate input for a Beverage or other type of receiving antenna. But what if you want to split one receiver antenna between several different receivers? This is a very common situation in contests; for example in a SO2R (single-operator 2-radio) setup. If you simply connect the Beverage coax in parallel to the two receivers operating on different frequencies, you don't really know what will happen—The input impedance of the receiver at the other frequency may be very low and may result in heavy swamping. Using a 3-dB splitter is technically OK, but you do lose 3 dB of signal. A neat solution I have used for some time now was developed by DL7AV and is shown in **Fig 3-11**. Three band filters are designed in such a way that the load impedance on the other frequencies are very high, effectively uncoupling the three band-outputs.

1.13. Intermodulation Created Outside the Receiver

If you hear what sounds like a spurious signals from a BC station in the ham bands, one way to tell if the product is occurring in your receiver is to insert an attenuator at the input of the receiver. If you observe a much greater drop in the garbage level than in the desired signals when attenuation is added, then you can bet the garbage comes from overload of your own receiver. If both the desired signals and the spurious drop the same amount with attenuation, then the generation of spurious signals is happening outside your receiver.

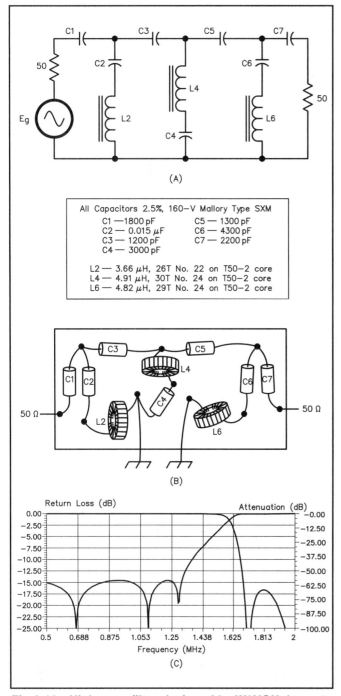

Fig 3-10—High-pass filter designed by W3NQN. Layout at A, schematic diagram at B and response curve at C.

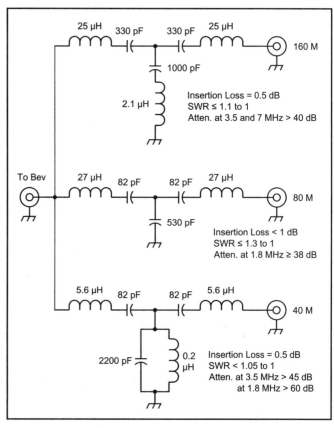

Fig 3-11—This 3-band Beverage splitter makes it possible to feed the signals from one Beverage to three receivers, operating on 160, 80 and 40 meters without minimal splitter loss (design by DL7AV).

I have witnessed this problem with aging Beverage antennas. Sometimes it is referred to as bad ground loops or even bad connections, but non-linearity caused by corrosion can create overload, cross modulation and intermodulation (in plain language—*mixing*). You need good connections in the system, even if you don't normally run transmitter power into a receiving antenna like a Beverage. If you suddenly hear all kinds of alien signals pop up in the band where they don't belong, it's time to go and check all the contacts in your receiving antenna system. Also check proper grounding of the coaxial feed line.

Such products can occur in poor electrical connections in cable TV (where aluminum cable is in contact with steel support strands), telephone wires, fences, towers and even in your own antennas. The mixing products are radiated by these inadvertent antennas into your receiving system. With broadband antennas such as Beverages, you may need to use high-pass filters or preselectors when operated in the vicinity of BC stations (as mentioned in Section 1.11).

1.14. Noise Blanker

A noise blanker, by nature of its principle of operation, only works on short-duration ignition-type pulses. Noise blankers detect strong noise pulses, and block (gate) the receiver's IF chain when these pulses are present. To detect these short pulses, we use wide roofing (first IF) filters, because narrow filters would lengthen and distort the noise pulses and make noise blanking impossible. Noise blankers are one of the reasons why modern transceivers use very wide (much too wide) first IF filters, leading to poor IMD performance on strong nearby signals. As the noise pulses are detected on our receiving frequency, rather than on any other frequency outside the busy amateur bands, noise blankers usually are ineffective when the band is fully loaded, such as during contests. Strong adjacent signals can gate the receiver, instead of noise pulses. Using a frequency outside the amateur bands to sense the noise, as was done in the Collins KWM-2 receiver more than 40 years ago, would be a solution to that problem.

It is well known that some top-end transceivers, such as the original Yaesu FT-1000D series, had a design flaw in the noise-blanker circuitry. This reduced the IMD performance significantly. In the FT-1000MP and MK 5 this problems exists when the NB gain pot is not set to minimum. In the FT-1000MP, if the gain control is set to CCW the NB has no effect on the receiver IMD. In the MK V if the menu is set to A=1 and B=1 (under software control), there is no IMD added by the noise blanker. There is a modification to overcome this flaw described on W8JI's web site (**www.w8ji.com/**). With Tom's mod installed in the FT-1000MP (or MK V), the A and B settings can be left anywhere and the IMD problem goes away if the NB switch is turned off. This is more convenient if you do regularly use the noise blanker.

1.15. Receiver Evaluations

A/B testing of radios is very tricky unless done at exactly the same time. On the low bands the type of noise in which we are listening for weak signals changes continuously. The tester's brain (doing the final decoding) may work differently (such as when you are tired) and many other circumstances can make results of so-called A/B testing vary quite a bit.

Of course, you would like to A/B-test all radios before buying one! However, results of such very subjective testing, done under totally uncontrolled circumstances, are just that— very subjective. It is not possible for most of us to perform exhaustive, laboratory-quality receiver tests ourselves. To minimize confusion, I have refrained from quoting test measurement data.

If you have more than a casual interest about how these relevant measurements are made, check W8JI's web site (**www.w8ji.com/receivers.htm**). Test methods have been described in the amateur literature (Ref 210, 211, 234 and 255). Hart, G3SJX, has published a series of excellent equipment evaluations in RSGB's *RadCom* (Ref 400-444). The reviews done by ARRL and published in *QST* and on their members-only Web site have made substantial progress as well. While the ARRL and RSGB do a good job of reviewing equipment, they publish somewhat useless wide-spaced data intermodulation dynamic range (see section 1.5).

Further, sometimes published values for two-tone dynamic range (IMD3-DR) and third-order intercept (IP3) don't make sense. This can happen where overload (and the consequent generation of distortion products) occurs in several different places in a receiver's front end simultaneously. So far as I'm concerned, the most fundamental measurement is the two-tone dynamic range, with test tones close to the desired frequency. You should recognize that it is impossible to actually measure third-order intercept, simply because the receiver saturates well before that point. While IP3 is a convenient, single number that is easy to remember, it is a theoretical number. You should be cautious when Equations 5 and 6 in Section 1.3 are not self-consistent. (Ref: *QEX* article by D. Smith, "Improved Dynamic-Range Testing," Jul/Aug 2002 or by U. Rohde, "Theory of Intermodulation and Reciprocal Mixing...," Jan/Feb and Mar/Apr 2003).

1.16. Adding Input Protection to Your RX Input Terminals

More and more of new transceivers have a separate receiver input for use with special receiving antennas such as Beverages. If those receiving antennas are installed very close to the transmit antenna, dangerously high voltages can destroy the input circuitry of the receiver. Since equipment manufacturers do not incorporate a suitable protective circuit, it may be wise to build one of your own.

Fig 3-12 shows a suitable protective circuit. A small relay shorts the input of the receiver during transmit. The

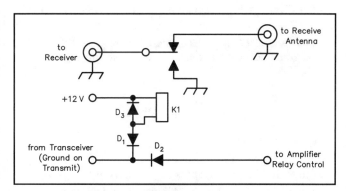

Fig 3-12—Receiver input protective circuit. Receiver input is grounded during transmit. D2, D2 and D3 are silicon diodes.

voltage for the relay usually be obtained from any 12-V source (usually from the transceiver itself), while the relay is activated by the amplifier control line. Two diodes make it possible to switch the amplifier and the protection circuit from the same line. It is clear that this circuit only protects your equipment from RF coming from the same transceiver. Where more than one transmitter is used (like in a multi-transmitter operation during contests), a different approach must be taken, such as using band-pass filters. It is also important to have a DC path to ground on the antenna jack. This can be a 2.5-mH choke or a 1-MΩ resistor.

1.17. Noise-Canceling Devices

When you are plagued with a local single-source manmade noise, you can often dramatically improve, or even eliminate it, using a so-called *noise-canceling device*. In a noise-canceling device, signals from two antennas (one is the regular receiving antenna, the second one is called the *noise source antenna*) are combined in such a way that the phase of the noise received on the noise antenna is of equal amplitude as on the normal receiving antenna, but exactly 180° out-of-phase. Details for such noise canceling devices can be found in Chapter 7 on Receiving Antennas.

1.18. In Practice

Now that you understand what makes a receiver good or bad for low-band DXing (and contesting) and after you study all the available equipment reviews, remember that what really counts is how the radio operates at your location, in your environment and with your antennas. You need to know how it satisfies your expectations and how it compares to the receiver you have been using. The easiest test is still to try the receiver when the band is really crowded, when signals are at their strongest. When you listen closely where it is relatively calm, you may hear weak crud that sounds like intermodulation or noise-mixing products. If you insert 10 or 20 dB of attenuation in the antenna input line and the crud is still there, there is a good chance that the crud is really being transmitted. (See the Sidebar "What About Spatter?") If the attenuation is raised and the crud goes away, it is likely that raising the intercept point by 10 or 20 dB would stop intermodulation created in the receiver itself.

1.19. Receiver Areas for Future Improvement

In the now 6-year old Third Edition of this book I wrote: "In the survey, which I sent out via Internet in early 1998, I asked: *'What are the main characteristics that would make a dream receiver?'* and *'What improvements would you like to see to the receiver you use now?'*" The top hits were:

- Better strong-signal handling capability with close signal spacing.
- Better selectivity (shape factor) and more selectivity choices.
- Lower VCO phase noise.
- Reduce wideband transmitted noise by using Band Pass Filters instead of Low Pass Filters in the PA output stage.
- Truly effective systems against manmade, as well as atmospheric noise.

Let's hope Yaesu, ICOM, Kenwood, Ten-Tec and other communication equipment designers read the following paragraphs about what they should do to please their most-demanding customers:

- Drastically improve IMD behavior for close-in spacing (down to 500 Hz); eg, by using roofing filters with final operational bandwidth (2.0 kHz on SSB and 500 Hz on CW). If necessary we will be happy to trade-in full HF-spectrum coverage (use a much lower first IF).
- Drastically improve the sideband noise of our oscillators, which will have a twofold benefit:
  - Improve the dynamic selectivity of the receiver.
  - Reduce the horrible noise sidebands on transmit.
- Develop systems that can effectively reduce or eliminate manmade or static noise (something radically different from noise blankers).
- Incorporate variable make and break CW shaping that tracks the CW speed so that we can have ideal wave-form shaping at any speed.
- Offer more and better (shape factor) IF filters; eg, 800 Hz to 1 KHz selectivity for CW
- Design and build quiet IF stages that don't hiss like an old steam train.
- Develop true diversity receiving systems with automatic selection (including computer controlled auto-tune noise canceling systems).
- Move DSP forward in the IF chain and write better and more user-friendly DSP software.
- Design and build quality audio stages, not just 1.5 W at 10% distortion that we have had pushed down our throat for years and years.
- Incorporate good front-end protection circuitry (for main as well as auxiliary antenna inputs).
- Make it possible to move the CW beat note as low as 200 Hz.

Did they read the Third Edition? Did they listen to their customers? Not to a large extent, I must say, except for Ten-Tec. Unfortunately it took Ten-Tec a long time as well. It is a little too early to assess the impact of the Ten-Tec ORION, but I know of many avid low-band DXers who have bought one or are waiting to buy one. As for me, I have been evaluating my ORION since mid-June, 2003. The essential lab tests (dynamic range, CW bandwidth, etc) were made in W8JI's lab. I have used this radio over many, many hours of listening, operating and contesting. The ORION is certainly the most revolutionary radio in concept and performance I have seen since the days when the first Drake radios hit the market. But it is a different kind of radio, and it takes time to learn how to operate and how to set all the parameters, since just about everything can be changed in software by the user. Since 80% of the radio is DSP, the manufacturer can actually update your radio by just sending you new firmware. Great!

Sometimes I have the impression that some of the design engineers of our transceivers never operate the radios themselves. Some of the control knobs we need to use often are tiny and in almost unreachable places on the front panel.

Six years ago I wrote: "Development of modern Amateur Radio equipment is largely market driven. If the marketers keep telling the designers they want more bells and whistles, that is what the user will get. If users tell the manufacturers they want better basic performance often enough, maybe

WHAT ABOUT SPLATTER?

By ARRL Senior Assistant Technical Editor, Dean Straw, N6BV

Once a reasonable level of performance is reached in terms of receiver sensitivity and selectivity, high dynamic range is the most important parameter for low-band DXers and contesters. After all, if your receiver itself generates garbage that covers up weak desired signals, you're not going to make QSOs. A dynamic range of 100 dB from the MDS to the level where third-order IMD products just start to appear is considered a very good figure of merit.

ON4UN has already chastised equipment manufacturers whose poorly engineered keying waveforms produce horrendous key clicks. But what about SSB voice operation? Having a receiver with a 100-dB dynamic range doesn't do you much good if the guy down the block with that huge signal drives his transmitter so hard that he splatters 100 kHz up and down the band! Well maybe 100 kHz is an extreme case. But we've all been infuriated by someone who moves perhaps 5 kHz below our frequency (on USB) and creates strong low-frequency buckshot—to the unmistakable cadence of "CQ contest, CQ contest." If he is S9 + 20 dB on-channel and his buckshot is S7 off-channel, his splatter is only suppressed 32 dB. That is not good enough.

Present-day specifications for amateur transmitters rarely look at IMD beyond the fifth-order products. In the USA, the Federal Communications Commission (FCC) regulates technical specifications for transmitters, including Amateur Radio transmitters, for which the specs are rather lax. Wn the other hand, FCC technical specifications for commercial transmitters (for example, Part 80 in the Maritime service) are the toughest in the world. The FCC commercial limit is: Suppression > 43 dB + 10 log (Average Power) for 11th-order or higher two-tone IMD products.

See **Fig 3-B**. For typical audio tones at 400 and 1800 Hz, the 11th-order IMD products are located about 7 kHz above and below the two main tones. For a 1500-W amplifier, this means that IMD products more than ±7 kHz away must be suppressed at least 75 dB below the *average* power (or 78 dB below *PEP*, the reference level usually used for Amateur transmitters). An IMD product that is 78 dB below an on-channel S9 + 20 dB signal would be less than S1. How many Amateur transmitters are that good?! (Just for reference, the same FCC limits for CW operation at 25 WPM requires 75 dB suppression of key clicks 300 Hz away from the carrier, resulting in "soft" keying. W1AW uses a commercial transmitter for their code-practice sessions and their code is perfectly copiable.)

Commercial transmitters can indeed meet such splatter and key-click specifications, so it isn't an impossible task!

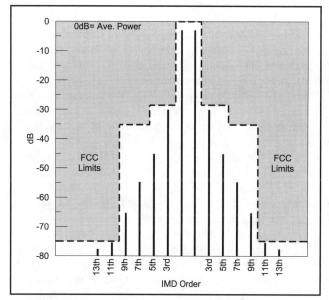

Fig 3-B—FCC commercial specifications for transmitted IMD. For 11th and higher-order IMD products, the required suppression for a 1500-W transmitter is 75 dB below average power (the FCC definition), or 78 dB below PEP (the convention used by amateurs).

Recently, Yaesu incorporated a "Class A" operating mode into their FT-1000MP MARK-V-FIELD model, where higher-level IMD products are very well suppressed on SSB. There are reports, however, that the amount of heat produced during contest operations using their Class-A mode is very high, leading to reliability problems.

When I discussed the topic of ultimate splatter performance with ON4UN, he suggested that many contesters prefer a wider transmitted signal—at least within reason. This is to keep interlopers away from their transmitting frequency. While this is a realistic view of the way a contester thinks, it doesn't absolve the manufacturers from the responsibility to produce really clean transmitters!

ON4UN suggests that really "dirty" signals in contests could have their scores reduced: "*We should have an independent jury noting the quality of the signals, and then giving a kind of multiplier to the score. Excellent signal: Multiplier 1; Bad signal: Multiplier 0.7; Very very bad signal: Multiplier = 0.4...*"

---

the designers will get the right message and we will see more progress toward better receiver performance." Maybe I was a little optimistic. We have told the Yaesus and the Kenwoods of this world what we want. There has been very little response from them. The little one on the list of ham radio equipment manufacturers (Ten-Tec) apparently got the message. In that respect radio development is indeed market driven: The market shouted out loud what it wanted, and at least one listened and heard us. If the claimed specs (at the time of writing the first units are being delivered) hold true, a drastic change in the market has to be envisaged.

2. TRANSMITTERS
2.1. Power

It should be the objective of every sensible ham to build a well-balanced station. Success in DXing can only be achieved if the performance of the transmitter setup is well-balanced with the performance of the receiving setup. It is true that you can only work what you can hear, but it is also true that you can only work the stations that can hear you. It is indeed frustrating when you can hear the DX very well but cannot make a QSO. There are some who cannot hear the DX, but they go so far as to make fictitious QSOs by "reading the Callbook." Fortunately those bad guys are rare.

A well-balanced station is the result of the combination of a good receiver, the necessary and reasonable amount of power and, most of all, the right transmitting and receiving antennas. It is, of course, handy to be able to run a lot of power for those occasions when it is necessary. In many countries in the world, amateur licenses stipulate that the minimum amount of power necessary to maintain a good contact should be used, while there is of course a upper limitation on the maximum power.

There are modes of communication where we have real-time feedback of the quality of the communication link, and that is AMTOR, PACTOR as well as other similar error-correcting digital transmission systems. In CW as well as SSB, we can only go by feeling and by reports received, and therefore we are most of the time tempted to run *power*.

There are a number of dedicated operators who have worked over 250 countries on 80 meters or 100 countries on 160 meters without running an amplifier. But a large majority of active low-band DXers run some form of power amplifier, and most of them run between 800 and 1500 W output.

2.2. Linear Amplifiers

Today the newest technologies are utilized in receivers, transmitters and transceivers to a degree that makes competitive home construction of those pieces of equipment out of reach for all but a few. Most high-power amplifiers still use vacuum tubes, however, and circuit integration as we know it for low-power devices has not yet come to the world of high-power amplifiers. At any major flea market you can buy all the parts for a linear amplifier. See **Fig 3-13**.

The amplifier builder will usually build more reserve into his design. He will have the option of spending a few more dollars on metal work and maybe on a larger power-supply transformer to have a better product that runs cool all the time and never lets him down. Maybe he will use two tubes instead of one, and run those very conservatively so that the eventual cost-effectiveness of his own design will be better than for a commercial black box.

Power amplifiers designers and builders now have their own reflector on the Internet, where very interesting information is exchanged between builders. To subscribe to the AMP reflector send a message to **amps-request@contesting. com** with "subscribe" in the body text.

A lot of very valuable information about building your own amplifier can be obtained at AG6K's Web Page: **www.vcnet.com/measures/**. AG6K described the addition of 160 meters to the Heath SB-220 (Ref 340), and in another article he covered the addition of QSK to the popular Kenwood TL-922 amplifier (Ref 338).

If you do consider building your own amplifier, make sure you carefully study *The ARRL Handbook*, as well as W8JI's website, which has lots of good information on this subject (**www.w8ji.com/Amplifiers.htm**).

2.3. Phone Operation

If you choose to play the DX game on phone (SSB), there are a few points to which you should pay great attention.

2.3.1. Microphones (and headphones)

Never choose a microphone because it looks pretty. Most of the microphones that match (aesthetically) the popular transceivers have very poor audio. Most dynamic microphones have too many lows and too few highs. In some cases the response can be improved by *equalizing* the microphone output. W2IHY developed an 8-band audio equalizer (plus noise gate that can produce excellent communications audio even from a mediocre microphone or a lousy voice! (See **w2ihy.com/Default.asp**). It also can turn a studio-quality microphone (such as the Heil Goldline mike) into an efficient DXing and contesting microphone.

In some cases the audio spectrum of a bad microphone can be drastically improved by changing the characteristics of the microphone resonant chamber. If the microphone has too many lows (which is usually the case), improvement can sometimes be obtained by filling the resonant chamber with absorbent foam material, or by closing any holes in the chamber (to dampen the membrane movement on the lower frequencies).

The most practical solution, however, is to use a microphone designed for communications service. A typical communications microphone should have a flat peak response between 2000 and 3000 to 4000 Hz, a smooth roll-off of about 7 to 10 dB from 2000 to 500 Hz and have a much steeper roll-off below 500 Hz. **Fig 3-14** shows typical response curves for the Heil HC4 and HC5 communications microphone elements (**www.heilsound.com/**). We should caution against overkill here too, however! We know that the higher voice frequencies carry the intelligence, while the lower frequencies carry the voice power. Therefore a good balance between the lows and highs is essential for maximum intelligibility combined with maximum power.

At this point I should mention that correct positioning of the carrier on the slope of the filter in the sideband-generating section of the transmitter is at least as important as the choice of a correct microphone. Therefore you should test your

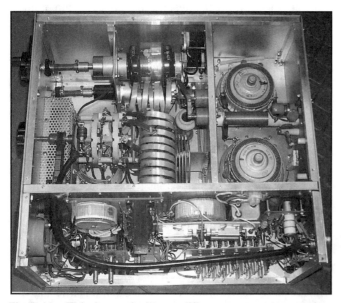

Fig 3-13—This home-built amplifier uses surplus parts obtained at a hamfest.

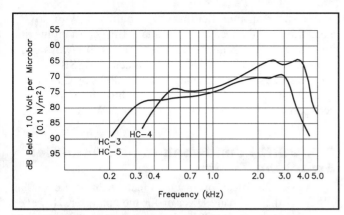

Fig 3-14—Typical responses of Heil communication microphones with a 2-kΩ resistive load. Note the sharp cut-off below 300 and 500 Hz.

equalized microphone system into a good-quality tape recorder before doing any on-the-air tests. Incorrectly positioned carrier crystals will also produce bad-sounding receive audio in a transceiver, since the same filter is (in most transceivers) used in both the transmit and receive chains.

One way of checking to see if the USB and LSB carrier crystals have been set to a similar point on the filter slopes is to switch the rig to a dead band, turn up the audio and switch from USB to LSB. The pitch of the noise will be a clear indication of the carrier position on the filter slope. The pitch should be identical on both sidebands. Most modern transceivers now allow tailoring of the AF bandpass curve through DSP. Some of the top range transceivers also make it possible to change the position of the carrier versus the filter curve through software programming.

As important as the *choice* of the microphone is how you *use* the microphone. Communications microphones are made to be held close to the mouth when spoken into. Always keep the microphone a maximum of two inches from your lips. A very easy way to control this is to use a headset/boom-microphone combination. Heil has various headset/microphone combinations, which can be equipped with either their HC4 or HC5 cartridge.

If you do not speak closely into the microphone, you will have to increase the microphone gain, which will bring the acoustics in your shack into the picture, and these are not always ideal. We often have a high background noise level because of the fans in our amplifiers. This background level—and the degree to which we practice close-talking into our microphone—that determines the maximum level of processing we can use.

2.3.2. Speech processing (clipping)

Speech processing should be applied to improve the intelligibility of the signal at the receiving station, not just to increase the ratio of average power to peak envelope power. After all, increased average power—together with the introduction of lots of distortion—will probably achieve little, or it may hurt your intelligibility. Although audio clippers can achieve a high degree of average power ratio increase, the generation of in-band distortion products raises the in-band equivalent noise power generated by harmonic and intermodulation distortion and in turn decrease the intelligibility (signal-to-distortion-and-noise ratio) at the receiving end.

RF clipping generates the same increase in the ratio of transmitted average power to PEP, but does not generate as much in-band distortion products. This basic difference eventually leads to a typical 8-dB improvement of intelligibility over AF clipping (Ref 322). Virtually all high-end current transceivers are equipped with RF or DSP speech processors.

Adjusting the speech-processor level seems to be a difficult task with some modern transceivers, if you judge from what we sometimes hear on the air. All modern transceivers have a compression-level indicator, which is very handy when adjusting the clipping level.

We already stated that the acoustics in the shack will be one of the factors determining the maximum allowable amount of speech clipping. By definition, a speech-clipped signal has a low dynamic range. In order not to be objectionable, the dynamic range should be kept on the order of 25 dB. This means that during speech pauses the transmitter output should be at least 25 dB down from the peak output power during speech. Let us assume we run 1400 W PEP output. A signal 25 dB down from 1400 W is just under 5 W PEP. Under no circumstances should our peak-reading wattmeter indicate more than 5 W peak (about 3 W average), or we will have objectionable background noise (Ref 305).

One way of getting rid of the background noise (from fans, etc), is to virtually extend the dynamic range of the audio by quieting the audio when the audio level drops below a certain threshold. The W2IHY audio equalizer described earlier has such a feature built-in, as does the Super Combo Keyer designed and built by ZS4TX (www.zs4tx.co.za/sck/). This keyer is multi-functional, and it includes an audio compressor and a noise gate that works very well. I have been using this unit very successfully for almost three years now. It also includes both a CW and a Voice keyer, each with 6 programmable memories. It has all that's needed for operating two radios in a SO2R (single-operator two-radio) contest setup, where the selection is controlled by the contesting software (compatible with *NA*, *CT*, *TR* and *Writelog*).

2.4. CW Operation

2.4.1. Keying waveform

In older generations of transmitters, you could adjust an RC network to change the leading and trailing edges of the keying waveform to ensure a clean CW signal without clicks. Today's transceivers are loaded with features, but judging from the lack of controls for the CW waveform, the designers of our present-day transceivers must not be serious CW operators.

Most transceivers seem to have the waveform shaping adjusted for 70 WPM. A state-of-the art transmitter should allow adjustment of wave shaping so that the rise and fall time are independent of speed. This shape cannot be obtained with a single, simple RC-network. It requires knowledgeable engineering to get the proper results. For example, you can adjust the Ten-Tec ORION's rise and fall times from 3 to 10 mS.

Six years ago I wrote in the Third Edition of this book: *"Let this be a message to the people who review the new transceivers to put emphasis on this issue, so that the designers wake up!"* What has happened since then? Little or nothing, except that serious operators became aware of the poor quality of the keying of many commercial transceivers. The

FT-1000 series transceiver, praised as being the best low-band rigs (and used by two-thirds of low-band DXers and contesters in my poll) is such an example, although not a lonely case! George Cutsogeorge, W2VJN, notes: *"The 'bad name' the MP has developed in recent times probably has something to do with the sheer number of radios in the field. In a major contest there are more MPs in use than all other radios together."*

For years we've told the designers at Yaesu that there was a serious problem, but succeeding FT-1000-style models all retained the same poor CW keying waveform shape. In the beginning of 2003 suddenly a modification was developed by Yaesu and applied to all new transceivers leaving the factory, although this was not announced to the public. Several people, including W8JI, tried the factory modification. The results were disappointing. W8JI, wrote to me: *"We are in very big trouble, because for every ten new radio's sold, at best one will get repaired correctly. This will eventually ruin the low bands for many years to come. Many people have horrible clicks and refuse to fix the radios, or use poor corrections. We need to put great pressure on Yaesu and other manufacturers to correct radios or we will slowly lose all pleasure on lowband CW."*

Both George, W2VJN, of International Radio (**www.qth.com/inrad/**) and Tom, W8JI, (**www.w8ji.com/keyclicks.htm**) dug into the FT-1000 problems on their own and came up with modifications that cure the problem. The bandwidth occupied in the W2JVN-modified FT-1000MP and the Ten-Tec ORION are very similar.

In truth, something should be done to improve key clicks in most popular radios, not only Yaesu's models. Most modern radios have worse keying characteristics than many rigs that are 30 or 40 years old. See "about-key clicks" on CD.

2.4.2. Harsh-sounding CW

The problem of harsh keying seems to go hand in hand with the problem of oscillator phase noise. Again, some 40-year old transmitters show better performance than today's radios. Sometimes you can often hear noise sidebands several kHz away!

2.4.3. Leading-edge spikes

Another common problem with several modern transceivers is that they generate a power surge on the leading edge of the first CW character. This surge is in some cases twice the level of a constant key-down signal. This causes increased transmitted garbage, sounding like key-clicks, and can trip the protective overdrive circuits of some commercial amplifiers.

In some transceivers this problem can be overcome by turning the RF OUT knob down to the point where the output power just begins to drop. But some transceivers use an internal ALC (automatic level control) loop that is controlled by the front-panel RF OUT knob. The attack time of an ALC is designed to be fast, but it isn't instantaneous. The delay before the ALC can automatically reduce the transmitter gain allows the initial spike to appear at the output.

2.4.4. Using ALC with an external amplifier

In a properly designed and operated station there is no need for external ALC between the amplifier and the transceiver. Controlling the output from a 200-W exciter by means of ALC from an amplifier that only requires 50 W of drive is a bad thing, again because of the attack time constant inevitably associated with an ALC circuit. This will always cause some overshoot on the first CW character or SSB syllable, showing up as extra sidebands—that is, clicks on CW or SSB splatter. In a good transmitter you can adjust the output power very precisely (in the Ten Tec ORION in 1-W increments, for example).

2.5. QSK, Semi Break-In and Amplifier Switching Timing

QSK (full break-in) is a nice feature, but not essential, either for the low-band DXer or for the contester. However, properly implemented QSK can be an asset to contesters and DXers. When calling, an operator can hear immediately when the DX transmits and can stop sending. This is beneficial to everyone on the frequency. QSK can help determine the DX station's pattern so that he can be called at the right time. Of course, good QSK used by two operators during a rag chew is really a pleasure.

If not properly designed and set up, however, QSK can be a disaster: It can generate severe key-clicks. It can ruin the antenna relay in the amplifier in no time, or cause component arcing and destruction of very expensive component (such as band switches) in the amplifier. Even stations operating semi-break-in on CW often exhibit poor timing and hot switching. In every JA-contest I seem to copy a lot of OA stations calling me—These are extreme cases where the entire first dot is missing. The same problem sometimes turns a W- station (USA) into a M-station (England).

If you don't want to ruin your amplifier, or ruin the bands with clicks and clacks for your neighbors, have a close look at the timings involved in your station. In every JA-contest I seem to copy a lot of OA stations calling me; these are extreme cases where the entire first dot is missing!

QSK (full break-in) is a nice feature, but it is not essential, either for the Low Band DXer or for the contester. However, properly designed QSK can be an asset to contesters and DXers. When calling a DX station the operator can hear immediately when he transmits and can stop sending. This is a benefit to all callers on the frequency. QSK can help determine the DX station's pattern so that he can be called at the right time to maximize returns. Of course, good QSK used by two operators during a rag chew is really a pleasure.

If not properly designed and set up, QSK can be disastrous: It can generate severe key-clicks, and it can ruin the antenna relay in the amplifier in no time, or cause component arcing and destruction of very expensive components (such as band switches) in the amplifier.

Even radios operating semi-break-in on CW can exhibit poor timing and hot switching, which can be avoided if the manufacturers obey the following general rules. The sequence of things happening on all **QSK** modes should be:

Make side:

- Appearance of signal input (key closure or data input).
- The transmitter immediately sends "on" signal to amplifier with minimum possible delay.
- Ideally the transmitter should have an adjustable RF-on delay (1 to 30 ms), after which it allows RF output to start

rising. If not adjustable, 15 to 20 mS is a must to accommodate amplifiers with slow relays.
- Wait for handshake signal (if handshake system is active), then deliver RF to the amplifier.

Break side:
- Data stops.
- RF output from transmitter stops with zero delay.
- After making sure envelope is just at zero, the amplifier keying line unkeys.

The sequence **on semi-break CW in or VOX** should be:

Make side:
- Appearance of TX Signal input (key closure, data input)
- Transmitter keys the amplifier relay line without any delay
- The transmitter should have adjustable RF-on delay (to make sure the amplifier relays are closed, see QSK mode) after which it allows RF output or checks for handshake if used. If no adjustable time, a minimum of 15-20 mS is required, and this may not be enough for some older amplifiers.

Break side:
- Data stops.
- RF envelope reaches zero.
- After independently adjustable OFF delay (0-1 second hang delay) amp unkeys. In better transmitters there is an independent adjustment for voice operation (VOX) and for semi-break-in CW. On CW this delay is advanced to a point where on semi-break on CW the amplifier relay does not clatter at the CW keying rate. It drops out after an adjustable delay.

On SSB the transmitter should have standard VOX adjustments (sensitivity, anti-VOX and hang-time) then generally follow this rule:

- The VOX trip and the amplifier immediately comes up
- After an adjustable TX Delay (can be same as CW or data) RF comes up (delay minimum of 15-20 mS)
- The VOX hang drops out only after the RF has reached zero, even at fastest hang setting

How can we know that the TX delay of 15 mS is enough? The best way is to listen for the transmitted signal quality of a second receiver. Listen around the transmitted signal and on its harmonics. Adjust the time delay for total cleanliness.

If the delay is longer than 20 mS it may fool very fast speed CW operators, and in that case it might be time to have a look at installing faster relays in the transmitter (for example, small vacuum relays).

2.6. DSP in the Transmitter

Most, if not all HF transceivers make extensive use of DSP technology. DSP is typically used to perform one of the following functions:
- DSP speech processing.
- DSP audio tailoring (nice velvet-like audio for a rag chew, and piercing sharp quality for the contest).
- CW make and break timing (hard or soft keying).
- DSP VOX control (delayed audio switching).
- Background noise elimination (to kill the noisy blower).

2.7. Signal-Monitoring Systems

You should have some means of monitoring the quality of your transmissions. All modern transceivers have some sort of built-in monitor system. The best ones are not mere audio output monitors, but instead monitor the directly detected SSB signal, so you can evaluate the adjustment of the speech processor. This feature allows the operator to check the audio quality and is particularly useful for checking for RF pickup into the microphone circuitry. A monitor-scope should also be mandatory in any amateur station. With a monitor scope you can:

- Monitor your output waveform (envelope).
- Check and monitor linearity of your amplifier (trapezoidal pattern).
- Monitor the keying shape on CW.
- Observe any trace of hot-switching on QSK.
- Check the tone of the CW signal (for power-supply ripple).
- Correctly adjust the speech processor.
- Correctly adjust the drive level of the exciter to optimize the make and the break waveform on CW and to avoid leading-edge overshoot.

I have been using a monitor scope at my stations ever since I got licensed almost 40 years ago, and without this simple tool I would feel distinctly uncomfortable when on the air (see **Fig 3-15**). Specific monitor 'scopes (eg, Yaesu, Kenwood) are rather expensive and have one distinct disadvantage: You must route the full output RF from the amplifier "through" the 'scope to tap off some RF, which is fed directly to the plates of the CRT. I decided to use a good second-hand professional one (Tektronix 2213, 20-MHz bandwidth), which cost less than a new monitor scope. You need to sample only a very small amount of RF to feed to the input of the scope. A small resistive power divider can be mounted at the output of the amplifier, from where millivolts of sampled RF can be

Fig 3-15—The Ten-Tec Orions are the new "battleships" at the ON4UN two-radio lowband DXing and contesting station. Between the two transceiver are all the antenna direction control units. Antenna selection is full automatic, controlled by the band data output from the transceiver. Two ACOM 2000 amplifiers complemented these radios.

routed to the scope through a small coaxial cable.

2.8. Transmitter Areas for Improvement

- Improve all intermodulation distortion products of the transmitter significantly.
- Noise sidebands (VCO noise) down to at least –135 dBc at 2-kHz separation.
- Easily and precisely adjustable power output (like the Ten-Tec ORION).
- No leading-edge power spikes on CW.
- Cure the key-click problems.
- SSB transmitter with 2.1-kHz bandwidth filters.
- Fully adjustable timing for QSK and semi break-in operation (to match amplifier characteristics).

3. CONCLUSION

If you're starting on the low bands and have a limited budget, then the best solution is to look for a decently priced second-hand transceiver with a decent reputation. The Kenwood TS-830 was one of the best low-band rigs in its day and would be a very good starting rig for any newcomer.

If you're not on a tight budget and you want only the best, there are a few choices. Judging by the popularity among by low-band DXers and contesters, you might look for a Yaesu FT-1000D transceiver. The later FT-1000 models (MP and MP MARK-V) have not brought any substantial improvements in basic performance, although they've added more bells and whistles.

The Ten-Tec ORION looks like it will be a serious competitor to the Japanese top-range transceivers. Yes, the ORION does not have the color display like the ICOM, but are you going to hear the DX better or work more stations in a contest thanks to a color display?

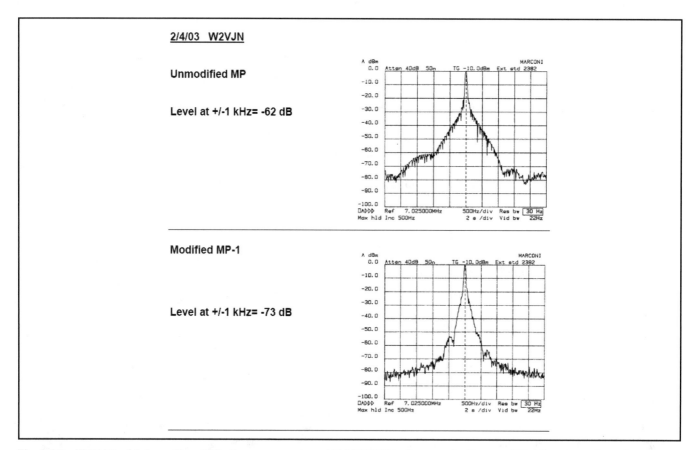

Fig 3-16—W2VJN of International Radio measured an FT-1000MP before and after modification to reduce key clicks. At spacings from the carrier of ±1 kHz, the modification reduced clicks by about 11 dB, a very significant amount.

CHAPTER 4

Antenna Design Software

Pierre Cornelis, ON7PC, an electronic engineer by education, has worked in several engineering jobs for the Belgian National TV and broadcasting company. He specialized in microwaves and is very familiar with antennas, both professionally and as an amateur. Pierre has lectured on antennas and antenna modeling before many radio clubs. He was an ideal partner for proofreading and counseling on these subjects. I thank my friend Pierre for his help and friendship.

When I talk at radio clubs, I usually ask how many in attendance have a PC and how many have Internet. In 2003 the answer is somewhere around 95%. The PC and the Internet connection have both become very important tools for the active radio amateur.

Most of us can hardly imagine what our daily life, what our hobby would be like, without our PC or the Internet. One of the few constants we see is change, and the increase of the rate of change! What today is state-of-the-art is outdated tomorrow.

For hams, computers are not only an excellent tool to gather information via Web sites, they are used for administrative tasks (logbooks, contests-logging, etc). In this chapter we'll look at how they can be used as design tools for circuits and antennas.

1. ANTENNA-MODELING PROGRAMS

Until not too long ago, predicting antenna performance was more a black art than a scientific or engineering activity, especially in Amateur Radio circles. Building full-size models or scale models and testing them on wide-open test sites were out of reach of most amateurs. This was when some of the old myths were born and the rat race for decibels was started.

What is modeling? It is evaluating the performance of a *system* that is governed by the laws of physics using a *model*. This may be a physical model (such as a scale model) or a mathematical model. Antenna-modeling programs are computer programs that via mathematics calculate and predict the performance (electrical, mechanical) of an antenna. Modeling is done in all branches of science. Modeling always has its limitations, partly because the model that we have to describe (enter into the program) can almost never be described in the same detail as the real thing (and especially its environment), and partly because of numerical limitations in the calculating code used. The final limitation is the operator, who enters the data and who interprets the results. In all cases a good deal of knowledge and experience in the field of antennas is required in order to draw the correct conclusions and take the right decisions during the process of modeling. Why do we want to model antennas?

- To understand how antennas work.
- To verify designs from literature.
- To optimize a design for your particular needs (frequency, height, application).
- To create a new design.

One of the first Yagi-modeling programs reported in the literature was written in 1965 by I. L. Morris for his PhD dissertation at Harvard University. Others (Mailloux, Thiele, Cheng and Cheng) have elaborated on this program to perform further analysis and optimization. Such a program was used by Hillenbrand, N2FB, to optimize Yagis. This program was later adapted for use on the IBM PC by Michaelis, N8TR (ex-N8ATR), and for Windows by Straw, N6BV. All the modeling programs described below are Windows based, except *ELNEC*, which is DOS based.

1.1. How Modeling Works

In an antenna model you must define the geometry of the antenna (all conductors, the feed points, the loads if any) as well as the environment in which the antenna works (free space, over perfect ground, over real ground, antenna height, etc). The basic concept is you need to describe all elements of the antenna (called *wires* in this context) by giving their X, Y and Z coordinates.

Once you have described all the elements (conductors, wires) geometrically, they will be split up into short *segments*. During modeling, the HF current in each segment is evaluated. The program calculates the self impedance and the mutual impedances for each of the segments. Then it computes the field created by the contribution from each segment. (I explain what mutual impedance is in Chapter 11 covering arrays.) Modeling can be done in free space, over perfect ground or over real ground.

This section of the book is not meant to be a tutorial on how to model. But it is hard to conceive that a serious lowbander would not, sooner or later, get involved in antenna modeling. After all, the low bands are the bands where we can still do a lot of home-antenna building and designing. That's what makes the low band so attractive to many.

You can learn the art of modeling by cut and try. The *EZNEC* manual is an excellent course by itself. If you are even more serious about it, have a look at the *ARRL Antenna Modeling Course* (**www.arrl.org/catalog/?item=8721** and **www.arrl.org/cce/courses.html**) and at the website of L.B. Cebik, W4RNL, at **www.cebik.com/**.

AC6LA's excellent website (**www.qsl.net/ac6la/ antmodaids.html**) has an abundance of interesting information about antenna modeling. A free *ARRL Antenna Modeling Course Aids* file listing details about the chapters and the large number of models used in the course can be downloaded from **www.qsl.net/ac6la/CourseAids.zip**.

Specific modeling issues, such as the required segment length, the segment length tapering technique, etc, are also covered in specific antenna chapters in this book (Verticals, Dipoles, Yagis and Quads) where relevant.

1.2. *MININEC*-Based Programs

MININEC (Mini Numerical Electromagnetic Code) was developed at the NOSC (Naval Ocean Systems Center) in San Diego by J. C. Logan and J. W. Rockway. The original *MININEC* was not a user-friendly program. Several people wrote pre- and post-processing programs to make *MININEC* (now at version 3.13) more user-friendly, in which the *MININEC* code is used as the core. For general antenna analysis that does not press its well-known limitations, *MININEC* is a highly competent code. It handles elements of changing diameter directly, and with segment-length tapering, can accurately model a wide range of antenna geometries.

1.2.1. *MININEC* limitations

The major limitation concerns calculations over real ground, which is limited to modeling far-field patterns. In the near field, *MININEC* assumes a perfectly conducting ground. Some of the consequences of this are that you cannot use *MININEC* to calculate the influence of radials on the feed-point impedance of a ground-mounted vertical. A quarter-wave vertical will yield a 36-Ω impedance over any type of ground. In reality the ground and the radials in the near field are important for collecting the return currents. This will influence the feed-point impedance and the efficiency of the antenna due to "lost return currents" in a poor ground. Radials can be specified with *MININEC*, but they will influence only low-angle reflection and attenuation in the far field. See Chapters 8 and 9 on dipole antennas and vertical antennas for details.

Further, *MININEC* reports the gain and the feed-point impedance of horizontally polarized antennas at low heights incorrectly. This is for horizontal antennas less than 0.25 wavelength above ground. For larger antennas the minimum height may be higher. At low heights the reported gain will be too high and the feed-point impedance too low. The shape of the radiation patterns will remain correct, however.

We thus are handicapped using *MININEC* on the low bands, where we often model antennas that are electrically close to the ground. For modeling antennas such as Yagis on higher-frequency bands, this is unlikely to be a problem because they are mounted higher than $1/4$ λ above ground. *MININEC* has other modeling problems with quads, which are detailed in the Chapter on Yagis and Quads.

In *MININEC* wires that are thicker than 0.001 λ may not be modeled accurately due to computational approximations in the code. While low-band antennas will not be affected, this limitation may be encountered when working on antennas for 10 meters and higher. These and other limitations are very well covered by R. Lewallen in "*MININEC*: The Other Edge of the Sword" (Ref 678) and on L.B. Cebik's (W4RNL) excellent Web site.

ELNEC (**www.eznec.com/**) is a DOS modeling program by Roy Lewallen, W7EL, based on *MININEC*. Note that W7EL doesn't actively market *ELNEC* any more.

ANTENNA MODEL (from Teri Software, **www.antennamodel.com/**) is a full-featured Windows version of *MININEC 3.13*. The core has virtually unlimited segment capacity for segments and uses improved algorithms to overcome many *MININEC* difficulties, fixing errors due to increasing frequency, angular junctions, wire junctions less than 28° and wires spaced closer than 0.23 λ. The program offers both 2D and 3D patterns and a variety of supplemental calculating features.

NEC4WIN95 (**www.orionmicro.com/**) is a Windows 95/98/NT 32-bit version of *MININEC*, using spreadsheet input page and pull-down boxes for other antenna parameters. 3D patterns are provided, as well as optimization routines. The user can vary the height of the antenna without invoking a complete recalculation of the matrix for faster results. There is a built-in loop correction feature allowing accurate modeling of square-loop antennas. The VM (virtual memory) version of the program permits almost unlimited numbers of segments in a model. L.B. Cebik, W4RNL, did an in-depth review: at **www.antennex.com/preview/Folder01/NEC4/n4w.htm**.

MMANA (by JE3HHT) is available as freeware from VK5KC's *MMHamsoft* website (**www.qsl.net/mmhamsoft/**). Based upon the *MININEC 3.13* core, the program offers a large segment (pulse) capacity and other advanced features, such as segment-length tapering, optimizing and network calculation, but it lacks some basic features, such as assigning a user-specified material conductivity or resistivity to the model wires, frequency compensation or close-wire compensation.

1.3. Programs Using the *NEC-2* Core

NEC is the full-fledged brother of *MININEC*, which means that *NEC* also employs the method-of-moments to model antennas. The original versions ran on mainframe computers only, and were accessible to professionals only. They had a very unfriendly user interface. In the last decade, however, a number of user-friendly *NEC*-based programs have been developed.

NEC-2, which is in the public domain, can model real ground in the near and fields. It does away with most of the limitations described above for *MININEC*. It can model antennas quite close to the ground, as well as radials above and even on the ground. (It cannot handle buried radials though.) *NEC-2* uses the Sommerfeld-Norton high-accuracy ground model to model horizontal wires close to the earth. One notable limitation of *NEC-2*, compared to *MININEC*, is its inability to model stepped-diameter wires (such as tapered Yagi elements, although this shortcoming has been overcome by some software providers using the *NEC-2* core. This problem has also been corrected in the newest version *NEC-4*, which also has the ability to model wires in the ground. I have frequently used *NEC* to model antennas where the limitation of *MININEC* would have made the results unreliable. I will review specific modeling issues when discussing those antennas (eg, Beverages, low Delta Loops, elevated radials, etc).

EZNEC Version 3 (**www.eznec.com/**) is written by Roy Lewallen, W7EL, who has been writing well-received modeling software for a long time. *EZNEC* offers 3D plots, 2D slicing, ground-wave output, direct entry for trap as well as for series and parallel R-L-C loads, stepped-diameter correction, and numerous short cuts for antenna-geometry modification. Standard *EZNEC Version 3* is restricted to 500 segments, while the *EZNEC Pro* version handles much larger arrays. **Fig 4-1** shows the "View Antenna" screen of a model representing 300-meter long Beverage antenna for 160 meters. This model uses two quarter-wave in-line terminations at each end (see Chapter 7).

NEC-Win Plus by Nittany Scientific (**www.nittany-scientific.com/**) is another popular Windows version of *NEC-2* that features spreadsheet-type input pages with design-by-equation capabilities. The program also offers stepped-diameter corrections. It provides 2D and 3D plots and antenna views and graphical outputs. *NEC-Win Pro* is the high-end versions of *NEC-Win Plus*. I have been told that the newest version (due end 2003) will include optimizing algorithms.

Antenna Solver (**www.gsolver.com/**) uses *NEC-2* Fortran translated into C++, with dynamic array allocation. The user interface, graphical editing features and data display capabilities allow analysis of antenna patterns for near, far and ground-wave fields, as well as currents and charge densities. A full-featured version of the program can be downloaded in the Demo mode for 30-day use, after which the purchase of a password will be needed to permanently enable the program.

The program *4nec2* by Arie (**4nec2@gmx.net/**) is another user-friendly shell wrapped around the standard *NEC-2* computing engine The *4nec2* package contains all the software to specify, calculate, evaluate and optimize your antenna system(for a single frequency or a band of frequencies). It is capable of modeling *NEC-2* files up to 11,000 segments. It also includes a Smith-chart display with integrated line-length calculator. A geometry-builder is included in the package. Best of all, it is freeware available at: **www.qsl.net/wb6tpu/swindex.html**.

Dimitry Fedorov, UA3AVR, (**ua3avr@au.ru**) wrote *NEC-2 for MMANA* (v 1.4). With this utility you can enjoy all the benefits of the *NEC-2* core while using modeling files created for *MMANA*. See **www.qsl.net/wb6tpu/swindex.html**.

1.4. Programs Using the *NEC-4* Core

The latest version of *NEC* is *NEC-4*, which overcomes most of the shortcomings with earlier *NEC-2* codes. *NEC-4* permits modeling of underground radial systems, elements of varying diameter sections, close-spaced parallel wires, as well as all the modeling capabilities of earlier versions of the code. While *NEC-2* is public-domain software, the copyright for *NEC-4* is held by Lawrence Livermore National Labs (**www.llnl.gov/**) and you must obtain a license to use either *NEC-4* or any of the other software packages that use the *NEC-4* core. The license at the time of writing is approx $800.00 and non-US citizens must apply for this license through their embassies.

EZNEC Pro, by Roy Lewallen, W7EL, (**www.eznec.com/**) can be bought with an option for *NEC-4*, if the purchaser can show a license for *NEC-4*. *EZNEC Pro* is also available for *NEC-2* (see above). *EZNEC Pro* imports and exports files in generic *.NEC format as well as *.EZ format.

GNEC (**www.nittany-scientific.com/**) is the *NEC-4* version of *NEC-Win Plus*. This program implements all or nearly all of the input "cards" of the complete *NEC-4* input deck. Output capabilities include 3D, polar plots and many rectangular (X-Y) graphs, as well as a large array of tabular reports. The spreadsheet and dialogue box interface is similar to *NEC-Win Pro*. Here too a license for the *NEC-4* core must be purchased separately.

If you are going to use a *NEC*-based program you should consult "The Unofficial Numerical Electromagnetic Code (NEC) Archives" at **www.qsl.net/wb6tpu/swindex.html**.

1.5. *MININEC* or *NEC*?

MININEC 3.13 shows its strength in the areas where *NEC-2* displays weaknesses—Stepped-diameter wire models mainly. It must be said, however, that for a large class of modeling tasks both *NEC* and *MININEC* are equally capable.

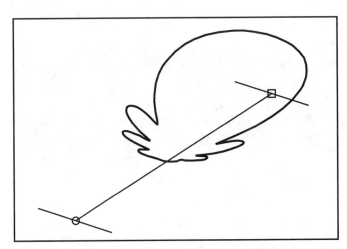

Fig 4-1—'View Antenna" screen in the *EZNEC 3* program of a model representing a 300-meter long Beverage for 160 meters, using 2 quarter-wave in-line terminations at each end.

1.6. Optimizing Programs.

With a regular *MININEC* or *NEC*-based program, you will have to spend quite some time if you want to optimize a design for a given parameter (whether that is gain, F/B or maybe impedance or SWR bandwidth). For this kind of application *optimizing programs* can be quite helpful. *YO* (Yagi Optimizer) by B. Beezley, K6STI, was the first optimizing program around, but it works only for monoband Yagi antennas. *YO* can optimize for any of the above mentioned characteristics or any weighted combination of those parameters. *AO* (Antenna Optimizer) is a similar program, but it works for any type of antenna. Both are based on *MININEC* and are no longer available nor supported by the author.

When using an optimizer you must be extremely cautious to keep an eye on all performance parameters, or to weigh different performance issues very carefully. Both *YO* and *AO* use what is called an "unconstrained local optimization." This means that the computer begins adjusting the antenna for user-defined variables until a performance maximum is obtained. The danger here is that the computer might reach a *local* maximum, even though there may be a better solution that remains undiscovered because it is too far from the starting point. Additionally, the computer might run away and give false results that are not physically possible to construct.

To date, the only optimization for *NEC* has been *NEC-OPT*, which was sold by Paragon Technologies. The software was very expensive and was purchased by only a few individuals and companies. Additionally the user interface was very difficult to use and could only be managed by a professional. Due to limited sales, this software is no longer sold.

You should understand that an optimizing algorithm is just a program that works purely on figures. Real optimizing must, to a large degree, come from brain of the antenna designer. You must first have a very good idea about what the final antenna should look like. For example, you must want a 5-element Yagi and then let the optimizer adjust the element lengths and spacings to achieve some desired goals: You cannot just tell the computer to design a "good antenna" for 20 meters, for example!

MultiNEC (**www.qsl.net/ac6la/index.html**) by AC6LA is listed here under optimizing programs because it can automate some repetitive steps in antenna modeling. You can start by building the model from scratch or by importing an existing model in one of several formats. Then you can run multiple test cases while letting the program make small changes to the model between runs. *MultiNEC* can be used by itself if you like, but it really shines when used in conjunction with an existing antenna modeling program. *MultiNEC* works with *EZNEC 3*, *EZNEC-M Pro*, *EZNEC/4 Pro* (*NEC-4* version), *NEC-Win Plus+*, *NEC-Win Pro*, *GNEC*, *Antenna Model* (Teri Software), and *4nec2*. I can import and use the data files from any of these programs and you can subsequently run these files on any (other) program listed above.

4nec2 is a *NEC-2* based program (freeware) that includes an optimization system for both a single frequency and a band of frequencies.

2. THE ON4UN LOW-BAND SOFTWARE

The nice thing about personal computers is that everyone can now handle the difficult mathematics pertaining to antennas and feed lines. All you need to do is understand the question... and the answers. The programs will do the hard mathematics for you and give you answers that you can understand. The theory of antennas and feed lines is not an easy subject.

We have all been brought up to know how much is 5 times 4. But nobody can tell off the top of his head how much $5 - j\,3$ times $12 + j\,12$ is. At least I cannot. When I started studying antennas and wanted not only to understand the theory, but also to be able to calculate things, I was immediately confronted with the problem of complex mathematics. While studying the subject I wrote a number of small computer programs to do complex-number calculations. They have since evolved to quite comprehensive engineering tools that should be part of the software library of every serious antenna builder. The software is on the CD in this book.

The NEW LOW BAND SOFTWARE is based on the original "Low Band DXing Software" I wrote in the mid 1980s, while preparing the original *Low Band DXing* book. The latest software (from the mid 1990s) is a very much

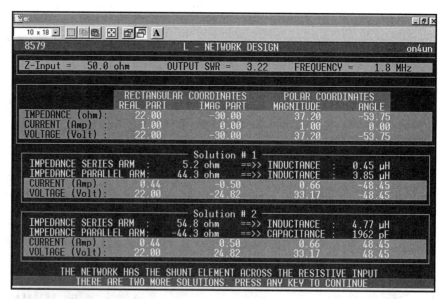

Fig 4-2—Screen capture of the L-network module of the ON4UN NEW LOW BAND SOFTWARE. This module is used very extensively in Chapter 11 for arrays. All relevant data (Impedance, Current and Voltage) are shown in both Cartesian (a + jb) as well as polar coordinates (A∠b°).

enhanced and much more user friendly. I wrote it under DOS using Q-Basic and it runs well in a DOS box on modern machines, even operating under Windows XP.

Each of the modules starts with a complete on-screen introduction, telling what the software is meant to do and how to use it. All propagation-related programs are integrated into a single module. There are many help screens in each of the modules.

2.1. Propagation Software

The propagation software module is covered in detail in Chapter 2. The module contains a low-band dedicated sunrise/sunset program and a gray-line program, based on a comprehensive database containing coordinates for over 550 locations, and which can be user changed or updated. The database can contain up to 750 locations.

2.2. Mutual Impedance and Driving Impedance

From a number of impedance measurements you can calculate the mutual impedance and eventually, knowing the antenna currents (magnitude and phase), you can calculate the driving impedance of each element of an array with up to 4 elements.

2.3. Coax Transformer/Smith Chart

The original software covered only ideal (lossless) cables. Now there are two versions of the program: for lossless cables and for real cables with losses. The real cable program will tell you everything about a feed line. You can analyze the feed line as seen from the generator (transmitter) or from the load (antenna). Impedance, voltage and currents are shown in both rectangular coordinates (real and imaginary parts) or in polar coordinates (magnitude and phase angle). You will see the Z, I and E values at the end of the line, the SWR (at the load and at the generator), as well as the loss—divided into cable loss and SWR loss.

A number of "classic" coaxial feed lines with their transmission parameters (impedance, loss) are part of the program, but you can specify your own cable as well. Try a 200-foot RG-58 feed line on 28 MHz with a 2:1 SWR and compare it to a ¾-inch Hardline with the same length and SWR, and find out for yourself that a "big" coax is not necessarily there just for power reasons. It makes no sense throwing away 2 or 3 dB of signal if you have spent a lot of effort building a top performance antenna. If you are going to design your own array, you will probably use this software module more than any other.

2.4. Impedance, Current and Voltage Along Feed Lines

Again, there are two versions of each module: loss-free and "real" cable.

2.4.1. Z, I and E listings

A coaxial cable, when not operated as a "flat" line (that is, it has an SWR greater than 1:1) acts as a transformer: The impedance, current and voltage are different at each point along the cable. You enter the feed-line data (impedance, attenuation data), the load data (impedance and current or voltage), and the program will display Z, I and E at any point of the cable.

2.4.2. Simultaneous voltage listing along feed lines

This module was written especially as a help for designing a KB8I (now K3LC) feed system for driven arrays. The program lists the voltage along feed lines, allowing the user to find points on the feed lines of individual array elements where the voltages are identical. These are the points where the feed lines can be connected in parallel (see Chapter 11 on arrays). This program is also helpful to see how high the voltage really rises on your feed line with a 4.5:1 SWR, for example.

2.5. Two- and Four-Element Vertical Arrays

These modules take you step-by-step through the theory and practical realization of a 2-element (cardioid) or 4-element (4-square) array, using the W7EL feed system. This tutorial and engineering program uses graphic displays to show the layout of the antenna with all the relevant electrical data. This unique module is extremely valuable if you want to understand arrays and if you want to build your own array with a feed system that really works.

2.6. The L Network

The L network is the most widely used matching network for matching feed lines and antennas. The module gives you all the L-network solutions for a given matching problem. The software also displays voltage and current at the input and output of the network, which can be valuable to assess component ratings in the network.

2.7. Series/Shunt Input L-Network Iteration

This module was written especially for use in the K2BT array-matching system, where L networks are used to provide a desired voltage magnitude at the input of the network, given

Fig 4-3—John, K9DX, using the SHUNT/SERIES impedance network module for designing the feed system of his 9-circle array (see Chapter 11).

an output impedance and output voltage. See Chapter 11 on phased arrays for details.

2.8. Shunt/Series Impedance Network

This is a simplified form of the L network, where a perfect match can be obtained with only a series or a shunt reactive element. It is also used in the modified Lewallen phase-adjusting network with arrays that are not quadrature fed (see Chapter 11 on vertical arrays).

2.9. Line Stretcher (Pi and T)

Line stretchers are constant-impedance transformers that provide a desired voltage phase shift. These networks are used in specific array feed systems (modified Lewallen method) to provide the required phase delay. See Chapter 11 on vertical arrays for details.

2.10. Stub Matching

Stub matching is a very attractive method of feed-line matching. This module facilitates matching a feed line to a load using a single stub placed along the transmission line. It is very handy for making a stub-matching system with an open-wire line feeding a high-impedance load (2000 to 5000 Ω).

2.11. Parallel Impedances (T Junction)

This module calculates the impedance resulting from connecting in parallel a number of impedances—Do you really want to calculate on your calculator what $21 - j\,34$ and $78 + j\,34$ ohms are in parallel?

2.12. SWR Value and SWR Iteration

2.12.1. SWR value

This calculates the SWR (for example, the SWR for a load of $34 - j\,12\,\Omega$ on a 75-Ω line). The mathematics are not complicated, but it's so much faster with the program (and error free!).

2.12.2. SWR iteration

This module was especially developed for use when designing a W1FC feed system for an array (using a hybrid coupler). See Chapter 11 on arrays for details.

2.13. Radiation Angle for Horizontal Antennas

This module calculates and displays the vertical radiation pattern of single or stacked antennas (fed in phase).

2.14. Coil Calculation

With this module you can calculate single-layer coils and toroidal coils. It works in both directions (coil data from required inductance, or inductance from coil data).

2.15. Gamma-Omega and Hairpin Matching

This is a simplified version of one of the modules of the YAGI DESIGN software (see information later in this chapter). Given the impedance of a Yagi and the diameter of the driven element (in the center), you can design and prune a gamma or omega or hairpin matches and see the results as if you were standing on a tower doing all the pruning and tweaking.

2.16. Element Taper

Antennas made of elements with tapering diameters show a different electrical length than if the element diameters had a constant diameter. This module calculates the electrical length of an element (quarter-wave vertical or half-wave dipole) made of sections with a tapering diameter. A modified W2PV tapering algorithm is used.

The NEW LOW BAND SOFTWARE is available from the author. See order form and details in the back of this book.

3. THE ON4UN YAGI DESIGN SOFTWARE

Like the NEW LOW BAND SOFTWARE, these programs were written under DOS using Q-Basic, and have not been changed for well over 10 years, with the exception of correcting 2 errors in two specific modules. The software runs well in a DOS box on a computer running the latest version of Windows, Windows XP. The Yagi software is on the CD.

Together with Roger Vermet, ON6WU, I have written a number of software programs dealing with both the electrical and the mechanical design of monoband Yagis. These programs were used for the Yagi designs presented in Chapter 13.

The 3-element 40-meter Yagi that I have been using since 1989, as well as all my other HF band Yagis, were designed using the YAGI DESIGN software. The 40-meter Yagi was instrumental in setting two all-time 40-meter European records in the 1992 ARRL CW and Phone contests. KS9K (now K4JA), before moving East from one of the top US Midwest contest stations, has been using designs from this software program for his entire antenna farm.

YAGI DESIGN is a multifunctional software package that takes the user through all the aspects of Yagi designing (mechanical as well as electrical). It is not a modeling program, but is based on a comprehensive database containing all the dimensional and performance data for 100 different HF Yagis (2 to 6 elements). The database contains approximately 20 "classic" reference designs by W6SAI, W2PV, N2FB, etc, but the majority are newly designed Yagis. Most of the new designs were verified by either modeling them on a scale frequency (72 MHz) or by making full-size HF-band models.

The YAGI DESIGN database has a Yagi for every application: From low to high-Q, contest, CW only, SSB only, narrow band, wide band, gain optimized, F/B optimized, etc. One of the software modules also allows you to create text (ASCII) input files for the *MN*, *AO* and *YO* modeling programs. This allows you to further change and manipulate any of the designs from the system database.

The mechanical design modules are based on the latest issue of the EIA/TIA-222-E standard, which is a much upgraded version of the older, well known EIA RS-222-C specification. The *cross-flow principle* is used to determine the effect of wind on a Yagi.

YAGI DESIGN consists of several modules, which are briefly described. Each time you leave a module, you can save the results in a work file that you can recall from any other module. You can also view the contents of the work file at any time, using the VIEW DATA FILES module.

3.1. The Analyze Module

Unless you are very familiar with the content of the database, it might take you a long time to browse through all the performance and dimensional data to choose a Yagi design. The main-menu option print database prints out the content of the entire database, either in a tabular format (only the key characteristics) or it can generate a full-blown data sheet for all the Yagis (with two designs per printed page. This represents a little booklet of 50 pages).

In the ANALYZE module you can specify some key characteristics, such as boom length (expressed in either wavelengths, feet or meters), minimum gain, minimum F/B, maximum Q factor, etc. The software will automatically select the designs that meet your criteria.

3.2. Generic Dimensions

Select the SELECT DESIGN module. After having chosen a proper design from the system database, the screen will display all the data relevant to this design—gain, F/B, impedance, etc, at the design frequency and 6 other frequencies spread up to ±1.5% of the design frequency.

You must now enter the design frequency (such as, 14.2 MHz). The screen now displays all the generic dimensions of the Yagi for the chosen design frequency. "Generic" means that the element lengths given in inches as well as centimeters are valid for an element diameter-to-wavelength ratio of 0.0010527. These are not the dimensions we will actually use to construct the Yagi, since the element will be made of tapered sections. The screen display also shows the amount of reactance that the driven element exhibits at the design frequency. The element positions along the boom are those that will be used in the final physical design.

3.3. Element Strength

Before we calculate the actual lengths of Yagi elements with tapering sections, we must first see which taper we might use. What are the required diameters and taper schedule that will provide the required strength at minimal cost, weight and element sag?

The ELEMENT STRENGTH module helps you build elements of the required strength at the minimum weight. Up to 9 sections of varying diameters can be specified (that's enough sections even for an 80-meter Yagi). Given the lengths (and overlap) of the different sections and the wall thickness entered from the keyboard, the program calculates the bending moments at the critical point of every section. The module lets you specify wind speeds and ice loading as well as a vibration-suppression internal rope and several types of aluminum material.

3.4. Element Taper

It's time now to calculate the exact length of the tapered elements. We follow the taper schedule we obtained with the ELEMENT STRENGTH module. An improved version of the well-known W2PV algorithm is used to calculate the exact length. A wide range of boom-to-element clamps (flat, square, L, rectangular, etc) can be specified. These clamps influence the eventual length of the tapered elements.

3.5. Mechanical Yagi Balance

This mechanical design module performs the following tasks.

3.5.1 Boom strength

This calculates the required boom diameter and wall thickness. An external sleeve (or internal coupler) can be defined to strengthen the central part of the boom. If the boom is split in the center, the sleeve or the coupler will have to take the entire bending moment. Material stresses at the boom-to-mast plate are displayed. Any of the dimensional inputs can be changed from the keyboard, resulting in an instantaneous display of the changed stress values.

3.5.2 Weight balance

Many of the newer computer-optimized Yagis have non-constant element spacing, and hence the weight is not distributed evenly along the two boom halves. The WEIGHT BALANCE module shifts the mast plate (attachment point) on the boom until a perfect weight balance is achieved. It is nice to have a weight-balanced Yagi when laboring to mount it on the mast!

3.5.3 Yagi wind load

This program calculates the angle at which the wind area and wind load are largest. In most literature the wind area and wind load are specified for a wind angle of 45° (wind blowing at a 0° angle blows along the boom; at 90° it blows right onto the boom). This is incorrect, because the largest wind load always occurs either with the boom broadside to the wind or with the elements broadside to the wind. With large low-band antennas, it is likely that the elements broadside to the wind produces the largest wind area. With higher-frequency long-boom Yagis having many elements (eg, a 5- or 6-element 10- or 15-meter Yagi), the boom is likely to produce more thrust than the elements. The wind load is calculated in increments of 5°, given a user-specified wind speed.

3.5.4. Torque balancing

Torque balance ensures that the wind does not induce any undue torque on the mast. This can only be achieved by a symmetrical boom moment. When the boom-to-mast plate is not at the center of the boom, a "boom dummy" will have to be installed to compensate for the different wind area between the two boom halves. The program calculates the area and the position of the boom dummy, if required.

3.6. Yagi Wind Area

Specifying the wind area of a Yagi is often a subject of great confusion. Wind thrust is generated by the wind hitting a surface that is exposed to that wind. The force is the product of the dynamic wind pressure multiplied by the exposed area, and with a so-called drag coefficient, which is related to the shape of the exposed body. The "resistance" to wind of a flat body is obviously different from the resistance of a round-shaped body. This means that if we specify or calculate the wind area of a Yagi, we must always specify the equivalent wind area for a flat plate (which should be the standard) or if the area is simply the sum of the projected areas of all the elements (or the boom). In the former case we must use a drag coefficient of 2.0 according to the latest EIA/TIA-222-E standard, while for (long and slender) tubes a coefficient of 1.2 is applicable. This means that for a Yagi consisting only of tubular elements, the flat-plate wind area will be 66.6% lower (2.0/1.2) than the round-element wind area. The WIND AREA module calculates both the flat-plate wind area and the round-element wind area of a Yagi.

3.7. Matching

The software provides three widely used matching systems: gamma, omega and hairpin. When choosing the gamma or omega system, you will be asked to enter the antenna power, as the program will calculate the voltage across and current through the capacitor(s) used in the system. If no match can be found with a given element length and diameter as well as gamma (omega) rod diameter and spacing (eg, very low radiation resistance and not enough capacitive reactance), then the program allows you to change the physical dimensions of the components (diameter of rod and rod-to-element spacing in order to change the system step-up ratio) or to shorten the element length to introduce some capacitive feed-point reactance. In all cases a match will be found.

With a hairpin match the procedure is even simpler. The program will tell you exactly how much you will have to shorten the driven element (from the length shown in the table under "generic dimensions") and how long the hairpin should be.

The program also computes the match for the matching components chosen over a total frequency range from ±1.5% of the design frequency, in 0.5% steps. These includes antenna impedance before matching, antenna impedance after matching and the SWR after matching.

3.8. Optimize Gamma/Omega

Maybe you would like to see if other dimensions (lengths, spacings, diameters) of your gamma (omega) system would result in more favorable matching-system components? Maybe you would like to "balance" the SWR curve? Most Yagis exhibit an asymmetric SWR curve, which means that the SWR rises faster above the design frequency than below. If you want to have the same SWR values on both band ends, it is obvious that the SWR cannot be 1:1 at the center frequency. The OPTIMIZE GAMMA/OMEGA module allows you to change any of the matching-system variables to see how the output impedance and the SWR change. You can also change from gamma to omega and vice versa. Changing the variables from the keyboard simulates tuning the Yagi in practice. The module is also very well suited for balancing the SWR over a given frequency range.

3.9. Feed-Line Analysis

When designing a Yagi, you must have a look at the feed line as well. It makes no sense to build an optimized long Yagi, where every inch of metal in the air contributes to gain (and F/B) and then to throw half of the boom length away by using a mediocre, lossy feed line.

The FEED LINE ANALYSIS module assesses the performance of the feed line when connected to the Yagi under design. The characteristics of the most current 50-Ω coaxial cables are part of the software (from RG-58 to $^7/_8$-inch Hardline), but you may specify your own (exotic) cable as well.

3.10. Rotating Mast Calculation

A weak point in many Yagi installations is the rotating mast. The MAST module calculates the stresses in the rotating mast for a mast holding up to ten stacked antennas.

3.11. Utilities
3.11.1. Make input files for YO, MN or AO

The popular Yagi modeling programs *YO* (Yagi Optimizer), *MN* (MININEC) and *AO* (Antenna Optimizer) by Beezley (K6STI) require input text files. The YAGI DESIGN software package contains a program, *FILE.EXE*, that automatically creates a text input file in the correct format for *YO*, *MN* or *AO*.

In the case of *MN* you can also specify a stack of two antennas that are identical (and fed in phase), or different (eg, a 15-meter and a 10-meter Yagi). In this way you can model any of the 100 designs of the database in either *YO* or *MN* without having to retype into text-input files where you're bound to make typing errors.

3.11.2. Your own database

If you'd like to add your own designs, the software package has provided an empty database that can contain up to 100 records (Yagis). The OWNDATA module is used to enter all the dimensional and performance data in the database.

The NEW YAGI DESIGN SOFTWARE is available from the author. See order form and details in the back of this book.

4. *PROFESSIONAL RF NETWORK DESIGNER* BY KM5KG

Grant Bingeman, KM5KG, is a professional broadcast-antenna engineer, who wrote a series of what we could call *utility programs*, similar to those in my software packages NEW LOW BAND SOFTWARE and YAGI DESIGN. These program can greatly ease some of the tedium of RF and antenna system design. *Professional RF Network Designer* is a versatile Windows program.

There are two versions of the program, an amateur and a professional version. You can obtain either through antenneX (**www.antennex.com/shopping.htm**). A free trial version

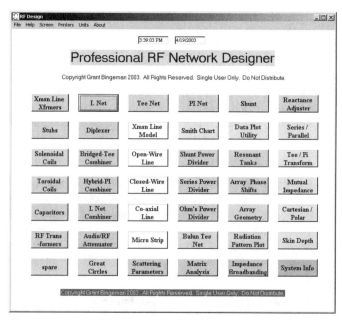

Fig 4-4—Opening screen of the *Professional RF Networks Designer* program by Grant Bingeman, KM5KG.

can be downloaded from **www.antennex.com/Sshack/rfnw/rfnw.html**. The same website has a very complete description of the program, so there is no need to repeat it in this publication.

With permission, I quote a short review by L.B. Cebik, W4RNL

"The buttons on the main screen are color coded by groups of related calculation sets. On the left are component calculations. The individual entries are unusually complete. For example, the capacitor entry not only provides calculations for standard 2-plate capacitors, but also concentric tubing capacitors as well. If you have never explored the relative frequency sensitivity of these two capacitor types, running some values over a large frequency span can be instructive. The middle of the upper-most row covers basic networks, while column 2 (counting from the left) provides entry into combiners and diplexers. The third column permits the user to custom design or analyze most forms of common transmission line configurations. The remaining columns below the top row provide an array of useful utilities, including Smith Chart analysis, Cartesian-to-polar (and back) conversions, and series/parallel tank circuit equivalencies

The individual calculation sets do not limit themselves to ideal lossless cases, but include all standard loss calculations as part of each exercise. There are a few special features worth noting in individual modules.

In all, Professional RF Network Designer *is a very useful tool for anyone designing or analyzing RF and antenna system circuitry, whether professional or amateur. Indeed, it is about the best of such tools that I have so far had a chance to sample or own. I highly recommend it. More importantly, I highly recommend that every purchaser spend a good bit of time with the program, sampling not only what features are available, but as well how networks operate. It only takes a systematic variation of the input variables of any module to acquire an appreciation and reasonable expectation for network variations with changing conditions. After this self-education will emerge a host of applications that we might not have previously imagined possible"*

CHAPTER 5

Antennas: General, Terms, Definitions

Lew Gordon, K4VX, needs no introduction to antenna designers and builders, nor to the contest community. I first met Lew through his excellent and still-popular *YagiMax* modeling software. A few years ago, when I evolved from an avid low-band DXer into an even more avid contester, Lew's contesting multi-op station in Missouri was an outstanding example of station and antenna design. It ranked with stations like W3LPL and K3LR. When I met Lew for the first time during WRTC (World Radio Team Championship) in San Francisco in the summer of 1996, I met a fine gentleman. When I asked Lew to godfather a few chapters of my new book, he immediately and enthusiastically accepted. Lew took care of Chapters 5 and 6 in this new book.

Lew graduated from Purdue University with a physics major. His professional career was as an RF systems engineer with the US Government. First licensed as W9APY in 1947, he has also held the calls WA4RPK and W4ZCY. Lew's antenna systems near Hannibal, Missouri, utilize a total of ten towers ranging from 50 to 170 feet in height. Although he professes to be mainly a contester instead of a DXer, his DXCC total stands at 349 confirmed.

Thank you, Lew, for your help and encouragement.

Agreeing on terms and definitions is important. Too many technical discussions seem to take place in the tower of Babble. First make sure you speak the same language; then speak. Before we get involved in a debate on what's the best antenna for the low bands (that must be the key question for most), we define what we want an antenna to do for us and how we will measure its performance.

Making antennas for the low bands is one area in Amateur Radio where home building can yield results that can substantially outperform most of what can be obtained commercially. All my antennas are homemade. Visitors often ask me, "Where do you buy the parts?" Or, "Do you have a machine shop to do all the mechanical work?" Very often I don't buy parts. And no, I don't have a machine shop, just run-of-the-mill hand tools. But my friends who are antenna builders and I keep our eyes open all the time for goodies that might be useful for our next antenna project. There is a very active swap activity between us. Among friends we have access to certain facilities that make antenna building easier. It's almost like we are a team, where each one of us has his own specialty.

Don't look at low-band antenna designing and building as a "kit project." You need some know-how, a good deal of imagination and inventiveness and often some organizational talent. But unlike the area of receivers and transmitters, where we homebuilders do not usually have access to custom-designed integrated circuits and other very specialized parts, we can build antennas and antenna systems using materials found locally.

A number of successful major low-band antennas are described in this book. These are not meant to be kits with step-by-step instructions, but are there to stimulate thinking and to put the newcomer to antenna building on the right track.

The antenna chapters of *Low-Band DXing* emphasize typical aspects of low-band antennas, and explain how and why some of the popular antennas work and what we can do to get the best results, given typical constraints. *The ARRL Antenna Book* (Ref 697) contains a wealth of excellent and accurate information on antennas.

1. THE PURPOSE OF AN ANTENNA
1.1. Transmitting Antennas

A transmitting antenna should radiate all the RF energy supplied to it in the desired direction, at the required elevation angle (directivity). We want to be loud; the issue is *gain*. We can do this by concentrating our RF in a given direction (in both the vertical and the horizontal planes).

1.1.1. Wanted direction
1.1.1.1 Horizontal directivity.

We learned in Chapter 1 (Propagation) that on the low bands, paths quite frequently deviate from the theoretical great-circle direction. This is especially so for paths going through or very near the auroral oval (such as West Coast or Mid-West USA to Europe). This is a fact we have to take into consideration for a fixed-direction antenna. For paths near the antipodes, signal direction can change as much as 180° (with every direction in-between) depending on the season. All this must be taken into account when designing an antenna system. Rotary systems, of course, provide the ultimate in flexibility so far as horizontal directivity is concerned.

I want to emphasize that the term *horizontal directivity* is really meaningless without further definition. Azimuthal directivity at a takeoff angle of 0° (perfectly parallel to the horizon) is of very little use, since practical antennas produce very little signal at a 0° wave angle over real ground. This issue is important when designing or modeling an antenna. It would be ideal to design an antenna that concentrates transmitted energy at a relatively low angle, while exhibiting the highest rejection off the back at a much higher angle (to achieve maximum rejection of stronger local signals, which as a rule come in at a much higher wave angle. Horizontal directivity should always be specified at a given elevation angle. An antenna can have quite different azimuthal directional properties at different elevation angles.

We will see further that a very low dipole radiates most of its energy directly overhead at 90° (zenith angle), and shows no directivity at high wave angles (60° to 90°). The same antenna, at the same height, shows a pronounced directivity (hardly any signal off the ends of the dipole) at very low wave angles, but hardly radiates at all at very low elevation angles. These issues must be very clear in our minds if we want to understand radiation patterns of antennas.

1.1.1.2. Vertical directivity

In the last few years a lot of modeling has been done using various propagation software packages. At ARRL HQ, D. Straw, N6BV, used *IONCAP* (Ionospheric Propagation Analysis and Prediction System) and *VOACAP* (a version of *IONCAP* upgraded by the Voice of America) to calculate elevation angles for various paths on the different amateur bands. *IONCAP* is based on a mass of propagation data collected over more than 35 years. **Table 5-1** shows the distribution of elevation angles on 40 and 80 meters for some typical DX paths, as does **Fig 5-1** in graphical form. This elevation-angle statistical information is derived from the data on the CD-ROM included with the 20th Edition of *The ARRL Antenna Book* (Ref 697).

Just after the 3rd Edition of *ON4UN's Low-Band DXing* went to press in 1999, N6BV discovered a bug in his elevation-statistics parsing software. Besides fixing the bug (which tended to emphasize medium-angle elevation angles), N6BV also elected to standardize on isotropic antennas (instead of dipoles and Yagis) in *VOACAP* so that the full range of possible elevation angles could be explored. This was done even though the lower angles (such as a 1° takeoff angle) would be very difficult to achieve with most real-world antennas (Ref 182).

The net result is that the range of elevation angles shown

Table 5-1

Range of Radiation Angles for 40 and 80 Meters for Various Paths

The values are averages across the complete sunspot cycle and across the seasons. The value between parentheses is the most common radiation angle (peak value in the distribution).

From	Path to	40 Meters	80 Meters
W. Europe (Belgium)	Southern Africa	1-18 (5)	1-17 (5)
	Japan	1-19 (3)	2-17 (3)
	Oceania	1-4 (1)	No Data
	South Asia	1-17 (4)	3-5 (4)
	USA (W1-W6)	2-33 (5)	1-35 (4)
	South America	1-17 (1)	1-12 (1)
USA East Coast	Southern Africa	1-16 (3)	3-4 (4)
	Japan	1-15 (1)	1-12 (5)
	Oceania	1-9 (1)	No Data
	South Asia	1-9 (1)	No Data
	South America	1-23 (5)	1-21 (10)
	Europe	1-38 (6)	1-31 (13)
USA Midwest	Southern Africa	1-8 (4)	No Data
	Japan	1-17 (2)	1-17 (1)
	Oceania	1-12 (3)	No Data
	South Asia	No Data	No Data
	South America	2-21 (4)	1-16 (4)
	Europe	1-29 (1)	1-34 (13)
USA West Coast	Southern Africa	1-4 (1)	No Data
	Japan	1-27 (5)	2-27 (10)
	Oceania	1-17 (2)	No Data
	South Asia	1-16 (4)	No Data
	South America	1-16 (6)	1-8 (1)
	Europe	1-21 (5)	1-23 (4)

in Table 5-1 and Fig 5-1 are now generally lower than the values shown in the 3rd Edition of *Low-Band DXing*. "No Data" means that there are no data available from the model. This does *not* mean that there is *no* possibility of propagation. On 80 and 40 meters propagation is possible from any point of the world to any other point of the world—given the right moment of the year and the right time of the day, under good propagation conditions —even though such propagation may not be statistically "significant." After all, low-band hams thrive on adversity and they love to pursue openings that are not shown in the statistics!

The elevation-angle distributions are based on statistical figures for various levels of solar activity over an entire solar cycle, and for various times and months. These distributions assume undisturbed geomagnetic conditions. There is anecdotal evidence that the prevailing elevation angles go higher during disturbed conditions. You will note that there is no statistical information for 160 meters, mainly because *IONCAP* and its derivatives do not explicitly take into account the Earth's magnetic field, which is crucially important on Top Band.

Because of the use of isotropic radiators in *IONCAP*, the range of elevation angles is limited only by the all the propagation "possibilities" and *not* by the antenna used at either the transmitting or receiving site. In other words, the charts assume a hypothetical antenna transmits and receives equally well at a 1° wave angle as it does at 10°, 20° or 30° angles. An isotropic antenna, of course, does not actually exist, although

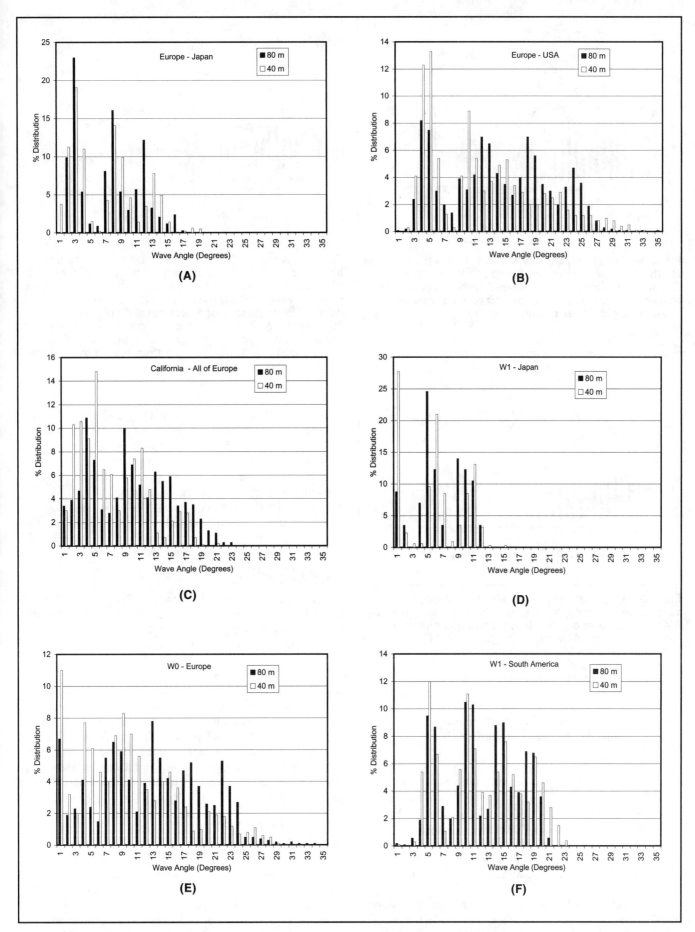

Fig 5-1—Distribution of wave angles (elevation angles) for a few common paths on 80 and 40 meters. Notice that the distribution is not a Gaussian one. This is because many mechanisms are involved that are totally unrelated.

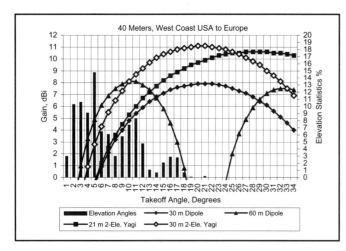

Fig 5-2—The statistical distribution of elevation angles for the 40-meter path from Europe to the US West Coast (San Francisco), compared with the elevation responses for horizontally polarized antennas at several heights over flat ground.

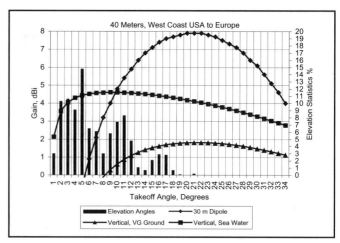

Fig 5-4—A comparison of horizontal versus vertical antennas for the 40-meter path from Europe to the US West Coast. At very low angles (less than about 10°) a quarter-wave vertical over saltwater would have a decided advantage over a horizontal dipole that is 30 meters high over flat ground. A quarter-wave vertical mounted over "very good" ground (typical of farmland in Belgium) would be stronger than the 30-meter high dipole at elevation angles below about 4°.

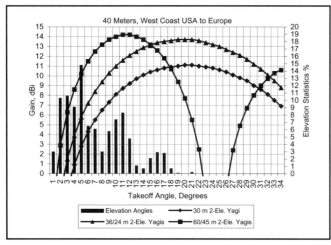

Fig 5-3—A comparison of the elevation responses versus elevation-angle statistics for the same path as Fig 5-2, but for more ambitious antennas mounted over flat ground. At very low angles (2° to 4°), you gain approximately 8 dB going from a single 30-meter high 2-element Yagi to a very high stack of identical Yagis at 60 and 45 meters. Ambitious, indeed!

a vertical over salt water or a high horizontal antenna over a sloping terrain can approach such performance.

1.1.1.3. 40 Meters

Now that we know the range of angles we need to cover, let's have a look at how we could do this. Wave angles of 1° to 20° (except for the path from the US East Coast to Europe, where the range extends to 30°) seem to be most common on 40 meters. Let's analyze how we might achieve this range over *flat terrain*. We'll take a look at three common types of antennas: A dipole, a 2-element Yagi and a λ/4 vertical over average ground.

To work at the lower angles, you need an impressively high horizontal antenna to match the wave angle distribution. In **Fig 5-2**, only the 60-meter high dipole comes relatively close to matching the statistics for the path from the US West Coast to all of Europe on 40 meters. **Fig 5-3** shows even better matches, but look at the heights involved. The stack of 2-element Yagis at 45 and 60 meters is at least 12 dB better than our 30-meter high dipole for wave angles of 5° and less!

What about verticals? **Fig 5-4** shows a single quarter-wave vertical, over very good ground (such as at ON4UN) with 100 λ/4 radials. This is still a poor match to the wave-angle distribution. Now, place that same vertical over salt water and see what happens. An almost perfect match results, even better than the stack of 2-element Yagis at 45 and 60 meters!

1.1.1.4. 80 Meters

Let's have a look at 80 meters. From Table 5-1 and Fig 5-1 you can see there is little difference in the overall range of elevation angles between 40 and 80 meters. Let us analyze the US East Coast to Europe path, where elevation angles extend up to approximately 35° on 80 meters.

The horizontal dipoles in **Fig 5-5** are relatively poor performers for this range of elevation angles, even for antennas at a height of 45 meters! Only a giant 2-element 80-meter Yagi at that height covers the low wave angles reasonably well down to about 5°. On this band verticals fare much better than on 40 meters (**Fig 5-6**). The single vertical over very good ground is better than the horizontal dipole at 24 meters, and covers the elevation angles almost as well as the 2-element Yagi at 45 meters. The λ/4 vertical over salt-water is unbeatable!

1.1.1.5. 160 meters

On Top Band most of us have the choice between an antenna that shoots straight up (a horizontal dipole or inverted-V dipole even at 30 meters in height will produce a 90° takeoff angle), and a vertical (it may be shortened or in the form of an inverted-L or T-antenna) that produces a good low radiation angle (20° to 40° depending on the ground quality). This means we have little chance to experience the differences in signal strength between different radiation angles. The

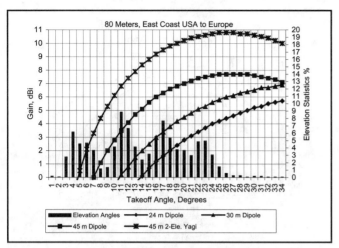

Fig 5-5— A comparison of the elevation responses versus elevation-angle statistics for the 80-meter path from Washington, DC, on the US East Coast, to Europe. Note the response for a gigantic 2-element 80-meter Yagi at 45 meters, a truly heroic antenna!

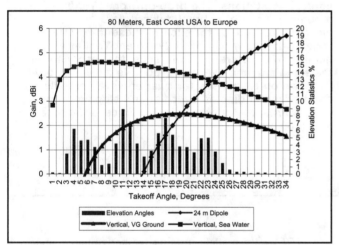

Fig 5-6—A comparison of horizontal and vertical antennas on 80 meters from the US East Coast to Europe. A quarter-wave vertical over saltwater is virtually unbeatable for angles lower than about 20°.

generated using a mathematical model, based on long-term observed propagation data. The wave angles are averages over many sunspot cycles, throughout the different seasons of the years and throughout the night (darkness path). Looking at the California-to-Europe angle distribution on 80 meters, we see that there is 3% chance that the angle is 1° and 1% chance as well that the angle is 20°. But what will the exact wave angle be tonight? The models give us good insight on the range of what is possible. They do not tell us anything about "when" a particular angle will occur. Fortunately our real-life antennas are not radiating at just one wave angle, but rather over a range of angles. The trick is to have an antenna or antennas where the range of actual radiating angles matches the range of statistically available wave angles as closely as possible. That way you cover all the possibilities.

What we also learn from the model is that propagation angles above 35° are rarely present on 40 and 80 meters under normal geomagnetic conditions, and that there is usually some sort of mechanism that supports propagation at very low wave angles. Does this come as a surprise? No. We all have heard, time after time, that vertical antennas on the beach radiating over salt water produce astonishingly strong signals on the low bands (Ref 183). There is also a lot of evidence about high-angle propagation near sunrise/sunset, where a high wave angle appears to be required to initiate ducting (see Chapter 1). Such "anomalies," which are by definition of short duration, are not included in the statistical data on which *IONCAP* and *VOACAP* are based.

1.1.2. The influence of sloping terrain

Where I live in Belgium, it's really, really flat. About 65 km from the coast, my QTH is 30 meters above sea level. It's flat as a pancake! But many low-band DXers live in hill country or even on mountaintops. It's not only saltwater locations that can do wonders—A mountaintop with the right slope and the right type of terrain pattern in the far field can also work wonders. In the mid 1980s I wrote a simple software program that could evaluate simple sloping terrains. That program is still part of the YAGI DESIGN SOFTWARE (see Chapter 4). Years later K6STI developed *TA* (Terrain Analysis) and N6BV developed *YT* (Yagi Terrain analysis) that ray trace over complex terrain using diffraction methods. The 20th Edition of *The ARRL Antenna Book* (Ref 697) now includes a full-blown Windows program called *HFTA* (High Frequency Terrain Analysis) by N6BV.

Let's have a look at some 40 and 80-meter antennas on "hilly terrains." **Fig 5-7A** shows the terrain for several prominent contest and DX stations. K1KI's QTH in Connecticut has a gentle slope that drops about 12 meters over the first 300 meters distance from the tower towards Europe. The impact of this downslope is nevertheless quite substantial and low takeoff angles are covered much better than over flat terrain, as shown in Fig 5-7B.

When he was in New Hampshire, N6BV's terrain sloped down 20 meters in the first 300 meters from the tower base and this too yielded a good improvement at low angles. The third example is the spectacular mountaintop QTH of YT6A, which features a very steep slope of almost 600 meters all the way down to the sea, some 2800 meters from his tower. The low-angle fill-in is quite spectacular! Ranko can hear signals arriving at 1° some 3 or 4 S units better than I can from my flat-terrain QTH.

antenna with a low wave angle would be the best in maybe 99% of the cases. Again, there are (even more than on 80 meters) exceptional cases where a high radiation angle is required to launch into a ducting mechanism, which around sunset or sunrise can produce much stronger signals than can be achieved using a low wave-angle antenna at these times.

Even if you have a very high horizontally polarized antenna (as does Tom, W8JI, with his 100-meter high inverted V), this does not mean it will perform as well as a vertical on 160 meters. The reason for that is explained in Chapter 1 (Section 3.4 and 3.5). Tom confirms that his very high dipole almost never equals his 4-square array, which uses quarter-wave verticals. The suspected mechanism only applies to 160 meters, because of the proximity of 1.8 MHz to the electron gyro frequency.

1.1.1.6. Conclusion, elevation angles

For 40 and 80 meters we have elevation-angle statistics

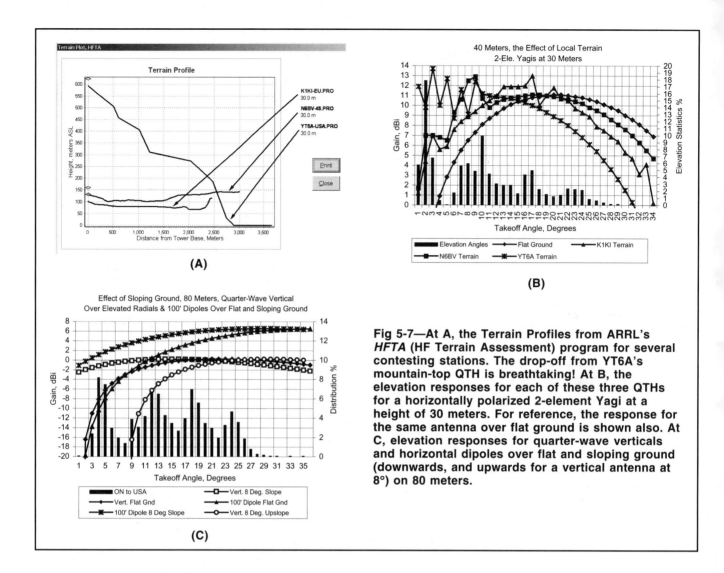

Fig 5-7—At A, the Terrain Profiles from ARRL's *HFTA* (HF Terrain Assessment) program for several contesting stations. The drop-off from YT6A's mountain-top QTH is breathtaking! At B, the elevation responses for each of these three QTHs for a horizontally polarized 2-element Yagi at a height of 30 meters. For reference, the response for the same antenna over flat ground is shown also. At C, elevation responses for quarter-wave verticals and horizontal dipoles over flat and sloping ground (downwards, and upwards for a vertical antenna at 8°) on 80 meters.

Fig 5-7C compares quarter-wave verticals with horizontal dipoles on 80 meters. In each case the terrain is either flat ground or ground with an 8° downslope, which is close to the YT6A terrain. A downslope in the direction of interest can materially aid low elevation angles for verticals as well as horizontals. One trace in Fig 5-7C is for a vertical antenna with ground sloping upwards at 8°, effectively blocking really low takeoff angles.

If you do live in a hilly country, you really should use terrain-modeling software to see the effects that real-world terrain has on the launch of HF signals into the ionosphere. You will have to make terrain data files for your particular QTH for all directions of interest. You can do this manually: Buy a detailed paper topographic map and note the terrain height at intervals (usually corresponding to the height contours on the topo map) along the direction of interest. You can also use *MicroDEM* (**www.nadn.navy.mil/Users/oceano/pguth/website/microdem.htm**), a sophisticated mapping and terrain-profiling program from the US Naval Academy. *MicroDEM* is supplied on the CD-ROM that comes with the 20th Edition of *The ARRL Antenna Book*. For the USA you can get the required electronic topographic maps from the USGS (US Geologic Survey) at **seamless.usgs.gov/**. All details can be found in the exhaustive manual that comes with the *HFTA* software program.

HFTA only models terrains for horizontally polarized antennas (for dipoles and 2 to 8-element Yagis). If you want to include a vertical over flat ground, you can model it (for example, with *EZNEC*). This is how the vertical patterns in Figs 5-4 and 5-6 were made.

1.2. Receiving Antennas

For a receiving antenna, the requirements are very different on the lower bands (80 and 160 meters). We expect the antenna to receive only signals from a given direction and at a given wave angle (directivity), and we expect the antenna to produce signals that are substantially stronger than the internally generated noise of the receiver, taking into account losses in matching networks and feeders. This means that the efficiency (see Section 2.5) of a receiving antenna is really not a requirement. The important asset of a good receiving antenna system is its *directivity*—the ability to be aimed in desired directions and to be switched in different directions rapidly. The ability to direct a null in a particular direction is often crucial.

In most amateur applications on the higher bands the transmitting antenna is used as the receiving antenna, and the transmitting requirements of the antenna outweigh typical receiving requirements. On the low bands, however, successful DXers most often use specialized receiving antennas, as we will see in Chapter 7 on Special Receiving Antennas. This is because most hams cannot build very directive (and efficient) transmit antennas, which are very large. It is possible, however, to build very effective directive receiving antennas that have poor efficiency, making them unsuitable for transmitting.

2. DEFINITIONS
2.1. The Isotropic Antenna

An *isotropic* antenna is a theoretical antenna of infinitely small dimensions that radiates equally well in all directions. This concept can be illustrated by a tiny light bulb placed in the center of a large sphere (see **Fig 5-8**). The lamp illuminates

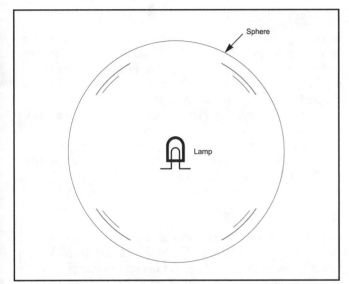

Fig 5-8—In this drawing the isotropic antenna is simulated by a small lamp in the center of a large sphere. The lamp illuminates the sphere equally well at all points.

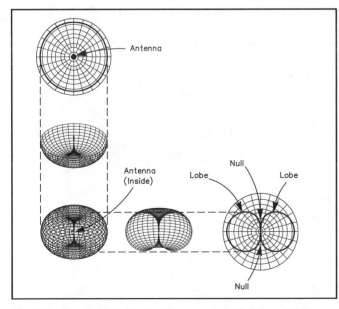

Fig 5-9—Vertical (left) and horizontal (right) radiation patterns as developed from the three-dimensional pattern of a horizontal dipole.

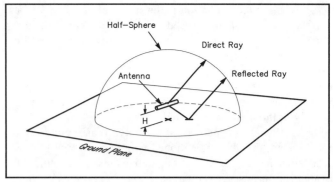

Fig 5-10—The effect of ground is simulated in a sphere by putting a plate (the reflecting ground plane) through the center of the sphere. Since the power in the antenna is now radiated in half the sphere's volume, the total radiated field in the half sphere is doubled. The ground reflection can add up to 6 dB of signal increase compared to free space. A smaller total gain is caused in practice, since part of the RF energy is absorbed in the poorly reflecting, lossy ground.

the interior of the sphere equally at all points. The isotropic antenna is often used as a reference antenna for gain comparison, expressed in decibels over isotropic (dBi). The radiation pattern of an isotropic antenna is a sphere, by definition. A dBi is no more and no less than a convenient abbreviation for power per unit area over the volume of a sphere.

2.2. Antennas in Free Space

Free space is a condition where no ground or any other conductor interacts with the radiation from the antenna. In practice, such conditions are approached only at VHF and UHF, where very high antennas (in wavelengths) are common. Also every real-life antenna has some degree of directivity. If is placed in the center of a large sphere, it will illuminate certain portions better than others. In antenna terms, the antenna radiates energy better in certain directions. A half-wave dipole has maximum radiation at right angles to the wire and minimum radiation off the ends. A half-wave dipole, in free space, has a gain of 2.15 dB over isotropic (2.15 dBi).

Radiation patterns are collections of all points in a given plane, having equal field strength. **Fig 5-9** shows the radiation pattern of a dipole in free space, as seen three dimensions and in two planes, the plane through the wire and the plane perpendicular to the wire.

2.3. Antennas over Ground

In real life, antennas are near the ground. We can best visualize this situation by cutting the sphere in Fig 5-8 in half, with a metal plate going through the center of the sphere. This plate represents the ground, a perfect electrical mirror. **Fig 5-10** shows what happens with an antenna near the ground: Direct and reflected waves combine and illuminate the sphere unequally at different points at different angles. For certain angles the direct and reflected waves are in phase and reinforce one another. The field is doubled, which means a power gain of 3 dB. In addition, we have only a half sphere to illuminate with the same power, and that provides another 3 dB of gain. This means that a dipole over perfect ground will have 6 dB of gain over a dipole in free space.

Over ground, radiation patterns are often identified as vertical (cutting plane perpendicular to the ground) or horizontal (cutting plane parallel to the ground). The latter is of very little use, since practical antennas over real ground produce no signal at a 0° wave angle. The so-called horizontal directivity should in all practical cases be specified as directivity in a plane making a given angle with the horizon, usually at the main takeoff angle.

Low-band antennas always involve real ground. With real ground, the above-mentioned gain of 6 dB will be lowered, since part of the RF is dissipated in the lossy ground. For evaluation purposes, we often specify *perfect ground*, a ground consisting of an infinitely large, perfect reflector.

Real grounds have varying properties, in both conductivity and dielectric constant. In this book, frequent reference will be made to different qualities of real grounds, as shown in **Table 5-2**.

2.4. Radiation Resistance

Radiation resistance (referred to a certain point in an antenna system) is the resistance, which if inserted at that point, would dissipate the same energy as is actually radiated from the antenna. In other words, radiation resistance is the total power radiated as electromagnetic radiation divided by the square of the current at some defined point in the system. This definition does not state where the antenna is being fed, however. There are two common ways of specifying radiation resistance:

- The antenna is fed at the current maximum: $R_{rad\ (I)}$
- The antenna is fed at the base, between the antenna lower end and ground: $R_{rad\ (B)}$

$R_{rad\ (I)} = R_{rad\ (B)}$ for verticals of ½ wavelength or shorter. $R_{rad\ (B)}$ is the radiation resistance used in all efficiency calculations for vertical antennas. Fig 9-10 in Chapter 9 shows the radiation resistance according to both definitions for four types of vertical antennas:

- A short vertical (< 90° high)
- A quarter-wave vertical
- A ³/₈- wave vertical (135° high)
- A ½-wave vertical

Radiation resistance is not the same as the feed-point impedance, since feed-point impedance consists of both radiation resistance and loss resistances, plus any reactance at the feed point.

2.5. Antenna Efficiency

The *antenna efficiency* of an antenna by itself located in free space is simply the ratio of power radiated from that antenna to the power applied to it. Any energy that is not radiated will be converted into heat in the lossy parts of the antenna. For a transmitting antenna, radiation efficiency is an important parameter. The efficiency of an antenna is expressed as follows:

$$\text{Efficiency} = R_{rad} / (R_{rad\ (B)} + R_{loss}) \quad \text{(Eq 1)}$$

where $R_{rad\ (B)}$ is the radiation resistance of the antenna as defined in Section 2.4, and R_{loss} is the total equivalent loss resistance of all elements of the antenna (resistance losses, dielectric losses, loading coils, etc). Loss resistance is normalized to the same point where R_{rad} was defined.

The *total efficiency* of an antenna setup is a rather different story. While antenna efficiency only considers the lossy parts of the antenna itself, total efficiency includes losses in its environment, including the ground. In other words, total efficiency takes into account all losses in the near field as well as in the far field (see Sections 2.6, 2.7 and 2.8).

2.6. Near Field

Depending on the physical dimension of the antenna the *radiating near field* (also called the *Fresnel* field) reaches out typically one or two wavelengths from simple wire antennas to many wavelengths in the case of long-boom Yagis on VHF and UHF. The relationship between magnetic and electric fields is a complex one in the near field. This is one of the reasons that we must not make antenna pattern measurements too close to the antenna. Antennas field-strength measurements should be done no less than a few wavelengths from the antenna.

Table 5-2
Conductivities and Dielectric Constants for Common Types of Earth

Surface Type	Dielectric Constant	Conductivity (S/m)	Relative Quality
Fresh water	80	0.001	
Salt water	81	5.0	Salt Water
Pastoral, low hills, rich soil, typ Dallas, TX, to Lincoln, NE areas	20	0.0303	Very Good
Pastoral, low hills, rich soil typ OH and IL	14	0.01	Good
Flat country, marshy, densely wooded, typ LA near Mississippi River	12	0.0075	
Pastoral, medium hills and forestation, typ MD, PA, NY, (exclusive of mountains and coastline)	13	0.006	
Pastoral, medium hills and forestation, heavy clay soil, typ central VA	13	0.005	Average
Rocky soil, steep hills, typ mountainous	12-14	0.002	Poor
Sandy, dry, flat, coastal	10	0.002	
Cities, industrial areas	5	0.001	Very Poor
Cities, heavy industrial areas, high buildings	3	0.001	Extremely Poor

With low-band antennas the ground will always be in the near field of our antennas, and losses in the near field will have to be considered. These losses will be discussed in detail in Chapter 9.

2.7. Induction Field

The *reactive near field* or *induction field* is a part of the near field, very close to the antenna where mutual coupling exists between conductors. This happens typically within a maximum of 0.5 wavelengths around the antenna.

2.8. Far Field

The *radiating far field* (or *Fraunhöfer* field) is the area around the antenna beyond the near field. This is where ground reflections for low-angle signals occur, which greatly interest us low-band operators. In the far field the power density is inversely proportional to the square of the distance from the antenna. Total energy is equally divided between electric and magnetic fields, and the relation is defined by: $E/H = Z_0 = 377\ \Omega$, the free-space impedance. See Chapter 9 for further discussion of far-field reflection losses.

2.9. Antenna Gain

The *gain* of an antenna is a measure of its ability to concentrate radiated energy in a desired direction (minus any losses in the antenna). Antenna gain is expressed in decibels, abbreviated dB. It tells us how much the antenna in question is better than a reference antenna, under defined circumstances. And that's where we enter the antenna-gain "jungle." Commonly, both the theoretical isotropic, as well as a real-world dipole, are used as reference antennas. In the former case the gain is expressed as dBi and in the latter as dBd.

But that's only part of the story. We can do a comparison in free space, or over perfect ground or over real ground. The only situation that makes a generic comparison possible is to compare antennas in free space. Gain in dBi in free space is what can always be compared; there is no inflation of gain figures by reflection. Very often manufacturers of commercial antennas will calculate gains including ground reflections—and often they will not mention this fact.

You might argue, "Why not use a real antenna, such as a dipole, as a reference, since the isotropic antenna is a theoretical antenna that does not exist, while a half-wave dipole does?" Comparing gains is really comparing the field strength of an antenna under investigation with that of our reference antenna. With an isotropic antenna the situation is clear. It radiates equally well in all directions and the three-dimensional radiation pattern is a sphere. What about the dipole as a reference? The gain of a half-wave, lossless half-wave dipole in free space over an isotropic is 2.15 dBi. But that does not mean that a real dipole has a gain of 2.15 dBi. It only means that the gain of a lossless dipole in free space (that's a theoretical condition as well, because nothing is really in free space) is 2.15 dB over an isotropic radiator. If we put the dipole over a perfect ground, it suddenly shows a gain of 8.15 dBi! You pick up 6 dB by radiating the power in half a hemisphere instead of a whole hemisphere, as in the theoretical case of free space. With less-than-perfect ground, part of the power will be absorbed in the ground and the ground-reflection gain will be less than 6 dB. It is clear that the only generic way of comparing antenna gains is in dBi, using an isotropic antenna as the only generic reference antenna not influenced by height or ground conditions. In this publication we will always quote gain figures in dBi—that is, referenced to an isotropic antenna in free space. (Ref 688).

2.10. Front-to-Back Ratio

Being a ratio (just like gain), we would expect front-to-back ratio to be expressed in decibels, which it is. The front-to-back ratio (F/B) is a measure expressing an antenna's ability to radiate a minimum of energy in the direction directly in the back of the antenna.

Free-space front-to-back ratio is always measured at a 0° wave angle. Over ground the F/B depends on the vertical radiation angle being considered. In most cases a horizontal radiation pattern over real ground is not really the pattern in the horizontal plane, but in a plane that corresponds to the main wave angle. If we look at the back lobe at that angle, it may be okay, but at the same time there may be a significant back lobe at a much different angle.

With the advent and the widespread use of modeling programs, especially some of the optimizer programs, the rat race started for the most ludicrous F/B figure. Let's not forget that mathematics is one thing, while antenna physics is another thing. It is possible to calculate an antenna exhibiting a F/B of 70 dB in a given direction, at a given wave angle. But that's all there is to it. One degree away the rejection may be down 40 or 50 dB. When you understand the physics behind all of this, it will be clear that F/B above a certain level (maybe 35 dB) is rather meaningless.

2.10.1 Geometric front-to-back ratio

In the past, front-to-back ratios were usually defined in the sense of a geometric front-to-back—the radiation 180° directly behind the front (0° lobe) of the antenna. We thus compare the "forward" power at the main forward radiation angle to the "backward" power radiated at the same wave angle in the backward direction.

The pattern of an antenna discriminates against unwanted signals coming from directions other than the front of the antenna. It is very unlikely that unwanted signals will be generated exactly 180° off the beam direction or at a radiation angle that is the same as the main forward lobe's radiation angle. Therefore, geometric F/B can be ruled out immediately as a meaningful way of defining the antenna's ability to discriminate against unwanted signals.

2.10.2 Average front-to-back (integrated front-to-back) ratio

The *average front-to-back ratio* can be defined as the average value of the front-to-back as measured (or computed) over a given back angle (both in the horizontal as well as the vertical plane). In the chapter on special receiving antennas (Chapter 7, Section 1.8 and 1.9) I use this concept for evaluating different antennas.

2.10.3. Worst-case front-to-rear ratio (F/R)

Another meaningful way to quantify the F/B ratio of an antenna is to measure the ratio of the forward power to the power in the "worst" lobe in the entire back of the antenna (from 90° to 270° azimuth). This is the standard used for example in *The ARRL Antenna Book* for Yagis and quads.

2.10.4. Front-to-back ratio and gain

Is there a link between gain and the front-to-back ratio of an antenna? Let's visualize a three-dimensional radiation pattern of a simple Yagi. The front lobe resembles a long stretched pear, while the back lobe (let's assume for the time we have a single back lobe) is a much smaller pear. The antenna sits where the stems of the two pears touch. The volume of the two pears (the total volume of the three-dimensional radiation pattern) is determined only by the power fed to the antenna. If you increase the power, the volume of the large as well as the small pear will increase in the same proportion. Let's take for definition of front-to-back ratio the ratio of the power radiated in the back versus the power radiated in the front. This means that the F/B ratio is proportional to the ratio of the volume of the two pears.

By changing the design of the Yagi (by changing element lengths or element positions), we change the size and the shape of the two pears. But so long as we feed the same power to it, the sum of the volumes of the two pears remains unchanged. It's as if the two pear-shaped bodies are connected with a tube, and are filled with a liquid. By changing the design of the antenna, we merely push liquid from one pear into the other. If the antenna were isotropic, the radiation body would be a sphere having the volume of the sum of the two pears.

Assume we have 100 W of power with 10% of this power applied to the antenna in the back-lobe. The F/B will be 10 × log (10 / 1) = 10 dB. Ninety percent of the applied power is available to produce the forward lobe.

Let's take a second case, where only 0.1% of the applied power is in the back lobe. The F/B ratio will be 10 × log (100 / 0.1) = 30 dB. Now we have 99.9% of the power available in the front lobe.

The antenna gain realized by having 99.9 watts instead of 90 watts in the forward lobe is 10 × log (99.9 / 90) = 0.45 dB. Pruning an antenna with a modest F/B pattern (10 dB) to an exceptional 30-dB value, gives us 0.45 dB more forward gain, provided that the extra liquid is used to lengthen the cone of the big pear.

The mechanism for obtaining gain and F/B is much more complicated than that described above. I am only trying to explain that optimizing an antenna for F/B does not necessarily mean that it will be optimized for gain. What is always true, however, is that a high-gain antenna will have a narrow forward lobe. You cannot concentrate energy in one direction without taking it away from other directions! We will see later that maximum-gain Yagis show a narrow forward lobe, but often a poor front-to-back ratio. This is the case with very high-Q, gain-optimized 3-element Yagis, for example.

Conclusion: There is no simple relationship between front-to-back ratio and gain of an antenna.

2.10.5. The importance of directivity

Directivity can be important for two very different reasons: With *transmit antennas* we want to have directivity because directivity is invariably linked to gain. What you take away in certain directions is added in other directions. We want gain because we want to be heard (to be strong), and that also implies the notion of efficiency.

With *receiving antennas* the story is different. We want to hear well above the noise (manmade, atmospherics, QRM, etc). The issue is one of signal-to-noise ratio, not just signal strength. While antenna efficiency is a secondary issue with receiving antennas, directivity is primary. That's why the concept of quantifying the directivity of an antenna was developed.

2.11. Directivity Merit Figure and Directivity Factor

For a receiving antenna directivity is the main concern. The average front-to-back (the peak forward lobe versus what happens in the back 180° degrees over the entire elevation angle range) gives a good indication of directivity. I used it in the Third Edition of this book to quantify some of the special receiving antennas in Chapter 7. I call the difference between the forward gain (at the desired wave angle, such as 20°) and the average gain in the back of the antenna, the *Directivity Merit Figure* (see Chapter 7, Section 1.10).

Tom, W8JI, (www.w8ji.com/) goes a step further and compares the forward-lobe gain to the average gain of the antenna in all directions (both azimuth and elevation). This figure does tells you not only how good the average front-to-back ratio is, but also how narrow your forward (wanted) lobe is (see Chapter 7, Section 1.11 on Special receiving antennas for more details). His merit figure is called *RDF* (Receiving Directivity Factor).

2.12. Standing-Wave Ratio

SWR is *not* a performance measure of an antenna! SWR is only a measure of how well the feed-point impedance of the antenna is matched to the characteristic impedance of the feed line. If a 50-Ω feed line is terminated in a 50-Ω load, then the impedance at any point on any length of the cable is 50 Ω.

If the same feed line is terminated in an impedance different from 50 Ω, the impedance will vary along the line. The SWR is a measure of the match between the line and the load. Changing the length of a feed line does not change the SWR on the line (apart from minute changes due to feed-line loss with longer lengths of line). What changes is the impedance at the input end of the line.

If changing the line length slightly changes the SWR reading on your SWR meter, then your SWR meter is not measuring correctly (many SWR meters fall into this category) or else you have stray common-mode current flowing on the shield of your feed line. A good test for an SWR meter is to insert short cable lengths between the end of the antenna feed line and the SWR meter (a few feet at a time). If the SWR reading changes significantly, don't expect correct SWR values.

If there are stray currents on the outside of the coaxial cable shield, a change in position on the line can indeed change the SWR reading (see Chapter 6). That's why we use a *balun* (*bal*anced-to-*un*balanced transformer) when feeding balanced feed points with a coaxial cable. In fact, current baluns (choke baluns) are a good idea to install on any coaxial feed line. You can insert a current balun (eg, a short length of coax equipped with a stack of 50 to 100 ferrite cores) at the SWR meter. If this balun changes the SWR value, RF currents are flowing on the outside of the coaxial cable.

Changing the feed-line length doesn't change the performance of the antenna. A feed line is an element that is *not* supposed to radiate. SWR on a feed line has no relation whatsoever to the radiation characteristics of the antenna. A perfect match between the line and the antenna results in a 1:1 SWR. What are the reasons we like a 1:1 SWR or the lowest possible SWR value?

- Showing a convenient 50-Ω impedance: Unless we want to use a transmission line as an impedance transformer, we would like all feed lines to show a 1:1 SWR. This would present the design load impedance of 50 Ω for solid-state transceivers.
- Minimizing losses: All feed lines have inherent losses. This loss is minimal when the feed line is operated as a flat line (SWR = 1:1) and increases when the SWR rises. On the low bands this will seldom be a criterion for desiring a very low SWR, because the nominal losses on the low frequencies are quite negligible, unless very long lengths are used.

For many hams, SWR is the only property they can measure. Measuring gain and F/B with any degree of accuracy is beyond the capability of most. That is why most hams pay attention only to SWR properties. The amount of SWR that can be tolerated on a line depends on:

- Additional loss caused by SWR—determined by the quality of the feed line. A high quality feed line can tolerate more SWR from an additional-loss point of view than a mediocre quality line.
- How much SWR the transceiver or linear amplifier can live with.
- How much power we will run into a line of given physical dimensions (for a given power, a larger coax will withstand a higher SWR without damage than a smaller one).

It must be said that a poor-quality line (a small-diameter cable with high intrinsic losses), when terminated in a load different from its characteristic impedance, will show at its input end a lower SWR value than if a good (low-loss, large-diameter) cable is used. Remember that a very long, poor (having high losses) coaxial cable (whether terminated, open or shorted at the end) will exhibit a 1:1 SWR at the input (a perfect dummy load) because of those losses.

From a practical point of view an SWR limit of 2:1 is usually sought after. From a loss point of view, it is clear that higher values can easily be tolerated on the low bands. Coaxial feed lines used in the feed systems of multi-element low-band arrays sometimes work with an SWR of 10:1!

You can always use an antenna tuner if the SWR is higher than the transceiver or the amplifier needs to work into (usually less than 2:1). Remember that the antenna tuner will not change the SWR on the line itself; it will merely transform the impedance existing at the line input and present the transceiver (linear) with a reasonable and more convenient SWR value. While this approach is valid on the low bands, I strongly suggest not using it on the higher frequencies, since the additional line losses caused by the SWR can become quite significant.

2.13. Bandwidth

The *bandwidth* of an antenna is the difference between the highest and the lowest frequency on which a given property exceeds or meets a given performance mark. This can be gain, front-to-back ratio or SWR. In this book, "bandwidth" refers to SWR bandwidth, unless otherwise specified. In most cases the SWR bandwidth is determined by the 2:1 SWR points on the SWR curve. In this text the SWR limits will be specified when dealing with antenna bandwidths. Many amateurs only think of SWR bandwidth when the term bandwidth is used. In actual practice, the bandwidth can refer to other properties at least as important, if not more important. Consider a dummy load, which has an excellent SWR bandwidth, but a very poor gain figure, since it does not radiate at all!

Bandwidth is an important performance criterion on the low bands. The relative bandwidth of the low bands is large compared to the higher HF bands. Special attention must be given to all bandwidth aspects, not only SWR bandwidth.

2.14. Q-Factor

2.14.1. The tuned circuit equivalent

An antenna can be compared to a tuned LCR circuit. The *Q factor* of an antenna is a measure of the SWR bandwidth of an antenna. The Q factor is directly proportional to the difference in reactance on two frequencies around the frequency of analysis, and inversely proportional to the radiation resistance and relative frequency change.

$$Q = \frac{F_0 \times (X1 - X2)}{2 \times R \times \Delta F} \quad \text{(Eq 2)}$$

where

$X1$ = reactance at the lower frequency
$X2$ = reactance at the higher frequency
R = average value of resistive part of feed-point impedance at frequencies of analysis ($R_{rad} + R_{losses}$)
ΔF = relative frequency change between the higher and the lower frequency of analysis

Example:

F_{low} = 3.5 MHz

F_{high} = 3.6 MHz

F_0 = 3.55

ΔF = 3.6 − 3.5 = 0.1

$R_{feed (Ave)}$ = 50 Ω

$X1$ = −20 Ω

$X2$ = +20 Ω

$$Q = \frac{3.55 \times (20 - (-20))}{2 \times 50 \times 0.1} = 14.2$$

It is clear that a low Q can be obtained through:

- A high value of radiation resistance
- High loss resistance
- A flat reactance curve.

An antenna with a low Q will have a large SWR bandwidth, and an antenna with a high Q will have a narrow SWR bandwidth. Antenna Q factors are used mainly to compare the (SWR) bandwidth characteristics of antennas.

2.14.2. The transmission-line equivalent

A single-conductor antenna (vertical or dipole) with sinusoidal current distribution can be considered as a single-wire transmission line for which a number of calculations can be done, just as for a transmission line.

2.14.2.1 Surge Impedance

The characteristic impedance of the antenna seen as a transmission line is called the *surge impedance* of the antenna. The surge impedance of a vertical is given by:

$$Z_{surge} = 60 \times \ln\left[\frac{4h}{d} - 1\right] \quad \text{(Eq 3)}$$

where
- h = antenna height (length of equivalent transmission line)
- d = antenna diameter (same units).

The surge impedance of a dipole is:

$$Z_{surge} = 276 \times \log\left[\frac{S}{d \times \sqrt{1 + \frac{S}{4h}}}\right] \quad \text{(Eq 4)}$$

where
- S = length of antenna
- d = diameter of antenna
- h = height of antenna above ground.

2.14.2.2 Q-factor

The Q-factor of the transmission-line equivalent of the antenna is given by:

$$Q = \frac{Z_{surge}}{R_{rad} + R_{loss}} \quad \text{(Eq 5)}$$

Example 1:

A 20-meter (66-foot) vertical with OD = 5 cm (1.6 inches), and $R_{rad} + R_{loss} = 45 \ \Omega$.

$$Z_{surge} = 60 \times \ln\left[\frac{4 \times 2000}{5} - 1\right] = 443 \ \Omega$$

Q = 443/45 = 9.8

Example 2:

A 40-meter (131-foot) long dipole, at 20 meters (66 feet) height is made of 2 mm OD wire (AWG 12). The feed-point impedance is 75 Ω.

$$Z_{surge} = 276 \times \log\left[\frac{4000}{0.2 \times \sqrt{1 + \frac{4000}{4 \times 2000}}}\right] = 1163 \ \Omega$$

Q = 1163 / 75 = 16.

CHAPTER 6

The Feed Line and The antenna

The feed line is the necessary link between the antenna and the transmitter/receiver. It may seem odd that I cover feed lines and antenna matching before discussing any type of antenna. I want to make it clear that antenna matching and feeding has no influence on the characteristics or the performance of the antenna itself (unless the matching system and/or feed lines also radiate). Antenna matching is generic, which means that any matching system can, in theory, be used with any antenna. Antenna matching must therefore be treated as a separate subject.

The following topics are covered:
- Coaxial lines; open-wire lines
- Loss mechanisms
- Real need for low SWR
- Quarter-wave transformers
- L networks
- Stub matching
- Wide-band transformers
- 75-Ω feed lines in 50-Ω systems
- Baluns
- Connectors

Before we discuss antennas from a theoretical point of view and describe practical antenna installations, let us analyze what matching the antenna to the feed line really means and how we can do it.

1. PURPOSE OF THE FEED LINE

The feed line transports RF from a source to a load. The most common example is from a transmitter to an antenna. When terminated in a resistor having the same value as its own characteristic impedance, a transmission line operates under ideal circumstances. The line will be *flat*—meaning that there are no standing waves on the line. The value of the impedance will be the same in each point of the line. If the feed line were lossless, the magnitude of the voltage and the current would also be the same along the line. The only thing that would change is the phase angle and that would be directly proportional to the line length. All practical feed lines have losses, however, and the values of current and voltage decrease along the line.

In the real world the feed line will rarely if ever be terminated in a load giving a 1:1 SWR. Since the line is most frequently terminated in a load with a complex impedance, in addition to acting as a transport vehicle for RF, the feed line also acts as a transformer. The impedance (also the voltage and current) will be different at each point along a mismatched line.

Besides transporting energy from the source to the load, feed lines are also used to feed the elements of an antenna array, whereby the characteristics of the feed lines (with SWR) are used to supply current at each element with the required relative magnitude and phase angle. This application is covered in detail in Chapter 11, Vertical Arrays and Chapter 7 (Receiving antennas).

2. FEED LINES WITH SWR

The typical characteristics of a line with SWR are:
- The impedance in every point of the line is different; the line acts as an impedance transformer. (While the impedances in a lossless line repeat themselves every half wavelength, the impedances in a real-world lossy line do not repeat exactly.)
- The voltage and the current at every point on the feed line are different.
- The losses of the line are higher than for a flat line.
- The phase shift in current and voltage is not linearly proportional to the line length. (Line length in degrees does not equal phase shift in degrees, except in very special cases such as for 90° long lines.)

Most transmitters, amplifiers and transceivers are designed to work into a nominal impedance of 50 Ω. Although they will provide a match to a range of impedances that are not too far from the 50-Ω value (eg, within the 2:1 SWR circle on the Smith Chart), it is generally a proof of good engineering and workmanship that an antenna on its design frequency, shows a 1:1 SWR on the feed line. This means that the feed-point impedance of the antenna must be *matched* to the characteristic impedance of the line at the design frequency. The SWR bandwidth of the antenna will be determined in the first place by the Q factor of the antenna, but the bandwidth will be largest if the antenna has been matched to the feed line (1:1 SWR) at a design frequency within that passband, unless special broadband matching techniques are employed. This means we want a low SWR for reasons of convenience: We don't want to be forced to use an antenna tuner between the transmitter and the feed line in order to obtain a match.

2.1. Conjugate Match

A conjugate match is a situation where all the available power is coupled from the transmitter into the line. In a conjugate match with lossless line, the impedance seen looking towards the load (a + jb) at a point in the transmission line is the complex conjugate of that seen looking towards the source (a − jb). A conjugate match is automatically achieved when we adjust the transmitter for maximum power transfer into the line. In transmitters or amplifiers using vacuum tubes, this is done by properly adjusting the common pi or pi-L network. Modern transceivers with fixed-impedance solid-state amplifiers do not have this flexibility, and an external antenna tuner will be required in most cases if the SWR is higher than 1.5:1 or 2:1. Many present-day transceivers have built-in antenna tuners that automatically take care of this situation.

But this is not the main reason for low SWR. The above reason is one of "convenience." The real reason is one of losses or attenuation. A feed line is usually made of two conductors with an insulating material in between. Open-wire feeders and coaxial feed lines are the two most commonly used types of feed lines.

2.2. Coaxial Cable

Coaxial feed lines are by far the most popular type of feed lines in amateur use, for one specific reason: Due to their coaxial (unbalanced) structure, all magnetic fields caused by RF current in the feed line are kept inside the coaxial structure. This means that a coaxial feed line is totally *inert* from the outside, when terminated in an unbalanced load (whether it has SWR or not). An unbalanced load is a load where one of the terminals is grounded. This means you can bury the coax, affix it to the wall, under the carpet, tape it to a steel post or to the tower without in any way upsetting the electrical properties of the feed line. Sharp bending of coax should be avoided, however, to prevent impedance irregularities and permanent displacement of the center conductor caused by cable dielectric heating and induced stresses. A minimum bending radius of five times the cable outside diameter is a good rule of thumb for coaxial cables with a braided shield.

Like anything exposed to the elements, coaxial cables deteriorate with age. Under the influence of heat and ultraviolet light, some of the components of the outer sheath of the coaxial cable can decompose and migrate down through the copper braid into the dielectric material, causing degradation of the cable. Ordinary PVC jackets used on older coaxial cables (RG-8, RG-11) showed migration of the plasticizer into the polyethylene dielectric. Newer types of cable (RG-8A, RG-11A, RG-213 and so on) use non-contaminating sheaths that greatly extend the life of the cable.

Also, coaxial cables love to drink water! Make sure the end connections and the connectors are well sealed. Because of the structure of the braided shield, the interstices between the inner conductor insulation and the outer sheath will literally suck up liters (quarts) of water, even if only a pin hole is present. Once water has penetrated cable with a woven copper shield, it is ruined. Here is one of the big advantages of the larger coaxial cables using expanded polyethylene and a corrugated solid copper outer conductor: Since the polyethylene sticks (bonds) to the copper, water penetration is impossible even if the outer jacket is damaged.

You should check the attenuation of your feed lines at regular intervals. You can easily do this by opening the feed line at the far end. Then feed some power into the line through an accurate SWR meter (such as a Bird wattmeter), and measure the SWR at the input end of the line. A lossless line will show infinite SWR (Ref 1321).

The loss in the cable at the frequency you do the measurement is given by:

$$\text{Loss (dB)} = \log \left[\frac{\text{SWR} + 1}{\text{SWR} - 1} \right] \quad \text{(Eq 1)}$$

The attenuation can also be computed using the graph in **Fig 6-1**. It is difficult to do this test at low frequencies because the low attenuation is such that accurate measurements are difficult. For best measurement accuracy the loss of the cable to be measured should be on the order of 2 to 4 dB (SWR between 2:1 and 4:1). The test frequency can be chosen accordingly. Use a professional type SWR meter such as a Bird wattmeter. Many cheaper SWR meters are inadequate.

It is obvious that this measurement can also be done using done of the popular Antenna Analyzers (eg, MFJ, Autek or AEA). Dave Hachadorian, K6LL, described a variant of the above method: "Plug your antenna into the feed line in the shack and tune it to a frequency where it shows a peak SWR. At this frequency, the antenna, whatever it is, will be a good approximation of an open or short circuit. The frequency will probably NOT be in the ham bands. Start at 30 MHz and work down." The same formula as above and the graph in Fig 6-1 apply to this technique too.

The advantage of K6LL's method is that you can actually

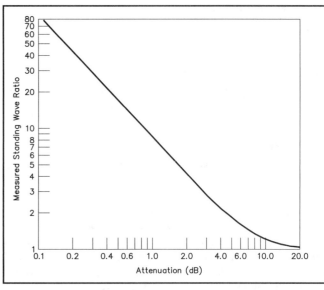

Fig 6-1—Cable loss as a function of SWR measured at the input end of an open or short-circuited feed line. For best accuracy, the SWR should be in the 1:1 to 4:1 range.

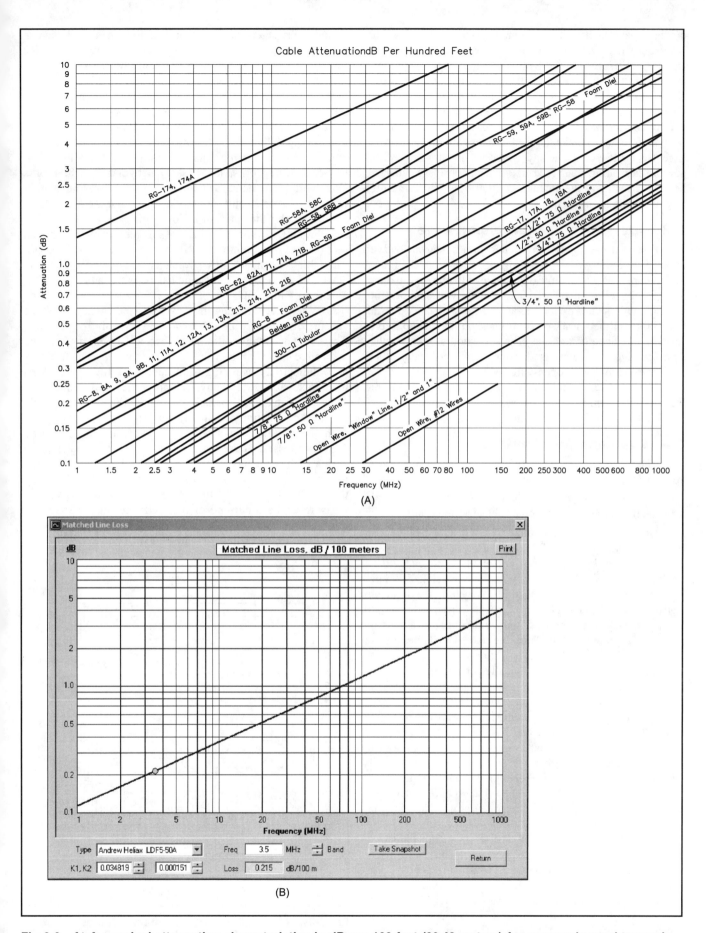

Fig 6-2—At A, nominal attenuation characteristics in dB per 100 feet (30.48 meters) for commonly used transmission lines. (*Courtesy of* The ARRL Antenna Book.) At B, attenuation vs frequency chart generated with AC6LA's *TLDetails* software for Andrews LDF5-50A Heliax. (Note that the attenuation shown here is per 100 meters.)

do the measurement without having to disconnect the feed line at the antenna. Using one of the popular SWR analyzers, you should make sure that the SWR measured at this worst frequency is at least 15:1 (equivalent to a line loss of 0.6 dB). K6LL points out that the impedance of most non-resonant antennas is several thousand ohms (SWR > 40). If the presence of an antenna does degrade the measurement at all, it will be in a direction to make the feed line loss appear higher than it really is. If you have any concerns about whether this method is making your feed line appear too lossy, you will have to disconnect the antenna. At that time you can do the measurement at any frequency.

2.3. Open-Wire Transmission Line

Even when properly terminated in a balanced load, an open-wire feeder will exhibit a strong RF field in the immediate vicinity of the feedline (try a neon bulb close to an open-wire feeder with RF on it!). This means you cannot "fool around" with open-wire feeders as you can with coax. During installation all necessary precautions should be taken to preserve the balance of the line: The line should be kept away from conductive materials. In one word, generally it's a nuisance to work with open-wire feeders!

But apart from this mechanical problem, open-wire feeders outperform coaxial feed lines in all respects on HF (VHF/UHF can be another matter).

2.4. The Loss Mechanism

The intrinsic losses of a feed line (coaxial or open-wire) are caused by two mechanisms:

- Conductor losses (losses in the copper conductors).
- Dielectric losses (losses in the dielectric material).

Dry air is an excellent insulator. From that point of view, an open-wire line is unbeatable. Coaxial feed lines generally use polyethylene as a dielectric, or polyethylene mixed with air (cellular PE or foam PE). Cables with foam or cellular PE have lower losses than cables with solid PE. They have the disadvantage of potentially having less mechanical (impact and pressure) resistance. Cell-flex cables using a solid copper or aluminum outer conductor are the top-of-the line coaxial feed lines used in amateur applications. Sometimes Teflon is used as dielectric material. This material is mechanically very stable and electrically very superior, but very expensive. Teflon-insulated coaxial cables are often used in baluns. (See Section 7.)

Coaxial cables generally come in two impedances: 50 Ω and 75 Ω. For a given cable outer diameter, 75-Ω cable will show the lowest losses. That's why 75 Ω is always used in systems where losses are of primary importance, such as CATV. If power handling is the major concern, a much lower impedance is optimum (35 Ω). The standard of 50 Ω has been created as a good compromise between power handling and attenuation.

Fig 6-2 shows typical matched-line attenuation characteristics for many common transmission lines. Note how the open-wire line outperforms even its biggest coaxial brother by a large margin. But these attenuation figures are only the "nominal" attenuation figures for lines operating with a 1:1 SWR.

TLDetails (see Section 2.4 and Fig 6-2A) is a freeware software program by Dan, AC6LA (**www.qsl.net/ac6la/tldetails.html**) that can generate beautiful attenuation vs frequency charts for any type of transmission line. *TLDetails*

includes characteristics for 49 built-in line types, and you can specify your own. For information concerning the K1 and K2 loss coefficients see **www.qsl.net/ac6la/bestfit.html**.

TLW (Transmission Line for Windows), by N6BV, is available from the ARRL as part of the CD that comes with the 20th Edition of *The ARRL Antenna Book*. *TLW* is a full-featured transmission line analysis program with beautiful graphic capabilities. It includes a design section for an antenna matcher (tuner), using four possible networks: high and low-pass L-networks, low-pass Pi networks and high-pass T-networks. The database contains transmission line characteristics of over 30 current types of lines, and the user can, in addition, enter the specs of his own line.

Frank Donovan, W3LPL, put together **Table 6-1**, which lists most of the commonly used coaxial cable types in the US. The table was made in two versions, one giving the classic atenuation/100 foot. The second list gives the cable length for 1 dB of attenuation.

When there are standing waves on a feed line, the voltage and the current will be different at every point on the line. Current and voltage will change periodically along the line and can reach very high values at certain points. The feed line uses dielectric (insulating) and conductor (mostly copper) materials with certain physical properties and limitations. The very high currents at peaks along the line are responsible for extra conductivity-related losses. The voltages associated with the voltage peaks will be responsible for increased dielectric losses. These are the mechanisms that make a line with a high SWR have more losses than the same line when matched. **Fig 6-3** shows additional losses caused by SWR. By the way, the losses of the line are the reason why the SWR we measure at the input end of the feed line (in the shack) is always lower than the SWR at the load.

An extreme example is that of a very long cable, having

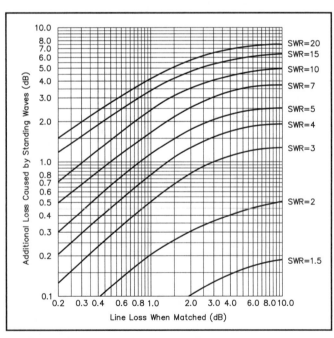

Fig 6-3—This graph shows how much additional loss occurs for a given SWR on a line with a known (nominal) flat-line attenuation. (*Courtesy of* The ARRL Antenna Book.)

Table 6-1

	Cable Attenuation (dB per 100 feet)									
MHz	1.8	3.5	7.0	14.0	21.0	28.0	50.0	144	440	1296
LDF7-50A	0.03	0.04	0.06	0.08	0.10	0.12	0.16	0.27	0.5	0.9
FHJ-7	0.03	0.05	0.07	0.10	0.12	0.15	0.20	0.37	0.8	1.7
LDF5-50A	0.04	0.06	0.09	0.14	0.17	0.19	0.26	0.45	0.8	1.5
FXA78-50J	0.06	0.08	0.13	0.17	0.23	0.27	0.39	0.77	1.4	2.8
3/4" CATV	0.06	0.08	0.13	0.17	0.23	0.26	0.38	0.62	1.7	3.0
LDF4-50A	0.09	0.13	0.17	0.25	0.31	0.36	0.48	0.84	1.4	2.5
RG-17	0.10	0.13	0.18	0.27	0.34	0.40	0.50	1.3	2.5	5.0
SLA12-50J	0.11	0.15	0.20	0.28	0.35	0.42	0.56	1.0	1.9	3.0
FXA12-50J	0.12	0.16	0.22	0.33	0.40	0.47	0.65	1.2	2.1	4.0
FXA38-50J	0.16	0.23	0.31	0.45	0.53	0.64	0.85	1.5	2.7	4.9
9913	0.16	0.23	0.31	0.45	0.53	0.64	0.92	1.6	2.7	5.0
RG-217	0.19	0.27	0.36	0.51	0.61	0.73	1.1	2.0	4.0	7.0
RG-213	0.25	0.37	0.55	0.75	1.0	1.2	1.6	2.8	5.1	10.0
RG-8X	0.49	0.68	1.0	1.4	1.7	1.9	2.5	4.5	8.4	17.8
	Cable Attenuation (Feet per dB)									
MHz	1.8	3.5	7.0	14.0	21.0	28.0	50.0	144	440	1296
LDF7-50A	3333	2500	1666	1250	1000	833	625	370	200	110
FHJ-7	2775	2080	1390	1040	833	667	520	310	165	92
LDF5-50A	2500	1666	1111	714	588	526	385	222	125	67
FXA78-50J	1666	1250	769	588	435	370	256	130	71	36
3/4" CATV	1666	1250	769	588	435	385	275	161	59	33
LDF4-50A	1111	769	588	400	323	266	208	119	71	40
RG-17	1000	769	556	370	294	250	200	77	40	20
SLA12-50J	909	667	500	355	285	235	175	100	53	34
FXA12-50J	834	625	455	300	250	210	150	83	48	25
FXA38-50J	625	435	320	220	190	155	115	67	37	20
9913	625	435	320	220	190	155	110	62	37	20
RG-217	525	370	275	195	160	135	90	50	25	14
RG-213	400	270	180	130	100	83	62	36	20	10
RG-8X	204	147	100	71	59	53	40	22	12	6

LDF7-50A is Andrew 1⅝" 50 Ω foam dielectric Heliax
LDF4-50A is Andrew ½" 50 Ω foam dielectric Heliax
LDF5-50A is Andrew ⅞" 50 Ω foam dielectric Heliax
FHJ-7 is an older version of Andrew 1⅝" 50 Ω foam dielectric Heliax
FXA78-50J is Cablewave ⅞" 50 Ω aluminum jacketed foam dielectric hardline
FXA12-50J is Cablewave ½" 50 Ω aluminum jacketed foam dielectric hardline
SLA12-50J is Cablewave ½" 50 Ω aluminum jacketed air dielectric hardline
FXA38-50J is Cablewave ⅜" 50 Ω aluminum jacketed foam dielectric hardline

a loss of at least 20 dB, where you can either short or open the end and in both cases measure a 1:1 SWR at the input. Such a cable is a perfect dummy load!

For a transmission line to operate successfully under high SWR, we need a low-loss feed line with good dielectric properties and high current-handling capabilities. The feeder with such properties is the open-wire line. Air makes an excellent dielectric, and the conductivity can be made as good as required by using heavy gauge conductors. Good-quality open-wire feeders have always proved to be excellent as feed-line transformers. Elwell, N4UH, has described the use and construction of homemade, low-loss open-wire transmission lines for long-distance transmission (Ref 1320). In many cases, the open-wire feeders are used under high SWR conditions (where the feeders do not introduce large additional losses) and are terminated in an antenna tuner. On the low bands the extra losses caused by SWR are usually negligible (Ref. 1319, 322), even for coaxial cables.

2.5. The Universal Transmission-Line Program

The COAX TRANSFOMER/SMITH CHART computer program, which is part of the NEW LOW-BAND SOFTWARE, is a good tool for evaluating the behavior of feed lines. Let us analyze the case of a 50-meter long RG-213 coax, feeding an impedance of 36.6 Ω (without a matching network). The frequency is 3.5 MHz.

Fig 6-4 shows a screen print obtained from the COAX TRANSFOMER/SMITH CHART module (using the program WITH cable losses). All the operating parameters are listed on the screen: impedance, voltage and current at both ends of the line, as well as the attenuation data split into nominal coax losses (0.61 dB) and losses due to SWR (0.03 dB). We also see the real powers involved. In our case we need to put 1734 W into the 100-meter long RG-213 cable to obtain 1500 W at the load, which represents a total efficiency of 86%. Note also the difference in SWR at the load

(1.4:1) and at the feed line end (1.3:1). For higher frequencies, longer cables or higher SWR values, this software module is a real eye-opener.

TLDetails (Transmission Line Details) mentioned above in Section 2.4 is a small standalone Windows program by AC6LA (**www.qsl.net/ac6la/tldetails.html**) that does exactly what my program does, and more. Dan wrote in an E-mail to me: *"When I was first playing with the transmission line equation several years ago, I remember comparing my results to several examples shown in your LOW BAND DXing book.* *I considered my code to be debugged when my results matched yours!"*

Another interesting software tool by AC6LA is *XLZIZL* (**www.qsl.net/ac6la/xlzizl.html**). This is an Excel application that analyzes the components of an antenna feed system, including transmission lines, stubs, baluns and tuners. Calculations include impedance transforms, SWR and reflection coefficient, power loss, voltage and current standing waves, stress on tuner components, network attenuation (S21), and return loss (S11). Analysis results are available in spreadsheet

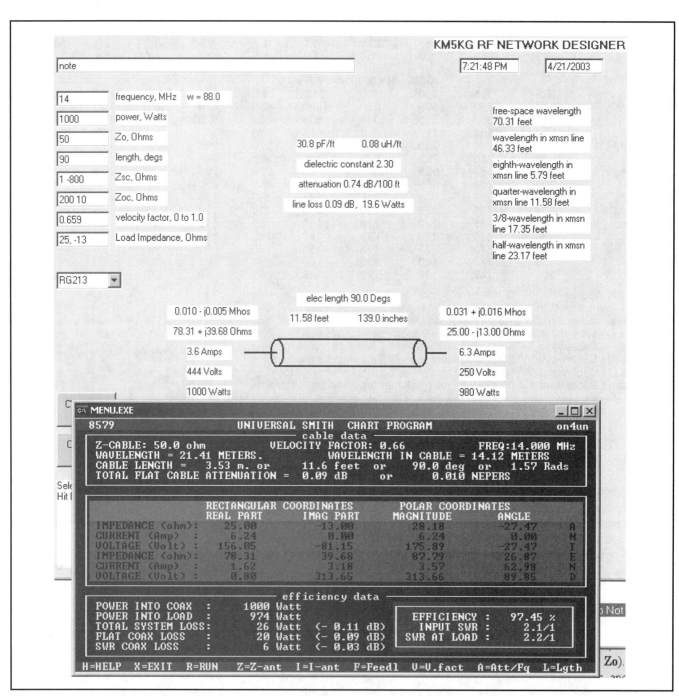

Fig 6-4—An overlay of two transmission-line programs. At the top, the *KM5KG RF NETWORK DESIGNER* program shows that how a load impedance of 25 − *j* 13 Ω is transformed through a 90°-long 50-Ω feed line (having a loss of 0.74 dB/100 feet at 14 MHz). At the bottom is shown ON4UN's UNIVERSAL SMITH CHART, a module of the **NEW LOW BAND SOFTWARE**. See text for details. Both programs calculate the impedance at the end of this (lossy) line as 78.31 + *j* 39.68 Ω.

format and in five different chart formats, including Smith charts.

You can also use the *Transmission Line Transformer* or the *Transmission Line Model* module from Grant Bingeman's (KM5KG) software *Professional RF Network Designer* (see Chapter 4). Fig 6-4B shows the screen result of Bingeman's program and the same using the author's program, both yielding exactly the same results.

2.6. Which Size of Coaxial Cable?

RG-213 will easily handle powers up to 2 kW on the low bands, even with moderate SWR. Is there any point in using "heavier" coax? 100 meters of RG213, when perfectly matched (SWR of 1:1) gives a loss of approx 2 dB/100 meters on 7 MHz, 1.2 dB/100 meters on 3.5 MHz and 0.8 dB/100 meters on 1.8 MHz. What's 0.8 dB? Do you have to worry about 0.8 dB? The answer is: You need not to worry about 0.8 dB. But you should worry about 0.8 dB here, 0.5 dB there and again 0.3 dB somewhere else. It's the sum of all these fractions of dB you need to worry about!

I use $7/8$-inch hardline on all my antennas, even on 160 meters (loss is approximately 0.25 dB /100 meters at 1.8 MHz). If your run is 100 meters long, you "gain" 0.55 dB over the same length of RG-213, which is a gain of 13% in power.

An additional reason for using hardline is that it is practically indestructible. With a solid copper shield, water ingress is impossible, and the black PE sheath used on these types of cables is perfectly UV resistant for lifetime! In addition this cable can often be obtained for less money than new RG-213 from Cell phone companies renewing their sites.

2.7. Conclusions

Coaxial lines are generally used when the SWR is less than 3:1. Higher SWR values can result in excessive losses when long runs are involved, and also in reduced power-handling capability. Many popular low-band antennas have feed-point impedances that are reasonably low, and can result in an acceptable match to either a 50-Ω or a 75-Ω coaxial cable.

In some cases we will intentionally use feed lines with high SWR as part of a matching system (eg, stub matching) or as a part of a feed system for a multi-element phased array. It is good engineering practice to use a feed line with the lowest possible attenuation—This employs the concept of cost versus performance, called in the USA getting the most "bang for the buck." We would like that cable to operate at a 1:1 SWR at the design frequency of our antenna system.

3. THE ANTENNA AS A LOAD

A very small antenna can radiate the power supplied to it almost as efficiently as much larger ones (see Chapter 9 on vertical antennas), but small antennas have two disadvantages. Since their radiation resistance is very low, antenna efficiency will be lower than it would be if the radiation resistance were much higher. Further, if short antennas use loading, the losses of the loading devices have to be taken into account when calculating antenna efficiency. On the other hand, if the short antenna (dipole or monopole) is not loaded, the feed-point impedance will exhibit a large capacitive reactance in addition to the resistive component.

You could install some sort of remote tuner at the antenna feed point to match the complex antenna impedance to the feed-line impedance. Then the matched feed line will no longer act as a transformer itself. Matching done with such a remote tuner results in a certain sacrifice in efficiency, especially for extreme impedance ratios. Transforming a very short vertical with a feed-point impedance of, say, $0.5 - j\,3000\,\Omega$ to a $50 + j\,0\,\Omega$ transmission line is a very difficult task, one that can't be done without a great deal of loss.

You can also supply power to an antenna point without inserting a tuner at the antenna's feed point. In this case the feed line itself acts as a transformer. In the above example of $0.5 - j\,3000\,\Omega$, an extremely high SWR would be present on the feed line. The losses in the transmission line itself will be determined by the quality of the materials used to make the feed line. In pre-WW II days, when coaxial cables were still unknown, everybody used 600 Ω open-wire lines, and nobody knew (or cared) about SWR. The transmission line is fed with a low-loss antenna tuner in the shack. What is a quality antenna tuner? The same qualifications for feed lines apply here: One that can transform the impedances involved, at the required power levels and with minimal losses.

Many modern unbalanced to unbalanced antenna tuners use a toroidal transformer/balun to achieve a relatively high-impedance balanced output. This principle is cost effective, but has its limitations where extreme transformations are required. The "old" tuners (for example, Johnson Matchboxes) are well suited for matching a wide range of impedances. Unfortunately these Matchboxes are no longer available commercially and are not designed to cover 160 meters.

4. A MATCHING NETWORK AT THE ANTENNA

Let's analyze a few of the most commonly used matching systems.

4.1. Quarter-Wave Matching Sections

For a given design frequency you can transform impedance A to impedance B by inserting a quarter-wave long coaxial cable between A and B having a characteristic impedance equal to the square root of the product A × B.

$$Z_{\lambda/4} = \sqrt{A \times B} \tag{Eq 2}$$

Example:

Assume we have a short vertical antenna that we wish to feed with 75-Ω coax. We have determined that the radiation resistance of the vertical is 23 Ω, and the resistance from earth losses is 10 Ω (making the feed-point resistance 33 Ω). We can use a $1/4$-wave section of line to provide a match, as shown in **Fig 6-5**. The impedance of this line is determined to be

$$\sqrt{33 \times 75} = 50\,\Omega.$$

Coaxial cables can also be paralleled to obtain half the nominal impedance. A coaxial feed line of 35 Ω can be made by using two parallel 70-Ω cables. Time Microwave (**www.timesmicrowave.com/**) offers a 35-Ω coaxial line (RG-83), which may be somewhat hard to find. This cable can, of course, be replaced with two paralleled 75-Ω coaxes.

You can parallel coaxial cable of different impedances to obtain odd impedances, which may be required for specific matching or feeding purposes. See **Table 6-2**. Make sure you use cable of exactly the same electrical length! Don't fool yourself—just because you parallel three identical cables the attenuation will not be one-third the attenuation of one cable. There is no change: Currents are now divided by the three cables, so all remains the same. Three cables in parallel will increase the power handling capability though.

One way to adjust ¼- or ½-wavelength cables exactly for a given frequency is shown in **Fig 6-6**. Connect the transmitter through a good SWR meter (ON4UN uses a Bird Model 43) to a 50-Ω dummy load. Insert a coaxial-T connector at the output of the SWR bridge. Connect the length of coax to be adjusted at this point and use the reading of the SWR bridge to indicate where the length is resonant. Quarter-wave lines should be short-circuited at the far end, and half-wave lines left open. At the resonant frequency, a cable of the proper length represents an infinite impedance (assuming lossless cable) to the T-junction. At the resonant frequency, the SWR will not change when the quarter-wave shorted line (or half-wave open line) is connected in parallel with the dummy load. At slightly different frequencies, the line will present small

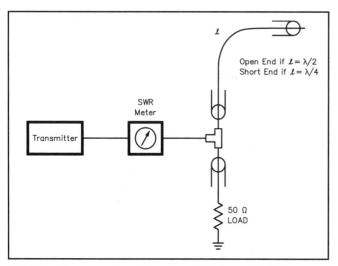

Fig 6-5—Example of a quarter-wave transformer, used to match a short vertical antenna (R_{rad} = 23 Ω, R_{ground} = 10 Ω, Z_{feed} = 33 Ω) to a 75-Ω feed line. In this case a perfect match can be obtained with a 50-Ω quarter-wave section.

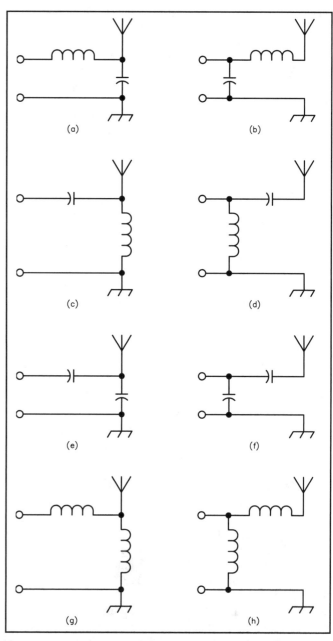

Fig 6-6—Very precise trimming of ¼ λ and ½ λ lines can be done by connecting the line under test in parallel with a 50-Ω dummy load and watching the SWR meter while the feed line length or the transmit frequency is changed. See text for details.

Fig 6-7—Eight possible L-network configurations. (After W. N. Caron, *ARRL Impedance Matching*.)

Table 6-2
Net characteristic impedance resulting from paralleling different coaxial cables.

Cables in Parallel	Net Impedance
75 Ω + 75 Ω	37.5 Ω
75 Ω + 50 Ω	30 Ω
50 Ω + 50 Ω	25 Ω
75 Ω + 75 Ω + 50 Ω	21.5 Ω
75 Ω + 50 Ω + 50 Ω	18.8 Ω
50 Ω + 50 Ω + 50 Ω	16.7 Ω

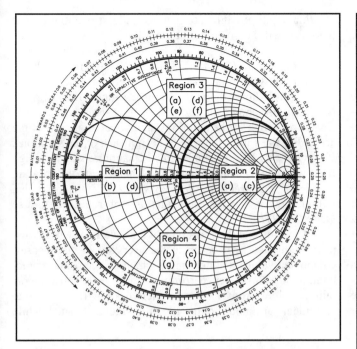

Fig 6-8—The Smith Chart subdivided in four regions, in each of which two or four L-network solutions are possible. The graphic solution methods are illustrated in Fig 6-9. (After W. N. Caron, *ARRL Impedance Matching*.)

values of inductance or capacitance across the dummy load, and these will influence the SWR reading accordingly. I have found this method very accurate, and the lengths can be trimmed precisely, to within a few kHz. Alternative methods are described in Chapter 11, Section 3.3.8.3.

Odd lengths, other than 1/4- or 1/2 wavelength, can also be trimmed this way. First calculate the required length difference between a quarter (or half) wavelength on the desired frequency and the actual length of the line on the desired frequency. For example, if you need a 73° length of feed line on 3.8 MHz, that cable would be 90° long on (3.8 × 90°/73°) = 4.685 MHz. The cable can now be cut to a quarter wavelength on 4.685 MHz using the method described above.

Some people use a dip oscillator, but this method isn't the most accurate way to cut a 90° length of feed line, and it often accounts for length variations of 2° or 3° (due to the inductance of the link use to couple to the GDO). You can also use a noise bridge and use the line under test to effectively short-circuit the output of the noise bridge to the receiver.

4.2. The L Network

The L network is probably the most commonly used network for matching antennas to coaxial transmission line. In special cases the L network is reduced to a single-element network, being a series or a parallel impedance network (just an L or C in series or in parallel with the load).

Fig 6-9—Design procedures on the Smith Chart for solution *a* through *h* as explained in Fig 6-8. (After W. N. Caron, *ARRL Impedance Matching*.) If you have a PC you can use the program *ARRL MICROSMITH* to quickly and easily calculate the matching values graphically on screen.

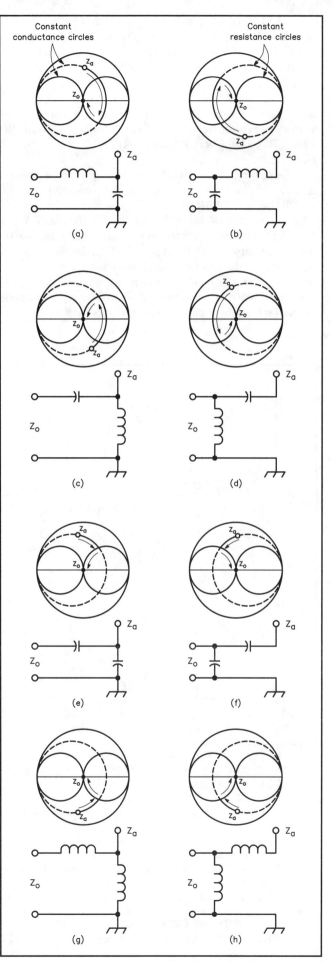

The Feed Line and the Antenna 6-9

The L network is treated in great detail by W. N. Caron in his excellent book *Antenna Impedance Matching* (an ARRL publication). Caron exclusively used the graphical Smith Chart technique to design antenna-matching networks. The book also contains an excellent general treatment of the Smith Chart and other basics of feed lines, SWR and matching techniques. Graphic solutions of impedance-matching networks have been treated by I. L. McNally, W1NCK (Ref 1446). R. E. Leo, W7LR (Ref 1404) and B. Baird, W7CSD (Ref 1402).

Designing an L network is something you can easily do using a computer program. I have written a computer program (L-NETWORK DESIGN) that will just do that for you. The program is part of the NEW LOW BAND SOFTWARE. So-called shunt-input L networks are used when the resistive part of the output impedance is lower than the required input impedance of the network. The series-input L network is used when the opposite condition exists. In some cases, a series-input L network can also be used when the output resistance is smaller than the input resistance (in this case we have four solutions). All possible alternatives (at least two, but four at the most) will be given by the program.

Other similar computer programs have been described in amateur literature (Ref 1441). The ARRL program *TLW* can design L-networks that take into account component losses.

Fig 6-7 shows the eight possible L-network configurations. **Fig 6-8** shows the four different regions of the Smith Chart and which of the solutions are available in each of the areas. **Fig 6-9** shows the way to design each of the solutions. If you have an IBM or compatible PC, another way to design L networks with an on-screen Smith Chart is with the program *ARRL MICROSMITH* by W. Hayward, W7ZOI. A detailed knowledge of the Smith Chart is not required to use *MICROSMITH*.

The choice of the exact type of L network to be used (low pass, high pass) will be up to the user, but in many cases, component values will determine which choice is more practical. In other instances, performance may be the most important consideration: Low-pass networks will give some additional harmonic suppression of the radiated signal, while

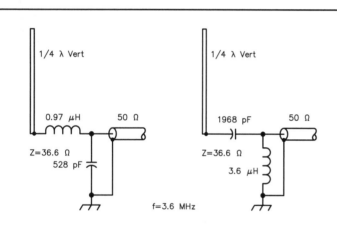

Fig 6-10—Design of an L-network to match a resonant quarter-wave vertical with a feed-point impedance of 36.6 Ω to a 50-Ω line. Note that in practice we must add the ground resistance to the radiation resistance to obtain the feed-point impedance. Therefore, in most cases the impedance of a quarter-wave vertical will be fairly close to 50 Ω.

```
7921                         L - NETWORK DESIGN                               on4un

Z-Input =    50.0 ohm          OUTPUT SWR =  1.37          FREQUENCY =    3.6 MHz

                      RECTANGULAR COORDINATES         POLAR COORDINATES
                      REAL PART      IMAG PART      MAGNITUDE       ANGLE
   IMPEDANCE (ohm) =    36.60          0.00           36.60          0.00
   CURRENT   (Amp) =     6.40          0.00            0.00          0.00
   VOLTAGE  (Volt) =   234.31          0.00          234.31          0.00
                              ─────── Solution # 1 ───────
   ┌─────────────────────────────────────────────────────────────────────────┐
   │ IMPEDANCE SERIES ARM   =  -22.1 ohm    ==>> CAPACITANCE =   1996 pF     │
   │ IMPEDANCE PARALLEL ARM =   82.6 ohm    ==>> INDUCTANCE  =   3.65 µH     │
   │ CURRENT  (Amp)    =     4.69        -2.84          5.48       -31.18    │
   │ VOLTAGE  (Volt)   =   234.31      -141.78        273.87       -31.18    │
   └─────────────────────────────────────────────────────────────────────────┘
                              ─────── Solution # 2 ───────
   ┌─────────────────────────────────────────────────────────────────────────┐
   │ IMPEDANCE SERIES ARM   =   22.1 ohm    ==>> INDUCTANCE  =   0.98 µH     │
   │ IMPEDANCE PARALLEL ARM =  -82.6 ohm    ==>> CAPACITANCE =    535 pF     │
   │ CURRENT  (Amp)    =     4.69         2.84          5.48        31.18    │
   │ VOLTAGE  (Volt)   =   234.31       141.78        273.87        31.18    │
   └─────────────────────────────────────────────────────────────────────────┘
           THE NETWORK HAS THE SHUNT ELEMENT ACROSS THE RESISTIVE INPUT
   X:EXIT     N:NEW RUN     R:Z-out    Z:Z-load    E:load volt   I:load curr    F:Freq
```

a high-pass filter may help to reduce the strength of strong medium-wave broadcast signals from local stations.

Some solutions provide a direct dc ground path for the antenna through the coil. If dc grounding is required, such as in areas with frequent thunderstorms, this can also be achieved by placing an appropriate RF choke at the base of the antenna (between the driven element and ground).

The L-NETWORK software module from the NEW LOW BAND SOFTWARE also calculates the input and output voltages and currents of the network. These can be used to determine the required component ratings. Capacitor current ratings are especially important when the capacitor is the series element in a network. The voltage rating is most important when the capacitor is the shunt element in the network. Consideration regarding component ratings and the construction of toroidal coils are covered in Section 4.2.1.2.

The user provides to the L-NETWORK software module:
- Design frequency
- Cable impedance
- Load resistance
- Load reactance

Fig 6-10 shows the screen display of a case where we calculate an L network to match $(36.6 - j\,0)\,\Omega$ to a 50-Ω transmission line. From the prompt line you can easily change any of the inputs. If the outcome of the transformation is a network with one component having a very high reactance (low C value or high L value), then we can try to eliminate this component all together. The SERIES NETWORK or SHUNT NETWORK programs will tell you exactly what value to use, and if the match is not perfect you may want to assess the SWR by switching to the SWR CALCULATION module of the NEW LOW BAND SOFTWARE to do that.

The same values can also be calculated with the L-network Module of Grant Bingeman's *Professional RF Network Designer* or with ARRL's *TLW*.

4.2.1. Component ratings

What kind of capacitors and inductors do we need for building the L networks?

4.2.1.1. Capacitors

The transmitter power as well as the position of the component in the L network will determine the voltage and current ratings that are required for the capacitor.
- If the capacitor is connected in parallel with the 50-Ω transmission line (assuming we have a 1:1 SWR), then the voltage across the capacitor is given by $E = \sqrt{P \times R}$. Assume 1500 W and a 50-Ω feed line.

$$E = \sqrt{1500 \times 50} = 274 \text{ V RMS}$$

The peak voltage is $274 \times \sqrt{2} = 387$ V peak.
- If the capacitor is connected between the antenna base and ground, we can follow a similar reasoning. But this time we need to know the absolute value of the antenna impedance. Assume the feed point impedance is $90 + j\,110\,\Omega$, where $R_r = 90\,\Omega$. The magnitude of the antenna impedance is:

$$Z_{ant} = \sqrt{90^2 + 110^2} = 142.1\,\Omega$$

The voltage across the antenna feed point is given by:

$$E = I \times Z_{ant} = \sqrt{\frac{P}{R_r}} \times Z_{ant}$$

$$= \sqrt{\frac{1500}{90}} \times 142.2 = 580 \text{ V RMS} = 820 \text{ V peak.}$$

- If the capacitor is the series element in the network, and if the parallel element is connected between the feed line and ground (transmitter side of the network), then the current through the capacitor equals the antenna feed current. Assume a new feed-point impedance of $120 + j\,190\,\Omega$. The magnitude of the antenna feed-point impedance is:

$$Z = \sqrt{120^2 + 190^2} = 225\,\Omega$$

Again assume 1500 W. The magnitude of the feed current is:

$$I = \sqrt{\frac{P}{Z}} = \sqrt{\frac{1500}{225}} = 2.58 \text{ A}$$

Assume the capacitor has a value of 200 pF and the operating frequency is 3.65 MHz. The impedance of the capacitor is:

$$X_C = \frac{10^6}{2\pi f\,C} = \frac{10^6}{2\pi \times 3.65 \times 200} = 218\,\Omega$$

where f is in MHz and C is in pF. The voltage across capacitor is:

$$E = I \times Z = 2.58 \times 218 = 562 \text{ V RMS or 795 V peak}$$

- If the capacitor is the series element in the L network and if the parallel element is connected between the feed point of the antenna and ground, then the current through the capacitor is the current going in the 50-Ω feed line. Assuming we have a 1:1 SWR in a 50-Ω feed line and a power level of 1500 W, the current is given by:

$$I = \sqrt{\frac{P}{Z}} = \sqrt{\frac{1500}{50}} = 5.48 \text{ A}$$

Assume the same 200-pF capacitor as above, whose impedance at 3.65 MHz was calculated to be 218 Ω. The voltage across the capacitor now is:

$$E = I \times Z = 5.48 \times 218 = 1194 \text{ V RMS or 1689 V peak}$$

When calculating required voltage ratings we must always calculate the peak value, while for currents we can use the RMS value. This is because the current failure mechanism is a thermal mechanism. In practice we should always use at least a 100% safety factor on these components. For the capacitors across low-impedance points, transmitting type mica capacitors can be used, as well as BC-type variables such as normally used as the loading capacitor in the pi network of a linear amplifier.

Table 6-3
Toroid Cores Suitable for Matching Networks

Supplier	Code	Permeability	OD (in.)	ID (in.)	Height (in.)	A_L
Amidon	T-40-A2	10	4.00	2.25	1.30	360
Amidon	T-400-2	10	4.00	2.25	0.65	185
Amidon	T-300-2	10	3.05	1.92	0.50	115
Amidon	T-225-A2	10	2.25	1.41	1.00	215

For series capacitors, only transmitting type ceramic capacitors (eg, doorknob capacitors) should be used because of the high RF current. For fine tuning, high-voltage variables or preferably vacuum variables can be used. I normally use parallel-connected transmitting-type ceramics across a low-value vacuum variable (these can usually be obtained at real bargain prices at flea markets).

4.2.1.2. Coils

Up to inductor values of approximately 5 µH, air-wound coils are usually the best choice. A roller inductor comes in very handy when trying out a new network. Once the computed values have been verified by experimentation, the variable inductor can be replaced with a fixed inductor. Large-diameter, heavy-gauge Air Dux coils are well suited for the application.

Above approximately 5 µH, powdered-iron toroidal cores can be used. Ferrite cores are not suitable for this application, since these cores are much less stable and are easily saturated. The larger size powdered-iron toroidal cores, which can be used for such applications, are listed in **Table 6-3**.

The required number of turns for a certain coil can be determined as follows:

$$N = 100 \times \sqrt{\frac{L}{A_L}} \quad \text{(Eq 2)}$$

where L is the required inductance in µH. The A_L value is taken from Table 6-3. The transmitter power determines the required core size. It is a good idea to choose a core somewhat on the large side for a margin of safety. You may also stack two identical cores to increase power-handling capability, as well as the A_L factor. The power limitations of powdered-iron cores are usually determined by the temperature increase of the core. Use large-gauge enameled copper wire for minimum resistive loss, and wrap the core with glass-cloth electrical tape before winding the inductor. This will prevent arcing at high power levels.

Consider this example: A 14.4-µH coil requires 20 turns on a T-400-A2 core. AWG 4 or AWG 6 wire can be used with equally-spaced turns around the core. This core will easily handle well over 1500 W.

In all cases you must measure the inductance. A_L values can easily vary 10%. It appears that several distributors (such as Amidon) sell cores under the name type number coming from various manufacturers and this accounts for the spread in characteristics.

When measuring the inductance of a toroidal core, it is important to do this on the operating frequency, especially when dealing with ferrite material. The impedance versus frequency ratio is far from linear for this type of material. Be careful when using a digital L-C meter, which usually uses one fixed frequency for all measurements (eg, 1 MHz). Accurate methods of measuring impedances on specific frequencies are covered in Chapter 11 (Arrays).

4.2.1.3. The smoke test

Two things can go wrong with the matching network:

- Capacitors and coils can flash over (short circuit, explode, vaporize, catch fire, burn up, etc) if their voltage rating is too low.
- Capacitors or coils will heat up (and eventually be destroyed after a certain time), if the current through the component is too high or the component's current carrying capability is too low.

In the second case excessive current will heat up either the conductor in a coil or the dielectric in capacitor. One way to find out if there are any losses in the capacitor, resulting from large RF currents, is to measure or feel the temperature of the components in question (not with power applied!) after having stressed them with a solid carrier for a few minutes. This is a valid test for both coils and capacitors in a network. If excessive heating is apparent, consider using heavier-duty components. This procedure also applies to toroidal cores.

4.3. Stub Matching

Stub matching can be used to match resistive or complex impedances to a given transmission-line impedance. The STUB MATCHING software module, a part of the NEW LOW BAND SOFTWARE, allows you to calculate *the position* of the stub on the line and *the length of the stub*, and whether the stub must be open or shorted at the end. This method of matching a (complex) impedance to a line can replace an L network. The approach saves the two L-network components, but necessitates extra cable to make the stub. The stub may also be located at a point along the feed line that is difficult to reach. **Fig 6-11** shows the screen of the computer program where we are matching an impedance of 36.6 Ω to a 50-Ω feed line. Note that between the load and the stub the line is not flat, but once beyond the stub the line is now matched. The computer program gives line position and line length in electrical degrees. To convert this to cable length you must take into account the velocity factor of the feed line being used.

4.3.1. Replacing the stub with a discrete component.

Stub matching is often unattractive on the lower bands because of the lengths of cable required to make the stub. The module STUB MATCHING also displays the equivalent component value of the stub (in either µH or pF). You can replace the stub with an equivalent capacitor or inductor, which is

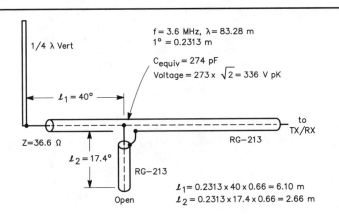

Fig 6-11—A 36.6-Ω resistive load is matched to a 50-Ω feed line using stub matching.

```
7921                    STUB MATCHING PROGRAM                           on4un

Freq:  3.6  Mhz            Z-line: 50.0 ohm                   SWR:    1.1

                 RECTANGULAR COORDINATES        POLAR COORDINATES
                 REAL PART     IMAG PART      MAGNITUDE      ANGLE
IMPEDANCE (ohm) =   36.60         0.00           36.60        0.00      A
CURRENT  (Amp)  =    6.40         0.00            6.40        0.00      N
VOLTAGE  (Volt) =  234.35         0.00          234.35        0.00      T
Posit.     IMPEDANCE           VOLTAGE         -------------STUB------------  IMP
Stub    Resis    React      Magnit    Angl    Imped   Value   Length   Type   ohm
 35     43.2     12.9       265.7     43.7   -157.9   280 pF   17.6    OPEN   45.8
 36     43.6     13.1       267.1     44.8   -157.7   280 pF   17.6    OPEN   46.5
 37     44.0     13.4       268.6     45.8   -157.8   280 pF   17.6    OPEN   47.2
 38     44.4     13.7       270.1     46.9   -158.0   280 pF   17.6    OPEN   48.0
 39     44.8     13.9       271.6     47.9   -158.5   279 pF   17.5    OPEN   48.7
 40     45.3     14.1       273.1     48.9   -159.2   278 pF   17.4    OPEN   49.4
 41     45.7     14.4       274.6     49.9   -160.0   276 pF   17.4    OPEN   50.1
 42     46.2     14.6       276.1     50.9   -161.1   274 pF   17.2    OPEN   50.9
 43     46.7     14.8       277.6     51.9   -162.3   272 pF   17.1    OPEN   51.6
 44     47.2     14.9       279.1     52.8   -163.8   270 pF   17.0    OPEN   52.3
 45     47.7     15.1       280.5     53.8   -165.4   267 pF   16.8    OPEN   53.0
 46     48.2     15.3       282.0     54.7   -167.3   264 pF   16.6    OPEN   53.7

H:HELP    X:EXIT    R:RUN    Z:Z-cable   F:Freq   I:Imp.load   C:Curr.load   V:Volt.load
```

then connected in parallel with the feed line at the point where the stub would have been placed. The same program shows the voltage where the stub or discrete element is placed. To determine the voltage requirement for a parallel capacitor, you must know the voltage at the load.

Consider the following example: The load is 50 Ω (resistive), the line impedance is 75 Ω, and the power at the antenna is 1500 W. Therefore, the RMS voltage at the antenna is:

$$E = \sqrt{P \times R} = \sqrt{1500 \times 50} = 274 \text{ V}$$

Running the STUB MATCHING software module, we find that a 75-Ω impedance point is located at a distance of 39° from the load. See **Fig 6-12** for details of this example. The required 75-Ω stub length, open-circuited at the far end, to achieve this resistive impedance is 22.2° (equivalent to 230 pF for a design frequency of 3.6 MHz). The voltage at that point on the line is 334 V RMS (472 V peak). Note that the length of a stub will never be longer than ¼ wavelength (either open-circuited or short-circuited).

4.3.2. Matching with series-connected discrete components.

In stub matching in a 50-Ω system, we look on a line with SWR for a point where the impedance on the line, together with the impedance of the stub (in parallel) will produce a 50-Ω impedance. A variation consists of looking along the line for a point where the insertion of a series impedance will yield 50 Ω. At that point the impedance will look like 50 + jX Ω or 50 − jY Ω. All we need to do is to put a capacitor or inductor in series with the cable at that point. A capacitor will have a reactance of X Ω or an inductor of Y Ω.

Example: Match a 50-Ω load to a 75-Ω line (same example as above). The software module IMPEDANCES, CURRENTS AND VOLTAGES ALONG FEEDLINES from the NEW LOW BAND SOFTWARE lists the impedance along the line in 1° increments, 2 starting at 1° from the load.

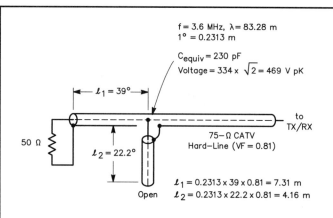

Fig 6-12—Example of how a simple stub can match a 50-Ω load to a 75-Ω transmission line. Note that between the load and the stub the SWR on the line is 1.5:1. Beyond the stub the SWR is 1:1.

```
7921                    STUB MATCHING PROGRAM                            on4un

Freq:   3.6  Mhz                Z-line:  75.0 ohm                SWR:    1.1

                    RECTANGULAR COORDINATES          POLAR COORDINATES
                    REAL PART     IMAG PART      MAGNITUDE      ANGLE
 IMPEDANCE (ohm) =    50.00         0.00           50.00         0.00      A
 CURRENT   (Amp) =     5.47         0.00            5.47         0.00      N
 VOLTAGE  (Volt) =   273.50         0.00          273.50         0.00      T
 Posit.    IMPEDANCE         VOLTAGE          ------------STUB------------ IMP
 Stub    Resis  React     Magnit   Angl    Imped    Value   Length   Type  ohm
  34     60.5   23.4      322.6    45.3   -180.0    246 pF   22.6   OPEN   67.8
  35     61.2   24.0      324.9    46.4   -180.2    245 pF   22.6   OPEN   69.2
  36     61.9   24.5      327.3    47.5   -180.7    245 pF   22.5   OPEN   70.6
  37     62.6   25.1      329.6    48.5   -181.4    244 pF   22.5   OPEN   71.9
  38     63.3   25.6      332.0    49.5   -182.3    243 pF   22.4   OPEN   73.3
  39     64.1   26.1      334.4    50.5   -183.4    241 pF   22.2   OPEN   74.7
  40     64.9   26.6      336.8    51.5   -184.8    239 pF   22.1   OPEN   76.0
  41     65.7   27.1      339.2    52.5   -186.4    237 pF   21.9   OPEN   77.4
  42     66.6   27.6      341.6    53.5   -188.2    235 pF   21.7   OPEN   78.7
  43     67.4   28.0      343.9    54.4   -190.2    232 pF   21.5   OPEN   80.0
  44     68.3   28.4      346.3    55.4   -192.5    230 pF   21.3   OPEN   81.3
  45     69.2   28.8      348.6    56.3   -195.0    227 pF   21.0   OPEN   82.5

 H:HELP   X:EXIT   R:RUN   Z:Z-cable   F:Freq   I:Imp.load   C:Curr.load   V:Volt.load
```

Somewhere along the line we will find an impedance where the real part is 75 Ω (see details in **Fig 6-13**). Note the distance from the load. In our example this is 51° from the 50-Ω load. The impedance at that point is 75.2 + j 30.7 Ω.

If we want to assess the current through the series element (which is especially important if the series element is a capacitor), we must enter actual values for either current or voltage at the load when running the program. Assuming an antenna power of 1500 W, the current at the antenna is:

$$I = \sqrt{\frac{P}{R}} = \sqrt{\frac{1500}{50}} = 5.47 \text{ A}$$

All we need to do now is connect an impedance of −30.7 Ω (capacitive reactance) in series with the line at that point. Also note that at this point the current is:

$$I = \sqrt{\frac{1500}{75.2}} = 4.46 \text{ A}$$

The software module SERIES IMPEDANCE NETWORK can be used to calculate the required component value. In this example, the required capacitor has a value of 1442 pF for a frequency of 3.6 MHz (see **Fig 6-14**). The required voltage rating (RMS) is calculated by multiplying the current through the capacitor times the capacitive reactance, which yields a value of E = I × Z = 4.46 × 30.7 = 136.9 V RMS = 193.6 V peak at 1500 W. As outlined above you need to take the peak value into consideration for a capacitor, and apply a safety factor of approximately two. The most important property of this capacitor is its current-handling capability, and we should use a capacitor that is rated approximately 10 A for the job.

In the case of a complex load impedance, the procedure is identical, but instead of entering the resistive load impedance (50 Ω in the above example), we must enter the complex impedance.

4.4. High-Impedance Matching Systems

Unbalanced high-impedance feed points, such as a half-

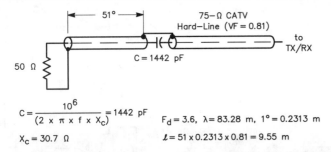

$$C = \frac{10^6}{(2 \times \pi \times f \times X_C)} = 1442 \text{ pF}$$

$X_C = 30.7 \ \Omega$

$F_d = 3.6, \ \lambda = 83.28 \text{ m}, \ 1° = 0.2313 \text{ m}$

$\ell = 51 \times 0.2313 \times 0.81 = 9.55 \text{ m}$

Fig 6-13—Example of how a series element can match a 50-Ω load to a 75-Ω transmission line. See text for details.

```
8579              LOSS FREE CABLE Z / I / E LISTING PROGRAM                on4un
     Z-CABLE: 75.0 ohm              STEP = 1.00 deg.              SWR =   1.50
   Length   Z-real      Z-imag    I-magnitude    I-angle    E-magnitude   E-angle
    0.00     50.0        0.0         1.0           0.0         50.0        0.0
   39.00     64.1       26.1         0.88         28.4         61.1       50.5
   40.00     64.9       26.6         0.88         29.2         61.6       51.5
   41.00     65.7       27.1         0.87         30.1         62.0       52.5
   42.00     66.6       27.6         0.87         31.0         62.4       53.5
   43.00     67.4       28.0         0.86         31.9         62.9       54.4
   44.00     68.3       28.4         0.86         32.8         63.3       55.4
   45.00     69.2       28.8         0.85         33.7         63.7       56.3
   46.00     70.2       29.2         0.84         34.6         64.2       57.2
   47.00     71.1       29.6         0.84         35.6         64.6       58.1
   48.00     72.1       29.9         0.83         36.5         65.0       59.0
   49.00     73.1       30.2         0.83         37.5         65.4       59.9
   50.00     74.2       30.4         0.82         38.5         65.8       60.8
   51.00     75.2       30.7         0.82         39.5         66.2       61.6
   52.00     76.3       30.9         0.81         40.5         66.6       62.5
   53.00     77.4       31.0         0.80         41.5         67.0       63.3
   54.00     78.6       31.1         0.80         42.5         67.4       64.2
   55.00     79.7       31.2         0.79         43.6         67.8       65.0
   56.00     80.9       31.2         0.79         44.7         68.2       65.8
   57.00     82.1       31.2         0.78         45.8         68.5       66.6
             ESC: STOP LISTING              ENTER KEY: RESUME LISTING
```

```
7921                    SERIES IMPEDANCE NETWORK (L OR C)                 on4un


                    RECTANGULAR COORDINATES        POLAR COORDINATES
                    REAL PART    IMAG PART       MAGNITUDE      ANGLE
IMPEDANCE (ohm) =     75.25        30.67            81.26       22.17      O
CURRENT   (Amp) =      3.44         2.83             4.46       39.46      L
VOLTAGE  (Volt) =    172.12       318.82           362.32       61.64      D

IMPEDANCE (ohm) =     75.25         0.00            75.25        0.00      N
CURRENT   (Amp) =      3.44         2.83             4.46       39.46      E
VOLTAGE  (Volt) =    259.04       213.25           335.52       39.46      W


      CAPACITANCE =   1442 pF           FREQUENCY =    3.60 MHz




      X = EXIT    R = RUN    Z = Z-load    E = E-load    I = I-load    F = Freq
```

Fig 6-14—Calculations of the value of the series element required to tune out the reactance of the load 75.248 + *j* 30.668 Ω. See text for details.

The Feed Line and the Antenna

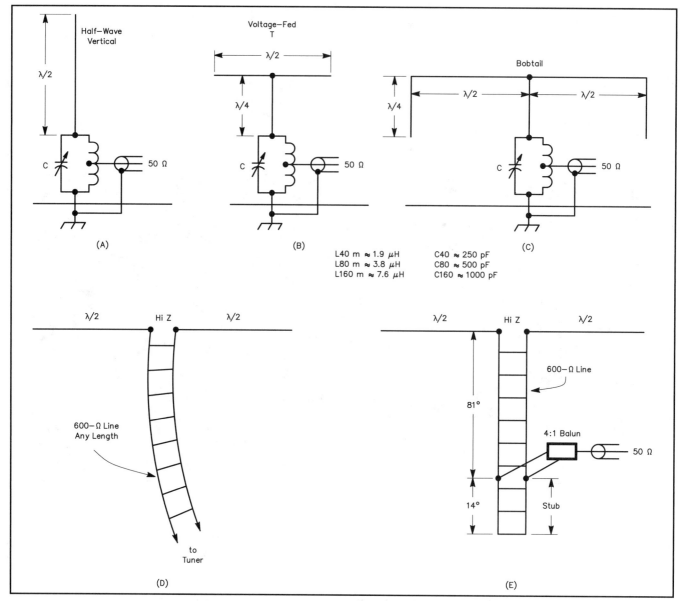

Fig 6-15—Recommended feed methods for high-impedance (2000 to 5000-Ω) feed points. Asymmetrical feed points can be fed via a tuned circuit. The symmetrical feed points can be fed via an open-wire line to a tuner, or via a stub-matching arrangement to a 4:1 (200 to 50-Ω) balun and a 50-Ω feed line.

wave vertical fed against ground, a voltage-fed T-antenna, the Bobtail antenna, etc, can best be fed using a parallel-tuned circuit on which the 50-Ω cable is tapped for the lowest SWR value. See **Fig 6-15**. Symmetrical high-impedance feed points, such as for two half-wave (collinear) dipoles in phase, the bi-square, etc, can be fed directly with a 600-Ω open-wire feeder into a quality antenna tuner (see Fig 6-15D).

Another attractive solution is to use a 600-Ω line and stub matching, as shown in Fig 6-15E. Assume the feed-point impedance is 5000 Ω. Running the STUB MATCHING software module, we find that a 200-Ω impedance point is located at a distance of 81° from the load. The required 600-Ω stub to be connected in parallel at that point is 14° long (X = 154 Ω). The impedance is now a balanced 200 Ω. Using a 4:1 balun, this point can now be connected to a 50-Ω feed line.

Let me sum up some of the advantages and disadvantages of both feed systems.

Tuned open-wire feeders:

- Fewest components, which means the least chance of something going wrong.
- Least likely loss.
- Very flexible (can be tuned from the shack).
- Open-wire lines are mechanically less attractive.

Stub matching plus balun and coax line:

- Coaxial cables are much easier to handle.

4.5. Wideband Transformers
4.5.1. Low-impedance wideband transformers

Broadband transformers exist in two varieties: The classic autotransfomer and the transmission-line transformer. The first is a variant of the Variac, a genuine autotransformer. The second makes use of transmission-line principles. What they have in common is that they are often wound on toroidal cores. It is beyond the scope of this chapter to go into details on this subject. More details can be found in Chapter 7 (Special Receiving Antennas), where such broadband transformers are commonly used to feed receiving antennas such as Beverages. *Transmission Line Transformers* by J. Sevick, W2FMI, is an excellent textbook on the subject of transmission-line transformers. It covers all you might need in the field of wide-band RF transformers.

4.5.2. High-impedance wideband transformer

If the antenna load impedance is both high and almost perfectly resistive (such as for a half wavelength vertical fed at the bottom), you may also use a broadband transformer such as is used in transistor power amplifier output stages. **Fig 6-16** shows the transformer design used by F. Collins, W1FC. Two turns of AWG 12 Teflon-insulated wire are fed through two stacks of 15½-inch (OD) powdered-iron toroidal cores (Amidon T50-2) as the primary low impedance winding. The secondary consists of 8 turns. The turns ratio is 4:1, the impedance ratio 16:1.

The efficiency of the transformer can be checked by terminating it with a high-power 800 Ω dummy load (or with the antenna, if no suitable load is available), and running full power to the transformer for a couple of minutes. Start with low power. Better safe than sorry. If there are signs of heating in the cores, add more cores to the stack. Such a transformer has the advantage of introducing no phase shift between input and output, and therefore can easily be incorporated into phased arrays.

5. 75-Ω CABLES IN 50-Ω SYSTEMS

Lengths of 75-Ω hardline coaxial cable can often be obtained from local TV cable companies. If very long runs to low-band antennas are involved, the low attenuation of hardline is an attractive asset. If you are concerned with providing a 50-Ω impedance, you need to use a transformer system. Transformers using toroidal cores (so called *ununs*) have been described (Ref 1307, 1517, 1518, 1521, 1522, 1523, 1524, 1525, 1526, 1527, 1528, 1829, 1830).

Ununs (Unbalanced to Unbalanced transformers) are really *autotransformers* and have been described for a wide range of impedance ratios. One application is as a matching system for a short, loaded vertical. If the short, loaded vertical is used over a good ground radial system, its impedance will be lower than 50 Ω. Ununs have been described that will match 25 Ω to 50 Ω, or 37.5 Ω to 50 Ω.

Ununs can also be used in array-matching systems to provide proper drive for various elements (see Chapter 7 and 11). Transformer systems can also be made using only coaxial cable, without any discrete components. If 60-Ω coaxial cable is available (as in many European countries), a quarter-wave transformer will readily transform 75 Ω to 50 Ω at the end of the hardline.

Carroll, K1XX, described the non-synchronous matching transformer and compared it to a stub-matching system (Ref 1318). While the toroidal transformer is broadbanded, the stub and non-synchronous transformers are single-band devices.

Compared to quarter-wave transformers, which need coaxial cable having an impedance equal to the geometric mean of the two impedances to be matched, the non-synchronous transformer requires only cables of the same impedances as the values to be matched (see **Fig 6-17**).

On the low bands (and even up to 30 MHz) the losses caused by using 75-Ω hard line in a 50-Ω system (50-Ω

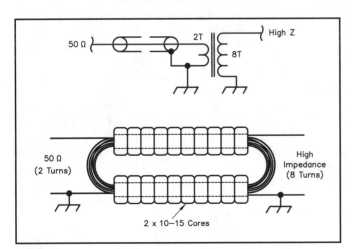

Fig 6-16—A wideband high-power transformer for large transformation ratios, such as for feeding a half-wave vertical at its base (600 to 10,000 Ω), uses two stacks of 10 to 15 half-inch-OD powdered-iron cores (eg, Amidon T502-2). The primary consists of 2 turns and the secondary has 8 turns (for a 50 to 800-Ω ratio). See text for details.

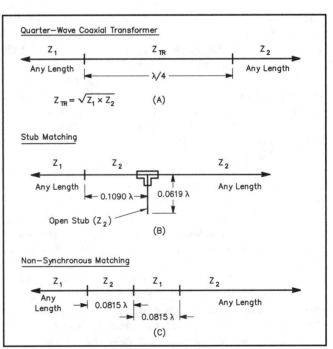

Fig 6-17—Methods of matching 75-Ω cables in 50-Ω systems. The quarter-wave transformer at A requires a cable having an impedance that is the geometric mean of the values being matched. The stub matching system at B and the non-synchronous matching system at C require only cables of the impedances being matched. The stub can be replaced with a capacitor or an inductor. All these matching systems are frequency sensitive.
Z_{TR}—60-Ω line.
Z_1—50-Ω line (or load).
Z_0—75-Ω line.

antenna and 50-Ω transceiver/amplifier) are generally negligible. A real problem is that 75-Ω feed line itself works as a transformer, and even when terminated with a perfect 50-Ω load, will show 100 Ω at the end of the line if the line is an odd multiple of quarter-waves long. This may cause problems for your linear amplifier. There is an easy solution to that problem, which is using ½-λ (of multiples of) lines. If you use a multiband antenna, make sure that the line is a number of half waves on all the frequencies used. For an antenna that works on 80 and 160 meters, make the coaxial line a multiple of half waves on 160 meters. Assuming a 75-Ω hardline with a Velocity Factor of 0.8, then the line should be 0.8 × (300/1.83)/2 = 65.6 meters, or any multiple thereof. You can trim the length by terminating the line with a 50-Ω load, and adjusting the length for minimum SWR on the highest frequency (in the above case, 3.66 MHz). Don't fool yourself though, in this case the SWR on the 75-Ω line is still 1.5:1, but the consequences are minimal so far as additional losses are concerned (because we use a feed line with intrinsic low losses) and are compensated for as far as the transformation effect is concerned, by using ½-λ lengths. To be fully correct the transformation is not a perfect 1:1 transformation with a real line, but close (1:1 is only with a lossless line).

6. THE NEED FOR LOW SWR

In the past many radio amateurs did not understand SWR. Unfortunately, many still don't understand SWR. Reasons for low SWR are often false and SWR is often cited as the single parameter telling us all about the performance of an antenna.

Maxwell, W2DU, published a series of articles on the subject of transmission lines. They are excellent reading material for anyone who has more than just a casual interest in antennas and transmission lines (Refs 1308-1311, 1325-1330 and 1332). These articles have been combined and, with new information added, published as a book, *Reflections II: Transmission Lines and Antennas* (WorldRadio Books). J. Battle, N4OE, wrote a very instructive article "What is your Real Standing Wave Ratio" (Ref 1319), treating in detail the influence of line loss on the SWR (difference between apparent SWR and real SWR).

Everyone has heard comments like: "My antenna really gets out because the SWR does not rise above 1.5:1 at the band edges." Low SWR is no indication at all of good antenna performance! It is often the contrary. The "antenna" with the best SWR is a quality dummy load. Antennas using dummy resistors as part of loading devices come next (Ref 663).The TTFD (Tilted Terminated Folded Dipole) and the B&W broadband folded dipole model BWD-18-30 are such examples. You should conclude from this that low SWR is no guarantee of radiation efficiency. The reason that SWR has been wrongly used as an important evaluation criterion for antennas is that it can be easily measured, while important parameters such as efficiency and radiation characteristics are more difficult to measure.

Antennas with lossy loading devices, poor earth systems, high-resistance conductors and the like, will show flat SWR curves. Electrically short antennas should always have narrow bandwidths. If they do not, it means that they are inefficient. In Chapter 5, Section 2.12. I explain further what are valid reasons for a low SWR.

7. THE BALUN

Balun is a term coming from the words *balanced* and *unbalanced*. It is a device we must insert between a symmetrical feed line (such as an open-wire feeder) and an asymmetric load (such as a ground-mounted vertical monopole) or an asymmetric feed line (for example, coax) and a symmetric load (such as a center-fed half-wave dipole). If we feed a balanced feed point with a coaxial feed line, currents will flow on both the outside of the coaxial braid (where we don't want them) and on the inside (where we do want them). Currents on the outside will cause radiation from the line.

Unbalanced loads can be recognized by the fact that one of the terminals is at ground potential. Examples: the base of a monopole vertical (the feed point of any antenna fed against real ground), the feed point of an antenna fed against radials (that's an artificial ground), the terminals of a gamma match or omega match, etc.

Balanced loads are presented by dipoles, sloping dipoles, delta loops fed at a corner, quad loops, collinear antennas, bi-square, cubical quad antennas, split-element Yagis, the feed points of a T match, a delta match, etc.

Many years ago I had an inverted-V dipole on my 25-meter tower and the feed line was just hanging unsupported alongside the tower, swinging nicely in the wind. When I took down the antenna some time later, I noticed that in several places where the coax had touched the tower in the breeze holes were burned through the outer jacket of the RG-213. Further, water had penetrated the coax, rendering it worthless. The phenomena of burning holes illustrates that currents (thus also voltages) are present on the coax if no balun is used. Such currents also create radiated fields, and fields from the feed line upset the field pattern from the antenna.

How much radiation there is from such a feed line depends on several factors, the main one being its length. In most cases the feed-line outer conductor will be (RF) grounded at the station. Assume the feed line is an odd number of quarter-waves long. In that case the impedance of the long wire (which is the outer shield of the feed line) will be very high at the antenna feed point, and hence the currents will be minimal, resulting in low unwanted radiation from it. If, however, the feed line is a number of half-waves long (and the outer shield grounded at the end), then we have a low-impedance point at the antenna end, consequently a large current can flow. In actual practice, unless the feed lines are a multiple of half-waves long, the impedance of the "long-wire" will be reactive, which in parallel with the resistive and low impedance of the real-antenna (at resonance) will result in a relatively small currents flowing on the outer shield of the coaxial feed line. The best answer is "take no chances" and to use a current balun, especially if you use (multiple of) half-wave long feed lines (see also Section 5).

Baluns have been described in abundance in the amateur literature (Refs 1504, 1505, 1502, 1503, 1515, 1519, and 1520 through 1530). In the simplest form a balun consists of a number of turns of coaxial cable wound into a close coil. In order to present enough reactance at the low-band frequencies, a fairly large coil is required.

Another approach was introduced by Maxwell, W2DU. This involves slipping a stack of high-permeability ferrite cores over the outer shield of the coaxial cable at the load terminals. In order to reduce the required ID of the toroids or beads, you can use a short piece of Teflon-insulated coaxial cable such as RG-141, RG-142 or RG-303. These have ODs of approximately 5 mm. A balun covering 1.8 to 30 MHz uses

50 no. 73 beads (Amidon no. FB-73-2401 or Fair-Rite no. 2673002401-0) to cover a length of approximately 30 cm (12 inches) of coaxial cable.

The stack of beads on the outer shield of the coax creates an impedance of one to several kΩ, effectively suppressing any current from flowing down on the feed line. Amidon beads type 43-1024 can be used on RG-213 cable. Ten to thirty will be required, depending on the lowest operating frequency. In general we can state that a choking effect of at least 1 kΩ is required for this *common-mode current arrestor* to be effective. This type of balun transformer is a true transmission line just like the beaded balun. But it can gain a much higher choking action from the transformation of the N turns power.

The two above approaches are called *current* or often *choke* baluns. They are called current-type baluns because even when the balun is terminated in unequal resistances, it will still force equal, opposite-in-phase currents into each resistance.

Current baluns made according to this principle are commercially available from Antennas Etc, PO Box 4215, Andover MA 01810; The Radio Works Inc, Box 6159, Portsmouth VA 23703. The Wireman, Inc, 261 Pittman Road, Landrum SC 29356 (**www.thewireman.com/**). This last supplier also sells a kit consisting of a length of Teflon coax (RG-141 or RG-303) plus 50 ferrite beads to be slipped over the Teflon coax at a very attractive price.

The traditional balun (for example, the well-known W6TC balun) is a *voltage balun*, which produces equal, opposite-phase voltages into the two resistances. With the two resistances we mean the two "halves" of the load, which are "symmetrical" with respect to ground (not necessarily in value!). If the load is perfect in common-mode balance and of a controlled impedance, a voltage-type balun is as good as a choke-type balun. But the choke-type balun is almost always much better in the real world.

The toroidal-core type voltage and current baluns are covered in *Transmission Line Transformers* by J. Sevick, W2FMI. **Fig 6-18** shows construction details for a the W6TC voltage-type balun designed for best performance on 160, 80 and 40 meters, as well as a current-type balun.

I have stated on several occasions that if the reading of an SWR meter changes with its position on the line (small changes in position, not affected by attenuation) this means the SWR meter is not functioning properly. The only other possible reason for a different SWR reading with position on the line is the presence of RF currents on the outside of the coax. For that reason it is common practice in professional SWR-measuring setups to put a number of ferrite cores on the coaxial cable on both sides of the measuring equipment.

We've touched upon three good reasons for using a balun with a symmetrical feed point:

- We don't want to distort the radiation pattern of the antenna.
- We don't want to burn holes in our coax.
- We want our SWR readings to be correct.

Are there good reasons to put a so-called current-balun on a feed-line attached to a asymmetrical feed point? Yes, there are. Assume a vertical antenna using two elevated-radials. The feed point is an asymmetric one, but the ends of the two radials are *not* the real ground, which usually is some distance below it. If we do not connect a current balun at the

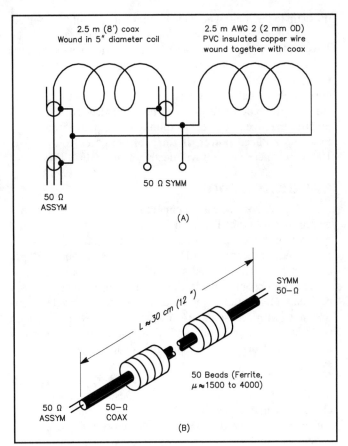

Fig 6-18—At A, details of a W6TC voltage-type balun for 160-40 meters, and at B, a current transformer for 160-10 meters. See text for details.

antenna feed point, antenna-return currents will flow on the outside of the coaxial feed line in addition to flowing in the elevated radials, which is not what we want with elevated radials (see also Chapter 9).

Is it harmful to put a current balun on all the coaxial antenna feed lines for all your antennas? Not at all. If the feed point is symmetric, there will be no current flowing and the beads will do no harm. As a matter of fact they may help reduce unwanted coupling from antennas into feed lines of other nearby antennas. A good thing is to use an RF-current meter (see Chapter 11) and check currents on the outside of any feed line while transmitting on any nearby (within ½ wavelength) antenna. These current should be zero; if not, they act as parasitically excited elements, which will influence the radiation pattern of your antenna.

How many ferrite beads (toroidal cores) are required on a coaxial cable to make a good current balun? From a choking impedance point of view you need at least 1 kΩ on the lowest operating frequency. The ferrite cores are not lossless, and depending on the mix used, they can be quite lossy. Where no power is involved (such as for solving EMC problems) this is never a problem. The total loss of the RF choke is then made up by the impedance of the inductance in series with the loss resistance. In other words, you have a low Q-coil. Where we use such ferrite cores to choke off potentially high RF currents (this is mostly the case with current baluns on transmitter feed lines), the resistive losses of the ferrites may actually heat those up to the point where they either become totally ineffective (permanently destroyed) or actually crack or explode! This problem can be avoided by using ferrite material that is not very lossy on the transmit frequency. In actual practice you can successfully combine two sorts of ferrite cores in a current choke balun: low resistive (high Q) cores at the "hot side" of the balun and lower-Q beads at the "cold side). In practice the touch-and-feel method is an adequate test method. First run reduced power. If some of the cores get warm at 100 W, chances are you will destroy them with a kW.

8. CONNECTORS

A good coaxial cable connector, such as a PL-259 connector, has a loss of less than 0.01 dB, even at 30 MHz, and typically 0.005 dB or less on the low bands. This means that for 1 KW of power you will have a heat loss of about 1 W per connector. Given the mass of a connector, and the heat-dissipating capacity of the cable, this will produce a hardly noticeable temperature increase. If you feel a connector getting hot (with "reasonable" power) on the low bands, then there is something wrong with that connector. You needn't avoid connectors for their high intrinsic losses, as claimed by some.

But when using connectors make sure they are well installed, and properly waterproofed. Despite what some may claim, N-connectors will easily take 5 kW on the Low bands, and over 2 kW on 30 MHz. N-connectors are intrinsically waterproof and the newer models are extremely easy to assemble (much faster than a PL-259). A PL-259 connector is not a constant-impedance connector, but that is not relevant on the low bands. It is, however, a connector that is difficult to waterproof without external means. I always use a generous amount of medical-grade petroleum jelly (Vaseline) inside the connector to keep moisture out. Some cheaper coax, as well as semi-air-insulated coax, may see the inner conductor retract or protrude after time. Such coaxial cables are best used with PL-259 connectors, where you can mechanically anchor the inner conductor in the connector by soldering. In an N-connector, the retracting inner conductor sometimes will retract the connector pin to the point of breaking the contact.

9. BROADBAND MATCHING

A steep SWR curve is due to the rapid change in reactance in the antenna feed-point impedance as the frequency is moved away from the resonant frequency. There are a few ways to try to broadband an antenna:

- Employ elements in the antenna that counteract the effect of the rapid change in reactance. The so-called "Double Bazooka" dipole is a well-known (and controversial) example. This solution is dealt with in more detail in the chapter on dipoles.
- Instead of using a simple L network, use a multiple-pole matching network that can flatten the SWR curve.

The second solution is covered in great detail in *Antenna Impedance Matching*, by W. N. Caron, published by the ARRL. *ANTMAT* is a computer program described in technical Document 1148 (Sep 1987) of the NOSC (Naval Ocean Systems Center). The document describing the matching methodology as well as the software is called "The Design of Impedance Matching Networks for Broadband Antennas." The computer program assists in designing matching networks to match antennas (such as small whip antennas) over a wide frequency range.

CHAPTER 7

Receiving Antennas

Tom, W8JI, is my antenna guru, my technical guru, and he's a great friend to have. I have read everything he's published in his Web pages or on the Topband Reflector at least a few times. Tom not only has the knowledge, he's got the expertise and he uses it. He uses it to build the best Topband station of the US South. His contesting and DXing results are the proverbial proof of the pudding. But what I admire most about Tom is his missionary approach: He wants technically better hams. The introduction on his Web site says it so well: *"The most important thing any of us can do to make the Web an asset is to help each other with review to insure technical accuracy IMPROVES with time! Let's work to make Internet a reliable source of information instead of a collection of folklore!! ... Like you, I also learn new things every day. As my knowledge improves, I revise technical articles. I'll note revision dates on articles with changes, if the changes affect technical content."*

No technical question is too difficult for Tom, not even godfathering the chapters on special receiving antennas and on phased arrays. I know Tom regretted he could only spend a hundred hours or more reviewing these chapters... It indeed is an honor for me to have him help me with the Fourth Edition of this book.

Some of the Questions I Will Try to Answer in This Chapter:

- Why do we need separate receiving antennas?
- What is noise? How to eliminate noise.
- Are Beverages so superior?
- Is a longer Beverage better?
- How about vertical receiving arrays?
- What's the correct way of feeding special receiving antennas?
- Can I do as well from my city lot as the big guns from their rural farm?
- Are Flags, Pennants and K9AY loops an alternative to Beverages?
- Why not receiving arrays with parasitic elements?
- When will we have receiving arrays with "active" antennas?

In an e-mail K9RJ wrote, *"The challenge of 160-meter (Low Band) DXing is receiving. It should be no surprise that the highest DXCC totals on this band are achieved only by those who have the space for good receiving antennas, or who live in a location where much of the DX is close by. I'm not aware of any exceptions to this. Thus, the greatest need is for creative development of low-noise directional receiving antennas or techniques such as active noise canceling that can be used to improve receiving capability."*

Not so long ago, any mention of "receiving antenna" usually invoked thoughts of "Beverage Antennas." The evolution in all technical fields is staggering, and it includes receiving antennas. Not that something spectacularly new has been invented, but our ability to communicate worldwide at leisure has improved drastically thanks to the Internet. Technical knowledge is spread more easily, and technical discussions have become accessible to nearly everyone interested.

1. INTRODUCTION TO RECEIVING ANTENNAS

This chapter no longer is a Beverage-only chapter. Unlike previous editions, it will not even start with Beverages. Readers have asked for more receiving antennas, so here they are! However, before we get into describing receiving antennas and antenna projects in detail, it is important to understand a few basics.

1.1. Why Separate Receiving Antennas?

Separate antennas are necessary because optimum receiving and transmitting have different requirements. For a transmit antenna, we want maximum possible field strength in a given direction (or directions) at the most useful elevation (wave) angles. We cannot tolerate unnecessary power loss in a transmit antenna, because any amount of transmitting loss decreases signal-to-noise ratio at the distant receiver. Antenna efficiency is an important issue for transmitting. It is obvious that for a given elevation angle and direction the highest gain antenna will deliver the strongest signal to the target area. We really do not care if we are being heard in other directions (areas) or not, we are only interested in the target direction.

Choosing a transmit antenna is a matter of properly positioned gain. Transmitting antennas require high directivity to achieve high gain, not directivity just for the sake of eliminating transmitting signal in unwanted directions. Tom, W8JI, at **www.w8ji.com/** adds to that: *"Takeoff angle is not important, what we actually need is maximum possible gain at the desired angle and direction. After all, we don't care where the peak is as long as the antenna we pick has more signal (gain) at the desired spot than other antenna choices!"*

A receiving antenna on the other hand has a different design priority. The goal is obtaining a signal that can be read comfortably, which means having the minimum possible amount of QRM and noise. The important issue when receiving is *signal-to-noise ratio* (S/N). The receiving antenna providing the best performance can and will be different under different circumstances, even at the same or similar locations. There is no such thing as a universal "best low-band receiving antenna."

Why doesn't the reciprocity law apply to signal-to-noise ratio as it applies to signal level? It's easiest to explain this with an example: Consider a high-band Yagi with 7-dBd gain, including ground-reflection gain. This antenna will improve the transmitted signal by 7 dB over a dipole, provided both have peak gain oriented to the target area. Does a 3-element Yagi with the same efficiency as a dipole improve reception S/N by the same amount as it improves transmission?

The answer is simple: Probably not! S/N will improve much more than the 7-dBd gain when very strong noise sources are located in a pattern null. If the null is –25 dBd, S/N can increase as much as 32 dB (+7 dBd signal to –25 dBd noise). Of course, the improvement will normally be less than 32 dB, since that is extreme.

If the noise comes from exactly the same direction as the desired signal, the Yagi's 7-dBd gain will not improve S/N at all. The Yagi will deliver equally increased signal and noise power, both being 7 dB stronger than the dipole.

The Yagi also might have decreased efficiency. This is actually very common, because of losses caused by increased element current. In reality, a 7-dBd Yagi often has more than 7-dBd directivity. If Yagi efficiency were only 50%, 7-dBd gain would require having 10-dB directivity increase. These are all reasons why gain does not determine receiving S/N improvement, and why the higher-gain antenna very often does not provide the best reception.

There is one predictable effect of gain. Signal levels will be increased by the amount of gain, both in transmitting and receiving. (Keep in mind that signal level is not the same as signal-to-noise ratio.) Continuing the example above and assuming perfect lobe alignment with the path, the Yagi's signal level will be 7 dB above the dipole in the target area. The distant receiver will always have 7-dB more S/N when the Yagi is used. This is true regardless of any S/N improvement we might or might not observe when receiving with the same Yagi. What counts for improving communications is the ratio of signal-to-noise on both receiving ends of the circuit. In practice this means there is no reciprocity "in readability"—reciprocity only applies to signal level. This is *not* "one-way" propagation, although it sometimes may cause people to think this is happening.

1.2. Gain Versus Directivity

Gain is a function of efficiency and directivity. High gain means an antenna has high directivity and reasonable efficiency. The increased field strength comes with a price. The extra energy found in the main lobe is energy that was removed from other directions (also see Chapter 5, Section 2.1.).

The answer to improved receiving can be the same as transmitting. Installing a highly directive transmit antenna results in high-performance receiving, so long as the antenna is not aligned with or installed near noise sources. Unfortunately, the physical size and height of efficient antennas—especially on 160 and 80 meters—often makes high-gain transmitting antennas prohibitively expensive.

Fortunately, high or even modest efficiency is not a direct requirement for directivity and receiving. This chapter will show it is possible to build relatively small receiving antennas that exhibit excellent directivity and greatly improve receiving, even though the antennas are useless for transmitting because of high losses and low gain.

Directivity is not the same as gain. It is possible to construct very directive antennas that actually have negative gain but that provide phenomenal receiving improvements. It is worth repeating: We need directivity—not gain—for a good receiving system.

The next question is: How much negative gain can we live with? The answer is fairly simple once an antenna is installed. If you can easily detect a background noise increase when a dummy load is removed and the antenna connected under the quietest operating conditions (usually winter daytime within a few hours of sunrise or sunset) with the narrowest IF filter selected, gain is OK! As Tom, W8JI, puts it with regard to preamps and matching devices in particular: *"Once you clearly hear external noise, amplifiers or impedance matching won't help. Just be sure you can hear noise at the quietest time you expect to operate."*

We learned in Chapter 3 (Section 1.2) that our present-day receivers have a large sensitivity margin when used with reasonably efficient antennas, especially considering the large amount of noise on the low bands (unless you live on a desert island or in the wilderness). Many receivers are sensitive enough to use with antennas having –10 to –20 dBi gain, depending on various factors. (See Section 1.2 in Chapter 3.) For the rest, we can always use a preamplifier to boost the signal to a more comfortable level.

In very quiet locations, with 250-Hz selectivity, a minimum discernable signal sensitivity of –140 to –145 dBm might be required while using narrow-pattern, low-efficiency receiving antennas. In suburban locations, –125 to –135 dBm sensitivity is often adequate. (See also Chapter 3.)

Very directional antennas and narrow selectivity reduce noise power, requiring less receiving system sensitivity to yield a satisfactory output S/N. Since noise power is proportional to bandwidth, a 250-Hz filter requires 10-dB more receiver sensitivity compared to the same system using a 2.5-kHz filter. Directivity has the same effect when noise is evenly distributed. A 3-dB increase in directivity for a given amount of antenna gain will provide 3-dB less noise power, and require a receiver sensitivity decrease of 3 dB. The key factor for the sensitivity required is if external noise from outside the antenna system clearly dominates the receiver noise at the narrowest selectivity being used.

1.3. Noise

We have covered the nature of noise and its intensity in different environments (urban, suburban, rural) in detail in Chapter 3, Section 1.2.4. What is noise? Noise is the sum of many signals, with most sources unintentional. We can distinguish three sorts:

- Noise generated by nature: noise from thunderstorms (static, QRN); precipitation static
- Noise generated by man: mostly from arcs or rapidly switched sources, such as power lines, switching power supplies, digital systems, electronic voltage controls such as dimmers or motor speed controls, defective doorbell transformers, lighting systems, electric fences, thermostats and so on.
- Noise generated by poorly designed, operated or maintained transmitters: CW clicks, sideband splatter, noise-sidebands, spurious oscillations and other transmitter defects.

When we consider how noise propagates or travels to our locations, we can distinguish:

- Near-field noise generated in the antenna system, or coupled directly to the antenna though induction or electric fields from nearby wiring. This near-field noise includes precipitation static, but is mostly man-made switching or sparking noise.
- Fresnel region noise generated outside the induction field area but before the antenna pattern is completely formed. This noise includes man-made noises, such as those from arcing high-voltage wiring or strong local static discharges.
- Noise propagated from the far field by groundwave or ionospheric propagation. This noise includes CW clicks, sideband splatter, noise-sidebands, lightning noise and other natural and man-made sources. It includes the sum of many hundreds of thousands of low-level noise sources, such as the accumulated noise from entire cities.

We usually refer to the sum of all *unidentifiable* noises as *band noise* or even *background noise*. We generally classify identifiable noise generated by intentional transmitters as QRM, although the end effects are largely the same as any other noise.

In quiet rural locations (away from polar regions) lower-frequency band noise is evenly distributed at all wave angles and directions whenever darkness surrounds the receiving location. Noise is only lacking in directions where propagation is very poor, or directions having a total lack of noise sources. We do not consider the sky as "quiet" on the lower bands, because the ionosphere reflects all types of very small noise sources from both nearby and distant sources. While the amount of noise from each source might be very small, the accumulated effect of innumerable noise sources is a smooth broadband hissing noise.

In some locations, QRM consistently arrives from well-defined directions. If you live in Western Europe, almost all QRM (transmitter generated noise) arrives from the East. In the very Northeast coast of Canada, QRM generally arrives from the mainland USA to the southwest. In many locations, QRM comes from many (if not all) directions, with nearly random distribution. You may want a different antenna pattern when DXing on relatively clear bands compared to patterns used during crowded contests.

1.4. Reducing Various Noise Types

Noise has exactly the same characteristics, so far as an antenna is concerned, as signals from *intentional* transmitters. There is no way to sort "good signals" from "bad noise" except through directional characteristics or *directivity* of the receiving antenna.

External noises can be eliminated or reduced only by the principle of phase opposition: Receive the noise with at least two different antennas (elements) and add the signals received from the elements in such a way that the sum is zero (equal amplitude and 180° out-of-phase). We can do this using arrays (groups of antennas) or using a special configuration where one antenna is a (usually small) noise pick-up antenna and the second one is the "regular" receiving antenna. In this case a so-called noise-canceller (such as the MFJ-1025) will combine the two signals to cancel a given noise signal (see Section 1.5).

Different types of noise are controlled through different methods. There are three primary sources of noise:

- Noise from thunderstorms
- Precipitation static
- Man-made noise

1.4.1. Noise From Thunderstorms

If a very active thunderstorm is local (directly overhead), noise is the least of our worries. We really should disconnect lightning-sensitive or inadequately protected equipment (before the storm) and stay away from the radios! If the storm is somewhere in the distance (usually covering a wide azimuth), an antenna with a very broad pattern null and extremely good front-to-storm-direction ratio will help (see Chapter 5, Section 2.10).

1.4.2. Precipitation Static

While often attributed to charged particles (such as water droplets) hitting an antenna, most precipitation static is actually caused by intense electric field gradients in the area surrounding the antenna. Such conditions commonly appear during inclement weather, when movement of particles or moisture causes concentrated areas of charges. The strong electric fields are responsible for noise-producing *corona discharges*. The noise comes from low-current corona discharges from sharp or protruding objects.

Sailors saw this effect on tall-masted ships, calling it *St Elmo's fire*. This noise generally builds slowly from a sizzle to a high-pitched whine and disappears with nearby lightning flashes. Lightning "equalizes" the potential difference between earth and nearby clouds, reducing the charge gradient and corona. Since this noise is generated in or very near the antenna, directivity is of no help.

Using an antenna at a lower height reduces corona current—The electric-field gradient is smaller close to the wide smooth surface of the earth. This is especially true when the low antenna is surrounded by taller structures. Round, smooth and insulated conductors are helpful, because they reduce voltage gradient and resulting corona discharges. Vertical antennas are particularly sensitive to precipitation static; they have pointed ends protruding upwards towards the oppositely charged sky. The corona also comes from the very high-impedance antenna end, which aids in coupling power into the receive system.

Beverages on the other hand, being near earth, will have fewer corona discharge problems. They also have low surge impedances. This means the low-current high-voltage arcs transfer very little noise power into the antenna. Beverages are thus quite resistant to precipitation static.

Quads are more resistant than Yagis because quads have long flat sides with blunt lower- impedance high-current areas towards the sky. Yagis have protruding high-impedance pointed ends. Low-current arcs are not only more likely to happen in Yagis, they are also better impedance matched to the antenna! Quads have a reputation for being "quiet antennas," but this only applies to corona. For all other noises quads are no better than any other antenna.

1.4.3. Man-Made Noise

Local man-made noise is received several ways. When the source is a modest distance (1 to 10 km) away, noise arrives by groundwave propagation. If noise comes from just outside or nearly outside the antenna's Fresnel zone, it can be eliminated with pattern nulls. The Fresnel-zone area is where the pattern is not fully formed. The zone is related to array size. It can extend a few kilometers with a very large array, particularly one using broadside elements on low frequencies. If the noise source cannot be eliminated using a directive antenna, we often make use of so-called *noise-cancellers* to solve the problem. If the noise source is from a single source we can define a few solutions.

1.4.3.1. Single-Point Radiation Far Source

Local noise arriving from one clear radiation point, even if multiple sources, can easily be nulled. The antennas need not be similar, but deep nulls require two antennas that both "hear" the noise. The sense antenna should be placed closer to and directly in-line with the noise source. The spacing can be nearly any distance, but λ/4 or more is always best. There must be a stable RF phase relationship between the noise received in the main receive antenna and the noise-sense antenna. Since local noise is received by surface or ground wave, the phase, polarization, and amplitude are constant. This allows a stable deep null to be obtained, using equipment such as the MFJ-1025 noise canceller.

1.4.3.2. Distributed Radiation Source

Noise from a single source or multiple sources can be fully nulled if the distance to the radiation area is large compared to the length of the radiating area. This is true even if the noise follows power lines and radiates from multiple points or is from multiple sources. The sense antenna must clearly and strongly pick up the noise. The ideal case is where the sense antenna is very close to the source and the signal antenna is a much larger distance away. If however the noise source is right on your street and the radiating power lines are in front of your house, it is likely that all of this happens in the near field of both the receive and sense antennas, and in that case nulling will be impossible.

In all cases the sense antenna should ideally hear only the noise and not the wanted signals, which means it must be fairly close to noise source. And the sense antenna should be fairly small.

1.4.3.3. Nearby Man-Made Noise

If the noise source is very close (in the near or *induction* field), it becomes difficult or impossible to eliminate noise through antennas arrays. In this case the problem must be tackled in a different way, either by eliminating the noise source or experimenting (trial and error) with various antennas. Using a portable receiver or a fox-hunting (DFing) receiver for 160 or 80 meters, local sources can be easily found. If the noise cannot be killed, such a single source noise, even in the near field, can often be completely nulled out provided the sense antenna is installed near the noise source, and the main receiving antenna is located farther from the noise. It's obvious, however, that the best solution in this case is to "kill" the noise source directly.

1.4.3.4. Propagated Noise

This noise generally sounds like a smooth hiss, even though it is coming from hundreds or thousands of raspy or harsh noise sources. Propagated noise is rarely, if ever, audible in urban areas on 160 meters, since it is masked by harsh local noises. Propagated noise is sometimes audible in quieter directions of suburban areas on 160 meters, but not in "noisy" groundwave directions or if a local dominant noise is present. Propagated noise is often responsible for the entire noise floor in remote rural areas. It is often possible to find the direction of strong band openings by looking for highest propagated noise, because the enhanced propagation can sum countless noise sources for many thousands of km! Unfortunately, as Tom, W8JI, says: "*Propagated noise reduces the advantage of super-quiet locations during the night.*" Hearing propagated noise is a good indicator of how quiet your location is and how good your receiving system is. For example, the winter season 160-meter daytime-to-nighttime noise level increase at W8JI has been measured at 15 dB. This is in the absence of thunderstorms within many thousands of miles.

While local man-made noise can often be nulled, propagated noise is another story. Canceling propagated noise only works with antennas of identical polarization and similar patterns. The antennas must be close to each other, so they receive signals in a constant phase and amplitude relationship, with no space diversity. However, propagated noise constantly changes phase, polarization and amplitude. Different types of antennas in a canceling system respond differently, making canceling impossible or very unstable. For any relief from such propagated noise, it must arrive from a significantly different direction than the desired signal.

1.4.3.5. QRM

CW clicks, splatter, noise sidebands, etc is usually called QRM, but it is just another form of noise. We do not deal with QRM any different than the way we deal with other propagated noises. If you are lucky, the QRM does *not* come from the same direction as the desired signal. If it does, there is very little you can do about such noise with your antennas.

1.5. Suppressing, Canceling and Noise Canceling

What is the principle of canceling and suppressing? It's really fairly simple. First, we must receive the unwanted signal with two different antennas. The main antenna would receive as much desired signal as possible. Ideally, the second antenna would hear only noise, with very little desired signal. The noise outputs of the antennas would then be adjusted so they are exactly equal, and the results combined exactly out-of-phase (180°). Total canceling would occur when these two conditions are met. If the noise antenna hears very little desired signal, all noise from the main source would be removed, without any change in desired signal level.

Noise cancellers are simple in theory. They allow adjustment of level, and rotation (or shift) of phase. When selecting a noise canceller, the following technical parameters are important:

- Low amplitude change with phase adjustment
- Wide amplitude range
- Wide phase range
- No loss
- Immunity to overload (good dynamic range).

Noise cancellers are most frequently used to cancel the noise from a single noise source. But, provided they are designed for it, they can also be used to feed elements of a receiving array, without the aim of canceling a specific noise source. (See Section 1.35.)

Homebrewers should exercise caution in selecting a noise-canceling circuit design. Some designs are very poor, having circuits that do not actually rotate or shift phase. It is impossible to shift phase in a simple transformer system, since transformers only invert phase. We cannot mix only 180° out-of-phase signals to obtain phase variation. L/C circuits, R/C circuits, or delay lines must be used.

The best noise-canceling circuits are bridge-type phasing systems. These circuits look much like a standard Wheatstone bridge, except a relatively high value reactance is substituted for at least one resistance. If such a circuit drives a high-impedance load, considerable phase shift can occur with minimal amplitude change.

1.6. Directive Receiving Arrays: How to Obtain a Null

Let's develop an antenna array that produces a null. In order to form a deep predictable null, we need two antennas with nearly identical patterns. Let's assume we will use two vertical antennas. In terms of physical size, the most efficient two-element combination is an end-fire array—This is where maximum radiation occurs in-line with the elements. The ideal spacing ranges from λ/4 downwards.

We must choose an optimum spacing between elements and select the phase delay where the signals are combined to null signals from unwanted directions. In other words, the outputs produced by these antennas are combined equally in amplitude and precisely 180° out-of-phase for signals arriving from unwanted directions. At the same time, we must be sure the null does not reduce the desired signal.

Since the two antenna elements are not at exactly the same physical point, an incoming wave takes a small time to travel between each antenna. Put another way, the two antenna elements receive the same signal, with slightly different phase differences for different directions of arrival. The exact phase difference depends on the distance between elements and the angle at which the signal arrives (both in the horizontal as well as the vertical plane). The largest phase difference occurs when signals arrive in-line with the elements.

Let us assume there are two vertical elements spaced λ/4 (90°). Refer to **Fig 7-1**. A signal coming from the right (in-line with the two antennas) at a low elevation angle will arrive at the second antenna (B) later than the first one (A). The phase difference will be 90° (since this is the physical separation). To completely cancel this signal you must combine the 90° shifted outputs (due to the physical separation) 180° out-of-phase. You can do this by connecting impedance-matched feed lines to both antennas, making the feed line to antenna B 90° electrically longer than the line to antenna A. If you connect these two lines together using a method that produces equal currents in each antenna, the signals coming from the back towards A will cancel out.

Signals from the front direction towards B add quite differently. As they arrive at A, they have a spatial delay of 90° at A. Since the feed line to B also delays phase 90°, the total phase shift is 0°. Signals from the direction of B are in phase and add together.

While many people use 90°-shift with 90° phasing, it is not the optimum phase delay. What amateur operator wants maximum nulling at a 0° elevation angle? Few signals arrive at (nearly) a zero elevation angle, except ground-wave signals that are perfectly in-line with the array.

We often need to place the null at higher angles, or move it slightly off the back. This not only increases usefulness of the null, it increases gain and directivity of the array.

Let's look at how to produce a null at a given wave angle, in this example 52.5°. Refer to **Fig 7-2**. Both antennas are still spaced spaced λ/4 apart. The rearward signal again arrives at antenna A before antenna B, but this time the difference will be shorter than 90°. A little trigonometry shows us that the spatial phase delay is now 90° × cosine (52.5°) = 55°. To create a null, we have to combine signals exactly out-of-phase, but the spatial delay is now 55°. The extra delay in the feed line

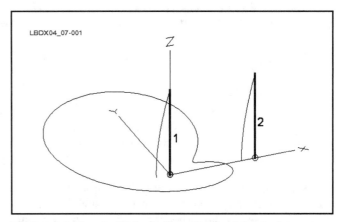

Fig 7-1—A signal arriving off the back of the 2-element end-fire array hits element A earlier than element B. See text about how the directivity pattern shown is obtained.

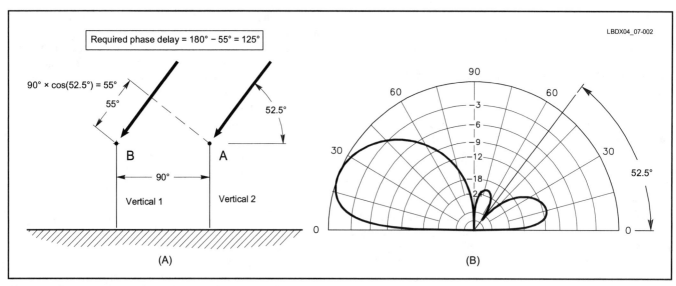

Fig 7-2—Development of a null at a given elevation angle in a 2-element end-fire array. See text for details.

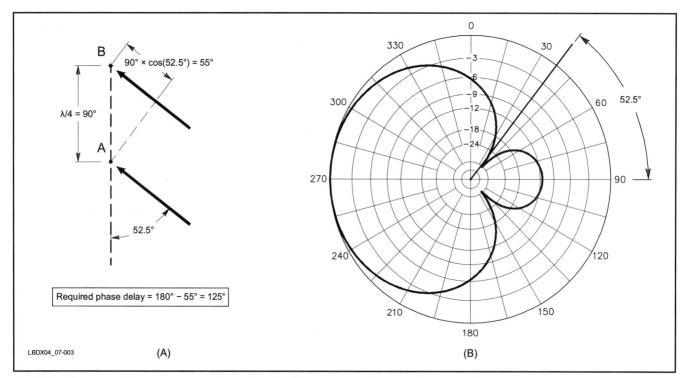

Fig 7-3—The same trigonometry applies when looking at the null angles in the horizontal plane. See text for details.

to element B becomes 180° – 55° = 125°. Signals from the front are now 125° – 55° = 70° out-of-phase. This actually does not decrease gain, because the transmitter power has to go someplace. What it does do is increase element currents, making the feed resistance of each element drop more than the case of 90° phasing. Each element actually changes reactance and resistance from mutual-coupling effects. The overall gain or directivity of this array actually increases slightly over the 90° phasing case. We have a better receiving and transmitting antenna.

In this λ/4 spacing example, we can obtain a null at any wave angle by changing phase delay between 90° (λ/4 long for 0° null angle) and 180° (λ/2 long for 90° null angle). This is very useful, and it also works for other element spacings if we use the proper phase delay ranges. (We must always be sure the feed and phasing system compensates for impedance changes.)

Null elevation is not the only pattern change. The array phasing change also moves the null in the azimuth plane. In **Fig 7-3** we can see the same geometry and trigonometry applies when examining the azimuth pattern at 0° elevation angle over a perfect ground. At a 0° wave angle, the 125° array phasing moves full cancellation to two points 52.5° either side of an imaginary line drawn through the line through the two antennas.

At different elevation angles, nulls are present in different directions. **Fig 7-4** shows a three-dimensional view of the pattern. The null actually forms a deep cone surrounding the back lobe. This much wider null greatly increases the area of zero response. Removing the response over this large area decreases noise pick up and improves gain and directivity of this array. It also gives two deep nulls along the ground and pulls in the sides of the pattern, decreasing antenna response to ground-wave noise.

The offset angle for maximum attenuation is identical in the horizontal and vertical plane. This is useful information. We can adjust an antenna for maximum attenuation at a 35° elevation by adjusting for maximum attenuation 35° offset from the back. It is not necessary to hire a helicopter!

In the foregoing examples, elements were spaced λ/4 (90°) apart. The same analysis can be used with other spacings, such as λ/8.

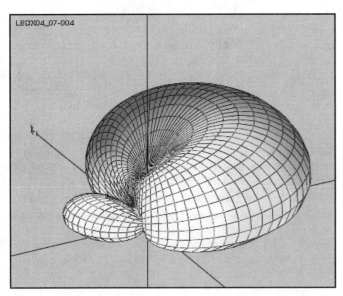

Fig 7-4—A 3D radiation pattern of the 2-element end-fire array (plotted by *Antenna Model*).

Table 7-1 shows phase delay for different null angles and element spacings (from 45° to 100°). Phase delay is given by $\phi = 180° - [S \times \cos(Na)]$ where S = spacing in degrees and Na = null angle in degrees.

1.7. Where Do We Put the Null?

We now understand how to move the null in an array, but haven't discussed the best position for the null. Since circumstances vary, there is no universal "best position." We always want to position the null to remove the maximum amount of accumulated noise power compared to the response at the desired signal elevation and azimuth angles. There are three distinct cases to consider:

- A location with desired signals arriving from a reasonably wide expanse of low noise, but with very high levels of noise covering a wide quadrant behind the array. An example of this would be a location in the suburbs of a high-noise city looking out over the very quiet ocean towards many DX contacts. Another case would be a noisy power distribution line going past a very quiet location. In this case the vast majority of noise (more than 15 dB higher than normal ambient in the forward direction) would always arrive from a well-defined relatively wide area, while desired signals arrive from a relatively low-noise wide forward area.
- The second case would be a quiet rural or somewhat quiet suburban location, with noise randomly distributed in all directions. This would be the case for most suburban and rural amateur stations, where widely distributed rearward local noise or point-source noise is not a major issue. This situation also applies where noise averages over time to be about the same from all directions, and isn't greatly stronger (greatly would be 15 dB or more) from any single area.
- The final case is where a strong single-source noise profoundly dominates all other noise arriving at the site. This may be typical of amateurs living in an area where all noise comes from an electrical substation, or a somewhat distant group of arcs or noise sources concentrated in one narrow direction.

In the first case, we should compare response in the rearward (or any other exceptionally noisy) area to the desired signal direction. The signal must be from a point in the relatively quiet front area of the antenna; the noise from a

Table 7-1
Required phasing angle φ as a function of spacing and notch angle

Null Angle | | | | | | | Element Spacing, Degrees | | | | | | | | |

Null Angle	23	35	45	50	60	70	80	90	100	110	120	130	140	150	160	170	180
0	158	145	135	130	120	110	100	90	80	70	60	50	40	30	20	10	0
5	158	145	135	130	120	110	100	90	80	70	60	50	41	31	21	11	1
10	158	146	136	131	121	111	101	91	82	72	62	52	42	32	22	13	3
15	158	146	137	132	122	112	103	93	83	74	64	54	45	35	25	16	6
20	159	147	138	133	124	114	105	95	86	77	67	58	48	39	30	20	11
25	160	148	139	135	126	117	107	98	89	80	71	62	53	44	35	26	17
30	161	150	141	137	128	119	111	102	93	85	76	67	59	50	41	33	24
35	162	151	143	139	131	123	114	106	98	90	82	74	65	57	49	41	33
40	163	153	146	142	134	126	119	111	103	96	88	80	73	65	57	50	42
45	164	155	148	145	138	131	123	116	109	102	95	88	81	74	67	60	53
50	166	158	151	148	141	135	129	122	116	109	103	96	90	84	77	71	64
55	167	160	154	151	146	140	134	128	123	117	111	105	100	94	88	82	77

noisy "problem" area. Good performance means we need a very broad null in the direction of the strong noise rather than the typical requirement of high directivity. We can identify this situation by changing antenna direction. If the noise shows about the same F/S and F/R as signals and is very clearly in one general direction (a F/R or F/S change on your "S" meter similar to that of regular signals), you should pay close attention to the Directivity Merit Figure (DMF), described below.

In the second case, noise averages to be within several dB from every direction. We need low average gain compared to point-gain at the specific elevation angle and azimuth of the desired signal. This translates to the need for high directivity.

In the final case, we are mostly concerned about maintaining a very deep null in one direction. Noise from one specific direction is hundreds of times stronger than normal band noise, ruining our DX. We need to totally remove it without removing the desired signal.

1.8. The Directivity Merit Figure (DMF)

In the Third Edition of this book, I ranked receiving antennas by calculating the average front-to-back, which was calculated at 120°, 140°, 160°, 180°, 200°, 220° and 240° azimuth and between 10° to 80° in elevation. This gives 7 × 8 = 56 gain figures, which were then compared to the maximum forward gain of the antenna.

Keeping the same basic approach, I elaborated on the idea and now calculate the average gain in the entire back azimuth half of the antenna, from 90° to 270°, and over the entire elevation range from 2.5° to 87.5°. Doing all of this at 5° increments means we consider 37 × 18 = 666 gain values. The average rearward gain now is the average of 666 values (watch out, you cannot average dB figures directly and have to compensate for area). We can now define a figure of merit for the directivity (front response to back half-hemisphere) as being the difference between the forward gain at an optimum wave angle (for example, 20°) and the average rearward gain.

For the 2-element array we developed above (with 90° spacing), this Directivity Merit Figure is 11.6 dB. The directivity merit figure (DMF) is the peak front lobe (at a specified elevation angle) gain versus the average back half-hemisphere gain.

This method of evaluating a receive antenna applies to a case where a dominant noise arrives from a relatively wide half-hemisphere. If the noise is evenly distributed in all directions (eg, in a very quiet location), the RDF ranking system discussed below should be used.

Let's examine DMF further. What is the DMF (at 20°) of a vertical antenna? By definition the vertical is an omnidirectional antenna. Does that means the DMF is 0 dB? No, its peak lobe (in all horizontal directions) is at an elevation angle of approximately 25° to 30°, but the antenna has good rejection at high elevation angles in all horizontal directions (**Fig 7-5**). The DMF of a single vertical (12 meters long at 1.83 MHz) is 3.8 dB (calculated for a 20° wave angle).

A word of caution: The non-professional versions of *EZNEC* and other software programs calculate patterns at an "infinite" distance from the antenna. If your software does not allow you to set a pattern distance, it almost certainly will not accurately evaluate groundwave signals. This means that anything less than perfect-ground causes vertically polarized,

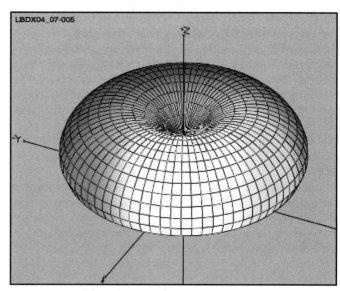

Fig 7-5—A 3D radiation pattern of a single vertical (plotted by *Antenna Model*).

zero-degree responses to incorrectly appear as zero. If groundwave or directly propagated noise exceeds skywave noise levels, vertically polarized antennas (including Beverages) may thus appear significantly better than they actually are for locally noisy locations. Despite that, these models work quite well at any location where skywave exceeds groundwave noise level.

What does a 11.6-dB DMF mean for our 2-element endfire array and what does the 3.8 dB DMF mean for our single vertical? It tells us the 2-element array will deliver 7.8 dB better S/N ratio, provided the dominant noise is skywave in the rearward area and provided that the desired signal arrives in the center of the forward lobe peak (at a 20° elevation angle). The test for this condition is simple. If you see a similar change in noise level to the change in signal level when the array is reversed, you should be using DMF to evaluate antennas. You should set the rear measurement window to the direction of strong noise.

If you have a strong single-point noise and if that noise arrives in a deep notch in your receiving antenna pattern, the S/N improvement may be much greater than expected. If noise arrives predominantly from a higher antenna response area, the S/N improvement will be proportionally less. Another important thing to consider: Signals almost never arrive from a single angle or direction. A range of angles is involved, and a single-angle evaluation does not fully represent the real world. (If it did, signals would not have nearly as much QSB!)

Many noise sources vary in direction, arrival angle and polarization tilt. The same is true for desired signals. Because of this, we really only are considering "average" results over time. Averages are not foolproof under every condition. There is a fable or story about a person who decided it was safe to wade across a river, because the depth averaged only 4 feet. Well, he never made it to the other bank.

How can we calculate the DMF of an antenna? First we model the antenna under generic reference conditions: The ground quality assumed is "average ground" (see Table 5-2 in Chapter 5), which means σ = 5.0 mS/m conductivity and a relative dielectric constant ε = 13. Using a modeling program, you calculate the far-field gain (at an infinite distance) between 90° and 270° azimuths, in 5° increments and between 2.5° and 87.5° elevation angles, also in 5° increments. These data are saved in a text file. Using a small dedicated software program I calculate the average of all these gain values. You cannot average dB figures directly, but must transform them into power ratios first. Next you have to compensate for area, since the physical length of 1° of azimuth near the zenith is very short compared to the length of 1° at 0° elevation. The length changes according the cosine rule. Finally, you must average these values and convert back to dB. This is the Average Back Half-Hemisphere gain. Next compute the difference between the forward gain at a 20° elevation angle and the Average Back Half-Hemisphere gain. The result is the DMF.

If we do this calculation for our end-fire array using 90° spacing and 12-meter long elements at 1.83 MHz with 105°

phasing, the DMF = 13.8 dB. Remember the concept of DMF assumes that the majority of the noise comes from the rear of the antenna and the forward lobe area is very quiet.

Fig 7-6 shows the radiation patterns for a 2-element end-fire array with λ/4 spacing and various phase angles. Note that the highest DMF is obtained for fairly high phasing angles. These phasing angles result in a good rejection at high elevation angles, although significant lobes at low angles are formed. On the average, though, higher phase angles achieve a better F/B, and thus a higher DMF. For higher phasing angles the forward lobe also gets narrower. Hence the Receiving Directivity Factor (RDF, discussed next) also becomes better for such angles.

1.9. The W8JI Receiving Directivity Factor (RDF)

W8JI has developed a similar measure to quantify receiving properties of antennas. While the Directivity Merit Factor (DMF) compares forward gain at the desired wave angle to the average gain in the rear half hemisphere, Tom's *Receiving Directivity Factor* (RDF) compares forward gain at a desired direction and elevation angle to average gain over the *entire sphere*. RDF includes all areas around and above the antenna, considering noise to be evenly distributed and aligned with the element polarization. Losses are factored out, and we find the directivity of the array. If noise, on average, is evenly distributed in *all directions* (including forward and side lobe areas) this method provides an accurate picture of receiving ability. (Keep in mind most antenna modeling programs used by amateurs calculate pattern at infinite distances and ignore groundwave response. RDF models, like DMF models, are not reliable when groundwave noise dominates skywave noise.)

For everything but an omnidirectional antenna, the RDF will be different from the DMF. You have to decide if your location has dominant skywave noise in the rearward area (DMF), or if skywave noise is evenly distributed on average (RDF). Do not compare RDF with DMF.

Calculating the RDF is very simple. First, you carefully model the antenna with a Windows version of *EZNEC* by plotting the 3D pattern. The main *EZNEC* window shows average gain at the very bottom. You normally use this average-gain figure (with all lossy antenna elements set to zero loss and in free-space or over perfect ground) so that you can isolate actual ground and element losses from possible deficiencies in the model itself. You must fix any model deficiencies before proceeding. Once you've determined that the model itself is OK, you can resume using lossy elements and real ground to calculate the average gain figure.

Now, you go to a two-dimensional elevation or azimuth pattern and select the desired elevation angle and/or azimuth of the desired signal with the gain cursor and note the gain. The difference between the overall average gain and gain at the desired direction and elevation angle is the RDF. The front lobe does not have to align with the desired signal. You can move the cursor around and look at the RDF for off-path signals.

For our 2-element end-fire array (with 90° spacing of 12-meter long elements at 1.83 MHz) with 105° phasing, RDF = 8.2 dB. See Fig 7-6. The RDF of a single vertical is 4.8 dB. If noise or interference is somewhat evenly distributed, the end-fire array will show 8.4 − 4.8 = 3.65 dB signal-

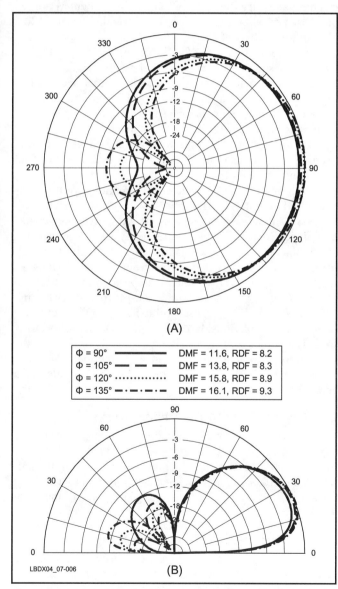

Fig 7-6—At A, azimuth patterns and at B, elevation patterns for 2-element end-fire array with λ/4 spacing and various phase angles. Solid line: φ = 90°; dashed line: φ = 105°; dotted line: φ = 120°; dashed-dotted line: φ = 135°. See Figs 7-2 and 7-3.

to-noise ratio improvement over a single vertical (this is at 20° elevation in the main lobe peak).

Throughout this chapter we will assess the quality of the antennas by calculating both the DMF as well as the RDF, in addition to the −3-dB (half-power) beamwidth. Also remember a few dB of improvement in S/N, while meaningless on strong signals, can make a profound difference in readability of signals near the noise level.

1.10. RDF or DMF?

Both evaluation systems have their merit. If you're in a location that's always very quiet, with no specific noise or QRM sources from a particular direction, then RDF is most meaningful. The exception would be if you always had grossly dominant noise (or QRM) only from one direction. The dominant noise would have to be so strong as to exceed distributed background noise by the null-depth ratio between an antenna selected by RDF compared to a F/R selection for a F/R selection to be valid, and would have to do so with some consistency.

For example, assume noise from a rearward quadrant was 20 dB higher than average noise from all other directions. Once the array had greater than 20 dB F/R ratio you could simply quit worrying about looking at F/R averages. Once the spot noise is down in the average noise, any additional depth is meaningless. At that point RDF takes over.

Another very important thing is when we work DX at local sunset or sunrise, the rearward area is looking into a zone of poor propagation. W8JI wrote *"At my very quiet QTH I see a 5-10 dB noise drop to the east at sunrise, and a 10-15 dB drop to the west near my sunset. This is because distant noise does not propagate in through the daylight areas. In this case, F/R is virtually meaningless and probably is "over considered" even in RDF. Another thing is when we look into an area of good propagation, noise is enhanced from that direction also. The same mechanisms that enhance noise propagation enhance signals, so we had better consider beamwidth (which RDF does)."*

RDF does not work well for local noise, but then nothing else will either. That's because EZNEC and other programs do patterns at "infinite" distance and do not show true response along the earth. They have no groundwave. If your modeling program does not have an input for distance, you can be sure it ignores groundwave. As such, there isn't an accurate model or method for those of us limited by local noise sources. RDF is exceptionally good for comparing similar antennas, such as a single Beverage to phased Beverages.

I think there are very few skywave noise cases where anything but RDF applies. Even while it is far from perfect, it is the best overall method. If you have a case like those of us in the SE USA do, where a certain land mass has frequent thunderstorms (Florida and S Georgia), then it might pay to always be sure to have a deep null over that area. Even so, I would never pick the antenna exclusively based on the ratio of average gain in the null area to gain in the desired direction.

At my QTH, antennas that have a poor 15 dB F/B hear just as well or better than antenna with huge 40 dB F/B ratios when the forward BW of the modest F/B arrays is narrower. The exception is summertime, when thunderstorms are off the rear. The worse thing you can do, over time, is go for extreme F/R at the expense of Half Power Beamwidth."

If you are in a less-ideal situation, it seems to me that you first have to take care of the noise/QRM that is predominant from one direction. At my QTH that is the East/South-East. A good F/B (good DMF) is essential. Once that has been taken care of, further noise reduction can only be achieved by narrowing the forward lobe beamwidth, provided you have the room to do it, because broadside arrays that narrow the forward lobe require a great deal of space! In a nutshell: have a look at both the DMF and the RDF figures, and understand what they mean.

1.11. Broadside Arrays

In end-fire arrays, we considered the case of two elements fed with different phasing. In a broadside array, radiation occurs in a direction perpendicular to the line through the elements. What happens if we feed the elements in-phase and vary the spacing? Since both antennas are fed in phase, maximum radiation is perpendicular to a line bisecting the elements regardless of spacing. Areas in-line with the elements are the "side" of the array. Fig 11-2 in Chapter 11 shows patterns at a 0° elevation angle.

At zero spacing, there would be no spatial phase delay. Signals arriving from the sides would be in-phase at both elements. As spacing is increased, a point is reached where elements are separated λ/2. Signals arriving at 0° (groundwave) from the sides will be delayed λ/2 in space, exciting each element 180° out-of-phase. Since the feed system has no element-to-element phase shift, zero-degree elevation angle signals arrive at the common point out-of-phase and completely cancel.

As you increase the distance between elements, the bi-directional lobe becomes progressively narrower, and the same time the vertical elevation angle at which the null occurs off the side, is lifted off the ground, which is what we really want. However, beyond λ/2 spurious lobes begin to appear. These spurious lobes increase in strength as spacing is increased, and may cause problems if they fall in noisy directions. It is obvious that with wider spacing, more directivity is obtained through narrowing of lobes. If noise arrives in roughly similar amounts from all directions, narrower lobes (more directivity) will translate into higher S/N ratios and higher RDF numbers.

1.11.1. Two-Element Broadside Arrays

Fig 7-7 shows the bi-directional radiation pattern for various spacings on 160 meters. At first glance you might think that 90- or 110-meter spacings give better overall directivity than 110 or 120 meters, but this is not so. The RDF peaks for approximately 110-meter spacing (0.67 λ), as shown in **Table 7-2**.

1.11.2. Four-Element Broadside Array

In a broadside array with more than two elements the current should taper away from the center. In a 4-element broadside array the outer elements should be fed with half the current of the center elements for best directivity. The pattern shown in **Fig 7-8** is for a 160-meter array with four elements spaced 110 meters apart. The −3-dB beamwidth is only 25° and the RDF is 12.4 dB.

Table 7-2

Spacing	90 m	100 m	110 m	120 m
–3-dB Angle	58°	52°	47°	43°
RDF	8.6 dB	9.1 dB	9.7 dB	9.4 dB

1.12. End-Fire/Broadside Combination

End-fire arrays can provide a single-direction pattern, and you can move null angles by changing the phase delay. The forward lobe of such short end-fire arrays is rather wide (Fig 7-6). A 2-element broadside array on the other hand is bi-directional, but can have a narrow lobe. Combining these two systems can produce huge benefits, and is the most space efficient way to obtain very high directivity. See **Fig 7-8**. The feed systems for these arrays are described later in Section 1.17.

If broadside and end-fire elements are used together, the properties combine to produce an array that has a good front-to-back ratio and a narrow forward lobe. Half-wave broadside

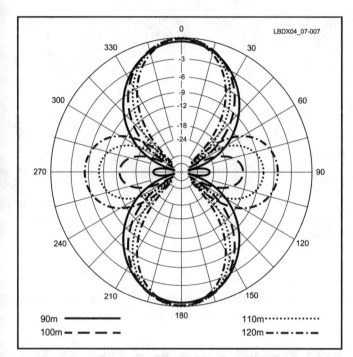

Fig 7-7—Azimuth patterns for various spacings on 160 meters at 20° elevation for 2-element broadside vertical array (two verticals fed in phase). Spacings for solid line = 90 meters; dashed line = 100 meters; dotted line: 110 meters; dashed-dotted line = 120 meters.

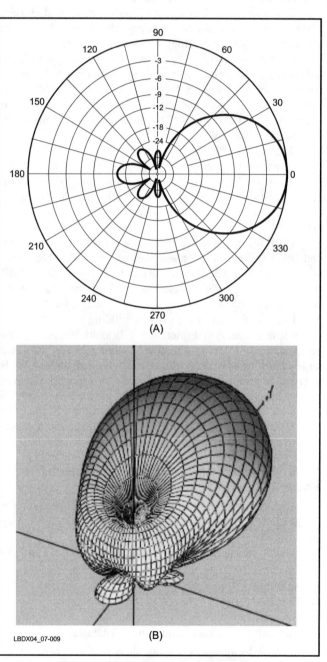

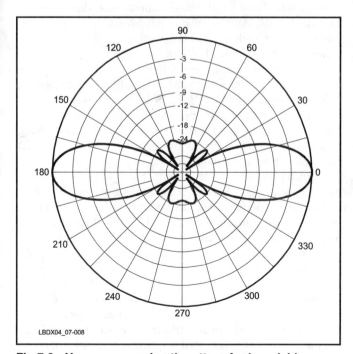

Fig 7-8—Very narrow azimuth pattern for broadside array consisting of four verticals fed in phase.

Fig 7-9—Azimuth and 3D radiation pattern for combination end-fire/broadside array, containing two (side-by-side) end-fire cells separated 198° (0.55 λ). (Plotting by *EZNEC* and *Antenna Model*.)

Receiving Antennas 7-11

spacing gives the best side-rejection at a 0° elevation angle but far from optimum directivity.

Noise-rejection is a three-dimensional problem. Wider spacing moves the nulls up off the ground and makes them more useful for distant QRM and noise. Wider spacing provides twice as many deep groundwave nulls and a noticeably narrower main lobe.

The elevation angle of the null center is given by: α = arc cos (λ/2 / Spacing). If we space the two "cells" a little wider than λ/2 we move the null up and form a cone reaching the ground, similar to the cone formed in end-fire arrays with larger phase lags. The same mechanism explained for end-fire arrays in Figs 7-2 and 7-3 applies to nulls in a broadside arrangement. **Fig 7-9** illustrates that spacings slightly larger than λ/2 cause small sidelobes to form.

A spacing of λ/2 is optimum only when the dominant noise is groundwave, the source being at least a few "broadside spacings" away (that is, it must be outside the Fresnel zone) from the antenna, and directly off the side at exactly 90°. If the null is moved to higher elevation angles, patterns will change from the two right-angle nulls to four nulls. A 25° to 30° side-null elevation is a good target angle if noise comes somewhat evenly from all directions. This is the angle that produces maximum directivity.

The plots in Fig 7-9 were generated using two end-fire groups (each 90° spacing, 105° phasing). These end-fire cells were then placed side-by-side with a separation of 198° (0.55 λ), placing additional side-nulls above the horizon and creating extra ground-level nulls. The side-null offset is arc cosine (180/198) = 25°.

DMF for this end-fire/broad-side array (with λ/4 spaced end-fire cells and 105° phasing, and 110-meter broadside spacing) is 19.1 dB, and the RDF is 12.7 dB; the –3-dB beamwidth is 46°. This is close to the optimum that can be achieved for a receiving pattern with four elements. End-fire/broadside combinations are used as building blocks for large multi-direction arrays (see Sections 1.29 and 1.30), or they can be expanded into super-directive receiving arrays. It was a large super-directive array of loop antennas that allowed W8JI to be the first station east of the western USA to work Japan in the presence of multiple extremely high-power pulse LORAN transmitters in the early 1970s. Pulse transmitters at 50 dB over S9 were taken to S2 with a custom blanker and a super-directive array.

1.13. Broadside Array Consisting of Four 2-Element End-Fire Cells

If you have lots of room (like 330 meters = 1000 feet) for broadside spacing, four end-fire cells will yield a razor-sharp pattern with a –3-dB beamwidth of approx 25°. The design parameters for the array whose pattern is shown in **Fig 7-10** are:

- **Spacing**: 110 meters
- **End fire cells**: ¼ λ spacing, 105° phase delay
- **Current taper**: 0.5, 1, 1, 0.5 (the outer element are fed with half the current of the inner elements).

The performance is nothing short of phenomenal, with a –3-dB beamwidth of 24.5°, RDF = 15.7 dB and DMF = 27.1 dB. More practical design of such arrays is covered in Section 1.23. See also Section 1.33 (parasitic receiving arrays),

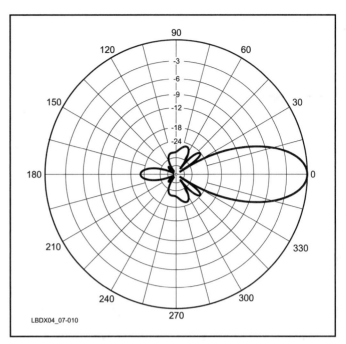

Fig 7-10—Azimuth pattern (at 20° elevation angle) for a broadside array consisting of four 2-element end-fire array cells. See text for details.

where a similar array was designed around four cells, where each cell consists of a 2-element parasitic array (driven element and reflector).

1.13.1. The –3-dB Forward Beamwidth

A narrow forward beamwidth is great, provided you know exactly where signals are coming from (remember the crooked or skewed paths from Chapter 1). You will need more receiving arrays if each array has a very narrow beamwidth.

We probably should try to define *narrow*. A 2-element end-fire array has a –3-dB beamwidth of somewhere between 110° and 180°, depending on spacing and phasing angle. Three or four such arrays will work over the entire 360° azimuth without serious holes in coverage.

The end-fire/broadside combination just described has a –3-dB beamwidth of less than 60° (similar to a 3-element Yagi antenna). It would take eight arrays to cover the entire azimuth without significant pattern holes. The –3-dB beamwidth is actually a serious limit when you are looking at signals close to the noise floor, because even one or two dB can make or break a contact.

Each person has to decide how much work they want to make very weak signal contacts. When you hear a station consistently working weak DX you cannot hear, he is either in a much better location (such as the edge of the ocean) or has taken the time to build very directional antennas and a wide enough variety of them to cover every possible condition and direction.

There are two essential characteristics of a receiving antenna: the RDF (DMF if strong noise is in one defined area) and the –3-dB beamwidth. Gain is meaningless so long as external noise is several dB stronger than the receiving system's internal noise at the quietest time of operation using the narrowest selectivity.

1.14. Modeling Limitations and Tolerances

Models really are shortcuts, where everything is assumed to be simple and perfect. Sources are ideal in current, power and phase. The ground in the model is both flat and homogeneous, and there are no unwanted feed-line currents. The model often has no transmission lines, with no SWR and phase-shift errors that would plague real-world systems. Models work with numbers that are 32 digits long, or longer if we choose!

Real-world antennas are often very different from models. In the real world everything is subject to tolerances. We should always keep this in mind as we examine antennas. If you model 2-element end-fire arrays with spacings closer than $1/4 \lambda$, you can obtain slightly better patterns. Spacings of $1/8 \lambda$ are often used for end-fire arrays, even in transmitting applications. (See **Fig 7-11**)

Note that the DMP peaks for $\phi \approx 155°$, while RDF is still higher at a 165° delay. This is mainly due to the further narrowing of the forward lobe, which overcompensates the worse F/R. **Fig 7-12** shows similar data for the $\lambda/16$-spaced

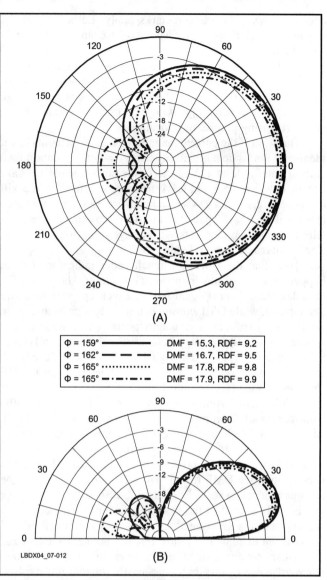

Fig 7-12—At A, azimuth patterns and at B, elevation patterns for 2-element end-fire array with $\lambda/16$ spacing and various phase angles. Solid line: $\phi = 159°$; dashed line: $\phi = 162°$; dotted line: $\phi = 165°$; dashed-dotted line: $\phi = 168°$.

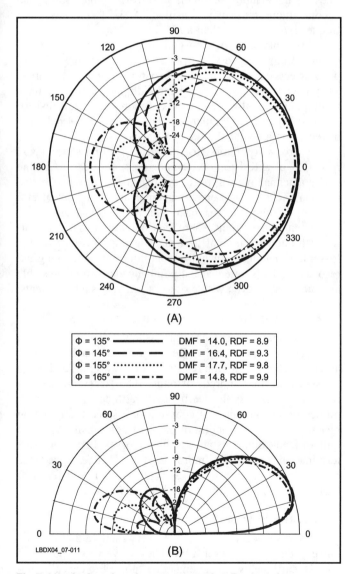

Fig 7-11—At A, azimuth patterns and at B, elevation patterns for two 2-element end-fire array with $\lambda/8$ spacing and various phase angles. Solid line: $\phi = 135°$; dashed line: $\phi = 145°$; dotted line: $\phi = 155°$; dashed-dotted line: $\phi = 165°$.

Table 7-3

$\lambda/4$ spacing, $\phi=120°$					
Ele 1, I (A)	1.0	1.0	1.0	1.0	1.0
Ele 2, I (A)	0.6	0.8	1.0	1.2	1.4
RDF (dB)	8.4	8.8	8.9	8.8	8.7
DMF (dB)	12.7	15.0	15.8	15.2	14.1
$\lambda/8$ spacing, $\phi=155°$					
Ele 1, I (A)	1.0	1.0	1.0	1.0	1.0
Ele 2, I (A)	0.6	0.8	1.0	1.2	1.4
RDF (dB)	8.3	9.7	9.8	9.7	9.0
DMF (dB)	14.9	16.9	17.8	17.0	12.8
$\lambda/16$ spacing, $\phi=162°$					
Ele 1, I (A)	1.0	1.0	1.0	1.0	1.0
Ele 2, I (A)	0.8	0.9	1	1.1	1.2
RDF (dB)	8.7	9.3	9.5	9.3	8.9
DMF (dB)	12.2	15.1	16.7	15.2	13.0

end-fire array. The same remarks apply. Let's look at the impact of variations in feed-point phase and feed current amplitude.

If the phase delay is not exactly what we intend in a model, the only influence is a change in the position nulls. Figures 7-6, 7-11 and 7-12 show us that this is not a large problem, except for arrays with extremely close spacing or arrays requiring precise null locations.

But what if the current magnitude in both elements is not identical? Let us do a sensitivity analysis. Let's maintain feed current amplitude at 1 A in the first element, while we vary current between 0.6 A and 1.4 A in the second element. We will do this exercise on a $\lambda/4$ spaced end-fire array, a $\lambda/8$ spaced array and on a $\lambda/16$ spaced array. All calculations were done with 12-meter long elements over average ground on 1.83 MHz. The results are given in **Table 7-3**.

Note that the DMF is a much better indicator of what happens in the back quadrisphere. This is logical because RDF includes average gain (signal and noise pick-up) over the entire hemisphere, while DMF considers unwanted signals and noise arriving only from the rear quadrisphere. These are two very different cases. If you live where noise arrives from all directions with somewhat similar signal levels over time, use RDF. If noise comes largely from the rear of the antenna, use DMF for comparisons (see Section 1.10).

With normal somewhat-even noise (within several dB) from most or all directions, S/N will not change a great deal with slight errors in element currents. There is a clear rule for considering F/R. If the ratio of arriving rearward noise to arriving forward noise approaches or exceeds the DMF, you will need to improve DMF to improve S/N ratio. If reversing the antenna produces a very clear noise increase of more than 10 or 20 dB, DMF should be the governing factor in array choice. If signals change a great deal but noise does not, use RDF and forget about extreme F/R ratios. F/R will not help a great deal.

Quarter-wave spaced arrays, even in situations where rearward noise is very high, can easily tolerate up to ±40% deviation from the nominal feed current without showing much S/N deterioration. The acceptable feed current amplitude tolerance for a spacing of $1/8 \lambda$ (145° phase) is ±20% where rearward noise is a problem.

What if you space the elements even closer? Data for the $\lambda/16$ spaced array ($\phi=162°$) in Fig 7-12 shows the range of feed current magnitudes providing good directivity is much narrower. You now have only have half the room for error. If you want to keep the DMF high (for concentrated rearward noise), you should keep the feed current tolerance to ±10%.

Close-spaced arrays are much less forgiving of errors than wider-spaced arrays. We can make an end-fire array quite small, and if properly designed and constructed, it will still perform well. There is no way, however, to reduce in size the width of the end-fire/broadside combination array described in Section 1.11. The large broadside spacing is required to reduce beamwidth and improve S/N. This is true for broadside arrays of all types, including Beverages.

1.15. Conclusions

1. In arrays requiring deep nulls, control of feed-current magnitude can be more important than control of phase angle.
2. Close-spaced arrays require accurate current and phase control

1.16. The Choice of Receiving Array Elements

It's very easy to look at pattern changes caused by current and phase variation in receiving arrays. We simply model the array in a program's wire table, enter the correct element phase and current ratios in a source menu and insert proper loads. If we experiment with the phase and ratio of currents, we can observe changes in pattern as the elements depart from optimum phase and current ratios.

If our feed lines were terminated in resistors having an impedance equal to the line's characteristic impedance, we could indeed adjust the line length to change element phase. In this ideal situation, where feed-line SWR is a perfect 1:1 ratio, phase shift is the same as electrical line length in degrees. Once we have a mismatch—that is, we have standing waves—the line no longer has a phase shift equal to line length, unless that line is an exact multiple of 90°. The higher the SWR becomes, the greater the phase error. Line lengths that are odd-multiples of $\lambda/8$ provide the worse SWR-related phase errors. Phase errors in each section of line will add, causing longer lines to have more accumulated error. Longer lines also become more frequency sensitive. A $3\lambda/4$ line has more frequency/phase error than a $\lambda/4$ line.

Elements showing constant impedance over wide frequency ranges are very desirable, especially if mutual-coupling effects can be eliminated. Resistors have these qualities. They have a very wide SWR bandwidth and show no effects from mutual coupling at spacings of more than a few resistor lengths! Unfortunately, there isn't much useful EM radiation or reception associated with small resistors.

There is a solution to the lack of EM radiation and reception of a resistor—A resistor does not need to be the *entire* antenna. We can make the resistor a large part of the antenna, including just enough antenna area to receive useful amounts of signal. Broadband phasing systems are easily implemented in systems where feed-point impedance is stabilized through intentional loss mechanisms. If we make the losses large enough to swamp out or dilute mutual coupling and resonance effects, antenna feed-point impedance remains stable and predictable, even with close-spaced, very short elements.

1.16.1. Naturally Lossy Elements

You can use an antenna that has low radiation resistance and high loss resistance. A Beverage is just such a natural antenna. It has a radiation resistance of a few ohms and a loss resistance in the hundreds of ohms. The large antenna-loss resistance caused by the nearby lossy earth below the antenna dilutes or swamps mutual coupling out, since the radiation resistance is a tiny fraction of the feed-point resistance. The bulk of feed-point resistance is due to losses. In addition, the termination resistor adds more loss and stabilizes impedance over wide frequency ranges (see Section 2.16). Non-resonant loops also meet these requirements (see Section 3).

1.16.2. Resistance-Swamped Elements

Short vertical elements have a very small radiation resistance, yet they are very sensitive. They are vertically polarized and the earth below the antenna does not try to cancel radiation. If you load a short vertical with significant resistance and then cancel the antenna's reactance, you can build an antenna with a wide SWR bandwidth. In addition, the high loss of the loading resistor swamps out mutual coupling effects. Element Q is almost totally defined by the ratio of loading reactance to loading resistance. With 200 Ω of reactance and 73 Ω of resistance, element Q is 200/73 = 2.7, more than enough to cover the widest amateur band (and then some). Higher resistance reduces Q and increases array bandwidth, but it also decreases sensitivity. Lower values of lumped reactance reduce Q. Top-loading with a large capacitance hat increases radiation resistance and sensitivity, and also reduces reactance and Q. The combination of a top-loading hat and 75-Ω feed system results in very wide bandwidth and a very stable array. The 160-meter arrays at W8JI are actually useable well into the AM BC band and far above 160 meters.

1.16.3. Active Antenna Elements

A third solution is to use active antennas as non-resonant elements. Each element consists of a fairly short vertical element (perhaps 3 meters high), with a semiconductor (FET) source-follower circuit. The source follower presents constant impedance to the feed line, isolating reactive components from the element. Of course the circuit needs to fulfill another number of criteria: It must withstand high RF level without damage from your own transmitting antennas, and it must not have intermodulation or harmonic distortion of signals while receiving. It also must withstand electrostatic fields and lighting discharges.

W8JI used active elements in the 1980s, when Tom lived near Cleveland, Ohio. Those elements used very expensive 1.5-dB NF 28-V FETs operating at 400 mA quiescent current, not something the casual experimenter would have available.

Beverages are easy to set up in arrays (broadside and end-fire; see Section 2.16). Use of short elements with resistor loading and inductors to cancel reactance is the current practice in vertical receiving arrays. Active antennas have not been widely described in literature or other publications, probably because of the technical or cost difficulties. W8JI is developing less-expensive active elements for such arrays. Hopefully the cost problems of earlier elements will be solved with a different approach.

1.17. Feeding the Elements of an Array

We need to combine array elements with carefully controlled amplitude and phase, but how do we achieve this? Transmission lines are ideal for moving radio frequency energy from antenna or array to the receiver, but at the same time they can act as delay lines. When improperly terminated, transmission lines become impedance transformers (see Chapter 5). A transmission line can be a very flexible and easy to adjust phasing line if we are careful to follow good engineering practices.

So far we have not said much about impedances. I have modeled arrays (on 160 meters) using generic elements, each 12-meters long and 40 mm in diameter. Let's have a look at the 4-element array (end-fire/broadside) with end-fire cells spaced at λ/4 and end-fire cell phasing at 105°.

Looking at **Table 7-4** we see an almost constant imaginary part, at –663 Ω. The real part (the radiation resistance) is very low, and varies quite a bit from –0.16 to +3.17 Ω. While the R_{rad} of a single vertical is 2.0 Ω, mutual coupling between the various verticals causes the wide variation in feed-point resistance, even negative resistances, once elements are combined in an array. (See Chapter 11 for a fully detailed explanation.)

The solution recently popularized by W8JI is matching the short element impedance to the feed line (preferably 75 Ω) by inserting a resistor and inductor in series with the feed. Reactance is cancelled by the loading inductor's reactance. The resistor is selected so total loss resistance, including ground loss and loading inductor ESR (equivalent series resistance), is approximately 72 Ω. Adding enough loss resis-

Table 7-4
Source Impedance for Various Vertical Arrays

	Ele 1	Ele 2	Ele 3	Ele 4
Single Vertical	2.0 – j 663 Ω			
2-Ele End-Fire array, λ/4 Spacing, 105°	0.67 – j 664 Ω	3.05 – j 662 Ω		
4-Ele End-Fire/Broadside, 90 m Spacing	–0.16 – j 664 Ω	3.17 – j 663 Ω	–0.16 – j 664 Ω	3.17 – j 663 Ω

Table 7-5
Resistor-Swamped Feed Impedance

	Ele 1	Ele 2	Ele 3	Ele 4
Single Vertical	74.0 + j 0.5 Ω			
2-Ele End-Fire Array, λ/4 Spacing, 105°	72.67 – j 0.5 Ω	75.05 – j 1.5 Ω		
4-Ele End-Fire/Broadside, 90-m Broadside Spacing	71.84 – j 0.5 Ω	75.17 + j 0.5 Ω	71.84 – j 0.5 Ω	75.17 + j 0.5 Ω

Table 7-6
SWR Values

	Ele 1	Ele 2	Ele 3	Ele 4
Single vertical	1.02			
2-ele end-fire array, λ/4 spacing, 105°	1.03	1.02		
4-ele end-fire/broadside, 90-m lateral spacing	1.04	1.01	1.04	1.01

tance in series with an inductor of + j 663.5 to equal 72 + j 663.5 Ω feed-point impedance are shown in **Table 7-5**. The resulting 75-Ω SWR is shown in **Table 7-6**.

The 75-Ω feed lines are terminated in very little radiation resistance, but high loss resistances. The large loss resistance swamps out mutual-coupling effects, stabilizing feed impedances, regardless of element phasing and spacing (within reason).

With all feed lines operating at very low SWR it becomes very easy to design phasing systems. When the lines are matched, feed-line phase delay equals feed-line electrical length, for any length of feed line. Additionally, current and voltage along any length of line are equal, except for attenuation through normal feed-line loss. See Chapter 11.

1.18. Sensitivity Analysis

Section 1.14 examined current magnitude and phase errors and how they affect directivity. We estimated the tolerable magnitude of phase and current error in simple end-fire arrays. The next step is learning how to achieve our goals. Section 1.17 also described methods of making element impedance more constant, and how to maintain very low feed-line SWR despite mutual coupling effects in end-fire arrays.

Section 1.16 clarified the important fact that phase delay equals feed-line length only if the line is flat (SWR = 1:1 or $Z_{load} = Z_0$) or a critical length (multiples of $1/4 \lambda$). We know current magnitude is the most critical parameter for null depth, while phase controls null placement

Let's have a look at what happens in a feed line. Using the "VOLTAGE, CURRENT, AND IMPEDANCE ALONG LINE" module of the *Low Band Software*, we can calculate important parameters along the line in step sizes we desire.

Table 7-7 shows some relevant data for a RG-6-type cable used on 1.83 MHz, and terminated in 75 – j 8 Ω.

- Column 1: Line length, in degrees
- Column 2: Current magnitude along the line, A
- Column 3: Current angle along the line, in degrees
- Column 4: Voltage magnitude along the line, V
- Column 5: Voltage angle along the line, in degrees
- Column 6: Normalized voltage angle, in degrees

The table tells us that we will have to put a voltage of 80.7 V magnitude in the line (logical, because we have attenuation). It also says that in order to have I = 1 A ∠0° at the end of the line, the voltage we need to enter in this 130°-long line will be 80.7/–127.6°. Along the 130°-long line the voltage will phase shift over 133.7°.

In our antenna feed systems we cut our feed lines as if line length is equal to phase shift, and that assumes SWR is 1:1. In Table 7-7 you now see that both the phase and the magnitude of the feed current will be slightly off from what it should be in the ideal case. How much can you tolerate? There are a large number of variables involved: You already know that feed-current magnitude is more important than the phase angle, since a slight shift in phase angle merely moves the nulls around in the back of the antenna, while incorrect magnitude will make it impossible to obtain full cancellation—at whatever wave angle.

Doing a complete sensitivity analysis involving all parameters is quite complex and beyond the scope of this book. However, to give you an idea, an SWR of 1.1:1 on the phasing lines can introduce phase-angle errors ranging from 0° to 6°, depending on the line length. Note that lines that are 90° long will always give a 90° phase shift between input voltage and output current, whatever the SWR.

The same 1.1:1 SWR can cause voltage magnitude errors of up to 8%, again depending on line length. Phase error actually peaks in lines 3λ/8 long, or other odd λ/8 multiple. Phase error is minimum in lines that are any multiple of 90°. The amount of deviation we can tolerate for current phase angle and current magnitude will greatly depend on the size of the array and the end use of the array. Wide-spaced arrays are much more tolerant than smaller spaced arrays.

Without going into further details, it is safe to state that you should try to design the array so that the SWR at the band edges is as low as possible, preferably less than 1.2:1. If SWR is high, it is also advisable to use element feed lines in electrical multiples of 90°. With spacings of less than λ/8 between elements, this issue becomes quite important.

In practice there are several things we can do to minimize phase and amplitude errors:

- Make as large an array as possible, without compromising directivity. If you want to build a receiving 4-square and you have room for a λ/8 spaced or larger array, do not build a λ/16-sided array!
- Make sure you have a stable ground system that does not change with weather and season. Long and short-term impedance and loss stability with climatic changes is very important.

Table 7-7
Voltage and Current Along a RG-6 Feed Line Terminated in 75 – j 8 Ω

Line Length	Current Magnitude	Current Angle	Voltage Magnitude	Voltage Angle	Normalized Voltage Angle
0°	1.0	0.0°	75.4	–6.1°	0.0°
10°	1.0	9.8°	74.2	4.0°	10.1°
20°	1.0	19.3°	73.1	14.4°	20.5°
30°	1.1	28.5°	72.2	25.2°	31.3°
40°	1.1	37.6°	71.8	36.2°	42.3°
50°	1.1	46.6°	71.9	47.3°	53.4°
60°	1.1	55.7°	72.4	58.4°	64.5°
70°	1.1	64.8°	73.4	69.2°	75.3°
80°	1.0	74.3°	74.7	79.2°	85.3°
90°	1.0	84.0°	76.2	89.9°	96.0°
100°	1.0	94.1°	77.7	99.7°	105.8°
110°	1.0	104.5°	79.0	109.2°	115.3°
120°	1.0	115.3°	80.1	118.5°	124.6°
130°	1.0	126.3°	80.7	127.6°	133.7°
140°	1.0	137.3°	80.9	136.6°	142.7°
150°	1.0	148.3°	80.6	145.7°	151.8°
160°	1.0	159.1°	80.0	154.9°	161.0°
170°	1.0	169.6°	79.0	164.4°	170.5°
180°	1.0	179.8°	77.9	174.1°	180.2°

Table 7-8

λ/4 Spacing, 12-meter Long Elements, F = 1.83 MHz
Spacing	90°	105°	120°	135°	150°	165°
Gain, dBi	−11.0	−11.1	−11.3	−11.8	−12.4	−13.2

λ/8 Spacing, 12-meter Long Elements, F = 1.83 MHz
Spacing	135°	145°	155°	165°	175°
Gain	−14.1	−14.9	−15.9	−17.2	−18.8

λ/16 Spacing, 12-meter Long Elements, F = 1.83 MHz
Spacing	159°	162°	165°	168°
Gain	−19.7	−20.3	−21.0	−21.7

- Carefully measure and adjust SWR of the elements. Makes sure the SWR at the band edges is low enough. Shoot for 1.2:1 SWR maximum at band edges, and use proper line length planning if SWR is higher.
- Measure feed current and feed angle at each element. This can be done quite easily as explained in Section 1.25.
- Checking element feed impedance regularly is a must. If it is not stable over time, you will have to add radials (and/or increase the lumped constant resistor value).

1.19. What About Gain? (Signal Output)

I intentionally have not given a single gain figure so far, since I have insisted that gain (array output) is not an important issue for receiving antennas. That is also why I left the dBi figures out of all plots. Let us now analyze gain figures for arrays using 12-meter long loaded elements. See **Table 7-8**.

As we will see later, the output of the λ/4 wave spaced array is similar to the output of a reasonably long Beverage antenna. Under normal circumstances, with feed-line losses of less than a few dB, you should not need a preamplifier, unless you are in a very quiet location and use narrow selectivity. With λ/8 spacing, a little amplification (see Section 6, covering preamplifiers) would be necessary, while λ/16 spacing requires at least 10 dB additional gain.

1.20. Feeding the End-Fire Array (Cross-Fire Feeding)

1.20.1. At 1.81 MHz

- Spacing = 20 meters = 43.5°, required φ = 180° − 43.5° = 136.5°
- We install the 136.5° long phasing line in the feed line to element B (leading current element)
- Delay to element A: L1°
- Inversion to element B: 180° − (180° − φ°) + L1° = L1° + φ
- Element B has the leading feed current, as required. (In practice the 180° inversion can be on either element, a useful tool for building multiple element arrays.)

1.20.2. At 3.5 MHz

- Spacing = 20 meters = 84°, required φ = 180° − 84° = 96° (for 0° elevation angle)
- The same phasing line is now: φ = 180° − (43.9° × 3.5/1.81) = 84° long.
- The phasing angle is 180° − 84° = 96° (180° from the inversion and 84° from the phasing line.)

We can do the same analysis looking at what happens with signals coming from the front of the antenna. We will see that for all frequencies below λ/2 element spacing, signals from the front will never be out of phase! (W8JI has a detailed explanation at **www.w8ji.com/crossfire_phasing.htm**.)

Using the same phasing-line length, the feed system maintains correct phase delay on both 160 and 80 meters. In actuality, phasing is correct from just above dc to the frequency where element spacing greatly exceeds 90°. This is a very unique phasing system.

The use of the phase-inversion transformer, and the fact that we put the delay line in the back element instead of the front element, results in subtraction of phase, causing the phase delay system to fully track with changes in frequency.

The phase-inversion transformer is identical to a regular 1:1 transformer, where input and output are "cross-connected." See **Fig 7-13**. W8JI recommends 73-material binocular cores for the job. He winds them with six passes (3 turns, 3 times through both holes) of #24 to #26 twisted-pair enameled wire, using Fair-Rite Products 2873000202 cores (about ½ inch square and ¼ inch thick 73 material). Others have also successfully used type 75 and type 43 material cores (FT114-75 or -43 toroids) for the same application.

At the feed coax in Fig 7-13 (To Rx) the impedance is now 37.5 Ω. For a perfect match to our feed line we should provide a small wideband-matching transformer. We can use the same Fair-Rite 2873000202 core. For a match from 37.5 to 75 Ω we use a primary of 3 turns (6 passes) primary and 4 turns (8 passes) secondary.

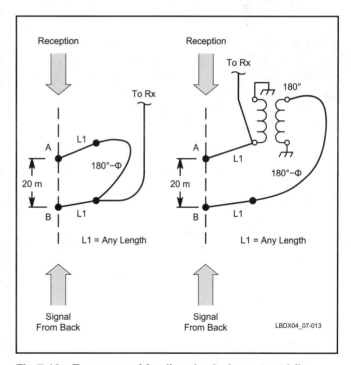

Fig 7-13—Two ways of feeding the 2-element end-fire array: the system on the left is good for one frequency, while the system on the right can be used with the same length of phasing cable over a very wide range of frequencies (easily two bands).

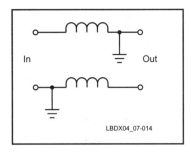

Fig 7-14—Phase-inversion transformer.

1.21. The Vertical Elements in Our Receiving Arrays

The issue is to make elements that have a very low Q over the entire band. Low Q means low SWR at band edges. Why do we want this? Because we use the feed lines to the elements as phasing lines, and to ensure proper phasing the line SWR must be very low, since only a 1:1 SWR means line length in degrees = phasing in degrees. Two parameters influence the variation of the impedance in an array as a function of frequency:

- The Q of the element itself
- The amount of mutual coupling in the array—Large arrays with wide element spacing have much less mutual coupling than small, narrow-spaced arrays.

This means we can live with higher-Q elements in a wide-spaced array as compared to a narrow-spaced array. This constitutes the limiting factor in small arrays: Low-Q elements have very low output. As we make our arrays smaller the output will drop, at the same time with bandwidth. Maybe most important of all is that small arrays are very critical to build and to adjust.

Let's examine a few types of short elements that can be used to build 80- and 160-meter vertical receiving arrays.

1.21.1. W8JI-Style Element (Umbrella Loading)

One of the nice things of this element is that you can easily build it to be resonant on 80 meters. This makes it a very attractive element for a 2-band array, since you will not need to load to resonance on 80. See **Fig 7-15**. If you want to model or build the element, first tune the element for 80 meters (3.65 MHz). Modeling was done with a 30-mm OD vertical tube and 2-mm loading wires. Exact length will depend of the diameter of the vertical tube and the size of the sloping wires.

The properties of the element on 160 meters are given in **Table 7-9**. The third column shows the impedance for the element loaded with 73.8 Ω in series with a coil of $X_L = 277$ Ω on 1.83 MHz. Note that when modeling an element with a loading coil on various frequencies, if you specify the loading element(s) as Laplace Transforms in *NEC-2*, the impedance of the coils is tracked on various frequencies. Gain on 1.83 MHz over average ground is –16.7 dBi.

We should not forget that all of this is modeling. In real life we have tolerances and extra unknowns and unstable parameters involved.

After having checked the behaviour of the W8JI-style top loaded element by itself, we will see how it behaves in an array. I modeled the element in a 2-element end-fire arry with 20-meter spacing in **Fig 7-16**, using 105° phase shift on 80 and 140° phase shift on 160 meters. The results are listed in **Table 7-10**. The SWR levels at the band edges are very acceptable. On 80 meters we can tolerate a little more SWR, with a little deviation from ideal phase shift, as I explained and calculated in Section 1.14.

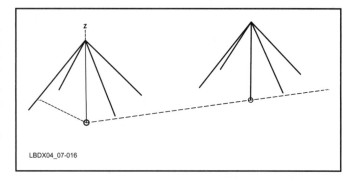

Fig 7-16—Two W8JI-style elements are also evaluated in an end-fire array (see text for details).

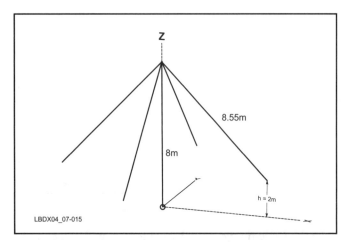

Fig 7-15—W8JI-style element with slanted top-loading wires. This element is resonant on 80 meters.

Table 7-9
W8JI-Loading (Umbrella)

Freq (MHz)	Zant	Zant-Loaded	SWR
1.81	1.2 – j 282 Ω	75 – j 7.8 Ω	1.11
1.83	1.2 – j 277 Ω	75 Ω	1.00
1.85	1.2 – j 272 Ω	75 + j 8 Ω	1.11
1.87	1.2 – j 267 Ω	75 + j 15.7 Ω	1.23

Table 7-10
W8JI-Loading (Umbrella) in a 2-Ele End-Fire Array

Freq (MHz)	Element	Ω	SWR
1.81	Ele 1	73.0 – j 7.9	1.12
	Ele 2	76.86 – j 6.6	1.09
1.83	Ele 1	73.0 – j 0.5	1.03
	Ele 2	76.8 + j 0.7	1.03
1.85	Ele 1	73.1 + j 6.7	1.10
	Ele 2	76.8 + j 8.0	1.11
1.87	Ele 1	73.1 + j 13.9	1.21
	Ele 2	76.86 + j 15.2	1.22

1.21.2. The K8BHZ Element

K8BHZ developed another form of top-loaded element for his HEX-array in Section 1.29. He uses just two flat-top top-capacity wires that run from one element to the next one in the circle containing the six elements. See **Fig 7-17**. The length of the top-hat wires is obviously half the spacing between the elements. As the wires are not 100% in-line (120° instead of 180°), horizontal radiation from these wires is not fully cancelled, but it is down just over 30 dB, which is acceptable.

The R_{rad} on 160 is a little higher than for the W8JI-element. As a consequence the output is a little higher (–14.9 dBi) on 160 meters. Logically, the bandwidth in the test configuration array (2-element end-fire) on 160 is a little bit less than with the W8JI element, as shown in **Table 7-11**.

1.21.3. Base-Loaded Elements

In some environments (in my front garden, for example) it is impossible to use top capacity and guy wires. I'm lucky enough to be able to put up four self-supporting verticals, using slender tapering 11-meter long elements, which are mounted on bases set in concrete. See **Figs 7-18** and **7-19**.

The results in **Table 7-12** show that an array with these unloaded elements will have a slightly narrower bandwidth on 160 meters than an array made with W8JI-style elements. Gain is –14.7 dBi.

Fig 7-18—Concrete base for the self-supporting 11-meter long elements used by the author. The concrete base goes down about 0.75 meters.

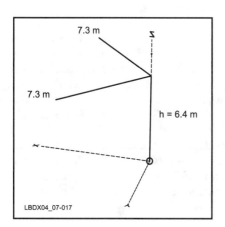

Fig 7-17—K8BHZ-style element for the Stone-HEX array.

Fig 7-19—Roger, ON6WU, working on one of the 11-meter long self-supporting elements of the array at ON4UN's QTH.

Table 7-11
K8BHZ Elements in End-Fire Array

Freq (MHz)	Zant	Zant + 73.5 Ω	SWR
1.81	1.6 – j 448	75.0 – j 11.3	1.16
1.83	1.6 – j 442	75.0	1.00
1.85	1.7 – j 436	75.1 + j 11.1	1.16
1.87	1.7 – j 430	75.2 + j 22.1	1.34

Table 7-12
Base-Loaded Elements in End-Fire Array

Freq (MHz)	Zant Ω	Zant + 75 Ω	SWR
1.81	1.7 – j 700	75.0 – j 9	1.19
1.83	1.8 – j 691	75.0 + j 0	1.00
1.85	1.8 – j 682	75 + j 9	1.19
1.87	1.8 – j 674	75 + j 17	1.25

1.21.4. Mechanical Construction of ON4UN 11-meter Long Elements

The bottom 6 meters of the 11-meter elements are made of steel pipe measuring 60.3-mm OD with a 3-mm wall thickness. Above that is a tapering element similar to a half element of a 20-meter Yagi (tapering from 35 mm OD to 20 mm OD). **Fig 7-20** shows the transition from a large-diameter tube to a much smaller one. Doughnut-like adapters are used, made of short lengths of aluminum tubing. Two stainless-steel bolts are driven through the element to secure everything.

Fig 7-20—Transition of the 60.3-mm OD steel pipe to the 35-mm OD aluminum element.

We see that very similar results can be obtained with all of the elements described above.

1.22. Other Array Configurations

There are more configurations than the 2-element end-fire and the 4-element end-fire/broadside arrays we have been studying. I examined these because they are the simplest and most forgiving, and I used them as a step-by-step exercise to show all the aspects involved.

Other arrays developed according to the guidelines and procedures explained above are:

- **Broadside** arrays and **broadside/end-fire** combination arrays
- The **Four-Square** array: Four elements set-up in a square, yielding an array that is switchable in four different directions (receiving directions along the diagonals going through the corners of the square).
- The **Hex array**: Six elements spaced 45° in a circle, yielding an array that is switchable in six different directions.
- The **Eight-Circle** array: Eight elements spaced 60° in a circle, yielding an array that is switchable in eight different directions.

We will now analyze these configurations one-by-one, and evaluate various versions. As with the 2-element end-fire, it is possible to develop these arrays on different scales. As is the case with the end-fire array, smaller arrays have slightly better theoretical performance, but are more difficult (less forgiving) to build, have a narrower bandwidth and substantially less output.

Many other configurations are possible, but once you understand the design procedure, you can design your own array.

1.23. Broadside Arrays and Broadside/End-Fire Arrays

In Section 1-11 we reviewed the 2- and the 4-element broadside array, which develop narrow patterns. How do we feed these arrays?

1.23.1. Two-Element Broadside Array

This bi-directional array has a −3-dB forward angle of 47° and an RDF of 9.7 dB. Feeding is extremely simple: Run two 75-Ω feed lines of identical length (if the impedance of the elements is 75 Ω) to a common point. See **Fig 7-21**. At that point connect them in parallel and use a 2:1 transformer to get the impedance back up to 75 Ω. For this transformer you can use a Fair-Rite binocular core (2873000202) with a primary of 6 passes (3 turns), and a secondary of 8 passes (4 turns; see also Section 2.8.1), yielding an SWR of 1.1:1 with 75-Ω loads.

1.23.2. Four-Element Broadside Array

The feed principle is the same. However, in this array, with a 3-dB opening angle of 25° and an RFD of 12.4 dB, we need to feed the outer elements with half the current of the center elements.

Ideally we would feed the four elements with coaxial cables of identical length, and not the outer elements via an extra length of 1 λ as shown in **Fig 7-22**. This would make sure that if you move away from the design frequency of the array, the four elements would still be fed perfectly in-phase. If the lengths to the outer elements are 1 λ longer, the four elements are fed in-phase only on the frequency where the extra cable is exactly 1 λ long, but as we move from this frequency the phase change will be much faster in the longer cables to the outer elements than in the cables going to the center elements. It is questionable though if the practical consequences are worth the extra 220 meters of cable though!

If you decide to feed all four antennas with cables of equal length, then you need to insert 3-dB pads (R1 = 28 Ω and R2 = 430 Ω), as shown in Fig 7-22. If you use the alternative, where you feed the outer elements with 1 λ extra feed line, you should take into account the loss caused by this extra feed line.

Table 7-13 shows data for commonly used 75-Ω cables. If you feed the elements with RG-59B or RG-59 foam cable, the extra loss due to the 1-λ long cable is 2.4 to 2.5 dB, which is close to what we need. With this cable we can simply leave out the attenuators. If you use RG-11-type cable, you would need to insert attenuators of 2.1 to 2.2 dB (R1 = 180 Ω, R2 = 680 Ω). Note the different velocity factors of the cables in

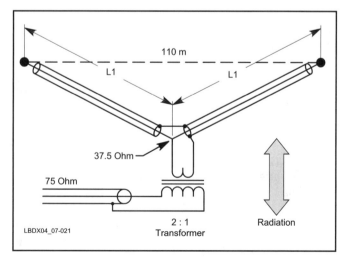

Fig 7-21—Feed system for a 2-element broadside array.

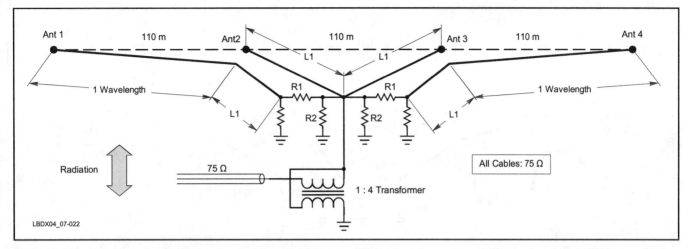

Fig 7-22—Feed system for the four element broadside array. The outer elements need to be fed with half the current of the inner elements to obtain maximum directivity. See text for details.

Table 7-13

Cable Type	Manufacturer	1-λ Length 1.83 MHz, m	Loss for 1 λ dB
RG-59B	Belden 8263	108	2.4
RG59-Foam	Belden 8212	128	2.5
RG-11	Belden 8161	109	0.9
RG11-foam	Belden 8213	128	0.8
½-in 75-Ω Harline	—	133	0.4

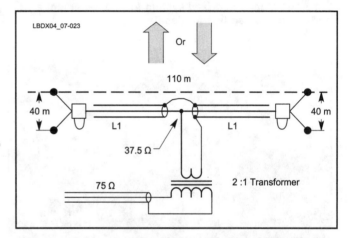

Fig 7-23—Suggested feed system for the broadside-end-fire array with two end-fire cells. Each cell is fed using the system described in Section 1.20.

Table 7-13, resulting in different physical lengths for 1 λ at 1.83 MHz.

The 4:1 impedance transformer (1:2 turns ratio) can be wound on a Fair-Rite binocular core (2873000202). I recommend a primary (37.5 Ω) of 6 passes (3 turns), and a secondary (75 Ω) of 12 passes (8 turns). See also Section 2.8.1.

1.23.3. Broadside/End-Fire Array With Two End-Fire Cells

Section 1.12 explained how this array works. Each of the end-fire cells is fed as described in detail in Section 1.20. The output impedance of the feed system for the individual cells is 75 Ω. We now run two lengths of 75-Ω cable to a common point, where we combine the feed lines and use another step-up transformer to bring the combined impedance of 37.5 Ω back up to 50 Ω (see Section 1.23.1). See **Fig 7-23**.

1.23.4. Broadside/End-Fire Array With Four End-Fire Cells

Feeding this array is a combination of what is explained previously. Ideally, we would run four feed lines of identical length to the end-fire cells, and use 3-dB attenuators to reduce the feed current in the outer cells by 50%. See **Fig 7-24**.

1.24. The Four-Square Receiving Array

Most of us undoubtedly know about the classic Four-Square, with λ/4 sides and elements fed in increments of 90° (quadrature). This configuration became popular because it was the first described in literature, and because it can, as a transmit antenna, easily be fed with a so-called *hybrid network* (see Chapter 11). See **Fig 7-25**.

But there is no reason why this 90°/90° (90° side dimension, 90° phase increment) would be magical or better than other configurations. I analyzed three types of Four-Squares:

- Large footprint Four-Square, with sides = λ/4
- Small footprint Four-Square, with sides = λ/8
- Very small footprint Four-Square, with sides = λ/16

Fig 7-26 shows the horizontal radiation patterns at a 20° elevation angle for Four-Squares with λ/4 side spacing, with various phasing-step increments. Changing the side-element phase angle from 90° to 130° (with rear-element phasing at twice that of side element) does the following:

- It narrows the forward lobe
- It increases the size of the side lobes

Note that the geometric F/B remains very high for all elevation angles. The RDF gets better as you increase the

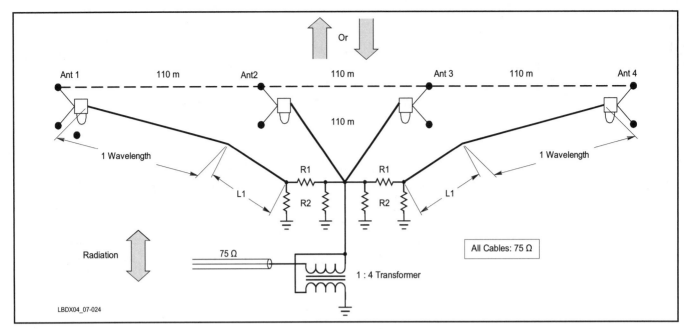

Fig 7-24—Suggested feed system for the 8-element array consisting of a 4-element broadside array using 2-element end-fire cells. See text for details

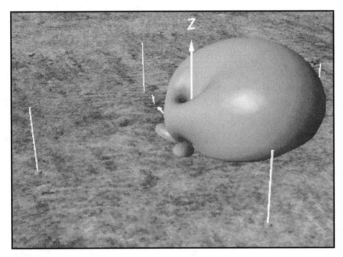

Fig 7-25—The Four Square has its main direction along the diagonals of the square.

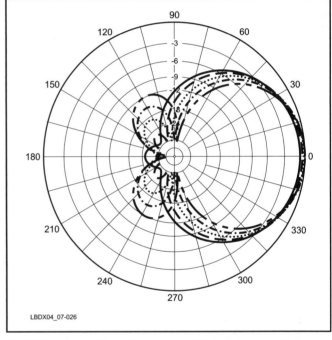

Fig 7-26—Azimuth patterns at a 20° elevation angle for various values of phasing step for a Four Square measuring $\lambda/4$ (side dimension). Solid line = 90°; dashed line = 100°; dotted line = 110°; dashed-dotted line = 120°; dotted-dotted-dashed line = 130°.

phase angle, simply because you substantially narrow the forward lobe. Looking only at the back (RDF), 90° phasing seems to be the best choice. Your choice of best phase angle should be dictated by whether or not you need to look at RDF or DMF. (See Sections 1.8, 1.9 and 1.10.) However, the two-dimensional patterns shown in **Table 7-14** can easily fool you! I would opt for 120° phasing.

Tables 7-14 also shows the performance data for a "medium-sized" Four-Square and for a "mini-sized" Four-Square. The horizontal radiation patterns for these Four-Squares are shown in **Figs 7-27** and **7-28**.

As for the large-sized Four-Square, the higher-angle steps result in better directivity and a narrower forward lobe. As we make the array smaller, however, its output (gain) drops. Compared to the large Four-Square ($\lambda/4$ sides), the medium-sized one has 7.5 dB less output, and the mini-size Four Square ($\lambda/16$ side) has 15 dB less, rather dramatic drops.

What we see in these Four Squares with various phasing steps is very similar to what we saw happening to the 2-element end-fire arrays. See Figs 7-6, 7-8 and 7-9.

Transmitting Four Squares have sometimes acquired a reputation for not being very good on receiving. The reason is that many transmit Four Squares have never been optimized

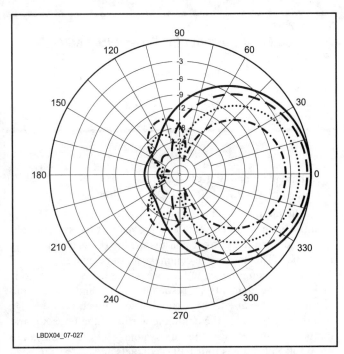

Fig 7-27—Azimuth patterns at a 20° elevation angle for various values of phasing steps for a Four Square measuring λ/8 (side dimension). Solid line = 130°; dashed line = 140°; dotted line = 150°; dashed-dotted line = 160°.

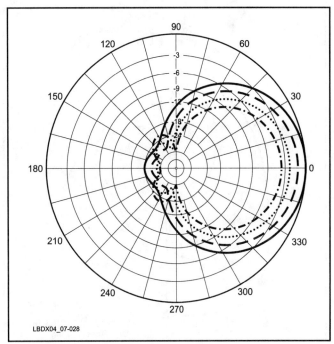

Fig 7-28—Azimuth patterns at a 20° elevation angle for various values of phasing steps for a Four Square measuring λ/16 (side dimension). Solid line = 157.5°; dashed line = 160°; dotted line = 162.5°; dashed-dotted line = 165°.

Table 7-14
Performance Data for 4-Square With λ/4 Side Dimension

Phasing Step	3-dB Angle Degrees	DMF dB	RDF dB
90°	100°	19.1	10.1
100°	92°	22.2	10.6
110°	86°	**24.8**	11.1
120°	81°	24.0	11.5
130°	74°	21.1	**11.9**

Performance Data for 4-Square With λ/8 Side Dimension

130°	96°	19.6	10.5
140°	86°	24.8	11.2
150°	76°	**25.9**	11.9
160°	66°	19.5	**12.2**

Performance Data for 4-Square With λ/16 Side Dimension

157.5°	89°	22.6	11.0
160°	85°	25.4	11.3
162.5°	80°	**27.6**	11.7
165°	76°	26.3	**12.0**

for best directivity. You can build a Four Square with element currents and magnitudes way off from what they should be and yet they will still show almost maximum gain, even though directivity might suffer seriously.

1.25. Feeding the Four-Square Array

Tom, W8JI, introduced the crossfire phasing feed system described here. It can be used with any array that requires feed currents with three or more different phase angles. This means it can also be used for the HEX-array described in Section 1.29. As explained in Section 1.20 it tracks frequency over a very wide range.

This system cannot be used for transmitting arrays, as it is based on the elements of the array showing essentially a constant and purely resistive impedance at all frequencies. Refer to **Fig 7-29**. The front element is fed without delay lines. The two center elements are fed like the reflector in the 2-element end-fire array, with (180°) delay lines from the inverted output of the phase inverter T1. The rear element of the array is fed by a phasing line that measures $2 \times (180 - \phi)$.

Let us check if the phase angles remain OK over a wide frequency spectrum. Assume the Four Square has elements that show 75-Ω feed impedance on both 3.5 and 1.83 MHz. We'll neglect the extra phase shift caused by L1, which is the same for all elements, and look at the phase delays. The required feed currents are:

1.25.1. On 160 meters

- $\phi = 150°$
- Element 1: 0°
- Element 2 and 3: $180° - (180° - 150°) = 180° - 30° = +150°$
- Element 4: $0° - 2 \times (180° - 150°) = -2 \times (30°) = -60° = +300°$
- Phasing line length to Elements 2 and 3: 30° long
- Phasing line length to Element 4: 60° long

1.25.2. On 80 meters

- $\phi = 120°$
- The phasing lines remain physically the same, but are now twice as long in degrees (60° and 120°)
- Element 1: 0°

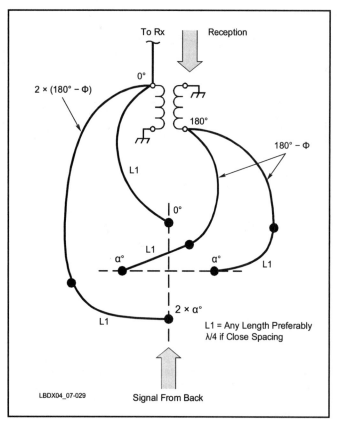

Fig 7-29—Feed system for the Four-Square array, based on the crossfire system. With one set of phasing cables the correct phasing is easily maintained for two adjacent bands (80 and 160 meters).

Table 7-15
Required Feed Currents for Cross-Fire Fed 4-Square

	160 meters	80 meters
Ele 1 (front element)	0°	0°
Ele 2 and Ele 3	150°	120°
Ele 4 (back element)	300°	240°

- Elements 2 and 3: $180° - 60° = +120°$
- Element 4: $-2 \times (180° - 120°) = +240°$

The required feed currents are listed in **Table 7-15**. The results obtained are correct. This proves that the system does track and keep the correct phase shift for (theoretically) any frequency. This means we can make a Four Square for 80 and 160 meters using the same feed system. All we have to do is make sure the elements are resonant for the bands we want to use it on (for example, by switching loading coils).

Fig 7-30 shows the complete wiring of the Four Square, including a direction-switching system. Only two relays are required. Relay 2 has four DPDT contacts and can be replaced by two relays each having two sets of DPDT contacts. The construction of the phase-inverting transformer is described in Section 1.20.

The 4:1 balun is wound exactly like the phase inverter, as shown in **Fig 7-31**. W8JI has designed a commercially available switching system for small receiving Four Squares. The product is available through DX Engineering (**www.dxengineering.com/**).

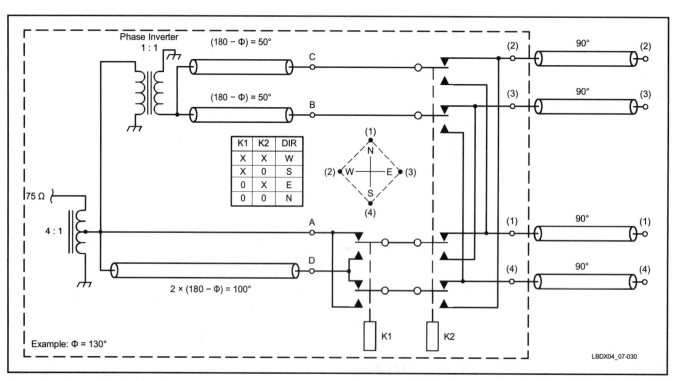

Fig 7-30—Complete wiring diagram of the Four-Square array. Inside the dashed lines is the phasing/switching system. The four 90° lines to the right run out to the array elements. All coax is 75 Ω.

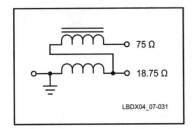

Fig 7-31—A 1:4 wideband balun transformer used to transform from 18.75 Ω (four 75 Ω lines) to 75 Ω (system impedance).

The crossfire phasing system maintains the correct phase over a wide frequency range. Tom also used this system in transmitting arrays as early as the mid-1970s. Despite trying to popularize this system verbally over the years, it wasn't until the advent of the Internet that the word got out.

1.26. A Word of Caution About Small Receiving (Mini) Four Squares

We have seen that small arrays, although mathematically the equivalent of their bigger brothers, are much more critical to build. Down to λ/8 spacing there should be no problems, if the array is built with sufficient care. Very small Four Squares (eg, λ/16 spacing) are even more sensitive to variations in feed angle and amplitude. The tolerances are small and extreme care must be taken to keep the operating parameters within strict limits.

In Figs 7-11 and 7-12, I showed responses of the 2-element end-fire array with λ/8 and λ/16 spacing, and explained issues with smaller arrays. The same concerns exist for the Four Square, but to an even higher degree.

1.27. Measuring and Tuning the Four-Square Receiving Array

The first thing you must do is to measure the impedance at the base of the elements. Make sure you have a good ground system. How do you make sure? Put a long wire on the ground as a single radial approximately 30 meters long. Measure the impedance of the (loaded and resonated) vertical with your antenna analyzer using a very short coax, maximum of 1 meter. Connect the radial wire to the ground at the antenna. If you see any appreciable change in readings on your antenna analyzer, you will need to improve your ground.

Make sure your coax sees 75 Ω at the design frequency. Measure the impedance at the band edges. It should preferably not be more than about 75 − j 8 Ω and 75 + j 8 Ω. If you use feed lines that are λ/4 long you can live with a little higher SWR without upsetting the directivity.

Make sure your feed lines are exactly the same length. It is not strictly necessary that the feed lines be λ/4 long, but this is highly recommended. Don't use poor quality coax to feed the array. Bury the feed lines in a plastic flexible cable duct about 50 cm deep.

Remember that element *current* (not voltage) sets radiation intensity in the fed elements. One interesting and useful property of a λ/4 feed line is that a constant voltage at the source end of the feed line produces constant a current magnitude equal to the input voltage divided by Z_0 of the line and a phase delay of the current compared to the input voltage of 90° at the other end of the feed line, regardless of the impedance at the load end of the line.

If you use quarter-wave feed lines you can simply measure voltage (using a vector voltmeter) at the end of those feed lines to know the antenna feed currents. This principle works with lines that are an odd multiple of quarter-waves long (λ/4, 3λ/4, 5λ/4, etc). But it is important to remember that feed lines that are 3λ/4 long will result in a narrower system bandwidth compared with feed lines that are λ/4 long. The reason is that, as you move away from the frequency where the lines are exact odd multiples of λ/4, the phase error will grow faster in the longer lines.

The only error is this measurement is due to feed-line loss, and on the low bands this loss is usually negligible for small cable lengths. Use good quality cable to reduce the error. The vector voltmeter (such as a surplus HP-8405) or a more expensive vector network analyzer is ideal for the purpose, but a good dual-trace oscilloscope can be pressed into service if the user is careful. Finally a permanent phase/amplitude indicator can be installed, as is explained below.

If all these precautions have been taken, you can do the final feed-current magnitude adjustments. See **Fig 7-32**, where I have added two T-attenuators, one in the feed line going to the front element and one going to the back element. It is likely that the total loss in the leg going to the middle elements (including the loss in the phase inverter) will be greater than the loss in the two other legs. The amount of adjustment is 0.3 to 1.0-dB steps. W8JI uses 2.2 Ω for the series resistors (R_s) and a 1000-Ω carbon variable for the parallel resistor (R_p). This is close enough to a 75-Ω match, and any mismatch created by imperfect attenuator values can be compensated by adjusting the attenuation. Remember that non-equal feed current magnitudes upset the F/B much more than slight deviations from the theoretical phase angle. This is why we fine-tune the current magnitudes with these attenuators.

If you have access to a RF vector voltmeter or similar equipment that let you accurately measure RF voltage magnitude and phase, you can measure the voltages at the ends of the λ/4 lines going to the elements, and adjust the T attenuators for identical voltage magnitude on the three lines.

But you can also build a simple piece of test equipment designed by Robye Lahlum, W1MK. Using two 90° hybrids shown in Fig 7-32, you can adjust the attenuator values for the deepest null at the output port (3) of the hybrid. You will, however, need a few watts of RF to do this. You can use your transceiver, followed by a good filter (eg, a W3NQN bandpass filter), as an RF power source. In Chapter 11 there is more detailed information on this adjustment procedure.

If you use this technique, you will also have to take certain steps before you start building the array. You will have to be able to insert a few watts of RF into the array, without burning out the loading coils or resistors at the base of your element. That means that miniature coils used on printed circuit boards are out of the question. Wind your own coils on powdered iron cores. Use loading resistors of sufficient wattage; connect several in parallel if necessary. You should build both the 1:1 phase inverter as well as the unun (4:1 impedance transformer) using sufficiently large cores, although for 5 W you do not need a huge core.

When using low RF power you can use a receiver instead of a detector and voltmeter for a null indicator. Make sure you have an extra attenuator in-line with the input of the receiver and make sure you apply the minimum amount of power necessary to obtain a clear null. Do the adjustment during the

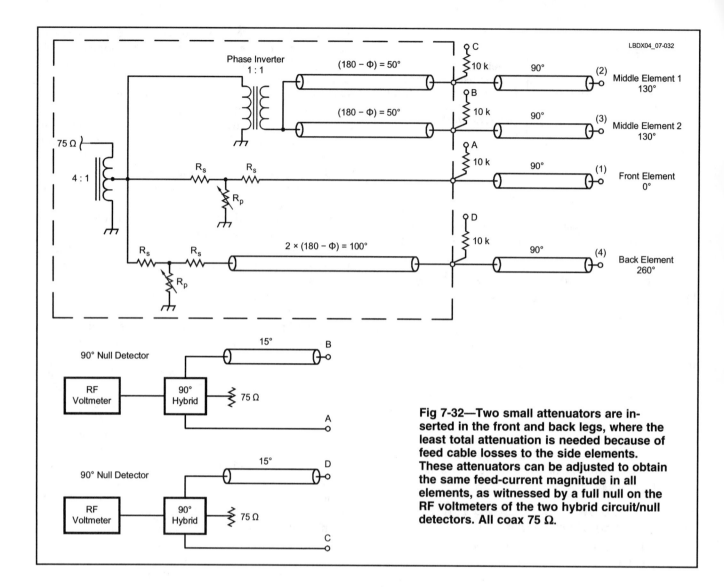

Fig 7-32—Two small attenuators are inserted in the front and back legs, where the least total attenuation is needed because of feed cable losses to the side elements. These attenuators can be adjusted to obtain the same feed-current magnitude in all elements, as witnessed by a full null on the RF voltmeters of the two hybrid circuit/null detectors. All coax 75 Ω.

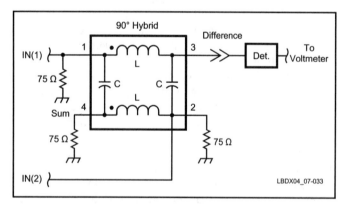

Fig 7-33—A 90° hybrid: if signals at IN(1) and IN(2) are exactly 90° out-of-phase and have the same amplitude, the output of the coupler to port 3 will be zero.

day when there are no signals on the band.

The hybrid coupler is a one-band device. The values of the components are: $L = Z_0/(2\pi F)$, where Z_0 = characteristic impedance (here 75 Ω), and F is in MHz. For 1.83 MHz the coil inductance is $L = 75/11.5 = 6.5$ µH and $C = 10^6/(4\pi F Z_0) = 580$ pF.

The coils must be wound in a bifilar fashion (the wires can be twisted). Note the phasing dots in **Fig 7-33**. Powdered-iron cores must be used. A T-50 core (red mix) has an A_L of 49 (meaning that 49 turns are required for 100 µH inductance). The required number of turns is given by $N = 100 [L/A_L]^{1/2}$ where L is expressed in µH, and A_L in µH/100 turns.

If the array requires 90°-phase shift between the elements ($\phi = 90°$), you need not add any additional phasing line to one of the inputs of the hybrid. In Figure 7-32, I inserted 15° long phasing lines, since the φ of this array is not 90° but 105°. Make sure you have the 75-Ω line termination resistors at ports 1, 4 and 2 of the hybrid. You can connect the hybrid to proper points (A, B, C and D) at the input of the λ/4 lines to the elements, with short but equal lengths of coax (one line will, in addition, have the extra line length as explained above, in case φ is not exactly 90°).

There is no real need for these circuits to be made for a 75-Ω system impedance; they can just as well be made for 50 Ω. In that case the extra phasing lines should be made with 50-Ω cable.

The two hybrid test circuits can be left in-line permanently. To test the array, inject a few watts of power, connect a receiver to the output port of the hybrid and adjust the T attenuator(s) for minimum signal. You can, of course, use a small dedicated detector/voltmeter instead of a receiver for this purpose (see Chapter 11).

1.28. The Mini Receiving Four Square at ON4UN

During the winter I can use the terrain you can see in the background of the picture in **Fig 7-34** to put up 12 Beverage antennas (the terrain is approximately 160 by 300 meters). Fortunately most of the DX on the low bands is worked in winter, although I have often put up a single Beverage across the cornfield in the summer when it was necessary to work a new one! You really should try to lay a Beverage wire on top of 2.5-meter tall corn: a unique experience!

To have some decent receiving capabilities also in the summertime, I decided to try a receiving Four Square in the small garden in front on the house. The arrays is about 40 meters from the 160-meter transmit antenna. This is really too close and requires the transmit antenna to be detuned while I'm listening. (See Section 6.6 in Chapter 9.) **Fig 7-35** shows my front garden in a picture taken from the top of one the other towers. The schematic for the switching employed at each of the duo-band 80/160-meter elements is shown in **Fig 7-36**. **Fig 7-37** shows the weatherproof plastic box used to hold the switching/matching/loading components for each vertical.

Long radials were not possible, so I put down about 18 radials per element, after installing two crossing bus-wires to which the ends of the inward-looking radials are soldered. See **Fig 7-38**. With a 1.5-meter long ground rod at each element, I estimated the equivalent ground loss resistance to be approximately 20 Ω.

It is quite easy to determine the ground loss resistance. Assume your element radiation resistance is 2 Ω, and the loading-coil equivalent loss is 2 Ω. We need to add 50-Ω series loading resistor to obtain a 1:1 SWR in a 75-Ω system,

Fig 7-34—You can see three of the elements of the mini Four Square in the front garden at ON4UN in this photo.

Fig 7-35—Radial layout used at ON4UN's mini Four Square. Each vertical has 18 short ground radials, which are soldered to a bus wire. A 1.5 meter long copperclad steel ground rod is used at each element.

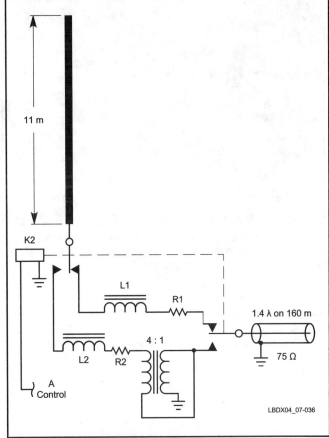

Fig 7-36—Duo-band switching system used by ON4UN. See text for details.

Receiving Antennas 7-27

Fig 7-37—A small plastic box houses the matching/loading (swamping) components. The relay switch the components for 80 and 160 meters.

Fig 7-38—Base of a vertical. All copper radials are soldered to a copper ring, and connected to the strap going to the 1.5 meter long ground rod.

so we can conclude that the ground loss resistance is $75 - 2 - 2 - 50 = 21\ \Omega$.

I wanted to use the small Four Square on both 80 and 160 meters. Using a loading resistor with a typical value of 50 to 60 Ω, to end up with a 75-Ω resistance (R1 + ground losses + losses in loading coil L1) this 11-meter long vertical gives adequate signal output (~ −15 dBi) on 160, with reasonable bandwidth (SWR approx 1.3:1 on 1810 and on 1850). The required loading coil has an inductance of ~ 60 µH.

On 80 meters I needed much more bandwidth, which means much more resistive loading. This is no problem as the 11-meter tall vertical is fairly long for this application and extra swamping is required to reduce the effects of mutual coupling. Instead of loading the antenna to a total resistance of 75 Ω, I loaded the antenna to 300 Ω, which resulted in an output of approximately −15 dBi, the same level as on 160 meters. On 80 meters I incorporate a small 4:1 broadband transformer, to bring the impedance down to 75 Ω. The 4:1 impedance transformer (1:2 turns ratio) can be wound on a Fair-Rite binocular core (2873000202) with a primary (300 Ω) of 6 turns (12 passes), and a secondary (75 Ω) of 3 turns (6 passes). See also Section 2.8.1.3). The bandwidth of this 80-meter element is extremely wide, with an SWR of 1.1:1 on 3.5 MHz, 1:1 on 3.65 MHz and 1.1:1 on 3.8 MHz. **Fig 7-36** shows the band-switching arrangement at the base of each element.

Using 18 short radials and a 1.5-meter ground rod at each vertical, the required value of the loading resistor for 160 meters (R1) was approx 50 Ω to achieve a 75-Ω feed impedance. For 80 meters a value of 270 Ω was used (R2).

Use metal-compositions resistors (not film resistors), such as Ohmite OY, OX resistors or equivalent in this application. I wound the coils on 1.3-inch OD powdered iron cores (T-130, RED mix, $A_L = 100$). The number of turns is given by the following equation:

$N = 100 \times [L/A_L]^{1/2}$, where L is expressed in µH, and A_L in µH/100 turns. A coil with an inductance of 50 µH requires: $100 \times [50/110]^{1/2} = 67$ turns.

After installing the elements, check the feed impedance on the design frequency. It should be as close as possible to 75 Ω. Also check the band edges. The SWR at the band edges must in no case be higher than 1.3:1.

1.29. Stone-HEX Array (K8BHZ)

Brian, K8BHZ, sent me details of a 6-element round array, which he calls a Stone-HEX (referring to Stonehenge). K8BHZ's hexagonal array is small: the six elements are located on a circle measuring only 30 meters in diameter. It can be switched in six directions, which is exactly what's needed with its −3-dB beamwidth of 78°. See **Fig 7-39**.

I discussed in great detail the disadvantages of close-spaced arrays previously. If there is no real need to make the array so small, I recommend making it about twice the size. This will result in a less critical design (and better bandwidth).

The radiation pattern and directivity figures of the larger array are identical to those of the small array. Only the output will be substantially higher. **Table 7-16** gives the design and performance data for the Stone-HEX array for three different sizes.

1.29.1. Feed System Principle

First read Sections 1.24, 1.25 and 1.26 before proceeding here. This antenna requires three different feed currents, just like the Four Square. These do not all have the same feed-current amplitude. This can easily be dealt with, however. Since the central elements require twice the current compared to the other elements, the phase-inverter transformer must be a 1:4 impedance transformer (1:2 voltage). The output of this transformer, if connected as shown in **Fig 7-40**, will provide a 180° shift and an output voltage twice the input voltage. This will result in double the feed current in the middle elements. At the input side of the transformer we see an impedance of

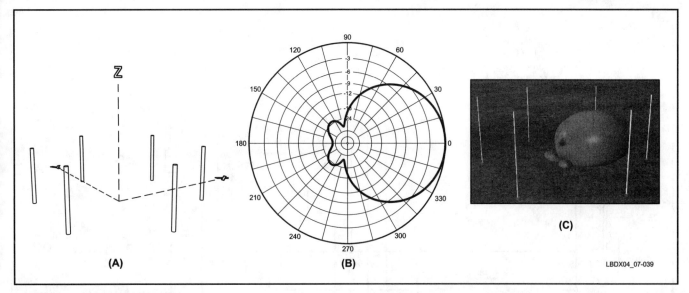

Fig 7-39—The Stone-Hex array has improved directivity over a Four Square and is switchable in six directions. The feed system is basically the same as used with the Four Square, since this array also requires three different feed currents. See text for details.

75/8 = 9.4 Ω. There are four more 75-Ω feed lines connecting to that point (75/4 = 18.75) so the total parallel impedance of 18.75 and 9.375 is 6.25 Ω.

L1 are equal line lengths, preferably λ/4 lines, which allow us to measure voltage at the end of these lines and know the currents at the element feed points. For the smaller Stone-HEX, with φ = 149°, the length of the phasing line to the central elements is (180° − 149°) = 31°. The length of the extra phasing line to the back elements is 0° − 298° = 360° − 298° = 62°. These are electrical lengths. When converting to coax length take into account the velocity factor of the coax.

The phase-inverter transformer can be wound as a 9:1 transmission-line type transformer (3:1 turns ratio). See also Sections 1.20 and 2.8.1.

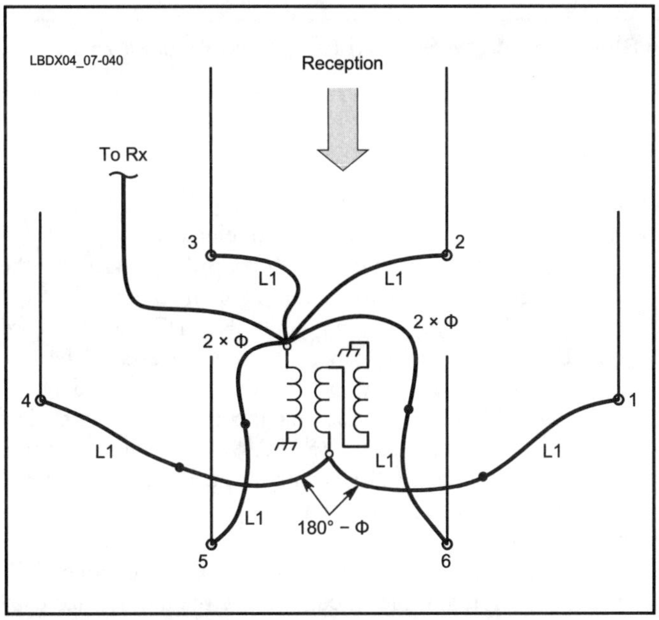

Table 7-16
Three Sizes of Stone-HEX Arrays

Circle Dia, m	φ Degrees	Gain Over One Ele, dB	3-dB Beamwidth	DMF dB	RDF dB
30	149°	6.7	78°	27.4	11.7
45	130°	6.5	80°	26.1	11.5
60	120°	6.7	75°	25.8	11.7

1.29.2. A Practical Stone-HEX Feed System With Direction Switching

You can use a Fair-Rite Products 2873000202 binocular core (about 1/2 inch square and 1/3 inch thick 73 material), with 1.5 turns (3 passes) on the low-Z side and 7.5 turns (15 passes) for the high-Z winding. You can also use type-75 and type-43 material cores (FT114-75 or -43 toroid) for the same application.

To perfectly balance the feed currents, insert T-attenuators at the points indicated as A, B, C and D in Fig 7-41. See Section 1.27 for details on these attenuators. The same 90° hybrids described in Section 1.27 can be used to tune the array. Four such circuits will be needed. One is connected between points A and E, with an extra phasing line going to E. The second circuit goes between C and F with the extra line going to F. The third goes between D and E with the extra line going to E and the final one goes between B and F with the extra coax going to point F.

The direction-switching system is quite a bit more complex than with a Four Square, where switching can be done with three SPDT relays. Here you need seven such relays (see **Fig 7-41**). Brian, K8BHZ uses a different concept for his feed/switching system (see **Fig 7-42**). Instead of using seven DPDT

Fig 7-40—Wiring of the feed system for the Stone-Hex-array. Notice that the phase shift transformer is also a 1:4 impedance transformer. See text for details.

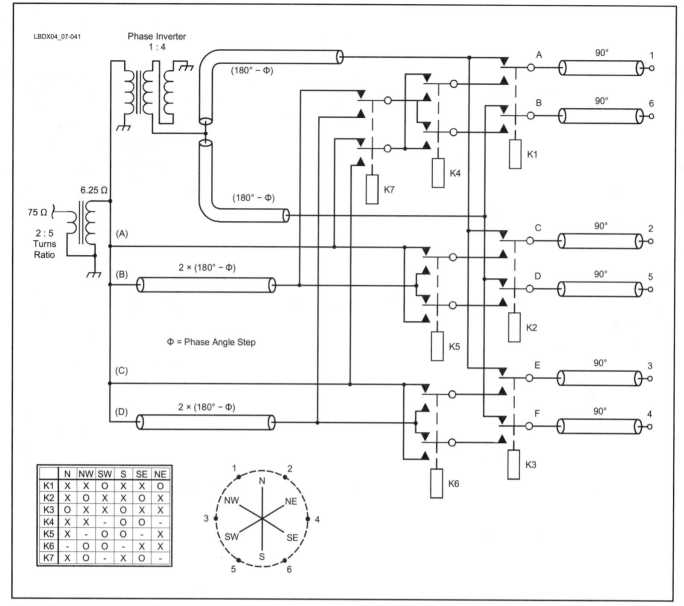

Fig 7-41—A practical feed/direction switching system for the Stone-HEX array. See text for details. All coax 75 Ω.

relays, he uses 18 small SPST reed relays in a matrix-switching configuration. Also, instead of using individual phasing lines going to the center and the back elements, he parallels the quarter-wave feed lines coming from each pair of those elements at the switching box (resulting in 37.5-Ω impedance at that point) and uses two parallel 75-Ω phasing line (of electrically identical length) to make up the required 37.5-Ω impedance phasing line.

This approach uses exactly the same length of phasing lines. Brian claims this is the better solution since a difference in phasing delay to the pairs of elements (center of back) are avoided. It must be clear, however, that to make a perfect 37.5-Ω phasing line of a given length, it also is essential that both 75-Ω lines have identical lengths. In my opinion both approaches are identical in all respects. In Brian's approach of using a matrix-type switching configuration it is, of course, advantageous to have only three input lines: One line with $E1 = 2 \times a \angle 0°$ (for the center elements) one line with $E2 = a \angle +\phi°$ (reflector elements) and one line with $E3 = a \angle -\phi°$ (director elements).

The three vertical "bus bar" lines in Fig 7-41 carry the feed voltages for the two center elements, the two front elements and the two back elements. K8BHZ uses standard 8-conductor rotator cable for switching the array. The two heavy wires carry +12 V dc for the relays and a shack switchable +12 V dc for feeding the remote preamp. The "Rotor" switch in the shack is a SP6T, with the center common contact going to the 12-V-dc ground. The six wires select the six directions. The Preamp ground return comes back through the coax feed-line shield.

Note also that the original K8BHZ six-circle Stone-HEX uses a 50-Ω system impedance, because the builder had easy access to 50-Ω cable. I recommend using 75-Ω feed lines to obtain better bandwidth. **Fig 7-43** is a panoramic photo of K8BHZ's Stone-HEX array.

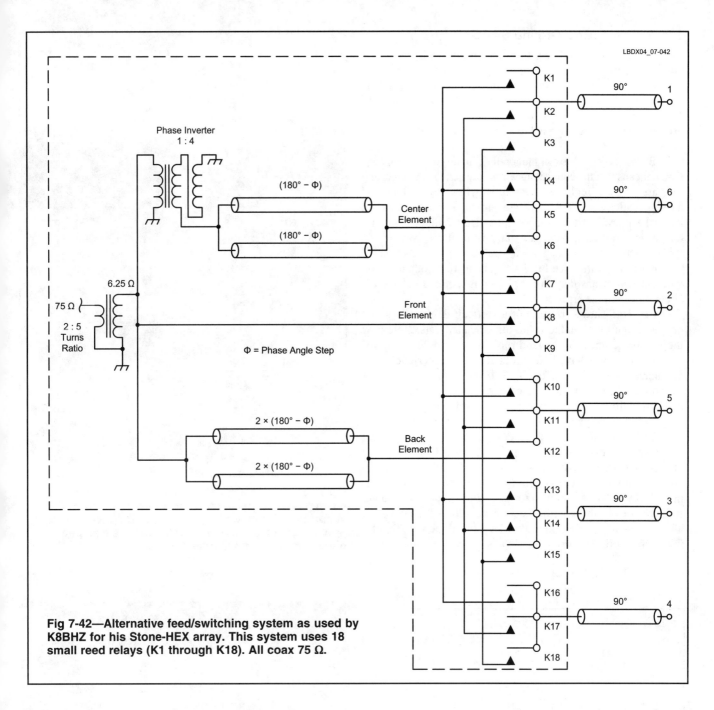

Fig 7-42—Alternative feed/switching system as used by K8BHZ for his Stone-HEX array. This system uses 18 small reed relays (K1 through K18). All coax 75 Ω.

Fig 7-43—The Stone-Hex array as set up at K8BHZ's QTH.

1.29.3. The K8BHZ Ground System

K8BHZ uses quite an unusual ground system for his array that relies on many judiciously located ground rods rather than an elaborate radial system. Brian had measured the ground loss for a single 2.4-meter long rod to be about 90 to 100 Ω. He came up with the idea of connecting multiple ground rods to lower this and to average out variations in that value.

Brian used the Moxon monopole/counterpoise symmetry, which means that the connecting wires between the three ground rods are an exact replica of the top-hat wires; just as long and directly underneath. This way the ground resistance at each vertical was brought down to about 30 Ω. He used three 2.4-meter long ground rods, one at the base of the vertical, and two more at the end of the 7.3-meter long radial wires running exactly beneath the top loading wires (see **Fig 7-44**). There are no other ground wires, which means that no extra room is required outside the circle on which the array is built.

Fig 7-45 shows the base of a vertical. Brian uses small printed-circuit type inductors to tune the elements. Note how the box is clamped to a ground rod with a large U-clamp.

Personally, I have had extremely poor experience with galvanized ground rods. In **Fig 7-46** you see my neighbor friend George holding a piece of rusted steel ground rod that had originally been galvanized. All that was left of a 150-cm long rod after 20 years in the ground was 30 cm long.

I would certainly only recommend copper-clad steel ground rods or better yet pure copper bar ground rods. You can clamp copper conductors onto copper-clad steel ground rods. After clamping, cover everything with a liquid rubber compound. If you solder a conductor to a solid copper-bar ground rod, cover everything with a heat-shrink tube with a hot-melt adhesive inside. Never expose solder joints done with regular 60/40 Sn/Pb to bare ground. Always cover the solder joints

Fig 7-45—K8BHZ uses 10-cm × 10-cm treated wooden fence posts, buried in concrete 1 meter in the ground to support his 6.4-meter long elements. The 2.4-meter long ground rod emerges above ground and is clamped directly to an aluminum die-cast box.

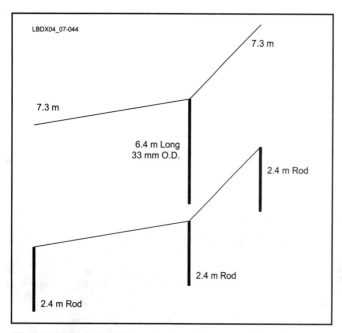

Fig 7-44—The K8BHZ ground system uses three 2.4 meter long ground rods connected with a wire that is an exact mirror image of the loading wires.

Fig 7-46—Remnants of a 1.5-meter galvanized ground rod after 20 years in Belgian soil. After seeing this I never used galvanized-steel ground-rod components again.

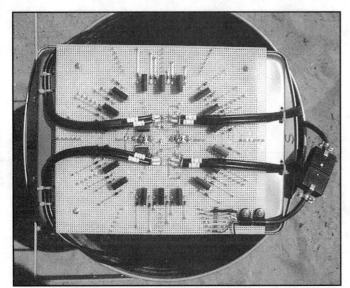

Fig 7-47—The phasing/matching/switching board made by Brian, K8BHZ.

Fig 7-48—A 10-gallon garbage can is used to house K8BHZ's phasing/switching circuitry, as well as the extra phasing-line coax, coiled up at the bottom of the can. The can is grounded using two 2.4 meter long ground rods.

liberally with a couple of layers of liquid rubber before burying.

Fig 7-47 shows the construction of the matching/phasing/switching board by K8BHZ, using a large vector board. Note the excellent RF layout, which results in minimum stray coupling. **Fig 7-48** shows the board in a 10-gallon garbage can used to protect it from the elements.

The direction-switching lines are the long outer conductors connected to the rotor cable. The symmetrical relay triads are self-explanatory; the 6-element connectors are directly below them. The short inner rings are for Front/Center/Rear, and the two transformers are visible, as are the phasing lines coming and going. The output connector is directly at the center. Brian also mentioned he made a second layout for a two-sided PCB with ground plane.

1.30. The Eight Circle Array (W8JI)

The Eight Circle array developed by Tom, W8JI, is undoubtedly one of the best performing vertical arrays, with directivity results equaling those of phased Beverages. In Section 1.12 I described the end-fire/broadside array. W8JI developed a direction-switchable array based on this four-element cell. **Fig 7-49** shows the design and the relationships between the end-fire spacing (EFS) and the broadside spacing (BSS).

A broadside spacing of 0.55 λ (90 meters on 1.83 MHz) results in the highest attenuation off the side for a 24° elevation angle (elevation angle = arc cosine (82/90), where 90 meters is the separation and 82 meters is $\lambda/2$ at 1.83 MHz. W8JI used a separation of 107 meters (0.65 λ), which results in a few more sidelobes but a substantially smaller forward beamwidth.

In this configuration the circle diameter is 97.4 meters and the element spacing in the end-fire cell is 37.25 meters. I calculated the DMF and RDF for various phase angles of the end-fire cells with a 37.25-meter spacing. The results are shown in **Table 7-17**.

The gain remains the same within 0.1 dB and is approximately 7.6 dB over a single element. The highest RDF and DMF are obtained with a phasing angle of approximately 120° to 130°. From the point of view of total noise these are the

Table 7-17
End-Fire Cells with 37.25-meter Spacing

Phase step (ϕ)	110°	120°	130°	140°
DMF (dB)	21.4	22.6	21.8	19.4
RDF (dB)	12.1	12.3	12.4	12.4
–3-dB Angle	56	55	54.5	54

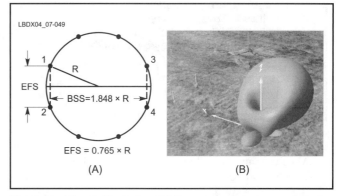

Fig 7-49—The 8-Circle is nothing more than a set of end-fire/broadside arrays, with the elements arranged at 45° intervals on a circle. Properly dimensioned (slightly more than $\lambda/2$ side-spacing), this configuration makes a super receiving array that can be switched in 45° intervals.

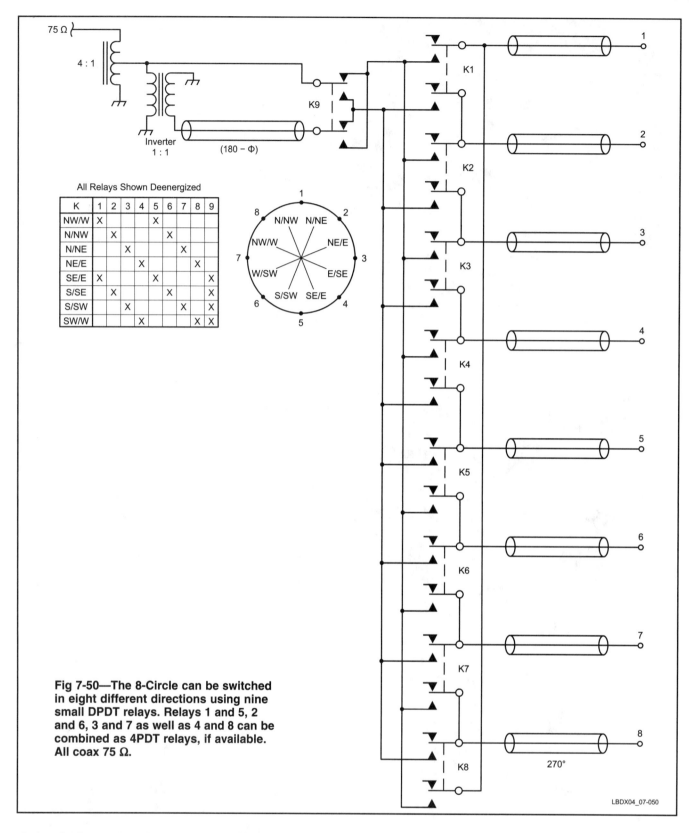

Fig 7-50—The 8-Circle can be switched in eight different directions using nine small DPDT relays. Relays 1 and 5, 2 and 6, 3 and 7 as well as 4 and 8 can be combined as 4PDT relays, if available. All coax 75 Ω.

choice phasing angles. If you have a lot of local ground-wave QRM coming in at very low angles, go for φ= 110° though for a better low-angle F/R.

1.30.1. Feed Currents

Elements 1 and 3: 1 $\angle 0°$ A

Elements 3 and 4: 1 $\angle +\lambda°$ A

With these feed current the array will be pointing North (in Fig 7-49). Note than only four of the eight elements are in use in any direction.

Contrary to what we saw with the 2-element end-fire, the Four Square and the Stone-HEX array, this array cannot be scaled to a smaller version and still maintain its excellent directivity. This is because scaling would change the broadside phasing distance, which must be a little over λ/2 for proper operation. This is also the reason why the array only works on one band.

It's interesting to compare this array with the Stone-HEX array. Gain-wise there is just over 1 dB difference, but who cares about gain for a receiving array? From a DMF point of view (looking at directivity in the back), there is very little difference between the two arrays. As expected the RDF is about 1 dB better with the Eight Circle because of its narrower forward lobe. The $\lambda/2$ broadside spacing of the two end-fire cells does the trick.

1.31. Feeding the Eight Circle

Remember how easy it was to feed the 2-element end-fire array? This is just as easy, because we have here two end-fire array groups fed in-phase. The switching too becomes quite easy, and a system using nine small DPDT relays does the job (**Fig 7-50**). If you can obtain 4PDT relays, you can do the job with five relays.

Because of the size of the array, $\lambda/4$ feed lines will not reach the center of the array. If you want to be able to measure voltage at the end of the feed lines to assess the current at the elements, you will have to install $3\lambda/4$ feed lines. If this is not considered a requirement, equal-length lines of any length reaching the center of the array may be used. $3\lambda/4$ feed lines will also give you somewhat wider performance bandwidth because of the current-forcing principle typical of lines that are multiples of $\lambda/4$ long.

The two front verticals are fed at 0° phase angle. The feed lines to these arrays are paralleled (Z_{tot} = 37.5 Ω) and connected to the primary of the inverter transformer. The feed lines to the back elements are also connected in parallel (37.5 Ω) and then fed to the secondary of the inverter transformer via the phasing line, which is made of two paralleled 75-Ω cables. At the input of the inverter transformer the impedance will be 37.5/2 = 18.75 Ω. A 4:1 unun transformer will match this to the 75-Ω feed line.

The 4:1 impedance transformer (1:2 turns ratio) can be wound on a Fair-Rite binocular core (2873000202) with a primary of 4 passes (2 turns), and a secondary of 8 passes (4 turns). (See also Section 2.8.1 and Fig 7-31).

The 1:1 inverter is shown in Fig 7-14. **Fig 7-51** shows the remote-control unit, in this example using two ganged 8-position switches. By making use of +12 V and –12 V and a bunch of diodes, you can limit the required number of conductors going to the remote unit to three. You could feed it through the coax, and make use of +12 Vdc, –12Vdc and 12 V ac, but I do not recommend this solution since it may induce noise in the line, and create havoc on your feed line if there is the slightest ingress of water even only in the connector.

If you want to experiment with various null angles, you could easily add a few relays and change the length of the delay line from 50° (gives 180 – 50 = 130° phase shift), over 60° (120° phase shift) to 70° (110° phase shift). By switching between these lines it is possible to position the nulls in the back of the array at various angles in both the horizontal and vertical planes.

1.32. The Eight Circle in Practice

To my knowledge, two receiving type Eight Circle arrays are in use at the time of writing, one at W8JI, who designed the array and the other at W8LRL. Wally presently has the highest DXCC country score in the world on Topband. Like W8JI,

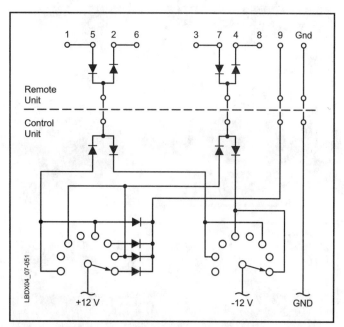

Fig 7-51—Wiring of the switch box at the receiver. As few as three wires can be used to switch the array if you use the shield of the coax as a ground line.

Fig 7-52—Wally, W8LRL, clearing the grass around one of the elements of his 8-Circle array. Like Tom, W8JI, he uses a 2 × 2-inch wooden support for the aluminum tube elements.

Fig 7-53—The professional grade printed-circuit board used in the 8-Circle remote unit box shown in Fig 7-50.

W8LRL has lots for acres for receiving antennas, and he finds the Eight Circle a good complement to the extended range of Beverages he runs on his very impressive 160-meter antenna farm. It's proven over and over again, that the most successful Topband DXers are those with the best receiving antennas in the quietest locations! See **Fig 7-52**.

W8JI developed a pc board that does all the switching and which is available from DX Engineering. He mounted it in a die-cast aluminum box. (**www.dxengineering.com/**). See **Figs 7-53** and **7-54**. W8LRL made his own matching/switching box and mounted it in a red plastic bucket. See **Fig 7-55**.

1.33. Parasitic Receiving Arrays

Why don't we see small parasitic receiving arrays? Can't we make our transmit arrays be excellent receiving arrays? Transmit arrays can be designed to have excellent directivity, but as a rule they are very large. For a transmit antenna efficiency is a major concern, unlike the case for receiving antennas.

Parasitic arrays require a high-Q to work well. Mutual coupling is essential, since the current in the parasitic element is obtained *only* through mutual coupling. Parasitic arrays, whether intended for receiving or transmitting, must be designed to have the lowest possible losses.

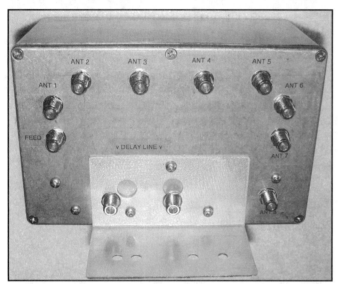

Fig 7-54—The 8-Circle remote switching box designed by W8JI and sold by DX Engineering. There are eight connectors for the coax going to the eight elements, two for the delay line and one for the coax to the receiver or preamplifier.

Do we need full-sized ($\lambda/4$) elements to have low-loss, high-Q elements? The answer depends on how good the ground system is. In other words, with a nearly perfect ground system (perfect in terms of near-field, where return currents are collected) shortened elements can be used. Jim, N7JW, and Al, K7CA, have designed various transmitting parasitic arrays for 160 meter that use 16.45-meter long elements, top loaded with two drooping wires (see **Fig 7-56**). This element is self-resonant slightly higher than 1.83 MHz, and has a radiation resistance of about 11.5 Ω. In the N7JW/K7CA transmitting arrays, a "nearly perfect" ground radial system is used, consisting of 120 40-meter long radials. The equivalent loss resistance is in the order of 1 Ω. This brings the feed-point impedance of the element to approximately 12.5 Ω.

In the array the element is tuned exactly to resonance on 1.83 MHz with a small coil at the base (0.1 µH). All the elements are made identical. **Fig 7-57** shows the element configuration. When the relay is energized, the bottom of the element is connected to ground via a coil of 0.5 µH, which turns it into a reflector (tuning the element to a self resonance of approximately 1812 kHz). With no voltage applied to the relay, the base of the element is resonated to 1.83 MHz with a coil of 0.1 µH and connected to a 4:1 wide-band transformer.

The basic idea of the transmitting array is the same as the concept of an Eight Circle, where eight elements are on a circle. Just like in the Eight-Circle receiving array described in Section 1.30, four-elements are active at a time (See Fig 7-49). The circle diameter is 94 meters, and the elements are spaced 36 meters in the circle. This makes for two broadside 2-element arrays, spaced 86.9 meters, while the elements of each cell are spaced 36 meters (0.2 λ).

Fig 7-55—Wally, W8LRL, designed and made his own matching/switching box, contained in a red plastic bucket, located at the center of his array.

The parasitic element in each cell is a reflector. The

Fig 7-56—At N7JW's QTH in St George, Utah, we see 8 elements in a line on top of a ridge.

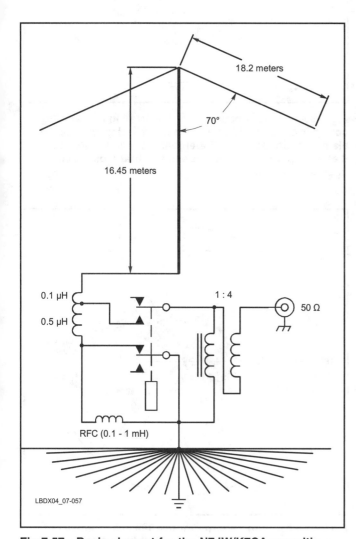

Fig 7-57—Basic element for the N7JW/K7CA parasitic array.

Table 7-18
Performance Data for 2- and 4-Element Parasitic Arrays

Configuration	Gain dBi	−3-dB Angle Degrees	RDF dB	DMF dB
2-Element	4.6	150°	8.5	12.8
4-Element	8.2	58°	11.9	20.3

performance data for two cells in a broadside array are given in **Table 7-18**. How does this N7JW/K7CA Eight Circle transmitting array compare to the W8JI Eight Circle receiving array in Section 1.30?

- The directivity (both RDF and DMF) are not quite as good as the for the all-driven array but they are still excellent.
- The parasitic array requires a "perfect" ground system.
- The parasitic array requires longer vertical elements (16.5 vs 6 meters).
- The parasitic array has a gain of +8.2 dBi vs −8.5 dBi for the all-fed receiving array. This makes it an excellent transmitting array, with 2.5 to 3 dB more gain than the well-known transmitting Four Square.

N7JW and K7CA went one step further and also built 160-meter receiving arrays that use four broadside 2-element cells. With a length of 261 meters and a width of 36 meters, this is a receiving array with a large footprint! See **Figs 7-58** and **7-59**. A 6-element array could also be made that would measure "only" 174 meters wide. And by the way, all these dimensions are without the radials extending 40 meters in all directions!

The performances of these receiving arrays are quite spectacular, as shown in **Table 7-19** and in **Fig 7-60**. With an RDF of 15.1 dB this array is almost 3 dB better than the all-

Table 7-19
Performance Data for 6- and 8-Ele Parasitic Arrays

Configuration	Gain (dBi) dBi	3 dB Angle Degrees	RDF dB	DMF dB
6-Element	10.0	37°	13.7	23.4
8-Element	11.3	28°	15.1	25.6

Table 7-20
Operational Bandwidth of 8-Element Array

Freq	Gain dBi	RDF dB	DMF dB	SWR
1810	10.8	14.9	30.3	1.6
1820	11.2	15.2	30.7	1.1
1830	11.3	15.1	29.8	1.1
1840	11.1	14.6	26.2	1.1
1850	10.8	14.2	24.0	1.1

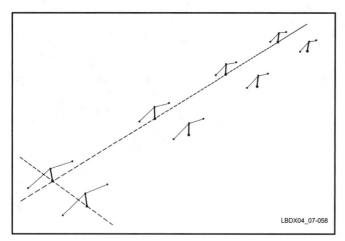

Fig 7-58—The 8-element parasitic array at N7JW/K7CA consists of four 2-element parasitic cells spaced just over λ/2 apart.

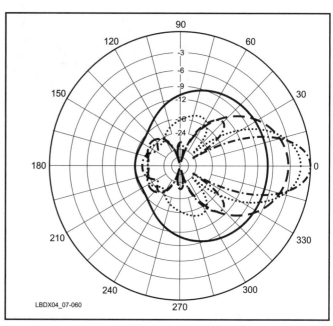

Fig 7-60—Azimuth patterns for the array in its various configurations. Solid line = 2 elements; dashed line = 4 elements; dotted line = 6 elements; dashed-dotted line = 8 elements. For gain, nose beamwidth and directivity figures, see Table 7-23.

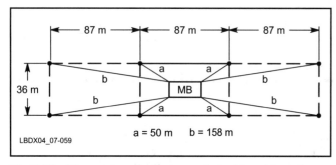

Fig 7-59—The center elements are fed with cables of equal length to the central matchbox (MB). The feed lines to the outer elements are 1 λ longer (108 meters for RG213) to keep the four cells fed in phase.

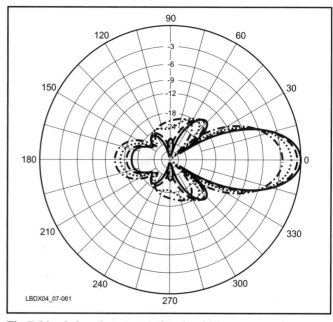

Fig 7-61—Azimuth patterns for the 8-element receiving array at various elevation angles. Solid line = 10°; dashed line = 20°; dotted line = 30°; dashed-dotted line = 40°.

fed transmitting Eight Circle. Note however that this is obtained through an extremely narrow forward lobe, where the −3-dB angle is only 28°.

In the 4-element array, both driven elements are fed with the same current. If you use two equal-length 50-Ω feed lines going to the centrally located matching and switching box, the combined impedance will be 25 Ω. This can be matched to the feed line by any convenient means, such as a 1:2 transformer, L-network or a 37.5-Ω quarter-wave transformer.

For four cells, the best directivity results from feeding the outer driven element with 80% of the current value used for the two center driven elements. This is automatically obtained by using a "lossy long line" to these elements. As all four elements need to be fed in-phase, you would feed the outer elements with feed lines that are 1 λ longer than the center driven elements. This is 108 meters of RG-213

(VF = 0.66), which has a loss of approx 0.9 dB. This means a current ratio of 1:1.11 or 90%, not exactly what we need but close enough. A little lossier coax would be even better. The same applies for the 6-element array, where you should feed the two outer elements with approximately 80% of the feed current of the center element.

The 6-element array requires the same current tapering across the cells (0.8:1.0:0.8) for best directivity. In the centrally located match-switchbox (MB) the feed lines to one side of the array are connected in parallel, as well as those to the other side. A simple relay switches directions by selecting which side to feed. At the same time, the elements on the back of the array are turned into reflectors by energizing the relays at their base (see Fig 7-57). A 1:4 wideband transformer (identical to the one used at the feed point of each element) is used to transform the 12.5 (four 50-Ω lines in parallel) back up to 50 Ω.

The directivity of this array is excellent over a wide range of elevation angles, as shown in **Fig 7-61**. Contrary to what you might expect, the operational bandwidth is also excellent, as shown in **Table 7-20**. The SWR bandwidth is very flat from just below 1820 to well over 1850 kHz, and the directivity figures remain remarkably high over about 50 kHz as well.

N7JW and K7CA built a remote antenna site some 35 miles from their QTH in the middle of the desert, where manmade noise is totally non-existent. Several such arrays are located at this remote site. The remote is operated via VHF links as a receiving site only, as there is no mains supply and all the equipment at the remote is powered by solar panels. The arrays can also used for transmit with high power in contests, when a generator and a 2-kW amplifier is brought to the site.

Jim and Al have Beverages up there as well, but they swear by the 8-element arrays. They claim its better performance is due to the cleaner vertical pattern, and to the fact that the elevation angle (at 25° to 30°) is substantially higher than for long (300-meter minimum) Beverages. See **Fig 7-61**.

The 8-element array get its directivity from the very narrow forward lobe. From Utah, looking into Europe, the back is California and the Pacific. Very little 160-meter activity originates from there, at least as compared to Europe. In Belgium, I have the entire continent of Central and Eastern Europe behind me, and I need my directivity mainly in the back, a very different situation. This goes to illustrate that what is best in one situation is not necessarily so in another one.

Going against the mainstream of using separate receiving and transmitting antennas and building these high performance TX/RX arrays has proven to be a winner for N7JW/K7CA, who are consistently either the loudest or the only stations heard in Europe from that part of the world. And it works on reception too, since I never have to call them twice. K7CA is shown next to one vertical in **Fig 7-62**. Part of the array is shown in **Fig 7-63** and a schematic of the directions covered is shown in **Fig 7-64**.

Fig 7-62—K7CA at the base of one of the array elements. Note the heavy copper ring to which all 120 radials are soldered. The two loading coils are wound on the black insulator, while the relay and the 1:4 transformer are placed in a plastic enclosure.

Fig 7-63—View of seven of the eight elements of one of the N7JW/K7CA arrays at their remote antenna farm in Utah.

1.34. Vertical Receiving Arrays Compared

Table 7-21 gives an overview of the characteristics of a single element (see Section 1.19) in the various receiving arrays examined so far. To obtain reasonable bandwidth the R_{rad} of an element should be kept low enough compared to the total equivalent series loss resistance in the feed impedance. If we use a 75-Ω system impedance, we generally will require elements with $R_{rad} < 2\ \Omega$ to provide adequate bandwidth. Elements with a higher R_{rad} will need to be loaded with a higher series resistance, as was done in Section 1.27. In **Table 7-22**, DMF and RDF are shown vs the forward lobe at a 20° elevation angle for a variety of arrays.

1.34.1. Gain Corrections for Other Elements

For other elements, the gain can be corrected using the gain figures from Table 7-21. If you use, for example, 6-meter instead of 12-meter elements, the gain will be 19.9 – 13.9 = 6 dB lower than what is shown in the table. If you use the W8JI-style elements you have to subtract 17.4 – 13.9 = 3.5 dB.

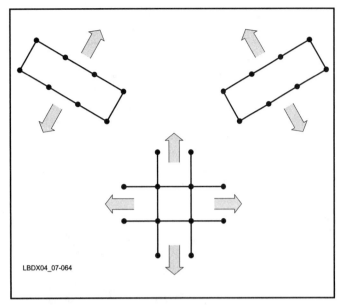

Fig. 7-64—Layout of the receiving arrays at N7JW, giving the choice for eight directions.

Table 7-21
Single Element Gain (at 20° Angle) and R_{rad} (Over Average Ground). Gain Calculated With a Series Resistance Totaling 75 Ω

Element	Reference	80-M Gain dBi	160-M Gain dBi	80-M Rrad Ω	160-M Rrad Ω
W8JI Element	Sect 1.20.1	–12	–17.4	5.3	1.2
K8BHZ Element	Sect 1.20.2	–10.1	–15.5	7.6	1.7
Base-loaded 6-m Ele	Sect 1.20.3	–14.8	–19.9	2.2	0.5
Base-loaded 9-m Ele	Sect 1.20.3	–11.1	–16.4	5.1	1.2
Base-loaded 12-m Ele	Sect 1.20.3	–8.3	–13.9	9.8	2.2

Table 7-22
DMF and RDF Comparisons, at 20° Takeoff Angle

Antenna	DMF dB	RDF dB	–3-dB Angle Degrees	Gain dBi	Reference
2 Ele End-Fire, λ/ Spacing, $\phi = 135°$	16.1	9.3	132	–12	Sect 1.8
2 Ele End-Fire, λ/8 Spacing, $\phi = 155°$	17.7	9.8	120	–16	Sect 1.14
2 Ele End-Fire, λ/16 Spacing, $\phi = 165°$	17.9	9.8	121	–21	Sect 1.14
4-Square, Side λ/4, $\phi = 120°$	24.0	11.6	86	–8	Sect 1.23
4-Square, Side λ/8= 140°	25.9	11.9	86	–15	Sect 1.23
4-Square, Side λ/16, $\phi = 160°$	27.6	12.0	85	–23	Sect1.23
6-Ele Stone-Hex, 305-m Dia	27.4	11.7	79	–19	Sect 1.28
8-Circle, $\phi = 120°$, Dia 0.594 λ	22.6	12.3	55	–8	Sect 1.29
Broadside 2-Ele Bidirectional	#	9.7	47	–10.5	Sect 1.11, 122.1
Broadside 4-Ele Bidirectional	#	12.4	25	–7.9	Sect 1.11, 1.22.2
Broadside/End-Fire 4-Ele	21.0	12.7	46	–4.7	Sect 1.12, 1.22.3
Broadside/End-Fire 8-Ele	26.6	15.7	24.5	0.7	Sect 1.12, 1.22.4

Note: Gain (dBi) on 1.8 MHz over average ground, using a 12-meter long non-top-loaded element, bottom loaded to achieve 75 Ω impedance.

#It does not make sense to calculate the DMF of a bidirectional antenna.

Table 7-23
N7JW/K7CA Arrays (Receive and Transmit Antennas)

Antenna Description	DMF dB	RDF dB	–3-dB Angle Degrees	Gain dBi	Reference
4-Ele Parasitic Array (N7JW)	20.3	12.9	58	+7.1	Sect 1.32
8-Ele Parasitic Array (N7JW)	25.6	15.1	28	+10.2	Sect 1.32

1.34.2. Transmitting Arrays With Outstanding Receiving Performance

While these antennas are not in a strict sense receiving antennas only, I will list their performance figures here in **Table 7-23**. The N7JW/K7CA array is covered in this chapter; the WØUN/K9DX Nine Circle is covered in Chapter 11.

1.35. Using a Noise-Canceling Bridge to Feed a Receiving Array

In Sections 1.4 and 1.5, I briefly touched on the subject of noise-canceling devices. A noise-canceling device is nothing but a phasing/combining system, in which you combine the output of two antennas and adjust the relative amplitude and phase for full cancellation of a particular signal.

The feed system for a receiving array can be seen as a noise-canceling bridge. For two signals to produce zero output when combined, they must be the same amplitude and 180° out-of-phase. In the receiving arrays described so far, identical elements were used (same polarization, same directivity pattern), close together (less than 1 or 2 λ apart to avoid space diversity effects). These are prerequisites for signals with identical magnitudes, when you consider signals (QRM, noise, etc) received via skywave. If you use different antennas to feed a noise bridge, the varying polarization and signal strength will make it impossible to get a stable null on skywave signals. As W8JI stated: *"Despite folklore... mixing a horizontal antenna with a vertical is real trouble on skywave circuits. Fading, averaged over a short time, actually increases..."*

The main difference between a directive array and a noise-canceling set-up can be summarized as follows:

- A directional array uses identical antenna elements in the array.
- A directional array is designed to provided directivity, not to null out a specific noise source. (The array has a narrow forward lobe in which we receive, and it suppresses signals as much as possible in all other directions, to increase RDF and DMF.)
- A noise-canceling set-up is intended to reduce/eliminate locally generated noise.
- A noise-canceling set-up normally uses a small-size noise-sensing antenna located near the noise source.

Instead of designing the feed system of a directive array for a fixed (optimized) phasing angle to produce the best possible DMF or RDF, we can bring the individual feed lines of the end-fire array and do the phasing "in-house" using a phasor-combiner, usually called a *noise bridge*. The MFJ-1025 and 1026 are commercial units that perform very well on the low bands. The advantage of such setup is that you can move the null of your array continuously around from inside the shack. They are great for nulling out a particular QRM source.

However, noise usually comes from many directions, the DMF or RDF of your receiving set-up is really much more important than being able to null out a signal from one specific direction. In addition, moving a null by varying the phase not only moves it in the horizontal plane, but also moves it at the same time in the vertical plane (see Section 1.6).

A noise-bridge is really the best instrument if you want to cancel out QRM received from nearby sources on groundwave. Still better, of course, is to kill the noise source itself. When used for such an application, any small vertically polarized sensing antenna can be used. The noise-sensing and receiving antennas can be very dissimilar, so long as the noise antenna hears the noise well. Ideally, the noise antenna should hear only the unwanted signal. Since both antennas receive the noise by groundwave, there are no variations like we normally experience on sky-wave signals.

W8JI has covered the MFJ-1025/1026 noise canceller in great detail on his Web page. Tom describes some modifications that can be made to improve its efficiency on 160 and 80 meters (**www.w8ji.com/mfj-1025_1026.htm**).

The noise canceller can also be used to enhance received signals. W8JI wrote on his Web site: *"Enhancing signals requires both antennas to have similar S/N ratios, and ideally they would have similar patterns. That's true if it is a Beverage and a Four Square, K9AY loops or Flags, two regular (so-called magnetic) loops, or whatever being mixed. Vertically polarized antennas mix well with other vertically polarized antennas. Since my Four Square has a similar pattern to a pair of echelon-staggered Beverages, they mix almost perfectly! Remember if you combine one poor S/N antenna with a good S/N antenna in an effort to enhance signals, you will almost certainly make the good antenna become worse! Mixing in a poorer S/N antenna is great for nulling local noise, but not peaking distant signals. I mix three Beverages with a Four Square, using any combo that works best. Sometimes the improvement is 15 dB or more. Sometimes it won't help, but mixing anything vertically polarized with my dipoles is almost always a real bust for signal enhancement."*

Mauri, I4JMY, who made his own noise-canceller, wrote on the Topband reflector: *"The null steerers are top performers to avoid overload by local huge signals on same or another band; ie, nearby contester running a pile-up or rag chewing even a few kHz apart."* This very true, and if you have a local station (within a few kilometers) on 160 meters running a kW, it's likely he'll be S9+40 or better, and you may hear him literally all over the band. The noise canceller can literally kill him, without having to go to prison!

I have successfully used the MJF noise canceller for getting rid of local groundwave noise. I use my transmitting vertical as the noise antenna, with a attenuator so that only the local noise was heard, while the main antenna was one of the Beverages. This worked very well. The built-in small 1-foot long sensing whip is really not much of a sense antenna, and I never could make good use of it.

It takes some understanding of what happens inside the black box before you can adjust the noise canceller properly. Dave, NR1DX, wrote: *"I first note the S-meter reading of the noise with the "main antenna gain" all the way up and the noise antenna gain all the way down. I then turn main control to zero. I turn up the gain of the noise antenna until I get the same S-meter reading. Next I turn the main antenna gain all the way back up and adjust the phasing control for a null. Careful tweaking of the noise gain and the phasing control then tames the beast. If I can't get an equal level through the noise-antenna side, I either live with the noise or try a different antenna as the noise antenna. Successful noise nulling is an art and it takes a good noise antenna plus patience to tune out a noise source as the controls can be quite sharp."*

2. AN INTRODUCTION TO BEVERAGE ANTENNAS

2.1. The Beverage Antenna: Some History

The Beverage antenna (named after Harold Beverage, W2BML) made history in 1921. In fact, a Beverage antenna was used in the first transatlantic tests on approximately 1.2 MHz. For many decades, the Beverage antenna wasn't used very much by hams, but in the last 25 years it has gained tremendous popularity with low-band DXers. The early articles on the Beverage antenna (Ref 1200-1204) are excellent reading material for those who want to familiarize themselves with this unique antenna.

A revealing interview with Dr. Harold H. Beverage and H.O. Peterson (interview by Norval Dwyer, done in 1968 and 1973) can be read at: **www.hard-core-dx.com/nordicdx/ antenna/wire/beverage/interview1.html**.

2.2. The Beverage Antenna Principles

Fig 7-65 shows the basic configuration of the *Beverage*

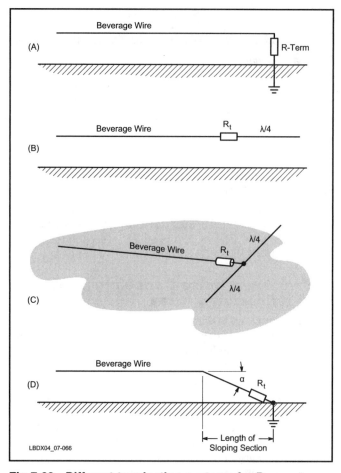

Fig 7-66—Different terminating systems for Beverage antennas. The version at A suffers from stray pickup because of the vertical down lead. At C, two in-line quarter-wave lines terminate the Beverage, by which the radiation from these lines is effectively canceled. This is not a very practical solution because of its area requirements. This configuration is used through the book for modeling Beverage antennas, however. At D, the method most widely used with real-world Beverages. Using a long sloping section is thought by many to reduce omnidirectional pickup.

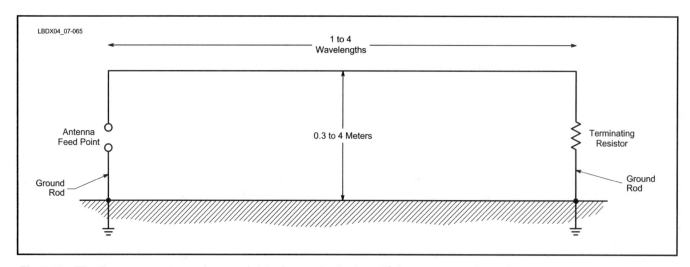

Fig 7-65—The Beverage antenna is a straight wire, typically 1 to 4 λ long, mounted parallel to the ground at a height of 0.01 to 0.03 λ.

antenna (also called the "wave antenna"). It consists of a long wire (typically 1 to several λ long) erected at a low height above the ground. The Beverage antenna has very interesting directional properties for an antenna so close to the ground, but it is relatively inefficient. This is why the antenna is primarily used only for reception on the amateur low bands.

The Beverage antenna can be thought of as an open-wire transmission line with the ground as one conductor and the antenna wire as the other. To achieve a unidirectional pattern, the antenna must be terminated at the far end in a resistor equal to the surge impedance of the antenna. See **Fig 7-66**. So-called *bi-directional Beverages* are covered in Section 2-12.

If the Beverage antenna is to be used on VLF (where it was originally used), the velocity of propagation in the two wires (one is the antenna conductor, the other one is its image in the earth) has to be different, so that the arriving wave front (at a 0° elevation angle for groundwave VLF signals) inclines onto the wire and induces an EMF in the wire. Therefore, the ground under the antenna must have rather poor conductivity for best performance.

A radio wave travels in the air with a velocity factor of 100%. In the antenna wire it travels at a slower speed, depending on its height above ground and the quality of the ground. As we make the wire antenna longer, an increasingly large phase difference exists along the wire between the wave in the air and the wave in the wire. When the difference becomes 90°, the EMF induced by the radio wave will start subtracting from the traveling wave in the wire instead of adding. This mechanism limits the length for maximum gain in this antenna.

On the amateur MF/HF low bands, the situation is different, because signals do not come in at 0° elevation angle. The elevation angle is typically above 0° and below 30° for DX signals on 160/80 meters. Here too we have to look at the phase difference between the wave in the air and the induced EMF wave in the antenna wire. It's clear that in the horizontal plane through the wire, the projected wavelength is now the wavelength in the air multiplied with the cosine of the elevation angle. The wave in the wire is now the faster one, at least for high enough elevation wave angles. For a given height and ground quality there is an elevation angle at which the VF and the elevation angle compensate one another and where both remain in phase all the time. Below this elevation angle the wave in the antenna is the slower one, and above this angle it is the slower one. This angle is also the elevation angle of the antenna, or the angle at which the gain is greatest. For signals coming in at higher elevation angles (and long enough antenna wires) the gain drops, goes through a minimum (180° phase difference), goes up again, etc, in a cyclic way.

This is the mechanism that forms the radiation pattern of the Beverage. See **Fig 7-67**. Notice the secondary lobes. The longer the antenna is the more nulls there are, as this phenomenon of adding/subtracting keeps repeating along the wire.

For the main elevation wave angle (the angle at which the main lobe peaks) there is a length beyond which the gain will drop. From Fig 7-67 you can deduce the theoretical maximum, as a function of practical parameters, such as the antenna height and ground quality, both of which together determine the velocity factor of the antenna. However, the velocity factor of the antenna is not the mechanism that limits the useful maximum length of a Beverage antenna. In our approach above, we assume that the wave in the air is homogeneous and linear in space, which in reality is not the case (re space diversity). The true mechanism that limits Beverage length is discussed in Section 2.4.1.

The velocity factor of a Beverage will vary typically from about 90% on 160 meters to 95% on 40 meters. These figures are for a height of 3.0 to 3.5 meters. At a 1-meter height the velocity factor can be significantly lower, depending on ground quality. BOGs (Beverages on ground) may have a velocity factor around 60%, which means that very short, very low Beverages can exhibit radiation patterns similar to higher Beverages that are much longer. A drawback of low Beverages is that their output is much lower. But may not be our main worry. After all, people trip on them, deer get tangled in them and critters eat them!

2.3. Modeling Beverage Antennas

2.3.1. Modeling with *MININEC*

I don't advise modeling Beverages with *MININEC* based programs (see also Chapter 4, Section 1.2) because *MININEC* assumes a perfect ground under the antenna. This has a number of consequences such as:

- *MININEC* works with a 100% velocity factor (because of the perfect ground under the antenna): The gain will increase with length, without reaching a maximum, which is not correct.
- *MININEC* will show an antenna impedance that is typically 10 to 20% lower than the actual impedance over real ground.
- *MININEC* patterns will show deep nulls in between the different lobes. This is not correct. The various lobes merge into one another due to real-ground conditions.
- Gains reported with *MININEC* are too high.

2.3.2. Modeling With *NEC*

We can do much more accurate modeling using modeling programs based on *NEC-2* (or *NEC-4*). See Chapter 4, Section 1.3. Using a *NEC-2* based program, the Beverage antenna's velocity factor is taken into consideration. Modeling various Beverage lengths will show the gain going through a maximum at about 5 to 8 λ, depending on height and ground quality.

NEC-2 is, however, well-known for predicting excessive gains for wires close to the ground. It underestimates loss along the antenna, which distorts the pattern. W8JI wrote: "*I measure as much as ~70% current loss in 500 feet, which is 3-dB loss.*" As gain is no issue with receiving antennas, this should not be a problem.

With *NEC-2* you cannot connect any part of the antenna to the ground. With *NEC-4* you can do this, but *NEC-4* is not released to the general public and a *NEC-4* license is still very expensive (see also Chapter 4, Section 1.4).

There is a simple way though to overcome this problem with *NEC-2*: We terminate the Beverage in our computer model using quarter-wave wires as ground terminations. If you use such a scheme in reality, you will have a single-band Beverage, since the termination wires are only λ/4, a dead short, on one frequency, and its odd multiples.

The correct configuration for modeling Beverages uses two λ/4 termination wires (in-line) at each end of the Beverage (see Fig 7-66C). Considering the current distribution,

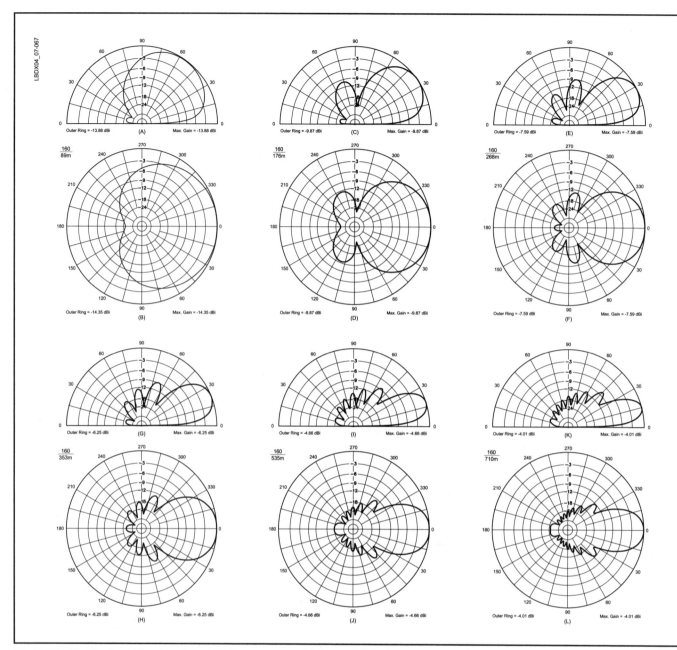

Fig 7-67—Vertical and horizontal radiation pattern of 2-meter high Beverage antennas over good ground, for different "cone of silence" antenna lengths for 80 and 160 meters.

these wires do not radiate in the far field, just like symmetrical radials or top-loading wires with verticals (see Chapter 9). We should not automatically extrapolate this to the real world. In the real world, in close vicinity to a ground that does not exhibit constant characteristics all along these wires, near-field losses will be different in different places, resulting in a imperfect cancellation of the radiation from the two λ/4 wires in the far field. A similar problem is discussed in Chapter 9, where we deal with verticals using just a few elevated radials.

In the real world, quarter-wave terminations are rarely used. If they are used, it would mostly be as a single wire, in-line with the antenna wire. Directivity is slightly better with these models than in the real world, where we use a vertical or a sloping download at both ends. This vertical download (or the vertical component of the sloping lead) is responsible for some omnidirectional signal pick up. The impact of the pick up from the down leads is further discussed in Section 2.5.5.

Most Beverage patterns described in this chapter were modeled using the *EZNEC* program (which uses the *NEC-2* computing core) and using the real "High Accuracy" ground

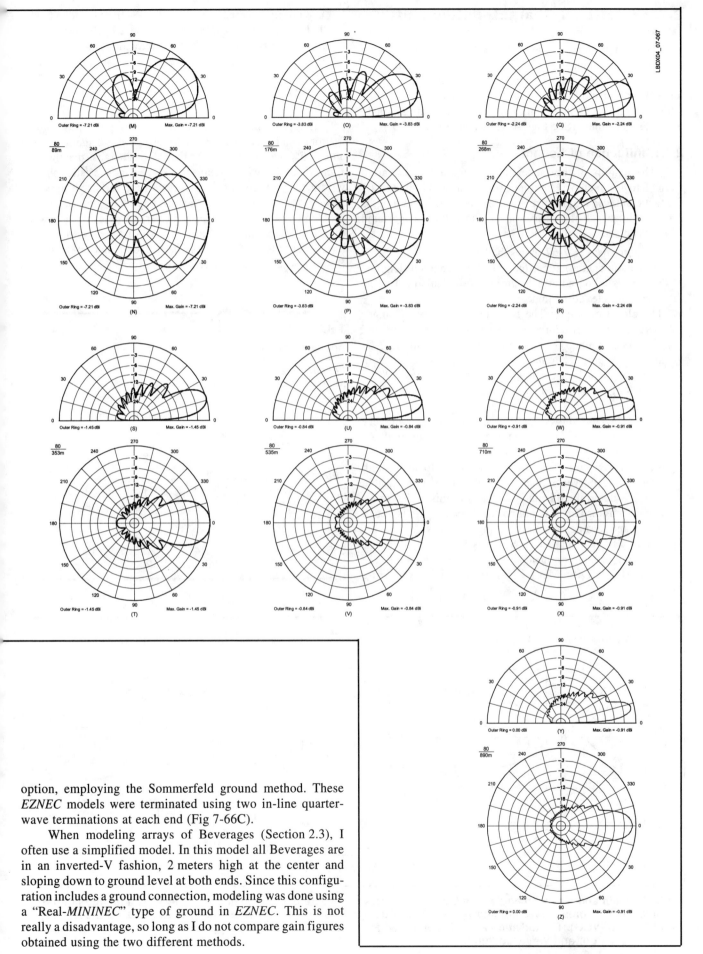

option, employing the Sommerfeld ground method. These *EZNEC* models were terminated using two in-line quarter-wave terminations at each end (Fig 7-66C).

When modeling arrays of Beverages (Section 2.3), I often use a simplified model. In this model all Beverages are in an inverted-V fashion, 2 meters high at the center and sloping down to ground level at both ends. Since this configuration includes a ground connection, modeling was done using a "Real-*MININEC*" type of ground in *EZNEC*. This is not really a disadvantage, so long as I do not compare gain figures obtained using the two different methods.

2.4. Directional Characteristics and Gain

Directivity is the name of the game with any receiving antenna. Forget about gain; since preamplifiers can do wonders! I analyzed a series of Beverage antennas (at 2 meters height) for 160 and 80 meters. I modeled these antennas over good ground (see Table 5-2 in Chapter 5), terminating them in a 600-Ω resistance. (See Fig 7-67 and also Section 2.5.)

2.4.1. Influence of Length

The lengths of the modeled Beverage antennas varied from 89 to 890 meters. The choice of lengths was indicated by the fact that these lengths happen to be the ideal target lengths for obtaining best F/B (see Section 4). The influence of ground quality is discussed in Section 2.4.4.

- The patterns: Fig 7-67 shows the horizontal and the vertical radiation pattern for different Beverage antenna lengths. The horizontal radiation patterns are calculated for the maximum elevation angle.
- The elevation angle: The radiation angle only changes marginally (a few degrees) between very poor ground and very good ground. The elevation angles shown in Fig 7-67 are for Average Ground. Note the large difference in elevation angle between long and short Beverages: 17° for a 3-λ long antenna and approximately 40° for a 1-λ long antenna!
- Gain: **Fig 7-68** shows gain curves vs length, over very poor ground, average ground and very good ground. As indicated above, the gain figures may be a little optimistic, a known flaw with *NEC-2* for antennas close to the ground. From these curves it may seem that maximum usable length is determined by the way gain diminishes beyond a certain length. This is not important because gain is not an important parameter with receiving antennas.
- Directivity: We learned in Sections 1.8 and 1.9 that the most important parameters for receiving antennas are DMF (Directivity Merit Figure) and RDF (Receiving Directivity Factor), as well as the −3-dB main-lobe angle. **Table 7-24** lists the DMF and RDF vs 20°-elevation-angle forward lobe and the −3-dB beamwidth angle shown in Fig 7-67 for 80 and 160 meters. Note that these are modeled values that assume there is no space diversity effect involved.
- Is longer really better?

It appears that you can build very long beverages (4 to 5 λ long) and get really superb directivity. Here again, models and reality may not always be the same. There is such thing as *space diversity*, which means that wave characteristics change with place. As long as you stay within a radius of approximately two wavelengths, this usually does not cause any problems. This is the reason why very large arrays and very long Beverages, may actually behave differently from what the model tells us. *"Longer is better"* does not hold true for Beverages (as for any large receiving array).

Table 7-24
80-meter Beverages

Length Meters	Gain dBi	−3-dB Angle Degrees	DMF dB	RDF dB
89	−8.6	90	15.4	9.3
176	−4.1	59	20.6	13.1
268	−2.2	43	23.6	15.0
353	−1.6	35	24.6	15.8
535	−2.3	26	23.8	15.4

160-meter Beverages

Length Meters	Gain dBi	−3-dB Angle Degrees	DMF dB	RDF dB
89	−15.9	122	11.7	6.5
176	−10.6	86	16.6	10.1
268	−7.8	66	21.3	12.2
353	−36.3	55	21.8	13.6
535	−4.8	40	24.5	15.3
710	−4.6	32	24.2	15.6

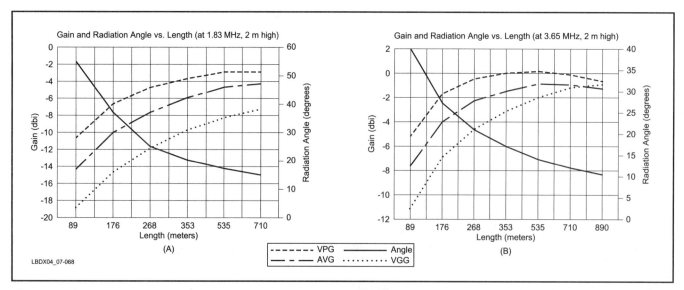

Fig 7-68—Gain and elevation angle for a 2-meter high Beverage antenna for 160 and 80 meters, as a function of the antenna length. Three curves are shows: over Very Poor Ground (VPG), over Average Ground (AVG), and over Very Good Ground (VGG). The radiation angle is computed for Average Ground. This angle only changes marginally between Very Poor and Very Good ground.

In real life the limit is not imposed by the velocity factor of the antenna (see Section 2.2), but by the space diversity. In modeling with *NEC*-based programs losses are definitely underestimated, as all Topbanders who have actually measured losses can confirm. But again, losses by itself are not really an issue.

Beyond a length of three wavelengths, little seems to be gained, as I have learned from real-life experience. If you look at the gain in DMF and RDF beyond three wavelengths, there is little to be gained there as well. And since we don't go by gain with receive antennas, we can safely conclude that three wavelengths is the maximum we want to use.

If you want to improve your receiving antenna beyond this point, you can go to staggered end-fire phased Beverages or broadside phased pairs (see Section 2.16). If you have enough space I would recommend a 268-meter long antenna as a best compromise for the two bands. This is 1.5 λ on 160 and 3.0 λ on 80 meters. However, a 176-meter long Beverage (1 λ on 160, 2 λ on 80 meters) is quite powerful as well.

These exact length figures are more symbolic than anything else. We will see in Section 2.4.2 that there are no such things as *magic* Beverage lengths. In a nutshell: A length between 160 to 270 meters seems to be optimal for Topband.

If you choose to use 300-meter long Beverages to cover all directions, and you want to use them on 80 and 160, you have to take into account the HPBW (half-power or −3-dB beamwidth) of 40° on 80 meters. This means you need to use at least nine Beverages to cover all directions equally well. At my QTH I have Beverages ranging from 170 to 300 meters long, and I have 12 of these spread out every 30°.

2.4.2. The "Cone-of-Silence" Length

Authors looking for F/B optimization as a function of antenna length developed the so-called "cone of silence" length. We now know that F/B does not mean much unless you would want to null out a specific local noise source. What we need to evaluate is DMF or RDF.

The concept of the cone of silence resulted in "sacred" Beverage antenna lengths, lengths that were supposedly better than others. It appears that the Front-to-Back ratio (geometric F/B) goes through maximum values for lengths that are a multiple of electrical half-wavelengths. This is logical, since it is pure trigonometry. However, the geometrical F/B is not very relevant, as explained in Section 1.7.

If you assess the overall directivity performance (DMF, RDF) of a Beverage, you come to the conclusion that there are no "special" lengths, provided the Beverage is properly terminated. I evaluated the DMFs and RDFs for Beverage antennas with lengths varying from 140 to 300 meters, using two different termination models: The "perfect model" using two T-shaped quarter-wave wire terminations (Fig 7-66C) and the "sloping" model, where from the middle of both antenna halves slope down to the ground level.

The results for 1.83 and 3.65 MHz are shown in Table 7-24 for both sloping-wire and T-termination models. When properly terminated for best F/B, DMR and RDF both increase monolithically with length, without appreciable bumps. RDF is a fairly linear curve, mostly determined by the forward lobe. The terminations varied between 425 and 525 Ω.

In the DMF curve there seems to be some kind of "wave" superposed on the curve, probably generated by the effect of the geometric F/B, which is largely undone in the RDF curve because of the impact of the forward lobe. We see that the wave tops out at about 160 and around 300 meters, which are the so-called "cone-of-silence" lengths.

2.4.3. Influence of Antenna Height

The general rule is as follows:

- Higher Beverages produce higher output
- Higher Beverages have larger side-lobes
- Higher Beverages have a higher elevation angle
- Higher Beverages have a wider 3-dB forward lobe

Fig 7-70 shows the elevation and azimuth patterns of a

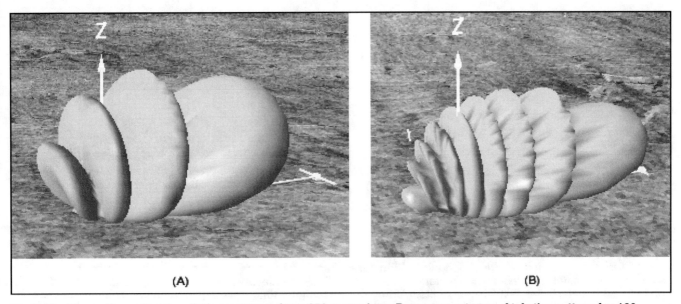

Fig 7-69—Three-dimensional radiation patterns for a 353-meter long Beverage antenna. At A, the pattern for 160 meters, where the antenna is 2 λ long. At B, the pattern for the same antenna on 80 meters (4 λ). Patterns generated with *4Nec2*.

**Table 7-25
Performance for 320-m Beverage at Various Heights**

1.83 MHz

Height	1 m	2 m	4 m	6 m
DMF (dB)	21.6	21.6	20.7	19.4
RDF (dB)	13.6	13.1	12.4	11.7
Opt Rterm	450 Ω	500 Ω	525 Ω	550 Ω

3.65 MHz

	1 m	2 m	4 m	6 m
DMF (dB)	23.3	24.2	23.6	22.2
RDF (dB)	15.2	15.5	15.2	14.7
Opt Rterm	450 Ω	500 Ω	525 Ω	550 Ω

320-meter long Beverage for 160 meters, at various heights (1, 2, 4 and 6 meters) over average ground. The horizontal pattern was calculated for a 20° elevation angle. The DMF and RDF figures for 160 meters are listed in **Table 7-25**. The variation in gain between 1 and 6 meters height is less than 3 dB on 160 meters.

What you see in the table is what you'd expect. The secondary lobes become more outspoken at greater heights, which reduces both the RDF as well as the DMF. The secondary lobes are present in the front half of the radiation pattern as well as in the back half.

On 160 meters a height of 4 meters seems to be still very good, and even 6 meters does not sacrifice much. What about using the same antenna on 80 meters? Fig 7-70 shows the

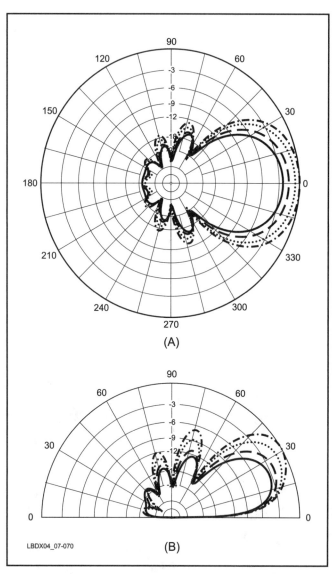

Fig 7-70—Elevation and azimuth radiation patterns for 320-meter long Beverage antenna on 160 meters over average ground, for various heights. Solid line = 1 meters; dashed line = 2 meters; dotted line = 4 meters; dashed-dotted line = 6 meters. See text for comments. (Although the patterns for the different heights are shown together, again I want to emphasize the differences in patterns, since gain is not important for receiving.)

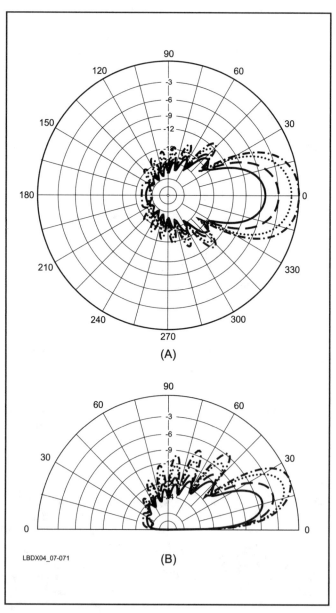

Fig 7-71—Elevation and azimuth radiation patterns for 320-meter long Beverage antenna on 80 meters over average ground, for various heights: solid line = 1 meter; dashed line = 2 meters; dotted line = 4 meters; dashed-dotted line = 6 meters.

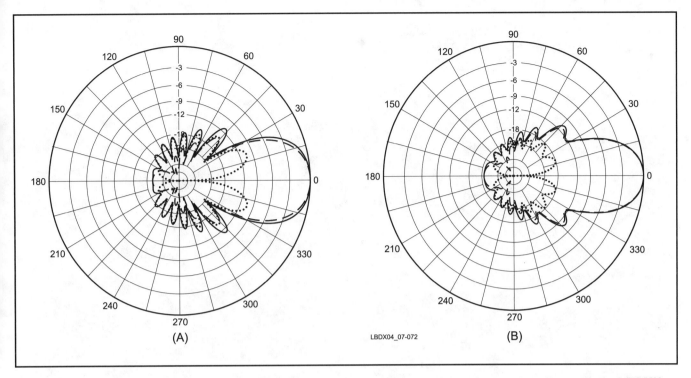

Fig 7-72—Azimuth radiation patterns for 6-meter high (A) and 1-meter high (B), 320-meter long Beverage at 3.65 MHz. The patterns show the total fields (solid lines) as well as the horizontal and vertical components (dotted and dashed lines). At the 6-meter height, the azimuth component is in certain directions approximately 10 dB stronger than at 1 meter high. This broadens the forward lobe.

story. Amazingly enough even at 4 meters the secondary lobes are down almost as much as they are at 1 meter in height, and even 6 meters, which is generally considered as being way too high for 80 meters, is still a very good Beverage antenna! The directivity figures for 80 meters are also given in Table 7-25.

You have to be careful about extending these model findings to real life. The model used in the configuration shown in Fig 7-66C uses two $\lambda/4$ wires in-line as terminations, which means there is no influence from omnidirectional pick-up from a vertical or sloping down lead (see Section 2.5.5). The high-angle lobes that appear with higher Beverages are due to the increasing horizontally polarized radiation component. **Fig 7-72** shows the azimuth patterns (both vertically and horizontally polarized components, plus total pattern) for the 320-meter long Beverage at 3.65 MHz for heights of 6 and 1 meters. The horizontal component is significantly more important at 6 meters than at 1 meter. **Fig 7-73** shows the whole situation in 3D, with the horizontally polarized component on the right of the total pattern for 80 meters at the top, and 160 meters at the bottom.

When elevated even higher, the Beverage will start behaving like a terminated long wire, not like a Beverage antenna. This increased high-angle response of a high Beverage is often used by those who don't believe in important path skewing, to explain "apparent" path skewing (see Chapter 1). Although I don't deny that reception of high-angle sidelobes may cause some confusion at times, the existence of direction skewing has been confirmed repeatedly through the use of other directive antennas, such as phased arrays, which do not have such high-angle secondary lobes.

If you suspect you are receiving signals from such high-angle sidelobes, you can usually verify this by switching to a high-angle antenna, such as a low dipole. As explained in the chapter on propagation this often can occur at sunrise or sunset (gray-line propagation) or during very disturbed conditions.

2.4.4.1. Height of Beverage Antennas: Conclusion

The height is not all that critical. Below 2 meters, Beverages can be a hazard for men and animals. If you must cross a driveway or small street, you can put your Beverage up to 6 meters high and still have a working Beverage. You can also slope the Beverage gently up from 2 meters to 6 meters to cross the obstacle without much harm at all. Tom, W8JI, writes on his Web page (www.w8ji.com): *"I've found very little performance difference with height, unless the Beverage is more than 0.05 λ high."*

If the Beverage can be constructed on terrain that is inaccessible to people, deer and other animals, then you can consider using Beverages at a height of 0.5 or 1 meters for added high-angle discrimination and reduced omnidirectional pick up. All of mine are about 2.2 meters high at the support post. Since I use fairly thin wire, mine sag quite a bit between the supports (to a height of about 2 meters), but I do seem to hear well nevertheless.

2.4.4. Influence of Ground Quality

The general mechanism is:
- The better the ground, the lower the output from the antenna (around 6 to 8 dB difference between very poor ground and very good ground). But even over very good ground Bev-

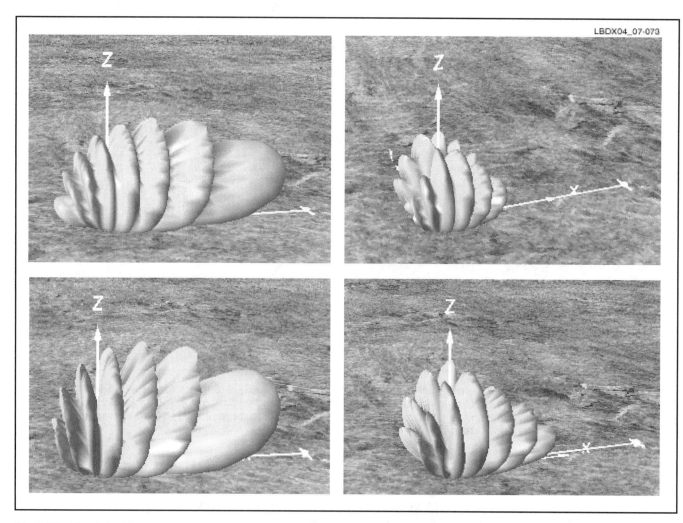

Fig 7-73—Top left: 3D pattern for 320-meter long Beverage on 80 meters. Top right: the horizontally polarized component. Note this is a typical radiation pattern of the many lobes perpendicular to the wire, which we know from (high) "long" long-wire antennas. Left bottom: 3D pattern for the same Beverage at 6 meters height. Note the slightly fatter and higher forward lobe, and the more outspoken secondary lobes. Right bottom: the horizontal component for the same 6-meter high Beverage. Note this component is much more important. (Patterns by *4Nec2*.)

erages have more than enough gain.
- The peak elevation angle changes only slightly with ground quality. For example, a 300-meter long Beverage peaks at 27° over Very Poor Ground. The response at 10° elevation is down 3.4 dB from the peak. Over Very Good ground, the lobe peaks at 29°, and the response at a 10° elevation angle is down 2.8 dB from the peak response.
- The poorer the ground quality, the less pronounced the nulls will be between the different lobes. This is similar to what we notice with horizontally polarized antennas over real ground.
- The directivity factors (DMF and RDF) of a Beverage antenna remains almost constant for grounds ranging from Very Good to Very Poor.
- The Beverage does not work at all over sea water. Its output is down 15 dB compared to the same antenna over poor ground and the main elevation angle is at 45°. This confirms the observations made by Ben Moeller, OZ8BV, that his Beverages near the sea never worked well at all. The beverages at VKØIR, erected over a saltwater marsh never worked either (as I told them would happen!).
- With good ground, the vertical ends do become much more important than over poor ground.

2.5. Terminating the Beverage Antenna

I have calculated the directivity patterns for a 160-meter long Beverage (over average ground, 2 meters high, wire: AWG #12) on both 160 and 80 meters. While the F/B changes for a given elevation angle, the RDF and DMF figures remain relatively constant, as shown in **Fig 7-74**, which shows that the 3.7-MHz geometric F/B peaks for a termination value between 400 and 500 Ω. For thinner wire (#20) these values will be somewhat higher. Unless you need to null out a local noise or QRM source right off the back of the antenna, the exact value of the termination resistance is far from critical. Varying the termination resistance just changes the position of the notches in the back of the Beverage. See **Fig 7-75**.

2.5.1. Beverage Impedance (Surge Impedance)

Over perfect ground the single wire Beverage impedance can be calculated using the formula of the single-wire trans-

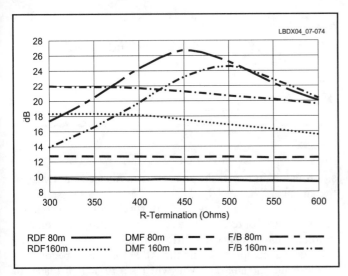

Fig 7-74—F/B, DMR and RDF for a 160-meter long Beverage antenna (2 meters high, #12 conductor, over AVG) terminated in a resistance between 300 and 600 Ω (for 80 and 160 meters).

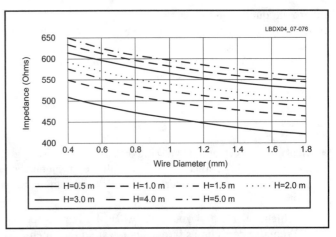

Fig 7-76—Characteristic impedance of a single-wire Beverage antenna for different conductor diameters and different antenna heights. The values are calculated for the single-wire feed line equivalent. In practice the values can be 10 to 30% higher, depending on the ground quality. See text for details.

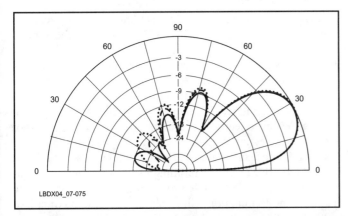

Fig 7-75—Changing the termination resistance of 160-meter long Beverage from 300 Ω (solid line) to 450 Ω (dashed line) to 600 Ω (dotted line) on 3.65 MHz merely changes the place of the nulls in the back of the antenna, and hardly impacts the total directivity (DMF and RDF) of the antenna.

mission line over ground:

$$Z = 138 \log\left(\frac{4h}{d}\right) \quad \text{(Eq 7-1)}$$

where

    h = height of wire
    d = wire diameter (in the same units).

The theoretical impedance values listed in **Fig 7-76** are calculated over perfect ground. They are useful for estimating the terminating resistor for a single-wire Beverage and for designing matching transformers and networks. Note that the impedance does not change drastically with height or wire size.

Over real ground the surge impedance appears to be higher than over perfect ground. This is because the electrical ground is not the surface of the ground, but a little deeper. The following correction figures can be used as compared to the impedance over perfect ground in Fig 7-76.

- Good ground: +12%
- Average ground: +20%
- Very poor ground: +30%

These termination values are for highest F/B at low elevation angle (450 Ω in Fig 7-74 on 80 meters). Very low Beverages (BOGs) do *not* have a very low impedance as is sometimes claimed. A BOG on the grass shows about a 300-Ω surge impedance because the actual ground is deeper than the surface of the soil.

2.5.2. Determining the Best Termination Resistance

With the correct terminating resistance the antenna current along the beverage shows no sinusoidal pattern but rather it shows an exponential decrease towards the termination (due to the attenuation). There are different ways to determine the best termination resistance value. The principle with all of them is to vary the resistance value for minimum standing waves on the antenna.

- Couple your antenna analyzer to the Beverage transformer and tune across the spectrum (for example, from 1 to 10 MHz). When you do this with the far end open-ended (or short-circuited) you will see large swings in impedance or SWR. The proper terminating resistance is the value for which the variation in impedance or SWR is least when tuning across the entire spectrum over which you want to use the Beverage. In this exercise the termination should *not* be adjusted to achieve the best SWR, but rather the *flattest SWR curve*.
- Excite the antenna with a small signal, and measure the current along the antenna with a clamp-on RF current meter or RF voltage meter. Adjust the termination resistance until the voltage or current has a uniformly smooth taper towards the far end (typically 25% to 50 % depending on

Receiving Antennas 7-51

ground quality and antenna length).
- Using your antenna analyzer, measure the feed-point impedance across a range of frequencies (perhaps 1.5 to 3 MHz) and note the lowest (Z_{min}) and the highest impedance value (Z_{max}). The Beverage surge impedance is then given by:

$$Z_{Bev} = \sqrt{Z_{max} \times Z_{min}}$$

- Measurements methods making use of a field-strength meter are useless, since pretty much all measurements are done in the induction (near) field, unless extreme care is taken!

Whichever method you pick, when the impedance (or SWR) remains constant as frequency is varied the antenna is properly terminated. A very small impedance change with frequency results in the best F/B at low elevation angles. Do not forget, however, that RDF and DMF are generally much more important than F/B and are hardly influenced by the value of termination resistor (see Fig 7-74).

2.5.3. Inductive and Capacitive Load Terminations

Take the example of a 200-meter long Beverage (2 meters high, over AVG ground, #20 wire) terminated in a resistor giving the best F/B: 525 Ω yields a 20-dB F/B. If you terminate the Beverage in a complex impedance of 475 + j 125 Ω, you obtain a much higher F/B, as **Fig 7-77** and **Table 7-26** show. Let's analyze the DMF and RDF for both cases:

Looking only at the F/B clearly leads us to the wrong—or rather to an incomplete—conclusion. All you do by changing from a resistive value to a complex load, is to move the nulls around in the back of the antenna, but that does not

Table 7-26
DMF and RDF for Resistive and Inductive Terminations

Termination	DMF	RDF	F/B
525 + j 0 Ω	17.5 dB	10.7 dB	20 dB
475 + j 125 Ω	17.2 dB	10.6 dB	44 dB

significantly influence the global directivity (RDF or DMF) of the antenna.

Unless you want to use your Beverage for nulling out one specific noise source in a very specific direction, the complex termination impedance has little or no added value. When you use such a complex termination the Beverage becomes a single-band antenna, and you will have to switch loads for different bands. The only sensible application for inductive terminations is with very short Beverages less than 0.5 λ long (see Section 2.13).

2.5.4. Ground Requirements at the Far End (Termination End)

If you have real soil (not rock), the ground system can consist of a ground rod. The RF resistance of the ground system at the far end (termination end) of the Beverage does not have to be very low, since even high ground resistance is effectively in series with the terminating resistance. A 1.5 meter (5/8-inch OD) copper clad steel rod will have an RF ground resistance ranging from ~50 to ~300 Ω depending on ground quality. We have seen that the termination resistor value is not critical (see Section 2.5). Let's assume you need a 425 Ω total termination resistance and that your ground rod is 100 Ω. You would require 325 Ω as a termination resistor. In most situations where you use a single ground rod, the actual resistance value of the termination will likely be 250 to a maximum of 400 Ω.

Do you want to know the RF ground resistance for the single ground rod at your particular QTH? Drive one rod in the ground, plus three more at about 2 meters distance around the first rod. Attach eight 25-meter long equally spaced wire radials to the ground system. Now you have a pretty decent ground of 20 Ω or better. Apply one of the procedures outlined in Section 2.5.2 to determine the optimum termination resistor value. Note this value (say, 415 Ω). Remove (not just disconnect) the radials and the extra three ground rods, and repeat the same procedure. Note the new value (for example, 300 Ω). The single ground rod RF earth resistance is thus: 415 + 10 − 300 = 125 Ω.

Where ground conductivity is really bad (Ref 1260) or where you cannot drive in a sufficiently long ground rod, the resistance of a single ground rod may actually be higher than the required terminating resistance. In that case you can install multiple ground rods, combined with a number of short radials, to bring the resistance down to an acceptable value. Do not use one or two long radials, but rather a large number of short radials forming a capacitance to earth. This was confirmed on the Internet by Greg, ZS5K, who wrote: "*With my 800 ft Beverage, I have found it desirable to have more than one rod, and use more inserted resistance, so that the total termination resistance is less dependent on the weather.*

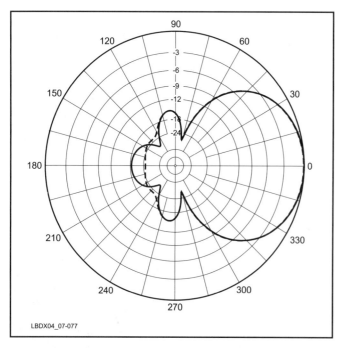

Fig 7-77—A complex termination impedance can be used to move the notch in the back of the antenna around (dashed line). It hardly affects the overall directivity (DMF and RDF) however.

I have found that with only one rod, things change quite noticeably with the seasons."

At the receiving end of the Beverage you should pay a little more attention to the ground system. This is explained in Section 2.8.

2.5.5. Vertical Down-Lead, Sloping Beverage or Quarter-Wave Terminations

A common way of terminating the single-wire Beverage is to connect the proper terminating resistor between the end of the Beverage and the earth. The systems using quarter-wave terminations (see Fig 7-66C) is only a good solution for modeling a Beverage. The reason is that these two quarter-wave wires (in-line), over real (not perfect) ground, with varying ground characteristics along their length, will pick up signals themselves. In a transmitting model they would radiate. The mechanism is similar to the one described in Chapter 9 dealing with low elevated radials and radiation from these radials.

Over the years, some have attributed magic properties to sloping terminations, since the vertical component of the down lead supposedly *disappears*. Many of us slope the beverage wire down from its nominal height to a ground stake, where the termination resistor is attached between the sloping antenna end and the ground. The ground stake serves two purposes, it is the electrical ground and the anchoring post for the Beverage wire.

There is nothing wrong by doing this, but such a sloping wire does not pick up fewer signals than a vertical wire of the same height. It doesn't matter whether you slope the last 20 meters of the Beverage or continue 18 meters of it horizontal and run 2 meters straight down. Either way you have 2 meters of vertical distance. A vertically-polarized wavefront arriving from the side will see only the 2-meter high vertical component of the sloping end, the same as with a straight vertical wire.

The best proof that a sloping wires works like a vertical wire is the fact that a Pennant antenna (or Delta-shaped loop; see Sections 3.4 and 3.5) works. The Pennant, despite having one end sloped and one end vertical, has nearly identical vertical sensitivity from both ends. This clearly proves that the sloped wire behaves almost exactly like a vertical of the same height, or else the Pennant would have a 0-dB F/B ratio.

Tom, W8JI explains: *"The vertical end-coupling of Beverages is a mostly non-problem that has been over rated, and the "cures" are mostly non-cures for a non-problem. One way to look at it is six feet of vertical drop is "six feet" of vertical drop, no matter how the "six foot drop" is spread around, and that isn't any big deal when the antenna is several hundreds of feet long. If that isn't reason enough not to worry, the entire antenna responds to vertical signals anyway...especially on groundwave!!*

Second, it is physically and electrically impossible to "shield" the vertical end lead no matter what scheme you employ. The antenna MUST always have the same net common mode current flowing to earth over the vertical lead distance to ground. The only way possible to prevent that effect is to move the entire ground system up to the element height, and that means work with a bulldozer making a large mound or installing a large counterpoise hanging in the air at antenna height. Only those options can prevent common mode current from flowing to earth!

Some Beverage books will give you the idea that a particular scheme does something to "shield" the vertical end wire, but it does not. On the outside of the tubing, we would measure exactly the same common-mode current as the end-current in the antenna. The vertical wire carrying current in the center of the tube induces an exactly equal opposing flow on the wall, that spills over at the open end and flows down the outside of the tube. Not that it matters, since that radiation is generally insignificant."

John, K9DX, did a well-controlled test to assess the difference between sloping terminations and terminations with a vertical down lead. He put up two widely separated 2-λ Beverages in parallel. The reference one had two in-line quarter-wave radials, which precluded any omnidirectional pick up. With this reference antenna a transformer with minimum capacitive coupling was required, since the antenna was now ungrounded. A binocular core transformer (see Fig 7-86) provided 30 dB of isolation.

The test antennas had either long sloping ends or simple vertical down leads. John concluded from extensive testing that there was absolutely no difference at all between sloping and vertical feeds. Shielding the vertical down lead also does *not* work. **Fig 7-78** shows the professional manner in which K9DX treats his sloping Beverage terminations.

All in all, John came to the conclusion that: *"... the vertical feed, as poor as it is, seems to hear as well as the raised feed. The noise picked up by the beverage in an omnidirectional noise environment appears dominated by the front lobe. If you want better signal to noise, reduce the beamwidth. Going from 15 to 30 dB off the sides or back won't do much."* All of this is, of course, considering "noise" coming in equal strength from all directions, and you're not

Fig 7-78—Sloping down lead termination for one of the Beverages at K9DX. Big 4 × 4-inch wooden poles are used as end posts, taking the stress of the copperclad steel Beverage wires. The supports along the Beverage are 1.5-inch plastic pipes slit over a rebar driven into the ground.

trying to reduce QRM and noise from a particular direction. In quiet location, such as at K9DX and W8JI, noise is evenly distributed in all directions and at all angles. Under these circumstances looking at the RDF is much more important than looking at nulls in specific directions. This does not mean that you cannot experience a tremendous improvement by going from 15 to 30-dB F/B. In other real-life situations where most of the noise is coming from the back—such as in Western Europe where most of the QRM, splatter and clicks, come from the east.

The sloping termination has been a subject of discussions and opinions for a long time. Let me add mine. Whether or not you use a perfectly vertical or a sloping termination wire, these wires pick up vertically polarized waves from all directions. We can envision in a Beverage with the two vertical termination wires separated by a number of half waves (considering the velocity factor of the antenna), the signals induced on both verticals will be out-of-phase when arriving at the receiving end. This may lead to so-called optimum lengths, where the ill effects of the vertical down leads are partially annihilated. (There still is the loss in the Beverage as a transmission line resulting in incomplete cancellation).

This only works in one direction, and for the 180° example given, assumes a wave angle coming at right angle to the Beverage. (The signal arrives at the two verticals in-phase, and will be added out-of-phase because of the delay in the Beverage acting as a transmission line.) So in this particular case the front-to-side ratio would be improved. For other lengths of Beverages, other directions will be nulled out. In other directions the noise may add and the directivity may go down. The canceling mechanism works along the same principles explained in Section 1.6.

Sloping ends can be considered as a number of short verticals, placed in slightly different locations (along the sloping end) so we cannot really talk about a single separation distance between the two sloping ends—It is smeared out. If you now use a Beverage that consists only of sloping ends (one central high post and two sloping wires, each being half the length of the vertical), you will have the disturbing effect of many very small verticals spread all along the Beverage. Depending on the direction of the noise, the contributing EMFs may add or subtract, and in a long Beverage all of this happens at the same time. The net effect is a general, small decrease in the overall directivity in all directions, without creating any specific nulls or maxima for any particular direction. I use this Beverage model (I call it the inverted-V shaped Beverage) when modeling Beverage arrays (Section 2.16) and find that the directivity patterns are very similar to those modeled using the T-terminations with two λ/4 in-line radials (Fig 7-66C).

If you add the above reasoning to the sensitivity analysis done by W8JI, you can conclude that whereas vertical feeds in almost all cases are not a *real* problem, long sloping ends because of the phase distribution are even better. Long sloping terminations can indeed help reduce the ill effects of the vertical distance involved in the down leads.

By using Beverages in a broadside or end-fire array configuration, the effects of vertical down leads can also be greatly compensated (see Section 2.16.3), at least for specific directions. A broadside/end-fire is even better.

2.5.6. Adjusting the Termination for Best F/B

Unless you want to null out a specific noise source that is present all the time, it makes no sense to tune a Beverage antenna. By adjusting the termination impedance (either as a pure resistor or as a complex impedance as explained in Section 2.5) you can put a null right in the direction of the noise source. The adjustment of the terminating impedance can easily be done using a small signal generator with a small whip antenna. This should be placed outside the near field of the antenna, a minimum of 2 λ from the far end of the Beverage.

For nulling out a local noise source, you will have to put the notch at close to a 0° elevation angle. **Fig 7-79** shows the radiation pattern at a 1° takeoff angle for a 200-meter long Beverage. The notch at exactly 180° results from a termination impedance of $525 + j\,115$ Ω. The notches at 135° and 225° (±45°) were obtained with a termination resistor of 500 Ω. This simple method allows us to move the notch (at a 1° elevation angle, which is ground wave) around over a total angle of 90°.

I must say, however, that using a second noise-sampling antenna and a noise canceller (as described in Section 1.35) is usually a simpler and better solution to the problem of eliminating a noise from a fixed source in the neighborhood. You will only eliminate the local noise and not all the signals coming from the direction of the noise, because the small pick-up antenna will not hear them well enough.

2.5.7. Terminating Resistors

The termination resistor must be a low-inductance resistor, which means that it cannot be a wire-wound resistor. In

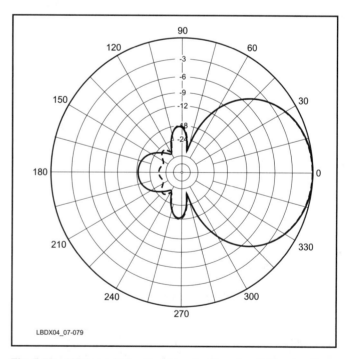

Fig 7-79—With an inductive termination (resistor + coil, shown with dashed line) we can move around the notch angle at 1°-elevation angle to suppress a local noise source arriving by ground wave. Response with just a resistive termination is shown by solid line.

principle, any wattage will do but if you have a Beverage close to a transmitting antenna (which is bad), you may want to use a few 1 or 2-W resistors in parallel. I use a single 2-W resistor and have never seen one discolored or burned up. If they do burn up, you either have your Beverages much too close to your transmit antenna or you run excessive power—or lightning has been causing such problems.

Any type of resistor can be killed by a long-term overload. In some areas, where lightning is frequent, inappropriate types of terminating resistors easily get blown out by lightning. Metal-film or carbon-film resistors are not good since they cannot withstand short overloads. Old type carbon-composition resistors are excellent, but they are difficult to find nowadays. Metallic-composition resistors, like standard Ohmite OX or OY resistors, (**www.ohmite.com/catalog/ox_oy_series.html**) will do the job as well. AB (Allen Bradley) still manufactures carbon-composition resistors but they are expensive when bought new (**www.welbornelabs.com/ab.htm**). K9DX uses a large fuse and clip for quick replacement when his terminating resistors are destroyed by lightning. See **Fig 7-80**.

2.5.8. Lightning Protection

If you live in an area with a high thunderstorm occurrence, the use of carbon-composition or metallic-composition resistors is a must. You can add further protection with small gas-discharge tubes connected across the resistor, or even by putting a pair of homemade small air-gap electrodes across the resistor. Riki, 4X4NJ, described such a homemade spark gap: *"The spark gap consists of heavy solid wires—about 10 AWG, soldered to teardrop terminals that are placed under the screws going to the antenna and ground. The ends of the wires are cut with "side cutters" leaving a nice "knife edge," and I position the two edges very close to each other. A piece of paper makes a nice "feeler gauge" for this purpose. It is very effective, most simple, and negligible cost."*

Bill, KØHA reported using *"Taiwan Semiconductor's SRYH-90L gas tube surge voltage protectors (the $1.85 CATV model, with high current capability)"* for the same purpose. DX Engineering (**www.dxengineering.com/**) sells high surge energy terminations.

2.6. Beverage Wire

A Beverage is a lossy antenna. There is no point in striving for the lowest possible conductor loss. Any type of wire that is mechanically suitable should do the job. I have used bare copper wire, enameled copper wire, PVC-insulated copper wire, with the copper having a diameter anywhere from 0.6 to 1 mm on my Beverages.

Soft-drawn copper wire cannot be tensioned to any great degree without stretching and it is not suitable for very long unsupported spans of wire. If you don't mind using many supports, however, soft wire may be used for the antenna. I use soft-drawn single strand copper wire of 1.0 mm OD (#18 AWG) for my winter-type Beverages, where the bamboo posts are separated by approximately 20 meters.

For permanent Beverages separated by distant posts I use bronze- and copper-clad steel wire (1.6 mm OD, #14), which makes it possible to cover 100-meter stretches without intermediate posts. I pull this tight with approximately 45-kg tension. Copper-clad steel is good, if you can get hold of it.

Insulated wire is potentially better than a bare conductor when the antenna wire touches branches or brushes against leaves in the wind. On the other hand, insulation may hide a broken conductor, unless the antenna wire is under significant tension.

Some people have used surplus telephone-pair wire. This wire is often available at hamfests. Don't try to separate the wires, just connect them in parallel at both ends. Using a pair as a Beverage has the advantage that you can check continuity from one end using an ohmmeter. All connections must be soldered, preferable with silver solder. Common Sn/Pb solder rots away after years in the outdoors. If you use regular 60/40 Sn/Pb solder, cover the solder joints with liquid rubber or a dollop of Vaseline.

Long Beverages made of thin soft-drawn copper wire can stretch in high wind. A solution is to use a well-tensioned bungee cord at the end of the Beverage.

Aluminum wire is widely used as electric-fence wire and makes for excellent Beverage wire. It is lightweight and strong and readily available from outlets selling farming equipment. It is commonly available in diameters from 1.15 mm (#17) to 1.65 mm (#14). See for example **www.tippertie.com/fencing/neverrust.asp**.

Earl, K6SE, uses galvanized-steel wire. Earl claims that the use of this high-loss resistance wire improves the F/B, since the signal from the reverse (unwanted) direction now travels down the lossy wire twice to reach the receiver—once toward the terminated end and then whatever signal is reflected from the termination must travel down the wire again toward the receiver. Signals coming from the forward (desired) direction must travel down the lossy wire only once.

Tom, W8JI, rightly points out: *"Steel fence wire would aggravate losses that already limit the benefits of using long*

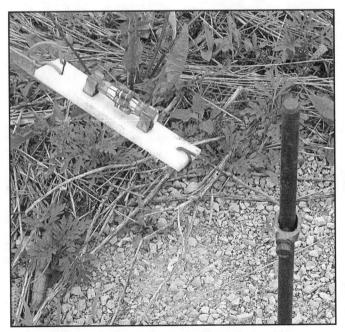

Fig 7-80—K9DX uses a fuse clip and parallel carbon resistors soldered across a (blown) fuse. This makes replacement, in case of lightning strike destruction very easy.

Beverage antennas. In a very long antenna, the small additional loss of steel fence wire might slightly reduce performance."

Please remember that one of the most important considerations in the selection of wire for a Beverage is maintenance!

2.7. Supports, Poles, Insulators and Other Hardware

You can use any convenient support for a permanently installed Beverage: tree trunks (use nail-type electric-fence insulators), wooden poles, metal posts, plastic tubes, etc. Electric fence hardware is cheap and suits the purpose. I always use the in-between posts only to support the wire, letting the wire freely slide through an insulator. This way tension will equalize in all stretches between supports, and if at one place the copper wire is stretched by accident this will smooth out over a longer stretch of wire.

If the wire breaks you will also see it at any point along the antenna. A long piece of wire is also much more difficult to break. Soft copper will typically stretch 25% before breaking! If someone accidentally walks into a taut, short span of wire, he will likely break it and may hurt himself.

A long span of wire will stretch without breaking and not hurt the "offender." Many Beverage users only put them up during the winter, when they can use their neighbors' land for the purpose. I have a good neighbor who is a farmer and who lets me use about 20 acres from the end of October to the end of March every year. I use 2.4-meter long bamboo sticks and support the wire in a simple loop made of electric tape. I do not wrap the Beverage wire around the stick, or around branches. This creates an inductance, which should be avoided, and it prevents the wire from freely moving. Bamboo supports are cheap ($25 per 100 at wholesale garden outlets) but do a wonderful job (see **Fig 7-81** and **7-82**). They last about four years before they rot at the bottom.

If you don't like the plastic tape loop, you can also slide a 10-cm long piece of 5/8-inch plastic tube over the top of the stick. Fix it with some electric tape. Make a slit in the tube with a hack saw, and just drop the wire into the slit. The same technique can be used when using slit 1.5-inch PVC pipe fitted over a rebar driven in the ground. (See **Fig 7-83**.)

Fig 7-82—Method for affixing the Beverage wire on the bamboo stick, using a loop made of electrical tape.

Fig 7-83—K9DX uses 1.5-inch plastic pipe for his Beverage wire supports. A 1-cm deep slot in which the Beverage wire is dropped is made in the pipe using a hacksaw.

I have a big box with old-fashioned egg-type ceramic insulators, with two holes. These are excellent for end-type insulators. If there is a lot of tension on the wire (in a permanent installation using copper-clad or bronze wire) be careful selecting plastic insulators. Some very thin plastic-compression insulators will actually cold-flow and allow the wire to pass through the insulation. Heavy-walled ceramic egg insulators are much more reliable.

2.8. Feeding the Beverage Antenna

In Section 2.5.1 I discussed the surge impedance of a Beverage. The Beverage feed system transforms the antenna impedance to the transmission-line impedance (usually 75 or 50 Ω) and transports the received signals to the receiver. The entire feed-system can be broken up into three parts:

- The feed-point transformer
- The antenna ground
- The feed line

Fig 7-81—2.4-meter long bamboo sticks are used to support the wintertime Beverages at ON4UN. This picture shows heavy ice-loading on the antenna, which makes the antenna visible.

The technical issues involved are:
- Good impedance matching
- Low loss
- Good common-mode suppression

2.8.1. The Beverage Feed-Point Transformer

The easiest way to match the Beverage (typically 450 to 600 Ω) to common coaxial cable (50 or 75 Ω) is to use a wideband transformer. Such transformers are usually wound on magnetic-material cores and the most common shapes of the cores used are the toroidal core and the binocular core. Such transformers are commercially available from various sources but can easily be homemade for a fraction of the price of commercial units.

2.8.1.1. Core Material

Several core materials can be used for this job. First of all, don't use just any core you find in your scrap box. Transformation ratio or SWR is not the only issue. I have seen transformers that showed a perfect SWR with the secondary terminated with 450 Ω—but they also exhibited a prohibitive loss of about 5 dB!

You can generally distinguish between two types of core material used at RF: *powdered iron* and *ferrite*. Ferrites have much higher permeability (up to 10,000) than powdered-iron materials (only up to 100), but ferrites are less stable at higher frequencies and saturate more easily. For wideband transformers used for receiving Beverages, ferrite cores are the most logical approach.

Another distinction you have to make is between a transmission-line type transformer and a regular transformer. Core materials that are excellent for transmission-line-type transformers are not necessarily the best for a regular transformer.

What is a transmission-line type transformer?

The transformers where we wind two, three or more wires in parallel to make sections of a transformer are transmission-line transformers. See **Fig 7-84A**. With transmission-line transformers, the core material loss tangent isn't critical. All we want is a high impedance.

What is a regular transformer?

Regular transformers are transformer where the primary and secondary are wound as separate windings on the core. They can be heavily coupled (one winding on top of the other) or wound to have minimum capacitive coupling by winding them on opposite sides of a toroidal core. With this type of transformer, loss tangent does become a factor. Regular transformers rely on magnetic coupling, while transmission-line transformers do not.

The more isolated the windings, the more critical the loss tangent becomes. This means that the loss-tangent issue is especially important with transformers where primary and secondary are separated to obtain minimum capacitive coupling.

Various ferrite-core materials have been used for Beverage transformers: Type-77, 75, 73, 43 and 71 materials are the most commonly used grades. The 77, 73 or 43 materials will require a smaller number of turns. Fewer turns reduces unwanted coupling from primary to secondary (in a non-

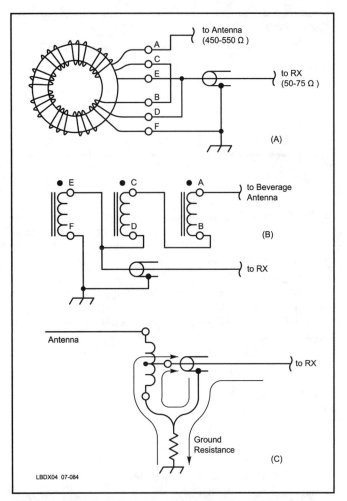

Fig 7-84—The transmission line transformer (A) consists of three parallel-wound wires, equally spaced across a toroid core. The proper connecting scheme (B) is shown for a 9:1 impedance transformation ratio. The enameled copper can be securely fixed to the core using tape or Q-dope. Antenna currents (C) on the outside of the feed line (acting as a sort of "on the ground Beverage") are routed into the inside of the coax via the common ground resistance and via the low-Z link of the transformer.

transmission-line type of transformer), but they saturate at high signal levels. They also work much better at higher frequencies, making the Beverage useful at 7 MHz or higher.

For decades I have used Ceramic Magnetics MN8CX high-permeability cores and I have not had a single failure or complaint. More recently I have become an enthusiast for the binocular-core transformers recommended by W8JI. These use type-73 material (see Section 2.8.1.3) and are much easier to wind than toroidal transformers.

2.8.1.2. Core Shape

Most Beverage transformers are probably wound on doughnut-shaped *toroidal* cores, although W8JI has been able to break the trend with his *binocular cores*. So far as the size of toroidal cores is concerned, 0.5-inch cores or smaller will do the job if the builder is not tempted to transmit with too much power on the Beverage.

I made several transformers using various types of cores and thoroughly tested them for insertion loss and SWR band-

Table 7-27
Manufacturer's Part Numbers for Toroid Cores

Core Type	Turns
MN8CX (doughnut)	4
FT50-75, FT82-75, FT114-75 (doughnut)	8
FT50-43, FT82-43, FT114-43 (doughnut)	8
Fair-Rite Products 2873000202 (binocular)	2 or 3*

*1 turn=2 passes

width. **Table 7-27** lists the required number of turns for a 50- or 75-Ω windings. This means that a 9:1 transformer made on a MN8CX core requires four turns for the primary; the Fair-Rite binocular requires two or three turns (six passes) for a low-Z primary.

W8JI highly recommends using type-73 material for receiving applications between 1 and 30 MHz. This material minimizes the turns count required, without introducing excessive loss. You can find a very thorough discussion on the subject of balun and transformer core selection on W8JI's website: www.w8ji.com/.

2.8.1.3. Winding Your Own Transmission-Line-Type Transformers

The Beverage and its feed coax are both unbalanced, so you can wind a transmission-line transformer using a trifilar winding. This gives a 9:1 impedance transformation to 50- or 75-Ω coax, which works well for Beverages mounted at the usual heights.

The schematic and the winding information for a 9:1 wideband transmission-line type transformer is shown in Fig 7-84. Note that the ground for both the coaxial feed line and the antenna are connected together in this transformer. The antenna and the feed line thus use a common ground, although this is often not the best configuration.

Refer to **Table 7-28**, which shows winding information for some common core materials and toroids. The same number of turns can be used for 50- or 75-Ω system impedances. Data are given for the frequency range 1.8 to 10 MHz.

Winding such a transformer is not difficult and the binocular is particularly easy. Take three lengths of 0.4-mm enameled wire (#26 or #28 AWG) and twist them together with a pitch of approximately two or three turns per cm. Use a small, variable-speed drill to get an even twist. Fix one end of the wires in a vice and the other end in the head of the drill. Make sure the tension in the three wires is identical and that the three wires run perfectly parallel (no wire crossings!). If you have access to enameled wire in different colors, then the task of identifying the wires is easy. Otherwise you will have to scrape off (or burn off) some insulation material and identify the three wires with an ohmmeter. Connect the wires exactly as shown in Fig 7-84. After soldering, cover the soldered wires with some shrink sleeving.

In the past I used mainly single MN8CX cores, as well as two stacked MN8CX cores, for my Beverage transformers. See **Fig 2-85**. Table 7-28 lists alternate core suppliers and **Table 7-29** shows the insertion loss and the SWR for transmission-line type transformers made with different types of cores. The measurements were made using an Rhode and Schwartz Network analyzer. Insertion loss was measured using two transformers back-to-back.

W8JI tested his binocular transformers (a single Fair-Rite Products 2873000202) using several methods and quoted a 0.45 dB insertion loss at 1.8 MHz, which is roughly the same as what I measured for a single MN8CX core. In previous sections of this chapter I have repeatedly said that gain of a receiving antenna is not the real issue. If you want to continue this logic, you should not start a good-better-best discussion over hundredths of a dB!

What is one turn? With toroidal cores you count one turn each time the wire goes through the core opening. A binocular core has two holes. See **Fig 7-86**. A wire though one hole is one pass. One complete turn is thus two passes through a binocular core. If you thread the wire in through one hole and bring it back through the other hole, you have one turn. With a binocular core the classic 9:1

You should insert two short pieces of thin-wall insula-

Table 7-28
Alternative Core Suppliers

Type	Alternative Supplier and Code
Fair-Rite Binocular 2873000202	Amidon BN 73-202 Ameritron 412-5250
Amidon FT50-75	Fair-Rite 5975000301 Ferroxcube 768T188/3E2A
Amidon FT82-75	Fair-Rite 5975000601 Ferroxcube 846T250/3E2A
Amidon FT114-75	Fair-Rite 59750001001 Ferroxcube 502T300/3E2A
Amidon FT50-43	Fair-Rite 5943000301 Ferroxcube 768T188/3D2
Amidon FT82-43	Fair-Rite 5943000601
Amidon FT114-43	Fair-Rite 59430001001

Fig 7-85—9:1 transmission-line Beverage transformers. From left to right, FT82 size core, two stacked MN8CX cores, FT114 core and FT50 core. Twisted wires were used on the smaller cores, while three parallel conductors were used on the larger cores.

Table 7-29
Transmission-Line Transformers With Different Cores

Type		1.8 MHz	3.65 MHz	7.1 MHz
[A]	Insertion Loss (dB)	0.50 dB	0.45 dB	0.45 dB
	SWR	< 1.1	<1.1	< 1.1
[B]	Insertion Loss	0.25 dB	0.25 dB	0.25 dB
	SWR	< 1.1	< 1.1	< 1.1
[C]	Insertion Loss	0.28 dB	0.20 dB	0.18 dB
	SWR	< 1.1	< 1.1	< 1.1
[D]	Insertion Loss	0.20 dB	0.21 dB	0.28 dB
	SWR	< 1.1	< 1.1	< 1.1
[E]	Insertion Loss	0.3	0.3	0.35
	SWR	< 1.1	< 1.1	< 1.1

[A]: Single MN8CX core (3 × 4 turns)
[B]: Stack of 2 MN8CX cores (3 × 4 turns)
[C]: Single FT114-43 core (3 × 8 turns)
[D]: Single FT114-75 core (3 × 8 turns)
[E]: Fair-Rite binocular 2873000202 (3 × 3 turns, equivalent to 3 × 6 passes)

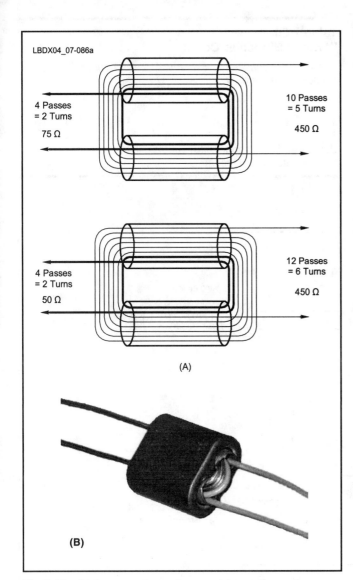

Fig 7-86—At A, correct way to count turns through a binocular-shaped core. At B, photo of finished transformer using binocular core (Fair-Rite Products 2873000202). Here, two turns were used for the low-Z primary and six turns for the high-Z secondary.

tion tubing that fits the ID of the binocular holes. This will prevent the enameled wire from damage scraping against the walls of the core.

2.8.1.4. Transformers With Isolated (Separate) Grounds

There are many circumstances when it is advisable to use a transformer with split primary and secondary windings. The outside of the coaxial feed line acts as a very low (on the ground) lossy long-wire. This is an unterminated Beverage, receiving signals and noise from directions other than those you are interested in. These signals cause currents to flow on the outside of the feed line (I_{tot}), which can be fed into the inside of the feed line through coupling via the common Earth resistance (I1, I3) as well as via the low-Z winding of the transformer (I2). See Fig 7-84C.

This is a common cause of spurious signal and noise reception, especially if long feed lines are involved. To prevent this, the Beverage and the coaxial feed line should be connected to different, well-separated grounds. This requires a transformer where the primary and the secondary grounds are galvanically isolated and where they exhibit the least possible capacitive coupling.

Using a transformer with a common ground requires a *very* low ground-resistance (*not* just a single ground rod) and a common-mode choke of some kind along the feed line. A far more effective way to reduce the common-mode coupling between primary and secondary to a minimum is to reduce the stray capacitive coupling between primary and secondary as much as possible. A transmission-line type transformer using the twisted wire method shown in Fig 7-84) with very tight coupling between primary and secondary is not the best solution in this respect.

Winding the primary and the secondary of a non transmission-type transformer on the opposite sides of a doughnut-shaped toroidal core will result in minimum capacitive coupling (**Fig 7-87**). This type of transformer has slightly more losses than the transmission-line type transformer and has a poorer high-frequency insertion loss because of the decreasing magnetic coupling efficiency of ferrite as the frequency increases.

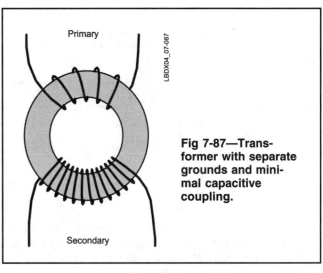

Fig 7-87—Transformer with separate grounds and minimal capacitive coupling.

On a classic toroidal core you can also wind separate primary and secondary windings, one on top of the other. This will ensure maximum coupling (and hence least loss) and better bandwidth, but also results in higher inter-winding capacitance.

A binocular core does not leave you a choice of winding techniques. The winding procedure shown in Fig 7-86 uses separate primary and secondary windings, but it is not possible to wind them in a specific way to minimize inter-winding capacitance. Tom, W8JI, wrote on this subject: *"As for spacing the windings or using a Faraday shield, neither are necessary or useful. I have typically measured about 10 pF or less of capacitance with one winding laid directly over the other. That is more than 9k-ohms of leakage reactance, and that would easily put any common mode well into the Beverage's noise floor."*

The Fair-Rite Products 2873000202 binocular core (type-73 material) requires only two turns (four passes) to three turns (six passes) for a low-Z winding (see Fig 7-86), so capacitive coupling is low. Four turns give marginally less loss on 1.8 MHz (a difference less than 0.1 dB, not worth worrying about), but a little more loss at 30 MHz, which is not really a worry either, since Beverages are not often used at 30 MHz. More turns means more capacitive coupling between primary and secondary.

Table 7-30 shows recommended winding data for binocular cores for various impedance transformations. W8JI addresses the issue of reception of noise and spurious signals due to common-mode problems in detail on his website (www.w8ji.com). Refer to **Fig 7-88** for how to connect this type of transformer to the feed line and to the antenna.

2.8.1.5. Checking Your Transformer's Performance

After winding a transformer it's a good idea to check its performance. Connect a 450-Ω non-inductive resistor (a small metal film will do) across the secondary and check the impedance or SWR using a Noise Bridge or an antenna analyzer (or still better yet, a Network Analyzer). With a well-made transformer the SWR should be less than 1.2:1 from 1.5 to > 10 MHz.

W8JI describes another testing procedure: *"Terminate the transformer with a resistor twice the normal secondary resistance and measure the SWR. Repeat with ½ the normal resistance. The SWR should be 2.0 ideally. Multiply the two SWR reading together and take the square root, if it comes out close to 2, you have a pretty good transformer."*

You can easily evaluate the insertion loss of a homemade transformer. Build two identical transformers and connect them back-to-back. Insert the back-to-back configuration in

Table 7-30
Winding Binocular Cores

Passes Primary	Passes Secondary	Primary Z Ω	Secondary Z Ω
4	10	75	450
6	16	75	533
4	12	50	450
6	20	50	550

Using Fair-Rite Products 2873000202 binocular core (where 1 turn = 2 passes)

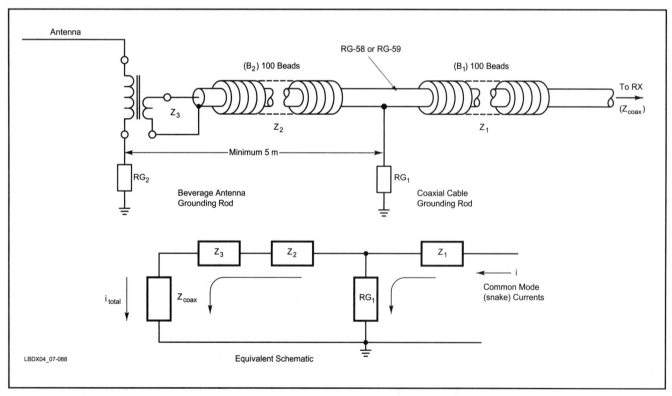

Fig 7-88—Method for connecting the feed line to the Beverage transformer, guaranteeing maximum suppression of common-mode signals induced on the outside of the feed line. See text for details.

the feed line to your transceiver. You should not be able to detect any change in signal strength, as two well-built transformers should exhibit no more than 0.5 dB of insertion loss, which is just about immeasurable without special test equipment. If you have access to professional test equipment you can do an actual insertion loss measurement in a 50-Ω system. A Network Analyzer will also give you accurate results.

2.8.2. Ground Systems at the Beverage Receiving End

In Section 2.5.4 I covered the ground requirements for the far-end termination. Requirements are a little different at the receiving end of a Beverage antenna. The Beverage is in essence an asymmetric (unbalanced) system, one terminal being the antenna wire, and the other being the ground. This second terminal is what we call the *antenna ground*. This ground can be a good ground (low loss, low resistance) or a bad ground. There are two issues:

- **Loss issue:** There will be signal loss due to high ground antenna ground resistance: the ground loss resistance is in series with the high-Z winding of the transformer.

  Assume we use just a short ground rod in poor ground. The RF ground resistance can be as high as 300 Ω. This ground resistance is in series with the secondary (high-Z; eg, 450 Ω) winding of the matching transformer. In this case the voltage loss would be: 20 log (450/450+300) = –4.4 dB.

  I've always said that gain is not the real issue of a receiving antenna, but that only holds true if you try to obtain gain at the sacrifice of the only important receiving antenna parameter, which is directivity. Using a better ground will not upset the antenna directivity.

- **Common-mode issue:** If the Beverage were perfectly unbalanced, that would mean there would be no possibility for common mode coupling. Since the ground is not perfect, any wire connected (physically or by capacitive or inductive coupling) to that ground will, through the presence of common-mode currents on that wire, induce RF voltages across the earth resistance. The greater the earth resistance (the poorer the ground) the higher will be induced voltages. If you use a common-ground transformer, the outside of the coax—which is just a long wire to the outside world—is directly connected to that poor ground and induces common-mode currents!

The only way to reduce the effect is to lower the earth resistance, or as explained in Section 2.8.1 you could use a transformer with separate primary and secondary windings. And even if you have separate windings there is always capacitive coupling between the primary and secondary. This is another common-mode path, but of relatively high impedance, typically 5 kΩ or more if the transformer is correctly designed and wound. Conclusion: a low-Z ground is a good idea here too!

For both reasons mentioned above, whenever possible you should provide a good RF ground at the feed point. In Section 2.5.4 I explained how to assess the equivalent ground resistance of a single ground rod. I would suggest that you try for a ground resistance that is less than 100 Ω. With very good ground a single ground rod might do the trick. But you might need multiple ground rods, spaced at least the length of the ground rods to lower the ground resistance. When you install rods too close together their effective *fields of influence* overlap and the end result is hardly much better than the first rod by itself.

Where you cannot use a ground rods because of rocky ground, you can use a large ground mat made of large strips of *chicken wire*, made into the shape of a cross measuring approximately 6 × 6 meters. Or you could use a large number of short interconnected radials laid on the ground. Do not use a small number of long radials, as they could upset the directivity of the Beverage.

If you use a transformer with a common ground for the primary and secondary windings you must improve the ground system (low resistance) to minimize common mode ingress into the feed line. Anyway, I do recommend that you avoid using common-ground transformers.

2.8.3. Is Your Feed Line Also an Antenna?

There are a few elementary tests that you can do to see if your feed line is behaving properly and is not picking up too much common-mode noise. If you short the coax's center conductor and shield at the Beverage feed point (temporarily disconnect the shield from the Beverage ground rod), the receiver should be completely dead. You can run this test on the AM broadcast band.

If you do hear signals, you've got some troubleshooting to do. Make sure the receiver itself has a good RF ground and is not just "grounded" through the shield of one or more coaxial feed lines. Make sure the coax is well-grounded where it enters the shack. Just relying on grounding at the receiver chassis is bad practice. It is quite common that the power mains feeding your receiver are carrying conducted RFI. This RFI is bypassed to the receiver chassis through any mains decoupling capacitors and the shield of the cable becomes the new path for this unwanted noise to leak via the cable shield all the way back to the feed point of the Beverage.

It is essential to feed the equipment at the shack through high-quality mains filters and ground these to an excellent RF ground. The bottom side of the operating table in my shack (which is 8 meters long) is completely covered with aluminum sheet. This represents a lot of capacitance and virtually zero inductance, which is just what you want! Quality mains filters are bolted directly to those sheets and the mains outlet to which the equipment is connected as well. The ground plane is connected with very short (less than 0.5 meter) low-inductance wide straps to long copper ground rods. Ground the feed lines to *another* good quality ground system where they leave the house of the shack. Ground rods for this ground should be a minimum of 5 meters from the ground rods grounding your equipment.

In really tough cases, you may need to eliminate common-mode signals on the shield of your coaxial feed line by using multiple, independent ground rods along the feed line near the antenna. Use a series of common-mode choke baluns between each ground rod. This forms a multi-section pi attenuator, making even modest choke impedances very effective. See Fig 7-88. Do not ground the far end of the cable and make sure you place your last coax ground at least 5 meters from your antenna ground. Use a stack of ferrite cores at the end of the feed line. This common-mode filter can also be made by winding 50 turns of miniature 50-Ω or 75-Ω coax on a stack of ferrite high-μ toroids.

I use a simple digital LC meter and use as many cores as

necessary to achieve an impedance of at least 1000 Ω on 160 meters (L = 130 µH). Make sure the LC meter actually operates near the frequency of interest (mine operates at 1 MHz). The exact ferrite core material is not critical for this purpose, but a high permeability is required. A low-Q situation is preferred (lots of resistance in the impedance number) to avoid resonance effects, but if the coax lays on the ground resonance effects are really excluded. A number of turns of small-diameter coax through a large high-µ ferrite core does the same job. Follow the wiring exactly in Fig 7-88. Install a stack of beads on the feed line at regular distances (make it a Snake antenna with even more losses). This is not practical if you use a large diameter feed line, of course.

In the most obnoxious cases of ground currents, it may be necessary to ground the shield of the feed line every 50 meters with independent ground rod—ground rods without any other connection. As an additional benefit, of course, lightning paths are disrupted by this method.

Bury your coax. *Never* suspend the feed line to a Beverage antenna in the air! Feed lines on the ground, or even better *in* the ground, will automatically choke off RF from the outside of the shield.

Do not run feed lines parallel to and in close proximity to elevated radials. Otherwise there will be field coupling from induction or radiation fields. Reduce coupling to other antennas (see also Section 2.9.1), towers, power lines, etc by careful placement of the antennas and by judicious feed-line routing.

All of these measures may not be necessary, but in stubborn cases the combination of all of these will guarantee a minimum amount of common-mode RF currents on the outside of your feed line. If your feed line is not properly decoupled it can upset the effective directivity pattern of your Beverage and turn a fantastic receiving antenna into a mediocre one.

Now you can take care of the transformer, as explained in Sections 2.8. Make sure the antenna ground (at the bottom of the transformer's high-Z winding) is grounded some distance away (at least 5 meters) from the closest ground rod grounding the feed line. This will introduce several thousand ohms of reactance in the common-mode signal path, as well as provide another path to earth for common-mode noise. Keep the connections to these grounds as far away from one another as possible.

Note that the primary (low-Z) winding of the transformer is *not* grounded but merely connected to the shield of the coax, which itself is grounded some distance away (minimum 5 meters). Fig 7-88 shows how the feed line should be connected to the transformer to achieve the greatest possible attenuation of signals picked up by the outside of the feed line. The stack of beads (B1) will form a high impedance (Z1 = typical 1500 Ω on 1.8 MHz for 100 stacked Wireman beads) for common-mode currents. In the equivalent schematic Z_1 and RG_2 (the ground resistance of the coaxial cable grounding rod) form a voltage divider. Assume this ground resistance (RG_2) is 100 Ω (that's a fairly good ground). This means that we have a voltage divider of 100:1100 or 1:11.1, which results in a signal attenuation of 20 log (1/11.1) = −20.8 dB.

The spurious signals left across RG_2 will now be fed *into* the feed line via another voltage divider made up by the impedance of the second bead stack (B_2) in series with the impedance of the low Z winding of the transformer (Z_3) and in series with the impedance of the coaxial cable (Z_k = 75 Ω, assuming the cable is terminated in its own characteristic impedance at the receiver end). This time we have a voltage divider made up by Z_2 = 1000 Ω (bead stack) + Z_3 = 250 Ω (transformer low Z winding) and Z_{coax} = 75 Ω. The attenuation of the voltage divider is 20 log (75/(1000+250+75)) = −25 dB. The total attenuation of this setup is about 46 dB for common-mode signals captured by the outside of the coaxial feed line. If that does not cure the problem, you can ground the coax at regular intervals or look for a different mechanism of common-mode signal ingress into your receiving setup.

Note that the stacks of ferrite cores across the coaxial feed line can be replaced by coiled-up lengths of coax. In order to achieve a choking impedance of approximately 1500 Ω on 160 meters, the inductance must be 125 µH. This requires a coil of coaxial cable measuring 30 cm in diameter and having 20 close-wound turns. Such a coil requires not less than 20 meters of coax, and two such coils are required in the system! If you use this approach, make sure the two coils are at right angles, to minimize coupling between them. Another alternative is to wind miniature type coaxial cable on a large high-permeability core. Five turns of miniature coax through a FT150A-F core (µ = 3000 and A_L = 5000) achieves an impedance of 1500 Ω (7 turns = 3 kΩ on 160 meters).

You should now have a "quiet" feed system to the Beverage. Connecting everything to the system except for the Beverage wire itself should yield no signals.

2.8.4. More About the Feed Line to the Beverage

As explained above, you should pay as much attention to the feed line as to the Beverage antenna itself. Bad feed-line practice can completely annihilate the directive properties of the antenna. The Beverage principles explained here apply equally well to all other low-signal receiving antennas described in this Chapter. The feed line issues are:

2.8.4.1. Attenuation

As your Beverage antennas will most likely be operated on relatively low frequencies, you need not use a feed line with the lowest possible loss, especially where all but the very longest feed-line runs are being considered. For runs up to 100 meters RG-58 (or 59) sized coax will be just fine. Table 7-13 shows the typical losses for common coaxial cables on 1.8, 3.5 and 7 MHz (or use *TLW*, the program that comes with the 20th Edition of *The ARRL Antenna Book*).

What we really require under the quietest circumstances (daytime) and using the narrowest receiver bandwidth is to be able to clearly hear the antenna noise over the internal receiver noise. If you are in a very quiet environment (way out in the country, no nearby power lines, or in the middle of the ocean on an uninhabited island) the surplus sensitivity that all modern receivers (transceivers) have may not be enough. In that case you will need a preamplifier. It will practically never be necessary to put the preamplifier at the antenna though.

2.8.4.2. Mechanical Properties

Running standard coax on the ground can be a problem for some. Many low banders report having lost small or medium-sized RG-type feed lines, as well as CATV drop-type feed lines due to animal bites. Small bites will usually not open or short the feed line, but will do enough damage to cause moisture migration and corrosion. There are three ways to prevent this from happening:

- Use CATV Hardline ($1/2$- or $5/8$-inch stuff is very sturdy, especially the *figure-8* type cable).
- Use quad shielded and flooded RG-6 type coax. It appears that the critters don't like the flooding compound.
- Bury the cables in a closed cable duct underground.

2.8.4.3. Availability

For short runs any coax you can buy at the local flea market will do, provided the shield or the inner conductor is not corroded (green and black for copper shields, dull and white for aluminum shields) from moisture ingress. Often, 75-Ω CATV-type coax leftovers (often lengths up to 100 meters!) can be found at reasonable prices from the local cable TV company. The flexible coax used for drop lines is good for anything but very long runs. Hardline is the ultimate choice for long runs, since it offers the lowest attenuation.

George, K8GG, uses $1/2$-inch 75-Ω CATV from his shack to the Beverage antenna park "headend," which is 1200 meters (yes, 4,000 feet) from the shack. From the headend he uses RG-59 flooded cable for all the connection runs to the remote antenna selector.

2.8.4.4. Coax Impedance

This is definitely the least important issue: It is totally irrelevant whether you use 75 or 50-Ω coaxial cable. The purist may want to design the matching transformer according to the impedance of the coax used but even that is a bit far fetched.

2.8.5.5. Shielding Effectiveness

It is important to use well-shielded coax to achieve a quiet feed system under all circumstances and on all frequencies. Some of the very cheap (non Mil-standard) coax has a very poor shield coverage factor (50% or so). This cable should *not* be used. Hardline is intrinsically the best choice, provided the solid shield is not broken. In many European countries CATV companies use so-called figure-8 Hardline (instead of lashing the coax to a messenger wire). Figure-8 Hardline structurally incorporates a support cable. Overhead Hardline often develop cracks and eventually break in the solid shield due to work hardening as they swing in the wind. In a CATV trunking network these breaks are responsible for radiating cables, which many of us have experienced. Watch out when buying *used* CATV Hardline, and always check the cable for visual shield damage as well as for electrical shield continuity.

2.9. Terrain and Layout Considerations

- How far should I keep the Beverage from my vertical transmit antenna?
- How close can I run two Beverages in parallel—with one shooting in the opposite direction?
- Can Beverages cross one another?
- How close can they cross?
- Can I run different Beverages from one feed point?
- How straight must a Beverage wire be?
- What if my terrain slopes up and down?

These are all very valid questions. And failing to understand what and why may turn a potentially wonderful antenna into a really lousy performer.

2.9.1. Proximity to Transmit Antennas

When they are located close to large, resonant transmit antennas, Beverages pick up noise and signals that are retransmitted by those transmit antennas. If you have a number of Beverages, and one or two are always noisier than the others, then there is a good chance those are the ones closest to your transmit antenna. Another indication is that a Beverage antenna is not much quieter than your vertical. It should be.

Is that noise really coming from your transmit antenna? A simple test is to connect a variable impedance like an antenna tuner to the transmit antenna's feed line and rotate it through all possible inductance and capacitance ranges while listening. By doing so you will actually change the resonant frequency of the antenna. If you hear any difference in noise level in the receiving antenna while doing this, you can be sure the transmit antenna is coupling noise into the receive antenna.

If you have the appropriate test equipment, a more

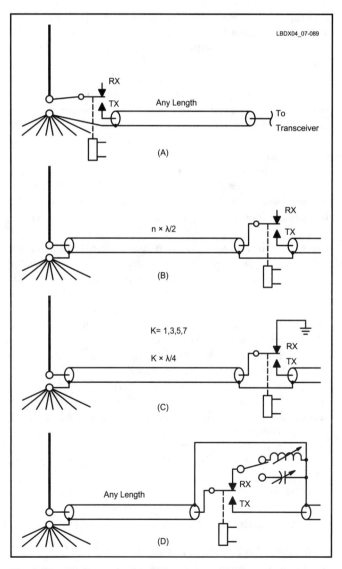

Fig 7-89—Various methods for decoupling a series-fed vertical. See text for details.

thorough approach is to measure the isolation between the two antennas by measuring power on one while feeding power to the other. If you detect any coupling you will need to decouple the transmit antenna, or else move your receive antenna further away.

If you have a series-fed λ/4 ground-mounted vertical, decoupling can be as simple as disconnecting the coax at the antenna, so that the vertical part of the antenna is now "floating." The antenna will now be resonant at twice the frequency and mutual coupling to the Beverage will be minimal. You can use relays at the base of the antenna (see **Fig 7-89A**), or you can do it at the end of the feed line. If the feed line has an electrical length equal to an odd number of quarter-waves, then the feed line should be short-circuited in the shack during reception (Fig 7-89B). If the feed line length is a multiple of half waves long, then the end of the feed line should be left open in the shack during transmit (Fig 7-89C).

If you don't know the feed line length, just terminate the feed line in a variable capacitor (such as a four-gang BC variable) or a variable inductor and tune these to obtain minimum noise level in the Beverage. This capacitor or inductor can then be switched across the end of the feed line during reception with a relay, as shown in Fig 7-89D. You can also use a series-resonant LC circuit with small values of C, or a parallel LC-circuit with large C and low L values, to do your testing. Replace these with a simple L or C once the right value has been determined.

If you have a shunt-fed vertical such as a loaded tower, just disconnecting the feed line may not help! Once you shunt feed a non-resonant structure, it becomes a resonant structure at the frequency to which it is tuned. To detune it you will need to turn a section of the tower into a parallel-resonant circuit, as explained in Section 3.10.

If you have a vertical with elevated radials, then you are really in trouble, unless you use only one radial. Systems with two or four verticals are very nasty. Two radials (more or less in-line) are a perfect half wave element, reradiating like crazy! Two more at right angles makes things even worse. **Fig 7-90** shows how the radiation pattern of a 268-meter long Beverage with and without a quarter-wave vertical with four elevated radials, 4 meters high. Note how a F/B of almost 30 dB falls to a mere 12 dB. A large number of elevated radials will eventually lose all its resonant characteristics and act like a disk. In this case current and power in each radial is a fraction of the total so radiation is miniscule.

With a 160-meter vertical using a single elevated quarter-wave radial, decoupling the feed line leaves only two λ/4 long wires (one vertical, one horizontal), which will do no harm, except on 80 meters, where they are λ/2 and hence resonant. Your Beverage will be compromised if it is used for 80 meters also!

Even if you don't notice an observable noise increase, the proximity of other antennas can hurt the directivity of your receive antenna. This may not always be easily detected just by listening to an antenna. A Beverage may still "out hear" a vertical, even if it is severely compromised by the proximity of another nearby antenna.

Conclusion, Nearby Transmitting Antennas
- Noise picked up by a large transmitting antenna can easily be coupled into a nearby receiving antenna.

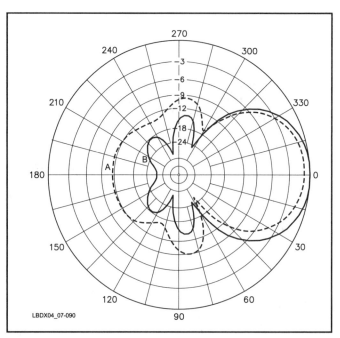

Fig 7-90—Influence of the λ/4 160-meter vertical with four elevated radials on the azimuth pattern of a 268-meter long Beverage (dashed line). In this example the Beverage comes within 10 meters of the vertical.

- To resolve this problem you can increase the distance between the antennas (usually a minimum of λ/2 depending whether or not the receive antenna is firing into the transmit antenna or not).
- Shift the resonant frequency of the transmit antenna as far away as possible from the operating frequency during receive.

If your Beverage is close to the transmit antenna, and if you have no apparent noise problem, this can be for one of the following reasons:

- Your transmit antenna is right in the null direction of the Beverage.
- You have an extremely quiet location (maybe Heard Island?). If there is no noise being picked up on the vertical, then it would be very hard to discern any change in the Beverage performance.

Under any circumstance it is better to keep any antennas separated as far as possible. This is also true for transmit antennas, of course. A situation where a wire (or more wires) are stuck randomly in the near field of an antenna ends up with results that are a matter of blind luck.

2.9.2. Parallel Beverages

Beverages are non resonant and exhibit very little mutual coupling. The rule of thumb is to keep two parallel Beverages separated by a distance equal to their height above ground. This separation guarantees a decoupling of at least 30 dB.

2.9.3. Crossing Beverages

Beverage antennas for different directions may cross each other if the wires are separated by at least 10 cm (4 inches). Beverage antennas should also not be run in close proximity to parallel conductors such as fences, telephone lines, power

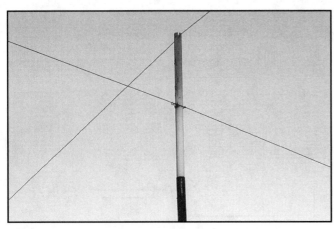

Fig 7-91—Crossing Beverages at W8JI. The top 25 cm of the support is a PVC electrical pipe.

lines (even if quiet) and the like. **Fig 7-91** show how W8JI does it.

2.9.4. Multiple Beverages From One Hub

Sometimes we need to run a long feed line to a central hub, from where other feed lines will run to different feed posts for single or multiple Beverages. This hub will simply switch the main feed line coming from the shack to the feed lines going out to the end posts. Make sure you always switch both the inner conductor and the shield of these feed lines. Do *not* connect the shields of all these feed lines together or to a common ground. If you want to ground them, do so at a distance of 5 meters from the switch box, to individual ground stakes that are well separated.

If you want to run various Beverage from one "feed post" there are several approaches to do this. Refer to **Fig 7-92**. Ranking from worst to best, we have the following methods:

- Keep the high-Z ground of the various transformers connected to a single ground rod at all times, and switch the hot-conductor of the low-Z point with a relay. The common ground resistance will mutually couple signal from one antenna into the other, as will the ground sides of the low-Z points of the antennas, which are connected together. Avoid this situation. See Fig 7-92A.
- Same as above, but now you switch both the outer and the inner conductor of the feed line. The feed line is grounded some distance away (5 meters or more), where a common mode filter is installed (see Section 2.8.3).
- Do the switching at the high-Z side of the transformer. Some purists argue that this increases crosstalk if a poor layout and bad wiring practices are used. But if you can minimize such crosstalk if you use a DPDT relay and wire both contacts in series, where the jumper between the two sets of contacts is grounded to minimize crosstalk. See Fig 7-92C.
- You could terminate any unused Beverage with a 450-Ω resistor to a separate ground, say 5 meters from the antenna ground. Terminating is better than grounding; you will have less current from the undesired direction flowing in the non-ideal ground system (Fig 7-92C), which, if close to the Receiver-transformer ground will couple into that ground circuit. It is obvious that you should not use the same ground for Fig 7-92A and Fig 7-92C. Use separate ground rods for Fig 7-92C, separated by at least 5 meters from the ground point.
- If at all possible I suggest using widely separated end posts for the Beverages, with at least 5 meters separation and separate grounds (Fig 7-92E). Avoid using the antenna ground of one antenna as the termination ground of another one that might be running in the opposite direction.

Little or no coupling occurs between the antenna wires, unless they are parallel and quite close together. I have not noticed any problem with antennas arriving at one point under a 30° angle.

2.9.5. Only Straight Wires?

A nice clear straight wire, without sags, not only looks great, it does hear better! Sags in the wire are just like sloping terminations (see Section 2.5) and help upset the directivity of the antenna. If the terrain is irregular, minor ups and downs in height or dips or valleys will not ruin your Beverage performance though. If your terrain is not flat, follow the contour of gradual slopes and go straight across ditches or narrow ravines without following the contour.

Roger, VE3ZI wrote: *"I have 12 Beverages, each about 800' long at 30 degree intervals. Some of them are relatively flat, some of them go up and down by about 100', some parts are across rock and some across swamp. I really can detect no difference in performance between any of them. I'm quite sure there is a difference, but my feeling is that it is so small as not to be relevant. Surely the real point is that Beverages work adequately even when gross compromises are made in their construction."*

2.10. Beverage Maintenance

It's always a good idea to log the measurements you've made on your Beverage antennas when you build it. Write down the terminating resistance that gave minimum standing waves on the antenna, as well as the SWR curve over the bands of interest. Next, do a checkup every year, and measure those two parameters. Don't forget to check your feed lines also (measure at least insulation resistance (far end open) and continuity (far end shorted).

2.11. BOG or Snake Antenna

What we now, often jokingly, call a *snake* antenna is in fact the shape of the very first Beverage antenna in history. The first antenna Harold Beverage used was simply several miles of wire laying on a sandy path on Long Island. Only later was the Beverage antenna refined to be an elevated wire with a resistive termination! Mr Beverage was the inventor of the snake antenna. The Snake antenna or *BOG* (Beverage on Ground) works because of common-mode excitation on the outside of a coaxial feed line. The entire shield picks up signal from the outside, and the on-the-ground-Beverage is simply a reverse-fed thick random wire lying on the ground.

The Snake antenna obviously gets its name from what it looks like: it is usually a coaxial cable just stretched out on the ground like a giant reptile. The antenna is said to have properties similar to those of a genuine Beverage antenna: It receives off the far end of the Snake, at least if we may believe all that's been published. DeMaw, W1FB, (Ref 1254) described different versions, and gave us some rather interesting material to read.

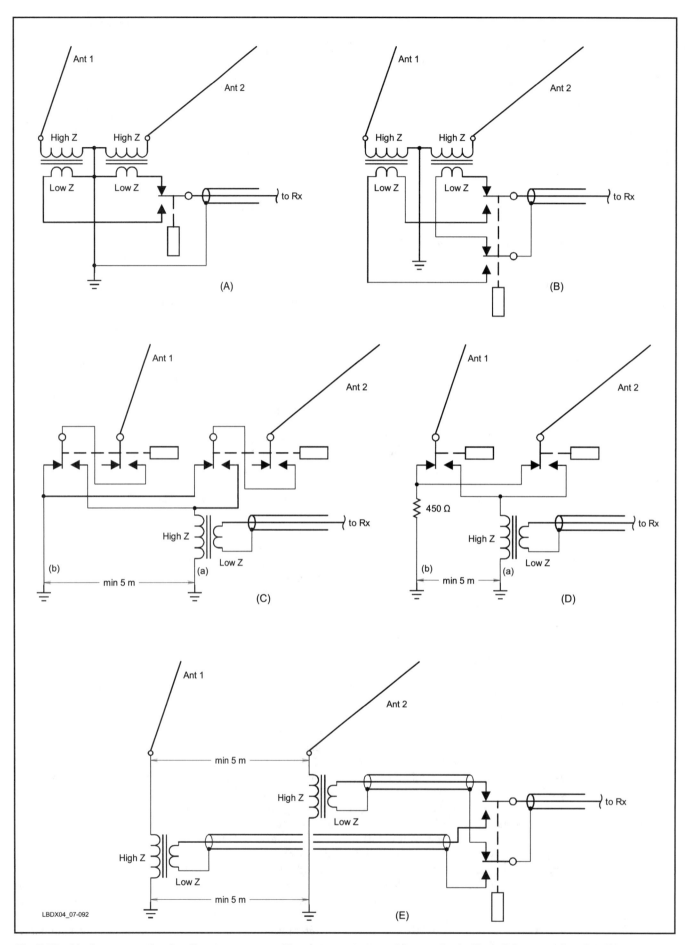

Fig 7-92—Various ways for feeding two or more Beverage antennas from a single "hub." See text for details.

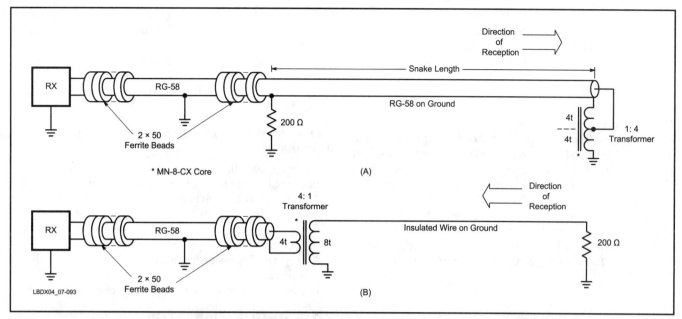

Fig 7-93—At A, the so-called Snake antenna, where the coax on the ground actually serves as feed line bringing the received signals off the back of the antenna to the receiver. Careful decoupling of the feed line is required (see text). At B, a simplified Snake using a single insulated wire on the ground. In this case reception is from the far-end of the antenna.

One type DeMaw described was short circuited at its end. He stated the cable will present a 50-Ω load at the receiver provided the line is lossy enough. It happens that at least 20 dB is required for a short circuited line to appear like a matched line at the other end. Well, that's going to have to be a very long Snake, if you remember that even a mediocre RG-8 cable exhibits 0.5 dB per 100 feet at 1.8 MHz. Yes, to get the 20-dB attenuation you would need a 4000-foot long Snake. DeMaw also said that the velocity factor of the coaxial line should be taken into account when constructing the antenna.

Let's leave the tales for what they are worth and look at the facts. Seen from the outside (by the radio wave we want to capture) the shield of the coax acts as a single fat wire antenna. Obviously everything important goes on *outside* the coax. The velocity factor of the coax has nothing to see with the coax acting as a fat wire antenna, contrary to what De Maw stated. The velocity factor of the coax relates to the coax acting as a transmission line, and then everything happens *inside* the coax.

Let us look at this very low fat Beverage. Most users report having measured an antenna impedance of approximately 300 Ω. It is that high because of the losses of the ground. Over a perfect conductor it would be as low as 50 Ω. The Snake has a very low velocity factor and of course, very low output, but it still retains essentially the same directivity characteristics as a Beverage at more usual heights.

Fig 7-93A shows a Snake antenna, where the coax on the ground also serves as feed line, bringing signals received off the back of the antenna to the receiver. Careful decoupling of the feed line is required. Fig 7-93B shows a simplified Snake using a simple insulated wire on the ground. In this case reception is off the far-end of the antenna.

At the end of the Snake in Fig 7-93A is a reflection transformer (200-300 Ω to 50 Ω). A similar reflection transformer is also used with a two-wire reversible Beverage (see also Section 2.15). At the near end of the Snake there is a terminating resistance, used to dissipate the waves received from the rear direction. This should be a 200 to 300-Ω resistor connected to a good RF ground. A common-mode choke, consisting of a number of ferrites on the cable, prevents the ground at the receiving end of the feed line from swamping the termination resistor. A number of ferrite beads on both sides of another good RF ground will do the job (see also Section 2.8.3). If you want to receive off the other end, you would construct your Snake as shown in Fig 7-93B. In this case we save some money and use insulated wire (rather than expensive coax) as the antenna.

At the near end the antenna wire can be connected via a 4:1 transformer (bifilar-wound transformer) to the inner conductor of the coaxial feed line. A transformer with separate primary/secondary grounds is required here. Here too, you will need to ground the end of the feed line, and a series of ferrite beads on the feed line will discourage any common-mode RF currents from messing up your Snake.

You can make a Snake like the two-wire antenna in **Fig 7-94B**, where you have two directions available with the flip of a switch in your shack. The only difference between this and the "high" Beverage antenna as described is that the antenna impedance is typically around 300 Ω instead of 450 to 600 Ω, that the velocity factor is much lower (typically 0.6) and its output is down about 10 to 15 dB.

A number of users have reported their BOG (Snake) to be much quieter than an equivalent Beverage antenna. You should be careful though not to compare a 150-meter long Beverage with a 150-meter long BOG. While a Beverage at a 2-meter height may show a velocity factor of about 95-98%, a BOG will show a VF figure of around 60%.

Riki, 4X4NJ, is an avid user of BOGs and he wrote: "*All of my 5 Beverages are BOG. I don't believe that this is a compromise antenna for terrain such as mine. My QTH is on* a rocky mountaintop with very poor soil conductivity. Of course there is capacitance to ground. Some of my BOGs are unterminated, and others are terminated. Those that are terminated use a 200 Ω resistor to a quarter wave wire just laying on the ground. I use the ¼ wave wire termination because of low ground conductivity and the impossibility to drive in a ground rod — they always hit underground rocks. My experience has been that the unterminated BOGs have excellent F/B. Receiving from the fed end direction is down at least 2 or 3 S-units. I can't properly explain why this is so, but I expect that it is because of the low impedance and losses. (The usual references predict only about a 3 dB difference). My gut feeling is that in my particular QTH there is no significant advantage in terminating the BOG. Besides its simplicity, this is a rather stealthy antenna. From my house with a small yard, the BOGs go out into the surrounding public area. Two of them cross roads, so I have buried the wire a few inches beneath the (gravel) road. I've also buried more extended length —about 100 feet in order to hide the wire from view."

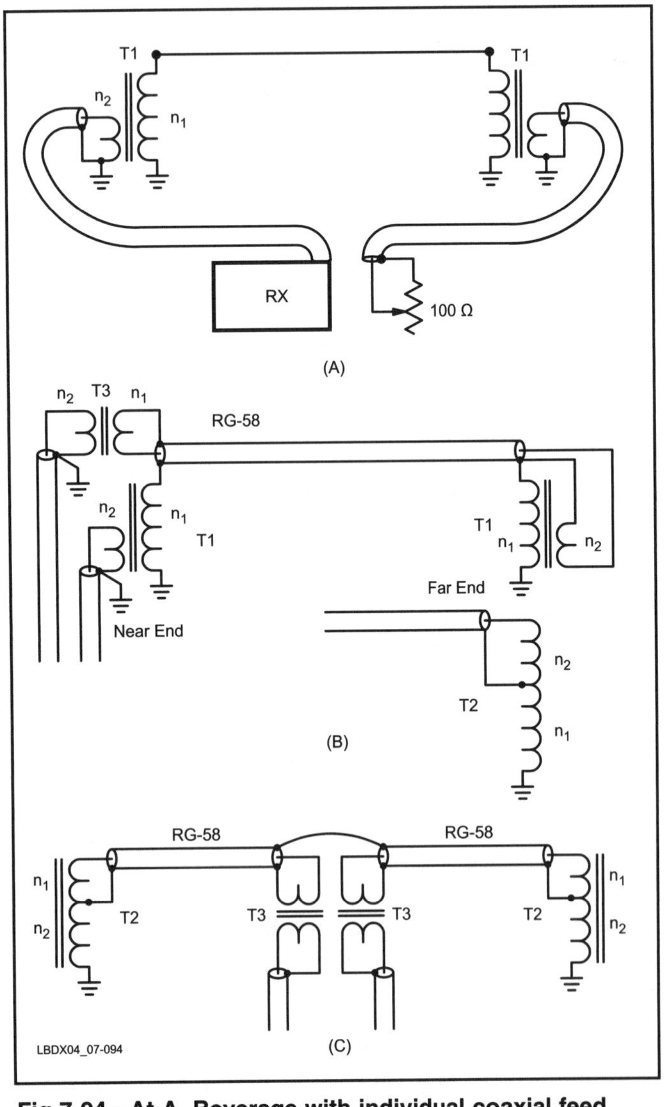

A BOG may also be an attractive receiving antenna for DXpeditions. But they do not work in close proximity to salt water nor over very good, conducting ground (such as a dry salt lake).

2.11.1. What to Remember About the Snake Antenna

- A piece of coax thrown on the ground, in any of the fashions described by DeMaw will not work like a Beverage antenna.
- A proper Snake requires good grounds at both ends.
- Feeding a Snake is even more delicate that feeding a regular Beverage, since signals are much weaker and common-mode currents can more easily do harm.
- A properly dimensioned Snake has the same pattern as a Beverage.
- The Snake will more than likely require a preamplifier.

2.12. Bi-Directional (Unterminated) Beverage Antenna

When the Beverage antenna is not terminated, the directivity will be essentially bi-directional. In real life the unterminated Beverage is not a true bi-directional antenna. The SWR and the loss on the antenna make it have some F/B. F/B is related to standing waves on the antenna; that is, to the reflection making it back to the receiver. **Fig 7-95** shows the horizontal directivity pattern for a 1-λ long unterminated Beverage antenna. Notice the slight attenuation from the back direction because of the extra loss of the reflected wave in the wire.

A method of switching the Beverage from unidirectional

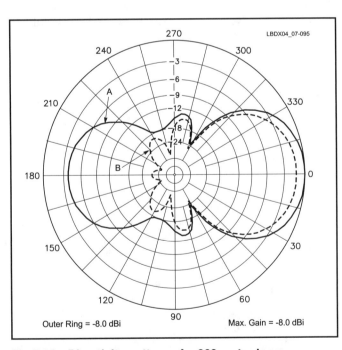

Fig 7-94—At A, Beverage with individual coaxial feed lines brought to both ends of the antenna. At B, an array where the antenna wire is a coaxial cable, which is also used to transport the signal from the "back direction" of the array to the front. At C, a variant, where the coaxial beverage is fed somewhere along its length. See text for details.

Fig 7-95—Directivity pattern of a 268-meter long unterminated Beverage antenna (solid line) at a height of 2 meters over good ground, compared to the same antenna when properly terminated (dashed line). Note the difference in gain from the back, resulting from the extra losses encountered by the reflected wave.

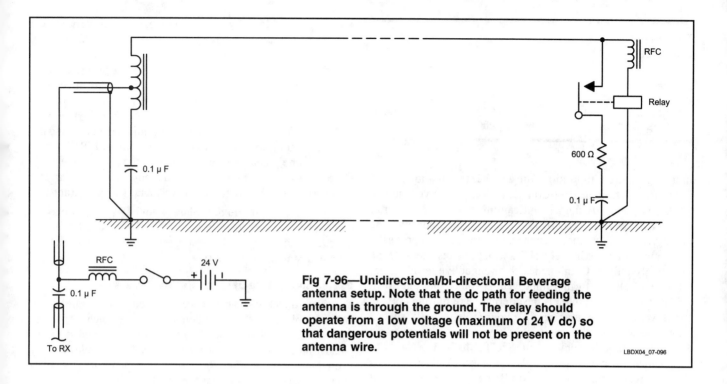

Fig 7-96—Unidirectional/bi-directional Beverage antenna setup. Note that the dc path for feeding the antenna is through the ground. The relay should operate from a low voltage (maximum of 24 V dc) so that dangerous potentials will not be present on the antenna wire.

to bi-directional is shown in **Fig 7-96**. A relay at the far end of the antenna wire is powered through the antenna wire and the earth return via an RF choke and blocking capacitors.

2.13. The Slinky Beverage and Alternatives

Carl, KM1H, had to explain to me what a slinky was. Apparently it's a child's toy, but I still don't know how children play with it... Anyhow, it looks like some Topbanders now also play with it.

"It is a continuously wound coil of about 7 cm in diameter and about 90 turns for each unit. I look at it as a helically wound Beverage that exhibits very low noise pickup and is still very efficient as compared to a regular wire Beverage."

Although Carl does not claim to be the inventor of the Slinky he says its initial use goes back to 1985 and started more as a joke and a dare as the result of an "after midnight" 160-meter SSB roundtable. At the same time he claims that the Slinky equals the performance of a Beverage of 1.5 to 2 times the length of the Slinky.

He uses the 7-cm diameter version and has connected five of them in series stretched over a total length of approximately 53 meters. This makes 8 turns per meter, which according to Carl gives a velocity of propagation (VP) of 0.58. A VP of 0.5 means the wave travels half the speed of free space, so the antenna looks twice as long as it is physically. A VP of 0.58 means a 55-meter long Slinky is 95 meters long electrically.

By the way, you can obtain exactly the same results, so far as a very low VP is concerned, by just laying your Beverage on the ground (see Section 2.11). You can't obtain a good F/B with a Beverage that is shorter than λ/2, unless you significantly reduce the velocity factor. This can be done in several ways:

- By using a R + L termination
- By adding inductance along its length
- By adding parallel capacitance to ground

You can, of course, use any combination of these methods. What we are trying to do with the loading is make a 50-meter long wire behave like an 80-meter long one (λ/2 on 3.6 MHz). This means we are shooting for a velocity factor of 50/80 = 63%.

Assume you have 50 meters in your garden. **Fig 7-97** shows the azimuth patterns for four configurations. The com-

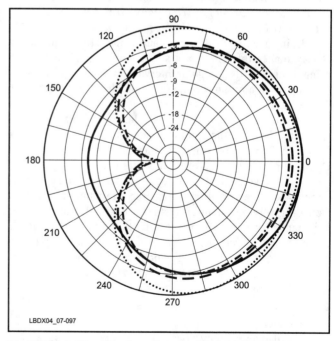

Fig 7-97—A 50-meter long beverage terminated in 450 Ω (solid line); same antenna terminated in 250 + j 280 Ω (dashed line); same length BOG (dotted line); same length with inductive loading in element (dashed-dotted line). See text for details. All models on 1.83 MHz.

Table 7-31
Short Beverages

Height	Termination, Ω	DMF	RDF
2 meters	450	4.5	4.0
2 meters	250 + j 250	6.0	4.5
0.1 meters	330 + j 200	7.8	5.6
2 meters, 4 Loading Coils	300 + j 200	6.0	4.6

mon Beverage configuration with a 450-Ω termination is very poor. If you terminate the Beverage with an inductive load you get a very good F/B and a significant improvement in overall directivity (especially DMF). We can further improve the directivity by laying the antenna on the ground. This drops the VF down to approximately 60%, which makes the antenna effectively longer. Combined with an inductive termination, DMF and RDF are further improved. If you cannot put down a BOG (Snake) then there is the alternative of putting coils in the Beverage (at regular height). Four coils, equally spaced along the wire, can do the trick. Or you may want to use a Slinky to obtain the same result. See **Table 7-31**.

Unless special steps are taken, Fig 7-97A clearly shows that a conventional 50-meter short Beverage is of little use. If discrete coils are used they should be spaced a maximum of λ/8 apart. Modeling shows good results with as few as four coils in a 50-meter long short Beverage. Coil inductance does not seem to be very critical, but a reactance of about 200 Ω seems indicated (approx 17 μH on 160 meters). The value is *not* critical. Have a go at it with your favorite modeling program.

If you want to use your Slinkies as antennas, W8JI offers the following advise on his website: *"Slinky users should be particularly cautious to extend the coils an optimum amount. With 1/2 λ of distance, you would want somewhat less than 1 λ of total conductor length. Too many turns-per-foot and the antenna will try to fire backwards, towards the feed point."*

Using discrete loading coils and an inductive termination clearly makes this antenna a single-band affair! Instead of using discrete coils at regular intervals, or a string of Slinkies (supported on a nylon rope), you could also use a number of ferrite beads on the Beverage wire.

It should be clear that if you have only 50 meters of space, the BOG is the right choice. It's easy to put up and almost invisible. You can cross the driveway if you put the wire in a plastic tube barely below ground.

How do you adjust the value of the termination complex impedance? Put a small signal source a few hundred meters from the receiving antenna (outside the near field) and tune the inductance for best F/B. You can do this for a zero azimuth angle (signal source in-line with the antenna) or for a given azimuth angle (say, 20°), in which case you put the signal source at the back of the antenna with a 20° offset to one of the sides. Remember:

- A random Slinky will most likely *not* work (unless you're very lucky!)
- A well-tuned Slinky always has less output than a full-sized Beverage of the same electrical length. The output is typically 3 to 5 dB down from its full-size λ/2 brother, but that should be no problem since we are talking about receiving antennas.
- A well-tuned Slinky or an inductively/capacitively-loaded short Beverage can have as good a F/B as a λ/2 properly terminated Beverage.
- If you have little room, try a BOG first. A 50-meter BOG with R+L termination is as good as a classic 80-meter long Beverage so far as directivity is concerned.
- Tune the loaded antenna or Slinky for best F/B (see above). This is a *must*. For a Slinky adjust the pitch and the termination impedance to obtain maximum F/B.
- The same precautions as outlined with regular Beverages so far as termination ground and feed-point ground apply, as well as coupling to nearby resonant transmit antennas.

Conclusion: If you don't have a lot of room, this may be something to try, especially the BOG. If you have 80 meters of space, also try two end-fire staggered short Beverages, with 30-meter stagger distance (see Section 2.16).

2.14. Two Directions From One Wire

Two directions can also be obtained from a single-wire Beverage by feeding coax to both ends of the antenna, with an appropriate matching network (**Fig 7-98A**). One end is terminated in the shack with an appropriate impedance, while the other is used for reception. (See Ref 1210.) The appropriate terminating impedance can be found as follows. First, determine the surge impedance of the Beverage using the method explained earlier in this chapter. Connect an impedance bridge across the high-impedance secondary of the matching transformer. Adjust the terminating impedance at the end of the feed line (inside the shack) until it is the same as the Beverage impedance.

In many cases, the impedance will be non-reactive and a simple resistor in the range of 50 to 75 Ω will provide the proper termination. An alternative method is to use a small signal source in the back of the antenna (remember the offset angle, see Section 1.6) and simply adjust a 200-Ω potentiom-

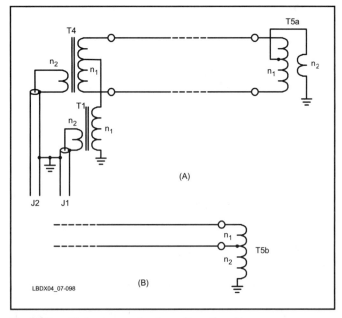

Fig 7-98—At A, open-wire two-direction Beverage antenna. This uses reflecting transformers to obtain both reflections and proper impedance matching (see text for details). At B, an autotransformer is used for the same purposes.

eter at the end of the second feed line for maximum front to back. You switch directions by interchanging the coax going to the receiver with the one going to the terminating resistor.

2.15. Two-Wire Switchable-Direction Beverage Antennas

Two-wire switchable Beverage antennas are covered in great detail by V. Misek, W1WCR (Ref 1206). I strongly recommend that you read this book if you want to get serious with open-wire type Beverage antennas.

2.15.1. Coaxial Cable as Feed Line and as Antenna Wire

We know that a coaxial feed line can act as a transmission line, where everything happens inside the cable between the center conductor and the inside of the shield. But it can also act as a single fat conductor, when you look at the coaxial cable "from the outside" (remember the Snake antenna in Section 2.11).

Fig 7-94B shows a small coax cable used as a Beverage antenna wire. The coax cable also serves as the feed line for the far end of the array. Transformer T1 is a classic 9:1 transformer. The reflection transformer at the far end provides the right impedance transformation to feed the RF via a good match to the inside of the coax. This transformer is identical to the transformer T1 used at the near-end. At the near end a 1:1 isolating transformer (T3) must be used.

The same antenna can also be fed in the center—or anywhere along its length—as sketched in Fig 7-94C. Here, a transformer similar to the one used at T2 in case Fig 7-94B is used at both ends. The coaxial antenna wire/feed line is opened at the appropriate point, where the isolating 1:1 transformers are used in both directions (T3). The design and construction of the transformers is covered in detail in Section 2.15.4.

2.15.2. Using Parallel-Wire Feed Line as Feed Line and Antenna Wire

Instead of using a small-diameter coaxial cable (which is relatively heavy and expensive), you can also use a parallel-conductor transmission line to achieve the same results. Using a parallel-conductor transmission line is somewhat more delicate since the only thing that stops the line (when used as a transmission line) from picking up unwanted signals is the balance of the system. If one conductor is closer to ground over the entire length of the line it unbalances the system.

This is why each of the parallel transmission line wires must be parallel to the ground. A line should not be constructed in a vertical fashion, using single support poles with one wire under the other. The closer the wire spacing of the line, the lower the impedance of the line, and the less critical installation-related unbalances become. The higher you install the line above ground, the less unbalance any unequal coupling to earth creates.

You can use a homemade open-wire feeder with a wire spacing of approximately 30 cm. Signals arriving off the near end the antenna will induce equal in-phase voltages in both wires. Because of the close spacing of the wires, there is no space-diversity effect. The end of the two wires are connected to a properly designed push-pull transformer (T4), which has the RF from the push-push (in-phase) signals on its center tap.

This signal in inductively coupled, via the n2:n1 windings in Fig 7-98A to the balanced secondary, which now feeds the balanced transmission line made up by the two parallel antenna wires. The two parallel wires act as an in-phase antenna wire and as out-of-phase transmission-line wires.

The signals reflected by the reflection transformer (T5) are routed to the near-end of the antenna, where they arrive at transformer T1. These push-pull signals at the primary are routed through the coax feed line to J2.

Signals arriving off the far end of this antenna, the normal receiving direction, arrive at transformer T4 in push-push (in-phase) mode and are available only at the center-tap of the secondary (n1) and not at the primary (n2). The 9:1 transformer (T1) is now connected to the center tap of T4, where the push-push signals from signal received off the far-end of the Beverage are available. These signals received from the far-end will not produce any output at the low-Z side of T4, provided a good balance is achieved in the transformer. Outputs from both directions are simultaneously available from outputs J1 and J2 of this system.

When only one of the feed lines is connected to a receiver, make sure the other feed line is terminated. A variable potentiometer in the shack can be helpful to optimize the F/B under all circumstances.

Reflection transformer T5 in Fig 7-98B must be built to have the correct transformation ratio from the transmission line impedance (for example, 710 Ω balanced for the open-wire line) to the antenna impedance (approximately 332 Ω if the two-wire Beverage is 2 meters high).

Using twin-lead transmission line is obviously a good choice, since it intrinsically guarantees better balance. When using twin lead it is appropriate the twist the line two or three turns per meter. If you use open-wire line, it is also a good idea to "flip" the wires every now and then along their run.

You can also make the reflection transformer as an autotransformer, as shown in Fig 7-98B. The same precautions explained in Section 2.8.3 apply regarding the intrusion of common-mode currents from the feed lines into your antenna system. Therefore use separate ground rods for the antenna and for the feed lines, and install common-mode filters on the feed lines as shown in Fig 7-88.

The design and construction of the transformers is covered in detail in Section 2.15.3. The push-pull impedance of the open-wire line is given by:

$$Z = 276 \log \frac{2S}{d} \qquad \text{(Eq 7-2)}$$

where

S = spacing between conductors
d = diameter of wires (in the same units).

See **Table 7-32**. The impedance of the parallel wires over ground is given by:

$$Z = 69 \log \left[\frac{4h}{d} \sqrt{1 + \frac{(2h)^2}{S}} \right] \qquad \text{(Eq 7-3)}$$

where

S = spacing between wires
d = diameter of wires
h = height of wires above ground (all in the same units)

Table 7-32
The Impedance of an Open-Wire Transmission Line Made of Parallel Conductors. This is the Impedance of the Two-Wire Transmission Line Used to Transport the RF Back From the Far End to the Feed Point.

Wire spacing	1.3-mm Wire #16 AWG	1.6-mm Wire #14 AWG	2-mm Wire #12 AWG
25 cm	713 Ω	688 Ω	661 Ω
30 cm	735 Ω	710 Ω	683 Ω

Table 7-33
Terminating Impedance for a Beverage Antenna Made of Two Parallel Wires (Open-Wire, Spaced 30 cm) and Three Types of Twinlead).

Height Meters	For 30-cm Open-Wire Line $Z_0 = 710$ Ω	For 50-Ω Twinlead Ω	For 300-Ω Twinlead Ω	For 75-Ω Twinlead Ω
0.5	252	341	365	393
1.0	295	383	407	435
2.0	332	424	449	476
3.0	357	449	473	500
4.0	375	466	490	518

See **Table 7-33**. For such a two-wire Beverage system to work properly, great care must be taken to the balance of everything that needs to be balanced, the open-wire transmission line and the transformer with push-pull windings. With open-wire transmission lines the balance to ground can be improved by periodically transposing the wires (say every λ/4 or less). This becomes more critical as the line gets longer. This is also why you should not use a two-wire Beverage with an open-wire transmission line where the conductors are mounted one above the other, although here too frequent conductor transpositions can reduce the imbalance effects (together with a "good" height above ground).

2.15.3. Designing the Transformers

The exact number of turns you need to wind your transformers depends on the permeability of the material used. Using cores with a high permeability has the advantage of requiring fewer turns, which means smaller capacitive coupling and a wider frequency coverage.

Table 7-34 lists the turn ratios for the transformers used with the two-wire Beverage antennas shown in Fig 7-94. In Table 7-34, three Beverage antenna impedances are listed: 450 Ω for a low Beverage (0.5 meters high), 525 Ω for a 2.0-meter high Beverage and 600 Ω for a 4-meter high Beverage. The procedure to determine the exact number of required turns is as follows:

- Choose the core to be used from Table 7-27.
- Note the numbers of turns required for the 50-Ω winding.
- Multiply the turn ratios from Table 7-27 by the number of turns taken from Table 7-34.

Example: You can make the transformer on binocular cores (using a single Fair-Rite 2873000202). For this case the ratios must be multiplied by a factor of two or three (1 turn = 2 passes).

Table 7-35 lists the turn ratios for the open-wire Beverages from Fig 7-98. Case A is for a two-wire Beverage made of 30-cm spaced open-wire feeders (Z = 710 Ω). Case B is for 450-Ω twin-lead. Case C is for 300-Ω twin lead, and Case D is for 75-Ω twinlead. The calculations were made for the antenna wire 2 meters above ground.

Commercially made transformers (all three) for the two-

Fig 7-99—This near-end receiving board, designed by W8JI, contains two transformers, the direction-switching relay and the termination resistor for the unused direction.

Table 7-34
Transformer Design for Low, Two-Direction Beverages. See Fig 7-94 for Cases.

Case	Zant	Zcoax	T1 n1	T1 n2	T2 n1	T2 n2	T3 n1	T3 n2
A	450	50	3.0	1.0				
	450	75	3.0	1.2				
	525	50	3.2	1.0				
	525	75	3.2	1.2				
	600	50	3.5	1.0				
	600	75	3.5	1.2	–	–	–	–
B	450	50	3.0	1.0	3.0	1.0	2.0	2.0
	450	75	3.6	1.2	3.6	1.2	2.4	2.4
C	450	50	–	–	3.0	1.0	2.0	2.0
	450	75	–	–	3.6	1.2	2.2	2.4

The numbers are turn ratios. See text for details.

Table 7-35
Winding Information for the Transformers Used in Fig 7-98. The Numbers are Turn Ratios. See Text for Details.

Case in Fig 7-98	Z_0 Coax	Zant	T4 n2	T4 n1	T1 n2	T1 n1	T5a n2	T5a n1	T5b n2	T5b n1
(a)	50	330	1.0	3.8	1.0	2.6	2.6	3.8	0.7	3.8
(a)	75	330	1.2	3.8	1.2	2.6	2.6	3.8	0.7	3.8
(b)	50	425	1.0	3.0	1.0	2.9	2.9	3.0	1.4	3.0
(b)	75	425	1.2	3.0	1.2	2.9	2.9	3.0	1.4	3.0
(c)	50	450	1.0	2.4	1.0	3.0	3.0	2.4	1.8	2.4
(c)	75	450	1.2	2.4	1.2	3.0	3.0	2.4	1.8	2.4
(d)	50	475	1.0	1.2	1.0	3.1	3.1	1.2	2.5	1.2
(d)	75	475	1.2	1.2	1.2	3.1	3.1	1.2	2.5	1.2

wire Beverage antenna are available from K1FZ (**www.qsl.net/k1fz/**) and from DX Engineering (**www.dxengineering.com/**). **Fig 7-99** shows the circuitry at the "near-end" of the two-wire beverage, where the feed line is connected. **Fig 7-100** shows the far-end reflecting transformer.

2.15.4. Constructing Your Own Transformers

I will briefly discuss the construction method of each of the transformers listed in Tables 7-34 and 7-35 and show typical performance data. The transformers were all built using MN8CX cores, but transformers made on other types of cores (see Section 2.8) should yield comparable results, as well as the binocular core (Fair-Rite 2873000202). The transformers wound on the binocular core have an extremely wide bandwidth (1 to over 30 MHz).

Transformer T1 can be wound as shown in **Fig 7-101A**. Transformer T2 (Fig 7-94) should be wound as shown in Fig 7-101B. The isolation transformer T3 (Fig 7-94) can be wound as shown in Fig 7-101C.

Fig 7-101D shows the winding layout for transformer T4 from Fig 7-98. When properly adjusted the suppression of the common-mode signal was measured as > 45 dB.

The open-wire reflection transformer (T5a and T5b from Fig 7-98) with separate primary and secondary windings is shown in Fig 7-101E. Fig 7-101F shows the equivalent transformer wound as an auto-transformer.

Fig 7-100—The far end reflecting transformer with its three terminals: two going to the two-wire Beverage and one going to ground.

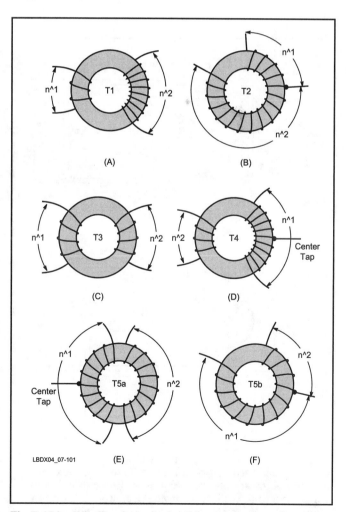

Fig 7-101—Winding layouts for the various transformers described in the text.

2.15.4.1. Housing the Transformers

To house the transformers I have used both aluminum and PVC-type diecast boxes of appropriate size. The PVC boxes are inexpensive and do not need feedthrough standoffs for the output to the Beverage antenna. A stainless-steel screw with nuts and bolts mounted right through the plastic wall is all that is needed.

Do not economize on coax connectors. One excellent way of waterproofing coaxial connections (and any others) is to squeeze plenty of medical-grade petroleum jelly into both connector parts before mating. Some petroleum jelly can be applied to the box seams to waterproof the whole assembly once the cover has been put on. Petroleum jelly is an excellent electrical insulator since it has a very low dielectric constant similar to Teflon and polyethylene. I have used the same material for protecting non stainless-steel antenna hardware. After many years of use in a humid and aggressive climate, the hardware still comes apart with no problems. **Fig 7-102** shows possible layouts for transformer units for single-wire and two-wire Beverage antennas.

2.15.5. Testing the Transformers

The transformer T1, T2, T3 and T4 can easily be tested for SWR by terminating the secondary with a terminating resistor and checking the SWR with an antenna analyzer or a noise bridge. The balance of push-pull transformer T4 can be tested by connecting the two secondary outputs together and feeding a small amount of RF from any low-power source to this connection. A perfectly balanced transformer should yield no output at the low-impedance secondary from this configuration. The physical symmetry of the transformer may be adjusted (slightly adjusting turn spacings on the toroidal core) while performing this test until the lowest possible signal output is achieved. The balance can be assessed by temporarily disconnecting one of the secondary leads and measuring the signal-strength difference on the receiver. Better than a 40-dB difference should be easily obtainable.

The insertion loss can checked as explained in Section 2.8. It should be less than 1 dB (typically ~ 0.5 dB) for all transformers.

John, W1FV describes a procedure for "tweaking" the reflection transformer: *"Adjust the turns ratio until the SWR in the reverse direction at the receiver end is flattest. By flattest I mean that SWR exhibits the smallest variations up and down over a wide range of frequencies, and not necessarily the lowest SWR. With an antenna analyzer like the AEA or MFJ, sweep the frequency from 1.8 to as high as 10 MHz and look for the SWR to change the least over this entire range, not just inside the amateur bands. This procedure usually gets you very close to the optimum."*

2.16. Arrays of Beverages

In the part of this chapter dealing with the principles of receiving antennas (Sections 1.6 through 1.12) I discussed the principles of broadside and end-fire arrays. Beverages make ideal elements for an antenna array. In both our vertical arrays as well as Beverages, broadband-phasing systems can easily be implemented because the feed-point impedance is stabilized through lossy mechanisms, such as series resistors in our vertical arrays or "natural" loss mechanisms in Beverages. If the losses are made large enough to swamp out or dilute mutual coupling and resonance effects the antenna feed-point impedance will remain stable and predictable, even when elements are end-fire phased with a unidirectional pattern and close spacing.

In Section 1.16 you learned that in receiving arrays using short verticals as elements, you should use heavy resistor loading (75 Ω) in combination with the very low radiation resistance (1-2 Ω) to achieve this goal. You can use lossy elements because all you need is to overcome the noise, which on the low bands is noise received by the antenna rather than internal receiver noise.

The radiation resistance of a Beverage is very low (1 to 2 Ω). The remainder of the Beverage surge impedance (300 to 500 Ω) is made up of dissipative losses in the ground under the antenna plus the termination resistance. With a single-rod ground termination these losses may be as high as 25% of the total system loss. This means that the overall loss resistance is typically 200 to 300 times higher than the Beverage's radiation resistance, making for a very stable feed-point impedance and very low mutual coupling in the feed-point impedance even with other closely spaced Beverages nearby.

Beverages also have the ideal characteristic of providing a relatively constant feed-point impedance over wide frequency range. This makes arrays of Beverages ideal candidates for wide-bandwidth phasing systems, eliminating complex switching systems that are required when large

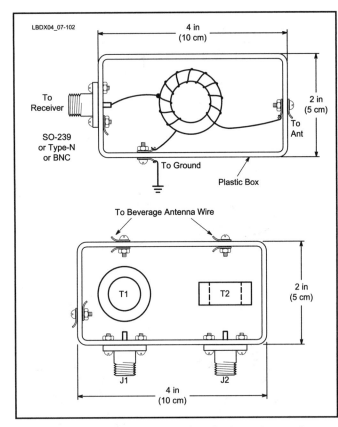

Fig 7-102—Suggested layouts for single and two-wire receiving end terminations. Plastic boxes are inexpensive and allow simple feedthrough systems to be used (screw and nut), while the toroidal transformers can be secured directly to the plastic material with silicone sealant.

verticals are used as array elements, since the verticals must be tuned at each operating frequency.

Can you really improve Beverages by phasing them? Theoretically you can achieve a narrow forward lobe (and hence a very high DMF) using a very long Beverage. To achieve the same 55° frontal lobe beamwidth as from an 8-Circle antenna (Section 1.30), you would need a 3-λ long Beverage (~550 meters on 160 meters), which poses two problems:

- Not many have that much room for such long Beverages.
- You run into space-diversity problems with this long an antenna, and you will never reach the results predicted by modeling.

The only way to achieve a similarly narrow forward beam and hence a very high RDF is to use a broadside array of Beverages, exactly the same as what we did with short verticals in Section 1.11. You can obtain excellent results with shorter Beverages (1 to 2 λ long) but these require a broadside spacing of a little over λ/2. There is no free lunch!

Improving the Front-to-Rear ratio (the DMF) is also quite simple. Just put up two Beverages in an end-fire arrangement, exactly the same way I discussed using short verticals. End-fire Beverage phasing can also help reduce or eliminate signals picked up from the vertical down leads (Section 2.5.5).

For calculating Beverage arrays I use a model where each beverage is set up as an inverted-V shape, with the highest point half-way (at 2 meters high), and sloped down to the ground at both ends. This is because with the T-shaped λ/4 terminations (see Fig 7-66C) in an array the termination wires would be too close together and mess up the modeling results.

2.16.1. Beverages and Mutual Coupling

Beverage antennas are very lossy and tightly coupled into the nearby Earth. As a result the mutual coupling between even very closely spaced Beverages is barely visible as impedance changes. Being non-resonant antennas emphasizes this characteristic. Elements on or near resonance exhibit the highest degree of mutual coupling (that's why Yagi antennas work!).

This can experimentally be confirmed this by putting some RF into one Beverage and measure the signal injected into an adjacent Beverage. Parallel Beverages spaced from each other by the same distance as their height above ground have about 30 dB of isolation.

I confirmed this by modeling also. I modeled two Beverages fed in-phase and monitored the feed impedance while changing the separation between the antennas.

2.16.2. Broadside Array of Beverages (Beverages in Phase)

In Section 1.11 I discussed how the broadside configuration (for example, spacing two antennas λ/2 apart) can give a substantial narrowing of the forward lobe, however, without improving the F/B since signals arrive from the back and from the front in-phase. Narrowing the forward lobe is especially beneficial when you cannot use long Beverages to achieve this result. If you only have room for a couple of short Beverages, but you have the room to space two of them λ/2 apart, you're in business for a super antenna. Unfortunately, if you are limited in space this is usually true in all directions, so broadside configurations are not a cure-all for everyone!

Fig 7-103 shows the radiation patterns at a 20° elevation angle for a single 160-meter long Beverage (1 λ) and broadside pairs spaced 40 and 90 meters at 1.83 MHz. **Table 7-36** lists the directivity data. As expected, the RDF improves considerably since we are narrowing the forward lobe, while the DMF changes less, except for the case of 90-meter spacing. Here the back lobe gets much narrower, although the geometrical F/B remains identical.

Note the improvement in directivity in Table 7-36—not in the back but at right angles and in the front, as witnessed by a narrower forward lobe. By definition, two poor λ/2 Beverages make a relatively good pair in a broadside combination. The elevation angle is not lowered, however, and a pair of short Beverages will still exhibit a much higher takeoff angle (48° for 80-meter long Beverages) than a single long Beverage (33° for 1-λ long and 26° for 2-λ long). See **Fig 7-105**.

Arrays of two broadside Beverages are not very common because of the wide spacing (λ/2 or wider) required for good

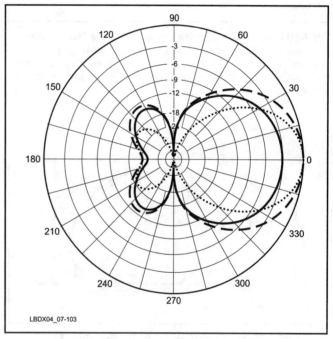

Fig 7-103—A single 160-meter long Beverage (solid line); two such Beverages in phase, side-by-side, spacing 40 meters (dashed line) and 90-meter spacing (slightly over λ/2, dotted line). Actual gain is irrelevant for receiving.

Table 7-36

Two Broadside 160-Meter Long Beverages at 1.83 MHz

	Single Antenna	Two, Spaced 40 meters	Two, Spaced 90 meters
DMF	19.0	19.5	21.3
RDF	10.2	10.6	11.9
–3-dB Angle	78°	69°	48°

Two Broadside 80-Meter Long Beverages at 1.83 MHz

DMF (dB)	11.1	–	14.4
RDF (dB)	7.3	–	9.6
–3-dB Angle	90°	–	48°

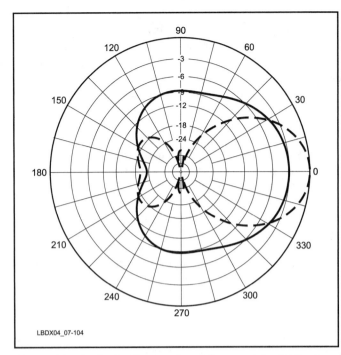

Fig 7-104—A single 80meter long Beverage (solid line) and two in a 90-meter-spaced broadside configuration (dashed line) on 1.83 MHz. Broadside makes the forward lobe narrow but does nothing for the F/B or the elevation angle.

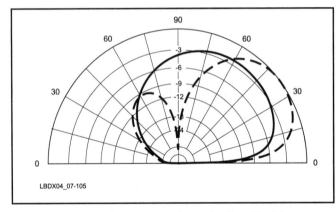

Fig 7-105—Elevation pattern (in main direction) for 80-meter long broadside pair (solid line) and for 160-meter long broadside pair (dashed line).

performance. That much space is not often found in the suburban environment where most hams live.

If you've got the space, however, a broadside configuration with a 0.67-λ spacing is a wonderful complement to an end-fire Beverage array (Section 2.16). This makes an end-fire/broadside combination a real winner, just as effective as an 8-Circle receiving vertical array (Section 1.30).

2.16.2.1. Feeding an In-Phase (Broadside) Array

Run equal-length coaxes to a common point, connect them in parallel and use a unun transformer to match the impedance to your feed line. DX Engineering sells a commercial splitter/combiner unit for this purpose (www.dxengineering.com/).

Large broadside spacings can give a narrow forward lobe. Through modeling it all seems very easy: Just increase the separation and keep feeding both in-phase. This way you can reduce the forward lobe's −3 dB angle from 48° for S = 90 meters to 35° for S = 135 meters and 27° for S = 180 meters. But that's just modeling. In real life there are two major problems associated with the use of wide-spaced arrays:

- The forward lobe becomes excessively narrow (30° and less), and you either have to be lucky that you're shooting in the right direction or you need dozens of such arrays to cover all azimuths.
- Space-diversity problems: Very often signals arrive at two very wide-spaced elements with different and constantly changing phases.

W8JI's experience is that up to 0.67 or 0.7 λ spacing the system is reliable and predictable, but over 0.8 λ the systems fall apart quickly. He wrote: *"Large spacings improve S/N under a wide range of conditions, but are more critical and often require constant re-adjustment of phase. I have to change phase between two Beverages spaced 500 feet as much as 130 degrees during the course of one opening at times!!! With wide spaced arrays, a phasing control is a MUST if you are going to use them night after night. Some nights phasing them will be useless or actually hurt because phase will be shifting so fast it will be impossible to add the desired signals. On calm stable propagation nights large arrays work exceptionally well, but on nights when we are having geomagnetic storms they are mostly useless!"*

2.16.3. The End-Fire Beverage Array

If you are after improving directivity in the rear of the antenna (improving the DMF), you can put up two Beverages fed as an end-fire phased array, just like what I described for short vertical elements in Section 1.6. The Beverages you wish to phase together need to be identical (same height and length).

Modeling shows that as you reduce the spacing (called *stagger* or *offset* for end-fire phased Beverages) you can maintain the same directivity, but your output (gain) drops. Gain is, of course, not a major issue with low-band receiving antennas. And as opposed to what happens with narrow-spaced verticals in an array, you do not have increased mutual coupling when the stagger distance is decreased. See **Fig 7-106**.

The stagger distance and the phasing angle (λ) are related to the null angle in exactly the same way as explained in Section 1.6. We can use the data in Table 7-1 for Beverages also. Assume you have two Beverages and you can stagger them 30 meters (66°) on 160 meters. You want to optimize the suppression for an offset null angle of, say, 30°. The phasing angle φ will need to be 125°. See **Table 7-37**.

Table 7-37
End-Fire Pair of 160-meter Long Beverages (30-m Stagger, 5-m Spacing, 1.83 MHz) Performance Vs Phasing

	1 Wire, Well Terminated	Two φ = 125°	Two φ = 140°
Gain (Ave Gnd) in dBi	−10.1	−8.4	−9.2
DMF (dB)	17.4	27.5	30.1
RDF (dB)	10.1	11.3	11.6

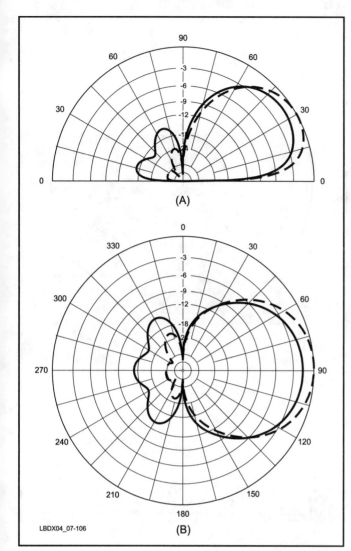

Fig 7-106—At A, elevation patterns for a "poorly terminated" 160-meter long Beverage (solid line) at 1.83 MHz compared to identical (poorly terminated) Beverage (dashed line), in an end-fire configuration (spacing: 5 meters, stagger: 30 meters, φ = 125°, R = 300 Ω). At B, comparison of azimuth patterns for same antennas. Note how both elevation and azimuth patterns have been cleaned up.

With end-fire phased Beverages the term *spacing* is normally used to indicate the lateral (side-by-side) spacing between two Beverages in an array. It is obvious that unlike a vertical array, we need to slightly side-position the Beverages, otherwise they would be on top of one another.

2.16.3.1. How Much Spacing?

In Section 2.16.1, I pointed out that Beverages hardly couple to one another. I have used spacings of 5 meters very successfully, and this can even be further reduced to 2 meters without any ill effects. Changing the spacing from 10 meters to 1 meter reduces the output by only 0.5 dB.

An interesting aspect of using two Beverages in an end-fire configuration is that the F/B of the individual elements is not so important to achieve a very good directivity (DMF). After all, we get good directivity (DMF) with two vertical elements, and they do not have any F/B by themselves! In an

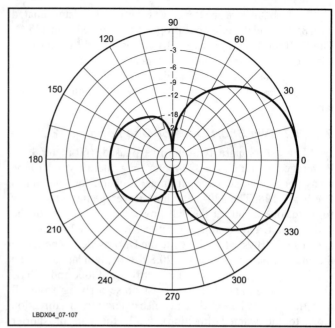

Fig 7-107—Azimuth pattern for the end-fire Beverage array in Fig 7-106, but where the two Beverages have different termination resistances (850 and 425 Ω) on purpose.

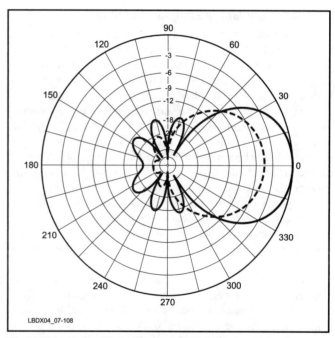

Fig 7-108—Azimuth pattern on 1.83 MHz for a single 320-meter Beverage (solid line); azimuth pattern for end-fire pair of 160-meter long Beverages, half the length (dashed line).

end-fire Beverage array it is important that the two elements have a similar F/B, as you are going to subtract the signals from both elements. You only get zero if you subtract two identical numbers. It does not matter what the numbers are though. Two Beverages with a 10-dB F/B can produce 30 to 40-dB nulls in an array, just as much as a pair where the individual elements already show 25 dB F/B.

Fig 7-107 shows the pattern for the same array as

Fig 7-106, but where one Beverage is purposely misterminated with 850 Ω and the other one properly terminated with 425 Ω. The DMF of this array is 16.9 dB versus 30.1 dB for the case with identical termination resistances (425 Ω). Obviously, you should take care to use similar and stable ground systems on both sides and to use exactly the same termination resistances.

2.16.3.2. Two Short Ones or One Long One?

Let us compare two end-fire phased 160-meter long Beverages with a single 320-meter long one. (I wonder how many people can actually put up a 320-meter long antenna!) See **Fig 7-108**. The 320-meter long Beverage has a narrower forward lobe of 60° at the half-power points, vs 70° for the end-fire pair. This results in a better RDF for the long Beverage of 13.2 dB vs. 11.6 dB, while the DMF is clearly better on the end-fire pair, at 30.1 dB vs 20.8 dB. In plain language, the single long Beverage has a narrower forward lobe and will be able to discriminate better against noise arriving from all directions (as well as QRM and noise coming from a direction close to the wanted direction), while the pair of shorter Beverages will give a much better discrimination against all QRM and noise coming off the back of the antennas. How narrow a forward lobe you can live with depends only on how many Beverages you can put up to cover all wanted directions.

2.16.3.3. How Much Stagger Distance?

If two Beverages (and the same holds true for verticals) are staggered λ/2 ("spaced" is the term for verticals), the array will need no extra phasing line to obtain a zero off the back at a zero wave angle, since signals at the end of equal-length feed lines are already out-of-phase.

To use a Beverage over a wide frequency range, you must examine the *stagger distances* you can use so that the system works correctly on all bands. The rule is simple: The stagger distance should be not larger than λ/2 at the highest frequency on which the antenna will be used. For 7, 3.5 and 1.8 MHz, the stagger distance should not be greater than 20 meters.

If you want to use the Beverages on 80 and 160 meters only, I would advocate using a stagger of 30 meters, which is a good compromise between size and gain. As with the vertical arrays, the gain drops as you reduce the stagger (spacing) distance.

Fig 7-109 shows azimuth and elevation patterns for end-fire Beverages of various lengths for 160 meters. In Section 1.6, I described how you can move the point of maximum rejection in both the horizontal as well as the vertical planes by changing the phase angle ϕ. If you consider the *average* rejection in the back (the DMF), changing the phase angle in a limited range will have little influence. **Fig 7-110** shows the 80-meter patterns in the back of a two 150-meter long Beverages in an end-fire array with 30 meters stagger and 5 meters spacing for various lengths. From the shack you could change the length of the phasing line using relays or a coaxial switch, but I have found this to be of no practical value, *unless* you would want to put a null directly in the direction where you have a noise source on groundwave. **Fig 7-111** shows the 3D pattern for the longest array in Fig 7-110. See **Table 7-38**.

John, K9DX, rightly points out that the success of two end-fire Beverages may to a great extent come from phasing out omnidirectional signals picked up by the vertical down

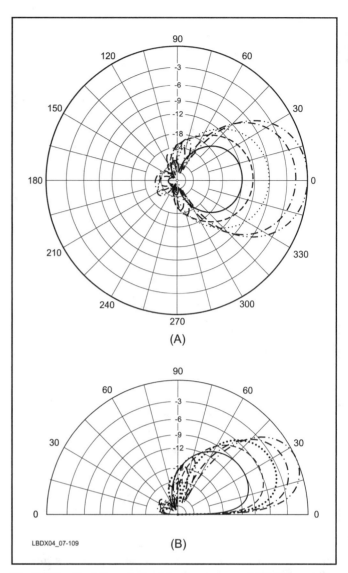

Fig 7-109—End-fire phased beverages on 1.83 MHz (5 meters spacing, 30 meters stagger, ϕ =140°), of various lengths: solid line = 100 meters; dashed line = 150 meters; 200 meters = dotted line; 250 meters = dashed-dotted line; 300 meters = dotted-dotted-dashed.

Table 7-38
Directivity of End-Fire Beverages
(Stagger = 30 meters, ϕ=140°) on 1.83 MHz

	100 m	150 m	200 m	250 m	300 m
DMF (dB)	24.6	30.6	29.0	33.0	33.8
RDF (dB)	10.3	11.4	12.4	13.1	13.9

Directivity of End-Fire Beverages
(Stagger = 30 meters, ϕ = 70°) on 3.65 MHz

	100 m	150 m	200 m	250 m	300 m
DMF (dB)	22.3	27.4	27.0	29.6	32.1
RDF (dB)	10.5	12.6	13.9	14.0	15.9

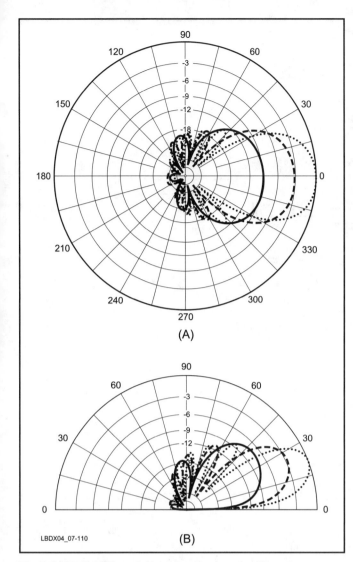

Fig 7-110—End-fire phased beverages on 80 meters (5 meters spacing, 30 meter stagger, $\phi = 70°$), of various lengths: solid line = 100 meters; dashed line = 200 meters; dotted line = 300 meters.

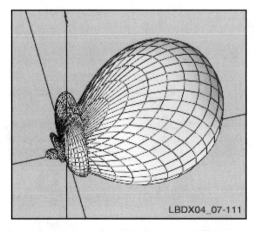

Fig 7-111—3D pattern of a 250-meter end-fire phased pair of Beverages on 160 meters. Note the extreme rejection of radiation in the back of the array. The DMF of this array is 34.2 dB (RDF = 12.2 dB). Pattern by *Antenna Model*.

Fig 7-112—Two of the many end-fire phased Beverages at W8LRL's super station.

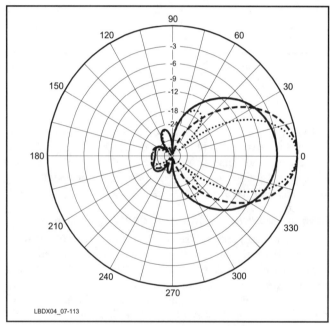

Fig 7-113—Two end-fire cells in a 1.83-MHz broadside combination: Solid line: a single cell; dashed line: two cells with 90-meter separation; dotted line: two cells with 135-meter separation. See text for details.

leads at both ends of the Beverage, or the vertical component in the sloping down leads, if used. **Fig 7-112** is a photo of some of W8LRL's end-fire Beverage arrays.

2.16.4. End-Fire Plus Broadside Combinations of Beverages

The ultimate array of Beverages is, like arrays of verticals, the end-fire/broadside combination. By putting two end-fire Beverage groups side-by-side (in-phase) you can further narrow the forward lobe and hence improve the RDF. **Fig 7-113** shows the radiation patterns of such an array, where each cell consists of two 160-meter long Beverages, spaced 5 meters,

Receiving Antennas 7-79

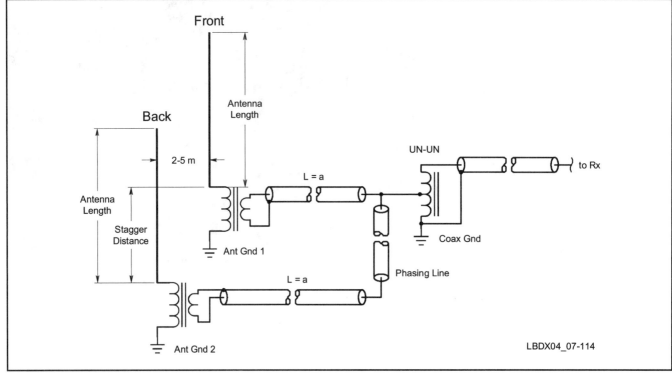

Fig 7-114—Feed system for the end-fire Beverages array (top view).

with a stagger distance of 30 meters and φ = 120°. A single cell is also shown in Fig 7-113 as reference. **Table 7-39** gives the numerical data. This is simply awesome! The broadside configuration, obviously, does not do anything to the geometric F/B (at 180°) of the array elements.

2.16.5. Feeding an End-Fire Beverage Array

Thanks to Tom, W8JI, who also brought to my attention the crossfire principle for feeding an end-fire array, feeding an array of end-fire Beverages has become very simple. Let us examine the design of an array for use on 160, 80 and 40 meters. Again, the stagger distance should not be more than λ/2 on the highest frequency (see Section 2.16).

For a spacing of 20 meters, and considering we want to lift the zero angle about 30° off the ground, we can calculate the design data using Table 7-1. In Section 1.19 I discussed the crossfire feeding principle for end-fire arrays. This principle can, of course, be applied to phased Beverages just as well as to phased verticals. From **Table 7-40** the length of the cross-fire phasing cable, expressed in meters, is the same for

Table 7-39
160-m Long Beverages in End-Fire/Broadside Combination (φ = 120°)

	Single End-Fire Cell	Cells Spaced 45 m	Cells Spaced 90 m	Cells Spaced 135 m
DMF (dB)	31.8	32	34	34.7
RDF (dB)	11.5	11.9	13.0	14.1
3-dB angle	70°	61°	46°	34°

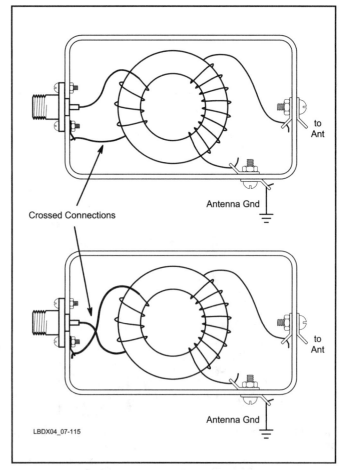

Fig 7-115—Here's how to make two phase-inverting Beverage transformers. Use a plastic box.

Table 7-40
Design Data for the 3-Band End-Fire Phased Beverage Array

	1.825 MHz	3.65 MHz	7.15 MHz
20-m spacing in °	45°	90°	180°
Phasing Angle	139°	98°	16°
Crossfire Phasing Cable Length	180 − 139 = 41°	180 − 98 = 82°	180 − 16 = 164°
Physical Length (VF = 0.66)	12.34 m	12.34 m	12.6 m

the three bands, because the system, as explained in Section 1.20, tracks frequency. Look at **Fig 7-114** and imagine A and B as feed points for two Beverages instead of two verticals. There is, of course, a slight offset (spacing) for the two parallel Beverages, which is not visible in Fig 7-114.

With a Beverage you do not even need an inverting transformer! Take a look at **Fig 7-115**. Make two such transformers, and invert the phase (either primary or secondary) in one of them. Run the phasing line length (12.34 meters in the table above) plus, for example, 5 meters to the antenna requiring the leading current—that is, the antenna in the back. Run the same length coax (5 meters) to the front element. Connect the two cables in a Tee and bingo! You may want to insert a unun at this point to match the combined impedance (25 Ω or 37.5 Ω) to the impedence of the feed line going to the receiver (50 or 75 Ω). See Fig 7-114 and **Fig 7-116**.

As we do use the coaxial cable to give us the correct phasing angle ϕ (line length = phase angle), we must make sure that the SWR on the line is *flat*. This means we have to pay a little more attention to matching the feed line to the phased Beverages compared to a system using just one Beverage.

Make sure you do everything in the proper sequence. First determine the best termination resistance for the Beverages by one of the methods outlined in Section 2.5. Make sure both termination grounds are of equal quality. Normally you should find the same resistance for two Beverages of equal length and spaced close together. If one turns out to be 350 Ω and the other 300 Ω, use 325 Ω on both. Next adjust the turns-ratio of your matching transformer for a 1:1 SWR at the impedance of the feed-phasing line you will use. You should be able to get it down to less than 1.1:1 on at least two bands. It may be a bit tricky to get this low on three bands, but set your priorities (160 and 80 meters, of course!).

Do *not* adjust the value of the termination resistances to improve the SWR on the feed lines. You should *only* adjust the turns-ratio of the matching transformer.

Of course all the precautions to avoid common-mode signal ingress into the feed lines apply for these systems as well as for single Beverages. Now you are all set for a unique experience in low-band listening delights!

If you are in a position where you can set up a broadside end-fire combination (Section 2.16.4), two identical feed system as outlined for the end-fire array should be used, and the feed lines to the feed systems as show in Fig 7-92 should be combined using equal-length feed lines (in-phase). Again, a unun transformer may be used to obtain a proper match.

2.16.6. The W8JI Parallelogram

Here is the W8JI's description of the antenna: "*Although it looks like a large horizontal loop, this antenna actually is an array of two ground-connection-independent end-fire staggered Beverages. The short end wires form two single-wire feed lines at each end. One end-wire is series terminated with*

Fig 7-116—A small unun transformer is used at the Tee junction of the feed lines coming from the two Beverages in the end-fire array. This makes for a 1:1 SWR on the feed line to the shack. The small waterproof box is elevated a few cm off the ground using small bamboo sticks.

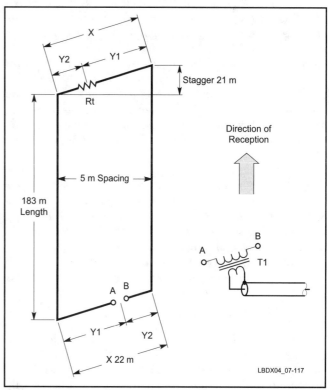

Fig 7-117—Top view of the layout of the W8JI Parallelogram array, which is basically a 2-element end-fire array of Beverages.

Receiving Antennas 7-81

exactly twice the resistance of a normal Beverage, while the short wire at the opposite end is the feed-point.

Stagger and spacing determines feed and termination location, the offset at the two ends being mirror images. A top view, looking down from straight above the antenna is given in **Fig 7-117**.

Notice the feed point is offset towards the termination end (front) of the antenna, and away from the null direction. This is typical for crossfire phased arrays. Crossfire arrays respond away from the delay line direction, exactly opposite conventional arrays. The termination is offset the same amount, but moves towards the feed-point end of the antenna.

The feed-point terminals, being floated (push-pull), provide 180-degree phasing between the two elements. The extra line length to the forward (left and front) element provides the "Stagger" delay. Consider the actual wire length of a "short side" called "X" (which is the same as Y1+Y2). This length is the same on both ends of the antenna. The difference between Y1 and Y2 must equal or be slightly less than stagger (S). To determine the offset of feed point and load:

1. *Measure length of the end-wire, length "X."*
2. *Measure stagger in the end-fire direction, "S"*
3. *(X − S)/2 = Y2*
4. *Y1 = X − Y2*

You may want to slightly offset the feed by making Y2 longer and Y1 shorter. This will move the null upwards, forming a cone. It is best to model the results. Let's review a system.

X = 22 meters
S = 21 meters
Y2 = (X − S)/2 = 1 meter
Y1 = 22 − 1 = 23 meters

The difference is 22 meters, and that is the phase delay= (22×1.83×360)/299.8 = 48.3°. We have a 180° shift at the push-pull feed point (between points A and B), so −48.3° rotates to +131.7°. We have a 131.7° lead in the forward element, with a 48.3° spatial array delay. In the forward direction (towards termination) phase is (−48.3°) + 131.7° = 83.4° out-of-phase. This results in nearly the voltage of one element alone, when the two element outputs of the long sides are summed. Towards the null (feed point) the phase is +48.3° + 131.7° = 180° for a sin 180° = 0 or zero voltage, a perfect 180° null.

On the second harmonic forward array feed system phase is −96.6° rotated to +83.4°. Spatial array phase is now 96.6°, or −96.6° towards termination. The result is (−96.6°)+ 83.4° = −13.2° in the forward direction. The result is nearly twice the voltage of one element in the forward direction. In the reverse direction, array phase is −96.6° − 83.4° or 180° out-of-phase. We once again have zero back-fire response.

The general pattern holds true for any length of S less than 180°, although grating lobes would make the pattern useless. This array, with 18-meter stagger S, is usable from about 5 MHz down to VLF."

Note that exactly the same analysis can be done with individually-fed end-fire arrays. You will also see that the shorter the stagger distance the lower the gain while the response off the back remains a perfect null, not necessarily in-line but at an offset angle (in elevation and azimuth). See **Fig 7-118**.

W8JI continues: *"Placing the feed point and termina-*

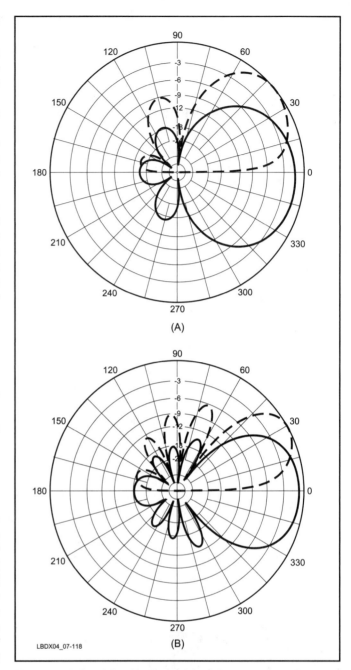

Fig 7-118—Combined azimuth/elevation patterns at 160 and 80 meters for the 150-meter long Parallelogram array. These are very similar to the patterns obtained by two individually fed end-fire Beverages.

tion centered on the end-wires is like using 180-degree phasing. Placing the feed point and termination at the stagger minus side length distance is like using S-180 degrees phasing on any frequency with one exception, the antenna fires towards the feed-point offset. This always results in a perfect backfire null, regardless of frequency.

If you wish to remove signals from the rear, the W8JI Parallelogram Array is a good broadband solution. It not only requires one transformer, one termination, and no ground systems at either end, it also is useful on several bands without switching anything!

Remember the above example is a reasonably short

Table 7-41
W8JI Parallelogram

	1.8 MHz	3.6 MHz
RDF (dB)	9.2	11.8
DMF (dB)	18.6	19.6
–3-dB Angle	87°	69°

antenna, performance improves as side-length increases. There is no reason why this antenna can't be used from a band where it is only $1/2$-λ long to bands where it is several wavelengths long, as long as you properly choose stagger and width."

Table 7-41 shows calculated performance data on 160 and 80 meters.

2.16.6.1 Feeding the W8JI Parallelogram

The feed point impedance is about 900 Ω, so a 16:1 transformer with minimal inter-winding capacitance should be used. The transformers described later in Section 3.8 (feed systems for elongated loops) are well-suited for this application.

2.17. Summing Up, Arrays of Beverage Antennas

- End-fire Beverages help in canceling out the omnidirectional pick-up caused by the vertical (or sloping) down leads at both ends of the transformers (this may be one of the main advantages of end-fire Beverages).
- Two 160-meter long Beverages in end-fire configuration allowed me to dig one layer deeper in the 160-meter noise. That's how I worked VP6DIA (Ducie Island) as only one of two Europeans on 160 meters to do so.
- The crossfire feed system is simple. (Thanks Tom, W8JI!). It works on up to three bands without any changes or switching.
- To obtain best results (which are just short of spectacular), make sure you terminate the antennas properly and have a very low SWR on the feed lines.
- If you are limited in length use two short beverages in an end-fire array (even two end-fire phased $\lambda/2$ Beverages are pretty good!)
- If you can put up a pair of $\lambda/2$ spaced Beverages, put up two groups of end-fire arrays rather than a broadside group of individual Beverages.
- You can't imagine how good an end-fire pair is unless you compared it side-by-side with a single one in the same direction. I have seen some spectacular improvements, like signals generally off the back dropping from S7 to S1 or less! Simply amazing.
- In a QTH where there is no direction where most of the noise comes from, it may pay off more to reduce the –3-dB forward angle by going to a broadside array, than trying to achieve the best possible front-to-rear with an end-fire array.

2.18. Beverages and Beverage Arrays Compared

In **Table 7-42** I compare DMFs and RDFs and the forward lobe at a 20° elevation angle for a variety of Beverages and Beverage arrays. All figures are for 1.83 MHz.

2.19. Summing Up, Beverage Antennas

I have been using Beverages since 1968. For me, Beverage antennas have undoubtedly been the key to working my last 50 countries on 80 meters. As far as 160 meters is concerned, I would not like to think what it would be like without them.

Since I started using end-fire phased Beverages, this has opened again a new world for me. Unfortunately, I cannot run broadside phased Beverages—the land available to me is much too small for that. I cannot not even run end-fire Beverage arrays in all directions, but I must say that my hearing to the US has been dramatically improved through the use of an end-fire pair. That is because most of the amateur-made noise on the bands (for example, during contests) comes from the SE, the opposite direction to the USA. Well, all of that has dropped another 20 dB since using the end-fire pair. It is just incredible. At first I though the antenna was "dead" as there seemed to be so little noise.

I use the Beverages all the time on 80 and 160 meters. Unfortunately, I do have to take down most of my Beverages

Table 7-42
Comparing Beverages and Beverage Arrays

Antenna description	DMF dB	RDF dB	–3-dB-Angle Degrees	Gain dBi
80-m Long Beverage	11.1	7.3	90	–16
160-m Long Beverage	19.0	10.2	78	–10
300-m Long Beverage	21.3	12.9	62	–5
Broadside 80-m Beverages, 90-m Spacing	14.4	9.6	48	–13.3
Broadside 160-m Beverages, 90-m Spacing	21.3	11.9	48	–7
Broadside 300-m Beverages, 90m Spacing	23.1	14.2	44	–2
80-m Long End-Fire Beverages, Stagger = 30 m, φ = 140°	20.0	9.7	77	–15.5
160-m Long End-Fire Beverages, Stagger = 30m, φ = 140°	30.1	11.6	69	–9
300-m Long End-Fire Beverages, Stagger = 30 m, φ = 140°	33.8	13.9	57	-4
160m Beverages in End-Fire/Broadside Array (*1)	34.0	13.0	46	-6.4
160-m Beverages in End-Fire/Broadside Array (*2)	34.7	14.1	34	-6.4

(*1): End-Fire Cell: φ = 140°, Stagger = 30 m, Broadside Spacing: 90 m
(*1): End-Fire Cell: φ = 140°, Stagger = 30 m, Broadside Spacing: 135 m

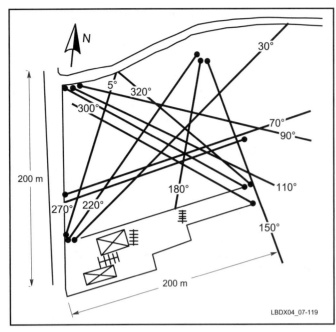

Fig 7-119—Beverages at ON4UN. The numbers are the target directions in degrees. Notice the phased Beverages to the US (300°). Twelve directions are covered, in approximately 30° steps.

Fig 7-120—View to the NE (Japan) across the fields where I have my Beverages in winter. In the front, 10 and 15-meter Yagis; to the right, the 160 meter λ/4 vertical.

I would like to thank Earl, K6SE, and Gary K9AY, for proofreading this next section.

3. THE FAMILY OF ELONGATED TERMINATED LOOPS

Loops have been around a long time. W8JI told me he'd been using arrays of loops to work Japan in the 1970s through 100-kW LORAN pulse emissions on 160 meters. These made it possible for Tom to be the first Eastern-USA station to work JA through the LORAN noise. In those days the JAs transmitted just above 1900 kHz, where all the East Coast LORAN pulse transmitters operated. Tom used an array consisting of no less than eight wire diamond elements in a series end-fire configuration, terminated at the far end. Tom told me it was Mr Topband himself, the much revered Stew Perry, W1BB, who told him that commercial antennas were also available using that principle as early as the 1960s, and that got Tom going! Nowadays, when we talk about small (terminated) loops, everyone thinks of EWEs, Flags, Pennants and such.

The EWE, the Pennant, Flag, Delta, Diamond, and the K9AY all operate on the same principle of the terminated loop. The differences are in the way ground is used, how they are fed, and their physical shape.

All these antennas basically produce a cardioid pattern. See **Fig 7-121**. The depth of the null and the null angle vary with the shape of the loop and the termination. The optimum termination value can vary with local ground conductivity. The directivity figures (DMF and RDF) are listed in **Table 7-43**. The directivity is very much the same as for a λ/2 Beverage that has been terminated for best directivity with a complex termination (see Table 7-38).

Most of the published designs are based on optimum

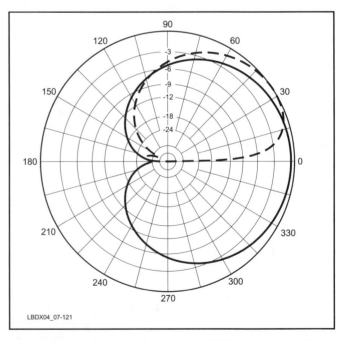

Fig 7-121—Radiation patterns of a EWE antenna (solid line = azimuth; dashed line = elevation). Whatever the dimension, whatever the shape (inverted U or inverted V), the shape of the pattern is always the same, provided the termination resistance is correct.

during the summer, which for me was "off-time" on the low bands. Summer is for bike riding, or writing a book... After some unsuccessful attempts using elongated loops such as Flags, I built a small receiving Four Square (see Section 1.28) in the small garden in the front of the house, so that I am not completely an alligator in the summer. I am very pleased with it, but it of course does not match the phased Beverages. With an 11-meter spacing the Four Square requires very careful adjustment to perform well on 160 meters. See **Fig 7-119** and **Fig 7-120**.

Table 7-43
Typical Directivity Figures for Elongated Terminated Loop Antennas

	DMF dB	RDF dB
Typical Elongated Loop	~11	~7.4

dimensions, but almost *any* loop size and shape, grounded or ground-independent, can be made to work. First, the feed point and termination must have significant separation—for example, between the two bottom corners of a triangle or square, or be separated by the ground connection (as in the K9AY loop) or by the two ground connections (as in the EWE). Then it needs the right termination to establish the current distribution and phase along the antenna conductors to create the directional pattern. The termination may be resistive, or it may be complex, requiring R+L or R+C to obtain the desired pattern.

While just about "any dimension" will give you a cardioid pattern for a specific termination resistance (impedance) over a more or less narrow frequency band, only the optimized designs combined bandwidth with performance and ease of terminating and feeding (with resistive impedances).

On this issue of the loop's size, K9AY wrote: "*Most of the terminated loop designs are "lowpass" in their characteristics. That is, they are quite consistent in feed point Z, termination and pattern below a cutoff frequency. For the K9AY loop, the published dimensions in Sep. '97 QST have that cutoff around 160 meters, which means that it behaves similarly in the AM broadcast band, or lower. A different termination is needed for optimum 80-meter performance, which is just above the "cutoff" frequency. The tradeoff is in the received signal levels. Smaller antennas, though more broadband, need more preamp gain. Larger antennas capture more signal. The voltage across the feed point is proportional to the area enclosed by the loop.*"

Fortunately all of these loops have been optimized (thanks to K9AY and K6SE) so that their cardioid null is constant over a wide frequency range (160, 80 and 40 meters for most designs) and this can be obtained with a simple resistive termination equal to the feed impedance of the antenna.

These elongated terminated loops are really a simple pair of verticals with a horizontal feed line, one being base-fed, the other being fed via the top wire. That provides crossfire phasing. When the element spacing is less than λ/4, this always fires towards the feed point end of the array. When properly terminated the loops show nearly constant current at all points on the loop. This constant current is achieved by terminating the far end of the loop in the same impedance as the feed point. Thus, the portions considered to be the vertical "elements" have equal current, with a phase equal to 180° plus the electrical length of the connecting wires, which is slightly more than the element spacing.

It is impossible to make a simple comparison between two phased bottom-fed verticals. Due to the terminating resistor the antenna works much more as a traveling-wave antenna than as a standing-wave antenna. The antenna current is nearly uniform or slowly tapers (due to loss from radiation and resistance) in the antenna system and contains no or greatly suppressed "end reflection." It is the terminating resistor that equalizes the currents. The conductor and radiation losses are compensated by slight changes in the value of the resistor. Therefore the antenna has no standing waves and can indeed be called a traveling-wave system. This is also why the antenna is very broadband, and why it exhibits a good F/B on both 160 and 80 meters.

In other words, the terminated loop is like two constant-current verticals with classic end-fire phasing. The only difference is the horizontal component in the radiation due to the radiating delay line (sloping or horizontal) of the terminated loops.

This is the operating principle for the entire family of elongated terminated loops. Sloping (inclined) wires used in some of the designs are just like a horizontal wire in combination with a vertical wire, and the same reasoning as above can be used.

3.1. Modeling Small Receiving Loops

K9AY, loop specialist par excellence, wrote on the Topband reflector: "*NEC-2 modeling programs are sufficiently accurate for modeling these loops, even for ground-connected antennas, but do not try the usual technique of increasing the number of segments until convergence is reached. When designing the original K9AY loops, I found that the model closely matched observed behavior with one segment for each 1½ feet of wire (on 160). Models for other loops behave similarly using this segmentation scheme.*"

There is nothing magic about the dimensions of a receiving antenna such as the K9AY Loop or the EWE. You can easily brew your own using a *NEC-2* based modeling program such as *EZNEC*. Just enter the dimensions and change the value of the terminating resistance until the best F/B is obtained. If you are looking for a well-balanced design, however, you will probably end up with the dimensions published in the original design.

I'd like to warn you once more that there is a difference between modeling and antenna building. Building a receiving loop (or any other receiving antenna for that matter) is not a mathematical exercise. I have seen umpteen radiation plots with high praises for a 67.37 dB F/B for a model. Great! Great what? Great mathematics, that's all. Look, the proud designer of this antenna needed a 937.735 Ω terminating resistance! Go to the radio shop, or call Allen-Bradley or Ohmite and ask them for such a resistor. To obtain that kind of phenomenal F/B the ground conductivity must be stable "like a rock." It should not vary more than 0.01 Ω. And the length of the antenna, should be measured to within 1 mm at worst...

There is a definite difference between a model and reality; I will say over and over. And I will refrain from publishing more digits of accuracy than make sense. We should think in terms of absolute and relative error all the time.

A little experimentation is normal after building the antenna according to the model. Also, once the antenna is working well enough to obtain 20 dB or more F/B, it takes careful measurements to discern any additional improvement.

3.2. The EWE Antenna

Floyd Koontz, WA2WVL, is the originator of the EWE. His publications (Ref 1263 and 1264) are must-reads on this novel receiving antenna. Incidentally, the curious name "EWE" came about because Floyd noted that his new antenna looked very much like an upside-down letter "U." He jokingly submitted a drawing of a female sheep—a ewe—to headline his February 1995 *QST* article, which he impishly titled with the triple entendre "Is This EWE for You?"

A EWE is small, requires little engineering, and can be designed to cover 80 and 160 meters. Moreover, the EWE is low profile and can be built for little money. In appearance, the EWE resembles a very short Beverage, though in fact it as an array of two short vertical antennas, where the horizontal wire is part of the (radiating) feed system. See **Fig 7-122**.

The horizontal wire is only about 5 to 15 meters long and is about 3 to 6 meters above ground. In other words, an EWE can fit in many tiny yards. The original description by WA2WVL uses 3-meter high "elements" with a spacing going from 4.5 to 18 meters, depending on the band (80 or 160 meters).

A good rule of thumb is to build an EWE where the length is twice the height. This rule can be followed for elements going from 3 to 6 meters in height, and will result in an optimum termination resistance of approximately 1000 Ω for both bands. The smaller the array, the easier it is to obtain a good deep null, but the lower the output level. The larger versions will be near the cut-off frequency, where a deep null cannot be obtained.

The output (gain) of the antenna depends on its size. The gain for 160 meters varies typically between −20 dBi and −30 dBi over average ground, and is about 9 dB higher on 80 meters. The inverted-V shaped EWEs looks quite attractive and can easily be made into a "hand-rotatable" receiving antenna: Just walk out in the garden and anchor the sloping wires to a different set of ground rods. It should be useful for DXpeditions as well.

All with all ground-based antennas a good ground will lower the takeoff angle (typically about 35° for average ground). As usual, you have to be cautious with the modeling results as they are calculated for a flat and perfectly homogenous ground.

The value of the terminating resistor is quite critical if you want to obtain a good notch in the back response. Since the ground resistance is part of the terminating resistance, a good low-impedance, stable ground is required. The ground resistance of a single ground rod may vary from as low as 50 Ω to as much as several hundred Ω. Although the terminating resistance is high (typically 1000 Ω), a ground consisting of more than one ground rod or a combination of a ground rod with various short radials is a good idea.

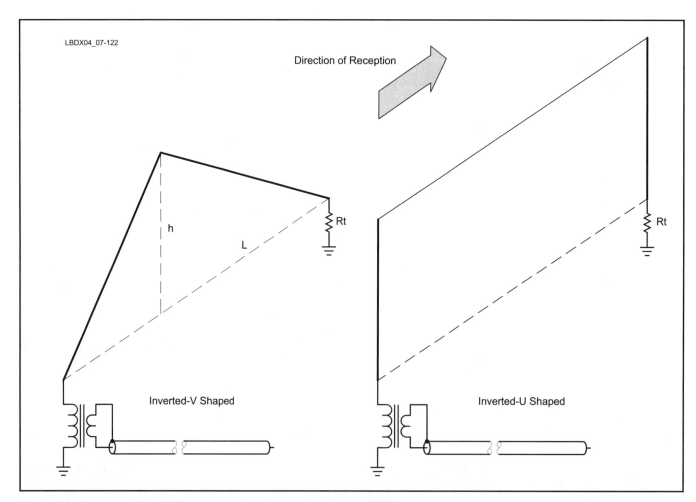

Fig 7-122—Layout of the EWE antenna. The inverted-V-shaped EWE requires one insulated support; the inverted-U-shaped EWE requires two base-insulated metallic supports that are part of the radiator.

Varying ground conductivity not only has an influence on the effective total termination resistance but also on the velocity factor of the various elements of the antenna. The velocity factor is critical in determining the current distribution necessary to obtain a cardioid pattern.

It has been reported that the best way to tune an array for best F/B is to find a medium-wave BC transmitter in the back of the antenna. Using a medium-wave signal during day time ensures that you will receive it on ground-wave, and as such have a constant and stable signal source. Tune the terminating resistor for a full notch, striving for a minimum of 25 dB. This value will be very close to the best value for 160 meters and close to what's best on 80 meters.

You should use broadcast stations between 1600 and 1700 kHz to do the null adjustment if you can. It is impossible to reliably adjust the termination resistance using signals arriving by skywave. You could also use a nearby local ham (in-line off the back) or set up a small signal generator off the back, using a small vertically polarized antenna, at a distance of at least 1 λ from the EWE (which means outside its near field).

The feed-point impedance of a properly terminated EWE usually has a reactive component, but the real part is usually in the 400 to 600-Ω range. Being a grounded antenna, the EWE has an advantage over an "elevated" elongated loop, in that problems with common-mode ingression at the antenna feed point are a little easier to solve.

The EWE antenna can be fed with a 1:9 transformer such as described in Section 2.8.1. Further, common-mode signals getting on the shield of the feed line can be avoided by applying the techniques outlined in Section 2.8.3. The features of the EWE can be summed up as:

- Easy to build, but requires good grounds and radials at both ends.
- Less critical in feeding compared to elevated loops.
- More sensitive to varying ground conditions compared to elevated loops.

3.3. The Rectangular Loop or Flag Antenna

A major drawback to the EWE is its extreme sensitivity to any change in the local soil conductivity. EA3VY was the first to come up with the idea of adding a ground wire between the bottom of the two verticals as an effective way to minimize the effect of different soil conductivities.

As pointed out in Section 3, many dimensions will work but only the optimized ones provide the best directivity over the largest frequency range with a single termination resistor. See **Fig 7-123**. These standard values developed by EA3VY and K6SE are 4.27 meter (14 feet) high by 8.84 meters (29 feet) wide, with a bottom wire height of 2 meters (6 feet) above ground. These dimensions result in an antenna that has excellent characteristics from 7.5 MHz down. K6SE says, *"The height by length ratio of the Flag is not a "happy medium" value. The correct dimensions were arrived at by designing the antenna to be broadband (ie, exhibit zero reactance at its feed point over a very wide frequency range). This required that the termination resistor value be equal to the feed-point resistance. To meet these criteria, the height and width dimensions had to be juggled until everything fit into place."*

This configuration has a gain of approximately –28 dBi

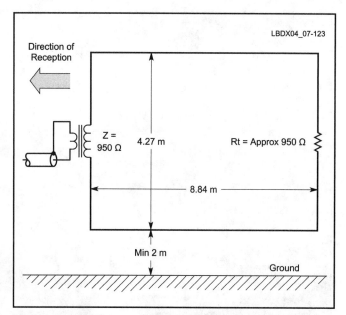

Fig 7-123—Optimized dimensions for the 160/80meter rectangular loop.

on 160 meters, –18 dBi on 80 meters and –10 dBi on 40 meters. The loop is terminated with 950 Ω and fed in the centers of opposite vertical sections. Its input impedance on 160 meters is 950 Ω with a small reactive component (10 to 50 Ω).

You can scale up the dimensions to get more signal output of the antenna. If you are not interested in using the antenna on 40 meters and if you have a lot of space, a flag twice the size of the K6SE/EA3VY design has a gain of –16 dBi on 160 meters, 12 dB more than the standard Flag. That is quite a difference and can be helpful overcoming problems with common-mode signal ingress (see Section 3.8). This scaling principle holds true for all types of elongated loop antennas.

A properly designed Flag exhibits a high F/B ratio over any type of soil and at virtually any height above ground without need to change the dimensions or termination value. The directivity figures (DMF and RDF) are given in Table 7-43.

K6SE, who has thoroughly tested all shapes and sizes of elongated terminated loop antennas, reports that the Flag antenna is probably the best of all configurations from the standpoint of being broadbanded and having gain. It has about 6 dB more gain than an equal-sized Pennant (see Section 3.4).

The shape and the size of this loop makes it attractive for a rotatable version. W7IUV describes such a design on his website (**www.qsl.net/w7iuv/**). **Fig 7-124** shows W7IUV's elegant loop. W7IUV says: *"The choice of boom material was easy; a friend had some chain link fence top rail laying in his horse pasture. I volunteered to clean it up for him. I wanted to use bamboo for the spreaders, but none could be found for free or even cheap. Several materials were experimented with before settling on wooden clothes poles. This material is almost as light as bamboo and is readily available in most lumberyards and home improvement stores. The spreaders were attached to the boom with square steel tubing welded to the ends of the boom. The tubing formed nice "sockets" for the wood poles to slip into. One through bolt holds it in place. If*

Fig 7-124—Photo of the W7IUV rotatable flag.

developed by K6SE and EA3VY. As explained earlier, other dimensions can be used, but these yield the highest F/B over the widest frequency range for a single value of termination resistance. This design can be used from 160 through 40 meters.

The correct termination resistor for the Pennant is about 900 Ω, which can be placed either at the point of the Pennant or in the center of the vertical section—the results are identical. The feed point is at the opposite end. Gain on 160 meters is about –34 dBi (which is 6 dB less than the Flag), –24 dBi on 80 meters and –16 dBi on 40 meters. The exact values depend on local ground conductivity. Note that the 6-dB lower signal level is proportional to the difference in area enclosed by the loop, which is half as much as the Flag: $1/2$ voltage = $1/4$ power (–6 dB).

welding is not your thing, consider making a spreader mounting plate from a square piece of aluminum about $12 \times 12 \times 1/4$ inch. Drill for U-bolts in appropriate places."

3.4. The Triangular Loop or Pennant

While optimizing the flag, K6SE and EA3VY developed the triangular-shaped loop. He named it the *Pennant* because its triangular shape resembles a flag pennant. The loop is fed in the center of the vertical section, while the load is situated where the sloping wires meet. It looks like the only limit as to shapes and dimensions of these kind of loops is your own imagination.

Fig 7-125 shows the dimensions of the optimized Pennant

3.5. The Delta-Shaped Loop

A EWE in the shape of a Delta loop has every bit as clean a pattern as the more classic ones. I think this antenna is really attractive for DXpeditions and for semi-rotatable setups. Just moving the base line around is all you have to do to change directions. During modeling it appeared that the best F/B was obtained with the terminating resistor mounted approximately 20% from the bottom corner of the Delta loop. The antenna is fed in the opposite bottom corner. See **Fig 7-126**.

The FO0AAA Clipperton Island DXpedition used this antenna in their operation. It is the only design that requires only one support and can be easily rotated manually, a very desirable feature for DXpedition use. The Clipperton team

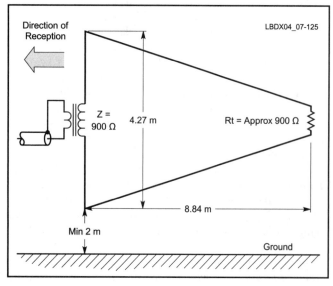

Fig 7-125—Layout and dimensions of the optimized Pennant antenna.

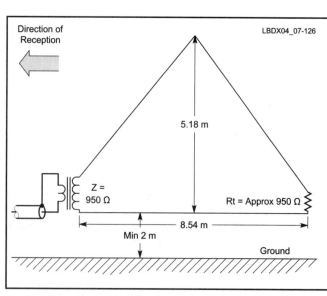

Fig 7-126—Layout and dimensions of the optimized Delta antenna.

found the antenna to be very successful and many other subsequent DXpeditions have also used the Delta configuration.

The optimum termination resistance for this design is also 950 Ω. The antenna was optimized for 160 meters and its output is about −33 dBi on 160 meters, −22 dBi on 80 and −13 dBi on 40 meters.

3.6. The Diamond-Shaped Loop

The diamond shaped loop derived from the flag, was developed thinking of a cubical quad element. The idea is to use the construction techniques and hardware for a 20-meter cubical quad element. See **Fig 7-127**.

It is not necessary to use an equilateral loop—it can be elongated if that is easier to do mechanically. The elongated version also seems to provide a wider bandwidth than the equilateral configuration, for a fixed termination resistance. The feed impedance and the termination resistance is approximately 1000 Ω in both cases and for the three lower bands.
Equilateral version:
L = H = 9 meters
Gain: −29 dBi on 160 meters; −18 dBi on 80 meters;
   −10 dBi on 40 meters.
Elongated version:
H = 7 meters, L = 10 meters
Gain: −31 dBi on 160 meters; −21 dBi on 80 meters;
   −12 dBi on 40 meters.

3.7. The K9AY Loop

Gary Breed, K9AY, described this loop very well in his Sep 1997 *QST* article (Ref 1265). This is another variant of the EWE, where the bottom wire of the loop is grounded in the center. See **Fig 7-128**. He pointed out in his article that the loop can really be any shape. The diamond shape K9AY used was dictated by practical construction considerations rather than anything else. All of the K9AY loops can be used on both 160 and 80 meters, although optimal performance may require slight adjustment of the terminating resistance, as K9AY

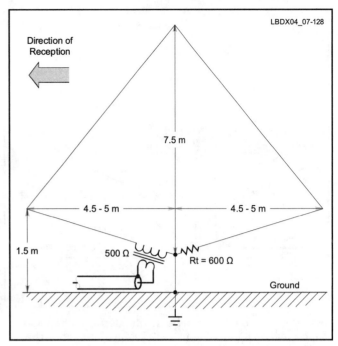

Fig 7-128—The K9AY loop.

pointed out in his article.

While in all the previously described loops the feed point and the termination are separated by distance (they are at opposite ends of the elongated loops), in the K9AY loop "separation" is achieved by grounding the loop between the adjacent feed and termination points. Having the feed and termination so close to one another makes it easy to switch directions from a single box.

Gary points out that the published dimensions in Sep 1997 *QST* have a cutoff around 160 meters, which means that it behaves similarly in the AM broadcast band, or lower. A different termination is needed for optimum 80-meter performance, which is above the cutoff frequency.

K9AY says: *"The shape of the loop affects the shape of the pattern. The delta/triangle shape for the K9AY loop was chosen for two reasons: 1) it only needs one support, and 2) it results in a null at about 30-40 degrees elevation, which is ideal for in-country QRM reduction. A rectangle will work, but the null will be at a lower angle. This would be good for knocking down neighborhood noises, but it will have less front-to-back on sky-wave arriving signals. Something close to a square is OK, but a rectangle that wider than its height will have its null at a very low angle."*

As with the other type of elongated terminated loops, the K9AY loop antenna can also be made smaller or larger by scaling it. The biggest tradeoff with small loops is in the received signal levels. Smaller antennas, though more broadband, will require more preamplifier gain. Larger antennas capture more signal, but are more difficult to build because they require more space, and they may not work on 7 MHz.

It is easiest to use an insulated mast for the loop. Make sure you use a good ground; for example, a 1.5-meter long ground rod. The ground rod is the antenna ground and should preferably not be used for grounding the shield of the coax, especially if the coax is not buried or placed on the ground, as I've recommended previously.

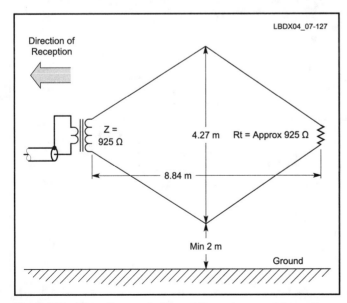

Fig 7-127—The diamond-shaped loop does not have to be exactly equilateral. The exact shape can be adjusted by the hardware you have available.

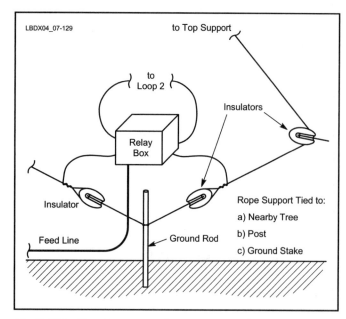

Fig 7-129—Two loops can be suspended from one mast, which makes it possible to switch four directions. The switch box can be mounted at the base of the mast.

You can successfully use a metal mast that is insulated from both the ground and from the antenna wires. The termination resistor varies from 500 to 600 Ω, and the feed-point impedance is around 500 Ω.

The K9AY loop can be adjusted just like any elongated loop, by varying the termination resistance for best F/B ratio. Since the highest rejection is at relatively high elevation angle, it is not possible to find a sharp null when testing on ground wave at an almost 0° elevation angle. On the other hand skywave signals are unstable in nature, and not very suitable for a nulling exercise at higher angles. But even if you do a null at groundwave angles by pruning the resistor value for maximum F/B (maybe only 10 to 15 dB), this will still result in a 35-dB notch in the same direction at higher elevation angles (40° to 50°).

The feed-point impedance is around 500 Ω, so a 9:1 transformer is indicated. I recommend a transformer with separate antenna and coax grounds using independent windings. You should run the coax under ground and do not ground the coax near the antenna.

The K9AY has a few advantages over the other loops that are above ground:

- Being grounded at the feed point, the problems with a balanced feed point don't exist.
- The feed line can be buried in the ground, which effectively keeps common-mode currents off the shield.
- The antenna is small and can easily be switched in four directions. See **Figs 7-129** to **7-131**.
- Because the "ground image" is part of the antenna, the signal levels with the K9AY loop are greater than with a Flag, Pennant or other ground-independent type with the same enclosed area.

There are commercially-made versions available. See **Fig-132**.

- Array Solutions, WXØB at **www.arraysolutions.com/**.
- Wellbrook Communication, UK at **www.wellbrook.uk.com/K9AY.html**.

3.8. Feeding Elongated Receiving Loops
3.8.1. Impedances

Optimized loops have a feed-point impedance that is essentially resistive over the design frequency range. The K9AY and the EWE have a feed-point impedance around 500 Ω. All the other loops (Flag, Pennant, Diamond, Delta) show about 950 Ω.

3.8.2. Symmetric—Asymmetric

In theory, the requirements for a EWE or K9AY transformer (like that for a Beverage) seem to be less onerous than for a Pennant or Flag, because the antenna operates in an unbalanced mode. If this is truly the case, there would be no other modes of operation to suppress, where the antenna can receive signals from all directions, filling in desired pattern nulls.

That is the ideal world, using an ideal ground. But its pretty likely that these antennas have less than perfect grounds. A few 1.5-meter long rods at the feed point of the antenna do not present a 0-Ω ground impedance, and hence the antenna is less than perfectly "unbalanced." As such, it is susceptible to common-mode ingress onto the outside of the feed line or onto any other conductor attached directly or indirectly to the ground system. That is why you should always use a transformer with separate primary and secondary windings, even

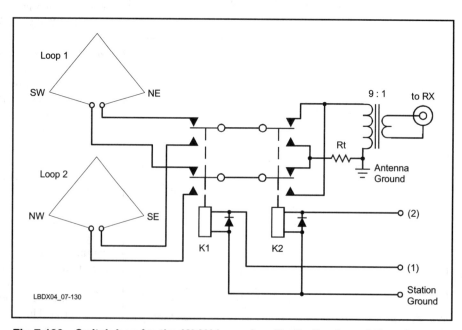

Fig 7-130—Switch box for the K9AY loop. A split-winding transformer has replaced the 9:1 transmission-line type transformer used in the original *QST* article.

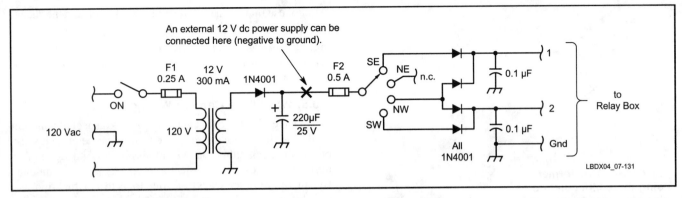

Fig 7-131—Direction-switching unit.

Fig 7-132—K9AY makes a commercial version of his K9AY loop.

with antennas that are nominally unbalanced.

With Flag/Pennant/Delta/Diamond antennas, which are fully balanced systems, preserving that balance is essential. For instance, do not use a metal box to house the transformer since it adds to the capacitive coupling from the loop to the feed line shield. A metal box may create imbalance unless very carefully positioned. The usual operating impedance of such loops is fairly high, and any added unbalanced capacitance can upset the voltage division between the loop terminals. This allows the loop to respond more to common-mode signals, where it acts like a short "longwire" antenna. Plastic boxes are cheaper and better in this case.

Although we consider the loop as fully balanced by itself, this really is not the case because one side of the antenna is physically closer to ground. Even a perfectly balanced feed transformer (in terms of capacitive balance) is not always a guarantee for a perfectly balanced *system*, because the antenna itself is not perfectly balanced.

This is also the reason why such loops should be well in the clear and everything around them should be as symmetrical as possible. The feed line should run away from the feed point along one of the axes of symmetry of the loop, preferably for a fairly long distance. This will preserve the symmetry, but on the other end it does create a long run of "in-the-air" coax, which again is like a nice "longwire" antenna. In this case the solution is to add a few common-mode chokes along the feed line so that the longwire effect is broken up.

We should indeed do everything we can to keep common-mode currents off the outside of the coax. The cable can radiate unwanted signals directly through the air and into the antenna itself, even if you ensure low capacitance between primary and secondary of your feed transformer. In other words, coupling can occur from induction and radiation fields even if perfect transformer isolation is obtained. As W8JI says: "*Just because we call it a feed line, does not mean it doesn't act like an antenna, one plate of a capacitor, or an inductor coupling through space to the antenna.*"

Once the cable drops down to the ground level, a series of high-μ cores on the coax, together with a good ground rod should keep common-mode currents on the remaining length of the feed line low. Once on—or preferably under—the ground, the feed line will carry very little common-mode signal.

3.8.3. The Transformer

The essential characteristics of a good transformer for a loop antenna are:

- Lowest possible capacitance coupling between primary and secondary windings.
- Low loss because signals are very low-level.
- Good SWR is *essential* if you want to phase such antennas (you want the phase angle to be the determined by the cable length!)

Let's have a look at doughnut-shaped toroidal cores and binocular cores. While binocular cores are an excellent choice for Beverages and EWEs, large-sized toroidal cores can achieve substantially less inter-winding capacitance, but with a sacrifice of a few tenths of a dB in insertion loss.

It is not a bad idea to use a large core, since it will give you less insertion loss vs inter-winding capacitance. Therefore a FT140-size core is not a luxury for a receiving antenna. K6SE recommends using the FT114-77 core and for 50-Ω coax suggests using four turns for the low-Z winding and 17 turns for the high-Z winding.

W8JI recommends the same type-77 mix ($\mu_i = 2000$) core. A FT82-77 toroid (OD is 0.825 inches and about 0.25 inches thick) requires five turns for a 50 or 75 Ω winding.

You can use a stack of three Fair-Rite binocular cores (2873000202) glued together with instant glue after you insert an insulating tube inside. Use #24 wire for the secondary and the same wire inside an insulating sleeve (to minimize capacitive coupling) for the primary. See **Fig 7-133**.

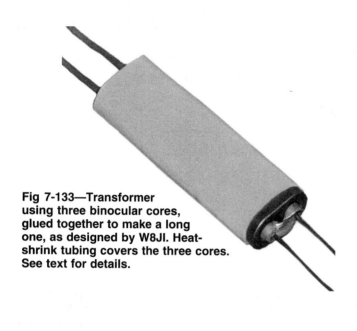

Fig 7-133—Transformer using three binocular cores, glued together to make a long one, as designed by W8JI. Heat-shrink tubing covers the three cores. See text for details.

Table 7-44
Winding Data for Binocular Transformer

Transformation	Low-Z	High-Z
500 Ω to 75 Ω	2 passes (1 turn)	5 passes
500 Ω to 50 Ω	2 passes (1 turn)	6 passes
950 Ω to 75 Ω	2 passes (1 turn)	7 passes
950 Ω to 50 Ω	2 passes (1 turn)	9 passes

Table 7-44 gives the proper winding data. Remember that going in one direction through a (stacked) hole is one pass, and two passes = one turn.

Some transformers are commercially available: the TR9X and TR16X (K9AY design available from Array Solutions, WXØB at **www.arraysolutions.com/**) have separate primary and secondary windings but are tightly coupled and not designed for minimum inter-winding capacitance. The TRX-16 is designed for much lower inter-winding capacitance than the TRX-9, but not so low as to compromise loss, which is less than 0.5 dB. The two models use different core sizes and different winding methods. The TRX-16 has 40 to 45 dB isolation between primary and secondary at 160 meters, as measured in a 50-Ω system. This is equivalent to a coupling capacitance between primary and secondary of about 12 to 15 pF.

3.8.4. Summing Up, Feeding Elongated Receiving Loops

- Make sure your loop is as physically balanced as possible.
- It's better to have the loop 5 meters high than 2 meters.
- Use a transformer with separate primary and secondary windings.
- Use a transformer with the lowest possible capacitive coupling between primary and secondary windings.
- Do not use a metal box to house the transformer.
- Run the feed line along the symmetry axis of the loop for several meters.
- Drop the coax down to ground, where you ground the shield to a good ground system. Insert a common-mode choke at that point (between the ground rod and the shack).

Finally, Tom, W8JI's advice is worth seriously considering: "*While you won't always see a difference with all these precautions, an ounce of prevention is worth it with low noise antennas.*"

3.8.5. What Else Can Go Wrong?

I have tried to cover every imaginable aspect of common-mode coupling from the feed line to a loop antenna. This kind of coupling can either make it impossible to find a good deep null off the back, or can inject a lot of trash into the antenna. But there are other ways for your loop to not function as it should.

If you cannot get a decent null, and you are not sure whether or not it's your feed line causing the problem, eliminate your feed line. Use a small battery powered receiver and a step-ladder (wooden!) and connect it to the loop with only a very short piece of coax. If you still cannot get a decent null, look for other conductors—most likely other antennas—coupling directly into your loop.

3.9. The Termination Resistance

Some users of receiving loops have found it useful to be able to adjust the termination resistor value from the shack. Some use a Vactrol VTL5C4 (75 Ω to 1.2 kΩ) or VTL5C2 (200 Ω to 5.5 kΩ), which is an opto-coupled variable resistance, where the resistance value is a function of the applied dc voltage. WA1ION covers this in detail on his web site: (**www.qsl.net/wa1ion/bev/bev_remote_term.htm**).

A word of caution: Using such a remote termination means more chances for common-mode problems! The routing of the control cable supplying the dc voltage—and its common-mode decoupling—is once more a critical issue (see Section 3.7).

3.10. Decoupling the Transmit Antenna

Resonant transmit antennas in the vicinity of the loops will make them worthless. You either need to decouple the transmit antenna (move its resonant frequency during receive) or move the receive loop at least λ/4 from the transmit antenna. And in any of these cases it's a good idea to erect a receiving loop by aiming it directly away from the vertical.

If you cannot separate your receiving antenna that far from the resonant transmit antenna, you will have to "decouple" it. This is not only true for receiving loops but for all types of receiving antennas. The lower the output of such antennas the more they will be sensitive to coupling. It has been my experience that Beverages should also be kept at least λ/4 away from a tall resonant transmit antenna.

The following information on tower detuning is copied, with permission, from W8JI's website. See **Fig 7-134**: "*We can minimize re-radiation by making an area or areas of the structure "electrically vanish." We often call this "de-tuning," even though it is more correctly electrical trapping or sectionalizing of a structure.*

Most structures or towers, when detuned, have a section adjusted to represent a parallel tuned circuit. Section A and B carry out-of-phase currents. Picture the current flowing upwards in A. It must then flow downwards in B. Since it is a closed loop, these out-of-phase currents are equal and flow in opposite directions at resonance. The result is radiation from

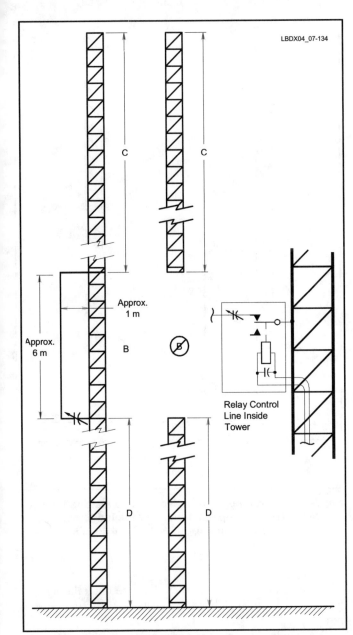

Fig 7-134—The decoupling effect of inserting a parallel-tuned circuit in a tower. Turning Section B of the tower into a loop tuned to the offending frequency makes that Section vanish altogether. See text for details.

sections A and B cancel each other. When section A and B are exactly resonant, sections D and C are isolated by a high impedance. The high impedance is caused by or related to the high current though the capacitor and the inductance of section A. When current is maximum, voltage drop is maximum.

This results in the electrical structure on the right, with section A and B removed. In effect, we have created a trap much like the trap in a dipole. As in the trap dipole, current is maximum in the trap at the trap's resonant frequency.

The condition of proper tuning occurs with maximum current in section B, NOT minimum current!! To electrically sectionalize the tower and isolate C and D (and minimize radiation from A) section B must be tuned for maximum current!

As either section C or D approach resonance by themselves, the tuning condition will change. This would occur when D is grounded and near $1/4\ \lambda$ or an odd multiple of $1/4\ \lambda$ long, or when C (with whatever is mounted on it) is self-resonant with section A removed!!

Under this condition, you would either need to sectionalize and detune C or D with additional detuning, or move the location of sections A and B to a new point that (when isolated) prevents resonance in C and/or D.

A few general rules apply. Pay attention to these guidelines to ensure best results:

- Never parallel-tune a large area. Certainly not an area over 3/16th wavelength long.
- The detuning "loop" must have a good solid connection to the structure being detuned. Don't connect the detuning wire out to a separate object or earth stake.
- We want to adjust for MAXIMUM current in section B, the exception being when that would cause resonance in C or D.
- We cannot have any electrically large structures or wires hanging from the tower in the area being detuned.
- Ideally any cables passing the detuning area should be grounded to the tower at the top and bottom of the detuning area, or pass through that area in the center of the tower or mast. At the very least, cable shields should be bonded to the tower at the top and bottom of the cable run and unshielded cables placed inside the tower.
- Tuning is fairly narrow. ~5% total BW is about all that can be expected in most cases, but this varies greatly with the system including distances to the other affected antennas and the amount of pattern distortion tolerated.
- I'm surprised cables are often not grounded at the top and bottom of tall towers, and that unshielded control cables are not passed through the inside of towers. Cables should always be treated that way for lightning protection if for no other reason!

Capacitor Size

The amount of capacitance and the voltage rating of the capacitor is not easy to predict. The size depends on unwanted power levels that excite the detuned structure, the electrical characteristics of the detuned structure, and the Q of the detuning section. Capacitance values will be fairly high with short sections on lower bands like 160, almost certainly in the range of a few thousand pF for eg, 6 m long sections. The exact value would depend heavily on dimensions of the A to B loop.

Voltages across the capacitor are generally not high, although they can be at times. The "loop Q" of A and B affects voltage, as does the amount of excitation and load presented by the impedances of C and D.

MFJ sells a clamp-on calibrated current meter that will not perturb the system. It is a cheap version of a current meter I designed. This is a calibrated meter with internal amplifier that measures current from a few mA to 3 amperes, not the non-calibrated RF-sniffer commonly sold. Some RF-sniffers, including those by MFJ, actually change the impedance and resonant frequency of the system because the pick-up transformers are not properly designed and terminated current transformers. Avoid loop-stick type current meters, since they measure ANY external field and can provide misleading results. Use a current meter that is directly inserted in line B, or clamps around line B with the closed core of a terminated current transformer. Use a meter that does NOT perturb the *system when removed!*

Lacking a current meter, it is possible to tune this system with a grid dip meter, by forming a small one or two turn coupling loop. As an alternative, the loop can be broken at any point near the capacitor and a MFJ-259 or similar antenna analyzer connected in series. Proper adjustment is at the point were minimum impedance occurs. If that impedance is not low, you probably are not effectively detuning the structure.

Multiple Stacked Antennas or Tall Structures

When multiple stacked antennas are used, especially on a fairly tall tower, it may be necessary to sectionalize multiple points. Individual sections between antennas can be resonant, or appear electrically long.

If the tower or structure or any part of the structure or tower becomes resonant when section A is tuned to present a high impedance, then we need to move section A or tune it to some condition other than maximum current (resonance). Adjustments under this condition can only be made two ways:

- *A sampling loop can be mounted on the structure $1/10\ \lambda$ or more above or below section B and adjusted for minimum terminal voltage*
- *Field strength of the pattern can be plotted, and the structure tuned for minimum pattern distortion*

Never detune an area that contains large Yagis or other electrically large objects, like long conductive guy lines, dipoles, or cables leaving the tower."

3.10.1. Switching the Decoupling Section During Transmit

If you are decoupling your transmitting antenna, you will have to open-circuit the decoupling loop during transmit. If you use high power a small vacuum relay is a good idea. Do not use a bulky slow relay, as you will need it to switch quickly, in pace with your amplifier keying line. Run the coil voltage through a cable on the inside of the tower and decouple it well at the base of the antenna. I use a ferrite rod with 20 turns of the small control cable on the rod. I added some good quality 0.01 µF decoupling caps to ground too. Again, see Fig 7-134.

The relay should be switched from the line that switches your amplifier—that means that the relay energizing voltage should come on approx 10 to 15 mS (depending how fast your relay is) before RF appears. The same rules that apply for switching amplifiers apply. Using this system full break QSK is not advised. If you do want to use QSK, you will probably have to switch off the trap system. See also Fig 3-12 in Chapter 3.

3.10.2. Gamma-Matched Tower

If you have a gamma matched tower you may consider just turning the section where the gamma-match systems is installed into a resonant loop, this way decoupling everything that's above the gamma attach point from ground. This way you only have to make a little switchbox located at the box containing the gamma series capacitor (see **Fig 7-135**).

If the tower is about 90° long the gamma-match attachment point will be about 8 meters high. Electrically longer and shorter gamma-matched towers will require a longer gamma match (see Chapter 9). Use a small and fast vacuum relay to switch the gamma rod to the feed line or the loop. The return

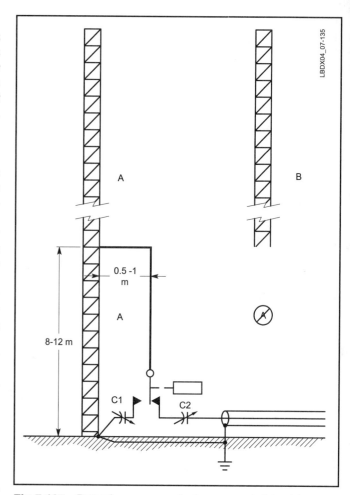

Fig 7-135—Detuning a grounded gamma-matched tower. C1 resonates the loop; C2 tunes out the inductive reactance of the gamma rod. See text for details.

of the loop at ground level should *not* be done through the ground. A heavy conductor or metal (aluminum) pipe is required.

At my location it was necessary to detune my 160-meter quarter-wave vertical to minimize coupling to my small receiving antenna, which is located less than λ/4 from the transmit antenna. I followed the tuning procedure described by W8JI above using an MFJ antenna analyzer. I tuned the loop with a 4-gang BC variable (2000 pF), which I later replaced with paralleled ceramic transmit-type capacitors.

3.10.3. Detuning Base-Insulated Towers

I described methods for detuning insulated towers in Section 2.9.1.

3.11. Arrays of Loops

Elongated terminated non-resonant loops make excellent candidates for wideband arrays, with elements either fed in-phase or in an end-fire configuration, using the crossfire phasing method. These are broadband, high-loss antennas. But the effects of mutual coupling are hardly visible on their feed-point impedances, similar to arrays of Beverages. Receiving loops can be used as elements in any of the array configurations described for verticals or Beverages.

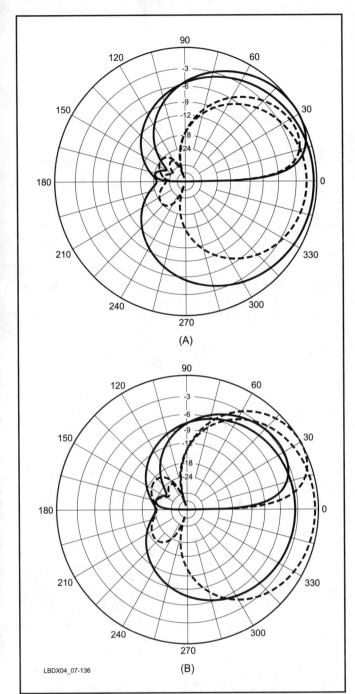

Fig 7-136—End-fire phased Flags, 5 meters spacing with 20-meter separation. At A, single Flag on 160 meters (solid line); phased Flags (ϕ = 160°, dashed line). At B, single Flag on 80 meters (solid line); phased Flags (ϕ = 140°, dashed line).

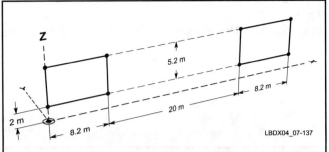

Fig 7-137—Layout of the 2-element end-fire Flag array.

Putting two Flags behind one another and feeding them end-fire with the appropriate phasing angle (see Table 7-1) can do wonders for directivity patterns (see **Figs 7-136** and **7-137**). These Flags were fed with ϕ = 160° on 160 and ϕ = 140° on 80 meters. **Table 7-45** shows directivity figures of phased loop arrays compared to single loops. Note that the DMF jumps up from about 10 dB to roughly 25 dB, quite a spectacular change. There is not much gain in RDF, since most of the improvements are in the back of the array. Note the substantial reduction in −3-dB beamwidth as well as the reduction in high-angle radiation because of the close spacing and the high ϕ angle. This high ϕ angle, of course, results in a relative loss of the array vs a single element, instead of what you might expect in terms of gain for an array. If you phase the system for gain, you would use a much smaller ϕ-angle, but would reap very little directivity improvement. Again: there is no free lunch!

This array can be fed using the crossfire principle as shown in Fig 7-114 for two Beverages. Here too you should take care that the SWR on the phasing lines is kept to less than 1.1:1. This can easily be achieved by adjusting the turns-ratio of the matching transformer.

F. Koontz, WA2WVL, described arrays of EWEs in one of his *QST* articles (Ref 1264). Two EWEs in broadside, spaced approximately λ/2, give the lobe-narrowing effect also seen with arrays using verticals and Beverages. There is no improvement in either back-lobe or high-angle behavior. We know from Section 1.11 that ~0.67-λ spacing results in much better directivity (RDF), since it lifts the elevation pattern off the ground.

The ultimate on-paper configuration is the four-element EWE array, being a combination of two side-by-side (in-phase) end-fire cells in an in-phase broadside array (similar to the array with vertical elements in Section 1.12 and the array of Beverages in Section 2.16.4). In this case the improvement from the end-fire cell is combined with the narrowing effect of

Table 7-45
Directivity Figures of Single and End-Fire Phased Flags

	Single Loop 160	Phased Loops 160	Single Loop 80	Phased Loops 80
Gain (dBi)	−29.0	−30.4	−18.5	−16.0
DMF (dB)	11.4	21.6	10.7	19.3
RDF (dB)	7.4	10.0	7.0	9.1
−3-dB Angle	147.0°	98.0°	156.0°	113.0°

Table 7-46
End-Fire and Broadside Combinations of Receiving Loops

	2 End-Fire Loops	2 Broadside Loops	4 End-Fire/ Broadside Loops	4 End-Fire/ Broadside Verticals Sec. 1.12	4 End-Fed/ Broadside Beverages Sec. 2.16.4
DMF (dB)	21.5	16.4	26.6	19.5	34.7
RDF (dB)	9.9	10.4	12.3	12.7	14.1
−3-dB Angle	98.0°	55.0°	50.0°	46.0°	34.0°

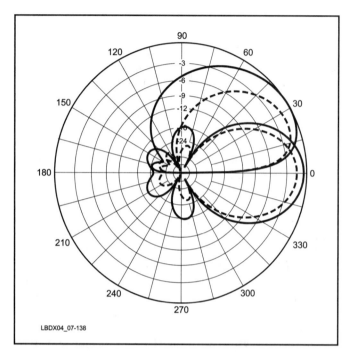

Fig 7-138—Broadside array of two Flags, separated 0.61 λ (solid lines) compared to broadside array (0.61 λ lateral spacing) of two flags cells, each cell containing two flags (dashed lines).

the broadside combination, yielding very good directivity figures (see **Figure 7-138** and **Table 7-46**).

Of course, once you start envisioning broadside arrays, you need a lot of room (100 meters spacing or more). We already know the disadvantages of a receiving antenna with a very narrow −3-dB forward beam angle: Where should you aim it? We should not forget that on the low bands signals often deviate 30° to 45° (and even more sometimes) from the theoretical great-circle direction. Of course, you can put up 12 such arrays. Did I start out saying that loops are receiving antennas suitable for being used on small properties? I think these giant arrays are more for modeling fun than anything else!

4. RANKING RECEIVING ANTENNAS

Elongated terminated loops (the EWE, Flag, K9AY, etc) are right at the bottom of the performance list, showing the least degree of directivity. But they are also so much better than having to listen on your vertical transmitting antenna. Ask anyone who has such a loop and who's never had a "big" receiving antenna. The merit of such receiving loops is a small footprint and their relative simplicity. We do, however, have

to take care of common-mode signal ingress into the feed system. Use the correct transformer, keep everything balanced, route the feed line correctly and decouple/ground the feed line to make it "dead" on the outside.

The best thing about Beverages is their simplicity. Even if they are not well engineered, and not installed or tested with care, they always seem to work "pretty well." Even the simplest forms of Snake antennas seem to work. But careful engineering and proper terminating can improve directivity quite substantially. Proper feeding and the use of good grounds and careful attention to common-mode problems are certainly areas where many existing Beverages can be improved.

Beverages are much simpler than arrays of verticals to engineer and to build, and the larger Beverage arrays outperform large vertical arrays by a significant margin. To be honest, I must admit that the end-fire broadside Beverage array takes 1.5 to 2 times as much room as the 8-Circle vertical array. Engineering, building and tuning a vertical array for receiving is a nice challenge for a technically inclined ham. Excellent directivity figures can be obtained on a relatively small footprint. A 2-element end-fire array may not take up much more area than a Flag but it will give you substantially better directivity.

Let's have a look at some directivity figures of vertical arrays and Beverages. All figures are for 1.83 MHz (modeled over "average" ground). From **Table 7-47**, which also lists the −3-dB forward angle, you can clearly see that in addition to the rejection off the back, the RDF figure also takes into account the narrowness of the forward lobe. Look at the 6-HEX and the 8-Circle, which both have an identical score for DMF (the directivity in the back), but where the 8-Circle scores a substantially higher RDF figure because of its narrow forward lobe (55° for the 8-Circle versus 78° for the 6-HEX).

5. SUMMING IT UP ON SPECIAL RECEIVING ANTENNAS

When I analyzed the results of the poll I ran among some 300 leading low-band DXers, I was amazed to see some of the top-scorers did *not* use Beverages or other special receiving arrays. After studying the data further, I found that those were the guys using directive antennas—arrays and/or Yagis on 80 meters.

Happy are the very few who categorically say: "I don't need a special receiving antenna. My transmit antenna is better than Beverages, even phased ones." I only know a few stations that can say that: K9DX on 160 and K4JA on 80 meters, both using a 9-Circle antenna (see Chapter 11).

I have received so many requests to write about small receiving antennas that would out-hear W8JI's Beverage

Table 7-47
Arrays of Short Verticals

	DMF	RDF	3-dB Angle	Gain dBi	Ref
2 Ele End-Fire λ/4 Spacing, φ = 135°	16.1	9.3	132	−12	Sect 1.8
2 Ele End-Fire λ/8 Spacing, φ = 155°	17.7	9.8	120	−16	Sect 1.14
2 Ele End-Fire λ/16 Spacing, φ = 165°	17.9	9.8	121	−21	Sect 1.14
4-Square Side λ/4, φ = 120°	24.0	11.6	86	−8	Sect 1.23
4-Square Side λ/8, φ = 140°	25.9	11.9	86	−15	Sect 1.23
4-Square Side λ/16, φ = 160°	27.6	12.0	85	−23	Sect1.23
6-Ele Stone-Hex 305-m Diameter	27.4	11.7	79	−19	Sect 1.28
8-Circle, φ = 120°, Diameter 0.594 λ	22.6	12.3	55	−8	Sect 1.29
Broadside 2 Ele Bidirectional	(#)	9.7	47	−10.5	Sect 1.11/1.2.2.1
Broadside 4 Ele Bidirectional	(#)	12.4	25	−7.9	Sect 1.11/1.22.2
Broadside/End-Fire 4 Ele	21.3	12.7	46	−4.7	Sect 1.12/1.22.3
Broadside/End-Fire 8 Ele	27.1	15.7	24.5	+0.7	Sect 1.12/1.22.4

Beverages and Arrays of Beverages

	DMF	RDF	3-dB Angle	Gain dBi	Ref
80-m Long Single Beverage	11.1	7.3	90	−16	Sect 2.16.2
160-m Long Single Beverage	19.0	10.2	78	−10	Sect 2.16.2
300-m Long Single Beverage	21.3	12.9	62	−5	−
Broadside 80-m Beverages, 90-m Spacing	14.4	9.6	48	−13.3	Sect 2.1.6.2
Broadside 160-m Beverages, 90-m Spacing	21.3	11.9	48	−7	Sect 2.1.6.2
Broadside 300-m Beverages, 90-m Spacing	23.1	14.2	44	−2	Sect 2.1.6.2
80-m Long End-Fire Beverages, Stagger = 30 m, φ = 140°	20.0	9.7	77	−15.5	−
160-m Long End-Fire Beverages, Stagger = 30 m, φ = 140°	30.1	11.6	69	−9	Sect 2.16.3
300-m Long End-Fire Beverages, Stagger = 30 m, φ = 140°	33.8	13.9	57	−4	Sect 2.16.3
160-m Beverages in End-Fire/Broadside Array (*1)	34.0	13.0	46	−6.4	Sect 2.16.4
160-m Beverages in End-Fire/Broadside Array (*2)	34.7	14.1	34	−6.4	Sect 2.16.4

Loops and Arrays of Loops

	DMF	RDF	3-dB Angle	Gain dBi	Ref
Elongated Terminated Loop (EWE, Flag, K9AY etc)	~11	7.5	~145	−29	Sect. 3
2 End-Fire Loops	21.5	9.9	89	−30	Sect 3.11

arrays or K9DX's 9-Circle. We should all know there is no free lunch in this cruel radio world! You now know the directivity figures (DMF and RDF) of the elongated loop antennas (Flags and such). That's it. You know you can do better with a small Four-Square array, but it takes a *lot* more engineering and building effort than to put up a Flag. Again, there is no free lunch. Even a 2-element end-fire array with vertical elements is substantially better than the elongated loops, because they have no high-angle radiation, while the loops have high-angle radiation from the horizontal or sloping wires. Clearly this is an incentive to build something better than such receiving loops. A 2-element end-fire pair does not have to be big. Six-meter tall elements (top loaded) and a spacing of 10-15 meters can set you on the road.

And, of course, you never can have too many good receiving antennas. Tom, W8JI, and Wally, W8LRL, are the proverbial proofs of the pudding. K4ISV's statement *"Can you imagine a fisherman going out with only one bait?"* makes a lot of sense. But I would immediately like to add that if you choose to have a bunch of antennas, do make sure they do not couple with one another or you will have a totally uncontrolled condition where anything can happen.

In Chapter 1 on propagation I mentioned a fairly typical phenomenon of high-angle propagation before sunset or after sunrise. Under such circumstances a low dipole will outperform a long Beverage because of its angle of radiation. A low dipole is definitely also useful in your gallery of receiving weapons. As Frank Donovan, W3LPL, has said: *"You can never have too many antennas."*

6. PREAMPLIFIERS

All special receiving antennas described in this book are low-gain antennas. Antennas with a nominal gain of as low as −15 dBi will normally not require a preamplifier, unless you live in a very quiet rural area and have very long, lossy feed lines. That means that Beverages, even with feed lines of many hundreds of meters long, will—as a rule—not require a preamplifier. Unless of course you like to see you're S-meter dancing up and down like a yoyo. But signal readability has nothing to see with dancing S-meter needles. It is only a question of signal-to-noise ratio.

But we have also seen receiving antennas such as loops with gains of −30 dBi, and that is pretty low and does require some signal boosting. How do you know if you need a preamplifier? The rule is simple. During the quietest moment of the day (usually at noon on 160 meters), with your receiver set at the narrowest bandwidth you normally use, can you hear the noise go up significantly when you switch from a dummy load to your receiving antenna? If you can, then you have enough gain in your system.

If you hear the receiver's internal noise and not band noise, then you need a preamplifier. You have a problem because:

- The antenna output might be extremely low (less than −15 dBi, for example).
- The feed line might be very long and/or lossy.
- You need to compensate for filter losses, splitter losses etc.

As a rule you should keep the signal level as low as possible to prevent chances of intermodulation and overload. This is also why our receivers have a front-end-attenuator, as well as a switchable preamplifier.

In most cases you can put the preamplifier in the shack. The signal loss in the feed line is a loss that affects both the signal and external noise. That means that the loss in the feed line does not affect the S/N ratio. In most circumstances moving the preamplifier to the antenna will not affect the system unless the feed line loss is so high that the noise floor is indeed established by the preamplifier and not the antenna.

All my Beverages are fed with ⅝-inch Hardline so the line losses are totally negligible, even though some of my feed lines are more than 300 meters long. My small receiving Four Square is very close to the house, so the coaxial losses using RG-6 are very low. I have use both a K9AY duo-band 80 and 160-meter preamplifier (model PRE1) and the DX Engineering push-pull preamplifier, both of which I can highly recommend. On 160, I can add another 10 dB of gain by using the very selective (band-pass filtered) preamplifier by Karl Braun. With all this gain I can hear band noise even during the quietest moments.

In the shack I can use the Braun SWF 1-40 (multiband) preselector preamplifier, which can give me up to 8 dB gain and a lot of selectivity, if required. A number of preamplifiers have been described in the literature (Ref 257, 267, 1232, 1251, 1254 and 1256).

Untuned preamplifiers using MMICs are not the right choice. MMICs have poor IMD performance and are also very poor with even-order harmonic distortion. They have a fixed gain, which is usually way too high anyhow. Last but not least, they withstand little abuse.

Larry, W7IUV, has published a 160-meter preamplifier design (**Fig 7-139**) with very good characteristics. With his permission I quote from his web site (**www.qsl.net/w7iuv/**):

"Here is a schematic of an amplifier that I have used for a number of years. It's based on a CATV transistor and gives pretty good IMD performance and a gain of about 20 dB. This circuit provides good performance from 100 KHz to at least 30 MHz.

Through the years I have modified the basic schematic for specialized applications. Recently I made a change to the bias and increased the idle current to around 75 mA. This causes the transistor to run a bit warmer, but also increased the third order intercept point. The "new" design measures 18 dB gain and +20 dBm power output for 1 dB compression. I was also able to run a third order intercept for this circuit and the OIP3 is reasonably good at +39 dBm. While the "old" circuit was good, I could just detect AM broadcast IMD products on a quiet band. With the "new" circuit, I can detect no IMD products any place on Topband when using any RX antenna I have. Remember the transistor will dissipate a little more power and really needs a small heat sink.

The 2N5109 transistor is the device recommended for this application. However, the 2N3866 is useable if the 2N5109 is not available. I also recommend that you shield the pre-amp very well and provide adequate de-coupling on the power supply leads. Noise from switching power supplies in local computers, as well as AM broadcast signals, can

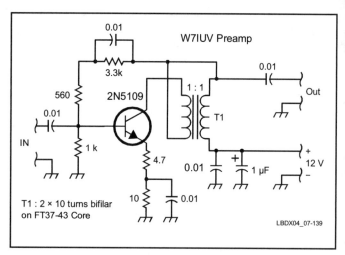

Fig 7-139—Schematic of the W7IUV 160-meter preamp, which has excellent IMD performance.

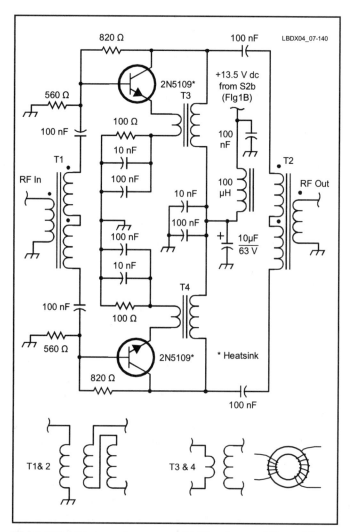

Fig 7-140—High-performance 12dB gain amplifier with excellent strong-signal handling characteristics, originally developed by Ulrich Rohde, KA2WEU, and modified by Sergio Cartoceti, IK4AUY. T1 and T2 are each 12 trifilar (twisted) turns of 0.35-mm (AWG #27) enameled wire on an FT-50-43 ferrite toroid core. T3 and T4 are each a 9-turn primary and a 2-turn secondary (0.35-mm AWG #27 enameled wire) on an FT37-43 core.

cause problems in overall performance if they get into the pre-amp power supply.

I have had a number of requests for vendors who could supply the 2N5109. Since I made my buy years ago, the vendor I used no longer stocks that part. However, Mouser Electronics (www.mouser.com) does carry it now."

Sergio, IK4AUY, described an excellent preamplifier as part of his article "A High Level Accessory Front End for the HF Amateur Bands" in Apr/March 2003 *QEX* (Ref 1267). The schematic of his preamplifier is shown in **Fig 7-140**. The performance data are just short of spectacular:

- 3rd order intercept (OIP3): +43 dBm
- 1 dB compression: +24.5 dBm
- Gain: 12 dB (1 to 30 MHz)
- Noise figure: 3.9 dB at 30 MHz

T1 and T2 are each 12 trifilar (twisted) turns of 0.35 mm (AWG#27) enameled wire on a FT50-43 ferrite toroidal core. T3 and T4 are nine turns primary and two turns secondary (same wire), wound on a smaller FT37-43 type of core. If a higher gain than 12 dB is required, you can easily connect two such preamplifiers in cascade.

Each transistor is biased for around 60 mA, so the total current drain is 120 mA at 13.5 V dc. All details on this preamplifier were graciously provided by Sergio Cartoceti, IK4AUY. See **Fig 7-141** and also **www.qsl.net/ik4auy/**.

Fig 7-141—The IK4AUY preamp, modified from the design by Ulrich Rohde.

Many commercial-grade preamplifiers are available, some better than others. The ones listed below have good to excellent characteristics. Some popular models behave very badly in the presence of strong signals.

K9AJ's excellent preamplifiers (model PRE1 covers 160 and 80, model PRE2 covers 160 through 40) as well as the ICE single-band preamplifiers) are available from Array Solutions-WXØB (**www.arraysolutions.com/**).

Advanced Receiver Research (**www.advancedreceiver.com/index1.html**) sells a 20-dB preamplifier (model PO.5-30/20VD) covering 0.5 to 30 MHz with a noise figure specified at 2.5 dB and excellent strong-signal-handling capability. Some measured performance data (source W8JI):

- Gain compression: 1 dB compression at 21.5 dBm output (spec = 22 dBm)
- 3rd order intercept point (OIP3) = + 30dBm
- Noise figure: 5.4 dB (spec = 2.5 dB)

Braun (Karl Braun, Deichlerstrasse 13, D-90489, Nuerenberg, Germany) makes a whole range of band-filtered mono-band preamplifiers (see Fig 3-14). The 160-meter model has better than 40 dB rejection at 1,5 MHz and at 2.4 MHz (passband is 1810 to 1910 kHz).

DX Engineering push-pull amplifier model RPA1: This new broadband amplifier, designed by W8JI and shown in **Fig 7-142**, has outstanding characteristics:

- Gain: 17 dB
- 1 dB compression: at +26 dBm (400 milliwatts output)
- 3rd order intercept point (OIP3): +43 dB
- Noise figure: 3.4 dB

For more details: **www.dxengineering.com/**. Tom, W8JI, has published measurement data on his web site: **www.w8ji.com/pre-amplifiers.htm**.

Fig 7-142—The DX Engineering push-pull amplifier designed by W8JI.

6.1. How to Assess the Relative Quality of Preamplifiers

W7ZOI introduced a figure of merit for receiver systems, which he called the RECEIVER FACTOR with bandwidth invariant parameters:

RF = IIP3 – NF

where
IIP3 is 3rd Order Intercept Point referenced to the input in dBm, NF is Noise Figure in dB

These are cumulative values for a system. The first is a measure of strong-signal performance, while the other defines weak signal behavior.

If the input intercept of an amplifier is known the intermodulation distortion is well defined for all input levels. We measure IMD at the output and if not otherwise specified IP3 for amplifiers is referenced to the output (OIP3).The difference is the simple stage gain:

IIP3 = OIP3 – G

where G= gain in dB

Anzac engineers called this many years ago the AMPLIFIER FACTOR (= IIP3 – NF) for a single-amplifier stage (the higher the number, the better, of course). If we apply this to the above-mentioned preamplifiers we get **Table 7-48**.

Table 7-48
Preamplifier Performance Characteristics

Preamplifier	OIP3 dBm	Gain dBm	IIP3 dB	NF dBm	AF = IIP3– NF dB
W7IUV	39	18	21	~5 (*)	~16 (*)
IK4AUY	43	12	31	3.9	27.1
Advanced Receiver R	30	20	10	5.4	4.6
DX Engineering (RPA1)	43	17	26	3.4	22.6
ICE 124A	14	16	−2	?	est. −7

(*) approximate value, not measured

CHAPTER 8

The Dipole

Klaus Owenier, DJ4AX, is an all-around amateur. Klaus teaches electronics and electromagnetics at the Ruhr-University Bochum. He was one of the first members of the world-famous Rhein-Ruhr DX Association (RRDXA), worldwide winner of the CQ-WW-DX-contest club competition for many years in a row. Klaus is an antenna expert and an excellent contester and CW operator. He has been a valuable and consistent presence during CQ Worldwide contests at OT*T for many years. Klaus volunteered to be my guide, counselor and helping hand. His critical analyses on dipole antennas have been very instrumental in the reworking of this chapter. Thank you, Klaus.

The first antenna most amateurs encounter is a dipole. I remember how, as a young boy, I put up my first 20-meter dipole between a second-floor window of our house and a nearby structure. It was fed with 75-Ω TV coax, and it worked—whatever that meant. For a while my whole antenna world was limited to a dipole. But there is more to dipoles.

Although we often think of dipoles as ½-λ long, center-fed antennas, this is not always the case. The definition used in this chapter is that of a center-fed radiator with a symmetrical sinusoidal standing-wave current distribution.

1. HORIZONTAL HALF-WAVE DIPOLE
1.1. Radiation Patterns of the Half-Wave Dipole in Free Space

The radiation pattern in the plane of the wire has the shape of a figure 8. The pattern in the plane perpendicular to the wire is a circle (see **Fig 8-1**). The three-dimensional representation of the radiation pattern is shown in the same figure and is a ring (torus). In free space the gain of this dipole over an isotropic radiator is 2.14 dB. This means that the dipole, at the tip of the ring where the radiation is maximum, has a gain of 2.14 dB compared to the theoretical isotropic antenna, which radiates equally well in all directions (its radiation pattern is a sphere).

1.2. The Half-Wave Dipole Over Ground

In any antenna system, the ground acts like an imperfect or lossy mirror that reflects energy. Assuming a perfect ground to simplify matters, we can apply the Fresnel reflection law, where the angles of incident and reflected rays are identical.

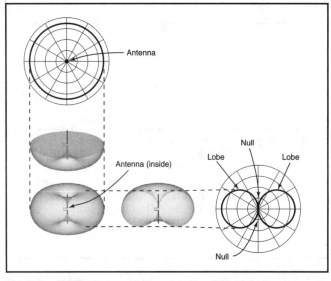

Fig 8-1—Radiation patterns as developed from the three-dimensional pattern of a half-wave dipole in free space. Upper left, vertical-plane pattern, and right, horizontal plane.

1.2.1. Vertical radiation pattern of the horizontal dipole

The vertical radiation pattern determines the *wave angle* of the antenna; the wave angle is the angle at which the radiation is maximum. Since obtaining a low angle of radiation is one of the main considerations when building low-band antennas, we will usually consider only the lowest lobe in case the antenna produces more than one vertical lobe. In free space, the radiation pattern of the isotropic antenna is a sphere. As a consequence, any plane pattern of the isotropic antenna in free space is a circle. In free space, the pattern of a dipole in a plane perpendicular to the antenna wire is also a circle. Therefore, if we analyze the vertical radiation pattern of the horizontal dipole over ground, its behavior is similar to an isotropic radiator over ground.

1.2.1.1 Ray analysis

Refer to **Fig 8-2**. In the vertical plane (perpendicular to the ground), an isotropic radiator radiates equal energy in all directions (by definition). Let us now examine a few typical rays. A and A′ radiate in opposite directions. A′ is reflected by the ground (A″) in the same direction as A. B″, the reflected ray of B′, is reflected in the same direction as B.

The important issue is the phase difference between A and A″, B and B″, etc. Phase difference is created by path-length difference (length is directly proportional to time, since the speed of propagation is constant), plus any phase shift at the reflection point itself. Horizontally polarized rays undergo a 180° phase shift when reflected from perfect ground. This can be simulated by feeding an image antenna with I′ = −I (see Fig 8-2).

If at a very distant point (in terms of wavelengths) the rays at points A and A″ are in phase, then their combined field strength will be at a maximum and will be equal to the sum of the magnitudes of the two rays. If they are out-of-phase, the resulting field strength will be less than the sum of the individual rays. If A and A″ are identical in magnitude and 180° out-of-phase, total cancellation will occur.

If the dipole antenna is at a very height (less than $1/4\lambda$), A and A″ will reinforce each other. Low-angle rays will be almost completely out-of-phase, resulting in cancellation, and thus there will be very little radiation at low angles. At increased heights, A and A″ may be 180° out-of-phase (no radiation at zenith angle), and lower angles may reinforce each other. In other words, the vertical radiation pattern of a dipole depends on the height of the antenna above the ground.

1.2.1.2 Vertical radiation pattern equations

The radiation pattern can be calculated with the following equation.

$$F_\alpha = \sin(h \sin \alpha) \qquad \text{(Eq 8-1)}$$

where

F_α = normalized field intensity at vertical angle α
h = height of antenna in degrees
α = vertical angle of radiation

One wavelength equals 360°. Eq 8-1 is valid only for perfectly reflecting grounds. For real ground the reflected wave must be multiplied by the complex reflection coefficient. This is shown in **Fig 8-3**; its total phase difference is then ≥ 180°, its magnitude ≤ 1. Eq 8-1 can be rewritten as follows:

$$H_1 = \frac{74.95}{f \sin \alpha} \qquad \text{(Eq 8-2)}$$

where

H_1 = height antenna in meters
f = frequency, MHz
α = vertical angle for which the antenna height is sought.

When more lobes are of interest, replace 90° with 270°

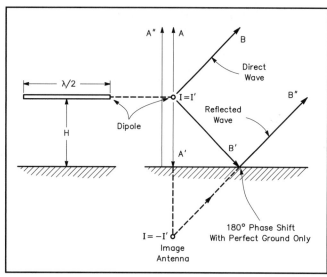

Fig 8-2—Reflection of RF energy by the electrical "ground mirror." The eventual phase relationship between the direct and the reflected horizontally polarized wave will depend primarily on the height of the dipole over the reflection ground (and to a small degree on the quality of the reflecting ground).

Table 8-1
Major Lobe Angles and Reflection Point for Various Dipole Antenna Heights

Antenna Height		40 Meters Angle Distance			80 Meters Angle Distance			160 Meters Angle Distance		
(ft)	(m)	(deg)	(ft)	(m)	(deg)	(ft)	(m)	(deg)	(ft)	(m)
60	18	36	83	25	90	0	0	90	0	0
80	24	26	163	50	54	58	18	90	0	0
100	30	20	266	81	40	118	36	90	0	0
120	36	17	391	119	33	187	57	90	0	0
140	42	15	540	148	28	268	82	77	31	9
160	48	13	710	217	24	362	110	59	97	30
180	54	-	-	-	21	467	142	49	154	47
200	60	-	-	-	18	584	178	43	213	66

for the second lobe, with 450° for third lobe, etc. If the nulls are sought, replace 90° with 180° for the first null, with 360° for the second null, etc.

Table 8-1 gives the major-lobe angles as well as reflection-point distances for heights ranging from 18 meters (60 feet) to 60 meters (200 feet) for 40, 80 and 160 meters.

1.2.1.3. Sloping ground locations

In many cases, an antenna cannot be erected above perfectly flat ground. A ground slope (Ref 630) can greatly influence the wave angle of the antenna. The RADIATION ANGLE HORIZONTAL ANTENNAS module of the NEW LOW BAND SOFTWARE calculates the radiation pattern of dipoles (or Yagis) as a function of the terrain slope. *TA* (Terrain Analyzer) from K6STI and *HFTA* (High Frequency Terrain Assessment) from N6BV are much more sophisticated programs that let you do basically the same thing. (See Section 1.1.2. in Chapter 5.)

Table 8-2 shows the influence of the slope angle on the required antenna height for a given wave angle on 80 meters. The table lists the required antenna height and the distance to the reflection point for a horizontally polarized antenna. A positive slope angle is an uphill slope. The results from this table can easily be extrapolated to 40 or 160 meters.

1.2.1.4. Antennas over real ground

Up to this point, a perfect ground has been assumed for most of the results presented. Perfect ground does not exist in practical installations, however. Perfect ground conditions are approached only when an antenna is erected over salt water.

Radiation efficiency and reflection efficiency

Contrary to the case with vertical antennas, a horizontal antenna does not rely on the ground to provide a return path for antenna currents. The physical "other half" takes care of that. This means that the ground will practically not play an important role in the radiation efficiency of the antenna. The radiation efficiency is related mainly to the losses in the antenna itself (conductor, insulator, loading coils, etc), although of course some of the total radiated energy can be dissipated in ground losses.

Both horizontally as well as vertically polarized antennas rely on the ground for reflection of the RF in the so-called Fresnel zone to build up the radiation pattern in combination with the direct wave, as shown in Fig 8-2. The efficiency of the reflection depends on the quality of the ground, and is called the *reflection efficiency*.

Reflection coefficient

The reflection from real ground is not like on a perfect

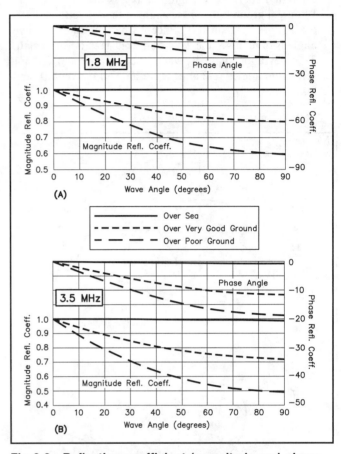

Fig 8-3—Reflection coefficient (magnitude and phase angle) of horizontally polarized waves over three types of ground: saltwater, average and very poor. See text for details.

Table 8-2
Slope Angle Versus Antenna Height at 3.5 MHz

Slope Angle (deg)	20° Wave Angle Height (ft)	20° Wave Angle Distance (ft)	30° Wave Angle Height (ft)	30° Wave Angle Distance (ft)	40° Wave Angle Height (ft)	40° Wave Angle Distance (ft)
35	-	-	-	-	906	10,364
30	-	-	-	-	430	2441
25	-	-	819	9367	275	1029
20	-	-	396	2249	201	553
15	768	8789	258	966	158	340
10	378	2146	192	528	131	227
5	251	937	153	329	113	161
0	189	520	129	224	100	120
−5	153	329	113	161	91	91
−10	131	227	102	121	85	72
−15	116	166	94	94	81	57
−20	107	127	89	75	79	45
−25	101	101	87	61	78	36
−30	97	91	86	49	78	28

mirror. The reflection coefficient is a complex number that describes the reflection from real ground:

- With a perfect mirror, all energy is reflected. There are no losses; the reflection coefficient magnitude is 1.
- With a perfect mirror, the phase of the reflected horizontal wave is shifted exactly 180° compared to the incoming wave.
- With real ground, part of the RF is absorbed, and the reflection coefficient magnitude is less than 1.
- With real ground, the phase angle of the reflection coefficient is greater than 180°. Except when the antenna wave angle is quite high, the deviation from 180° is very small. This deviation typically varies between 0° and 25° for reflection angles (equal to wave angles) between 0° and 90°.
- The magnitude of the reflection coefficient, which becomes smaller as the ground quality becomes poorer, is the reason that the dipole over real ground shows less gain than over perfect ground.

The reflection coefficient is a function the wave angle. The smaller the wave angle, the closer the reflection coefficient magnitude will approach 1. This explains why the loss with a dipole (poor ground vs perfect ground) is higher at high angles (for example, at the zenith) than at low angles. See **Fig 8-4**.

The fact that the dipole over poor ground seems to have a lower radiation angle than over perfect ground is because at lower angles there is less loss. In other words, over poor ground it just has less loss at low angles than at high angles.

The filling in of the deep notch at a 90° wave angle for the

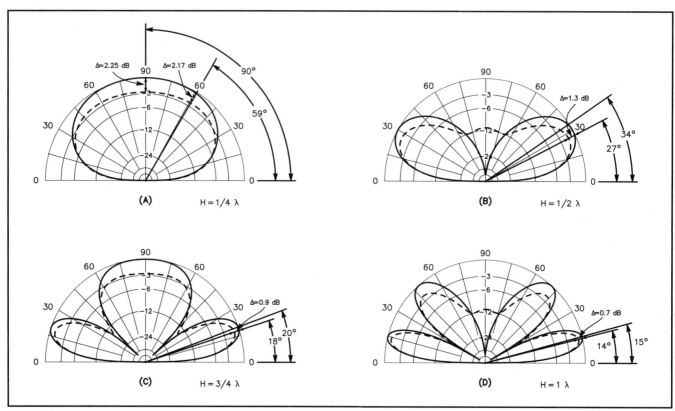

Fig 8-4—Vertical radiation patterns over two types of earth: saltwater (solid line in each set of plots) and very poor ground (broken line in each set of plots). The wave angle as well as the gain difference between saltwater and poor ground are given for four antenna heights.

Table 8-3
Relative gain (vs dipole over perfect ground) and wave angle (max vertical radiation angle) for $\frac{1}{2}$-λ dipoles at heights of $\frac{1}{4}\lambda$ and $\frac{1}{2}\lambda$.

	Height = $\frac{1}{4}\lambda$		Height = $\frac{1}{2}\lambda$	
	Rel. Loss (dB)	Wave Angle (deg)	Rel. Loss (dB)	Wave Angle (deg)
Perfect Ground	0	90	0	30
Saltwater	−0.05	90	−0.01	30
Very Good Ground	−0.57	71	−0.16	29
Average Ground	−1.23	62	−0.52	28
Very Poor Ground	−2.17	53	−1.21	26

dipole at ½ λ (and 1 λ) (Fig 8-4B and D) is because the reflected wave is considerably attenuated and phase shifted and can no longer cancel the direct wave. Note that changing the height of the antenna could compensate for the effect of the additional phase shift.

Again refer to Fig 8-3 showing the reflection coefficient (magnitude and phase) for a horizontally polarized wave. The information is for a horizontally polarized antenna over seawater, average ground and very poor ground, for both 160 and 80 meters. Eq 8-1 multiplied by this complex reflection coefficient gives the vertical radiation pattern over real ground.

Radiation patterns

Fig 8-4 shows vertical patterns of a horizontal half-wave dipole over both near-perfect ground (salt water) and desert, the two extremes. **Table 8-3** lists the wave angle and the relative loss for a half-wave dipole over five different types of ground and for two antenna heights. Note that for a dipole at ½ λ, the peak wave angle drops from 30° over seawater to 26° over desert. At the same time there is a radiation loss of 1.21 dB.

For an antenna at ¼ λ height (Fig 8-4A), maximum radiation occurs at 90° over a perfect conductor. Over very poor ground (desert), the maximum radiation is at 59°. This is not because more RF is concentrated at this lower angle, but only because more RF is being dissipated in the poor ground at the 90° angle than at 59° (the reflection coefficient is much lower at 90° than at 59°). The difference, however, between the radiation at 90° and at 59° is very small (0.08 dB). The difference in radiated power at 90° between salt water and a desert type of reflecting ground is 2.25 dB. Since 90° is a radiation angle of little practical use, the relatively high loss

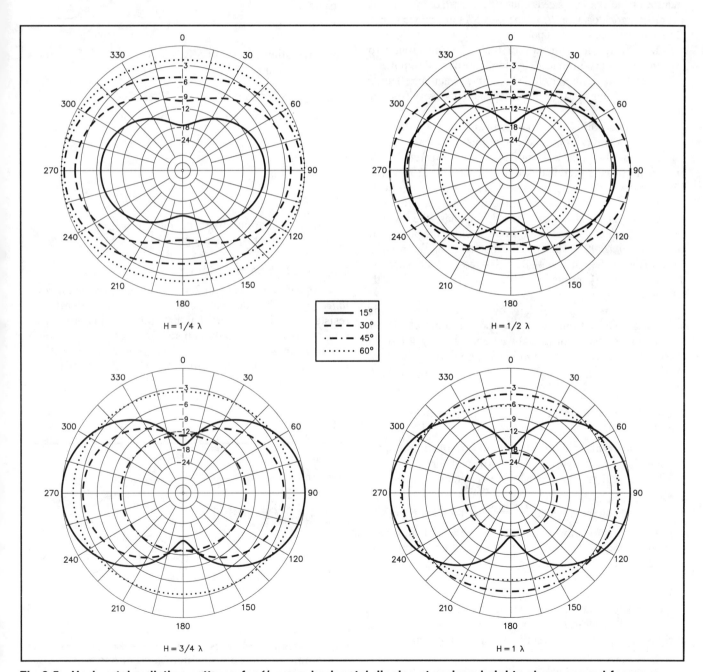

Fig 8-5—Horizontal radiation patterns for ½-wave horizontal dipoles at various heights above ground for wave angles of 15°, 30°, 45° and 60° (modeled over good ground).

at the zenith angle does not really bother us.

With a vertical antenna, poor ground results in loss at low angles, but with horizontal antennas the loss due to poor ground occurs at high angles. Notice that for a height of $1/2\ \lambda$ (Fig 8-4B), the sharp null at a 90°-elevation angle for perfect ground has been degraded to a mere 12-dB attenuation over desert-type ground.

Conclusion

We can conclude that the effects of absorption over poor ground are pronounced with low horizontally polarized antennas and become less pronounced as the antenna height is increased. Artificial improvement of the ground conditions by the installation of ground wires is only practical if one wants maximum gain at a 90°-wave angle (zenith) from a low dipole ($1/8$ to $1/4\ \lambda$ height). This can be done by burying a number of wires ($1/2$ to $1\ \lambda$ long) underneath the dipole, spaced about 60 cm apart, or by installing a parasitic reflector wire ($1/2\ \lambda$ long plus 5%) just above ground (2 meters high) under the dipole.

Improving the efficiency of the reflecting ground for low-angle signals produced by high horizontal dipoles is impractical and yields very little benefit. The active reflection area can be as far as 10 or more wavelengths away from the antenna!

Horizontal dipoles, unlike verticals, do not suffer to a great extent from poor ground conditions. The reason is that for horizontally polarized signals, when reflected by the ground, the phase shift remains almost constant at 180° (within 25°), whatever the incident angle of reflection (equal to the wave angle) may be. For verticals, the phase angle varies between 0° and 180°. For vertical antennas, the *pseudo-Brewster angle* is defined as the angle at which the phase shift at reflection is 90°. This means that there is no pseudo-Brewster angle with horizontally polarized antennas such as a dipole, because there never will be a 90° phase shift at the reflection point.

The effects are proved daily by the fact that on the low bands big signals from areas with poor ground conditions (mountainous, desert, etc) are always generated by horizontal antennas, while from areas with fertile, good RF ground, we often hear big signals from verticals and arrays made of verticals.

1.2.2. Horizontal pattern of horizontal half-wave dipole

The horizontal radiation pattern of a dipole in free space has the shape of a figure 8. The horizontal directivity of a dipole over real ground depends on two factors:
- Antenna height
- The wave angle at which we measure the directivity

Fig 8-5 shows the horizontal directivity of half-wave horizontal dipoles at heights of $1/4$, $1/2$, $3/4$ and $1\ \lambda$ over average ground. Directivity patterns are included for wave angles of 15° through 60° in increments of 15°. At high angles a low dipole shows practically no horizontal directivity. At low angles, where it has more directivity, the low dipole hardly radiates at all. Therefore, it is quite useless to put two dipoles at right angles for better overall coverage if those dipoles are at low heights.

At heights of $1/2\ \lambda$ and more, there is discernible directivity, especially at low angles. **Fig 8-6** shows the three-dimen-

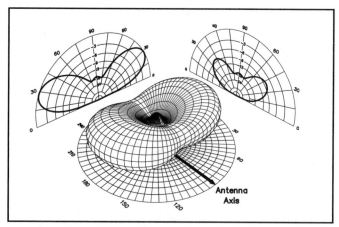

Fig 8-6—Three-dimensional representation of the radiation patterns of a half-wave dipole, $1/2\ \lambda$ above ground.

sional radiation pattern of a half-wave dipole at $1/2\ \lambda$ above average ground.

1.3. Half-Wave Dipole Efficiency

The radiation efficiency of an antenna is given by the equation

$$\text{Eff} = \frac{R_{rad}}{R_{rad} + R_{loss}} \qquad \text{(Eq 3)}$$

where

R_{rad} = radiation resistance, ohms
R_{loss} = loss resistance, ohms

1.3.1. Radiation resistance

As defined in Section 2.4 in Chapter 5, *radiation resistance* (referred to a certain point in an antenna system) is the resistance, which if inserted at that point, would dissipate the same energy as is actually radiated from the antenna. *Radiation resistance* is a fictional resistance. For a half-wave dipole at or near resonance, the radiation resistance is equal to the real (resistive) part of the feed-point impedance, assuming a perfectly lossless antenna system.

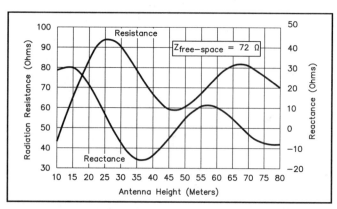

Fig 8-7—Radiation resistance and feed-point reactance of a dipole at various heights. Calculations were done at 3.65 MHz using a 2-mm OD conductor (AWG #12 wire) over good ground.

The relationship of the radiation resistance and reactance of a half-wave dipole to its height above flat ground is shown in **Fig 8-7**. The radiation resistance varies between 60 and 90 Ω for all practical heights on the low bands. For determining the reactance, the dipole was dimensioned to be resonant in free space (72 Ω). The resonant frequency changes with half-wave-dipole height above ground. Where the reactance is positive, the dipole appears to be too long, and too short where the reactance is negative.

1.3.2. Losses

The losses in a half-wave dipole are caused by:

- RF resistance of antenna conductor (wire)
- Dielectric losses of insulators
- Ground losses

Table 8-4 gives the effective RF resistance for common conductor materials, taking skin effect into account. The resistances are given in ohms per kilometer. The RF resistance values in the table are valid at 3.8 MHz. For 1.8 MHz the values must be divided by 1.4, while for 7.1 MHz the values must be multiplied by the same factor. The RF resistance of copper-clad steel is the same as for solid copper, since the steel core does not conduct any RF at HF. The dc resistance is higher by 3 to 4 times, depending on the copper/steel diameter ratio. The RF resistance at 3.8 MHz is 18 times higher than for dc (25 times for 7 MHz, and 13 times for 1.8 MHz). Steel wire is not shown in the table; it has a much higher RF resistance. Never use steel wire if you want good antenna performance.

1.3.2.1. Dielectric losses in insulators

Dielectric losses are difficult to assess quantitatively. Care should be taken to use good quality insulators, especially at the high-impedance ends of the dipole. Several insulators can be connected in series to improve the quality.

1.3.2.2. Ground losses

Reflection of RF at ground level coincides with absorption in the case of non-ideal ground. With a perfect reflector, the gain of a dipole above ground is 6 dB over a dipole in free space. The field intensity doubles, since the same power is now radiated in a half sphere instead of a full sphere; double field intensity means 4 times power, which equals 6 dB gain.

Real ground is never a perfect reflector. Therefore some RF will be dissipated in the ground. The effects of power absorption in the real ground have been covered in section 1.2.1.4. and illustrated in Fig 8-4 and Table 8-3.

Attempting to improve ground conductivity for improved performance is a common practice for vertical antennas. You can also improve ground conductivity under horizontal dipoles, although it is not quite as easy, especially if your are interested in low-angle radiation and if the antenna is physically high. From Table 8-1 you can find the distance from the antenna to the ground-reflection point. For the major low-angle lobe this is 36 meters (118 feet) away from an 80-meter dipole 30 meters (100 feet) high. Consequently, this is the place where the ground conductivity must be improved. Because of the horizontal polarization of the dipole, any wires that are laid on the ground (or buried in the ground) should be laid out parallel to the dipole. They should preferably be at least 1 λ long.

However, in view of the small gain that can be realized, especially with high antennas and for low wave angles, it is very doubtful that such improvement of the ground is worth all the effort! The only really worthwhile improvement will be obtained by moving to the seacoast or to a very small island surrounded by salt water. Don't forget that the quality of the reflecting ground with horizontal antennas is of far less importance than with vertical antennas.

The efficiency of low dipoles (¼ λ high and less), which essentially radiate at the zenith angle (90°), can be improved by placing wires under the antenna running in the same direction as the antenna.

1.4. Feeding the Half-Wave Dipole

In general, half-wave dipoles are fed in the center. This, however, is not a must. The Windom antenna is a half-wave dipole fed at approximately ⅙ from the end of the half-wave antenna, with a single-wire feed line. It has been proved (and can be confirmed by modeling) that careful placing of the feed point results in a perfect symmetrical and sinusoidal current distribution in the antenna (Ref 688). The disadvantage of the single-wire fed antenna (Windom) is that the feed line does radiate, and as such distorts the radiation pattern of the dipole. Belrose described a multiband "double Windom" antenna using a 6:1 balun and coaxial feed line in the above-mentioned publication.

1.4.1. The Center-Fed Dipole

The feed point of a center-fed dipole is symmetrical. The antenna can be fed with an open-wire transmission line if it is to be used on different frequencies (eg, as two half-waves in phase on the first harmonic frequency). It can also be fed with a coaxial feed line using a balun. The balun is mandatory to avoid upsetting the radiation pattern of the antenna. Baluns are covered in detail in the chapter on transmission lines. A current-type balun, consisting of a stack of high-permeability ferrite beads slipped over the coaxial cable at the load, is

---

Table 8-4
Resistance of Various Types of Wire Commonly Used for Constructing Antennas

Wire Diameter	Copper dc (Ω/km)	Copper 3.8 MHz (Ω/km)	Copper-Clad dc (Ω/km)	Copper-Clad 3.8 MHz (Ω/km)	Bronze dc (Ω/m)	Bronze 3.8 MHz (Ω/km)
2.5 mm (AWG 10)	3.4	61	8.7	61	4.5	81
2.0 mm (AWG 12)	5.4	97	13.8	97	7.2	130
1.6 mm (AWG 14)	8.6	154	22.0	154	11.4	206
1.3 mm (AWG 16)	13.6	246	35.0	246	18.2	326
1.0 mm (AWG 18)	21.7	391	55.6	391	29.0	521

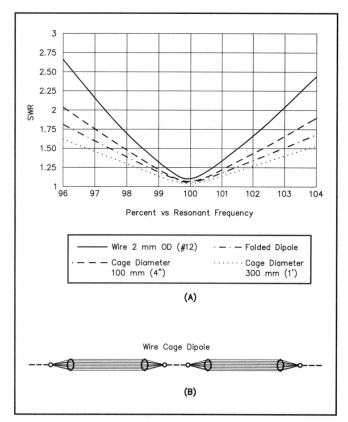

Fig 8-8—At A, SWR plots for 3.75-MHz half-wave dipoles in free space of various conductor diameters. The total bandwidth of the 80-meter band (3.5 to 3.8 MHz) is 8%. The 100-mm (4-inch) and 300-mm (12-inch) diameter conductors can be made as a cage of wires, as shown at B. Note that the SWR bandwidth of a folded dipole is substantially better than for a straight dipole. The spacing between the wires of the folded dipole does not influence the bandwidth to a large extent.

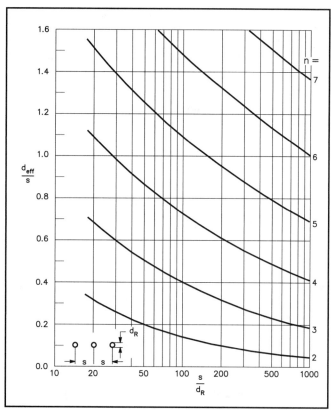

Fig 8-10—Normalized effective diameter d_{eff}/D_R for a flat multi-wire conductor, made out of n conductors (diameter d_R) spaced uniformly with a spacing S. (Source: *Kurze Antennen*, by Gerd Janzen, IBSN 3-440-05469-1.)

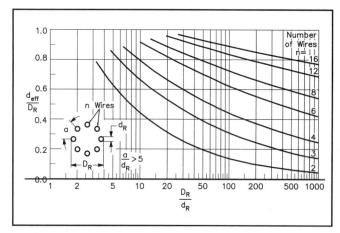

Fig 8-9—Normalized effective diameter d_{eff}/D_R for a wire-cage conductor, made out of n conductors (diameter D_R). Example: a wire cage made out of 6 wires of 2-mm diameter, spaced equally on a circle measuring 20 cm in diameter, had an equivalent diameter of a single solid conductor of $0.62 \times 20 = 125$-mm diameter. (Source: *Kurze Antennen*, by Gerd Janzen, IBSN 3-440-05469-1.)

recommended. The exact feed-point impedance of a horizontal dipole can be found from Fig 8-7.

Bandwidth

The SWR bandwidth of a full-size half-wave dipole is determined by the diameter of the conductor. **Fig 8-8A** shows the SWR curves for dipoles of different diameters. Large conductor diameters can be obtained by making a so-called *wire-cage* (Fig 8-8B). I used a wire-cage approach on my 80-meter vertical, with 6 wires forming a 30-cm (12-inch) diameter cage. **Fig 8-9** shows the effective equivalent diameter of such a cage conductor as a function of the number of wires making up the cage.

Instead of using a wire cage you can also use a configuration consisting of a number of identical wires in a plane. **Fig 8-10** shows the effective equivalent diameters of such a flat multi-wire configuration. (Source: *Kurze Antennen*, by Gerd Janzen, ISBN 3-440-05469-1). Example: Three parallel wires, each measuring 2 mm OD and equally spaced 5 cm, have an effective equivalent diameter of a solid conductor of $50 \times 0.65 = 32.5$ mm.

A folded dipole shows a substantially higher SWR bandwidth than a single-wire dipole. A folded dipole for 80 meters, made of AWG #12 wire, with a 15-cm (6-inch) spacing between the wires, will cover the entire 80-meter band (3.5 to 3.8 MHz) with an SWR of approximately 1.75:1, as compared to 2.5:1 or more for a straight dipole.

1.4.2. Broadband dipoles

Instead of decreasing the Q factor of the antenna, you can also devise a system to compensate the inductive part of the impedance as you move away from the resonant frequency of the antenna. The "double Bazooka dipole" is probably the best-known example of such an antenna, although it is rather controversial. In this antenna, part of the radiator is made of coaxial cable, connected in such a way as to present shunt impedances across the dipole feed point when moving away from the resonant frequency.

F. Witt, AI1H, designed a better broadband dipole antenna in detail (Ref 1012). **Fig 8-11** shows the dimensions of Witt's *80-Meter DX-Special* antenna, which has been dimensioned for minimum SWR at both the CW as well as SSB end of the 3.5 to 3.8-MHz band. Another innovative broadbanding technique was described by M. C. Hatley, GM3HAT (Ref 682).

R. Sevens, N6LF, described an 80-meter folded broadband dipole using the principle of the open-sleeve antenna (Ref 1014). **Fig 8-12A** shows the layout of this folded-dipole, where Severns inserted another nearly half-wave long wire between the legs of the folded dipole. The resulting SWR curve is shown in Fig 8-12B.

Of course, there is no reason why you couldn't apply switched inductive or capacitive loading devices, such as described in detail in the chapters on verticals and large loop antennas, although this is seldom done in practice.

1.4.3. Does a resonant dipole radiate better than a dipole off-resonance?

No, an infinitely short dipole would radiate as well as a

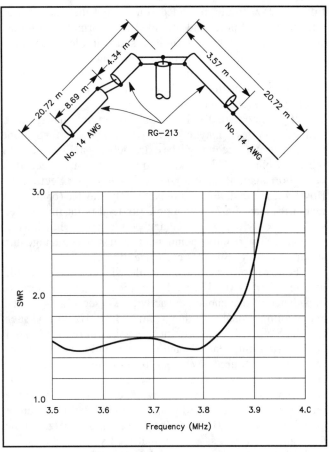

Fig 8-11—Dimensions and SWR curve of the "80-Meter DX Special," a design by F. Witt, AI1H.

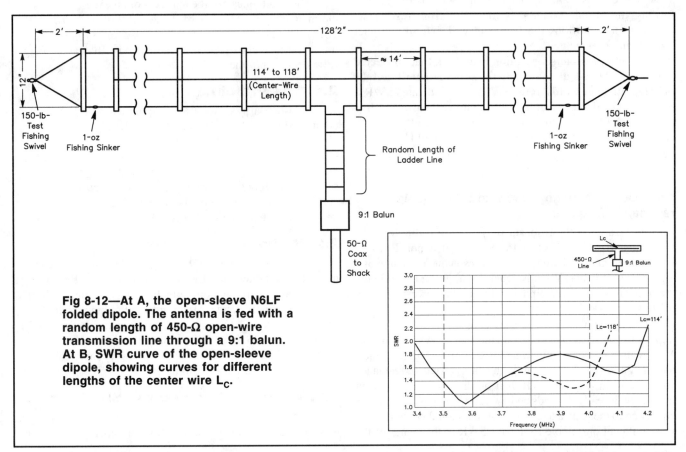

Fig 8-12—At A, the open-sleeve N6LF folded dipole. The antenna is fed with a random length of 450-Ω open-wire transmission line through a 9:1 balun. At B, SWR curve of the open-sleeve dipole, showing curves for different lengths of the center wire L_c.

full-size ½-λ dipole, provided you can get the same power into the short dipole, and provided the (normalized) losses are the same. Such an infinitely short dipole is called a *Hertzian* dipole. It has a constant current distribution and therefore a slightly different radiation pattern than a half-wave dipole. But that should not concern us, since it is a theoretical antenna anyhow.

If you have a really short dipole, the radiation resistance will be very low, maybe a few ohms, and the feed-point impedance at the center will be extremely capacitive (several thousand ohms). This makes it quite difficult to feed this very short antenna with a good efficiency (see Section 2). However, if the antenna is only slightly shorter (or longer) than a resonant half-wave dipole, its feed point impedance will vary only a few percentage points from what it is at resonance. The reactive components will still be manageable so far as feeding this off-resonance dipole.

For example, let's take a single-wire center-fed dipole tuned for 3.65 MHz. Let's assume it is at a height where its impedance at resonance is exactly 50 Ω (see Fig 8-7). The antenna impedance at both 3.5 and at 3.8 MHz will be such that the SWR will be approximately 2:1 (still referred to our 50-Ω system impedance), and this is mainly caused by the reactive component. Whether or not this non-resonant antenna will radiate as much power as its resonant counterpart, depends exclusively on how much loss there is in the feed system, now required to match a complex impedance: $40 + j\ 70\ \Omega$ at 3.5 MHz and $60 + j\ 70\ \Omega$ on 3.8 MHz. On the low bands, you can safely say that feed systems will show negligible losses when operated with SWRs below 2:1 or even 3:1.

Summarizing, the off-resonance dipole will radiate just as well as the resonant dipole. The only issue is the ease of feeding this dipole when it is far away from resonance. Under such conditions the feed line will exhibit a higher SWR than at resonance.

A widespread misconception is to think that "reflected power" (reflected at the load, the antenna) will not be radiated. In other words, if the SWR meter indicates SWR = 3:1 or 50% reflected, that 50% of the power is *wasted*. In a lossless feed line system, all power will eventually be radiated, whatever the SWR. Our only concern should be a small increase in additional attenuation in a real-world feed line due to SWR (see Chapter 6).

1.4.4. Does frequency of lowest SWR equals resonance frequency?

Is it true that the resonant frequency of a dipole is the frequency where the SWR is lowest? No, it is not. But is it important to know where is the exact resonant frequency of a dipole? No, it is not. What is generally important is to know the frequency where the dipole will cause the lowest SWR on the feed line.

Let me explain it with an example: A dipole has an impedance of 70 Ω at resonance and thus shows an SWR of 1.4:1 with 50-Ω coax. Somewhat lower in frequency, a combination of a lower resistive part with some capacitive reactance could result in a lower SWR than on the antenna's resonant frequency. It really depends on how fast the reactance changes compared to the resistance. But all of this should not bother us; we should cut our dipole for lowest SWR in the center of the (portion of the) band we want to cover. Whether or not this is the dipole's resonant frequency is irrelevant.

1.5. Getting the Full-Size Dipole in Your Backyard

The ends of the half-wave dipole can be bent (vertically or horizontally) without much effect on the radiation pattern or efficiency. The tips of the dipoles carry little current; hence, they contribute very little to the radiation of the antenna.

Bending the tips of a dipole is the same as "end loading" the dipole (equal to top-loading with verticals). The folded tips can be considered as capacitive loading devices. For more details see Section 2 in Chapter 9 on Vertical Antennas.

N. Mullani, KØNM, calculated the gain and the impedance of a half-wave dipole with its end hanging down vertically (Ref 691). He concluded that with horizontal lengths as short as 40% of full-size, the trade-offs are rather insignificant, being only about 0.6 dB in gain, and some reduction of SWR bandwidth. Bending the end may actually somewhat improve the match to a 50-Ω feed line, depending on antenna height. KØNM concludes: "Don't be afraid to bend your dipole antennas if you are cramped for space." (See also Chapter 14, Antennas for the Small Garden.)

2. THE SHORTENED HALF-WAVE DIPOLE

On the low bands, it is sometimes impossible to use full-size radiators. This section describes the characteristics of short dipoles, and how they can be successfully deployed. Short dipoles are often used as elements in reduced-size Yagis (see the chapter on Yagis) or to achieve manageable dimensions whereby the antenna can be fit into a city lot.

Short antennas are the subjects of an excellent book (in German language) by Gerd Janzen, DF6SJ/VK2BJZ (Ref 7818). This book is highly recommended for anyone who does not fear a formula and a graph, and who really wants to dig a little deeper into the subject.

2.1. The Principles

You can always look at a dipole as two back-to-back connected verticals, except that the "vertical" elements are no longer vertical. Instead of having the ground make the mirror image of the antenna (this is always the case with quarter-wave monopole verticals), we supply the mirror half ourselves in a dipole. All principles about radiation resistance and loading of short verticals, as explained in Chapter 9 on vertical antennas, can be directly applied to dipoles as well.

2.2. Radiation Resistance

The radiation resistance of a dipole (made of an infinitely thin conductor) in free space will be twice the value of the equivalent vertical monopole. For instance, the R_{rad} for the half-wave dipole made of an infinitely thin conductor is approximately 73.2 Ω, which is twice the value of the quarter-wave vertical (36.6 Ω). Over ground, the infinitely thin horizontal dipole's radiation resistance will vary in a similar way as the full-size half-wave dipole (see Fig 8-7).

2.3. Tuning or Loading the Short Dipole

Loading a short dipole consists of bringing the antenna to resonance. This means eliminating the capacitive reactance component in the feed-point impedance. Different loading methods yield different values of radiation resistance.

It is not necessary, however, to load a shortened antenna to resonance in order to operate it. You could connect a feed line to it, directly or via a matching network, without tuning out the capacitive reactance. Therefore you can consider a dipole together with its feed line as a *dipole system*, and analyze the system of a short dipole to see what the alternatives are. Sometimes this situation is referred to as a dipole with "tuned feeders."

There are different ways to operate the short dipole system:

- Tuned feeders
- Matching at the dipole feed point
- Coil loading to tune out the capacitive reactance
- Linear loading
- Capacitive end loading
- Combined loading methods

2.3.1. Tuned feeders

Tuned feeders were common in the days before the arrival of coaxial feed lines. Very low-loss open-wire feeders can be made. The *Levy* antenna is an example of a short dipole fed with open-wire line. In a typical configuration its overall length is $1/4\ \lambda$. This antenna (R_{rad} = approximately 13 Ω and X_C = approximately $-j\ 1100$ Ω), can be fed with open-wire feeders (450-600 Ω) of any length into the shack, where we can match it to 50 Ω with an antenna tuner. An outstanding feature of this approach is that the system can be "tuned" from the shack via the antenna tuner, and is not narrow banded, as is the case with loaded elements.

Let us calculate the losses in such a system. The losses of the antenna can be assumed to be zero (provided wire elements of the proper size and composition are used). The loss in a flat open-wire feeder is typically 0.01 dB per 100 feet at 3.5 MHz. The SWR on the line will be a mind-boggling value of 280:1 (this value was calculated using the program SWR RATIO, which is part of the NEW LOW BAND SOFTWARE. The additional line loss due to SWR for a 30-meter (100 foot) long line will be in the order of 1.5 dB (Ref 600, 602). On a line with standing waves, the impedance is different at every point. Changing the feeder length slightly can produce more manageable impedances, ones the tuner can cope with more easily. This can be done with the software module COAX TRANSFORMER/SMITH CHART or IMPEDANCE, CURRENT AND VOLTAGE ALONG FEED LINES. A good antenna tuner should be able to handle this matching task with a loss of less than 0.2 dB. The total system loss depends essentially on the efficiency with which the tuner can handle the impedance transformation. The typical total loss in the system should be less than 2.0 dB.

2.3.2. Matching at the dipole feed point

You could, of course, install a matching network at the dipole feed point, although this will be highly impractical in most cases. In the case of a vertical antenna this solution is practical, since the feed point is at ground level.

2.3.3. Coil loading

Loading coils can be installed anywhere in the short dipole halves, from the center to way out near the end. Loading near the end will result in a higher radiation resistance, but will also require a much larger coil, and hence introduce more coil losses.

2.3.3.1. Center loading

The inductive reactance required to resonate the Levy dipole from the previous example is approximately +1100 Ω. To achieve this, two 550-Ω (reactance) coils must be installed in series at the feed point. We should be able to realize a coil Q (quality factor) of 300. With good care 500 to 600 can be achieved as well (Ref 694).

$R_{loss} = 1100/600 = 1.83$ Ω

The total equivalent loss resistance of the two coils is 3.66 Ω. The antenna efficiency will be:

Eff = 13/(13 + 3.66) = 78 %

The equivalent power loss is $-10 \log (0.78) = 1.08$ dB.

The feed-point resistance of the antenna is 13 + 3.66 = 16.7 Ω at resonance. This assumes negligible losses from the antenna conductor (heavy copper wire). If the use of coaxial feed lines is desired, an additional matching system will be needed to adapt the 15-Ω balanced feed-point impedance to the 50- or 75-Ω unbalanced coaxial cable impedance. This example was calculated assuming free-space impedances. Over real ground the impedances can be different, and will vary as a function of the antenna height.

Another way to determine the necessary inductance is to model the antenna using a *MININEC* or *NEC-2* program (eg, *ELNEC* or *EZNEC*). Let us work out the example of the λ/4 long dipole using *EZNEC 3*.

Input data:

f = 1.83 MHz
h = 25 m
L_{ant} = λ/4 (half size)

Procedure:

Find a full-sized dipole length resonant at 1.83 MHz, using AWG #14 copper wire. (This turns out to be 80.1 m.) (Do not model the dipole at any lower height, if you are using a *MININEC*-based program, since the results will be erroneous.) It makes little difference what type of earth you model, but do not model the dipole in free space, since this will give incorrect results. A 1/4-wave-long dipole is half the above length: 40.05 m. Model the dipole again: The impedance is $11.5 - j\ 1122$ Ω.

The required center loading coil has a reactance of 1122 Ω. Assuming a loading-coil Q of 300, the total equivalent loss resistance is 3.74 Ω. The feed point resistance becomes 11.5 + 3.74 = 15.24 Ω.

Note that the figures obtained by this method confirm the numbers discussed earlier.

Matching to the feed-line impedance

One way of matching this impedance to a 50-Ω feed line is to use a quarter-wave transformer. The required impedance of the transformer is

$Z_0 = \sqrt{15.24 \times 50} = 27.6$ Ω

We can construct a feed line of 25 Ω (that's close) by paralleling two 50-Ω feed lines. Don't forget you need a 1:1 balun between the antenna terminals and the feed line.

Another attractive matching scheme used by a number of commercial manufacturers of short 40-meter Yagis is to use a single central loading coil, on which we install a link at the center. The link turns are adjusted to give a perfect match to the feed line.

Comparing losses

A good current-type balun should account for much less than 0.1 dB of loss. The loading coils (Q factor = 300) give a loss of 1.3 dB. Including 30 meters (100 feet) of RG-213 (with 0.23 dB loss), the total system loss can be estimated at 1.63 dB. The resulting efficiency is very close to the result obtained with open-wire feeders.

There are certain advantages and disadvantages to this concept, however. An advantage is that coaxial cable is easier to handle than open-wire line, especially when dealing with rotatable antenna systems. The high Q of the coils will make the antenna narrow-banded as far as the SWR is concerned. In the case of the open-wire feeders, retuning the tuner will solve the problem. With coaxial feed line you may still need a tuner at the input end if you want to cover a large bandwidth, in which case the extra losses due to SWR in the coaxial feed line may be objectionable.

Another disadvantage is that the loading-coil solution requires two more elements in the system—the coils. Each element in itself is an extra reliability risk, and even the best loading coils will age and require maintenance.

Instead of modeling the antenna with *EZNEC*, we can calculate the required loading coils as follows:

Length of the dipole = 22.5 m
f = 3.8 MHz
Wire diameter = 2 mm OD (AWG #12)
Antenna height = 20 m

A full-size dipole length (with a 2.5% shortening factor) is 38.5 m. We first calculate the surge impedance of the transmission-line equivalent of the short dipole using the equation:

$$Z_S = 276 \log \left[\frac{S}{d\sqrt{1 + \frac{S}{4h}}} \right] \quad \text{(Eq 8-4)}$$

where
S = dipole length = 2250 cm
d = conductor diameter = 0.2 cm
h = dipole height = 2000 cm

Thus, $Z_S = 1103\ \Omega$ and the electrical length of the 22.5-m long dipole is:

$$\ell = 180° \left(\frac{22.5}{38.5} \right) = 105.2°$$

The reactance of the dipole is given by:

$$X_L = Z_S \cot \frac{\ell}{2} = 1103 \cot 52.6° = 843\ \Omega$$

Separate *MININEC* calculations show a reactance of 908 Ω, which is within 7% of the value calculated above. The required inductance is:

$$L = \frac{X_L}{2\pi \times f}$$

where L is in µH and f is in MHz. For 3.8 MHz:

$$L = \frac{843}{2\pi \times 3.8} = 35.3\ \mu H$$

There are two ways of loading and feeding the shortened dipole with a centrally located loading coil:

- Use a single 35.3 µH loading coil and link couple the feed line to the coil. This method is used by Cushcraft for their shortened 40-meter antennas.
- A 35.3-µH loading coil can be opened in the center where it can be fed by a 1:1 balun.

2.3.3.2. Loading coils away from the center of the dipole

The location of the loading devices has a distinct influence on the radiation resistance of the antenna. This phenomenon is explained in detail in the chapter on short verticals.

Clearly, it is advantageous to move loading coils away from the center, provided the benefit of higher radiation resistance is not counteracted by higher losses in the loading device. It appears, however, that in practice there is very little difference.

When loading coils are placed farther out on the elements, the required coil inductance increases. With increasing values of inductance, the Q factor is likely to decrease, and the equivalent series losses will increase.

I have calculated a case where the 22.5-meter long dipole (for 3.8 MHz) from Section 2.3.3.1 was loaded with coils at different (symmetrical) positions along the half-dipole elements. In all cases I assumed a Q factor of 300.

The results of the case are shown in **Fig 8-13** for a dipole

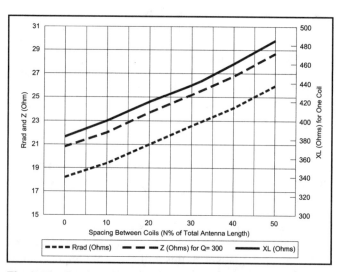

Fig 8-13—Design data for a 22.5-meter shortened dipole (F = 3.8 MHz, wire diameter = 2.5 mm), showing R_{rad}, Z_{feed} and required loading coil reactance X_L as a function of the spacing between the loading coils. Where the spacing between the coils is zero (center loading) the coil reactance is twice the value shown (2 × 386 Ω).

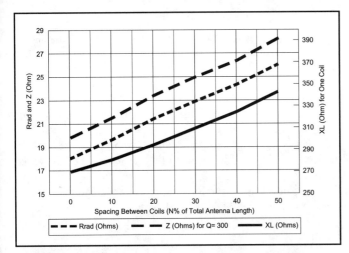

Fig 8-14—Design data for a 22.5-meter shortened dipole (F= 3.8 MHz, wire diameter = 25 mm), showing R_{rad}, Z_{feed} and required loading coil reactance X_L as a function of the spacing between the loading coils. Where the spacing between the coils is zero (center loading) the coil reactance is twice the value shown (2 × 269 Ω).

made with 2.5-mm-diameter wire and **Fig 8-14** for a dipole with an average conductor diameter of 25 mm. The charts include the reactance value of the required loading coils, the radiation resistance (R_{rad}), and the feed-point impedance at resonance (Z). The radiation efficiency is given by R_{rad}/Z. Over the entire experiment range the efficiency remains practically constant at 88% for 2.5 mm conductor diameter and 91% for 25 mm conductor diameter. You can also use a wire-cage type dipole (see Fig 8.8) to achieve an effective conductor diameter of 250 mm, yielding an efficiency of 95%. **Table 8-5** shows the influence of the coil Q and the effective wire diameter on the radiation efficiency of a shortened dipole.

The fact that the efficiency does not change much by moving the coils out on the elements means that the advantage we gain from an increased radiation resistance by moving the coils out on the dipole halves is balanced out by the increased ohmic losses of the higher coil values. In the experiment I assumed a constant Q of 300, which may not be realistic, as it is likely that the Q of lower-inductance coils will be higher than higher-inductance ones.

In Table 8.5 we see the influence of dropping the Q to 100 (pretty lousy) and raising it to 600 (excellent). The spread between the minimum wire diameter combined with the worst coil (Q = 100) and the maximum wire diameter with the best coil (Q = 600) is from 72% to 98%, which means a difference of 1.4 dB in signal strength.

Another marked advantage of using the large-diameter conductor is a substantially increased SWR bandwidth. The loaded 22.5-meter long dipole made of 2.5-mm diameter wire has a 2:1 SWR bandwidth of 50 kHz on 75 meters. The same dipole made out of a wire cage (6 wires in a circle with a 300-mm diameter, yielding an effective diameter of 250 mm) has a 2:1 SWR bandwidth of 100 kHz.

I did the calculation above in free space. Over real ground the radiation resistance (and Z) will vary to a rather large extent as a function of the height (see Fig 8-7). With the large diameter dipole (effective 25 mm diameter) we need only a reactance of 310 Ω to center load the 22.5-meter long dipole for 3.8 MHz. This is compared to 765 Ω for a wire dipole of the same length with 2.5 mm diameter.

Calculating the loading coil value

The method for calculating the loading coil value is described in detail in Sections 2.1.3 and 2.6.8 of the chapter on vertical antennas. In short the procedure is as follows:

- Calculate the surge impedance of the wire between the loading coil and the center of the antenna (Z_{S1})
- Calculate the surge impedance of the wire between the loading coil and the tip of the antenna (Z_{S2})
- Calculate the electrical length of the inner length (coil to center) = ℓ_1
- Calculate the electrical length of the tip (coil to tip) = ℓ_2
- Calculate the reactance of the l_1 part using: $X = Z_{S1} \tan(\ell_1)$
- Calculate the reactance of the inner part of the half dipole using: $X_1 = + j Z_{S1} \times \tan(\ell_1)$

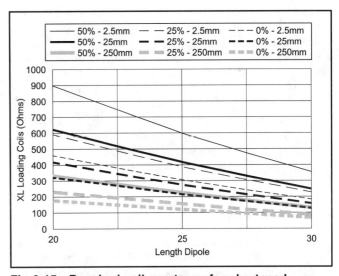

Fig 8-15—Required coil reactance for shortened dipoles with effective conductor diameters of 2.5, 25 and 250 mm (design frequency = 3.8 MHz) as a function of total antenna length (varying from 20 to 30 meters) and varying loading coil location (0% = center loaded, 25% means coils spaced 25% of total length, and 50% means coils spaced 50% of total length).

Table 8-5
Antenna radiation efficiency as a function of loading coil Q and conductor diameter

Conductor	Q=100	Q=300	Q=600
Diam 2.5 mm	72 %	88 %	94 %
Diam 25 mm	78 %	91 %	96 %
Diam 250 mm	84 %	95 %	98 %

- Calculate the reactance of the tip using: $X_2 = -j Z_{S2} / \tan(\ell_2)$
- Add the reactances (the sum will be a negative value, such as −1000 ohms). The loading coil will have a reactance with the same absolute value.

It is much faster to use a *NEC-2*-based modeling program, such as *EZNEC*, to calculate the elements of a short dipole. Results obtained with *EZNEC* match the results obtained by the above procedure.

Fig 8-15 shows required loading coils reactances for various combinations of antenna length, wire diameter and coil position. Note that for 0% spacing (center loading), the center-loading coil has twice the value indicated in the graph.

Conclusion:

Use loading coils with the highest possible Q and reduce the Q-factor of the antenna (the rate at which the reactive component of the impedance changes with frequency) by using a large diameter using a cage-type construction. The exact position of the coil does not significantly influence the antenna efficiency and is far from critical.

2.3.4 Linear loading

In the commercial world, we have seen *linear loading* used on shortened dipoles and Yagis for 40 and 80 meters. Linear-loading devices are usually installed at or near the center of the dipole. The required length of the loading device (in each dipole half) will be somewhat longer than the difference between the quarter-wave length and the physical length of the half-dipole. The farther away from the center that the loading device will be inserted, the longer the "stub" will have to be. The stub must run in parallel with the antenna wire if we want to take advantage of any radiation from the stub itself (see the chapter on vertical antennas).

Example: A short dipole for 3.8 MHz is physically 28 meters (91.9 feet) long. The full half-wavelength is 39 meters. The missing electrical length is 39 − 28 = 11 meters (36.1 feet). It is recommended that each linear-loading device be made 30% longer than half of this length:

L = 11/2 + 30% = 5.5 × 1.3 = 7 m

Trim the length of the loading device until resonance at the desired frequency is reached. When constructing an antenna with linear-loading devices, make sure the separation between the element and the folded linear-loading device is large enough, and that you use high-quality insulators to prevent arc-over and insulator damage. If directivity is not an issue you can hang the linear loading "stubs" vertically from the dipole.

Modeling the linear loaded dipole

Modeling antennas that use very close-spaced conductors (such as a linear-loading device that looks like a stub made of open-wire transmission line) is very tricky. I would not recommend trying this with a *MININEC*-based modeling program.

If linearly loaded dipoles are used as elements of an array it is very important that the linear loading devices run horizontally (in-line with the element) and not at angle. If run at an angle there will be vertically polarized radiation from the linear loading conductors, and this will mess up the directivity of the antenna.

2.3.5. Capacitive (end) loading

Capacitive loading has the advantage of physically shortening the element length at the end of the dipole where the current is lowest (least radiation), and without introducing noticeable losses (as inductors do). End-loaded short dipoles have the highest radiation resistance, and the intrinsic losses of the loading device are negligible. Thus, end or top loading is highly recommended.

Top loading, and the procedures to calculate the loading devices, are covered in detail in the vertical antenna chapter (Chapter 9) in Sec. 2.1.2 and 2.6.3. An example will best illustrate how capacitive loading can be calculated. A shortened dipole will be loaded for 80 meters. The physical length of the dipole is 18.75 meters (approximately a 40% shortening factor).

S = 18.75 m
d = 0.2 cm
h = 20 m

We calculate the surge impedance from Eq 8-4:

Z_S = 1084 Ω.

The antenna length to be replaced by a disk is:

t = 90° × 40% = 36°

This means we must replace the outside 36° of each side of the dipole with a capacitive hat (a half-wave dipole is 180°). The inductive reactance of the shorted transmission-line equivalent) is given by:

$$X_L = +j \frac{Z_S}{\tan t} = +j \frac{1084}{\tan 36°} = +j\, 1492 \, \Omega$$

A capacitive reactance of the same value (but opposite sign) will resonate the equivalent transmission line. The required capacitive reactance is $X_C = -j\, 1492 \, \Omega$.

The capacitance at 3.8 MHz is:

$$C = \frac{10^6}{2\pi \times f \times X_C} = 28.1 \, pF$$

The required diameter of the hat disk is given by

D = 2.85 × C

where

D = hat diameter in cm
C = the required capacitance in pF

In our example, D = 2.85 × 28.1 = 80.1 cm. The above formula to calculate the disk diameter is for a solid disk. A practical capacitive hat can be made in the shape of a wheel with at least eight large-diameter spokes. This design will approach the performance of a solid disk. For ease in construction, the spokes can be made of four radial wires, joined at the rim by another wire in the shape of a circle.

In its simplest form, capacitive end loading will consist of bending the tips of the dipole (usually downward) to make the antenna shorter. By doing so we create extra capacitance between those two tips, which will load the antenna and make it electrically longer. These wire tips have an approximate capacitance of 6 pF/m. For the above example, where a loading capacity of 28.1 pF is required at each dipole end, vertically drooping wires of 28.1/6 = 4.7 m would be required.

2.3.6. Combined methods

Any of the loading methods already discussed can be employed in combination. It is essential to develop a system that gives you the highest possible radiation resistance and that employs a loading technique with the lowest possible inherent losses.

Gorski, W9KYZ, (Ref 641) has described an efficient way to load short dipole elements by using a combination of linear and helical loading. He quotes a total efficiency of 98% for a 2-element Yagi using this technique. This very high percentage can be obtained by using a wide copper strap for the helically wound element, resulting in a very low RF resistance. Years ago, Kirk Electronics (W8FYR, SK) built Yagis for the HF bands, including 40 meters, using fiberglass elements wound with copper tape.

2.4. Bandwidth

The bandwidth of a dipole is determined by the Q factor of the antenna. The antenna Q factor is defined by:

$$Q = \frac{Z_S}{R_{rad} + R_{loss}}$$

where

Z_S = surge impedance of the antenna
R_{rad} = radiation resistance
R_{loss} = total loss resistance.

The 3-dB bandwidth can be calculated from:

$$BW = \frac{f_{MHz}}{Q}$$

The Q factor (and consequently the bandwidth) will depend on:

- The conductor-to-wavelength ratio (influences Z_S).
- The physical length of the antenna (influences R_{rad}).
- The type, quality, and placement of the loading devices (influences R_{rad}).
- The Q factor of the loading device(s) (influences R_{loss}).
- The height of the dipole above ground (influences R_{rad}).

For a given conductor length-to-diameter ratio and a given antenna height, the loaded antenna with the narrowest bandwidth will be the antenna with the highest efficiency. Indeed, large bandwidths can easily be achieved by incorporating pure resistors in the loading devices, such as in the Maxcom dipole (Ref 663). The worst-radiating antenna one can imagine is a dummy load, where the resistor is a loading device while the radiating component does not exist. Judging by SWR bandwidth alone, the dummy load is a wonderful "antenna," since a good dummy load can have an almost flat SWR curve over thousands of megahertz!

2.5. The Efficiency of the Shortened Dipole

Besides the radiation resistance, the RF loss resistance of the shortened-dipole conductor is an important factor in the antenna efficiency. Refer to Table 8-4 for the RF loss resistances of common wire conductors used for antennas. For self-supporting elements, aluminum tubing is usually used. Both the dc and RF resistances are quite low, but special care should be taken to ensure that you make the best possible electrical RF contacts between parts of the antenna. Some makers of military-specification antennas go so far as to gold plate the contact surfaces for low RF resistance!

As a rule, loading coils are the most lossy elements, and capacitive end loading should be employed if possible. In its simplest shape a capacitively end-loaded short dipole consists of a horizontal short dipole (say, 0.3 λ long), with vertical wires hanging down from each end. These wires carry little current and hence contribute only marginally to radiation, but their capacity effect lowers the resonance of the shortened dipole. Watch out—the end of these wires carry very high voltage and must be kept out of reach of humans and animals.

Linear loading is also a better choice than inductive loading, unless uttermost care is taken to construct loading coils with Qs in the range of 400 to 600. But you should remember that the linear loading stubs do radiate. If it is not in the plane of the element, this will upset directivity if the shortened element is part of an array such as a loaded Yagi. See Chapter 13.

It is very important to minimize the contact losses at any point in the antenna, especially where high currents are present. Corroded contacts can turn a good antenna into a radiating dummy load. These aspects are covered in more detail in the chapter on vertical antennas.

3. LONG DIPOLES

Provided the correct current distribution is maintained, long dipoles can give more gain and increased horizontal directivity compared to a half-wave dipole. The "long" antennas discussed in this paragraph are not strictly dipoles, but arrays of dipoles. They are the double-sized equivalents of the "long-verticals" covered in the chapter on verticals. The following antennas are covered:

- Two half-waves in phase
- Extended double Zepp

3.1. Radiation Patterns

Center-fed dipoles can be lengthened to approximately 1.25 λ to achieve increased directivity and gain without introducing objectionable side lobes. **Fig 8-16** shows the horizontal radiation patterns for three antennas in free space: a half-wave dipole, two half-waves in phase (also called *collinear dipoles*), and the extended double Zepp, which is a 1.25 λ long. Further lengthening of the dipole introduces major secondary lobes in the horizontal pattern unless phasing stubs are inserted to achieve the correct phasing between the half-wave elements.

As we know, a half-wave dipole has 2.14-dB gain over an isotropic antenna in free space. It is interesting to overlay the patterns of the two long dipoles on the same diagram, using the same dB scale. The extended double Zepp beats the dipole with almost 3 dB of gain. Note, however, how much more narrow the forward lobe on the pattern has become. This may be a disadvantage in view of varying propagation paths. The two-half-waves antenna is right between the dipole and the extended double Zepp, with 1.5-dB gain over the half-wave dipole.

Fig 8-17 and **Fig 8-18** show the horizontal radiation patterns for the half-waves-in-phase dipole and the extended double Zepp at various heights and wave angles. As with the half-wave dipole, the vertical radiation pattern depends on the height of the antenna above ground.

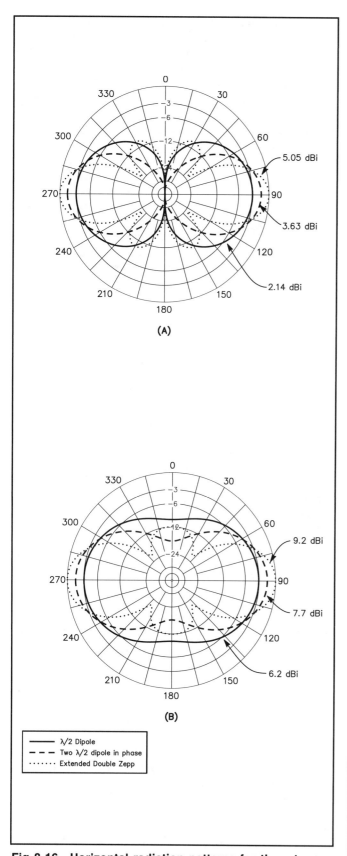

Fig 8-16—Horizontal radiation patterns for three types of dipoles: The half-wave dipole, the collinear dipole (two half waves in phase) and the extended double Zepp. At A, the radiation patterns at a 0° wave angles with the antennas in free space. At B, the patterns at a 37° wave angle with the antennas $3/8\ \lambda$ above good-quality ground. Notice the sidelobes apparent with the extended double Zepp antenna.

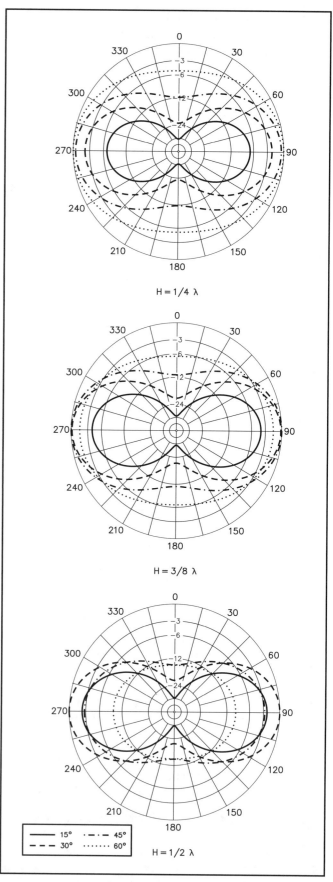

Fig 8-17—Horizontal radiation patterns for collinear dipoles (two half-waves in phase) for wave angles of 15°, 30°, 45° and 60°. As with a half-wave dipole, directivity is not very pronounced at low heights and at high wave angles.

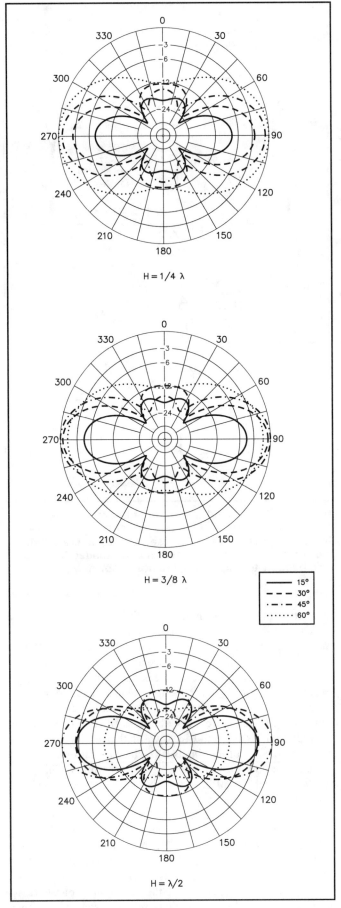

Fig 8-18—Horizontal radiation patterns for the extended double Zepp for wave angles of 15°, 30°, 45° and 60°.

3.2. Feed-point Impedance

The charts from Chapter 9 (Figs 9-8, 9-9, 9-11 and 9-12) can be used to estimate the feed-point impedances of long dipoles. The values from the charts that are made for monopoles must be doubled for dipole antennas.

The antennas can also be modeled with *NEC-2* or *MININEC*, but care should be taken with the results of long dipole antennas less than 0.25 λ above ground with *MININEC*. Remember, too, that except for free space, *MININEC* always reports the impedance for the antenna above a perfect ground conductor, but if the height is specified as mentioned above, the results should be reasonably close to the actual impedance over real ground. *NEC-2*-based programs using the Sommerfeld-Norton ground model are accurate at any height above 0.001 wavelength.

Since center-fed long antennas are not loaded with lossy tuning elements that would reduce their efficiency, long dipoles can have efficiencies very close to 100% if care is taken to use the best materials for the antenna conductor.

3.3. Feeding Long Dipoles

The software module COAX TRANSFORMER/SMITH CHART from the NEW LOW BAND SOFTWARE is an ideal tool for analyzing the impedances, currents, voltages and losses on transmission lines. The STUB MATCHING module can assist you in calculating a stub-matching system in seconds. In any case, we need to know the feed-point impedance of the antenna. Measuring the feed-point impedance is quite difficult, since you cannot use an impedance bridge unless it is specially configured for measuring balanced loads.

3.3.1. Collinear dipoles (two half-wave dipoles in phase)

The impedance at resonance for two half-waves in phase is several thousand ohms. With a 2-mm OD conductor (AWG #12), the impedance is approximately 6000 Ω on 3.5 MHz. The shortening factor in free space for this antenna is 0.952. The SWR bandwidth of the two half-waves in phase is given in **Fig 8-19**. The antenna covers a frequency range from 3.5 to 3.8 MHz with an SWR of less than 2:1.

The antenna can be fed with open-wire feeders into a tuner, or via a stub-matching system and balun as shown in Chapter 6 Fig 6-15. Using tuned feeders with a tuner can, of course, ensure a 1:1 SWR to the transmitter (50 Ω) at all times.

3.3.2. Extended double Zepp

The intrinsic SWR bandwidth of the extended double Zepp is much narrower than for the collinear dipoles. For an antenna made out of 2-mm OD wire (AWG #12) and with a total length of 1.24 λ, the feed-point impedance is approximately $200 - j\,1100$ Ω. The SWR curve (normalized to R_{rad} at the design frequency) is given in Fig 8-19. For lengths varying from 1.24 to 1.29 λ, the radiation resistance will vary from 200 to 130 Ω (decreasing resistance with increasing length).

The exact length of the antenna is not critical, but as we increase the length, the amplitude of the sidelobes increases. The magnitude of the reactance will depend on the length/diameter ratio of the antenna. An antenna made of a thin conductor will show a large reactance value, while the same antenna made of a large-diameter conductor will show much less reactance.

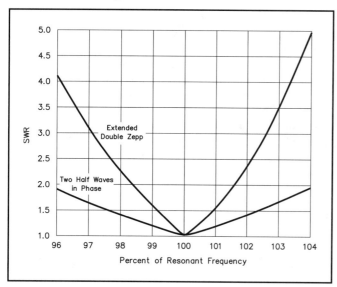

Fig 8-19—SWR curves for an extended double Zepp and for two half waves in phase (collinear array). The calculation was centered on 3.65 MHz using a conductor of 2-mm OD (AWG #12), and the results normalized to the radiation resistances. The SWR bandwidth of the collinear array is much higher than for the double extended Zepp.

The impedance of the extended double Zepp also changes with antenna height, as with a regular half-wave dipole. For the 1.24-λ long extended double Zepp, the resistive part changes between 150 and 260 Ω, and settles at 200 Ω at very high heights.

In principle, we can feed this antenna in exactly the same way as the collinear, but since the intrinsic bandwidth is much more limited, it is better to feed the antenna with open-wire lines running all the way into the shack and to an open-wire antenna tuner.

3.4. Three-Band Antenna (40, 80, 160 Meters)

Refer to the three-band antenna of **Fig 8-20**. On 40 meters the antenna is a collinear array (two half-waves in phase) at a height of 24 meters (80 feet). On 80 meters, it is a half-wave dipole. For 160, we connect the two conductors of the open-wire feeders together, and the antenna is now a flat-top loaded vertical (T antenna). The disadvantage is that we must install a switchable tuning network at the base, right under the antenna. Since the antenna is a vertical on 160 meters, its performance will largely depend on the quality of the ground and the radial system. Some slope away from vertical can, of course, be allowed in the feed line.

4. INVERTED-V DIPOLE

In the past, the inverted-V shaped dipole has often been credited with almost magical properties. The most frequently claimed special property is a low radiation angle. Some have more correctly called it a *poor man's dipole*, since it requires only one high support. Here are the facts.

4.1. Radiation Resistance

The radiation resistance of the inverted-V dipole changes with height above ground (as does a horizontal dipole) and as

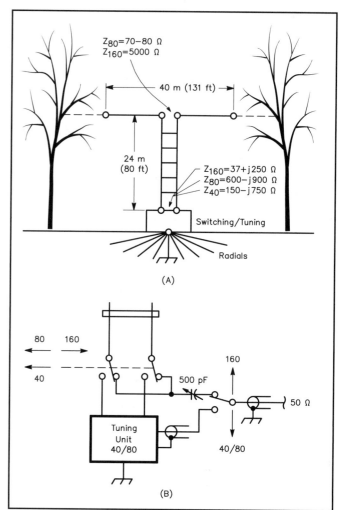

Fig 8-20—Three-band antenna configuration (40, 80 and 160 meters). On 40 meters the antenna is a collinear (two half waves in phase); on 80 meters a half-wave dipole; and on 160 meters a top-loaded T vertical. The bandswitching arrangement is shown at B.

a function of the apex angle, which is the angle between the legs of the dipole. Consider the two apex-angle extremes. When the angle is 180°, the inverted-V becomes a flattop dipole, and the radiation resistance in free space is 73 Ω. Now take the case where the apex angle is 0°. The inverted-V dipole becomes an open-wire transmission line, a quarter-wavelength long and open at the far end. This configuration will not radiate at all—the current distribution will completely cancel all radiation, as it should in a well-balanced feed line. The input impedance of the line is 0 Ω, since a quarter-wave stub open at the end reflects a dead short at the input. This zero-angle inverted-V will have a radiation resistance of 0 Ω and consequently will not radiate at all.

I modeled a range of inverted-V dipoles with different apex angles at different apex heights. This was done using *NEC-2*. **Fig 8-21** shows the radiation resistance as a function of the apex angle for a range of angles between 90° and 180° (a flattop dipole). The curve also shows the physical length that produces resonance, where the feed point is purely resistive. Decreasing the apex angle raises the resonant frequency of the inverted-V.

Fig 8-22 shows the feed-point resistance and reactance

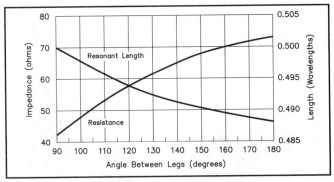

Fig 8-21—Radiation resistance (resistance at resonance) of the inverted-V dipole antenna in free space as a function of the angle between the legs of the dipole (apex angle). Also shown is the physical length (based on the free-space wavelength) for which resonance occurs.

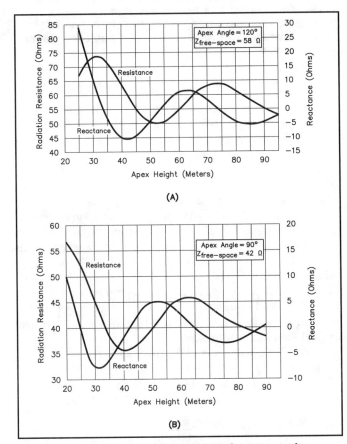

Fig 8-22—Impedance (feed-point resistance and reactance) of inverted-V dipoles as a function of height above ground. Analysis frequency is 3.75 MHz, with a 2-mm OD wire (AWG #12). Resistances at resonance are: 120° apex angle, 58 Ω; 90° apex angle, 42 Ω. *NEC-2* was used for these calculations, since *MININEC* is unreliable for impedance at low heights.

Fig 8-23—Radiation patterns for an inverted-V dipole with an apex angle of 90°. For comparison, the radiation pattern of a horizontal dipole is included in each plot, on the same dB scale. The horizontal pattern is shown for the main wave angle (28° for the straight dipole and 32° for the inverted V). The height of the inverted V is the height at its apex, which is the same height as the flattop horizontal dipole.

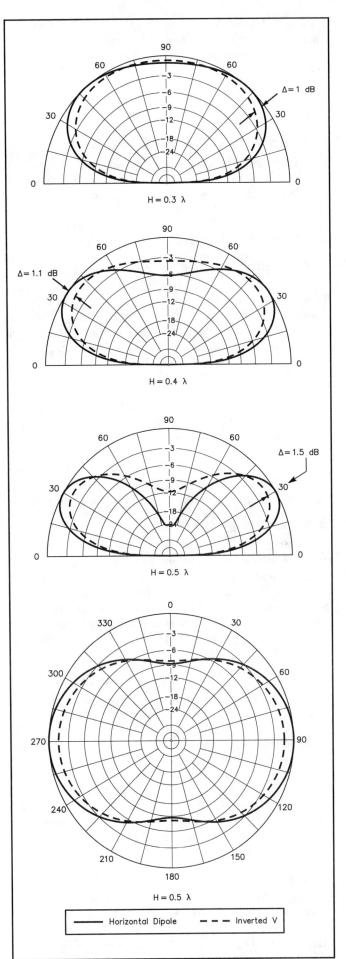

The Dipole   8-19

for inverted-V dipoles with apex angles of 120° and 90°. The antennas were first resonated in free space. Then the reactances were calculated over ground at various heights. Notice that the shape of both curves is similar to the shape of the straight dipole curve in Fig 8-7. Bringing the inverted-V closer to ground lowers its resonant frequency. This is a fairly linear function between 0.25 λ and 0.5 λ apex height.

4.2. Radiation Patterns and Gain

Previous paragraphs compare the inverted-V to a straight dipole at the same apex height. It is clear that the inverted-V is a compromise antenna when compared to the straight horizontal dipole. At low heights (0.25 to 0.35 λ), the gain difference is minimal, but at heights that produce low-angle radiation the dipole performs substantially better.

The 90° apex angle inverted-V dipole

Fig 8-23 shows the vertical and horizontal radiation patterns for inverted-Vs with a 90° apex angle at different apex heights. Modeling was done over good ground. For comparison, I have included the radiation pattern for a straight dipole at the same apex height. In the broadside direction, the inverted-V dipole shows 1 to 1.5 dB less gain than the flattop dipole and also a slightly higher wave angle.

The 120° apex angle inverted-V dipole

The flat-top dipole is 0.6 dB better than the inverted V at a height of 0.4 λ; 0.7 dB at 0.45 λ and 0.8 dB at 0.5 λ. In addition, the wave angle for the horizontal dipole is slightly lower than for the inverted V, at approximately 3° for heights from 0.35λ to 0.5λ. The difference is not spectacular but it is clear that the inverted-V dipole has no magical properties.

4.3. Antenna Height

In many situations it will be possible to erect an inverted-V dipole antenna much higher than a flat top dipole, in most cases because there is only one high support structure available. In this respect the high inverted V can be superior to a low horizontal dipole. The inverted V loses compared to a flattop dipole at the same apex height, but not everyone has two such high supports. And if they do, are they in the right direction?

4.4. Length of the Inverted-V Dipole

The usual formulas for calculating the length of the straight dipole cannot be applied to the inverted-V dipole. The length depends on both the apex angle of the antenna and the height of the antenna above ground. Fig 8-22 shows the feed-point impedance for inverted Vs of different configurations at different heights.

Closing the legs of the inverted-V in free space will increase the resonant frequency. On the other hand, the antenna will become electrically longer when closer to the ground due to the end-loading effect of the ground on the inverted-V ends.

4.5. Bandwidth

Fig 8-24 shows the SWR curves for three inverted-V dipoles with different apex angles: 90°, 120° and 180° (flattop dipole), for a conductor diameter of 2 mm (AWG #12) and a frequency of 3.65 MHz. As expected, the SWR bandwidth

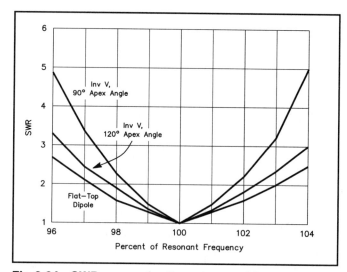

Fig 8-24—SWR curves for three types of free-space half-wave dipoles: The horizontal (flattop) dipole, and inverted-V dipoles with apex angles of 120° and 90°. Each curve is normalized to the feed-point resistance at resonance.

decreases with decreasing apex angle. The computed figures are for free space. The SWR values in Fig 8-24 are normalized figures. This means that the SWR at resonance is assumed to be 1:1, whatever the actual impedance (resistance) at resonance is. In practice, the SWR will almost never be 1:1 at resonance because the line impedance will be different from the feed-point impedance (see the impedance chart in Fig 8-7).

Over ground, the reactive part of the impedance remains almost the same value as in free space, after you have re-resonated the inverted V at the center frequency. This means that the SWR bandwidth will be largest for heights where the radiation resistance is highest. For the inverted-V dipole this is at an apex height of approximately 0.35 to 0.4 λ. Practically speaking, it means that for an apex height of 0.3 λ to 0.5 λ, the SWR curve will be somewhat flatter over ground than in free space.

The SWR bandwidth of the inverted-V can be increased significantly by making a folded-wire version of the antenna. The feed-point impedance of the folded-wire version is four times the impedance shown in Figs 8-20 and 8-21.

5. VERTICAL DIPOLE

The half-wave vertical is covered in detail in the chapter on vertical antennas. Whereas in that chapter we consider the half-wave vertical mainly as a base-fed antenna, we can of course use a dipole made of wire and feed it in the center. This is what we usually call a *vertical dipole*. In many practical cases a wire half-wave vertical will not be perfectly vertical, but will generally slope away from a tall support such as a tower or a building. Sloping half-wave verticals are covered in Section 6.

5.1. Radiation Patterns

Whether the half-wave vertical is base fed or fed in the center, the current distribution is identical, and hence the radiation pattern will be identical. Radiation patterns are shown in **Fig 8-25** when the lower end is near the ground. Over saltwater the half-wave vertical can yield 6.1-dBi gain, which drops to about 0 dBi over good soil. As with all verticals, it is mainly the quality of the ground in the Fresnel zone that determines how good a low-angle radiator the vertical dipole will be (see Section 3 of the chapter on verticals). Half-wave verticals produce excellent (very) low-angle radiation when erected in close proximity to saltwater. As a general-purpose DX antenna the vertical dipole may, however, produce too low an angle of radiation for some nearby DX paths.

Raising the half-wave vertical higher above the ground introduces multiple lobes. **Fig 8-26** shows the patterns for a half-wave center-fed vertical with the bottom $1/8\ \lambda$ above ground. Note the secondary lobe, which is similar to the lobe we encountered with the horizontally polarized extended double Zepp.

I also modeled a half-wave vertical on top of a rocky island with very poor ground, 250 meters (820 feet) above sea level, and some 100 meters (330 feet) from the sea. **Fig 8-27** shows the layout and the radiation pattern. Superimposed on the pattern are the patterns for the same antenna at sea level, as well as over very poor ground. Note that the extra height does not give any gain advantage over sea level but the extra height does help low-angle rays shoot across the poor ground (the rocky island) and find reflection at sea level some 250 meters below the antenna.

5.2. Radiation Resistance

The radiation resistance of a vertical half-wave dipole, fed at the current maximum (the center of the dipole), is shown in **Fig 8-28** as a function of its height above ground. The impedance remains fairly constant except for very low heights. No current flows at the tips of the dipole, and hence the small influence of the height on the impedance, except at very low heights where the capacitive effect of the bottom of the antenna against ground lowers the resonant frequency of the antenna.

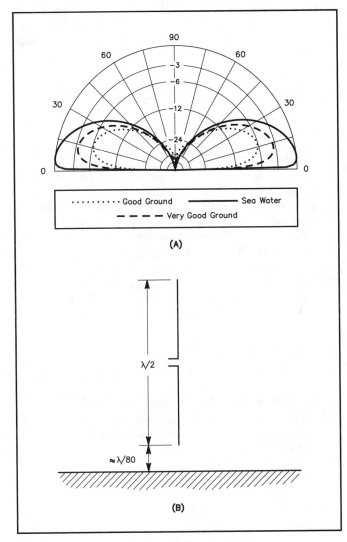

Fig 8-25—At A, vertical radiation patterns over various grounds for a vertical half-wave center-fed dipole with the bottom tip just clearing the ground, as shown at B. The gain is as high as 6.1 dBi over ground. The feed-point impedance is 100 Ω.

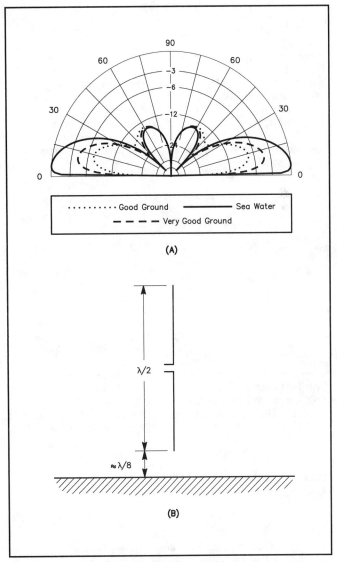

Fig 8-26—At A, vertical radiation patterns of the half-wave vertical dipole with the bottom tip $1/8\ \lambda$ off the ground, as shown at B.

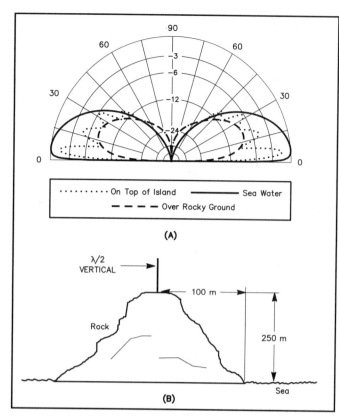

Fig 8-27—At A, the serrated radiation pattern of a half-wave vertical overlooking a slope of very poor ground (an island with volcanic soil) next to the ocean, as shown at B. Because of the antenna height above the sea, multiple lobes show up in the pattern. The radiation patterns of the half-wave vertical at sea level and the pattern over very poor ground are superimposed for comparison.

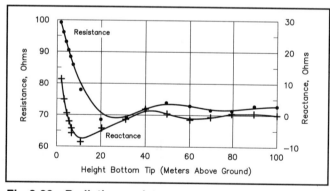

Fig 8-28—Radiation resistance and reactance of the half-wave vertical as a function of height above ground. The height is taken as the height of the bottom tip. Calculations are for a design frequency of 3.5 MHz.

5.3. Feeding the Vertical Half-Wave Dipole

There are two main approaches to feeding a vertical half-wave dipole:
- Base feeding against ground (voltage feeding)
- Feeding in the center (current feeding)

Base feeding is covered in Sect. 4.4 of Chapter 6 on matching and feed lines. In most cases you will use a parallel

Fig 8-29—View from the top of I8UDB's tower in Naples. Such an awesome view needs no comment.

tuned circuit on which the coax feed line is tapped. If the vertical is made using a sizable tower, the base impedance may be relatively low (600 Ω), and a broad-band matching system as described in Sect. 4.5.2 in Chapter 6 on matching and feed lines (the W1FC broadband transformer) may be used.

A center-fed vertical dipole must be fed in the same way as a horizontal dipole. It represents a balanced feed point, and can be fed using open-wire line to a balanced tuner, or via a balun to a coaxial feed line (see Sect. 1.4.).

6. SLOPING DIPOLE

Sloping half-wave dipoles are used very successfully by a number of stations, especially near the sea. FK8CP is using a half-wave sloper on 160 meters, with the end of the antenna connected about 15 meters above sea level, less than 50 meters from the saltwater. I8UDB is using a sloper on 160 from his mountaintop location near Naples, where electrical ground is nonexistent, but where the sea is only 100 meters away and a few hundred meters below the antenna. See **Fig 8-29**.

The half-wave sloper radiates a signal with both horizontal and vertical polarization components. Unless it is very high above the ground (such as I8UDB), you need not bother with the horizontal component. Low-angle radiation will be produced only by the vertical component. All modeling in this section was done on 80 meters, over a very good ground.

6.1. The Sloping Straight Dipole

Due to the weight of the feed line, a sloping dipole will seldom have two halves in a straight line. Let us nevertheless analyze the antenna as if it does.

Radiation patterns

Fig 8-30 shows the radiation patterns of sloping half-wave dipoles for apex angles of 15°, 30° and 45° over three types of ground (poor, good and sea). For the dipole with a 45° slope angle I include the pattern showing the vertical and the horizontal radiation separately (**Fig 8-31**).

It is obvious that the steeper the slope, the less horizontal radiation component there will be. High-angle radiation is only due to the horizontal radiation component. For the verti-

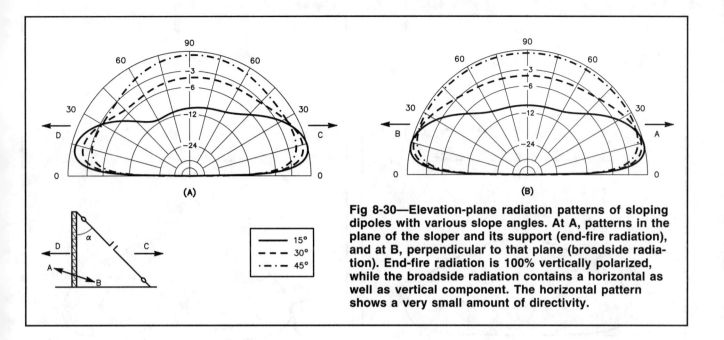

Fig 8-30—Elevation-plane radiation patterns of sloping dipoles with various slope angles. At A, patterns in the plane of the sloper and its support (end-fire radiation), and at B, perpendicular to that plane (broadside radiation). End-fire radiation is 100% vertically polarized, while the broadside radiation contains a horizontal as well as vertical component. The horizontal pattern shows a very small amount of directivity.

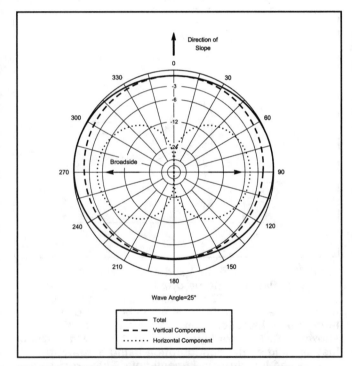

Fig 8-31—Azimuth-plane radiation pattern for the sloping dipole with a 45° slope angle, taken at a 25° wave angle. Patterns for the vertical and horizontal components of the total are also shown. The directivity is very limited. Actually, the sloping dipole radiates best about 70° either side of the slope direction.

cal component the same rules apply as for the half-wave vertical: In order to exploit the intrinsic very low-angle capabilities, you must have an excellent ground around the antenna. Don't forget, the Fresnel zone (the area where the reflection at ground level takes place) can stretch all the way out to 10 wavelengths or more from the antenna.

Fig 8-31 shows the horizontal pattern for a sloping dipole with a 45° slope angle. The sloper is almost omnidirectional, but radiates best broadside (perpendicular to the plane going through the sloper and the support). In the end-fire direction (in the plane of the sloper and its support), it has less than 1-dB F/B at an elevation angle of 25°. The antenna radiates a little better in the direction of the slope. The fact that it radiates best in the broadside direction is due to the horizontal component, which only radiates in the broadside direction.

Impedance

The radiation resistance of the sloping dipole with the bottom wire $1/80$ λ above ground (1 meter for an 80-meter antenna) varies from 96 Ω for a 15° slope angle to 81 Ω for a 45° slope angle.

6.2. The Bent-Wire Sloping Dipole

Most real-life sloping half-wave dipoles have a bent-wire shape, because of the weight of the feed line. **Fig 8-32** and **Fig 8-33** analyze a sloping vertical with a slope angle of 20° for the top half of the antenna, and slope angles of 40° and 60° respectively for the bottom half of the dipole. Using a 60° slope angle reduces the height requirement for the support.

The sloping dipole with a relatively horizontal bottom quarter-wave wire yields almost the same signal as the straight sloping dipole. It is important to keep the top half of the sloping dipole as vertical as possible. Analysis shows the angle of the bottom half of the antenna is relatively unimportant.

Feed point

Is the feed point of such a bent sloping dipole a symmetrical feed point? Not strictly speaking. If you use such an antenna, don't take any chances. It does not hurt to put a current balun at a load even when the load is asymmetric. Use a current-type choke balun to remove any current from the outside of the coaxial cable. A coiled coax or a stack of ferrite beads is the way to go (see the chapter on feed lines and antenna matching).

The Dipole   8-23

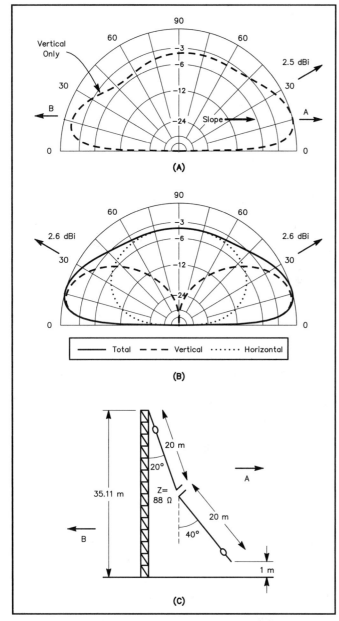

Fig 8-32—At A, "end-fire" and at B, "broadside" vertical radiation patterns of a bent-wire half-wave sloper for 3.6 MHz. The horizontal and vertical components of the total pattern are also shown at B. The bottom 0.25-λ section slopes at an angle of 40°, as shown at C. Modeling is done over very good ground.

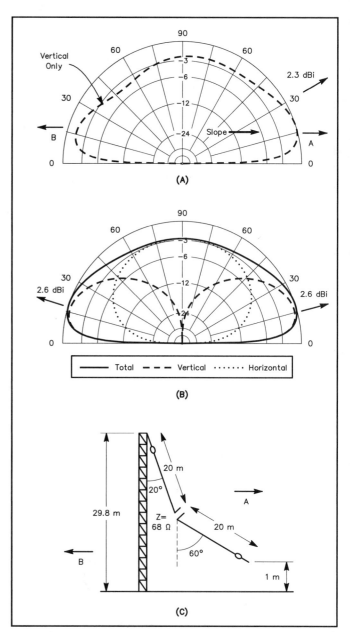

Fig 8-33—At A, "end-fire" and at B, "broadside" vertical radiation patterns of a bent-wire half-wave sloper for 3.6 MHz. The horizontal and vertical components of the total pattern are also shown at B. The bottom 0.25-λ section slopes at an angle of 60°, as shown at C. This configuration and that of Fig 8-32 are just as valid as the configuration using a straight sloper. The loss in gain is negligible. This arrangement requires less support height than that of the straight sloper or that of Fig 8-32.

6.3. Evolution into the Quarter-Wave Vertical

We can go one step further and bring the bottom quarter-wave all the way horizontal. If the top half were fully vertical, we now would have a quarter-wave vertical with a single elevated radial. This configuration is described in detail in the Chapter 9 on vertical antennas (see Fig 9-18).

To transform the half-wave sloper into a quarter-wave vertical, we first replace the sloping bottom half of the antenna with two wires, now called *radials*. Both radials are "in line" and slope toward the ground, as shown in **Fig 8-34C**. A and B of Fig 8-34 show the radiation patterns for this configuration.

Note that the high-angle radiation has been attenuated some 10 dB, and we pick up 0.5 to 0.8 dB of gain. The little horizontally polarized radiation left over is, of course, caused by the sloping radials. The configuration shows gain in the direction of the sloping wire of approximately 0.4 dB.

Next we move the radials up, so they are horizontal, and move the antenna down so the base is now 5 meters (16 feet) above ground (Fig 8-34E). All the horizontal radiation is gone, and the gain has settled halfway between the forward

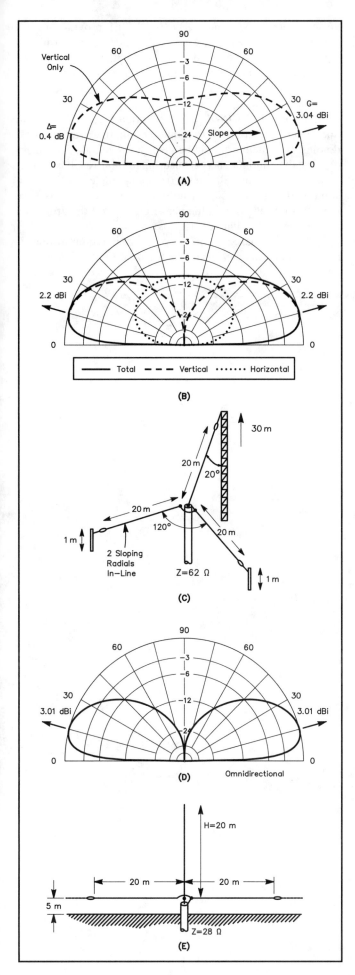

Fig 8-34—Transition from a sloping dipole to a 0.25-λ vertical with two radials. At C, the bottom half of the dipole is replaced by two 0.25-λ wires sloping to the ground; the resulting patterns are shown at A and B. At E the radials are lifted to the horizontal, with the resulting pattern at D. This changes eliminates all the horizontal radiation component that was originated by the sloping wires. Analysis frequency: 3.65 MHz.

and the backward gain of the previous model, which is to be expected. We now have a quarter-wave vertical with two radials, which is how the original ground plane was developed (see Sect.1.3.3 of the chapter on verticals).

The quarter-wave vertical with two radials definitely has an asymmetrical feed point. The feed line is exposed to the strong fields of the antenna and often is run on the ground under the two radials. So you should fully decouple the feed line from the feed point by using a current-type balun (coiled coax or stack of ferrite beads).

6.4. Conclusion

Some 6.1-dBi gain can be obtained with a half-wave vertical only over nearly perfect ground (such as saltwater). Even over very good soil, the half-wave vertical will not be any better than a quarter-wave vertical (3-dBi gain). This means that unless you are near the sea, you may as well stick with a quarter-wave vertical. The sloping vertical (make the sloping wire as vertical as possible) with two radials (5 meters high for 3.6 MHz) will produce as good a signal as a half-wave vertical or sloping half-wave vertical over very good ground. It will, however, only require a 25 meter support instead of a 35 or 40 meter support for a half-wave 80-meter vertical dipole.

7. MODELING DIPOLES

MININEC or *NEC-2*-based modeling programs are well suited for modeling dipoles. Straight dipoles can be accurately modeled with a total of 10 to 20 pulses. Inverted-V dipoles require more pulses, depending on the apex angle, to obtain accurate impedance data. **Table 8-6** shows impedance data for a straight dipole, and **Table 8-7** for an inverted-V dipole as a function of the pulses, wires and segments. An inverted-V with a 90° apex angle requires at least 50 equal-length segments for accurate impedance data. By using the TAPERING technique (see the chapter on Yagi and quad antennas), accurate results can be obtained with a total of only

Table 8-6
MININEC Pulses Versus Calculated Impedance for a Straight Dipole Antenna

Pulses	Impedance
5	$71 - j14$
10	$67 - j26$
20	$68 - j28$
30	$68.5 - j28$
50	$68.6 - j27.3$
80	$68.7 - j27.1$
100	$68.7 - j27.0$

Table 8-7
MININEC Pulses Vs Calculated Impedance for an Inverted-V Dipole Antenna

Pulses	Impedance
5	$43.6 - j\,23.7$
10	$44.3 + j\,10.3$
20	$44.6 + j\,28.4$
30	$44.6 + j\,34.3$
50	$44.7 + j\,38.1$
80	$44.7 + j\,39.8$
100	$44.8 + j\,42.0$
20 tapered, min 0.4 m, max 3.0 m	$44.2 + j\,36.1$
26 tapered, min 0.3 m, max 2.0 m	$44.4 + j\,37.6$
26 tapered, min 0.4 m, max 2.0 m	$44.4 + j\,36.3$
28 tapered, min 0.4 m, max 2.0 m	$44.4 + j\,38.9$
46 tapered, min 0.2 m, max 1.0 m	$44.2 + j\,40$

26 segments. *ELNEC* and *EZNEC* by W7EL provide an automatic feature for generating tapered segment lengths, which is a great asset when you model antennas with bent conductors.

Knowing the exact impedance is important only if you want to calculate the exact resonant length (or frequency) of a dipole, or if the dipole is part of an array. To obtain reliable results using a *MININEC*-based program the dipoles should not be modeled too close to ground. For half-wave horizontal dipoles, the antenna should be at least $0.2\,\lambda$ high. For longer dipoles, the minimum height ensuring reliable results is somewhat higher. Vertical dipoles and sloping dipoles (with a steep slope angle) can be modeled quite close to the ground, as there is very little radiation in the near-field toward the ground (a dipole does not radiate off its tips).

A *NEC-2*-based modeling program is required if accurate gain and impedance data are required for dipoles close to ground.

CHAPTER 9

Vertical Antennas

THANKS DJ2YA

Uli Weiss, DJ2YA, is an all-around radio amateur. His more-than-casual interest and in-depth knowledge of antenna matters and his eminent knowledge of the English language (Uli teaches English at a German "Gymnasium") has made him one of the few persons who could successfully translate the Low Band DXing book into the German language without any assistance from the author. It also makes him a very successful antenna builder and contest operator. Uli was, with Walter Skudlarek, DJ6QT, cofounder of the world-renowned RRDXA Contest Club, which has lead the CQ-WW Club championships for many years.

Uli has been an editor, helping hand and supporter for this chapter on vertical antennas. Thank you for your help, Uli.

The effects of the earth itself and the artificial ground system (if used) on the radiation pattern and the efficiency of vertically polarized antennas is often not understood. They have until recently not been covered extensively in the amateur literature.

The effects of the ground and the ground system are twofold. Near the antenna (in the *near field*), you need a good ground system to collect the antenna return currents without losses. This will determine the *radiation efficiency* of the antenna.

At distances farther away (in the *far field*, also called the *Fresnel zone*), the wave is reflected from the earth and combines with the direct wave to generate the overall radiation pattern. The absorption of the reflected wave is a function of the ground quality and the incident angle. This mechanism determines the *reflection efficiency* of the antenna.

Vertical monopole antennas are often called *ground-mounted verticals*, or simply verticals. They are, by definition, mounted perpendicular to the earth, and they produce a vertically polarized signal. Verticals are popular antennas for the low bands, since they can produce good low-angle radiation without the very high supports needed for horizontally polarized antennas to produce the same amount of radiation at low takeoff angles.

1. THE QUARTER-WAVE VERTICAL
1.1. Radiation Patterns
1.1.1. Vertical pattern of vertical monopoles over ideal ground

The radiation pattern produced by a ground-mounted quarter-wave vertical antenna is basically one-half that of a half-wave dipole antenna in free space. The dipole is twice the physical size of the vertical and has a symmetrical current distribution. A vertical antenna is frequently referred to as a "monopole" to distinguish it from a dipole. The radiation pattern of a quarter-wave vertical monopole over perfect ground is half of the figure-8 shown for the half-wave dipole in free space. See **Fig 9-1**.

The relative field strength of a vertical antenna with sinusoidal current distribution and a current node at the top is given by:

$$E_f = k \times I \left[\frac{\cos(L \sin \alpha) - \cos L}{\cos \alpha} \right] \qquad (Eq\ 9\text{-}1)$$

where

k = constant related to impedance
E_f = relative field strength
α = elevation angle above the horizon
L = electrical length (height) of the antenna
I = antenna current

This equation does not take imperfect ground conditions into account, and is valid for antenna heights between 0° and 180° (0 to λ/2). The "form factor" inside the square brackets containing the trigonometric functions is often published by itself for use in calculating the field strength of a vertical antenna. If used in this way, however, it appears that short verticals are vastly inferior to tall ones, since the antenna length appears only in the numerator of the fraction.

Replacing the current I in the equation with the term

$$\sqrt{\frac{P}{R_{rad}+R_{loss}}}$$

gives a better picture of the actual situation. For short verticals, the value of the radiation resistance is small, and this term largely compensates for the decrease in the form factor. This means that for a constant power input, the current into a small vertical will be greater than for a larger monopole.

The radiation resistance R_{rad} does not determine the current—the sum of the radiation resistance and the loss resistance(s) does. With a less-than-perfect ground system and short, less-than-perfect loading elements (lossy coils used with short verticals), the radiation can be significantly less than the case of a larger vertical (where R_{rad} is large in comparison to the ground loss and where there are no lossy loading devices).

Interestingly, short verticals are almost as efficient radiators as are longer verticals, provided the ground system is good and there are no lossy loading devices. When the losses of the ground system and the loading devices are brought into the picture, however, the sum $R_{rad} + R_{loss}$ will get larger, and as a result part of the supplied power will be lost in the form of heat in these elements. For instance, if $R_{rad} = R_{loss}$, half of the power will be lost. Note that with very short verticals, these losses can be much higher.

1.1.2. Vertical radiation pattern of a monopole over real ground

The three-dimensional radiation pattern from an antenna is made up of the combination of the direct wave and the wave resulting from reflection from the earth. The following explanation is valid only for reflection of vertically polarized waves. See Chapter 8 on dipole antennas for an explanation of the reflection mechanism for horizontally polarized waves.

For perfect earth there is no phase shift of the vertically polarized wave at the reflection point. The two waves add with a certain phase difference, due only to the different path lengths. This is the mechanism that creates the radiation pattern. Consider a distant point at a very low angle to the horizon. Since the path lengths are almost the same, reinforcement of the direct and reflected waves will be maximum. In case of a perfect ground, the radiation will be maximum just above a 0° elevation angle.

1.1.2.1. The reflection coefficient

Over real earth, reflection causes both amplitude and phase changes. The reflection coefficient describes how the incident (vertically polarized) wave is being reflected. The reflection coefficient of real earth is a complex number with magnitude and phase, and it varies with frequency. In the polar-coordinate system the reflection coefficient consists of:

- The magnitude of the reflection coefficient: It determines how much power is being reflected, and what percentage is being absorbed in the lossy ground. A figure of 0.6 means that 60% will be reflected and 40% absorbed.
- The phase angle: This is the phase shift that the reflected wave will undergo as compared to the incident wave.

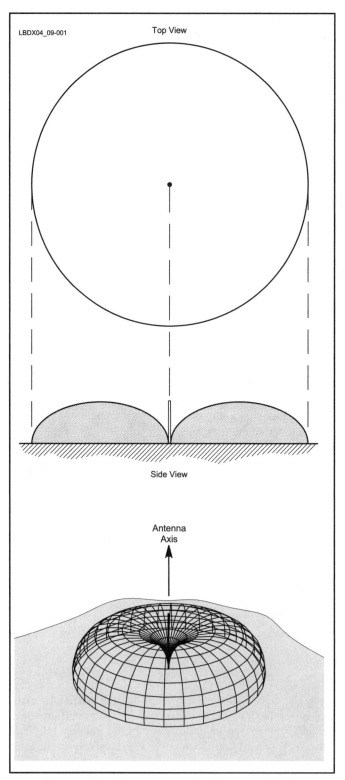

Fig 9-1—The radiation patterns produced by a vertical monopole over perfect ground. The top view is the horizontal pattern, and the side view is the vertical (elevation plane) pattern.

Over real earth the phase is always lagging (minus sign). At a 0° elevation angle, the phase is always −180°. This causes the total radiation to be zero (the incident and reflected waves, which are 180° out-of-phase and equal in magnitude, cancel each other). At higher elevation angles, the reflection phase angle will be close to zero (typically −5° to −15°, depending on the ground quality).

1.1.2.2. The pseudo-Brewster angle

The magnitude of the vertical reflection coefficient is minimum at a 90° phase angle. This is the reflection-coefficient phase angle at which the so-called *pseudo-Brewster wave* angle occurs. It is called the pseudo-Brewster angle because the RF effect is similar to the optical effect from which the term gets its name. At the pseudo-Brewster angle the reflected wave changes sign. Below the pseudo-Brewster angle the reflected wave will subtract from the direct wave. Above the pseudo-Brewster angle it adds to the direct wave. At the pseudo-Brewster angle the radiation is 6 dB down from the perfect ground pattern (see **Fig 9-2**).

All this should make it clear that knowing the pseudo-Brewster angle is important for each band at a given QTH. Most of us use a vertical to achieve good low-angle radiation.

Fig 9-3 shows the reflection coefficient (magnitude and phase) for 3.6 MHz and 1.8 MHz for three types of ground. Over seawater the reflection-coefficient phase angle changes from −180° at a 0° wave angle to −0.1° at less than 0.5° wave angle! The pseudo-Brewster angle is at approximately 0.2° over saltwater.

1.1.2.3. Ground-quality characterization

Ground quality is defined by two parameters: the dielectric constant and the conductivity, expressed in milliSiemens per meter (mS/m). Table 5-2 in Chapter 5 shows the characterization of various real-ground types. The table also shows five distinct types of ground, labeled as very good, average, poor, very poor and extremely poor. These come from Terman's classic Radio Engineers' Handbook, and are also used by Lewallen in his *ELNEC* and *EZNEC* modeling programs. The denominations and values listed in Table 5-2 are the standard ground types used throughout this book for modeling radiation patterns. In the real world, ground characteristics are never homogeneous, and extremely wide variations over short distances are common. Therefore any modeling results based on homogeneous ground characteristics will only be as accurate as the homogeneity of the ground itself.

1.1.2.4. Brewster angle equation

Terman (*Radio Engineers' Handbook*) publishes an equation that gives the pseudo-Brewster angle as a function of the ground permeability, the conductivity and the frequency. The chart in **Fig 9-4** uses the Terman equation. Note especially how saltwater has a dramatic influence on the low-angle radiation performance of verticals. In contrast, a sandy, dry ground yields a pseudo-Brewster angle of 13° to 15° on the low bands, and a city (heavy industrial) ground yields a pseudo-Brewster angle of nearly 30° on all frequencies! This means that under such circumstances the radiation efficiency

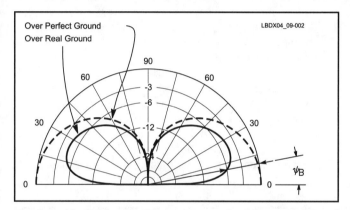

Fig 9-2—Vertical radiation patterns of a λ/4 monopole over perfect and imperfect earth. The pseudo-Brewster angle is the radiation angle at which the real-ground pattern is 6 dB down from the perfect-ground pattern.

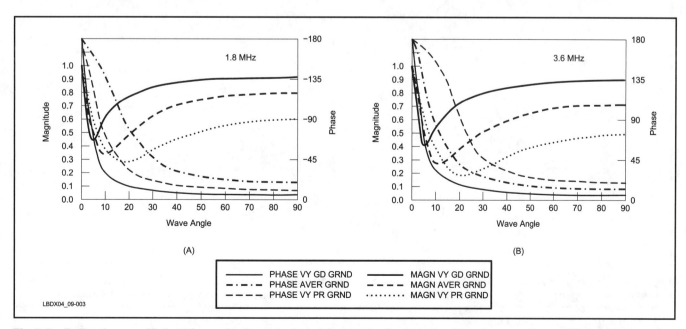

Fig 9-3—Reflection coefficient (magnitude and phase) for vertically polarized waves over three different types of ground (very good, average, and very poor).

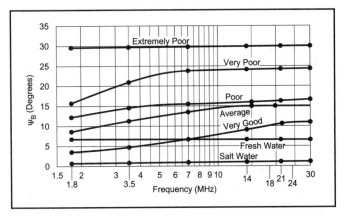

Fig 9-4—Pseudo-Brewster angle for different qualities of reflecting ground. Note that over salt water the pseudo-Brewster angle is constant for all frequencies, at less than 0.1°! That's why vertical antennas located right at the saltwater shore get out so well.

for angles under 30° will be severely degraded in a city environment.

1.1.2.5. Brewster angle and radials

Is there anything you can do about the pseudo-Brewster angle? Very little. Ground-radial systems are commonly used to reduce the losses in the near field of a vertical antenna. These ground-radial systems are usually 0.1 to 0.5 λ long, too short to improve the earth conditions in the area where reflection near the pseudo-Brewster angle takes place.

For quarter-wave verticals the Fresnel zone (the zone where the reflection takes place) is 1 to 2 λ away from the antenna. For longer verticals (such as a half-wave vertical) the Fresnel zone extends up to 100 wavelengths away from the antenna (for an elevation angle of about 0.25°).

This means that a good radial system improves the efficiency of the vertical in collecting return currents and shielding from lossy ground, but will not influence the radiation by improving the reflection mechanism in the Fresnel zone. Of course you could add 5 λ long radials, and keep the far ends of these radials less than 0.05 λ apart by using enough radials. But that seems rather impractical for most of us! In most practical cases radiation at low takeoff angles will be determined only by the real ground around the vertical antenna.

Conclusion

This information should make it clear that a vertical may not be the best antenna if you are living in an area with very poor ground characteristics. This has been widely confirmed in real life—Many top-notch DXers living in the Sonoran desert or in mountainous rocky areas on the West Coast swear by horizontal antennas for the low bands, at least on 80 meters, while some of their colleagues living in flat areas with rich fertile soil, or even better, on such a ground near the sea coast, will be living advocates for vertical antennas and arrays made of vertical antennas.

On Topband another mechanism enters into the game—the effect of power coupling (see Chapter 1, Section 3.5), which makes a vertically polarized antenna the better antenna in most places away from the equator (eg, North America and Europe) due to the influence of the Earth's magnetic field. In addition,

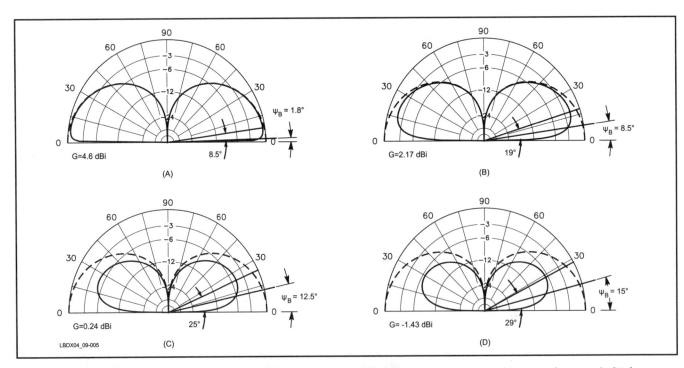

Fig 9-5—Vertical-plane radiation patterns of 80-meter λ/4 verticals over four standard types of ground. At A, over saltwater. At B, over very good ground. At C, over average ground. At D, over very poor ground. In each case using 64 radials, each 20 meters long. The perfect ground pattern is shown in each pattern as a reference (broken line, with a gain of 5.0 dBi). This reference pattern also allows us to calculate the pseudo-Brewster angle. The patterns and figures were obtained using the *NEC-4* modeling program. (*Modeling was done by R. Dean Straw, N6BV.*)

horizontally polarized antennas producing a low radiation angle on 160 meters are out of reach for all but a few, who have antenna supports that are several hundred meters high!

1.1.2.6. Vertical radiation patterns

It is important to understand that gain and directivity are two different things. A vertical antenna over poor ground may show a good wave angle for DX, but its gain may be poor. The difference in gain at a 10° elevation angle for a quarter-wave vertical over very poor ground, as compared to the same vertical over sea-water, is an impressive 6 dB. **Fig 9-5** shows the vertical-plane radiation pattern of a quarter-wave vertical over four types of "real" ground:

- Seawater
- Excellent ground
- Average ground
- Extremely poor ground

The patterns in Fig 9-5 are all plotted on the same scale.

1.1.2.7. Vertical radiation patterns over sloping grounds

So far all our discussions about radiation patterns assumed we have perfectly homogeneous flat ground stretching for tens of wavelengths around the antenna. In Section 1.1.2 of Chapter 5, I discussed the influence of sloping terrain on vertical radiation patterns of antennas on the low bands. **Fig 9-6** shows that a terrain that slopes downhill in the direction of the target is as helpful for vertical antennas as it is for horizontally polarized antennas. On the other hand, an upwards-sloping terrain works the other way!

1.1.3. Horizontal pattern of a vertical monopole

The horizontal radiation pattern of both the ground-mounted monopole and the vertical dipole is a circle.

1.2. Radiation Resistance of Monopoles

The IRE definition of radiation resistance says that radiation resistance is the total power radiated as electromagnetic radiation, divided by the *net* current causing that radiation.

The radiation resistance value of any antenna depends on where it is fed (see definition in Chapter 6, Section 3). I'll call the radiation resistance of a vertical antenna at a point of current maximum as $R_{rad(I)}$ and the radiation resistance of a vertical antenna when fed at its base as $R_{rad(B)}$. For verticals greater than one quarter-wave in height, these two are not the same. Why is it important to know the radiation resistance of our vertical? The information is required to calculate the efficiency of the vertical:

$$\text{Eff} = \frac{R_{rad}}{R_{rad} + R_{loss}}$$

The radiation resistance of the antenna plus the loss resistance R_{loss} is the resistive part of the feed-point impedance of the vertical. The feed-point resistance (and reactance) is required to design an appropriate matching network between the antenna and the feed line.

Fig 9-7 shows $R_{rad(I)}$ of verticals ranging in electrical height from 20° to 540°. (This is the radiation resistance referred to the current maximum.) The radiation resistance of a vertical shorter than or equal to a quarter wavelength and fed at its base [thus $R_{rad(I)} = R_{rad(B)}$] can be calculated as follows:

$$R_{rad} = \frac{1450\,h^2}{\lambda^2} \qquad \text{(Eq 9-2)}$$

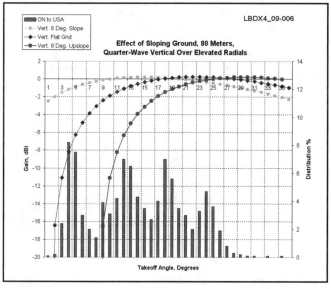

Fig 9-6—The bar graph represents the distribution of the wave angles encountered on 80 meters on a Europe to USA path. Modeling was done over good ground. The wave angles are shown for a λ/4 vertical over flat ground, over an uphill slope of 8° and over a downhill slope of 8%. The downhill slope is very helpful when it comes to very low angles.

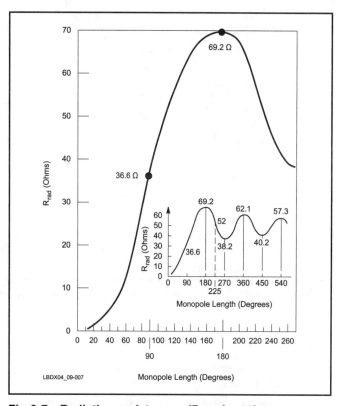

Fig 9-7—Radiation resistances ($R_{rad(I)}$, at the current maximum) of monopoles with sinusoidal current distribution. The chart can also be used for dipoles, but all values must be doubled.

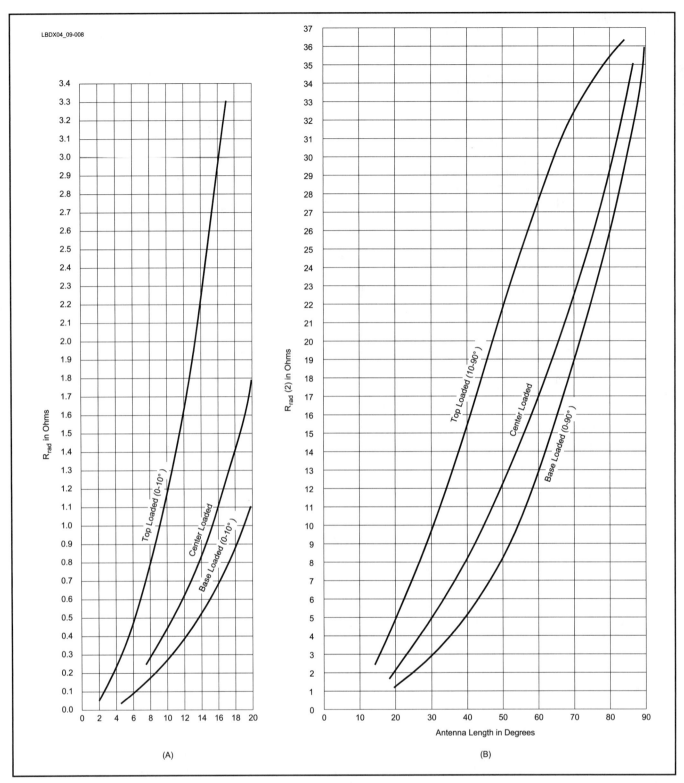

Fig 9-8—Radiation resistance charts (R_{rad}) for verticals up to 90° or λ/4 long. At A, for lengths up to 20°, and at B, for greater lengths.

where
    h = effective antenna height, meters
    λ = wavelength of operation, meters (= $300/f_{MHz}$)

The *effective height* of the antenna is the height of a theoretical antenna having a constant current distribution all along its length. The area under this current distribution line is equal to the area under the current distribution line of the "real" antenna. Equation 2 is valid for antennas with a ratio of antenna length to conductor diameter of greater than 500:1 (typical for wire antennas).

For a full-size, quarter-wave antenna the radiation resistance is determined by:

Current at the base of the antenna = 1 A (given)
Area under sinusoidal current-distribution curve =

1 A × 1 radian = 1 A ×180/π = 57.3 A-degrees
Equivalent length = 57.3° (1 radian)
Full electrical wavelength = 300/3.8 = 78.95 meters
Effective height = (78.95 × 57.3)/360 = 12.56°

$$R_{rad} = \frac{1450 \times 12.56^2}{78.95^2} = 36.6 \, \Omega$$

The same procedure can be used for calculating the radiation resistance of various types of short verticals.

Fig 9-8 shows the radiation resistance for a short vertical (valid for antennas with diameters ranging from 0.1° to 1°). For antennas made of thicker elements, **Fig 9-9** and **Fig 9-10** can be used. These charts are for antennas with a constant diameter.

For verticals with a tapering diameter, large deviations have been observed. W. J. Schultz describes a method for calculating the input impedance of a tapered vertical (Ref 795). It has also been reported that verticals with a large diameter

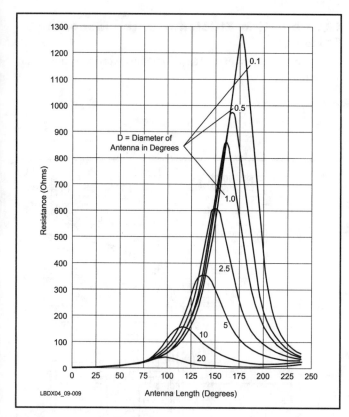

Fig 9-9—Radiation resistances for monopoles fed at the base. Curves are given for various conductor (tower) diameters. The values are valid for perfect ground only.

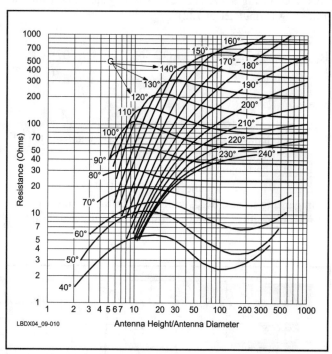

Fig 9-10—Radiation resistances for monopoles fed at the base. Curves are given for various height/diameter ratios over perfect ground.

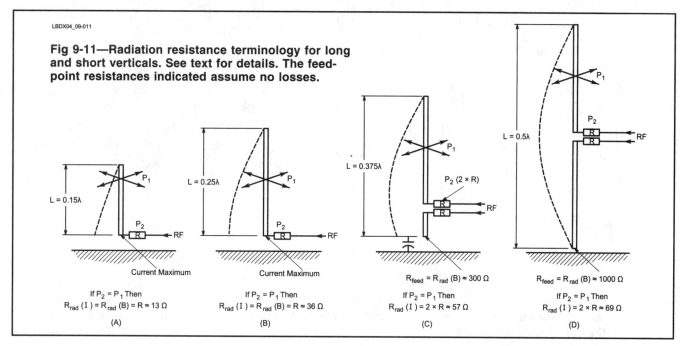

Fig 9-11—Radiation resistance terminology for long and short verticals. See text for details. The feed-point resistances indicated assume no losses.

Vertical Antennas 9-7

exhibit a much lower radiation resistance than the standard 36.6-Ω value. A. Doty, K8CFU, reports finding values as low as 21 Ω during his extensive experiments on elevated radial systems (Ref 793). I have measured a similar low value on my quarter-wave 160-meter vertical (see Section 6.5.) Section 1.2 shows how to calculate the radiation resistance of various types of short verticals.

Longer vertical monopoles are usually not fed at the current maximum, but rather at the antenna base, so that $R_{rad(I)}$ is no longer the same as $R_{rad(B)}$ for long verticals in Figs 9-9 and 9-10. (Source: Henney, *Radio Engineering Handbook*, McGraw-Hill, NY, 1959, used with permission.) $R_{rad(I)}$ is illustrated in **Fig 9-11**. The value can be calculated from the following formula (Ref 722):

$$R_{rad(I)} = \varepsilon - 0.7\,L + 0.1\left[20 \sin(12.56637 L - 4.08407)\right] + 45$$

(Eq 9-3)

where

ε = the base for natural logarithms, 2.71828 .
L = antenna length in radians (radians = degrees × π/180°
= degrees divided by 57.296).
The length must be greater than π/2 radians (90°).

Fig 9-11C shows the case of a 135° (3λ/4) antenna. Disregarding losses, $R_{rad(B)} = R_{feed} \approx 300\,\Omega$, but the value of 2R, the theoretical resistance at the maximum current point, will be lower (57 Ω). If P1 (radiated power) = P2 (power dissipated in 2R), then $R_{rad(I)} = 2R$.

These values of $R_{rad(I)}$ are given in Fig 9-6, while $R_{rad(B)}$ can be found in Figs 9-8 and 9-9. **Fig 9-12** and **Fig 9-13** show the reactance of monopoles (at the base feed point) for varying antenna lengths and antenna diameters (Source: E. A. Laport, *Radio Antenna Engineering*, McGraw-Hill, NY, 1952, used by permission.)

1.3. Radiation Efficiency of the Monopole Antenna

The radiation efficiency for short verticals has been defined as

$$\text{Eff} = \frac{R_{rad}}{R_{rad} + R_{loss}}$$

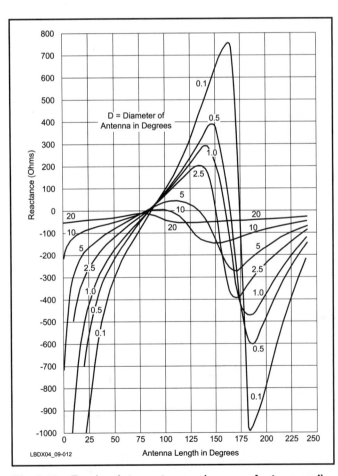

Fig 9-12—Feed-point reactances (over perfect ground) for monopoles with varying diameters.

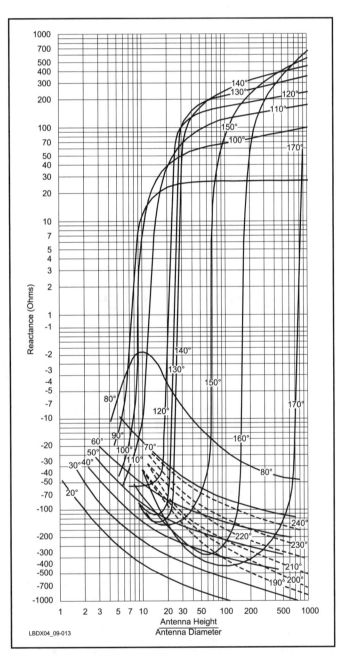

Fig 9-13—Feed-point reactances (over perfect ground) for monopoles with different height/diameter ratios.

For the case of any vertical, short or long, when fed at its base this equation becomes

$$\text{Eff} = \frac{R_{rad}}{R_{rad(B)} + R_{loss}} \quad \text{(Eq 9-4)}$$

The loss resistance of a vertical is composed of:
- Conductor RF resistance
- Parallel losses from insulators
- Equivalent series losses of the loading element(s)
- Ground losses part of the antenna current return circuit
- Ground absorption in the near field

1.3.1. Conductor RF resistance

When multisection towers are used for a vertical antenna, care should be taken to ensure proper electrical contact between the sections. If necessary, a copper braid strap should interconnect the sections. Rohrbacher, DJ2NN, provided a formula to calculate the effective RF resistance of conductors of copper, aluminum and bronze:

$$R_{loss} = (1 + 0.1\,L)\left(f^{0.125}\right)\left(0.5\frac{1.5}{D}\right) \times M \quad \text{(Eq 9-5)}$$

where
L = length of the vertical in meters
f = frequency of operation in MHz
D = conductor diameter in mm
M = material constant (M = 0.945 for copper, 1.0 for bronze, and 1.16 for aluminum)

1.3.2. Parallel losses in insulators

Base insulators often operate at low-impedance points. For monopoles near a half-wavelength long, however, care should be taken to use high-quality insulators, since very high voltages can be present. There are many military surplus insulators available for this purpose. For medium and low-impedance applications, insulators made of nylon stock (turned down to the appropriate diameter) are excellent, but a good old Coke bottle may do just as well!

1.3.3. Ground losses

Efficiency means: How many of the watts I deliver to the antenna are radiated as RF. Effectiveness means: Is the RF radiated where I want it? That is, at the right elevation angle and in the right direction. Your antenna can be very efficient but at the same time be very ineffective. Even the opposite is possible (killing a mouse with an A-bomb).

A large number of articles have been published in the literature concerning ground systems for verticals. The ground plays an important role in determining the efficiency as well as effectiveness of a vertical in two very distinct areas: the near field and the far field. Losses in the near field are losses causing the radiation efficiency to be less than 100%.

- I^2R **losses**: Antenna return currents travel through the ground, and back to the feed point, right at the base of the antenna (see Fig 9-41). The resistivity of the ground will play an important role if these antenna RF return currents travel through the (lossy) ground. Unless the vertical antenna uses elevated radials, the antenna return current will flow through the ground. These currents will cause I^2R losses. Even for elevated radials, return currents can partially flow through the ground if a return path exists (can be by capacitive coupling if raised radials are close to ground). With a small elevated system, loss increases with any RF ground path at the antenna base, including the path back by the coax shield. This why the feed line should be decoupled for common modes at the antenna feed point with an elevated radial system.

- **Absorption losses**: The conductivity and the dielectric properties of the ground will play an important role in absorption losses, caused by an electromagnetic wave penetrating the ground. These losses are due to the interaction of the near-field energy-storage fields of the antenna (or radials) with nearby lossy media, such as ground. These types of losses are present whether elevated radials are used or not. The radials should shield the antenna from the lossy soil and distribute the field evenly around the antenna. Most often elevated radials don't help much here, since they normally aren't dense enough to make an effective screen. Four radials are far from a screen! The field is concentrated near the radials, and other areas are directly exposed to the antenna's induction fields.

In the far field (efficiency and effectiveness issues):

- Up to many wavelengths away, the waves from the antenna are reflected by the ground and will combine with the direct waves to form the radiation at low angles, the angles we are concerned with for DXing. The reflection mechanism, which is similar to the reflection of light in a mirror is described in Section 1.1.2. The real part of the reflection coefficient determines what part of the reflected wave is absorbed. The absorbed part is responsible for Fresnel-zone reflection losses (efficiency).

- The ground characteristics in the Fresnel zone will also determine the low-angle performance of the vertical, and this is an effectiveness issue.

The effect of ground in these two different zones has been well covered by P. H. Lee, N6PL (Silent Key), in his excellent book, *Vertical Antenna Handbook*, p 81 (Ref 701). The next section will cover these and various other aspects of the subject.

2. GROUND AND RADIAL SYSTEM FOR VERTICAL ANTENNAS: THE BASICS

2.0.1. Ground-plane antennas

We all know that a VHF vertical antenna usually employs four radials as a "ground-plane," hence its popular name. But in fact, two radials would do the same job. All you need with a λ/4 vertical radiator is a λ/4 wire connected to the feed-line outer conductor in order to have an RF ground at that point. The radial provides the other terminal for the feed line to "push" against. Unless the feed line is radiating, you will have exactly the same current into the radial (system) as you have in the form of common-mode current exciting the vertical. That is the "push against" effect of the radials. This is also how the antenna return currents are collected.

But if you have only one radial, this radial would radiate a horizontal wave component. Two λ/4 radials in a straight line have their current distributed in such a way that radiation from the radials is essentially canceled in the far field, at least in an ideal situation. This is similar to what happens with top-wire loading (T antennas). Using three wires (at 120° intervals) or four radials at right angles does the same also.

It was George Brown himself, Mr 120-buried-radials, who invented elevated resonant radials. He invented the ground-plane antenna. The story goes that when Brown first tried to introduce his ground-plane antenna it had only two radials, but he had to add two extra radials because few of his customers believed that with only two radials the antenna would radiate equally well in all directions! In the case of a VHF ground plane mounted at any practical height above ground, there is no "poor ground" involved and all return currents are collected in the form of displacement currents going through the two, three or four radials.

The VHF case is where detrimental effects of real ground are eliminated by raising the antenna high above ground, electrically speaking. There are no I^2R losses, because the return currents are entirely routed through the low-loss radials. There also are no near-field absorption losses, since the real ground is several wavelengths away from the antenna. Third, on VHF/UHF we are not counting on reflection from the real earth to form our vertical radiation pattern; we are not confronted by losses of Fresnel reflection in the far field either. In other words, we have totally eliminated poor earth.

2.0.2. Verticals with an on-ground (or in-ground) radial system

The other approach in dealing with the poor earth is going to the other extreme—bring the antenna right down to ground level, and, by some witchcraft, turn the ground into a perfect conductor. This is what you try to do in the case of grounded verticals.

You can put down radials, or strips of "chicken wire" to improve the conductivity of the ground, and to reduce the I^2R losses as much as possible. This mechanism is well-known. You can also measure its effect: You know that as you gradually increase the number and the length of radials, the feed-point impedance is lowered, and with a fairly large number of long radials (for example, 120 radials, $\lambda/2$ long) you will reach the theoretical value of the radiation resistance of the vertical. In the worst case, when no measures are taken to improve ground conductivity, losses can be incurred that range from 5 to well over 10 dB with $\lambda/4$ long radiators, and much higher with shorter verticals.

The other mechanism—absorption by the lossy earth—is less well-known in amateur circles. This is partly because you cannot directly measure its effects (see also Section 2.4), as you can for I^2R losses. But the effect is nevertheless there and can result in 3 to 6 dB of signal loss, if not properly handled. For a ground-plane antenna you can improve the situation by moving the near field of the antenna well above ground. For a vertical with its base less than about $3\lambda/8$ above ground you can screen (literally hiding) the lossy ground from the near field of the antenna.

This means that in the case of buried or on-the-ground radials, their number and length must be such that the ground underneath is effectively made invisible to the antenna. It has been experimentally established that for a $\lambda/4$ vertical you must use at least $\lambda/4$-long radials, in sufficient number so that the tips of the radials are separated no more than 0.015 λ (1.2 meters on 80 meters and 2.4 meters on 160 meters). This means approximately 100 radials to achieve this goal. With half that number, you will lose approximately 0.5 dB due to near-field absorptive losses—This is RF "seeping" through an imperfect ground screen. In real life, taking good care of the I^2R losses with buried radials also means taking good care of the near-field absorption losses.

2.0.3. Verticals with a close-to-earth elevated radial system

In some cases it is difficult or impossible to build an on-the-ground radial system that meets this requirement, in most cases because of local terrain constraints. In this case a vertical with a radial system *barely* above ground may be an alternative. The question is: how good is this alternative and how should we handle this alternative? With radials at low height (typically less than 0.1 λ above ground) you still must deal with effectively collecting return currents and with absorption losses in the real ground.

It is clear that if you raise an almost-perfect on-ground radial system higher above ground should yield an almost-perfect elevated-radial system. The perfect on-ground system would consist of 50 to 100 $\lambda/4$-long radials. In fact, the screening effect that is good for radials laying directly *on* the lossy ground, will be even better if the system is raised somewhat above ground. That the screening of such a dense radial system is close to 100% effective was witnessed by Phil Clements, K5PC, who reported on the Internet that while walking below the elevated radial system (120 elevated radials) of a BC transmitter in Spokane, Washington, he could hardly hear the transmitted signal on a small portable receiver. The question, of course, is: Do we really need so many elevated radials, or can we live with many less? This question is one of the topics that I deal with in detail in Section 2.2 on elevated radial systems.

When dealing with the antenna return currents, it is clear that simple radial systems (in the most simple form a single radial) can be used. This has proven true for ages in VHF and UHF ground planes. The only issue here is the possible radiation of these radials in the far field, which could upset the effective radiation pattern of the antenna. This will also be dealt with in Section 2.2.

2.1. Buried Radials

Dr Brown's original work (Ref 801) on buried ground-radial systems dates from 1937. This classic work led to the still common requirement that broadcast antennas use at least 120 radials, each at least 0.5 λ long.

2.1.1. Near-field radiation efficiency

The effect of I^2R losses can be assessed by measuring the impedance of a $\lambda/4$ vertical, as a function of the number and length of the radials. Many have done this experiment. **Table 9-1** shows the equivalent loss resistance computed by deducting the radiation resistance from the measured impedance.

2.1.2. Modeling buried radials

Antenna modeling programs based on *NEC-3* or later can model buried radials. These programs address both the I^2R losses and the absorption losses in the near field, plus of course any far-field effects. These powerful new tools can be dangerous. They would make you believe you can now model everything, and that there is no need for validation. In the real

Table 9-1
Equivalent Resistances of Buried Radial Systems

Radial Length (λ)	Number of Radials				
	2	15	30	60	120
0.15	28.6	15.3	14.8	11.6	11.6
0.20	28.4	15.3	13.4	9.1	9.1
0.25	28.1	15.1	12.2	7.9	6.9
0.30	27.7	14.5	10.7	6.6	5.2
0.35	27.5	13.9	9.8	5.6	2.8
0.40	27.0	13.1	7.2	5.2	0.1

Table 9-2
Wave Angle and Pseudo-Brewster Angle for Ground-Mounted Vertical Antennas Over Different Grounds.

The Wave angle and the Pseudo Brewster angle are essentially independent of the radial system used, unless the radials are several wavelengths long.

Band/Ground Type	Wave Angle	Pseudo-Brewster Angle
80 meters		
Very Poor Ground	29°	15.5°
Average Ground	25°	12.5°
Very Good Ground	17°	7.0°
Sea Water	8.5°	1.8°
160 meters		
Very Poor Ground	28°	14.5°
Average Ground	23°	11°
Very Good Ground	19.5°	8.5°
Sea Water	8.5°	7.0°

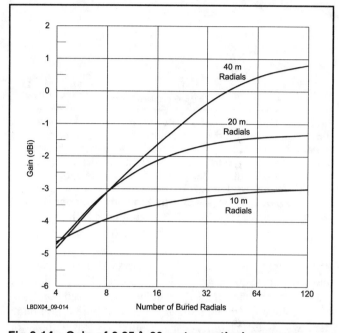

Fig 9-14—Gain of 0.25-λ 80-meter vertical over very poor ground as a function of radial length and number of radials. For short (10-m long) radials there is not much point in going above 16 radials. With 20-m radials you are within 0.5 dB of maximum gain with 32 radials. If you want maximum benefit from 0.5-λ radials (40 m), 120 radials are for you.

world, mainly due to the non-homogeneous nature of the ground surrounding our antennas, the slight variations we sometimes see from modeling results (many authors would rank modeled ground systems by quoting gains specified to a 1/100 of a dB!) are totally meaningless. At best modeling under such circumstances indicates a trend. Let's have a look at these trends.

R. Dean Straw, N6BV, ran a large number of models using *NEC-4* for me (*NEC-4* is not available to non-US citizens). Separate computations were done for 80 and 160 meters. The radiators were λ/4 long and the radials were buried 5 cm in the ground. The variables used were:

- Ground: very poor, average, very good
- Radial length: 10, 20 and 40 meters (for 80 meters), and 10, 40 and 80 meters (for 160 meters)
- Number of radials: 4, 8, 16, 32, 64 and 120.

We computed the gain, the elevation angle and the pseudo-Brewster angle. Although we ordinarily talk about λ/4 buried radials, buried radials by no means must be resonant. A λ/4 wire that is resonant above ground, is no longer resonant in the ground—not even on or near the ground. Typically for a wire on the ground, the physical length for λ/4 resonance will be approximately 0.14 λ and the exact length depending on ground quality and height over ground.

Quarter-wave radials, in the context of buried radials, are wires measuring λ/4 over ground (typically 20 meters long on 80 meters and 40 meters on 160 meters).

The gains of the modeling are shown in **Figs 9-14** through **9-19**. The wave angle as well as the Brewster angle are almost totally independent of the radial system in the near field. The values are listed in **Table 9-2**.

When modeling the antenna over poor ground using only four buried radials, it was apparent that the gain was slightly higher using 15-meter long radials rather than 20 meter or even 40-meter long radials (the gain difference being 0.7 dB, quite substantial). It happens that the resonant length of a λ/4 radial in such lossy ground is 10 to 15 meters (and not ≈ 20 meters as it would be in air). In case of a small number of radials, there is hardly any screening effect, and antenna return currents flow back through lossy, high-resistance earth to the antenna, as well as through the few radials. There are

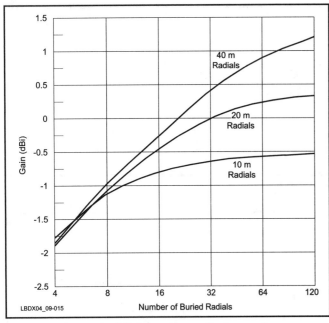

Fig 9-15—Gain of a λ/4 80-meter vertical over average ground, as a function of radial length and number of radials. Note that for 10-meter long radials there is practically no gain beyond about 52 radials. For quarter wave radials there is little to be gained beyond 104 radials, and the difference between 26 λ/4 radials and 104 λ/4 radials is only 0.5 dB. These are exactly the same number N7CL came up with by experiment (see Section 2.1.3).

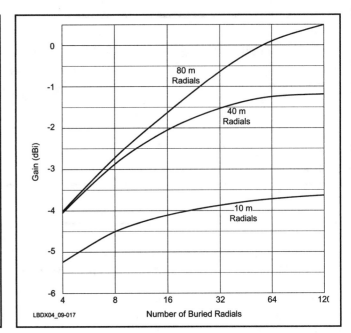

Fig 9-17—Gain of λ/4 160-meter vertical over very poor ground as a function of radial length and number of radials. Note that 10-meter radials, no matter how many, are really too short for 160 meters.

two parallel return circuits: a low-resistance one (the radials) and a high-resistance one (the lossy ground). If the radials are made resonant, their impedance at the antenna feed point will be low, thereby forcing most of the current to return through the few radials. If the impedance is high (such as with 20- or 40-meter long radials), a substantial part of the return currents can flow back through the lossy earth. (See Section 2.1.3.)

The same phenomenon is marginally present with radials in average ground as well, but has disappeared completely in good ground. These observations tend to confirm the mechanism that originates this apparent anomaly. All of this is of no real practical consequence, since four radials are largely insufficient, in whatever type of ground (except saltwater).

We also modeled radials in seawater. As expected, one radial does just as well as any other number. All we really need

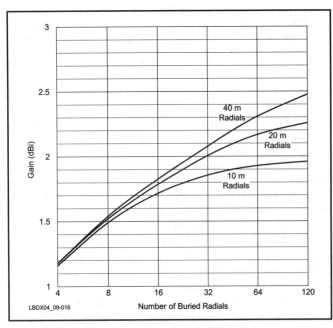

Fig 9-16—Gain of λ/4 80-meter vertical over very good ground as a function of radial length and number of radials.

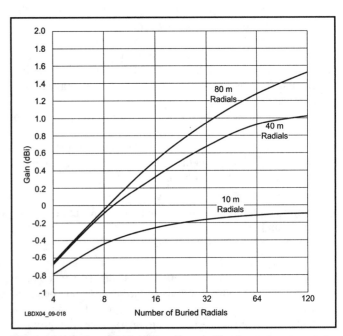

Fig 9-18—Gain of λ/4 160-meter vertical over average ground as a function of radial length and number of radials.

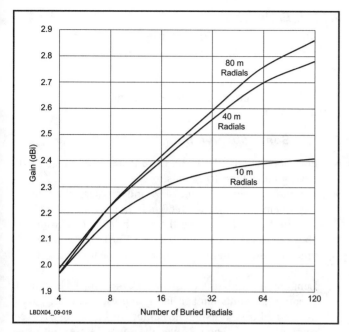

Fig 9-19—Gain of λ/4 160-meter vertical over very good ground as a function of radial length and number of radials. The λ/2 radials are really a waste over very good ground.

Table 9-3
Optimum Length Versus Number of Radials

Number of Radials	Optimum Length (λ)
4	0.10
12	0.15
24	0.25
48	0.35
96	0.45
120	0.50

This table considers only the effect of providing a low-loss return path for the antenna current (near field). It does not consider ground losses in the far field, which determine the very low-angle radiation properties of the antenna.

is to connect the base of the vertical to the almost-perfect conductor (and screen) that the seawater represents. See **Fig 9-20** for a fantastic saltwater location.

Years ago Brian Edward, N2MF, modeled the influence of buried radials (Ref 816), and discovered that for a given number of radial wires, there is a corresponding length beyond which there is no appreciable efficiency improvement. This corresponds very well with what we find in Figs 9-14 through 9-19. Brian found that this length is (maybe surprisingly at first sight) nearly independent of earth conditions. This indicates that it is the screening effect that is more important than the return-current I^2R loss effect. Indeed, the effectiveness of a screen only depends on its geometry and not on the quality of the ground underneath. **Table 9-3** shows the optimum radial length as a function of the number of radials. This was also confirmed through the experimental work by N7CL (see Section 2.1.3).

Conclusion

To me, the results obtained when modeling verticals using buried radials with NEC-4 seem to be rather optimistic, but the trends are clearly correct. Take the example of an 80-meter vertical over average ground: going from a lousy eight 20-meter long radials to 120 radials would only buy you 1.4 dB of gain, which is less than what I think it is in reality. In very good ground that difference would be only 0.7 dB!

There has been some documented proof that NEC-4 does not handle very low antennas correctly, and that the problem is a problem associated with near-field losses (see Section 2.2.2). Maybe this same limitation of NEC-4 causes the gain figures calculated with buried radials to be optimistic as well. The future will tell. No doubt further enhancements will be added to future NEC releases, which may well give us gain (loss) figures that I would feel more comfortable with for verticals with buried radials.

2.1.3. How many buried radials now, how long, what shape?

When discussing radial lengths, I usually talk about λ/4 or λ/8 radials. Mention of a λ/4 radial leads most of us to think of a 20-meter long radial on 80 meters. A wire up in the air at heights

Fig 9-20—XZ0A had an ideal location for far-field reflection efficiency: Saltwater all around. Four Squares were used on 80 and 160 meters, resulting in signals up to S9+20 dB in Europe on Topband, quite extraordinary from that part of the world.

where you normally have an antenna has a velocity factor (speed of travel vs speed of light) of about 98%. When you bring that same wire close to ground, the velocity factor starts dropping rapidly below a height of about 0.02 wavelength. On the ground, the velocity factor is on the order of 50-60%, which means that a radial that is physically 20 meters long is actually a half-wave long electrically! (See also Fig 9-32.)

If you use just a few on-the-ground radials over poor ground, the radials may act like they are somewhat resonant. The resonance vanishes if you have many radials or if the ground is good to excellent. For these cases it is best to use radials that are an *electrical* quarter-wave long. On 80 meters you should use 10-meter long radials, and on 160 meters you should use 20-meter long radials if you are only using a few (up to four). But that's bad practice anyhow: Four is far too few radials.

As soon as you use a larger number of equally spread radials the resonance effect disappears, and the radials form a disk, which becomes a screen with no resonance characteristics. In this case we no longer talk about length of radials but about the diameter of a disk hiding the lossy ground from the antenna.

Assume we have 1 km of radial wire and unrestricted space. How should we use it? Make one radial that is 1000 meters long, or 1000 radials that are 1 meter long? It's quite obvious the answer is somewhere in the middle.

2.1.3.1. Early work

Brown, Lewis, and Epstein in the June 1937 Proceedings of the IRE published measured field strength data at 1 miles (versus number and length of radials). Measurements were done at 3 MHz. The measured field strength was converted to dB vs the maximum measured field strength (for 113 radials of 0.411 λ).

2.1.3.2. Some observations

- For short radials (0.137 λ), there is negligible benefit in having more than 15 radials.
- For radial lengths of 0.274 λ and greater, continuous improvement is seen up to 60 radials. Note that doubling the number and doubling the length of radials from the above case (15 short radials of 0.137 λ) only gains 1 dB greater field strength, with four times the total amount of wire.
- Lengthening radials 50% from 0.274 λ to 0.411 λ and keeping the same number hardly represents an improvement (0.24 dB). Raising the number to 113 radials represents a gain of 0.66 dB over the second case, but uses nearly three times as much wire.

From these almost 70-year-old studies, we can conclude that 60 quarter-wave long radials is a cost-effective optimal solution for amateur purposes. The following rule was experimentally derived by N7CL and seems to be a very sound and easy one to follow. Put radials down in such a way that the distance between their tips is not more than 0.015 λ. This is 1.3 meters for 80 meters and 2.5 meters on 160 meters.

The circumference of a circle with a radius of $\lambda/4$ is $2 \times \pi \times 0.25 = 1.57 \lambda$. At a spacing of 0.015 λ at the tips, this circumference can accommodate 1.57/0.015 = 104 radials. With this configuration you are within 0.1 dB of maximum gain over average to good ground. If you space the tips 0.03 λ you will lose about 0.5 dB.

For radials that are only $\lambda/8$ long, a 0.03-λ tip spacing requires 52 radials. Here too, if you use only half that number, you will give up another 0.5 dB of gain. In general, the number that N7CL came by experimentally, closely follow those from Brown, Lewis and Epstein. Let us apply this simple rule to some real-world cases:

Example 1

Assume your lot is 20 by 20 meters and that you want to install a radial system for 80 and 160 meters. Draw a circle that

Fig 9-21—The Battle Creek Special that made Heard Island available on 160 for over 1000 different stations. Ghis, ON5NT, is not holding up the antenna; it is very well capable of standing up by itself. The antenna was located near the ocean's edge, on saltwater-soaked lava ash.

Table 9-4
From Brown, Lewis and Epstein
Signal Strength vs Length of Radials in Wavelengths

Number Radials	Length 0.137 λ)	Length 0.274λ)	Length 0.411λ)
2	−4.36	−4.36	−4.05 dB
15	−2.4	−1.93	−1.65 dB
30	−2.4	−1.44	−0.97 dB
60	−2.0	−0.66	−0.42 dB
113	−2.0	−0.51	0 dB (Ref)

fits your lot. This circle has a radius of
$\sqrt{20^2}/2 = 14$ meters

On each 20-meter side of your lot you would space the ends evenly by 1.3 meters. This means you can fit 16 radials on your property. The longest will be 14 meters; the shortest will be 10 meters long. The average radial length is 12 meters. You can install a total of 16 (radials) × 4 (sides) × 12 meters (average length) = 768 meters of radial wire, with a total of 64 radials. A radial system using 32 evenly spread radials, and using only 385 meters of wire, would compromise your system by about 0.5 dB.

In actual practice, when laying radials on an irregular lot where the limits are the boundaries of the lot, the most practical way to make best use of the wire you have is just walk the perimeter of the lot and start a radial from the perimeter (inward toward the base of the antenna) every 0.015 λ (1.3 meters for 80 meters or 2.5 meters for 160) as you walk along the perimeter.

Example 2

You have only 500 meters of wire and space is not a problem. How many radials and how long should they be to be used on both 80 and 160 meters?

The formula to be used is:

$$N = \frac{\sqrt{2 \times \pi \times L}}{A} \quad \text{(Eq 9-6)}$$

where

N = number of radials
L = total wire length available
A = distance between wire tips (1.3 meters for 80, 2.5 meters for 160, or twice that if 0.5 dB loss is tolerated).

For this example use L = 500 meter, A = 1.3 meters, and you calculate:

$$N = \frac{\sqrt{2 \times \pi \times 500}}{1.3} = 43 \text{ radials.}$$

Each radial will have a length of 500/43 = 11.6 meters.

You could also use A = 2.6 meters, in which case you wind up with 22 radials, each 18 meters long. However, the first solution will give you slightly less loss.

For a given length of wire, it is better to use a larger number of short radials than a smaller number of long radials, the limit being that the tips should not be closer than 0.015 λ.

Example 3

How much radial wire (number and length) is required to build a radial system (for a λ/4 vertical) that will be within 0.1 dB of maximum gain. How much to be within 0.5 dB?

The answer to the first question is 104 radials, each λ/4 long. The total wire length for 80 meters is: 2080 meters (4000 meters for 160). With 52 radials, each λ/4 long, you are within 0.5 dB of maximum gain. This translates to 1000 meters of radial wire required for 80 meters and 2000 meters for 160 meters.

Example 4

I can put down 15-meter long radials in all directions. How many should I put down, and how much radial wire is required?

The circumference of a circle with a radius of 15 meters is: 2 × π × 15 = 94.2 meters. With the tips of the radials separated by 1.3 meters, we have 94.2/1.3 = 72 radials. In total I would use 72 × 15 = 1080 meters of radial wire. There is no point in using more than 72 radials.

2.3.1.3. K3NA's work

In private correspondence ("Effects of Ground Screen Geometry on Verticals"), Eric Scace, K3NA, explained a simple rule of thumb he derived from an extensive modeling study he conducted using *NEC-4.1*. His conclusions are applicable for radials up to 3λ/8 in length:

- Measure R, the real component of the feed-point impedance.
- Double the number of radials.
- Measure R again.
- Continue doubling the number of radials until R changes by less than 1 Ω.

K3NA's detailed modeling study to evaluate the effectiveness of various radial configurations was similar to what N6BV did years ago for the Third Edition of this book. The main difference between the two studies is that K3NA calculated the gain versus the total amount of radial wire used for different configurations. He calculated the "sky Gain" (Gsky) to assess the quality of the radial system. Gsky is the total power radiated to the entire sky, covering all elevation angles, all azimuths. K3NA was concerned with two aspects: the *efficiency* issue, which is related to the task of collecting return currents in the vicinity of a lossy ground and doing so with the smallest possible losses. (See definitions in Section 1.3.3.) The second issue is that of *effectiveness*, which means putting the radiated power where we want it. For a single

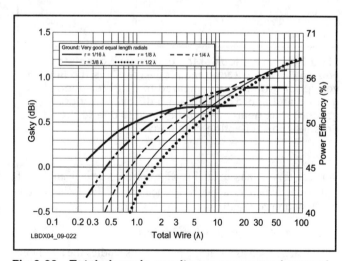

Fig 9-22—Total sky-gain results over very good ground for various radials systems using standard radials, shaped as the spokes of a wheel. The graph shows that with small amounts of wire, many short radials are the answer. It also tells us that 10 λ of radial wire used to make 80 λ/8 radials is only 0.2 dB down from 30 λ of radial wire used to make 120 λ/4 radials.

Vertical Antennas 9-15

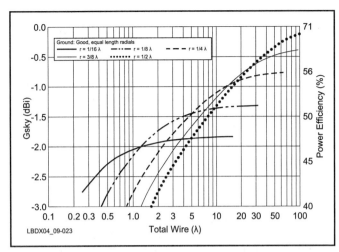

Fig 9-23—The same graph as in Fig 9-22 but for good ground. Unless you only have 4 λ of wire, λ/8 radials are really too short; λ/4 radials are just fine for up to about 20 λ of wire (this about 3.3 km or 10,000 ft of radials on 160 meters). Notice that this study also shows that there is little to be gained beyond approximately 100 λ/4 radials. 300 λ/2 radials only gain about 0.7 dB (a power increase of only 20%) over 100 λ/4 radials—not really a whole lot!

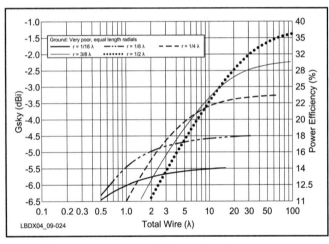

Fig 9-24—Have a look at the gain axis: No matter what you do (lots of λ/2 long radials), −1.3 dBi is the limit for very poor ground (as compared to 0 dBi for good ground and approximately + 1.5 dB over very good ground). The nearly 3-dB difference is due to the Fresnel-zone reflection efficiency.

vertical this means obtaining appropriate vertical angles of radiation, which is actually formed in the far field by the combination of the direct and the reflected waves.

2.3.1.3.1. Over very good ground

K3NA used as a starting point in his studies the available quantity of radial wire. For up to 3 λ of available wire, the most efficient solution is to use λ/16 radials, even if there is space for longer ones. Beyond 48 radials, he found hardly any improvement. This confirms what we show in Fig 9-16. Not everything in his study is however a perfect match the modeling results done several years ago by N6BV.

Figs 9-22 through 9-24 show the results for very good ground, good ground and very poor ground respectively. These confirm that any improvement in efficiency by improving the radial system improves radiation at all elevation angles equally. For regular-shaped radials laid out as the spokes of a wheel K3NA came to the conclusion that N7CL's rule of thumb, which says to separate the tips of the radials by no more than 0.015 λ, is confirmed by modeling, at least for radials up to λ/4 in length.

2.3.1.3.2. Other configurations

K3NA also investigated the possibility of using radials that split out along their way: fork-shaped radials. He found out that for a given amount of available wire, these fork-type radials do not perform any better than regular straight radials.

A third alternative he examined was alternating long (λ/4) and short (λ/8) radials. Here too this radial geometry reduces Gsky compared to a system using the same total length of radial wire used as uniform-length straight radials.

Eric went on to assess the performance of ground screens in square and triangular grids. Here again, for a given amount of radial wire, the performance did not meet that of a classical radial configuration.

Looking at all these very detailed modeling results you must ask yourself: "Is it really like this in real life?" We are playing with very minute changes in inputs and obtaining even smaller changes in results. Can you really trust these models? Earth is a very difficult thing to model, and it is very non-homogeneous.

It's obvious that we should be conscious of trends, and the modeling results confirm the trends revealed by N7CL's experimental work. There's an even simpler rule: Put in as many radials as you can, until you feel satisfied. If you think you can do better, do better. If you think "this is as far as I can go," be happy with it!

Tom, W8JI, wrote this interesting observation for the Topband reflector: *"Even a very small limited space antenna like an inverted L will do very well if some effort is put into the ground system. My friend K8GIJ was always within a few dB of my signal (I used a ¼ λ vertical tower with 100 radials), and all he had was a 15 by 100 ft back yard! But then Harold filled his small yard with radials, and even tied the fences and everything else in to his ground system."* So, you guys on a city size lot, there is no reason not to be loud on 160 meters.

Of course, to be able to *hear* as well as Tom, W8JI, is another challenge.

2.1.4. Two-wavelength-long radials and the far field

Everything that happens in the near field determines the radiated field strength in the far field. Radials, screens, and I^2R losses have very little influence on the radiation pattern of the vertical, except maybe at very high angles, which don't interest us anyhow. Any method of improving ground conductivity in the near field (up to λ/4 from the base of a λ/4 vertical) improves the entire radiation pattern, not just favoring certain radiation angles more than others.

In the far field, however, ground characteristics greatly influence the low-angle characteristics of a vertical antenna. For λ/4 verticals the area where Fresnel reflection occurs starts about 1 λ from the antenna and extends to a number of wavelengths.

For current collecting and near-field screening there is really no point in installing radials longer than λ/4. With 104 such radials you are within 0.1 dB of what is theoretically possible. The Brown rule (120 radials, 0.5-λ long) shoots for less than 0.1 dB and has some extra reserve built in.

If you want to influence the far field and pull down the radiation angle somewhat, or reduce the reflection loss, then we are talking about radials that are about 2 λ long. For this you would need a terrain measuring 660 × 660 meters (43 hectares or 100 acres) for Topband, which is hardly practical, of course.

The only practical way to influence the far-field reflection efficiency and effectiveness is to install your vertical in the middle of saltwater. In that case you will have a peak radiation angle of between 5 and 10° and a pseudo-Brewster angle of less than 1°! The elevation pattern becomes very flat, showing a −3-dB beamwidth ranging from 1 to 40°. All this is due to the wonderful conductivity properties of saltwater. No wonder such a QTH does wonders!

Tom Bevenham, DU7CC (also SM6CNS), testified: *"At my beach QTH on Cebu Island, I use all vertical antennas standing out in salt water. Also, at high tide, water comes all the way underneath the shack. On Topband, I use a folded monopole attached alongside a 105-ft bamboo pole. This antenna is a real winner. I use not much of a ground system, only a few hundred feet of junk wire at sea bottom. At the other QTH, less than half a mile from the beach, the same antennas with ground radials don't work at all."*

Of course, we have all heard how well the over-saltwater vertical antennas perform. I remember the operation from Heard Island (VK0IR) for one. The Battle Creek Special (see Section 6.6) was standing with its base right in the saltwater.

2.1.5. Ground rods

Ground rods are important for a good dc ground, which is necessary for adequate lightning protection, even if ground rods contribute very little to the RF ground system. If you use a series-fed (insulated-base) vertical, a lightning arrestor spark gap with a good dc ground is a good idea. In addition, you can install a 10 to 100-kΩ resistor or an RF choke between the base of the antenna and the dc ground to drain static charges.

2.1.6. Depth of buried radials

C. J. Michaels, W7XC (Silent Key), calculated the depth of penetration of RF current in different types of ground. He defined the depth of penetration as the depth at which the current density is 37% of what it is at the surface. On 80 meters he calculated a depth of penetration of 1.5 meters for very good ground. For very poor ground the depth reaches 12 meters!

From the point of view of I^2R loss, you can bury the radials "deep" without any ill effects. However, from near-field screening effect point of view, we need to have the radial system above the lossy material.

Bob Leo, W7LR, in Ref 808 reports that burying the radials a few inches below the surface does not detract from their performance. Al Christman, K3LC (ex KB8I), confirmed this when modeling his elevated radial systems using *NEC-4.1*. He found a difference of only hundredths of a dB between burying radials at 5 cm or 15 cm. I would not bury them much deeper though. The sound rule here is "the closer to the surface, the better."

2.1.7. Some practical hints
2.1.7.1. Local ground characteristics

It is impossible to make a direct measurement of ground characteristics. The most reliable source of information about local ground characteristics may be the engineer of your local AM broadcast station. The so-called "full proof-of-performance" record will document the average soil conductivity for each azimuth out to about 30 km (20 miles). But unfortunately this is hardly what you need to know. What you need is the ground characteristics in a circle with a λ/4 radius around the base of your vertical! In your modeling program you plug in a single set of values that supposedly characterize your ground. In the real world, the soil around an antenna is virtually never homogeneous—and almost always not even remotely close to homogeneous. Real-world earth is a widely varying mix of moisture, as well as different types of soil. Because of this, any model that treats the earth as a uniform medium will not be accurate. Verification by field-strength measurement is the only way to know for sure what's going on!

2.1.7.2. Radial bus-bar/low-loss connections

There are two good ways to collect the currents in the many radials at the base of the vertical. You could use a radial plate (see **Fig 9-25**) and use stainless-steel hardware to connect the radials. Using solder lugs and stainless steel hardware makes it possible to disconnect the radials so that individual radial-current measurements can be made.

Another method is to make a heavy gauge bus-bar made of a large diameter copper ring, and solder all (copper) radials to the bus (see **Fig 9-26**).

Fig 9-25—The stainless-steel radial plate made by DX-Engineering with 64 holes drilled around its perimeter. All stainless-steel hardware is provided to make a quality radial-connecting system using crimped lugs at the ends of the radials.

Fig 9-26—W8LRL uses a copper tube (about 10 mm in diameter) bent in a circle, to which he solders all his radials. For a permanent installation this is probably the best way to go, provided you use the correct solder or protect your 60/40 Sn-Pb solder joints with liquid rubber.

2.1.7.3. Soldering/welding radial wires

Tin-lead (Sn-Pb), which is often used to solder copper wires, will deteriorate in the ground and may be the source of bad contacts. Therefore you should silver-solder all copper radials, or even better yet, weld the radials. Information about CADWELD welding products from The RF Connection in Maryland is available on their Web page: **www.therfc.com/**.

If you decide to use regular 60/40 tin-lead solder, cover all soldered joints with several layers of liquid rubber, so that the acidity of the ground cannot reach the solder joint.

2.1.7.4. Sectorized radial systems

Very long radials (several wavelengths long) in a given direction have been evaluated and found to be effective for lowering the wave angle in that direction, but seem to be rather impractical for just about all amateur installations. A similar effect occurs when verticals are mounted right at the saltwater line. Similar in result to a sectorized radial system is the situation where an elevated radial system is used with only one radial (see Section 2.2.3).

2.1.7.5. Radial wire material

Use copper wire if at all possible. Galvanized-steel wire is a not good, as it has poor conductivity and will rust away in just a few years in wet acidic ground. Aluminum is OK as far as conductivity is concerned but aluminum gradually turns to a white powder as it reacts with the soil. Soldering aluminum wire is not easy, and crimp-on lugs are the only way to go, if you decide to use aluminum.

2.1.7.6. Radial wire gauge

When less than six radials are used, the gauge of the wires is important for maximum efficiency. The heavier the better—#16 wires are certainly no luxury when only a few buried radials are used. With many radials, wire size becomes unimportant since the return current is divided over a large number of conductors. DXpeditions using temporary antennas often take a small spool of #24 or #26 (0.5 or 0.4-mm diameter) enameled magnet wire. This is inexpensive and can be used to establish a very efficient RF ground system.

2.1.7.7. Bare or insulated wire?

Experience has shown that you can use insulated as well as bare copper wire for buried radials. L.B. Cebik, W4RNL, posted a short paper on this issue on his very informative web site. (**www.cebik.com/ir.html**). The *NEC-4* modeling program finds no noticeable difference between insulated and bare buried radials. This relates to the capacitive coupling between the radial wire and the earth around it. Experience is what counts, and the modeling program gives the correct answer on this issue!

2.1.7.8. A radial plow

Installing radials can be quite a chore. Hyder, W7IV, (Ref 815) and Mosser, K3ZAP, (Ref 812) have described systems and tools for easy installation of radials.

2.1.7.9. Radials on the ground

Radials can also be laid on the ground (instead of being buried in the ground) in areas that are suitable. A neat way of installing radials in a lawn-covered area is to cut the grass really short at the end of the season (October), and lay the radials flat on the ground, anchored here and there with metal hooks (clothespins, doll pins, gutter nails or fencing staples). By the next spring, the grass will have covered up most of the wires, and by the end of the following year the wires will be completely covered by the grass. This will also guarantee that your radials are "as close as possible" to the surface of the ground, which is ideal from a near-field screening point of view!

2.1.7.10. Radials and salt water

The conductivity of saltwater is excellent. But you should also remember that the skin depth of saltwater is very limited, and you better keep that in mind when you install radials "in" salt water. Throwing radials in salt water and letting them sink to the bottom is like installing radials "under" a copper plate: not much use! It seems best to have many short radials dangling from the base of the antenna into the saltwater or better yet, have a few copper plates extending into the salt water to ensure a large contact surface with the saltwater.

If your antenna is exposed to the tide it seems like a good idea to have a floating device with large copper fins extending under the device in the saltwater. If the area gets dry at low tide, you should also have regular radials lying on the ground. Over saltwater, two in-line elevated radials make a very valid alternative.

2.2. Elevated Radials

With elevated VHF or UHF ground-plane antennas the three or four radials are more an electrical *counterpoise* (a 0-Ω connection point high above ground), than a ground plane. The ground is so far away that any term including the word "ground" is really not applicable. The radials of such antennas radiate in the near field (radiation from the radials only cancels in the far field), but they do not suffer near-field absorption losses in the ground, because of their relative height above ground.

Using a small number of elevated radials does not prevent the antenna and its radials from coupling heavily to the feed line and from inducing common-mode currents onto the feed line. There will be substantial feed-line radiation unless you isolate the feed line from such common-mode currents. (See also Section 2.2.12.)

Such HF and VHF/UHF ground planes have been in use for many years. Studies that were undertaken in the past several years, however, are concerned with vertical antennas using radials at much lower heights, typically 0.01 to 0.04 λ above ground. That there is still quite a bit of controversy on this subject is no secret to insiders. It appears that a number of real-life results do confirm the current modeling results, while others do not. The jury is still out. I will try to represent both views in this book.

A. Doty, K8CFU, concluded from his experimental work (Ref 807 and 820) that a $\lambda/4$ vertical using an elevated counterpoise system can produce the same field strength as a $\lambda/4$ vertical using buried bare radials. The reasoning is that in the case of an elevated radial or counterpoise system, the return currents do not have to travel for a considerable distance through high-resistance earth, as is the case when buried radials are used. His article in April 1984 *CQ* also contains a very complete reference list of just about every publication on the subject of radials (72 references!).

Frey, W3ESU, used the same counterpoise system with his Minipoise short low-band vertical (Ref 824). He reported that connecting the elevated and insulated radial wires together at the periphery definitely yields improved performance. If a counterpoise system cannot be used, Doty recommends using insulated radials lying right on the ground, or buried as close as possible to the surface.

Quite a few years after these publications, A. Christman, R Redcliff, D. Adler, J. Breakall and A. Resnick used computer modeling to come to conclusions which are very similar to the findings brought forward after extensive field work by A. Doty. The publication in 1988 by A. Christman, K3LC (ex KB8I), has since become the standard reference work on elevated radial systems (Ref 825), work that has stirred up quite a bit of interest and further investigation.

The results from Christman's study were obtained by computer modeling using *NEC-GSD*. It is interesting to understand the different steps he followed in his analysis (all modeling was done using average ground):

1. Modeling of the $\lambda/4$ vertical with 120 buried radials (5-cm deep). This is the 1937 Brown reference. (See Section 2.1.3.)
2. The $\lambda/4$ vertical was modeled using only four radials at different radial elevations. For a modeling frequency of 3.8 MHz, Christman found that 4.5 meters was the height at which the four-radial systems equaled the 120-buried-radial systems so far as low-angle radiation performance is concerned.
3. Christman's studies also revealed that as the quality of the soil becomes worse, the elevated radial system must be raised progressively higher above the earth to reach performance on par with that of the reference 120-buried-radial vertical monopole. If the soil is highly conductive, the reverse is true.

The elevated-radial approach has become increasingly popular with low-band DXers since the publication of the above work, and it appears that elevated radials represent a viable alternative to digging and plowing, especially where the ground is unfriendly for such activities.

It is important to critically analyze the elevated-radial concept and therefore to understand the mechanism that governs the near-field absorptive losses (see Section 1.3.3) connected with elevated radials. In the case of an elevated-radial system these near-field losses can be minimized in only three ways:

1. By raising the elevated radials as high as possible (move the near field of the antenna away from the real lossy ground).
2. By installing many radials, so that these radials screen the near fields from "seeing" the underlying lossy earth.
3. By improving ground conductivity of the real ground below the raised radials.

Although the experts all agree on the mechanisms, there appears to be a good deal of controversy about the exact quantification of the losses involved (see Section 2.2.1). Incidentally, an elevated radial system does not imply that the base of the vertical must be elevated from the ground. The radials can, from ground level, slope up at a 45° angle to a support a few meters away, and from there run horizontally all the way to the end. It is a good idea to keep the radials high enough so no passersby can touch them. This is also true when radials are quite high. In an IEEE publication (Ref 7834) it was reported that significantly better field strengths were obtained with elevated radials at 10-meter height than at 5-meter height. In both cases the radials were sloping upward at a 45° angle from the insulated base of the vertical at ground level.

2.2.1. Modeling vs measuring? Elevated vs ground radials

The performance of an elevated radial system can be assessed by either computer modeling or by real-life testing and field-strength measuring. It would of course be ideal if the results from modeling and field-strength (FS) measurements match.

Al Christman used *NEC-4* to study the influence of the number of elevated radials and their height on antenna gain and antenna wave angle (Ref 7825) and came to the conclusion that if the height of the radials is at least 0.0375 λ (3 meters on 80, 6 meters on 160) there is very little gain difference between using four or up to 36 radials. He also concluded that the gain of antennas with an elevated radial system compared in gain to the same antenna with about 16 buried radials. Incidentally, the modeling also showed that for buried $\lambda/4$ radials the difference in gain between 16 radials and 120 radials is only about 0.74 dB. When raising the elevated radials to a height of 0.125 λ (20 meters on 160), the gain actually approached the gain of a vertical with 120 buried radials. The publication of these results (1988) gave a tremendous impetus in the use of elevated-radial systems.

In another study, Jack Belrose, VE2CV (Ref 7821 and 7824) also concluded that there was a good correlation between measured and computed results. In this study Belrose used a $\lambda/4$ vertical, as well as $\lambda/4$ (resonant) radials.

A good correlation between the modeled results and FS

measurements was established in several study cases. One of them was an extremely well-documented case, with thousands of FS measurements, which matched very well the figures obtained with modeling (*NEC-4*). Belrose's studies revealed that radials should be at least 0.03-λ high (2.5 meters on 80 meters, 5 meters on 160 meters) to avoid excessive near-field absorption ground losses, especially so if fewer than eight radials are used. With a large number of radials (>16) the radials can be much lower.

Another well-documented case was reported in a technical paper delivered by Clarence Beverage (nephew of Harold Beverage) at the 49th NAB Broadcast Engineering Conference entitled: "New AM Broadcast Antenna Designs Having Field Validated Performance." The paper covered antenna tests done in Newburgh, NY, under special FCC authority. The antenna system consisted of a tower 120 feet in height with an insulator at the 15-foot level and six elevated radials a quarter wavelength in length spaced evenly around the tower and elevated 15 feet above the ground. The system operated on 1580 kHz at a power of 750 W. The efficiency of the antenna was determined by radial field-intensity measurements (in 12 directions) extending out to distances up to 85 km. The measured RMS efficiency was 287 mV/m (normalized) to 1 kW at 1 km, which is the same measured value as would be expected for the tower above with 120 buried radials.

In a number of other cases however, it was reported that field-strength measurements indicated a discrepancy of 3 to 6 dB with the *NEC-4* computed results. Tom Rauch, W8JI, published the following measured results:

Number of Radials	On the ground	Elevated 0.03 λ
4	–5.5 dB	–4.3 dB
8	–2.7 dB	–2.4 dB
16	–1.3 dB	–0.8 dB
32	–0.8 dB	–0.7 dB
60	Reference (0 dB)	–0.2 dB

Calculations with *NEC-4* show a difference of only about 2 dB going from 4 to 60 buried radials, which is 3.5 dB less that Rauch's experiment showed. The 5 dB he found inspired the following comment: *"Consider that going from a single vertical to a four square only gained me 5 dB! I got almost that just by going from four radials to 60 radials."*

Eric Gustafson, N7CL, reported (on the Topband reflector) that several experiments comparing signal levels of a ground mounted λ/4 vertical with 120 radials with those produced by the same radiator with an elevated radial system (using a few radials) have been done a number of times by various researchers for various organizations ranging from the broadcast industry and universities to the military. He reported that the results of these studies always have returned the same results: The correctly sized, sufficiently dense screen is superior to four resonant radials in close proximity to earth. The quantification of the difference has varied. The largest difference Eric personally measured during research for the military was 5.8 dB, the smallest difference 3 dB. The latter one was measured over really good ground, being a dry salt-lake bed (measured conductivity approximately 20 mS during the test). It is clear that the quality of the ground plays a very important role in the exact amount of loss.

For those who would like to duplicate these tests, understand that you cannot do these tests on one and the same vertical, switching between elevated radials to ground-mounted radials, *unless* you remove (physically) the ground-mounted radials when you use the elevated ones. If not, you have an elevated radial system *plus* a screen, effectively screening the near fields from the underlying real ground.

It seems to me that elevated-radial systems are indeed a valid alternative for buried ones, especially if buried ones are not possible or very difficult to install for whatever reason. Even the broadcast industry now uses elevated-radial systems quite extensively and successfully where local soil conditions make it impossible to use the classic 120 buried λ/2 radials. It must be said though that most of these systems use more than just a few radials. I also know of many amateur antenna systems successfully using elevated radial systems. Whether they get optimum performance or lose maybe 2 to 5 dB because of near-field absorption losses, is hard to tell. As a matter of fact, there is still the possibility of improving the ground conductivity under the elevated radial system. More on that in Section 2.2.13.

The discrepancy between measured and modeled gain figures has been recognized by a number of expert *NEC* users. All of the current modeling programs have flaws, but most are known and can be compensated for by experienced users. It seems to me that modeling of very low wires even with current versions of *NEC-4* may be affected by such a flaw.

We should also recognize that the total losses due to mechanisms in the near field can amount to much more than 5 dB. Antenna return-current losses (sometimes also called "connection" losses) can amount easily from 10 to even 40 dB over poor ground. These losses can, however, easily be mastered with elevated radials and reduced to zero. The remaining 4 or 5 dB, accountable to near-field absorption losses, are indeed somewhat more difficult to deal with using elevated radials.

2.2.2. Modeling vertical antennas with elevated radials

As mentioned before only *NEC*-based programs can model antennas with elevated radials close to ground. Roy Lewallen's *EZNEC* program (using the *NEC-2* engine) incorporates the "high-accuracy" (NEC Sommerfeld) ground model, which should be accurate for low horizontal wires down to 0.005-λ high (about 2.7 feet on 160 meters).

Still, many cases have been reported indicating a difference of up to 6 dB in gain for antennas very close to ground. A similar flaw was already present in *NEC-2* and has been documented by John Belrose, VE2CV, who compared the experimentally obtained results, published by Hagn and Barker in 1970 ("Gain Measurements of a Low Dipole Antenna Over Known Soil") with the *NEC-2* predictions. At 0.01 λ above ground, *NEC-2* showed 5 dB more gain than the actual measured values.

All of this goes to say that modeling software is a mathematical tool. Most modeling programs have well-known, but also sometimes little-known or barely documented limitations. Field-strength measurements are the real thing (eating is the proof of the pudding). But we should be thankful for having access to antenna-modeling programs. They have undoubtedly helped the non-professionals to gain an enormous amount of insight they would miss without these tools. It is the role of the professional, and the experts to show non-expert users how to use them correctly, and make corrections if necessary.

2.2.3. How many elevated radials?

Through antenna modeling, K3LC (ex KB8I), calculated (for 80 meters), the $\lambda/4$ antenna gains for elevated radial heights of 5, 10, 15, 20, 25 and 30 meters, while varying the number of $\lambda/4$ radials between 4 and 36 (Ref 7825). According to these calculations, at a height of 4.5 meters (which is roughly what I have) it made less than 0.1 dB of difference between 4 and 32 radials, and this was within 0.3 dB of a buried radial system using 120 quarter-wave radials. These results were confirmed by Jack Belrose, VE2CV (Ref 7821) also through antenna modeling.

Eric Gustafson, N7CL, in a well documented e-mail addressed to the Topband reflector, explained that for a $\lambda/4$ vertical radiator, a radial system with 104 $\lambda/4$-long radials (resulting in wire ends separated not more than 0.015 λ at their tips) achieves 100% shielding effectiveness. His experimental work (radials about 5-meters high) further indicates that the screening effectiveness of a $\lambda/8$-long radial system does not improve above 52 radials. See Fig 9-14, where we note that the experimental work by N7CL confirms the modeling results. Beyond 104 $\lambda/4$ radials there hardly is any increase in gain, and the same is true beyond 52 radials that are $\lambda/8$ long. This means that the shielding effectiveness of the $\lambda/8$ radial system with 52 radials by itself is 100%, but that some loss will be caused by near fields "spilling over" the screen at its perimeter. (In other words, the screen is dense enough, but not large enough.) Using just 26 $\lambda/4$-long radials, you will typically lose about 0.5 dB due to near-field absorption losses in the ground.

N7CL goes on to say that a $\lambda/4$ vertical with only four elevated radials can indeed produce the same signal as a ground-mounted vertical with 120 radials $\lambda/4$ long, provided that:

1. The base of the vertical is at least $3\lambda/8$ high.
2. Or that the quality of the ground under the elevated radials has been improved so that it acts as an efficient screen, preventing the nearby field from interacting with the underlying lossy ground.

Unless such measures are effectively taken, N7CL calculated that the extra ground absorption losses can be as high as 5 or 6 dB. Loss figures of this order have been measured in a number of cases (eg, by Tom Rauch, W8JI) reported on the Topband Reflector (see Section 2.2.1).

2.2.3.1. Conclusion

According to the *NEC*-based modeling results, there should be no point in using more than four elevated radials. With four radials over good ground the gain of a $\lambda/4$ monopole is –0.1 dBi. Two such radials gives an average of –0.15 dBi (+0.14 and –0.47 dBi due to slight pattern squeezing). One elevated radial gives a gain of +1.04 dBi in the direction of the radial, and –2.3 dBi off its back, resulting in an integrated gain of 0.65 dBi. These optimistic figures drove many people to use four elevated radials on their verticals, convinced that they would be as loud as their neighbors using 120 buried radials. Over the years, though, the enthusiasm for elevated radials seems to have somewhat settled down, and many have returned to the old-fashioned large numbers of radials on the ground, at least where feasible.

The *NEC*-based modeling programs are overly optimistic when it comes to dealing with near-field absorption losses. Three or four elevated radials over a poor ground, in my humble opinion, can never be as good as 120 ground-mounted (or elevated for that matter) radials. There is simply no free lunch! If you need to use an elevated radial system, maybe it's not a bad idea after all to use 26 radials, which according to N7CL would put you within 0.5 dB of the Brown standard.

2.2.4. Radial layout

If you use a limited number of elevated radials (two, three or four), a symmetrical layout is necessary for the radiation from radials to cancel "as much as possible" in the far field. One radial is not symmetrical, but two and more are symmetrical, provided the radials are spread out evenly over 360°. When using more than four radials the exact layout as well as the exact radial length becomes of little importance about creating high-angle radiation.

2.2.4.1. Only one radial

In his original article on elevated radials (Ref 825) Christman showed the model of a $\lambda/4$ vertical using a single elevated radial. This pattern shown in **Fig 9-27** is for a radial height of 0.05 λ over average ground. He showed this vertical, with a single elevated radial, as having (within a minor fraction of a dB) the same gain in its favored direction as a ground-mounted vertical with 120 buried radials.

Note however that the pattern is non-symmetrical. The radiation favors the direction of the radial, resulting in a 3 to 4-dB F/B over average ground. Modeling the same vertical over very good ground results in much less directivity, and over saltwater the antenna becomes perfectly omnidirectional.

I expect that it is sufficient to install radials on the ground under the antenna to improve the properties of the ground in the near field of the antenna to a point where the directivity, due to the single radial, is reduced to less than 1 dB. The slight directivity can be used to advantage in a setup where one would have a vertical with four radials, which are then connected one at a time to the vertical antenna. Another application (Ref 7824) is where the vertical is part of a fixed array, and where you make use of the initial directivity of each element to provide some added directivity (see **Fig 9-28**).

The single radial does not only create some horizontal

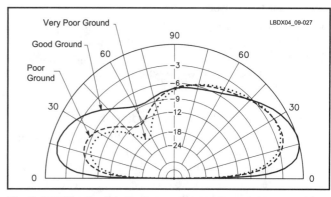

Fig 9-27—Vertical radiation pattern of a quarter-wave vertical with one horizontal $\lambda/4$ radial at a height of 0.05 λ over different types of ground.

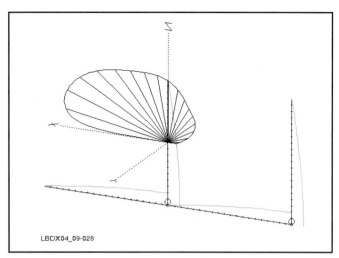

Fig 9-28—Two λ/4 verticals are used in an end-fire configuration (see Chapter 10), producing a cardioid pattern. By placing the single radial in the forward direction of the array, some additional gain can be achieved. This technique makes it impossible to switch directions.

directivity, it also introduces some high-angle radiation, caused by the radiation from the single radial. If two or more radials are used, they can be set up in such a way that the horizontal radiation of these radials is effectively canceled. Notice from Fig 9-27 that most of the high-angle pattern energy is at or near 90°.

If you are looking for maximum low-angle radiation (which is normally the case for DXing), using only one radial is not the best choice, especially if the antenna is going to be used for reception as well. In a contest-station environment, however, creating some high-angle radiation, to give some "presence" on the band with locals can be desirable. If separate directive low-angle receiving antennas (eg, Beverages) are used, using a single radial on a vertical may well be a logical choice. I am using a single 5-meter high elevated radial on my 80-meter Four Square (radials pointing out of the square). At the same time I have a decent shielding effect on the real ground because of the more than 200 radials for the 160-meter vertical, which supports the 80-meter wire Four Square (see Chapter 11).

A vertical with a single radial can also be a logical choice for a DXpedition antenna (over saltwater or over a ground screen) for two reasons:

1. Ease of adjusting resonance from the CW to the phone end of the band, by just lengthening the radial.
2. Extra gain by putting the radial in the wanted direction (toward areas of the world with high amateur population density).

2.2.5. How high should the radials be?

The *NEC*-modeling results, published by Christman, K3LC (ex KB8I), indicate that radials above a height of approximately 0.03 λ achieve gains within typically 0.2 dB of what can be achieved with 64 buried radials. In other words, there is no point in raising the radials any higher than 6 meters on 160 or 3 meters on 80 meters.

Measurements done by Eric Gustafson, N7CL, however,

tell us a totally different and very logical story. To prevent the near fields created by the radial currents from causing absorption losses in the underlying ground, the radials must be high enough so that the near fields do not touch ground. With up to six radials, this is between λ/8 and λ/4. Below λ/8 the losses are very considerable (if no other screen is available). For amateur purposes with four radials, a minimum height of λ/4 would be a reasonable limit to use. The minimum height decreases as the density of the radial screen is increased. With a density of about 100 quarter-wave long radials (in which case the distance between the tips of the radials is 0.015 λ) the radial plane can be lowered all the way *onto* the ground without incurring significant near-field absorption loss. This is shown in Fig 9-15, where beyond 100 radials there is little to gain. At a height of about 0.03 λ, 26 radials will result in an absorption loss of not more than 0.5 dB, according to N7CL.

Conclusion: If you want to play it extra safe, and if you have the tower height, get the radials up as high as possible and add a few more. Having more radials will make their exact length much less critical as well. Another solution that I have used is to put radials and chicken-wire strips on the ground to achieve an "on-the-ground screen" in addition to your small number of elevated radials (see Section 2.2.13).

It all is very logical. Get away from the lossy ground by raising the radials higher above the ground or hide the lossy ground with a dense screen using many radials.

2.2.6. Why quarter-wave radials in an elevated radial system?

In modeling it is quite easy to create perfectly resonant quarter-wave radials. Why do we want them to be exactly λ/4 long? Let's examine this issue. What we really want is the vertical plus the radials to be resonant, not because this would make the antenna radiate better, but only because that makes it easier to feed the antenna.

Dick Weber, K5IU, found through a lot of measuring and testing of real-life verticals with elevated radials that using λ/4-long elevated radials has a certain disadvantage. In his models he used four radials (one per 90° of azimuth) because he wanted the radiation from these radials to be completely canceled: no pattern distortion and no high-angle horizontally polarized radiation. He found out though that this is very

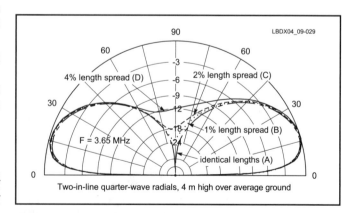

Fig 9-29—Vertical radiation patterns (over good ground) for a λ/4 long 80-meter vertical, with two in-line radials 4 meters high, for various radial lengths around λ/4. See text for details.

difficult, if not impossible, to achieve in the real world. Of course, λ/4 radials works fine on a computer model, since you can define four radials that have exactly the same electrical length. But this is not always the case in the real world. One radial will always be, perhaps by only a minute amount, electrically longer or shorter than another one. And therein lies the problem. We want these four radials all to carry exactly the same current, in order for the radiation to balance out.

The real question is how important are equal currents in the radials? I modeled several cases of intentional radial current imbalance. **Fig 9-29** shows the vertical radiation pattern of a λ/4-vertical (F = 3.65 MHz), with two elevated radials, 4 meters high. Pattern A is for two radials showing no reactance (both perfectly 90°, which can never be achieved in real life). For pattern B, I have intentionally shortened one radial about 20 cm (approximately 1% of the radial length). This introduced a reactance of $-j\,8\,\Omega$ for this radial. One radial now carried 62% of the antenna current, the other the remaining 38%. Over good ground this imbalance causes the horizontal pattern to be skewed about 0.6 dB (an inconsequential amount), but we see a fill-in of the high-angle rejection (around 90° elevation) that we would expect to have when the currents are really equal. Pattern C is for a case where one radial is 20 cm too short, and the other one 20 cm too long (reactance $-j\,8\,\Omega$ and $+j\,8\,\Omega$). In this case the relative current distribution was very similar as in the first case (63% and 36%). The horizontal pattern skewing was the same as well. Pattern D is for a rather extreme case where radials differ 80 cm in length ($+j\,16\,\Omega$ and $-j\,16\,\Omega$). Current imbalance has now increased to 76% versus 24%.

I did a similar computer analysis for a vertical using four elevated radials. In this case, I did the analysis over three different types of ground: good ground, very good ground and seawater (ideal case).

Fig 9-30 shows the results of these models. Case A is for equal currents in the four radials (theoretical case); case B is for radials showing reactances of $+j\,8\,\Omega$, $0\,\Omega$, $-j\,8\,\Omega$, and $+j\,10\,\Omega$. The relative current distribution in the four radials was: 51%, 39%, 5% and 5%, which are values very similar to what has been measured experimentally by K5IU. Pattern C shows a rather extreme imbalance with radial reactances of $-j\,16\,\Omega$, $0\,\Omega$, $+j\,16\,\Omega$ and $+j\,8\,\Omega$ (a total length spread of 4% of the nominal radial length). In this case the relative currents in the radials are 54%, 28%, 8% and 10%. Plot 1 is for the antenna over good ground, Plot 2 over very good ground, and Plot 3 over sea water.

Note that the pattern deformation depends to a very high degree on the quality of the ground under the antenna! Over seawater the current imbalances practically cause no pattern deformation at all. The horizontal pattern squeeze is at maximum 1.6 dB over good ground, and 0.6 dB over very good ground, computed at the main elevation angle.

From this it appears that in addition to using a few (typically less than 10) elevated radials, it is a good idea to improve the ground conductivity right under the radials by installing a ground screen using radials there as well. This is for two different reasons: To form a screen hiding the lossy ground from the antenna, and to reduce the effect of high-angle radiation from the radials.

You should understand that if you have enough elevated

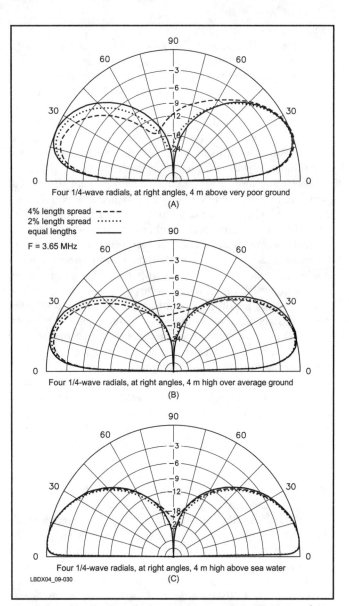

Fig 9-30—Vertical radiation patterns of an 80-meter λ/4 vertical with four elevated radials (4 meters high) over various types of ground. Patterns are for: (A) average ground, (B) very good ground and (C) saltwater. See text for details.

radials any variation in the exact electrical length will *not* result in high-angle radiation or pattern squeezing. With 16 radials, length variations of ±1.5%, and angular variations of ±5° (not evenly spaced in azimuth), the effect is of no consequence, resulting in horizontally polarized radiation components down > 40 dB). The radials now form a screen that no longer shows resonance, just like the case with radials on the ground.

You also need a large number of elevated radials to avoid excessive near-field losses. You can kill two birds with one stone with a raised radial system using at least 16 radials.

Dick Weber, K5IU, measured many real-life installations with either two, three or four elevated radials, and it was not uncommon to find one radial taking 80% of the antenna current, one radial 20% and the other two almost zero! The recorded variations in radial currents were used to calculate

Vertical Antennas 9-23

the patterns shown in Fig 9-30.

The question now is whether or not you can live with the high-angle fill in, (mostly around the 90° elevation angle) and slight pattern-squeeze (typically not more than 1 dB).

If you want maximum low-angle radiation, and if you don't want to lose a fraction of a dB, and if you don't want to put up a few more radials, then equal-radial currents may be for you. Or maybe you would like some high-angle radiation? Maybe you are not using your vertical or vertical array for reception, and you want some high-angle radiation? If you are a contest operator, this is a good idea (you want some local presence as well). In that case, don't bother with equal radial currents, maybe just one radial is the answer for you, as I did.

However, even a small number of radials that are laid out perfectly symmetrically and that carry identical currents are no guarantee of 100% cancellation of the horizontal high-angle radiation in the far field. Slight differences in ground quality under the radial wires (or environment, trees, bushes, buildings) can result in different near-field absorption losses under radials that would otherwise carry identical RF currents. The result will be incomplete cancellation of their radiated fields in the far field. Measuring radial currents does not, indeed, tell you the full story!

It is interesting though to understand why slight differences in radial lengths can cause such large differences in radial current. A λ/4 radial is equivalent to an open-circuited λ/4 transmission line that uses the ground as the second conductor. This acts like a dead short at its resonant frequency. When this short is connected in parallel with another λ/4 radial, it's like connecting a short circuit across another short circuit, and then expecting that both shorts will take exactly the same current.

We have similar situations in electronics when we parallel devices such as power transistors in power supplies, or when we parallel stubs to reject harmonics on the output of a transmitter. If one stub gives us 30 dB of attenuation, connecting a second one right across the first one will increase the attenuation by 3 dB at the most. If we take special measures (λ/4 lines at the harmonic frequency) between the two stubs, then we get greater attenuation (almost double that of the single stub, an additional 6 dB). **Fig 9-31** shows the equivalent schematic of the situation using λ/4 radials.

2.2.6.1. Conclusions

1. For elevated radial systems using two, three or four (resonant) λ/4 radials, slight differences in electrical length cause radial current imbalances, resulting in some high-angle radiation as well as some pattern squeezing, especially over less than very good ground. However, even perfectly balanced currents are not a 100% guarantee for zero high-angle radiation (due to unequal near-field ground losses under different radials).
2. Starting with eight radials (or more) the influence of unequal radial current on the generation of high-angle radiation is almost nonexistent. If you are greatly concerned about a little high-angle radiation, you should simply increase the number of elevated radials to eight.
3. Adding a good ground screen under the antenna totally annihilates the effects of unequal radial currents, and in addition it will raise the gain of the antenna by up to 5 dB!
4. By the way, you need not to concern about any of these issues with a classic in (or on) the ground radial system using 60 radials.

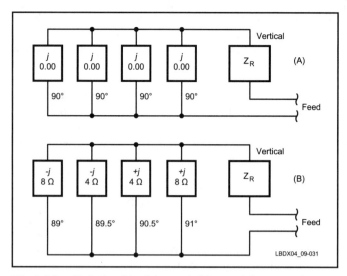

Fig 9-31—At A the ideal (not of this world) case where all four radials are exactly 90° long. They all are a perfect short and exhibit *zero* reactance. At B the real life situation, where it now is clear that in this circuit, where the current divides into four branches, these currents are now very unequal.

2.2.7. Making quarter-wave radials of equal length

Despite all of that, it's nice to know how you can make λ/4 radials of identical electrical length! In the past, one of the standard methods of making resonant radials, was to connect them as a (low) dipole and prune them to resonance. It is evident that resonance does not mean that both halves of the dipole have the same electrical length, even if both halves are the same physical length. One half could exhibit $+j\,20\,\Omega$ reactance, while the other half could exhibit a so-called conjugate reactance, $-j\,20\,\Omega$. At the same time the dipole would be perfectly resonant.

Nevertheless, there is a more valid method of constructing radials that have the same electrical length. Whether these are perfect λ/4 radials is not so important, we can always tune out any remaining reactance with a small series coil or a capacitor (if too long). This method is as follows:

- Model the length of the vertical to be λ/4 at the design frequency.
- Put up an elevated vertical of the computed length.
- Use one of the charts in **Fig 9-32** to determine the theoretical radial length. Note that the length is very dependent on radial height.
- Connect one radial.
- Trim the radial to bring the vertical to resonance.
- Disconnect the radial.
- Put up the second radial in line with number one.
- Trim this second radial for resonance.
- If you use four radials, do the same with the remaining two radials.

Then connect all radials to the vertical and check its resonant frequency. It is likely that the vertical will no longer be resonant at the design frequency. Is it necessary to have the

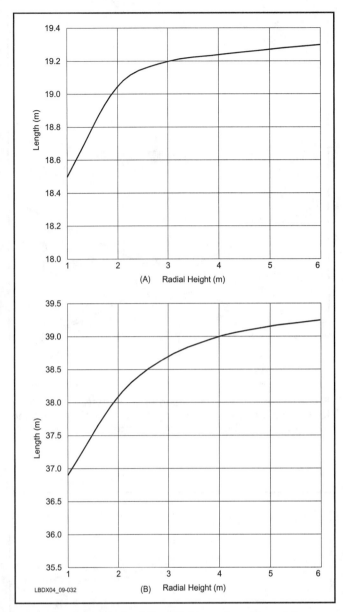

Fig 9-32—Length of a λ/4 radial as a function of the height above ground. For 80 meters at A; for 160 meters at B.

vertical at exactly λ/4? No, but if you want, here are two procedures to make the antenna plus radials perfectly resonant on your design frequency:

2.2.7.1. First method

This requires changing the length of the vertical to bring the system to resonance. Do not change any radial length, but change the length of the vertical to achieve resonance at the desired frequency.

2.2.7.2. Second method

Change all radials in length by exactly the same amount (all together, not one at a time) until you establish resonance. Neither of these two methods guarantees that both the radial system and the vertical are exactly a quarter wavelength, they only guarantee that both connected together are resonant.

Again, it is totally irrelevant whether both are 90° long or not. It is not unusual that radials of different physical length result in identical electrical lengths. This is mainly due to the variation of ground conductivity, which can vary to a wide degree over small distances. Other causes are coupling to nearby conductors.

On the other hand, radials of exactly the same electrical length are still no guarantee for identical radial current because of near-field losses being different under different radials (see Section 2.2.6).

2.2.8. The K5IU solution to unequal radial currents

D. Weber, K5IU, inspired by Moxon (Ref 693, pages 154-157 in the First Edition, pages 182-185 in the Second Edition, and Ref 7833) installed radials shorter than λ/4 and tuned the radial assembly to resonance with a coil. It appears that slight changes in electrical length of these "short" radials have little influence on the current in the various radials (Ref 7822 and 7823).

Weber's modeling studies showed that radial lengths between 45° and 60° and between 115° and 135° resulted in minimum creation of high-angle radiation from unequal electrical radial lengths. When using radials longer than 90° the system can be tuned to resonance using a series capacitor, which is easier to adjust than a coil and which also has intrinsically less losses (see **Fig 9-33**). The purist may even use a motor-driven (vacuum) capacitor, which could be used to obtain an almost perfect SWR anywhere in the band.

I would suggest, however, not to shorten the radials to less than approximately 60°-70° if not really necessary. It is clear that we cannot indefinitely shorten radials, and expect to get the same results. If that were true we should all use two in-line loaded mobile whips on our 160-meter tower as a radial (current collecting) system. T. Rauch, W8JI, put it very clearly on the Topband reflector: "*The last thing in the world*

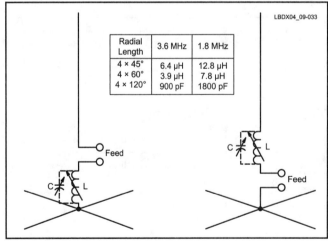

Fig 9-33—When radials shorter than 90° are used, the system must be tuned to resonance using a coil. With radials longer than 90° the tuning element is a capacitor. Typical values for the tuning elements are also shown. The feed line can be connected in two different ways: Between the tuning element and the radiator or between the tuning element and the radials. The result is exactly the same. In both cases, a coaxial feed line connected to the feed point *must* be equipped with a current balun.

I'd want to do is concentrate the current and voltage in smaller areas. Resonant radials, or especially shortened resonant radials, concentrate the electric and magnetic fields in a small area. This increases loss greatly. The ideal case is where the ground system carries current that evenly, and slowly, disperses over a large physical area, and has no large concentrated electric fields from high voltage." This is clearly another plea for the classic, multi-radial ground system. I did some modeling myself using *EZNEC* and found that:

- The fewer the radials, the greater the current imbalance due to length variations.
- The worse the ground quality the greater the impact of current imbalance on the radiation pattern.
- Starting with 16 radials, the effect of current imbalance is totally gone, even with 90° radials.

2.2.8.1. Conclusion

You can solve the problem of high-angle radiation by using a larger number of radials (for example, 16) or by improving the ground quality under the radials by installing a ground screen, at the same time yielding less near-field ground-absorption losses!

2.2.9. Should the vertical be a quarter-wave?

From a radiation point of view, neither a vertical with a buried-radial ground system nor one with an elevated-radial system necessarily must be resonant. We usually make these resonant because it makes feeding the antenna easier.

A buried ground-radial system is a non-resonant, low-impedance system. Over such a ground system the vertical is usually made resonant (90° long electrically), to have a non-reactive feed-point resistance. Verticals somewhat longer than λ/4 (usually about 3λ/8) can be tuned to resonance using a series capacitor. Although most 3λ/8 verticals use ground-mounted radials, the same can be done with a 3λ/8 vertical

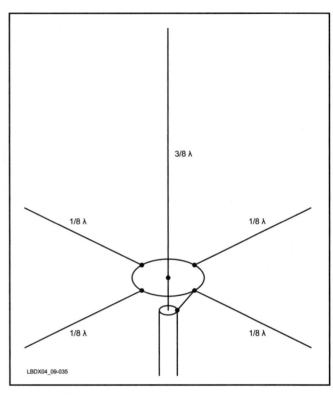

Fig 9-35—A 3λ/8 vertical used in conjunction with 45° long radials does not require any series coil to tune the antenna, hence losses are minimized.

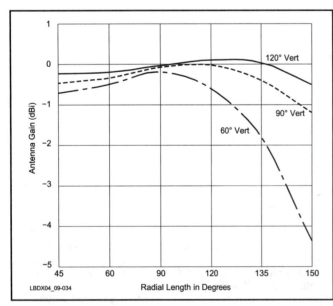

Fig 9-34—Gain as a function of radial length for verticals measuring 60°, 90° and 120° over average ground (all using four elevated radials about 0.012-λ high) as calculated by K5IU.

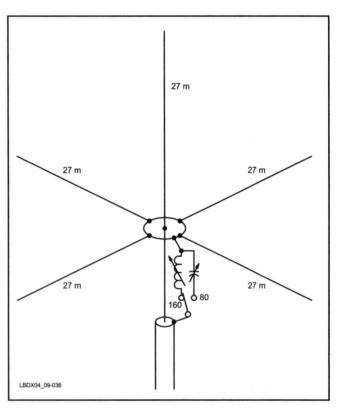

Fig 9-36—A 27-meter long vertical with 27-meter long radial makes an excellent antenna for both 80 and 160. Band switching only requires the switching of the loading element from a coil (160 meters) to a capacitor (80 meters).

using elevated radials.

Remember that with a small number of radials (up to about 10), the length of each of these radials is critical and the radial system has a resonant character that is more pronounced as the number of radials is reduced. This means that if you use only a few radials, you can adjust their length to change the resonant frequency of the vertical. With a large enough number of radials the system becomes non-resonant (like a ground screen) and changing radial lengths has no influence on the resonant frequency of the antenna system. See **Fig 9-34**.

Using this concept we can envisage a $3\lambda/8$ vertical to be used in conjunction with, say, $\lambda/8$ long radials. A 3.75-MHz vertical designed according to these principles is shown in **Fig 9-35**. The combination of a $3\lambda/8$-long radiator and $\lambda/8$-long radials does not require a coil to tune the antenna. The radiator length shown for a wire element whose diameter is 2 mm is 26.9 meters long. With four 10-meter long radials, the feed impedance is exactly 52 Ω, an excellent match for 50-Ω feed line.

The same vertical can be turned into an 80/160-meter vertical using 27-meter long radials (60° on 160 meters and 120° on 80 meters) as shown in **Fig 9-36**. The total system length on 160 meters is 60° + 60° = 120°, which is less than 180° ($\lambda/2$); hence a coil is required to resonate the antenna. On 80 meters, the total length is 120° + 120° = 240°, which is longer than $\lambda/2$; hence, a capacitor is required.

2.2.10. Elevated radials on grounded towers
2.2.10.1. The N4KG antenna

T. Russell, N4KG, an eminent low-band DXer, described a method of shunt feeding grounded towers in conjunction with elevated radials (Ref 7813 and 7832). His tower uses a TH7DX triband Yagi as top loading to make it about 90° long with respect to the feed point (see **Fig 9-37**). It is important to find the attachment point of the radials on the tower whereby the part of the tower above the feed point becomes resonant in conjunction with the radials. Russell installed 10 $\lambda/4$ radials and moved the ring to which these radials were attached up and down the tower until he found the system in resonance. This point was 4.5 meters above ground.

John Belrose, VE2CV, analyzed N4KG's setup using *NEC-4* (Ref 7821). He simulated the connection to earth of the tower (at the base) by using a 5-meter long ground rod (a decent dc ground). It is obvious that RF current is flowing through the tower section below the feed point. This current causes the gain of the antenna to be somewhat lower than that of a $\lambda/4$ base-fed tower. Belrose calculated the difference as 0.8 dB.

A typical configuration like the one described by N4KG will yield a 2:1 SWR bandwidth of 100 to 150 kHz. There are several approaches to broadband the design. Sam Leslie, W4PK, designed a system where he uses two sets of two radials, installed at right angles. One set is cut to resonate the system at the low end of 80 meters (CW band) and the other at the phone end. The SWR curve has two dips now, one on 3.5 and the other on 3.8 MHz.

Another approach is to design the antenna for resonance on 80-meter CW, and tune it to resonance in the SSB portion by inserting a capacitor between the feed line and the

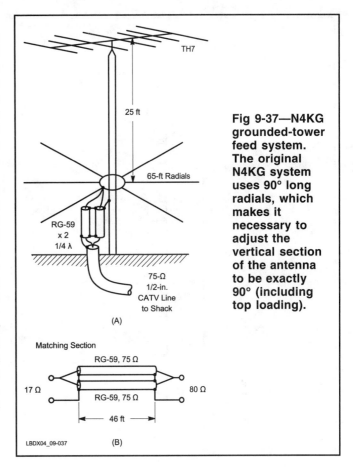

Fig 9-37—N4KG grounded-tower feed system. The original N4KG system uses 90° long radials, which makes it necessary to adjust the vertical section of the antenna to be exactly 90° (including top loading).

radials or the vertical conductor (tuning out the inductive reactance on 3.8 MHz).

2.2.10.2. Decoupling the tower base from the real ground

It is possible to minimize the loss by decoupling the base of the vertical from ground. Methods of doing so were described by Moxon (Ref 693 and 7833). **Fig 9-38** shows the layout of a so-called *linear trap* that turns the tower section between the feed point and ground into a high impedance, effectively isolating the antenna feed point from the dc-ground rod. The trap is constructed as follows:

- Connect a shunt arm about 50 cm in length to the tower, just below the antenna feed point.
- Connect a drop wire, parallel with the tower, from the end of the arm to ground level and connect it back to the base of the tower. This forms a loop.
- Insert a variable capacitor in the drop wire (wherever convenient).
- Excite the vertical antenna (above the linear stub) with some RF.
- Use an RF current probe (such as a Palomar type PCM1) and tune the capacitor for maximum current in the drop wire.
- You're done!

The loop tower + drop wire + capacitor now form a parallel-resonant circuit at the operating frequency. This ensures that no RF currents can flow through the bottom tower section to the lossy ground.

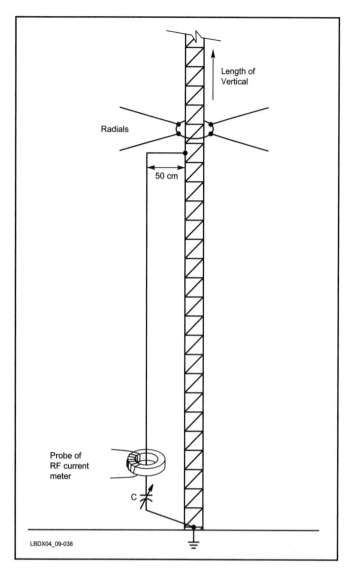

Fig 9-38—The grounded-tower section below the antenna feed point can be made a resonant linear trap, which inserts a high impedance between the antenna feed point and the bottom of the tower. Tune capacitor for maximum current in the loop.

2.2.10.3. Summing up

Using grounded towers with an elevated radial system can readily be done. The principles are simple:

- The vertical (top loaded or not) together with the radial system must be resonant
- Use the largest number of radials you can accommodate to obtain a ground-shielding effect.
- Provisions must be taken for minimum RF return current to flow in the ground. The section of the tower below the feed point should thus be decoupled.

2.2.11. The N4KG reverse-feed system

Russell feeds his design in Fig 9-37 in an unconventional way, with the center of the coax going to the radials, and the outer shield going to the vertical part. He claims this prevents arcing through from the braid of the coax to the tower. Tom coils up his parallel 75-Ω coax inside the tower leg, and that forms an RF choke. I would strongly suggest *not* to tape the coax (or the coiled coax) to the leg of the tower, especially when a linear trap is installed, since there may be a rather steep RF voltage gradient on that leg. I would keep the coax a few inches from all metal, and route it in the center *inside* the tower. In addition to the coiled coax I would certainly use a current balun made of a stack of ferrites, installed beyond the λ/4 transformer toward the transmitter. Whether or not the braid or the inner conductor goes to radials is irrelevant if a good current balun is used.

2.2.12. Practical design guidelines, elevated radials with grounded towers

If you have a grounded tower and you want to use it with an elevated radial system with four radials, you can proceed as follows:

1. Define the height where you want to have the radials. You might start at 6 meters. Convert to degrees (360° = 300/F_{MHz}) and 6 meters = 13° on 160 meters. If you have enough physical tower height, put the radials as high as possible, since this helps reduce the near-field absorption losses from the ground.
2. Define the electrical length of the tower. Let us assume you have a 30-meter tower with a 5-element 20-meter Yagi on top. From Fig 9-84 we learn that this tower has an electrical length of about 123°.
3. The electrical length of the tower above the radial attaching point is 123° – 13° = 110°.
4. Cut four radials to identical electrical length as explained in Section 2.2.7.
5. Whether or not you will require a coil or a capacitor to tune the system to resonance depends on the *total* length of the antenna vertical part *plus* radials. If the length is greater than 180°, a capacitor will be required. An inductor will be required if the total length is less than 180°. Assume for this example that you use 120° long radials, so that the total antenna length is 110° + 120° = 230°. A series capacitor will be required to tune the system to resonance.
6. Measure the impedance at resonance using an antenna analyzer. If necessary use an unun or a quarter-wave transformer (or other suitable impedance matching system) to get an acceptable match to your feed-line impedance.
7. Install the linear trap on the tower section under the feed point and tune the loop to resonance by adjusting the loop variable capacitor (see procedure above).
8. You are all done!

Fig 9-39 shows the final configuration of the antenna we designed above. It is obvious that the tower must use non-conducting guys, or if steel guy wires are used they must be broken up in short lengths so that they do not interfere with the vertical antenna.

Finally, here's some perspective. Maybe it's not such a good idea after all to have elevated radials on your grounded tower because it makes things more complicated. You need a linear trap to decouple the bottom of the tower from the real ground and you need to have radials above ground. Maybe 10 or 20 radials on the ground would do the job just as well. The real reason I can see for elevated radials on a grounded tower is when that tower is electrically too long (for example, > 140° rather than 90°). For this case you can shorten the tower

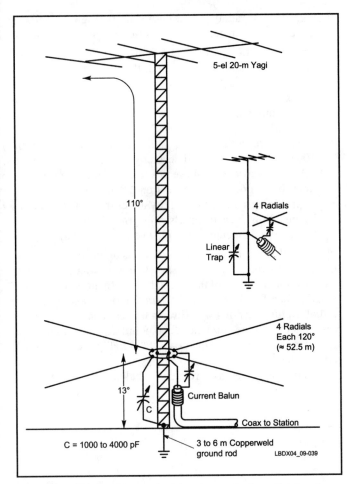

Fig 9-39—Design example of a grounded vertical using an elevated-radial system (see text for details).

If you have the space, and a potential 4 to 5 dB is worth the expense and effort to you, by all means provide a ground screen. In the case you do *not* want to use the screen for antenna current collecting, the screen does not have to have the shape of radial wires. A net of copper wires, with a mesh density measuring less than approx. $0.015\ \lambda$ (1 meter on 80; 2 meters on 160), or even $0.03\ \lambda$ if you are willing to sacrifice maybe 0.5 dB, is all that is needed to provide an effective near-field screen. Make sure that the crossing copper wires make good and permanent electrical connections at their joints (see Section 2.1.7).

If you use but one elevated radial, you may want to increase the ground net density in the area under that radial. In principle the screen should have a radius of $\lambda/4$ (for a $\lambda/4$ vertical), but a screen measuring only $\lambda/8$ in radius will typically be about 0.3 dB down from a $\lambda/4$ radius ground screen. Of course the saltwater environment shown in **Fig 9-40** makes for a virtually "perfect" ground screen, even though only two elevated radials were used!

For over five years now, I have very successfully used $\lambda/4$ verticals in my Four-Square array, each using a single $\lambda/4$ radial at about 5-meters in height. Judging an antenna's performance by the DX worked with it certainly makes no sense. But judging the same antenna's performance by the repetitive results obtained in world-class DX contests, may be

electrically using an elevated radial system. Watch out, however, if the radial system is fairly high above ground, because the vertical radiation pattern becomes different from that of a ground-mounted vertical.

2.2.13. Elevated radials combined with radial screen on the ground

All publications I have seen so far on the subject of elevated radials use either one of the modeling standard grounds (Average, Good, etc—see Table 5-2 in Chapter 5), or they have been done over whatever type of ground that happened to be there where the tests were run.

The modeling I have done suggests that improving the ground right under the vertical and its elevated radials can increase the system gain, especially if only one to four elevated radials are used (see Section 2.2.3 and Fig 9-27). For the case of a single radial or when using $\approx 90°$ long radials, improving the ground quality right under the antenna can greatly reduce horizontally polarized high-angle radiation and can increase the antenna gain. This can be accomplished by putting down radials or ground screens on the lossy ground.

It is important to understand that these on-the-ground radials (or screen in whatever shape) should *not* be galvanically connected in any way to the elevated radials in any way. They should be connected to nothing, since we don't want any antenna return currents to flow in the ground.

Fig 9-40—The Titanex V160E antenna on the beach at 3B7RF (St Brandon Island). Note the two elevated radials about 2 meters above salt water. The combination of one or two elevated radials with a perfect ground underneath is hard to beat.

a good indication indeed about whether the antenna works well or not. Operated over ground that is literally swamped with copper wire, I have never scored less than a first or second place for Europe in the ARRL International DX Contest (single-band 80 meters), both CW and SSB and that is in 18 contests since 1994. In addition, I set a new European record with that antenna. Taking into account that my QTH is certainly not the best for working Ws (Normandy or the UK West Coast are better places), this means that such a vertical—even with a single elevated radial—can be a top performer.

2.2.14. Avoiding return currents through the soil

Fig 9-41 shows the vertical antenna return paths for different radial configurations. Fig 9-41A shows the case where a simple ground rod is used, where the antenna return currents have to travel entirely through the lossy soil. This reduces the radiation efficiency of the vertical to a very high degree, because of the I^2R ground losses. Burying radials in the ground can greatly reduce the losses as the return currents can now travel, to a great extent (depending on the number and the length of the radials) through the low-loss radial conductors in the ground, as Fig 9-41B shows.

Fig 9-41C shows two radials elevated above ground. There are now two current return paths: the lossless path through the two radials and a lossy path through the soil.

We can minimize the currents in this parasitic path by:

- Raising the radials high above ground: Once the radials are a few meters above ground, the capacity to the lossy soil is rather small.
- Using fewer radials: More radials means more capacitance, thus more current in the ground and hence more ground losses.
- Using more radials: More radials means a better screen. 100 radials, λ/4 long will perfectly screen the earth underneath the vertical. (This seems to contradict the previous item, but it doesn't—see Section 2.3.).
- Improving ground conductivity under the elevated radials by installing buried radials or a ground screen (not galvanically connected to the elevated radials, though!).

Another important issue is currents on the outside of the coaxial feed line. Fig 9-41D shows how unwanted currents can flow on the shield of the coaxial cable. In this situation, the coaxial feed line is just another conductor, a random-length radial. Return currents will flow in that conductor unless it is disconnected at the antenna's feed point. The question is now how can we disconnect the coaxial "radial" wire and not the coaxial feed line?

You must insert a current choke balun at the antenna feed

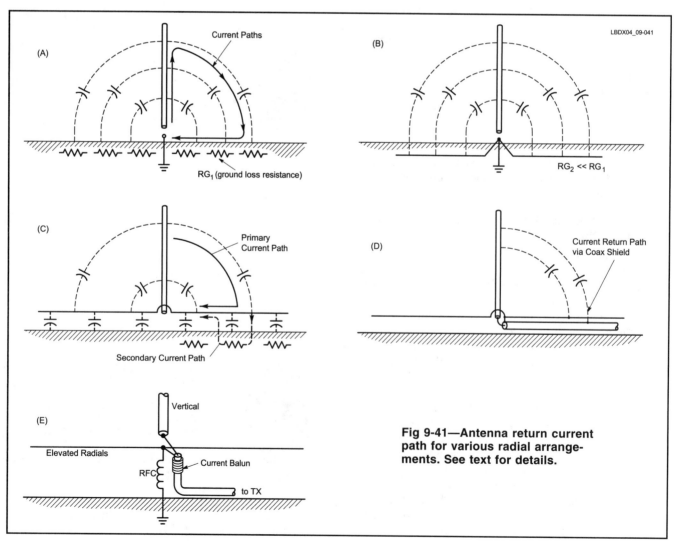

Fig 9-41—Antenna return current path for various radial arrangements. See text for details.

point (see Fig 9-41E). The high impedance the current balun presents to any currents on the outside of the coax shield effectively suppresses common-mode currents on the cable. Several types of current baluns are described in Chapter 6, Section 7. If you are forced to use (for layout reasons) $3\lambda/4$ feed lines in a Four-Square array, you will wind up with a lot of surplus coax length. Wind it all up in a coil and mount it as close as possible to the antenna feed point. This makes an excellent choke balun. It is always better to run the coax on or preferably in the ground, rather than supported on poles at a certain height, to prevent coupling and parasitic currents on the outer shield.

It also makes common sense to provide a dc ground for the common radial points. You can do this by connecting an RF choke (100 µH or more) between the radial common point and a safety ground rod below the antenna feed point, as shown in Fig 9-41E.

If you use only a few radials each of them can radiate considerable near-field energy. They can induce currents on the feed line beyond where the choke balun has been inserted at the feed point. Burying the feed line can improve this situation. Feed lines supported off the ground are very sensitive to this kind of coupling. If you use only two radials, run the feed line at right angles to the two in-line radials. In other words, keep the feed line away from the near fields of the radials.

When using a number of elevated radials (eg, > 20), it is unnecessary to use a current balun since the screening effect of the radials will be sufficient to prevent common-mode antenna-return currents of any significant magnitude to flow on the coax outer shield.

2.2.15. Elevated radials in vertical arrays

When a vertical is used as an element in an array, an additional parameter arises when choosing the ideal radial length, at least if you are concerned about reducing horizontally polarized high-angle radiation of the array to a minimum. Careful layout of the radials is very important. Never run radials belonging to two different array elements in parallel. Design your layout such that coupling is minimized.

Zero coupling is of course achieved by using buried radials, terminated in bus bars where radials of adjacent elements meet one another. (See Chapter 13, Section 9.10). I should point out that if you use four 90° long radials on each element of an array, and have them laid out in such a manner that coupling does not exist between radials of adjacent elements, it may be just as good to use a single radial!

2.3. Buried or Elevated, Final Thoughts

It is clear, and it has been proven over and over in the real world, that an elevated radial system at a relatively low height is a valid alternative for a system of buried radials, if there is a good reason you can't put down a decent radial system in or on the ground. If you use only a small number of radials, perhaps 1 to 8, their task will be almost exclusively to efficiently collect the return currents of the vertical, and you will have to suffer substantial near-field losses in the ground, up to 5 dB. With a larger number the screening effect becomes important and near-field ground losses can be reduced by making use of the screening effect of a large number of radials. Elevated radials can have advantages such as:

- Providing the possibility of installing a decent ground system under very unfriendly circumstances, such as over rocky ground.
- More flexibility in matching, since the real ground is not resonant. An elevated radial system using only a few radials—maximum of four—can be made inductive or capacitive, which may be an asset in designing a matching system.

For using elevated radials I would propose the following guidelines:

- Put the radials up as high as possible.
- Use as many radials as possible, since this makes the radial system non-resonant.
- If you use a small number (< 16), install a ground screen.

If you have the space and if the ground is not too unfriendly, I would suggest you use buried radials however.

2.4. Evaluating the Radial System

Evaluating means measuring antenna field strength (FS), or measuring certain parameters for which we know the

Fig 9-42—Walter Skudlarek, DJ6QT, inspecting some of the radials used on the 160-meter vertical at ON4UN. Half of the radials are buried (where the garden is), and half are just lying on the ground in the back of the garden behind the hedge (where the XYL can't see the mess from the house!). In total, some 250 radials are used, ranging in length from 15 to 75 meters.

Vertical Antennas 9-31

correlation with radiated FS. You cannot truly evaluate an antenna just by modeling it. You can develop, design and predict performance by modeling, but you cannot evaluate the actual performance of the antenna on a computer. However, there are some indirect measurements and checks that can and should be done:

2.4.1. Evaluating a buried-radial system

The classic way to evaluate the losses of a ground system is to measure the feed-point resistance of the vertical while steadily increasing the number of radials. The feed-point resistance will drop consistently and will approach a lower limit when a very good ground system has been installed. Be aware, however, that the intrinsic ground conductivity can vary greatly with time and weather, so it is recommended that you do such a test over a short time frame to minimize the effects of varying environmental factors on your tests (Ref 818, 819).

Peter Bobeck, DJ8WL, (now a Silent Key) performed such a test on his 23-meter long top-loaded (T) antenna. He added 50-meter long radials (on the ground) while measuring the feed-point impedance and found the following:

No. of radials	2	5	8	14	20	30	50
Impedance, Ω	122	66	48	39	35	32	29

Incidentally, eight radials look like a perfect match to 50-Ω coax, but the system efficiency for that case was below 50%!

Don't be surprised if the impedance gets lower than 36 Ω with a full-size λ/4 vertical. It first surprised me when I measured about 20 Ω for my 160-meter full-size λ/4 vertical made with a freestanding tower, but that was because of its very large effective diameter.

For calculating antenna efficiency, you can use the values from Table 9-1 that lists the equivalent resistance of buried radial systems in good-quality ground. For poor ground, higher resistances can be expected, especially with only a few radials.

Measuring the impedance of a vertical and watching it decrease as you add radials tells us nothing about the near-field absorption ground losses. It only gives us an indication of the I^2R losses that determine return-current collecting efficiency.

Periodic visual inspections of the radial system for broken wires and loose or corroded connections, etc will assure continued efficient operation. **Fig 9-42** shows DJ6QT examining the radials of the ON4UN 160-meter vertical. If you bury the radials, it is a good idea to make them accessible anyhow just where they connect to the bus bar. This way you can periodically check with a snap-on current meter if the radial still carries any current on transmit. If it doesn't, maybe the radial is broken at a short distance from the connection point.

2.4.2. Evaluating an elevated-radial system

Whether you have 1, 2 or 16 elevated radials, if these radials are the only antenna-current return paths (that is, the elevated radials are *not* connected to the lossy ground), the measured real part of the antenna impedance will not change. There is no gradual decrease of feed-point impedance as you increase the number of radials.

Measuring the antenna impedance does *not* give you any indication of near-field absorption ground losses. The only test you can perform on an elevated radial system is to measure the radial current, although this has little, if any, correlation

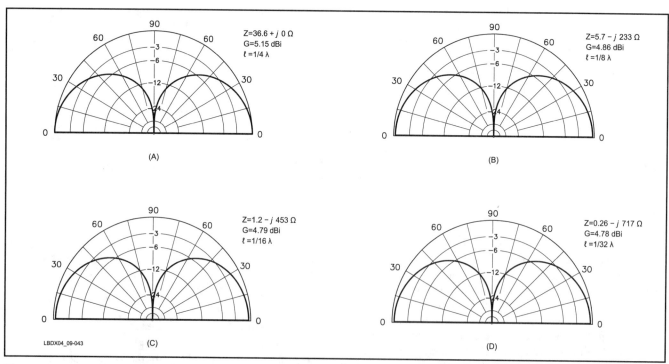

Fig 9-43—Elevation-plane radiation patterns and gain in dBi of verticals with different heights. The 0-dB reference for all patterns is 5.2 dBi. Note that the gain as well as the shape of the radiation patterns remain practically unchanged with height differences. The patterns were calculated with *ELNEC* over perfect ground, using a modeling frequency of 3.5 MHz and a conductor diameter of 2 mm. At A, height = λ/4. At B, height = λ/8. At C, height = λ/16. At D, height = λ/32.

with low-angle field strength. Nevertheless, when using only a few radials (2 to 8) it is a good idea to check the radial currents, and to make sure they are similar (± a few percent of one another).

Do regular inspections of your current balun. I would recommend to periodically measure its effectiveness by checking its inductance. This should be measured at the operating frequency.

3. SHORT VERTICALS

We usually consider verticals as being *short* if they are physically shorter than λ/4. Short verticals have been described in abundance in the amateur literature (Ref 771, 794, 746, 7793 and 1314). Gerd Janzen published an excellent book on this subject, *Kurze Antennen* (in German). Unfortunately, this was completely based on antenna modeling, where in my opinion real-world measured results are greatly lacking (Ref 7818).

The radiation pattern of a short vertical is essentially the same as that for a full-size λ/4 vertical. **Fig 9-43** shows the vertical radiation patterns of a range of short verticals over perfect ground, calculated using *ELNEC*. Notice that the gain is essentially the same in all cases (the theoretical difference is less than 0.5 dB).

If those short verticals over perfect ground are in essence almost as good as their full-size (λ/4) counterparts, why aren't we all using short verticals? A short monopole exhibits a feed-point impedance with a resistive component that is much

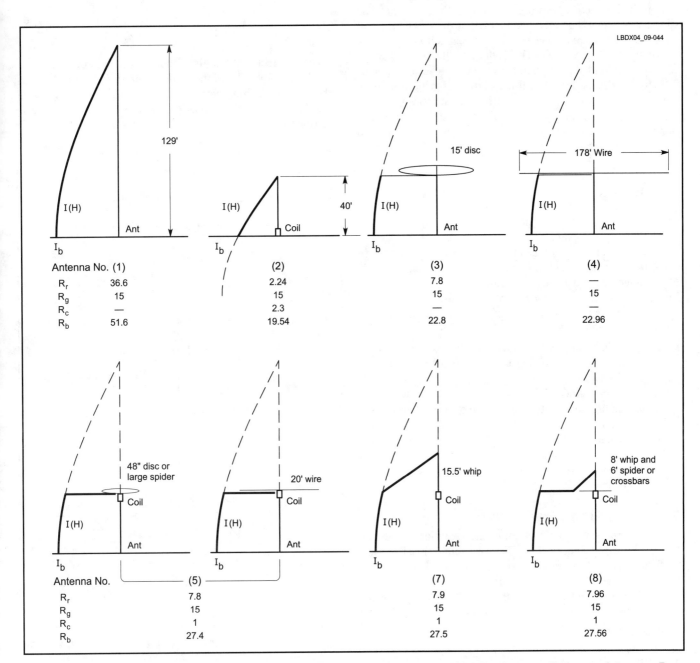

Fig 9-44—The antennas described in the text are shown with their current distributions, radiation resistances R_r, assumed ground loss resistance R_g, coil loss R_c (if any), total base input resistance R_b, base current I_b for 1000-W input to the antenna, and finally radiating efficiency in % (*Source*: "Evaluation of the Short Top Loaded Vertical" by W7XC, QST *March 1990*.)

smaller than 36.6 Ω and a reactive component that is highly capacitive. These two factors can make a short vertical more difficult to handle than a bigger one. To feed a short vertical with low losses using a coaxial feed line, you must first get rid of the reactive part and increase the real part of the feed impedance up to 50 Ω. This requires *loading* and *matching* the vertical and these can greatly impact efficiency.

Short verticals can be loaded to be resonant at the desired operating frequency in different ways. Various loading methods will be covered in this section, and the radiation resistance for each type will be calculated. Design rules will be given, and practical designs are worked out for each type of loaded vertical. Different loading methods will be compared in terms of efficiency.

Loading a short vertical means canceling the reactive part of the impedance to bring the antenna to resonance. The simplest way is to add a coil at the base of the antenna, a coil with an inductive reactance equal to the capacitive reactance shown by the short vertical. This is the so-called *base-loading method*. **Fig 9-44** shows a number of classic loading schemes for short verticals, along with the current distribution along the antenna. Remember from Section 1.2 that the radiation resistance is a measure of the area under the current-distribution curve. Also remember from Section 1.3 that the radiation efficiency is given by:

$$\text{Eff} = \frac{R_{rad}}{R_{rad} + R_{loss}}$$

The real issues with short verticals are *efficiency* and *bandwidth*. Let us examine these issues in detail. With short verticals the numerator of the efficiency formula decreases in value (smaller R_{rad}), and the term R_{loss} in the denominator is likely to increase (losses of the loading devices such as coils). This means we have two terms, which tend to decrease the efficiency of loaded verticals. Therefore maximum attention must be paid to these terms by

- Keeping the radiation resistance as high as possible (which is *not* the same as keeping the feed-point impedance as high as possible).
- Keeping the losses of the loading devices as low as possible. Maximum radiation resistance occurs when current integrated over the vertical section is as high as possible, which means maximum current mid-height in the vertical section. With very short verticals the current distribution is almost constant and the exact position of the maximum becomes irrelevant.

3.1. Radiation Resistance

The procedure for calculating the radiation resistance was explained in Section 1.2, where we found that for a λ/4 vertical made with a very small size conductor is 36.6 Ω. (See Fig 9-44). We will now analyze the following types of short verticals, all of which are about 30% of full-size quarter-wave (approximately 12 meters high on 160 meters) or 27.5° long:

1. Base loaded.
2. Top loaded.
3. Center loaded.
4. Base plus top loaded.
5. Linear loaded.

3.1.1. Base loading

The radiation resistance can be calculated as defined in Section 1.2. A trigonometric expression that gives the same results, is given below (Ref 742).

$$R_{rad} = 36.6 \times \frac{(1 - \cos L)^2}{\sin^2 L} \quad \text{(Eq 9-7)}$$

where L = the length of the monopole in degrees (1 λ = 360°).

According to Eq 9-7, the radiation resistance of the base-loaded vertical (electrical length = 27.5°) is 2.2 Ω. (See Fig 9-44.)

J. Hall, K1TD, derived another equation (Ref 1008):

$$R_{rad} = \frac{L^{2.736}}{6096} \quad \text{(Eq 9-8)}$$

where L = electrical length of the monopole in degrees.

This simple equation yields accurate results for monopole antenna lengths between 70° and 100°, but should be avoided for shorter antennas. A practical design example is described in Section 3.6.1.

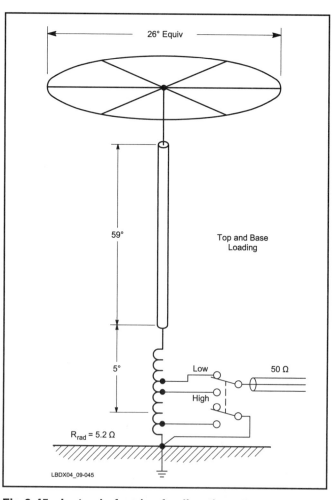

Fig 9-45—Instead of series-feeding the antenna, we can look for a tap on the coil that gives 50 Ω. The coil serves two purposes: Some base loading and also impedance matching. Using a DPDT relay you could make provisions for a perfect 50-Ω match on two frequencies; eg, on CW and on phone.

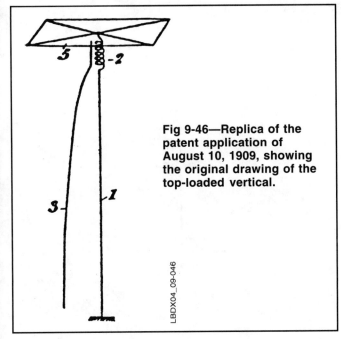

Fig 9-46—Replica of the patent application of August 10, 1909, showing the original drawing of the top-loaded vertical.

3.1.2. Top loading

The patent for the top-loaded vertical was granted to Simon Eisenstein of Kiev, Russia, in 1909. **Fig 9-46** is a copy of the original patent application, where you can see a combined loading coil plus top-hat loading configuration. The resulting current distribution is also shown.

The tip of the vertical antenna is the place where there is no current, and maximum voltage. This is the place where *capacitive* loading is most effective, and *inductive* loading (loading coils) is least effective. In some cases, inductive loading is combined with capacitive top loading. Top loading is achieved by one of the following methods (see **Fig 9-47**):

- **Capacitance top hat**: In the shape of a disk or the spokes of a wheel at the top of the shortened vertical. Details of how to design a vertical with a capacitance hat are given in Section 3.6.3.
- **Flat-top wire loading (T antenna)**: The flat-top wire is symmetrical with respect to the vertical. Equal currents flowing outward in both flat-top halves essentially cancel the radiation from the flat-top wire. For design details see Section 3.6.4.

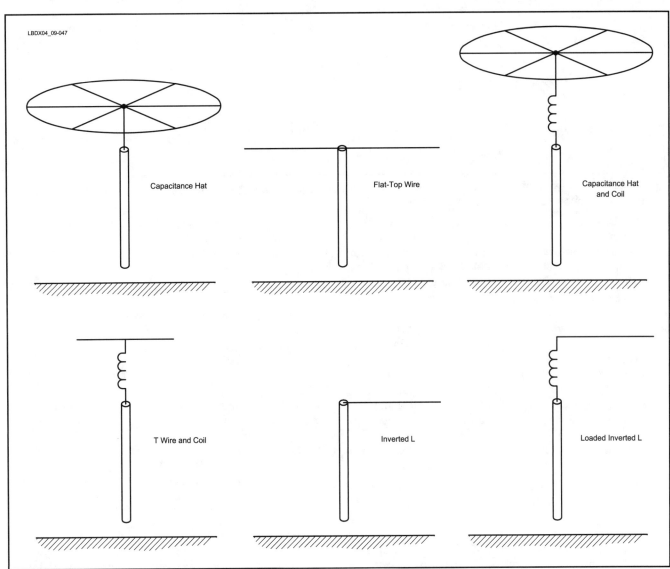

Fig 9-47—Common types of top loading for short verticals. The inverted L and loaded inverted L are not true verticals, since their radiation patterns contain horizontal components.

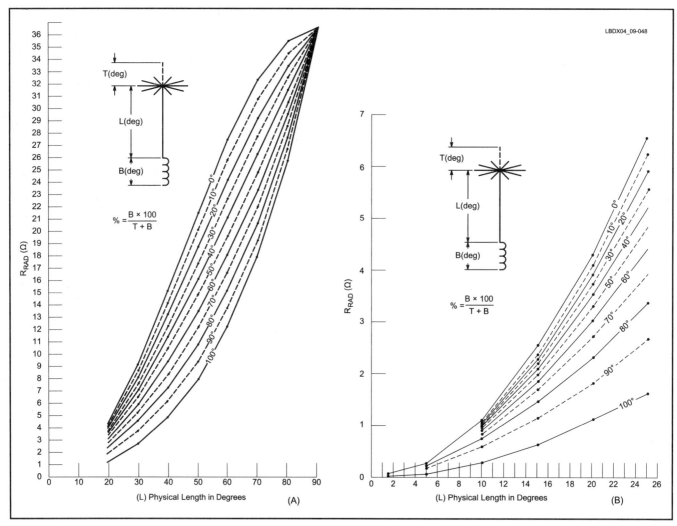

Fig 9-48—Radiation resistances of a monopole with combined top and base loading. Use the chart at B for shorter monopoles to obtain better accuracy.

- **Coil with capacitance hat**: In many instances a loading coil is used in combination with a capacitance hat to load a short monopole. This may be necessary, as otherwise an unusually large capacitance hat may be required to establish resonance at the desired frequency.
- **Coil with flat-top wire**: This loading method is similar to the coil with capacitance hat (see Section 3.6.5 for design example).
- **Inverted L**: This configuration is not really a top-loaded vertical, since the horizontal loading wire radiates along with the vertical mast to produce both vertical and horizontal polarization. Inverted-L antennas are covered separately in Section 7.
- **Coil with wire**: This too is not really a loaded short vertical, but a form of a loaded inverted L.

For calculating the radiation resistance of the top-loaded vertical, it is irrelevant which of the above loading methods is used. For a given vertical height, all achieve the same radiation resistance. However, when we deal with efficiency (where both R_{rad} and R_{loss} are involved) the different loading methods may behave differently because of different loss resistances.

The radiation resistance can be calculated as defined in Section 1.2. A trigonometric expression with the same results is given below (Ref 742 and 794):

$$R_{rad} = 36.6 \times \sin^2 L \qquad (Eq\ 9\text{-}9)$$

where L is the length of the vertical monopole in degrees.

The 27.5° short monopole with pure end loading (Fig 9-44) has a radiation resistance of

$$R_{rad} = 36.6 \times \sin^2 27.5° = 7.8\ \Omega$$

The radiation resistance of top-loaded verticals can be read from the charts in **Fig 9-48**. For top-loaded verticals, use only the 0% curves.

3.1.3. Center loading

The center-loaded monopole of Fig 9-44 is loaded with a coil positioned along the mast. The antenna section above the coil is often called the *whip*.

- Length of mast below the coil = 27.5°
- Length of whip above the coil = 3° (4.7 meters on 1.9 MHz)

The radiation resistance can be calculated as defined in Section 1.2. A trigonometric expression that gives the same results is shown below (Ref 42 and 7993):

$$R_{rad} = 36.6 \times \left(1 - \sin^2 t2 + \sin^2 t1\right) \quad \text{(Eq 9-10)}$$

where
t1 = length of vertical below loading coil (27.5°)
t2 = 90° − length of vertical above loading coil (the whip, 3°) = 87°

Using this formula, R_{rad} is calculated as = 7.9 Ω. Note that R_{rad} is essentially the same as the other top loaded schemes. The whip is often used in mobile antennas to fine-tune the antenna to resonance.

3.1.4. Combined top and base loading

Top and base loading are quite commonly used together, as shown in Fig 9-45. Top loading is often done with capacitance-hat loading, or even more frequently in the shape of two or more flat-top wires. If a wide frequency excursion is required (eg, 3.5 to 3.8 MHz), you can load the vertical to resonate at 3.8 MHz using the top-loading technique. When operating on 3.5 MHz, a little base loading is added to establish resonance at the lower frequency.

A trigonometric expression for R_{rad} is given below (Ref 742 and 7993):

$$R_{rad} = 36.6 \times \frac{(\sin t1 - \sin t2)^2}{\cos^2 t2} \quad \text{(Eq 9-11)}$$

where
t1 = electrical height of vertical mast
t2 = electrical length provided by the base-loading coil

In our example shown in Fig 9-45, t1 = 59° and t2 = 5°

$$R_{rad} = 36.6 \times \frac{(\sin 59° - \sin 5°)^2}{\cos^2 5°} = 21.9 \, \Omega$$

By replacing some of the top loading by base loading, the radiation resistance has only dropped a few tenths of an ohm. Fig 9-44 shows the radiation resistance for monopoles with combined top and base loading. The physical length of the antenna (L) plus top loading (T) plus base loading (B) must total 90°. The calculation of the required capacitance and the dimensions of the capacitance hat are explained further in Section 3.6.2.

When the antenna has a large capacitance hat compared to the distributed capacitance of the structure, there is no reason to put the coil high on the structure. Current distribution will be essentially the same no matter where you put the coil, even when the antenna is *far* from self-resonance with just the hat. We can simply use a large hat and put a coil at the base, where it can do double-duty for impedance matching and loading, and we can reach it easily for adjustment, as shown in Fig 9-45.

3.1.5. Linear loading

Linear loading is defined as replacing a loading coil at a given place in the vertical with a linear-loading section, which resembles a shorted stub, at the same place in the vertical. This places the two conductors of the loading device in parallel with the radiating element. Due to the current *not* being out-of-phase in the loading device, the device will radiate. The R_{rad} of the antenna will be slightly higher than if we were using a loading coil in the same place.

This linear-loading technique described above is used on the Hy-Gain 402BA shortened 40-meter beam, where linear loading is used at the center of the dipoles. It is also used successfully on the KLM 40 and 80-meter shortened Yagis and dipoles, where linear loading is applied at a certain distance from the center of the elements, but where the linear loading devices were *not* parallel to the elements, introducing some unwanted radiation. This reduced the directional characteristics of the antenna.

In recent years the better Yagi designs for 80 meters have employed optimized high-Q loading coils rather than linear-loading devices, with great success (see Chapter 13).

3.2. Keeping the Radiation Resistance High

As stated before, this is *not* the same as keeping the feed-point impedance high! Using any kind of transformers, such as folded elements or any other type of matching systems do *not* change the radiation resistance. The rule for keeping the radiation resistance as high as possible is simple:

1. Use as long a vertical as possible (up to 90°).
2. Use top-capacitance loading rather than center or bottom loading. Fig 9-48 gives the radiation resistance for monopoles with combined base and top loading. The graphs clearly show the advantage of top loading.

The values of R_{rad} given in these figures can be used for antennas with diameters ranging from 0.1° to 1° (360° = 1 λ). J. Sevick, W2FMI, (Ref 818) obtained very similar results experimentally, while the values in the figures mentioned above were derived mathematically.

For a given physical size, the way to maximize efficiency is to make current as large and uniform as possible over the maximum available vertical distance. The solution is to end-load the antenna with a large hat or some other form of termination that does *not* return to earth. The only thing fancy shunt tuning schemes or multiple drop wires do is to make the feed line see a new impedance.

Top loading with sloping wires is attractive from a mechanical point of view. Sloping loading wires do add capacitance, but only marginally increase R_{rad}, because of the shielding effect of the sloping wires around the vertical. In Chapter 7, we saw how W8JI uses sloping top-hat wires in his 8-circle receiving array, but bear in mind that in this receiving antenna and the designer is not after a larger R_{rad} but rather is trying to lengthen the vertical electrically.

3.3. Keeping Losses Associated with Loading Devices Low

- **Capacitance hat**: The losses associated with a capacitance hat are negligible. When applying top-capacitance loading, especially on 160 meters, the practical limitation is likely to be the size (diameter) of the top hat. Therefore, when designing a short vertical it is wise to start by dimensioning the top hat.
- **T-wire top loading**: This method is lossless, as with the capacitance hat. It may not always be possible, however, to have a perfectly horizontal top wire. Slightly drooping of top-loading wires is just as effective, and when used in pairs (each wire of a pair being in-line with the second

wire) the radiation from these loading wires is negligible.
- **Linear-loading**: W8JI measured the Q of typical linear loading devices and found an amazing low figure of between 50 and 100, while loading coils of moderate quality easily reach an unloaded Q of 200 and well-designed and optimized coils may reach a Q of well over 400. Tom, W8JI remarks: *"For example, the Q of a 400 ohm reactance with a #14 folded wire stub is much less than 100. I can easily obtain a Q of 300 with the same size wire in a conventional coil."*
- **Loading coil**: Even large loading coils are intrinsically lossy. The equivalent series loss resistance is given by:

$$R_{loss} = \frac{X_L}{Q} \qquad (Eq\ 9\text{-}12)$$

where

X_L = inductive reactance of the coil
Q = Q (quality) factor of the coil

Base loading requires a relatively small coil, so the Q losses will be relatively low, but the R_{rad} will be low as well. See Section 3.6 for practical design examples with real-life values.

Top loading requires a large-inductance coil, with correspondingly larger losses, while in this case the R_{rad} is much higher.

As mentioned above, unloaded Q factors of 200 to 300 are easy to obtain without special measures. Well-designed and carefully built loading coils can yield Q factors of up to 800 (Ref 694 and 695). W8JI, wrote: *"The most detailed and accurate loading inductor text readily available to amateurs appears in the chapter "Reactive Elements and Impedance Limits" in Kuecken's book "Antennas and Transmission Lines" (Ref 696). I've measured hundreds of inductors. A typical B&W Miniductor or Airdux coil of #12 wire operated far from self-resonance with a form factor of 2:1 L/D has a Q in the 300 range. Optimum Q almost always occurs with bare wire space wound one turn apart, but optimum L/D can range from 0.5 to 2 or more depending on how far below self-resonance you operate the inductor and what is around the inductor and how big the conductors in the coil are.*

Large optimal edge-wound or copper tubing coils can get into the Q ~800 range. I've never in my life seen an inductor of reasonable reactance above that Q, and very few make it that high."

3.4. Short-Vertical Design Guidelines

From the above considerations we can conclude the following:
- Make a short vertical physically as long as possible.
- Make use of top loading (capacitance hat or horizontal T wires) to achieve the highest radiation resistance possible.
- Use the best possible radial system.
- Design and build your own loading coils with great care (high Q).
- Take extremely good care of electrical contacts, contacts between antenna sections, between the antenna and the loading elements. This becomes increasingly important when the radiation resistance is low.

Though you may be able to build small verticals with low

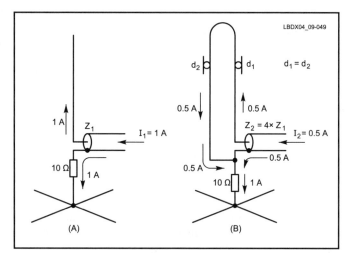

Fig 9-49—The same net current flows in the ground system, whether an open or a folded element is used. This is clearly illustrated for both cases. See text for details.

intrinsic losses, it may not always be possible to improve the losses in the ground-return circuit (radials and ground) to a point where a small loaded vertical achieves good efficiency. Small loaded verticals will often be imposed by area restrictions, which may also mean that an extensive and efficient ground (radial) system may be excluded. Keep in mind that with short loaded verticals, the ground system is even more important than with a full-size vertical.

It is a widespread misconception that vertical antennas don't require much space. Nothing is farther from the truth. Verticals take a lot of space! A good ground system for a short vertical takes much more space than a dipole, unless you live right at the coast, over saltwater, where you might get away with a simple ground system. By the way, it is the saltwater that allows a short loaded verticals to produce such excellent signals on many DXpeditions. Remember VK0IR (Heard Island) and ZL7DK (Chatham Island), just to name a couple of them.

3.4.1. Verticals with folded elements

Another common misconception is that folded elements increase the radiation resistance of an antenna, and thus increase the system efficiency. However, the radiation resistance of a folded element is not the same as its feed-point resistance.

A folded monopole with two equal-diameter legs will show a feed-point impedance with the resistive part equal to $4 \times R_{rad}$. The higher feed-point impedance does not reduce the losses due to low radiation resistance, however, since with the folded element the lower feed current now flows in one more conductor, totaling the same loss. In a folded monopole, the same current ends up flowing through the lossy ground system, resulting in the same loss whether a folded element is used or not.

This is illustrated in **Fig 9-49**. In the non-folded situation in Fig 9-49A it is clear that the total 1 A current flows through the 10-Ω equivalent ground-loss resistance. The ground loss is $I^2 \times R = 10$ W. Figure 9-49B shows the folded-element situation. In this example equal-diameter conductors are assumed; hence the feed impedance is four times the impedance of the single-conductor-equivalent vertical, and the current is half the value of the same antenna with a single conductor. Thus, 0.5 A flows in the folded-element wire and from the feed point down to the 10-Ω resistor. There is another 0.5 A coming down the folded wire and also going to the top of the 10-Ω resistor. In the ground system through the 10-Ω ground loss resistor, we have a total current of 1 A flowing, the same as with the unfolded vertical. The loss is again $I^2 \times R = 10$ W.

In other words, the impedance transformation of the folded monopole also transforms the ground loss part of the equation in the same way as it does for the radiation resistance, and there is no net improvement. It is just another form of transformer and is no different than adding a toroidal step-up transformer at the base of a regular monopole.

Although the folded monopole does not gain anything in efficiency due to the impedance transformation it does have some advantages. The impedance transformation will result in a higher impedance that might be more easily matched by a more efficient network than would be required by a plain monopole. The folded monopole has some advantages in lightning protection due to the possibility of dc grounding the structure. And the folded monopole may have a wider bandwidth due to the larger effective diameter of the two conductors (see also Chapter 8, Section 1.4.1).

Fig 9-50 shows the effective normalized diameter of two parallel conductors, as a function of the conductor diameters and spacing (from *Kurze Antennen*, by Gerd Janzen, ISBN 3-440-05469-1). A folded element consisting of a 5-cm OD tube and a 2-mm OD wire (d1/d2 = 25), spaced 25 cm has an effective round conductor diameter of 0.6×25 = 15 cm.

3.5. SWR Bandwidth of Short Verticals

3.5.1. Calculating the 3-dB bandwidth

One way of defining the Q of a vertical is:

$$Q = \frac{Z_{surge}}{R_{rad} + R_{loss}} \quad \text{(Eq 9-13)}$$

Z_{surge} is the characteristic impedance of the antenna seen as a short single-wire transmission line. The surge impedance is given by:

$$Z_{surge} = 60\left[\ln\left(\frac{4h}{d}\right) - 1\right] \quad \text{(Eq 9-14)}$$

where

h = antenna height (length of transmission line)
d = antenna diameter (transmission-line diameter)
and where values for h and d are in the same units

The 3-dB bandwidth is given by:

$$BW_{3dB} = \frac{f}{Q} \quad \text{(Eq 9-15)}$$

where f = the operating frequency.

Example:

Assume a top-loaded vertical 30 meters high, with an effective diameter of 25 cm and a capacitance hat that resonates the vertical at 1.835 MHz.

Using Eq 9-14: $Z_{surge} = 310$ Ω
The electrical length of the vertical is:

$$\frac{1.835}{300 \times 0.96} \times 30\,m \times 360° = 68.8°$$

Using Eq 9-7: $R_{rad} = 31.8$ Ω
Assume: $R_{ground} = 10$ Ω (an average ground system).
Using Eq 9-13:

$$Q = \frac{310}{31.810} = 7.42$$

Using Eq 9-15: $BW_{3dB} = \frac{1.835}{7.42} = 0.247\,MHz$

3.5.2. The 2:1 SWR bandwidth

A more practical way of knowing the SWR bandwidth performance is to model the antenna at different frequencies, using eg, *MININEC* or *EZNEC*. The Q of the vertical is a clear indicator of bandwidth. Antenna Q and SWR bandwidth are discussed in Chapter 5, Section 3.10.1.

Table 9-5 shows the results obtained by modeling full-size quarter-wave verticals of various conductor diameters. Both the perfect as well as the real-ground case are calculated. The vertical with a folded element clearly exhibits a larger SWR bandwidth than the single-wire vertical. Note that with a tower-size vertical (25-cm diameter), both the CW as well as the phone DX portions of the 80-meter band are well covered. If a wire vertical is planned (eg, suspended from trees), the

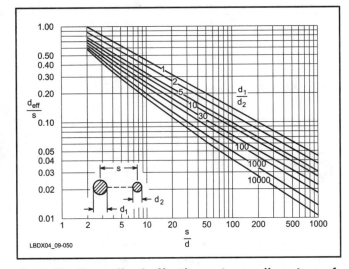

Fig 9-50—Normalized effective antenna diameters of a folded dipole using two conductors of unequal diameter, as a function of the individual conductor diameters d1 and d2, as well as the spacing between the two conductors (S). (*After Gerd Janzen,* **Kurze Antennen**)

Table 9-5
Quarter-Wave Verticals on 80 Meters
Z_t, SWR_t and Q_t indicate the theoretical figures assuming zero ground loss.
Z_g, SWR_g and Q_g values include an equivalent ground resistance of 10 Ω.

Diameter Vertical		2 mm (0.08")	40 mm (1.6")	250 mm (10")
3.5 MHz	$Z_t =$	31.6 − j 31.4	31.4 − j 23.5	31.1 − j 16.7
	$Z_g =$	41.6 − j 35.9	41.4 − j 23.5	41.1 − j 16.7
	$SWR_t =$	2.8:1	2.0:1	1.7:1
	$SWR_g =$	2.2:1	1.7:1	1.5:1
3.65 MHz	$Z_t =$	35.9	35.9	35.9
	$Z_g =$	45.9	45.9	45.9
	$SWR_t =$	1:1	1:1	1:1
	$SWR_g =$	1:1	1:1	1:1
3.8 MHz	$Z_t =$	40.0 + j 35.5	40.9 + j 24.5	41.1 + j 16.6
	$Z_g =$	50.0 + j 35.5	40.9 + j 24.5	51.1 + j 16.6
	$SWR_t =$	2.5:1	1.9:1	1.6:1
	$SWR_g =$	2.1:1	1.7:1	1.4:1
All	$Q_t =$	12.1	8.1	5.6
	$Q_g =$	9.5	6.4	4.4

Table 9-6
Verticals with 40-mm OD for 80 Meters
Z_t, SWR_t and Q_t are the values for a 0-Ω ground resistance. Z_g, SWR_g and Q_g relate to an equivalent ground resistance of 10 Ω.

Frequency		λ/8 Long (9.9 m) (28.4 ft)	3λ/16 Long (12.6 m) (41.3 ft)
3.5 MHz	$Z_t =$	5.37 − j 340	9.3 − j 237
	$Z_g =$	15.37 − j 340	19.3 − j 237
	$SWR_t =$	15.7:1	6.0:1
	$SWR_g =$	3.6:1	2.7:1
3.65 MHz	$Z_t =$	5.9 − j 319	10.3 − j 217
	$Z_g =$	10.5 − j 319	20.3 − j 217
	$SWR_t =$	1:1	1:1
	$SWR_g =$	1:1	1:1
3.8 MHz	$Z_t =$	6.47 − j 299	11.4 − j 198
	$Z_g =$	16.47 − j 299	21.4 − j 198
	$SWR_t =$	12.3:1	4.9:1
	$SWR_g =$	3.3:1	2.4:1
All	$Q_t =$	42	23
	$Q_g =$	15	12

folded version is to be preferred. Matching can easily be done with an L network.

It is evident that loaded verticals exhibit a much narrower bandwidth than their full-size λ/4 counterparts. With short verticals, the quality of the ground system (the equivalent loss resistance) plays a very important role in the bandwidth of the antenna. **Table 9-6** shows the calculated impedances and SWR values for short top-loaded verticals. The same equivalent ground resistance of 10 Ω used in Table 9-5 has a very drastic influence on the bandwidth of a very short vertical. Note the drastic drop in Q and the increase in bandwidth with the 10-Ω ground resistance.

Two factors definitely influence the SWR bandwidth of a vertical of a given length: the conductor diameter and the total loss resistance. We only want to increase the conductor diameter to increase the bandwidth where possible. If you want to use the loss resistance to increase the bandwidth, you might as well use a dummy load for an antenna. After all, a dummy load has a large SWR bandwidth and the worst possible radiating efficiency!

If you use a coil for loading a vertical (center or top loading), you can see that for a given antenna diameter, the bandwidth will decrease as the antenna is shortened and the missing part is partly or totally replaced by a loading coil. Then with more shortening, the bandwidth will begin to increase again as the influence of the equivalent resistive loss in the coil begins to affect the bandwidth of the antenna.

If you measure an unusually broad bandwidth for a given vertical design, you should suspect a poor-quality loading coil or some other lossy element in the system. (Or did you forget a ground system?)

3.6. Designing Short Loaded Verticals

Let us review some practical designs of short loaded verticals (Ref 794).

3.6.1. Base coil loading

Assume a 24-meter high vertical with an effective diameter of 25 cm, which you can use as a 3λ/8 vertical on 80 meters. You can also resonate it on 160 meters using a base-mounted loading coil (**Fig 9-51**). The electrical length

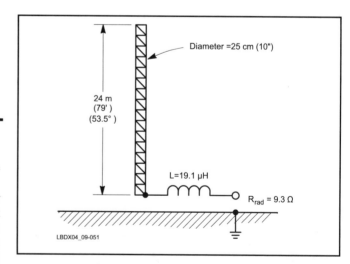

Fig 9-51—Base-loaded tower for 160 meters. See text for details on how to calculate the radiation resistance as well as the value of the loading coil. The loss resistance is effectively in series with the radiation resistance. With 60 λ/8 radials over good ground, the feed-point impedance will be approximately 20 Ω and the radiation efficiency about 50%.

on 160 meters is 53.5°. Calculate the surge impedance of the short vertical using Eq 9-14:

$$Z_{surge} = 60\left[\ln\left(\frac{4 \times 2400}{25}\right) - 1\right] = 297\,\Omega$$

3.6.1.1. Calculate the loading coil

The capacitive reactance of a short vertical is:

$$X_C = \frac{Z_{surge}}{\tan t} \qquad \text{(Eq 9-16)}$$

where t = the electrical length of the vertical in degrees (24 meters is 53.5°).

In this example, $X_C = \dfrac{297\,\Omega}{\tan 53.5°} = 220\,\Omega$

Since X_L must equal X_C,

$$L = \frac{X_L}{2\pi \times f} = \frac{220}{2\pi \times 1.83} = 19.1\,\mu H$$

Let us assume a Q factor of 300, which is easily achievable:

$$R_{loss} = \frac{X_L}{Q} = \frac{220\,\Omega}{300} = 0.73\,\Omega$$

This value of loss resistance is reasonably low, especially when you compare it with the value of R_{rad} calculated using Eq 9-7:

$$R_{rad} = 36.6 \times \frac{(1 - \cos 53.5°)^2}{\sin^2 53.5°} = 9.3\,\Omega$$

ELNEC also calculates R_{rad} as 9.3 Ω. The radiation resistance is effectively in series with the ground-loss resistance. Assuming 60 λ/8 radials over good ground, the estimated equivalent loss resistance is about 10 Ω, meaning the feed-point impedance will be approximately 20 Ω. The efficiency will be 50%. The quality of the ground system (its equivalent loss resistance, see Table 9-1) determines the antenna efficiency much more than the loading device.

3.6.2. Capacitance-hat loading

Consider the design of a 30-meter vertical that will be loaded with a capacitance hat to resonate on 1.83 MHz. The electrical length of the 30-meter vertical is 67°. We must replace the missing 23° of electrical height with a capacitance hat (**Fig 9-52**).

First we calculate the surge impedance of the short vertical using Eq 9-14, assuming that the vertical's diameter is 25 cm. The surge impedance is:

$$Z_{surge} = 60\left[\frac{(4 \times 3000)}{25} - 1\right] = 310\,\Omega$$

Notice that the conductor diameter has a great influence

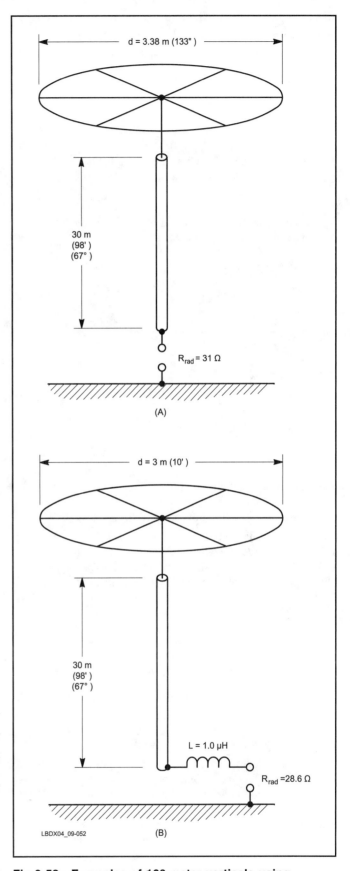

Fig 9-52—Examples of 160-meter verticals using capacitance hats. At A, the hat is dimensioned to tune the vertical to resonance at 1830 kHz. The antenna at B uses a capacitance hat of a given dimension, and resonance is achieved by using a small amount of base loading.

on the surge impedance. The same vertical made of 5-cm tubing would have a surge impedance of 407 Ω.

The electrical length of the capacitance top-hat is calculated:

$$X_C = \frac{Z_{surge}}{\tan t} \quad \text{(Eq 9-17)}$$

where

X_C = reactance of the capacitance hat (Ω)
t = electrical length of the top hat = 23°
Z_{surge} = 310 Ω

Eq 9-17 has the same form as Eq 9-15, but the definitions of terms are different.

$$X_C = \frac{310\ \Omega}{\tan 23°} = 730\ \Omega$$

$$C_{pF} = \frac{10^6}{2\pi \times f \times X_C} = \frac{10^6}{2\pi \times 1.82 \times 730} = 119\ \text{pF}$$

3.6.2.1. Capacity of a disk:

The approximate capacitance of a solid-disk-shaped capacitive loading device is given by (Ref 7818):

$$C = 35.4 \times D \quad \text{(if } D < h/2\text{)} \quad \text{(Eq 9-18)}$$

where

C = hat capacitance (in pF)
D = hat diameter (in meters)
h = height of disk above ground (in meters)

The capacitance of a solid disk can be achieved by using a disk in the shape of a wheel, having eight (large diameter) to 12 (small diameter) radial wires (Ref 7818). The capacitance of a single horizontal wire, used as a capacitive loading device is given by (Ref 7818):

$$C = k \times L \quad \text{(Eq 9-19)}$$

where

k = 10 pF/m for thick conductors (L/d < 200)
k = 6 pF/m for thin conductors (L/d > 3000)
C = hat capacitance (in pF)
L = length of wire (in meters)

3.6.2.2. Capacity of loading wires

If two loading wires are used at right angles to the vertical, the k-factors become approximately 8 pF/meter for thick conductors and 5 pF/meter for thin conductors. If the loading wires are not horizontal, they must be longer to achieve the same capacitive loading effect.

The capacitance of a sloping wire is given by:

$$C_{slope} = C_{horizontal} \times \cos \alpha \quad \text{(Eq 9-20)}$$

where

$C_{horizontal}$ = capacity of the horizontal wire
α = slope angle (with a horizontal wire α = 0°)

The required diameter of the disk need to achieve the 119 pF top-loading capacity is:

$$D = \frac{119}{35.4} = 3.4\ \text{meters}$$

Using a wire, the total required length of the (thin) wire is:

$$L = \frac{119}{6} = 19.8\ \text{meters}$$

This wire can be in the shape of a single horizontal or gently sloping wire; it can be the total length of the two legs of a T-shaped loading wire (horizontal or slightly sloping), or it can be the total length of four wires as shown in **Fig 9-53**.

The disk of a capacitance hat has a large screening effect to whatever is located above the disk. If there is a whip above a large disk, the lengthening effect of the whip may be largely undone. The same effect exists with towers loaded with Yagis. It is mainly the largest Yagi that determines the capacitance to ground. The capacitance hat in effect makes one plate of a capacitor with air dielectric; the ground is the other plate.

3.6.3. Capacitance hat with base loading

Consider the design of the same 30-meter vertical with a 3-meter diameter solid-disk capacitance hat for 1.83 MHz as shown in Fig 9-52B. The effective diameter of the vertical is again 25 cm. We know that this hat will be slightly too small to achieve resonance at 1.83 MHz. We will add some base loading to tune out the remaining capacitive reactance at the base of the vertical. This can be referred to as *fine tuning* the antenna. The coil will normally merge with the coil of an L network that might be used to match the vertical to the feed line.

The capacitance of a solid-disk hat is given by Eq 9-18:
C = 35.4 × D

In this example, C = 35.4 × 3 = 106 pF. The capacitive reactance of the hat at 1.83 MHz is:

$$\frac{10^6}{2\pi \times 1.83 \times 85} = 820\ \Omega$$

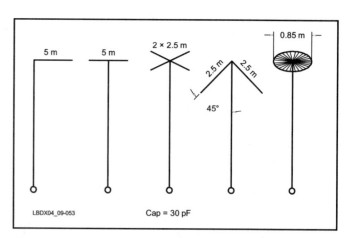

Fig 9-53—Capacitance hats can have various shapes, such as a disk, one or two wires, forming an inverted L or a T with the vertical. The lengths indicated are approximate values for a capacity of 30 pF.

Next we calculate the surge impedance:

$$Z_{surge} = 60\left[\ln\left(\frac{4 \times 3000}{25}\right) - 1\right] = 310\ \Omega$$

The electrical length of the capacitance top-hat is calculated using Eq 9-17, rewritten as:

$$\tan t = \frac{Z_{surge}}{X_C} \text{ or } t = \arctan\left(\frac{Z_{surge}}{X_C}\right)$$

$$t = \arctan\left(\frac{310}{820}\right) = 20.7°$$

For a thinner radiator, the electrical length of the hat would be higher, since Z_{surge} would be larger. The electrical length of our example vertical radiator is 67°, and the top-hat capacitance is 20.7°. Since the sum of the two is 87.7°, another 2.3° of loading is required to make a full 90°. Let us calculate the required loading coil for mounting at the base of the short vertical.

We must first calculate the surge impedance of the vertical with its capacitance top hat. The surge impedance was calculated above as 310 Ω. The capacitive reactance is calculated using Eq 9-16:

$$X_C = \frac{Z_{surge}}{\tan t} = \frac{310\ \Omega}{\tan 87.7°} = 12.4\ \Omega$$

Since X_L must equal X_C:

$$L = \frac{X_L}{2\pi \times F} = \frac{12.4}{2\pi \times 1.83} = 1.1\ \mu H$$

The coil can be calculated using the program module available in the NEW LOW BAND SOFTWARE. Let's see what the equivalent series loss resistance of the coil will be to assess how the base-loading coil influences the radiation efficiency of the system. We will assume a coil Q of 300. Using Eq 9-12 we calculate:

$$R_{loss} = \frac{X_L}{Q} = \frac{12\ \Omega}{300} = 0.04\ \Omega$$

This negligible loss resistance is effectively in series with the ground-loss resistance. Calculate the radiation resistance using Eq 9-11:

$$R_{rad} = 36.6 \frac{(\sin t1 - \sin t2)^2}{\cos^2 t2}$$

$$= 36.6 \frac{(\sin 67° - \sin 2.3°)^2}{\cos^2 2.3°}$$

$$= 28.4\ \Omega$$

With an equivalent ground resistance of 10 Ω, the efficiency of this system (Eq 9-4) is:

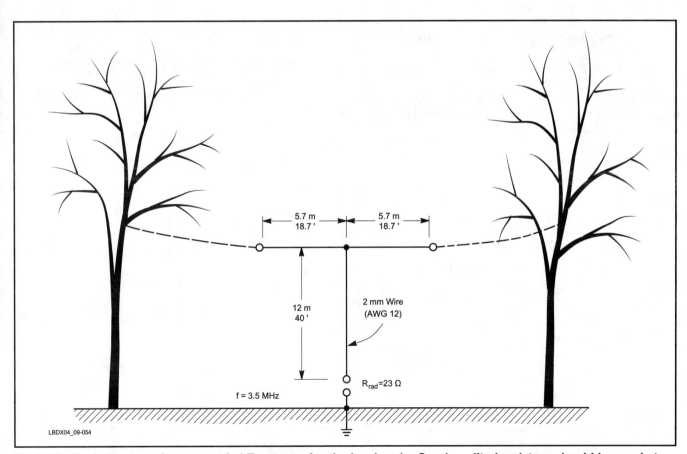

Fig 9-54—Typical setup of a current-fed T antenna for the low bands. Good-quality insulators should be used at both ends of the horizontal wire, as high voltages are present.

$$\text{Eff} = \frac{R_{rad}}{R_{rad}+R_{loss}} = \frac{28.4}{28.4+10+0.04} = 74\%$$

3.6.4. T-wire loading

If the vertical is attached at the center of the top-loading wire, the horizontal (high-angle) radiation from this top wire will be effectively canceled in the far field. The capacitance of a top-loading wire of small diameter is about 6 pF/meter for horizontal wires (see Chapter 8, Section 2.3.5). The total T-wire length is roughly twice the length of the missing portion of the vertical needed to make it into a λ/4 antenna.

Fig 9-54 shows a typical configuration of a T antenna. Two existing supports, such as trees, are used to hold the flattop wire. Try to keep the vertical wire as far as possible away from the supports, since power will inevitably be lost in the supports if close coupling exists.

Fig 9-55 shows a design chart derived using the *ELNEC* modeling program. The dimensions can easily be extrapolated to other design frequencies. In practice, the T-shaped loading wires will often be downward-sloping loading wires. In this case the radiation resistance will be slightly lower due to the vertical component from the downward-sloping current being in opposition with the current in the short vertical. Sloping loading wires will also be longer than horizontal ones, to achieve the same capacity (see Section 3.6.2.2 and Eq 9-20).

3.6.5. Capacitance hat plus coil

Often it will not be possible to achieve enough capacitance hat loading with practical structures, so additional coil loading may be required. If the hat is large enough to dwarf the distributed capacitance of the vertical, you can place a high-Q loading coil anyplace in the vertical and efficiency will remain essentially unchanged.

Let's work out an example of a 1.8-MHz antenna using a 12-meter mast, 5-cm OD, with a 1.2-meter diameter capacitance hat above the loading coil (**Fig 9-56**). The electrical length of the mast is 26.3° and the capacitance of the top hat, by rearranging Eq 9-18, is:

$$C = 35.4 \times D = 35.4 \times 1.2 = 42.5 \, pF$$

$$X_C = \frac{10^6}{2\pi \times 1.8 \times 42.5} = 2080 \, \Omega$$

The surge impedance of the vertical mast is calculated using Eq 9-14:

$$Z_{surge} = 60\left[\ln\left(\frac{4 \times 1200}{5}\right) - 1\right] = 352 \, \Omega$$

Let us analyze the vertical as a short-circuited transmission line with a characteristic impedance of 352 Ω. The input impedance of the short-circuited transmission line is given by:

$$Z = X_L = +jZ_0 \tan t \qquad \text{(Eq 9-21)}$$

where

Z = input impedance of short-circuited line
Z_0 = characteristic impedance of the line (352 Ω)
t = line length in degrees

Thus, $Z = +j \, 352 \times \tan(26.3°) = +j \, 174 \, \Omega$.

This means that the mast, as seen from above, has an inductive reactance of 174 Ω at the top. The capacitive reactance from the top hat is 2080 Ω. The loading coil, installed at the top of the mast, must have an inductive reactance of 2080 − 174 Ω = 1906 Ω.

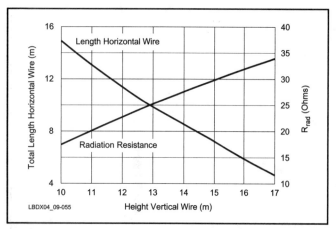

Fig 9-55—Design chart for a wire-type λ/4 current-fed T antenna made of 2-mm OD wire (AWG #12) for a design frequency of 3.5 MHz. For 160 meters the dimension should be multiplied by a factor of 1.9.

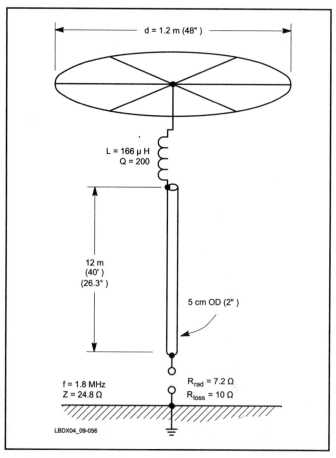

Fig 9-56—Top-loaded vertical for 160 meters, using a combination of a capacitance hat and a loading coil. See text for details.

$$L = \frac{1906}{2\pi \times 1.8} = 169 \; \mu H$$

Assuming you build a loading coil of such a high value with a Q of 200, the equivalent series loss resistance is:

$$R_{loss} = \frac{1906 \; \Omega}{200} = 9.5 \; \Omega$$

Using Eq 9-7, calculate the radiation resistance of the 12-meter long top-loaded vertical:
$R_{rad} = 36.6 \times \sin^2 26.3° = 7.2 \; \Omega$.

Notice that if you want to use the loss resistance of the (top) loading coil for determining the efficiency (or the feed-point impedance) of the vertical, you must transpose the loss resistance to the base of the vertical. This can be done by multiplying the loss resistance of the coil times the square of the cosine of the height of the coil. In our example the loss resistance transposed to the base is:

$$Loss_{base} = Loss_{coil} \times \cos^2 h = 9.5 \times \cos^2 26.3° = 7.6 \; \Omega$$
(Eq 9-22)

Assuming a ground loss of 10 Ω, the efficiency of the antenna is:

$$Eff = \frac{7.6}{7.6 + 10 + 7.4} = 29\%$$

If there were no coil loss, the efficiency would be 42%. This brings us to the point of power-handling capability of the loading coil.

3.6.5.1. Power dissipation of the loading coil

Let us determine how much power is dissipated in the loading coil, calculated as in Section 3.6.6 for an input power to the antenna of 1500 W. The base feed impedance is the sum of R_{rad}, R_{ground} and R_{coil}. The sum is 7.2 + 10 + 7.6 = 24.8 Ω.
The base current is:

$$I_{base} = \sqrt{\frac{1500}{24.8}} = 7.8 \; A$$

The resistance loss of the loading coil is 7.6 Ω. The current at the position of the coil (26.3° above the feed point) is:

$I_{coil} = 7.8 \times \cos 26.3° = 7 \; A$

The power dissipated in the coil is: $I_{coil}^2 \times R_{coil} = 7.0^2 \times 9.5 = 465$ W. This is an extremely high figure, and it is unlikely that we can construct a coil that will be able to dissipate this amount of power without failing (melting!). In practice, we will have to do one of the following things if we want the loading coil to survive:

- Run lower power. For 100 W of RF, the power dissipated in the coil is 31 W; for 200 W it is 62 W; for 400 W it is 124 W. Let us assume that 150 W is the amount of power that can safely be dissipated in a well-made, large-size coil. A maximum input power of 482 W can thus be applied to the vertical, where the assumed coil Q is 200.
- Use a coil of lower inductance and use more capacitive loading (with a larger hat or longer T wires). To allow a power input of 1500 W, and assuming a ground loss of 10 Ω and a coil Q of 200, the maximum value of the loading coil for 150-W dissipation is 42.1 μH. This value is verified as follows (the intermediate results printed here are rounded):

The reactance of the coil is = $2 \times \pi \times 1.8 \times 42.1 = 476 \; \Omega$.

The R_{loss} of the coil is $\frac{476 \; \Omega}{200} = 2.4 \; \Omega$.

Transposed to the base, $R_{loss} = 2.4 \times \cos^2 (26.3°) = 1.9 \; \Omega$.

$$I_{base} = \sqrt{\frac{1500}{7.2 + 1 + 1.9}} = 12.2 \; A$$

This current, transposed to the coil position, is 8.9 × cos 26.3° = 7.9 A.

$P_{coil} = 7.9^2 \times 2.4 = 150$ W.

This is only about 20% of the value of the original 168-μH inductance needed to resonate the antenna at 1.8 MHz. This smaller coil will require a substantially larger capacitance hat to resonate the antenna on 160 meters. T wires would also be a good way to tune the antenna to resonance.

- Make a coil with the largest possible Q. If we change the coil with a Q of 200 in the above example to 300 and run 1500 W, then the maximum coil inductance is 63.1 μH. The calculation procedure is identical to the above example.

The reactance of the coil is = $2 \times \pi \times 1.8 \times 63.1 = 714 \; \Omega$.
The R_{loss} of the coil is 714/300 = 2.4 Ω.
Transposed to the base, = $2.4 \times \cos^2 (26.3°) = 1.9 \; \Omega$

$$I_{base} = \sqrt{\frac{1500}{7.2 + 10 + 1.9}} = 8.9 \; A$$

This current, transposed to the coil position, is 8.9 × cos 26.3° = 7.9 A.

$P_{coil} = 7.9^2 \times 2.4 = 150$ W.

This means that an increase of Q from 200 to 300 allows us to use a loading coil of 63.1 μH instead of 42.1 μH, resulting in the same power being dissipated in the coil. As you can see, the inductance needed is inversely proportional to the Q for a constant power dissipation in the coil.

Notice that the ground-loss resistance again has a great influence on the power dissipated in the loading coil. Staying with the same example as above (Q = 300, L = 63.1 μH), the power loss in the coil for a ground-loss resistance of 1.0 Ω (an excellent ground system) is:

$$I_{base} = \sqrt{\frac{1500}{7.2 + 1 + 1.9}} = 12.2 \; A$$

$P_{coil} = (12.2 \times \cos 26.3°)^2 \times 2.4 = 284$ W.

The better the ground system, the more power will be dissipated in the loading coil. C. J. Michaels, W7XC, investigated the construction and the behavior of loading coils for 160 meters (Ref 797). In the above examples we assumed Q factors of 200 and 300. (See also Ref 694 and 695.) How can we build loading coils having the highest possible unloaded Q? Michaels came to the following conclusions:

- For coils with air dielectric, the L/D (length/diameter) ratio should not exceed 2:1.
- For coils wound on a coil form, this L/D ratio should be 1:1.
- Long, small-diameter coils are not good.
- The highest Q that can be achieved for a 150-µH loading coil for 160 meters is approximately 800. This can be achieved with an air-wound coil (15-cm long by 15-cm diameter), using 35 turns of AWG #7 (3.7-mm diameter) wire, or with an air-wound coil (30-cm long by 15-cm diameter, wound with 55 turns of AWG #4 (5.1-mm diameter) wire.
- Coil diameters of 10 cm wound with AWG #10 to #14 wire can yield Q factors of 600, while coil diameters of 5 cm wound with BSWG #20 to #22 will not yield Q factors higher than approximately 250. These smaller wire gauges should not be used for high-power applications.

You can use some common sense and simple test methods for selecting an acceptable plastic coil-form material:

- High-temperature strength: Boil a sample for ½ hour in water, and check its rigidity immediately after boiling while still hot.
- Check the loss of the material by inserting a piece inside an air-wound coil, for which the Q is being measured. There should be little or no change in Q.
- Check water absorption of the material: Soak the sample for 24 hours in water and repeat the above test. There should be no change in Q.
- Dissipation factor: Put a sample of the material in a microwave oven, together with a cup of water to load the oven. Run the oven until the water boils. The sample should not get appreciably warm.

3.6.6. Coil with T wire

A coil with T-wire configuration at the top of the vertical is essentially the same as the one just described in Section 3.6.5. For a capacitance hat we would normally adjust the resonant frequency by pruning the value of the loading coil or by adding some reactance (inductor for positive or capacitor for negative) at the base of the antenna. For a T-loading wire system it is easier to tune the vertical to resonance by adjusting the length of the T wire.

You can also fine tune by changing the "slope" angle of the T wires. If the T wires are sloped downward the resonant frequency goes up, but also the radiation resistance will drop somewhat. **Fig 9-57** shows two examples of practical designs. For the guyed vertical shown in Fig 9-57B, changing the slope angle by dropping the wires from 68° (ends of T wires at 12-meter height) to 43° (ends at 9-meter height) raises the resonant frequency of the antenna from 1.835 kHz to 1.860 kHz. Note, though, that with this change the radiation resistance drops from 10.1 Ω to 8.3 Ω.

The larger the value of the coil, the lower the efficiency will be, as we found previously. The equivalent loss resistance

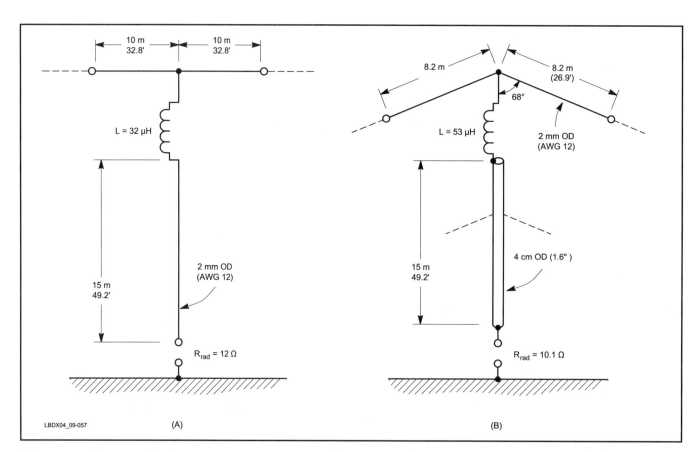

Fig 9-57—Practical examples of combined coil and flat-top wire loading. At A, a wire antenna with a loading coil at the top of the vertical section (no space for longer top-load wires). At B, a loaded vertical mast (4-cm OD) where two of the top guy wires, together with a loading coil, resonate the antenna at 1.835 MHz. The remaining guy wires are made of insulating material (eg, Kevlar, Phillystran, etc).

of the coil and the transposed loss resistance required to determine the efficiency and the feed impedance of the vertical can be calculated as shown in Section 3.6.5. Again, you should avoid having a coil of more than 75 µH of inductance.

3.6.7. Coil with whip

Now we consider a vertical antenna loaded with a whip and a loading coil, as shown in **Fig 9-58**. Let's work out an example for 160 meters:

Mast length below the coil = 18.16 meters = 40°
Mast length above the coil (whip) = 4.54 meters = 10°
Design frequency = 1.835 MHz
Mast diameter = 5 cm
Whip diameter = 2 cm

Calculate the surge impedance of the bottom mast section using Eq 9-14:

$$Z_{surge} = 60\left[\ln\left(\frac{4 \times 1816}{5}\right) - 1\right] = 377 \ \Omega$$

Looking at the base section as a short-circuited line with an impedance of 377 Ω, we can calculate the reactance at the top of the base section using Eq 9-17 rearranged:

$$Z = X_L = +j\ 377 \times \tan 40° = +j\ 316\ \Omega$$

Calculate the surge impedance of the whip section, again using Eq 9-14:

$$Z_{surge} = 60\left[\ln\left(\frac{4 \times 454}{2}\right) - 1\right] = 349 \ \Omega$$

Let us look at the whip as an open-circuited line having a characteristic impedance of 349 Ω. The input impedance of the open-circuited transmission line is given by:

$$Z = X_C = -j\frac{Z_0}{\tan t} \qquad \text{(Eq 9-23)}$$

where

Z_0 = characteristic impedance (here = 349 Ω)
t = electrical length of whip (here = 10°)

The reactance of the whip is:

$$Z = X_C = -j\frac{349}{\tan 10°} = -j\ 1979$$

Sum the reactances:

$X_{tot} = +j\ 316\ \Omega - j\ 1979\ \Omega = -j\ 1663\ \Omega$

This reactance is tuned out with a coil having a reactance of $+j\ 1663\ \Omega$:

$$L = \frac{X_L}{2\pi \times f} = \frac{1663}{2\pi \times 1.835} = 144\ \mu H$$

Assuming you build the loading coil with a Q of 300, the equivalent series loss resistance is

R_{loss} = XL/Q = 1663/300 = 5.5 Ω.

The coil is placed at a height of 40°. Transpose this 5.5-Ω loss to the base using Eq 9-22:

$R_{loss@base}$ = 5.5 Ω × cos² (40°) = 3.2 Ω

Calculate the radiation resistance using Eq 9-10:

R_{rad} = 36.6 × (1 − sin² 80° + sin² 40°)² = 16 Ω

Assuming a ground resistance of 10 Ω, the efficiency of this antenna is:

$$\text{Eff} = \frac{16}{16+10} = 55\%$$

I modeled the same configuration using *ELNEC* and found the following results:

Required coil = 1650 Ω reactance = 143 µH
R_{rad} = 20 Ω

The R_{rad} is 25% higher than what we found using Eq 9-10. This formula uses a few assumptions, such as equal diameters for the mast section above and below the coil, which is not the case in our design. This is probably the reason for the difference in R_{rad}.

3.6.8. Sloping loading wires

Using top loading in the shape of a number of wires

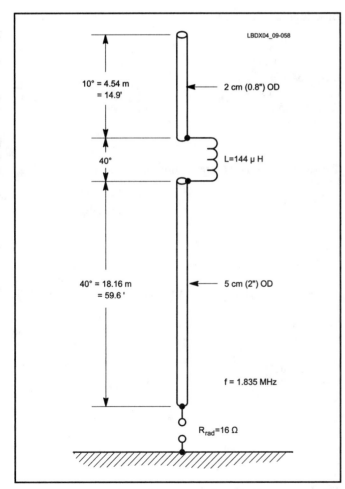

Fig 9-58—Practical example of a vertical loaded with a coil and whip. The length and diameter of the whip are kept within reasonable dimensions that can be realized on top of a loading coil without guying.

radially extending from the top of the vertical is, together with the disk solution, by far the most efficient way to load a short vertical. Often though, we slope these wires down at an angle, lacking suitable supports to erect them horizontally. In this configuration the radiation resistance will be lower due to the vertical component from the downward-sloping current being in opposition with the current in the short vertical. Sloping loading wires must also be longer than horizontal ones to achieve the same capacity (see Section 3.6.3 and Eq 9-18). The reduction in R_{rad} results in an inevitable reduction in efficiency, given the same ground loss resistance. With a lossless ground (such as saltwater), there is no reduction in efficiency.

Mauri, I4JMY published on the Topband reflector some modeling results using a 9-meter long vertical with four 20-meter long top hat wires. He calculated the efficiency, assuming a ground loss resistance of 5 Ω, which is for a fairly elaborate ground system (see Table 9-1).

- Horizontal hat wires: R_{rad} = 5.5 Ω; Z_{feed} = 10.5 Ω; Eff = 48%, Ref = 0 dB
- Hat wires sloping down to 4.5 meters: R_{rad} = 3.2 Ω; Z_{feed} = 8.2 Ω, Eff = 39%, −1.8 dB
- Hat wires sloping down to 1.5 meters: R_{rad} = 2.0 Ω; Z_{feed} = 7 Ω, Eff = 28%, −4.7 dB
- Hat wires sloping down to 0.3 meters : R_{rad} = 1.6 Ω; Z_{feed} = 6.6 Ω, Eff = 24%, −6 dB

If you were using the same vertical with a base-loading coil (see procedure in Section 3.6.1.), you would have R_{rad}= 1.2 Ω, required loading coil reactance ~ 900 Ω, which, assuming a Q of 300, means a coil loss resistance of 3 Ω. Total efficiency of this setup (assuming the same 5 Ω ground loss) is 1.2/(1.2+3+5) = 13%. From this perspective, even the last of the above solutions with four top hat wires sloping down almost to the ground has double the efficiency compared to base coil loading, or a relative gain of 3 dB! In addition, the sloping-hat-wire solution presents a higher feed resistance, which makes it somewhat easier to match with low losses.

Mauri rightfully adds *"I'd keep the ends of the sloping hat wires as high as I could. I would also keep the antenna impedance slightly capacitive using a smaller hat than required. This I'd do in order to use a coil at the antenna base that would serve both to resonate the antenna and to act as a step-up autotransformer."* (This is shown in Fig 9-45.) You should not forget either that the solution with the sloping hat wires has considerably more bandwidth than when using a large loading coil.

How steep a slope angle can be tolerated? Preferably not more than approximately 45°. Four top hat wires sloping at a 45° angle reduce R_{rad} to 50% already. Tom, W8JI, summed it all up nicely by saying: *"Any vertical you can build without a hat, I can build better with one.... even if I have to fold the hat down."*

An important mechanical issue: Top hats on verticals must be pulled out as tight as possible. If not, they will blow around in the wind, or sag a lot with ice and your resonant point will blow and sag with them.

The same remarks on down-sloping top hat wires also apply to an inverted-L antenna. See Section 7.

3.6.9. Using modeling programs

In this section on short verticals I used equations for the transmission-line equivalent for an antenna. You can, of course, obtain the same information by modeling these antennas with a modeling program such as *EZNEC*. In this age of antenna modeling, I thought it was a good idea to use simple math and trigonometry to understand the physics and to calculate the numbers.

3.6.10. Comparing different loading methods

To see how different loading methods work, let's compare verticals of identical physical lengths over a relatively poor ground. Where you cannot erect a full-size vertical, you probably won't be able to put down an elaborate radial system either, so we'll use a rather high ground resistance in this comparative study. The study is based on the following assumptions:

- Physical antenna length = 45° ($\lambda/8$)
- L = 20.5 meters
- Design frequency = 1.83 MHz.
- Antenna diameter = 0.1° on 160 meters = 4.55 cm
- Ground-system loss resistance = 15 Ω.

Quarter-wave full size (reference values):
R_{rad} = 36 Ω
R_{ground} = 15 Ω
$R_{ant\ loss}$ = 0 Ω
Z_{feed} = 51 Ω
Eff = 71%
Loss = 1.5 dB

Base loading, $\lambda/8$ size:
R_{rad} = 6.2 Ω
R_{ground} = 15 Ω
Coil Q = 300
L_{coil} = 34 µH
$R_{coil\ loss}$ = 1.3 Ω
Z_{feed} = 22.5 Ω
Eff = 28%
Loss = 5.6 dB

Top-loaded vertical (capacitance hat or horizontal T wire, $\lambda/8$ size):
R_{rad} = 18 Ω
R_{ground} = 15 Ω
Z_{feed} = 33 Ω
Eff = 55%
Loss = 2.6 dB

Top-loaded vertical (coil with capacitance hat at top, $\lambda/8$ size):
R_{rad} = 18 Ω
R_{ground} = 15 Ω
Diameter of capacitance hat = 3 meters
L_{coil} = 37 µH
Coil Q = 200
$R_{coil\ loss}$ = 2.1 Ω
$R_{coil\ loss}$ transposed to base = 1 Ω
Z_{feed} = 34 Ω
Eff = 53%
Loss = 2.8 dB

Top-loaded vertical (coil with whip, $\lambda/8$ size):
R_{rad} = 12.7 Ω
R_{ground} = 15 Ω
Length of whip = 10° (4.55 meters on 1.83 MHz)

L_{coil} = 150 μH
Coil Q = 200
$R_{coil\,loss}$ = 8.6 Ω
$R_{coil\,loss}$ transposed to base = 5.8 Ω
Z_{feed} = 33.5 Ω
Eff = 38%
Loss = 4.2 dB

3.6.11. Conclusions

With an average to poor ground system (15 Ω), a λ/8 vertical with capacitance top loading is only 1.1 dB down from a full-size λ/4 vertical. Over a better ground the difference is even less. If possible, stay away from loading schemes that require a large coil.

4. TALL VERTICALS

In this section we'll examine verticals that are substantially longer than λ/4, especially their behavior over different types of ground. Is a very low elevation angle computed over ideal ground ever realized in practice?

First of all, you need to ask whether you really need very low elevation angles on the low bands. A very low incident angle grazes the ionosphere for a long distance increasing loss. More hops with less loss from a sharper angle can actually decrease propagation loss. We saw in Chapter 1 that relatively high launch angles are actually a prerequisite to allow a "duct" to work on 160 meters, typically at sunrise. On 160 meters, we can state that the antenna with the most gain at the lowest elevation angle under almost all circumstances will produce the strongest signal.

In this section I will dispel a myth that voltage-fed antennas do not require an elaborate ground system. In fact, long verticals require an even better radial system and an even better ground quality in the Fresnel zone to achieve their low-angle and gain potential compared to a λ/4 vertical.

In earlier sections of this chapter, I dealt with short verticals in detail, mostly for 160 meters. On higher frequencies, electrically taller verticals are quite feasible. A full-size λ/4 radiator on 80 meters is approximately 19.5 meters in height. Long verticals are considered to be λ/2 to 5λ/8 in length. Verticals that are slightly longer than a quarter-wave (up to 0.35 λ) do not fall in the *long vertical* category.

4.1. Vertical Radiation Angle

Fig 9-59 shows the vertical radiation patterns of two long verticals of different lengths. These are analyzed over an identical ground system consisting of average earth with 60 λ/4 radials. A λ/4 vertical is included for comparison.

Note that going from a λ/4 vertical to a λ/2 vertical drops the maximum-elevation angle from 26° to 21°. More important, however, is that the −3-dB vertical beamwidth drops from 42° to 29°. Going to a 5λ/8 vertical drops the elevation angle to 15° with a −3-dB beamwidth of only 23°. But notice the high-angle lobe showing up with the 5λ/8 vertical. If we make the vertical still longer, the low-angle lobe will disappear and be replaced by a higher-angle lobe. A 3λ/4 vertical has a radiation angle of 45°.

Whatever the quality of the ground, the 5λ/8 vertical will always produce a lower angle of radiation and also a narrower vertical beamwidth. The story gets more complicated, though, when you compare the efficiency of the antennas.

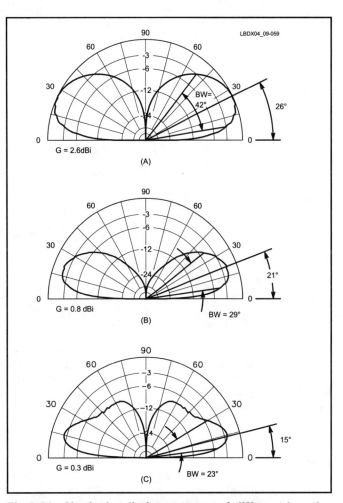

Fig 9-59—Vertical radiation patterns of different-length verticals over average ground, using 60 λ/4 radials. The 0-dB reference for all patterns is 2.6 dBi. At A, λ/4 vertical. At B, λ/2 and at C, 5λ/8.

4.2. Gain

I have modeled both a λ/4 as well as a 5λ/8 vertical over different types of ground, in each case using a realistic number of 60 λ/4 radials. Fig 9-5 shows the patterns and the gains in dBi for the quarter-wave vertical, and **Fig 9-60** shows the results for the 5λ/8 antenna.

Over perfect ground, the 5λ/8 vertical has 3.0 dB more gain than the λ/4 vertical at a 0° elevation angle. Note the very narrow lobe width and the minor high-angle lobe (broken-line patterns in Fig 9-60).

Over saltwater the 5λ/8 has lost 0.8 dB of its gain already; the λ/4 only 0.4 dB. The 5λ/8 vertical has an extremely low elevation angle of 5° and a vertical beamwidth of only 17°. The λ/4 has an 8° take off angle, but a 40° vertical beamwidth.

Over very good ground, the 5λ/8 vertical has now lost 5.0 dB; the λ/4 only 1.9 dB. The actual gain of the λ/4 in other words equals the gain of the 5λ/8! Note also that the high-angle lobe of the 5λ/8 becomes more predominant as the quality of the ground decreases.

Over average ground the situation becomes really poor for the 5λ/8 vertical. The gain has dropped 7.3 dB, and the secondary high-angle lobe is only 4 dB down from the low-

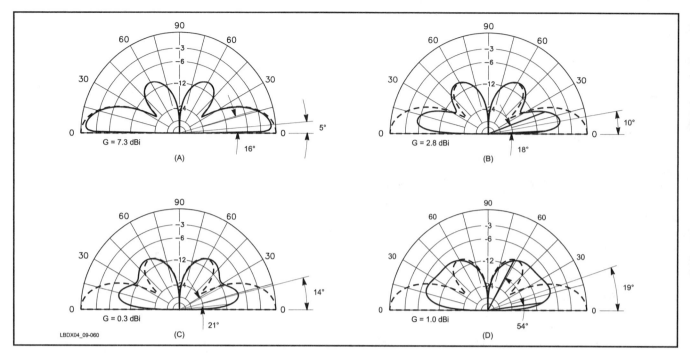

Fig 9-60—Vertical radiation pattern of the 5λ/8 vertical over different types of ground. In all cases, 60 λ/4 radials were used. The theoretical perfect-ground pattern is shown in each case as a reference (broken line, with a gain of 8.1 dBi). Compare with the patterns and gains of the λ/4 vertical, modeled under identical circumstances (Fig 9-5). At A, over saltwater. At B, over very good ground. At C, over average ground. At D, over very poor ground.

angle lobe. The λ/4 vertical has lost 2.6 dB versus ideal ground, and now shows 2.0 dB *more* gain than the 5λ/8 vertical!

Over very poor ground the 5λ/8 vertical has lost 6.6 dB from the perfect-ground situation, while the λ/4 vertical has lost only 3.0 dB. Note that the 5λ/8 vertical seems to pick up some gain compared to the situation over average ground. From Fig 9-60 you can see this is because the radiation at lower angles is now attenuated so much that the radiation from the high-angle lobe at 60° becomes dominant. Note also that the level of the high-angle lobe hardly changes from the perfect-ground situation to the situation over very poor ground. This is because the reflection for this very high angle takes place right under the antenna, where the ground quality has been improved by the 60 λ/4 radials.

This must come as a surprise to most. How can we explain this? An antenna that intrinsically produces a very low angle (at least in the perfect-ground model) relies on reflection at great distances from the antenna to produce its low-angle radiation. At these distances, radials of limited length do not play any role in improving the ground. With poor ground, a great deal of the power that is sent out at a very low angle to the ground-reflection point is being absorbed in the ground rather than being reflected (see also Section 1.1.2). For Fresnel-zone reflections the long vertical requires a better ground than the λ/4 vertical to realize its full potential as a low-angle radiator.

4.3. The Radial System for a Half-Wave Vertical

Here comes another surprise. A terrible misconception about voltage-fed verticals is that they do not require either a good ground or an extensive radial system.

4.3.1. The near field

If you measure the current going into the ground at the base of a λ/2 vertical, the current will be very low (theoretically zero). With λ/4 and shorter verticals, the current in the radials increases in value as you get closer to the base of the vertical. That's why, for a given amount of radial wire, it is better to use many short radials than just a few long ones.

With voltage-fed antennas, however, the earth current will increase as you move away from the vertical. Brown (Ref 7997) calculated that the highest current density exists at approximately 0.35 λ from the base of the voltage-fed λ/2 vertical. Therefore it is even more important to have a good radial system with a voltage-fed antenna such as the voltage-fed T or a λ/2 vertical. These verticals require longer radials to do their job efficiently compared to current-fed verticals.

4.3.2. The far field

In the far field, the requirement for a good ground with a long vertical is much more important than for a λ/4 vertical. I have modeled the influence of the ground quality on the gain of a vertical by the following experiment.

- I compared three antennas: a λ/4 vertical, a voltage-fed λ/4 T (also called an inverted ground plane) and a λ/2 vertical.
- I modeled all three antennas over average ground.
- I put them in the center of a disk of perfectly conducting material and changed the diameter of the disk to determine the extent of the Fresnel zone for the three antennas.

The results of the experiment are shown in **Fig 9-61**. Let us analyze those results.

- With a conducting disk λ/4 in radius (equal to a large number of λ/4 radials) the λ/4 current-fed vertical is

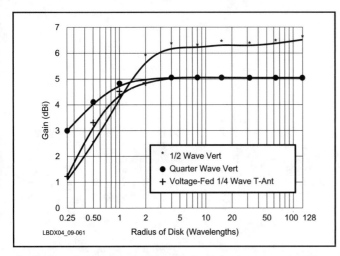

Fig 9-61—Gain of three types of verticals over a perfectly conducting disk of varying radius. The ground beyond the disk is of good quality. This means that the λ/2 vertical requires 0.6-λ radials to perform as well as the λ/4 vertical with λ/4 radials. Be aware that the radiation angle of the λ/2 vertical will be much lower, however.

almost 2 dB better than the voltage-fed λ/4 and the λ/2 vertical.
- The λ/4 vertical remains better than the other antennas up to a disk size of 1.5-λ diameter. This means that over good ground you must be able to put out radials at least 2-λ long with a λ/2 vertical before it shows any gain over the λ/4 current-fed vertical.
- The voltage-fed λ/4 vertical (voltage-fed T) equals the current-fed λ/4 for a disk size of at least 2 λ in diameter. This is because the current maximum is at the top of the antenna, which means that for a given elevation angle, the Fresnel zone (where the main wave hits the ground to be reflected) is much farther away from the base of the vertical than is the case with a λ/4 current-fed vertical. In other words, there is no advantage in using such a voltage-fed λ/4 antenna.
- For both the voltage and the current-fed λ/4 vertical, the Fresnel zone is situated up to 4 λ away from the vertical. For the λ/2 vertical, the Fresnel zone stretches out to some 100 λ!

4.4. In Practice

On 40 meters, a height more than λ/4 (10 meters) should be easy to install in most places. In many cases it will be the same vertical that is used as a λ/4 vertical on 80 meters.

I have been using a 5λ/8 vertical for 40 meters for more than 20 years with good success. With Beverage receiving antennas it has always been a relatively good performer. Now that I have been using a 3-element Yagi at 30 meters for a few years, I know that the vertical solution was far from ideal.

Earl Cunningham, K6SE's, experience confirms this: "*I used a grounded ½-λ vertical in the Houston/Gulf Coast area where the soil conductivity is abnormally high. It was a super performer. The same vertical here in the desert (Palmdale, CA) was a ho-hum performer, even with a much more extensive ground radial system.*"

A similar testimony comes from Tom Rauch, W8JI, who wrote: "*...I had the same results using BC arrays on 160 meters. The 250-ft to 300-ft verticals stunk; my ¼-λ vertical would beat them. I find the same effect on 80 meters.*"

Figs 9-9 and 9-12 give the base resistance, $R_{Rad(B)}$, and feed-point reactance for monopoles as a function of the conductor diameter in degrees, and in Figs 9-10 and 9-13 as a function of the antenna length-to-diameter ratio. The graphs are accurate only for structures with rather large diameters (not for single-wire structures) and that have *uniform* diameters. A conductor diameter of 1° equals 833/f (MHz) in mm.

5. MODELING VERTICAL ANTENNAS

ELNEC as well as other versions of *MININEC* are well suited to do your own vertical antenna modeling, as is the *NEC-2*-based *EZNEC* program. Be aware, however, that all *MININEC*-based antenna modeling programs assume a perfect ground under the antenna base for computing the impedance of the antenna. You cannot use these programs to assess the efficiency of the vertical, where I have defined efficiency as:

$$\text{Eff} = \frac{R_{rad}}{R_{rad} + R_{loss}}$$

MININEC will show the influence of the reflecting ground in the far field that creates the low-angle radiation pattern of the vertical antenna. If you want to include the losses of the ground, you can insert a resistance at the feed point, having a value equivalent to the assumed loss resistance of the ground (see Table 9-1).

5.1. Wires and Segments

In modeling terminology, a *wire* is a straight conductor and is part of the antenna. A *segment* is a part of a wire. Each wire can be broken up into a number of segments, usually all with the same length. Each segment has a different current. The more segments a wire has, the closer the current (pulse) distribution will come to the actual current distribution. There are limits, however.

- Many segments take a lot of computing time.
- Each segment should be at least 2.5 times the wire diameter (according to *MININEC* documentation).

There is no general rule about the minimum number of segments that should be used on a wire. There is only the cut-and-try rule, where you gradually increase the number of segments and look for the point where no further significant changes in the results are observed. This is commonly called *convergence testing* (see also Chapter 4).

Fig 9-62 shows an example of a straight vertical for 80 meters (19 meters long). This antenna consists of a single wire. To evaluate the effect of the segment length, I varied the number of segments in the wire from five to 150. Gain and pattern are very close to modeling with only five segments compared to 150. For impedance calculations, at least 20 sections are required for a reasonably accurate result. The table in Fig 9-62 also shows an example of too many segments for a vertical measuring 250 mm in diameter. As the segment length becomes very short in comparison to the wire diameter, the results become erroneous.

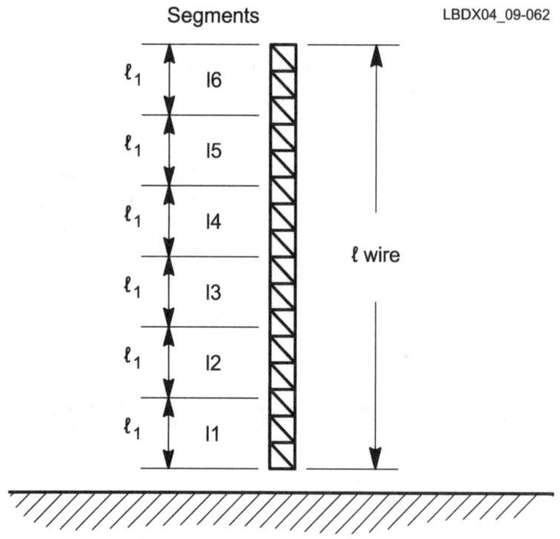

Number of Segments	Segment Length (mm)	2 mm OD Wire (AWG No. 12)	250 mm Mast Diam
5	3800	43.1 − j10.9	36.0 + j3.0
10	1900	34.4 − j11.6	37.0 + j4.1
20	950	34.6 − j11.2	37.5 + j5.2
30	630	34.6 − j11.0	37.7 + j5.8
50	380	34.7 − j10.8	38.0 + j6.2
70	240	34.7 − j10.8	38.0 + j6.3
100	190	34.7 − j11.1	34.5 − j13.0
150	130		34.0 − j14.0

Fig 9-62—*MININEC* analysis of a straight 19-meter vertical antenna shown in the drawing. The analysis frequency is 3.8 MHz. *MININEC* impedance results are shown as a function of the number of segments in the table. Note that for reliability with a "fat" (200-mm) vertical, the maximum number of segments (in this case segments = pulses) is 70. The *MININEC* documentation states that the segment length should be greater than 2.5 times the wire diameter (2.5×200 mm = 500 mm). In this particular case errors occur when the segment length is smaller than the wire diameter.

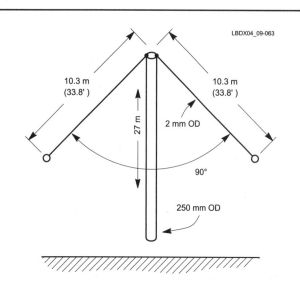

Fig 9-63—Impedances calculated by *MININEC* for a top-loaded 1.8-MHz vertical, using a 250-mm OD mast and two 2-mm OD slant loading wires. The segment lengths are stated in mm. A large number of segments on all wires always gives more reliable results, provided the segment length is not very different. Judicious choice of segment length on the different wires can also yield very accurate results with a smaller number of total segments. To obtain accurate impedance results using *MININEC*, the wire sections near the acute-angle wire junctions must be short.

	Vertical Mast				**Slant Wires**		
No. of Segments	Segment Length (min.)	Segment Length (max.)	No. of Segments	Segment Length (min.)	Segment Length (max.)	Total Pulses	Impedance
3	9000	9000	3	3765	3765	9	9.0 − j184.0
5	5400	5400	5	2260	2260	1	14.2 − j107.0
10	2700	2700	10	1130	1130	3	16.2 − j83.5
20	1350	1350	20	565	565	6	16.6 − j87.1
30	900	900	30	437	437	9	16.7 − j77.0
40	625	625	40	282	282	120	16.6 − j76.5
50	540	540	50	226	226	150	16.8 − j76.3
10	2700	2700	5	2260	2260	15	16.8 − j78.6
6	4500	4500	2	5650	5650	10	16.8 − j80.4
5	5400	5400	2	5650	5650	9	16.7 − j80.7
35	770	770	2	5650	5650	39	14.9 − j103.0

f = 1.8 MHz
length of vert mast = 27 m
slant wire = 10.3 m

5.2. Modeling Antennas with Wire Connections

When the antenna consists of several straight conductors, things become more complicated. **Fig 9-63** shows the example of a 27-meter vertical tower (250-mm OD), loaded with two sloping top-hat wires, measuring 2-mm OD (AWG #12).

The standard approach is to use three wires, one for each of the three antenna parts, and divide the three conductors into a number of segments (which are equal length inside each wire).

To obtain reliable results, you must make sure that the lengths of the segments near the junctions of wires are similar. The table in Fig 9-63 shows the impedance obtained for the top-loaded vertical with different numbers of segments. A large number of segments on the vertical mast (eg, 35 segments, which results in a segment length of 770 mm), together with a small number of segments on the sloping wires, give an unreliable result, while a good result is obtained with a total of just nine segments if the lengths are carefully matched. The segment *tapering* technique, described in Chapter 14 on Yagis and quads, can also be used to minimize the number of segments and improve the accuracy of the results.

5.3. Modeling Verticals Including Radial Systems

MININEC does not analyze antenna systems with horizontal wires close to the ground. Therefore, modeling ground systems as part of the antenna requires the *NEC* software. *NEC-2*, or software such as *EZNEC*, which uses the *NEC-2* engine, can model radials over ground. There seem, however, documented cases (models verified against real-world measurements) of *NEC-2* giving very optimistic results (sometimes up to nearly 6 dB too high gain). *NEC-3* and *NEC-4* can model buried radials (see Chapter 4, Section 1.4), but apparently still show optimistic gain values for wires very close to ground. The results of modeling buried radial systems (see Sections 2.1.2 and 2.1.3) are largely confirmed by N7CL's experimental work. While we can't be absolutely sure real gain figures match modeled numbers within fractions of a dB, nonetheless the trends are certainly correct.

5.4. Radiation at Very Low Angles

Most modeling programs most amateurs use show zero radiation at zero elevation and very little at low angles, unless over salt water. How can we hear ground-wave signals even over average ground? We should not, according to what the model tells us. Experiments show that in real life the very low-angle performance of vertical is better than these modeling programs tell us.

5.5. Measurements, Verifications, Real Life

It is beyond the reach of almost all amateurs to do real-life experiments with low band antennas. The reasons are many. Verifications of modeling results can only be done by few, because of lack of test equipment and most of all the necessary acres... On the other hand, modeling involves mathematics, and a computer can show us results expressed in fractions of a dB. Some models were never verified, and we suspect that the error could be many dBs... Modeling most often does seem to make sense if you compare one model to another model, but you should not automatically conclude that the results apply directly to the real world!

6. PRACTICAL VERTICAL ANTENNAS

A number of practical designs of verticals for 40, 80 and 160 meters are covered in this section, as well as dual and triband systems. A number of practical matching cases are solved, and the component ratings for the elements are discussed. All the L networks have been calculated using the L-NETWORK DESIGN module from the NEW LOW BAND SOFTWARE.

6.1. Single-Band Quarter-Wave Vertical for 40, 80 or 160

Large-diameter conductors are used for various reasons, such as increasing the bandwidth (by increasing the D/L ratio) or simply for mechanical reasons. The effective diameter of wire cages and flat multi-wire configurations is covered in Chapter 8 in Section 2.4.

Often, triangular tower sections are used to make vertical antennas. The effective equivalent diameter of a tower section is shown in **Fig 9-64**. A tower section measuring 25-cm wide, with vertical tubes measuring 2.5-cm diameter, has an equivalent diameter of $0.7 \times 25 = 17.5$ cm.

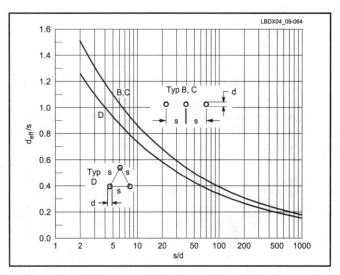

Fig 9-64—Normalized (round solid conductor) effective diameter of a triangular tower section as a function of the vertical tube diameter (d) and the tower width (s). The graph for three parallel conductors is also given (curve BC). (After Gerd Janzen, *Kurze Antennen*.)

Table 9-7

λ/4 Resonance for Vertical as Function of Length/Diameter

Length/Diameter Ratio	Shortening Factor (%)
5000	97.3
2500	97.1
1000	96.8
500	96.2
250	95.7
100	94.6
50	93.4

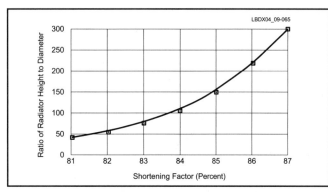

Fig 9-65—Graph showing the amount a λ/4 vertical must be shortened for resonance as a function of the length-to-diameter ratio.

The length of a resonant full-size quarter-wave vertical depends on its physical diameter. See **Table 9-7**, which shows the physical shortening factor of a λ/4 resonant antenna as a function of the ratio of antenna length to antenna diameter. The required physical length is given by:

$$L = \frac{74.95 \times p}{f_{MHz}}$$

where
 L = length (height) of the vertical in meters
 SF = correction factor (from **Fig 9-65**)
 f_{MHz} = design frequency in MHz.

Quarter-wave verticals are easy to match to 50-Ω coaxial feed lines. The radiation resistance plus the usual earth losses will produce a feed-point resistance close to 50 Ω.

If you don't mind using a matching network at the antenna base, and if you can manage a few more meters of antenna height, the extra height will give you increased radiation resistance and higher efficiency. The feed-point impedance can be found in the charts of Figs 9-9, 9-10, 9-12 and 9-13.

Consider the following examples (see **Fig 9-66**):

Example 1:
Tower height = 27 meters
Tower diameter = 25 cm
Design frequency = 3.8 MHz
From the appropriate charts or through modeling we find:

R = 185 Ω
X = + j 215 Ω

Let's assume we have a pretty good ground radial system, with an equivalent ground resistance of 5 Ω. We calculate the matching L network with the following values:

Z_{in} = 190 + j 215 Ω
Z_{out} = 50 Ω

The values of the matching network were calculated for 3.8 MHz. The two matching-network alternatives (low and high-pass) are shown in Fig 9-66A. The low-pass filter network gives a little additional harmonic suppression, while the high-pass assures a direct dc ground for the antenna and some rejection of medium-wave broadcast signals.

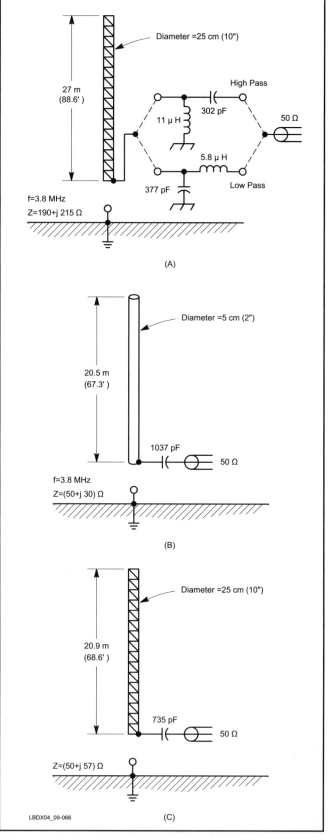

Fig 9-66—Three different 80-meter verticals that are no longer than λ/4, together with their matching networks. Designs at B and C are dimensioned in such a way that the resistive part of the impedance is 50 Ω, in which case the matching network consists of only a series capacitor. The difference between B and C is the diameter of the vertical. In all cases a perfect ground (zero loss) is assumed.

Example 2:

This time we are setting out to build a vertical that is a little longer than a λ/4, so the resistive part of the feed-point impedance at the design frequency will be exactly 50 Ω. This matching network will consist of a simple series capacitor to tune out the inductive reactance of the feed-point impedance. We use *ELNEC* to design two models:

- Vertical mast diameter = 5 cm, length = 20.5 meters, Z = 50 + j 39 Ω (see Fig 9-64B). The matching network consists of a series capacitor with a reactance of 39 Ω at 3.8 MHz. The value of the capacitor is:

$$C = \frac{10^6}{2\pi \times 3.8 \times 39} = 1073 \text{ pF}$$

- Vertical mast diameter = 25 cm, length = 20.9 meters, Z = 50 + j 57 Ω (see Fig 9-64C). In this case the series-matching capacitor has a value of 735 pF.

Note that for the above examples we assumed a zero ground loss. The values of the series-matching capacitor can also be calculated using the SERIES IMPEDANCE module of the NEW LOW BAND SOFTWARE package.

What type of capacitor should you use? The current requirement can be calculated as follows: $I = \sqrt{P/R}$. For 1.5 kW, $I = \sqrt{1500/50} = 5.48$ A. For voltage, $E = \sqrt{PR} = \sqrt{1500 \times 50} = 274$ V. These are RMS values, so you should use components that are built to withstand two to three times these values (a vacuum variable is recommended).

6.1.1. Mechanical design

I don't want to give many detailed mechanical designs, listing materials, tubing diameter, etc since their availability is different in every country. Guy Hamblen, AA7QZ/2, described an attractive 80/75-meter design that uses 12-foot long aluminum tubing sections ranging from 1.5-inch OD to 0.875-inch OD. He also describes the installation details (Ref 7819).

If you consider making a vertical with a rather long un-guyed top section, you can use the ELEMENT STRENGTH MODULE of the ON4UN YAGI DESIGN SOFTWARE. Using the software you can design a Yagi element with a length equal to twice the length you need for the non-guyed top section of the vertical. Because this top section, unlike the Yagi half-element, will not be loaded by its own weight (causing the sag in a Yagi element), the vertical section will have an added safety factor.

Fig 9-68 shows the design for an 80-meter vertical using 4 and 3-inch aluminum irrigation tubing, as designed by Steve Kelly, K7EM. The verticals are mounted on 6×6-inch pressure treated lumber. The total length of each post is 12 feet, of which 4 feet is in the ground. The arms that hold the verticals in place are made from 2×6-inch lumber. Steve used ³/₈-inch threaded rod to bind the arms to the posts and a ¹/₂-inch threaded rod goes through the base of each vertical (see **Fig 9-69**). The ¹/₂-inch rod acts as a hinge for raising and lowering and is insulated from the vertical with PVC tubing. Steve recently replaced the 2×6-inch lumber with ¹/₂-inch thick Plexiglas sheet. The bottom ends of the 4-inch tubing is insulated by 4-inch (inside diameter) PVC pipe. Steve mentions splitting this pipe lengthwise, heating it with a special PVC bending blanket and then sliding it over the 4-inch irrigation tubing.

6.2. Top-Loaded Vertical

The design of loaded verticals has been covered in great detail in Section 3.6. Capacitive top loading using wires (usually slightly sloping) are quite easily constructed from a mechanical point of view. It is more difficult to insert a husky loading coil in a vertical antenna. In addition, because of their intrinsic losses, loading coils are always a second choice when it comes to loading a vertical.

A wire-loaded vertical for 160 meters is described in Section 6.4 as part of an 80/160-meter duoband system. Inverted-L antennas, which are a specific form of top-loaded verticals, are the subject of Section 7.

6.3. Duo-Bander for 80/160 Meters

Full-size, λ/4 verticals (40-meters tall on 160 meters) are out of reach for most amateurs. Often an 80/160-meter duoband vertical will be limited to a height of around 30 meters. This represents an electrical length of 140° at 3.65 MHz and 70° on 160 meters. We can determine R and X from Figs 9-9 and 9-11 or through modeling:

80 meters: Z = 280 + j 278 Ω
160 meters: Z = 17 − j 102 Ω

Fig 9-67—Ninety-foot irrigation-pipe vertical at W7LR in Montana, which is used for both 80 and 160 meters.

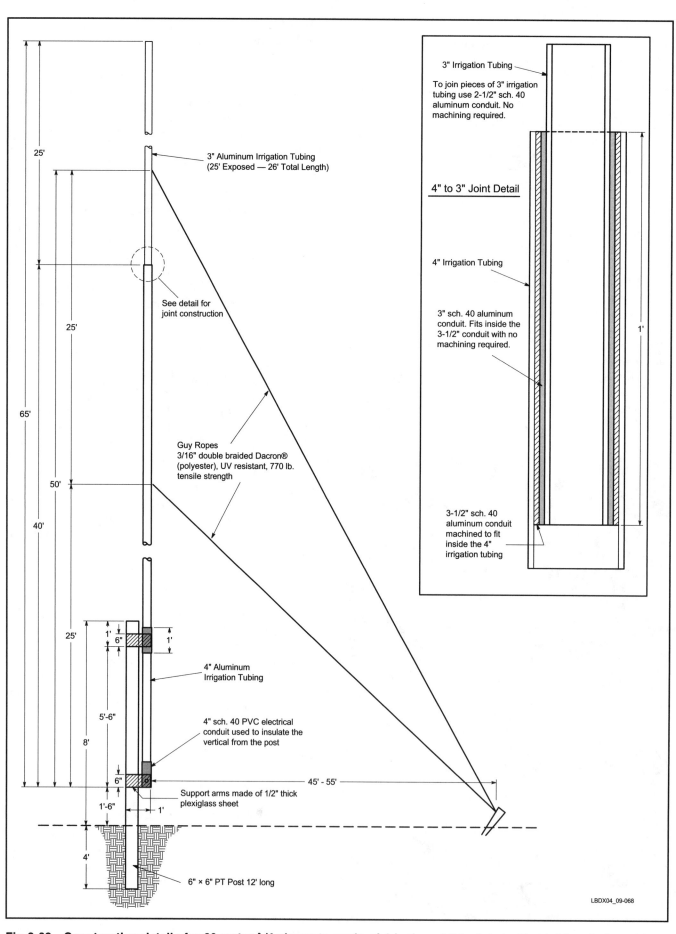

Fig 9-68—Construction details for 80-meter λ/4 elements made of 4-inch and 3-inch irrigation tubing designed by K7EM.

Fig 9-69—Base of one of the irrigation-pipe 80-meter verticals constructed by K7EM.

We use L-networks to match these impedances to our feed line. It's a good idea to use your antenna analyzer to check the impedances on 80 and 160. They should be close to those mentioned above. Remember that the magnitude of the reactive part depends on the effective diameter of the antenna (large diameter antennas exhibit less reactance). Use the L-networks section of the NEW LOW BAND SOFTWARE to calculate the component values. **Fig 9-70** shows the antenna configuration together with the switchable matching system.

6.4. 80/160 Top-Loaded Vertical with Trap

Traps are frequency-selective devices incorporated in radiating elements to adapt the electrical length of the element depending on the frequency at which the element is being used.

Commercial multiband antennas make frequent use of traps. Home-made antennas use the technique less often. There are two types of commonly used traps:

- Isolating traps
- Shortening/lengthening traps

For details about traps and modeling antennas, visit: **www.cebik.com/trap.html**.

6.4.1. Isolating traps

An isolating trap is a parallel-tuned circuit that presents a high impedance at the design frequency, effectively decoupling the "outer" section of the radiator from the "inner" section. A good isolating trap meets the following specifications:

- It represents a high impedance on the design frequency.
- It represents as low a Q as possible, together with the high impedance.
- It represents as low a series inductance as possible on the frequencies where the trap is not resonant (minimize inductive loading), unless you want to use the trap off resonance as a loading device, which could shorten or lengthen the element. The LC circuit off resonance acts as an L or a C, depending which side of resonance you are.

Traps have been described in literature is several configurations:

- Regular LC parallel-tuned circuits
- Resonant circuits with the coil created by a so-called linear loading device (for example, see KT34 Yagis)
- Traps made with coaxial cable, where the capacitance of the cable is used as capacitor, and where the cable shield acts as the coil in the resonant circuit.

Losses are the main issue with traps. An ideal trap will have an infinite parallel (shunt) loss resistance (R_p). Tom, W8JI investigated different traps and found the following results.

- Copper tubing and vacuum cap: R_p = 300,000 Ω
- 60-pF doorknob and #10 Airdux coil: R_p = 250,000 Ω
- 100-pF doorknob and #12 Airdux coil: R_p = 99,850 Ω
- Mosely TA33: R_p = 79,000 Ω
- Cushcraft A3 R_p = 76,270 Ω
- Coax RG-58/U R_p = 17,800 Ω
- Teflon-insulated semi-rigid copper tubing type coax: R_p = 45,000 Ω.

W8JI adds: "*Stubs, linear-loading, and coaxial-cable*

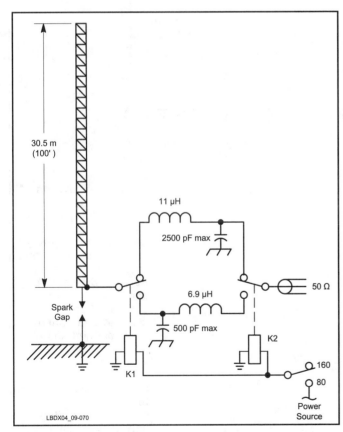

Fig 9-70—Two-band (80 and 160-meter) vertical system using L-networks to match the 50-Ω feed line. The values of the L networks were calculated assuming an equivalent ground-loss resistance of 10 and 5 Ω respectively on 160 and 80 meters.

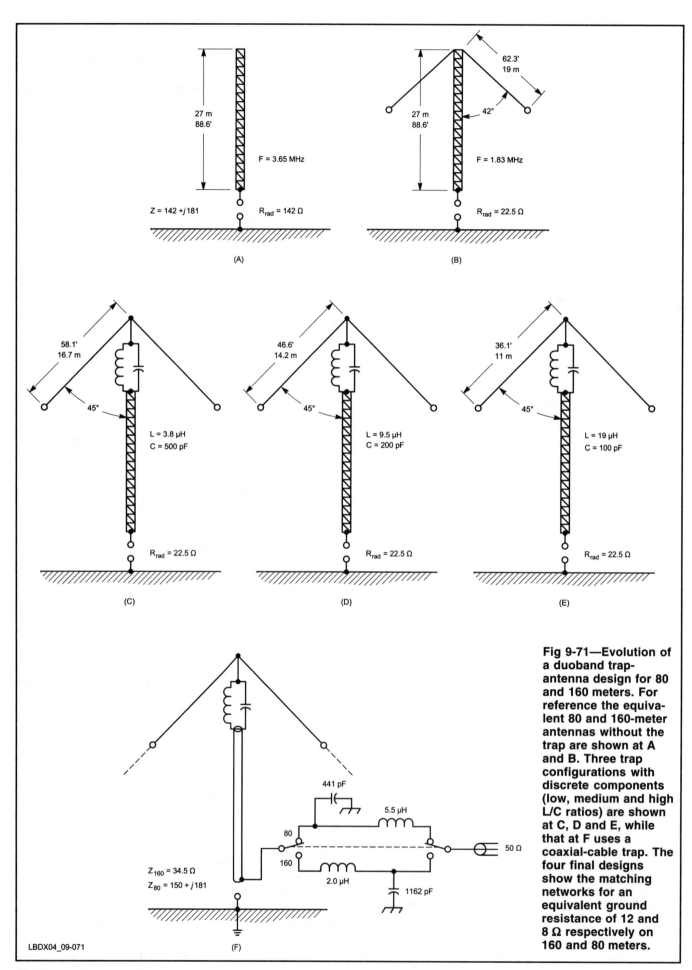

Fig 9-71—Evolution of a duoband trap-antenna design for 80 and 160 meters. For reference the equivalent 80 and 160-meter antennas without the trap are shown at A and B. Three trap configurations with discrete components (low, medium and high L/C ratios) are shown at C, D and E, while that at F uses a coaxial-cable trap. The four final designs show the matching networks for an equivalent ground resistance of 12 and 8 Ω respectively on 160 and 80 meters.

capacitors generally have very low Q compared to other systems. Expect about 1.5 dB or so loss for a coaxial trap in an Inverted L or vertical at the trapped frequency. Loss using a small Airdux coil and a doorknob capacitor would be less than 0.25 dB. Loss on other than the exact trapping frequency are insignificant with all types of traps."

The exception would be traps using very low C and high inductive reactance, which may have significant loss at the pass frequency. This type of trap would be the high-inductance coils used to isolate antenna sections while, substantially reducing length on the non-trap frequency (as sometimes used in small Yagis). Those traps can seriously degrade the pass-frequency performance.

The example in **Fig 9-71** shows a 27-meter vertical mast, measuring 25 cm in effective diameter. We want to use this mast on 80 meters and load it to resonance on 160 meters using two flat-top wires. The trap at the top of the vertical will isolate the loading wires from the mast when operating on 80 meters; it will have to be resonant on 80 meters. Let us design a system that covers 3.5 to 3.8 MHz and see what the performance will be when compared to two monoband systems using basically the same configuration.

Resonance of the trap at 3.65 MHz can be obtained with an unlimited number of L/C combinations:

$$F_{res} = \frac{10^3}{2\pi\sqrt{LC}}$$

where f is in MHz, L in µH and C in pF.

Let us evaluate three different L/C ratios that resonate at 3.65 MHz:

L = 3.8 µH, C = 500 pF (X = 87 Ω)
L = 9.5 µH, C = 200 pF (X = 218 Ω)
L = 19 µH, C =100 pF (X = 436 Ω)

Use standard capacitor values (doorknob or high-quality ceramic-transmitting type capacitors) and adjust the coil turns to obtain the desired resonant frequency, as measured with a Grid Dip meter. The winding data for the coil can be calculated using the COIL module of the NEW LOW BAND SOFTWARE.

As standards of comparison we'll use the stand-alone 27-meter tower (no trap, no flat-top wires) on 80 meters, and the same 27-meter tower with two sloping flat-top wires on 160 meters. The dimensions of the five different configurations are shown in Fig 9-71.

The impedance at 1.835 MHz (resonance with the loading wires, Fig 9-71B) is 22.5 Ω. At 3.65 MHz (Fig 9-71A) the feed-point impedance is 142 + j 181 Ω.

Fig 9-72 shows the influence of the slope angle of the top wires on the resonant frequency. The values are for a 27-meter vertical with two sloping loading wires, 19 meters long.

The three different trap solutions (different L/C ratios) have a significant influence on the SWR-bandwidth behavior of the antenna. The influence of the L/C ratio is the opposite on 80 meters from what it is on 160. The low-L, high-C solution (3.8 µH and 500 pF) yields the highest bandwidth on 160, and the lowest bandwidth on 80 meters. The opposite is also true. With L = 19 µH and C = 100 pF, the SWR curve on 80 is almost as flat as for the reference antenna (just the 27-meter vertical with no trap nor loading wire).

The bandwidth results for the different designs are shown in **Fig 9-73** for both 80 and 160 meters. I have calculated the theoretical bandwidth, excluding the ground losses, as well as the practical bandwidth, including ground losses.

Solution two (9.5 µH and 200 pF, Fig 9-71D) is certainly an excellent compromise if both bands are to be treated with equal attention. This antenna should be matched at the bottom using L-networks (one for each band). Alternatively, you could use a 2:1 unun on 160 meters, but you would still need an L-network on 80 meters.

6.4.2. Shortening/lengthening traps

If the isolating trap principle were to be used on a triband antenna, it would require two isolating traps. Three-band trap Yagis of the early sixties indeed used two traps on each element half, the inner one being resonant on the highest band, the outer one on the middle band. Modern trap-design Yagis only use a single trap in each element half to achieve the same purpose. Y. Beers, WØJF, wrote an excellent article covering the design of these traps (Ref 680). In this design, the trap is not resonant on the high-band frequency, but somewhere in between the low and the high band. In the balanced design described by Y. Beers, the frequency at which the trap is resonant is the geometrical mean of the two operating frequencies (equal to the square root of the product of the two operating frequencies). For an 80/160-meter vertical, the trap would be resonant at:

$$f = \sqrt{1.83 \times 3.85} = 2.65 \text{ MHz}$$

On a frequency below the trap resonant frequency the trap will show a positive reactance (it acts like an inductor), while above the resonant frequency the trap acts as a capacitor. A single parallel-tuned circuit can be designed that inserts the necessary positive reactance at the lowest frequency and negative reactance at the highest frequency. In the balanced design the absolute value of the reactances is identical for the two bands; only the sign is different. There are five variables

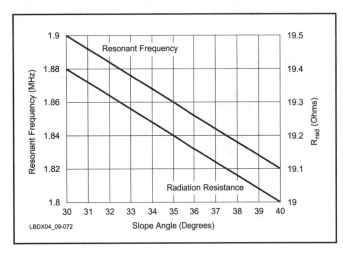

Fig 9-72—This chart shows the variation in resonant frequency and radiation resistance for a 27-meter vertical with 19-meter long sloping top-load wires. A change in slope angle of 30° to 40° shifts the resonant frequency by 80 kHz.

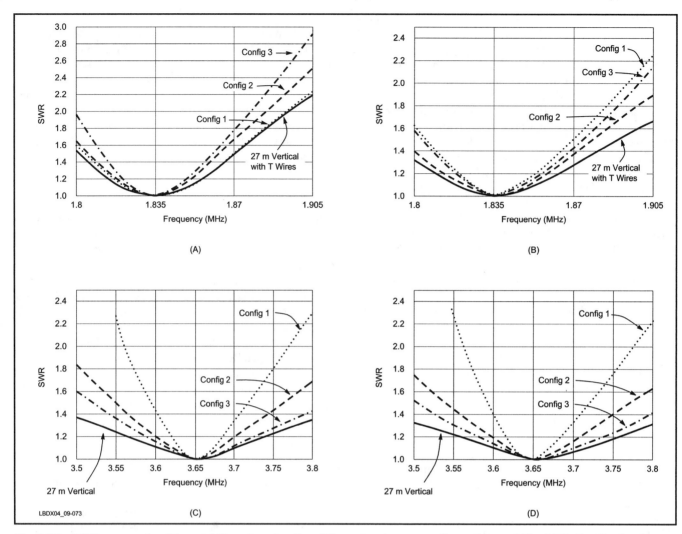

Fig 9-73—SWR curves for 80 and 160 meters for the different antenna configurations of Fig 9-71. Config 1 refers to Fig 9-71C, Config 2 to Fig 9-71D, and Config 3 to Fig 9-71E. The curves are plotted for perfect ground (at A and C) as well as for real ground (at B and D). Note the increased bandwidth on 160 meters caused by the ground losses and the low R_{rad}. On 80 meters the ground losses have almost no influence due to the high R_{rad}.

involved in the design of such a trap system: the two operating frequencies, the trap resonant frequency, the total length and the L/C ratio used in the trap parallel circuit. The design procedure and the mathematics are covered in detail in the above-mentioned article.

It's clear that the "isolating" trap we designed in Section 6.4.1 was really a lengthening trap as we designed it to be a trap slightly lower than 80 meters.

6.5. The Self-Supporting Full-Size 160-Meter Vertical at ON4UN

A full-size λ/4 vertical antenna for 160 meters is just about the best transmitting antenna you can have on that band, with the exception of an array made of full-size or top-loaded verticals. I use a 32-meter triangular self-supporting tower, measuring 1.8 meters across at the base, and tapering to 20 cm at the top. I knew that the taper would make the tower electrically shorter than if it had a constant diameter, so that had to be accounted for. On top of the tower I mounted a 7-meter-long mast. It is steel at the bottom and aluminum at the top, tapering from 50-mm OD to 12 mm at the top.

To make up for the shortening due to the tower taper, I knew I had to install a capacitance hat somewhere near the top of the tower. The highest point I could do this was at 32.5 meters. I decided to try a disk with a diameter of 6 meters, because I had 6-meter-long aluminum tubing available. Two aluminum tubes were mounted at right angles, the ends being connected by copper wire to make a square. **Fig 9-74** shows my vertical.

I hoped I would come close to an electrical λ/4 on 160 meters, and fortunately the antenna resonated on exactly 1830 kHz. In the beginning I had the tower insulated at the base, and was able to measure its impedance, approximately 20 Ω, while I expected 36 Ω with a 0-Ω earth-system resistance. Such a low radiation resistance has been reported in the literature, and must be due to the large tower cross-section. Originally I suspected mutual coupling with one of two other towers (or both), but decoupling or detuning those towers did not change anything.

Fig 9-75 shows the radiation resistance of a λ/4 vertical over a radial system consisting of 60 λ/4 radials, measured as a function of the diameter of the vertical. You can see that for a height/diameter ratio of 44 (eg, a self-supporting tower with a diameter of 1 meter operating at 1.83 MHz) the radiation resistance should be about 20 Ω. The classic 36-Ω figure applies for a very thin conductor!

After a series of unsuccessful attempts to use the vertical on 80 meters, I grounded the tower and shunt fed it using a gamma match for 160 meters and dropped the idea of using this "much-too-long" tower on 80 meters. A tap 8-meters high, and a 500-pF series capacitor provided a 1:1 SWR on Topband, and a 2:1 SWR bandwidth of 175 kHz. The gamma wire is approximately 1.5 meters from the tower. This vertical really plays extremely well. I use quite an extensive radial system, consist-

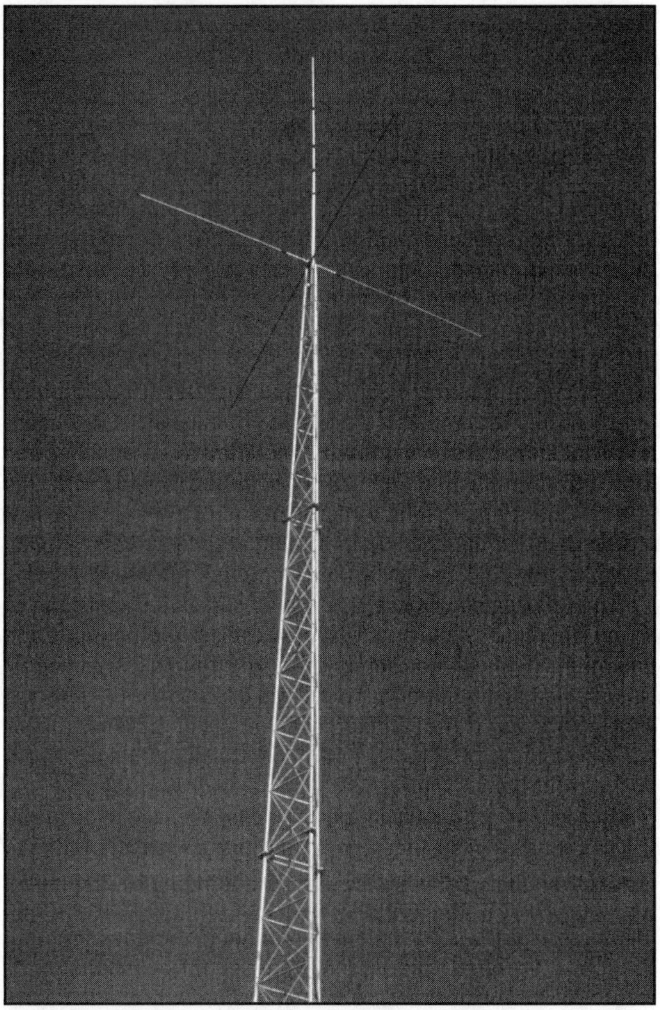

Fig 9-74—Self-supporting 39.5-meter λ/4 vertical for 160 meters at ON4UN. The base is 1.8-meters wide and the tower tapers to just a few inches at the top. The tower is shunt fed with a gamma match and also serves as a support for an 80-meter Four-Square array made of λ/4 verticals, supported from sloping catenary lines running from the 160-meter tower.

Fig 9-76—Giving out new countries and chasing new countries on 160 meters are not the only hobbies for Rudi, DK7PE, (left) and ON4UN, who are ready to go on a bike trip. In the background is the base of ON4UN's 160-meter vertical showing the cabinet that houses the matching circuitry for the 160-meter vertical and the Four-Square 80-meter system.

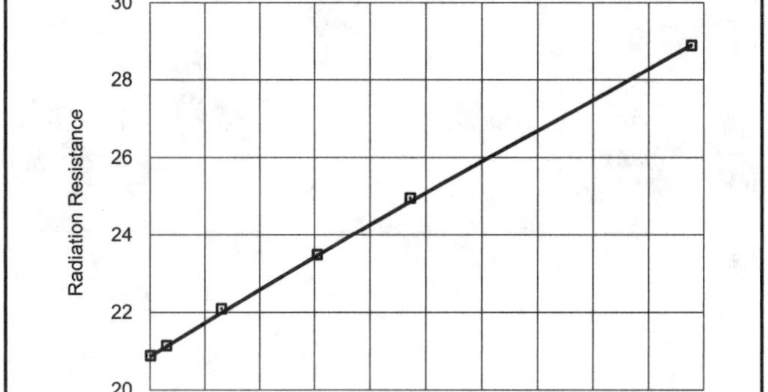

Fig 9-75—The feed-point resistance of a resonant λ/4 vertical over 60 λ/4 radials, as a function of the conductor diameter. Verticals made of a large-diameter conductor, such as a tower, exhibit much lower feed-point resistances than encountered with wire verticals.

Fig 9-77—The base of ON4UN's 160-meter vertical and the cabinet housing both the 80-hybrid-coupler Comtek and WX0B Lahlum networks for the 80-meter Four Square. R. Vermet, ON6WU, with his professional antenna measuring setup, tunes the 80-meter vertical. An HP network analyzer is used that directly produces a Smith Chart.

ing of approximately 250 radials ranging from 18 to 75 meters in length. The tower now also supports a Four Square sloping λ/4 vertical array as described in the chapter on vertical arrays.

Fig 9-76 shows the vertical's base and the cabinet housing the series capacitor for the 160-meter gamma match (as well as the hybrid coupler for the 80-meter Four-Square array). I obtained detailed feed-point information for the ON4UN vertical with the assistance of ON6WU and his professional-grade test equipment (HP Network analyzer). See **Fig 9-77**.

6.6. The Battle Creek Special Antenna

Everyone familiar with DX operating on 160 meters has heard about the Battle Creek Special and its predecessor, the Minooka Special. These antennas are transportable verticals for operating on the low bands. The Minooka Special (Ref 761) was designed by B. Boothe, W9UCW, for B. Walsh, WA8MOA, to take on his trips to Mellish Reef and Heard Island many years ago.

Basically the antennas were designed to complement a triband Yagi on DXpeditions to provide excellent six-band coverage for the serious DXpeditioner. The original Minooka was a 40 through 160-meter antenna, using an L network for matching and an impressively long 160-meter loading coil near the top. WØCD built a very rugged and easily transportable version of the Minooka Special, but soon found out that the slender loading coil simply melted when the antenna was taking high power for longer than a few seconds. No wonder! It was more than 100-cm long with a diameter of only 27 mm. Michaels, W7XC, later calculated the Q factor of the coil to be around 20! That's an equivalent loss resistance of 100 Ω!

WØCD improved the antenna both mechanically and electrically. Instead of developing a better loading coil, he simply did away with the delicate part, and replaced the loading coil with a loading wire. His design uses two sloping wires, one for 80 meters and one for 160, which now makes it really an inverted L, but nothing would prevent you from using a T-shaped loading wire as described in Section 3.6.4.

The new design, named the Battle Creek Special, takes 1.5 kW of RF on SSB or CW without any problem for several minutes. For continuous-duty digital modes the RF output should not exceed 600 W. An 80-meter trap isolates the loading wires for 80 and 160.

The section below the 40-meter trap is 9.75 meters long, which makes it a full-size quarter-wave on that band. The SWR bandwidth is less than 2:1 from 7 to 7.3 MHz.

On 80 meters the 15 meters of tubing below the 80-meter trap, together with the loading wire, make it an inverted L. The antenna will cover 3.5 to 3.6 MHz with an SWR of less than 2:1. On 3.8 MHz the antenna is "too long," but a simple series capacitor of 200 to 250 pF will reduce the SWR to a very acceptable level (typically 1.3:1).

On 160 meters the entire vertical antenna plus the top-loading wire make it a λ/4 L antenna. The SWR is typically 2:1 over 20 kHz, indicating a feed-point impedance of approximately 25 Ω (depending to a large extent on the quality of the radial system).

There are several ways to obtain a better match to the feed line. WØCD uses an unun with a 2:1 impedance ratio (see also Chapter 6 on Feed Lines and Matching). The unun is switched in the circuit on 160 meters, and out of the circuit on 80 and 40 meters. Under certain circumstances it can be even advantageous to use the unun on the higher bands as well. The unun is an unbalanced-to-unbalanced wideband toroidal transformer (Ref 1521 and 1522). WØCD actually built a 9:4 (2.25:1) balun, and removed the top turn to get an exact 2:1 ratio.

Another alternative is to use an L network. A simple tunable L network that has been especially designed for matching "short" 160-meter loaded verticals is shown in **Figs 9-78** and **9-79**. The L network was made by ON7TK and has been traveling around the world on various DXpeditions (A61, 9K2, FOØC, etc).

The Battle Creek Special uses high strength aluminum tubing, 6061-T6 alloy, in sizes ranging from 2 inches to 1 inch (5 to 2.5 cm). The guy lines are 2.4-mm Dacron double-braided rope with a rating of 118 kg breaking strength. Wind survival rating is 160 k/hr assuming proper guy-rope anchors.

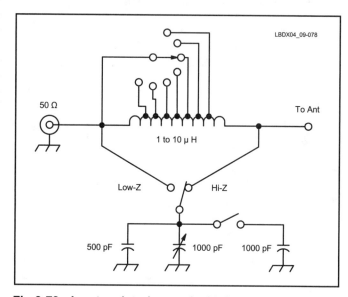

Fig 9-78—L network to be used with Inverted-L antennas and other loaded 160-meter verticals. With the component values shown, impedances in the range 20 − *j* 100 to 100 + *j* 100 Ω can easily be matched on 160 meters.

Fig 9-79—The L network of Fig 9-80 is contained in a small plastic housing. This particular unit was built by ON7TK and used on several DXpeditions (A61, 9K2, FOØC).

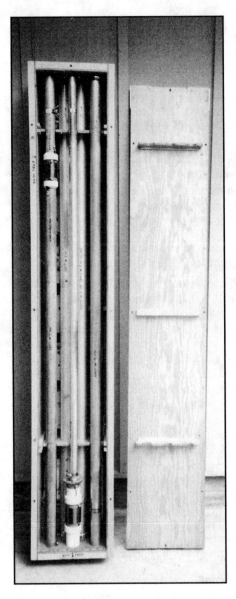

Fig 9-80—The wooden crate containing the Battle Creek Special antenna, a three-band (40, 80 and 160-meter) vertical. The wooden crate is especially designed to ensure safe transportation of the antenna to the most remote parts of the world. It contains all the antenna parts and accessories, such as guy ropes, anchors, hinged base plate, radial wires, etc.

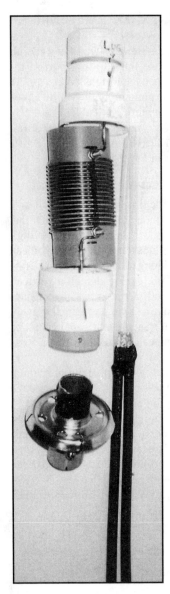

Fig 9-81—The new 80-meter trap for the Battle Creek Special. The original coaxial-cable traps have been replaced with regular LC traps, the capacitor being made by short pieces of coaxial cable. See text for details.

Fig 9-82 (below)—This version of the famous Battle Creek vertical is about 18 to 19 meters tall. If resonance on 80 cannot be achieved where wanted (if the support is not high enough), an 80-meter loading wire (typically 1 to 2 m long) can be added as shown. Slope this wire at right angle to the 40 and 160-meter loading wires if possible. The sloping angle of the 160-meter loading wire has a pronounced influence of the impedance of the antenna. You can use this to tune the antenna on 160. Keep the wire as horizontal as possible, however, it's better is to trim the length of the wire to tune it.

It is guyed four ways at three levels so the side guy ropes act as a hinge allowing it to be "walked up" by one person.

The original traps were coaxial-cable traps using RG-58, but they ran too hot with power levels over 800 W. Instead of changing to Teflon coax the designers decided to switch to regular L/C traps with the inductor made of #10 wire and the capacitor made from some lengths of RG-213 with 100 pF/m. The coaxial capacitors fit inside the aluminum mast sections. A single open-ended coax stub of about 90-cm length (90 pF) is used for the 40-meter trap and two parallel-connected pieces of coax of approximately 120-cm length each (240 pF total) are used for the 80-meter trap.

W0CD recommends using at least 30 radials, each of 20-meters length. I consider this a bare minimum. The Battle Creek Special is not for sale, but is available for loan to DXpeditions to rare countries. Interested and qualified DXpeditioners should contact W8UVZ for further details. The antenna was used at Bouvet on 80 and 160 meters in 1989/90, and during the DXpeditions to ZS0Z, 7P8EN, 7P8BH, G4FAM/3DA, 3Y5X, 5X4F, ZS8IR, XR0Y, VP8SGP, YK0A, 8Q7AJ, VK0IR, ZS9Z, V51Z, P40GG, CY9AA, ZS6EX, ZS6NW, AH0/AC8W, AL7EL/KH9, XF4DX, AH1A, 3Y0PI,

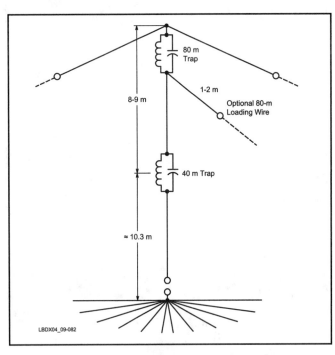

9MØC, J37XT, K5VT/JT, VK9LX, ZK1XXP, 3B7RF and many other locations with great success.

The entire antenna, with its base, guy-wires and radials is packed in a strong wooden case for safe transport to the remotest DXpedition spot. The package weighs 30 kg (66 lb). **Fig 9-80** shows the wooden crate containing the Battle Creek Special. **Fig 9-81** shows the 80-meter coax-cable trap.

For construction details, visit: **www.ok1rr.com/view.php?cisloclanku=2004122518**. A wire-type Battle Creek special vertical is shown in **Fig 9-82**.

6.7. A Very Attractive 160-Meter Vertical

Remember that a short vertical is as good as a full-size vertical if the losses in the system are zero. That means that your loading system has no losses (which means top loading). It also means that your ground losses are zero, which you can come close to if you use 100 λ/4 long radials or if you operate over saltwater. K7CA and N7JW developed such a vertical, which they both use in their 160-meter arrays (see Chapter 11), but which is very attractive as a single vertical as well. See **Fig 9-83**. The design set out to achieve R_{rad} = 12.5 Ω, so that a simple λ/4 coax transformer can be used to match a 50-Ω feed line.

The vertical is 20-meters tall, and is made of aluminum tubing, top loaded with two in-line-sloping top hat wires, each approx 18 meters long and sloping at an angle of about 55°. The tips end up at a 10-meter height. This has a resonant frequency of 1830 kHz and the desired feed-point impedance of 12.5 Ω. All you need to do is use a 4:1 broadband transformer (2:1 turns ration on an appropriate ferrite core) to have a perfect 50-Ω match. This short antenna has a 2:1 SWR bandwidth of 60 kHz and a 1.5:1 SWR bandwidth of 40 kHz, more than decent!

If you want play in the top league of the 160-meter DXers, make no compromises in your ground system! This is also the ultimate performer above saltwater, where just two elevated radials in a gull-wing configuration would do the job. This would be quite an attractive design for DXpeditions where the antenna can be set up over saltwater.

6.8. Using the Beam/Tower as a Low-Band Vertical

The tower supporting the HF antennas can often make a very good loaded vertical for 160 meters. A 24-meter tower with a triband or monoband Yagi, or a stack of Yagis, will exhibit an electrical length between 90° and 150° on 160 meters. These are lengths that are very attractive for low-angle work on 160.

6.8.1. The electrical length of a loaded tower

You can use **Fig 9-84** to determine the electrical length of a tower loaded with a Yagi antenna. The chart shows the situation for a tower with an effective diameter of 30 cm, loaded with five different types of Yagis, ranging from a 3-element, 20-meter Yagi to a 3-element 40-meter full-size Yagi. These figures are for Yagis that have their elements electrically connected to the boom, using so-called "plumber's delight" construction. An antenna like a KT-34 will show little capacity loading, because all elements are insulated from the boom. More on this below.

A 24-meter tower, loaded with a 5-element, 20-meter Yagi, will have an electrical length of 103° on 1.825 MHz. The effect of capacitance top loading depends to a great extent on the diameter of the tower under the capacitance hat. The capacitance hat (the Yagis) will have a greater influence with "slim" towers than with large-diameter towers. If you increase the tower diameter to 60 cm, this will shorten the electrical length between 4° and 7° (4° for the tower loaded with the 3-element, 20-meter Yagi and 7° for the tower loaded with the

Fig 9-83—20-meter tall vertical with an eagle sitting on top in the desert location of N7JW's remote-antenna setup in Southern Utah.

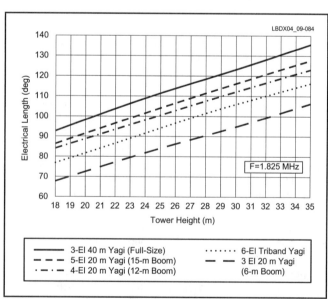

Fig 9-84—Electrical length of a tower loaded with a Yagi antenna. The chart is valid for 160 meters (1.825 MHz) and a tower equivalent diameter of 30 cm. For a larger tower diameter, the electrical length will be shorter (4° to 7° for a tower measuring 60 cm in diameter).

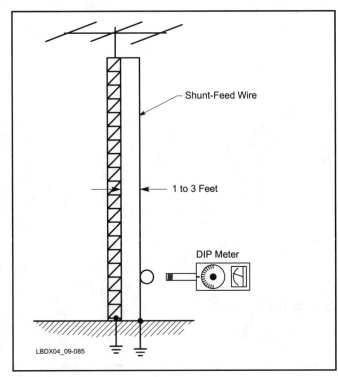

Fig 9-85—A method of "dipping" a tower with a shunt-feed wire connected to the top.

40-meter, 3-element full-size Yagi). W. J. Schultz, K3OQF, published the mathematical derivation of the shunt-fed top-loaded vertical (Ref 7995).

There are neither data nor formulas available for calculating the exact electrical length of a tower loaded with multiple Yagis. The best way to find out is to attach a drop wire to the very top of the tower (turn it into a folded element) and grid dip the entire structure, as shown in **Fig 9-85**. You could, of course, also model the entire structure, but that seems a rather tedious task.

If your tower, top loaded with a Yagi, is still a little short, you may want to add some extra wires from the top of the tower sloping down to increase the top loading. You might use part of the top set of guy wires, for example.

Towers with stacked Yagis are more difficult to assess. Basically it's the bottom Yagi that determines the capacity, as this Yagi *hides* the Yagis above it, especially if they are nearby and smaller.

6.9.2. Yagis with elements insulated from the boom

There are two problems associated to having Yagis with insulated elements on a tower for use on the lower bands:

- The insulated elements will only add little top loading
- Possible arcing of the Yagi insulating parts and destruction of baluns

Some Yagis use fiberglass or PVC for insulating their elements from the boom. While these are good enough for the Yagi, where the voltage between the center of the floating elements and the boom are low, voltages in case of a top-loaded tower may be very high in these same places. The highest RF voltage always occurs at the farthest end of an antenna from the feed point. For a shunt-fed tower with an HF Yagi with insulated elements used for top loading this highest RF voltage point would be at the ends of the Yagi's boom (eg, the 20-meter reflector). In most cases you can simply ground the center of the elements to the boom.

PA3DZN had to do this with his KTX-34 tribander, and after having grounded all elements, he could not detect any change in performance of the Yagi, but his arcing problem was solved. John, K9DX, reported that he burned the feed line off his KLM 40-meter beam when it flashed over to the boom. Grounding all the elements solved the problem. Although direct grounding of the elements should in most cases not upset the functioning of the Yagi, you could connect the elements to the boom using RF inductors having a few hundred ohms reactance on the lowest band the Yagi is used on.

You should be careful grounding the center of the driven element, because that might upset the matching system. Although insulated from the boom, the driven element usually acts as if it were grounded to the boom on 160 meters because of the feed line's coupling to the tower on its way down to the ground. Therefore, the driven element already fully adds to the top-loading and there is no reason to "ground" the driven element to the boom when you shunt feed the tower on 160 meters.

Certain types of baluns used on Yagis can be destroyed by shunt feeding the tower. If you use a balun using ferrite material (eg, W2DU, Hy-Gain, or Force-12) you may have to decouple 160-meter RF from reaching the balun, which is not easily done at a high-impedance point, near the end of the 160-meter antenna (see also Section 6.8.8). Plumber's-delight Yagis (all elements connected to the boom and using a Gamma or Omega match) are the ideal solutions for Yagi-loaded towers.

6.8.3. Measuring the electrical length

A second and very practical method of determining the resonant length of a tower system was given by DeMaw, W1FB (Ref 774). A shunt-feed wire is dropped from the top of the tower to ground level. What you want to do is turn a grounded single-conductor vertical into a folded-element vertical, where you now can easily do measurements in the drop wire. Attach a small 2-turn loop between the end of the wire and ground and couple this loop to the grid dip meter (Fig 9-85).

The lowest dip found then is the resonant frequency of the tower/beam. The electrical length at the design frequency is given by:

$$L \text{ (in degrees)} = 90° \frac{f_{design}}{f_{resonant}}$$

Therefore, if $f_{resonant} = 1.6$ MHz and $f_{design} = 1.8$ MHz, then $L = 101°$.

6.8.4. Gamma and omega matching

There are many approaches to matching a loaded, grounded tower. Three popular methods are:

- Slant-wire shunt feeding (Section 6.8.5)
- Folded-monopole feeding (Section 6.8.6)
- Gamma or omega-match shunt feeding.

Gamma and omega-matching techniques are widely used on loaded towers. The design of gamma matches has often

Table 9-8
Gamma-Match Data for a Shunt-Fed Tower with 50-cm Gamma-Wire Spacing

Tower electrical height = 100 degrees
Tower diameter = 250 mm (10 inches)

	1.730	1.765	1.800	1.835	1.870	1.905	1.940 MHz
Gamma-wire diameter = 2 mm (AWG 12); tap height = 19.5 m (64.0 ft)							
R	80.6	66.8	56.5	50.0	43.3	39.2	36.0
X	+330	+338	+350	+363	+377	+392	+407
SWR	2.0	1.7	1.3	1.0	1.4	2.0	2.8
Gamma-wire diameter = 10 mm (0.4 in.); tap height = 19.8 m (65.0 ft)							
R	82.9	68.6	58	50	44.8	40.6	37.6
X	+250	+257	+267	+278	+291	+303	+316
SWR	1.9	1.6	1.3	1.0	1.3	1.8	2.4
Gamma-wire diameter = 50 mm (2 in.); tap height =20.0 m (65.6 ft)							
R	80.8	66.9	56.9	50	44.3	40.3	37.3
X	+164	+171	+179	+188	+198	+208	+218
SWR	1.8	1.5	1.2	1.0	1.3	1.6	2.1
Gamma-wire diameter = 250 mm (10 in.); tap height =20.2 m (66.3 ft)							
R	78.8	65.5	56	50	44	41	38.3
X	+75	+82	+90	+98	+105	+113	+121
SWR	1.8	1.5	1.2	1.0	1.2	1.5	1.8

been described in the literature (Ref 1401, 1414, 1421, 1426 and 1441).

Fig 9-86 shows the height of the gamma-match tap, as well as the value of the gamma capacitor for a range of antenna lengths varying from 60° to 180°. The chart was developed using a gamma wire of 10-mm diameter. There are three sets of graphs, for three different wire spacings (0.5, 1.0 and 1.5 meters).

It is a fairly common misconception to think that the tower must be resonant to be able to match it correctly to 50 Ω at the desired frequency. This isn't true. However, there are some advantages to having the tower (with its top loading) resonant near the desired frequency:

- A resonant tower makes it possible to design an efficient shunt-feed system.
- A tower resonant slightly off the desired frequency can exhibit a broader SWR bandwidth because there are now two dips in the SWR curve, one caused by the resonant frequency of the tower and another one caused by the pulling of the Gamma or Omega match of the resonance to a slightly different frequency.
- The 50-Ω tap point is closest to the ground with a λ/4 resonant tower (only about 8 meters high on 160 meters).
- The 2:1 SWR bandwidth is better than with a non ¼-wave shunt-fed vertical (see also Section 6.8.4.3).
- The RF voltage across the gamma capacitor will be lowest.

However, it is not necessary to strive for λ/4-wave resonance. Virtually any size vertical can be successfully shunt-fed and will perform well.

6.8.4.1. Close spacing versus wide gamma spacing

The wider the spacing, the shorter the gamma wire needs to be. Shorter gamma wires will naturally show less inductive reactance, which means that the series capacitor must be larger in value.

Electrically very long verticals will require a tap that is 20 to 30 meters up on the tower. The required series capacitor will be small in value (typically 100 to 150 pF). There will be a very high voltage across capacitors of such small value.

In case the required gamma-wire length shown in Fig 9-85 appears to be longer than the physical length of the tower, you will need an omega match (see Section 6.8.4).

6.8.4.2. Influence of gamma-wire diameter

The gamma-wire diameter has little influence on the

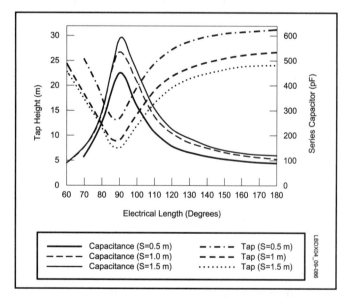

Fig 9-86—Tap height and values of the gamma series capacitor for a shunt-fed tower at 1.835 MHz. The tower diameter is 250 mm, and the gamma wire has a diameter of 10 mm. Three sets of curves are shown, for three spacings (S). The spacing is the distance from the wire to the tower center.

Table 9-9
Gamma-Match Data for a Shunt-Fed Tower with 150-cm Gamma-Wire Spacing

Tower electrical height = 100 degrees
Tower diameter = 250 mm (10 inches)

	1.730	1.765	1.800	1.835	1.870	1.905	1.940 MHz
Gamma-wire diameter = 2 mm (AWG 12); tap height = 11.9 m (39.0 ft)							
R	86.8	71	59	50	43.7	38.8	35
X	+226	+229	+236	+244	+253	+262	+272
SWR	1.8	1.5	1.2	1.0	1.2	1.6	2.1
Gamma-wire diameter = 10 mm (0.4 in.); tap height = 12.0 m (39.4 ft)							
R	87.8	71.7	59.8	50	44.4	39.5	25.7
X	+179	+181	+187	+195	+203	+212	+220
SWR	1.8	1.5	1.3	1.0	1.2	1.6	2.0
Gamma-wire diameter = 50 mm (2 in.); tap height =12.0 m (39.4 ft)							
R	87	71	59	50	44.3	39.5	35.8
X	+120	+132	+137	+144	+152	+159	+166
SWR	1.8	1.5	1.2	1.0	1.2	1.5	1.8
Gamma-wire diameter = 250 mm (10 in.); tap height =11.9 m (39.0 ft)							
R	85.2	69.6	58.3	50	44	39.4	36
X	+79	+82	+87	+93	+100	+1-6	+112
SWR	1.8	1.5	1.2	1.0	1.2	1.5	1.7

length of the gamma wire (position of the tap on the tower). A larger diameter wire will require a somewhat shorter gamma wire. The wire diameter has a pronounced influence, however, on the required gamma capacitor. It also has some influence on the SWR bandwidth of the antenna system, but less than most believe.

6.8.4.3. SWR bandwidth

Tables 9-8 and **9-9** show the feed-point impedance and the SWR versus frequency for a vertical of 100° electrical length, fed with a gamma match. A spacing of 50 cm is used in Table 9-8, and 150-cm spacing in Table 9-9. Wire diameters of 2 mm (AWG #12), 10 mm, 50 mm and 250 mm are included. The 2-mm (#12) wire is certainly not responsible for a narrow bandwidth. It does not seem worth using a "wire cage" gamma-wire to improve the bandwidth.

For loaded towers that are much longer than 100°, the bandwidth behavior is quite different. The longer the electrical length of the vertical, the narrower the SWR bandwidth. **Table 9-10** shows the feed-point impedance and the SWR for a vertical of 150° electrical length, fed with a gamma-match and a gamma-wire of both 10 mm and 250 mm OD. In contrast with the effect on the shorter vertical (100°), the wire diameter now has a pronounced influence on the bandwidth. The 10-mm wire yields a 70-kHz bandwidth; the 250-mm wire cage almost 130 kHz. As can be seen from the impedance values listed in Table 9-9, it is the large variation in reactance that is responsible for the steep SWR response. This can be overcome using a motor-driven variable capacitor. The 150° long antenna with a 10-mm-OD gamma wire shows an SWR of less than 1.3:1 over more than 200 kHz, if a variable capacitor with a tuning range of 100 to 175 pF is used. A high-voltage (eg, 10 kV) vacuum variable is a must.

This simple way of obtaining a very flat SWR does not apply to shorter verticals (90° to 110°), where a much larger variation in the resistive part of the feed-point impedance is

Table 9-10
Gamma-Match Data for a Shunt-Fed Tower with 150-cm Gamma-Wire Spacing

Tower electrical height = 150 degrees
Tower diameter = 250 mm (10 inches)

	1.730	1.765	1.800	1.835	1.870	1.905	1.940 MHz
Gamma-wire diameter =102 mm (0.4 in.); tap height = 25.9 m (65.0 ft)							
R	43.1	45.2	47.7	50	54	58	62.7
X	+567	+597	+628	+661	+697	+736	+778
SWR	6.0	3.5	1.9	1.0	2.0	3.7	6.3
Gamma-wire diameter = 250 mm (10 in.); tap height = 24.8 m (81.4 ft)							
R	41.5	43.8	46.6	50	54	58.5	64
X	+286	+303	+320	+340	+362	+384	+409
SWR	3.2	2.2	1.5	1.0	1.5	2.2	3.2

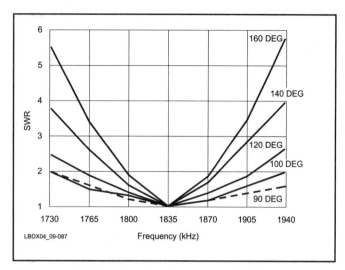

Fig 9-87—SWR curves for gamma-fed towers using a 10-mm OD gamma wire and a spacing of 50 cm, for electrical tower lengths varying from 90° to 160°. The SWR of the longer vertical can be tuned over a wide bandwidth using a motor-driven variable series capacitor.

responsible for the SWR. **Fig 9-87** shows SWR plots for gamma-fed towers of varying electrical length, using a 10-mm OD gamma wire, spaced 150 cm from the tower.

6.8.4.4. Adjusting the gamma-matching system

The easiest way to fine tune the gamma-matching system is to vary the spacing of the gamma wire. This changes the resistive part of the feed-point impedance. Then you can tune out the inductive reactance using the series Gamma capacitor for a 1:1 SWR.

Example:

For a vertical 100° electrical long and with a tower diameter of 250 mm, we install the tap at 14 meters. At that point the spacing is 1 meters. Changing the spacing at ground level has the following influence:

Spacing = 0.5 meters: $Z = 38 + j\ 206\ \Omega$
Spacing = 0.75 meters: $Z = 44.8 + j\ 298\ \Omega$
Spacing = 1.0 meters: $Z = 49.3 + j\ 211\ \Omega$
Spacing = 1.25 meters: $Z = 56.4 + j\ 213\ \Omega$
Spacing = 1.5 meters: $Z = 61.5 + j\ 214\ \Omega$

This demonstrates how fine tuning can easily be done on the gamma-matching system.

6.8.4.5. Using the omega matching system

If you can tune your tower using a gamma, I would not advise using an omega system. The omega match requires one more component, which means additional losses and additional chances for a component breakdown. It is possible, however, to use a gamma-rod (wire) length that is up to 50% shorter than the length shown in Fig 9-86 when you use an omega match. An Omega system is similar to a Gamma system except that a parallel capacitor is connected between the bottom end of the gamma wire and ground.

The 100°-long vertical requires a 14-meter long gamma wire, with 100-cm gamma-wire (OD 10 mm) spacing. If we shorten the gamma wire to 8 meters, the transformed impedance becomes $14.1 + j\ 127\ \Omega$. This can be matched to 50 Ω using an L network. One of the solutions of this L network consists of two capacitors: the well-known parallel and series capacitor of the omega-matching system. To calculate the omega-system, use the following procedure:

- Model the vertical with the shorter gamma rod. Make sure you use enough segments (pulses). For 160 meters, segment lengths of 100 cm gives good results. Note the input impedance, which will be lower than 50 Ω and inductive.
- Use the L NETWORK module of the NEW LOW BAND SOFTWARE to calculate the capacitance of the parallel and the series capacitor.

In our example above, the 8-meter-long gamma wire requires a parallel capacitor of 369 pF and a series capacitor of 323 pF.

Table 9-11
Omega-Match Data for a Shunt-Fed Tower with 50-cm Gamma-Wire Spacing

Tower electrical height = 140 degrees
Tower diameter = 250 mm (10 in.)

	1.730	1.765	1.800	1.835	1.870	1.905	1.940 MHz
Gamma-wire diameter = 2 mm (AWG 12); tap height = 24.0 m (78.7 ft)							
R	16.0	16.3	16.7	17.2	17.9	18.6	19.3
X	+514	+535	+552	+579	+603	+629	+650
With parallel capacitor of 62 pF added							
R	37.4	40.7	44.8	50	56.8	65.3	74.7
X	+785	+845	+910	+986	+1073	+1178	+1287
With fixed series capacitor of 88 pF added							
R	37.4	40.7	44.8	50	56.8	65.3	74.7
X	−261	−180	−95	0	+106	+228	+348
SWR	38.0	17.9	5.9	1.0	5.8	17.9	35
With variable series capacitor, 50 to 125 pF (adjusted to cancel inductive reactance)							
R	37.4	40.7	44.8	50	56.8	65.3	74.7
X	0	0	0	0	0	0	0
SWR	1.3	1.2	1.1	1.0	1.1	1.3	1.5

If you have a physically short tower with a lot of loading, it may be that the required tap height is greater than your tower height. In this case an omega match is the only solution (if you have already tried a larger spacing).

Example:

See **Fig 9-89**. The tower was grid-dipped and the electrical length turned out to be 140°. The physical height is 24 meters. Fig 9-86 shows a required gamma-wire length of 30 m for a 2-mm OD gamma wire and a 50-cm spacing. In this case we will connect the gamma wire at the top of the tower (h = 24 meters). Using *NEC-2*, we calculate the feed-point impedance as $Z = 17.2 + j\,579\,\Omega$. From the L NETWORK software module, the capacitor values are calculated as $C_{par} = 62$ pF, $C_{series} = 88$ pF. Note that these very low-value capacitors will carry very high voltages across their terminals with high power.

This L network is a high-Q network. Table 9-10 lists the impedances at the end of the gamma wire before and after transformation by the capacitors of the omega-match. Note the very narrow bandwidth of this high-Q matching system. If we adjust the omega-capacitors for a 1:1 SWR on 1835 kHz, the 2:1 bandwidth will be typically only 20 kHz. If we make the series capacitor adjustable (60 to 120 pF), we can tune the antenna to an SWR of less than 1.5:1 over more than 200 kHz.

6.8.4.6. Conclusion

If after modeling or actually measuring your loaded tower it turns out to be longer than about 140°, you might consider using an elevated feed system, with lots of elevated radials (see Section 2.2.10). Within reason, the longer the section below the feed point, the easier to decouple this section. If you keep raising the base of the vertical, however, you will achieve a vertical radiation pattern that is not ideal for most DX work.

If you have an electrically long gamma-matched vertical (more than about 110° high), you can use a large-diameter cage-type gamma wire and a large wire-to-tower spacing. Making the series capacitor remotely tunable will certainly make the antenna much more broadbanded. Do not shorten the gamma wire unless required because of the physical length of the tower.

Fig 9-89 shows the correct wiring of both the gamma and omega-matching networks on a loaded tower. Notice the correct connection of the shunt capacitor in the case of the omega match. Make sure the ground wires all have very low resistance and the lowest possible reactance. Use flat solid-copper strip if possible. Do *not* use flat braided strip, which has high RF losses, contrary to popular belief.

The same principles can, of course, be applied to 80 meters, although it is probable that a tower of reasonable height, loaded with a Yagi antenna, will result in too long an antenna for operation on that band.

6.8.4.7. A few practical hints

All cables leading to the tower and up to the rotator and antennas should be firmly secured to a tower leg, on the inside of

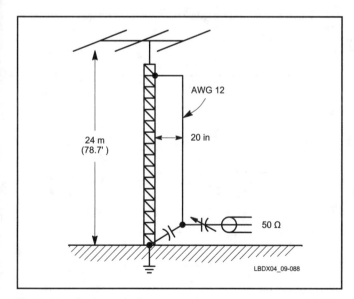

Fig 9-88—A shunt-fed tower using an omega matching system. The tower is electrically 140° long. An omega match is required, since the tower is physically too short to accommodate a gamma match with a 2-mm gamma wire. Table 9-11 lists the impedances at the end of the gamma wire before and after transformation by the capacitors of the omega-match system.

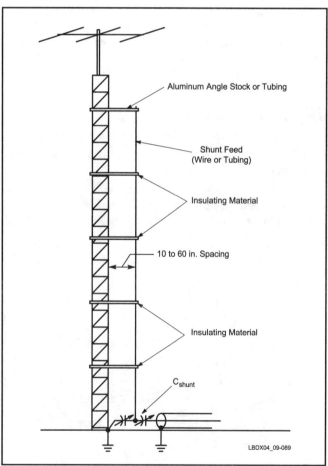

Fig 9-89—The omega-matching system (a gamma match with an additional shunt capacitor) adds a great deal of flexibility to the shunt-fed-tower arrangement. To maintain maximum bandwidth make the gamma wire as long as possible. If the antenna is electrically longer than 120°, a variable series capacitor will make it possible to obtain a very low SWR over a wide bandwidth.

the tower. All leads from the shack to the tower base should be buried underground to provide sufficient RF decoupling. If there is RF on some of the cables, you should coil up a length of the cable running up your shunt-fed or series-fed tower to make a common-mode choke at the tower base. A coil consisting of 50 turns on a 10-cm diameter form yields about 100 µH of inductance. If you still detect some RF on these cables entering your shack, install another coil at that point or put some ferrite beads on the cable. Coaxial cables running down the tower should preferably be grounded right at the base of the vertical.

Take care to ensure good electrical continuity between the tower sections, and between the rotator, the mast and the tower. Again, use flat-strip copper conductors, not the woven battery-connecting flat strips, which are good only for dc.

A gamma rod can be supported with sections of plastic pipe, attached to the tower with U bolts or stainless-steel radiator hose clamps. If the tower is a crank-up type, heavy, insulated copper wire can be used for the gamma element.

6.8.5. The slant-wire feed system

The slant-wire feed system is very similar to a gamma feed system. The feed wire is attached at a certain height on the tower and slopes at an angle to the ground, where a series capacitor tunes out the reactance. The advantage of this system is that a match can be obtained with a lower tap point, which makes it possible to avoid using an omega match on physically short towers. The disadvantage is that the slant-wire feed also radiates a horizontally polarized component. The slant-wire feed system can easily be modeled using a computer program, just like the gamma and omega-matching systems.

6.8.6. Folded monopoles

Folded antennas have the following advantages:
- Higher bandwidth due to a larger effective antenna diameter
- Higher feed-point impedance.

Fig 9-90 shows how you can manipulate the wire diameter and spacing to obtain up-transformation ratios ranging from two to well over 10. (Source: *Kurze Antennen*, by Gerd Janzen, ISBN 3-440-05469-1). The configuration of the two-wire folded monopole is shown in the same figure. One leg is grounded, while the antenna is fed between the bottom of the other leg and ground.

The effective diameter of multi-wire elements can be calculated from the chart shown in Fig 9-50. Three-wire folded-element configurations allow even higher transformation ratios, as can be seen in **Fig 9-91**. The effective antenna diameter (which determines the bandwidth of the antenna) is given in Fig 9-64 for the various configurations. The configurations are also shown in the same figure.

6.8.7 Modeling shunt-fed towers

MININEC or *NEC-2* can be used for modeling the gamma, omega and slant-wire matching systems on shunt-fed grounded towers. Satisfactory results are obtained using the following guidelines:

- The horizontal wire connecting the gamma wire to the tower has one segment (the length of the segment is the spacing from the gamma wire to the tower).
- Use approximately the same segment lengths on all wires of the antenna.
- All segments should have the same length; this is determined by the length of the horizontal wire connecting the gamma-wire to the tower.
- Do not try to model the capacitance top load. It is much easier to first grid-dip the tower (see Fig 9-85), calculate the electrical length of the loaded tower and then use an equivalent straight tower to do the gamma-match modeling.

Example:

A tower dips at 1.42 MHz. The required operating frequency is 1.835 MHz. The electrical length is:

$$L \text{ (in degrees)} = 90° \frac{1.835}{1.42} = 116°$$

The physical length of a λ/4 tower (equivalent diameter = 250 mm) is 39 meters. The equivalent tower length for 116° is:

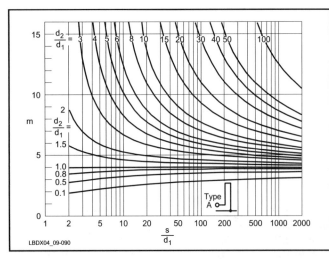

Fig 9-90—Transformation ratio (m) of a two-wire folded monopole, as function of the wire spacing (S/d1) and the ratio of the conductor diameter (d2 = diameter of the grounded conductor, d1 = diameter of the fed conductor). (*After Gerd Janzen,* Kurze Antennen.)

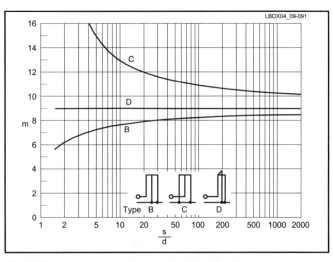

Fig 9-91—Transformation ratio (m) of a three-wire folded monopole, as function of the spacing between the wire (S/d) and the configuration B, C or D. In this case three conductors of equal diameter are assumed. (*After Gerd Janzen,* Kurze Antennen.)

$$39 \, \text{m} \times \frac{116°}{90°} = 50.3 \text{ meters}$$

Now model a vertical with a diameter of 250 mm and of 50.3 meters length. According to Fig 9-86, the tap will be at a height of between 17 and 25 meters, depending on the wire spacing.

6.8.8. Decoupling antennas at high-impedance points

When shunt feeding a tower that supports various antennas, including wire antennas, you may wish to decouple the wire antennas from the vertical radiator. Otherwise, the wire antennas will act as top loading to the vertical. For relatively short towers, the extra loading may be welcome, but in other cases the loaded vertical may become too long with the additional loading of large wire antennas, such as a 160-meter or an 80-meter inverted V.

These wire antennas are usually installed at the top of the tower, at a high-impedance point. This makes decoupling of the wire antennas more difficult. Conventional common-mode current baluns are not suitable, since they do not have enough inductance to effectively decouple the antenna. Such baluns can lead to unexpected changes in the feed-point impedance of the loaded vertical while transmitting. When first transmitting the SWR may be normal, but soon the ferrite material used in the current balun will heat to the point where the Curie temperature is reached, resulting in a sudden drop in magnetic susceptibility of the ferrite material. The balun will no longer represent enough impedance, causing the dipole to load the tower with a change in SWR as a result.

For effective decoupling in this application, a high-impedance balun is required. This balun is rather like a trap—a parallel-tuned resonant circuit—tuned to the frequency of the loaded vertical. The RF currents that flow from the vertical to the dipole (which we want to decouple) are common-mode currents, which means they flow only on the outside of the coaxial feed line of the inverted V dipole.

The trap is made by winding a single-layer coil of coax onto a suitable form, and resonating the coil with a suitable capacitor. Jim Jorgenson, K9RJ, made such a trap for 160 meters. It consisted of 21 turns of RG-213, wound close spaced on a PVC pipe 10-cm in diameter. The coil is about 33 cm long, and the measured inductance of this coil is 33 µH. The coil is held in place on the form by drilling close-fitting holes at an angle through the PVC pipe and passing the coax through these holes into the interior of the pipe at both ends. At one end the shield and the inner connector are separated and connected to stainless steel eyebolts that are used to connect the two legs of the inverted V antenna. At the other (bottom) end the coax passes out through a standard PVC end cap and a PL-259 connector is attached at that end.

The capacitance needed to resonate this coil on 1830 kHz is about 200 pF. You could use a quality transmitting-type ceramic capacitor, but a suitable capacitor can be made from a short piece of coax. RG-213 coax has a capacitance of 100 pF/m, which means that an open-ended piece of RG-213, 2 meters long will resonate the coil on 160 meters. The resonant frequency of the trap can easily be measured using a grid-dip meter. **Fig 9-92** shows the layout of the trap and **Fig 9-93** shows it deployed on the tower. To tune the trap, you can deliberately make the coaxial-line capacitor too

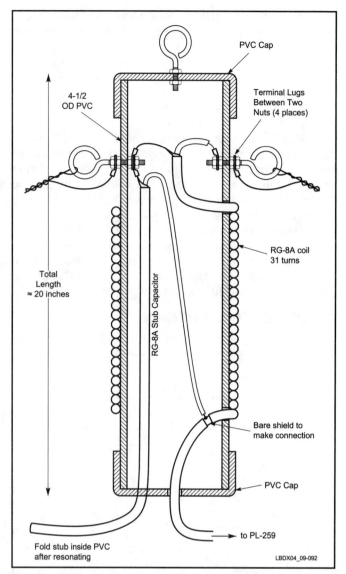

Fig 9-92—Construction details of the K9RJ decoupling trap.

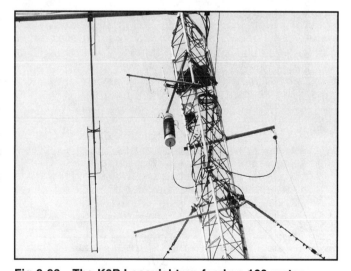

Fig 9-93—The K9RJ coaxial trap feeds a 160-meter inverted V hanging from the top of a tower, which is also used as a shunt-fed vertical for 160 meters. The entire construction looks somewhat like an oversized center insulator or balun.

Vertical Antennas 9-71

long, and then cut small pieces at a time until resonance is obtained at the desired frequency. The stub capacitor can then be folded inside the PVC tube before putting on the bottom end cap. Jim reports that since he has been using this balun there has been no change of the shunt-fed tower impedance, with the 160-meter inverted V attached. This proves that the trap is now fully decoupling the inverted V from the vertical.

Whether or not this concept is good enough to decouple the inverted V-antenna (in this case) from the tower depend on where along the tower this system is installed. If it is installed right at the top of the tower with no additional top loading, the impedance at that point is very high, meaning extremely high voltages at that point. This system worked for a 160-meter inverted-V mounted at the 80-foot level on K9RJ's 100-foot shunt-fed and top-loaded tower. It probably would work well enough for most typical shunt-fed and top loaded tower installations on 160 meters, but not when the antenna is installed right at the top of an unloaded tower.

7. INVERTED-L ANTENNAS

The ever-so popular inverted-L is analyzed in this section and a few practical designs, such as the well-known "AKI Special," are given particular attention.

The inverted L is a popular antenna, especially on 160 meters. These antennas are not truly verticals, as part of the antenna is horizontal and thus radiates a horizontally polarized component. We often form the wrong mental picture of what actually happens because most antenna modeling programs only express the field in two distinct polarizations. We wrongly picture two distinct fields. The actual field is the vector sum of the two fields, and is a single polarization wave with a tilt and a distinct total null at 90-degrees from the peak response.

Most inverted Ls are of the λ/4 variety, although this does not necessarily need to be the case. The vertical portion of an inverted L can be put up alongside a tower supporting HF antennas. In such a setup one must take care that the tower plus HF antenna does not resonate near the design frequency of the inverted L. Grid dip your supporting tower using the method shown in Fig 9-85. If it dips anywhere near the operating frequency, maybe you should shunt feed the tower instead of using it as a support for an inverted L.

If you choose do the inverted L, you can detune the tower to make sure the highest possible current flows in the parallel-tuned structure (see Section 3.10 in Chapter 7).

The longer the vertical part of the antenna, the better the low-angle radiation characteristics of the antenna and the higher the radiation resistance (see **Fig 9-94**). The horizontal

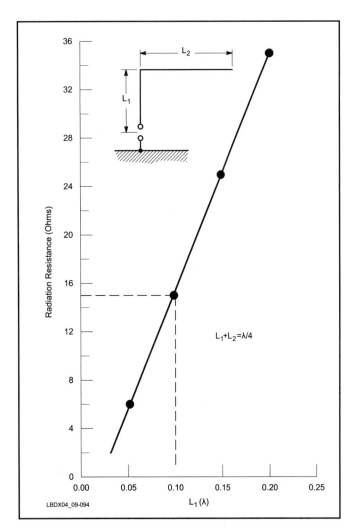

Fig 9-94—Radiation resistance of an inverted-L antenna as a function of the lengths of the horizontal wire versus the vertical conductor size.

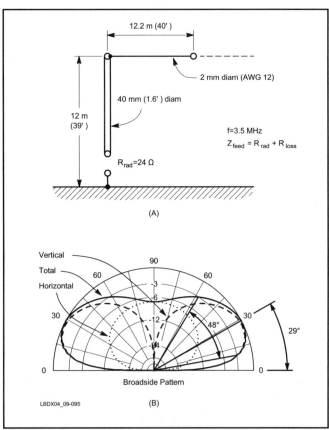

Fig 9-95—At A, a 3.5-MHz inverted L with a 12-meter vertical mast. The vertical radiation pattern is shown at B. The pattern has both vertically and horizontally polarized components and these components are also plotted at B. The pattern is generated over average ground, using 60 λ/4 radials. Note that the angle of maximum radiation is 29°, not bad for a DX antenna.

part of the antenna accounts for the high-angle radiation that the antenna produces but this normally is low, since the bulk of the radiation comes from the bottom part of the antenna, where the current is highest. Since it is a top-loaded monopole, an inverted L still requires a good ground system.

Fig 9-95 shows the vertical and horizontal radiation patterns for a practical design of an inverted-L antenna for 3.5 MHz, one having a 12-meter vertical mast. Notice how the vertical part of the antenna takes care of the low-angle radiation, while the horizontal part gives high-angle output. The radiation pattern shown is for the direction perpendicular to the plane of the inverted L.

An inverted L is also an attractive solution for the operator who wants to use an 80-meter vertical antenna as a support for a 160-meter antenna (**Fig 9-96A**). The easiest solution is to insert a trap at the top of the 80-meter vertical. The exact L/C ratio is not important, but it influences the length of the loading wire and the SWR behavior of the antenna on both 80 and 160 meters. See also Figs 9-71 and 9-73.

A second alternative, shown in Fig 9-96B, uses an 80-meter trap to isolate the horizontal part of the 160-meter inverted-L antenna when operating on 80 meters (Ref 659). The trap can be a coaxial-cable trap as explained in Section 6.4.

The inverted L has been extensively described in amateur literature as a good antenna for producing a low-angle signal on Topband (Ref 798 and 7994). The Battle Creek Special, described in Section 6.6 is an example of an inverted L (on 80 and 160 meters).

7.1. The AKI Special

The AKI Special is another DXpedition-style inverted L, as used by Aki Nago, JA5DQH, during his operations on 160 meters from several rare DX spots. From Kingman Reef (May 1988), Nago used the inverted L shown in **Fig 9-97**. The vertical part is made of a 12-meter aluminum mast, which is extended by an 8-meter-long fiberglass fishing rod, to which a

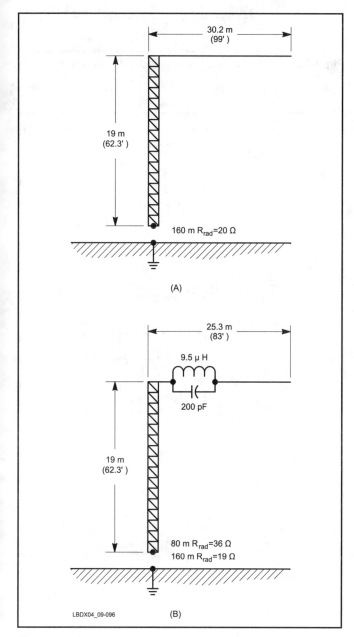

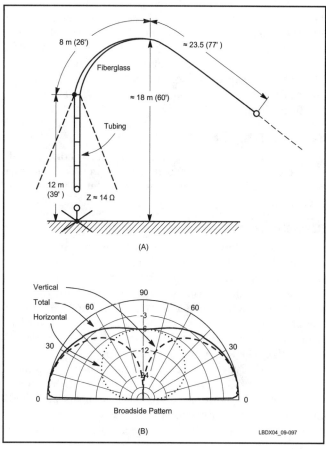

Fig 9-96—At A, an inverted-L antenna for 160 meters, using a 19-meter vertical tower. To cover both 80 and 160 meters, a trap can be installed at the top of the tower as shown at B. With the trap installed, the loading wire is shorter, because the trap shows a positive reactance (loading effect) on 160 meters. See also Figs 9-71 and 9-73.

Fig 9-97—The "AKI Special," a typical DXpedition type 160-meter inverted L. A collapsible fiberglass fishing rod (available in Europe in lengths of up to 12 meters) is used on top of a 12-meter aluminum mast. A #12 wire is attached to the rod, and slopes to a distant point to make the sloping (horizontal) part of the antenna. The radiation pattern is over saltwater. (That's where the island DXpeditioners put these antennas.)

Vertical Antennas 9-73

copper wire has been attached. From the tip of the (bent) fishing rod, the sloping wire extends another 23.5 meters, to be terminated with a fishing line supported by a 3-meter pole at some distance. Aki used about 800 meters of radials running into the Pacific Ocean. He used a very similar 160-meter antenna successfully from Palmyra during the same DXpedition trip in 1988, and during a more recent DXpedition to Ogasawara by JA5AUC. The excellent signals from VKØIR (1996) on 80-meter SSB were also produced with an AKI-type inverted L, using two elevated radials, above a large number of ground wires (not connected to the radials or the feed system). The calculated radiation resistance of this antenna is approximately 14 Ω. The main radiation angle (over sea water) is 10°, but due to the relatively long horizontal (sloping) wire, the radiation at higher angles is only slightly suppressed.

7.1.1. Tuning procedure

When cutting the length of the sloping wire, cut it at first 2 meters too long. Put up the antenna, and connect one of the popular antenna analyzers (MFJ, AEA or Autek) between the bottom of the antenna and the ground system. Adjust the length of the sloping top wire for minimum SWR. Now read the resistance value off the scale of your analyzer. If it is between 35 Ω and 70 Ω, the SWR will be pretty acceptable (1.5:1) and you may want to feed the antenna directly with 50-Ω feed line. From the difference between the R value and the calculated 14-Ω radiation resistance, you can calculate the effective ground-loss resistance of the ground radial system. If the feed-point impedance is above 50 Ω, you really need to improve the radial system. At 50 Ω the efficiency would be 14/50 = 28%. Any value higher than 50 Ω indicates an even lower efficiency. If you want a perfect match you can use an L network or an unun (Ref 1522).

8. THE T ANTENNA

The current-fed T antenna is a top-loaded short vertical, as covered earlier in this chapter. The voltage-fed T antenna is given special attention here, as well as different top-loading structures.

8.1. Current-Fed T Antennas

T-wire loading (flat-top wire) is covered in detail in Sections 3.6.4 and 3.6.5 when dealing with top loading of short verticals. The advantage of the horizontal T-wire loading system over the inverted-L system is that the top-wire does not contribute to the total radiation pattern. **Fig 9-98** shows a practical current-fed T design, where a 12-meter long vertical is loaded with a horizontal top-load wire to achieve resonance at 3.5 MHz. The R_{rad} of this design is approximately 23.5 Ω.

Fig 9-55 gives a design chart for λ/4 T antennas. If there is not enough room for a single flat-top wire, two wires (or any number of wires positioned in equal increments on a 360° circle) can be used. If you use two in-line wires the length of the wire will be about 60% of the length of a single wire.

8.2. Voltage-Fed T Antennas

Voltage-fed T antennas are loaded vertical antennas with a current minimum at ground level. A specific case consists of a quarter-wave vertical, loaded with a half-wave top wire. **Fig 9-99** shows the configuration of this antenna and the current distribution. In this case, the impedance at the base of the antenna is high and purely resistive. The current maximum

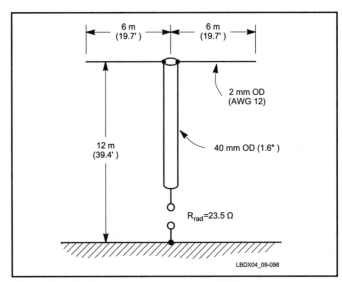

Fig 9-98—T-wire loaded current-fed λ/4 antenna for 3.5 MHz.

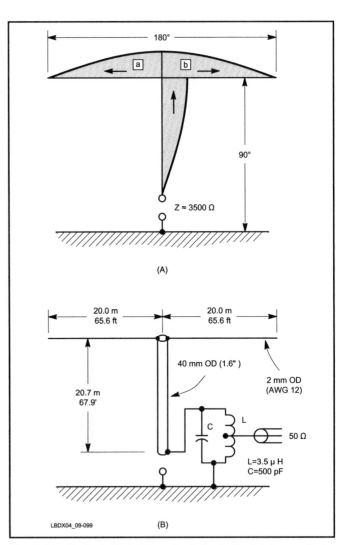

Fig 9-99—Voltage-fed 80-meter λ/4 vertical (also called inverted vertical), using a λ/2 long top-loading wire. The T wire has a twofold function—providing a low impedance at the top of the vertical, and having a configuration whereby horizontally polarized radiation is essentially canceled (area A = area B, hence no radiation). C—500 pF. L—3.5 μH.

is at the antenna top. The antenna is sometimes called an inverted vertical, as it has its current maximum at the top. In theory, the current in both halves of the flat-top wire is such that radiation from that wire is zero. (In practice there is a very small amount of horizontal radiation.) The disadvantage of this construction is that the antenna requires a very long flat-top wire. Fig 9-99 also shows the dimensions for such a vertical for a practical design on 3.5 MHz.

Hille, DL1VU, dramatically improved the T antenna by folding the λ/2 flat-top section in such a way that the radiation from the flat-top section is effectively suppressed. **Fig 9-100** shows the configuration of this antenna. It can easily be proved that the area under the current distribution line for the central part (which is λ/12 long) is the same as the area for the remaining part of the loading device (which is λ/6 long). Because of the way the wires are folded, the radiation from the horizontal loading device is effectively canceled.

The latest design of a T-type top-load by Hille requires only a single λ/4 flat top. To cancel all possible horizontal radiation from this flat-top wire, the λ/4 is folded back as shown in **Fig 9-101**. Notice that the top load is asymmetrical.

A single quarter-wave flat top acts as a short circuit at the top of the vertical, the same way that radials provide a low-impedance attachment point for the outer conductor of the coax feed line in the case of a ground-plane antenna.

Hille also described a vertical with a physical length of only 0.39 λ, using the λ/4-long top-load wire configuration described above (Ref 7991). This antenna produces the same field strength as a 5λ/8 (0.64-λ) vertical antenna.

The T antenna can also be seen as a Bobtail Curtain antenna with the two vertical end sections missing. As such, this antenna is a poor performer with respect to the Bobtail antenna, where the directivity and gain is obtained through the use of three vertical elements.

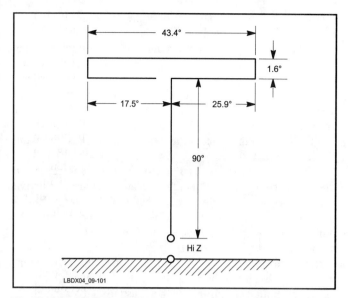

Fig 9-101—Voltage-fed T antenna with a λ/4 long top load, arranged in such a way that there is no radiation from the flat-top section.

8.2.1. Feeding the antenna

The voltage-fed T antenna can best be fed by means of a parallel tuned circuit (see Fig 9-99). You can either tap the coax on the coil for the lowest SWR point or tap the antenna near the top of the coil. Either method is valid.

8.2.2. The required ground and radial system

The ground and radial requirements are identical to those required for a λ/2 vertical (see Section 4.3).

8.3. Close Spaced Short Vertical and Reduced Losses

If you are in a situation where you cannot put down a good radial system (such as 100 λ/4 radials) for your short verticals, but you can erect several of those verticals close together (eg, with λ/16 spacing), this can be a way of improving the efficiency of your antenna.

John, W1FV, wrote: *"I've been doing this for years on 160 with three 60-foot verticals (actually my 80-meter vertical system) spaced 35 feet apart. When fed in-phase, the feed-point radiation resistance at each vertical is around 18 ohms without a top hat. A single 60-foot vertical system has a radiation resistance of around one third of that. When the total system resistive loss (ground loss plus other component losses) is high (much bigger than 5-6 ohms), the efficiency of the three vertical system would be improved as much as a factor of three (5 dB) over a single vertical. For two in-phase verticals, the improvement would be around 3 dB. When the system loss starts out low and the single vertical efficiency is pretty good, there is obviously less to be gained, but that's also true when other loading schemes are used with short verticals."*

This concept of using close-spaced in-phase verticals dates from 1920 (described in Jasik's *Antenna Engineering Handbook*, 1st Edition, page 19-9). Ground loss remains constant for a given area of ground system and antenna, because the sum of currents from each vertical flowing into that fixed

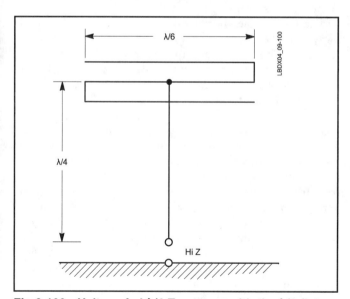

Fig 9-100—Voltage-fed λ/4 T antenna with the λ/2 flat-top wire folded to have a total span of only λ/6. The current distribution in the folded top load is such that radiation from the top load is effectively canceled. The advantage of this design over the original voltage-fed T antenna is that it requires a much shorter top-load space.

size ground system remains exactly the same no matter how many verticals are added.

The only improvement occurs when multiple antennas are far enough apart so that return currents at the base of one vertical are not influenced by currents from another vertical. This means that each antenna must have a small ground system area, well separated from the ground systems of the other verticals. The end result, however, is no better than making a single large ground system the exact size of the sum of the small ground systems.

To have an efficiency advantage in such a configuration there must be loss present in the verticals, including resistive ground loss (for example, a very small radial system). Resistive loss is proportional to current squared, so reducing current for constant power reduces loss. Since the drive currents to the verticals are split equally (in the ideal case) between N verticals, the current per vertical is 1/N times the current that would flow in one of the verticals by itself. This neglects the effects of mutual coupling, which are usually rather insignificant between short monopoles. This means the loss per vertical is $(1/N)^2$ times the loss of the single vertical. Since there are N verticals, the net system loss is N times $(1/N)^2$, or just 1/N times the loss of one vertical. This can be a significant improvement over a single vertical that would otherwise be lossy or inefficient by itself.

Tom, W8JI pointed out a possible application: *"Where this would help is when a driveway would be in the middle of an area, and you couldn't cross the driveway with radials. You could build an antenna consisting of two verticals, with one on either side of the driveway, and separate "half" ground systems on either side that are not connected. In this example efficiency would be identical to a single vertical in the middle of the driveway with a full radial system that covers exactly the same physical area, but you can still have a driveway."*

Another application of the principle would be where you would use four 80-meter verticals forming an 80-m Four Square, each vertical using a radial system designed for 80 meters, and where you would feed all four verticals in phase. That system acts like a single vertical of the same height placed in the exact middle of the 80-meter array. That would be better than feeding only one element at the edge of the ground system. But it gains nothing over a single vertical loaded the same way with the same area ground system, except convenience. W8JI pointed out: *"A Four Square works the same way. The center two elements combine to effectively make one element in the middle of the array. That is why we can feed a four-square with a 1:1:1:1 current ratio when a three-element array requires a 1:2:1 ratio! The center two elements (being in-phase) form one "radiation fat" element."*

W8JI concluded saying: *"If it were a 160-only array, he almost certainly would be better off putting the same effort into a single vertical and one big ground system covering the same overall area. RCA found this to be true in an actual test at a VLF station, where they initially used multiple antenna elements over multiple distributed grounds to obtain the same 1/N efficiency as described above. When they pulled the multiple elements and the multiple independent ground systems out and replaced everything with a conventional system of radials filling the same area, efficiency actually went up a considerable amount (and they got rid of many maintenance headaches). Tom also points out that in recent tests (on VLF) the Air Force did at Marion the conclusion was a normal large radial ground system resulted in considerably less ground loss than had previously been obtained with a combination of multiple verticals using independent smaller ground systems, with a complex overhead distribution and equalizing system."*

9. LOCATION OF A VERTICAL ANTENNA

Let's tackle the often-asked question, "Will a vertical work in my particular location?" Verticals for working DX on the low bands are certainly not space-saving antennas but to the contrary, require a lot of space and a good ground. Many low-band DXers have wondered why some verticals don't work well at all, while others work "like gangbusters." The poor performers generally have the poor locations. To repeat, a vertical is not a space-saving antenna! A good vertical takes a lot of real estate. In addition, it must be real estate with a good RF ground!

The standard for buried radials is that for best radiation efficiency you need 120 λ/2 radials. This means that for 80 meters, you need about an 80×80-meter lot in which to place all the radials. The radials are there to provide a low-resistance return path for the antenna current to achieve good efficiency.

The area beyond the ends of the radials is at least as important, because that's where the low-angle reflection at ground level takes place (the Fresnel zone). This is where the reflection efficiency is determined.

Up to λ/2 away from the vertical, most of the reflection will take place that is responsible for the 25° radiation (main angle) of a typical λ/4 vertical over average ground. Therefore, beyond this point, a clear path should be available for these low-angle rays to obtain maximum low-angle radiation. It is clear that for even lower angles of radiation, the ground at even greater distances becomes important. As explained earlier, this is even more so with "long" verticals (eg, λ/2 vertical), where the Fresnel reflection takes place up to 100 λ away from the antenna (for wave angles down to 0.25°).

To avoid excessive absorption verticals should be kept at least λ/4 away from residential houses. This means, for instance, that at a point 60 meters from a 3.5-MHz antenna, the maximum height of a structure should be limited to 9.1 meters. What about trees closer in? Trees can be reasonably good conductors and can be very lossy elements in the near field of a radiator. A case has been reported in the literature where a λ/4 vertical with an excellent ground system showed a much lower radiation resistance than expected. It was found that trees in the immediate area were coupling heavily with the vertical and were causing the radiation resistance of the vertical to be very low. Under such circumstances of uncontrolled coupling into very lossy elements, far from optimum performance can be expected. Of course, if the trees are short in relation to the quarter wavelength, it is reasonable to assume that the result of such coupling will be minimal.

Even though neighboring (lossy) structures such as trees may not be resonant, they will always absorb some RF to an unknown degree. Other objects that are likely to affect the performance of a vertical are nearby antennas and towers. Mutual coupling can be considered the culprit if the radiation resistance of the vertical is lower than expected. Another way of checking for coupling with other antennas is to alternately open and short-circuit the suspected antenna feed lines while watching the SWR or the radiation resistance of the vertical antenna. If there is any change, you are in trouble. Checking for resonance of towers has been described in Section 7.

It may come as a surprise that a vertical is so demanding of space. Most amateur verticals are not anywhere near ideal, yet good performance can still be obtained from practical setups. But the builder of a vertical should understand which factors are important for optimum performance, and why.

10. 160-METER DXPEDITION ANTENNAS

I have talked at great length with well-known DXpeditioners who have been especially successful on the low bands. I'd like to share the following rules with candidate DXpeditioners with respect to the low bands.

If you're on an island, erect the station on that side of the island where you will have the most difficult propagation path or where you are facing the most stations (eg, if you are on an island in the South Indian Ocean, try a shore on the northwest side of the island, looking into both Europe and North America). By all means erect the antenna very close to saltwater, or over (or in) saltwater. This will help you lower the pseudo-Brewster angle, and will ensure a good low-angle take off.

Unless you have a very tall support of at least 30 meters, use a vertical. Good choices are the Battle Creek Special, the BC Trapper, the AKI Special, the Titanex V160E or any inverted L, for which you should try to make the vertical part as long as possible. The vertical section should be at least 15-meters tall, 12 meters being an absolute minimum for 160 meters. If there are some trees, you may try to climb a tall tree, and use a collapsible fiberglass fishing rod (they exist in 12-meter lengths) to extend the effective support height. Use as many radials as you can, and let them run into the salt water. Very thin wire is just fine if you use many (current is shared by the many wires). A small spool of #28 enameled copper magnet wire can hold a lot of wire and takes little space. Equally as good is to use two in-line elevated radials. These make switching from the CW to the phone band very simple by merely adjusting the length of the two radials.

The Titanex verticals are very special in that they are made of an aluminum-titanium alloy that is very strong and extremely lightweight. The model V160E vertical is a 26.7-meter long vertical that weighs only 7.5 kg. See **Fig 9-102**. The maximum section length is only 2.1 meters, and the total antenna can easily be erected by two to three persons. This, as well as the low weight, makes this a very attractive antenna for DXpeditions. The guy wires are 2-mm Kevlar, and guys are placed at heights of 6, 9, 12, 15 and 18 meters. The upper 8 meters of the vertical swings freely in the wind. With a total length of 26 meters, this antenna has a very respectable radiation resistance of 12 Ω, which is 50% higher than that of the Battle Creek Special (which is 10 meters shorter!). The antenna is 3λ/8 on 80, and 5λ/8 on 40. Also on 40 this should make it a killer antenna if erected over saltwater. The V80E vertical measures only 20-meters tall, which is good for a R_{rad} of about 8.5 Ω, which is similar to the R_{rad} of the Battle Creek Special. Titanex also provides a three-band relay-switched matchbox providing a 1:1 SWR on the three low bands. More info at **www.qth.com/titanex**. The Titanex antennas are expensive mechanical marvels but they have been used

Fig 9-102—The Titanex V160E antenna on the beach on VK9CR, surrounded by beautiful coconut trees. This 26.8-meter-long special DXpedition vertical weighs only 7.5 kg and disassembles into 2.1-meter long sections, ideal for traveling!

extremely successfully during a number of expeditions; eg, VK9CR, VK9XY, C56CW, FW2OI, S21XX, P29VXX, DL7FD/HR3, K7K, K4M, T31BB, 9M0C, TJ1GB, ZL7DK, YJ0ADJ, FO0FI, FO0FR, and 3B7RF.

Don't bother putting up a Beverage near the sea; it won't work well. The VK0IR guys did not believe me. They put one up; it never worked. Anyhow, it's unlikely you will have to deal with a lot of local QRM or man-made noise, which makes the use of directive antennas pretty senseless. If you do need directivity, try an EWE or a K9AY loop, which are receiving antennas that actually work better over good ground.

If there is a tall support, you may want to use a sloping half-wave vertical, especially if you are near the sea (see Chapter 8 on dipole antennas). The sloping vertical builds up its image as far as 100 λ away from the antenna. If there is no saltwater nearby and ground conductivity is poor, use a high support for an inverted-V dipole. Don't try an inverted V or any other horizontally polarized antenna at a height of 15 meters or less. All you will get is very high-angle radiation.

Here is a hint from Rudi, DK7PE: If you are on a DXpedition in a country with a substantial tourist business, choose the tallest hotel (Hilton, Sheraton or Intercontinental hotels usually do well in this respect). Slope a dipole from the

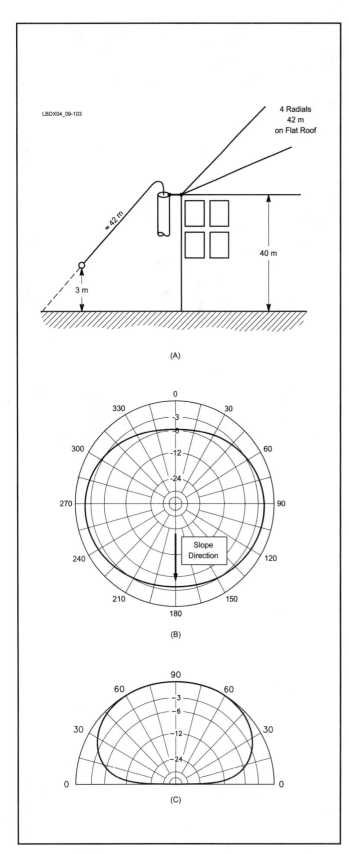

Fig 9-103—Configuration and radiation patterns of the inverted λ/4 sloper antenna used by 9M2AX. The azimuth pattern is shown at B for an elevation angle of 30°, and at C is the elevation pattern. (The elevation pattern is taken in the 90-270° direction as displayed in the azimuth pattern.) Note the relative high amount of high-angle radiation. Using just two radials in-line would improve this situation considerably.

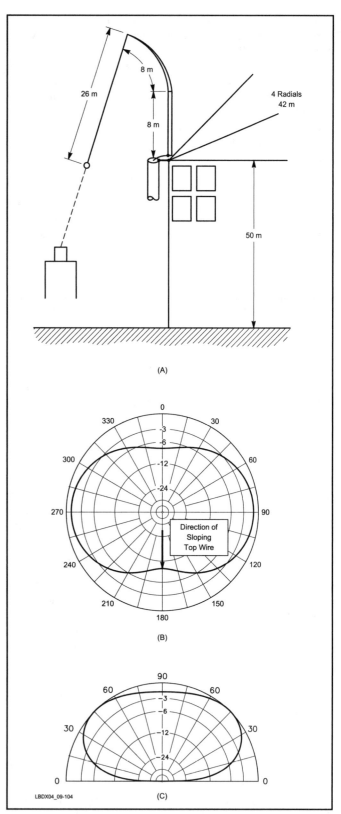

Fig 9-104—Configuration and radiation patterns of the inverted λ/4 sloper antenna used by 9M2AX during his expedition from East Malaysia (Sarawak) as 9M8AX. The azimuth pattern is shown at B, and the elevation pattern at C. (The elevation pattern is taken in the 90-270° direction as displayed in the azimuth pattern.) The antenna was installed on the edge of a 50-meter high flat roof. Four λ/4 long radials were laid on the roof. The metal mast plus the fiberglass rod are 16 meters long. The sloping wire was adjusted for minimum SWR at resonance.

top of the building to some distant point and let the feed line come to your room, which can be a few stories below the roof. Make the dipole as vertical as possible. This is by far the best antenna if you are in such a situation.

DK7PE proved it during his operation from D2CW (August 1992) where he had his sloping dipole attached some 60 meters above street level, facing north, and within 1 λ of the South Atlantic Ocean. Rudi's signals were always S9 in Europe on 160 meters. During his more recent operation from Ethiopia and Eritrea (9F2CW), he proved it again. Rudi's total antenna system for his DXpeditions (covering 160 through 10 meters) can be packed in a small handbag. The RG-58 cable takes up 80% of the volume. The antenna consists of precut lengths of flexible insulated wire, with small insulators and a variety of alligator clips that let him change bands. On the higher bands he can configure the wire into a 2-element Yagi.

R. E. Tanaka, 9M2AX, well-known 160-meter operator from the Far East, sent me the sketches of the antennas he is using in 9M2 as well as when he operated from 9M8AX. The antennas Ross was using can be put up at any tall hotel, and should be excellent suggestions for 160-meter DXpeditioners. **Figs 9-103** and **9-104** show the layouts of the two antenna setups and their radiation patterns. The radial system covering only one quadrant (90°) results in a significant high-angle radiation component with the 9M2AX version. The low-angle radiation is very pronounced as well. From modeling, the "inverted sloping wire vertical" from the 9M2 QTH has a feed-point impedance of about 75 Ω. The 9M8AX configuration is an inverted L with a sloping flat-top. The calculated impedance from modeling is nearly 60 Ω. This antenna has better low-angle radiation than the 9M2AX version, which is normal. In order to eliminate the high-angle radiation for the 9M2AX version, it would be necessary to install just two radials (in-line with one another), so that the radiation from these wires would be canceled. The radials are *not* there to provide a ground plane, but are merely serving to provide a low-impedance point to which to connect the outer shield of the feed line. One λ/4 radial would serve that purpose, but would radiate a lot of horizontal component. Two radials in-line would provide a low impedance point just as well, but would not radiate a high-angle horizontal component.

11. BUYING A COMMERCIAL VERTICAL

I sincerely hope that this chapter on verticals has incited you to build your own antenna. You cannot believe how much more satisfaction you get out of using something you made or designed yourself, rather than going to the store, opening your wallet and then playing the appliance-type ham.

Anyhow, if you choose not to make your own, here are a few rules to help you select your new low-band commercial vertical:

1. Most, if not all companies advertising their products, largely exaggerate the performance, especially if it's something different from a straightforward vertical.
2. A short vertical with a large bandwidth means there are a lot of losses. With short antennas a large bandwidth is a direct measure of its poor efficiency (lots of losses).
3. The efficiency of a vertical is in the first place determined by the physical length of the vertical (and the ground system, which you will have to install yourself anyhow).
4. Only top loading is efficient.
5. Verticals with coil loading are bound to be inefficient. An 8-meter long vertical with center-coil loading is bound to be a very poor performer on 160 meters as a transmitting antenna.
6. To be a reasonable performer a minimum physical vertical length of about 14 meters is needed on 160 meters.
7. Good hardware (stainless steel, good finishing, etc) are no guarantee for a good antenna.
8. A fancy feed system or folded elements that claims to reduce losses and increase efficiency are a total fallacy.
9. A producer of a 160-meter vertical who prescribes using only a few 10-meter-long radials does not know what he is talking about.
10. Advertisers bragging that their product is bought by government agencies are not proving anything. Remember the Maxcom "dummy load" antenna-matching network used extensively by the armed forces?
11. An advertiser specifying his 8-meter long 160-meter vertical has 75% efficiency, without specifying the ground radial system is telling you stories.
12. Advertisers selling their product by telling how many new countries one of their customers has worked with it, are… Well, you know. Maybe, with a good homemade vertical he would have worked double the number of new countries! Not very scientific advertising, anyhow.

Spending nearly $500 for a 9-meter long radiator is a heck of a lot of money. You could buy some simple aluminum tubing (TV-type push-up mast, about $70), some copper wire to make a number of top-loading wires (add another $10), some nylon guy rope (another $10), maybe an (empty) Coke bottle for an insulator (free), and you have exactly the same for about 20% of the price of the commercial thing. It won't work any better, but at least you won't feel like you've been robbed. And spending nearly $400 for an 8-meter-long 160-meter vertical, with a slim (and thus very lossy) loading coil, is even worse, of course.

Amateur Radio is a technical hobby. It is true that the progress of microelectronics has made it very difficult for the average ham to do much home designing and home building in the field of receivers and transmitters. Building antennas is one of the few fields where we can, ourselves, through our own knowledge, understanding and expertise, do as well and usually much better than the commercial companies. Let's grab this opportunity with both hands, and build our own vertical for the low bands. This will give you the ultimate kick, I promise you!

CHAPTER 10
Large Loop Antennas

The delta-loop antenna is a superb example of a high-performance compromise antenna. The single-element loop antenna is almost exclusively used on the low bands, where it can produce low-angle radiation, requiring only a single quarter-wave high support. We will see that a vertically polarized loop is really an array of two phased verticals, and that the ground requirements are the same as for any other vertically polarized antenna.

This means that with low delta loops, the horizontal wire will couple heavily to the lossy ground and induce significant losses, unless we have improved the ground by putting a ground screen under the antenna. (See Chapter 9, Section 1.3.3 and Section 2.) I have seen it stated in various places that delta loops don't require a good ground system. This is as true as saying that verticals with a single elevated radial don't require a good ground system.

Loop antennas have been popular with 80-meter DXers for more than 30 years. Resonant loop antennas have a circumference of 1 λ. The exact shape of the loop is not particularly important. In free space, the loop with the highest gain, however, is the loop with the shape that encloses the largest area for a given circumference. This is a circular loop, which is difficult to construct. Second best is the square loop (quad), and in third place comes the equilateral triangle (delta) loop (Ref 677).

The maximum gain of a 1-λ loop over a $\lambda/2$ dipole in free space is approximately 1.35 dB. Delta loops are used extensively on the low bands at apex heights of $\lambda/4$ to $3\lambda/8$ above ground. At such heights the vertically polarized loops far outperform dipoles or inverted-V dipoles for low-angle DXing, assuming good ground conductivity.

Loops are generally erected with the plane of the loop perpendicular to the ground. Whether or not the loop produces a vertically or a horizontally polarized signal (or a combination of both) depends only on how (or on which side) the loop is being fed.

Sometimes we hear about horizontal loops. These are antennas with the plane of the loop parallel to the ground. They produce horizontal radiation with takeoff angles determined, as usual, by the height of the horizontal loop over ground.

1. QUAD LOOPS

Belcher, WA4JVE, Casper, K4HKX, (Ref 1128), and Dietrich, WA0RDX, (Ref 677), have published studies comparing the horizontally polarized vertical quad loop with a dipole. A horizontally polarized quad loop antenna (**Fig 10-1A**) can be seen as two short, end-loaded dipoles stacked $\lambda/4$ apart, with the top antenna at $\lambda/4$ and the bottom one just above

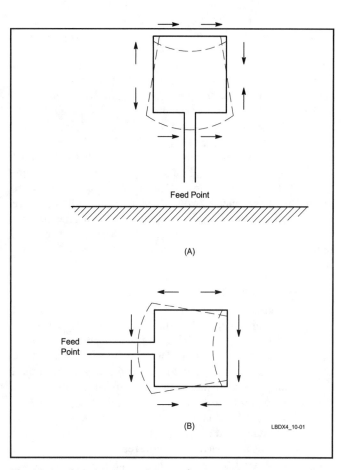

Fig 10-1—Quad loops with a 1-λ circumference. The current distribution is shown for (A) horizontal and (B) vertical polarization. Note how the opposing currents in the two legs result in cancellation of the radiation in the plane of those legs, while the currents in the other legs are in-phase and reinforce each other in the broadside direction (perpendicular to the plane of the antenna).

Large Loop Antennas   10-1

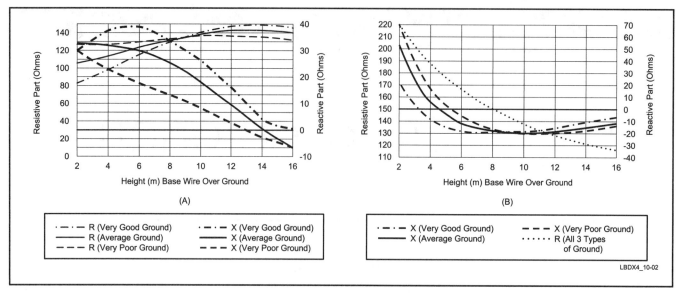

Fig 10-2—Radiation resistance and feed-point resistance for square loops at different heights above real ground. The loop was first dimensioned to be resonant in free space (reactance equal to zero), and those dimensions were used for calculating the impedance over ground. At A, for horizontal polarization, and at B, for vertical polarization. Analysis was with *NEC* at 3.75 MHz.

ground level. There is no broadside radiation from the vertical wires of the quad because of the current opposition in the vertical members. In a similar manner, the vertically polarized quad loop consists of two top-loaded, $\lambda/4$ vertical dipoles, spaced $\lambda/4$ apart. Fig 10-1B shows how the current distribution along the elements produces cancellation of radiation from certain parts of the antenna, while radiation from other parts (the horizontally or vertically stacked short dipoles) is reinforced.

The square quad can be fed for either horizontal or vertical polarization merely by placing the feed point at the center of a horizontal arm or at the center of a vertical arm. At the higher frequencies in the HF range, where the quads are typically half to several wavelengths high, quad loops are usually fed to produce horizontal polarization, although there is no specific reason for this except maybe from a mechanical standpoint. Polarization by itself is of little importance at HF (except on 160 meters! See Chapter 1), because it becomes random after ionospheric reflection.

1.1. Impedance

The radiation resistance of an equilateral quad loop in free space is approximately 120 Ω. The radiation resistance for a quad loop as a function of its height above ground is given in **Fig 10-2**. The impedance data were obtained by modeling an equilateral quad loop over three types of ground (very good, average and very poor ground) using *NEC*. *MININEC* cannot be used for calculating loop impedances at low heights (see Section 2.9).

The reactance data can assist you in evaluating the influence of the antenna height on the resonant frequency. The loop antenna was first modeled in free space to be resonant at 3.75 MHz and the reactance data was obtained with those free-space resonant-loop dimensions.

For the vertically polarized quad loop, the resistive part of the impedance changes very little with the type of ground under the antenna. The feed-point reactance is influenced by the ground quality, especially at lower heights. For the horizontally polarized loop, the radiation resistance is noticeably influenced by the ground quality, especially at low heights. The same is true for the reactance.

1.2. Square Loop Patterns
1.2.1. Vertical polarization

The vertically polarized quad loop, Fig 10-1B, can be considered as two shortened top-loaded vertical dipoles, spaced $\lambda/4$ apart. Broadside radiation from the horizontal elements of the quad is canceled, because of the opposition of currents in the vertical legs. The wave angle in the broadside direction will be essentially the same as for either of the vertical members. The resulting radiation angle will depend on the quality of the ground up to several wavelengths away from the antenna, as is the case with all vertically polarized antennas.

The quality of the reflecting ground will also influence the gain of the vertically polarized loop to a great extent. The quality of the ground is as important as it is for any other vertical antenna, meaning that vertically polarized loops close to the ground will not work well over poor soil.

Fig 10-3 shows both the azimuth and elevation radiation patterns of a vertically polarized quad loop with a top height of 0.3 λ (bottom wire at approximately 0.04 λ). This is a very realistic situation, especially on 80 meters. The loop radiates an excellent low-angle wave (lobe peak at approximately 21°) when operated over average ground. Over poorer ground, the wave angle would be closer to 30°. The horizontal directivity, Fig 10-3C, is rather poor, and amounts to approximately 3.3 dB of side rejection at any wave angle.

1.2.2. Horizontal polarization

A horizontally polarized quad-loop antenna (two stacked short dipoles) produces a wave angle that is dependent on the height of the loop. The low horizontally polarized quad (top at 0.3 λ) radiates most of its energy right at or near zenith angle (straight up).

Fig 10-4 shows directivity patterns for a horizontally

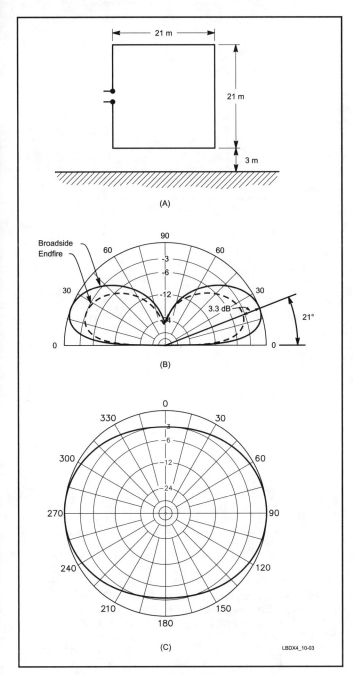

Fig 10-3—Shown at A is a square loop, with its elevation-plane pattern at B and azimuth pattern at C. The patterns are generated for good ground. The bottom wire is 0.0375 λ above ground (3 meters or 10 feet on 80 meters). At C, the pattern is for a wave angle of 21°.

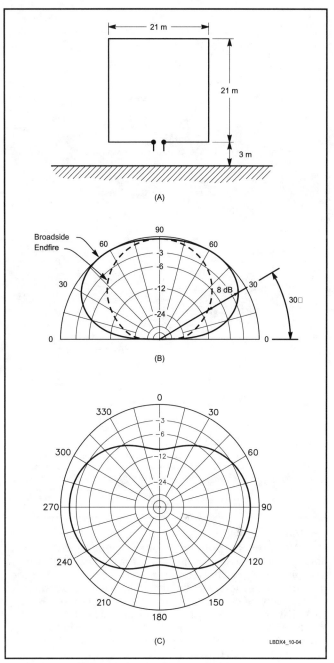

Fig 10-4—Azimuth and elevation patterns of the horizontally polarized quad loop at low height (bottom wire 0.0375 λ above ground). At an elevation angle of 30°, the loop has a front-to-side ratio of approximately 8 dB.

polarized loop. The horizontal pattern, Fig 10-4C, is plotted for a takeoff angle of 30°. At low wave angles (20° to 45°), the horizontally polarized loop shows more front-to-side ratio (5 to 10 dB) than the vertically polarized rectangular loop.

1.2.3. Vertical versus horizontal polarization

Vertically polarized loops should be used only where very good ground conductivity is available. From **Fig 10-5A** we see that the gain of the vertically polarized quad loop, as well as the wave angle, does not change very much as a function of the antenna height. This makes sense, since the vertically polarized loop is in the first place two phased verticals, each with its own radial. However, the gain is drastically influenced by the quality of the ground. At low heights, the gain difference between very poor ground and very good ground is a solid 5 dB! The wave angle for the vertically polarized quad loop at a low height (bottom wire at 0.03 λ) varies from 25° over very poor ground to 17° over very good ground.

I have frequently read in Internet messages that a delta loop has certain advantages over a vertical antenna (or arrays of vertical antennas) since the loop antenna does not require

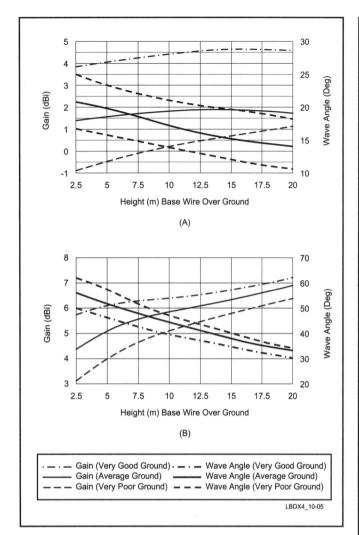

Fig 10-5—Radiation angle and gain of the horizontally and the vertically polarized square loops at different heights over good ground. At A, for vertical polarization, and at B, for horizontal polarization. Note that the gain of the vertically polarized loop never exceeds 4.6 dBi, but its wave angle is low for any height (14 to 20°). The horizontally polarized loop can exhibit a much higher gain provided the loop is very high. Modeling was done over average ground for a frequency of 3.75 MHz, using *NEC*.

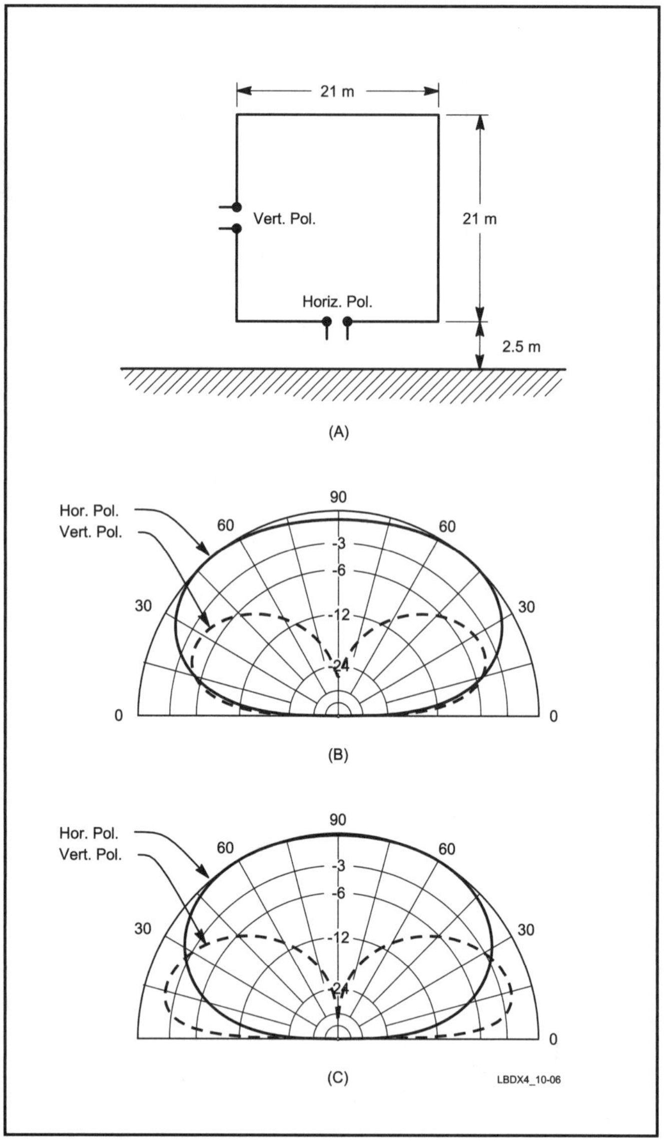

Fig 10-6—Superimposed (same dB scale) patterns for horizontally and vertically polarized square quad loops (shown at A) over very poor ground (B) and very good ground (C). In the vertical polarization mode the ground quality is of utmost importance, as it is with all verticals. See also Fig 10-14.

any radials. This statement is really quite misleading—It is like saying that a vertical with elevated radials does not require any radials. Indeed, in a delta loop (and a quad loop), the "element" that takes care of the return current is part of the antenna itself just like with a dipole!

With a horizontally polarized quad loop the wave angle is very dependent on the antenna height, but not so much by the quality of the ground. At very low heights, the main wave angle varies between 50° and 60° (but is rather constant all the way up to 90°), but these are rather useless radiation angles for DX work.

As far as gain is concerned, there is a 2.5-dB gain difference between very good and very poor ground, which is only half the difference we found with the vertically polarized loop. Comparing the gain to the gain of the vertically polarized loop, we see that at very low antenna heights the gain is about 3-dB better than for the vertically polarized loop. But this gain exists at a high wave angle (50° to 90°), while the vertically polarized loop at very low heights radiates at 17° to 25°.

Fig 10-6 shows the vertical-plane radiation patterns for both types of quad loops over very poor ground and over very good ground on the same dB scale. For more details see Section 2.3.

1.3. A Rectangular Quad Loop

A rectangular quad loop, with unequal side dimensions, can be used with very good results on the low bands. An impressive signal used to be generated by 5N0MVE from Nigeria with such a loop antenna. The single quad-loop element is strung between two 30-meter high coconut trees, some 57 meters apart in the bush of Nigeria. 5N0MVE fed the loop in the center of one of the vertical members. He first tried to feed it for horizontal polarization but he says it did not work well. The vertical and the horizontal radiation patterns for this quad loop over good ground are shown in **Fig 10-7**. The horizontal directivity is approximately 6 dB (front-to-side ratio).

Even in free space, the impedance of the two varieties of this rectangular loop is not the same. When fed in the center of a short (27-meter) side, the radiation resistance at resonance is 44 Ω. When fed in the center of one of the long (57-meter) sides, the resistance is 215 Ω. Over real ground the feed-point impedance is different in both configurations as well; depending on the quality of the ground, the impedance varies between 40 and 90 Ω.

1.4. Loop Dimensions

The total length for a resonant loop is approximately 5 to 6% longer than the free-space wavelength.

1.5. Feeding the Quad Loop

The quad loop feed point is symmetrical, whether you feed the quad in the middle of the vertical or the horizontal wire. A balun should be used. Baluns are described in Chapter 6 on matching and feed lines.

Alternatively, you could use open-wire feeders (for example, 450-Ω line). The open-wire-feeder alternative has the advantage of being a lightweight solution. With a tuner you will be able to cover a wide frequency spectrum with no compromises.

2. DELTA LOOPS

Just as the inverted-V dipole has been described as the poor man's dipole, the delta loop can be called the poor man's quad loop. Because of its shape, the delta loop with the apex on top is a very popular antenna for the low bands; it needs only one support.

In free space the equilateral triangle produces the highest gain and the highest radiation resistance for a three-sided loop configuration. As we deviate from an equilateral triangle toward a triangle with a long baseline, the effective gain and the radiation resistance of the loop will decrease for a bottom-corner-fed delta loop. In the extreme case (where the height of the triangle is reduced to zero), the loop has become a half-wavelength-long transmission line that is shorted at the end, which shows a zero-Ω input impedance (radiation resistance), and thus zero radiation (well-balanced open-wire line does not radiate).

Just as with the quad loop, we can switch from horizontal to vertical polarization by changing the position of the feed point on the loop. For horizontal polarization the loop is fed either at the center of the baseline or at the top of the loop. For vertical polarization the loop should be fed on one of the sloping sides, at λ/4 from the apex of the delta. **Fig 10-8** shows the current distribution in both cases.

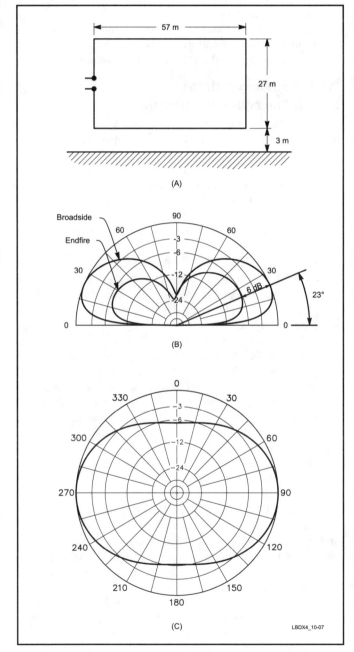

Fig 10-7—At A, a rectangular loop with its baseline approximately twice as long as the vertical height. At B and C, the vertical and horizontal radiation patterns, generated over good ground. The loop was dimensioned to be resonant at 1.83 MHz. The azimuth pattern at C is taken at a 23° elevation angle.

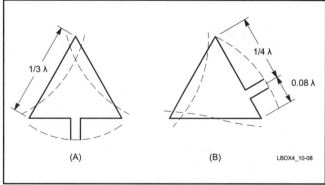

Fig 10-8—Current distribution for equilateral delta loops fed for (A) horizontal and (B) vertical polarization.

2.1. Vertical Polarization
2.1.1. How it works

Refer to **Fig 10-9**. In the vertical-polarization mode the delta loop can be seen as two sloping quarter-wave verticals (their apexes touch at the top of the support), while the baseline (and the part of the sloper under the feed point) takes care of feeding the "other" sloper with the correct phase. The top connection of the sloping verticals can be left open without changing anything about the operation of the delta loop. The same is true for the baseline, where the middle of the baseline could be opened without changing anything. These two points are the high-impedance points of the antenna. Either the apex or the center of the baseline must be shorted, however, in order to provide feed voltage to the other half of the antenna. Normally, of course, we use a fully closed loop in the standard delta loop, although for single-band operation this is not strictly necessary.

Assume we construct the antenna with the center of the horizontal bottom wire open. Now we can see the two half baselines as two λ/4 radials, one of which provides the necessary low-impedance point for connecting the shield of the coax. The other radial is connected to the bottom of the second sloping vertical, which is the other sloping wire of the delta loop.

This is similar to the situation encountered with a λ/4 vertical using a single elevated radial (see Chapter 9 on vertical antennas). The current distribution in the two quarter-wave radials is such that all radiation from these radials is effectively canceled. The same situation exists with the voltage-fed T antenna, where we use a half-wave flat top (equals two λ/4 radials) to provide the necessary low-impedance point to raise the current maximum to the top of the T antenna.

The vertically polarized delta loop is really an array of two λ/4 verticals, with the high-current points spaced 0.25 λ to 0.3 λ, and operating in phase. The fact that the tops of the verticals are close together does not influence the performance to a large degree. The reason is that the current near the apex of the delta is at a minimum (it is current that takes care of radiation!).

Considering a pair of phased verticals, we know from the study on verticals that the quality of the ground will be very important as to the efficient operation of the antenna:

This does not mean that the delta loop requires radials. It has two elevated radials that are an integral part of the loop and take care of the return currents. The presence of the (lossy) ground under the antenna is responsible for near-field losses, unless we can shield it from the antenna by using a ground screen or a radial system, which should not be connected to the antenna.

As with all vertically polarized antennas the quality of the ground within a radius of several wavelengths will determine the low-angle radiation of the loop antenna.

2.1.2. Radiation patterns
2.1.2.1. The equilateral triangle

Fig 10-10 shows the configuration as well as both the broadside and the end-fire vertical radiation patterns of the vertically polarized equilateral-triangle delta loop antenna. The model was constructed for a frequency of 3.75 MHz. The baseline is 2.5 meters above ground, which puts the apex at 26.83 meters. The model was made over good ground. The delta loop shows nearly 3 dB front-to-side ratio at the main wave angle of 22°. With average ground the gain is 1.3 dBi.

2.1.2.2. The compressed delta loop

Fig 10-11 shows an 80-meter delta loop with the apex at 24 meters and the baseline at 3 meters. This delta loop has a long baseline of 30.4 meters. The feed point is again located λ/4 from the apex.

The front-to-side ratio is 3.8 dB. The gain with average ground is 1.6 dBi. In free space the equilateral triangle gives a higher gain than the "flat" delta. Over real ground and in the vertically polarized mode, the gain of the flat delta loop is 0.3 dB better than the equilateral delta, however. This must be explained by the fact that the longer baseline yields a wider separation of the two "sloping" verticals, yielding a slightly higher gain.

For a 100-kHz bandwidth (on 80 meters) the SWR rises to 1.4:1 at the edges. The 2:1-SWR bandwidth is approximately 175 kHz.

Bill, W4ZV, used what he calls a "squashed" delta loop very successfully on 160 meters. The apex is 36-meters high and Bill claims that this configuration actually has improved gain over the equilateral delta loop, which can indeed be verified by accurate modeling. The antenna was also fed a λ/4 from the apex, using a λ/4 75-Ω matching stub. Bill says that this loop can actually be installed on a 27-meter tower by pulling the base away from the tower. By pulling the base away about 8 meters from a tower, you can actually use a full-wave delta on a 24-meter high tower, with very little trade-off.

2.1.2.3. The bottom-corner-fed delta loop

Fig 10-12 shows the layout of the delta loop being fed at one of the two bottom corners. The antenna has the same

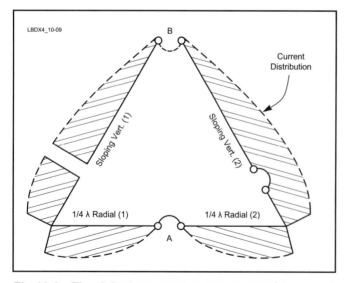

Fig 10-9—The delta loop can be seen as two λ/4 sloping verticals, each using one radial. Because of the current distribution in the radials, the radiation from the radials is effectively canceled.

apex and baseline height as the loop described in Section 2.1.2.2. Because of the "incorrect" location of the feed point, cancellation of radiation from the base wire (the two "radials") is not 100% effective, resulting in a significant horizontally polarized radiation component. The total field has a very uniform gain coverage (within 1 dB) from 25° to 90°. This may be a disadvantage for the rejection of high-angle signals when operating DX at low wave angles.

Due to the incorrect feed-point location, the end-fire radiation (radiation in line with the loop) has become asymmetrical. The horizontal radiation pattern shown in Fig 10-12D is for a wave angle of 29°. Note the deep side null (nearly 12 dB) at that wave angle. The loop actually radiates its maximum signal about 18° off the broadside direction.

All this is to explain that this feed-point configuration (in the corner of the compressed loop) is to be avoided, as it really deteriorates the performance of the antenna.

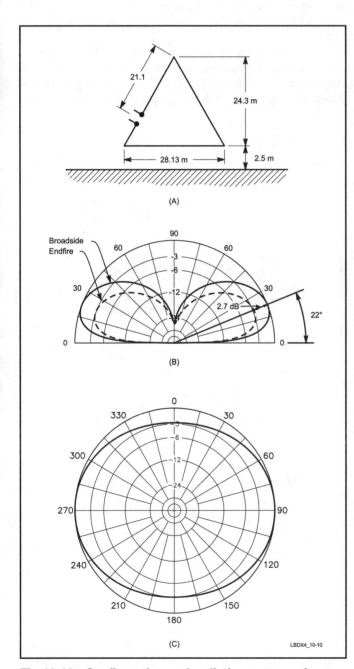

Fig 10-10—Configuration and radiation patterns for a vertically polarized equilateral delta loop antenna. The model was calculated over good ground, for a frequency of 3.8 MHz. The elevation angle for the azimuth pattern at C is 22°.

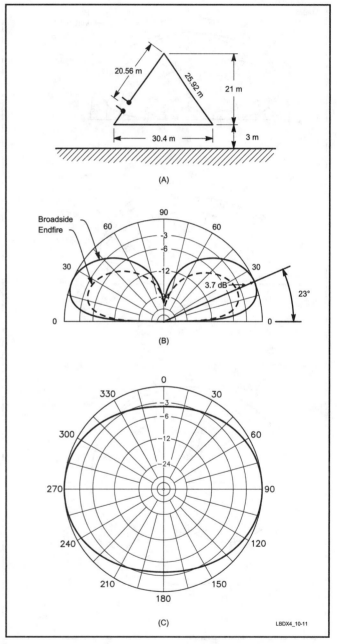

Fig 10-11—Configuration and radiation patterns for the "compressed" delta loop, which has a baseline slightly longer than the sloping wires. The model was dimensioned for 3.8 MHz to have an apex height of 24 meters and a bottom wire height of 3 meters. Calculations are done over good ground at a frequency of 3.8 MHz. The azimuth pattern at C is for an elevation angle of 23°. Note that the correct feed point remains at λ/4 from the apex of the loop.

2.2. Horizontal Polarization

2.2.1. How it works

In the horizontal polarization mode, the delta loop can be seen as an inverted-V dipole on top of a very low dipole with its ends bent upward to connect to the tips of the inverted V. The loop will act as any horizontally polarized antenna over real ground; its wave angle will depend on the height of the antenna over the ground.

2.2.2. Radiation patterns

Fig 10-13 shows the vertical and the horizontal radiation patterns for an equilateral-triangle delta loop, fed at the center of the bottom wire. As anticipated, the radiation is maximum at zenith. The front-to-side ratio is around 3 dB for a 15 to 45° wave angle. Over average ground the gain is 2.5 dBi.

Looking at the pattern shape, one would be tempted to say that this antenna is no good for DX. So far we have only spoken about relative patterns. What about real gain figures from the vertically and the horizontally polarized delta loops?

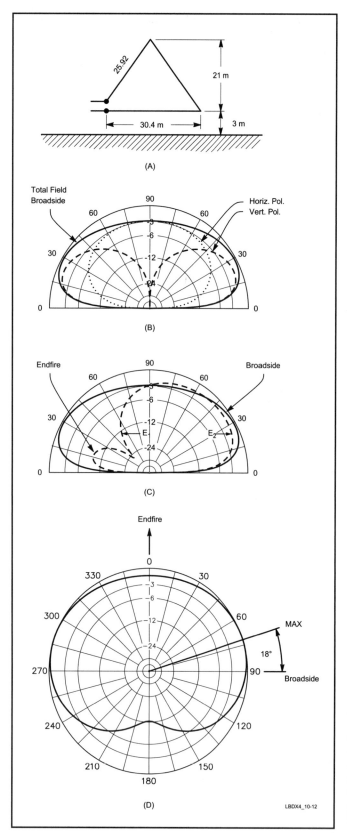

Fig 10-12—Configuration and radiation patterns for the compressed delta loop of Fig 10-11 when fed in one of the bottom corners at a frequency of 3.75 MHz. Improper cancellation of radiation from the horizontal wire produces a strong high-angle horizontally polarized component. The delta loop now also shows a strange horizontal directivity pattern (at D), the shape of which is very sensitive to slight frequency deviations. This pattern is for an elevation angle of 29°.

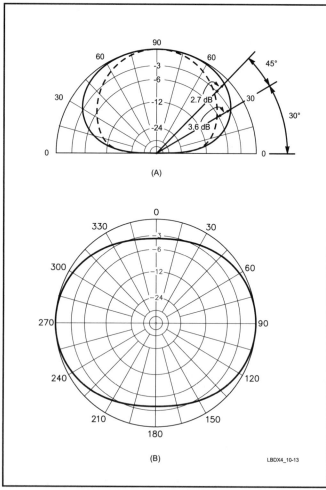

Fig 10-13—Vertical and horizontal radiation patterns for an 80-meter equilateral delta loop fed for horizontal polarization, with the bottom wire at 3 meters. The radiation is essentially at very high angles, comparable to what can be obtained from a dipole or inverted-V dipole at the same (apex) height.

2.3. Vertical Versus Horizontal Polarization

Fig 10-14 shows the superimposed elevation patterns for vertically and horizontally polarized low-height equilateral-triangle delta loops over two different types of ground (same dB scale). *MININEC*-based modeling programs cannot be used to compute the gain figures of these loops, since impedance and gain figures are incorrect for very low antenna heights.

2.3.1. Over very poor ground

The horizontally polarized delta loop is better than the vertically polarized loop for all wave angles above 35°. Below 35° the vertically polarized loop takes over, but quite marginally. The maximum gain of the vertically and the horizontally polarized loops differs by only 2 dB, but the big difference is that for the horizontally polarized loop, the gain occurs at almost 90°, while for the vertically polarized loop it occurs at 25°.

One might argue that for a 30° elevation angle, the horizontally polarized loop is as good as the vertically polarized loop. It is clear, however, that the vertically polarized antenna gives good high-angle rejection (rejection against local signals), while the horizontally polarized loop will not.

2.3.2. Over very good ground

The same thing that happens with any vertical happens with our vertically polarized delta: The performance at low angles is greatly improved with good ground. The vertically polarized loop is still better at any wave angle under 30° than when horizontally polarized. At a 10° radiation angle, the difference is as high as 10 dB. We have learned, in Chapter 5, that on the low bands very low angles (down to just a few degrees) are often involved, and in that respect the vertically polarized delta over good ground is far superior.

2.3.3. Conclusion

Over very poor ground, the vertically polarized loops do not provide much better low-angle radiation when compared to the horizontally polarized loops. They have the advantage of giving substantial rejection at high angles, however.

Over good ground, the vertically polarized loop will give up to 10-dB and more gain at low radiation angles as compared to the horizontally polarized loop, in addition to its high-angle rejection. See Fig 10-14B.

2.4. Dimensions

The length of the resonant delta loop is approximately 1.05 to 1.06 λ. When putting up a loop, cut the wire at 1.06 λ, check the frequency of minimum SWR (it is always the resonant frequency), and trim the length. The wavelength is given by

$$\lambda = \frac{299.8}{f \text{ (MHz)}} \quad \text{(Eq 1)}$$

2.5. Feeding The Delta Loop

The feed point of the delta loop in free space is symmetrical. At high heights above ground the loop feed point is to be considered as symmetrical, especially when we feed the loop in the center of the bottom line (or at the apex), because of its full symmetry with respect to the ground.

Fig 10-15 shows the radiation resistance and reactance for both the horizontally and the vertically polarized equilateral delta loops as a function of height above ground. At low heights, when fed for vertical polarization, the feed point is to be considered as asymmetric, whereby the "cold" point is the point to which the "radials" are connected. The center conductor of a coax feed line goes to the sloping vertical section. Many users have, however, used (symmetric) open-wire line to feed the vertically polarized loop (eg, 450-Ω line).

Most practical delta loops show a feed-point impedance between 50 and 100 Ω, depending on the exact geometry and coupling to other antennas. In most cases the feed point can be reached, so it is quite easy to measure the feed-point impedance using, for example, a good-quality noise bridge connected directly to the antenna terminals. If the impedance is much higher than 100 Ω (equilateral triangle), feeding via a 450-Ω open-wire feeder may be warranted. Alternatively, you could use an unun (unbalanced-to-unbalanced) transformer, which can be made to cover a very wide range of impedance ratios (see Chapter 6 on feed lines and matching). With somewhat compressed delta loops, the feed-point impedance is usually between 50 and 100 Ω. Feeding can be done directly with a 50 or 70-Ω coaxial cable, or with a 50-Ω cable via a 70-Ω quarter-wave transformer (Zant = 100 Ω).

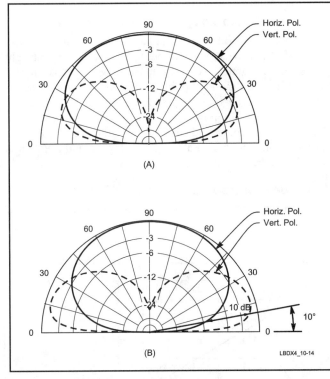

Fig 10-14—Radiation patterns of vertically and horizontally polarized delta loops on the same dB scale. At A, over very poor ground, and at B, over very good ground. These patterns illustrate the tremendous importance of ground conductivity with vertically polarized antennas. Over better ground, the vertically polarized loop performs much better at low radiation angles, while over both good and poor ground the vertically polarized loop gives good discrimination against high-angle radiation. This is not the case for the horizontally polarized loop.

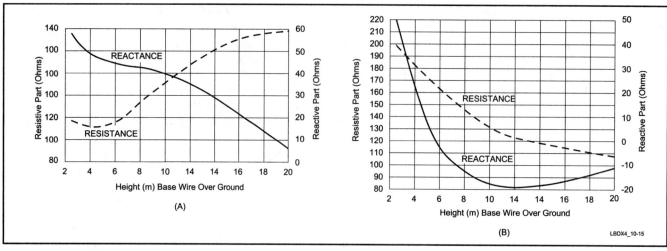

Fig 10-15—Radiation resistance of (A) horizontally and (B) vertically polarized equilateral delta loops as a function of height above average ground. The delta loop was first dimensioned to be resonant in free space (reactance equals zero). Those dimensions were then used for calculating the impedance over real ground. Modeling was done at 3.75 MHz over good ground, using *NEC*.

To keep RF off the feed line it is best to use a balun, although the feed point of the vertically polarized delta loop is not strictly symmetrical. In this case, however, we want to keep any RF current from flowing on the outside of the coaxial feed line, as these parasitic currents could upset the radiation pattern of the delta loop. A stack of toroidal cores on the feed line near the feed point, or a coiled-up length of transmission line (making an RF choke with an impedance of approximately 1000 Ω) will also be useful. For more details refer to Section 7 of Chapter 6 on feed lines and matching.

2.6. Gain and Radiation Angle

Fig 10-16 shows the gain and the main-lobe radiation angle for the equilateral delta loop at different heights. The values were obtained by modeling a 3.8-MHz loop over average ground using *NEC*.

Earl Cunningham, K6SE, investigated different configurations of single element loops for 160 meters, and came up with the results listed in **Table 10-1** (modeling done with *EZNEC* over good ground). These data correspond surprisingly well with those shown in Fig 10-16 (where the ground was average), which explains the slight difference in gain.

2.7. Two Delta Loops at Right Angles

If the 4 to 5 dB front-to-side ratio bothers you, and if you have sufficient space, you can put up two delta loops at right angles on the same tower. You must, however, make provisions to open up the feed point of the antenna not in use, as well as its apex. This results in two non-resonant wires that do not influence the loop in use. If you would leave the unused delta loop in its connected configuration, the two loops would influence one another to a very high degree, and the results would be very disappointing.

2.8. Loop Supports

Vertically polarized loop antennas are really an array of

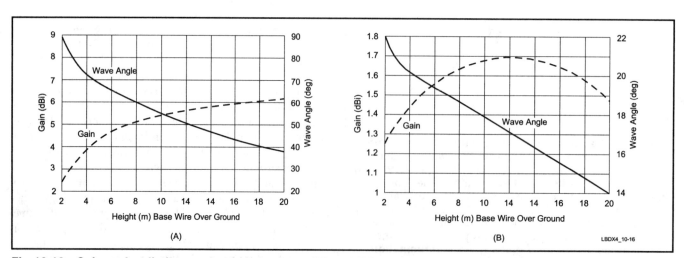

Fig 10-16—Gain and radiation angle of (A) horizontally and (B) vertically polarized equilateral delta loops as a function of the height above ground. Modeling was done at 3.75 MHz over average ground, using *NEC*.

Table 10-1
Loop Antennas for 160 Meters

Description	Feeding Method	Gain dBi	Elevation Angle
Diamond loop, bottom 2.5 m high	In side corner	2.15 dBi	18.0°
Square loop, bottom 2.5 m high	In center of one vertical wire	2.06 dBi	20.5°
Inverted equilateral delta loop (flat wire on top)	Fed λ/4 from bottom	1.91 dBi	20.9°
Regular equilateral delta loop	Fed λ/4 from top	1.90 dBi	18.1°

two (sloping) verticals, each with an elevated radial. This means that if you support the delta loop from a metal tower, this tower may well influence the radiation pattern of the loop if it resonated anywhere in the vicinity of the loop. You can investigate this by modeling, but when the tower is loaded with Yagis, it is often difficult to exactly model the Yagis and their influence on the electrical length of the tower.

The safest thing you can do is to detune the tower to make sure the smallest possible current flows in the structure. The easiest way I found to do this is to drop a wire from the top of the tower, parallel with the tower (at 0.5 to 1.0 meter distance) and terminate this wire via a 2000 pF variable capacitor to ground. Next use a current probe (such as is shown in Fig 11-17 in the chapter on Vertical Arrays) and adjust the variable capacitor for *maximum* current while transmitting on the delta loop. The capacitor can be replaced with a fixed one (using a parallel combination of several values, if necessary) having the same value. This procedure will guarantee minimum mutual coupling between the loop and the supporting tower.

2.9. Modeling Loops

Loops can be modeled with *MININEC* or *NEC-2* when it comes to radiation patterns. Because of the acute angles at the corners of the delta loop, special attention must be paid to the length of the wire segments near the corners. Wire segments that are too long near wire junctions with acute angles will cause pulse overlap (the total conductor will look shorter than it actually is). The wire segments need to be short enough in order to obtain reliable impedance results. Wire segments of 20 cm length are in order for an 80-meter delta loop if accurate results are required. To limit the total number of pulses, the segment-length tapering technique can be used: The segments are shortest near the wire junction, and get gradually longer away from the junction. *ELNEC* as well as *EZNEC* have a special provision that automatically generates tapered wire segments (Ref 678).

At low heights (bottom of the antenna below approx. 0.2 λ), the gain and impedance figures obtained with a *MININEC*-based program are incorrect. The gain is too high, and the impedance too low. This is because *MININEC* calculates using a perfect ground under the antenna. Correct gain and impedance calculations at such low heights require modeling with a *NEC*-based program, such as *EZNEC*. All gain and impedance data listed in this chapter were obtained by using such a *NEC*-based modeling program.

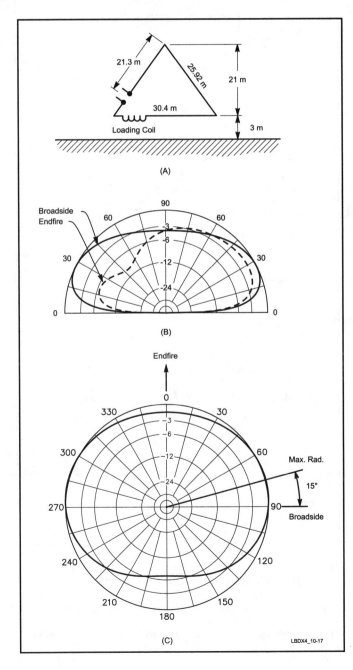

Fig 10-17—To shift the resonant frequency of the delta loop from 3.75 MHz to 3.55 MHz, a loading coil (or stub) is inserted in one bottom corner of the loop, near the feed point (A). This has eliminated the reactive component, but has also upset the symmetrical current distribution in the bottom wire. Vertical patterns are shown at B, and the horizontal pattern is shown at C for a 27° elevation angle. As with the loop shown in Fig 10-12, high-angle radiation (horizontal component) has appeared, and the horizontal pattern exhibits a notch in the endfire direction. Maximum radiation is again slightly off from the broadside direction.

3. LOADED LOOPS
3.1. CW and SSB 80-Meter Coverage

An 80-meter delta loop or quad loop will not cover 3.5 through 3.8 MHz with an SWR below 2:1. There are two ways to achieve a wide-band coverage:

1) Feed the loop with an open-wire line (450 Ω to a matching network).
2) Use inductive or capacitive loading on the loop to lower its resonant frequency.

3.1.1. Inductive loading

For more information about inductive loading, you can refer to the detailed treatment of short verticals in Chapter 9 on vertical antennas. There are three principles:

- The required inductance of the loading devices (coils or stubs) to achieve a given downward shift of the resonant frequency will be minimum if the devices are inserted at the maximum current point (similar to base loading with a vertical). At the minimum current point the inductive loading devices will not have any influence. This means that for a vertically polarized delta loop, the loading coil (or stub) cannot be inserted at the apex of the loop, nor in the middle of the bottom wire.
- Do not insert the loading devices in the radiating parts of the loop. Insert them in the part where the radiation is canceled. For example, in a vertically polarized delta loop, the loading devices should be inserted in the bottom (horizontal) wire near the corners.
- Always keep the symmetry of the loop intact, including after having added a loading device.

From a practical (mechanical) point of view it is convenient to insert the loading coil (stub) in one of the bottom corners. **Fig 10-17A** shows the loaded, compressed delta loop (with the same physical dimensions as the loop shown in Fig 10-11), where we have inserted a loading inductance in the bottom corner near the feed point.

A coil with a reactance of 240 Ω (on 3.5 MHz) or an inductance of 10.9 μH will resonate the delta on 3.5 MHz. The 100-kHz SWR bandwidth is 1.5:1. Note again the high-angle fill in the broadside pattern (no longer symmetrical baseline configuration), as well as the asymmetrical front-to-side ratio of the loop.

Although a well-designed and well made loading coil (see Chapter 9, Section 3.3) can have a much higher Q than a linear loading stub, it sometimes is easier to quickly tune the loop with a stub. The 10.9-μH coil can be replaced with a shorted stub. The inductive reactance of the closed stub is given by:

$$X_L = Z \tan \ell \quad \text{(Eq 2)}$$

where

Z = characteristic impedance of stub (transmission line)
ℓ = length of line, degrees
X_L = inductive reactance

From this,

$$\ell = \arctan\left(\frac{X_L}{Z}\right) \quad \text{(Eq 3)}$$

In our example:
X_L = 240 Ω
Z = 450 Ω
ℓ = arctan (240/450) = 28°

Assuming a 95% velocity factor for the transmission line, we can calculate the physical length of the stub as follows:

$$\text{Wavelength} = \frac{299.8}{3.5} = 85.66 \text{ meters (for } 360°\text{)}$$

$$\text{Physical Length} = 85.66 \text{ meters} \times 0.95 \times \frac{28°}{360°} = 6.33 \text{ meters}$$

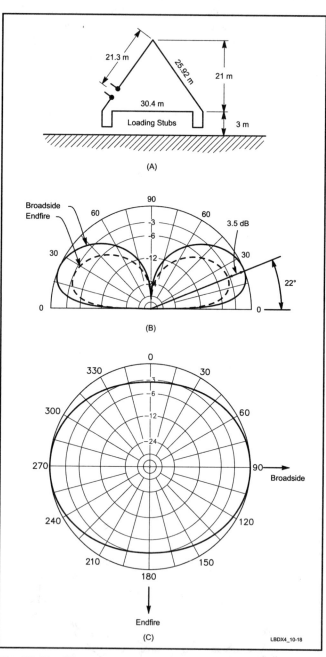

Fig 10-18—The correct way of loading the delta loop is to insert two loading coils (or stubs), one in each bottom corner. This keeps the current distribution in the baseline symmetrical, and preserves a "clean" radiation pattern in the horizontal as well as the vertical plane. The horizontal pattern at C is for an elevation angle of 22°.

Parts B and C of Fig 10-17 show the radiation patterns resulting from the insertion of a single stub (or coil) in one of the bottom corners of the delta loop. The insertion of the single loading device has broken the symmetry in the loop, and the bottom wire now radiates as well, upsetting the pattern of the loop.

This can be avoided by using two loading coils or stubs, located symmetrically about the center of the baseline. The example in **Fig 10-18A** shows two stubs, one located in each bottom corner of the loop. Each loading device has an inductive reactance of 142 Ω. For 3.5 MHz this is:

$$\frac{142}{2\pi \times 3.5} = 6.46\,\mu H$$

A 450-Ω short circuited line is 3.96 meters long (see calculation method above). The corresponding radiation patterns in Fig 10-18 are now fully symmetrical, and the annoying high-angle radiation is totally gone. The 100-kHz SWR bandwidth is 1.45:1. The 2:1 SWR bandwidth is 170 kHz.

Fig 10-19 shows the practical arrangement that can be used for installing the switchable stubs at the two delta-loop bottom corners. A small plastic box is mounted on a piece of epoxy printed-circuit-board material that is also part of the guying system. In the high-frequency position the stub should be completely isolated from the loop. Use a good-quality open-wire line and DPDT relay with ceramic insulation. The stub can be attached to the guy lines, which must be made of insulating material. If at all possible, make a high-Q coil, and replace the loading stub with the coil!

3.1.2. Capacitive loading

You can also use capacitive loading in the same way that we employ capacitive loading on a vertical. Capacitive loading is to be preferred over inductive loading because it is essentially lossless. Capacitive loading has the most effect when applied at a voltage antinode (also called a voltage point).

This capacitive loading is much easier to install than the inductive loading, and requires only a single-pole (high-voltage!) relay to switch the capacitance wires in or out of the circuit. *Keep the ends of the wires out of reach of people and animals, as extremely high voltage is present.*

Fig 10-20 shows different possibilities for capacitive

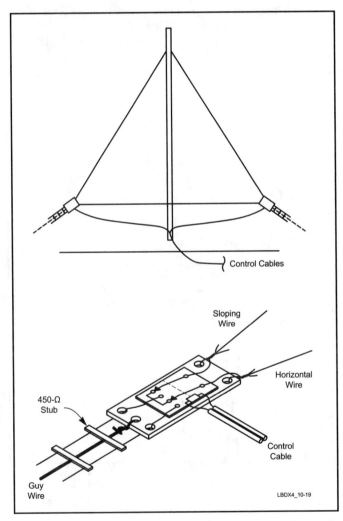

Fig 10-19—Small plastic boxes, mounted on a piece of glass-epoxy board, are mounted at both bottom corners of the loop, and house DPDT relays for switching the stubs in and out of the circuit. The stubs can be routed along the guy lines (guy lines must be made of insulating material). The control-voltage lines for the relays can be run to a post at the center of the baseline and from there to the shack. Do not install the control lines parallel to the stubs.

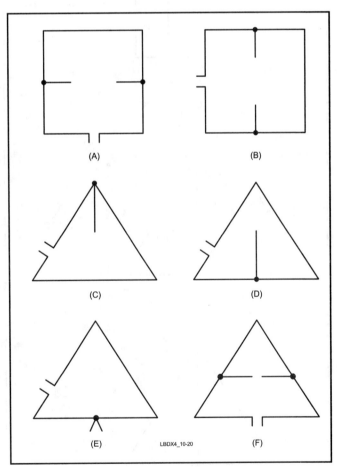

Fig 10-20—Various loop configurations and possible capacitive loading alternatives. Capacitive loading must be applied at the voltage maximum points of the loops to have maximum effectiveness. The loading wires carry very high voltages, and good insulators should be used in their insulation.

loading on both horizontally and vertically polarized loops. If installed at the top of the delta loop as in Fig 10-20C, a 9-meter long wire inside the loop will shift the 3.8-MHz loop (from Fig 10-11) to resonance at 3.5 MHz. For installation at the center of the baseline, you can use a single wire (Fig 10-20D), or two wires in the configuration of an inverted V (Fig 10-20E). Several wires can be connected in parallel to increase the capacitance. *(Watch out, since there is very high voltage on those wires while transmitting!)*

The same symmetry guidelines should be applied as explained in Section 3.1 to preserve symmetrical current distribution.

3.1.3. Adjustment

Once the loop has been trimmed for resonance at the high-frequency end of the band, just attach a length of wire with a clip at the voltage point and check the SWR to see how much the resonant frequency has been lowered. It should not take you more than a few iterations to determine the correct wire length. If a single wire turns out to require too much length, connect two or more wires in parallel, and fan out the wire ends to create a higher capacitance.

3.1.4. Bandwidth

By using one of the above-mentioned loading methods and a switching arrangement, a loop can be made that covers the entire 80-meter band with an SWR below 2:1.

3.2. Reduced-Size Loops

Reduced-size loops have been described in amateur literature (Refs 1115, 1116, 1121, 1129). **Fig 10-21** shows some of the possibilities of applying capacitive loading to loops, whereby a substantial shift in frequency can be obtained. G3FPQ uses a reduced-size 2-element 80-meter quad that makes use of capacitive-loaded square elements as shown in Fig 10-21A. The fiberglass spreaders of the quad support the loading wires.

It is possible to lower the frequency by a factor of 1.5 with this method, without lowering the radiation resistance to an unacceptable value (a loop dimensioned for 5.7 MHz can be loaded down to 3.8 MHz). The triangular loop can also be loaded in the same way, although the mechanical construction

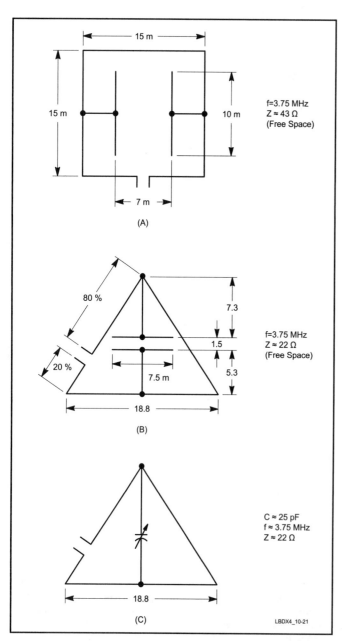

Fig 10-21—Capacitive loading can be used on loops of approximately ²/₃ full size. See text for details.

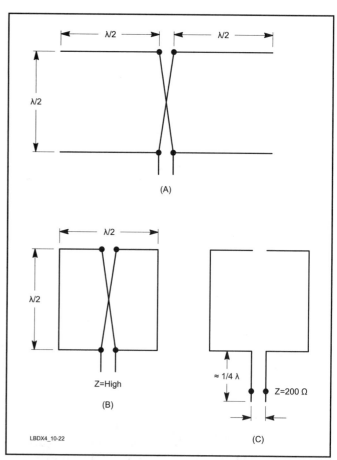

Fig 10-22—The bi-square antenna is a lazy-H antenna (two λ/2 collinear dipoles, stacked λ/2 apart and fed in phase), with the ends of the dipoles bent down (or up) and connected. The feed-point impedance is high and the array can best be fed via a λ/4 stub arrangement.

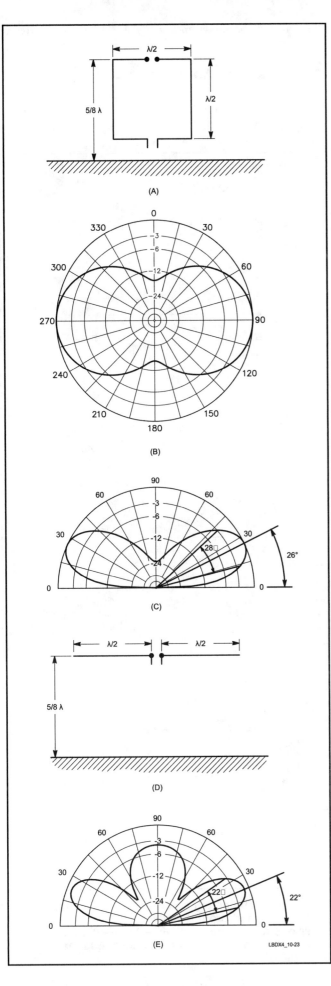

may be more complicated than with the square loop. See Fig 10-21B.

In principle, we can replace the parallel wires with a (variable) capacitor. This would allow us to tune the loop. The example in Fig 10-21C requires approximately 30 pF to shift the antenna from 5.7 to 3.8 MHz. Beware, however, that extremely high voltages exist across the capacitor. It would certainly not be over-engineering to use a 50-kV or higher capacitor for the application.

4. BI-SQUARE

The bi-square antenna has a circumference of 2 λ and is opened at a point opposite the feed point. A quad antenna can be considered as a pair of shortened dipoles with λ/4 spacing. In a similar way, the bi-square can be considered as a lazy-H antenna with the ends folded vertically, as shown in **Fig 10-22**. Not many people are able to erect a bi-square antenna, as the dimensions involved on the low bands are quite large.

In free space the bi-square has 3-dB gain over two λ/2 dipoles in phase (collinear), and almost 5 dB over a single λ/2 dipole. Over real ground, with the bottom wire λ/8 above ground (10 meters for an 80-meter bi-square), the gain of the bi-square is the same as for the two λ/2 dipoles in phase. The bottom two λ/2 sections do not contribute to low-angle radiation of the antenna.

The bi-square has the advantage over two half-waves in phase that the antenna does not exhibit the major high-angle sidelobe that is present with the collinear antenna when the height is over λ/2. **Fig 10-23** shows the radiation patterns of the bi-square and the collinear with the top of the antenna 5λ/8 high. Notice the cleaner low-angle pattern of the bi-square. Of course you could obtain almost the same result by lowering the collinear from 5λ/8 to λ/2 high!

The bi-square can be raised even higher in order to further reduce the wave angle without introducing high-angle lobes, up to a top height of 2 λ. At that height the wave angle is 14°, without any secondary high-angle lobe. With the top at 5λ/8, the takeoff angle is 26°.

To exploit the advantages of the bi-square antenna, you need quite impressive heights on the low bands. N7UA is one of the few stations using such an antenna, and he produces a most impressive signal on the long path into Europe on 80 meters. With a proper switching arrangement, the antenna can be made to operate as a full-wave loop on half the frequency (eg, 160 meters for an 80-meter bi-square).

The feed-point impedance is high (a few thousand ohms), and the recommended feed system consists of 600-Ω line with a stub to obtain a 200-Ω feed point. By using a 4:1 balun, a coaxial cable can be run from that point to the shack. Another alternative is to run the 600-Ω line all the way to the shack into an open-wire antenna tuner.

Fig 10-23—The bi-square antenna (A) and its radiation patterns (B and C). The azimuth pattern at B is for an elevation angle of 25°. At D, two half waves in phase and at E, its radiation pattern. Note that for a top-wire height of 5λ/8, the bi-square does not exhibit the annoying high-angle lobe of the collinear antenna.

Large Loop Antennas 10-15

5. THE HALF LOOP

The *half loop* was first described by Belrose, VE2CV (Ref 1120 and 1130). This antenna, unlike the half sloper, cannot be mounted on a tall tower supporting a quad or Yagi. If this was done, the half loop would shunt-feed RF to the tower and the radiation pattern would be upset. This can be avoided by decoupling the tower using a $\lambda/4$ stub (Ref 1130).

The half loop as shown in **Fig 10-24** can be fed in different ways.

5.1. The Low-Angle Half Loop

For low-angle radiation, the feed point can be at the end of the sloping wire (with the tower grounded), or else at the base of the tower (with the end of the sloping wire grounded).

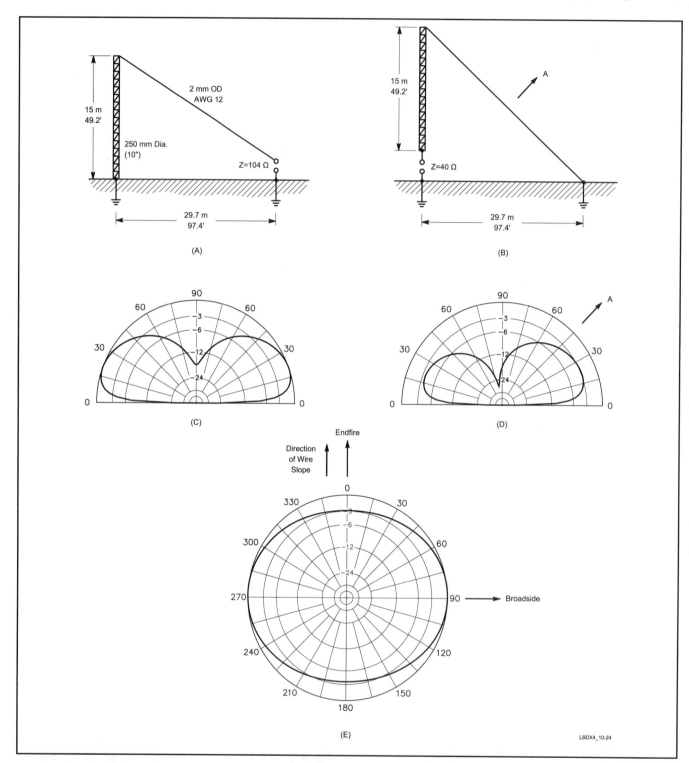

Fig 10-24—Half-loop antenna for 3.75 MHz, fed for low-angle radiation. The antenna can be fed at either end against ground (A and B). The grounded end must be connected to a good ground system, as must the ground-return conductor of the feeder. Radials are essential for proper operation. Note that while the feed-point locations are different, the radiation patterns do not change. C shows the broadside vertical pattern, E is the end-fire vertical pattern, and E is the azimuth pattern for an elevation angle of 20°.

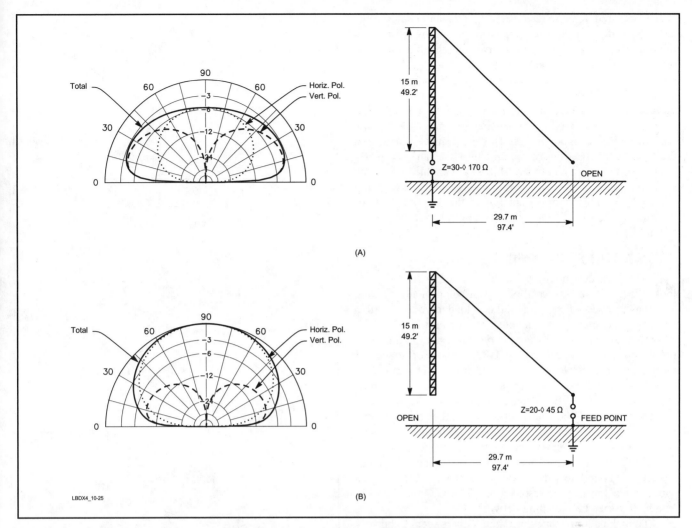

Fig 10-25—High-angle versions of the half delta loop antenna for 3.75 MHz. As with the low-angle version, the antenna can be fed at either end (against ground). The other end, however, must be left floating. The two different feed points produce different high-angle patterns as well as different feed-point impedances.

The radiation pattern in both cases is identical. The front-to-side ratio is approximately 3 dB, and the antenna radiates best in the broadside direction (the direction perpendicular to the plane containing the vertical and the sloping wire).

There is some pattern distortion in the end-fire direction, but the horizontal radiation pattern is fairly omnidirectional. Most of the radiation is vertically polarized, so the antenna requires a good ground and radial system, as for any vertical antenna. As such, the half loop does not really belong to the family of large loop antennas, but as it is derived from the full-size loop, it is treated in this chapter rather than as a top-loaded short vertical.

The exact resonant frequency depends to a great extent on the ratio of the diameter of the vertical mast to the slant wire. The dimensions shown in Fig 10-24 are only indicative. Fine-tuning the dimensions will have to be done in the field.

5.2. The High-Angle Half Loop

The half delta loop antenna can also be used as a high-angle antenna. In that case you must isolate the tower section from the ground (use a good insulator because it now will be at a high-impedance point) and feed the end of the sloping wire. Alternatively, you can feed the antenna between the end of the sloping wire and ground, while insulating the bottom of the tower from ground. Using the same dimensions that made the low-angle version resonant no longer produces resonance in these configurations.

Fig 10-25 shows the low-angle configurations with the radiation patterns. Note that the alternative where the end of the slant wire is fed against ground produces much more high-angle radiation than the alternative where the bottom end of the tower is fed. In both cases, the other end of the aerial is left floating (not connected to ground).

Dimensional configurations other than those shown in the relevant figures can be used as well, such as with a higher tower section and a shorter slant wire. If you move the end of the sloping wire farther away from the tower, you will need to decrease the height of the tower to keep resonance, and the radiation resistance will decrease. This will, of course, adversely influence the efficiency of the antenna. If the bottom of the sloping wire is moved toward the tower, the length of the vertical will have to be increased to preserve resonance. When the end of the sloping wire has been moved all the way to the base of the tower we have a λ/4 vertical with a folded feed system. The feed-point impedance will depend on the spacing and the ratio of the tower diameter to the feed-wire diameter.

Compared to a loaded vertical, this antenna has the advantage of giving the added possibility for switching to a high-angle configuration. For a given height, the radiation resistance is slightly higher than for the top-loaded vertical, whereby there is no radiation from the top load. The sloping wire in this half-loop configuration adds somewhat to the vertical radiation, hence the increase (10% to 15%) in R_{rad}.

Being able to feed the antenna at the end of the sloping wire may also be an advantage: This point may be located at the transmitter location, so the sloping wire can be directly connected to an antenna tuner. This would enable wide-band coverage by simply retuning the antenna tuner. Switching from a high to a low-angle antenna in that case consists of shorting the base of the tower to ground (for low-angle radiation).

6. THE HALF SLOPER

Although the so-called half sloper of **Fig 10-26A** may look like a half delta, it really does not belong with the loop antennas. As we will see, it is rather a loaded vertical with a specific matching system and current distribution.

Quarter-wave slopers are the typical result of ham ingenuity and inventiveness. Many DXers, short of space for putting up large, proven low-angle radiators, have found their half slopers to be good performers. Of course they don't know how much better other antennas might be, as they have no room to try them. Others have reported that they could not get their half sloper to resonate on the desired frequency (that's because they gave up trying before having found the proverbial needle in the haystack). Of course resonating and radiating are two completely different things. It's not because you cannot make the antenna resonant that it will not radiate well. Maybe they need a matching network?

To make a long story short, half slopers seem to be very unpredictable. There are a large number of parameters (different tower heights, different tower loading, different slope angles, and so forth) that determine the resonant frequency and the feed-point impedance of the sloper.

Unlike the half delta loop, the half sloper is a very difficult antenna to analyze from a generic point of view, as each half sloper is different from any other. Belrose, VE2CV, thoroughly analyzed the half sloper using scale models on a professional test range (Ref 647). His findings were confirmed by DeMaw, W1FB (Ref 650). Earlier, Atchley, W1CF, reported outstanding performance from his half sloper on 160 meters (Ref 645).

I have modeled an 80-meter half sloper using *MININEC*. After many hours of studying the influence of varying the many parameters (tower height, size of the top load, height of the attachment point, length of the sloper, angle of the sloper, ground characteristics, etc), I came to the following conclusions:

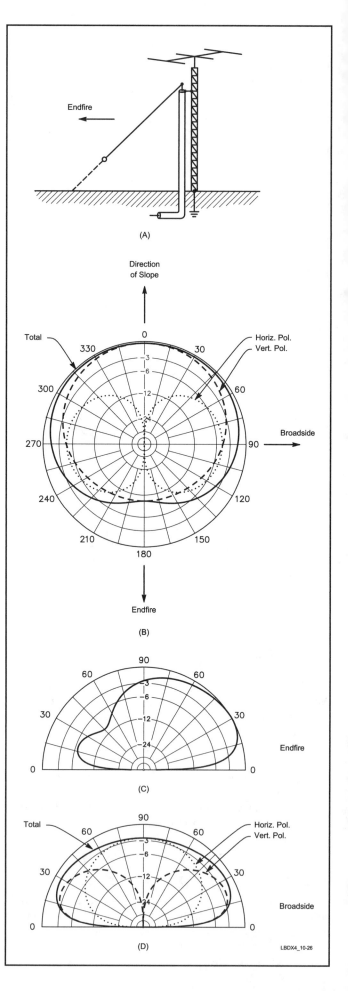

Fig 10-26—At A, a half-sloper mounted on an 18-meter tower that supports a 3-element full-size 20-meter Yagi. See text for details. At B, the azimuth pattern for a 45° elevation angle with vertically and horizontally polarized components, and at C and D, elevation patterns. The antenna shows a modest F/B ratio in the end-fire direction at a 45° elevation angle.

- The so-called half sloper is made up of a vertical and a slant wire. Both contribute to the radiation pattern. The radiation pattern is essentially omnidirectional. The low-angle radiation comes from the loaded tower, the high-angle radiation from the horizontal component of the slant wire. The antenna radiates a lot of high-angle signal (coming from the slant wire).
- Over poor ground the antenna has some front-to-back advantage in the direction of the slope, ranging from 10 to 15 dB at certain wave angles. Over good and excellent ground the F/B ratio is not more than a few dB.

An interesting testimony was sent on Internet by Rys, SP5EWY, who wrote *"Well, I had previously used my tower without radials and the half sloper favored the South, with the wire sloping in that direction, by at least 1 S-unit. Later I added 20 radials and since then it seems to radiate equally well in all directions."*

In essence the half sloper is a top-loaded vertical, which is fed at a point along the tower where the combination of the tower impedance and the impedance presented by the sloping wire combine to a 50-Ω impedance (at least that's what we want). The sloping wire also acts as a sort of *radial* to which the *other* conductor of the feed line is connected (like radials on a vertical to push against). In other words, the sloping wire is only a minor part of the antenna, a part that helps to create resonance as well as to match the feed line. Belrose (Ref 647) also recognized that the half sloper is effectively a top-loaded vertical. Fig 10-26 (B through D) shows the typical radiation patterns obtained with a half-sloper antenna.

While modeling the antenna, it was very critical to find a point on the tower and a sloper length and angle that give a good match to a 50-Ω line. The attachment point on the tower need not be at the top. It is not important how high it is, as you are not really interested in the radiation from the slant wire.

Changing the attachment point and the sloper length does not appreciably change the radiation pattern. This indicates that it is the tower (capacitively loaded with the Yagi) that does the bulk of the radiating. As the antenna mainly produces a vertically polarized wave, it requires a good ground system, at least as far as its performance as a low-angle radiator is concerned.

From my experience in spending a few nights modeling half slopers, I would highly recommend any prospective user to first model the antenna using *EZNEC*, which is great for such a purpose and which has the most user-friendly interfaces for multiple iterations.

There is an interesting analysis by D. DeMaw (Ref 650). DeMaw correctly points out that the antenna requires a metal support, and that a tree or a wooden mast will not do. But he does not emphasize anywhere in his study that it is the metal support that is responsible for most of the desirable low-angle radiation. DeMaw, however, recognizes the necessity of a good ground system on the tower, which implicitly admits that the tower does the radiating. DeMaw also says, *"The antenna is not resonant at the operating frequency,"* by which he means that the slant wire is not a quarter-wave long. This is again very confusing, as it seems to indicate that the slant wire is the antenna, which it is not. Describing his on-the-air results, DeMaw confirms what we have modeled: Due to the presence of high-angle radiation, it outperforms the vertical for short and medium-range contacts, while the vertical takes over at low angles for real DX contacts.

To summarize the performance of half slopers, it is worthwhile to note Belrose's comment, *"If I had a single quarter-wave tower, I'd employ a full-wave delta loop, apex up, lower-corner fed, the best DX-type antenna I have modeled."*

Of course, a delta loop still has a baseline of approximately 100 feet (on the 80-meter band), which is not the case with the half sloper. But the half sloper, like any vertical, requires radials in order to work well. It may look like the half sloper has a space advantage over many other low-band antennas, but this is only as true as for any vertical.

CHAPTER 11

Phased Arrays

Hardly any active ham needs an introduction to John Brosnahan, WØUN. Another American antenna guru, John retired as president of Alpha/Power, Inc, the Colorado-based manufacturer of top-notch power amplifiers. He is now relocating to the hill country of south-central Texas, moving all of his Colorado antennas and towers to his new 126-acre (50-hectare) site. Although the new site lacks a lot in ground conductivity, it should more than make up for that with its lack of man-made noise and no tower regulations! Although I had met John eye-to-eye on a number of occasions at Dayton Hamventions over the years it was during WRTC 1996 in San Francisco that we got to know each other better, and I have followed his moves in Amateur Radio ever since.

John is a research physicist by education who has spent his career on the electrical engineering side of remote-sensing instrumentation. From 1973 to 1978 he was the engineer for the University of Colorado's radio-astronomy observatory, designing receivers and antenna arrays for HF and VHF radio astronomy. Since then he has been founder and president of two companies that design and build HF and VHF radar systems for remote sensing of the atmosphere and ionosphere. He has designed and built arrays all the way from 80 dipoles at 2.66 MHz, which covered 40 acres (16 hectares), to a 12,288-dipole array at 49 MHz. He has also built numerous Yagi arrays, including a 768-element array at 52 MHz to a 500-element array at 404 MHz.

When I asked him, John immediately volunteered to review the chapters on arrays for the low bands, as he did in the previous edition. Thank you, John, for your help, your input and also your friendship!

When John had finished his proofreading he wrote to me: "*I loved this chapter. Excellent balance of the whole range of typical ham phased arrays, with a lot of very solid practical information and enough new stuff to make it worth the money to buy the new edition.*"

Corresponding with Robye Lahlum on the issue of L-networks for feeding arrays was more than interesting. Robye developed the mathematics for feeding the arrays in a Lewallen fashion, but with no limitation of phase angle. You can design a network for any array in a minute with the mathematics Robye provided for this book. Robye also developed a novel test setup that allows us to adjust the components of the L-network until we are right on the nose.

In addition, Robye has proven to be a very meticulous and thorough proofreader! I am proud having your contributions in my new book, *Robye*.

John Battin, K9DX, was the first to dare building a 9-Circle array, which I described in the Third Edition of this book. It's been a very enlightening experience for me discussing various issues of array feeding with John, and I am extremely grateful to him for letting me share his experience with the readers of this book. Thank you John.

Phased Arrays 11-1

If you want gain and directivity on one of the low bands and if you live in an area with good or excellent ground, an array made of vertical elements may be the answer, provided you have room for it. Arrays made with vertical elements have the same requirements as single vertical antennas so far as ground quality is concerned. Before you decide to put one up, take the time to understand the mechanism of an array with all-fed elements.

In this chapter I cover the subject of arrays made of elements that, by themselves, have an omnidirectional horizontal radiation pattern; that is, vertical antennas.

1. RADIATION PATTERNS
1.1. How the Pattern is Formed

In Chapter 7 we explored in great detail how the radiation pattern of an array was formed.

1.2. Directivity Over Perfect Ground

Fig 11-2 shows a range of radiation patterns obtained by different combinations of two monopoles over perfect ground and at a 0° elevation wave angle. These directivity patterns are classics in every good antenna handbook.

1.3. Directivity Over Real Ground

Over real ground there is no radiation at a 0° elevation angle. All the effects of real ground, which were described in detail in Chapter 9 on verticals, apply to arrays of verticals.

1.4. Direction of Firing

The rule is simple: An array always fires in the direction of the element with the lagging feed current.

1.5. Phase Angle Sign

Phase angles are a relative thing, which means you can put your *reference* phase angle of 0° anywhere in the array. We will stick to our own convention of assigning the 0° phase angle to the back element of an array. This means that the feed currents in all other elements will carry a negative sign.

2. ARRAY ELEMENTS

In principle, you can use verticals of length longer than λ/4 (electrically) for building arrays, but in that case the various feed systems described in this chapter do not apply. However, the whole range of verticals described in Chapter 9 can be used as elements for these vertical arrays, provided they are base-fed and are not longer than λ/4 electrically.

Quarter-wave elements have gained a reputation for giving a reasonable match to a 50-Ω line, which is certainly true for single vertical antennas. In this chapter we will learn the reason why quarter-wave resonant verticals do not have a resistive 36-Ω feed-point impedance when operated in arrays (even assuming a perfect ground). Quarter-wave elements still remain a good choice, since they have a reasonably high radiation resistance. This ensures good overall efficiency. On 160 meters, the elements could be top-loaded verticals, as described in Chapter 9.

The design methodology for arrays given in Section 3, as well as all the designs described in Section 4, assume that all the array elements are physically identical, with a current distribution that is the same on each element. In practice this means that only elements with a length of up to λ/4 should be used. Remember, the patterns given in Section 4 do not apply if you use elements much longer than λ/4. They certainly do not apply for elements that are λ/2 or 5λ/8 long. If you want to use long elements, you will have to model the design using the particular element lengths (Ref 959). This may be a problem if you want to use shunt-fed towers carrying HF beams as elements for an array. With their top loads, these towers are electrically often much longer than λ/4.

3. DESIGNING AN ARRAY

The radiation patterns shown in Fig 11-2 give a good idea what can be obtained with different spacings and different phase delays for a 2-element array. For arrays with more elements there are a number of popular classic designs. Many of those are covered in detail in this chapter. A good array should meet the following specifications:

- High gain (you want to be loud)
- Good directivity (F/B, forward beamwidth, especially if you will not be using separate receiving antennas)
- Ease of feeding
- Ease of direction switching

3.1. Modeling Arrays

If you feed the elements of an array at a current maximum (which is usually the case with base-fed elements not longer than λ/4 long) the RF drive has to be specified as a current. If you feed at a voltage maximum (which we hardly ever do in vertical arrays), the RF drive will need to be specified as a voltage. We normally define currents for the RF sources since it is the current (magnitude and phase) in each element that determines the radiation pattern in such an array. Therefore currents, rather than voltages, should be specified. All modern modeling programs allow you to define "sources" as current sources or as voltage sources.

You can do initial pattern assessments using a *MININEC*-based program, although the latest versions of some programs no longer include such a computing engine (eg, *EZNEC 4.0*) because modern computers allow fast analysis with *NEC*-engines. Also, if you want to do some modeling that includes the influence of a radial system plus the influence of a poor ground, the full-blown *NEC* program is required. Studies on elevated radial systems and on buried radials require *NEC* (*NEC-3* or better yet, *NEC-4* if buried radials are involved).

In this chapter, the influence of the loss introduced by an imperfect radial system has been included in the form of an equivalent loss resistance in series with each element feed point (most model used 2 Ω).

3.2 About Polar and Rectangular Coordinates

We will be going into detail on various issues and aspects of arrays and will be talking impedances all the time. It's a good idea to review a few basics.

- **Complex impedance**: A complex impedance is an impedance consisting of a real part (resistive part) and an imaginary part (reactive part).
- **Complex number**: A complex impedance is represented by a complex number.
- **Complex number representation**: While a real number can be represented as a point on a line, a complex number

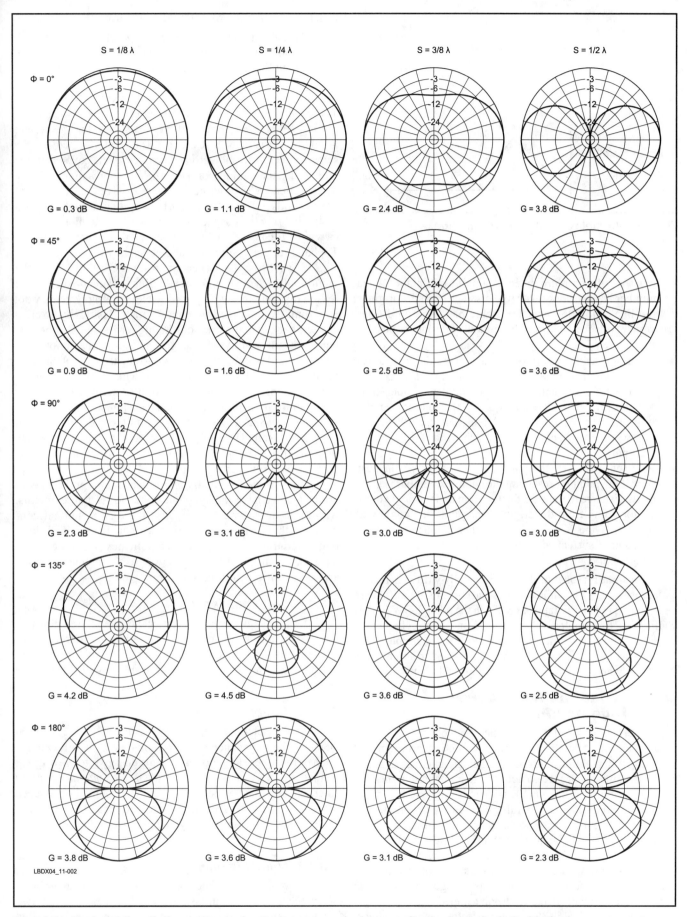

Fig 11-2—Horizontal radiation patterns for 2-element vertical arrays (both elements fed with the same current magnitude). The elements are in the vertical axis, and the top element is the one with the lagging phase angle. Patterns are for 0° elevation angle over ideal ground. *(Courtesy* The ARRL Antenna Book.)

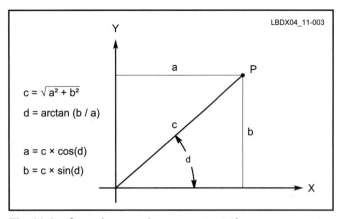

Fig 11-3—Complex number representation.

must always be represented as a point in a plane. A real number has one coordinate (the distance from the origin on the line) while the complex figure has two coordinates, which are necessary to unambiguously define its position in a plane.

- **Rectangular coordinates**: In a rectangular coordinate system, which in a plane consists of an X and a Y-axis, the X and the Y coordinates define the complex number. If the X-value is a and the Y-value equals b, the complex number is written as $a + jb$. The "j" indicates that the figure following is the Y-coordinate, which stands for the imaginary part.
- **Polar coordinates**: In a polar-coordinate system the position of the point representing the complex number is given by its distance to the coordinate origin and the angle of the vector going from the origin to the point, the angle with respect of the X-axis. The complex number in a polar coordinate system is written as $c\ \underline{/d°}$ where c = vector length and d = angle.

For some reason impedances are usually written in rectangular form as $a + jb$, while voltages and currents are most often represented in polar notation as $c\underline{/d°}$. In the NEW LOW BAND SOFTWARE, complex values of Z, I and E are always expressed in both coordinate systems. With some simple trigonometry we can always convert from one system to another (see **Fig 11-3**, where conversion formulas are included).

3.3. Getting the Right Current Magnitude and Phase

There is a world of difference between designing an array on paper or with a computer modeling program and realizing it in real life. With single-element antennas (a single vertical, a dipole, etc) we do not have to bother about the feed current (magnitude and phase), as there is only one feed point anyway. With phased arrays things are vastly different.

First, we must decide which array to build. Once we do this, the problem will be how to achieve the right feed currents in all the elements (magnitude and phase angle). When we analyze an array with a modeling program, we notice that the feed-point impedances of the elements change from the value for a single element. If the feed-point impedance of a single quarter-wave vertical is 36 Ω over perfect ground, it is almost always different from that value in an array because of mutual coupling.

3.3.1. The effects of mutual coupling

Until about 10 to 15 years ago, few articles in Amateur-Radio publications addressed the problems associated with mutual coupling in designing a phased array and in making it work as it should. Gehrke, K2BT, wrote an outstanding series of articles on the design of phased arrays (Refs 921-925, 927). These are highly recommended for anyone who is considering putting up phased arrays of verticals. Another excellent article by Christman, K3LC (ex-KB8I) (Ref 929), covers the same subject. The subject has been very well covered in the 15th and later editions of *The ARRL Antenna Book*, where R. Lewallen, W7EL, wrote a comprehensive contribution on arrays. Today, Tom Rauch, W8JI, is a good teacher on principles and practical aspects of arrays in his excellent website (**www.w8ji.com/**), and his advice in these matters on the Topband reflector are much appreciated by all.

If we bring two (nearly) resonant circuits into the vicinity of each other, mutual coupling will occur. This is the reason that antennas with parasitic elements work as they do. Horizontally polarized antennas with parasitically excited elements are widely used on the higher bands. On the low bands the proximity of the ground limits the amount of control the designer has on the current in each of the elements. Arrays of vertical antennas, where each element is fed, overcome this limitation, and in principle the designer has an unlimited control over all the design parameters. With so-called *phased arrays*, all elements are individually and physically excited by applying power to the elements through individual feed lines. Each feed line supplies current of the correct magnitude and phase.

There is one frequently overlooked major problem with arrays. As we have made up our minds to feed all elements, we too often assume (incorrectly) there is no mutual coupling or that it is so small that we can ignore it. Taking mutual coupling into account complicates life, as we now have two sources of applied power to the elements of the array: parasitic coupling plus direct feeding.

3.3.1.1. Self-impedance

If a single quarter-wave vertical is erected, we know that the feed-point impedance will be $36 + j\,0\ \Omega$, assuming resonance, a perfect ground system and a reasonably thin conductor diameter. In the context of our array we will call this the *self-impedance* of the element.

3.3.1.2. Coupled impedance

If other elements are closely coupled to the original element, the impedance of the original element will change. Each of the other elements will couple energy into the original element and vice versa. This is often termed *mutual coupling* since each element affects the other. The coupled impedance is the impedance of an element being influenced by one other element and it is significantly different from the self-impedance in most cases.

3.3.1.3. Mutual impedance

The mutual impedance is a term that defines unambiguously the effect of mutual coupling between a set of two antenna elements. Mutual impedance is an impedance that cannot be measured. It can only be calculated. The calculated mutual impedances and driving impedances have been extensively covered by Gehrke, K2BT (Ref 923).

3.3.1.4. Drive impedance

To design the correct feed system for an array, you must know the drive impedances of each of the elements, as well as the correct current magnitude and angle needed to feed the element(s).

3.3.2. Calculating the drive impedances

Mutual impedances are calculated from measured self-impedances and drive impedances. Here is an example: We are constructing an array with three $\lambda/4$ elements in a triangle, spaced $\lambda/4$ apart. We erect the three elements and install the final ground system, making the ground system as symmetrical as possible. Where the buried radials cross, we terminate them in a bus. Then the following steps are carried out:

1. Open-circuit elements 2 and 3. Opening an element will effectively isolate it from the other elements in the case of quarter-wave elements. (When using half-wave elements the elements must be grounded for maximum isolation and open-circuited for maximum coupling.)
2. Measure the self-impedance of element 1 (= Z11).
3. Ground element 2.
4. Measure the coupled impedance of element 1 with element 2 coupled (= Z1,2).
5. Open-circuit element 2.
6. Ground element 3.
7. Measure the coupled impedance of element 1 with element 3 coupled (= Z1,3).
8. Open-circuit element 3.
9. Open-circuit element 1.
10. Measure the self-impedance of element 2 (= Z22).
11. Ground element 3.
12. Measure the coupled impedance of element 2 with element 3 coupled (= Z2,3).
13. Open-circuit element 3.
14. Ground element 1.
15. Measure the coupled impedance of element 2 with element 1 coupled (= Z2,1).
16. Open-circuit element 1.
17. Open-circuit element 2.
18. Measure the self-impedance of element 3 (= Z33).
19. Ground element 2.
20. Measure the coupled impedance of element 3 with element 2 coupled (= Z3,2).
21. Open-circuit element 2.
22. Ground element 1.
23. Measure the coupled impedance of element 3 with element 1 coupled (= Z3,1).

This is the procedure for an array with 3 elements. The procedures for 2 and 4-element arrays can be derived from the above.

As you can see, measurement of coupling is done for pairs of elements. At step 15, you are measuring the effect of mutual coupling between elements 2 and 1, and it may be argued that this had already been done in step 4. It is useful, however, to make these measurements again to recheck the previous measurements and calculations. Calculated mutual couplings Z12 and Z21 (see below) using the Z1,2 and Z2,1 inputs should in theory be identical, and in practice should be within an ohm or so. The self-impedances and the driving impedances of the different elements should match closely if the array is to be made switchable.

The mutual impedances can be calculated as follows:

$$Z12 = \pm\sqrt{Z22 \times (Z11 - Z1,2)}$$

$$Z21 = \pm\sqrt{Z11 \times (Z22 - Z2,1)}$$

$$Z13 = \pm\sqrt{Z33 \times (Z11 - Z1,3)}$$

$$Z31 = \pm\sqrt{Z11 \times (Z33 - Z1,3)}$$

$$Z23 = \pm\sqrt{Z33 \times (Z22 - Z2,3)}$$

$$Z32 = \pm\sqrt{Z22 \times (Z33 - Z3,2)}$$

It is obvious that if Z11 = Z22 and Z1,2 = Z2,1, then Z12 = Z21. If the array is perfectly symmetrical (such as in a 2-element array or in a 3-element array with the elements in an equilateral triangle), all self-impedances will be identical (Z11 = Z22 = Z33), and all driving impedances as well (Z2,1 = Z1,2 = Z3,1 = Z1,3 = Z2,3 = Z3,2). Consequently, all mutual impedances will be identical as well (Z12 = Z21 = Z31 = Z13 = Z23 = Z32). In practice, the values of the mutual impedances will vary slightly, even when good care is taken to obtain maximum symmetry.

Because all impedances are complex values (having real and imaginary components), the mathematics involved are difficult. The MUTUAL IMPEDANCE AND DRIVING IMPEDANCE software module of the NEW LOW BAND SOFTWARE will do all the calculations in seconds. No need to bother with complex algebra. Just answer the questions on the screen.

Fig 11-4 shows the mutual impedance to be expected for quarter-wave elements at spacings from 0 to 1.0 λ. The resistance and reactance values vary with element separation as a damped sine wave, starting at zero separation with both signs positive. At about 0.10 to 0.15-λ spacing, the reactance sign changes from + to –. This is important to know in order to assign the correct sign to the reactive value (obtained via a square root).

Gehrke, K2BT, emphasizes that the designer should actually measure the impedances and not take them from tables. Some methods of doing this are described in Ref 923. The published tables show ballpark figures, enabling you to verify the square-root sign of your calculated results.

After calculating the mutual impedances, the drive impedances can be calculated, taking into account the drive current (amplitude and phase). The driving-point impedances are given by:

$$Zn = \frac{I1}{In} \times Zn1 + \frac{I2}{In} \times Zn2 + \frac{I3}{In} \times Zn3 ... + \frac{In}{In} \times Znn$$

where n is the total number of elements. The number of equations is n. The above formula is for the nth element. Note

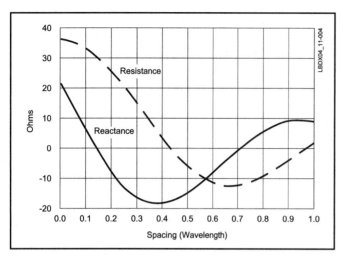

Fig 11-4—Mutual impedance for two λ/4 elements. For shorter vertical elements (length between 0.1 λ and 0.25 λ), you can calculate the mutual impedance by multiplying the figures from the graph by the ratio $R_{rad}/36.6$ where R_{rad} = the radiation resistance of the short vertical.

$Z1 = 55.8 + j\,19.8\,\Omega$ for the $-90°$ element.
$Z2 = 24.8 - j\,19.8\,\Omega$ for the $0°$ element

We have now calculated the impedance of each element of the array, the array being fed with the current (magnitude and phase) as set out. We have used impedances that we have measured; we are not working with theoretical impedances.

The 2 EL AND 4 EL VERTICAL ARRAYS module of the NEW LOW BAND SOFTWARE is the perfect tool to guide you along the design of an array. You can enter your own values or just work your way through using a standard set of values.

3.3.3. Modeling the array

With the latest *NEC-3* or *NEC-4* based software, you can include a buried radial system, but for the design and evaluation of arrays, a *MININEC*-based modeling program, or even better a *NEC*-based program using a *MININEC*-type of ground (such as provided in *EZNEC*) will work, so long as we realize that we must add some equivalent series resistance to account for the ground-losses of the radial system. To simulate the

also that Z12 = Z21 and Z13 = Z31, etc.

The above-mentioned program module performs the rather complex driving-point impedance calculations for arrays with up to 4 elements. The required inputs are:

1. The number of elements.
2. The driving current and phase for each element.
3. The mutual impedances for all element pairs.

The outputs are the driving-point impedances Z1 through Zn.

3.3.2.1. Design example

Let us examine an array consisting of two λ/4 long verticals, spaced λ/4 apart and fed with equal magnitude currents, with the current in element 2 lagging the current in element 1 by 90°. This is the most common (though not necessarily the best) end-fire configuration with a cardioid pattern.

Self impedance

The quarter-wave long elements of such an array are assumed to have a self-impedance of 36.4 Ω over perfect ground. A nearly perfect ground system consists of at least 120 half-wave radials (see Chapter 9). For example, a system with only 60 radials may (depending on the ground quality) show a self-impedance on the order of 40 Ω.

Coupled impedance

We measured $37.5 + j\,15.2\,\Omega$.

Mutual impedance

The mutual impedances were calculated with the above-mentioned computer program: $Z12 = Z21 = 19.76 - j\,15.18\,\Omega$. From the mutual impedance curves in Fig 11-4 it is clear that the minus sign is the correct sign for the reactive part of the impedance.

Drive impedances

The same software module calculates the drive impedances (also called feed-point impedances) of the two elements:

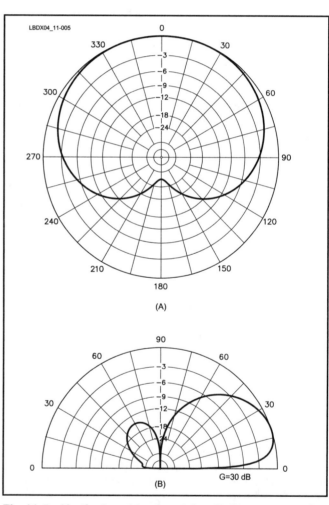

Fig 11-5—Vertical and horizontal radiation patterns for the 2-element cardioid array, spaced 90° and fed with 90° phase difference. The pattern was calculated for very good ground with a radial system consisting of 120 radials, each 0.4 λ long (the equivalent ground resistance is 2 Ω). The gain is 3.0 dB compared to a single vertical over the same ground and radial system. The horizontal pattern at A is for an elevation angle of 19°.

effect of a radial system consisting of 60 quarter-wave radials I inserted 4 Ω in series with the feed point of each antenna.

Modeling the cardioid antenna over *MININEC*-type ground with 4-Ω loss resistance included in each element, *EZNEC* comes up with the following impedances:

$Z1 = 55.0 + j\ 22.7\ \Omega$
$Z2 = 26.5 - j\ 19.5\ \Omega$

These are close to the values worked out with the NEW LOW BAND SOFTWARE, which were based on measured values of coupled and self impedances. The vertical and the horizontal radiation patterns for the 2-element cardioid array are shown in **Fig 11-5**.

3.4. Designing a Feed System

The challenge now is to design a feed system that will supply the right current to each of the array elements. As we now know the current requirements as well as the drive-impedance data for each element of the array, we have all the required inputs to design a feed system.

Each element will need to be supplied power through its own feed line. In a driven array each element either gets power, or it possibly delivers power into the feed system. During calculations we will sometimes encounter a negative feed-point impedance, which means the element is actually delivering power into the feed network. If the element impedance is zero, this means that the element can be shorted to ground. It then acts as a parasitic element.

Eventually all the feed lines will be connected to a common point, which will be the common feed point for the entire array. You can only connect feed lines in parallel if the voltages on the feed lines (at that point) are identical (in magnitude and phase)—the same as with ac power!

Designing a feed system consists of calculating the feed lines (impedance and length) as well as the component values of networks used in the feed system, so that the voltages at the input ends of the lines are identical. It is as simple as that.

The ARRL has published the original (1982) work by Lewallen, W7EL, in the last five editions of *The ARRL Antenna Book*. This material is a must for every potential array builder. However, there are other feed methods than the Lewallen method. Various feed systems are covered in the following sections of this book:

- Christman method
- Using flat lines
- Cross-fire principle
- Lewallen (quadrature fed arrays)
- Lewallen/Lahlum (any phase angle, any current ratio)
- Collins (hybrid coupler)
- Gehrke (broadcast approach)
- Lahlum/Gehrke (non current-forcing, L-network)

3.4.1. The wrong way

In just about all cases, the drive impedance of each element will be different from the characteristic impedance of the feed line. This means that there will be standing waves on the line. This has the following consequences:

- The impedance, voltage and current will be different in each point of the feed line.
- The current and voltage phase shift is not proportional to the feed line length, except for a few special cases (eg, a half-wave-long feed line).

This means that if we feed these elements with 50-Ω coaxial cable, we cannot simply use lengths of feed line as phasing lines by making the line length in degrees equal to the desired phase delay in degrees. In the past we have seen arrays where a 90° long coax line was inserted in one of the feed lines to an element to create a 90° antenna current phase shift. Let us take the example of the 2-element cardioid array (as described above) and see what happens (see **Fig 11-6**).

We run two 90° long coax cables to a common point. Using the COAX TRANSFORMER/SMITH CHART software module of the NEW LOW BAND SOFTWARE, we calculate the impedances at the end of those lines (I took RG-213 with 0.35 dB/100 feet attenuation at 3.5 MHz). Using round figures, the array element feed impedances, including 2 Ω of equivalent ground loss resistance, are:

$Z1 = 51 + j\ 20\ \Omega$
$I1 = 1\ A\ /\!\!-90°$

From $E = Z / I$ we can calculate (don't worry, the software does it for you):

$E1 = 54.8\ /\!\!-68.6°\ V$

and

$Z2 = 21 - j\ 20\ \Omega$
$I2 = 1\ /0°\ A$
$E2 = 29\ /\!\!-43.6°\ V$

At the end of the 90° long RG-213 feed lines the impedances (and voltages) become:

$Z1' = 42.81 - j\ 16.18\ \Omega$
$E1' = 50.89\ /0.39°\ V$
$I1' = 1.11\ /21.09°\ A$

and

$Z2' = 63.1 + j\ 56.94\ \Omega$
$E2' = 50.37\ /89.61°\ V$
$I2' = 0.59\ /47.54°\ A$

If we make the line to the lagging element 180° long (90° plus the extra 90° to obtain an extra 90° phase shift), we end up with:

$Z1'' = 51.18 + j\ 18.64\ \Omega$
$E1'' = 56.42\ /110.77°\ V$
$I1'' = 1.04\ /90.76°\ A$

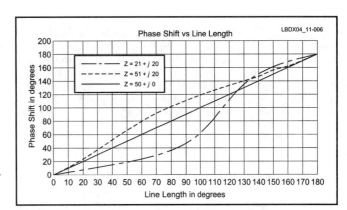

Fig 11-6—Graph showing the current phase shift in a 50-Ω line (RG-213, on 80 meters), as a function of the load impedance. The loads shown are those for a 2-element cardioid array. Note that the phase shift does *not* equal line length, except when the line is terminated in its own characteristic impedance!

analyzing this phenomenon. Look at the values of voltage and current as you scan along the line, and remember we want the right current phase shift and we want the same voltage where we connect the feed lines in parallel.

If you have such a feed system, do not despair. Simply by shortening the phasing line from 90° to 70°, you can obtain an almost perfect feed system. (See **Fig 11-7**.)

Watch out, if you want to use this system, make sure you have the same feed impedances as in the model above. How? By calculating the drive impedances as outlined in Section 3.3.2, or by carefully modeling your array, making sure you take into account all the small details!

3.4.2. Christman (K3LC) method

In the Christman, K3LC (ex-KB8I) method (Ref 929), we scan the feed lines to the different elements looking for points where the voltages are identical. If we find such points, we connect them together, and we are all done! It's really as simple as that. Whatever the length of the lines are, provided you have the right current magnitude and phase at the input ends of the lines, you can always connect two points with identical voltages in parallel. That's also where you feed the entire array.

Christman makes very clever use of the transformation characteristics of the feed lines. We know that on a feed line with SWR, voltage, current and impedance are different in every point of the line. The questions are now, "Are there points with identical voltage to be found on all of the feed lines?" and "Are the points located conveniently; in other words, are the feed lines long enough to be joined?" This has to be examined case by case.

It must be said that we cannot apply the Christman method in all cases. I have encountered situations where identical voltage points along the feed lines could not be found. The software module, IMPEDANCE, CURRENT AND VOLTAGE ALONG FEED LINES, which is part of the NEW LOW BAND SOFTWARE, can provide a printout of the voltages along the feed lines. The required inputs are:

- Feed-line impedance.
- Driving-point impedances (R and X).
- Current magnitude and phase.

Continuing with the above example of a 2-element configuration (90° spacing, 90° phase difference, equal currents, cardioid pattern), we find:

E1 = (155° from the antenna element) = 47.28 /86.1°V
E2 = (84° from the antenna element) = 47.27 /85.9°V

Notice on the printout that the voltages at the 180° point on line 1 and at the 90° point on line 2 are not identical (see Section 3.4.1), which means that if you connect the lines in parallel in those points, you will not have the proper current in the antennas.

We need now to connect the two feed lines together where the voltages are identical. If you want to make the array switchable, run two 84° long feed lines to a switch box, and insert a 155° – 84° = 71° long phasing line, which will give you the required 90° antenna-current phase shift. **Fig 11-8** shows the Christman feed method.

Of course the impedance at the junction of the two feed lines is not 50 Ω. Using the COAX TRANSFORMER/SMITH CHART software module, we calculate the impedances at the

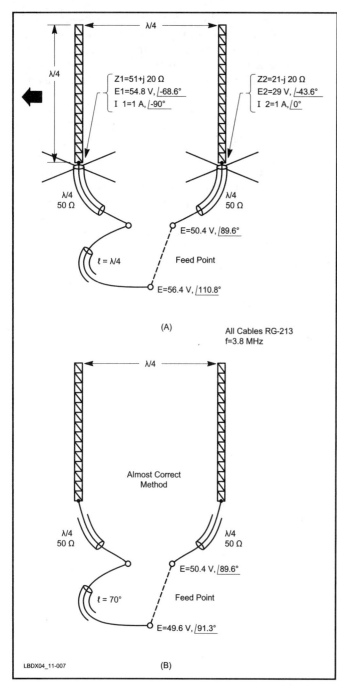

Fig 11-7—At A, the incorrect way of feeding a 2-element cardioid array (90° phase, 90° spacing). Note that the voltages at the input ends of the two feed lines are not identical. In B we see the same system with a 70° long phasing line, which now produces almost correct voltages. The F/B ratio of existing installations will jump up by 10 or 15 dB, just by changing the line length from 90° to 70°.

Note that E2′ and E1″ are not identical. This means we cannot connect the lines in parallel at those points without upsetting the antenna current (magnitude and phase). From the above voltages we see that the extra 90° line created an actual current phase difference of 90.76° – 21.09° = 68.67°, and *not* 90° as required.

The software module IMPEDANCES, CURRENTS AND VOLTAGES ALONG FEED LINES is ideally suited for

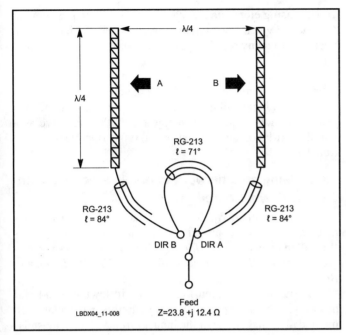

Fig 11-8—Christman feed system for the 2-element λ/4-spaced cardioid array fed 90° out-of-phase. Note that the two feed lines are 84° long (not 90°), and that the "90° phasing line" is actually 71 electrical degrees in length. The impedance at the connection point of the two lines is 23.8 + j 12.4 Ω (representing an SWR of 2.3:1 for a 50-Ω line), so some form of matching network is desirable.

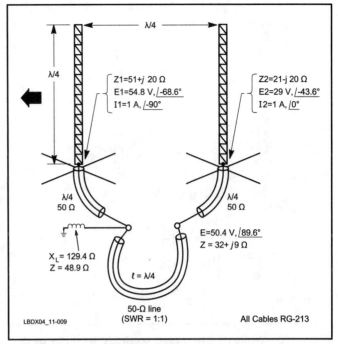

Fig 11-9—Adding a shunt coil with a reactance of +129.4 Ω at the end of the λ/4 feed line going to the front element turns the impedance at that point into 48.9 Ω, very close to 50 Ω. Now we can insert a 50-Ω delay line and be assured that the phase shift equals the line length.

input ends of the two lines we are connecting in parallel:

Z1end = 39 + j 12 Ω
Z2end = 50 + j 52 Ω

The software module PARALLEL IMPEDANCES calculates the parallel impedance as 23.8 + j 12.4 Ω. This is the feed-point impedance of the array. You can use an L network, or any other appropriate matching system to obtain a more convenient SWR on the 50-Ω feed line.

3.4.3. Using flat lines (SWR ~ 1:1) with "length = phase shift"

Let's go back to Fig 11-7. The impedance at the end of the quarter-wave line going to the front element is 42.81 − j 16.18 Ω. Maybe we can turn it in a purely resistive impedance of convenient value by connecting a reactance in parallel. Using the SHUNT/SERIES IMPEDANCE NETWORK module of the LOW BAND SOFTWARE, we can easily calculate the required parallel impedance to make it a purely resistive impedance. In this case it appears that putting an inductance of +129.4 Ω in parallel at that point, turns the impedance to 48.9 Ω, very close to 50 Ω. Let's do that, and now connect a quarter-wave phasing line from that point to the end of the quarter-wave line coming from the back element. As the line now operates with an SWR of very close to 1:1, phase difference equals line length, and we have exactly what we want.

Fig 11-9 shows the layout of this system. If you want more phase shift, eg 120° to lift the notch off the ground (see Chapter 7) you simply make the phasing line 120° long. Note however that the element feed impedances shown are for 90° phase shift and that those are slightly different when you change the elevation angle.

It is obvious that such method can only be applied when you are lucky to find an impedance (after tuning out the reactance by a parallel element) that matches an existing feed-line impedance. You can, of course, use parallel feed line to obtain low impedances, and actually connect feed lines of different impedances in parallel (25 Ω = two 50-Ω lines in parallel; 30 Ω = a 50-Ω and a 75-Ω line in parallel; 37.7 Ω = two 75-Ω in parallel).

3.4.4. The Cross Fire (W8JI) principle

In a "standard" array, for example as shown in Section 3.4.2 and 3.4.3, the feed line goes to the back element, and the front element is fed via a phasing line. Let us analyze what happens in such a design when we change frequency away from the nominal design frequency. Assume we have a 2-element end-fire array, spaced exactly λ/4 (90°) and with exactly 90° phase shift (this is by far not the best arrangement!). Our notch elevation angle will be 0° (see Chapter 7). If we increase the frequency by 5%, the spacing becomes 94.9° and the phasing becomes also larger (if the lines are relatively flat also about 5% longer). But, in order to maintain the zero notch angle at ground level, we need the phasing line to be *shorter* by about 5%. This mechanism limits the usable bandwidth in such arrays. In simple 2-element arrays this usually is not a problem, but in more complex arrays using four or more elements it can become a key design factor.

Tom Rauch, W8JI, pointed out that we can also use the cross-fire principle feed method, where we feed the array at the front element using a phase inverter (a 180° transformer) and feed the back element with a phasing line that is complementary in length to the required phasing angle (see Chap-

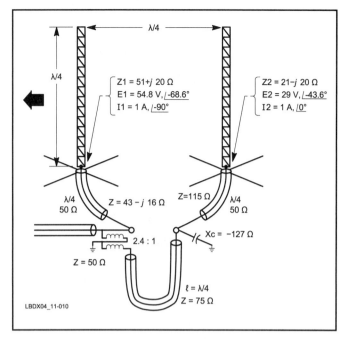

Fig 11-10—While all other feed methods feed the back element directly and provide phase delay via coaxial cable or a network to the front element, the cross-fire feeding system does the opposite. It makes uses of a 180° phase-inverter transformer to achieve a feed system that guarantees that the phase delay remain correct when the frequency is changed. See text for details.

ter 7). In this case the phase-shift transformer produces a 180° shift over a wide frequency range. At a frequency that is 5% higher than the design frequency, the phase shift produced by the phasing line becomes about 95° long. Subtracting this value from the 180° phase shift obtained by the transformer, the phase difference becomes 85° at the higher frequency. With this cross-fire principle the tracking is achieved, which is exactly what we want.

In **Fig 11-10** we see that we will have to put the phasing line in the feed line going to the back-element. The impedance at the end of the quarter wave line to the elements is $63.1 + j\, 56.94\, \Omega$. Using the SHUNT/PARALLEL IMPEDANCE section from the NEW LOW BAND SOFTWARE program, we find that a parallel capacitor with an impedance of $-127\, \Omega$ will turn the impedance into $115\, \Omega$, not exactly a common coaxial cable impedance. But what if we used a quarter-wave 75-Ω feed line for achieving a 90° phase shift? This will work but because the antenna impedance is not the same as the load impedance, the typical quarter-wave impedance transformation will occur. The impedance at the end of the line will be $(75 \times 75)/115 = 49\, \Omega$. This means that there will be a voltage transformation of $115/49 = 2.3:1$. In this particular setup, we will need to use a 180°-phase-shift transformer that has a transformation ratio (turns ratio) of 2.3:1 if we want to end up with equal current magnitudes at both elements.

Would you ever want to go through this procedure to achieve tracking? No, because tracking is limited anyhow by the variation in element feed impedances as you change frequency. This principle holds very well, however, when you are using elements that show little or no change in feed impedance when the frequency is changed, which is what occurs with many receiving antennas, as explained in Chapter 7.

This principle can also be used with complex arrays (4 elements and more) to achieve better bandwidth. Such designs are far from being "plug and play" and are explained for the reader to understand the principle rather than to serve as a building kit! For an application of this principle see Section 4.7.2.

3.4.5. Using an L network to obtain a desired shift
3.4.5.1. Current Forcing:

Roy Lewallen, W7EL, uses a method that takes advantage of the specific properties of quarter-wave feed lines (Lewallen calls it *current-forcing*). This method is covered in great detail by W7EL in recent editions of *The ARRL Antenna Book*.

A quarter-wave feed line has the following wonderful property, which is put to work with this particular feed method: The magnitude of the input current of a λ/4 transmission line is equal to the output voltage divided by the characteristic impedance of the line. It is independent of the load impedance. In addition, the input current lags the output voltage by 90° and is also independent of the load impedance.

3.4.5.2. Using a simple L-network to obtain the right phase shift

The method of using an L-network to obtain the proper phase shift was also introduced by Lewallen, W7EL. The original Lewallen feed method could only be applied to antennas fed in quadrature, which means antennas where the elements are fed with phase differences that are a multiple of 90°. Later the L-network technique approach was made more flexible, and the equations were made available where you could calculated the L-network for arrays where the L-network feeds more than one element, as well as for arrays where the current magnitude is not the same in all elements. Robye Lahlum, W1MK, worked out the following equations for such an L-network, including any arbitrarily chosen phase angle (no longer only multiples of 90°).

Let's have a look at **Fig 11-11**. This is an example of a

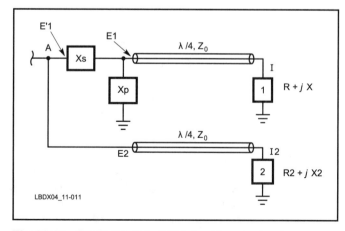

Fig 11-11—Basic layout of the L-network phasing system developed by R. Lewallen, W7EL, and enhanced for any phase angle by R. Lahlum, W1MK.

2-element array, where element 2 is fed directly, and element 1 is fed through an L-network. Both elements are fed through quarter-wave current-forcing feed lines—although this is not strictly necessary as explained in Section 3.4.9—but it makes measuring and tuning easier.

Voltage E1, at the end of the feed line going to element 1 is transformed in the L-network to E1′. The transformation is:

$$E1' \rightarrow k \times E1 \underline{/\theta°} \qquad \text{(Eq 11-1)}$$

The k factor is related to the transformation's magnitude and the desired phase shift is represented by the angle θ. Obviously, we want to connect the input of the L-network (where the voltage is E1′) to the input of the quarter-wave feed line going to element 2, where the voltage is E2.

We can connect those two points together, if the voltages in those points are identical. In other words if:

$$E2 = k \times E1 \ \underline{/\ \theta°} \qquad \text{(Eq 11-2)}$$

The condition for this to apply is:

$$X_S = \frac{-\sin\theta \times Z_0^2}{n \times k \times R} \qquad \text{(Eq 11-3)}$$

$$X_P = \frac{X_S}{\left[\dfrac{n \times X \times X_s}{Z_0^2} - 1 + \dfrac{\cos\theta}{k}\right]} \qquad \text{(Eq 11-4)}$$

Theta (θ) is the desired difference between the current phase angle at the element fed through the L-network and the phase angle at the input of the network. The phase angle is responsible for a time delay, and q must be negative. If necessary subtract 360° to obtain a negative value. Make sure you do not invert signs! Follow the examples given to understand the procedure.

The letter k is the ratio of the current supplied to the element in the branch fed through the L-network (in this case it is feed current magnitude of element 1), versus the current in the element fed directly (in this case, element 2).

The letter n is the number of identical elements (with identical feed currents) that are fed through the branch containing the L-network (see **Fig 11-12**) for a case where two elements are fed with identical current magnitudes and phases.

Z_0 is the characteristic impedance of the quarter-wave (or 3λ/4 or 5λ/4, etc) current-forcing feed lines.

R is the real part of the feed-point impedance of one of the identical element(s).

X is the imaginary part of the feed-point impedance ($Z = X + j\,X$).

X_S is the impedance of the series element in the L-network.

X_P is the impedance of the parallel (shunt) element in the L-network.

These apply under all circumstances where you feed the elements via current-forcing feed lines. The impedance R + j X is *not* the impedance at the end of the feed line but the feed-point impedance of an antenna element.

The equations do not work for 0° or 180°, but for 0° you do not need a phase-shifter and for 180° we have the choice between a half-wave long feed line or a 180°-phase-reversal

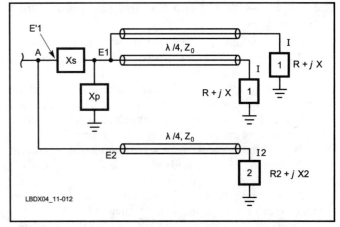

Fig 11-12—In this particular case the L-networks feeds two elements with identical feed currents. All you need to do is enter n = 2 in the *Lahlum.xls* spreadsheet.

transformer (see Section 4.15.7).

Note that in these equations no consideration was given to the losses in the feed lines nor in the network. Under most real-life conditions these losses are small on the low bands. We can, however, do the calculation including cable losses as well (see Sections 3.4.5.4).

Also assuming no feed-line losses, we can easily calculate the input impedance at the input side of the L-network. The parallel input impedance components at the input of the network are:

$$R_{par} = \frac{Z_0^2}{k^2 \times n \times R} \qquad \text{(Eq 11-5)}$$

$$X_{par} = \frac{X_S}{1 - k \times \cos\theta} \qquad \text{(Eq 11-6)}$$

These values must be converted to their equivalent series-input impedances:

$$R_{ser} = \frac{R_{par} \times X_{par}^2}{R_{par}^2 + X_{par}^2} \qquad \text{(Eq 11-7)}$$

and

$$X_{ser} = \frac{R_{par}^2 \times X_{par}}{R_{par}^2 + X_{par}^2} \qquad \text{(Eq 11-8)}$$

This parallel-to-series calculation can also be done using the using the RC/RL transformation module of the NEW LOW BAND SOFTWARE program.

3.4.5.3 The Lahlum.xls spreadsheet tool

I wrote an *Excel* spreadsheet (*Lahlum.xls*) that is on the CD-ROM bundled with this book. This tool allows you to calculate the values of the L network, as well as the resulting input impedance of the branch with the L-network. Usage is simple and self-explanatory. The spreadsheet uses the formulas shown in Section 3.4.5.2.

3.4.5.4. Two-element end-fire array in quadrature feed

In this example in **Fig 11-13** for a 2-element end-fire array from Section 3.4.1, the L-network goes to one element (in a Four-Square it may drive two elements), so enter 1 for the number of elements. Z_0 is the characteristic impedance of the quarter-wave line going from the L-network to the element(s). R and X are the real and the imaginary values of the impedance of the element at the end of that line (in our case R = 51 and X = +20). For the moment enter k = 1, meaning that the current magnitude in the elements is identical. Use theta = (−90) − (0) = −90°.

As explained above, the formulas used in the spreadsheet assume no cable loss. If you want to calculate the L-network values and include cable loss, you can calculate the impedance at the end of the current-forcing feed line, using the COAX TRANSFORMER/SMITH CHART module of the NEW LOW BAND SOFTWARE, and use the option "with cable losses." You can also use a transmission-line program such as ARRL's *TLW*. Once you know the impedance at the end of the feed lines, you can calculate the L-network component values using the second part of the spreadsheet (called: "For system NOT USING current-forcing, or if using "real" quarter-wave lines").

The first part of the *Lahlum.xls* spreadsheet calculates without taking into account cable losses. **Fig 11-14** shows the feed network for the case where cable losses are included. Note that the difference in L-network values is very small. In most cases the lossless calculation will suffice. In most of the examples in this chapter, thus, we will use lossless calculations.

For a 2-element end-fire array we normally feed the back element directly, with the exception of feeding using the cross-fire principle (see Section 3.4.4.). We can, however, feed the front element directly and the back element with a phase shift. In the case of quadrature feeding, this is +90°, which equals +90 − 360 = −270°. We can achieve the −270° phase shift by designing an L-network to do just that, or we can do this using an L-network that takes care of −90°, followed by a half-wave of feed line, for another 180°. When *Lahlum.xls* is used with θ = 270°, the resulting L-network values are 352.0 pF and 104.7 μH. The inductance required is rather high, which is not desirable. If however we replace −270° with −90°, and add a half-wave feed line at the input of the L-network, we end up with much more attractive component values of 687.2 pF and 5.0 μH.

In many of the phased arrays described in this chapter, the rear element has a very low feed impedance, often with a negative value for the series resistance. At the end of the λ/4 current-forcing feed line, the impedance becomes very high. If we design a feed system that includes an L-network in this branch, we will very often end up with extreme component values. If the reactances are very high, the Q will be high and bandwidth very low. In many cases we will see reactances change from high positive values to high negative values with just a small change in frequency. This situation must be

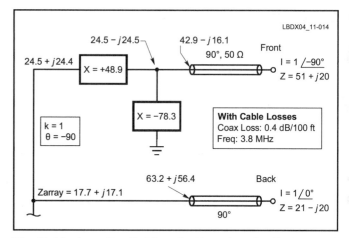

Fig 11-14—Lewallen/Lahlum feed system for the 2-element cardioid array from Fig 11-10. Calculations were done including cable losses. Note the minute difference between the lossless and the "real-world" calculation results in Fig 11-13.

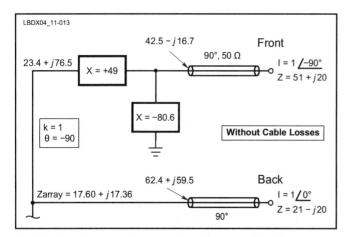

Fig 11-13—Lewallen/Lahlum feed system for the 2-element cardioid array from Fig 11-10. Calculations were done assuming zero cable losses.

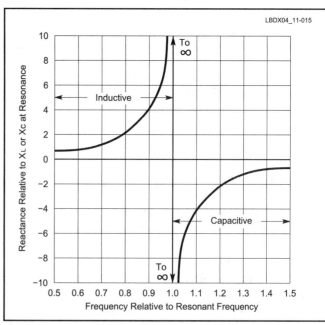

Fig 11-15—This demonstrates that the change in reactance is much greater near resonance than far away.

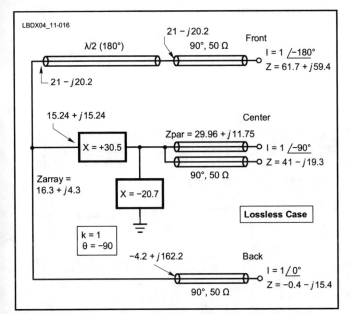

Fig 11-16—Feed system for the quadrature-fed Four-Square, using 50-Ω current-forcing feed lines.

with $X = -39.1$ Ω, the feed impedance becomes 39.1 Ω, which results in an acceptable 1.25:1 SWR for 50-Ω coax. Using a parallel capacitor with a reactance of −78 Ω, results in a feed impedance of 78 Ω, giving a good match to a 75-Ω feed line, if you'd like to use that.

3.4.5.5. Calculation of array feed impedance

There are two ways of doing this: without losses and with losses. In most cases the lossless way will suffice, but I will explain both ways.

3.4.5.6. Without losses: Using the Lahlum.xls spreadsheet:

See Section 3.4.5.2 for the formulas, but the top part of the spreadsheet does all the work. Let's do it, step by step:

$R_{par} = Z_0^2/(n \times R \times k^2)$

and

$X_{par} = X_S/(1 - k \times \cos \theta)$

In this case $k = 1$ and $\theta = -90°$ so the formula becomes:

$X_{par} = Z_0^2/R$

and

$X_{par} = X_S$

Using the figures from the above example we have:

$R_{par} = (50 \times 50)/51 = 49$ Ω

and $X_{par} = 49$ Ω

These values must be converted to their series-equivalent input impedances using the following formulas:

$R_{ser} = (R_{par} \times X_{par}^2)/(R_{par}^2 + X_{par}^2) = (49 \times 49 \times 49)/(49 \times 49 + 49 \times 49) = 24.5$ Ω

$X_{ser} = (R_{par}^2 \times X_{par})/(R_{par}^2 + X_{par}^2) = (49 \times 49 \times 49)/(49 \times 49 + 49 \times 49) = 24.5$ Ω

The transformation from parallel to serial impedance (and vice versa) can also be calculated using the RC/RL transformation module of the LOW BAND SOFTWARE program. Now we connect this impedance in parallel with $62.4 + j\,59.5$ Ω. The result is $17.60 + j\,17.36$ Ω.

3.4.5.7. Including losses:

$Z1 = 51 + j\,20$ Ω
$Z2 = 21 - j\,20$ Ω

Using the COAX TRANSFORMER/SMITH CHART module of the NEW LOW BAND SOFTWARE, we calculate the transformed impedances at the end of 90° long feed lines (Vf = 0.66, attenuation = 0.3 dB/100 feet, at F = 3.8 MHz):

$Z1' = 42.95 - j\,16.1$ Ω
$Z2' = 63.2 + j\,56.4$ Ω

Now $-j\,80.6$ Ω in parallel with $42.5 - j\,16.7$ Ω $= 24.49 - j\,24.53$ Ω. This is in series with $+j\,49$ Ω, yielding $24.49 + j\,24.53$ Ω. Now, we connect this impedance in parallel with $63.2 + j\,56.4$ Ω and the result is $17.67 + j\,17.12$ Ω.

This calculation includes cable losses but not the losses from the L-network components. Note that this values is very close to what we calculated in the lossless case.

3.4.5.8. Tutorial:

The 2 EL AND 4 EL VERTICAL ARRAYS module of the NEW LOW BAND SOFTWARE is a tutorial and engi-

---

avoided. Therefore it is always best to feed the rear element directly, and the center and front elements via L-networks.

Fig 11-15 shows how the reactance near resonance abruptly changes from inductive to capacitive, and also demonstrates that the relative change in reactance is much greater in that area than farther away. It's a good rule of thumb to design a network where the absolute value of the component reactances are not larger than about 250 Ω.

It's a good idea always to work out all the alternative feed systems. You can do these exercises with 50 and with 75-Ω cable. And if there are phasing angles involved that are larger than 180°, you can use a half-wave coax cable to the 180° part (see **Fig 11-16**). For each alternative, look at the total array feed impedance and at the L-network component values.

3.4.5.5. Using a different Z_0 (current-forcing feed-line impedance)

The example of Figs 11-13 and 11-14 results in a relatively low array feed impedance (~ $17.6 + j\,17.4$ Ω). We could do the same exercise using 75-Ω feed lines, and we will see a higher feed impedance. How much higher? Robye, W1MK, pointed out a simple rule-of-thumb (not 100% correct but a close indicator):

$Z(\text{feed–75 Ω}) = Z(\text{feed–50 Ω}) \times [75/50] \times 2 = Z(\text{feed–50 Ω}) \times 2.25$ (Eq 11-9)

In our example the estimated (lossless) impedance, according to this rule is $2.25 \times (17.6 + j\,17.4) = 39.6 + j\,39.15$ Ω. If we do the detailed calculations, the feed impedance, using 75-Ω element feed lines, turns out to be: $39.6 + j\,39.1$ Ω, which confirms the simple rule above.

In this particular case it would certainly be better using 75-Ω element feed lines. Note that in many arrays using 75-Ω feed lines achieves an overall network drive impedance closer to 50 Ω than is the case when using 50-Ω lines.

The $39.6 + j\,39.1$ Ω can be matched pretty well to either a 50-Ω or a 75-Ω feed line to the shack using a series capacitor neering program that takes you step by step through the design of a 2-element cardioid type phased array (and also the famous Four-Square array, which I'll describe later in this chapter). The results as displayed in that program will be slightly different from the results shown here, since the software uses lossless feed lines.

3.4.5.9. The quadrature-fed Four-Square

Let's assume we have obtained the following feed impedance values through modeling a Four-Square array:

Z1 = 61.7 + j 59.4 Ω (at the front element, fed with a −180° current phase angle)

Z2 = Z3 = 41 − j 19.3 Ω (the center elements, both fed with a −90° current phase angle)

Z4 = −0.4 − j 15.4 Ω (at the 0° element, the back element)

Note that the −0.4-Ω resistive part of the feed impedance Z4 means that the antenna is not *taking* power from the feed network, bur rather *delivering* power to it. This is excess power due to mutual coupling to the other elements. Note also that in a lossless calculation such a negative (usually very low) value will show up as a negative (high) value at the end of the λ/4 feed line. If, however, the nominal value is low, and the cable attenuation is taken into consideration, a small negative R-value at the antenna can turn up a high positive R-value at the other end. This is due to the effect of cable loss.

Note also that in this array, as is the case in most multi-element arrays, the SWR of the feed line going to the back element is *very* high, which normally causes a lot of additional power loss due to SWR. But in this case, the power flow is so small into the feed line to the back element that it does not matter much. High SWR, but no power flow, results in very little watts being lost. If you look at the resistive part of the equivalent-parallel resistance (several thousand Ω) at the end of the λ/4 feed line, any reduction in the exact value due to losses would cause very little increase in input power to get the same current to flow into the loads. This means that you can use the lossless model to calculate the feed system.

As explained for the 2-element end-fire array we can design the feed system in different ways. The most common approaches are:

- Feeding the back element directly, the front element via a 180° phase shift line (λ/2) and the central elements via a L-network "from the back element," all of this with 50-Ω, λ/4 feed lines. See **Fig 11-16**.
- Identical as above, but with 75-Ω feed lines. See **Fig 11-17**.

There is no absolute need to feed the back element directly and the middle and front via a phasing system. You could feed the center elements directly and the front with a −90° phasing system (L network) and the back element with a −270° phasing system. In a third alternative you could feed the front directly, the center elements with −270° phase shift and the back with −180° phase shift.

Each solution will have different L-network component values and a different array input impedance and different values for the L-network components. You can then select the network with the most manageable network component values and the most attractive feed impedance (avoid values below 10 Ω).

It's a good idea always to work out all the alternative feed systems. In the case of a Four-Square array you can use either

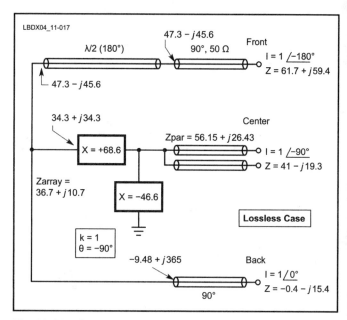

Fig 11-17—Same feed system but using 75-Ω feed lines. This results in a significantly higher feed impedance, which is desirable.

the branch to the back element as the *reference* branch, which is fed directly, or the branch to the center element or even the branch to the front element. You can do these exercises with 50-Ω and with 75-Ω cable. And if there are phasing values involved that are larger than 180°, you can use a half-wave coax cable to the 180° part (see Fig 11-16).

Table 11-1 and **Table 11-2** show the *Lahlum.xls* results for a 75-Ω and for a 50-Ω system impedance, if we apply R = 41, X = −19.3, n = 2 and λ = −90°. It is obvious that the 75-Ω solution is the better one, since it results in a much more convenient array feed impedance.

3.4.5.10. The example of a three-in-line end-fire array with binomial current distribution.

You can also use the Lewallen feed system in arrays using different current magnitudes on each of the elements. Using parallel cables is not the right solution with the Lewallen method. The formulas, as given above, assume that the quarter-wave feed lines to all the elements in the array have the same impedance. If the current magnitude of the element fed through the L network needs to be different from the magnitude of the current to the other elements in the array, the appropriate current can be achieved by specifying the correct k-value in the *Lahlum.xls* spreadsheet or in the formulas from Section 3.4.5.

Let's work out the example of the 3 elements in-line array, each spaced λ/4, fed in 90° increments, but with the center element fed with double the current magnitude. See **Fig 11-18**.

Front element: Z1 = 76.1 + j 51 Ω
Center element: Z2 = 26.3 − j 0.4 Ω
Back element: Z3 = 15 − j 22.6 Ω

In the *Lahum.xls* spreadsheet in **Table 11-3** we enter k = 2, which means that the element(s) fed through the L-networks will have twice the current magnitude as the

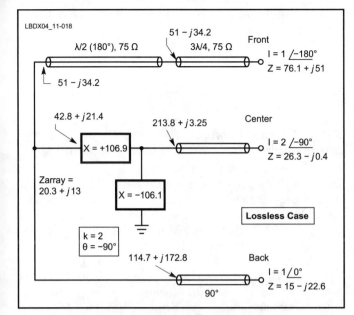

Fig 11-18—Classic configuration with direct feed to the back element. Specifying k = 2 in the *Lahlum.xls* spreadsheet allows us to double the feed current magnitude without having to resort to paralleled feed lines.

reference element in the array.

Fig 11-18 uses 75-Ω feed lines and results in an array feed impedance of 20.3 + j 13 Ω. Using 50-Ω feed lines, the array impedance would be approx 2.25 times lower, which is certainly not the best solution! Hence 75 Ω is recommended.

If we had included losses, the real part of the feed impedance would have been slightly higher, since you need more driving power into the feed system to get the same amount of radiated power.

3.4.5.10.1 Calculating the array input impedance

In order to prove that the real part of the input impedance would indeed be higher, we will carry out a calculation in a real-world environment. Note that in order to reach the center of the array (which is necessary if you want to switch directions) you will need $3\lambda/4$ feed lines, as the element physical spacing is $\lambda/4$.

Let's do some impedance calculations using the applicable modules of the NEW LOW BAND SOFTWARE. Using the COAX TRANSFORMER/SMITH CHART module we first calculate the impedances at the end of our $3\lambda/4$ feed lines ($5\lambda/4$ feed line to front element). As explained earlier, a lossless calculation will do

Front element: $Z1 = 76.1 + j\,51\,\Omega \rightarrow 79.1 + j\,27.3\,\Omega$
Center element: $Z2 = 26.3 - j\,0.4\,\Omega \rightarrow 180.7 + j\,2.2\,\Omega$
Back element: $Z3 = 15 - j\,22.6\,\Omega \rightarrow 128.9 + j\,134.9\,\Omega$

I used 75-Ω coax (Vf = 0.8) with a loss of 0.2 dB/100 feet for the calculation (design frequency = 1.8 MHz). Next we calculate the parallel impedance caused by the parallel reactance Xp1, using the module PARALLEL IMPEDANCES (T-JUNCTION).

Xp1 calculates as $-j\,106\,\Omega$ in parallel with $180.7 + j\,2.2\,\Omega$, which gives $46.8 - j\,79.2\,\Omega$. Adding $+j\,106.9\,\Omega$ in series yields: $46.8 + j\,27.78\,\Omega$.

Table 11-1
Lahlum.xls spreadsheet results (see Fig 11-17):

INPUT DATA

n elem	2.00
Z	75.00 ohm
R⁰	41.00 ohm
X	−19.30 ohm
k	1.00
theta	−90.00 deg
freq	3.80 MHz

RESULTS

Xs	68.60 ohm
Xp	−46.64 ohm
Series elem	2.9 µH
Par elem	898.4 pF
Rpar	68.60 ohm
Xpar	68.60 ohm
Rser	34.30 ohm
Xser	34.30 ohm

Table 11-2
Lahlum.xls spreadsheet results (see Fig 11-16):

INPUT DATA

n elem	2.00
Zo	50.00 ohm
R	41.00 ohm
X	−19.30 ohm
k	1.00
theta	−90.00 deg
freq	3.80 MHz

RESULTS

Xs	30.49 ohm
Xp	−20.73 ohm
Series elem	1.3 µH
Par elem	2021.4 pF
Rpar	30.49 ohm
Xpar	30.49 ohm
Rser	15.24 ohm
Xser	15.24 ohm

Table 11-3
Lahlum.xls spreadsheet results (see Fig 11-18):

INPUT DATA

n elem	1.00
Zo	75.00 ohm
R	26.30 ohm
X	−0.40 ohm
k	2.00
theta	−90.00 deg
freq	3.80 MHz

RESULTS

Xs	106.94 ohm
Xp	−106.13 ohm
Series elem	4.5 µH
Par elem	394.8 pF
Rpar	53.47 ohm
Xpar	106.94 ohm
Rser	42.78 ohm
Xser	21.39 ohm

For the back element we have an impedance of $128.9 + j\,134.91\ \Omega$ at the end of the $3\lambda/4$ feed line. For the front element we have an impedance of $79.1 + j\,37.3\ \Omega$ at the end of the $5\lambda/8$ feed line.

All three are in parallel, so $Z_{tot} = 24.5 + j\,14.9\ \Omega$. This is what we expected. Compared to the value we calculated without losses ($20.3 + j\,13$) the feed impedance goes a little higher when losses are included.

3.4.5.11. Input impedance at the input side of the L-network

If we neglect the effect of losses in the feed lines, we can also calculate the input impedance using the R_{ser} and X_{ser} values from the *Lahlum.xls* worksheet. $R_{ser} = 42.78\ \Omega$ and $X_{ser} = 21.39\ \Omega$

These values are somewhat lower than those calculated considering cable losses ($46.8 + j\,27.78\ \Omega$) The difference is relatively high because in this case we are using 270° feed lines, which represents more loss. But for all practical purposes the lossless calculations are adequate.

3.4.5.12. Using Lahlum's formulas for desired phase angles—the modified Lewallen method

So far we have used $\theta = 90°$ in the generic formulas shown in Section 3.4.5. Robye Lahlum, W1MK, developed the formulas that allow us to use the L-network to obtain a phase shift other than 90° with different current magnitudes, and he decided to share them with me for publication in this book, for which I am very grateful!

As we will see in Section 5 it appears that we can significantly improve the performance of a Four-Square by not feeding the element in quadrature (in 90° steps) and with equal current magnitudes. Jim Breakall, WA3FET developed such an optimized version of a Four-Square array.

In **Fig 11-19** the back element is the reference element, with $\theta = 0°$ and $k = 1$. The two center elements are fed with a phase angle of $-111°$ and a current magnitude ratio of $k = 0.9$, the front element with $\theta = -218°$ and $k = 0.872$. In this example I used the following feed impedances for a full-size quarter-wave spaced Four-Square, including 2 Ω ground-loss resistance:

Z-front element: $36.6 + j\,69.4\ \Omega$
Z-center-elements: $33.1\ \Omega$
Z-back element: $5.7 + j\,3.5\ \Omega$

The component values are computed in the *Lahlum.xls* spreadsheet. See **Table 11-4** and **Table 11-5**, based on a 75-Ω cable impedance. If I had used 50-Ω feed lines, the array impedance would have been approx 2.25 times lower than shown in Fig 11-19 or approx $28 - j\,2.2\ \Omega$.

As explained above, the formulas used in the spreadsheet assume zero-loss transmission lines. In most cases this will give a result accurate enough to tell you what the approximate value of the components of the L-network will be. You can however also do the exercise including cable losses. See Section 3.4.9 and **Fig 11-20**, which shows but one of the many alternative solutions that can be calculated using the *Lahlum.xls* calculation tool.

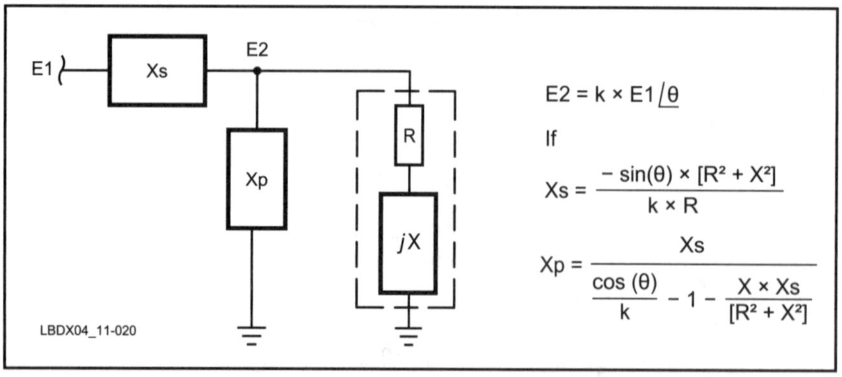

3.4.5.13. Array impedance:

From the spreadsheet we find the input impedance to both L-networks:

Z (to center elements): $30.17 + j\,47.48\ \Omega$
Z (to front element): $18.58 - j\,58.4\ \Omega$

Let's now use the COAX TRANSFORMER/SMITH CHART module of the NEW LOW BAND SOFTWARE to calculate the impedances at the end of the $\lambda/4$ feed line going to the back element (fed without phasing):

Back element: $Z = 5.7 + j\,3.5\ \Omega$. At the other end of the current-forcing feed

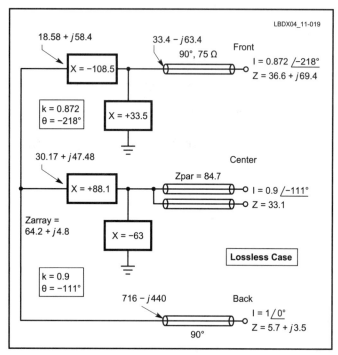

Fig 11-19—Lahlum/Lewallen feed network for a $\lambda/4$-spaced Four-Square array using current-forcing with 75-Ω, $\lambda/4$ feed lines and where the back element is directly fed.

Fig 11-20—Equations for calculating the L-network components needed to produce a desired phase shift θ, based upon the feed-point impedances (R = real part, X = reactive part). (These equations do not use lossless current-forcing feed lines that are odd multiples of $\lambda/4$, although that option is available in the upper portion of the *Lahlum.xls* spreadsheet.) See text for details.

Table 11-4
Lahlum.xls spreadsheet results (see Fig 11-19):

INPUT DATA
n elem	2
Zo	75.00 Ω
R	33.10 Ω
X	0.00 Ω
k	0.90
theta	−111.00°
freq	3.80 MHz

RESULTS
Xs	88.14 Ω
Xp	−63.04 Ω
Series elem	3.7 µH
Par elem	664.7 pF
Rpar	104.90 Ω
Xpar	66.65 Ω
Rser	30.17 Ω
Xser	47.48 Ω

Table 11-5
Lahlum.xls spreadsheet results (see Fig 11-19):

INPUT DATA
n elem	1
Zo	75.00 Ω
R	36.60 Ω
X	69.40 Ω
k	0.87
theta	−218.00 deg
freq	3.80 MHz

RESULTS
Xs	−108.51 Ω
Xp	33.46 Ω
Series elem	386.2 pF
Par elem	1.4 µH
Rpar	202.12 Ω
Xpar	−64.31 Ω
Rser	18.58 Ω
Xser	−58.40 Ω

line: $Z = 716 - j\,440$ Ω. All three in parallel (calculated with the PARALLEL IMPEDANCES module of the NEW LOW BAND SOFTWARE: $Z_{tot} = 64.2 - j\,4.8$ Ω.

3.4.6. Collins (W1FC) hybrid-coupler method

Fred Collins, W1FC, developed a feed system similar to the Lewallen system in that it uses current-forcing λ/4 feed lines to the individual elements. There is one difference, however. Instead of using an L network, Collins uses a quadrature *hybrid coupler*, shown in **Fig 11-21**.

The hybrid coupler divides the input power (at port 1) equally between ports 2 and 4, with theoretically no power output at port 3 if all four port impedances are the same. When the output impedances are not the same, power will be dissipated in the load resistor connected to port 4. In addition, the phase difference between the signal at ports 2 and 4 will not be different by 90° if the load impedance of these ports is not real or, if complex, they do not have an identical reactive part.

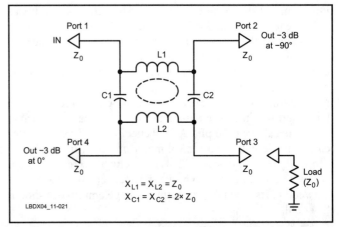

Fig 11-21—Hybrid coupler providing two −3 dB outputs with a phase difference of 90°. L1 and L2 are closely coupled. See text for construction details.

We will examine whether or not this characteristic of the hybrid coupler is important to its application as a feed system for a quadrature-fed array.

3.4.6.1. Hybrid coupler construction

The values of the hybrid coupler components are:

XL1 = XL2 = 50-Ω system impedance
XC1 = XC2 = 2 × 50 Ω = 100 Ω

For 3.65 MHz the component values are:

$$L1 = L2 = \frac{X_L}{2\pi f} = \frac{50}{2\pi \times 3.65} = 2.18\,\mu H$$

$$C1 = C2 = \frac{10^6}{2\pi f X_C} = \frac{10^6}{2\pi \times 3.65 \times 100} = 436\,pF$$

When constructing the coupler, you should take into account the capacitance between the wires of the inductors, L1 and L2, which can be as high as 10% of the required total value for C1 and C2. The correct procedure is to first wind the tightly coupled coils L1 and L2, then measure the inter-winding capacitance and deduct that value from the theoretical value of C1 and C2 to determine the required capacitor value. For best coupling, the coils should be wound on powdered-iron toroidal cores. The T225-2 (µ = 10) cores from Amidon are a good choice for power levels well in excess of 2 kW. The larger the core, the higher the power-handling capability. Consult Table 6-3 in Chapter 6 for core data. The T225-2 core has an A_L factor of 120. The required number of turns is calculated as:

$$N = 100\sqrt{\frac{2.18}{120}} = 13.4\,turns$$

The coils can be wound with AWG #14 or AWG #16 multi-strand Teflon-covered wire. The two coils can be wound with the turns of both coils wound adjacent to one another, or the two wires of the two coils can be twisted together at a rate of 5 to 7 turns per inch before winding them (equally spaced) onto the core.

At this point, measure the inductance of the coils (with an

impedance bridge or an LC meter) and trim them as closely as possible to the required value of 2.09 µH for each coil. Do *not* merely go by the calculated number of turns, since the permeability of these cores can vary quite significantly from production lot or one manufacturer to another. Moving the windings on the core can help you fine-tune the inductance of the coil. Now the interwinding capacitance can be measured. This is the value that must be subtracted from the capacitor value calculated above (436 pF). A final check of the hybrid coupler can be made with a vector voltmeter or a dual-trace oscilloscope. By terminating ports 2, 3 and 4 with 50-Ω resistors, you can now fine-tune the hybrid for an exact 90° phase shift between ports 2 and 4. The output voltage amplitudes should be equal.

3.4.7. Gehrke (K2BT) method

Gehrke, K2BT, has developed a technique that is fairly standard in the broadcast world. The elements of the array are fed with randomly selected lengths of feed line, and the required feed currents at each element are obtained by the insertion of discrete component (lumped-constant) networks in the feed system. He makes use of L networks and constant-impedance T or pi phasing networks. The detailed description of this procedure is given in Ref 924.

The Gehrke method consists of selecting equal lengths (not necessarily 90°lengths) for the feed lines running from the elements to a common point where the array switching and matching are done. With this method, the length of the feed lines can be chosen by the designer to suit any physical requirements of the particular installation. The cables should be long enough to reach a common point, such as the middle of the triangle in the case of a triangle-shaped array.

As this method is rarely used in amateur circles, I decided not to describe it in detail in this edition of the book (but it was covered in all previous editions). This method however has the tremendous merit that it was the first one described in amateur literature that was technically 100% correct.

3.4.8. Lahlum (W1MK)/Gehrke(K2BT)

The Lahlum/Lewallen method described in sect 3.4.6 can be applied with any coax feed length—the length does not necessarily have to be λ/4 or odd multiples of λ/4. While the use of current-forcing is a very desirable feature there are situations where you might not care to use current forcing. For example, the use of the array on multiple bands with the use of the same coax feed for both bands. The Lahlum/Lewallen method is suitable in this situation.

I called this system the "Lahlum/Gehrke" system, since it uses the mathematics developed by Robye Lahlum, W1MK, and follows more or less the principle of Gehrke's original methods, where arbitrary lengths of feed lines were used to the elements.

In this case we will first have to calculate the impedances at the end of the feed lines; eg, using the COAX TRANSFORMER/SMITH CHART module of the NEW LOW BAND SOFTWARE. The formulas involved are given in Fig 11-20. R and jX are the impedance values of the feed impedances of the antenna elements, transformed by the coaxial feed line.

In the situation explained in Section 3.4.6, k is the ratio of the feed currents when we use current-forcing feed lines. In this application however, k = E1/E2, the ratio between the *voltages* at the end of the feed lines. These feed lines are not necessarily 90° long—or odd multiple thereof— and they do not even have to be of equal length. θ is again the phase shift caused by the L-network. It is the phase angle difference between the voltages at the end of the two equal-length feed lines. More precisely it is the difference between the voltage phase angle at the output of the L-network and the phase angle at the input of the network. θ *must* be negative. If necessary subtract 360° to obtain a negative value. **Fig 11-22** shows the principle.

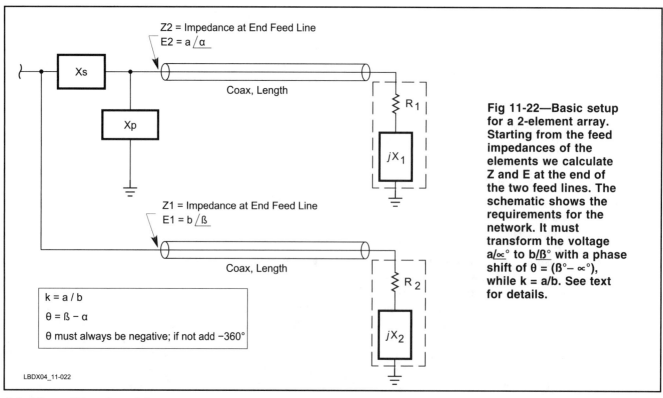

Fig 11-22—Basic setup for a 2-element array. Starting from the feed impedances of the elements we calculate Z and E at the end of the two feed lines. The schematic shows the requirements for the network. It must transform the voltage a/∝° to b/ß° with a phase shift of θ = (ß°− ∝°), while k = a/b. See text for details.

This non-current-forcing feed systems is an elegant solution where you want to built two-band arrays; for example, covering 80 meters with wide spacing (approx λ/4) and 160 meters with close (λ/8) spacing. Let's work out an example for a 2 element, λ/8 spacing case, where the phase shift is −135°. Through antenna modeling we obtain the following element impedance values:

Back element:

I_{back} = 1 /0° A

Z_{back} = 13 − j 21 Ω

Front element:

I_{front} = 1/−135° A

Z_{front} = 18 + j 23 Ω

Using the COAX TRANSFORMER/SMITH CHART module of the NEW LOW BAND SOFTWARE, the values at the end of a 38.4° long feed line are calculated. (Note: It's *not* necessary that both feed lines be of equal length, unless of course you want to switch directions.). I used a frequency of 1.83 MHz, using real cable (RG-213, 0.2 dB loss/100 feet). We now need to look at the voltage at the end of the feed lines, since we need to connect them in parallel (equal voltages required!). The transformed values are:

At end feed line to back element:

Eb' = 18.12 /54.04° V

Zb' = 12.07 + j 12.13 Ω

At end feed line to front element:

Ef' = 51.23 /−61.24° V

Zf' = 61.07 + j 69.94 Ω

We need to insert an L network in either the feed line to the front or to the back element. This L-network has to perform the followings two tasks:
- Perform the required phase shift
- Perform the required voltage transformation so that the input voltage to the L-network is identical to the voltage at the end of the other feed line (so that we can connect them in parallel).

3.4.8.1. Solution 1

See **Fig 11-23**. We put the L-network in the feed-line going to the

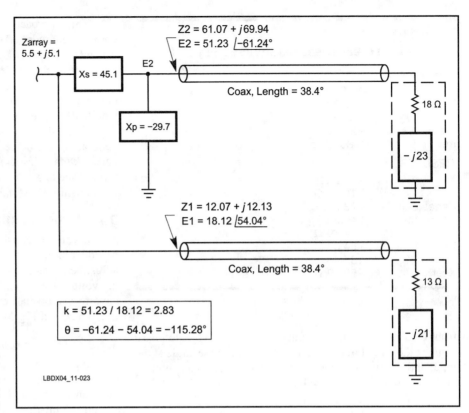

Fig 11-23—First solution for a 2-element end-fire array (λ/8 spacing). See text for details.

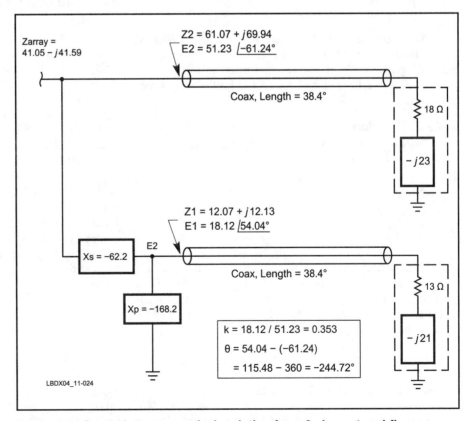

Fig 11-24—Second, more practical, solution for a 2-element end-fire array (λ/8 spacing). See text for details.

Table 11-6
Lahlum.xls spreadsheet results (see Fig 11-24):

INPUT DATA
R	12.07 Ω
X	12.13 Ω
k	0.353
theta	−244.80°
freq	3.80 MHz

RESULTS
Xs	−62.2 Ω
Xp	−168.2 Ω
Series elem	673.9 pF
Par elem	249.2 pF
Rpar	194.69 Ω
Xpar	−54.06 Ω
Rser	13.94 Ω
Xser	−50.19 Ω

---

front element. θ is the difference between the voltage phase angle at the output of the L-network and the phase angle at the input of the network. λ must be negative. If necessary subtract 360° to obtain a negative value.

θ = (−61.24) − (54.04) = −115.28°
k = ratio of the voltage magnitudes at the end of the feed lines: k = 51.23/18.12 = 2.83

We can plug these values in the formulas shown in Fig 11-20, or better yet use the special spreadsheet tool *Lahlum.xls*. This tool allows you to calculate the values of the L network directly. For this example (see **Fig 11-23**):

X_{ser} = 45.11 Ω
X_{par} = −29.7 Ω

An impedance of −29.7 Ω in parallel with 61.07 − j 69.94 Ω gives 10.07 − j 36.34 Ω. Adding the series reactance of 45.11 Ω gives 10.07 + j 8.76 Ω. Paralleling this impedance with 12.07 + j 12.13 Ω gives 5.5 + j 5.1 Ω for the array's feed impedance.

3.4.8.2. Solution 2

See **Fig 11-24** and **Table 11-6**. The L-network is in the feed line going to the back element:

θ = (54.04) − (−61.24) = + 115.28 = (−360+115.28) = −244.72°
k = 18.12/51.23 = 0.353
X_{ser} = −62.2 Ω
X_{par} = −168.2 Ω

Note that this requires two capacitors, rather than a capacitor and an inductor, for the L-network. −168.2 Ω in parallel with 12.07 + j 12.13 Ω gives 13.94 + j 12.00 Ω. Adding the series reactance of −62.2 Ω gives 13.93 − j 50.2 Ω. Paralleling this impedance with 61.07 + j 69.94 Ω gives 47.04 − j 41.59 Ω for the array feed impedance

Both solutions are valid, the only difference is the resulting input impedance. In Solution 1 the resulting input impedance is very low (5.5 + j 5.1 Ω). Solution 2 yields an array feed impedance that is much closer to 50 Ω (47 − j 41 Ω), and the use of a series inductor would give an almost perfect match to 50-Ω cable.

This approach to solving the problem of obtaining the correct amplitude and phase shift using coax feeds of any length is similar to the method of Gehrke, K2BT, however it results in much fewer circuit elements. Solving this same problem using Gehrke's method would result in the need for six or seven elements (see *Low Band DXing*, Editions 1, 2 or 3), all of which would affect the amplitude/phase relationships.

Using the Lahlum/Lewallen approach, four elements in general would be required. Two of them would be an L-network matching the array input impedance to the feed-line impedance and only two of them affect the amplitude/phase relationship, thus making it much easier to adjust.

3.4.8.3. Adjusting the network values

If you do *not* use current-forcing (feed lines that are λ/4 or odd multiples thereof), you cannot use the testing and adjustment procedure as described in Section 3.6.2. (measuring voltages at the end of the feed lines). In this case you will have to use a small current probe at the elements (see Section 3.5.5. and Fig 11-29).

3.4.8.4. Other applications of the software

While the calculation procedures described in Section 3.4.5 assume current-forcing feed lines without losses, you can use the above procedure to take actual losses into account. You first need to calculate the impedances at the end of the current forcing feed lines, using the COAX TRANSFORMER/SMITH CHART module of the NEW LOW BAND SOFTWARE (option "with cable losses") and then use these values as input date for the *Lahlum.xls* spreadsheet.

3.4.9. Choosing a feed system

Until Gehrke published his excellent series on vertical arrays, it was general practice to simply use feed lines as phasing lines, and to equate electrical line length to phase delay under all circumstances. We now know that there are better ways of accomplishing the same goal (Ref Section 3.3.1).

Fortunately, as Gehrke states, these vertical arrays are relatively easy to get working. **Fig 11-25** shows the results of an analysis of the 2-element cardioid array with deviating feed currents. The feed-current magnitude ratio as well as the phase angle are quite forgiving so far as gain is concerned. As a matter of fact, a greater phase delay (eg, 100° versus 90° will increase the gain by about 0.3 dB. The picture is totally different so far as F/B ratio is concerned. To achieve an F/B of better than 20 dB, the current magnitude as well as the phase angle need to be tightly controlled. But even with a "way off" feed system it looks like you always get between 8 and 12 dB of F/B ratio, which is indeed what we used to see from arrays that were incorrectly fed with coaxial phasing lines having the electrical length of the required phase shift.

3.4.9.1. Collins (hybrid coupler) system

We know that the perfect 90° phase shift with identical antenna feed-current magnitudes can never be obtained with this system because the hybrid is never terminated in its design impedance (50 Ω) but rather in different complex feed impedances of the elements of the array.

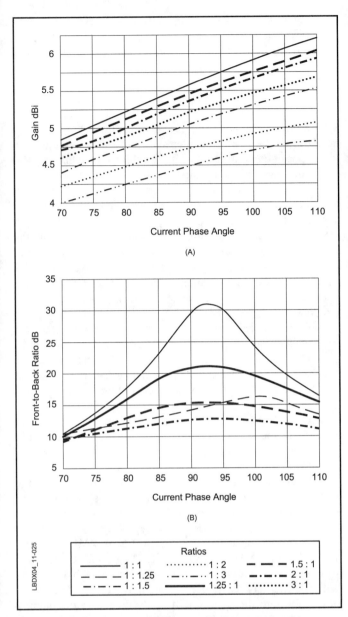

Fig 11-25—Calculated gain and the front-to-back ratio of a 2-element cardioid array versus current magnitudes and phase shifts. Calculations are for very good ground at the main elevation angle. The array tolerates large variations so far as gain is concerned, but is very sensitive so far as front-to-back ratio is concerned.

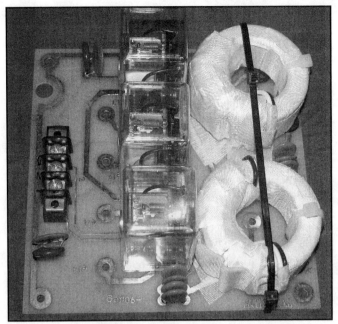

Fig 11-26—The internal works of the Comtek hybrid coupler: PC board showing two large toroidal cores: one is used in the hybrid coupler, while the second one serves to make a 180° phase-reversal transformer (used instead of a 180° phasing line). Note also the three heavy-duty relays for direction switching.

3.4.9.1.1. Performance of the hybrid coupler

I have tested the performance of a commercially made hybrid coupler (Comtek, see Section 3.4.6 and **Fig 11-26**). First the coupler was tested with the two load ports (ports 2 and 4) terminated in a 50-Ω load resistor. Under those conditions the power dissipated in the 50-Ω dummy resistor (port 3) was 21 dB down from the input power level. This means that the coupler has a directivity of 21 dB under ideal loading conditions (equivalent to 12 W dissipated in the dummy load for a 1500-W input). The results were identical for both 3.5 and 3.8 MHz. The input SWR under the same test conditions was approximately 1.1:1 (a 25-dB return loss).

I also checked the hybrid coupler for its ability to provide a 3-dB signal split with a 90° phase-angle difference. When the two hybrid ports were terminated in a 50-Ω load I measured a difference in voltage magnitude between the two output ports of 1.7 dB, with a phase-angle difference of 88° at 3.8 MHz. At 3.5 MHz the phase-angle difference remained 88°, but the difference in magnitude was down to 1.2 dB. Theoretically the difference should be 0 dB and 90°. A 1.2-dB difference means a voltage or current ratio of k = 0.87.

The commercially available hybrid coupler system from Comtek Systems (**comtek4@juno.com**) uses a toroidal-wound transmission line to achieve a 180° phase shift over a wide band-width. The phase transformer consists of a bifilar-wound conductor pair, where wire A is grounded on one end and wire B on the other end of the coil. The other two ends are the input and output connections, whereby the voltages are shifted 180° in phase. This approach eliminates the long (λ/2) coax that is otherwise required for achieving the 180° phase shift and it is broadbanded as well.

I measured the performance of this "compressed" 180° phasing line. Using a 50-Ω load, the output phase angle was −168°, with an insertion loss of 0.8 dB. With a complex-impedance load the phase shift varied between −160° and −178°. Measurements were done with a Hewlett-Packard vector voltmeter. The hybrid coupler was also evaluated using real loads in a Four-Square array.

After investigating the components of the Comtek hybrid-coupler system, I evaluated the performance of the coupler (without the 180° phase inverter transformer), using impedances found at the input ends of the λ/4 feed lines in real arrays as load impedances for ports 2 and 4 of the coupler. Let us examine the facts and figures for our 2-element end-fire cardioid array.

The SWR on the quarter-wave feed lines to the two elements (in the cardioid-pattern configuration) is not 1:1. Therefore, the impedance at the ends of the quarter-wave feed lines will depend on the element impedances and the characteristic impedances of the feed lines. We want to choose the feed-line impedances such that a minimum amount of power is dissipated in the port-3 terminating resistor.

The impedances at the end of the 90°-long real-world feed lines (λ/4 RG-213 with 0.35 dB/100 feet on 80 meters) are:

$Z1' = 42.81 - j\,16.18\ \Omega$
$Z2' = 63.1 - j\,56.94\ \Omega$

These values are reasonably close to the 50-Ω design impedance of our commercial hybrid coupler. With 75-Ω feed lines the impedance would be:

$Z1' = 95.11 - j\,35.88\ \Omega$
$Z2' = 141.05 - j\,125.4\ \Omega$

It is obvious that for a 2-element cardioid array, 50 Ω is the logical choice for the feed-line impedance. This can be different for other types of arrays. The basic 4-element Four-Square array, with λ/4 spacing and quadrature-fed, is covered in detail in Section 4.7. A special version of the Four-Square array is analyzed in detail in Section 6.

3.4.9.1.2. Array performance

Although the voltage magnitudes and phase at the ends of the two quarter-wave feed lines are not exactly what is needed for a perfect quadrature feed, it turns out that the array only suffers slightly from the minor difference. The incorrect phase angle will likely deteriorate the F/B, but the gain will remain almost the same as with the nominal quadrature driving conditions, which again, do not result in optimum gain nor directivity (see also Fig 11-25).

3.4.9.1.3. Different design impedance

We can also design the hybrid coupler with an impedance that is different from the 50-Ω quarter-wave feed-line impedance in order to realize a lower SWR at ports 2 and 4 of the coupler. The load resistor at port 3 must of course have the same ohmic value as the hybrid design impedance. Alternatively we can use a standard 50-Ω dummy load with a small L network connected between the load and the output of the hybrid coupler.

With the aid of the software module SWR ITERATION, you can scan the SWR values at ports 2 and 4 for a range of design impedances. The results can be cross-checked by measuring the power in the terminating resistor and alternately connecting 50-Ω and 75-Ω quarter-wave feed lines to the elements. A practical design case is illustrated in Section 4.7.1.2.

By choosing the most appropriate feed-line impedance as well as the optimum hybrid-coupler design impedance, it is possible to reduce the power dissipated in the load resistor to 2% to 5% of the input power. Whether or not reducing the lost power to such a low degree is worth all the effort may be questionable, but covering the issue in detail will certainly help in better understanding the hybrid coupler and its operation as a feed system for a phased array with elements fed in quadrature.

3.4.9.1.4. Bottom line

The Collins feed method (with the hybrid coupler) is only applicable in situations where the elements are fed in quadrature relationship (in increments of 90°). We also must realize that the hybrid-coupler system does *not* produce the exact phase-quadrature phase shift unless some very specific load conditions exist (resistive loading or loading with identical reactive components on both ports).

Fortunately most of the quadrature-fed arrays are quite lenient, tolerating a certain degree of deviation from the perfect quadrature condition. We know however that the quadrature feed configuration is not the best configuration, and 0.6 up to 1 dB more gain and better directivity (narrower forward lobe) can be obtained with other phasing angles and different current magnitudes (see Section 4.7.2 and Section 6.8.4).

Over the years the Collins method has become the most popular feed method, clearly because it is a "plug and play" type solution, which works most of the time! The tradeoff for this is that you are not getting peak performance, such as can be obtained with a properly adjusted Lahlum/Lewallen feed system.

One advantage with the Collins system, however, is that essentially the same (but compromised) front-to-back ratio can be achieved over the entire band (3.5 to 4.0 MHz on 80 meters).

Watch out though and don't make the error to judge the operational bandwidth of the hybrid-coupler system by measuring the SWR curve at the input of the coupler. The coupler will show a very flat SWR curve (typically less than 1.3:1) under *all* circumstances, even from 3.5 to 4 MHz or from 1.8 MHz to 1.9 MHz on 160 meters. The reason is that, away from its design frequency, the impedances on the hybrid ports will be extremely reactive, resulting in the fact that nearly all power fed into the system will be dissipated in the dummy resistor. It is typical that an array tuned for element resonance at 3.8 MHz will dissipate 50% to 80% of its input power in the dummy load when operating at 3.5 MHz. The exact amount will depend on the Q factor of the elements. On receive, the same array will still exhibit excellent directivity on 3.5 MHz, but its gain will be down by 3 to 7 dB from the gain at 3.8 MHz, since it is wasting 50% to 80% of the received signal as well into the dummy resistor.

It is clear that the only bandwidth-determining parameter is the power wasted in the load resistor. So stop bragging about your SWR curves, but let's see your dummy-load power instead! The hybrid coupler has the drawback of wasting part of the transmitter power (and receive power as well, but that's probably much less relevant) in the dummy-load resistor. Ten percent power loss may not seem a lot, but on 160 meters, where signals are often riding on or in the noise, 10% of power, which equals 0.5 dB, can be meaningful.

3.4.9.2 Christman system

The Christman method makes maximum use of the transformation characteristics of coaxial feed lines, thus minimizing the number of discrete components required in the feed network. This is an attractive solution, and should not scare off potential array builders. For a 2-element cardioid array this is certainly a good way to go. Of course, you need to go through the trouble of measuring the impedances.

With arrays of more elements, it is likely that identical voltages will only be found on two lines. For the third line, lumped-constant networks will have to be added. In such case the Lewallen or Lahlum/Lewallen method is preferred.

3.4.9.3. Lewallen and Lahlum/Lewallen systems
3.4.9.3.1. The quadrature Lewallen system

The Lewallen feed system has been used very successfully by many array builders, especially those that want no compromises and only care for peak performance. The system can produce the right phase angle and feed current magnitude for any load impedance, and one can adjust ("tune") the values of the L-network to obtain the desired values.

Lewallen, W7EL, published in the last several issues of *The ARRL Antenna Book* a number of L-network values for the 2-element cardioid and the 4-element square arrays, which a builder can use for building the L network without doing any measuring.

3.4.9.3.2. Any phase angle with Lahlum's approach

With Lahlum's introduction in this book of the extra feature that allows you to program any phase angle at any feed current magnitude, the enhanced Lahlum/Lewallen system should be considered as the best engineering choice, and should attract all those who want nothing but the best. Lahlum made the mathematics and the calculation method for this fully flexible system available to all home-builders.

It is interesting however to see that the only commercially available feed-system according to the Lewallen feed system (**www.arraysolutions.com**) in fact already was using an approach that seems to be similar if not identical to the Lahlum system. See **Fig 11-27**.

Array Solutions advertises two versions of their Four-Square feed system. One is the quadrature system (0°, –90°, –180°), the other one is called the "optimized version" with phase angles of 0°, –111° and –218°, with unspecified feed-current magnitudes. In the optimized version, the phase in the front element could be made longer than 180° (obtained through a λ/2 feed line) with the addition of a small L-network, which is exactly what is done in Lahlum's solution. Array Solutions tunes all of its feed systems for the desired feed current (magnitude and phase angle).

This system, which employs two L-networks, is "fully adjustable," which is a great advantage. Using quarter-wave (or 3λ/4 feed lines) to your array elements, you can measure the voltage (magnitude and phase) at the start of these lines, and tune the L-networks elements until you obtain exactly what you want. A simple procedure to do that is outlined in Section 3.6.

3.4.9.3.3. My experience

After having used the hybrid system for a number of years I installed a feed system according the Lahlum/Lewallen system, manufactured by Array Solutions, as shown in Fig 11-27. In Section 3.5 I cover some test equipment I used for tuning the array. See also Chapter 7. When properly tuned,

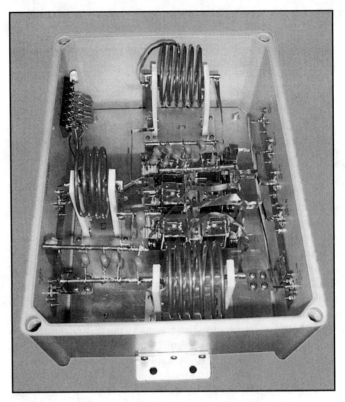

Fig 11-27—Lahlum/Lewallen feed system for a Four-Square built by Array Solutions (WXØB). The unit includes an L-network for a perfect match to the feed line as well as an omnidirectional position.

using the right test equipment, you can expect a little better performance from this system compared to the hybrid-coupler system.

The design parameters for my particular Four-Square (using one elevated radial, as described in Section 6) were:

Front element: I = 1.5/–220° A
Center elements: I = 1/–111° A
Back element: I = 0.85/0° A

These feed currents give about 0.6 dB more gain than a perfectly working quadrature feeding solution and the directivity is much enhanced (see also Section 3). I used a vector voltmeter to measured the voltages at the start of the λ/4 feed lines, and transformed to the feed currents at the elements. The measurement was confirmed used the method described in Section 3.6. A multi-channel scope brought further confirmation. The design phase and amplitude are obtained through carefully adjusting the network.

3.4.9.3.4. Bottom line

I went into great detail in the foregoing sections to explain step by step how you can calculate the Lahlum/Lewallen feed system and build one yourself. The procedure is simple:

- Model the planned array as accurately as possible.
- Use the spreadsheet program (*Lahlum.xls*) to calculate the L-network components.
- Use the NEW LOW BAND SOFTWARE to calculate the array feed impedance.

This is all pretty straightforward. Once you understand, you can calculate any array in less than 10 minutes! Make sure you calculate the L network component values based on real antenna impedances and not 50 Ω. This would yield incorrect values.

3.5. Measuring and Tuning

3.5.1. Can I put up an array without any test equipment?

None of the arrays described in this chapter can be built or set-up without any measuring. The simplest array uses a quadrature configuration, which makes it possible to use a hybrid coupler for obtaining the required phase shift within most often acceptable tolerances. Even in that case, the elements of the array will have to be tuned to proper resonance. Use an SWR meter to trim the elements to resonance. Don't forget to decouple the "other" elements. Just assume the point of lowest SWR is the resonant frequency (which is not quite true), and you will be close enough for a 2, 3 or 4 element array fed (in quadrature) with a hybrid coupler. The only other thing you should measure in such an array is the power dumped in your hybrid termination resistor. This should never be more than about 10% of the power going into the hybrid. If the power is high, try 75-Ω, λ/4 feed lines instead of 50-Ω lines, or vice versa. OK, so far we have not needed any special test equipment!

In order to obtain maximum directivity from an array, it is essential that the self-impedances of the elements be identical. Measurement of these impedances requires special test equipment, and the method explained in Section 3.6 is recommended. Equalizing the resonant frequency can be done by changing the radiator lengths, while equalizing the self-impedance can be done by changing the number of radials used. If you start putting down perfectly identical and symmetrical radial systems, you will likely get very similar values for the resistive part of the various elements. If you cannot easily get equal impedances, you will have to suspect that one or more of the array elements are coupling into another antenna or conducting structure. Take down all other antennas that are within λ/2 from the array to be erected. Do not change the length of one of the radiators to get the equal values for the resistive parts of the elements. The elements should all have the same physical height (within a few percent).

3.5.2. Can I cut my λ/4 feed lines without special test equipment?

Yes you can, but first a word of warning: Never go by the published velocity-factor figures, certainly not when you are dealing with foam coax. There are several valid methods for cutting λ/4 or λ/2 cable lengths.

You can simply use your transceiver, a good SWR bridge and a good dummy load to cut your phasing lines. Maybe the accuracy will not be as good as with other methods described later, but it is totally feasible. Connect your transmitter through a good SWR meter (a Bird 43 is a good choice) to a 50-Ω dummy load. Insert a coaxial-T connector at the output of the SWR bridge. See **Fig 11-28**.

If you need to cut a quarter-wave line (or an odd multiple of λ/4), first short the end of the coax. Make sure it is a good short, not a short with a lot of inductance. Insert the cable in the T connector. If the cable is a quarter-wave long, the cable

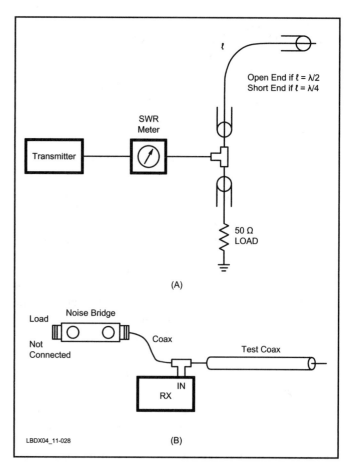

Fig 11-28—At A, very precise trimming of λ/4 and λ/2 lines can be accomplished by connecting the line under test in parallel with a 50-Ω dummy load. Watch the SWR meter while the line length or the transmitter frequency is being changed. At B, alternative method uses a noise bridge and a receiver. See text for details.

end at the T connector will show as an infinite impedance and there will be no change at all in SWR (will remain 1:1). If you change the frequency of the transmitter you will see that on both sides of the resonant frequency of the line, the SWR will rise rather sharply. For fine tuning you can use high power (eg, 1 kW) and use a sensitive meter position for measuring the reflected power. I have found this method very accurate, and the cable lengths can be trimmed very precisely.

Make sure the harmonic content from your transmitter is very low. It's a good idea to use a good low-pass or bandpass filter between the generator (transmitter) and the T-connector. A W3NQN bandbass-filter (see Chapter 15, Section 6.3) is ideal for this purpose.

3.5.3. Is there a better way to cut the λ/4 lines?

Yes, there are more accurate ways:
- Using a noise bridge (two methods are described in Section 3.6.)
- Using your antenna analyzer
- Preferred method: using the W1MK 6-dB hybrid and detector/power meter (see Section 3.6.5.)

11-24   Chapter 11

3.5.4. What about arrays using the other feed systems (Christman/Lewallen Lahlum)?

In this case we do need to measure the *self impedance* of the elements. This means you need some test equipment.

- You can use your MFJ or AEA antenna analyzer, but their precision is not always very good.
- Much better is to use the W1MK method described in Section 3.6.4.
- Best is to use a professional network analyzer or the VNA (Vector Network Analyzer) described is Section 3.6.9.
- Or use a good old-fashioned Impedance Bridge (eg, General Radio) as described in Section 3.6.10.

You should not only measure the self impedances, but you should try to make them equal, as explained in Section 3.5. Once you have measured the self impedances of all elements, you can calculate the feed impedances, as explained in Section 3.3. Check if the values you calculated are in the same ballpark as the results you obtained through modeling.

If you use a Christman feed method you should now look for points on both feed lines where the voltages are identical (see Section 3.4.2). If you use a Lewallen/Lahlum feed system, you can now calculate the value of the L networks(s) using the *Lahlum.xls* spreadsheet tool, as explained in Sections 3.4.5.

3.5.5. How can I measure that the values of the feed-current magnitude and phase angle at the elements are what I really want?

It is essential to be able to measure the feed current to assess the correct operation of the array. A good-quality RF ammeter is used for element-current magnitude measurements and a good dual-trace oscilloscope to measure the phase difference. The two inputs to the oscilloscope will have to be fed via identical lengths of coaxial cable.

Fig 11-29 shows the schematic diagram of the RF current probes for current amplitude and phase-angle measurement. Details of the devices can be found in Ref 927. D. M. Malozzi, N1DM, pointed out that it is important that the secondary of the toroidal transformer always sees its load resistor, as otherwise the voltage on the secondary can rise to extremely high values and can destroy components and also the input of an oscilloscope if the probe is to be used with a scope. He also pointed out that it is best to connect two identical load resistors at each end of the coax connecting the probe to the oscilloscope. Both resistors should have the impedance of the coax. Make sure the resistors are non-inductive, and of adequate power rating. It is not necessary to do your measurement with high power (nor advisable from a safety point of view).

3.5.6. Are there other methods that are more accurate?

Measuring voltage magnitude and phase is easier than measuring current magnitude and phase. We learned in Section 3.4.5 that λ/4 feed lines have this wonderful property called current-forcing. The property allows us to measure voltage at one end of a λ/4 cable to tell us the current at the other end of that cable. This means we make our feed lines quarter-wave (or 3-quarter-wave), and measure the voltage at the end of the feed lines where they all come together.

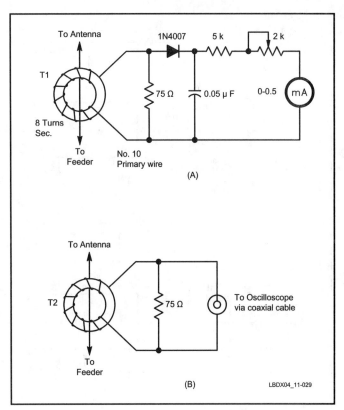

Fig 11-29—Current amplitude probe (at A) and phase probe (at B) for measuring the exact current at the feed point of each array element. See text for details.

T1, T2—Primary, single wire passing through center of core; secondary, 8 turns evenly spaced. Core is ½-in. diameter ferrite, A_L = 125 (Amidon FT-5061 or equivalent).

3.5.7. How do I measure magnitude and phase of these voltages?

The HP Vector-voltmeter (model HP-8405A) is an ideal tool, provided you can find one that has a probe in good condition. Surplus HP-8405As very often have defective probes!

3.5.8. Do I really need such lab-grade test-equipment?

No, a very attractive, simple and inexpensive, but very accurate, test method is described in Section 3.6.2.

3.6. Test Equipment and Test Procedures for Array Builders

3.6.1. Dual channel RF detector/wattmeter (by W1MK)

Various test methods described in this chapter require a sensitive null detector. In most cases a receiver can be used, but a small dedicated and calibrated (in dBm) test instrument is a real asset for any ham who wants to venture into array building.

Robye Lahlum, W1MK, built a dual-channel detector/wattmeter (a modified W7ZOI design), using two AD8307 logamps that give him a sensitivity of better than −70 dBm.

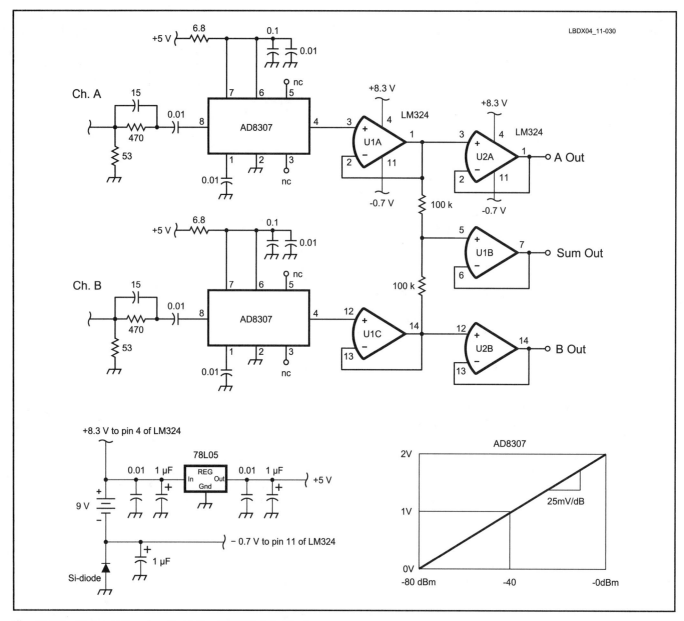

Fig 11-30—Schematic circuit of the W1MK detector/power-meter circuit. First connect one input and adjust the RF drive for 2 V output. Then the components of the LC circuit(s) are adjusted until the sum output (A + B) reads minimum.

The schematic is shown in **Fig 11-30**. In this circuit we see two identical detector/amplifiers, with three outputs: one for channel A, one for channel B and one for the sum of channel A and B. This comes in very handy if we when adjusting a Four-Square array using the Lewallen/Lahlum feed methods using two independent L-networks (see Section 3.6.2).

The output of all three ports varies between 0 and 2 V, where 2 V equals 0 dBm and 0 V equal –80 dBm. The maximum sensitivity is about –75 dBm and it has a bandwidth of approximately 500 MHz.

The circuit shown in **Fig 11-31** makes it possible to read the power in dBm on the scale of the DVM used as indicator.

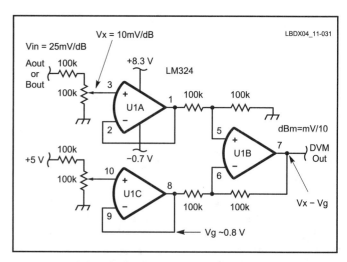

Fig 11-31—With this additional circuit, the output reading becomes easy to interpret: –50 dBm = –500 mV, and 0 dBm equals 0 mV. If you use a digital voltmeter as an output device, a reading of 0.375 V means a signal of –37.5 dBm.

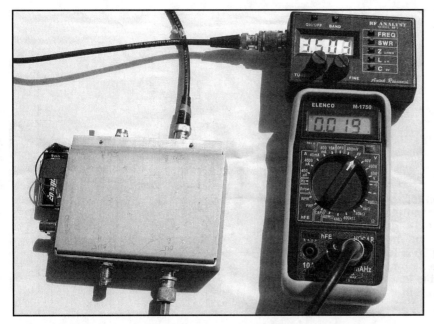

Fig 11-32—W1MK's array alignment setup. An Autek Research RF-1 is used as the RF generator. On the left the dual-channel RF wattmeter described in Fig 11-30 and 11-31. The DVM is used as a digital readout.

The scaling is as follows: Power in dBm = mV/10. Example:

Power in = –50 dBm → –500 mV
Power in = –35 dBm → –350 mV
Power in = 0 dBm → 0 mV

Fig 11-32 shows W1MK's test setup in action on 80 meters, with an Autek RF-1 used as a signal generator.

3.6.2 A hybrid-coupler phase-measuring circuit

The hybrid coupler as used in the W1FC feed systems can be used as the heart of a simple but very effective phase-measuring device for quadrature-fed arrays. If two voltages of identical magnitude but 90° out-of-phase are applied, the bridge circuit will be fully balanced and the output is null. The design also comes from Robye Lahlum, W1MK (Ref 968). **Fig 11-33** shows the hybrid in a simple test circuit for a quadrature-fed Four-Square. After having built the hybrid for the test circuit (see Fig 11-21), use the layout described in **Fig 11-34** to test the hybrid.

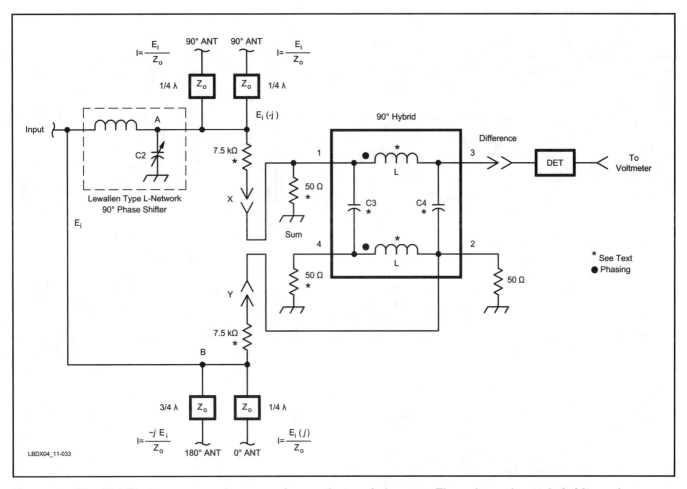

Fig 11-33—The W1MK phase-measuring setup for quadrature-fed arrays. The unit employs a hybrid coupler as used in the Collins feed system for arrays. The unit can be left permanently in the circuit if the voltage dividing resistors are of adequate wattage. See text for details.

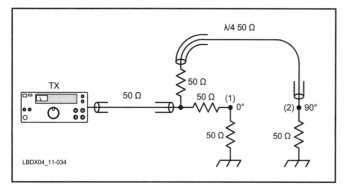

Fig 11-34—In this phase-calibration system for the quadrature tester, RF voltage from the transmitter is divided down with two 50-Ω series resistors (to ensure a 1:1 SWR), routed directly to a 50-Ω lead, and through a 90°-long 50-Ω line (RG-58) to the second 50-Ω load. For a frequency of 3.65 MHz, the cable has a nominal length of 44.49 feet (13.56 meters). The cable length should be tuned using the method described in Chapter 6 on feed lines and matching.

Note that the principle can be used with phase angles differences other than 90° as well. Let's work with an example. **Fig 11-35** shows the WA3FET Four-Square, described in Section 4.7. The elements are fed via λ/4 feed lines, which means we can measure the voltages at the end of these lines to determine the currents at the antenna feed point (current equals voltage divided by feed line impedance).

Using a voltage divider (with a high enough dividing ratio so as not to disturb the impedance involved), we sample some voltages at those points and bring them with equal length coaxial cables to our hybrid-coupler test setup. Three possibilities exist:

- Assume first that the array is fed in 90° increments (quadrature feeding). The sampled voltage at the end of our probe lines will be 90° out-of-phase and the output of the hybrid coupler will be zero.
- Assume that we are feeding with 90° phase shift but with slightly unequal current magnitudes. In this case we need to compensate for that with a calibrated attenuator in the probe line at the hybrid coupler input. It is essential that the probe coaxial cables are terminated in their characteristic impedances so that line length equals phase shift.
- Assume the array is not fed in 90° current increments, but with a phase difference of 111° (such as between the center elements and the back element in the WA3FET Four-Square). All we need to do in that case is insert an additional line length of (111−90) = 21° in the line going

Fig 11-35—Some RF is sampled at the end of the λ/4 lines going to the antenna elements. This is fed via RG-58 voltage sampling lines of equal length to the measuring equipment. Short line lengths and small attenuators can be inserted to compensate for non-quadrature setups and unequal drive currents. The schematic of the 90° hybrid is given in Fig 11-21. Section 3.4.6 explains how to calculate Xs1, Xp1, Xs2 and Xp2. V is a detector, which can be the detector/wattmeter described in Section 3.6.1 or a receiver. BPF is a bandpass filter.

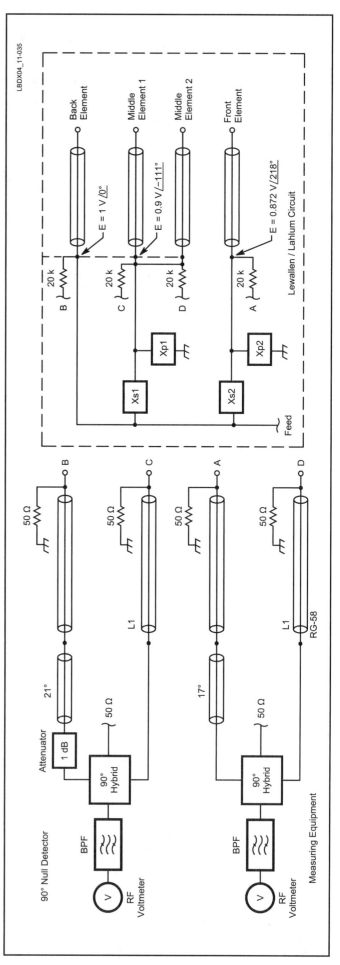

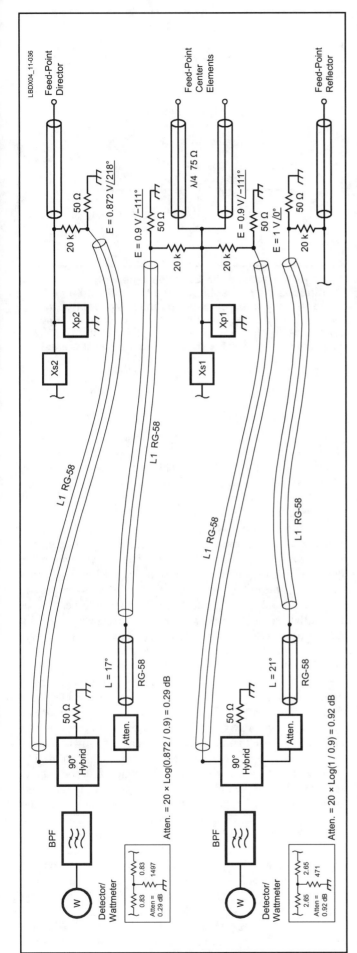

Fig. 11-36—Detailed schematic of the test setup for the WA3FET optimized Four-Square.

to the element with the leading phase, so that the net result again is 90°. See **Fig 11-36**.

In the same example the phase difference between the center elements and the director is −107°, hence we need an additional line length in the measuring set up of 17°. When measuring between points A and C, we need to insert a 1-dB (a 0.89:1 voltage ratio) attenuator in the line to point B to compensate for the unequal drive currents. The value of the sampling resistors depends on the power you want to do the testing with, and the detector's sensitivity.

3.6.2.1 Discussion on required signal levels, BC interference, and detector sensitivity.

Ideally we would want to be able to do some testing with an antenna analyzer (eg, the MFJ-259B) as a signal source, and using a small Detector/Wattmeter as described in Section 3.6.1. This way we can work on the antenna with really portable equipment. This should do for initial tuning even if you are not able to get a null better than 30 dB. As a final touch up, you can always use the station transmitter as a signal source for doing final alignment.

- What are the limiting factors?
- BC signals or even broadband noise.
- Detector sensitivity (noise figure)
- Available testing power

W1MK says that when he starts a measurement session, he first measures the level of background signals or noise on the antenna. For that you simply connect the detector/wattmeter to the antenna you will be testing. A broadband noise level of −35 dBm for 80 meters and even more on 160 is not uncommon, and in some case can be much higher (10 or 20 dB higher!). These values will of course be different in different locations.

Adding a band-pass filter (BPF) in front of the broadband detector should drop the meter readings signficantly. The values, of course, will be different for different locations. For example, W1KM experiences very high levels (−45 dBm) even with a BPF in front of the detector due to strong BC interference levels. In most situations the majority of the power hitting the detector is from out-of-band signals and if not filtered out by a selective circuit will reduce the amount of null that can be obtained. If the interference is inside the BPF, you can apply more power, or use a receiver to provide more selectivity.

For minimum measurement error a sampling resistor value of 20 kΩ is recommended. This means that the sampled signal will be approx −52 dB down from the applied power. If we apply power with the MFJ-259, the level will be +13−52 = approximately −40 dBm.

If we use the detector/wattmeter described in Section 3.6.1 (which has a maximum sensitivity of −75 dBm) and if we are not limited by BC signals, we can see a null down as far as −35 dB. This is not bad for a starter! An S9+40 signal represents −32 dBm, which means that the sensitivity of the detector/wattmeter matches pretty well with the level of a S9+40 signal, and even with such strong broadcast signals you will be able to see nulls of approx −30 to −35 dB.

In case of very stubborn noise/interference problems you can, of course, use your receiver as a null-detector. It has surplus sensitivity and should have enough selectivity to reject offending signals.

Your ability to obtain a deep null with a simple detector/wattmeter will always be either noise limitation (the internal noise or the noise figure of the detector/wattmeter) or interference limitation. If it is out-of-band interference, a BPF will help. If the interference is *on* your desired testing frequency you can move the test frequency slightly, or even better apply more power.

You might use 10-kΩ sampling resistors, if sensitivity is a problem but that is the limit—It is better to use higher testing power. A simple testing procedure is the following:

- Always use a bandpass filter at the input of the detector/wattmeter.
- Start you session with a portable source, such as the MFJ-259 antenna analyzer.
- Adjust the L-network values for maximum null. You should be able to obtain a null of at least −30 dB.
- If you are satisfied with a 30-dB null, now use your exciter as a signal source and apply 10 Watt (+40 dBm). This about 27 dB better than the MFJ-259, which means that under the same circumstances you now will be able to see a null down to 50 dB.

For fine trimming the phase and amplitude you must be able to fine adjust both the series and the parallel reactances of the L-network. A variable capacitor is an obvious choice for fine trimming. You can make the equivalent of a variable inductor with a little trick. For example, if the networks requires a coil with a reactance of +50 Ω, make a coil with double the reactance (100 Ω or 4.2 µH at 3.8 MHz) and connect in series a variable capacitor with (at maximum capacitance) a reactance of −50 Ω or less. If you use −25 Ω (1675 pF at 3.8 MHz), the series connection of the two elements will now yield a continuously variable reactance (at 3.8 MHz) of +25 (or less) to +75 Ω. See **Fig 11-37**.

The nice feature of such a test setup is that you can leave it permanently connected. Make sure that your sampling resistors are of high wattage if you run high power. Using 20-kΩ sampling resistors and running 1500 W the resistors dissipate 3.75 W, so two 40-kΩ, 2-W resistors in parallel is adequate.

The sampled power level going into the hybrid is −50 to −60 dB down from the transmit power, which puts it in the 1 to 10-mW (0 to +10 dBm) level for 1000 W (= +60 dBm) transmit power. A 40-dB null would show up as −30 to −40 dBm on your detector/wattmeter in the shack.

A −30 dBm level is 7 mV in 50-Ω. If you just want a kind of alarm system that tells you when things are really wrong, a simple germanium diode detector and a sensitive analog microamp meter (eg, 50 µA full scale) could be used.

Don't expect to have enough nulling sensitivity with this setup to properly adjust the L-network components. For that you need the sensitive wattmeter in Fig 11-30. To avoid overdriving the detector-wattmeter you should provide a 10/20/30-dB step attenuator when running high power.

3.6.3. Measuring antenna resonance

The true resonant frequency is the frequency where the

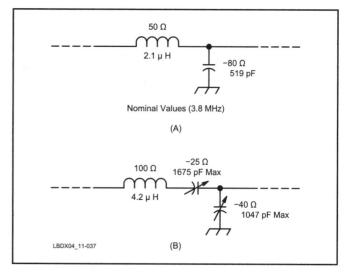

Fig 11-37—To make the Lewallen L-network continuously adjustable, replace the coil with a coil of twice the required value and connect a capacitor in series. The net result will be a continuously variable reactance. With the values shown, the nominal +50-Ω reactance is adjustable from +75 to +25 Ω (and less). The two capacitors can be motor driven to make the phase-shift network remotely controllable.

reactive part of the impedance equals zero. You can use one of the common antenna analyzers (see Section 3.6.8.), but their accuracy is not always the best, at least not when compared to the method described below.

W1MK uses the detector/power meter described in Section 3.6.1, together with a so-called *6-dB hybrid* to measure the resonant frequency, as well as the $R_{rad} + R_{loss}$ of the antenna very accurately. The circuit is very simple. See Fig 11-37. It boils down to a resistive bridge, where the detector has an asymmetric input is fed via a balun. The circuit is similar to the old "Antennascope" described 50 years ago in many handbooks, except that W1MK now uses a very sensitive null detector. This allows him to achieve a very deep null and to determine the exact resonant frequency. The signal source is not very critical and a typical antenna analyzer such as MFJ-259B should do.

If you cannot achieve a very deep null, BC band signals or the harmonic content of the signal generator may be a problem. Insert a band-bass filter between the generator and the bridge. The W3NQN bandpass filters are the best in this application. I use them between the exciter and my amplifier, so they are always available for such an application. The 50-Ω, non-inductive resistors must be matched if you want to read the value of the antenna total resistance from the potentiometer scale. T1 is a little balun that can be wound on a FairRite Products 2873000202 core (or similar). Use twisted-pair enameled wire (#24 to #26) to wind six passes (= 3 turns, = 3 times through both holes) on the binocular core.

Robye, W1MK, points out that he made provisions allowing him to actually measure the value of the variable resistor, using his digital multimeter, which allows him to get very accurate results.

Connect the antenna to the ANT terminal and adjust the frequency of the generator and the value of the potentiometer until the deepest null is reached. This will be at the antenna's resonant frequency. The value that you read off the potentiometer is the sum of R_{rad} and R_{loss} of the antenna.

3.6.4. Measuring antenna impedance using the W1MK 6-dB hybrid

Although the 6-dB hybrid (or Antennascope) described in Section 3.6.2 is merely is a resistive bridge circuit that can only be nulled when terminated in a purely resistive load, we can still use it to make accurate impedance measurements.

What we need to do is tune out the reactive part of the antenna impedance before it is connected to the bridge. We can do this simply by connecting a coil or a capacitor of the appropriate value in series. (Alternatively you could put the reactance in series with the 100-Ω potentiometer). Once this is done you can read the real part of the antenna impedance from the calibrated potentiometer scale on the 6-dB hybrid. See **Fig 11-38**.

The imaginary part of the impedance is the conjugate value (just change the sign) of the value of the series coil or capacitor used to tune out the reactance. **Fig 11-39** shows the schematic for a unit I built around a beautiful 5 × 4000 pF BC variable with built-in 91:1 gear reduction. In combination with a Groth turns counter, it is possible (after calibration against a laboratory grade instrument) to read off the capacitance over the entire range with an accuracy of a few pF!

With S1 in position a, you can obtain C values from about 100 pF to 6000 pF, which means capacitances ranging from −5.5 Ω to > −500 Ω on 80 meters and −11 Ω to > −1000 Ω on 160 meters. Of course S2 or S2 and S3 will need to be closed for the lower values. If needed, we can always add extra capacitors to obtain even lower values.

With S1 in position b (S1 and S3 open), you can obtain reactance values going from a few ohms to −170 Ω on 160 meters and up to −340 Ω on 80 meters. If that is not enough, we can put S1 in position C, where these values are doubled.

Ideally, this unit should be calibrated using a professional-grade network analyzer or impedance bridge. Once this is done you are all set with a very accurate impedance measurement set-up for antennas.

3.6.5. Using the 6-dB hybrid to make λ/4 lines, λ/2 lines or multiples thereof.

The 6-dB hybrid circuit described in Section 3.6.2 makes an ideal piece of test equipment for cutting stubs. Refer again to Fig 11-38. You can trim λ/4 long lines by leaving the far end open-circuited. For trimming λ/2 lines you can do it with the far end open-circuited on a frequency that is twice the design frequency. You can, of course, also use shorted (at the far end) lines, but make sure the short is a zero inductance short! It is easier to make a perfect open-circuit than a perfect short-circuit. Here is the procedure:

1. First short the "CABLE MEAS PORT" connector, preferably with a coaxial short (not just a wire loop, since that is *not* a very good short at RF).
2. Adjust the generator (antenna analyzer) to the desired frequency, where the feed line will be a short.
3. Next adjust the potentiometer for maximum notch (minimum power as detected by the W1MK detector/power meter). The value should be approximately 10 Ω.
4. Connect the stub, whose far end has been shorted for λ/2 or open-circuited for λ/4, to the "CABLE MEAS PORT" connector.
5. Tune the generator frequency and find the frequency of deepest null while slightly changing the value of the potentiometer for the best null.
6. I hope you started with a stub that was too long! Now cut off short lengths at a time, taking care to preserve a good dead short at the end with no inductance for λ/2, until you are right on the dot.

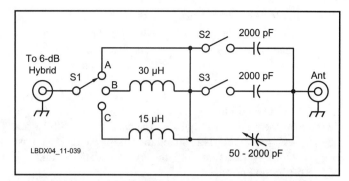

Fig 11-39—You can always change a complex feedpoint impedance of an antenna to a pure resistive impedance (which means bringing it to resonance) by adding the appropriate value of reactance in series. This simple circuit allows you add a wide range of positive, as well as negative, reactances to do this. See text for details.

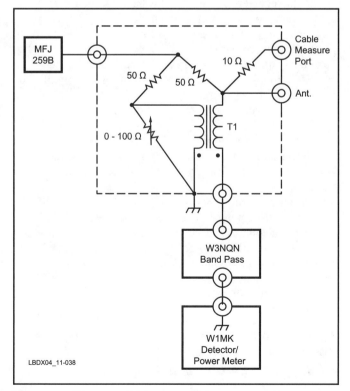

Fig 11-38—The 6-dB hybrid is the heart of the measuring setup for determining antenna resonance. See text for details.

3.6.5.1. Measuring cable loss with this set up

A lossless λ/4-long cable open-circuited at its end represents a dead short at the other end. The low resistance valued measured is a measure of the loss of the cable. If the cable were truly lossless, the potentiometer setting would be 10 Ω, the value of the series resistor going to the "CABLE MEAS PORT." If the potentiometer reads (10 + a) Ω for bridge balance at the resonant frequency, the cable loss = 8.69 × a × Z_0 where Z_0 = characteristic impedance of the cable. You must very accurately measure both the 10 Ω resistance and the value of the potentiometer!

Example: Z_0 = 50 Ω, a = 1 Ω (that is, the potentiometer is 11 Ω). The Matched Loss (dB) = 8.69 × 1/50 = 0.17 dB. The results are very accurate for a loss up to 5 dB. See **Table 11-7**.

3.6.6. Noise bridges

Commercially available noise bridges will almost certainly not give the required degree of accuracy, since rather small deviations in resistance and reactance must be accurately recorded. A genuine impedance bridge is more suitable. But with care, a well-constructed and carefully calibrated noise bridge may be used.

Several excellent articles covering noise bridge design and construction have been published, written by Hubbs, W6BXI; Doting, W6NKU (Ref 1607); Gehrke, K2BT (Ref 1610); and J. Grebenkemper, KI6WX (Ref 1623); D. DeMaw, W1FB (Ref 1620); and J. Belrose, VE2CV (Ref 1621). These articles are recommended reading material for anyone considering using a noise bridge in array design and measurement work.

The software module RC/RL TRANSFORMATION part of the NEW LOW BAND SOFTWARE is very handy for transforming the value of the noise-bridge capacitor, connected in parallel with either the variable resistor or the unknown impedance, first to a parallel reactance value and then to an equivalent reactance value for a series-LC circuit. This enables the immediate computation of the real and imaginary parts of the series impedance equivalent, expressed in "A + j B" form. Noise bridges are frequently used to cut quarter-wave or half wave transmission lines:

3.6.6.1. Using the noise bridge as a noise source only

If you have a noise bridge such as the Palomar bridge, you can use it as a wide-band noise source, without using the internal bridge. Instead you will connect the line to be trimmed across the output of the noise bridge and trim the length until the noise level on the receiver is reduced to zero. Switch off the receiver AGC to make the final adjustments (see Fig 11-28B). Tune the receiver back and forth across the frequency to determine the frequency of maximum rejection quite accurately. In this method λ/4 lines should be open-circuited at the end, and λ/2 lines should be short circuited.

3.6.6.2. The K4PI method.

Another method consists in using the noise bridge not only as a noise generator, but also as a bridge. Here is the procedure Mike Greenway, K4PI, uses with great success:

First put a really good RF short at the "UNKNOWN" terminals of the bridge. Using the XL/XC control and the RESISTANCE knobs alternatively, null the noise in the receiver. This is an important step. Keep increasing RF and AF gain and moving the bridge controls to obtain the lowest noise hiss you can. If you do it correctly you will get to the point where the receiver will sound almost dead.

Now, treat λ/4 wave sections as a λ/2 section because the λ/4 method shows too broad a reading. Prepare the short at the end of the coax by removing some of the outer plastic sheath. Push back some of the shield and remove some center-conductor insulation. Pull the shield back and squeeze it onto the center conductor and apply some solder. This makes a good RF short.

Switch the receiver (detector) to AM with the AGC off. The receiver must then be tuned to the area you are expecting to find the null. Connect the coax to the "UNKNOWN" terminals taking care not to touch the XL/XC and RX settings. Now use the RESISTANCE knob to null the noise along with tuning the receiver up and down the band for the lowest noise point. If you do everything right and listen very carefully you can get a null on an 80-meter λ/4 line being checked around 7300 kHz (where the stub is λ/2 long) to within 5 - 8 kHz. Take the center of that spread as the true null frequency.

Here too, if you have problems getting a deep null, you may want to try a bandpass filter (eg, W3NQN) between the noise bridge and the receiver. See **Figure 11-40**.

3.6.7. Network analyzers

Professional network analyzers are, in principle, ideal tools for measuring impedances. There are various types on the market, and second hand you may be able to get a system with an analyzer and generator for between $1000 and $2000. When measuring antennas on 80 and 160 meters, it is important that you do these measurements during day time, because during the night the average signal power on the band is so great that this background noise will cause erroneous readings on the equipment. With good quality equipment, one can adjust the generator power level to overcome this problem to a certain degree.

3.6.8. Measuring antenna impedance using one of the popular antenna analyzers
3.6.8.1. The AEA CIA-HF antenna analyzer

The AEA-CIA-HF analyzer is a one-port network analyzer with limited capabilities. It measures impedances (and of course SWR) by a swept-frequency method, over a range

Table 11-7
Conversion for 50 and 75-Ω systems

a (Ω)	Loss (dB) for Z_0 = 50 Ω	Loss (dB) for Z_0 = 75 Ω
1	0.17	0.11
2	0.34	0.22
3	0.52	0.34
4	0.69	0.46
5	0.87	0.58
6	1.02	0.70
7	1.22	0.81
8	1.39	0.93
9	1.56	1.04
10	1.74	1.16

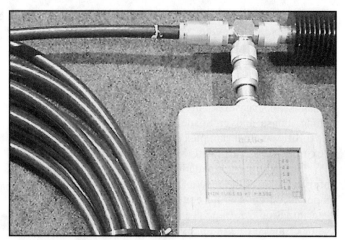

Fig 11-40—The AEA CIA-HF showing the SWR curve and the stub frequency (frequency of minimum SWR). The stub was initially cut for exactly 3.5 MHz using and R&S network analyzer. The text under the graph reads "MIN SWR 1.01 at 3.500 MHz."

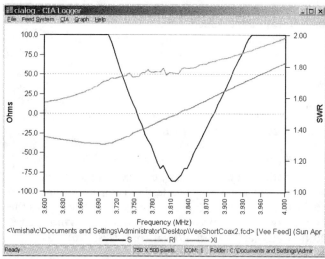

Fig 11-41—Screen shot of the software developed by W8WWV for the CIA-HF analyzer. See text for details.

that you can set between 0.4 and 54 MHz. The nice thing is that it is portable, and can operate from built-in batteries. However the power consumption is pretty high and it's a good idea to run it from a small 12-V power supply in the house or from a small lead-battery on a shoulder strap when in the field. When using it to measure antenna impedances, I have found it quite useful on all bands, down to 40 meters, and sometimes 80 meters. See **Fig 11-40**. On 160 meters, signals picked up from the broadcast band are too strong and mess up the readings, even during daytime.

The challenge will be for someone to come up with filters that will eliminate the BC interference, without causing any impedance transformation in the measuring range. This is quite a challenge. Another solution would be to have a higher output, but that may conflict with the FCC regulations on this subject.

The little screen on the unit does not show much detail, and when you use it in the shack the use a PC with the appropriate control and display software is recommended. AEA (**www.aea-wireless.com/cia.htm**) has such software, called "Via Director". The VIA HF is similar to the CIA-HF but has slightly extended frequency range.

Greg, W8WWV also developed similar software for the CIA-HF, It can be downloaded from his website at **www.seed-solutions.com/gregordy/Software/cialog.htm**. This web page describes the software, and near the bottom there is a link to download the self-extracting program that installs the software. See **Fig 11-41** for a screen shot of the graph produced by W8WWV's software.

AEA now also has an improved version of the CIA-HF, called the VIA-Bravo, which goes all the way up to 200 MHz. The VIA-Bravo provides greater accuracy in all complex measurements including 0.01° phase-angle resolution at lower angles. The unit, however, is very expensive.

3.6.8.2. The MFJ-259B antenna analyzer

The MFJ-259B antenna analyzer is different from the older MFJ-259. It uses a microprocessor and four voltage detectors in a bridge to directly measure reactance, resistance and VSWR. With so much information available, uses are limited mostly by your imagination and technical knowledge.

The main application for antenna builders is its capability of measuring SWR (also in terms of reflection loss). It will also measure the resistive part and the absolute value of the reactive part of a complex impedance. The MFJ-259B isn't smart enough give the sign of the reactive part without some minor help. You must vary the frequency slightly and watch the reactance change to determine the sign of the reactance and the type of component required to resonate the system. If adjusting the frequency slightly higher increases reactance (X), the load is inductive and requires a series capacitance for resonance. If increasing frequency slightly reduces reactance, the load is capacitive and requires a series inductance for resonance. This general rule works with most antennas, but not necessarily all of them.

The designers of the unit have added a "transparent filter" to cope with the problems of strong signals messing up low-level reflected-power readings in the vicinity of broadcast transmitters or during night time measurements on the low bands. This accessory includes an adjustable notch filter and selective bandpass filter. This handy accessory allows the MFJ-259B to be used on large low band antennas, even if the antenna is located in the area of a broadcast transmitter.

At first blush the major difference between the MFJ unit and the AEA unit is the fact that the MFJ-unit does not generate a spectral display of the units measured. It is basically a single-point (one frequency) measurement system.

3.6.8.3. Cutting stubs with antenna analyzers

All of the popular antenna analyzers can be used in this application. The method consists of connecting a 50-Ω dummy load to the analyzer via a T connector. The transmission line or stub is connected in parallel with the dummy load. The antenna analyzer is then adjusted for the frequency with the lowest SWR ratio. For an open-circuited cable this is at the frequencies where the cable is λ/2 or multiple thereof. For a short-circuited stub this is for a length of λ/4 or any odd multiple thereof. The AEA CIA-HF Analyzer has a nice feature where it can calculate the frequency of lowest SWR and show it on the screen.

Using the AEA CIA-HF Analyzer, it is possible to determine very accurately the frequency of minimum SWR (which equals the frequency where the stub is resonant). When measuring a stub that was cut for 3.5 MHz using the R&S network analyzer, the average of a number of measurements gave 3.492 MHz for the stub-resonant frequency, which agrees within 0.2%, an excellent figure.

3.6.8.4. Which antenna analyzer?

On the subject of analyzer measurement accuracy, W8WWV did some elaborate testing comparing some of the popular units. The detailed information is available on his site: (**www.seed-solutions.com/gregordy/Amateur% 20Radio/Experimentation/EvalAnalyzers.htm**).

3.6.9. N2PK VNA (Vector Network Analyzer)

Genuine network analyzers are expensive, even second hand, but they provide much better accuracy than the present day Antenna Analyzers. At the time this Fourth Edition goes to press, it appears that a Vector Network analyzer developed by N2PK will be a valid replacement for these expensive instruments, and provide results with comparable accuracy. See **Fig 11-42**.

This unit is capable of both transmission and reflection measurements from 0.05 to 60 MHz, with about 0.035-Hz frequency resolution and over 110 dB of dynamic range. Its transmission-measurement capabilities include gain/loss magnitude, phase and group delay. Its reflection-measurement capabilities include complex impedance and admittance, complex reflection coefficient, VSWR and return loss.

Unlike other impedance measuring instruments that infer the sign of the reactance (sometimes incorrectly) from impedance trends with frequency, a VNA is able to make this determination from data at a single frequency. This is a direct result of measuring the phase as well as the magnitude of an RF signal at each test frequency.

N2PK (**users.adelphia.net/~n2pk/**) impressed all of us hams looking for an affordable network analyzer by the level of documentation that he has made available for anyone wanting to build a unit. And the performance is *much* better than anything else you might be able to build or buy for the amount of money you will spend building his VNA. However, building a VNA is not for a first-time kit builder, although there are interest groups supporting potential builders (**www.seed-solutions.com/gregordy/Amateur%20Radio/ Experimentation/N2PKVNA/N2PKVNA.htm**).

Even better news is that a "plug and play" commercial version may be available soon. The basic VNA unit works all the way up to 60 MHz, so it's got plenty of range for the low-band enthusiast. Comparing the measurements of the VNA with a top grade network analyzer shows that it is *very* close to a professional instrument.

3.6.10. The good old impedance/admittance bridge

All the above-mentioned instruments have the same intrinsic problem of suffering from alien-signal overload when measuring large antennas on the low bands, especially 160 meters, where BC signals are likely to cause false readings unless clever computer algorithms are used to compensate for them. The only other way to overcome this problem is to measure with more generator power, which to a degree is possible with professional-grade network analyzers. Or in the worst of cases, you may have to resort to a good old General Radio bridge, driven by a signal source of sufficient level. This method, of course, lacks the flexibility of a real frequency-sweeping network analyzer.

3.7. Mutual-Coupling Issues

3.7.1. Too little mutual coupling where you want it

When we set up an array, we need to calculate the mutual impedance from the measurements of the self impedance and the coupled impedance (see Section 3.3.1). The normal procedure is to first measure the self impedance, and then couple one element at a time and measure the coupled impedance.

If you measure little or no difference between the self impedance and the coupled impedance, then have a look at the value of the self impedance. It is likely that the resistive part of the impedance is much higher than the impedance you have calculated by modeling the antenna. For example, if you use inverted-L elements with λ/8 vertical portions, you should

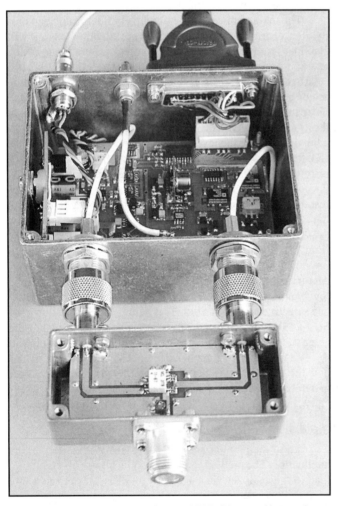

Fig 11-42—The N2PK-designed VNA (Vector Network Analyzer) constructed by G3SEK. This is all the hardware needed! A PC does the control and interface, of course. It is very likely that soon a commercial version will be available.

expect a self impedance of approximately 17 Ω over a perfect ground. If you measure 50 Ω, it means that you have an equivalent loss resistance of 33 Ω! With so much loss resistance you will see—even with very close coupling such as an array with λ/8 spacing—only a little difference between self impedance and coupled impedance. Such an array will still show the proper directivity, but its gain will be way down. In the above example the gain will be down 4 to 5 dB from what it would be over an excellent ground system. So, if you see no effect of mutual coupling where you *should* see it, suspect you have large losses involved somewhere.

3.7.2. Unwanted mutual coupling.

There are cases where you don't want to see the effect of mutual coupling. But they are there and you want to control them. If you happen to have towers (or other metal structures or antennas) within λ/4 of one of the elements of an array, you may induce a lot of current into that tower by mutual coupling. The tower acts as a parasitic element, which will upset the radiation pattern of the array and also change the feed impedances of the elements and the array. To eliminate the unwanted effect from the parasitic coupling proceed as follows:

- Decouple all the elements of the array, with the exception of the element closest to the suspect parasitic tower. For quarter-wave element decoupling, this means lifting the elements from ground.
- Measure the feed-point impedance of the vertical under investigation.
- If a suspect tower is heavily coupled to one of the elements of the array, a substantial current will flow in it. Probe the current by one of the methods described by D. DeMaw, W1FB (Ref *W1FB's Antenna Notebook*, ARRL publication, 1987, p 121) and shown in **Fig 11-43**. If there is an appreciable current, you will have to *detune* the tower. Methods for detuning a tower are given in detail in Chapter 7 (Section 3.7.2).
- After detuning the offending tower, measure the feed-point impedance of the vertical again. If you have properly detuned the parasitic tower, you will likely see a rise in impedance and a shift in resonant frequency.
- Reconnect the whole array and fire in the direction of the parasitic tower.
- Check the current in the parasitic tower and if necessary make final adjustments to minimize the current in the tower. You can use high power now to be able to tune the tower very sharply. In general the tuning will be quite broad, however.
- You now have made the offending tower invisible to your array.

3.8. Network Component Dimensioning

When designing array feed networks using the computer modules from the NEW LOW BAND SOFTWARE, you can use absolute currents instead of relative currents. The feed currents for the 2-element cardioid array (used so far as a design example) have so far been specified as $I1 = 1\ /\!\!-\!90°$ A and $I2 = 1\ /0°$ A. The feed-point impedances of the array are:

$Z1 = 51 + j\ 20\ \Omega$
$Z2 = 21 - j\ 20\ \Omega$

With 1 A antenna current in each element, the total power taken by the array is $51 + 21 = 72$ W. If the power is 1500 W, the true current in each of the elements will be:

$$I = \sqrt{\frac{1500}{72}} = 4.56\ A$$

Using this current magnitude in the relevant computer program module COAXIAL TRANSFORMER will now show the user the real current and voltage information all through the network design phase. The components can be chosen according to the current and voltage information shown.

If you plan to build your own Lahlum/Lewallen network, it's a good idea to stick to air-wound coils (have a look at Fig 11-27) for inductances up to approx 5 µH. Above this value you will have to revert to toroidal cores. Ferrite cores should not be used in this application since they tend to be unstable under certain circumstances. Only use powdered-iron cores. The red cores (mix 2) are a good choice for both 160, 80 and 40 meters. How large a core do you need to use? The rule is never to wind more than a

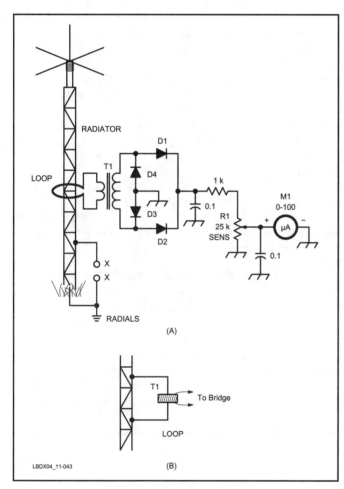

Fig 11-43—Current–sampling methods for use with vertical antennas, as described by DeMaw, W1FB. Method A requires a single-turn loop of insulated wire around the tower. The loop is connected to a broadband transformer, T1. A high-mu ferrite toroid, as used with Beverage receiving antennas (see Chapter 7 on special receiving antennas), can be used with a 2-turn primary and 2 to 10-turn secondary, depending on the power level used for testing.

Table 11-8
Maximum inductance for a single layer winding, as a function of wire diameter

Type	A_L	#10	#12	#14	#16	#18	#20
T106-2	135	3.9	6	9	15	22	35
T157-2	140	12	18	24	47	68	110
T200-2	120	16	25	40	65	95	153
T200A-2	218	29	46	73	119	172	278

single layer. **Table 11-8** gives you the maximum inductance that you can get with a given wire size (AWG #) for a given core.

Example: Assume you need a reactance of +800 Ω. On 1.83 MHz that represents 69 µH. You may marginally make it on a T157-2 core with #18 wire. In most cases where such high values of inductance are involved, current through the coil will be very small and #18 enameled wire would be just fine. Only in cases were inductances of between 10 and 15µH are required, I would use A T200 or T200A core with #10 or even #8 wire.

4. POPULAR ARRAYS

Whereas in previous editions I described in detail how various feed systems can be applied to various arrays, I decided to describe mainly two feed systems (with one exception) in detail:

- The hybrid-coupler method (plug and play), where applicable for quadrature feeding
- The Lewallen/Lahlum feed method, which allows the most flexibility.

All arrays were modeled using *NEC-2* over "good ground"(conductivity 5 mS/m, dielectric constant 13), with an extensive radial system that accounts for an equivalent-series-loss resistance of 2 Ω for each element. The element feed-point impedances shown include this 2 Ω of loss resistance. If you want to calculate your feed system for different equivalent-ground-loss resistances, apply the following procedure:

- Take the values from the array data (see further). The resistive part includes 2 Ω of loss resistance. If you want the feed-point impedance with 10 Ω of loss resistance, just add 8 Ω to the resistive part of the feed-point impedance shown in the array data. The imaginary part of the impedance remains unchanged.
- Follow the feed-system design criteria as shown, but apply the new feed-point impedance values.

I did the modeling using a wire diameter of 200 mm (approximating a Rohn 25 tower) for the vertical element, and the elements were adjusted to resonance, with all other elements decoupled, meaning floating.

The gain is expressed in dBi (over good ground as specified above). For each array we also calculated the directivity, expressed in RDF (Receiving Directivity Factor) and in DMF (Directivity Merit Figure). See Chapter 7.

In many arrays you will see a negative impedance, in most cases for the "back" element of the array. Again, the negative impedance merely means that the feed network is not supplying power to that element but rather taking power from that element. The different modules of the NEW LOW BAND SOFTWARE as well as the *Lahlum.xls* spreadsheet program handle these negative values without problems.

All Lahlum/Lewallen feed networks are calculated without taking into consideration the effects of cable losses. These effects are quite small on the low frequency bands, if good cables are used. Only with very long cable lengths (eg, 3λ/4 current-forcing feed lines plus a 180° phasing line, losses can be significant. I made several calculations between ideal case (no losses) and the real-world case, and the differences of the L-networks values were well within the typical tuning range of the components. When you take into account the losses, the feed impedance of the network will be slightly higher (typically a few percent).

4.1. Two-Element End-Fire Arrays

The principles of operation of the 2-element end-fire array were explained in detail in Chapter 7. Most of us probably think of a λ/4 spaced array, where the elements are fed 90° out-of-phase, but this is not necessarily the best solution. If you want to use 90° phase shift, for instance because you want to use a hybrid coupler to feed the array, then a spacing of about 105° achieves just marginally better DMF than 90°. Staying with quarter-wave spacing, a phase difference of

Table 11-9
Main Data for a Range of 2-Element End-Fire Arrays

Spacing	Phase	Gain (dBi)	3-dB BW dB	RDF ele.	DMF ele.	Zfront	Zback
105°	−90°	4.24	177	8.13	13.1	55 + j 13	19 − j 13
90°	−90°	4.23	177	8.11	12.3	53 + j 18	21 − j 19
90°	−105°	4.72	159	8.70	14.4	48 + j 21	17 − j 14
90°	−110°	4,87	154	8.88	15.1	46 + j 23	16 − j 15
75°	−120°	5.05	146	9.14	15.6	37 + j 25	14 − j 15
60°	−135°	5.20	135	9.50	17.2	24 + j 24	12 − j 15
45°	−145°	5.00	131	9.57	16.6	14 + j 19	10 − j 17

about 105° is recommended, in which case you can no longer feed the array with a hybrid coupler. The larger the array the better the bandwidth, and this shows in the element impedances. Small arrays, such as those λ/8 spacing, give excellent directivity but the element feed impedances become low, causing drop in gain or a given ground system and small bandwidths over which directivity will hold.

4.1.1. Data, 2-element end-fire array

Table 11-9 shows the main data for a range of 2-element end-fire arrays. The first impression is that 60° spacing with 135° phase shift is best, but note the relatively low feed impedance, which means narrower bandwidth than for a wider-spaced array. The gain figures are over average ground ($\varepsilon = 13$ and $\sigma = 5$ mS).

4.1.2. Feed systems, 2-element end-fire array

Several feed methods were illustrated with a 2-element end-fire array in Section 3.4.

4.1.2.1. Christman feed, 2-element end-fire array

See Section 3.4.2. This approach uses a minimum of components, but since it does not use current-forcing feed lines you cannot measure voltage to determine the feed current. This means you either need to be able to measure feed current (not so easy to do accurately), or you need to do some precise element-impedance measurements (coupled and uncoupled), calculate the mutual coupling and from there figure the actual feed impedances. (You can use the module MUTUAL IMPEDANCE AND DRIVING IMPEDANCE from the NEW LOW BAND SOFTWARE.)

Fig 11-7 shows how you can switch the array in the two end-fire directions. When both elements are fed in-phase the array will have a bi-directional broadside pattern (see Section 4.2) with a gain of 1 dB over a single vertical. The front-to-side ratio is only 3 dB. The feed impedance of two quarter-wave-spaced elements fed in-phase is approximately $57 - j\,15\,\Omega$, assuming an almost-perfect ground system with 2-Ω equivalent-ground-loss resistance. Notice that both elements have the same impedance, which is logical since they are fed in-phase.

We can easily add the broadside direction (both elements fed in-phase) by adding a switch or relay that shorts the 71° long phasing line, as shown in **Fig 11-44**. L networks can be designed to match the array output impedance to the feed line. Don't forget that you need to measure impedances to calculate the line lengths that will give you the required phase shifts. Merely going by published figures will not get you optimum performance!

4.1.2.2. Hybrid-coupler feed, 2-element end-fire array

See Section 3.4.6. When you buy a commercial hybrid coupler, you don't really need to do any impedance measurements. All you will have to do is trim the elements to resonance (decoupled from one another!). Commercial hybrid couplers are made to accommodate Four-Square arrays, and normally use four relays to do the direction switching. For a 2-element end-fire array, a much simpler switching system, using a single DPDT relay will do the job if only the two end-fire directions are required.

In this case you can delete K1 and its associated wiring from the schematic shown in **Fig 11-45**. On the low-bands any 10-A relay will do. If you want the bi-directional broadside pattern as well, two relays and an L-C network are needed.

4.1.2.3. Lewallen feed, 2-element end-fire array

The application of the Lewallen feed method for the 2-element end-fire array was described in detail in Section 3.4.5 and is shown in Fig 11-45. Two-element end-fire arrays are commonly used in a broadside/end-fire combination to increase directivity and gain.

Using the Lewallen feed system, you can adjust the L-network values to obtain the proper feed current magnitude and phase shift, using the simple test method and equipment developed by Robye, W1MK, and described in Section 3.6.

4.2. The 2-Element Broadside Array

If you feed two elements in-phase, they will produce a broadside (radiation in a direction perpendicular to the line connecting the two elements) bidirectional figure-eight pattern, provided the spacing is wide enough. The array with 90° spacing is often used as a "third" direction with an end-fire array and gives about 1 dB gain over a single vertical.

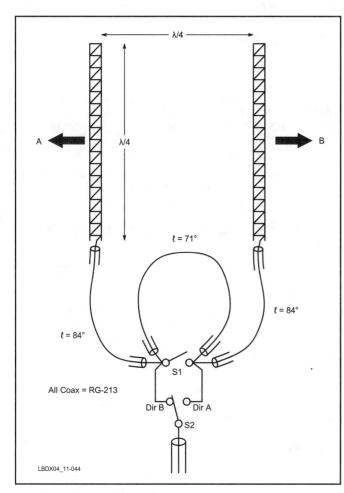

Fig 11-44—The 2-element vertical array (λ/4 spacing) can be fed in-phase to cover the broadside directions. I added switch S1 to the Christman feed system as described in Fig 11-8. When S1 is closed, both antennas are fed in-phase, resulting in bi-directional broadside radiation.

4.2.1. Data, 2-element broadside arrays

Narrow spacing yields a wide forward pattern. When we reach $\lambda/2$ spacing, and up to about $5\lambda/8$ spacing, the forward lobe is at its narrowest without excessive sidelobes. At $\lambda/2$ spacing the rejection off the side is maximum at zero elevation angle. Increasing the spacing lifts the maximum rejection off the ground, resulting in better directivity and higher gain (by way of narrower forward lobe).

See **Table 11-10**. Gain is over average ground, and includes the effect of a 2-Ω equivalent-ground-loss resistance in each element.

4.2.2. Feed systems, 2-element broadside arrays

As the elements are fed in-phase, you can feed them with equal-length feed lines to a common point where you parallel the ends of the feed lines. In principle the array can be fed with two feed lines of any equal lengths. Feeding via $\lambda/4$ or $3\lambda/4$ feed lines, however, has the advantage of forcing equal currents in both elements, whatever the difference in element impedances might be. I therefore advise people to feed the array via two $3\lambda/4$ feed lines. Quarter-wave feed lines are too short (due to the coax's velocity factor) to reach the center of the array.

Using the COAX TRANSFORMER/SMITH CHART and the PARALLEL IMPEDANCES modules of the NEW LOW BAND SOFTWARE program, you can easily calculate the feed impedance of this antenna. Let's work out an example of a broadside array with 193° spacing:

$$Z_{elem} = 28 - j\,12\;\Omega$$

Assume loss-free cables: The impedance at the end of $3\lambda/4$-long current-forcing feed lines ($Z_0 = 50\;\Omega$) is: $75.4 + j\,32.3\;\Omega$. Paralleling the two feed lines yields: $Z = 38.7 + j\,6.1\;\Omega$.

Run the SHUNT/SERIES IMPEDANCE NETWORK MODULE and find out that by putting a reactance of $-109\;\Omega$ (a capacitor) in parallel with this impedance, transforming it into 45 Ω, an almost perfect match for the 50-Ω feed line.

4.3. Three- and Four-Element Broadside Arrays

If more than two elements are used in a broadside combination (all in-line and fed in-phase), the current magnitude should taper off towards the outside elements to obtain the best directivity and gain. This current distribution is what is called the *binomial current distribution*. Multi-element broadside arrays are also covered in Chapter 7 on receiving arrays.

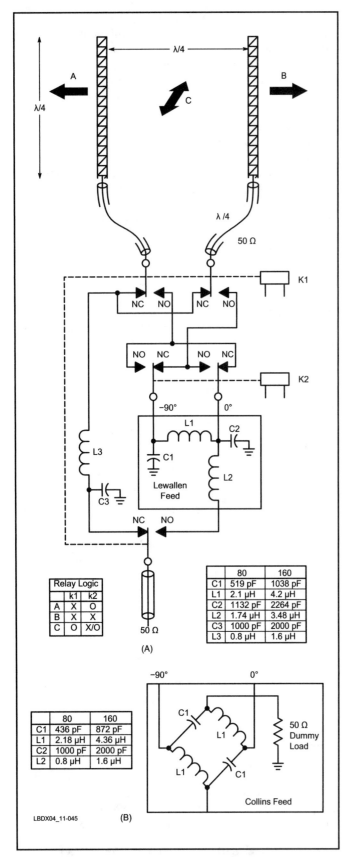

Fig 11-45— The 2-element vertical array ($\lambda/4$ spacing) can be fed in-phase to cover the broadside directions. Two feed methods are shown: At A, the Lewallen feed method, and at B, the Collins hybrid-coupler method. In both these cases relay K1 chooses between the end-fire and the broadside configurations. Relay K2 switches directions in the end-fire position.

Table 11-10
Data for 2-Element Broadside Arrays

Spacing	Gain (dBi)	3-dB Beamwidth	Impedance
90°	2.31	—	56 – j 17
135°	3.52	90°	41 – j 19
180°	4.91	64°	30 – j 14
193°	5.26	59°	28 – j 12
208°	5.59	55°	27 – j 9
225°	5.81	50°	25 – j 5

4.3.1. Data, 3- and 4-element broadside arrays

The radiation pattern is similar to what is shown in Fig 11-5, only the patterns get narrower and the gain increases as we use more elements. See **Table 11-11**.

4.3.2 Feed systems, 3- and 4-element broadside arrays

4.3.2.1. Feed systems, 3-element broadside array

If we design the array with $\lambda/2$ spacing between the elements, the feed lines will need to be $3\lambda/4$ long if we want to follow the current-forcing principle. To obtain double the feed current magnitude in the center element, we need to feed the central element with two parallel feed lines. Using 75-Ω coax for the feed lines we have at the end of those feed lines:

Outer elements: $144 + j\,182\ \Omega$
Center element: $38 + j\,17\ \Omega$

Connected in parallel we obtain an array feed impedance of: $27 + j\,16\ \Omega$, which we can easily match with an L-network to 50 Ω.

4.3.2.1. Feed systems, 4-element broadside array

Here too, if we want to use current-forcing feed lines, we will need to use $3\lambda/4$ feed lines to the center elements and $5\lambda/4$ feed lines to the outer elements. This involves a lot of coax. If instead of spacing the elements $\lambda/2$ we space them $0.8 \times \lambda/2$ (0.8 being the velocity factor of foam coax), we will reach out with $\lambda/4$ feed lines to the center elements and $3\lambda/4$ lines to the outer elements. To maintain good directivity and well-suppressed side lobes for this particular case, the current magnitude distribution along the elements is 1:2:2:1. There is some loss in gain vs the $\lambda/2$-spaced array (6.8 vs 7.2 dBi), and the 3-dB bandwidth is now 42°. The feed impedances are: $31 - j\,23\ \Omega$ for the outer elements and $36 - j\,24\ \Omega$ for the center elements. To obtain a relatively high total-array feed impedance, it is best to use 75-Ω current-forcing feed lines. We need to run two cables in parallel to the two central elements and single feed lines to the outer elements. The impedance at the end of those feed lines are:

Outer elements: $117 + j\,87\ \Omega$
Inner elements: $27 + j\,18\ \Omega$

All connected together, the impedance is $11 + j\,7.5\ \Omega$. We can match this to a 50-Ω feed line with an L-network.

4.4. The 3-Element End-Fire Arrays

We have covered the 2-element end-fire arrays in Section 4.1. Just as we have 2- and 3-element Yagis, we can have 2- and 3-element end-fire arrays.

As we have seen with 2-element end-fire arrays, there is nothing sacred about spacing or phase angles. It is true, of course, that an array with quadrature feeding (phasing angles that are in 90° steps) with identical current magnitudes have

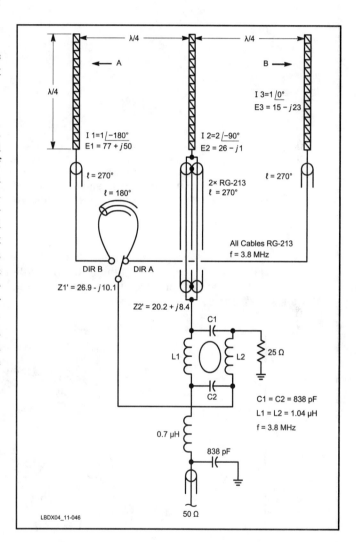

Fig 11-46—Feed system for the 3-in-line broadside array with binomial current distribution and quadrature phase currents. The center element is fed via two parallel 75-Ω feed lines to obtain double the feed-current magnitude. The current-forcing method ensures that variations in element self-impedances have minimum impact on the performance of the array.

Table 11-11
Data, 3- and 4-Element Broadside Arrays

Array	Element Feed Currents (1)	Gain dBi	3-dB Beamwidth	Element Feed Impedances (Ω)
3 ele	1, 2, 1	6.33	46°	$25 - j\,19$; $31 - j\,14$; $25 - j\,19$
4 ele	1, 3, 3, 1	7.21	37°	$23 - j\,23$; $29 - j\,16$; $29 - j\,16$; $23 - j\,22$
4 ele (2)	1, 2, 2, 1	6.80	42°	$31 - j\,23$; $36 - j\,24$; $36 - j\,24$; $31 - j\,23$

(1) Current magnitude
(2) Element spacing = 0.4 λ (see text)

THE λ/4-SPACED ARRAY—END-FIRE AND BROADSIDE

Feed-Current Phasing in an End-Fire Array

For a 2-element array spaced 90° (λ/4), varying the phase of the feed current can be used to not only increase the gain, but also to shift the position of nulls in the rearward direction. **Fig A** shows the physical layout of two λ/4 verticals.

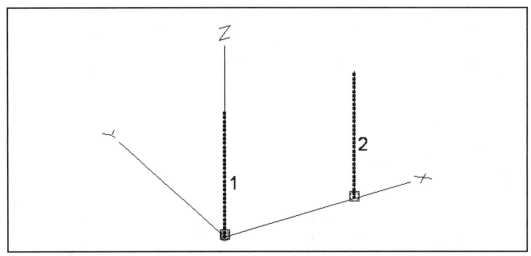

Fig A—Quarter-wave long vertical elements are positioned on the X-axis. For this example of end-fire operation, element number 1 is fed at 0° phase angle, while element number 2 is fed at either a −90° or a −110° phase angle. Both feed currents have the same magnitude. For broadside operation, both elements are fed with equal-amplitude currents at the same 0° phase angle.

Fig B illustrates how the azimuth patterns change for this array with end-fire feed-current phases of 0° (broadside operation), −90° and −105°. In the 0° phase configuration, the gain decreases compared to the end-fire configurations as the pattern becomes closer to "omnidirectional" but the peak gain is rotated 90° from the peak for the end-fire array—hence the name "broadside."

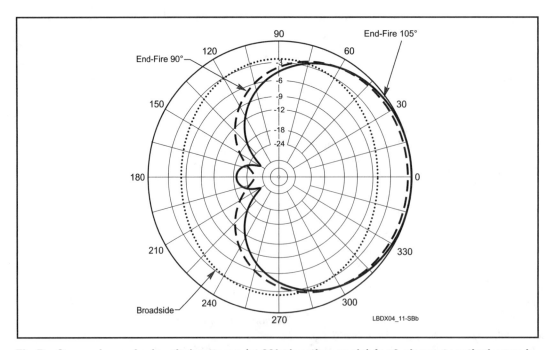

Fig B—Comparison of azimuthal patterns (at 20° elevation angle) for 2-element vertical array in Fig A, operated end-fire at phases of −90° and −105°. Also shown is response of the array operated at 0° phase, which is the broadside feed configuration. Each element is physically spaced λ/4 from the other. The end-fire peak is along the line between the two elements and is greater than the broadside peak, which is perpendicular to the line between the elements.

Fig C shows the elevation-plane patterns for the two end-fire and one broadside arrangements for the λ/4-spaced 2-element vertical array.

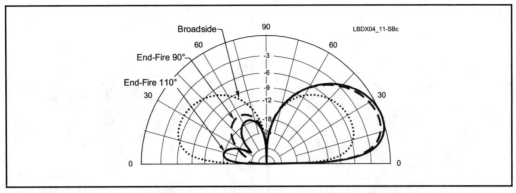

Fig C—Comparison of elevation-plane patterns for end-fire and broadside arrays shown in Fig A. The 110° phase shift used in the end-fire array puts a null at about a 40° rearward elevation angle and achieves a much better overall directivity in the back compared to a quadrature (90°) phase shift. Of course, the end-fire gain is also higher than the more "omnidirectional" gain of the broadside array.

Changing the Element Spacing for Broadside Operation

If the physical spacing between the elements in a 2-element array operated in broadside is varied, the gain will increase with increasing spacing beyond 90° (λ/4). However, more than a spacing of about 225° (5λ/8) results in objectionable sidelobes in the azimuth-plane pattern, as illustrated in **Fig D**. The gain is largest and the sidelobe pattern is cleanest at 193° (0.536 λ) physical spacing.

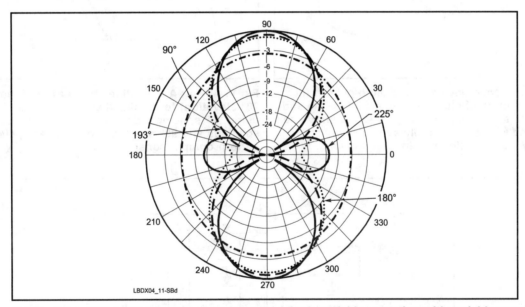

Fig D—Azimuthal pattern (at 20° elevation angle) for broadside operation with variable spacing between the two elements. Note the sizeable sidelobe that appears for the 225° (5λ/8) case.

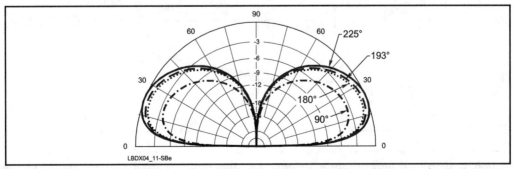

Fig E—Elevation-plane patterns for different physical spacings between the 2-elements in a broadside array.

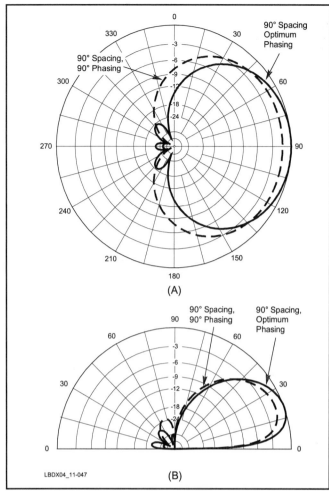

Fig 11-47—At A, solid line shows azimuth pattern (at 20° elevation) for quadrature-fed, 3-element in-line end-fire array, with spacings of λ/4 (Fig 11-46). Dashed line is for array fed with optimized phase angles and amplitudes. At B, elevation pattern comparisons.

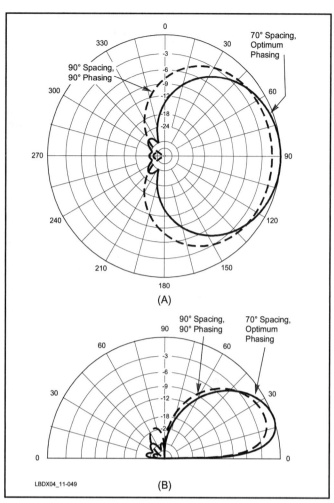

Fig 11-49—At A, solid line shows azimuth pattern (at 20° elevation) for Lahlum/Lewallen feed-optimized array using 70° spacings. Dashed line is reference with 90° spacings and 90° and 180° phasing. At B, elevation pattern comparisons.

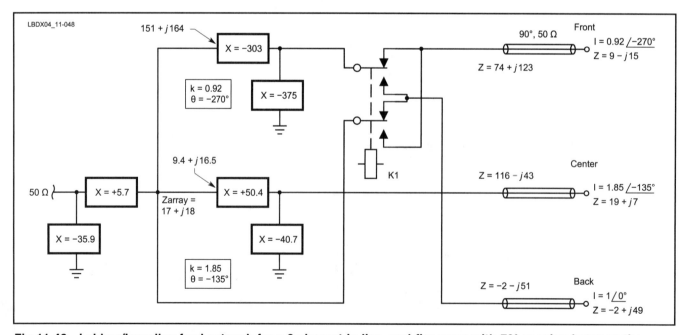

Fig 11-48—Lahlum/Lewallen feed network for a 3-element in-line, end-fire array with 70° spacing between the elements. This element phasing was chosen to be able to use λ/4 current forcing feed lines (Vf = 0.8). Direction switching is included.

11-42 Chapter 11

Table 11-12
Data, 3-element broadside arrays

Spacing	Feed Currents	Gain dBi	Beamwidth 3-dB	RDF dB	DMF dB	Feed Impedances (Back, Mid, Front)
90°	1, 0°; 2, –90°, 1, –180°	5.28	143°	9.24	17.9	$15 - j\,23$; $26 - j\,1$; $77 + j\,50$
90°	1, 0°; 1.75, –125°; 0.9, –250°	6.59	106°	10.9	27.9	$11 - j\,14$; $26 + j\,9$; $30 + j\,60$
70°	1, 0°; 1.85, –135°; 0.92, –270°	6.57	98°	11.2	27.8	$9 - j\,15$; $19 + j\,7$; $-2 + j\,49$
45°	1, 0°; 1.9, –150°; 0.95, –300°	5.19	91°	11.5	27.5	$5 - j\,17$; $11 + j\,1$; $-18 + j\,11$

a certain attraction, since they make it possible to use the hybrid coupler (Collins) feed system.

4.4.1 Data, 3-element end-fire arrays

Note the negative impedance for the 70°-spacing case in **Table 11-12**. This happens frequently in multi-element arrays for the element in the back, especially at close spacings.

Note that with close spacing, especially at λ/4, the feed impedances become very low, which results in small bandwidth, critical tuning and less gain (R_{rad} becomes small while R_{loss} remains constant at 2 Ω).

4.4.2 Feed systems, 3-element end-fire arrays
4.4.2.1. Hybrid-coupler feed, 3-element end-fire arrays

The λ/4-spaced non-optimized version of this array can be fed with a hybrid coupler. **Fig 11-46** shows the feed system and direction switching and **Fig 11-47** shows the horizontal and vertical radiation patterns. As we need double the current in one of the elements of such an array, all we need to do is to run a coaxial cable with half the impedance of the coax feeding the other elements. In other words, the feed line to the center element will consist of two parallel-connected feed lines.

The transformed impedance for the center element (now being fed via a 270° long 25-Ω line) is $20.2 + j\,8.4$ Ω. The impedance at the end of the feed line going to the front elements is: $22.8 - j\,18.4$ Ω. To the back element: $49.8 + j\,76.3$ Ω (all calculated with the COAX TRANSFORMER/SMITH CHART software module). In parallel, those two give: $26.9 - j\,10.1$ Ω.

Notice that both impedances result in a low SWR in a 25-Ω system. The performance of the coupler will be very good if we design the hybrid coupler with a nominal impedance of 25 Ω. The values of the coupler components are:

$X_{L1} = X_{L2} = 25\ \Omega$; $X_{C1} = X_{C2} = 2 \times 25 = 50\ \Omega$

4.4.2.2. Lahlum-Lewallen feed, 3-element broadside array

The array with 70° spacing between the elements has the advantage of not requiring 3λ/4 current-forcing feed lines if we use coaxial lines with a velocity factor of 0.8.

Fig 11-48 shows the feed network, including the direction switching using a DPDT relay K1. **Fig 11-49** shows the radiation patterns for this feed-optimized array. Here, 50-Ω feed lines were used since they prevent the components in the L-network to the front element from having too high an impedance. A similar network can be calculated for other spacings and phase angles, using the *Lahlum.xls* spreadsheet and the appropriate NEW LOW BAND SOFTWARE modules. The procedure to adjust the L-network values is covered in Sections 3.6.1 and 3.6.2.

4.5. A Bidirectional End-Fire Array

Assume we have a 2-element broadside array with λ/2

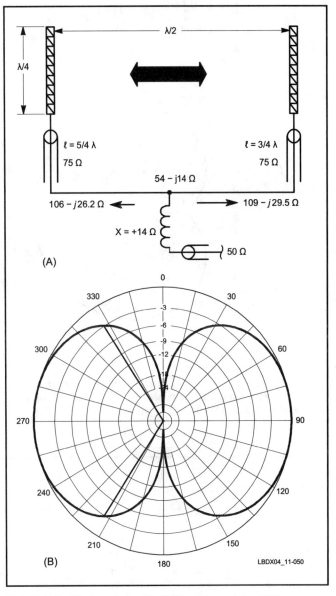

Fig 11-50—Horizontal radiation pattern (at a 20° elevation) for the 2-element out-of-phase, end-fire array with λ/2 spacing. Elements in 90°-270° plane.

spacing. How can we cover the 90° off directions? This can be done by feeding the two elements 180° out-of-phase, which also results in a bidirectional pattern but with a much broader lobe (beamwidth of 116° vs 64° in broadside) and less gain (3.5 dBi vs 4.9 dBi). See **Fig 11-50**.

4.5.1. Data, bidirectional end-fire array

Spacing: $\lambda/2$
Feed currents: I1 = 1 $\underline{/0°}$ A; I2 = 1 $\underline{/-180°}$ A

Feed point impedance: Z1 = Z2 = 45 + j 14 Ω
Gain (over average ground): 3.51 dBi

4.5.2. Current-forcing feed system, bidirectional end-fire array

We will run a 270°-long (3λ/4) feed line to the element with the leading current, and a 450°-long (5λ/4) feed line to the element with the lagging feed current. With the lines being odd multiples of λ/4 long, we can use the current-forcing principle. A 90° and a 270°-long feed line are physically too short for the array, since the elements are spaced λ/2. To preserve symmetry, the T junction where the lines to the elements join must be located at the center of the array.

The impedances at the end of the feed lines can be calculated with the COAX TRANSFORMER software module. Using 75-Ω coax and zero losses we have: Z1' = Z2' = 114 – j 35 Ω. The combined impedance is 57 – j 17.5 Ω.

If we do the calculation including cable losses (there is a lot of cable in the two feed lines), assuming 0.2 dB/100 feet at 1.8 MHz and Vf = 0.66, we would have a feed impedance of 54 – j 14 Ω, which is a good match. In both cases we can tune out the negative reactance with a small series coil, and end up with a feed impedance very close to 50 Ω. See Fig 11-50.

4.6. Triangular Arrays

The original description by D. Atchley, W1CF, was for a 3-element array, where the verticals were positioned in an equilateral triangle with sides measuring 0.29 λ, or 104°. (Ref 939 and 941). The original version of the array used equal current magnitude in all elements. Later, Gehrke, K2BT, improved the array by feeding the two back elements with half the current of the front element. This very significantly improved the directivity of the array.

We can operate a triangle array in two different configurations:

- Beaming off the top of the triangle. The top corner (the front element) is fed with a phase delay vs the two bottom-line verticals, which are fed with the reference phase angle (0°)
- Beaming off the bottom of the triangle. In this case the bottom-corner elements are fed by the current with a phase delay vs the top vertical (the back element), which is fed with the reference phase angle of 0°.

In both cases the solitary element is usually fed with twice (or slightly less) the current magnitude when compared to the two non-solitary elements of the triangle, which are fed with the same current magnitude. Being a triangle,

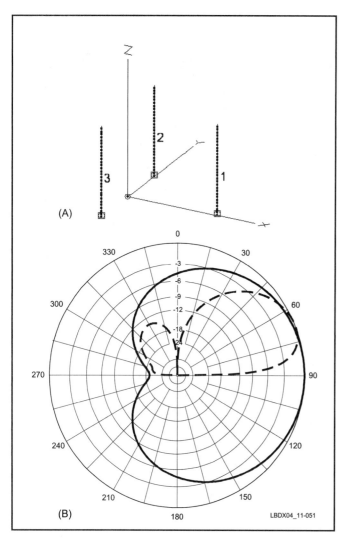

Fig 11-51—Triangular array with 0.29-λ spacings between elements. Azimuth plot is at 20° elevation angle. See Table 11-13.

Table 11-13
Triangular Array Data

Side	Config	Feed Current	Gain dBi	BW	RDF dB	DMF dB	Feed Impedance Front, Back
0.29λ	A	2, –90°; 1, 0°; 1, 0°	4.87	150°	8.68	13.7	55 + j 19; 13 – j 36 (2x)
0.29λ	A	1.8, –110°; 1, 0°; 1, 0°	5.47	129°	9.40	16.0	53 + j 17; 13 – j 21 (2x)
0.29λ	B	2, 0°; 1, –90°; 1, –90°	5.01	146°	8.82	14.3	87 + j 0 (2x); 18 – j 9
0.29λ	B	1.8, 0°; 1, –110°; 1, –110°	5.56	126°	9.49	16.3	76 + j 9 (2x); 14 – j 2

Side = side dimension in degrees (90° = λ/4)
Config A = shooting of the top of the triangle, B = shooting off the base

each array can be switched in three directions. Three directions fire off the top of a triangle, the other three off the bottom-line of a triangle. This means that a triangular array can be made switchable in six directions. All directions have the same gain (within 0.1 dB) and a very similar radiation pattern.

As expected, the performance (gain, beamwidth, directivity) is somewhere between the 2-element end-fire array and the Four-Square array. See **Table 11-13** and **Fig 11-51**.

4.6.1. Feed systems, triangular arrays

If we use quadrature feeding through a hybrid coupler, we are confronted with a practical switching problem. We need to feed the solitary element with double the feed current, which means with two paralleled feed lines. This means that a bunch of relays will be required to switch the extra feed line in parallel, depending on the direction.

If you want to erect a triangle array, you should opt for the current-optimized Lahlum/Lewallen version, where you can achieve the double feed current magnitude by simply dimensioning the L-network components correctly. **Fig 11-52** shows the Lahlum/Lewallen feed networks for both triangle configurations.

Fig 11-53 shows the direction switching for the array. As the feed impedances are different for the "A" and the "B" directions, we need two phasing networks. To do the direction switching we need a small matrix of SPST relays plus a seventh relay with three inverting contacts. This may seem complicated but using the two L-networks makes it possible to adjust the values to obtain the exact feed currents required. The measuring set up as described in Sections 3.6.1 and 3.6.2 should be used to make the adjustments.

4.7. The Four-Square Array

In 1965 D. Atchley, (then W1HKK, later W1CF, now a Silent Key), described two arrays that were computer modeled, and later built and tested with good success (Refs 930, 941). Although the theoretical benefits of the Four-Square were well understood, it took a while before the correct feed methods were developed that could guarantee performance on a par with the theory.

The Four-Square is in fact similar to a 3-in-line

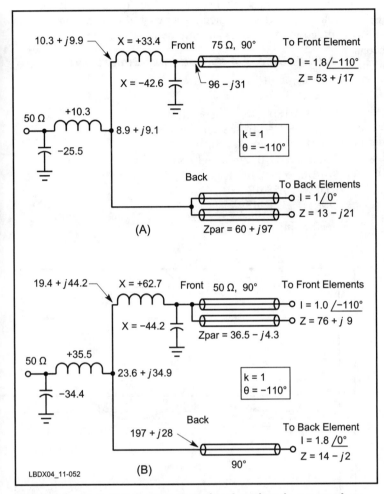

Fig 11-52—At A, the feed system for the triangle array when firing off the top of the triangle. At B, the feed system when firing off the base line of the triangle. If you want six directions, you will need a switching system that selects the proper network, as shown in Fig 11-53.

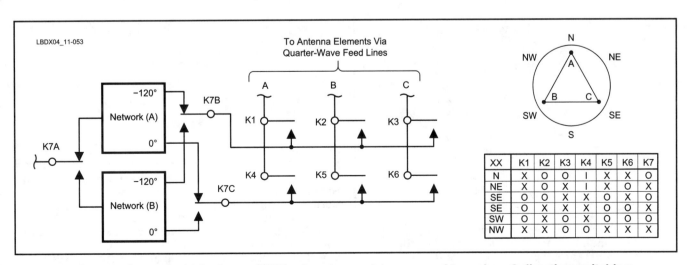

Fig 11-53—Seven relays, of which six are SPST relays in a matrix, are used to make a 6-direction switching network/feed system. The networks are shown in Fig 11-52.

end-fire array—the center two elements are fed in-phase and act as one common element. If all four elements have equal current, the total center-element current (for both in-phase elements together) is twice the current at each end. The required 1:2:1 current distribution as explained in Section 4.4 is satisfied.

The Four-Square can be switched in four quadrants. Atchley also developed a switching arrangement that made it possible to switch the array directivity in increments of 45°. The second configuration consists of two side-by-side cardioid arrays. This antenna is discussed in detail in Section 4.8.

The practical advantage of the extra directivity steps, however, does not seem to be worth the effort required to design the much more complicated feeding and switching system, since the forward lobe is so broad that switching in 45° steps makes very little difference. It is also important to keep in mind that the more complicated a system is, the more failure-prone it is.

4.7.1. Quadrature-fed, λ/4-spaced Four-Square

Placement of elements is in a square, spaced λ/4 per side. All elements are fed with equal currents. The back element is fed with the reference feed current angle of 0°, the two center elements with −90° phase, and the front element with −180° phase difference.

Fig 11-54 shows the radiation patterns for this array. The direction of maximum signal is along the diagonal from the rear to the front element. An array always radiates in the direction of the element with the lagging current.

4.7.1.1. Data, λ/4-spaced Four-Square array

Dimension of square side: λ/4.
Feed currents:

$I1 = 1 \angle{-180°}$ (front element)
$I2 = I4 = 1 \angle{-90°}$ (center elements)
$I3 = 1 \angle{0°}$ (back element)
Gain: 6.67 dBi over good ground
3-dB beamwidth: 98°
RDF = 10.58 dB
DMF = 21.02 dB

Feed-point impedances:

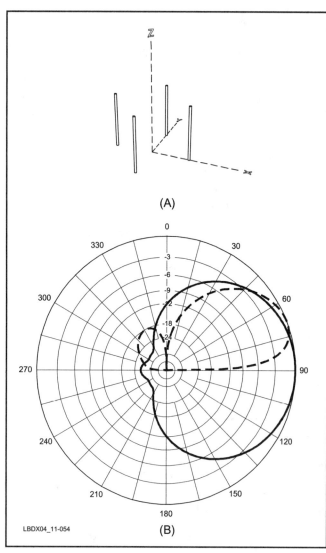

Fig 11-54—Radiation patterns (horizontal at a 20° elevation angle) for a typical quadrature-fed Four-Square array. Notice the important back lobe at relatively high elevation angles (about 60°).

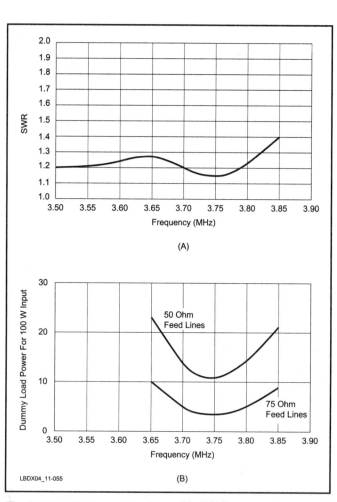

Fig 11-55—SWR and dissipated-power curves for a Four-Square array tuned for operation in the 3.7 to 3.8-MHz portion of the 80-meter band. Note that the dissipated power is much lower with 75-Ω feed line than with the 50-Ω feed line. The SWR curves for both the 50- and the 75-Ω systems are identical. The curve remains very flat anywhere in the band, but it is clear that the power dissipated in the load resistor is what determines a meaningful bandwidth criterion for this antenna.

$Z1 = 62.3 + j\,53.4\ \Omega$
$Z2 = Z4 = 40.5 - j\,19\ \Omega$
$Z3 = -0.3 - j\,15.2\ \Omega$

4.7.1.2. Feed system, quadrature–fed Four-Square

4.7.1.2.1. Hybrid-coupler Collins feed, quadrature-fed Four-Square

Since the antenna is fed in quadrature, a hybrid feed system is possible (see Section 3.4). We can feed the array with either 50 or 75-Ω, $\lambda/4$ current-forcing feed lines. Using 75-Ω feed lines generally results in less power being dumped in the load resistor if the hybrid network is designed for a system impedance of 50 Ω. On the antenna design frequency it should be possible to dump no more than 1% to 5% (–20 to –13 dB) of the transmit power in the dummy load. A 200-W dummy load should normally be sufficient for 1.5 kW power output into the antenna. See **Fig 11-55**. It's not a bad idea, however, to have a bigger one. In case of malfunction of the antenna much more power can be dumped into the load! Many operators measure the power dumped in the dummy and have an indicator in the shack.

4.7.1.2.2. Lewallen feed, quadrature-fed Four-Square

In Section 3.4.5 we see the detailed calculation of the Lewallen feed system (LC-network) using the *Lahlum.xls* spreadsheet. The Lewallen feed method for this array is worked out in great detail in *The ARRL Antenna Book*, where L-network values are listed for a range of feed-line impedances and ground systems.

4.7.2. WA3FET-optimized Four-Square array

Jim Breakall, WA3FET, optimized the quarter-wave-spaced Four-Square array to obtain higher gain and better directivity. Fig 11-54 shows that the original Four-Square exhibits a big high-angle backlobe (down only 15 dB at 120° in elevation). By changing the feed current magnitude and angle to the various elements you can change the size and the shape of the backlobes as well as the width of the front lobe. Full optimization is a compromise between optimization in the elevation and the azimuth planes. With Breakall's optimization, the gain of the array goes up by 0.6 dB. At least as important is a significant gain in directivity (RDF and DMF).

4.7.2.1 Data, WA3FET-optimized Four-Square

Dimension of square side: $\lambda/4$
Feed currents: I1 = 0.872 /–218° A (front)
I2 = I4 = 0.9 /–111° A
I3 = 1 /0° A (back)
Gain: 7.25 dBi
3-dB beamwidth: 85°
RDF = 11.4 dB
DMF = 24.4 dB
Feed-point impedances:
$Z1 = 37.5 + j\,57.7\ \Omega$ (front)
$Z2 = Z3 = 30.8 - j\,7.0\ \Omega$ (center)
$Z4 = 6.0 - j\,3.4\ \Omega$ (back)

4.7.3. Lahlum/Lewallen feed system, quadrature-fed Four-Square

In Section 3.4 I covered in detail the design of the Lahlum/Lewallen feed system for this array. Note that we

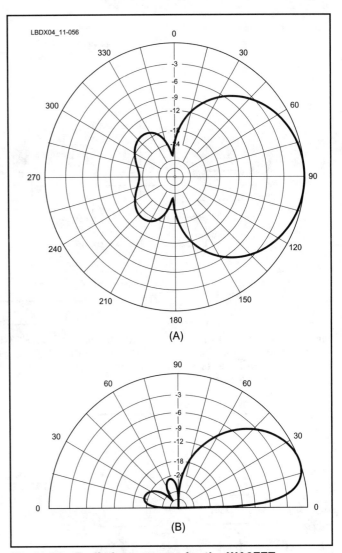

Fig 11-56—Radiation patterns for the WA3FET-optimized Four-Square, where the high-angle back lobe has been reduced substantially. Net result is 0.7 dB more gain and increased directivity.

lengthened all elements an equal amount to obtain a non-reactive impedance in the center two elements, resulting in slightly different component values and impedances.

4.7.4. W8JI cross-fire feed, Four-Square

W8JI's 4-square has the following configuration:

Dimension of square side: $\lambda/4$
Feed currents: I1 = 1 /–240° (front element)
I2 = I4 = 1 /–120° (center elements)
I3 = 1 /0° (back element)

If you model this configuration you find:

Gain: 7.45 dBi (0.8 dB better than quadrature-fed)
3-dB beamwidth: 79°
RDF = 11.78 dB
DMF = 17.4 dB
Feed-point impedances:
$Z1 = 27 + j\,56\ \Omega$
$Z2 = Z4 = 24\ \Omega$
$Z3 = 6.6 + j\,3\ \Omega$

The impedance of 24 Ω (at 1.83 MHz) was obtained by

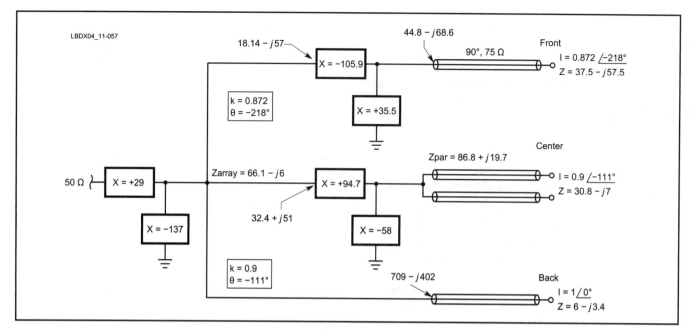

Fig 11-57—Lahlum/Lewallen feed circuit for the WA3FET-style Four-Square, with optimized phase angle and drive current magnitudes. In Fig 11-19 slightly different element impedances were used. Note that the variation of the L-network components are well within the normal tuning range. The feed impedances are also within a few percent of one another.

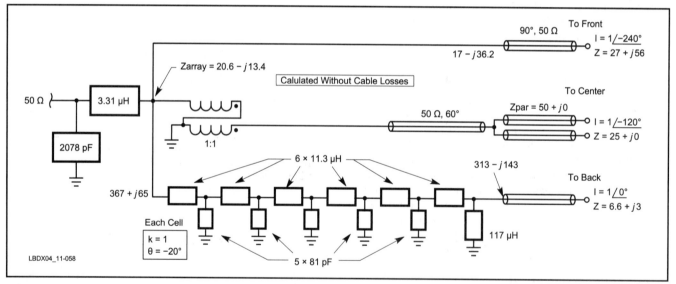

Fig 11-58—Feed system used by W8JI for his 160-meter Four-Square, which uses 120° phasing angle increments. See text for details.

tuning the solitary elements for resonance at 1.818 MHz.

Fig 11-58 shows the feed system developed by Tom, W8JI. The cross-fire principle means that we feed the array from the front element, and use a 180° phase-reversal transformer to feed the other elements (see Section 3.4 and also Chapter 7).

With respect to the front element, the required phase shift to the center elements is +120°, which is equal to +120−360 = −240°. Note that the parallel impedance of the λ/4 current-forcing feed lines to the two center elements is very close to 50 Ω. This means that we will be able to obtain any desired phase shift by using a 50-Ω feed line of a length (in degrees) equal to the required phase shift.

Note the 180° phase-reversal transformer, which takes care of −180° of the required 240° phase shift. The remaining 60° is obtained through a 50-Ω coax 60° in length. The feed system for the front and the center elements is extremely broad-banded, as phase shift remains constant with changing frequency, because of the cross-fire principle.

The "bad boy" is the back element. We can feed it with an L-network and use the *Lahlum.xls* spreadsheet to calculate the components. It is obvious that this branch will be the bottleneck for bandwidth. You could develop two L-networks, one for each band section of interest, and switch them, however.

Tom, W8JI, used what he calls "an artificial transmission

line using lumped components, composed of multiple L/C sections to simulate a transmission line with a characteristic impedance matching the rear element." Tom quotes the following advantages: "*Q is low, making phase-shift much less frequency critical. and I can easily tweak delay-line characteristics with a few adjustments to optimize the array null.*"

How do you calculate such an artificial line? You can consider it as a series connection of a number of L-networks—which we know from the Lewallen/Lahlum principle. Here is how to calculate the components of the "artificial transmission line" using *Lahlum.xls*:

- First calculate the impedance at the end of the λ/4 feed line to the back element: $313 - j\,143\ \Omega$.
- Next use this impedance as an input for R and X in the second part of the spreadsheet (called "for non-current forcing").
- For the regular Lahlum network composed of a single L-network cell, we would enter a required phase shift of $(+240 - 360) = -120°$, and end up with a parallel cap of 293 pF and a series coil of 28.5 μH (all calculated for 1.83 MHz). The input impedance into the L-network would be $94.6 + j\,163.8\ \Omega$.
- But you can specify, for example, a required phase shift of $-20°$. In that case this L-network cell will require a parallel coil of 117 μH and a series coil of 11.3 μH. Look now at the input impedance and note it is $367 + j\,65\ \Omega$.
- All we need to do now is, using the same spreadsheet, calculate another five L-networks, around an output impedance of $367 + j\,65\ \Omega$, each for a 20° phase shift. Each of these L-network cells has a parallel capacitor of 81 pF and a series coil of 11.3 μH, and the input impedance of this artificial line (consisting of six L-network cells), is $367 + j\,65\ \Omega$.

W8JI has experimented a lot with this system, and notes: "*Because the current is low, components can be modest sized. The end result is more bandwidth, more stability and less loss than a simple one-stage network.*" It is not strictly necessary to use six cells, of course, but the greater the number, the better the bandwidth.

Another way to calculate the "artificial transmission line" is to first tune out the reactance of the impedance at the end of the λ/4 feed line. A parallel coil of 77 μH, which represents $+828\ \Omega$ reactance, will turn the impedance into 378 Ω. Now we can use the PI-LINE STRETCHER module from the NEW LOW BAND SOFTWARE to calculate cells that each give 20° phase shift and for a characteristic impedance of 378 Ω. The values of the components are identical.

4.7.4.1. Other applications, cross-fire principle

Single L-networks can be replaced with multiple-section networks to improve bandwidth. This is especially true where high-impedances are encountered, which is most frequently the case with the "back element" of an array.

4.7.4.1.1. Where can we apply this cross-fire principle?

To be able to feed an element through a 1:1 (180°) transformer and a coaxial phasing line (whose length is 180° minus the required phase delay), you need to be able to achieve a pure resistive impedance at the end of the current-forcing feed line. You can shorten/lengthen the element somewhat so that the feed-

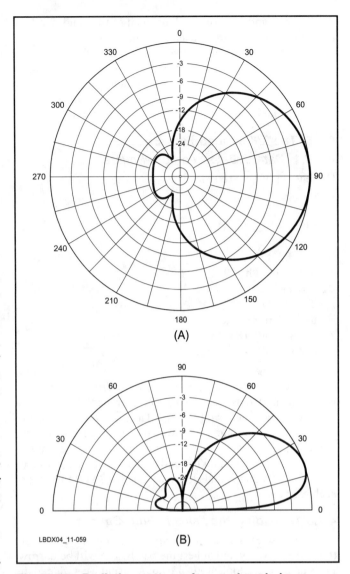

Fig 11-59—Radiation patterns for a reduced-size (λ/8 side) Four-Square, which exhibits even better directivity than the larger varieties but with slightly less gain and less bandwidth.

point impedance at the element is purely resistive.

Next, you can select the impedance of the current forcing feed lines (usually 75 or 50 Ω), and see if any combination turns out to be a good one. Good ones are: 25 Ω (two 50 Ω in parallel), 30 Ω (50 and 75 Ω in parallel) 37.5 Ω (two 75 Ω in parallel), 50 Ω and 75 Ω.

It might be better to make an element somewhat non-resonant, so that with the addition of a parallel reactance (coil or capacitor) we end up near one of the above-mentioned impedances.

4.7.4.2. Conclusion, cross-fire principle

It has to be rather a lucky shot if you can apply this principle. It is clear that the cross-fire principle may track frequency a little better than the other feed methods, and this is certainly so with receiving arrays (and phased Beverages) where the element impedances hardly change with frequency.

If you need bandwidth, it seems to me that the easiest solution is to provide switchable L-networks; that is, one for the CW end and one for the Phone end of the band, while at the same time changing the length of the current-forcing feed lines. (You would add some extra length for the lower frequency) and retune the elements for resonance.) That is certainly a guarantee for peak performance and at the same time it gives you the ability to prune the array for optimum performance, even at the sacrifice of some bandwidth.

4.7.5. The λ/8-spaced Four-Square

4.7.3.1. Data, λ/8-spaced Four-Square

Dimension of square side: λ/8
Feed currents:
$I1 = 1 \angle -270°$ A
$I2 = I3 = 1 \angle -135°$ A
$I4 = 1.1 \angle 0°$ A
Gain: 5.85 dBi
3-dB beamwidth: 89°
RDF = 11.3 dB
DMF = 25.0 dB
Feed-point impedances:
$Z1 = -11.3 + j\, 18.7\ \Omega$
$Z2 = Z3 = 18.4 - j\, 5.6\ \Omega$
$Z4 = 1.3 - j\, 11.8\ \Omega$

This small-footprint Four-Square sacrifices 1.4 dB of gain compared to its optimized big brother, but it has every bit as good or even better directivity. The main disadvantage of this design is the much narrower bandwidth. Note that the reduction in gain is to a large extent due to the lower impedances of the elements, taking into account that we inserted an equivalent-ground-radials loss resistance of 2 Ω at the base of each element.

4.7.5.1. Feeding the "small" Four-Square

Because of the very low impedances involved, feeding this array is tricky and at best the bandwidth will be narrow. Let's take a close look at **Fig 11-60**. As usual we will feed the back element directly. Note that the negative impedance we've become accustomed to is large, which indicates very heavy mutual coupling, obviously due to the proximity of the elements involved. The center elements have reasonable impedance values, which translates into normal L-network components in the center branch. The branch to the front element is very peculiar. The real part of the impedance of the front element (including 2-Ω ground losses) is 1.3 Ω. This means that this element is taking almost no power at all. If we do the calculating of the Lahlum-network we will some up with an "extreme" value for the series element in the network (−4755 Ω), which represents 18 pF at 1.8 MHz. It is clear that due to stray capacity this is a impossible value. The value X1 (for the parallel element of the L-network) is the Lahlum-network value. Let's see what would be the value of a parallel impedance that turns $51.9 + j\, 471\ \Omega$ into a pure resistance. Using the HUNT-SERIES IMPEDANCE NETWORK module of the NEW LOW BAND SOFTWARE, it appears that it is −477 Ω, and that the resistive impedance at that point is 4,326 Ω, a high value as expected.

At this point is appears to be much simpler to turn the front element into a parasitic element and not feed it at all. The parasitic element can now be tuned by simply tuning the parallel reactance (a capacitor), which in this case has the value X2 = −477 Ω. Note also that X2 is almost the same as X1. If the front element was taking no power at all, these two value would have been identical.

In practice, you can just leave out the series element of the L-network in the front element branch. If you can, stay away from arrays with such close coupling and such low impedances. They mean critical alignment, high Q and low bandwidth!

4.7.6. Direction switching for Four-Square arrays

Fig 11-61 shows a direction-switching system that can be used with all Four-Square arrays. The front element (in the direction of firing) will be fed with the most lagging feed angle (−180° for quadrature feeding); the back element will be with the zero reference feed angle. "Mid", Back" and

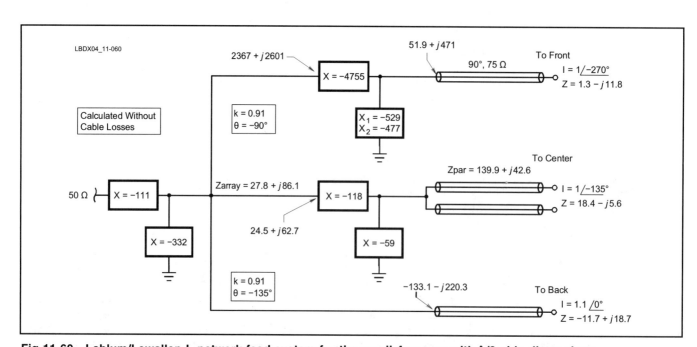

Fig 11-60—Lahlum/Lewallen L-network feed system for the small 4-square with λ/8 side dimensions.

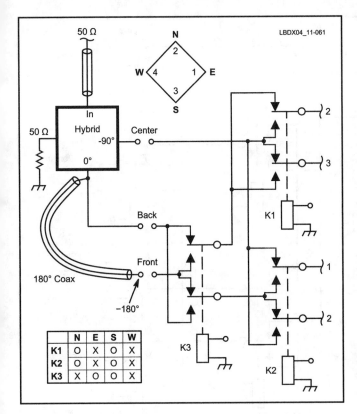

Fig. 11-61—This direction-switching system shown can be used with λ/8 Four-Square arrays, with whatever phasing circuitry is used.

"Front" go to the corresponding points in the feed circuits for a Lahlum/Lewallen circuit.

For a quadrature-fed array, a hybrid can be used, wired as shown in Fig 11-61. The Comtek Hybrid Coupler includes a 180° phase-inversion transformer. In that case "Front" is directly connected to that transformer output, without using the 180° phase-shift coax shown in the figure.

4.8. Four-Square Array with 8 Directions

A Four-Square usually has a −3-dB forward lobe beamwidth of 85° to 100° (depending on spacing and phasing), so four directions can quite adequately cover all azimuths. Some people think they need more. Admittedly, more than four directions may be advantageous for moving the nulls, which may be advantageous for nulling out QRM and noise from certain directions.

In the half-angle intermediate position, the four verticals are fed like two side-by-side (broadside is not an adequate term—"narrow-side" would be better) end-fire cells, spaced only λ/4. **Fig 11-62** shows the radiation patterns for a Four-Square using side-by-side, end-fire feeding.

4.8.1. Quadrature-fed, 8-direction Four-Square

4.8.1.1. Data, quadrature-fed 8-direction Four-Square

Side of square: λ/4
Feed currents:
$I_4 = I_3 = 1 \angle{-90°}$ A (front)
$I_1 = I_2 = 1 \angle{0°}$ A (back)

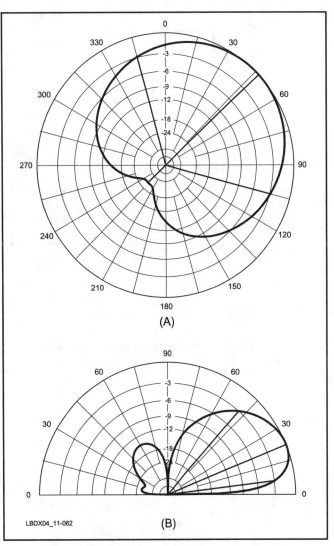

Fig 11-62—Layout and radiation patterns (horizontal at 20° elevation angle) for the close-spaced side-by-side, end-fire cell configuration, used in an "intermediate" direction in a Four-Square. Compare with the patterns of a conventional Four-Square in Fig 11-54.

Gain: 5.51 dBi (6.67 dBi in regular Four-Square configuration)
3-dB beamwidth: 123°
RDF = 9.22 dB
DMF = 14.7 dB
Feed-point impedances:
$Z_4 = Z_3 = 88.5 + j\, 6.6\, \Omega$ (front)
$Z_1 = Z_2 = 17.7 - j\, 36.1\, \Omega$ (back)

4.8.1.2. Feed system, quadrature-fed 8-direction Four-Square

If you use quadrature feeding, you can feed the array in all eight directions using a hybrid coupler feed system. **Fig 11-63** shows a possible direction-switching and feed system for the quadrature-fed Four-Square with hybrid coupler, including the four additional "mid-direction" firing directions.

4.8.2. Optimized feeding, 8-direction Four-Square

For spacing of λ/4 between the elements, if we increase the phase shift to 105°, we get somewhat higher gain and directivity. For that case we need to feed the array with a Lewallen feed

Phased Arrays 11-51

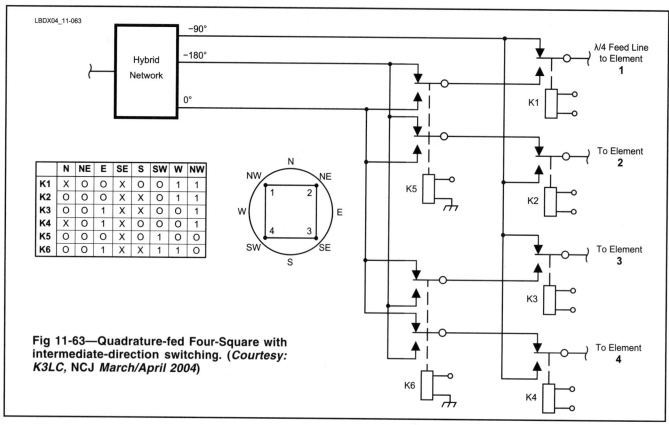

Fig 11-63—Quadrature-fed Four-Square with intermediate-direction switching. (*Courtesy: K3LC, NCJ March/April 2004*)

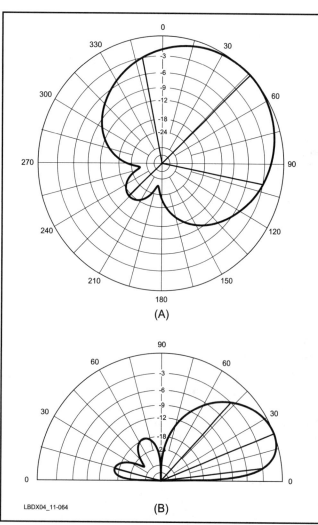

Fig 11-64—Radiation patterns (horizontal at 20° elevation angle) for the optimized side-by-side, end-fire cells, showing improved high-angle rejection off the back due to 105° phase-shift feed at intermediate angles.

system (L-network). If you decide to add four more directions to an optimized Four-Square, this is the way to go.

4.8.2.1. Data, optimized 8-direction Four-Square

Side of square: $\lambda/4$
Feed currents: I4 = I3 = 1 $\underline{/-105°}$ A (front)
I1 = I2 = 1 $\underline{/0°}$ A (back)
Gain: 5.92 dBi (optimized regular Four-Square configuration: 7.25 dBi)
3-dB beamwidth: 113°
RDF = 9.72 dB
DMF = 16.5 dB
Feed-point impedances:
Z4 = Z3 = 81.8 + j 15 Ω (front)
Z1 = Z2 = 13.2 – j 26.3 Ω (back)

4.8.2.2. Feed system, optimized 8-direction Four-Square

Fig 11-65 shows the Lahlum/Lewallen feed system for the Four-Square in its intermediate directions. Note that we use 75-Ω feed lines, which results in a relatively high feed impedance for the array. In this case a simple coil in parallel with the input (X = +113.9 Ω) will turn the feed impedance into 52 Ω, a perfect match!

If we choose to feed the back elements directly and the front elements through a 105° L-network delay circuit, we will have L-network components of lower impedance (Xp =

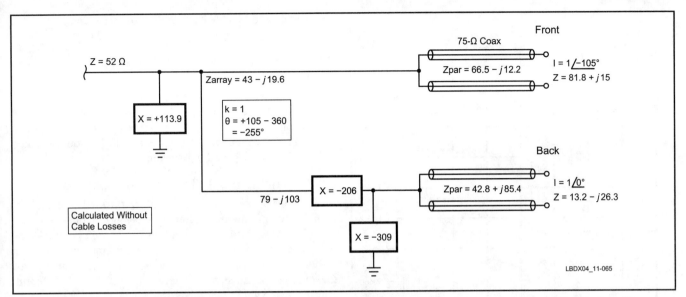

Fig 11-65—Lahlum/Lewallen feed circuit for a Four-Square working as two closely spaced end-fire cells with optimized phasing, shooting along the directions of the side of the square.

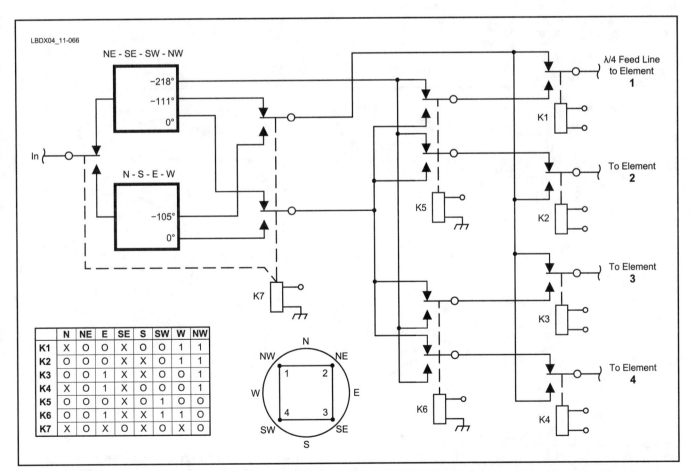

Fig 11-66—Direction-switching for an optimized Four-Square (WA3FET) in an intermediate direction. Two feed networks are used, selected by relay K7. The feed network for the main directions (NE, SE, SW and NW) is shown in Fig 11-65.

Phased Arrays 11-53

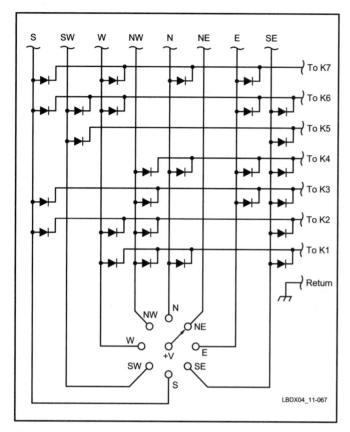

Fig 11-67—Diode matrix for switching the Four-Square in eight directions. This circuit can be applied to both Figs 11-63 and 11-66. When used with Fig 11-63 you do not use the relay line "to K7".

−61.4 Ω and X_s = +66.4 Ω), but the disadvantage is that the array impedance is relatively low at 16 + j 24 Ω.

4.8.2.3 Direction switching, optimized 8-direction Four-Square

Fig 11-66 shows a possible direction switching method, similar to the one used in Fig 11-63, with the difference that there is one extra relay (K7) for switching the two networks. In **Fig 11-67** we see the truth table from Fig 11-66 translated to a diode-matrix switching system.

4.8.3. Conclusion, 8-direction Four-Square

When you are happy with the quadrature-fed Four-Square, adding the intermediate directions is only a question of a slightly more complicated direction-switching system. As the hybrid coupler will see vastly different impedances when switching from the "full" directions to the "half" directions, the input impedance of the network will also be different, making fast switching impossible, unless you add two L-networks, which would be switched automatically. With the optimized versions things become a bit more complicated as you will require two different feed/phasing circuits.

It is questionable if all these efforts are worthwhile. Having been a user of a Four-Square with just four directions for over 10 years now, I have never felt the urge of adding four more directions. Take that for whatever you think it is worth!

4.9. The Broadside/End-Fire Array

This array has a spacing in the end-fire cells of 45 meters (on 160 meters) and a broadside spacing of 90 meters. The broadside spacing should be a minimum of 80 meters and can

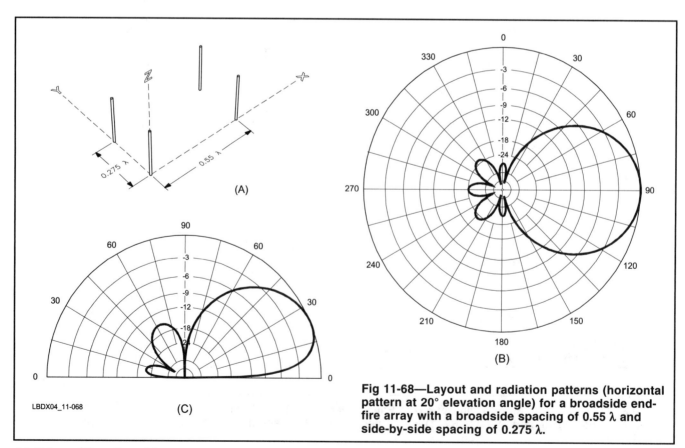

Fig 11-68—Layout and radiation patterns (horizontal pattern at 20° elevation angle) for a broadside end-fire array with a broadside spacing of 0.55 λ and side-by-side spacing of 0.275 λ.

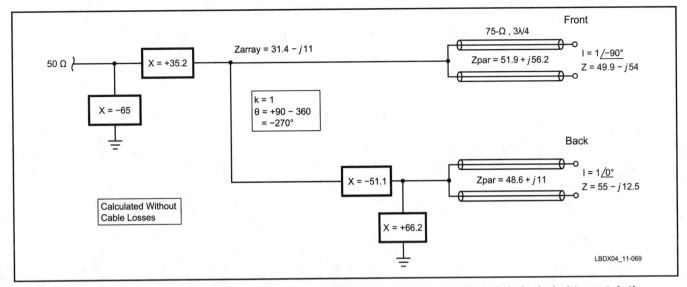

Fig 11-69—Lewallen feed system (L-network) for the array in Fig 11-68. An L-network is included to match the feed system input impedance to 50 Ω. The alternative, where we feed the back elements directly and the front elements via an L-network, results in a much lower feed impedance (19 + j 12 Ω).

be as much as 125 meters, achieving better rejection at high elevation angles but poorer directivity at low angles.

The end-fire cells could also have much smaller spacing, in which case a different phasing would be required; for example, 135° to 145° for λ/4 spacing. The bandwidth would however suffer from the small spacing. The array that was calculated is fed with 90° phase shift. These end-fire broadside combinations are also used in the 8-Circle array (Section 4.14).

4.9.1. Data, broadside/end-fire array

Broadside spacing: 0.55 λ
End-fire cell separation: 0.275 λ
End-fire cell phase: 90°
Feed currents:
Element 1 and Element 2 (back): I = 1 /0° A
Element 3 and Element 4 (front): 1 /–90° A
Gain = 8.43 dBi
3-dB forward angle: 57.4°
RDF = 12.11 dB
DMF = 20.8 dB
Feed impedances:
Z(Element 1) = Z(Element 2) = 55 – j 12.5 Ω (back)
Z(Element 3) = Z(Element 4) = 38.6 + j 11 Ω (front)

4.9.2. Feed system, broadside/end-fire array

This array, which is quadrature-fed (in 90° increments) can use a hybrid coupler. I designed a Lewallen LC-network type coupler, which has the advantage of being able to tune the array. The network was designed around 75-Ω current-forcing feed lines, which results in higher impedances than when using 50-Ω lines. Note that you need to use 270° long feed lines because of the physical separation of the two end-fire cells. **Fig 11-69** shows the Lewallen feed system for the array.

Direction switching is very simple, all you need is a single DPDT relay to invert the paired feed lines to the front and to back (points indicated as Fr and Bk in Fig 11-69).

4.10. The 5-Rectangle Array

One way of getting radiation "off the side" from an end-fire broadside array (see Section 4.5) with a decent pattern is by adding a fifth radiator right in the center of the rectangle. **Fig 11-70** shows the configuration and the radiation patterns obtained.

4.10.1. Data, 5-Rectangle array

Length of rectangle: 0.55 λ
Width of rectangle: 0.275 λ
Feed currents: Element 1 and Element 2 (back):
I = 1 /0° A
Element 5 (center): I = 3 /–90° A
Element 3 and Element 4 (front): I = 1 /–180° A
Gain = 5.92 dBi
3-dB forward angle: 122°
RDF = 9.82 dB
DMF = 19.4 dB
Feed impedances:
Z (Element 1) = Z (Element 2) = 23.3 – j 37.3 Ω (back)
Z (Element 5) = 35.4 + j 1.8 Ω (center, middle)
Z (Element 3) = Z (Element 4) = 130 + j 30 Ω (front)

4.10.2 Feed system, 5-Rectangle array

The Lahlum/Lewallen LC-network feed system shown in **Fig 11-71** feeds the central element directly. The front and back elements are fed via LC networks, the front element with a –90° phase shift, the back element with a phase shift of +90°, which equals –270°.

4.10.3. Direction switching, 5-Rectangle array

As the 5-Rectange is a configuration that will only be used together with an end-fore/broadside array for giving right-angle coverage, I designed a direction-switching system that switches both configurations. In **Fig 11-72** North and South are the high-gain directions, and East and West the "fill-in" directions.

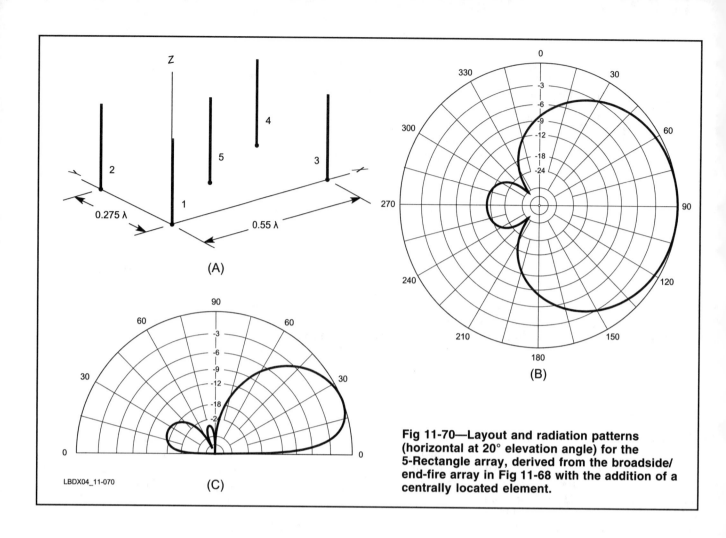

Fig 11-70—Layout and radiation patterns (horizontal at 20° elevation angle) for the 5-Rectangle array, derived from the broadside/end-fire array in Fig 11-68 with the addition of a centrally located element.

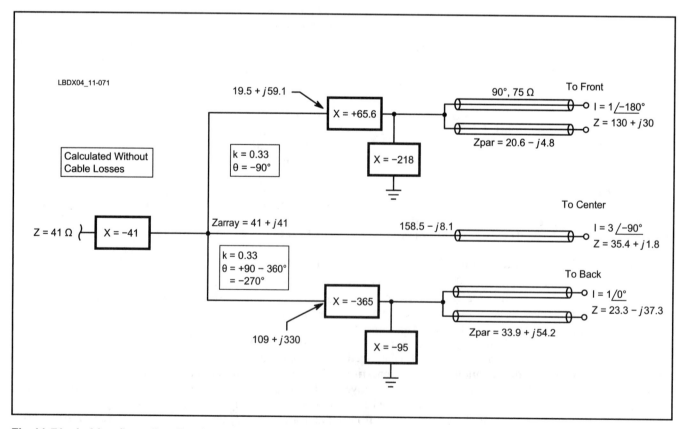

Fig 11-71—Lahlum/Lewallen Feed system for the 5-Rectangle.

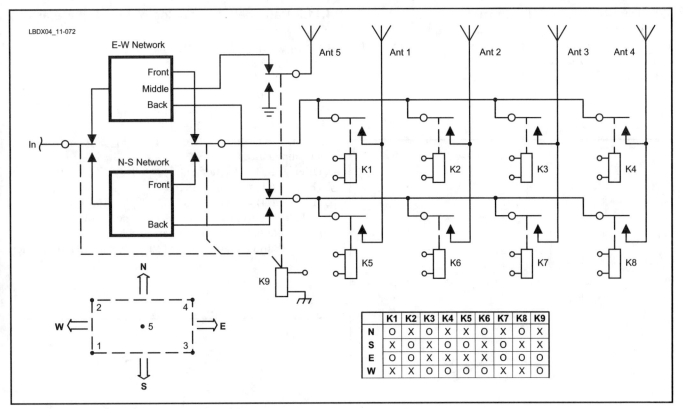

Fig 11-72—Direction-switching for the 5-Rectangle, as fill-in directions for a broadside/end-fire array described in Section 4.7.

Fig 11-73—Russian-made small vacuum relays having single make/break contacts. These are well-suited for a relay matrix.

You will need to build two networks and the right feed network will be selected by K9. The selection of the elements utilizes a small relay matrix consisting of eight relays. **Fig 11-73** shows a Russian-made vacuum relay that I use for switching.

4.11. Five-Square Array

The 5-Square is a modified Four-Square as shown in **Fig 11-74.** The side of square is 0.3 λ. The array can be made to cover eight directions. Shooting along the X and the Y axis the array has one reflector (Element 5), one director (Element 4) and three elements (1, 2 and 3) fed in phase, although with slightly different current magnitudes.

4.11.1. Data, "diagonal" operation of Five-Square array

Gain = 7.67 dBi (and that is 1 dB better than a classic Four-Square)
3-dB forward angle: 78°
RDF = 11.92 dB
DMF = 23.0 dB
Feed currents:
Element 5 (back): I = 1.25 /0° A
Element 1 (center): I = 1 /–125° A
Element 2 = Element 3 (outside center): 1 /–125° A
Element 4 (front): I = 1.25 /–255° A
Feed impedances:
Z (Element 5) = 2.5 – j 2.9 Ω (back)
Z (Element 1) = 45.4 – j 6.5 Ω (center, middle)
Z (Element 2) = Z (Element 3) = 37.2 – j 1.7 Ω (outside center)

Phased Arrays 11-57

Z(Element 4) = 28.5 + j 61 Ω (front)

Using the same physical layout we can shoot along the bisector of the X-Y axis, adding another four directions to the array. In this configuration we have two directors (Elements 4 and 2) and 2 reflectors (Elements 3 and 5).

4.11.2 Data, half-way angles of Five-Square array

Gain = 6.2 dBi
3-dB forward angle: 102.4°
RDF = 10.52 dB
DMF = 18.5 dB
Feed currents:
Element 3 = Element 5 (back): I = 0.6 $\underline{/0°}$ A
Element 1 (center): I = 2 $\underline{/-110°}$ A
Element 4 = Element 2 (front): I = 0.7 $\underline{/-230°}$ A
Feed impedances:
Z (Element 3) = Z (Element 5) = 8.8 − j 32.5 Ω (back)
Z (Element 1) = 22.8 + j 5.1 Ω
Z (Element 4) = Z (Element 2) = 23.4 + j 68.1 Ω (front)

In this configuration the performance is substantially the same as obtained by a Four-Square à la WA3FET. You can get a better F/B by changing the feed currents but will have a much

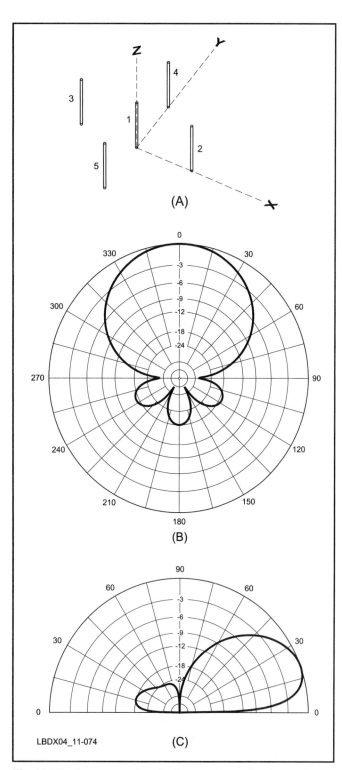

Fig 11-74—Layout and radiation patterns (horizontal at 20° elevation angle) for the Five-Square array operating in the diagonal fashion. The excellent directivity is mainly derived from its relatively narrow forward lobe (typically 20-25° less than for a Four-Square).

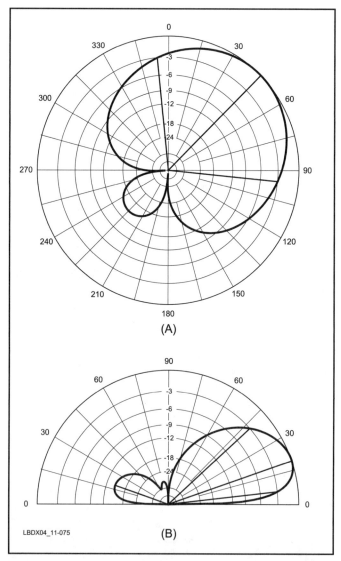

Fig 11-75—When shooting along the sides of the square, the array operates with one center element with two directors and two reflectors. The 3-dB forward lobe angle is roughly the same as for a classic Four-Square, and its gain is similar to a Four-Square optimized by WA3FET.

wider forward lobe, decreasing the RDF. By playing with a modeling program you can fine-tune this array to your liking.

It is obvious that the "main" directions are those along the X and Y-axes with 1.6 dB more gain and 1.4 dB more RDF (directivity), mainly obtained through a substantially narrower 3-dB forward beamwidth (78° vs 102°). In both configurations the array can be fed using the Lahlum/Lewallen feed system.

A nice thing about this Five-Square is that you don't really need five towers—You can hang the center element from some nylon or Dacron catenary cables strung between four towers. If the center element is slightly shorter because of the sag of the catenary cables, just top load it with four cross wires, running in the direction of the support cables.

4.11.3. Feed system, Five-Square array
4.11.3.1. Diagonal operation, Five-Square array

Fig 11-76 and **Fig 11-77** show two feed systems developed according the Lahlum/Lewallen system, using the "Lahlum.xls" spreadsheet. If you want to make the array cover eight directions, you will have to install both networks, and switch them in and out of the circuit according to the direction used.

4.11.4. Switching directions, Five-Square array

Fig 11-78 shows the direction switching, accomplished

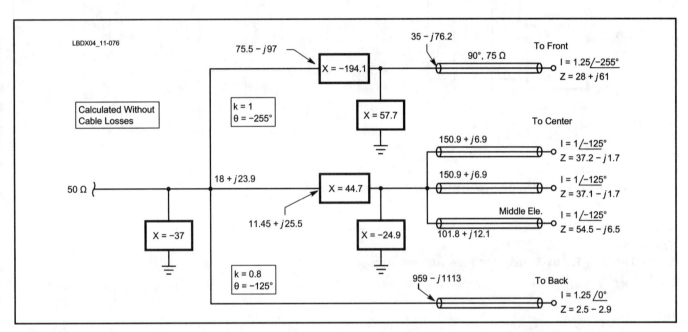

Fig 11-76—Lahlum feed network for the Five-Square array shooting diagonally across the square in one of its main directions.

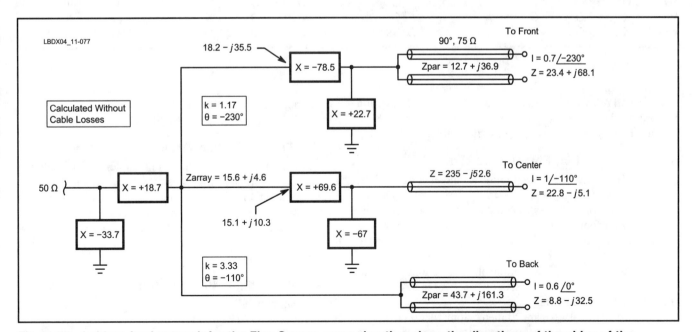

Fig 11-77—Lahlum feed network for the Five-Square array shooting along the directions of the sides of the square in the "intermediate" directions.

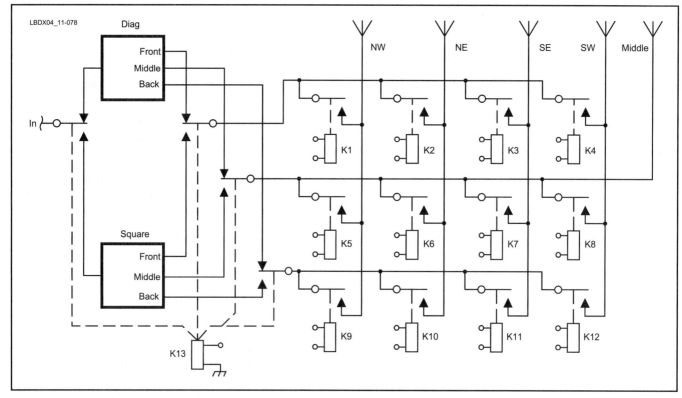

Fig 11-78—Direction-switching system for the Five-Square array.

Table 11-14
Truth Table of Relay Matrix for Five-Square Array

	K1	K2	K3	K4	K5	K6	K7	K8	K9	K10	K11	K12	K13
N	X	X	O	O	O	O	O	O	O	O	X	X	X
NE	O	X	O	O	X	O	O	X	O	O	O	X	O
E	O	X	X	O	O	O	O	O	X	O	O	X	X
SE	O	O	X	O	O	X	O	X	X	O	O	O	O
S	O	O	X	X	O	O	O	O	X	X	O	O	X
SW	O	O	O	X	X	O	X	O	O	X	O	O	O
W	X	O	O	X	O	O	O	O	O	X	X	O	X
NW	X	O	O	O	O	X	O	X	O	O	X	O	O

with a small relay matrix. According to the selected direction the appropriate drive network is selected with relay K13. **Table 11-14** below shows the relay truth table.

4.11.5. Conclusion, Five-Square array

For 160 meters, and including 40-meter-long radials, this array requires an area measuring 107 by 107 meters, (1.15 hectares, or ~ 3 acres). With 40-meter long radials, the antenna has a foot print that is only 20% larger than for the classic Four-Square (with λ/4 sides), yet produces 2 dB more gain and has the possibility of switching in eight directions. Clearly a winner! If you care for the eight directions you will, of course, need two different sets of L-networks to establish the required phased shifts.

With 120 radials that are 20 meters long for all elements, the footprint is 0.8 hectares (approximately 2 acres), and the trade for gain will be marginal (a fraction of a dB). Obviously for 80 meters, the required real estate is four times smaller. **Fig 11-79** is a photograph of NO8D's Five-Square array.

4.12. The 6-Circle Array

The 6-Circle was described in Chapter 7 as a receiving antenna. I developed two transmit versions, one having a circle diameter of 80 meters and one smaller version with 60-meter diameter (for 1.83 MHz). Still smaller versions (eg, 30-meters diameter) work very well as receiving arrays, but have low feed impedances and narrow bandwidth when used as transmit antennas.

4.12.1. Array data (40-meter radius), 6-Circle array

Circle diameter: 80 meters
Frequency: 1.83 MHz
Gain = 7.8 dBi
3-dB forwards angle: 74.6°

Fig 11-79—160-meter Five-Square at NO8D.

RDF = 11.63 dB
DMF = 24.2 dB
Feed currents:
Element 5 and Element 6 (back): 1 /0° A
Element 1 and Element 2 (center): I = 2 /–90° A
Element 3 and Element 4 (front): I = 1 /–180° A
Feed impedances:
Z (Element 5) = Z (Element 6) = 5.3 – j 15.4 Ω (back)
Z (Element 1) = Z (Element 2) = 27.9 – j 11.6 Ω
Z (Element 4) = Z (Element 3) = 129 + j 48.8 Ω (front)

This version with a circle radius of 40 meters is quadrature fed. We can feed it with a hybrid coupler (see Section 3.4.6) but since the feed current in the center elements is twice as high as for the other elements we need to feed them with two paralleled feed lines.

If you decide to feed this array using a hybrid coupler, use 3λ/4-long feed lines of 75-Ω rather than 50-Ω impedance, which will result in less power lost in the hybrid's dummy load. One issue with an 80-meter diameter 6-Circle array is that λ/4-long feed lines do not reach the center of the array. This requires 3λ/4 lines, with all the attendant drawbacks (more loss, more cost and much less bandwidth).

For this reason I developed a slightly smaller 6-Circle, which has a radius of 30 meters. This accommodates λ/4 feed lines using foam coaxial cables with VF of approx 0.8.

4.11.2 Array data (30-meter radius), 6-Circle array

Circle diameter: 60 meters
Frequency: 1.83 MHz
Feed currents:
Element 5 and Element 6 (back): 1 /0° A
Element 1 and Element 2 (center): I = 2 /–110° A
Element 3 and Element 4(front): I = 1 /–225° A
Gain = 7.6 dBi
3-dB forward angle: 78.2°
RDF = 11.72 dB
DMF = 25.5 dB
Feed impedances:
Z (Element 5) = Z (Element 6) = 4.3 – j 16.1 Ω (back)
Z (Element 1) = Z (Element 2) = 25.2 – j 7.7 Ω
Z (Element 4) = Z (Element 3) = 52 + j 93 Ω (front)

This smaller version of the 6-Circle is no longer quadra-

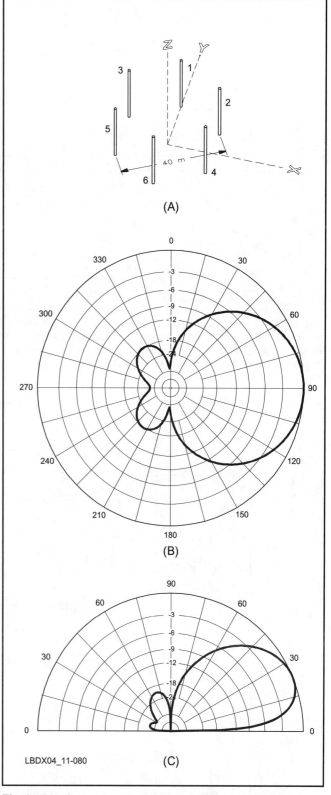

Fig 11-80—Layout and radiation patterns (horizontal pattern at 20° elevation angle) for the 6-Circle array at 3.7 MHz. The patterns remain almost identical for different sizes of the array. Larger arrays will show higher impedances and will be somewhat easier to feed. They also have somewhat more bandwidth. Larger arrays will require 3λ/4 current-forcing feed lines, which have definite disadvantages when it comes to bandwidth. If you have a diameter of 30 meters, λ/4 feed lines will reach if you use foam coax with Vf ~ 0.8.

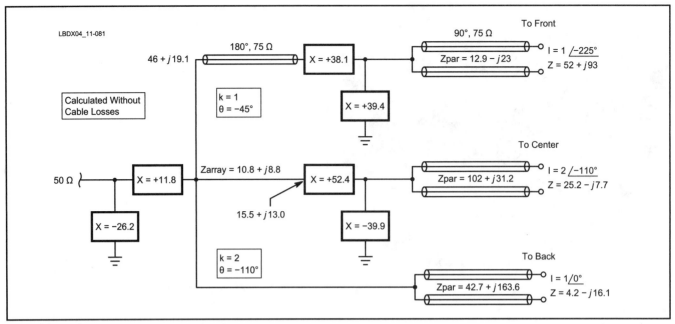

Fig 11-81—Feed system for a "small" 6-Circle (radius = 30 meters). In this array the back elements are fed directly, the center elements via an L-network and the front elements via a 180° line plus L-network. This results in the most realizable network-components reactances, but a low feed impedance. Other solutions result in a higher feed impedance but require high L-network component reactances.

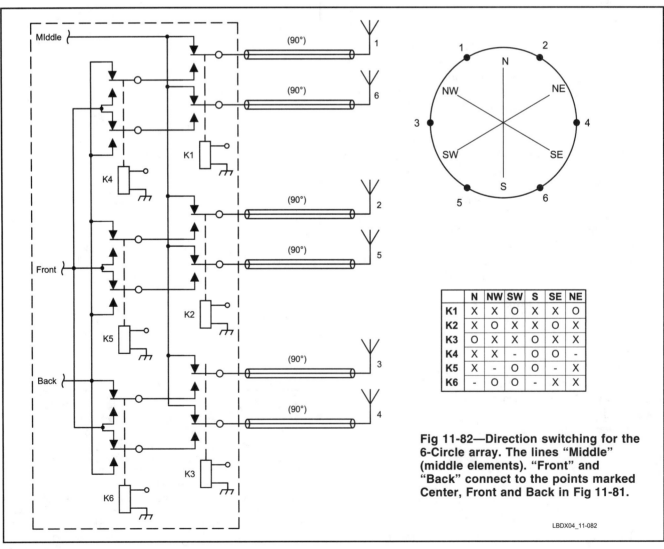

Fig 11-82—Direction switching for the 6-Circle array. The lines "Middle" (middle elements). "Front" and "Back" connect to the points marked Center, Front and Back in Fig 11-81.

11-62   Chapter 11

ture-fed and requires the Lewallen/Lahlum feed system. Using this feed system there is no need for parallel feed lines, since the double feed current magnitude can be obtained through a proper L-network circuit (see Section 3.4.5).

4.12.3. Feed system, 6-Circle array

The Lahlum/Lewallen feed system in **Fig 11-81** feeds this smaller array. Here too, 75-Ω current-forcing feed lines are used to achieve a higher input impedance (which is only about 16 Ω if using 50-Ω feed lines).

4.12.4. Direction switching, 6-Circle array

Fig 11-82 shows the direction switching circuitry for the 6-Circle array. Six DPDT relays are required.

4.12.5. Conclusion, 6-Circle array

On 160 meters, using 40-meter long radials, this array requires an area measuring 140 by 140 meter, approximately 2 hectares or 5 acres. It is a very nice performer, with excellent directivity and it should appeal to all those who want a top-notch array but just do not have the real estate for a 9-Circle (see Section 4.15).

With 120 radials, each 20-meters long, on each of the verticals the foot print is reduced to only 1 hectare (2.5 acres), with a loss in gain of just a fraction of a dB.

4.13. The 7-Circle Array

I developed a 7-Circle array using the same 60-meter circle diameter. This makes it possible to use $\lambda/4$ feed lines (with foam coax, Vf ~ 0.8) on 160 meters. This is identical to the 6-Circle but has another element in the center of the circle. It did not seem to be possible to improve the performance of the 6-Circle though.

4.13.1. Data, 7-Circle array

Gain = 7.61 dBi
3-dB forward angle: 80.4°
RDF = 11.74 dB
DMF = 26.3 dB
Feed currents:
Element 5 and Element 6 (Back): 0.8 /0° A
Element 1, Element 2, Element 7 (middle): I = 1 /–120° A
Element 3 and Element 4(front): I = 0.73 /–250° A
Feed impedances:
Z (Element 5) = Z (Element 6) = 11.5 – j 21.1 Ω (back)
Z (Element 1) = Z (Element 2) = 35.3 – j 2.6 Ω (outer middle elements)
Z (Element 7) = 40.5 – j 1.5 Ω (middle, center element)
Z (Element 4) = Z (Element 3) = 13 + j 88 Ω (front)

Feed system: The same schematic as used for the 6-Circle (Section 4.12) can be used, as the seventh element needs no switching when changing directions.

4.13.2. Conclusion, 7-Circle array

Nothing seems to be gained with the extra element in the middle.

4.14. The 8-Circle Array

In an 8-Circle array (see Chapter 8), we only use four elements at a time. The 8-Circle consists of two broadside cells, spaced about 0.65 λ, each cell consisting of a 2-element

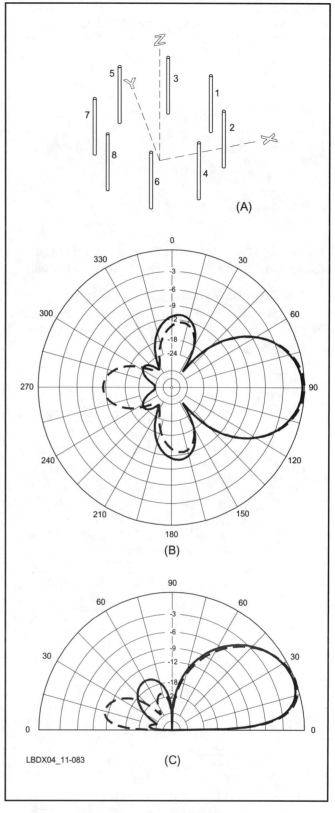

Fig 11-83—At A, layout of the 8-Circle array. At B, comparison of azimuthal patterns at 20° elevation for phase shifts of 90° (solid line) and 120° (dashed line) in the end-fire cells. At B, elevation pattern comparison. Note good F/B at low angles for 90° phasing but large sidelobe at 60° elevation. For 120° phasing the F/B is better at high angles. The large azimuthal sidelobes for both phases are due to 0.65-λ spacing, which also gives the narrow forward lobe.

end-fire array.

Using a side-by-side separation of 0.65 λ, you can obtain the narrowest possible forward beamwidth without creating excessive sidelobes. At 1.83 MHz, this results in a circle diameter of 115.8 meters and a separation between the two elements of the end-fire cell of 44.3 meters.

4.14.1. Data, quadrature-fed 8-Circle array
Gain = 8.95 dBi
3-dB forward angle: 47.2°
RDF = 12.86 dB
DMF: 21.8 dB
Feed currents:
Element 5 and Element 6 (back): 1 /0° A
Element 3 and Element 4(front): I = 1 /–90° A
Feed impedances:
Z (Element 5) = Z (Element 6) = 12.1 – j 6.3 Ω (back)
Z (Element 4) = Z (Element 3) = 38.6 + j 5.6 Ω (front)

4.14.2. Optimized phasing, 8-Circle array
When feeding the cells with a larger phase angle (120° vs 90°) we can increase the gain and the RDF, although the F/B at low angles suffers. The data now are:
Gain = 9.2 dBi
3-dB forward angle: 46.0°
RDF = 13.29 dB
DMF = 21.8 dB
Feed currents:
Element 5 and Element 6 (Back): 1 /0° A
Element 3 and Element 4(front): I = 1 /–120° A
Feed impedances:
Z (Element 5) = Z (Element 6) = 10.8 + j 1.1 Ω (back)
Z (Element 4) = Z (Element 3) = 33.8 + j 11.5 Ω (front)

4.14.3. Feed system, 8-Circle array
When fed in quadrature (90° phase shift) the array can be fed using a hybrid coupler. Using a Lewallen L-network feed system gives the advantage of being able to adjust the network components to obtain the desired phase shift and current magnitude in the front element.

Fig 11-84 shows the networks calculated with the *Lahlum.xls* spreadsheet. Both the 90° phase angle solution (A) and the 120° phase angle alternative (B) are shown. 75-Ω feed lines are recommended in order to achieve a high enough network input impedance. **Fig 11-85** shows a direction-switching system that can be used with the 8-Circle array.

4.14.4. Discussion, 8-Circle array
Of all large high-performance arrays, the 8-Circle is certainly the easiest one to build, as there are only two phase angles involved. The directivity is excellent, especially the RDF, the narrow forward lobe being responsible of this. With this array you will hardly need separate receiving antennas. Unfortunately, you will need at least 4 hectares (~ 10 acres) of real estate for this array on 160 meters, or 1 hectares (2.5 acres) on 80 meters!

4.15. The 9-Circle Array
The 9-Circle was originally developed by John Brosnahan, WØUN. John Battin, K9DX, has a 160-meter array at his remote station near Chicago and is planning to build an 80-meter version during the summer of 2004. The other existing 80-meter 9-Circle is at the QTH of Paul Hellenberg, K4JA, in Virginia.

4.15.1 Data, 160-meter 9-Circle array
Size = 128 meters diameter
Gain = 9.05 dBi over Good Ground. (Over Very Good Ground (ε = 20 and conductivity = 30 mS), the gain is 11.0 dBi.)
3-dB forward angle: 58.5°
RDF = 12.99 dB
DMF = 31.7 dB
Feed currents:
Element 6 (back, tip): I = 1 /0° A
Element 5 and Element 7 (back, side-by-side):
    I = 1.66 /–90° A
Element 4 and Element 6 (middle, outer): I = 1 /–180° A
Element 1 (middle, center): I = 3 /–180° A
Element 3 and Element 9 (front, side-by-side):
    I = 1.66 /–270° A

Element 2 (front, tip): I = 1 /–360° A

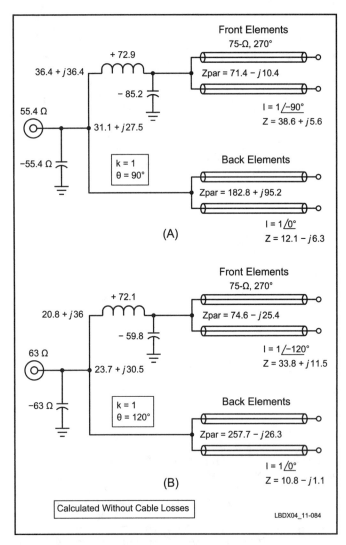

Fig 11-84—Lahlum/Lewallen feed systems for the 8-Circle. At A, the values for a 90° phase shift. At B, the values when using a 120° phase shift.

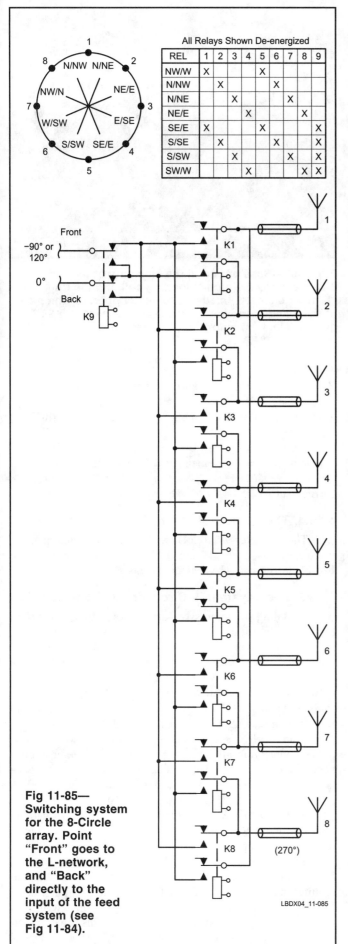

Fig 11-85—Switching system for the 8-Circle array. Point "Front" goes to the L-network, and "Back" directly to the input of the feed system (see Fig 11-84).

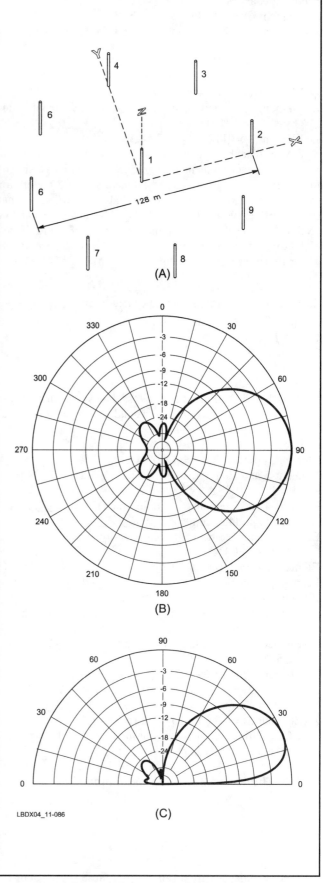

Fig 11-86—Layout of the 9-Circle, along with the horizontal (at a 20° elevation angle) and vertical radiation patterns obtained with the larger 9-circle, which measures 128 meters in diameter.

Phased Arrays   11-65

Feed impedances:
Z (Element 6) = −27.4 + j 2.2 Ω (back)
Z (Element 5) = Z (Element 7) = 12.3 − j 12 Ω (back, side-by-side)
Z (Element 4) = Z (Element 8) = 58.7 − j 26 Ω (middle, outer)
Z (Element 1) = 36.7 + j 3.1 Ω (middle, center)
Z (Element 3) = Z (Element 9) =75.9 − j 2.3 Ω (front, side-by-side)
Z (Element 2) = 112.5 + j 157 Ω (front, tip)

Note that the impedances include 2-Ω equivalent loss resistance in each element. I also modeled a smaller version of the 9-Circle, with a diameter of 80 meters vs 128 meters for the "big" one above. As expected the gain is down somewhat (by 0.7 dB), as well as the 3-dB forward beamwidth (69.6° vs 58.5°). This resulted in a somewhat lower RDF (12.39 vs. 12.99 dB). The behavior in the back is almost as spectacular as for its larger brother, resulting in a DMF of not less than 31 dB!

4.15.2. Data, Small 9-Circle array

Size = 80 meters diameter
Gain = 8.15 dBi
3-dB forward angle: 69.6°
RDF = 12.39 dB
DMF = 31.0 dB
Both RDF and DMF are nothing less than spectacular. Over very good ground (ε = 20 and conductivity = 30 mS), the gain is 9.9 dBi.
Feed currents:
Element 6 (back, tip): I = 1 /0° A
Element 5 and Element 7 (back, side-by-side):
    I = 1.4 /−100° A
Element 4 and Element 6 (middle, outer): I = 1 /−200° A
Element 1 (middle, center): I = 3 /−200° A
Element 3 and Element 9 (front, side-by-side):
    I = 1.4 /−300° A
Element 2 (front, tip): I = 1 /−40° A
Feed impedances:
Z (Element 6) = −18 − j 7.5 Ω (back)
Z (Element 5) = Z (Element 7) = 10.9 − j 20.1 Ω (back, side-by-side)
Z (Element 4) = Z (Element 8) = 54.9 − j 11.1 Ω (middle, outer)
Z (Element 1) = 27.6 + j 2.2 Ω (middle, center)
Z (Element 3) = Z (Element 9) = 54.2 + j 48.8 Ω (front, side-by-side)
Z (Element 2) = −77 + j 104.7 Ω (front, tip)

Calculations are done including an equivalent ground loss resistance of 2 Ω in each element.

4.15.3. The K9DX 9 Circle near Chicago

John, K9DX, built his 9-Circle using 27-meter long elements (Titanex 160HD). He used 120 quarter-wave long radials on each element and the tradeoff caused by these shorter element is nil. John tuned the element with a high-Q coil at the bottom of each element. See **Fig 11-87**. Of course, the feed impedances are different from those shown above.

K9DX supplied these impedances (including the loading coil, which has an inductive reactance of 120 Ω):

Z (Element 6) = −16.6 − j 1.0 Ω (back)

Fig 11-87—Base of one of the elements of the K9DX 9-Circle. Note the high-Q loading coil and the ring (1-meter diameter) made of 10 mm copper, to which all of the 120 quarter-wave radials are connected. K9DX uses 1⁵/₈-inch coax for the 3λ/4 feed lines.

Z (Element 5) = Z (Element 7) = 5.9 − j 7 Ω (back, side-by-side)
Z (Element 4) = Z (Element 8) = 30.3 − j 15 Ω (middle, outer)
Z (Element 1) = 18.7 + j 0 Ω (middle, center)
Z (Element 3) = Z (Element 9) = 52.2 + j 48.8 Ω (front, side-by-side)
Z (Element 2) = 53.7 − j 127 Ω (front, tip)

4.15.4. The K9JA 9-Circle in Virginia

The 80-meter 9-Circle at K4JA uses full-sized quarter-wave elements, made of 5-cm OD aluminum tubing elements. **Fig 11-88** shows K4JA's elegant array.

4.15.5. Feed system, 9-Circle array

Modeling an impressive array like a 9-Circle is one thing. Building it and making it to work like it does on the computer screen is a totally different thing! So far as designing a feed system, there are many roads that lead to Rome.

In the design shown in **Fig 11-89** I made use of two L-networks and three transformers (one 9:1 impedance ratio transformer and two 1:1 transformers). Note that this is not the feed system that K9DX is actually using.

John, K9DX uses a motor-driven rotary switch with heavy contacts on nine ceramic wafers. Each wafer is connected to one element. This system ensures minimum inductance, but is expensive if you need to buy the switch new. It is available from Multi-tech Industries, (**www.multi-tech-industries.com/**). See **Fig 11-91**. K4JA changes directions on his 80-meter 9-Circle using small vacuum relays mounted to an aluminum plate, as shown in **Fig 11-92**.

The feed system in Figs 11-89 and 11-90 uses only two L-networks and no additional coaxial cables to obtain the required phasing angles. The center element (with a relative feed current of 3) is fed directly from the input terminals through a 180° phase-reversal transformer. This is better than a 180°-long piece of coax because it is frequency-indepen-

Fig 11-88—Arial picture of the K9DX 9-Circle array. The circular terrain, including the radials that extend 40 m from the array itself, has a diameter of 208 m (two soccer fields long!). The brown stripes are the feed line trenches that had not grown over when the picture was taken.

dent, and if well-made has little loss and a phase delay of a little bit less than 180° (see Section 4.15).

The two elements that are fed in phase with the center element, but with 1/3 of the current magnitude, are also fed via an identical 180° phase-inverter transformer, and through a 9:1 impedance ratio transformer (which is a 3:1 voltage ratio, and a 3:1 turns ratio). This ensures that these elements get three times less feed current compared to the center element.

The front and the back elements are 360° out-of-phase, which means that they are in-phase. Since they have a feed current magnitude of 1, they can be connected directly to the output (low Z side) of the 9:1 transformer.

So far we have the three center elements fed at –180°, the front element at –360° and the back element at 0°. This is a very simple setup using only broad-band transformers.

We will now design two L-networks that take care of the proper phase angle and magnitude for feeding the remaining elements—two directors and two reflectors. These are all fed at a relative current magnitude of 1.66. For this we use the *Lahlum.xls* spreadsheet. The inputs of the L-networks are connected to the line that feeds the center element, which has a phase angle of –180°. The two directors require a phase angle of –270°, so will require a θ of –90°. The k factor is 1.66/3 = 0.55. The two reflectors require a phase angle of –90°, which means that for this L-network θ = –90 –(–180) = + 90° – 360° = –270°. The same k factor applies (0.55).

Fed this way, the array appears to have a total array feed impedance of 19 Ω, which is quite acceptable. Using 75-Ω cable (see Fig 11-90) for the current-forcing feed lines, we obtain a feed impedance of 40.4 + j 19.7 Ω. With just a paral-

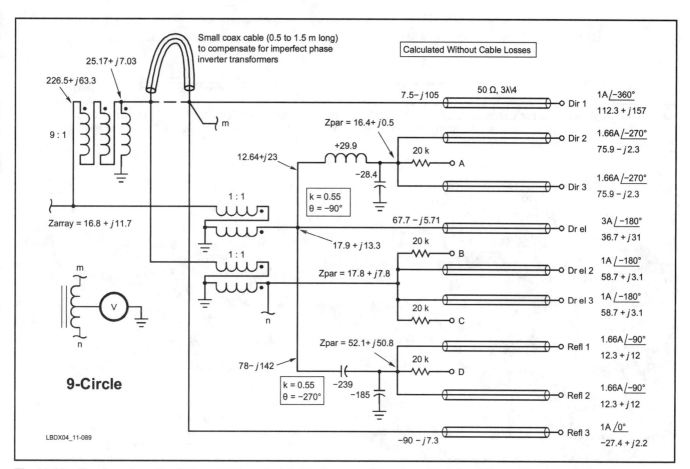

Fig 11-89—Feed system for the quadrature-fed 9-Circle array based on 50-Ω feed lines. The calculations were done without taking cable losses into account. See text for details.

Phased Arrays 11-67

lel capacitor (reactance = −102.6 Ω), the feed impedance turn into exactly 50 Ω (lucky strike!). With "real cables" the impedance will be a little higher, maybe 52 to 55 Ω, depending on cable losses.

It turns out that the values calculated for the 2 L-networks are very normal and quite manageable. To adjust the components to obtain the right phase angle and feed current magnitude, we can use the hybrid-coupler adjustment system developed by W1MK (see Section 3.6.2).

In Figs 11-89 and 11-90 we see four voltage divider resistors (20 kΩ) installed. All we need to do is connect a hybrid coupler between points A and B and another one

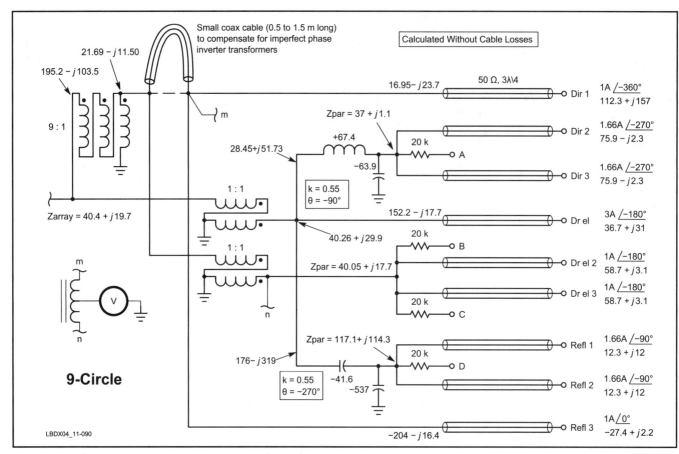

Fig 11-90—Feed system for the quadrature-fed 9-Circle array based on 75-Ω feed lines. The calculations were done without taking cable losses into account. With real-world lines that have line losses the array feed impedance will be very close to 50 Ω.

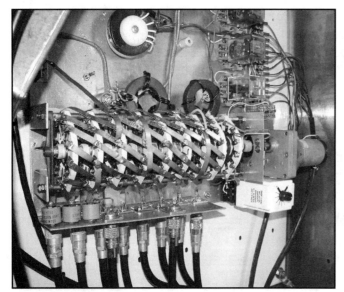

Fig 11-91—A motor-driven rotary switch is used for direction switching at K9DX.

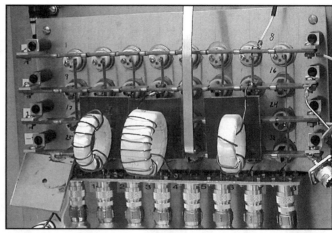

Fig 11-92—K4JA uses a matrix of 8 by 4 small vacuum relays to do direction switching. This has inherently more inductance, but if you use the adjustable L-networks it should be possible to tune out the effect of the stray inductances in the wiring.

between C and D. As the required phase angle difference is 90°, we need no extra lengths of coax to correct for non-quadrature phase angles. We will have to provide some attenuation in the legs going to the points B and C though. The voltage ratio is 3/1.66 = 1.81, or in dB: 20 × log 1.81 = 5.14 dB. We need two 50-Ω attenuators of 5.14 dB. A T-attenuator using 15-Ω resistors in the series branches and 82-Ω in the parallel branch will be very close.

If we use the detector/wattmeter designed by W1MK, we can adjust the values of the four components until we get a good null on the summed output. Bingo!

4.15.6. Transformer construction, 9-Circle array

K9DX, has gained a lot of first-hand experience building RF transformers required in this array. John winds most of his transformers with RG-303 single-shield Teflon coax. For the 9:1 unit he takes the shield from another piece of coax and slides it over the Teflon insulation to get three conductors. The core material used is 61 material, permeability of 125, 2.4-inch OD (A_L = 171).

Not all of us have access to professional equipment to measure loss, impedance ratio and phase delay, so John developed a quick way of checking the transformer's performance. He terminates the transformer with the impedance it will normally see, then opens and shorts the output terminals. The input impedance with an open-circuit must go up by 10 times, and the impedance with a short-circuit must go down by 10 times. If it meets these requirements, it will likely measure OK on sophisticated test equipment!

4.15.7. 180° phase-shift transformers, 9-Circle array

K9DX also found that unless the transformer is allowed to be lossy (when wound with too few turns), the resulting delay will be 180° plus the delay inherent in the length of coax or wire involved. The usual result is in the 183° to 185° range. **Fig 11-93** is a photograph of one of K9DX's 180° phase-inverting transformers. But there is a way to compensate for this imperfection.

If we can introduce a similar amount of "extra phase" shift (3° to 5°) in the branches not fed via the 180° transformer(s) we can compensate for the imperfect phase-reversal transformers. A short piece of coax (0.5 to 1.5 meters long, the exact length will depend upon the SWR on the line) has this compensating effect. To adjust the short coax length to obtain 180° phase delay in the system, we can use a small push-pull (balanced) transformer. Connect the balanced inputs (m and n in Figs 11-89 and 11-90), and adjust the length for minimum voltage between the center tap and ground. You can use the detector-wattmeter described in Section 3.6 as an RF voltmeter.

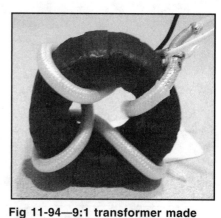

Fig 11-94—9:1 transformer made by John, K9DX. The transformer has four trifilar turns. The trifilar wiring is made by using a small Teflon coax equipped with a second shield. See text for details.

K9DX commented: "Th*is 'over 180° problem' is one of the reasons I will probably stick to 180° coaxial lines in my 80 meter system. The lines can be adjusted to hit the delay right on the nose. Of course the downside is that their length must be changed between phone and CW, which adds more complexity."*

A 9:1 wide-band transformer built by K9DX is shown in **Fig 11-94**. **Fig 11-95** shows a wire-connection diagram for this transformer.

4.15.8. Bandwidth, 9-Circle array

A major issue in designing a feed network is bandwidth. It is relatively easy to adjust the phase and magnitude of the antenna currents at one frequency, but depending on the feed system used, things can fall apart rapidly when the frequency is changed.

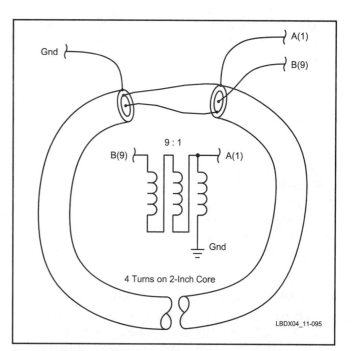

Fig 11-95—Construction of 9:1 transformer. See text for details.

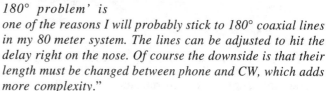

Fig 11-93—K9DX made a 180° 1:1 phase-reversal transformer wound using 13 turns of RG-303 coax on a two side-by-side, 2-inch OD #61 cores.

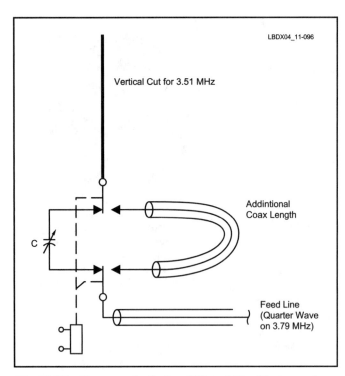

Fig 11-96—To cover both the CW and the Phone ends on 80 meters, you must re-resonate the elements and change the length of the current-forcing feed lines.

You can always make a feed system that does exactly what you engineered, at least on the frequency you engineered it for! If you set out to build an array like a 9-Circle, there is no room for compromises. Compromising to improved bandwidth is inevitably a losing battle. K9DX says that his 160-meter 9-Circle holds perfect directivity (down 30 to 40 dB off the side and in the back) over approx 30 kHz (1.8% bandwidth), which is adequate for that band.

On 80 meters, trying to cover the CW end and the Phone end (8% relative bandwidth) in one network is just *not* possible. The antenna elements become the wrong lengths, the phase shifting networks (coaxial or L-networks) move far away from their design center, and the current-forcing feeds destroy the amplitude and phase relationships. Note that is holds true not only for the 9-Circle but for all phased arrays that we want to operate on both the CW- and the phone end of the 80-meter band.

The only good solution on 80 meters is to resonate the elements on both ends of the band (for example, with a small loading coil on the CW end, or with a series capacitor to tune a CW element to become a phone-band element). See **Fig 11-96**. We then adjust the $3\lambda/4$ (or $\lambda/4$ in case of "smaller" arrays) current-forcing lines to the exact lengths needed, by inserting extra cable lengths when operating at the CW end of the band.

With a quarter-wave current-forcing feed line, going from 3.5 to 3.8 MHz with a line cut for 3.8 MHz introduces

Table 11-15
Summary, Phased Arrays

Array Type	Gain (1) dBi	RDF dB	DMF dB	3-dB BW, Deg	Footprint (1)	Footprint (2)	Reference
Single vertical	0.9	—	—	—	0.16	0.4	—
2-ele end-fire quadrature	4.2	8.1	12.3	177°	0.32	1.0	4.1.1.
2-ele end-fire optimized	4.0	8.9	5.0	154°	0.28	0.9	4.1.1
2-ele end-fire $\lambda/8$ spacing	5.0	9.6	16.6	131°	0.24	0.8	4.1.1.
3-in-line quadrature	5.3	9.2	17.9	143°	0.50	1.3	4.4.1
3-in-line optimized	5.6	11.2	27.8	98°	0.40	1.1	4.4.1
Triangle wide, quadrature	5.0	8.8	14.3	150°	—	—	4.6
Triangle optimized	5.8	9.8	16.9	129°	—	—	4.6
Four-Square, quadrature	6.7	10.6	21.0	98°	0.64	1.4	4.7.1
Four-Square WA3FET	7.3	11.4	24.4	85°	0.64	1.4	4.8.2
Four-Square, half dir. quadrature	5.5	9.2	14.7	123°	0.64	1.4	4.8.1
Four-Square half dir. optimized	5.9	9.7	16.5	113°	0.64	1.4	4.8.2
Broadside/end-fire, 0.55 λ spacing	8.4	12.1	20.8	58°	1.2	2.1	4.9
5-Rectangle	5.9	9.8	19.4	122°	1.2	2.1	4.10
5-Square	7.7	11.9	23.0	78°	0.8	1.7	4.11
5-Square, half direction	6.2	10.5	18.5	102°	0.8	1.7	4.11.2
6-Circle 60-m diameter	7.6	11.7	25.5	78°	1.0	2.0	4.12.2
6-Circle 80-m diameter	7.8	11.6	26.3	75°	1.5	2.6	4.12.1
8-Circle optimized	9.2	13.3	21.8	58°	2.4	3.8	4.14.2
9-Circle 128-m diameter	9.1	13.0	31.7	70°	2.8	4.3	4.15
9-Circle 80-m diameter	8.15	12.4	31.0	70°	1.5	2.5	4.15.2

Gain (1): in dBi over good ground (ε = 13, conductivity = 5 mS/m)
Gain (2): in dBi over Very Good Ground (ε = 30, conductivity = 30 mS/m)
Footprint (1): footprint in hectares with 20-meter long radials (1 hectare ~ 2.5 acres)
Footprint (2): footprint in hectares with 40-meter long radials

a phase shift error of no less than 7°, if the line is flat. With 3λ/4 current-forcing lines the phase error becomes three times that much on a flat line, or 21°. On a feed line with SWR (they all have SWR, sometimes very high such as on the "back" or the "front" elements of the array, the phase angle error for a 3λ/4 line cut for 3.5 MHz and operating on 3.8 MHz can be as much as 100° or 150°! This make it totally impossible to make the antenna work on both ends of the band.

4.15.9. Current-forcing feed lines, 9-Circle array

One solution for an 80-meter array is to resonate the elements on the CW end, and re-resonate them in the SSB band with a relay and a series capacitor. In addition you can add a short piece of coax to the quarter wave feed lines to make an exact λ/4-feed line also on 3.5 MHz. This is even more important if you are using 270°-long feed lines, where things move three times as fast off target as compared to using 90° feed lines.

4.15.10. Conclusion, 9-Circle array

The 9-Circle is a low-band array most of us can only dream of. It's an interesting subject where you let your imagination go and design your own feed system. There are numerous alternatives, and they all have pros and cons.

Building and owning a 9-Circle array is not for everyone: a 160-meter version requires over 4 hectares (~ 10 acres) of real estate! It also requires more-than-average knowledge in antenna matters and in electronics to design the feed system, to build it and to adjust it.

4.16. Summary, Phased Arrays

Table 11-15 gives a performance overview of the arrays covered in this chapter. Gain is given over "Average Ground." RDF and DMF are also listed. Interesting information is the footprint information, which is given for the array with 20-meter long radials, as well as for 40-meter long radials. The figures apply for a 160-meter antenna. For an 80-meter array, the footprints are four times smaller.

5. ELEMENT CONSTRUCTION
5.1. Mechanical Considerations

Self-supporting λ/4 elements are easy to construct on 40 meters. On 80 meters it becomes more of a challenge, but self-supporting elements are feasible even with tubular elements when using the correct materials and element taper. Tubular full-size λ/4 elements for 80 meters are available commercially from Arrays Solutions (**www.arraysolutions.com/**) as well as from Titanex (**www.titanex.de/**). See **Fig 11-97**. Lattice-type construction is commonly used, with tapering-diameter aluminum tubing

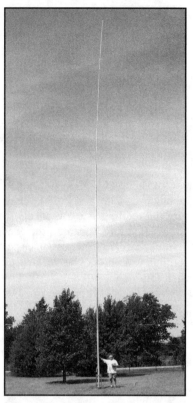

Fig 11-97—Self supporting full-size 80-meter λ/4 elements are commercially available from WX0B at Array Solutions.

at the top. On Topband, most quarter-wave vertical radiators are guyed towers. Rohn 25 tower is often used. See **Fig 11-98**. As it is highly recommended to series feed the elements of an array, the elements should be insulated from ground. This poses additional mechanical constraints on construction.

I used the ELEMENT STRESS ANALYSIS module of the YAGI DESIGN software (see Chapter 4 on software) to develop self-supporting elements for 40 and 80 meters that withstand high wind loads. As the element is vertical, there is no loading of the element by its own weight, which means that the same element in a vertical position will sustain a higher wind load than in a horizontal position. When using the ELEMENT STRESS ANALYSIS module, you can create this condition by entering a near-zero specific weight for the material used.

It will, however, be much easier if you plan to have at least one level where the vertical can be guyed. This will typically lower the material cost for constructing a wind-survival vertical by a factor of three or more. Finally, the element construction that is best for your project will be dictated to a large extent by material availability.

Needless to say, guying materials need to be electrically transparent guy wires (Kevlar, Phillystran, Nylon, Dacron, etc) or metallic guy wires broken up into small nonresonant lengths with egg-type insulators. Refer to *The ARRL Antenna Book* (Chapter 22), which covers this aspect in great detail.

Fig 11-98—Roger Vermet, ON6WU, installing the base insulator on the Rohn 25 tower sections used for the 160-meter Four-Square at A61AJ in Dubai.

All the array data in this chapter are for λ/4-full-size elements. Electrically longer elements are to be avoided at all times, since such elements would be no longer fed at a current maximum and design becomes more complicated. It is not crucial, however, to use full-size elements. Top-loaded elements that are physically ²/₃ full-size length can be used without much compromise. Make sure, however, that all elements in the array use the same amount of top-loading. If not, special precautions have to be taken (see Section 6.5). If guyed aluminum-tubing elements are used, the top set of guy wires can be used to load the element (see Chapter 9 on vertical antennas). If the array must cover 3.8 MHz as well as 3.5 MHz, a small inductance can be inserted at the base of each vertical (make sure the loading coils are identical!) to establish resonance for all elements at 3.5 MHz.

The main cause of failure with guyed aluminum tubing elements is *buckling*. This usually happens when four conditions are met:

1. Distance between guying points is too long.
2. Thin-wall flimsy aluminum material (easy bending).
3. Too much vertical load on the mast (too much guy pulling).
4. Too much wind (bending in between guying points, eventually turning into buckling).

5.2. Shunt Versus Series Feeding

Shunt feeding the elements of an all-fed array is to be avoided in just about all cases. The matching system (gamma match, omega match, slant-wire match, etc) introduces additional phase shifts that are difficult to control. Such phase shifts will mess up the correct feed current in the antenna elements.

Only with arrays where all the feed impedances are identical can shunt feeding be applied successfully. The feed impedances of all elements of an array will be identical only when all the elements are fed in-phase (or 180° out-of-phase). Shunt feeding may be considered for such arrays if the vertical elements as well as the matching systems are identical (including the values of any capacitors or inductors used in the matching system).

If you feel tempted to use your tower loaded with HF antennas as an element of an array, be aware that you might be trying to achieve the impossible.

- The loaded tower may be electrically quite long, which could very well be a hindrance to achieve the required directivity (see Section 2).
- You will be forced to use shunt feeding, which is just about uncontrollable, especially if all elements are not strictly identical (which will hardly ever be the case with such loaded towers).

Loaded towers are just great for single verticals, but are more than a hassle in driven arrays.

5.3. Loaded Elements

It is not always possible to use full size λ/4 elements, and provided you install a very low-loss ground system, full-size elements are not really required. On 160 meter many arrays have been built with inverted L elements, although T-loaded elements will produce much better directivity.

5.3.1. Inverted-L elements

The inverted-L vertical is described in Chapter 9, Section 7. For a single vertical, where we really do not expect much directivity at all, the inverted-L is a good antenna. In an array however, where you really are mostly after gain *and* directivity, the horizontally polarized high-angle component radiating from the flat-top section of the inverted-L is a problem. T-loading, with horizontal or even sloping top-loading wires arranged symmetrically, will reduce the high-angle radiation from these loading wires to almost zero.

5.3.2. T-Loaded Vertical Elements

If the central tower is not high enough to support full-size λ/4 verticals from the sloping support wires, these verticals can be top-loaded with a sloping top-wire. The top-loading wires can be part of the support system, as shown in **Fig 11-99**. The vertical elements are loaded with sloping top-wires to resonate them at 3.8 MHz. The sloping support wires have the property of not producing any horizontally polarized signal, provided the lengths on both sides of the vertical are the same.

As long as the vertical wire is not shorter than about λ/8, the loaded verticals will produce the same results as full-size verticals, with only some reduction in bandwidth. Section 7.5 describes a 160-meter Four-Square with vertical elements that are not longer than 18.5 meters.

6. ON4UN 4-SQUARE ARRAY WITH WIRE ELEMENTS

6.1. The Mechanical Concept

An 80-meter 4-square takes a lot of room to put up, not to speak about one on 160 meters! I have installed a somewhat special version of the 80-meter Four-Square around my full-size λ/4 160-meter vertical. This design has become very popular since it was first published. From the top of the vertical I run four 6-mm (¹/₄-inch) nylon ropes in 90° increments, to distant support poles. These nylon ropes serve as support cables from which I suspend the four verticals. A single radial is directed away from the center of the square

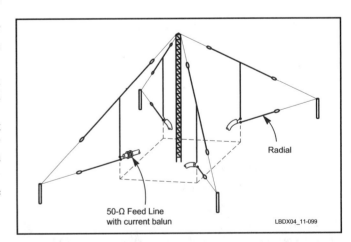

Fig 11-99—If enough space is available, you can run four catenary cables from the top of the support tower to four supports mounted in a square. These catenaries support the verticals and their loading structures, if any. Sloping top-loading wires as shown here exhibit no horizontal radiation component, provided the length is the same on both sides of the vertical.

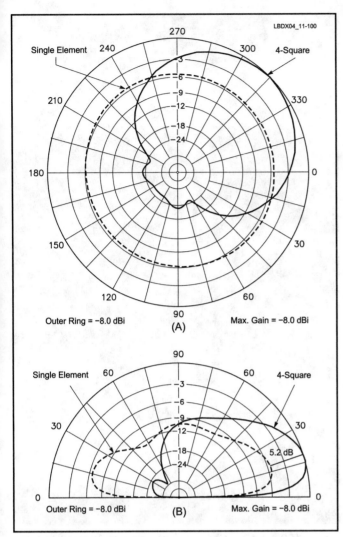

Fig 11-100—Horizontal and vertical radiation patterns of the ON4UN Four-Square array with one elevated radial. Also shown is the pattern of a single vertical element. Both are modeled with a single radial per element, but over an extensive buried-radial system, 5 meters (17 feet) below the radial over very good ground. The buried radials are installed like spokes from the center of the square.

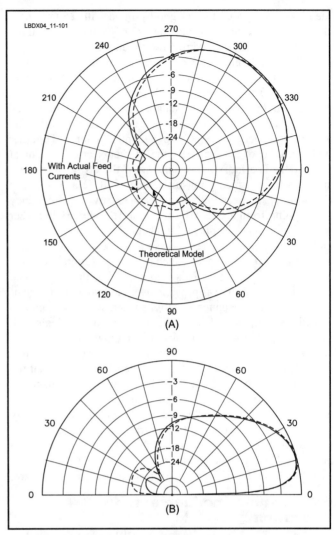

Fig 11-101—Vertical and horizontal patterns (at a 20° elevation angle) for the theoretical currents and the actual currents at each element of the Four-Square array (see Fig 11-100). Note that although there are some significant current deviations (phase angle and magnitude) from the theoretical values, the array suffers only very slightly from these differences. These patterns were calculated using *EZNEC*.

(where the 160-meter vertical is located). See **Fig 11-100**.

Since my support tower was high enough, I could manage full-size 80-meter vertical elements, with the feed point and the single radial 5 meters above ground. In this setup the single radial serves three purposes:

- It provides the necessary low-impedance connection for the feed line outer shield.
- It helps to establish the resonance of the antenna (which is not the case with a large number of radials or buried radials, where the resonance is mainly determined by the length of the vertical member).
- It provides some high-angle radiation. We can debate whether or not this is wanted, but in my particular case I wanted a fair amount of high-angle radiation as well, in order to be able to use the array successfully in contests, where shorter range contacts are also needed.

Just a single radial, without an extra ground screen on or in the ground will make you lose up to 6 dB of maximum achievable gain, depending on the quality of the ground below. I strongly advocate the use of extra radials or a ground screen under the verticals (see Chapter 9). In my particular case, there are some 250 radials (20 to 60 meters long) under my array, basically serving as the radial system for the 160-meter vertical that supports the array. With this extensive radial system, my 80-meter array exhibits a low-angle F/B of 20 to 25 dB, and still a very reasonable amount of directivity at relatively high angles (20 dB F/B at 60°).

The gain of this array at low angles is only 0.4 dB less than the gain of a Four-Square over a system using a perfect ground system. The slight drop is due to the power radiated at higher angles. The gain is 5.1 dB over a single element, which is very substantial.

Fig 11-101 shows the horizontal and vertical radiation patterns for my array, as well as that of a single element over

Phased Arrays   11-73

identical good ground (very good ground with 250 radials). Both the single vertical and the Four-Square use a single elevated radial for each vertical in the model.

The bottom ends of the four vertical wires are supported by steel masts that are located on the corners of a square measuring 20 meters, with the 39-meter tall 160-meter vertical right in the center of the square. The masts can be folded over for easy access to each element's feed point. The 80-meter vertical elements are 19.5 meters long. Together with a radial of 18.7 meters, the elements are resonant at 3.75 MHz. The individual elements of the array were measured to have a feed-point resistance of 40 Ω at resonance (3.75 MHz). I measured the impedance over a frequency range going from 2 to 5 MHz using an HP network analyzer with a Smith Chart display. Mutual coupling to other antennas and surrounding structures shows up on the Smith Chart as a kink or a dip in the impedance chart of one or more elements at a specific frequency. It is important that the impedance curves be as near alike as possible over the frequency range of interest, if the impedance variations when switching antenna directions are to be kept at a minimum. Section 3.3 deals with the problem of eliminating unwanted mutual coupling.

A word of caution: if the central supporting tower is a base-insulated tower, tuned to 160 meters, make sure that the tower is effectively grounded when you use the 80-meter Four-Square. If it is left floating, the central element can act as a half-wave element on 80 meters and can interfere heavily with the array. Grounding the central tower can be done in several ways, as discussed in Chapter 7 on special receiving antennas. The grounded 160-meter resonant tower in the center does not influence the performance of the Four-Square.

6.2. Loading the Elements for CW Operation

To make the antenna cover the CW end of the band as well, I use a stub (or linear loading section), inserted in the radial at the feed point, to shift the resonance of the elements to 3.5 MHz. A small box is mounted on top of each mast. All connections (to the vertical element, radial, and feed lines) are made inside this box. The box also contains a vacuum relay that can switch the stub in and out of the circuit. The stub is supported by stand-off insulators along the metal support mast (see **Fig 11-102**).

The calculated reactance of the stub is +130 Ω. Using 3-mm-OD (AWG #9) copper wire with a spacing of 20 cm, the length of the stub turned out to be 2.25-meters long to lower the resonant frequency to 3.505 MHz. The same stub, when shortened to 75 cm resonates the element at 3.65 MHz. A nice feature is that the resonant frequency can be changed anywhere between 3.5 and 3.75 MHz by using a movable shorting bar across the stub. This way, you can create different operating windows on 80 meters. A relay can be used to switch the 3.65-MHz shorting bar in and out of the circuit, making the window selection remotely controlled. The direction control box contains a three-position lever switch, which selects the three band segments.

6.3. The λ/4 Feed Lines

Each element is fed via an electrical λ/4 of coaxial feed line with a current balun (50 stacked ferrite beads on a short length of small-diameter Teflon coax) at the feed point. The

Fig 11-102—A 10 × 10 × 3 cm plastic box is mounted on the top of the 5-meter support for each of the elevated wire verticals. Inside the box, the vertical wire and the single radial are connected to the feed line, which is equipped with a stack of 50 ferrite cores to remove any RF from flowing on the outside of the feed line. The box also houses the relay that switches the stub in and out of the circuit to lower the operating frequency to 3.5 MHz. The stub can be seen running along the steel support mast.

feed lines were cut to be λ/4 at 3.75 MHz. If a perfect 90° phase shift is desired at 3.5 MHz, the feed lines can be lengthened by a 1-meter long piece of coax (VF = 66%).

6.4. Wasted Power

In the case of a quadrature-fed Four-Square with 90° side dimensions, the use of 50-Ω feed lines results in combined feed-line impedances that load the ports of a 50-Ω hybrid coupler heavily (22 + j 19 Ω and 23 + j 17 Ω). This results in up to 11% of the power being dissipated in the dummy load resistor. With 75-Ω feed lines, these impedances are much higher (51 + j 37 Ω and 48 + j 43 Ω), which results in much less power being dissipated in the load resistor (4%). Again, the main parameter that determines the operational bandwidth of an array fed with a hybrid coupler is the amount of power being dissipated in the load resistor.

6.5. Gain and Directivity

Section 3.6.2 explains that the feed current in the elements can be assessed by measuring the voltage at the end of the quarter-wave feed lines going to the elements. When I initially built the array, I used a vector voltmeter to measure the voltages.

With the current-forcing method employed, the relative element feed-current requirement for equal-magnitude in quadrature phase is voltages with equal magnitudes at the ends of the $\lambda/4$ feed lines. (Here, $E = Z_0 \times I$ or $E = 50 \times V$ for a 50-Ω line). There was deviation from the theoretical values. As expected, the voltage magnitudes and phase angles were not exactly perfect. The voltage magnitude varied as much as 1.7 dB (41 V versus 50 V theoretically), while the phase angle was up to 13° off from the theoretical value for the 50-Ω feedline case. In a pleasant surprise, even the relatively important deviations for the 50-Ω impedance case influenced the directivity pattern and gain only very marginally.

Using 75-Ω, $\lambda/4$ (or $3\lambda/4$) feed lines does not make this a 75-Ω system. In this particular case I am still using a hybrid coupler with a 50-Ω nominal design impedance. The 75-Ω cables are used only because they transform the element feed-point impedances to more suitable values, resulting in less power dissipation in the dummy resistor.

6.6. Construction

In the original ON4UN layout, the Comtek Systems 50-Ω hybrid coupler (including a 180° phase-shift transformer) and the hybrid-coupler load resistor are located in a cabinet mounted at the base of the 160-meter vertical, in the center of the 80-meter Four-Square. The Comtek unit was removed from its normal housing and the PL-259 hardware was replaced with N connectors. The cabinet also contains the relay that switches the feed line between the 160-meter vertical and the 80-meter Four-Square. In 2004 I modified the cabinet to include a Lahlum/Lewallen system manufactured by Array Solutions. Both can be instantly selected from the shack. See **Fig 11-103**.

In order to know at all times how much power is being dissipated in the dummy load, I added a small RF detector to the dummy-load resistor and fed the dc voltage into the shack, where the relative power is displayed on a small moving-coil meter mounted on the homemade direction-switching box. The box also contains the switch to select the subbands, mentioned above. In addition, a level-detector circuit is included, using an LM-339 voltage comparator, which turns on a red LED if the dissipated power goes above a preset value. **Fig 11-104** shows the schematic of the system and **Fig 11-105** shows the actual switch box.

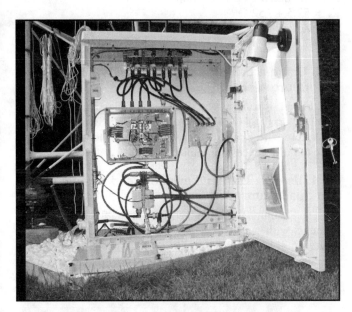

Fig 11-103—Cabinet located at the base of the 160-meter vertical, housing the Comtek hybrid-coupler feed system and directional switching circuitry for the Four-Square array, along with a Lahlum/Lewallen feed system built by Array Solutions. The feed lines to the elements are 75-Ω ½-inch cable (Vf = 0.88) and cut for the system to be $\lambda/4$ on 3.775 MHz. In the cabinet a short length (0.97 meters) of RG-11 is added to make the feed lines exactly $\lambda/4$ on 3.51 MHz. The 180° delay line for the Array Solutions box is made of ⅞-inch 50-Ω hard line. It too is switched to be an exact half wave on 3.51 MHz or 3.885 MHz with a 3-meter section of 1-inch hard line (coiled up under the relay box).

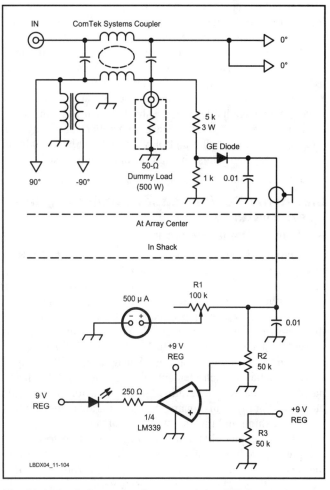

Fig 11-104—Schematic diagram of the RF detector and voltage comparator used to monitor the RF into the hybrid terminating resistor. The LED will switch on if the voltage coming from the detector is higher than the preset voltage supplied by the potentiometer R3. R1 adjusts the sensitivity of the indicator, and the R2 sets the alarm level.

Fig 11-105—Array-direction switch box at ON4UN, with 1 direction switch for 40; 2 for 80. A switch selects the Comtek or Lahlum/WX0B system. Relative power and alarm circuit (Fig 11-104) is for Comtek. A lever switch selects one of three band segments on 80.

6.7. Using the Lewallen Phasing Feed System

As explained earlier, the Lahlum/Lewallen system has two design advantages: There is no power lost in a dummy resistor and you can design the system for any phase angle and feed current magnitude. Non-quadrature feeding (eg, the configuration discussed in Sections 4.7.2 and 6.8.4) can give 1.0 dB of extra gain. Considering the power loss in the dummy load, 1.5 to 2 dB can be gained by going from a quadrature hybrid system to a Lahlum/Lewallen system.

In 2004 I rebuilt the 80-meter Four Square to use either feed system, switchable from the shack. When you consider a Lahlum/Lewallen system, please remember that a hybrid coupler system is a "plug-and-play" system, while a Lahlum/Lewallen system must be tuned and requires some test equipment.

6.8. Array Performance: Comparing the Feed Systems

6.8.1 Using the hybrid coupler

Judging the array performance by measuring its SWR is totally meaningless (see Section 3.4). With a hybrid coupler, my old array showed an SWR of less than 1.3:1 over the entire 80-meter band, wherever the resonances of the elements were.

Based on a wasted-power bandwidth criterion, however, the bandwidth of this array is 100 kHz without tuning for individual band segments. Fig 11-55 showed the wasted-power curve for my original Four-Square array with a single elevated radial at each element. The steepness of the dissipated-power curve is determined by the Q factor of the array elements. For my Four-Square, the elements are made of wire, so the Q is high and the bandwidth is narrow.

Also the fact that I use a single radial instead of a comprehensive buried radial system adds to the sharpness of the curve. While changing the frequency away from the design frequency, the single radial (just like the vertical element) will introduce reactance into the feed-point impedance, which would not be the case with a buried radial system.

Practically speaking, this array was by far the best antenna I had ever had on 80 meters. On-the-air tests continuously indicated that the signal strength on DX was ranked with the best signals from the European continent. As far as directivity is concerned, it is clear that the array had a nice wide forward lobe, and that the relative loss half-way between two adjacent forward lobes was hardly noticeable (typically 2 dB). Long-haul DX very often reported, "You are S9 on the front and not copyable off the back." Even on high-angle European signals there was always a good deal of directivity with this array (typically 15 dB).

However, it was not a good receiving antenna when compared to the range of Beverages I have at my QTH. One of the reasons, of course, is the single elevated radials, which causes a big bulge in the vertical radiation pattern at high angles. As explained before, this was done on purpose, so that the antenna would radiated a reasonably strong signal at high angle as well, which is a real asset when working contests.

6.8.2. ON4UN Four-Square array, data for quadrature feed

Design frequency: 3.7 MHz
Length of verticals: 18.7 meters (2-mm OD wire)
Length of radials: 21.2 meters
Height of feed point/radial: about 5 meters
Feed currents:
$I1 = 1 \underline{/-180°}$ A
$I2 = I3 = 1 \underline{/-90°}$ A
$I4 = 1 \underline{/0°}$ A
Gain: 5.2 dBi (over good ground)
3-dB beamwidth: 102°
F/B: 17 dB
RFD (receiving directivity factor): 9.08 dB
DMF: 13.3 dB
The calculated feed-point impedances are:
$Z1 = 52.5 + j\ 52\ \Omega$
$Z2 = Z3 = 34 + j\ 0\ \Omega$
$Z4 = 7.5 - j\ 2.5\ \Omega$

6.8.3 Using a Lahlum/Lewallen feed system in a quadrature feed.

Fig 11-106 shows the Lahlum/Lewallen network for the 80-meter array at ON4UN, when the Four-Square is fed in quadrature. These calculations were done without cable losses since the differences are minute.

6.8.4. Using a Lahlum/Lewallen feed system in a non-quadrature feed

The Lahlum/Lewallen feed system enables us to feed the elements with whatever feed currents and phase angles we want. A half hour using *EZNEC* resulted in a design that uses the same geometrical layout of elements but that represents a worthwhile improvement over the equal-current/quadrature-phasing configuration.

6.8.4.1. Array data

Design frequency: 3.7 MHz
Length of verticals: 18.7 meters (2-mm-OD wire)
Length of radials: 21.2 meters
Height of feed point/radial: about 5 meters
Feed currents: $I1 = 0.7 \underline{/0°}$ A
$I2 = I3 = 1 \underline{/-110°}$ A
$I4 = 1.4 \underline{/-220°}$ A
Gain: 6.2 dBi—This is 1 dB better than the same array fed in quadrature!
3-dB beamwidth: 88°

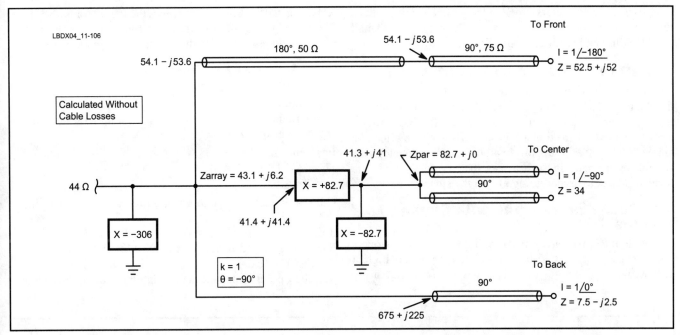

Fig 11-106—The 4-element wire Four-Square at ON4UN was modeled in *EZNEC* and the impedance data used to design a quadrature-fed Lahlum/Lewallen feed system. The front element is fed via a 180° long phasing line, which could be replaced by a 1:1 phase-inverting transformer, as is done in the Comtek hybrid coupler.

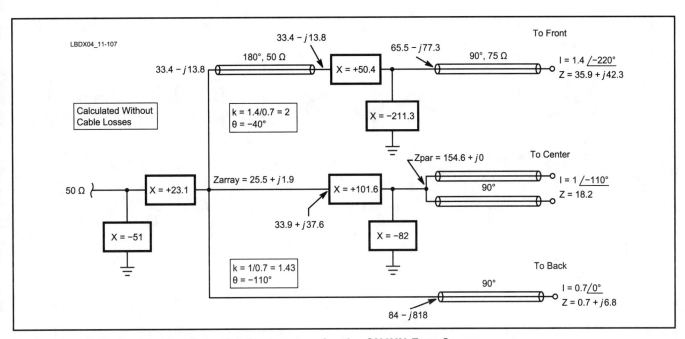

Fig 11-107—Optimized Lahlum/Lewallen feed system for the ON4UN Four-Square.

F/B: 21 dB
RFD: 9.57 dB
DMF: 14.4 dB
The calculated feed-point impedances are:
Z1 = 0.7 + j 6.8 Ω
Z2 = Z3 = 18.2 + j 0 Ω
Z4 = 55.9 + j 42.3 Ω

Fig 11-107 shows the new Lahlum/Lewallen feed system designed for this array. The values of the two LC network components can be adjusted using the test setup described in Section 3.6.2.

6.8.5. T-Loaded vertical elements

If the central tower is not high enough to support full-size quarter-wave verticals from the sloping support wires, these verticals can be top-loaded by a sloping top-wire. The top-loading wires can be part of the support system, as shown in Fig 11-99. The vertical elements are loaded with sloping top-wires to resonate at 3.8 MHz. The sloping-support wires have the property of not producing any horizontally polarized signal, provided the lengths on both sides of the vertical are the same.

As long as the vertical wire is not shorter than ²/₃ full size (approximately 15 meters), the loaded verticals will produce the same results as the full-size verticals, with only a small reduction in bandwidth and gain.

7. ARRAYS OF SLOPING VERTICALS

In the chapter on dipoles, I describe a vertical half-wave dipole, as well as a sloping half-wave dipole and its evolution into a quarter-wave vertical with one radial. Sloping verticals are well suited for making a Four-Square array from using a single, tall tower as a support. In all these arrays the elements should be arranged in such a way that the feed points are located on a square measuring λ/4 on the side.

7.1. Four-Square Array with Sloping λ/2 Dipoles

A Four-Square array made of four λ/2 dipole slopers requires a support that is about 0.36 λ tall, 55 meters high at 1.84 MHz. The four feed points make a square measuring λ/4 on each side. **Fig 11-108** shows the layout for a 160-meter system. Fig 11-108B shows a side view of a single dipole, which includes the droop due to weight of the feed line.

7.2. Array data, Four-Square with Sloping λ/2 Dipoles

Quadrature feeding does not result in a very good pattern. Through modeling using *EZNEC* I came up with the following design for 1.8 MHz:

Feed currents: I1 = 1 /0° A
I2 = I3 = 1 /–139° A
I4 = 1 /–263° A

Gain: 4.56 dBi over good ground (about 4 dB over a single vertical)
3-dB beamwidth: 87° (vs 108° if quadrature fed!)
RDF (receiving directivity factor): 8.06 dB
DMF: 11.9 dB

The feed-point impedances calculated including a grounded

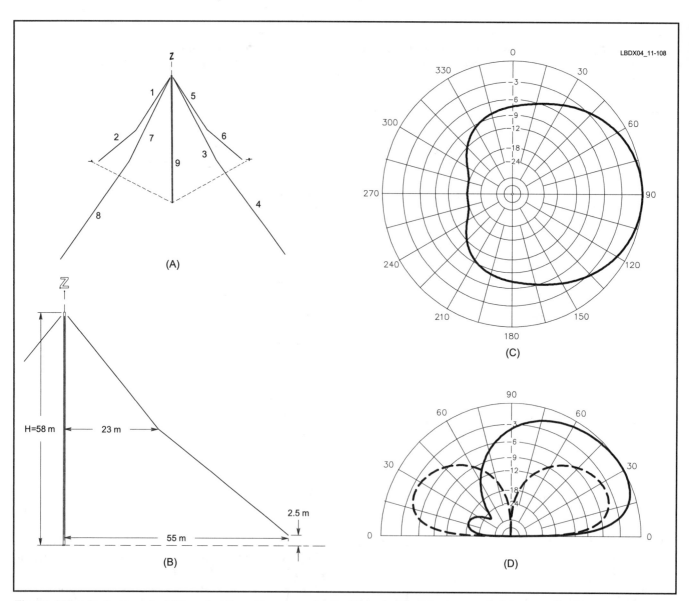

Fig 11-108—At A, layout of 80-meter Four-Square using half-wave dipoles suspended from a single tall tower. At B, layout of one of the sloping dipoles, showing the droop due to the weight of the feed line. At C and D, horizontal and vertical radiation patterns for this array. The vertical pattern is compared with that for a single vertical.

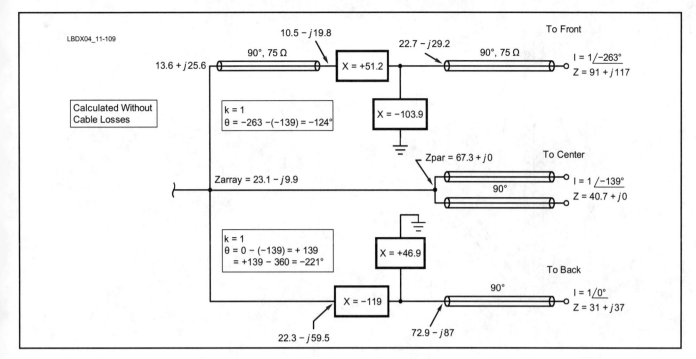

Fig 11-109—Lahlum/Lewallen feed network for the Four-Square with bent half-wave dipoles shown in **Fig 11-108A**.

central support tower are:

Z1 = 31 + j 37
Z2 = Z3 = 40.7 + j 0 Ω
Z4 = 91 + j 117

Fig 11-108B and 108C show the horizontal and vertical radiation patterns. For comparison, the vertical pattern of a single vertical is also included. **Fig 11-109** shows a feed method using the Lahlum/Lewallen feed system. The array shows a fair amount of high-angle radiation, due to the horizontal radiation component originated by the sloping wires. That's also why the RDF is pretty poor (8.06 dB vs 11.04 dB for the K8UR-antenna described in Section 7.3 below).

7.3. The K8UR Sloping-Dipole Four-Square Array

D. C. Mitchell, K8UR, described his 4-element sloping array in Ref 975. He uses half-wave sloping dipoles (sometimes called *slopers*) where the bottom half is sloped back toward the tower. This eliminates much of the high-angle radiation, as most of the horizontal component is now canceled due to the folding of the elements. The four feed points form a square measuring λ/4 on each side. See **Fig 11-110A**.

The original design was fed in quadrature. We now know that much better directivity can be obtained with phasing angles that are much larger than 90°. Modeling turned out a design that has excellent properties:
Feed currents:

I1 = 1 $\underline{/0°}$ A
I2 = I3 = 1 $\underline{/-137°}$ A
I4 = 1 $\underline{/-263°}$ A

Gain: 6.3 dBi (over average ground)
3-dB beamwidth: 88.4° (120° if quadrature fed!)
RDF = 11.07 dB
DMF = 22.7 dB

The feed-point impedances calculated with a grounded 58-meter support tower are:
Z1 = 24.5 + j 83 Ω
Z2 = Z3 = 36.3 Ω
Z4 = −17.8 + j 12 Ω

Fig 11-111 shows the radiation patterns. For comparison, the vertical pattern of a single quarter-wave vertical is also included in B. Mitchell used a Collins-type, quadrature-feed system, although he replaced the 180° phasing line with a hybrid-type network, taking care of the required 180° phase shift. These hybrid couplers are now commercialized by Comtek Systems. Fig 11-111 shows a possible Lahlum/Lewallen feed system for this array.

Mike Greenway, K4PI, developed an innovative way for switching the dipoles of his K8UR-type array from the phone end of the band to the CW end of the band. See **Fig 11-112**.

If you have a tall tower, a Four-Square sloping array à la K8UR is a good way to go, with a clean pattern, excellent gain, good RDF and DMF because there is no high-angle radiation. This array lends itself to a feed design according to the cross-fire principle (see Section 3.4.4 and 4.7.4), where we end up with exactly 75 Ω for Z_{par}, using the two feed lines to the center elements in parallel (see **Fig 11-113**).

I inserted a 1:1 (180° phase-shift) transformer in the center branch. The required phase shift vs the front element (the element being fed directly) is −137 − (−263) = +126°. The coax has to take care of: −180 + 126 = −54°. The back element needs to be fed with a phase angle of +263° vs the front element, and that is the same as +263 − 360 = −97°.

John Brosnahan, WØUN, pointed out that because of the ends of the dipoles being close together (near the tower), mutual coupling is very high. This also shows in the high negative impedance of the back element, which means there is a lot of power coupled into this element by mutual coupling, power, that is fed back into the network. This results in a high-

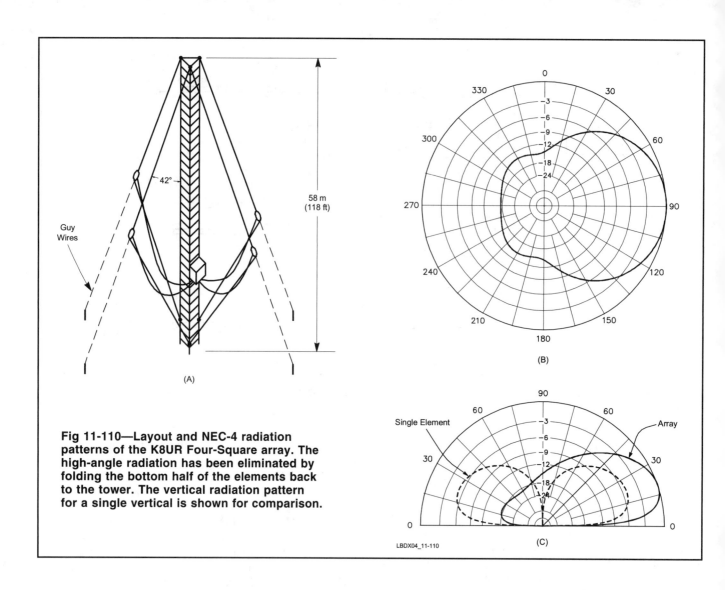

Fig 11-110—Layout and NEC-4 radiation patterns of the K8UR Four-Square array. The high-angle radiation has been eliminated by folding the bottom half of the elements back to the tower. The vertical radiation pattern for a single vertical is shown for comparison.

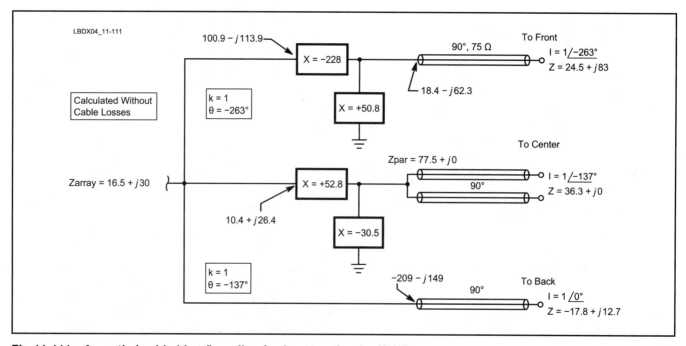

Fig 11-111—An optimized Lahlum/Lewallen feed system for the K8UR array.

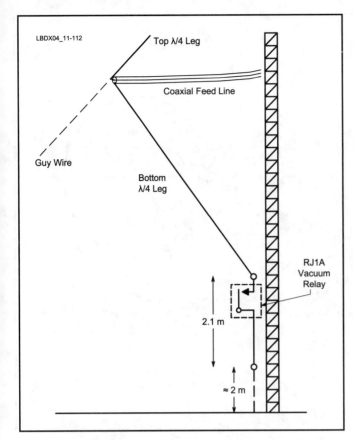

Fig 11-112—Method for switching a K8UR-style dipole from 3.8 to 3.5 MHz. The bottom end of the dipole is lengthened with a piece of wire (which can be called capacitive loading) to lower its resonant frequency to 3.5 MHz. K4PI uses a 2.1-meter long wire, spaced about 0.5 meters from the tower. The relay must be able to withstand high voltage. Mike uses RJ1A vacuum relays.

Q system and a narrow SWR and pattern bandwidth.

7.3.1. Conclusion, K8UR array

If you care about bandwidth, and are not just a CW operator, I would opt for two switchable L-networks in the Lahlum/Lewallen feed system for the K8UR element configuration. I'd adjust each for peak performance in the CW and SSB portions of the 80-meter band instead of using all sorts of tricks to try to improve operational bandwidth. But I find it interesting to analyze these cases, one by one, as they do improve our understanding in these matters.

7.4. The Four-Square Array with Sloping Quarter-Wave Verticals

If you cannot run long catenary cables to support wire vertical elements, you can stick some insulating booms made of fiberglass or aluminum broken up with insulators at the top of your tower. See **Fig 11-114**.

7.4.1. Array data

F = 3.75 MHz
Length sloping wires: 19.86 meters
Side of square: 20 meters
Feed currents:
$I1 = 1 \angle 0°$ A
$I2 = I3 = 1 \angle -120°$ A
$I4 = 0.9 \angle -235°$ A
Gain: 6.04 dBi
3-dB beamwidth: 87°
RDF: 10.55 dB
DMF: 22.4 dB
Feed-point impedances:
$Z1 = 24.5 + j\,60\ \Omega$
$Z2 = Z4 = 25.6 + j\,5\ \Omega$
$Z3 = -7 + j\,5\ \Omega$

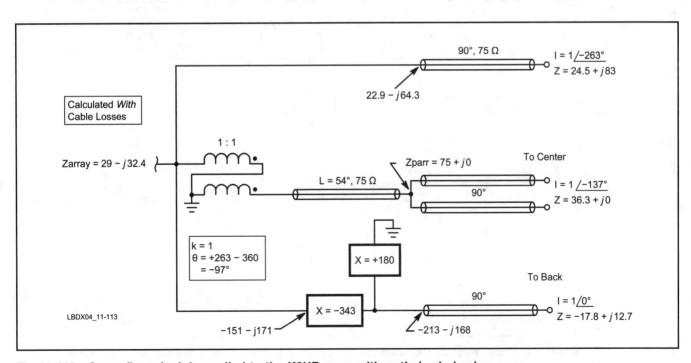

Fig 11-113—Cross-fire principle applied to the K8UR array with optimized phasing.

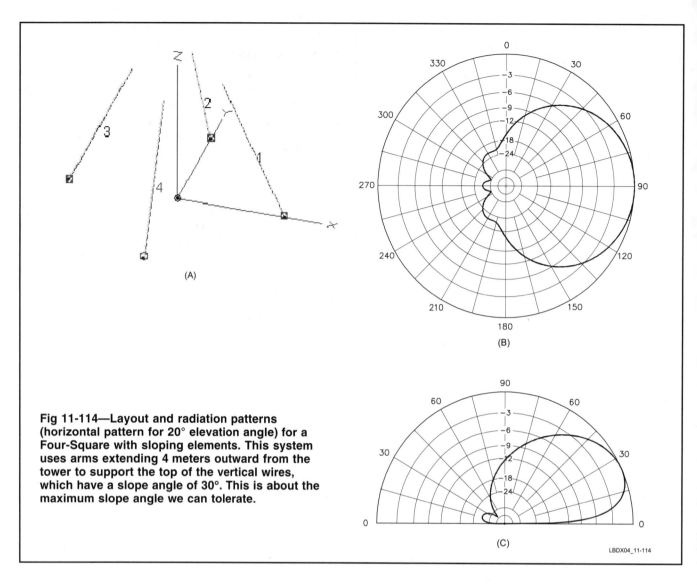

Fig 11-114—Layout and radiation patterns (horizontal pattern for 20° elevation angle) for a Four-Square with sloping elements. This system uses arms extending 4 meters outward from the tower to support the top of the vertical wires, which have a slope angle of 30°. This is about the maximum slope angle we can tolerate.

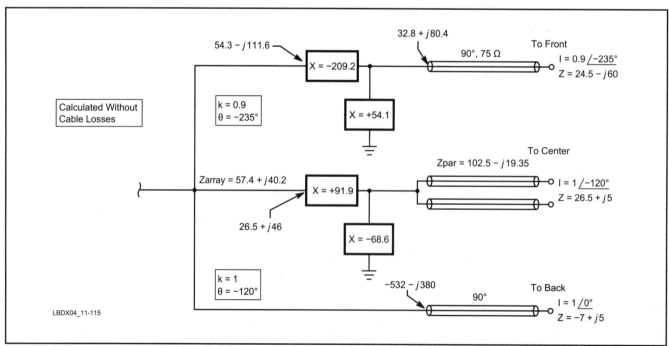

Fig 1-115—One possible Lahlum/Lewallen feed method for a Four-Square with sloping elements shown in Fig 11-114.

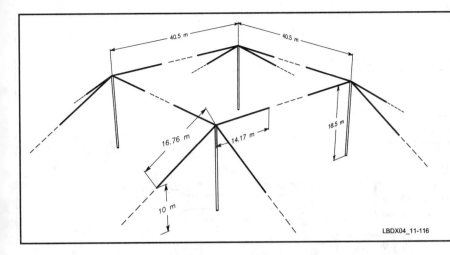

Fig 11-116—This reduced-height Four-Square measures $\lambda/4$ along the side of the square and it uses $\lambda/8$-high elements. The loading wires are symmetrical and don't contribute high-angle radiation. Some fine tuning of the resonant frequency of the elements can be done by changing the height of the ends of the sloping loading wires (nominally 10 meters high).

This array is very attractive if you have one central support and if you cannot run long catenary sloping wires to support the wire verticals. Make sure the central support does not upset the radiation pattern. If it is grounded, it may be too close to resonance and you may have to decouple it (see Chapter 7). **Fig 11-115** shows a Lahlum/Lewallen feed system for this array.

If you want to operate this array on both the CW and the phone end of the band, it is best first to model the antenna at 3.51 MHz. For that case the sloping vertical wires should be about 20.9 meters long (assuming 2 mm OD copper conductors). On 3.775 MHz the elements are too long and need to be shortened by a series capacitor of about 600 pF. A relay can switch the capacitor in and out of the circuit. At the same time the current-forcing feed line can be lengthened on 3.5 MHz to maintain its proper ($\lambda/4$) length. See also Fig 11-96.

It is clear that the same configuration can be used in 160 meters, where everything will be approximately twice the size.

7.5. An Attractive 160-Meter Four-Square With 18.5-Meter Tall Verticals

One does *not* need to use full-size $\lambda/4$ long vertical radiators to build a high-performance array. Using short verticals has a few consequences though:

1. Lower feed point impedances
2. Meaning that an excellent radial system is even more important
3. A low-loss top-loading configuration is essential

Top loading means wire loading. While inverted-L elements can be used, they will radiate a lot at high angles and completely destroy the RDF and DMF directivity figures of the array. Top-loading wires need to be arranged in such a way that far-field cancellation occurs for all radiation from the loading wires.

The design shown in **Fig 11-116** is a good example of such a design. Compared to a Four-Square with full-size (39-meter long) elements, the quadrature fed version of this array with 18.5-meter long elements (46% of full-size) gives up only 0.5 dB in gain, assuming an identical radial system is used (with 2-Ω equivalent loss resistance).

7.5.1. Quadrature fed

Dimension of square side: $\lambda/4$
Elements: 18.5 meters tall with top loading wires.

Feed currents:
$I1 = 1\ \underline{/-180°}$ A (front element)
$I2 = I4 = 1\ \underline{/-90°}$ A (center elements)
$I3 = 1\ \underline{/0°}$ A (back element)
Gain: 6.10 dBi (Average Ground) and 7.73 dBi (Very Good Ground)
3-dB beamwidth: 100°
RDF = 10.20 dB
DMF = 19.1 dB
Feed-point impedances:
$Z1 = 27.4 + j\,36.2\ \Omega$
$Z2 = Z4 = 19.5 - j\,1.2\ \Omega$
$Z3 = -1.8 - j\,1.4\ \Omega$

7.5.2. Feeding the array

As this array is quadrature-fed you can feed it with a hybrid coupler (eg, Comtek). A Lahlum/Lewallen feed system is shown in **Fig 11-117**.

7.5.3. Optimized feed, Short Four-Square

We can get slightly more gain and better directivity with the WA3FET feed-current values. **Fig 11-118** shows the radiation patterns for this array.
Dimension of square side: $\lambda/4$
Feed currents:
$I1 = 0.872\ \underline{/-218°}$ A (front element)
$I2 = I4 = 0.9\ \underline{/-111°}$ A (center elements)
$I3 = 1\ \underline{/+0°}$ A (back element)
Gain: 6.58 dBi (Average Ground) and 8.22 dBi (Very Good Ground)
3-dB beamwidth: 87°
RDF = 11.09 dB
DMF = 24.8 dB
Feed-point impedances:
$Z1 = 16.5 + j\,29.9\ \Omega$
$Z2 = Z4 = 14.3 - j\,1.5\ \Omega$
$Z3 = 0.9 - j\,0.2\ \Omega$

In this configuration you must use an L-network feed system (Lahlum/ Lewallen). **Fig 11-119** shows one of the possible solutions.

8. RADIAL SYSTEMS FOR ARRAYS
8.1. Buried Radials

Radials of the elements of an array cross each other. It is

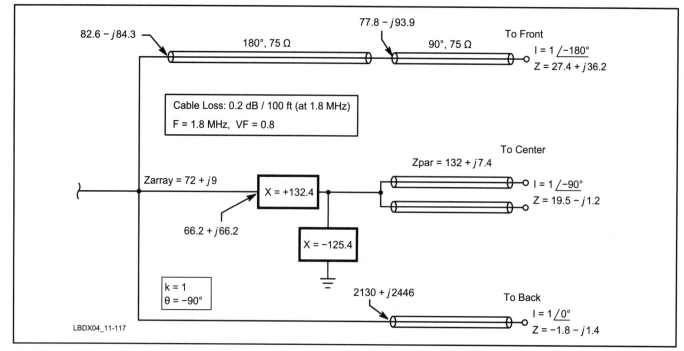

Fig 11-117—Lahlum/Lewallen feed system for the quadrature-fed Four-Square with λ/8 top loaded elements.

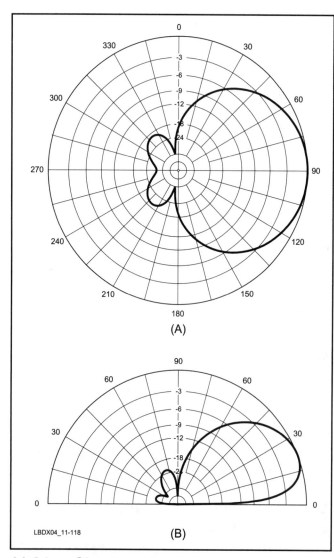

standard procedure to install a bus (AWG #6 or #8 copper strip) half way between the elements and to connect the radials to this "bus." The radials can be any wire size, if many are used. The size will be dictated more by mechanical strength than current-carrying capability. **Fig 11-120** shows various typical radial layouts for a 2-element cardioid array, a 3-element triangular array and for the classic Four-Square array.

8.2. Elevated Radials

When a large number of elevated radials are used on each of the elements of an array, these radials become non-resonant, and they can be connected to a bus system in exactly the same way as shown in **Fig 11-120**.

When only a few radials are used (typically 1 to 4 radials), the situation is very different. In this case the radials can couple heavily with adjacent (especially parallel) radials from other elements and can upset the directivity of the array, and create uncontrolled and unwanted high-angle radiation from the elevated radials. **Fig 11-121** shows a few possible layouts that try to minimize the coupling.

9. CONCLUSION

Now that we have powerful modeling programs available (eg, *EZNEC*), I would like to encourage everyone to try to develop an array that fits his or her own requirements. The dBi gain is *not* the gain you should expect if you change from a single vertical to an array. Over average ground (dielectric constant = 13, conductivity = 5 mS) a single element has a gain in the order of 1 dBi, so if your modeling program tell you the array has 6 dBi gain, expect about 5 dB gain over a single vertical.

During your modeling sessions, make sure you use the

Fig 11-118—Radiation patterns (horizontal for 20° elevation angle) for the optimized 160-meter Four-Square with λ/8-long top loaded elements.

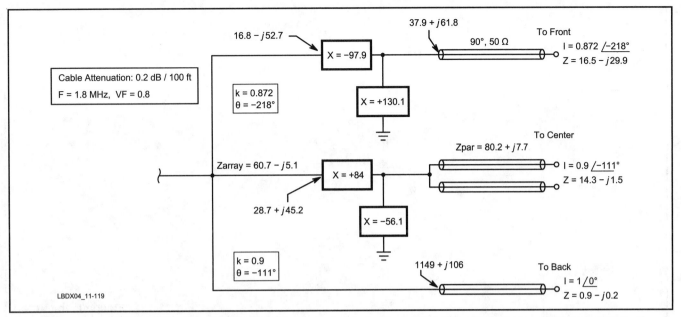

Fig 11-119—Lahlum/Lewallen feed system for the optimized short-element Four-Square for 160 meters. This time 50-Ω feed lines turned out to be the better choice.

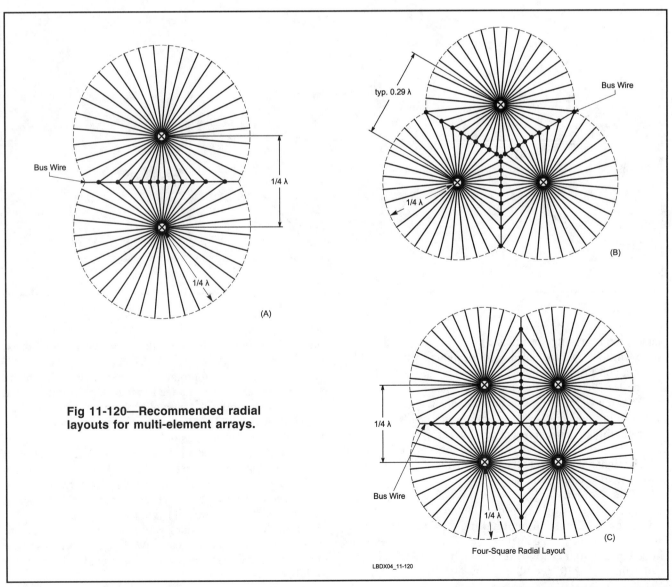

Fig 11-120—Recommended radial layouts for multi-element arrays.

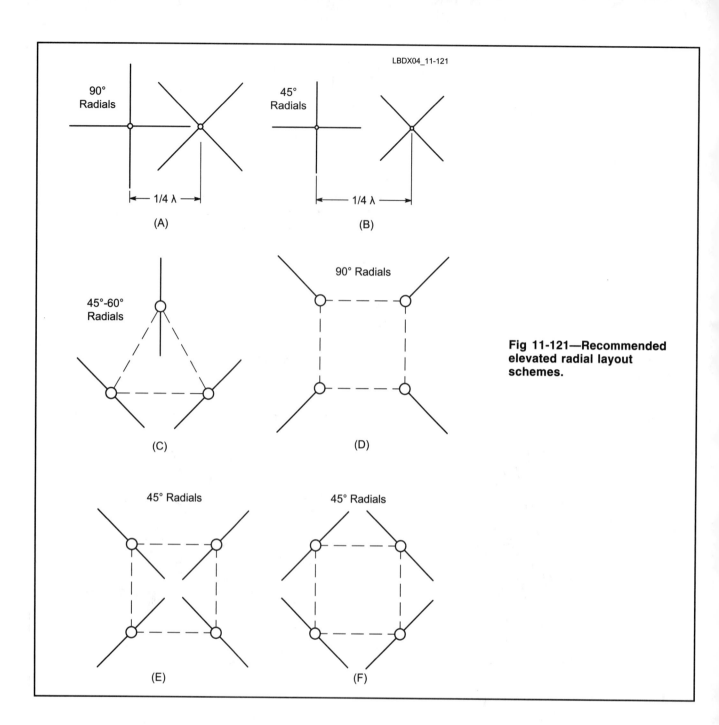

Fig 11-121—Recommended elevated radial layout schemes.

same ground quality specifications in all circumstances. If you want to model the antenna for designing your own feed system, don't forget to include the equivalent loss resistance for the-less-than perfect ground radial system.

Let me warn the readers once again: Antenna modeling is one thing and practical antenna design can be a totally different thing. You can model complex arrays with staggering characteristics, but which may be difficult to build and which might not a project for a newcomer with little technical expertise to attempt to build.

Don't forget you have to feed the array, by practical means, with real-world feed lines. After having done your initial calculations without taking cable losses into account, give it a final go with losses. Especially if extreme impedances (high or low) are involved and relatively long coax runs (eg, 270° feed lines) you may be surprised how much the difference between the ideal world (no losses) and the real world can be!

If you have to compromise a little in order to be able to use a much easier-to-make feed-system (eg, quadrature fed versus the requirement for exotic phase angles or current magnitudes), I would advise the compromise, unless you have the required measuring setup.

A simple test system developed by W1MK makes it possible to tune each L-network feed system for top-notch performance. There no longer is a need for vector volt-meters. You can now get top performance just using very affordable test instruments that you can build yourself. Go for it!

If designing and building your own feed system does not attract you, know that you can nowadays buy systems using the Lahlum/Lewallen approach commercially (from Array Solutions).

CHAPTER 12

Other Arrays

Chapter 11 on phased arrays only covered arrays made of vertical (omnidirectional) radiators. You can, of course, design phased arrays using elements that, by themselves, already exhibit some horizontal directivity; eg, horizontal dipoles.

Even at relatively low heights (0.3 λ), arrays made of horizontal elements (dipoles) can be quite attractive. Their intrinsic radiation angle is certainly higher than for an array made of vertical elements, but unless the electrical quality of the ground is good to excellent, the horizontal array may actually outperform the vertical array even at low angles.

The vertical radiation angle (wave angle) of arrays made with vertical elements (typical λ/4 long elements) depends only on the quality of the ground in the Fresnel zone. Radiation angles range typically from 15° to 25°. The same is true for arrays made with horizontally polarized elements, but we have learned that reflection efficiency is better over bad ground with horizontal polarization than it is with vertical polarization (see Chapter 9, Section 1.1.2 and Chapter 8, Section 1.2.1.1).

The elevation angle for antennas with horizontally polarized elements basically depends on the height of the antenna above ground. For low antennas (with resulting high elevation angles), the quality of the ground right under the antenna (in the near field) will also play a role in determining the elevation angle (see Chapter 8, Section 1.2.1.4). But as DXers, we are not interested in antennas producing wave angles that radiate almost at the zenith.

Over good ground, a dipole at λ/4 height radiates its maximum energy at the zenith. Over average ground, the wave angle is 72°. The only way to drastically lower the radiation angle with an antenna at such low height is to add another element.

If we install a second dipole at close spacing (eg, λ/8), and at the same height (λ/4), and feed this second dipole 180° out-of-phase with respect to the first dipole, we achieve two things:

- Approximately 2.5 dB of gain in a bidirectional pattern.
- A lowering of the elevation angle from 72° to 37°!

At the zenith angle the radiation is a perfect null, whatever the quality of the ground is. This is because, at the zenith, the reflected wave from element number 1 (reflected from the ground right under the antenna) will cancel the direct wave from element number 2. The same applies to the reflected wave from element number 1 and the direct 90° wave from element number 2. All the power that is subtracted from the high angles is now concentrated at lower angles. Of course there also is a narrowing of the horizontal forward lobe. Example: A λ/2 80-meter dipole at 25 meters has a –3 dB forward-lobe beamwidth of 124° at an elevation angle of 45°. The 2-element version, described above, has a –3-dB angle of 95° at the same 45° elevation angle. The impedance of the two dipoles has dropped very significantly to approximately 8 Ω.

Fig 12-1 shows the elevation angles for three types of antennas over average ground: a horizontal dipole, two half waves fed 180° out-of-phase (spaced λ/8), and a 2-element Yagi. From this graph you can see that the only way to achieve a reasonably low radiation angle from a horizontally polarized antenna at low height of λ/3 or less is to add a second element. The 180° out-of-phase element lowers the radiation angle at lower antenna heights (below 0.35 λ) significantly more than a Yagi or a 2-element all-fed array. It also has the distinct advantage of suppressing all the high-angle radiation, which is not the case with the Yagis or all-fed arrays.

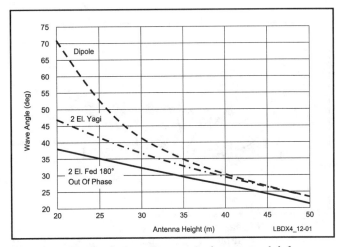

Fig 12-1—Vertical elevation angle (wave angle) for three types of antennas over average ground: a half-wave dipole, a 2-element parasitic Yagi array and two close-spaced half-wave dipoles fed 180° out-of-phase. Note the remarkable superiority of the last antenna at low heights. The graph is applicable for 80 meters.

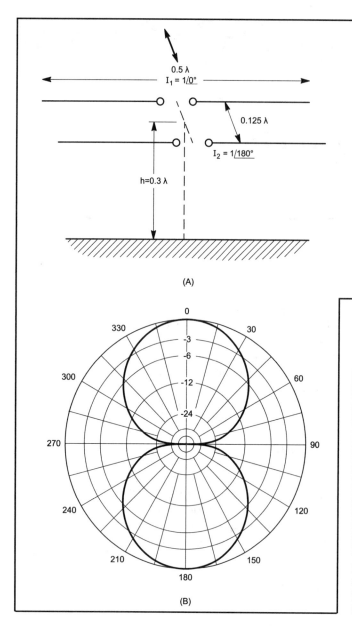

(A)

(B)

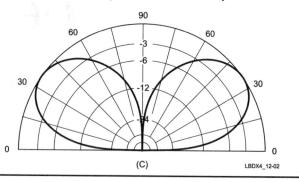

(C)

Fig 12-2—Configuration and radiation patterns of two close-spaced half-wave dipoles fed 180° out-of-phase at a height of 0.3 λ above average ground. The azimuth pattern at B is taken for an elevation angle of 36°. Note in the elevation pattern at C that all radiation at the zenith angle is effectively canceled out (see text for details).

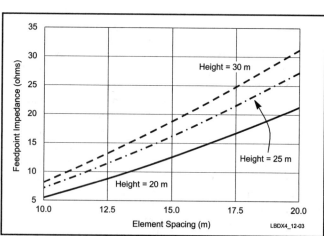

Fig 12-3—Feed-point impedance of the 2-element close-spaced array with elements fed 180° out-of-phase, as a function of spacing between the elements and heights above ground. The design frequency is 3.75 MHz.

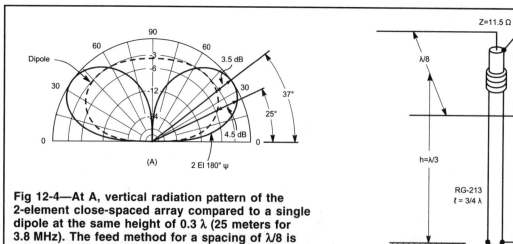

Fig 12-4—At A, vertical radiation pattern of the 2-element close-spaced array compared to a single dipole at the same height of 0.3 λ (25 meters for 3.8 MHz). The feed method for a spacing of λ/8 is shown at B. The feed-point impedance is about 100 Ω at the junction of the λ/4 and the 3λ/4 50-Ω feed lines. A λ/4 long 70-Ω feed line can be used to provide a perfect match to a 50-Ω feed line.

1. TWO-ELEMENT ARRAY SPACED λ/8, FED 180° OUT-OF-PHASE

The vertical and the horizontal radiation patterns of the 2-element array are shown in **Fig 12-2**. As the antenna elements are fed with a 180° phase difference, feeding is simple. The impedances at both elements are identical. **Fig 12-3** gives the feed-point impedance of the elements as a function of the spacing between the elements and the height. Within the limits shown, spacing has no influence on the gain or the directivity pattern. Very close spacings give very low impedances, which makes feeding more complicated and increases losses in the system. A minimum spacing of 0.15 λ is recommended.

Compared to a single dipole at the same height, this

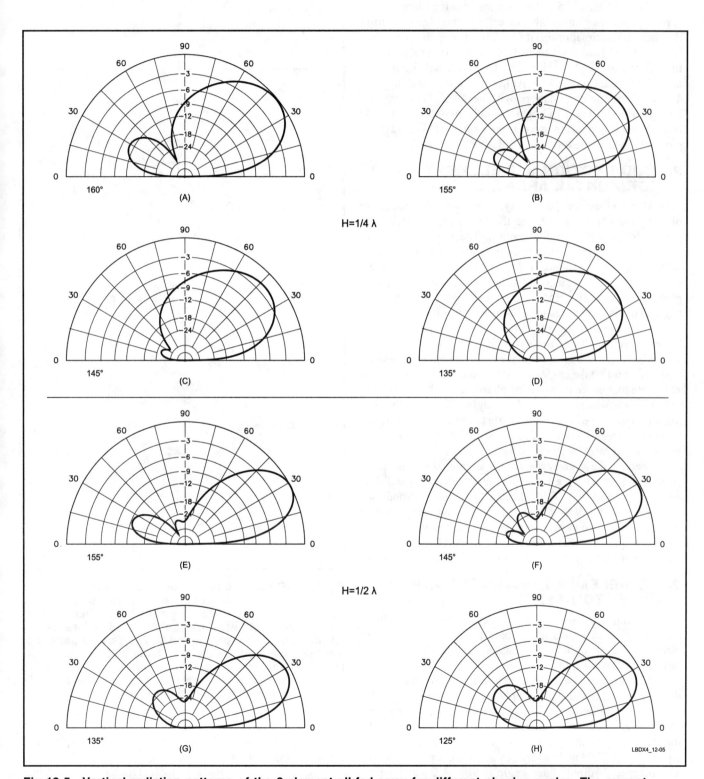

Fig 12-5—Vertical radiation patterns of the 2-element all-fed array for different phasing angles. The current magnitude is the same for both elements. All patterns are plotted to the same scale. Patterns are shown for antenna heights of λ/4 (at A through D) and λ/2 (at E through H). A—160° phase difference. B and E—155° phase difference. C and F—145° phase difference. D and G—135° phase difference. H—125° phase difference.

Other Arrays   12-3

antenna has a gain of 3.5 dB at its main elevation angle of 37°, and of 4.5 dB at an elevation angle of 25° (see **Fig 12-4**).

Feeding the array is done by running a λ/4 feed line to one element, and a 3λ/4 feed line to the other element. The feed point at the junction of the two feed lines is approximately 100 Ω for an element spacing of 0.125 λ. A λ/4 long 75-Ω cable will provide a perfect match to a 50-Ω feed line.

You will have a 5:1 SWR on the two feed lines, so be careful when running high power! Another feed solution that may be more appropriate for high power is to run two parallel 50-Ω feed lines to each element, giving a feed line impedance of 25 Ω. In this case the SWR will be a more acceptable 2.2:1 on the line. At the end of the feed lines (λ/4 and 3λ/4) the impedances will be 54 Ω. The parallel combination will be 27 Ω, which can be matched to a 50-Ω line through a quarter-wave transformer of 37.5 Ω (two parallel 75-Ω cables) or via a suitable L network.

2. UNIDIRECTIONAL 2-ELEMENT HORIZONTAL ARRAY

Starting from the above array, we can now alter the phase of the feed current to change the bidirectional horizontal pattern into a unidirectional pattern. The required phase to obtain beneficial gain and especially front-to-back ratio varies with height above ground. At λ/2 and higher, a phase difference of 135° produces a good result. At lower heights, a larger phase difference (eg, 155°) helps to lower the main wave angle. This is logical, as the closer we go to the 180° phase difference, the more the effect of the phase radiation cancellation at high angles comes into effect.

Fig 12-5 shows the vertical radiation patterns obtained with different phase angles for a 2-element array at λ/4 and λ/2. Note that as we increase the phase angle, the high-angle radiation decreases, but the low-angle F/B worsens. The higher phase angle also yields a little better gain. For antenna heights between λ/4 and λ/2, a phase angle of 145° seems a good compromise.

Feeding these arrays is not so simple, since the feed-current phase angles are not in quadrature (phase angle differences in steps of 90°). For a discussion of feed methods see Chapter 11 on vertical arrays. *Current forcing* using a modified Lewallen feed system seems to be the best choice.

The question that comes to mind is, "Can we obtain similar gain and directivity with a parasitic array?" Let's see.

3. TWO-ELEMENT PARASITIC ARRAY (DIRECTOR TYPE)

Our modeling tools teach us that we can indeed obtain exactly the same results with a parasitic array. A 2-element director-type array produces the same gain and a front-to-back ratio that is even slightly superior.

As a practical 2-element parasitic-type wire array, I have developed a Yagi with 2 inverted-V-dipole elements. **Fig 12-6** shows the configuration as well as the radiation patterns obtained at a height of 25 meters (0.3 λ on 80 meters). To make the array easily switchable, both wire elements are made equally long (39.94 meters for a design frequency of 3.8 MHz). The inverted-V-dipole apex angle is 90°. A 25-meter high mast or tower is required. At that height we need to install a 10-meter long horizontal support boom, from the end of which we can hang the inverted-V dipoles. The gain is 3.9 dB versus

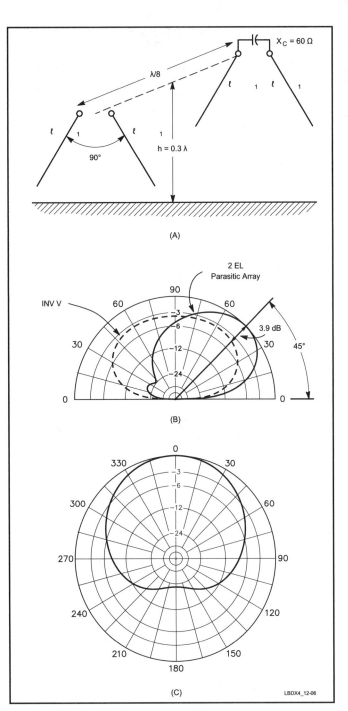

Fig 12-6—Configuration and calculated radiation patterns for the 2-element parasitic array using inverted-V dipole elements. The array is installed with an apex angle of 90°, at a height of 0.3 λ (25 meters for 3.8 MHz). Element spacing is λ/8. The vertical pattern of a single inverted-V dipole is included at B for comparison. At C, the azimuth pattern is shown for an elevation angle of 45°. The gain at the 45° peak elevation angle is 3.9 dB over the single inverted-V dipole.

an inverted-V dipole at the same height, measured at the main elevation angle of 45°.

A loading capacitor with a reactance of −j 65 Ω produces the right current phase in the *director*. The radiation resistance of the array is 24 Ω. To make the array easily switchable, we run two feed lines of equal length to the

elements. From here on there are two possibilities:
- We use a length of coax feed line to provide the required reactance of $-j\,65\,\Omega$ at the element.
- We use a variable capacitor at the end of a $\lambda/2$ feed line. The theoretical value of the capacitor is:

$$\frac{10^6}{2\pi \times 3.8 \times 65} = 644\,\text{pF}$$

Now we calculate the length of the open feed line that exhibits a capacitance of 644 pF on 3.8 MHz. The reactance at the end of an open feed line is given by:

$$X = Z_C \times \tan(90 - L) \qquad \text{(Eq 12-1)}$$

where
Z_C = characteristic impedance of the line
L = length of the line in degrees

This can be rewritten as

$$L = 90 - \arctan\frac{X}{Z_C} \qquad \text{(Eq 12-2)}$$

In our case we need $X = -60\,\Omega$. Thus,

$$L = 90 - \arctan\frac{60}{50} = 39.8°$$

The physical length of this line is given by:

$$L_{\text{meters}} = \frac{833 \times V_f \times \ell}{1000 \times F_q} \qquad \text{(Eq 12-3)}$$

where
V_f = velocity factor (0.66 for RG-213)
F_q = design frequency
ℓ = length in degrees

$$L_{\text{meters}} = \frac{833 \times 0.66 \times 39.8}{1000 \times 3.8} = 5.76\,\text{meters}$$

Fig 12-7 shows the feed and switching arrangements according to the two above-mentioned systems.

4. TWO-ELEMENT DELTA-LOOP ARRAY (REFLECTOR TYPE)

Using the same support as described above (a 10-meter long boom at 25 meters), we can also design a 2-element delta-loop configuration. If the ground conductivity is excellent, and if we can install radials (a ground screen), the 2-element delta-loop array should provide a lower angle of radiation and comparable gain to the 2-element inverted-V-dipole array described in Section 3.

4.1. Two-Element Delta Loop with Sloping Elements

Since the low-impedance feed point of the vertically polarized delta loop is quite a distance from the apex, and as most of the radiation comes from the high-current areas of the antenna, we can consider using delta-loop elements that are sloping away from the tower. We could not do this with the inverted-V, 2-element array, since the high-current points are right at the apex.

In this example I have provided a boom of 6 meters length at the top of the support at 25 meters. From the tips of the boom we slope the two triangles so that the base lines are

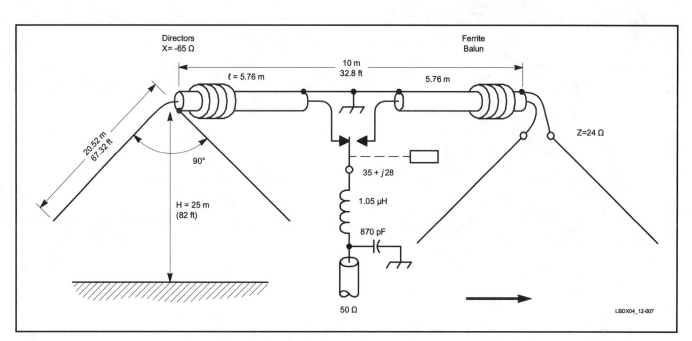

Fig 12-7—Feeding arrangement for the 2-element parasitic array shown in Fig 12-6. Two lengths of RG-213 run to a switch box in the center of the array. The coax feeding the director is left open at the end, producing a reactance of $-j\,65\,\Omega$ (equivalent to 644 pF at 3.8 MHz) at the element feed point. The radiation resistance of the 2-element array is 29 Ω. An L network can be provided to obtain a perfect match to the 50-Ω feed line. A current type of balun (eg, stack of ferrite beads) *must* be provided at both element feed points.

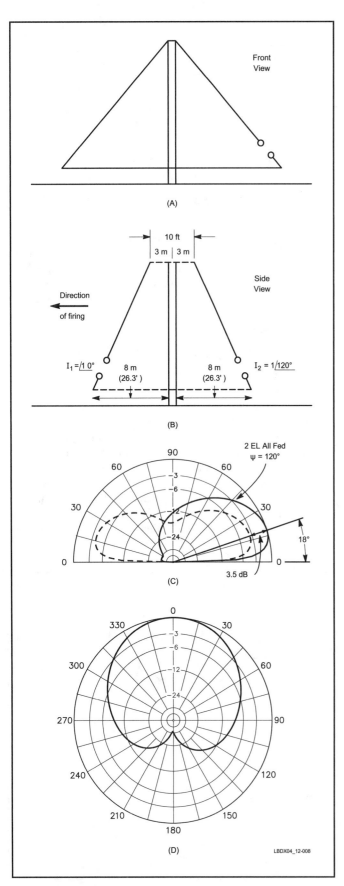

Fig 12-8—Configuration and radiation patterns of a 2-element delta-loop array, using sloping elements. The elements are fed with equal-magnitude currents and with a phase difference of 120°. The horizontal pattern at D is for an elevation angle of 18°.

now 8 meters away from the support and approximately 2.5 meters above the ground.

Fig 12-8 shows the radiation pattern obtained with the array when the loops are fed with equal current magnitude and with a phase difference of 120°. Note the tremendous F/B at low angles (more than 45 dB!). Gain over a single-element loop is 3.5 dB. The wave angle is 18° over a very good ground. One of the problems is, of course, the feed system for an array that is not fed in quadrature.

Fig 12-9 shows the radiation patterns for the 2-element array with a parasitic reflector. The gain is the same as for the all-fed array and 3.4 dB over a single delta-loop element. The parasitic array shows a little less F/B at low angles, as compared to the all-fed array (see Fig 12-8), but the difference is slight.

As with the 2-element dipole array, my personal preference goes to the parasitic array, since the all-fed array is not fed in quadrature, which means that the feed arrangement is all but simple (it requires a modified Lewallen feed system). The obvious feed method for the 2-element parasitic array uses two equal-length feed lines to a common point mid-way between the two loops. A small support can house the switching and matching hardware.

As with the 2-element inverted-V array, we use two loops of identical length, and use a length of shorted feed line to provide the required inductive loading with the reflector element. The length of the feed line required to achieve the required 140° inductive reactance is calculated as follows:

$$X_L = Z_C \times \tan \ell \quad \text{(Eq 12-4)}$$

where

X_L = required inductance
Z_C = cable impedance
ℓ = cable length in degrees

This can be rewritten as

$$\ell = \arctan \frac{X_L}{X_C} \quad \text{(Eq 12-5)}$$

or

$$\ell = \arctan \frac{140}{75} = 61.8°$$

The physical length is given by

$$L_{meters} = \frac{833 \times V_f \times \ell}{1000 \times F_q} \quad \text{(Eq 12-6)}$$

where

L_{meters} = length, meters
ℓ = length in degrees
V_f = velocity factor of the cable
F_q = design frequency, MHz

We use foam-type RG-11 ($V_f = 0.81$), because solid PE-

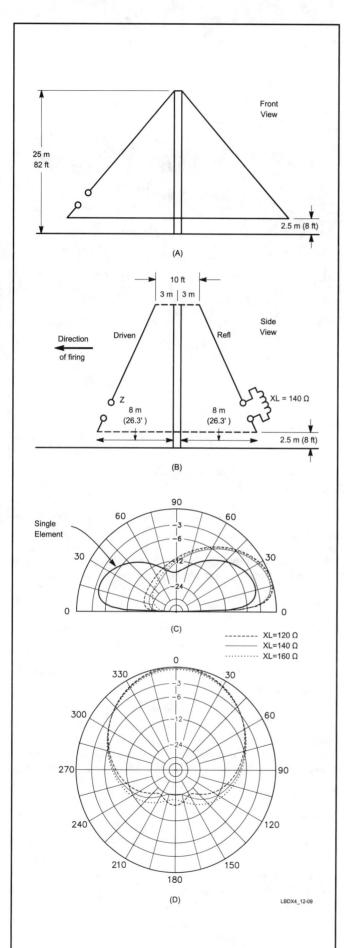

Fig 12-9—Radiation patterns for the 2-element delta-loop array having the same physical dimensions as the all-fed array of Fig 12-8, but with one element tuned as a reflector. In practice both triangles are made equal size, and the required loading inductance is inserted to achieve the phase angle. Patterns shown are for different values of loading coils (X_L = 120, 140 and 160 Ω). The feed-point impedance of the array will vary between 80 and 150 Ω, depending on the ground quality.

type coax (Vf = 0.66) will be too short to reach the switch box.

$$L_{meters} = \frac{833 \times 0.81 \times 61.8}{1000 \times 3.8} = 10.97 \text{ meters}$$

Fig 12-10 shows the feed line and the switching arrangement for the array. Note that the cable going to the reflector must be short-circuited. The two coaxial feed lines must be equipped with current-type baluns (a stack of ferrite beads).

The impedance of the array varies between 75 Ω and 150 Ω, depending on the ground quality. If necessary, the impedance can easily be matched to the 50-Ω feed line using a small L network. This array can be made switchable from the SSB end of the band to the CW end by applying the capacitive loading technique as described in Chapter 10.

Since this array was published in the Second Edition of this book, I have received numerous comments from people who have successfully constructed it.

5. THREE-ELEMENT DIPOLE ARRAY WITH ALL-FED ELEMENTS

A 3-element phased array made of λ/2 dipoles can be dimensioned to achieve a very good gain together with an outstanding F/B ratio. Three elements on a λ/4 boom (giving λ/8 spacing between elements) can yield nearly 6 dB of gain at the major radiation angle of 38° over a single dipole at the same height (over average ground).

A. Christman, KB8I, described a 3-element dipole array with outstanding directional and gain properties. (Ref 963.) I have modeled a 3-element inverted-V-dipole array using the same phase angles. The inverted-V elements have an apex angle of 90°, and the apex at 25 meters above ground. The radiation patterns are shown in Fig 12-11.

The elements are fed with the following currents:
$I1 = 1 \angle -149°$ A
$I2 = 1 \angle 0°$ A
$I3 = 1 \angle 146°$ A

With the antenna at 25 meters above ground and elements that are 39.72-meters long (design frequency = 3.8 MHz), the element feed-point impedances are:
$Z1 = -36 + j\,24.5$ Ω
$Z2 = 12.3 + j\,25$ Ω
$Z3 = 7.6 - j\,12.2$ Ω

If you are confused by the minus sign in front of the real part of Z1, it just means that in this array, element number 1 is actually *delivering* power into the feed system, rather than taking power from it. This is a very common situation with driven arrays, especially where close spacing is used.

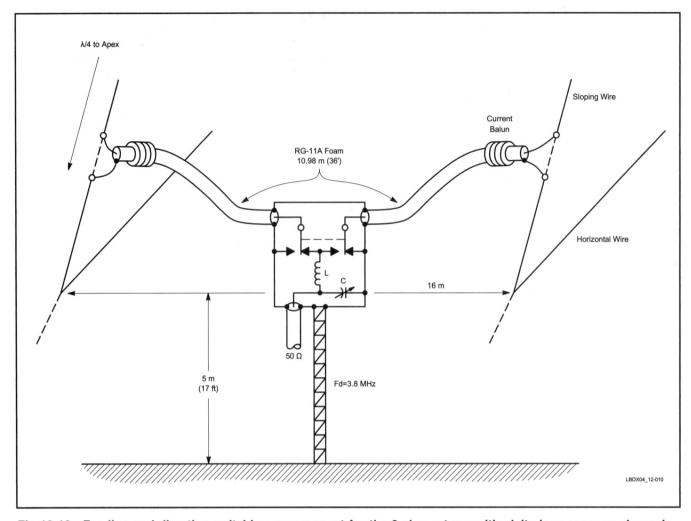

Fig 12-10—Feeding and direction-switching arrangement for the 2-element parasitic delta-loop array as shown in Fig 12-9. The length of the 75-Ω feed lines going from the feed points to the switch box is 61.8°. For 3.8 MHz, and using foam-type coax (Vf = 0.81), this equals 10.98 meters. The spacing between the elements at the height of the feed points is about 5 meters. Note that the feed line to the reflector needs to be short-circuited. A simple L network provides a perfect match for a 50-Ω feed line.

A possible feed method consists of running three λ/4 lines to a common point. Current forcing is employed: We use 50-Ω feed lines to the outer elements, and two parallel 50-Ω lines to the central element. The method is described in detail in Chapter 11 on vertical arrays.

It is much easier to model such a wonderful array and to calculate a matching network than to build and align the matching system. Slight deviations from the calculated impedance values mean that the network component values will be different as well. There is no method of measuring the driven impedances of the elements. All you can do in the way of measuring is use an HF vector voltmeter and measure the voltages at the end of the three feed lines. The voltage magnitudes should be identical, and the phase as indicated above (E1, E2 and E3). If they are not, the values of the networks can be tweaked to obtain the required phase angles. Good luck!

We have seen that we can just about match the performance of a 2-element all-fed array with a parasitic array. We will see that the same can be done with a 3-element array.

6. THREE-ELEMENT PARASITIC DIPOLE ARRAY

The model that was developed has a gain of 4.5 dB over a single inverted V-element (at the same height) for its main elevation angle of 43°. The F/B ratio is just over 20 dB, as compared to just over 30 dB with the all-driven array. At the same antenna height (0.3 λ), the radiation angle of the 3-element parasitic was also slightly higher (43 Ω) than for the 3-element all-fed array (38 Ω), modeled over the same average ground.

Fig 12-11 shows the superimposed patterns for the all-driven and the parasitic 3-element array (for 80 meters at 25 meters height). Note that the 3-element all-fed has a better rejection at high angles. This is because the currents in the outer elements have a greater phase shift (versus the driven element) than in the parasitic array. These phase shifts are:

Reflector:
    All-driven array: −149°
    Parasitic array: −147°

Director:
All-driven array: +147°
Parasitic array: +105°

This demonstrates again that, with an all-driven array, we have more control over all the parameters that determine the radiation pattern of the array. Like the 2-element array described in Section 3, the 3-element array is also made using three elements identical in length. The required element reactances for the director and reflector are obtained by inserting the required inductance or capacitance in the center of the element. In practice we bring a feed line to the outer elements as well. The feed lines are used as stubs, which represent the required loading to turn the elements into a reflector or director.

The question is, which is the most appropriate type of feed line for the job, and what should be its impedance? **Table 12-1** shows the stub lengths obtained with various types of feed lines. The length of the open-ended stub serving to produce a negative reactance (for use as a director stub) is given by:

$$\ell° = 90 - \arctan \frac{X_C}{Z_C} \quad \text{(Eq 12-7)}$$

For the short-circuited stub serving to produce a positive reactance (for the reflector), the formula is:

$$\ell° = \arctan \frac{X_L}{Z_C}$$

- From **Table 12-1** we learn the 450-Ω stub requires a very long length to produce the required negative reactance for the director (17.28 meters).

Table 12-1
Required Line Length for the Loading Stubs of the Parasitic Version of the 3-Element Array of Fig 12-11

Z_C Ω	VF	Length, Degrees	Length Meters	Length Feet
Director				
50	0.66	42.3	6.12	20.08
75	0.66	53.75	7.77	25.49
100	0.95	83.03	8.85	29.04
450	0.95	83.03	17.28	56.70
Reflector				
50	0.66	52.53	7.58	13.39
75	0.66	40.91	5.91	13.39
100	0.66	33.02	4.77	15.65
450	0.95	8.22	1.71	5.61

Other data:
Design frequency = 3.8 MHz, wavelength = 78.89 meters
Director X_C = –55 Ω
Reflector X_L = +65 Ω

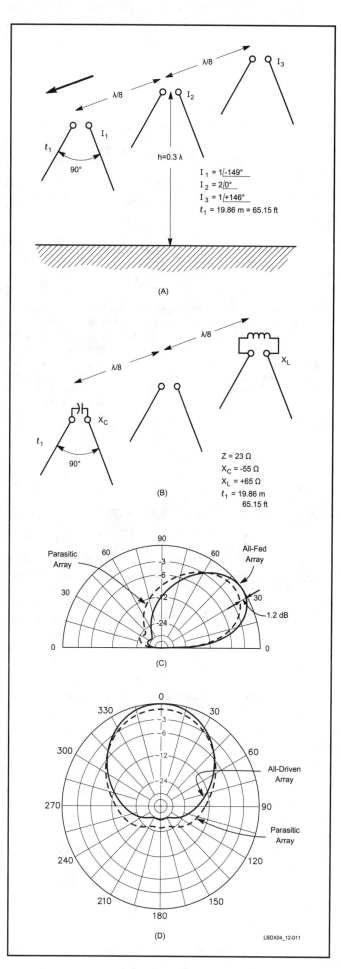

Fig 12-11—Configuration and radiation patterns for two types of 3-element inverted-V-dipole arrays with apexes at 0.3 λ. At both C and D, one pattern is for the all-fed array and the other for an array with a parasitic reflector and director. The all-fed array outperforms the Yagi-type array by approximately 1 dB in gain, as well as 10 dB in F/B.

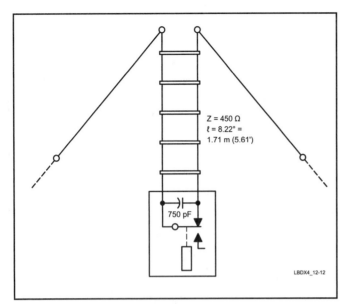

Fig 12-12—The 3-element parasitic type inverted-V dipole array is made with elements that have exactly the same length. The required element loading is obtained by inserting the required capacitance or inductance in the center of these elements. This is obtained by using stubs, as shown here. With a 450-Ω transmission line we require only a short 1.71-meter long piece of short-circuited line to make a stub for the reflector. For the director we connect a 750-pF capacitor across the end of the open-circuit line. This can be switched with a single-pole relay, as explained in the text.

- When made from 50-Ω or 75-Ω coax, we obtain attractive short lengths. The disadvantage is that you need to put a current balun at the end of the stubs to keep any current from flowing on the outside of the coax shield.
- A third solution is to use a 100-Ω shielded balanced line, made of two 50-Ω coax cables. The lengths are still very attractive, and you no longer require the current balun.
- A final solution is to use the 450-Ω transmission line for the reflector (1.71 meters long) and to load the line with an extra capacitor to turn it into a capacitor. I assumed a velocity factor of 0.95 for the transmission line. You must check this in all cases (see Chapter 11 on vertical arrays). The capacitive reactance produced by an open-circuited line of 1.71 meters length at 3.8 MHz is:

$$X_L = 450 \tan(90° - 8.22°) = +j\,3115\ \Omega$$

This represents a capacitance value of only:

$$\frac{10^6}{2\pi \times 3.8 \times 95} = 13.4\ \text{pF}$$

The required capacitive reactance was $-j\,55\ \Omega$, which represents a capacitance value of

$$\frac{10^6}{2\pi \times 3.8 \times 95} = 762\ \text{pF}$$

This means we need to connect a capacitor with a value

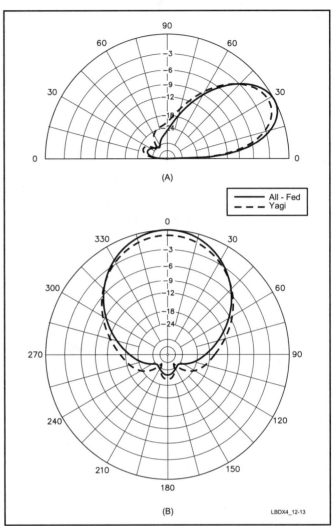

Fig 12-13—Radiation pattern of the 3-element inverted-V type array at a height of λ/2. Note that the all-fed array still outperforms the Yagi-type array, but with a smaller margin than at a height of 0.3 λ (Fig 12-11). To produce an optimum radiation pattern, the values of the loading impedances were different than those for a height of 0.3 λ. See text for details.

of 762 − 13.4 = 750 pF across the end of the open stub. This last solution seems to be the most flexible one. A parallel connection of two transmitting-type ceramic capacitors, 500 pF and 250 pF, will do the job perfectly. If you want even more flexibility you can use a 500-pF motor-driven variable in parallel with a 500-pF fixed capacitor. This will allow you to tune the array for best F/B.

The practical arrangement is shown in **Fig 12-12.** From each outer element we run a 1.71-meter long piece of 450-Ω line to a small box mounted on the boom. The box can also be mounted right at the center of the inverted-V element, whereby the 1.71-meter transmission line is shaped in a large 1-turn loop. The box houses a small relay, which either shorts the stub (reflector) or opens, leaving the 750-pF capacitor across the line.

Is the relative "inferiority" of the parasitic array due to the low height? In order to find out I modeled the same antennas at λ/2 height. **Fig 12-13** shows the vertical and the horizontal radiation patterns for the all-driven and parasitic-array versions of the 3-element inverted-V array at this height. Note that the all-driven array still has 0.9 dB better gain than the parasitic array. The F/B is still a little better as well, although the difference is less pronounced than at lower height. The optimum pattern was obtained when loading the director with a −50-Ω impedance and the reflector with a +30-Ω impedance. The gain of the all-fed array is 5.7 dB versus a dipole at the same height (at 28° elevation angle). For the 3-element parasitic array, the gain is 4.8 dB versus the dipole at its main elevation angle of 29°.

In looking at the vertical radiation pattern it is remarkable again that the all-driven array excels in F/B performance at high angles. Notice the "bulge" that is responsible for 5 to 10 dB less F/B in the 35°-50° wave-angle region.

It must be said that I did not try to further optimize the parasitic array by shifting the relative position of the elements. By doing this, further improvement could no doubt be made. This, of course, would make it impossible to switch directions, since the array would no longer be symmetrical.

6.1. Conclusion

All-fed arrays made of horizontal dipoles or inverted-V dipoles always outperform the parasitic-type equivalents in gain as well as F/B performance. As they are not fed in quadrature, it is elaborate or even "difficult" to feed them correctly.

The parasitic-type arrays lend themselves very well for remote tuning of the parasitic elements. Short stubs (open-ended to make a capacitor, or short-circuited to make an inductor) make good tuning systems for the parasitic elements. Switching from director to reflector can easily be done with a single-pole relay and a capacitor at the end of a short open-wire stub.

The same 3-element array made of fully horizontal (flat top) dipoles exhibits 1.0 dB more gain than the inverted-V version at the same apex height.

7. DELTA LOOPS IN PHASE (COLLINEAR)

Two delta loops can be erected in the same plane and fed with in-phase currents to provide gain and directivity. In order to obtain maximum gain, the loops must be separated about λ/8, as shown in **Fig 12-14**. In this case the two loops, fed in phase exhibit a gain of almost 3.5 dB over a single loop! The array has a front-to-side directivity of at least 15 dB, not negligible. The impedance on a single loop is between 125 and 160 Ω. Each element can be fed via a 75-Ω λ/2 feed line. At the point where they join the impedance will be 60 to 80 Ω. The radiation patterns and the configuration are shown in Fig 12-14.

This may be an interesting array if you have two towers with the right separation and pointing in the right direction. As with all vertically polarized delta loops, the ground quality is very important as to the efficiency and the low-angle radiation

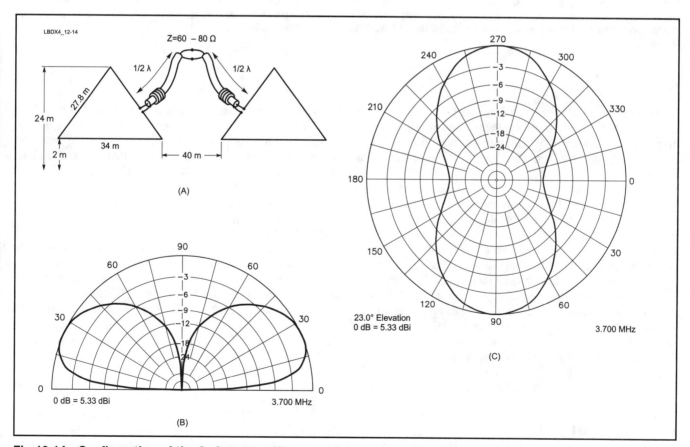

Fig 12-14—Configuration of the 2-element collinear delta-loop array with 10-meter spacing between the tips of the deltas. This array has a gain of 3.0 dB over a single delta loop. The loops are fed λ/4 from the apex on the sloping wire in the center of the array (see text for details). The pattern at C is for an elevation angle of 18°.

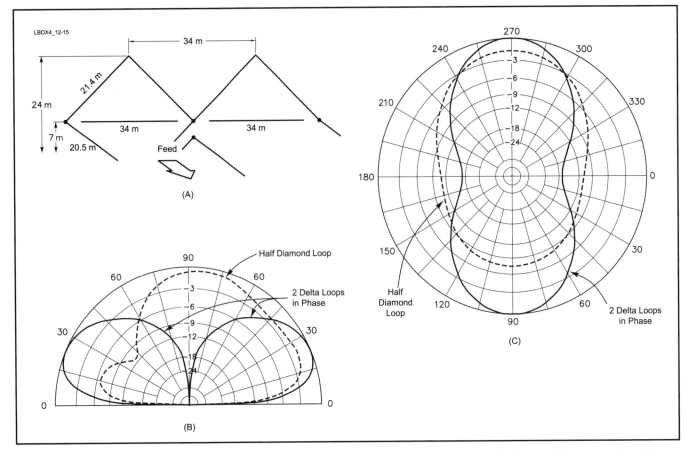

Fig 12-15—There is some similarity between the half-diamond loop, described by VE2CV and shown at A and two delta loops in phase. Overlays of the vertical (B) and horizontal (C) patterns show, however, that the 2-element delta loop has better high-angle discrimination, in addition to almost 1 dB more gain.

of the array (see Chapter 10 on large loops).

Putting the loops closer together results in a spectacular drop in gain. Loops with touching tips only exhibit approximately 1-dB gain over a single element—they're not worth the effort!

In one of his articles on elevated radials, John Belrose, VE2VC, mentioned the half-diamond loop, which has a significant resemblance to the delta loop (Ref 7824). I modeled this array and compared it to the 2-element delta loop shown in Fig 12-14. **Fig 12-15** shows both the horizontal and the vertical radiation pattern of both antennas in overlay. The 2-element delta has almost 0.7 dB more gain and has excellent high-angle rejection, while the half-diamond loop has some very strong high-angle response, which is of course due to the way the radials are laid out, resulting in zero high-angle cancellation. The extra gain that was thought to be achieved by laying radials in one direction, is apparently more than wasted in high angle radiation. It seems that the two in-phase delta loops are still, by far, the best choice.

8. ZL Special

The *ZL Special*, sometimes called the *HB9CV*, is a 2-element dipole array with the elements fed 135° out-of-phase. This configuration is described in Section 2. It is the equivalent of the vertical arrays described in Chapter 11.

These well-known configurations make use of a specific feeding method. The feed points of the two elements are connected via an open-wire feed line that is crossed. The crossing introduces a 180° phase shift. The length of the line, with a spacing of λ/8 between the elements, introduces an additional phase shift of approximately 45°. The net result is 180° + 45° = 225° phase shift, lagging. This is equivalent to 360° − 225° = 135° leading.

Different dimensions for this array have been printed in various publications. Correct dimensions for optimum performance will depend on the material used for the elements and the phasing lines. Jordan, WA6TKT, who designed the ZL Special entirely with 300-Ω twin lead (Ref 908), recommends that the director (driven element) be $447.3/f_{MHz}$ and the reflector be $475.7/f_{MHz}$, with an element spacing of approximately 0.12 λ.

Using air-spaced phasing line with a velocity factor of 0.97, the phasing-line length is 119.3°. This configuration of the ZL Special with practical dimensions for a design frequency of 3.8 MHz is given in **Fig 12-16**, along with radiation patterns. As it is rather unlikely that this antenna will be made rotatable on the low bands, I recommend the use of open-wire feeders to an antenna tuner. Alternatively, a coaxial feed line can be used via a balun.

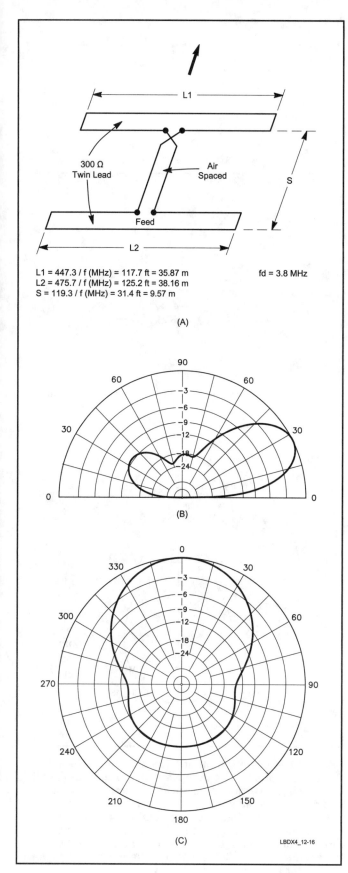

Fig 12-16—The ZL Special (or HB9CV) antenna is a popular design that gives good gain and F/B for close spacing. Radiation patterns were calculated with *ELNEC* for the dimensions shown at A, for a height of λ/2 above average ground. The horizontal pattern at C is for an elevation angle of 27°.

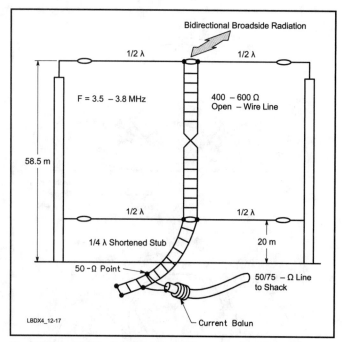

Fig 12-17—Typical Lazy-H configuration for 80 meters. The same array can obviously be made for 40 meters with all dimensions halved.

9. LAZY H

The Lazy-H antenna is an array that is often used by lowbanders that have a bunch of tall towers, where they can support Lazy-Hs between them. **Fig 12-17** shows a typical Lazy-H layout for use on 80 meters. Such a 4-element Lazy-H has a very respectable gain of about 11 dBi over average ground, as shown in **Fig 12-18**. Its gain at a 20° elevation angle is nearly 4 dB above a flat-top dipole at the same height, and 1.7 dB over a collinear (two λ/2 waves) at the same height. The outstanding feature of the Lazy-H is however, that the 90° (zenith) radiation, which is very dominant with the dipole and the collinear, is almost totally suppressed. This makes it a good DX-listening antenna as well!

The easiest way to feed the array is shown in Fig 12-18. A λ/4 open-wire line, shorted at its end, is probed to find the low-impedance point (50 or 75 Ω). Fine adjustment of the length of the line and the position of the tap make it possible to find a perfect resistive 50 or 75-Ω point. One of the popular antenna analyzers is a valuable tool to find the exact match. The same antenna can be used for both ends of the 80-meter band, all that is required is a different set of values for the length of the λ/4 stub and the position of the tap. This can be achieved with some rather simple relay switching.

10. BOBTAIL CURTAIN

The Bobtail Curtain consists of three phased λ/4 verticals, spaced λ/2 apart, where the center element is fed at the base, while the outer elements are fed via a horizontal wire section between the tips of the verticals. Through this feeding arrangement, the current magnitude in the outer verticals is half of the current in the center vertical. The current distribution in the top wire is such that all radiation from this horizontal section is effectively canceled. The configuration as well

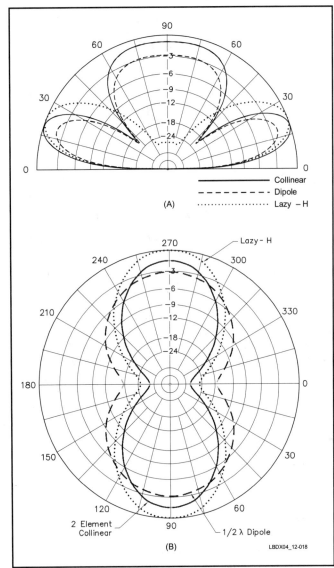

Fig 12-18—Vertical and horizontal radiation patterns of the 80-meter Lazy H shown in Fig 12-17 compared to the patterns of a flat-top dipole and a 2 × λ/2 collinear at the same height (over average ground).

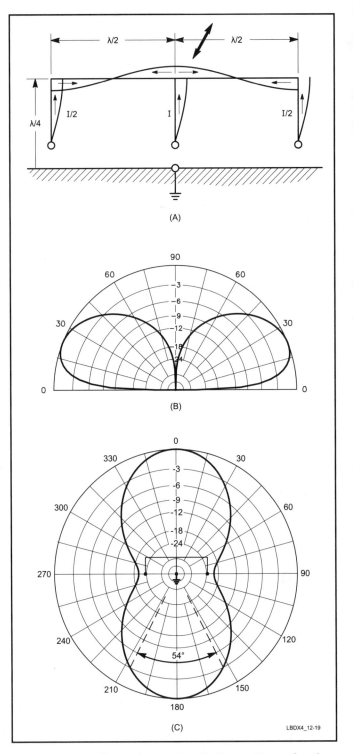

Fig 12-19—Configuration and radiation patterns for the Bobtail curtain. This antenna exhibits a gain of 4.4 dB over a single vertical element. The current distribution, shown at A, reveals how the three vertical elements contribute to the low-angle broadside bidirectional radiation of the array. The horizontal section acts as a phasing and feed line and has no influence on the broadside radiation of the array. The horizontal pattern at C is for an elevation angle of 22°.

as the radiation patterns are shown in **Fig 12-19**.

The gain of this array over a single vertical is 4.4 dB. The −3-dB forward-lobe beamwidth is only 54°, which is quite narrow. This is because the radiation is bidirectional. K. Svensson, SM4CAN, who published an interesting little booklet on the Bobtail Array, recommends the following formulas for calculating the lengths of the elements of the array.

Vertical radiators: $\ell = 68.63/F_{MHz}$
Horizontal wire: $\ell = 143.82/F_{MHz}$

where

F_{MHz} = design frequency, MHz
ℓ = length, meters

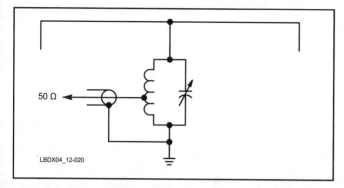

Fig 12-20—The Bobtail Curtain is fed at a high-impedance point with a parallel-tuned circuit, where the coax is tapped a few turns from the cold end of the coil. The array can be made to operate over a very large bandwidth by simply retuning the tuned circuit.

The antenna feed-point impedance is high (several thousand ohms). The array can be fed as shown in **Fig 12-20**. This is the same feed arrangement as for the voltage-fed T antenna, described in Chapter 9 on vertical antennas. In order to make the Bobtail antenna cover both the CW as well as the phone end of the band, it is sufficient to retune the parallel resonant circuit. This can be done by switching a little extra capacitor in parallel with the tuned circuit of the lower frequency, using a high-voltage relay.

The bottom ends of the three verticals are very hot with RF. You must take special precautions so that people and animals cannot touch the vertical conductors.

Do not be misled into thinking that the Bobtail Array does not require a good ground system just because it is a voltage-fed antenna. As with all vertically polarized antennas, it is the electrical quality of the reflecting ground that determines the efficiency and the low-angle radiation of the array.

11. HALF-SQUARE ANTENNA

The Half-Square antenna was first described by Vester, K3BC (Ref 1125). As its name implies, the Half-Square is half of a Bi-Square antenna (on its side), with the ground making up the other half of the antenna (see Chapter 10 on large loop antennas). It can also be seen as a Bobtail with part of the antenna missing.

Fig 12-21 shows the antenna configuration and the radiation patterns. The feed-point impedance is very high (several thousand ohms), and the antenna is fed like the Bobtail. The gain is somewhat less than 3.4 dB over a single λ/4 vertical. The forward-lobe beamwidth is 68°, and the pattern is essentially bidirectional. There is some asymmetry in the pattern, which is caused by the asymmetry of the design: The current flowing in the two verticals is not identical. As far as the required ground system is concerned, the same remarks apply as for the Bobtail antenna.

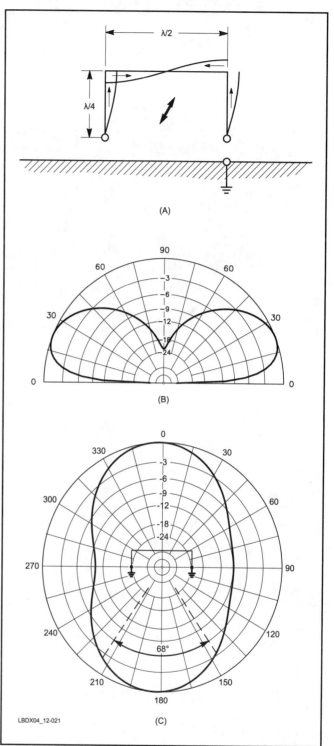

Fig 12-21—Configuration and radiation patterns of the Half-Square array, with a gain of 3.4 dB over a single vertical. The antenna pattern is somewhat asymmetrical because the currents in the vertical conductors are not identical. The azimuth pattern at C is for an elevation angle of 22°.

CHAPTER 13

Yagis and Quads

Fig 13-1—Tim Duffy, K3LR, a well-known contester and superstation builder from Western Pennsylvania.

Tim Duffy, K3LR, needs no introduction to readers of this book. The way Tim runs his Dayton Antenna Forum and his own super contest station tells a lot about the man. He is thorough, well organized, punctual, and a super host on top of it all! No wonder there's a long line of operators who want to operate from K3LR in the big contests!

Tim is Senior Vice President and Chief Technical Officer of Dobson Communications Corporation (the 9th largest Cellular Telephone company in the USA). He has been employed in the broadcast and wireless engineering discipline for over 28 years. Tim is a graduate of The Pennsylvania State University and has been a licensed amateur radio operator for over 32 years. He currently maintains his large 9-tower, 12-operating position multi-multi station in Western Pennsylvania and experiments with large high-gain contest antennas. Tim took the time to review this chapter on Yagis and Quads, for which I am very thankful.

On the higher HF bands, almost all dedicated DXers use some type of rotatable directional antenna. Directional antennas produce gain to be better heard. They also show directivity, which is a help when listening. Yagi and cubical-quad antennas are certainly the most popular antennas on those bands.

On the low bands, rotatable directive antennas are large. Forty-meter Yagis and quads—even full-size—exist in greater numbers these days. On 80 meters there are only a few full-size Yagis and quads, while reduced-size Yagis and quads are a little more common. They seem to come and go, and are rather difficult to keep in the air. On 160 meters, rotatable Yagis still belong to dreamland.

I have had the chance to operate a 3-element full-size quad, as well as a 3-element full-size Yagi, on 80 meters, and I must admit that it is only when you have played with such monsters that you appreciate what you are missing without them. The same is even more true on 40 meters, where full-size Yagis and quads appear in ever-growing numbers. Until the day I had my own full-size 40-meter Yagi, I always considered 40 as my worst band. Now that I have the full-size Yagi, I think it has become my "best" band.

Much of the work presented in this chapter is the result of a number of major antenna projects that were realized with the help of R. Vermet, ON6WU, who has been a most assiduous supporter and advocate in all my antenna work.

1. ARRAYS WITH PARASITIC ELEMENTS

In Chapter 11 on vertical arrays I discuss arrays of antennas, where each antenna element is fed with a dedicated feed line. During the analysis of these arrays I noticed that elements sometimes exhibit a negative impedance, which means that these elements do not draw power from the feed line but actually deliver power into the feed system.

In such a case mutual coupling has already supplied enough (or too much) current into the element. Negative feed-point impedances are typical with close-spaced arrays, where the coupling is heavy.

Parasitic arrays are arrays where (most often) only one element is fed, and where the other elements obtain their feed current only by mutual coupling with the various elements of the array. To obtain the desired radiation pattern and gain, feed-current magnitudes and phases need to be carefully adjusted. This is done by changing the relative positions of the elements and by changing the lengths of the elements. The exact length of the *driven element* will not influence the pattern nor the gain of the array; it will only influence the feed-point impedance.

Unlike with driven arrays, you cannot obtain just any specific feed-current magnitude and angle. In driven arrays you "force" the antenna currents, which means you add (or subtract) feed current to the element current already obtained through mutual coupling. You could, for example, make a driven array with three elements in-line where all elements have an identical feed current. You cannot make a parasitic array where the three elements have the same current phase and magnitude. Arrays with parasitic elements are limited in terms of the current distribution in the elements.

2. QUADS VERSUS YAGIS

It is not my intention to get into the debate of quads versus Yagis. But before I tackle both in more depth, let me clarify a few points and kill a few myths:

- For a given height above ground, the quad does *not* produce a markedly lower radiation angle than the Yagi. The vertical radiation angle of a horizontally polarized antenna in the first place depends on the height of the antenna above ground.
- There is a very slight difference (perhaps a few degrees, depending on actual height) in favor of the quad, as there is some more squeezing of the vertical plane due to the effect of the stacked two horizontal elements that make a horizontally polarized cubical quad (Ref 980).
- For a given boom length, a quad will produce slightly more gain than a Yagi. This is logical since the aperture (capture area) is larger. The principle is simple: Everything being optimized, the antenna with the largest capture area has the highest gain, or can show the highest directivity.
- Yagis as a rule are easier to build and maintain. A Yagi is two-dimensional, and the problems involved with low-band antennas are simplified by an order of magnitude. Problems of wire breaking are nonexistent with Yagis. Large Yagis are also easier to handle and to install on a tower than large quads.
- There are other factors that will determine the eventual choice between a Yagi or a quad, such as material availability, maximum turning radius (the quad takes less rotating space) and, of course, personal preference.

3. YAGIS

There have been a number of good publications on Yagi antennas. Until about 20 years ago, before we all knew about the effect of tapered elements, the W6SAI/W2LX *Beam Antenna Handbook* was in many circles considered the Yagi "bible." I built my first Yagi based on information from this work.

Dr Jim Lawson, W2PV (SK), wrote a very good series on Yagis back in the early 1980s. Later the ARRL published his work in the excellent book, *Yagi Antenna Design* (Ref 957). Lawson explained how he scientifically designed a winning contest station, based on high-level engineering work.

Lawson was the first in amateur circles to methodically study the effect of tapered elements. He came up with a tapering algorithm that is still widely called the *W2PV algorithm*. It calculates the correct electrical length of an element as a function of the length and diameters of individual tapered sections.

3.1. Modeling Yagi Antennas

We now have very sophisticated modeling software available for Yagis, most of them based on the method of moments. See Chapter 4 to see what's available. Here are some things you should keep in mind:

- Make sure you know exactly what you want before you start: maximum boom length, maximum gain, maximum directivity, large SWR bandwidth, etc.
- Always model the antenna first in free space.
- Always model the antenna on a range of frequencies (eg, 7.0 to 7.3 MHz), so you can assess the SWR, gain and F/B of the design over the whole band.
- Make sure the feed-point impedance is reasonable (it can be anything between 18 Ω and 30 Ω).
- When the array is optimized and meets your requirements in free space, you should repeat the exercise over real ground at the actual antenna height, usually using a *NEC-2*-derived program such as *EZNEC*.
- If the antenna is stacked with other antennas, include the other antennas in the model as well. This is especially so when considering stacking Yagis for the same band. F/B may be totally ruined due to stacking. Stacks need to be optimized as stacks!
- If you consider making a Yagi with loaded elements, first model the full-size equivalent. When applying the loading devices, don't forget to include the resistance losses and possible parasitic capacitances or inductances.
- If you are about to model your own Yagi using loading devices, such as linear-loading stubs or capacity-loading wires, you should be very careful. The best approach is to first model the antenna using all wires of the same diameter. This should prove the feasibility of the concept. Next, you should determine the resonant frequencies of the individual elements, by removing other elements from the model. These resonant frequencies are excellent guides for the actual tune-up of the antenna.

3.2. Mechanical Design

Making a perfect electrical design of a low-band Yagi is a piece of cake nowadays with all the magnificent modeling software available. The real challenge comes when you have to turn your model into a mechanical design! When building a mechanically sound 40-meter Yagi, there is no room for guesswork. Don't ever take anything for granted when you are building a very large antenna. If you want your beam to survive the winds and ice loading you expect, you *must* go through a fair amount of calculations. The same holds true for an 80-meter Yagi, of course, but with the magnitude squared!

Physical Design of Yagi Antennas, by D. Leeson, W6QHS, (Ref 964) covers all aspects of mechanical Yagi design. Leeson uses the "variable area" principle to assess the influence of wind on the Yagi. The book unfortunately does not give any design examples of practical full-size 40 or for 80-meter Yagis. The only low-band antenna covered is the Cushcraft 40-2CD, a shortened 2-element 40-meter Yagi. Leeson's modification to strengthen the Cushcraft 40-2CD has become a classic, and is a must for everyone who has this antenna and who does not want to see it ripped to piecesin a storm.

Over the years standards dealing with mechanical issues for towers and antennas have evolved. The well-known EIA RS-222 standard has evolved from 222-C through suffix D and eventually to the RS-222-E standard. While the earlier versions of Leeson's software that he supplied with his book were based on C, the latest versions are now based on E. The E-version (and also ASCE 74) treats wind statistics and force on cylindrical elements more realistically than C and D, and the difference shows up in the question of forces on cylinders at an angle to the wind. This affects boom strength and rotating torque. The article by K5IU (Ref 958) uses the E approach, as well as the ON4UN LOW BAND SOFTWARE modules dealing with boom strength and torque balancing.

Curt Andress, NI6W (now K7NV) wrote an interesting software package that addresses all of the mechanical issues concerning antenna strength. *YS* (Yagi Stress) is easy to use, has lots of data about materials and tubing in easy-to-access form. For information contact **K7NV@contesting.com**. A free trial download can be obtained from WXØB's website at **www.arraysolutions.com/Products/yagistress.htm**.

All of these tools deal with static wind-load models. The question, of course, is how reliable all these models are in a complex aerodynamic situation. As Leeson puts it, ". . . but we're not dealing with mathematical models when the wind is roaring through here at 134 mi/h. Either model (C or E) results in booms that break upward in the wind if you ignore vertical gusting..." In particular locations, such as hilltop QTHs, there may be vertical updraft winds that can break a boom unless three-way boom guys are used. But these are rather extreme conditions, not the run-of-the mill situations.

The real proof of the pudding is in the building of big antennas, and even more so keeping them up year after year. The mathematics involved in calculating all the structural aspects of a low-band Yagi element are rather complex. It is a subject that is ideally suited for computer assistance. Together with my friend R. Vermet, ON6WU, I have written a comprehensive computer program, YAGI DESIGN, which was released in early 1988 and updated a few times since.

In addition to the traditional electrical aspects, YAGI DESIGN tackles the mechanical-design aspects. This is especially of interest to the prospective builder of 40 and 80-meter Yagi antennas. While Yagis for the higher HF bands can be built "by feel," 40 and 80-meter Yagis require much closer attention if you want these antennas to stay up.

The different modules of the YAGI DESIGN software are reviewed in Chapter 4 on low-band software. This book is not a textbook on mechanical engineering, but a few definitions are needed in order to better understand some of the formulas I use in this chapter.

3.2.1. Terms and definitions

Stress: Stress is the force applied to a material per unit of cross-sectional area. Bending stress is the stress applied to a structure by a bending moment. Shearing stress is the stress applied to a structure by a shearing moment. The stress is expressed in units of force divided by units of area (usually expressed in kg/mm^2 or lb/inch^2).

Breaking Stress: The breaking stress is the stress at which the material breaks.

Yield Stress: Yield stress is the stress where a material suddenly becomes plastic (non-reversible deformation). The yield stress to breaking stress ratio differs from material to material. For aluminum the yield stress is usually close to the breaking stress. For most steel materials the yield stress is approximately 70% of the breaking stress. Never confuse breaking stress with yield stress, unless you want something to happen that you will never forget.

Elastic Deformation: Elastic deformation of a material is deformation that will revert to the original shape after removal of the external force causing the deformation.

Compression or Elongation Strain: Compression strain is the percentage change of dimension under the influence of a force applied to it. Being a ratio, strain is an abstract figure.

Shear Strain: Shear strain is the deformation of a material divided by the couple arm. It is a ratio and thus an abstract figure.

Shear Angle: This is the material displacement divided by the couple arm. As the angles involved are small, the ratio is a direct expression of the shear angle expressed in radians. To obtain degrees, multiply by $180/\pi$.

Elasticity Modulus: Elasticity modulus is the ratio stress/strain as applied to compression or elongation strain. This is a constant for every material. It determines how much a material will deform under a certain load. The elasticity modulus is the material constant that plays a role in determining the sag of a Yagi element. The elasticity modulus is expressed in units of force divided by the square of units of dimension (unit of area).

Rigidity Modulus: Rigidity modulus is the ratio shear-stress/strain as applied to shear strain. The rigidity modulus is the material constant that will determine how much a shaft (or tube) will twist under the influence of a torque moment (eg, the drive shaft between the antenna mast and the rotator). The rigidity modulus is expressed in units of force divided by units of area.

Bending Section Modulus: Each material structure (tube, shaft, plate T-profile, I-profile, etc) will resist a bending moment differently. The section modulus is determined by the shape as well as the cross-section of the structure. The section modulus determines how well a particular shape will resist a bending moment. The section modulus is proper to a shape and not to a material. The bending section modulus for a tube is given by:

$$S = \pi \times \frac{OD^4 - ID^4}{32 \times OD} \qquad \text{(Eq 13-1)}$$

where

OD = outer diameter of tube
ID = inner diameter of tube

The bending section modulus is expressed in units of length to the third power.

Shear Section Modulus: Different shapes will also respond differently to shear stresses. The shear stress modulus determines how well a given shape will stand stress deformation. For a hollow tube the shear section modulus is given by:

$$S = \pi \times \frac{OD^4 - ID^4}{16 \times OD} \qquad \text{(Eq 13-2)}$$

where

OD = outer diameter of tube
ID = inner diameter of tube

The bending section modulus is expressed in units of length to the third power.

3.3. Computer-Designed 3-Element 40-Meter Yagi at ON4UN

Let us go through the design of a very strong 3-element full-size 40-meter Yagi. This is not meant to be a step-by-step description of a building project, but I will try to cover all the critical aspects of designing a sound and lasting 40-meter Yagi. The Yagi described also happens to be the Yagi I have been using successfully over the past several years on 40 meters (it has brought several new European records in major contests on 40 meters). The design criteria for the Yagi are:

- Low Q, good bandwidth, F/B optimized.
- Survival at wind speeds up to 140 km/h with the elements broadside to the wind.
- Maximum ice load 10 mm at 60 km/h wind.
- Lifetime greater than 20 years.
- Boom length 10.7 meters maximum (only because I happened to have this boom)

3.3.1. Selecting an electrical design

Design number 10 from the database of the YAGI DESIGN software program meets all the above specifications. **Fig 13-2** shows a copy of the *TLW* main screen for my 40-meter Yagi. While I could have selected another design with up to 0.5 dB more gain, I selected this design because of its excellent F/B pattern and wide bandwidth for SWR, gain and F/B.

I mounted this Yagi 5 meters above my 20-meter Yagi (design number 68 from the database), 30 meters above ground. The combination of both antennas was modeled once more over real ground at the final height using a *MININEC*-based modeling program, to see if there would be an important change in pattern and gain due to the presence of the second antenna. The performance figures (gain, F/B) and directivity pattern of the 40-meter Yagi changed very little at the 5-meter stacking distance.

3.3.2. Principles of Mechanical Load and Strength Calculations for Yagi Antennas

R. Weber, K5IU, brought to our attention (Ref 958) that the variable-area method, commonly employed by most Yagi manufacturers, and used by many authors in their publications as well as software, has *no* basis in science, nor is there any experimental evidence for the method.

The variable-area method assumes that the direction of the force created by the wind on an element is always in line with the wind direction, and that the magnitude is proportional to the area of the element as projected onto a plane perpendicular to the wind direction (proportional to the sine of the wind angle).

The scientifically correct method of analyzing the wind-force behavior, called the "cross-flow" principle, says that the direction of the force due to the wind is *always* perpendicular to the plane in which the element is situated and that its magnitude is proportional to the square of the sine of the wind angle.

Fig 13-3 shows both principles. It is easy to see that the cross-flow principle is the correct one. The experiment described by K5IU can be carried out by anyone, and should convince anyone who has doubts: "Take a 1-meter long piece of aluminum tubing (approximately 25 mm in diameter) for a car ride. One person drives, while another sits in the passenger seat. The passenger holds the tube in his hand and puts his arm out the window positioning the tube vertically. The tube is now perpendicular to the wind stream (wind angle = zero). It is easy to observe a force (drag force) that is *in-line* with the

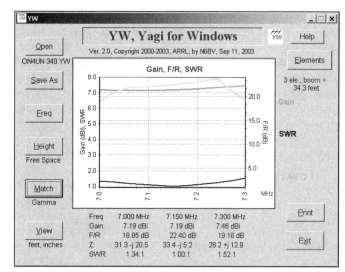

Fig 13-2—Free-space performance data for full-sized 3-element Yagi design number 10 from the YAGI DESIGN software suite. This was created by the YW (Yagi for Windows) program.

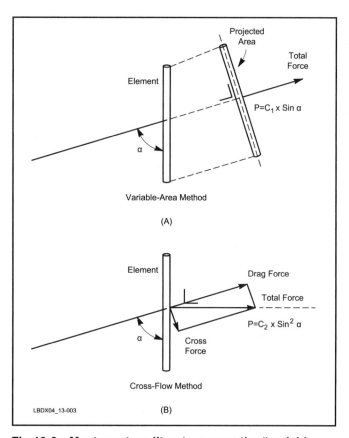

Fig 13-3—Most amateur literature uses the "variable area" method shown at A for calculating the effect of wind on an element. The principle says that the direction of the force created by the wind on an element is *always* in-line with the direction of the wind, which is clearly incorrect. If this were correct, no plane would ever fly! The "cross-flow" principle, illustrated at B, states that the direction of the force is *always* perpendicular to the element, and is the resultant of two components, the drag force and the cross force (which is the lifting force in the case of an airplane wing). See text for details.

wind (and at the same time perpendicular to the axis of the tube). The passenger now rotates the tube approximately 45°, top end forward. The person holding the tube will now clearly feel a force that pushes the tube *backward* (drag force), but at the same time tries to *lift* (cross force) the tube. The resulting force of these two components (the drag and cross force) is a force that is *always* perpendicular to the direction of the tube. If the tube is inclined with the bottom end forward, the force will try to push the tube downward."

This means that the direction of the force developed by the wind on an object exposed to the wind is not necessarily the same as the wind direction. There are some specific conditions where the two directions are the same, such as the case where a flat object is broadside to the wind direction. If you put a plate (1 meter^2) on top of a tower, and have the wind hit the plate at a 45° angle, it will be clear that the push developed by the wind hitting the plate will not be developed in the direction of the wind, but in the direction perpendicular to the plane of the flat plate. If you have any feeling for mechanics and physics, this should be fairly evident.

To remove any doubt from your mind, D. Weber states that Alexandre Eiffel, builder of the Paris Eiffel tower, used the cross-flow principle for calculating his tower. And it still stands there after more than 100 years.

Now comes a surprise: Take a Yagi, with the wind hitting the elements at a given wind angle (forget about the boom at this time). The direction of the force caused by the wind hitting the element at whatever wind angle, will always be perpendicular to the element. This means that the force will be in-line with the boom. The force will not create any bending moment in the boom; it will merely be a compression or elongation force in the boom. All of this, of course, provided the element is fully symmetrical with respect to the boom.

This force in the boom should not be of any concern, as the boom will certainly be strong enough to cope with the bending moments caused by wind broadside to the boom. These bending moments in the boom at the mast attachment plate are caused only by the force created by the wind on the boom only (by the same "cross-flow" principle) or any other components that have an exposed wind area in-line with the boom.

If the mast-to-boom plate is located in the center of the boom, the wind areas on both sides of the mast are identical, and the bending moments in the boom on both sides of the mast (at the boom-to-mast plate) will be identical. This means there is no *mast torque*. If the areas are unequal, mast torque will result. This mast torque puts extra strain on the rotator, and should be avoided. Torque balancing can be done by adding a *boom dummy*, which is a small plate placed near the end of the shorter boom half, and which serves to reestablish the balance in bending moments between the left and the right side of the boom.

This may seem strange since intuitively you may have difficulty accepting that the extreme case of a Yagi having one element sitting on one end of a boom would not create any rotating torque in the mast, whatever the wind direction is. Surprisingly enough, this is the case. You cannot compare this situation with a weathervane, where the boom area at both sides of the rotating mast is vastly different. It is the vast difference in boom area that makes the weathervane turn into the wind.

Fig 13-4 shows the situation in theory, and what's likely to happen in the real world. At A and B the wind only sees the element (the boom is not visible), and if the element is fully symmetrical with respect to the boom, there will be no torque moment at the element-to-boom interface. Hence this is a fully stable situation. At C the situation where the boom is facing the wind is shown. The element is invisible now and as the boom is supposed to be wind-load balanced, the boom by itself creates no torque at the boom-mast interface. At D we see that the cross-flow principle only creates a force in-line with the boom. This means that this example still guarantees a well-balanced situation, and the structure will not rotate in the wind.

But let's be practical. The wind blowing on the long flexible elements of a Yagi will make the elements bend slightly, as shown in Fig 13-4E. In this case now it is clear that the pressure induced by the wind on side (a) of the element will be much greater than on side (b) as side (a) now faces the wind much more than side (b). In this case the antennas will tend to rotate in the sense indicated by the arrow.

Taking all of this into account it seems to be a good idea not only to try to achieve full boom (area) symmetry but full element (area) symmetry as well. Leeson came to the conclusion that he prefers to balance in the element plane by offset-

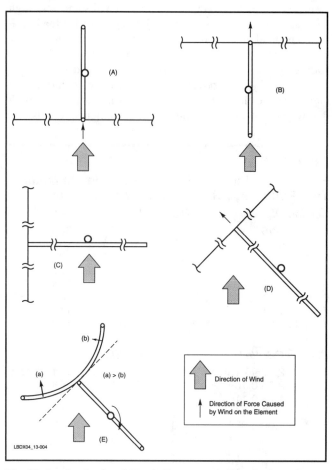

Fig 13-4—Analysis of the influence of the wind on the mast torque for a single element sitting on the end of a boom. In all cases, A through D, no mast torque is induced. Only in case E, where the element is deformed by the wind, will mast torque be induced. See text for details.

ting the element ensemble to eliminate the need for a torque balancing element, then using a vane (boom torque compensating plate) on the now unbalanced boom. If offsetting the element ensemble creates an important weight imbalance, this can always be compensated for by inserting some form of weight in the boom near one tip.

Not adding extra dummy elements seems to be a good idea, as in dynamic situations (wind turbulence) these may actually deteriorate the situation rather than improve it. Since in principle the Yagi elements do *not* contribute to the boom moments, and therefore not to the mast torque, it makes no sense to create dummy elements to try to achieve a torque-balanced Yagi.

The MECHANICAL YAGI BALANCE module of the YAGI DESIGN software addresses all the issues as explained above and uses the cross-flow principle. It uses latest data from the latest EIA/TIA-222-E specification, which is somewhat different from the older EIA standard RS-222-C.

3.3.3. Element strength calculation

While it is standard procedure to correct boom sag using truss cables, element sag must be controlled to a maximum degree by using the properly designed tapered sections for making the element. Guyed elements are normally only used with 80-meter Yagis. Unguyed 40-meter full-size tubular elements (24 meters long) can be built to withstand very high wind speeds, as well as a substantial degree of ice loading.

The mathematics involved are quite tedious, and a very good subject for a computer program. Leeson (Ref 964) addresses the issue in detail in his book, and he made a spreadsheet type of program available for calculating elements. As the element-strength analysis is always done with the wind blowing broadside to the elements, the issues of variable area or cross-flow principle don't have to be taken into consideration.

The ELEMENT STRENGTH module of the YAGI DESIGN software is a dedicated software program that allows the user to calculate the structural behavior of Yagi elements with up to nine tapering elements. This module operates in the English measurement system as well as in the metric system (as do all other modules of the integrated YAGI DESIGN software). A drag factor of 1.2 is used for the element calculations (as opposed to 0.66 in the older RS-222-C standard).

Interactive designing of elements enables the user to achieve element sections that are equally loaded. Many published element designs show one section loaded to the limit, while other sections still exhibit a large safety margin. Such unbalanced designs are always inefficient with respect to weight, wind area and load, as well as cost.

Each change (number of sections, section length, section diameter, wind speed, aluminum quality, ice load, etc) is immediately reflected in a change of the moment value at the interface of each taper section, as well as at the center of the element. When a safe limit is exceeded, the unsafe value will blink. The screen also shows the weight of the element, the wind area, and the wind load for the specified wind speed.

It is obvious that the design in the first place will be dictated by the material available. Material quality, availability and economical lengths are discussed in Section 3.3.6 where **Table 13-2** shows a range of aluminum tubing material commonly available in Europe.

A 40-meter Yagi reflector is approximately 23-meters long. This is twice the length of a 20-meter element. Designing a good 40-meter element can be done starting from a sound 20-meter element, which is then lengthened by more tapered sections toward the boom, calculating the bending stresses at each section drop.

When designing a Yagi element you must make sure that the actual bending moments (LM_t) at all the critical points match the maximum allowable bending moments (RM) as closely as possible. LM_v is the bending moment in the vertical plane, created by the weight of the element. This is the moment that creates the sag of the element. LM_t is the sum of LM_v and the moment created by the wind in the horizontal plane. Adding those together may seem to create some safety, although it can be argued that turbulent wind may in actual fact blow vertically in a downward direction.

The reflector element for my 40-meter Yagi uses material with metric dimensions available in Europe. The design was done for a maximum average wind speed of 140 km/h, using F22 quality (Al Mg Si 0.5%) material. This material has a yield strength of 22 kg/mm^2 (31,225 lb/inch^2). For material specifications see Section 3.3.6.

All calculations are done for a static condition. Dynamic wind conditions can be significantly different, however. The highest bending moment is at the center of the element. Inserting a 2-meter long steel tube (5 or 7-mm wall) in the center of the center element will not only provide additional strength but also further reduce the sag.

Whether 140 km/h will be sufficient in your particular case depends on the following factors:

- The rating of the wind zone where the antenna is to be used. The latest EIA/TIA-222-E standard lists the recommended wind speed by county in the US.

Table 13-1

Element Design Data for the 3-Element 40-Meter Yagi Reflector, Driven Element and Director

Section	OD/Wall	Dir.	Dr. Ele.	Refl.
1	60/5	300	300	300
2	50/5	285	285	285
3	35/2	60	85	84
4	30/2	60	112	100
5	25/1.5	135	135	176
6	15/1	60	80	82
7	12/1	111	80	113
Total length (cm)		1011	1077	1150

Section	OD/Wall	Dir.	Dr. Ele.	Refl.
1	2.375/0.154	144	144	144
2	2.00/0.109	55	66	66
3	1.25/0.11	34	42	50
4	1.00/0.11	30	30	30
5	1.00/0.058	30	38	42
6	0.625/0.11	18	15	21
7	0.625/0.058	28	30	34
8	0.50/0.058	60	63	65
Total length (inches)		399	428	452

Note: This design assumes a boom diameter of 75 mm (3 inches) and U-type clamps to mount the element to the boom (L = 300 mm, W = 150 mm, H = 70 mm). Availability of materials will be the first restriction when designing a Yagi antenna.

- Whether modifiers or safety factors are recommended (see EIA/TIA-222-E standard).
- Whether you will expose the element to the wind or put the boom into the wind (see Section 3.3.4).
- Whether you have your Yagi on a crank-up tower, so that you can nest it at protected heights during high wind storms.

Fig 13-5 shows the 3-element full-size 40-meter Yagi placed 5 meters above my 5-element 20-meter Yagi, which has a similar taper design. Note the very limited sag on the elements. The telescopic fits are discussed in Section 3.3.7. **Figs 13-6** and **13-7** show the section layout of the 40-meter reflector element, calculated for both metric and US (inch) materials.

3.3.3.1. Element sag

Although element sag is not a primary design parameter, I included the mathematics to calculate element sag in the ELEMENT STRENGTH module of the YAGI DESIGN software. While designing, it is interesting to watch the total element sag. Minimal element sag is an excellent indicator of a good mechanical design. Too much sag means there is somewhere along the element too much weight that does not contribute to the strength of the element. The sag of each of the sections of an element depends on:

- The section's own weight.
- The moment created by the section(s) beyond the section being investigated (toward the tip).
- The length of the section.
- The diameter of the section.
- The wall thickness of the section.
- The elasticity modulus of the material used.

The total sag of the element is the sum of the sag of each section. The elasticity modulus is a measure of how much a material can be bent or stretched without inducing permanent deformation. The elasticity modulus for all aluminum alloys is 700,000 kg/cm^2 (9,935,000 lb/inch^2). This means that an element with a stronger alloy will exhibit the same sag as an element made with an alloy of lesser strength.

The 40-meter reflector designed above has a calculated

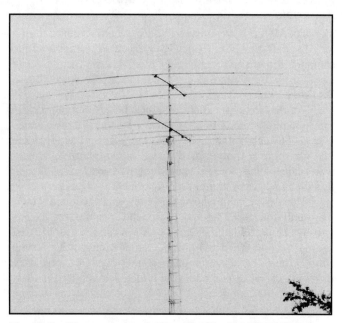

Fig 13-5—Three-element 40-meter Yagi at ON4UN. The Yagi is mounted 5 meters above a 5-element 20-meter Yagi with a 15-meter boom, at a height of 30 meters. Note the very limited degree of element sag, which is proof of a good physical design.

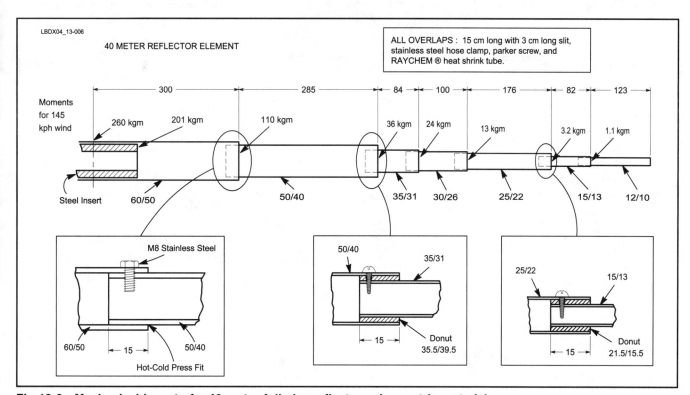

Fig 13-6—Mechanical layout of a 40-meter full-size reflector using metric materials.

sag of 129.5 cm, not taking into account the influence of the steel insert (coupler). The steel coupler reduces the sag to approximately 91 cm. These are impressive figures for a 40-meter Yagi. With everything scaled down properly, the sag is comparable to that of most commercial 20-meter Yagis. After mounting the element, the total element sag was that calculated by the software.

3.3.3.2. Alternative element designs using US materials

The US design is made by starting from standard tubing lengths of 144 inches. **Tables 13-3** and **13-4** list some of the standard dimensions commonly available in the US. The availability of aluminum tubes and pipes is discussed in Section 3.3.6.

For the two larger-diameter tubes, I used aluminum pipe. The remaining sections are from the standard tubing series with 0.058-inch wall thickness. From the design table we see that for some sections I used a wall thickness of 0.11 inch, which means that we are using a tight-fit section of 1/8-inch less diameter as an internal reinforcement.

The design table shows that the center sections would marginally fail at a 90-mi/h design wind speed. In reality this will not be a problem, since this design requires an internal coupler to join the two 144-inch center sections. This steel coupler must be strong enough to take the entire bending moment. The section modulus of a tube is given by Eq 13-1:

$$S = \pi \times \frac{OD^4 - ID^4}{32 \times OD}$$

where

S = section modulus
OD = tube outer diameter
ID = tube inner diameter

The maximum moment a tube can take is given by:

$$M_{max} = YS \times S \qquad (Eq\ 13\text{-}3)$$

where

YS = yield strength of the material
S = section modulus as calculated above

or

$$M_{max} = YS \times \pi \times \frac{OD^4 - ID^4}{32 \times OD} \qquad (Eq\ 13\text{-}4)$$

The yield strength varies to a very large degree (Ref 964 p 7-3). For different steel alloys it can vary from 21 kg/mm² (29,800 lb/inch²) to 50 kg/mm² (71,000 lb/inch²).

A 2-inch OD steel insert (with aluminum shimming material) made of high-tensile steel with a YS = 55,000 lb/inch² would require a wall thickness of 0.15 inches to cope with the maximum moment of 19.622 inch-lb at the center of the 40-meter reflector element.

Note that the element sag (42.1 inches with a 2×40-inch-long steel coupler) is very similar to the sag obtained in the previous metric design example. It is obvious that for an optimized Yagi element (and for a given survival wind speed), the element sag will always be the same, whatever the exact

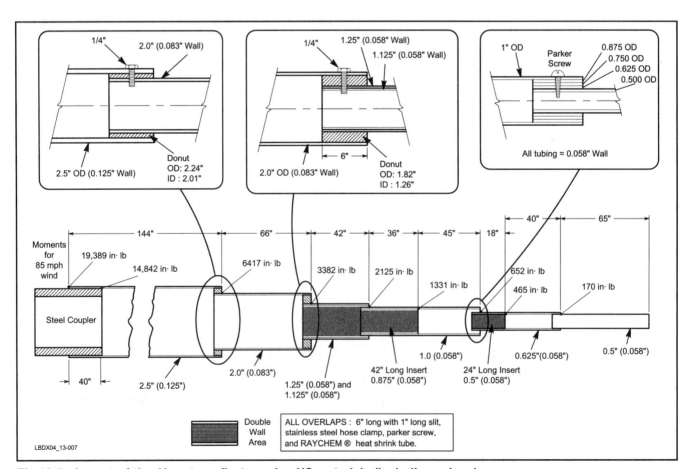

Fig 13-7—Layout of the 40-meter reflector using US materials (inch dimensions).

taper scheme may be. In other words, a good 40-meter Yagi reflector element, designed to withstand a 140 km/h (87 mi/h) wind should not exhibit a sag of more than 40 inches (100 cm) when constructed totally of tubular elements. More sag than that proves it is a poor design.

3.3.3.3. The driven element and the director

Once we have designed the longest element, we can easily design the shorter ones. We should consider taking the "left over" lengths from the reflector for use in the director. The lengths of the different sections for the 3-element Yagi number 10 from the YAGI DESIGN database, according to the metric and US systems, are shown in **Table 13-1**. Typically, if the reflector is good for 144 km/h, the director and the driven element will withstand 160 to 170 km/h.

3.3.3.4. Final element tweaking

Once the mechanical design of the element has been finalized, the exact length of the element tips will have to be calculated using the ELEMENT TAPER module of the software. You can also use a modeling programs such as *EZNEC* or *YW* and enter all the tapered sections directly.

3.3.4. Boom design

Now that we have a sound element for the 40-meter Yagi, we must pay attention to the boom. When the wind blows at a right angle to the boom, the maximum pressure is developed on the boom area. At the same time, the loading on the Yagi elements will be minimum. There is no intermediate angle at which the loading on the boom is higher than at a 90° wind angle, when the wind blows broadside onto the boom.

3.3.4.1. Pointing the Yagi in the wind

We all have heard the question, "Should I point the elements into the wind, or should I point the boom into the wind?" The answer is simple. If the area of the boom is smaller than the area of all the elements, then put the boom perpendicular to the wind. And vice versa.

Let me illustrate this with some figures for a 40-meter Yagi. Calculations are done for a 140 km/h wind, with the boom-to-mast plate in the center of the boom. The figures below were calculated in the MECHANICAL YAGI BALANCE module of the YAGI DESIGN software.

Zero-degree wind angle (wind blowing broadside to the elements):
- Boom moment in the horizontal plane: Zero
- Thrust on tower/mast 323 kg
- Maximum bending moment in the elements

90° wind angle (wind blowing broadside to the boom):
- Boom moment 114 kg-m
- Thrust on tower/mast: 87 kg
- Minimum bending moment in the elements

In this case it is obvious that we should at all times try to put the boom perpendicular to the wind during a storm with high winds. For calculating and designing the rotating mast and tower, I recommend, however, that you take into account the worst-case wind pressure of 323 kg.

What about a long-boom HF Yagi? For a 6-element 10-meter Yagi, putting the elements perpendicular to the wind would be the logical choice. But relying on the exact direction of the Yagi as a function of wind direction is a dangerous practice and I don't want to encourage this. This does not mean that in case of high winds you couldn't take advantage of the best wind angle to relieve load on the Yagi, mast or tower, but what is gained by doing so should only be considered as extra safety margin only!

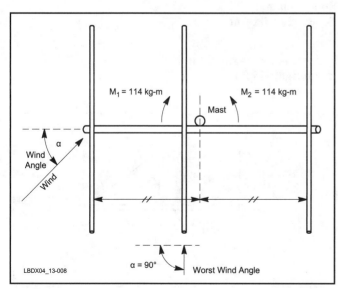

Fig 13-8—Boom moments in the horizontal plane as a result of the wind blowing onto the boom and the elements. The forces produced by the wind on the Yagi *elements* do not contribute to the boom moment; they only create a compression force in the boom (see text). The highest boom moments occur when the wind blows at a 90° angle, broadside to the boom.

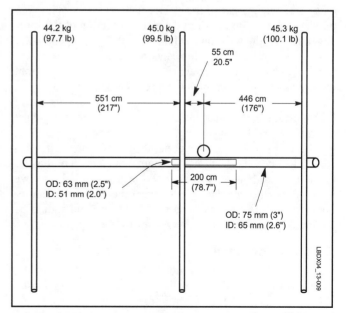

Fig 13-9—Weight-balanced layout of the 3-element 40-meter Yagi, showing the internal boom coupler. The net weight, without a match box (containing the Gamma or Omega matching capacitors) and without the boom-to-mast plate is 183 kg.

3.3.4.2. Weight balancing

In **Fig 13-8**, I assumed that the mast is at the physical center of the boom. As the driven element is offset toward the reflector, the Yagi will not be weight balanced. A good physical design must result in a perfect weight balance, since it is extremely difficult to handle an unbalanced 40-meter monster on a tower when trying to mount it to the rotating mast. The obvious solution is to shift the mast attachment point in such a way that a perfect balance is achieved.

The MECHANICAL YAGI BALANCE module of my software calculates weight-balancing for a Yagi. It automatically calculates the area of the required boom dummy plate (see Section 3.3.4), to reestablish torque balance. Components taken into account for calculating the weight balance are:

- The Yagi elements
- The boom
- The boom coupler (if any)
- The boom dummy (see Section 3.3.4.3 below)
- The match box (box containing gamma/omega matching components).

Fig 13-9 shows the layout that produces perfect weight balance. In our example I have assumed no match box. Slightly offsetting the driven element of the 3-element Yagi avoids the conflict between the location for the mast and for the driven element attach point.

3.3.4.3. Yagi torque balancing

The cause of mast torque has been explained in Section 3.3.2. If the bending moment in the boom on one side of the mast is not the same as the bending moment at the other side of the mast, we have a net *mast torque*. One moment is trying to rotate the mast clockwise, while the other tries to rotate the mast counterclockwise.

Only when the boom areas on both sides of the mast are identical will the Yagi be perfectly torque-balanced. The wind area of the elements and their placement on the boom do *not* play any role in the mast torque, as the direction of the force developed by the wind on an element is always perpendicular to the element itself, which means in-line with the boom. As such, element wind area cannot create a boom moment, but merely loads the boom with compression or elongation.

It is the mast torque that makes an antenna *windmill* in high winds. A good mechanical design must be torque-free at all wind angles. During our weight-balancing exercise earlier, we shifted the mast attachment point somewhat to reestablish weight balance. This causes the boom moments on both sides of the mast to become different. To reestablish balance, we mount a small *boom dummy plate* near the end of the shorter boom half. This plate has an area of 133 cm^2 and should be mounted 50 cm from the reflector for torque-balance.

3.3.4.4. Boom moments

I calculated the boom moments after torque-balancing and found that the boom bending moments have increased slightly, from 114 kg-m for the "non-weight-balanced Yagi" to 120 kg-m after weight balancing and adding the boom dummy. This is a negligible price to pay for having a weight-balanced Yagi.

The software calculate everything related to the boom design. The material stresses are computed for the coupler, as well as for the boom. The boom stress is only meaningful if the boom is not split in the center. With a split boom it is the coupler that takes the entire stress.

Even for a 140-km/h wind, the stresses in the boom are low. But as we will likely point the boom into the wind in windstorms (Section 3.3.4.1), we should build in a lot of safety. Also, as mentioned before, the 140-km/h does not include any safety factors or modifiers, as may be prescribed in the standard EIA/TIA-222.

To me, it is proof of *poor* engineering to design a boom that needs support guys to make it strong enough to withstand the forces from the wind and the bending moments caused by it. If guy wires are employed to provide the required strength, guying will have to be done in both the horizontal as well as the vertical plane. Guy wires can be used to eliminate boom sag. This will only be done for cosmetic rather than strength reasons.

Three-way guying may be necessary where vertical gusts can be expected (hilltop QTHs) to prevent the boom from dancing up and down due to vertical up drafts.

3.3.4.5. Boom sag

The boom as now designed will withstand 140-km/h winds, with a good safety factor. The same boom, however, without any wind loading will have to endure a fair bending moment in the vertical plane, caused by the weight of the elements and the weight of the boom itself.

Fig 13-10 shows the forces and dimensions that create these bending moments. The weight moments were obtained earlier when calculating the Yagi weight balance.

Weight moments to the "left" of the mast:
Element no. 1: –226.6 kg-m
Element no. 2: –24.5 kg-m
Boom left: –37.1 kg-m
Boom insert left: –4.2 kg-m
Boom dummy: –0.3 kg-m
Total: –292.7 kg-m

Weight moments to the "right" of the mast:
Element no. 3: 243.6 kg-m
Boom right: 45.2 kg-m

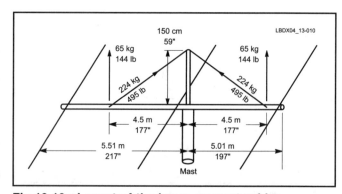

Fig 13-10—Layout of the boom-support cables (trusses) with the forces and tensions involved. The truss cables are not installed to provide additional strength to the boom; they merely support the boom in order to compensate for the sag from the weight of the elements on the boom.

Boom insert right: 4.2 kg-m
Total: 293 kg-m

The weight moment to the left of the mast is the same as to the right of the mast since the Yagi is weight-balanced. Here comes another surprise: The boom is loaded almost *three times* as much by weight loading in the vertical plane (293 kg-m) than it is by wind loading at 140 km/h in the horizontal plane (120 kg-m).

The maximum allowable bending moment for the boom steel insert with a diameter of 63 mm and 6-mm wall is 619 kg-m as calculated with Eq 13-2 for a material yield strength of 20 kg/mm^2. This steel coupler has a safety factor of *two* as far as the weight-loading in the vertical plane is concerned. Boom stress by weight will usually be the condition that will specify the size of the boom with large low-band Yagis using heavy elements.

The boom, using the above calculated coupler, does not require any guying for additional strength. However, the high weight loading of the very long elements sitting at the end of the boom halves will cause a very substantial sag in the boom. For my 40-meter beam the sag amounts to nearly 65 cm, which is really excessive from a cosmetic point of view. A sag of 10 cm is due to the boom's own weight and 55 cm is due to the weight of the elements at the tips of the boom.

Fig 13-11—Details of the tension-equalizing system at the top of the support mast, where the two boom-support trusses are attached. The triangular-shaped plate can rotate freely around the 10-mm bolt, which serves to equalize the tensions in the two guy wires. See text for details.

Again, I consider it a proof of good engineering to eliminate sag by supporting the boom using truss cables. The two boom halves are supported with two sets of dual parallel guy wires attached on the boom at a point 4.5 meters from the mast attachment point. The guy wires are supported from a 1.4-meter high support mast made of a 35 mm OD stainless steel tube, which is welded to the boom-to-mast plate. See Fig 13-10.

The weight that is supported is given by the previously calculated moment divided by the distance of the cable attachment point to the mast attachment point. Assuming the two boom halves are hinged at the mast, each support cable would have to support the total weight as shown above, divided by the sine of the angle the truss support cable makes with the boom.

Leeson (Ref 964) covers guyed booms well in his book. In the case above we are *not* guying the boom to give it additional strength, we do it only to eliminate sag. Guying a boom is not a simple problem of moments, but a problem of a compressed column, where the slenderness of the boom and the compression force caused by the guy wire (usually in three directions) come into the picture. In our case these forces are so low that we can simplify the model as done above. In the above case we implicitly assumed that the boom has enough lateral strength (which we had calculated). For solving the wire-truss problem we assume that the boom is a "nonattached" cantilever. The fact that the boom is attached introduces an additional safety factor.

If a single steel cable is used, a 6-mm ($^1/_4$-inch) OD cable is required to safely support this weight. I use *two* cables of 4-mm (0.16-inch) OD Kevlar (also known as Phillystran in the US). I use this because it was available at no cost, and it does not need to be broken up with egg insulators (Kevlar is a fully dielectric material which has the same breaking strength as steel and the same elongation under load). Note that turnbuckles may prove to be the weak link in the system and stainless-steel turnbuckles can be very expensive. If two parallel cables are used, a tension equalizer must be used to ensure perfect equal stress in both cables. In the case of two truss cables without equalization, one of the cables is likely to take most of the load.

Let me go into detail why I use two parallel support guys.

Fig 13-12—The element-to-boom mounting system as used on the ON4UN 40-meter Yagi.

Fig 13-13—The Omega matching system and plastic "drainpipe" box containing the two variable capacitors. Note also the boom-to-mast mounting plate made of 1-cm thick stainless steel. The boom is attached to this plate with eight U bolts and double saddles.

Fig 13-11 shows the top of the support mast, on which two triangular-shaped stainless-steel plates are mounted. These plates can pivot around their attachment point, which consists of a 1-cm diameter stainless-steel bolt. The two guy wires are connected with the correct hardware (very important—consult the supplier of the cable!) at the base of these triangular pivoting plates. The pivoting plates now serve a double purpose:

- They equalize the tension in the two guy wires.
- They serve as a visual indicator of the status of the guy wires.

If something goes wrong with one of the support wires, the triangular plate will pivot around its attachment point. At the same time the remaining support (if properly designed) will still support the boom, although with a greatly reduced safety factor.

To install the support cables and adjust the system for zero or minimum boom sag if you don't use turnbuckles, place the beam on two strong supports near the end of the boom so as to induce some inverse sag in the boom. Lift the center of the boom to control the amount of inverse sag. Now adjust the position of the boom attachment hardware to obtain the desired support behavior.

Make sure you properly terminate the cables with thimbles. The loads involved are not small, and improper terminations will not last. This is especially true when Kevlar rope is used.

3.3.5. Element-to-boom and boom-to-mast clamps

With an element weighing well over 40 kg, attaching such a mast at the end of a 5-meter arm must be done with great care. The forces involved when we rotate the Yagi (start and stop) and when the beam swings in storm winds are impressive.

After an initial failure, I designed an element-to-boom mounting system that consists of three stainless-steel U-channel profiles (50-cm long) welded together. The element is mounted inside the central channel profile using four U bolts with 12-mm wide aluminum saddles. Four double-saddle systems are used to mount the unit onto the boom (see **Fig 13-12**). U bolts must be used together with saddles and you must use saddles on both sides. The bearing strength of U bolts is far too low to provide a durable attachment under extreme wind loads without saddles on both sides. Never use U bolts made of threaded stainless-steel rods directly on the boom; if they can move but a hair, they become like perfect files that will machine a nice groove in the boom in no time!

At the center of the boom I mounted a 60-cm wide, 1-cm thick stainless-steel plate to which the 1.5-meter long support mast for the boom guying is welded. The boom is bolted to the boom-to-mast plate using eight U bolts with saddles matching the 75-mm OD boom (see **Fig 13-13**). On the tower, this plate is bolted to an identical plate (welded to the rotating mast) using four 18-mm OD stainless-steel bolts.

3.3.6. Materials

In the metric world (mainly Europe), aluminum tubes are usually available in 6-meter sections. **Table 13-2** lists dimensions and weights of a range of readily available tubes. Aluminum tubing in F22 quality (Al Mg Si 0.5%) is readily available in Europe in 6-meter lengths. The yield strength is 22 kg/mm^2.

Tables 13-3 and **13-4** show a range of material dimensions that are available in the US. *The ARRL Antenna Book* also lists a wide range of aluminum tubing sizes. Make sure you know which alloy you are buying. The most common aluminum specifications in the US are:

6061-T6: Yield strength = 24.7 kg/mm^2
6063-T6: Yield strength = 17.6 kg/mm^2
6063-T832: Yield strength = 24.7 kg/mm^2
6063-T835: Yield strength = 28.2 kg/mm^2

Economical Lengths

When designing the Yagi elements, a maximum effort should be made to use full fractions of the 6-meter tubing lengths, in order to maximize the effective use of the material purchased. A proper section overlap is 15 cm (6 inches). The effective net lengths of fractions of a 600-cm tube are 285, 185, 135, 85 and 60 cm.

In the US, aluminum is available in 12-ft lengths. The effective economical cuts (excluding the 6-inch overlap) are 66, 42, 30, 22.8 inches, etc.

3.3.7. Telescopic Fits

You can make well-fitting telescopic joints as follows: With a metal saw, make two slits of approximately 30-mm length into the tip of the larger section. To avoid corrosion, use plenty of Penetrox (available from Burndy) or other suitable contact grease when assembling the sections. A stainless-steel hose clamp will tighten the outer element closely onto the inner one (with shimming material in-between if necessary). A stainless-steel Parker screw will lock the sections lengthwise. For large diameters and heavy-wall sections, a stainless steel 6 or 8-mm bolt is preferred in a pre-threaded hole.

Table 13-2

Dimensions and Weight of Aluminum Tubing in F22 Quality

OD mm	Wall mm	Weight g/m	OD mm	Wall mm	Weight g/m
10	1	76	40	1.5	489
12	1	93	44	2	541
13	1	103	48	1.5	603
14	1	110	50	5	1923
15	1	127	50	2	820
19	1.5	227	52	1.5	654
20	1.5	235	57	2	940
22	2	339	60	5	2350
22	1.5	261	60	3	1460
25	2.5	477	62	2	1040
25	2	398	70	5	2757
25	1.5	298	70	3	1718
28	1.5	336	80	5	3181
30	3	687	80	4	2579
30	2	484	84	2	1385
32	1.5	387	90	5	3605
35	2	564	100	5	4029
36	1.5	438	100	2	1676
40	5	1495	110	5	4485
40	2	644			

Metric tube sections do not provide as snug a telescoping fit as do the US series with a 0.125-inch-diameter step and 0.058-in. wall thickness. At best there is a 1-mm difference between the OD of the smaller tube and the ID of the larger tube. A fairly good fit can be obtained, however, by using a piece of 0.3-mm-thick aluminum shimming material. The slit, hose clamp, Parker screw and heat-shrink tube make this a reliable joint as well.

Sometimes sections must be used where the OD of the smaller section is the same as the ID of the larger section. To achieve a fit, make a slit approximately 5 cm (2 inches) long in the smaller tube. Remove all burrs and then drive the smaller tube inside the larger to a depth of 3 times the slit length (eg, 15 cm). Do this after heating up the outer tube with a flame torch and cooling down the inner tube in ice water. The heated-up outer section will expand, while the cooled-down inner section will shrink. Use a good-sized plastic hammer and enough force to drive the inner tube quickly inside the larger tube before the temperature-expansion effect disappears. A solid unbreakable press fit can be obtained. A good Parker screw or stainless-steel bolt (with pre-threaded hole) is all that's needed to secure the taper connection.

Under certain circumstances a very significant drop in element diameter is required. In this case a so-called *doughnut* is required. The doughnut is a 15-cm long piece of aluminum tubing that is machined to exhibit the right OD and ID to fill up the gap between the tubes to be fit. Often the donut can be made from short lengths of heavy-wall aluminum tubing.

I always cover each taper-joint area with a piece of heat-shrinkable tube that is coated with hot-melt glue on the inside (Raychem, type ATUM). This protects the element joint and keeps the element perfectly watertight.

3.3.8. Material ratings and design conditions

All the above calculations are done in a static environment, assuming a wind blowing horizontally at a constant speed. Dynamic modeling is very complex and falls out of the scope of this book. If all the rules, the design methodology and the calculating methods as outlined above and as used in the mechanical design modules of the YAGI DESIGN software are closely followed, a Yagi will result that will withstand the forces of wind, even in a normal dynamic environment, as has been proved in practice. My 40-meter Yagi was designed to be able to withstand wind speeds of 140 km/h, according to the EIA/TIA-222-E standard. The 140-km/h wind does *not* include any safety factors or other modifiers.

The most important contribution of all the above calculations is that the stresses in all critical points of the Yagi are kept at a similar level when loading. In other words, the mechanical design should be well-balanced, since the system will only be as strong as the weakest.

Make sure you know exactly the rating of the materials you are using. The yield stress for various types of steel and especially stainless steel can vary with a factor of three! Do not go by assumptions. Make sure.

3.3.9. Element finishing

As a final touch I always paint my Yagi beams with three layers of transparent metal varnish. It keeps the aluminum nice and shiny for a long time.

3.3.10. Ice loading

Ice loading greatly reduces an antenna's wind-survival speed. Fortunately, heavy ice loading is not often accompanied by very high winds, with an exception for the most harsh environments (near the poles).

Although we are almost never subject to ice loading here in Northern Belgium, it is interesting to evaluate what the performance of the Yagi would be under ice loading conditions. **Table 13-5** shows the maximum wind survival speed and element sag as a function of radial ice thickness. As the ice thickness increases, the sections that will first break are the tips. The reflector of our metric-design element will take up to 16 mm of radial ice before breaking. At that time the sag of the tips of the reflector element will have increased from 100 cm without ice to approximately 500 cm with ice. If the Yagi must be built with heavy ice loading in mind, you will have to start from heavier tubing at the tips. The ELEMENT STRENGTH module will help you design an element meeting your requirements in only a few minutes.

3.3.11. Material fatigue

Many have observed that light elements (thin-wall, low wind-survival designs) will oscillate and flutter under mild wind conditions. Element tips can oscillate with an amplitude

Table 13-3
List of Currently Available Aluminum Tubing in the US

OD inches	Wall inches	Weight lb/foot	OD inches	Wall inches	Weight lb/foot
0.250	0.058	0.04	1.625	0.058	0.34
0.375	0.058	0.07	1.750	0.058	0.36
0.500	0.058	0.10	1.750	0.083	0.51
0.625	0.058	0.12	1.875	0.058	0.39
0.750	0.058	0.15	2.000	0.065	0.45
0.875	0.058	0.18	2.000	0.083	0.59
1.000	0.058	0.20	2.000	0.125	0.83
1.125	0.058	0.23	2.500	0.065	0.59
1.250	0.058	0.26	2.500	0.083	0.74
1.375	0.058	0.28	2.500	0.125	1.06
1.500	0.058	0.31	3.000	0.065	0.71
1.500	0.065	0.34	3.000	0.125	1.30
1.500	0.083	0.43			

Table 13-4
List of Currently Available Aluminum Pipe in the US

OD	Wall	OD	Wall
1.050	0.113	1.900	0.109
1.050	0.154	1.900	0.145
1.315	0.133	1.900	0.200
1.315	0.179	2.375	0.065
1.660	0.065	2.375	0.109
1.660	0.109	2.375	0.154
1.660	0.140	2.375	0.218
1.660	0.191	2.875	0.203
1.900	0.065	2.875	0.276

of well over 10 cm. Under such conditions a mechanical failure will be induced after a time. This failure mechanism is referred to as *material fatigue*.

Element vibrations can be prevented by designing elements consisting of strong heavy-wall sections. Avoid tip sections that are too light. Tip sections of a diameter of less than about 15 mm are not recommended, although they may be difficult to avoid with a large 40-meter Yagi. Through the entire length of each element I run an 8-mm nylon rope, which lies loosely inside the element. This rope dampens any self-oscillation that might start in the element.

At both ends, the rope is fastened at the element tips by injecting a good dose of silicone rubber into the tip of the element and onto the end of the rope. The tip is then covered with a heat-shrinkable plastic cable-head cover with internal hot-melt glue. And at both ends of the element you must drill a small hole (3 mm) at the underside of the element about 5 cm from the tip of the element to drain out any condensation water that may accumulate inside the element.

Make sure the rope lays loosely inside the element. The method is very effective, and not a single case of fatigue element failure has occurred when these guidelines were followed. A simple test consists of trying to hand excite the elements into a vibration mode. Without internal rope this can usually be done quite easily. You will become really frustrated trying to get into an oscillation mode when the rope is present. Try for yourself!

3.3.12. Matching the Yagi

The only thing left to do now is to design a system to match the antenna feed-point impedance (28 Ω) to the feed-line impedance (50 Ω). The choice of the omega match is obvious:

- No need for a split element (mechanical complications).
- No need to adjust the length of a gamma rod.
- Fully adjustable from the center of the antenna.

Only a true "plumber's delight" construction, with no floating or insulated elements, can guarantee proper operation as a top loading device (eg for 160 meter) on top of a tower. With floating elements the insulation may flash over and be destroyed when the Yagi is near or at the top of the tower, and cause destruction of the beam and erratic functioning as a loading device on 160 meters.

The two Omega-match capacitors are mounted in a housing made of a 50-cm long piece of plastic drainpipe (15-cm OD), which is mounted below the boom near the driven element (Fig 13-13). This is a very flexible way of constructing boxes for housing Gamma and Omega capacitors. The drain pipes are available in a range of diameters, and the length can be adjusted by cutting to the required length. End caps are available that make professional-looking and perfectly watertight units.

The design of the Omega match is described in detail in Section 3.10.2. **Fig 13-14** shows the SWR curve of my 40-meter Yagi. The 1.5:1 SWR bandwidth turned out to be 210 kHz.

3.3.13. Tower, mast, mast bearings, drive shaft and rotator

If you want a long-lasting low-band Yagi system, you must pay attention :

- The tower.
- The rotating mast.
- The mast bearings.
- The rotator.
- The drive shaft.

3.3.13.1. The tower

Your tower supplier or manufacturer will want to know the wind area of your antenna. Or maybe you have a tower that's good for 2 meters^2 of top load. Will it be okay for the 40-meter antenna?

Specifying the wind area of a Yagi is an issue of great confusion. Wind-thrust force is generated by the wind hitting a surface exposed to that wind. The thrust is the product of the dynamic wind pressure multiplied by the exposed area, and with a so-called *drag coefficient*, which is related to the *shape* of the body exposed to the wind. The resistance to wind of a flat-shaped body is obviously different (higher!) than the

Table 13-5
Ice Loading Performance of the 40-Meter Beam

Radial		Ice		Max Wind Speed	Sag	
mm	inch	kph	mph		cm	inch
2.5	0.1	116	72		132	52
5.0	0.2	96	60		183	72
7.5	0.3	79	49		242	95
10	0.4	64	40		310	122
12.5	0.5	47	29		386	152
15	0.6	25	15		435	171
16	0.63	0	0		Break	

Note: As designed, the Yagi element will break with a 16-mm (0.63 inch) radial ice thickness at zero wind load, or at lower values of ice loading when combined with the wind. The design was not optimized to resist ice loading. Optimized designs will use elements that are overall thicker, especially the tip elements.

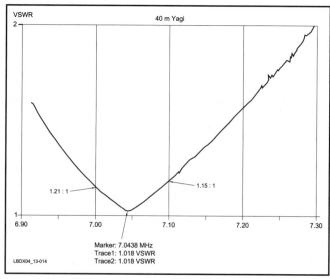

Fig 13-14—SWR curve of the ON4UN 40-meter Yagi, as shown on the screen of a PC running the Alpha Network Analyzer.

resistance of a ball-shaped or tubular-shaped body.

This means that if we specify or calculate the wind area of a Yagi, we must always specify if this is the equivalent wind area for a flat plate (which really should be the standard) or if the area is simply meant as the sum of the projected areas of all the elements (or the boom, whichever has the largest projected area; see Section 3.3.4.1).

In the former case we must use a drag coefficient of 2.0 (according to the latest EIA/TIA-222-E standard) to calculate the wind load, while for an assembly of long and slender tubes a coefficient of 1.2 is applicable. This means that for a Yagi consisting only of tubular elements (Yagi elements and boom), the flat-plate wind area will be 66.6% lower (2.0/1.2) than the round-element wind area.

The 40-meter Yagi, excluding the boom-to-mast plate, the rotating mast and any match box, has a flat-plate equivalent wind area of 1.65 meters^2. As the projected area of the three elements is 2.5 times larger than the projected area of the boom, the addition of the boom-to-mast plate and the match box will not change the wind load, which for this Yagi is only determined by the area of the elements. The round-element equivalent wind area for the Yagi is 2.74 meters^2.

The wind thrust generated by this Yagi at a wind speed of 140 km/h is 302 kg. This is for 140 km/h winds, without any safety margins or modifiers. Consult the EIA/TIA-222-E standard or your local building authorities to obtain the correct figure you should use in your specific case.

Let me make clear again that the thrust of 302 kg is only generated with the elements broadside to the wind. If you put the boom into the wind, the loading on the tower will be limited to 90 kg. However, I would not advise using a tower that will take less than 300 kg of top load. Consider the margin between the boom in the wind and the elements in the wind as a safety margin.

3.3.13.2. The rotating mast

Leeson (Ref 964) covered the issue of masts very well. Again, what you use will probably be dictated in the first place by what you can find. In any case, make sure you calculate the mast. My 3-element 40-meter beam sits on top of a 5-meter long stainless-steel mast, measuring 10 cm in diameter with a wall thickness of 10 mm. This mast is good for a wind load of 579 kg at the top. I calculated the maximum wind load as 302 kg. At the end of a 5-meter cantilever the bending moment caused by the beam is 1510 kg-m. Knowing the yield strength of the tube, we can calculate the minimum required dimensions for our mast using Eq 13-2.

$$M_{max} = YS \times \pi \times \frac{10^4 - 8^4}{32 \times 10} = YS \times 58$$

where YS = yield strength.

The stainless-steel tube I used has a yield strength of 50 kg/mm^2.

$M_{max} = 5000 \times 58 = 290,000$ kg-cm

It appears that we have a safety factor of 75% versus the moment created by the Yagi (1510 kg-m). I have not included the wind load of the mast itself, but the safety margin is more than enough to cover the bending moment caused by the mast.

In my installation I welded plates on the mast at the heights where the beam needs to be mounted. These plates are exact replicas of the stainless-steel plates mounted on the booms of the Yagis (the boom-to-mast coupling plates). When mounting the Yagi on the mast, you do not have to fool around with U bolts; the two plates are bolted together at the four corners with 18-mm-OD stainless-steel bolts.

One word of caution about stainless-steel hardware. Do not tighten stainless-steel bolts as you would do with steel bolts. Stainless-steel bolts gall when over tightened and are very difficult to remove later. It is always wise to use a special grease before assembling stainless-steel hardware. Also, where safety is a concern, use one normal bolt, doubled up with a special safety self-locking bolt (with plastic insert). Between the two plates a number of stairs have been welded in order to provide a convenient working situation when installing the antennas.

3.3.13.3. The mast bearings

The mast bearings are important parts of the antenna setup. Each tower with a rotating mast should use two types of bearings:
- The thrust bearing—it should take axial weight as well as a radial load.
- The second bearing should only take a radial load.

The thrust bearing should be capable of safely bearing the weight of the mast and all the antennas. The thrust-bearing assembly must be waterproof and have provisions for periodic lubrication. **Fig 13-15A** shows the thrust collar being welded on the stainless-steel mast inside my top tower section. Notice the stainless-steel housing of the thrust bearing. The bearing is a 120-mm ID, FAG model FAG30224A (T4FB120 according to DIN ISO 355). In my tower the thrust bearing is 2 meters below the top of the tower.

My tower's second bearing is mounted right at the top of the tower and consists of a simple 10-cm long nylon bushing with approximately 1-mm clearance with the mast OD. Note that the thrust bearing could instead be at the top, with the radial bearing at the lower point. The choice is dictated by practical construction aspects.

The mast and antenna weight should never be carried by the rotator. In my towers I have the rotator sitting at ground level, with a long drive shaft in the center of the self-supporting tower. The drive shaft is supported by the thrust bearing near the top of the tower. The fact that the heavy drive shaft hangs in the center of the tower adds to the stability of the tower. I can easily replace the rotator. The coupling between the rotator and the drive shaft is a cardan axle from a heavy truck, as shown in Fig 13-15B.

3.3.13.4. The rotator

I would not dare to suggest using one of the commercially available rotators with antennas of this size. Use a prop-pitch or a large industrial-type worm-gear reduction with the appropriate reduction ratio and motor. For example, the Prosistel "Big Boy" rotator is available from Array Solutions at **www.arraysolutions.com/**.

3.3.13.5. The drive shaft

The drive shaft is the tube connecting the rotating mast with the rotator. The drive shaft must meet the following specifications:

- It must act as a torque absorber when starting and stopping the motor. This effect can be witnessed when you start the rotator and the antenna actually starts moving a second later. This relieves a lot of stress on the rotator. Leeson (Ref 964) uses an automotive transmission damper as a torque absorber spring.
- The drive shaft should not have too much spring effect, so that the antenna points in the right direction even in high winds. If there is too much springiness, excessive swinging of the antenna could damage the antenna. The acceleration and the forces induced by the swinging of the elements could induce failure at the element-to-boom mounts.

The torque moment will deform (twist) the drive shaft (hollow tube). The angle over which the shaft is twisted is directly proportional to the length of the shaft. In practice, we should not allow for more than ±30° of rotation under the worst torque moment.

In an ideal world, the Yagi is *torque-balanced*, which means that even under high wind load there is no mast torque. In practice nothing is less true: Wind *turbulence* is the reason that the large wind capture area of the Yagi always creates a large amount of momentary torque moment during wind storms.

When rotation is initiated, the inertia of the Yagi induces twist in the drive shaft. The same is true after stopping the rotator, when the antenna overshoots a certain degree before coming back to its stop position.

In practice you will have to make a judicious choice between the length of the drive shaft and the size of the shaft. Using a long drive shaft and the rotator at ground level has the following advantages (in a non-crank-up, self-supporting tower):

- No torque induced on the tower above the point where the rotator isolated.
- Motor at ground level facilitates maintenance and supervision.
- Long crank shaft works as torsion spring and takes torque load off the motor.
- The disadvantage is that you will need a sizable shaft to keep the swinging under control.

3.3.13.6. Calculating the drive shaft

It is difficult, if not impossible, to calculate the torque moment caused by turbulent winds. I have estimated the momentary maximum torque moment to be 3 times as high as the torque moment on one side of the boom, as calculated before for a wind speed of 140 km/h. This is 360 kg-m. This means that the wind turbulence momentarily causes the antenna to rotate in only one direction, and that we disregard the forces trying to rotate the antenna in the opposite direction. In addition, I added a 200% safety factor. I use this figure as the maximum momentary torque moment to calculate the requirements for the drive shaft. I have not found any better approach yet, and it is my practical experience that this approach is a fair approximation of what can happen under the worst circumstances with peak winds in a highly turbulent environment.

Assumed momentary maximum torque moment T = 360,000 kg-mm, calculate the section shear modulus (Z):

$$Z = \pi \times \frac{D^4 - d^4}{16 \times D}$$

Assume the following:
D = 8 cm
d = 6.5 cm
Z = 56.7 cm^3

Calculate the shear stress (ST):
ST = T/Z
where
Z = modulus of section under shear stress
T = applied torque moment
ST = 36,000 kg-cm/56.7 cm^3 = 635 kg/cm^2
 = 6.35 kg/mm^2

This is a low figure, meaning the tube will certainly not break under the torque moment of 36,000 kg-cm. Calculate

(A)

(B)

Fig 13-15—At A, the thrust bearing for the 100-mm OD-mast inside the top section of the 24-meter tower at ON4UN. At B, base of the self-supporting 25-meter tower (measuring 1.5 meters across), with the prop-pitch motor installed 1 meter above the ground. The drive shaft is coupled to the prop-pitch motor via a cardan axle from a heavy truck. Having the motor at ground level facilitates service, and takes torque load off the tower. In addition, the long drive shaft acts as a shock (momentum) absorber, greatly reducing strain on the motor.

the maximum twist angle (TW). The twist angle of the shaft is directly proportional to the shaft length. In my case the rotator is 21 meters below the lower bearing, which makes the shaft 21 meters long. The critical part of the whole setup is the shaft-twist angle under maximum mast torque, where:

T = applied torque (360,000 kg-mm)
L = length of shaft (21,000 mm)
G = rigidity modulus of the material = 8000 kg/mm^2
J = section modulus × radius of tube =
　　56,700 mm^3 × 40 mm = 2,268,000 mm^4
TW = 0.44 radians = 25°

A twist angle of 25° is an acceptable figure. The twist should in all cases be kept below 30° to keep the antenna from excessively swinging back and forth in high winds. It is clear that the same result could be obtained with a much lighter tube, provided it was much shorter in length.

3.3.14. Raising the antenna

A 3-element full-size 40-meter Yagi, built according to the guidelines outlined in the previous paragraphs, is a "monster." Including the massive boom-to-mast plate, it weighs nearly 250 kg (500 lbs). A few years ago I befriended a man who has his own crane company. He has a whole fleet of hydraulic cranes that come in very handy for mounting large antennas on towers.

Fig 13-16 shows the crane arm extended to a full 48 meters, maneuvering the 40-meter Yagi on top of the 30-meter self-supporting tower. With the type of boom-to-mast plates shown in Fig 13-13, it takes but a few minutes to insert the four large bolts in the holes at the four corners of the plates and get the Yagi firmly mounted on the mast.

3.3.15. Conclusion

Long-lasting, full-size low-band Yagis are certainly not the result of improvisation. The 40-meter Yagi I've described here has been up for over 15 years now, without any repairs. Long lasting Yagis, especially for the low bands are the result of a serious design effort, which is 90% a mechanical engineering effort. Software is now available that will help design mechanically sound, large low-band Yagis. This makes it possible to build a reliable antenna system that will out-perform anything that is commercially available by a large margin. It also brings the joy of home-building back into our hobby, the joy and pride of having a no-compromise piece of equipment.

3.4. A Super-Performance, Super-Lightweight 3-Element 40-Meter Yagi

Nathan Miller, NW3Z, designed a very novel and attractive 3-element Yagi that was featured in *QST* (Ref 979). See **Fig 13-17**. It weighs only a tiny fraction of the battleship described in Section 3.3. The NW3Z antenna can be turned with a run-of-the-mill good-quality rotator. The antenna is based on a similar 2-element design by Jim Breakall, WA3FET.

This 3-element Yagi also uses the principle of instantaneous pattern reversal, which I described about 7 years ago in a previous edition of this book (see also Section 3.5). Basically this Yagi uses two directors, symmetrically located with respect of the driven element. By using small relays an inductance is inserted in the middle of the parasitic element to turn it into a reflector to make instantaneous direction switching possible. You must have experienced this feature in order to fully appreciate it!

3.4.1. Electrical performance

I modeled the antenna using *EZNEC*, both in free space as well as over real ground. The dimensions shown in **Fig 13-18** are very close to those published by NW3Z. In free space the antenna exhibits 7.34 dBi gain, with a feed impedance of 40.5 Ω, using a loading inductance with a reactance of 138 Ω as shown in the *QST* article. The gain is more than 7.1 dBi, across the whole 40-meter band. The F/B performance in free space is illustrated in **Fig 13-19**.

This is a fairly low-Q antenna, yielding a feed-point impedance of nearly 50 Ω, which means that the antenna is split fed with a current balun. The computer SWR values are shown in **Table 13-6**.

I also modeled the antenna over real ground, at a height of 21 meters. We learned in Chapter 5 that for most DX paths an elevation angle between 10° and 15° seems to be optimum. If the angle is 15°, a Yagi at 0.6 λ will lose about 1.5 dB compared to its brother at 1 λ, but it will have much better high-angle rejection. The high antenna rejects a signal at a wave angle of 60° in the forward direction by about 8 dB. The antenna at 0.6 λ will reject the same signal, about 18 dB!

This antenna is within reach of many and it can perform quite outstandingly at a 0.5 to 0.6-λ height. **Fig 13-20** shows

Fig 13-16—The ON4UN 40-meter Yagi is lowered on top of the rotating mast at a height of 30 meters using a 48-meter hydraulic crane.

the radiation patterns for 7.05 MHz. Note that in order to obtain best F/B at that frequency, the reactance of the loading coil should be changed from 138 Ω to 130 Ω. The SWR curve becomes a little steeper, especially on the low side.

3.4.2. Mechanical design

The prototype was made using two types of aluminum tubing: The 2.5 and 2.25-inch-OD tubing is an extruded 6061-T6 alloy with 0.125-inch walls. All other tubing is 6063-T832 with 0.058-inch wall. The parasitic elements are made of #10 aluminum plated steel wire. Copper-clad steel wire or bronze wire would also be appropriate.

Table 13-6
SWR Performance of the WA3FET/NW3Z Yagi Modeled in Free Space, Using X_L = 138 Ω.

7.0 MHz	7.05 MHz	7.1 MHz	7.2 MHz	7.3 MHz
1.4:1	1.2:1	1.1:1	1.3:1	1.7:1

Table 13-7
SWR Performance of the WA3FET/NW3Z Yagi at 21 Meters Over Average Ground, Using X_L = 132 Ω.

7.0 MHz	7.05 MHz	7.1 MHz	7.2 MHz	7.3 MHz
1.6:1	1.3:1	1.2:1	1.3:1	1.8:1

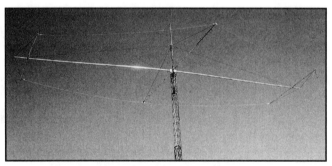

Fig 13-17—The 3-element 40-meter NW3Z Yagi is mounted on a 21-meter crankup tower at the Penn State University Dept of Electrical Engineering research facility at Rock Springs.

The boom, for which the taper schedule is shown in **Fig 13-21** weights only 9 kg. The entire Yagi weighs well under 50 kg, which makes this really a super lightweight 3-element full-size 40-meter antenna!

3.4.3. The parasitic element supports

Miller used an aluminum spreader (1.5-inch OD) tubing broken up with fiberglass rods to minimize loading of the director. Also, where the element supports are attached to the driven element, he uses a 30-cm long fiberglass rod, to keep the metal of the support far enough from the driven element. A valid alternative would of course be to use fiberglass poles along the entire length.

3.4.4. Truss wiring

Because of the additional cross-arm and parasitic-wire

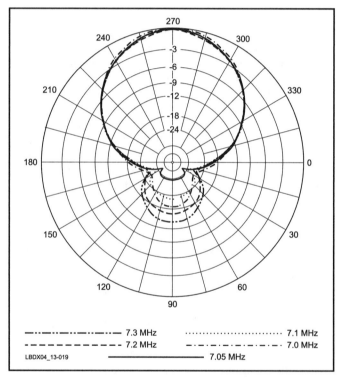

Fig 13-19—Horizontal radiation pattern in free space for the NW3Z/WA3FET 40-meter Yagi. The F/B is 20 dB or better from 7.0 to 7.1 MHz, and still a usable 17 dB at 7.2 MHz.

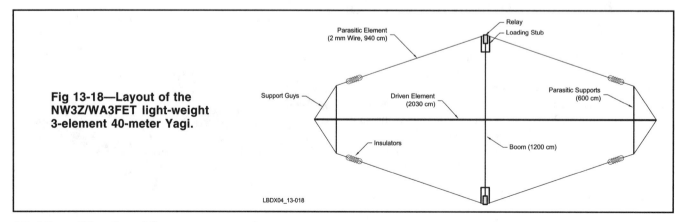

Fig 13-18—Layout of the NW3Z/WA3FET light-weight 3-element 40-meter Yagi.

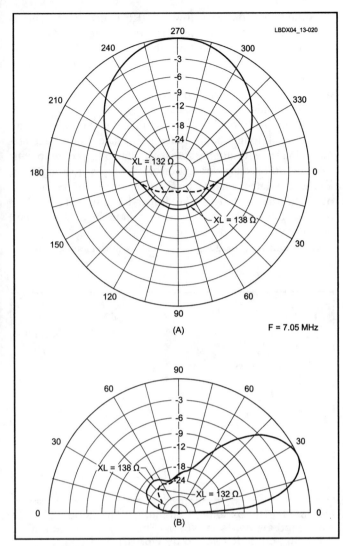

Fig 13-20—At A, horizontal radiation pattern at 21-meter height over average ground for the NW3Z/WA3FET 40-meter Yagi. The patterns are for 7.05 MHz. Reducing the value of the loading reactance from 138 to 132 Ω improves the F/B performance.

weight loading on the tips, the full-size driven element requires a supporting truss. The antenna uses Phillystran (PVC coated Kevlar rope) for the purpose. The boom is also guyed. Both sets of guy wires are attached to a support about 2 meters above the antenna. Horizontal support guys are used from the driven-element tip to the ends of the parasitic supports to counter the tension in the parasitic wires, as shown in **Fig 13-22**.

3.4.5. Tuning the Yagi

The shorted-stub loading reactance of 132 to 139 Ω represents an inductance of 3 to 3.1 µH. You can achieve an unloaded Q of more than 500 with a well-designed coil compared to a Q of about 100 with a linear-loading stub. So I designed a high-Q coil for this application:

Required inductance: 3.1 µH
Coil diameter: 7.5 cm (3 inches)
Coil length: 11.3 cm (4.5 inches)
Number of turns: 8 air-wound
Conductor: 6mm (1/4 inch) copper tubing

To tune the coil to the required exact value, just stretch or squeeze the turns. For a constant diameter (7.5 cm) and for 8 turns the inductance will vary from 2.8 µH to 3.8 µH by

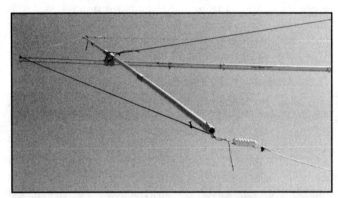

Fig 13-22—The parasitic-wire, cross support, parasitic elements, horizontal and vertical supports in place on the NW3Z/WA3FET Yagi.

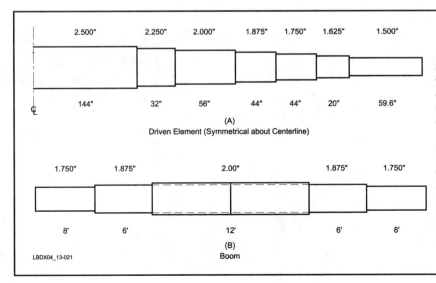

Fig 13-21—Taper schedule for the driven element (A) and boom (B) of the NW3Z/WA3FET Yagi. The driven-element taper schedule is quite different from an ordinary Yagi element taper, since the element supports the 6-meter (20-foot) long cross bars at the end, which in turn support the ends of the parasitic wires. In this design the driven element is more like a boom, while the boom can be much lighter because it only supports the centers of the parasitic wires. The driven element weighs about 22 kg.

changing the coil length from 11.3 cm (4.5 inches) to 15 cm (6 inches).

The *NEC* model shows that the reflector is resonant on 6.7 MHz, and the director on 7.0 MHz by themselves. The best way to make sure that the parasitic elements are resonant on 6.7 MHz would be to feed the elements temporarily with $\lambda/2$ feed lines and cut them for zero reactance using a network analyzer or an antenna analyzer. While doing this the driven element and the second parasitic element must be left "open." Adjusting the loading coil or stub can be done the same way: Connect the feed line in series (not in parallel!) with the loading coil.

3.4.6. Feeding the Yagi

The original design uses a very simple split-element feed, since the antenna impedance is around 40 Ω. This requires a split driven element. A fiberglass rod used as a element insert can be used for the purpose. When direct feed is used, a choke balun is required. Alternatively, you could use any of the other matching systems described in Section 3.8.

3.4.7. Conclusion

Considering that the NW3Z antenna only trades 0.2 dB of forward gain vs my heavy-weight 3-element Yagi, and given its additional feature of instant direction reversal, this antenna is one of the most interesting designs that has been published for a long time, and deserves great popularity. When will we see the first 80-meter version of this design?

3.5. Loaded 80-meter Yagi designs

Full-size 80-meter Yagis are obviously not for everyone. The investment is very important, and they are, let it be said, difficult to keep up. If carefully designed and well-made, Yagis with shortened elements can perform almost as well as a full-size Yagi. A three-element Yagi with shortened element can be made to have just as good a directivity as its full-sized brother. It may, however, give up a fraction of a dB in gain. Let me review some interesting designs that have appeared recently in literature or on the Internet.

3.5.1. Loading coils instead of linear-loading

Until the mid 1990s it was commonly understood that linear-loading devices (such as used for many years by KLM, Force 12 and M-Squared) were the best solution for low-loss loading of shortened low-band Yagis. Linear loading was assumed to have very low losses. Until then, for some strange reason, coil loading was generally assumed to be a lossy affair. Was that another tall tale?

As a consequence of a lot of experimenting and modeling in the last years our knowledge in this matter has improved a great deal, we now know that coil loading can actually be much better than linear-loading.

David Padrick, W6ANR (ex-KC7LU and WB6IIS), decided to analyze loading coils in detail and found out that most commercial coils did indeed exhibit poor unloaded Q. But David also came to the conclusion that it was not all that hard to make coils with Qs of 650 or even more. (Ref 694). However, David found that the dimensions are critical, and high-Q coils able to handle high power ended up being quite large! (See **Figs 13-23** and **13-24**.)

A high-Q coil is not enough by itself. It is equally important to make very low loss RF connections. This is where many commercial designs failed in the past. Good RF connections mean connections that are wide with respect to their length! RF conductor also are crucial. The location of the coils on the elements is important (see Chapter 7). With very high-Q coils we can afford putting them out further on the elements, which increases the radiation resistance without introducing additional losses, provided the Q remains high.

Padrick found that commercially made Yagis using elements loaded with sloping stubs exhibit two sorts of problems: Inferior F/B because of radiation off the sloping loading stubs, and accumulated resistive losses of loading stubs. He pointed out that the length of the linear-loading wire is 2 to 3 times longer than the wire or tubing in a high-Q coil providing the same degree of loading.

In addition, the gauge of a loading stub is usually only a

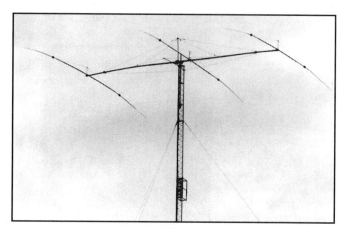

Fig 13-23—Three-element 80-meter Yagi using shortened elements at W6KW. The Yagi uses the high-Q loading coils developed by W6ANR. See text for details. The 55-meter high antenna sits on a knoll surrounded by flat terrain. To work on the antenna the platform visible along the tower is unfolded, and the Yagi is tilted 90° so that the center of the element is easily accessible from the platform.

Fig 13-24—W6KW holding two of the coils for his new 80-meter Yagi, as developed by W6ANR. These loading coils measure 7 inches in diameter and exhibit an unloaded Q of nearly 700!

fraction of what's used for high-Q coils. Poor connections to the element and the relays and jumpers used to switch band-segments add to these problems, and all of these critical items have been found to deteriorate over time.

Changing from a linear-loading stub to a coil actually yields a net increase in R_{rad}. More recently Tom, W8JI, tested linear-loading stubs and equivalent inductance coils. He found it was possible to build coils with Qs up to 800 (Ref 694), but also that the linear stubs never exceeded 100 (see Chapter 9)!

Peter Dalton, W6KW's new Yagi uses elements that are approximately 66% of full-size, and he has incorporated the loading coils out at about 55% of the total element length. Loading coils with an inductance of 17 µH were required to resonate the elements. See **Fig 13-25**. The final fine tuning of the parasitic elements was done by varying the length of the element tips. A Q of 650, as quoted by W6ANR, means a series loss resistance of only 0.624 Ω per coil. If you want to know all the details of loading coil design visit www.w8ji.com/loading_inductors.htm.

I calculated the influence of the Q factor on the gain and directivity. The influence on gain can be seen in **Table 13-8**. In a typical 3-element coil-loaded Yagi, with the reflector spaced closer than the director and tuned for best F/B, the current magnitude in the reflector is almost the same as for the driven element, while the director only carries 15 or 20% that much current. As a consequence, the influence of coil losses on antenna gain is the same for the driven element and the reflector, but much less for the director. In other words, if you have a lossy coil, better put it in the director element.

The Q factor has very little impact on directivity. An antenna that has been optimized for an excellent F/B with no coil losses, may actually show even marginally better directivity when small losses (1 or 2 Ω) are introduced! High losses reduce the mutual coupling to a degree that proper current magnitude and phase can no longer be set up in the elements to achieve a good F/B. As far as directivity is concerned in the model I used, it was possible to achieve a little deeper null with Qs of 200, compared to 400. This is actually irrelevant and only says something about the model, not about what can be achieved with a real antenna.

So far as directivity is concerned, you can say that there is very little to be gained by going for Q factors above 150 or 200, *but* from the point of view of total antenna gain, the higher the Q, the higher the gain. The difference in antenna gain between a Q of 700 and a Q of 175 is about 1 dB, which is certainly not negligible—that is like losing 30% of your power!

3.5.1.1. How to make high-Q coils

The pictures of the loading coils made by K7ZV in Figs 13-24 and 13-25 give you part of the answer. Heavy gauge copper-tubing conductors (6 mm or ¼ inch) and good low-loss contacts. But there is more. There are not only series losses involved, but also parallel losses.

Any leakage current between adjacent turns of the coil causes parallel losses. If we keep the Q high (800), it means that we have an equivalent series resistance (for an inductive reactance of 400 Ω) of 400/800 = 0.5 Ω. This is the equivalent of a parallel loss resistance of 400 × 800 = 320 kΩ. If we allow dirt, smoke deposits, etc, to accumulate on the surface of the bare copper tubing of an unprotected coil, we can expect parallel losses to drop well below 320 kΩ, especially when it gets wet. Therefore the coil must be properly pro-

Fig 13-25—The W6ANR loading coil, wound using ¼-inch copper tubing on a grooved ABS coil form, measuring 10 inches long and 7 inches in diameter. Note the husky, large-area contact clamp used to connect the coil to the element. The ¼-inch copper tubing is covered with a plastic head-shrink tube to protect it from surface contamination and surface leakage.

Fig 13-26—Forty-meter very high-Q loading coil developed by W6ANR. Not that contact blocks where the heavy gauge special enameled copper wire (AWG #8 or 3.3-mm OD) attaches to the element. The contact block is actually welded to the element to minimize loss resistance.

Table 13-8
Influence of Coil Losses on Antenna Gain

Loss R	0 Ω	0.5 Ω	1.0 Ω	1.5 Ω	2.0 Ω	2.5 Ω	3.0 Ω
Q factor	—	800	400	333	200	160	133
Gain, dB	0 dB	−0.34	−0.66	−0.97	−1.27	−1.55	−1.82

Reference: 12.3 dBi

tected. One neat way to protect the copper tubing used to wind the coil with a heat-shrink tube of a material that has a good dielectric for HF and is UV-resistant (consult Raychem for such material).

Later K7ZV models used very heavy AWG 3 gauge enameled copper wire. It doesn't pay to reduce the serial losses to almost zero if you don't take care of the parallel losses too. K7ZV, has a high-precision machine shop to construct the coils used at W6KW. He made clever use of readily available materials and machined them into real jewels of coil forms. In Fig 13-23 you can see the loading coils on W6KW's 3 element 80-meter Yagi. **Fig 13-26** shows a 40-meter loading coil also developed by W6ANR.

If you are interested in a low-band antenna using these High-Q coil assemblies contact W6ANR (**w6anr@pcmagic.net**). They are available for conversion of KLM and M-Squared 80-meter antennas only. For new antennas, contact M-Squared for the "W6RJ" version of the 80M3 designed by W6ANR for use with these coil assemblies. This antenna includes all the aluminum required to build the array. It does not come with coil assemblies or Phillystran needed for boom and element guying.

On his superb web-site (**www.qsl.net/ve6wz/**) Steve Babcock, VE6WZ, describes the design and construction details for his 2-element 80-meter Yagi, which he uses from his city lot on a 28-meter crank up tower to produce the most dominant signal from VE6-land into Europe. He explains the construction of his coils in great detail at **www.qsl.net/ve6wz/coil.htm**. See **Fig 13-27**.

Steve used *Coil.EXE* (a DOS program from Brian Beezley, K6STI), to design his low-loss loading coils. He found that not only conductor size and coil dimensions were important, but that form loss plays a significant role in determining the unloaded Q. You must be very careful with the program, however, as it sometimes predicts unreasonably high Qs. The final coil measures 15 cm in diameter and 15 cm long, and uses 15.5 turns of 3/16-inch (~ 5mm) copper tubing. The final physical coil design at VE6WZ is mostly air core, but uses two black ABS strips to give the required mechanical rigidity to the coil. Steve estimated a final Q of around 500 to 1000 (a loss of 0.7 to 1.5 Ω). To prevent copper corrosion and to ensure maximum surface insulation (especially important near the ABS strips), Steve painted the copper winding with red electrical Varnish (Q dope).

Steve also took all possible precautions to minimize contact and connection resistance. Notice in Fig 13-27 how the wide and thick aluminum strip is used as mechanical support and electrical connection as well. Steve also used a redundant connection made of an insulated wire.

3.5.1.2. Other big guns on 80 meters using shortened Yagis

Rich, K7ZV, lives in Oregon, close to the California border, in a county with lots of small mountains. One of those is his mountaintop. From his house the terrains slopes down in all directions. This is a real dream QTH, although there must be an important degree of signal scattering from the many other hilltops in just about all directions. Rich says it does not make much difference whether he has his antenna retracted to 16 meters or at its full 40 meters of height—The effective height is impressive in both cases. Fully retracted the antenna is about 40 ft. high. His 3 element is based on the "W6RJ" version of M2 aluminum, obviously with W6ANR/K7ZV loading coils.

Using *Yagistress* (by K7NV), K7ZV redid the mechanical design of the standard 80-M3 80-meter Yagi by M2, which was very similar to the original KLM design. The original boom was lengthened from 17.5 to 20 meters and strengthened as well. The new design uses a totally different boom and element-guying system, which allows tipping the boom vertical to access the director or reflector without the need for a crane. This makes assembly, installation and maintenance much easier. The elements are different as well, and while maintaining a similar taper schedule they are much stronger, with double-wall tubing throughout most of the of the span. With guying just inboard of the coils at the center of the half element they are also much stronger.

Rich's most impressive tool is his bucket truck, which is permanently parked at the base of the antenna tower. Need to work on the antenna? Need to measure something? Just start the truck, up goes the basket and within minutes you can reach any part of the antenna, at a height where it still works!

Rich's 3 element Yagi has been designed for 3.8 MHz but can be switched to the CW end of the band by inserting the appropriate loading coils in the 3 elements. **Fig 13-28** shows the center lengthening and hairpin coil for the driven element, in a plastic box together with switching relays.

Bob Ferrero, W6RJ, eminent low band DXer and contester and owner of HRO, was formerly K6AHV. Bob is well-known in the world of 80-meter DXers. He built his dream station on a 1000-meter high mountaintop about 70 km from his home on the east side of the San Francisco Bay. No neighbor within maybe 5 or 10 km—no noise, just nature and the sky.

In **Fig 13-29** you can see his 40-meter tall microwave tower, topped with the heaviest duty tower that US TOWERS makes. When fully extended, this can make the 3 element 80-meter Yagi almost disappear in the sky…

The crank-up tower and tilting boom allow all antenna work to be done from the large platform at the 40-meter level on the microwave tower. What an amazing sight! The tower and the shack sit right on top of the ridge, sloping quite steeply in most directions. This makes the effective height of the antennas *very* high. Bob told me it hardly makes any differ-

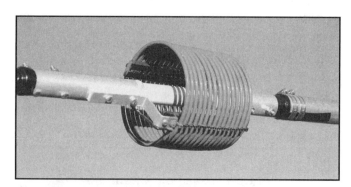

Fig 13-27—Steve Babcock, VE6WZ, also makes his own high-Q loading coils. The coils are air-wound except for the two lateral strips of insulating material to provide mechanical rigidity. Note the redundant connection made with insulated wire.

Fig 13-28—K7ZV's driven element feed system. Relays add in loading inductors to switch from phone to CW band segments. Center coil is the hairpin matching inductor.

Fig 13-30— OH2BH's 3-element 80-meter Yagi was flown by helicopter from the assembly area to the tower, which is in the middle of the woods about 1 km away.

ence whether he had the crank up extended or not. I think it does not make a bit of difference, given his QTH!

Since winter 2003/ 2004 Bob now also runs a wire Four-Square for 160 meters from the top of the tower, and that does the trick for him on Topband. He also operates the station on his mountaintop from his home by UHF remote control.

World famous Martti Laine, OH2BH, is another owner of a 3-element 80-meter Yagi according to the W6ANR design with K7ZV coils. See **Fig 13-30**. The antenna was built using the "W6RJ" version of M2 aluminum, but was heavily reinforced with three-dimensional truss wires on the elements, to help them cope with Scandinavian wind and ice loads. Martti can also operate his station in the middle of the woods from his Helsinki downtown QTH using ADSL telephone lines.

Fig 13-29— 3-element shortened 80-meter Yagi of W6ANR design with K7ZV coils at W6RJ's mountaintop QTH.

Steve Babcock, VE6WZ, was obviously inspired by the work done by W6ANR and K7ZV when he set out to build a 2-element 80-meter Yagi that would fit into his city lot on a 28-meter crankup tower. Steve told me, "*You must also know that it was your book that inspired me to build it.*" Steve has a Web site where he describes the design and construction of his antenna in great detail (**www.qsl.net/ve6wz/).** Steve also said, "*This 2-element Yagi is by far the best homebrew antenna I have ever built. It has substantially exceeded my expectations! I have built many, many other homebrew antennas over the years, but none have ever performed so well.*" Whereas K7ZV uses his bucket truck to get access to his antenna for measuring and experimenting, Steve uses the roof of his house as a work platform. His heavy-duty 28-meter crankup tower is just adjacent to his house. From the roof he has access to the whole antenna. To achieve the results Steve does, his antenna must be very carefully tuned. It certainly is not a plug-and-play design. But Steve is usually the first, if not the only one, I hear in Europe when the band decides to open up from his Northerly location. It is amazing what Steve does from a residential area.

Charley, WØYG, is another addict of large loading coils. After a disappointing experience with a Yagi using linear loading stubs, he rebuilt his antenna with W6ANR/K7ZV loading coils. He now says, "*If you doubt the efficiency of this design listen some time in an 80-meter pileup. The guys who really rule the roost use those W6ANR converted beams. They are all over the west coast, some in the interior of the States and some in Europe. The design is practical, tuning is a snap and results are, well, fantastic!*"

3.5.2. W7CY 2-element 80-meter Yagi

Rod Mack, W7CY, developed an interesting 2-element, capacitively end-loaded 2-element Yagi for 80 meters. He published pictures and some raw dimensions on the Internet Web pages at **www.ulio.com/ants.html**. See **Fig 13-31**). He claims in excess of 20 dB F/B, which is in line with other antennas using capacitively coupled element tips, similar to the Moxon-type Yagis popularized by L. B. Cebik, W4RNL.

I modeled W7CY's array using *EZNEC* and noticed some very interesting things. The layout of the array is shown in **Fig 13-32**. It uses a 24.4-meter long boom. At both ends of the boom he mounted 11-meter long spreaders, which serve as capacitive loading devices for the sloping elements. The center of the two elements is supported by an 11-meter boom, which is 3.6 meters above the main boom. This means that the central part of the elements is an inverted-V.

Rod stated that the capacitive loading spreaders are, of course, insulated from the boom, since they carry very high voltages at their ends. This is where he tunes the array by varying their length. Modeling demonstrates that the distance between the tips of the loading devices near the boom is a very critical item! By varying this distance, you can control the coupling between the two elements to a fine degree. Again, this is similar to the "Moxon" type of Yagi design.

For an element spacing of 11 meters, the ideal current relationship in the two elements is 1:1 for current magnitude and 125° for phase shift. You can model this easily by "planting" two verticals spaced 11 meters (on 3.8 MHz) and feeding them that way. If the tips are too close together you will have too much current in the reflector. With parts near the element tips so close together you can create a great deal of capacitive coupling.

I developed a system by which I loaded the sloping elements in two different ways (both capacitively): By changing the length of the horizontal loading wires (the support structure), and by adding some vertical aluminum tubing at the same point. By judiciously weighing the ratio of these two capacitive loading devices, I arrived to a point where the required current ratios in both elements were obtained. At that point the F/B was over 24 dB, together with a feed-point impedance of very close to 50 Ω.

I could not obtain these results by loading the elements with only the horizontal "spreader" loading tubes; they gave me too much coupling between the two elements, as their tips came too close together. I found out by inserting a resistor in the reflector, which reduced the current magnitude with a sacrifice of gain. The same result was obtained by spacing the tips of the horizontal "loading wires" further apart.

The array has a very nice bandwidth as well. **Fig 13-33** shows the radiation patterns for the array over a span of 40 kHz, without retuning the reflector. It is of course possible to tune the reflector by installing a variable capacitor in the center of the element and changing its capacitance as you move around on the band. By doing so the same F/B can be achieved (20 to 25 dB) anywhere in the band. Even without doing any retuning, this array has an SWR of less than 2:1 over more than 150 kHz. **Fig 13-34** shows some essential array data for a frequency range of 3750 to 3890 kHz.

3.5.2.1. Duplicating the W7CY antenna

First of all, it is important to know that the dimensions

Fig 13-31—Rod, W7CY, stands proudly on the boom of his 2-element Moxon-type loaded 80-meter Yagi.

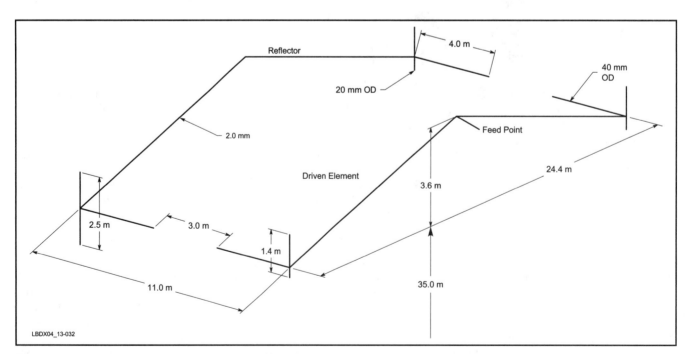

Fig 13-32—Approximate dimensions of the W7CY 2-element 80-meter capacitively coupled Yagi.

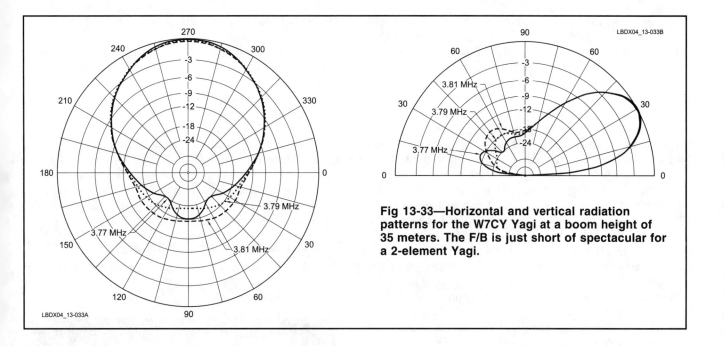

Fig 13-33—Horizontal and vertical radiation patterns for the W7CY Yagi at a boom height of 35 meters. The F/B is just short of spectacular for a 2-element Yagi.

shown in Fig 13-32 are ballpark figures. These are by no way "build-and-forget" dimensions. I modeled this antenna with several modeling programs: *AO* (*MININEC* based), *ELNEC* (*MININEC*), *EZNEC2* (*NEC-2* based) and *EZNEC-PRO* (*NEC-4.1* based). Although all of these programs achieved essentially the same radiation patterns after fine-tuning, these results were all obtained with slightly different dimensions. The main reason for this is the inability of some of these programs to handle wires with vastly different diameters. The problem lies in modeling the capacitance hat, which is made out of aluminum tubing, while the rest of the element is made of a much thinner wire. As an example, with the dimensions optimized for 3.79 MHz using *NEC-2*, the frequency shifted down approximately 80 kHz using *NEC4-1*. The dimensions shown in Fig 13-32 are the results of modeling with the *NEC4-1* engine, which is supposed to give the most accurate results in this case.

Whereas these models may not give us the exact lengths for a precise operating frequency, they give us a good idea of what can be achieved so far as directivity is concerned. Further, it is very important to model in order to determine the resonant frequency of the parasitic reflector and the driven element. We can use this information to tune the array in real life.

The models tell us that for an array optimized for 3.79 MHz, the reflector is resonant on 3.80 MHz, and the driven element by itself is resonant on 3.94 MHz. This clearly shows what mutual coupling does!

You must determine the resonant frequency of the driven element and the director with the other element removed from the model. For example, I decoupled the other element by inserting a load of R = 9999 Ω and X = 9999 Ω in the center of the element. Armed with this information, here is how we can tune the actual array:

1. Build the array according to the dimensions of your model. Be prepared, however, to change the dimensions of the loading devices.
2. Cut a feed line that is λ/2 at the resonant frequency of the reflector (in our case 3.8 MHz). For RG-213 cable the length is (0.66 × 299.8) / (3.8 × 2) = 26.03 meters. If you cannot reach the end of the feed line, you can use a full wavelength feed line as well (52.06 meters).
3. Connect this feed line to the reflector, raise the antenna to final height, decouple the driven element (leave the center open) and now adjust the loading devices symmetrically on both sides until you get resonance on 3.8 MHz. That takes care of tuning the reflector. Remove the feed line and close the reflector.
4. Now connect the feed line to the driven element, and prune the length of the loading devices for minimum SWR at the design frequency (3.79 MHz).

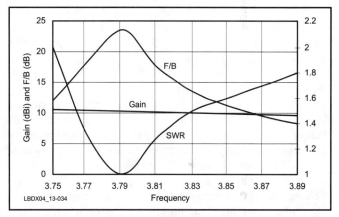

Fig 13-34—Gain over average ground, F/B and SWR for the W7CY 2-element shortened 80-meter Yagi as a function of frequency.

3.5.3. The K6UA 2-element 80-meter Yagi

Dale Hoppe, K6UA, must have been around 80-meter DXing almost as long as the band has been there. Some old timers may remember Dale as "W6 Very Strong Signal." With his beautiful hilltop QTH on an avocado plantation, not only avocados grow well, but also antennas!

Although from this way-above-average QTH almost any antenna would work, Dale has been an avid antenna experimenter and builder. The latest of his designs is a 2-element shortened 80-meter Yagi, which he described in *CQ* Magazine (Ref 978). It is clear that the tower and the boom, which I have seen at Dale's place for ages, were what set him on his way to develop the array (see **Fig 13-35**).

The boom of the array is a 22-meter long triangular tower, 30-cm wide. Fiberglass vaulting poles, measuring 4.5-meters long were mounted at both ends of the boom, providing the 9-meter spacing between the driven element and the reflector. The 22-meter long horizontal elements are loaded at both ends by loading stubs, as shown in **Fig 13-36**. The linear-loading stubs also serve as vertical bracing for the vaulting poles.

Dale reports raising and lowering the antenna about five times and cutting the length of the vertical trim wires to tune the array (see Ref 978). Tuning the reflector is quite critical, and changes of a few cm can make an important difference in antenna Q. If the reflector is too short, the feed-point impedance will be much lower than 50 Ω and the bandwidth will be very narrow. When properly tuned, the array exhibits a gain, F/B and SWR pattern as shown in **Fig 13-37**. The antenna has a fairly high bandwidth above its design frequency, and the gain remains fairly constant as well. Depending on the wave angle considered the F/B is >20 dB over approximately 50 kHz. This is quite a good figure for a 2-element Yagi.

This array is very similar to the W7CY array, the only difference being a slightly shorter elements, closer spacing and partial inductive loading of the elements. As with the W7CY array, the position of the tips of the loading elements facing one another (the tips of the stubs), determine the degree of coupling of the 2 elements in the array. The amount of coupling is quite critical to obtain maximum F/B ratio. In the model I used, a physical spacing of approximately 4 meters between the tips of the loading stubs gave the best results.

Instead of using aluminum-tubing capacitance hats, which cause a modeling problem due to the vast difference in diameter between the wires in the antenna (2 mm) and the tubes, I decided to keep the original K6UA approach and use "dangling" wires to tune the array. The length of these wires can be trimmed to change the frequency of the elements. To keep them more or less in place in the breeze, you could hang small weights at the end of those wires. The dimensions given in Fig 13-32 were obtained by modeling through *EZNEC* (*NEC-2* engine).

One way of tuning the reflector is to watch the SWR about 30 kHz below the design frequency and adjust the reflector (shorten it) until the SWR is about 2:1 (see **Fig 13-38**). Tuning of the driven element to obtain lowest SWR at the design frequency is the last step in tuning the array. The antenna can be fed directly with a 50-Ω feed line via a choke balun.

3.5.3.1. Alternative tuning method

Raising the Yagi repeatedly in order to tune the antenna for best F/B may not be the most attractive job. There is an alternative way though, that brings you an additional advantage. I intentionally lengthened the reflector a substantial degree, and made the vertical tuning wires about 1 meter longer (6.1 vs 5.13 meters). Now we have a reflector that is way too long and we can electrically tune it to where we want it, by simply inserting a capacitor in the center of the element. In the model case I achieved 23 dB F/B on *any* frequency between 3.75 MHz and 3.87 MHz, simply by adjusting the value of the capacitor from 663 pF on 3.75 MHz and 354 pF on 3.87 MHz (see **Table 13-9**)! **Fig 13-39** shows the vertical radiation patterns obtained at various frequencies in that range.

The same approach for remotely tuning the reflector can of course be used on the W7CY 2-element Yagi. In conclusion, this design is a good example of what can be achieved using locally available materials and a good deal of knowledge, insight and imagination. Listen for Dale's big signal on 80 meters!

3.6. Horizontal Wire Yagis

Yagis require a lot of space and electrical height in

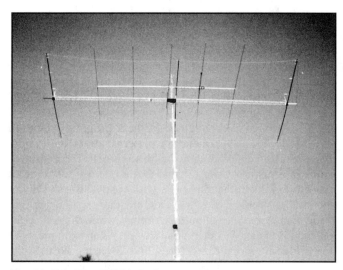

Fig 13-35—The K6UA 2-element 80-meter array mounted on a Telrex rotating pole, just under a 5-element 20-meter Yagi, which is dwarfed in comparison.

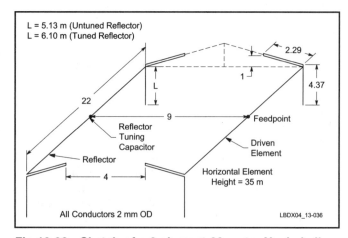

Fig 13-36—Sketch of a 2-element 80-meter Yagi similar to K6UA's design. The dimensions are approximate, and were used to calculate the patterns of the array. The wires of the loading stubs in this model are separated 20 cm.

order to perform well. Excellent results have been obtained with fixed-wire Yagis strung between high apartment buildings, or as inverted-V shaped or sloping elements from catenary cables strung between towers. There are few circumstances, however, where supports at the right height and in a favorable direction are available. When using wire elements, it is easy to determine the correct length of the elements using a *MININEC* or a *NEC*-derived modeling program (eg, *EZNEC*). Wire Yagis have been described in detail in Chapter 12 (Other Arrays).

3.7. Vertical Arrays with Parasitic Elements (Vertical Yagis)

Do vertical arrays with parasitic elements work on the low bands? If you are a Top Bander, look in your log for KØHA. He's either there long time before anyone else from his area, or he's there all by himself, or he's there much stronger than anyone else. Bill Hohnstein, KØHA, swears by his vertical Yagis. His farm grows vertical parasitic arrays in all sorts and sizes (see **Fig 13-40**).

There is no need to use full-size elements for putting

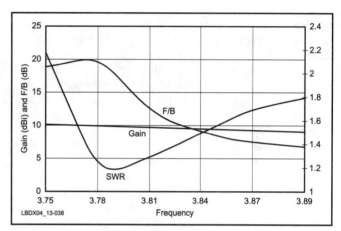

Fig 13-38—Gain, F/B and SWR for the 2-element K6UA 80-meter Yagi over average ground.

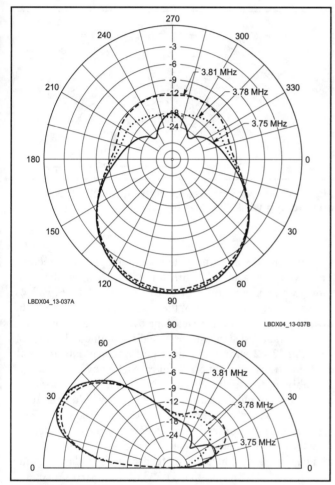

Fig 13-37—Radiation patterns for the K6UA Yagi. The F/B is 23 dB at the design frequency for an elevation angle of 30°.

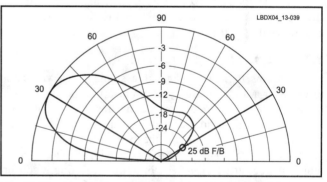

Fig 13-39—Vertical radiation pattern for the K6UA 2-element 80-meter Yagi, showing the directivity that can be obtained across 150 kHz of bandwidth by remotely tuning the reflector with a variable capacitor.

Table 13-9

Values of the Capacitive Reactance (X_C) and Corresponding Capacitance (pF) to Tune the K6UA for Best F/B Across a Wide Spectrum. Curve SWR (1) is for the Array Tuned for Best F/B at >23 dB (R_{rad} About 37 Ω). SWR (2) is for the Array Adjusted for 50-Ω Impedance (F/B Approximately 15 dB).

Freq (MHz)	3.75	3.78	3.81	3.84	3.87	3.9
X_C (Ω)	−64	−77	−90	−103	−116	−129
C (pF)	663	547	464	403	354	317
SWR (1)	2.0	1.5	1.3	1.5	1.9	>2.0
SWR (2)	1.4	1.1	1.2	1.6	2.0	>2.0

Fig 13-40—Bird's eye view from the driven element of the 160-meter array at K0HA. The first director, on the left of the picture, is hiding from the second director. On the right a line of elements for the 80-meter array aims at Europe. It also appears that Bill enjoys some of the best ground conductivity around. No wonder he's loud!

together effective and efficient arrays. Bill uses a shunt-fed 32-meter tower as the driven element, while his parasitic elements are approximately 26-meters high, and top loaded.

It is obvious that in a parasitic array, neither the feed method nor the exact electrical length affect the performance of the array. Shunt or series feeding can be used without preference. The elements should, however, not be much longer than λ/4.

I will take you on a little tour of some of the classic parasitic arrays, and point out what you should watch for if you want to build one. A modeling program seems to be essential, as you probably will be using existing towers as part of the antenna, and you will need to do some specific modeling. Watch out that you understand what the modeling program tells you, and be aware of what it does *not* tell you.

3.7.1. Turning your tower guy wires into parasitic elements

Several good articles have been published on this subject (Refs 981, 982 and 983). I recommend reading those if you plan to try one of these antennas.

3.7.1.1. One sloping wire

It seems logical to think of a sloping guy wire as a reflector or a director. But, you can also use the sloper as a driven element, and use the tower as a parasitic element! This last case may not be so practical, since in many cases it will probably not be possible to tune the tower to the exact required length. The tower could be tuned by changing its length, or by tuning it—eg, at its insulated bottom by inserting a coil or capacitor to ground.

I analyzed the case of a 40-meter high tower (25 cm equivalent effective diameter) with a 2-mm-OD wire measuring 40.5 meters, sloping from the top of the tower with a length of insulated rope to the ground point, 27 meters away from the tower base. This is an appropriate distance for the guy-wire anchor points for a 40-meter high tower. The top of the sloping wire is 4.7 meters from the tower. These dimensions are valid only for conductors of the diameter indicated. The resonant frequency of the vertical conductor by itself is 1.78 MHz, and the sloping wire is resonant at 1.822 MHz. These data make it possible to duplicate the array with conductors of different diameters. All you have to do is to dip the wires to the listed frequencies.

With the tower fed, the wire acts as a perfect reflector, giving 3.4-dB gain over the tower by itself, and a useful 17 dB of F/B at 1.83 MHz. With the tower grounded, and feeding the

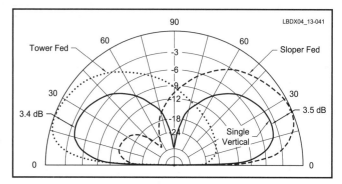

Fig 13-41—Vertical radiation patterns for the tower and one sloping wire. See text for details.

sloping wire, the array now shoots in the opposite direction. The tower now acted as a director, and the gain is about the same (3.5 dB), with a F/B of 15 dB on 1.83 MHz. **Fig 13-41** shows the vertical patterns of these arrays, as compared to the tower by itself. Modeling was done over average ground, and a perfect radial system was assumed for both conductors.

Fig 13-42 shows the main performance data for both configurations. In most practical cases, however, you would probably try to use the sloping wire as either a director or as a reflector, in which case the sloping wire would need to be tuned with a reactive element (L or C). See **Fig 13-43**. This specific case is interesting as it demonstrates that perfect directivity and pattern reversal can be obtained without need of a capacitor or inductor. The feed-point impedance, in the case of the fed tower, is $43 + j\,62\,\Omega$ on 1.83 MHz. When the sloping wire is fed, its feed impedance on 1.83 MHz is $43 + j\,62\,\Omega$. In both cases a small L-network (or just a series capacitor, if you can live with about 1.3:1 SWR at the design frequency) should be used to match the antenna to a 50-Ω feed line.

Using the same sloping wire (dimensions, placement) I tuned it by a series capacitor ($X_C = -j\,50\,\Omega$) for best performance. With less than 4 dB of F/B the gain was 2.9 dB over the vertical by itself. Not a very spectacular result. This was also reported by J. Stanley, K4ERO (Ref 982). The gain obtained is also almost 1-dB less than for the reflector case. Note that the results of this configuration are far inferior to those obtained when feeding the sloping wire and using the tower as a director (see Fig 13-41).

Most of the 40-meter tall towers used as 160-meter λ/4 verticals are probably guyed in four directions. That means

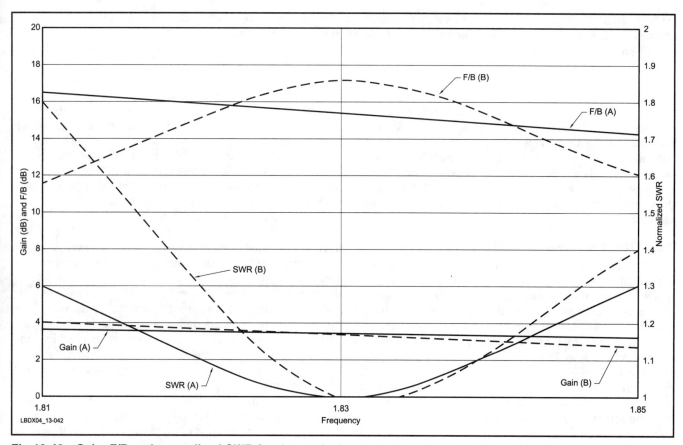

Fig 13-42—Gain, F/B and normalized SWR for the vertical and sloping wire array. Case A is the array with the sloper being fed, and the tower acting as a reflector. Case B is the tower being fed with the sloper acting as a reflector. See text for details.

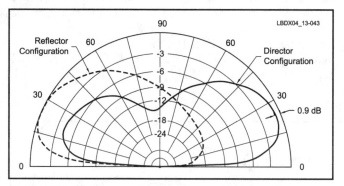

Fig 13-43—The same sloping wire tuned as a reflector and as a director. The director configuration yields a poor F/B and mediocre gain.

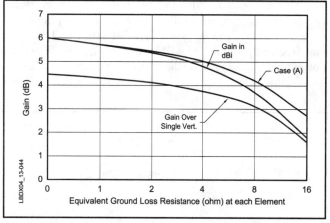

Fig 13-44—Gain of the 3-element guy-wire array as a function of the equivalent ground-loss resistance. Case A is for a driven element with a fixed 1-Ω loss resistance.

that we probably can hang four sloping wires from the top. This is the next case I investigated.

3.7.1.2. A reflector and a director

Continuing with the same physical configuration (the guy cables being anchored approximately 29 meters from the base of the 40-meter tower), the combination of using both a director as well as reflector was obvious. This combination can typically boost the gain another 1 dB, but has the disadvantage of reducing the F/B by about 6 dB. Tuning the director with a series reactance of $-j\,70\,\Omega$ is a compromise situation that does not yield maximum gain, but still yields a more or less acceptable F/B ratio. This compromise was described in detail by Christman (Ref 389). The resonant frequencies of the parasitic elements, when fully decoupled from one another and from the driven element are: director: 1.952 MHz, reflector: 1.822 MHz.

It is clear that for all of the above arrays it is important to have a good ground radial system, not only for the driven element but also for the parasitic elements. **Fig 13-44** shows

Yagis and Quads 13-29

the gain of the array as a function of the equivalent ground loss resistance. Case A is for a radiator with an almost-perfect ground radial system (= 1 Ω) but for varying ground loss resistances at the parasitic elements. Whereas 1-Ω ground systems yield a gain of 5.6 dBi for the array, the gain drops to 3.67 dBi if all elements have an equivalent ground loss of 8 Ω. If we have a 1-Ω loss resistance for the radiator, but a rather mediocre loss resistance of 8 Ω for the parasitic elements, the gain drops to 4.13 dBi. This shows that there really is very little room for a poor ground system under the parasitic elements too. It is obvious that it makes no difference whatsoever how the driven element is fed, series or parallel. **Fig 13-45** shows the radiation patterns for three types of slant-wire parasitic arrays.

3.7.2. Three-element vertical parasitic array

The arrays I analyzed in the last section all showed rather substantial high-angle radiation, which is caused by the horizontal component of the sloping parasitic wires. To improve on that situation we can try to bring the sloping elements as vertical as possible. If you do not use the parasitic-element wires as guy wires for the tower, you can consider the configuration shown in **Figs 13-46**.

The four cross arms mounted at the top of the tower support four sloping wires. The tower itself is a quarter-wave vertical. The bases of the parasitic elements are 0.125 λ away from the driven element. All four sloping wires are dimensioned to act as directors. When used as a reflector, a parasitic element is loaded with a coil at the bottom. The two sloping elements off the side are left floating. This array has a very respectable gain of 4.5 dB over a single vertical. At the main elevation angle the F/B ratio is an impressive 30 dB, as can be seen from the patterns in Fig 13-46.

You can "grid dip" the sloping wires to tune them. Make sure the driven element as well as the other three sloping wires are left floating when dipping a parasitic element. The resonant frequency should be 4.055 MHz. (f_{design} = 3.8 MHz). You can, of course also dip the wires with the loading coil in place to find the resonant frequency of the reflector. Again, all other elements must be fully decoupled when dipping one parasitic element. The resonant frequency for the reflector element is 3.745 MHz (f_{design} = 3.8 MHz).

If you want to totally eliminate the horizontal high angle radiation component, you can hang top-loaded elements from catenary cables, as shown in **Fig 13-47**. In this 160-meter (f_{design} = 1.832 MHz) version the parasitic elements are top loaded with sloping T-shaped wires. These wires can be supported along the catenary support cable or may be an intrinsic part of the support structure. Such slanted top-loading wires do not produce far-field horizontal radiation because they are symmetrical with respect to the vertical wire.

To make this an array that can be switched in four directions, you should slope four catenary cables at 90° increments from the top of the driven-element tower. With the appropriate hardware you can connect the parasitic elements directly to ground to serve as a director, to ground via a loading coil to serve as a reflector. Or you can leave the element floating, with the unused elements off the side of the firing direction. It's a good idea to provide a position in your switching system to have all elements floating, in which case you will have an omnidirectional antenna. This may be of interest for testing the array, or for taking a quick listen around in all directions.

The example I analyzed uses 23-meter-long vertical parasitic elements. Each is top-loaded with a 19.72-meter long sloping top wire that is part of the support cable. As the length of the top-loading wire is the same on both sides of the loaded vertical member, there is no horizontally polarized radiation from the top-loading structure.

The four parasitic elements are dimensioned to be resonant at 1.935 MHz. The same procedure as explained above for the 80-meter array can be used to tune the parasitic elements. When used as a reflector, the elements are tuned to be resonant at 1.778 MHz. This can be done by installing an inductance of 3.65 µH (reactance = 42 Ω at 1.832 MHz) between the bottom end of the parasitic element and ground.

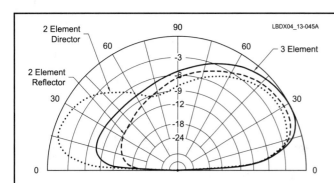

Fig 13-45—Vertical and horizontal radiation patterns (over excellent ground) of three types of slant-wire parasitic arrays: reflector parasitic, director parasitic and reflector plus director parasitic. These patterns are valid for parasitic elements spaced 0.18 λ from the driven element (at their bases), which appears to be a typical situation for guy wires on a 40-meter guyed tower. The horizontal patterns are for an elevation angle of 20°.

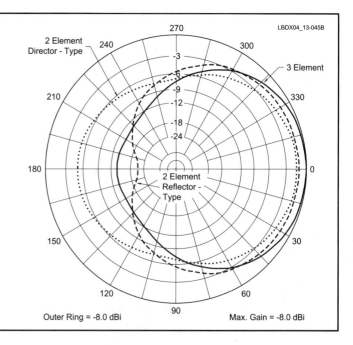

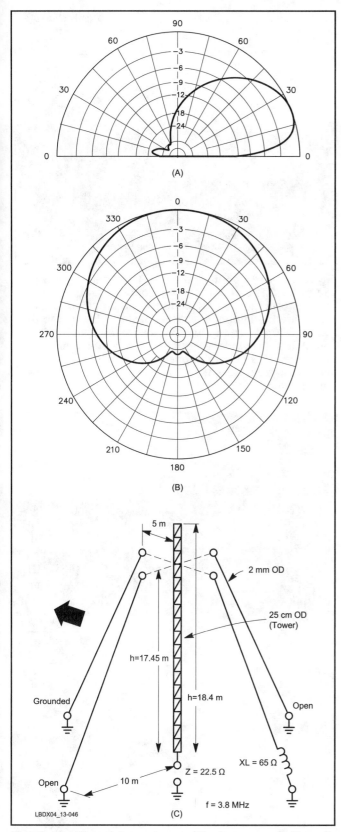

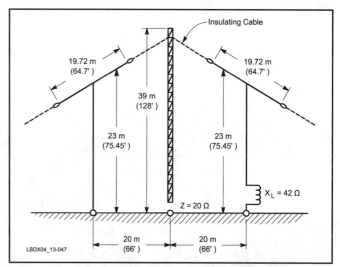

Fig 13-47—This 160-meter 3-element parasitic array produces 4.8 dB gain over a single vertical, and better than 25 dB F/B over 30 kHz of the band. With such an array there is no need for Beverage receiving antennas! The drawing shows only two of the four parasitic elements. The two other elements are left floating.

Fig 13-46—Three-element parasitic array, consisting of a central support tower with two support cross-arms mounted at 90° near the top. Two of the sloping wires are left floating, a third one is grounded as a director, and the fourth one is loaded with a coil to act as a reflector. The azimuth pattern at B is taken for a takeoff angle of 22°. Radials are required on all five ground points but they have been omitted on this drawing for clarity.

The radiation resistance of this array is around 20 Ω, and it has a gain of 4.8 dB over a single full-size vertical. **Fig 13-48** shows the radiation patterns of this array. The bandwidth behavior is excellent. The array shows a constant gain over more than 50 kHz and better than 25 dB F/B over more than 30 kHz. When tuned for a 1:1 SWR at 1.832 MHz, the SWR will be less than 1.2:1 from 1.820 to 1.850 kHz. This really is a winner antenna, and it requires only one full-size quarter-wave element, plus a lot of real estate to run the sloping support wires and the necessary radials.

The same principle with the sloping support wires and the top-loaded parasitic elements could, of course, be used with the 80-meter version of the 3-element vertical parasitic array. Tim Duffy, K3LR, made an almost exact copy of this array, after initially having used inverted Ls for the parasitic elements. Tim recognized that these inverted-L elements introduced a fair amount of horizontally polarized high-angle radiation. For a single-element vertical, this may be of very little importance, but for a parasitic element of an array, this will greatly reduce the directivity of the array, especially at high angles This can be important if the array is also used for receiving. Tim reported changing from inverted-L shaped elements to the sloping-T shaped elements and reported that the T-shaped elements work much better.

At K3LR the parasitic directors were resonated at 1.903 kHz, and the loading coils were 4.0 μH with a vertical length of 19.58 meters and a sloping-T-shaped top hat of 17.78 meters (all made of #12 copperweld wire). The array is matched to a 75-Ω coaxial feed line with an L network (see **Fig 13-49**). The measured SWR is 1.3:1 on 1.8 MHz, 1:1 on 1.83 MHz and 1.3:1 on 1.85 MHz. K3LR reports about 5 dB of gain and 30 dB of F/B at the design frequency (1.83 MHz). At 1.82 MHz the measured F/B is still 25 dB and at 1.84 MHz Tim measured 15 dB.

Tim has an omnidirectional mode, where he floats all parasitic elements. See **Fig 13-50**. Tim can't run many Bever-

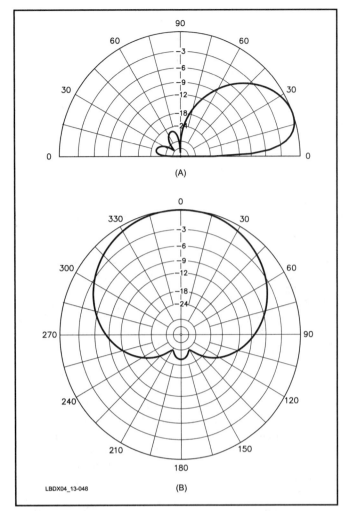

Fig 13-48—Horizontal and vertical radiation patterns for a 3-element vertical parasitic array for 160 meters. The azimuth pattern at B is for an elevation angle of 20°. Note the excellent pattern and F/B for the array.

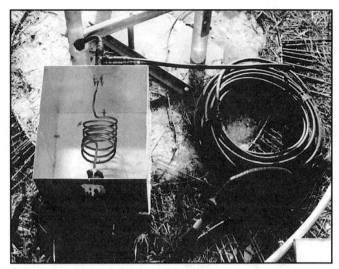

Fig 13-49—Feed point of the K3LR vertical Yagi for 160 meters. The aluminum box contains an L network to transform the array impedance (about 20 Ω) to the 75-Ω Hardline impedance. The coaxial cable coil to the right is a λ/4 shorted-stub that serves as a static drain and almost provides attenuation of the 80-meter harmonic. This is very important at a multioperator contest station! Note the 120 radials connected to the annular ring.

Fig 13-50—A plastic food box contains the Air Dux loading coil and a small relay to switch the coil in and out of the circuit at each of the four parasitic elements. The box is covered by an inverted plastic trash container. The base of each parasitic element is also equipped with a λ/4 coaxial stub and complemented with no less than 120 radials.

age antennas at his location. He has just one 1200 footer on Europe. He appreciates the excellent directivity for his 160-meter array on receive. Don't forget that in order to make such an array work correctly, you need an impressive radial system under each of the elements. K3LR uses not less than 15 km of radials in this system!

A recommended radial layout, which K3LR uses, is shown in **Fig 13-51**. Note that radials are even more important for a parasitic array than for an all-driven phased array. With parasitic arrays the gain seems to suffer even more quickly from poor ground systems, so a good radial system is mandatory.

Incidentally, computer modeling indicates that elevated radials do not work well with parasitic arrays. No matter what kind of elevated radial system I modeled (different numbers of radials, varied lengths, and different orientations), the result was a badly distorted pattern. This is logical in view of the influence of the capacitive coupling of the raised resonant radials. Compare this situation with what was experienced with the top-loaded 80-meter Yagis designed by W7CY and K6UA. With phased arrays using current-forcing methods, the feed method itself is responsible for overcoming the effects of mutual coupling due to the proximity of the wires.

3.7.3. The N9JW – K7CA array with parasitic elements

In chapter 7, Section 1.32 I covered "Parasitic receiving arrays." Since you cannot make parasitic receiving arrays with lossy elements, such arrays are equally good transmitting arrays! The various parasitic arrays that were built in the Utah desert really put N7JW and K7CA in the front row when the show is on to Europe on Topband.

3.7.4. The K1VR/W1FV Spitfire Array

Fred Hopengarten, K1VR, and John Kaufmann, W1FV, designed a somewhat novel 3-element parasitic array, which they called the "Spitfire Array." **Fig 13-52** shows the layout of the antenna. John, W1FV, described the array as a 3-element parasitic array with a vertical tower as the driven element and two sloping-wire parasitics, one a director and one a reflector. He adds that the parasitic elements are *not* grounded and do *not require a separate radial system of their own.* Rather the parasitic wires are folded around to achieve the required λ/2 wave resonances. The tower driven element does have its own radial system.

The antenna was modeled with *EZNEC*, using the *MININEC* ground analysis method (the *NEC-2* ground analysis method cannot be used because the driven element is a grounded element). Over average ground, the model shows a gain of 4.8 dB over a single vertical at a takeoff angle of 23°.

The antenna has a substantial amount of high-angle radiation and its pattern resembles that of a EWE antenna (see **Fig 13-53**). The high angle radiation is mainly caused by the radiation of the bottom half of the sloping half-wave parasitic elements. You can consider the bottom half of each parasitic element as a single radiating radial that is bent backward toward the driven element.

To change directions, you use a relay to switch in or out an additional wire segment on the lower horizontal portion of each parasitic to change from director to reflector operation. It is important to point out that this switching happens at high voltage points, and a well-insulated vacuum relay is certainly no luxury.

A model for four switching directions can be constructed by adding an identical set of parasitic wires (for a total of four wires) oriented at right angles to the original two. Only two wires at a time are active. The other two are detuned so they don't couple. Simply grounding them appears to accomplish this.

3.7.4.1. Critical analysis

If you compare this array with the classic 3-element parasitic array as described in Section 3.7, you will notice that the main difference is that this Spitfire array claims *not to require radials* for the parasitic elements. We learned in Chapter 9 about ground losses and radial systems for vertical antennas. I also pointed out that the Spitfire array has been modeled with a *MININEC*-based modeling program, which means that a perfectly conducting ground is assumed in the near field of the antenna. This is certainly not true in real life.

The bottom half of the sloping parasitic elements are very close to ground, and undoubtedly will cause a great deal of near-field absorption losses in the lossy ground, unless the ground is hidden from these low wires by an effective ground screen. The model used to develop this antenna does not take any of this into account. What does that mean? It does *not* mean that the antenna will not be able to give good directivity. But it means that the quoted gain figures are probably several dB higher than what can be accomplished in real life, if no radial or ground screen system is used that effectively screens the lossy ground under the array from the antenna. This could be accomplished by using extra long radials on the driven element that extend at least λ/8 beyond the tips of the parasitic elements. This would mean radials that are at least 60-meters long, with their tips separated not more than 0.015 λ (see Chapter 11). This means that 157 radials, each 60-meters long, fulfill this requirement. Only under these circumstances will we achieve the same gain as with ground-mounted quarter-wave parasitic elements, each using their own elaborate radial system (à la K3LR).

I modeled the same antenna, but using λ/4 grounded

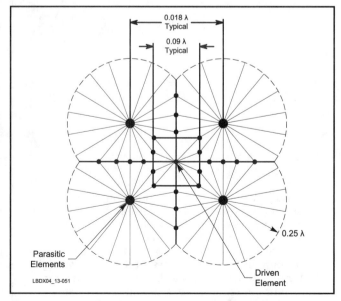

Fig 13-51—Radial layout used at K3LR on his 3-element 160-meter parasitic array. A total of 15 km of wire is used for radials in this array!

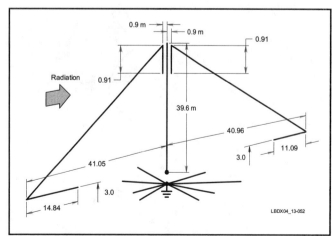

Fig 13-52—Layout and dimensions for the 160-meter Spitfire array.

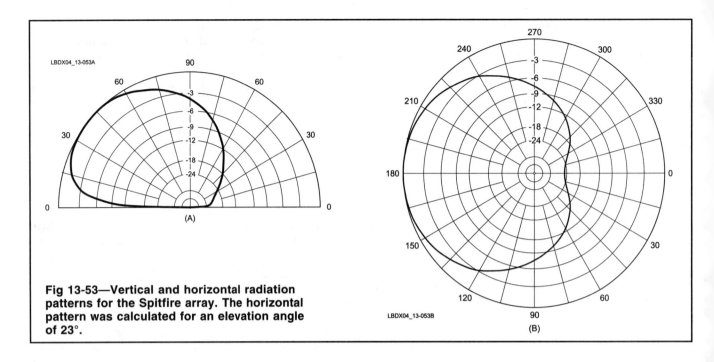

Fig 13-53—Vertical and horizontal radiation patterns for the Spitfire array. The horizontal pattern was calculated for an elevation angle of 23°.

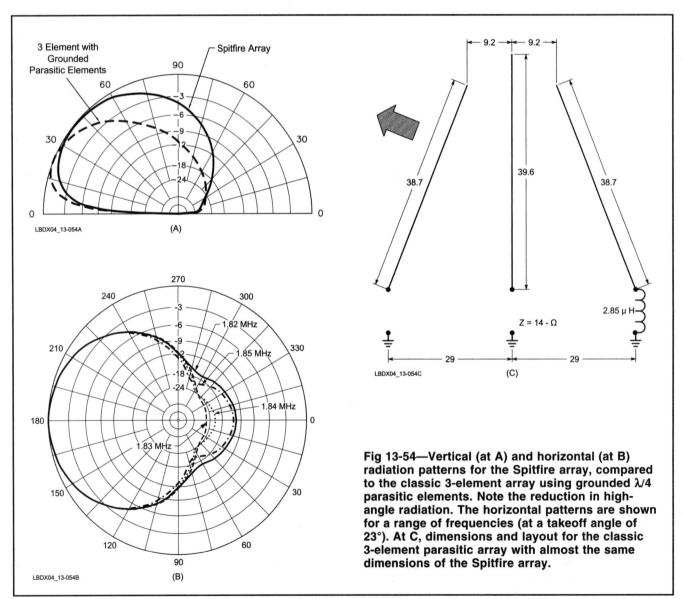

Fig 13-54—Vertical (at A) and horizontal (at B) radiation patterns for the Spitfire array, compared to the classic 3-element array using grounded λ/4 parasitic elements. Note the reduction in high-angle radiation. The horizontal patterns are shown for a range of frequencies (at a takeoff angle of 23°). At C, dimensions and layout for the classic 3-element parasitic array with almost the same dimensions of the Spitfire array.

13-34   Chapter 13

parasitic elements, and maintained the same average spacing from the tower. The sloping parasitic elements are both 38.7-meters long, spaced 9.2 meters from the tower at the top and 29 meters from the tower at the bottom. The tops of the parasitic elements are at 33 meters, which is 6 meters below the top of the 39-meter tower. The directors can be tuned to become reflectors by loading them with a coil with an inductance of 2.85 μH. This classic antenna has 0.7 dB more gain than the Spitfire over a perfect ground, and has a F/B and bandwidth that is comparable to what's been calculated for the Spitfire antenna. Most important is that the antenna does not show the high-angle radiation associated with the Spitfire array (see **Figs 13-54** and **13-55**).

3.7.4.2. Conclusion

Modeling tools are fantastic, but they are tools. Each tool has its limitations. As users of these tools, we should be aware of their limitations and know how to handle them. Elevated radials, half-wave parasitic elements, voltage-fed verticals, etc, do not have any magic properties. They do not make real ground vanish. It's still there, and if it's close to any radiating wires, it will cause losses, what we call the *near-field absorption losses*.

Modeling programs based on *MININEC* use a perfect near-field ground, which means that the results from those models do not take into account these real-world losses. If you would use a *NEC-4* based modeling program, where you can ground the driven element, you would likely still arrive at gain figures that are too high. This is because of a widely recognized flaw in *NEC* that results in too-low near-field losses for wires that are close to ground (see Chapter 9).

All parasitic vertical arrays with grounded elements suffer from the drawback that the real gain rapidly diminishes when the resistive connection loss of an imperfect ground system is considered. This can easily be modeled on a *MININEC*-based program by inserting a small resistor in series with the elements at their connection to ground. Not having a direct ground connection for the parasitic elements (as in the Spitfire) does not mean that there are no ground-related losses. The losses here are the near-field absorption losses, associated with low-to-the-ground wires, and these cannot be properly modeled with today's modeling tools. This does not detract from the fact that they are there, and can account for several dB of signal loss!

The Spitfire is an array that has its merits. The extra high-angle radiation may be an asset under certain circumstances, like in contesting where some extra local presence is welcome. Potential builders should know that a good ground screen is as essential with this antenna as it is for an array using grounded near-quarter-wave parasitic elements. Sorry, but again, there is simply no free lunch!

3.8. Yagi matching systems

The matching systems for Yagis I describe in this section are not only valid for the low bands. They work on higher frequencies just as well. I will cover the concept, design and realization of various popular matching systems used with Yagis, including:

- Gamma match
- Omega match
- Hairpin match
- Direct feed

My YAGI DESIGN software contains modules that make it possible to design these matching systems with no guesswork.

3.8.1. The Gamma match

In the past, Gamma-match systems have often been described in an over-simplifying way. A number of homebuilders must have gone half-crazy trying to match one of W2PV's 3-element Yagis with a Gamma match. The reason for that is the low radiation resistance in that design, coupled with the fact that the driven-element lengths were too long as published. The driven element of the 3-element 20-meter W2PV Yagi has a radiation resistance of only 13 Ω and an inductive reactance of + 18 Ω at the design frequency for the published radiator dimensions of 0.489661 λ (Ref 957). Yagis with such low radiation resistance and a positive reactance cannot be matched with a Gamma (or an Omega) match. It is necessary to shorten the driven element to introduce the required capacitive reactance in the feed-point impedance!

Yagis with a relatively high radiation resistance, say 25 Ω, or with some amount of capacitive reactance, typically –10 Ω, can easily be matched with a whole range of Gamma-match element combinations.

Fig 13-56 shows the electrical equivalent of the gamma match. Z_g is the element impedance to be matched. The gamma match must match the element impedance to the feed-line impedance, usually 50 Ω. The step-up ratio of a Gamma match depends on the dimensions of the physical elements (element diameter, Gamma-rod diameter and spacing) making up the matching section. **Fig 13-57** shows the step-up ratio as a function of the driven-element diameter, the gamma rod diameter and spacing between the two.

The calculation involves a fair bit of complex mathematics, but software tools have been made available from different sources to solve the Gamma-match problem. The YAGI

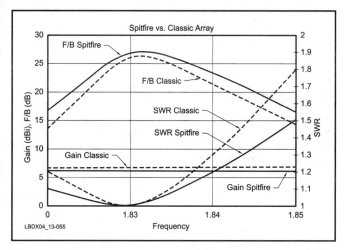

Fig 13-55—Performance characteristics of the Spitfire array compared to a classic 3-element parasitic array of essentially the same dimensions (Fig 13-54). Note that the gain for the Spitfire can only be achieved with an extensive radial or ground-screen, which is mandatory to prevent several dB of near-field ground-absorption losses of the half-wave elements that are very close to ground.

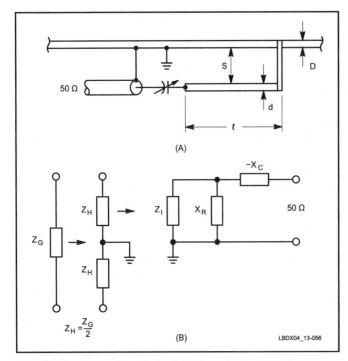

Fig 13-56—Layout and electrical equivalent of the Gamma match.

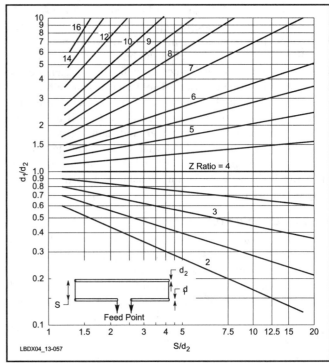

Fig 13-57—Step-up ratio for the Gamma and Omega matches as a function of element diameter (d2), rod diameter (d1) and spacing (S). (*After The ARRL Antenna Book.*)

DESIGN software addresses the problem in one of its modules (MATCHING SYSTEMS), as does *YW* (Yagi for Windows, supplied with late editions of *The ARRL Antenna Book*).

To illustrate the matching problems evoked above, I have listed the gamma-match element variables in **Table 13-10** for a Yagi with R_{rad} = 25 Ω, and in **Table 13-11** for a Yagi with R_{rad} = 15 Ω.

Table 13-10 shows that a Yagi with a radiation resistance of 25 Ω can easily be matched with a wide range of Gamma-match parameters, while the exact length of the driven element is not at all critical. It is clear that short elements (negative reactance) require a shorter Gamma rod and a slightly smaller value of Gamma capacitor.

Table 13-11 tells the story of a high-Q Yagi with a radiation resistance of 15 Ω, similar to the 3-element W2PV or W6SAI Yagis. If such a Yagi has a "long" driven element, a match cannot be achieved, not even with a step-up ratio of 15:1. With this type of Gamma (step-up = 15:1), the highest positive reactance that can be accommodated with a radiation resistance of 13 Ω is approximately +12 Ω. In other words, it is simply impossible to match the 13 + j 18-Ω impedance of the W2PV 3-element 20-meter Yagi with a Gamma match without first reducing the length of the driven element.

The first thing to do when matching a Yagi with a relatively low radiation resistance is to decrease the element length to introduce capacitive reactance, perhaps –15 Ω in the driven-element feed-point impedance. How much shortening is needed (in terms of element length) can be derived from **Fig 13-58**. Table 13-11 shows that an impedance of 15 – j 15 Ω can be easily matched with step-up ratios ranging from 5 to 8:1.

Several Yagis have been built and matched with Gamma systems, calculated as explained above. When the reactance of the driven element at the design frequency was exactly known, the computed rod length was always right on. In some cases

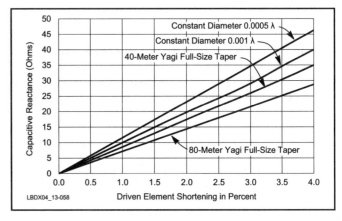

Fig 13-58—Capacitive reactance obtained by various percentages of driven-element shortening. The 40-meter full-size taper is the taper described in Table 13-1. The 80-meter taper is that for a gigantic Yagi using latticed-tower elements, varying from 42 cm at the boom down to 5 cm at the tips.

the series capacitor value turned out to be smaller than calculated. This is caused by the stray inductance of the wire connecting the end of the gamma rod with the plastic box containing the gamma capacitor, and the wire between the series capacitor and the coaxial feed line connector. The inductance of such a wire is not at all negligible, especially on the higher frequencies. With a pure coaxial construction, this should not occur.

A coaxial gamma rod is made of two concentric tubes, where the inner tube is covered with a dielectric material, such as heat-shrink tubing. The length of the inner tube, as

Table 13-10
Gamma-Match Element Data for a Yagi with a Radiation Resistance of 25 Ω

Rod d	S	Step up Rat.	−20 Ω L	C	−15 Ω L	C	−10 Ω L	C	−5 Ω L	C	0 Ω L	C	+5 Ω L	C
0.5	5.0	5.28	118	350	123	502	138	614	171	734	231	396	317	734
	4.0	5.42	131	342	135	488	151	592	184	700	255	376	331	700
	3.0	5.65	152	332	155	468	172	562	207	656	267	349	351	654
	2.5	5.83	169	324	172	452	189	540	224	634	285	332	369	624
0.38	5.0	5.87	119	322	121	450	133	536	158	618	203	328	269	618
	4.0	6.08	132	314	133	434	145	514	171	588	216	311	281	584
	3.0	6.43	153	302	154	412	165	482	192	548	238	288	302	558
	2.5	6.71	170	292	169	396	188	462	208	520	255	273	319	520
0.25	5.0	6.75	120	290	119	394	128	458	147	516	181	270	230	516
	4.0	7.07	133	282	131	374	140	430	158	482	192	251	239	482
	3.0	7.62	154	268	151	356	160	408	179	452	213	236	262	452
	2.5	8.06	172	258	167	340	175	398	195	428	230	223	278	428

Design parameters: D = 1.0; Z_{ant} = 25 Ω; Z_{cable} = 50 Ω. The element diameter is normalized as 1. Values are shown for a design frequency of 7.1 MHz. L is the length of the Gamma rod in cm, C is the value of the series capacitor in pF. The length of the Gamma rod can be converted to inches by dividing the values shown by 2.54.

Table 13-11
Gamma-Match Element Data for a Yagi with a Radiation Resistance of 15 Ω

Rod d	S	Step up Rat.	−20 Ω L	C	−15 Ω L	C	−10 Ω L	C	−5 Ω L	C	0 Ω L	C	+5 Ω L	C
0.5	5.0	5.28	93	410	92	586	116	1180	—	—	—	—	—	—
	4.0	5.42	103	400	102	566	121	1074	—	—	—	—	—	—
	3.0	5.65	120	386	117	538	136	948	—	—	—	—	—	—
	2.5	5.83	134	376	131	518	123	874	—	—	—	—	—	—
0.38	5.0	5.87	94	372	91	514	130	860	—	—	—	—	—	—
	4.0	6.08	104	362	101	494	113	996	206	3906	—	—	—	—
	3.0	6.43	121	346	117	466	128	716	208	1680	—	—	—	—
	2.5	6.71	136	334	130	446	140	666	210	1306	—	—	—	—
0.25	5.0	6.75	96	334	91	442	99	660	147	1268	—	—	—	—
	4.0	7.07	106	322	101	424	107	614	152	1060	376	1268	—	—
	3.0	7.62	123	304	117	396	122	556	161	864	309	1188	—	—
	2.5	8.06	138	292	131	376	135	518	172	766	295	982	—	—

Design parameters: D = 1.0; Z_{ant} = 15 Ω; Z_{cable} = 50 Ω. The element diameter is normalized as 1. Values are shown for a design frequency of 7.1 MHz. C is expressed in pF; L in cm (divide by 2.54 to obtain inches). Note there is a whole range where no match can be obtained. If sufficient negative reactance is provided in the driven-element impedance (with element shortening), there will be no problem in matching Yagis even with a low radiation resistance.

well as the material's dielectric and thickness, determine the capacitance of this coaxial capacitor. Make sure to properly seal both ends of the coaxial gamma rod to prevent moisture penetration.

Feeding a symmetric element with an asymmetric feed system has a slight impact on the radiation pattern of the Yagi. The forward pattern is skewed slightly toward the side where the gamma match is attached, but only a few degrees, which is of no practical concern. The more elements the Yagi has, the less the effect is noticeable.

The voltage across the series capacitor is quite small even with high power, but the current rating must be sufficient to carry the current in the feed line without warming up. For a power of 1500 W, the current through the series capacitor is 5.5 A (in a 50-Ω system) The voltage will vary between 200 and 400 V in most cases. This means that moderate-spacing air-variable capacitors can be used, although it is advisable to over-rate the capacitors, since slight corrosion of the capacitor plates normally caused by the humidity in the enclosure will derate the voltage handling of the capacitor.

3.8.2. The Omega match

The Omega match is a sophisticated Gamma match that uses two capacitors. Tuning of the matching system can be done by adjusting the two capacitors, without having to adjust the rod length. **Fig 13-59** shows the Omega match and its electrical equivalent. Comparing it with the Gamma electrical equivalent of Fig 13-56 reveals that the extra parallel capacitor, together with the series capacitor, now is part of an L network that follows the original Gamma match.

Again, the mathematics involved are complex, but the MATCHING section of the YAGI DESIGN software will do

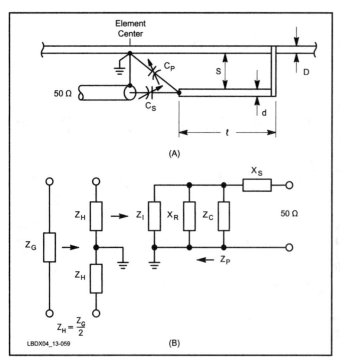

Fig 13-59—Layout and electrical equivalent of the Omega match.

the job in a second. From a practical point of view the Omega match is really unbeatable. The ultimate setup consists of a box containing the two capacitors, together with dc motors and gear-reductions. **Fig 13-60** shows the interior of such a unit using surplus capacitors and dc motors from a flea market. This system makes the adjustment very easy from the ground, and is the only practical solution when the driven element is located away from the center of the antenna.

The remarks given for the Gamma capacitors as to the required current and voltage rating are valid for the Omega match as well. The voltage across the Omega capacitor is of the same magnitude as the voltage across the Gamma capacitor, usually between 300 and 400 V, with a current of 2 to 4 A for a power of 1500 W.

3.8.3. The hairpin match

The feed-point impedance of a Yagi driven element that is about $\lambda/2$ long consists of a resistive part (the radiation resistance) in series with a reactance. The reactance is positive if the element is longer and negative if the element is shorter than the resonant length. In practice, resonance never occurs at a physical length of exactly $0.5\,\lambda$, but always at a shorter length. With a hairpin matching system we deliberately make the element short, meaning that the feed-point impedance will be capacitive. **Fig 13-61** shows the electrical equivalent of the hairpin matching system.

If we connect an inductor across the terminals of a short driven element, we can now consider the series capacitor and the parallel inductor to be the two arms of an L network. This L network can be dimensioned to give a 50-Ω output impedance. The parallel inductor is commonly replaced with a short length of short-circuited open-wire feed line having the shape of a hairpin, and hence the matching system's name.

The radiation resistance of the feed-point impedance

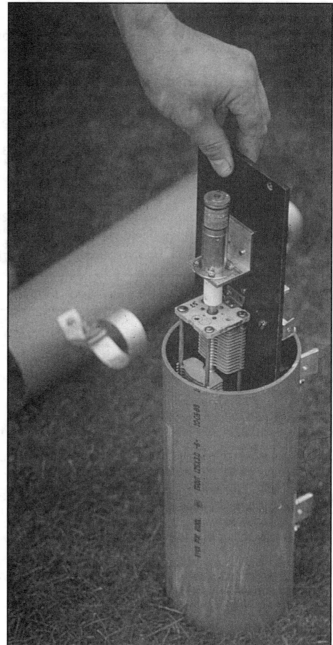

Fig 13-60—Motor-driven Omega matching unit. The two capacitors with their dc motors and gear boxes are mounted in-line on a piece of insulating substrate material. This can then be slid inside the housing, which is made of stock lengths of PVC water drainage pipe. The PVC pipe is easily cut to the desired length. The round shape of the housing also has an advantage so far as wind loading is concerned.

changes only slightly as a function of length if the element length is varied plus or minus 5% around the resonant length. The change in the reactance, however, is quite significant. The rate of change will be greatest with elements having smaller diameters (see Fig 13-58). The required hairpin reactance is given by:

$$X_{hairpin} = 50 \times \sqrt{\frac{R_{rad}}{50 - R_{rad}}} \qquad \text{(Eq 13-5)}$$

The formula for calculating the size of the shunt reactance depends on the shape of the inductor. There are two common types:
- A hairpin inductor
- A beta-match inductor

The hairpin inductor is a short piece of open-wire transmission line. The boom is basically outside the field of the transmission line. In practice the separation between the line and the boom should be at least equal to twice the spacing between the conductors of the transmission line. The characteristic impedance of such a transmission line is given by:

$$Z_C = 276 \times \log \frac{2 \times SP}{D} \quad \text{(Eq 13-6)}$$

where

SP = spacing between wires
D = diameter of the wires

In the so-called *beta-match*, the transmission line is made of two parallel conductors with the boom in between. This is the system used by Hy-Gain. The characteristic impedance of such a transmission line is given by:

$$Z_C = 553 \times \log \frac{2 \times SP}{D} \quad \text{(Eq 13-7)}$$

where

SP = spacing of wire to center of boom
D = diameter of wire
DB = diameter of boom

Z_C is the characteristic impedance of the open wire line made by the two parallel conductors of the hairpin or beta-match. The length of the hairpin or beta-match is given by:

$$\ell° = \arctan \frac{X_{hairpin}}{X_C} \quad \text{(Eq 13-8)}$$

where

$\ell°$ is the length of the hairpin expressed in degrees
arctan is the inverse tangent

To convert to real dimensions (assuming a velocity factor of 0.98 for a transmission line with air dielectric):

$$L_{cm} = \ell° \times \frac{81.6}{f} \quad \text{(Eq 13-9)}$$

where

f = design frequency, MHz

The required driven element reactance is given by:

$$X_C = -\frac{4_{rad} \times 50}{X_{hairpin}} \quad \text{(Eq 13-10)}$$

Table 13-12 shows the required values of capacitive reactance in the driven element, as well as the required

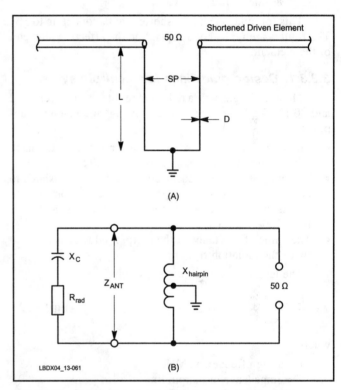

Fig 13-61—Layout and electrical equivalent of the hairpin match.

Table 13-12
Required Capacitive Reactance in Driven Element and in Hairpin Inductance

R_{rad} Ω	Antenna Reactance Ω	Inductance Hairpin Ω	Length hairpin (cm) (SS = 10D)	
			3.65 MHz	7.1 MHz
10.0	−20.0	25.0	89	46
12.5	−21.6	28.9	103	53
15.0	−22.9	32.7	117	60
17.5	−23.8	36.7	130	67
20.0	−24.5	40.8	144	74
22.5	−24.9	45.2	160	82
25.0	−25.0	50.0	177	91
27.5	−24.9	55.3	194	100
30.0	−24.5	61.2	216	111

Note: The feed-point impedance is 50 Ω. To obtain the hairpin length in inches, divide values shown by 2.54.

reactance for a range of radiation resistances. (For a hairpin with S = 10D as in the table, Z = 359 Ω.) The question now is how long the driven element must be to represent the required amount of negative reactance ($-X_C$). Fig 13-58 lists the reactance values obtained with several degrees of element shortening. Although the exact reactance differs for each one of the listed element diameter configurations, you can derive the following formula from the data in Fig 13-58.

$$X_C = -Sh \times A \qquad \text{(Eq 13-11)}$$

where

Sh = shortening in % versus the resonant length
X_C = reactance of the element in Ω
A = 8.75 (for a 40-meter full-size Yagi) or 7.35 (for an 80-meter full-size Yagi)

This formula is valid for shortening factors of up to 5%. The figures are typical and depend on the effective diameter of the element.

3.8.3.1. Design guidelines for a hairpin system

Most HF Yagis have a radiation resistance between 20 Ω and 30 Ω. For these Yagis the following rule-of-thumb applies:

- The required element reactance to obtain a 50-Ω match with a hairpin is approximately– 25 Ω. (Table 13-12).
- This almost constant reactance value can be translated to an element shortening of approximately 2.8% compared to the resonant element length for a 40-meter Yagi, and 3.5% for an 80-meter Yagi.
- The value of reactance of the hairpin inductor is equal to twice the radiation resistance.

The length of the hairpin is given by:

$$\ell = \frac{9286 \times R_{rad}}{f \times Z} \qquad \text{(Eq 13-12)}$$

where

f = design frequency, MHz
Z = impedance of hairpin line
ℓ = length in cm

The impedance of the hairpin line for a range of spacing-to-wire diameter ratios is shown in **Table 13-13**.

The real area of concern in designing a hairpin matching system is to have the correct element length that will produce the required amount of capacitive reactance. As we have an split element, we can theoretically measure the impedance, but this can be impractical for two reasons:

- The impedance measurement must be done at final installation height.
- The average ham does not have access to measuring equipment that can measure the impedance with the required degree of accuracy. A run-of-the-mill noise bridge will not suffice, and a professional impedance bridge or network analyzer is required.

Let us examine the impact of a driven element that does not have the required degree of capacitive reactance.

Table 13-14 shows the values of the transformed impedance and the resulting minimum SWR if the reactance of the driven element was off +5 Ω and –5 Ω versus the theoretically required value for an R_{rad} of 20 Ω. An error in reactance of 5 Ω either way is equivalent to an error length of 0.5% (see **Table 13-15**). In other words, an inaccuracy of 0.5% in element length will deteriorate the minimum SWR value from 1:1 to 1.25 or 1.3:1.

The mounting hardware for a split element will always introduce a certain amount of shunt capacitance at the driven-element feed point. This must be taken into account when designing a hairpin- or beta-match system (see example in Section 3.8.3.3).

3.8.3.2. Element loading and a hairpin

The length of the driven element that produces zero reactance at the design frequency is called the resonant length. This length also depends on the element diameter (in terms of wavelength). If any taper is employed for the construction of the element, the degree of taper will have its influence as well. Finally, the resonant length will differ with every Yagi design. This is caused by the effect of mutual coupling between the elements of the Yagi. For elements with a constant diameter of approximately 0.001 λ, the resonant-frequency length will usually be between 0.477 λ and 0.487 λ. The exact value for a given design can be obtained by modeling the Yagi or by

Table 13-13

Hairpin Line Impedance as a Function of Spacing-to-Diameter Ratio

S/D Ratio	Impedance, Ω
5	193
7.5	325
10	359
15	408
20	442
25	469
30	491
35	510
40	525
45	539
50	552

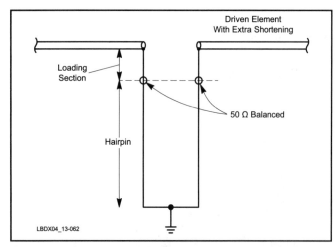

Fig 13-62—Layout of hairpin match combined with element loading, whereby the 50-Ω point is found along the hairpin at some distance from the element.

obtaining it from a reliable database.

If the exact resonant length is not known, then it is better to make the element somewhat too short, after which the element can be electrically lengthened by loading it in the center with a short piece of transmission line. Adjusting the amount of loading can often be done more easily than adjusting the element tips, especially if the driven element is located on the boom away from the tower.

The loading can be done using a short length of open-wire line. The short length of line can have the same wire diameter and spacing as used for the hairpin, usually between 300 Ω and 450 Ω. The layout and the electrical equivalent of this approach is shown in **Fig 13-62**. The transmission line acts as a loading device between the element feed point and the 50-Ω tap, and as the matching inductor beyond the 50-Ω tap (the hairpin). Another method of changing the electrical length of the driven element is described in Section 3.10.3.5, where a parallel capacitor is used to shorten the electrical length of the driven element.

Let us examine the impedance of the antenna feed point along a short 200-Ω to 450-Ω transmission line:

- The value of the resistive part will remain almost constant (change negligible).
- The value of the capacitive reactance will decrease by X ohms per degree, where X is given by:

$$\frac{X}{\text{degrees}} = Z \times 0.017 \quad \text{(Eq 13-13)}$$

The change in reactance per unit of length is:

$$\frac{X}{\text{cm}} = \frac{Z \times f \times 0.204}{1000} \quad \text{(Eq 13-14)}$$

where
f = frequency, MHz
Z = characteristic impedance of the line made by the two parallel wires of the hairpin or beta- match (see Eq 13-6 and Table 13-13)

Eqs 13-13 and 13-14 are valid for line lengths of 4° maximum. The line length required to achieve a given reactance shift X is given by:

$$L_{cm} = \frac{X \times 4900}{Z \times f} \quad \text{(Eq 13-15)}$$

This formula is valid for values of X of 25 Ω maximum.

Example: Let us assume that we start from a 20 − j 30 Ω impedance, and we need to electrically lengthen the driven element to yield an impedance of 20 − j 24.5 Ω (see Table 13-12). The design frequency is 7.1 MHz.

The required reactance difference is X = 30 − 24.5 = 5.5 Ω. The required 359-Ω-line length is:

$$\ell = \frac{5.5 \times 1929}{359 \times 7.1} = 10 \text{ cm}$$

The length of the hairpin section can be determined from Table 13-12 as 111 cm. This means that we can electrically load the element to the required length by adding an extra piece of hairpin line. The length of this line will be only a few inches long. In this case the 50-Ω tap will not be at the element but at a short distance on the hairpin line. The length of the hairpin matching inductor will remain the same, but the total transmission-line length will be slightly longer than the matching hairpin itself.

To adjust the entire system, look for the 50-Ω point on the line by moving the balun attachment point while at the same time adjusting the total length of the hairpin. The end of the hairpin shorting bar is usually grounded to the boom.

Design Rule of Thumb: The transmission-line loading device can be seen as part of the driven element folded back in the shape of the transmission line. For a 359-Ω transmission line (spacing = 10 × diameter), the length of the loading line will be exactly as long as the length that the element has been shortened. In other words, for every inch of total element length you shorten the driven element, you must add an equivalent inch in loading line. This rule is applicable only for 359-Ω lines and for a maximum length of 406/f cm, where f = design frequency in MHz. For other line impedances the

Table 13-14
Values of Transformed Impedance and SWR for a Range of Driven-Element Impedances

Driven Element Impedance	Hairpin Reactance, Ω	Resulting Impedance	SWR (vs 50 Ω)
20 − j 20	40.8	40 − j 0.81	1.25:1
20 − j 24.5	40.8	50 − j 0	1.00:1
20 − j 25	40.8	51.2 + j 0.26	1.02:1
20 − j 30	40.8	64.5 − j 5.92	1.32:1

Table 13-15
Capacitive Reactance Obtained by Various Percentages of DrivenElement Shortening

Shorten Element	Diameter in wavelengths 0.0010527	0.0004736	Light Taper	Heavy Taper
0%	0 Ω	0 Ω	0 Ω	0 Ω
0.5	−4.8	−5.5	−4.6	−4.8
1.0	−9.6	−11.1	−9.1	−9.7
1.5	−14.3	−16.5	−13.6	−14.3
2.0	−19.1	−22.2	−18.2	−19.2
2.5	−23.8	−27.5	−22.7	−23.7
3.0	−28.6	−32.8	−27.2	−28.6
3.5	−33.5	−38.2	−31.7	−33.5
4.0	−38.5	−43.5	−36.2	−38.3

calculation as shown above should be followed.

Example

A driven element is resonant at 7.1 MHz with a length of 2200 cm. We want to shorten the total element length by 25 cm, and restore resonance by inserting a 359-Ω loading line in the center. The length of the loading line will be approximately 25 cm.

3.8.3.3. Hairpin match design with parasitic element-to-boom capacitance

As explained in Section 3.4, it is virtually impossible to construct a split element without any capacitive coupling to the boom. With tubular elements a coaxial-type construction is often employed, which results in an important parasitic capacitance.

The Hy-Gain Yagis, which use a form of coaxial-insulating technique to provide a split element for their Yagis, exhibit the following parallel capacitances:

- 205BA: 27 pF
- 105BA and 155BA: 10 pF

Let's work out an example for a Yagi designed at 7.1 MHz. We model the driven element to be resonant at that design frequency. Assume the resonant length is 1985 cm.

The capacitance introduced by the split-element mounting hardware for a 40-meter element is 300 pF per side (I have measured this with a digital capacitance meter). Do not forget to measure the capacitance without the full element attached. The reactance of this capacitor is:

$$X_C = \frac{10^6}{2\pi \times 7.1 \times 300} = 75 \, \Omega$$

The capacitance across the feed point is 150 pF (two 300-pF capacitors in series):

$$X_C = 150 \, \Omega$$

Using the SHUNT NETWORK module of the NEW LOW BAND SOFTWARE, we calculate the resulting impedance of this capacitor in parallel with 28-Ω impedance at resonance as:

$$Z = 27.1 - j\,5 \, \Omega$$

The required inductance of the hairpin (using Eq 13-5) will be:

$$X_{hairpin} = 50 \times \sqrt{\frac{R_{rad}}{50 - R_{rad}}} = \sqrt{\frac{27.1}{50 - 27.1}} = 54.30 \, \Omega$$

Assume we are using a hairpin with two conductors with spacing = 10 × diameter. The impedance of the line is given by Eq 13-6 as:

$$Z_C = 276 \times \log \frac{2 \times SP}{D} = 276 \times \log(20) = 359 \, \Omega$$

The length of the hairpin is given by Eq 13-8 as:

$$\ell = \arctan \frac{X_{hairpin}}{Z_C} = \arctan \frac{54.3}{359} = 8.6°$$

The length in cm is given by Eq 13-9 as:

$$\ell = \ell° \times \frac{81.6}{f} = 8.6 \times \frac{81.6}{7.1} = 98.5 \, cm$$

For an impedance of 27.1 Ω we need an reactance (using Eq 13-10) of:

$$X_C = -\frac{R_{rad} \times 50}{X_{hairpin}} = \frac{27.1 \times 50}{54.3} = -24.95 \, \Omega$$

This means we have to add another 19.95 Ω of negative reactance to our driven element. This can be done by shortening the element approximately 2.2% (see Fig 13-58), which amounts to

$$1985 \times \frac{2.2}{100} = 44 \, cm \text{ or 22 cm on each side.}$$

Instead of shortening the element 22 cm on each side, we could shorten it, say, 42 cm on each side, which will now give us some range to fine tune the matching system. The 22 cm we have shortened the driven element on each side will be replaced with an extra 20-cm length of transmission line at the feed point. The 50-Ω point will now be located some 20 cm from the split driven element. The hairpin will extend another 98.5 cm beyond this point. Tuning the matching system consists of changing the position of the 50-Ω point on the hairpin as well as changing the length of the hairpin.

If you use "wires" to connect the split driven element to the matching system, you must take the inductance of this short transmission line into account as well.

3.8.3.4. Designing a hairpin match with the YAGI DESIGN software

You can use the MATCHING SYSTEMS module in the YAGI DESIGN software to design a hairpin. From the prompt line you can change any of the input data, which will be immediately reflected in the dimensions of the matching system. The value of the "parasitic" parallel capacitance can be specified, and is accounted for during the calculation of the matching system.

3.8.3.5. Using a parallel capacitor to fine-tune a hairpin matching system

A parallel capacitance of reasonable value across the split element only slightly lowers the resistive part of the impedance, while it introduces an appreciable amount of negative reactance.

Example

A capacitor of 150 pF in parallel with an impedance of 28 + j 0 Ω (at 7.1 MHz) lowers the impedance to 27.1 – j 5 Ω. This means that instead of fine-tuning the matching system by accurately shortening the driven element to obtain the required negative reactance, you can use a variable capacitor across the driven element to electrically shorten the element. This is a very elegant way of tuning the hairpin matching system "on the nose." The only drawback is that it requires another vulnerable component. This method is an alternative fine-tuning method to the configuration where the length of the driven element is altered by using a short length of transmission line.

3.8.4. Direct feed

If the driven element is split (not grounded to the boom), you can also envisage a direct-feed system. Most 3-element Yagis do not present a 50-Ω feed-point impedance, unless you design it with a low Q and trade matching ease for some forward gain. Three-element Yagis will typically show feed-point impedances varying between 18 Ω and 30 Ω. This means we really do have to use some kind of system to match the Yagi impedance to the feed-line impedance. In addition, if we want to use a feed system for an 80-meter Yagi, which has to cover both the CW and the SSB end of the band, we also will have to deal with the reactances involved.

3.8.4.1. Split element direct feed system with series compensation

You can often use a quarter-wave transformer to achieve a reasonable match between the Yagi impedance and a 50-Ω feed line. There are two solutions. For a feed-point impedance lower than 25 Ω you can use a λ/4 length of line with an impedance of 30 Ω. This can be made by paralleling a 50-Ω, λ/4 cable with a 75-Ω, λ/4 cable. The cable can be rolled up into a coil measuring about 30 cm in diameter, which will serve as common-mode choke balun.

For impedances between 25 and 30 Ω the required impedance for a quarter-wave transformer is 37.5 Ω, made by paralleling two 75-Ω, λ/4-λ cables. An example of such a matching system is given in **Fig 13-63**.

Let us analyze the case of a direct feed for the 80-meter Yagi described in Section 3.3. The real part of the impedance of the Yagi is around 28 Ω, which is easy to match to a 50-Ω feed line through a 37.5-Ω quarter-wave transformer made with two parallel 75-Ω cables.

There are different approaches to handling the inductive part of the impedance at the opposite end of the band. You could dimension the driven element to be resonant on 3.8 MHz and tune out the capacitive reactance (approximately 75 Ω) by using a coil in series with the coaxial feed line. The other alternative is to dimension the driven element for resonance in the CW band and then tune out the inductive reactance on 3.75 MHz using a series capacitor. In an example I dimensioned the driven element for resonance on 3.55 MHz, and calculated the value of the series capacitor to achieve resonance on 3.75 MHz. The capacitor value is 900 pF. This matching system is extremely simple, and will guarantee maximum bandwidth as well. Again, the quarter-wave 37.5-Ω transformer can be coiled up and serve as a choke balun.

Fig 13-64 shows the SWR curves for this feed arrangement. Note that with such an arrangement it is impossible to have a good SWR in the middle of the band.

3.8.4.2. Split element direct feed system with parallel compensation

The direct-feed system with parallel compensation does not require a quarter-wave transformer as it provides a good match directly to a 50-Ω impedance. In the case of the driven element of an 80-meter Yagi, you could dimension the driven element for resonance in the middle of the band at 3.65 MHz. In that case the reactance of a typical 3-element Yagi such as the antenna developed in Section 3.3, exhibits about $-j\,35\,\Omega$ at 3.5 MHz and $+j\,35\,\Omega$ at 3.8 MHz. The reactance can be tuned out by a parallel coil or capacitor (see **Fig 13-65**). A coil

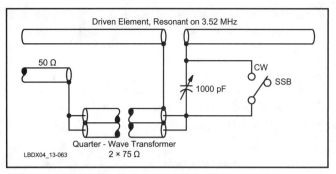

Fig 13-63—Split-element matching system for the 80-meter Yagi. The driven element is tuned to resonance in the CW end of the band. On phone (3.8 MHz) the inductive reactance is tuned out by a simple series capacitor of 560 pF. A relay can short out the capacitor on CW. A quarter-wave 37.5-Ω transformer made of two parallel 75-Ω coaxes may be coiled up to serve as a choke balun, representing a 50-Ω impedance at its end.

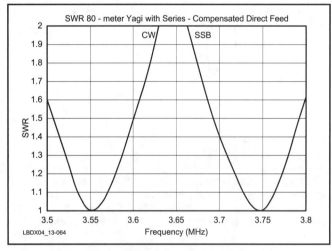

Fig 13-64—SWR curves for the split-element feed method with series compensation and a 37.5-Ω quarter-wave transformer.

with an inductance of 2.2 μH will tune the driven element to resonance on 3.55 MHz and yield an resistance of 50 Ω (a lucky coincidence!). Likewise, a capacitor with a value of 700 pF will tune the element to resonance on 3.8 MHz, also with a resistance of 50 Ω.

The value of these components can easily be calculated using the SHUNT IMPEDANCE NETWORK module of the NEW LOW BAND SOFTWARE. With a simple relay you can switch either the coil or the capacitor in parallel with the feed point, and obtain a fine matching system for either the CW or the SSB end of the band. **Fig 13-66** shows the SWR curves obtained with such an arrangement.

4. QUADS
4.1. Modeling Quad Antennas
4.1.1. MININEC-Based programs

Modeling quad antennas with *MININEC*-based programs requires very special attention. To obtain proper results the number of wire segments should be carefully chosen. Near the corners of the loop, the segments must be short enough not to introduce a significant error in the results. Segments as short as 20 cm must be used on an 80-meter quad to obtain reliable impedance results on multi-element loop antennas. Years ago, when most modeling programs still used a *MININEC* core, W7EL developed a *taper technique* in his *ELNEC* software, where segments automatically got progressively shorter when coming to a corner. See **Fig 13-67**.

4.1.2. NEC-based programs

NEC-based programs do not exhibit the above problem, and no special precautions have to be taken to obtain correct results.

4.2. Two-element full-size 80-meter quad with a parasitic reflector

Fig 13-68 shows the configuration of a 2-element 75-meter quad on a 12-meter boom, and **Fig 13-69** shows the radiation patterns. The optimum antenna height is 35 meters for the center of the quad. Whether you use the square or the

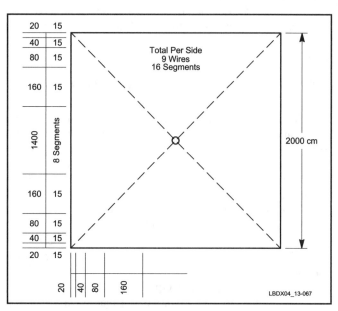

Fig 13-67—Tapering of segment lengths for *MININEC* analysis. See Table 13-16 for the results with different tapering arrangements. With the segment-length-taper procedure shown here, the result with a total of just 56 tapered segments is as good as for 240 segments of identical length.

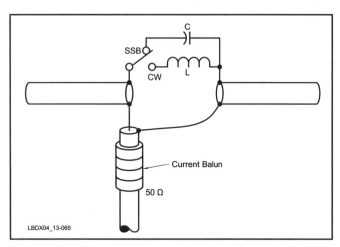

Fig 13-65—Direct-feed system for the driven element of an 80-meter array with parallel compensation, which makes it possible to obtain a good SWR in both the CW and SSB sections of the band. See text for details.

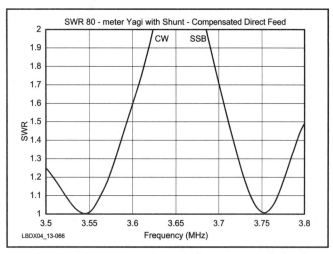

Fig 13-66—SWR curves for the 80-meter 3-element Yagi using parallel compensation. The driven element, initially tuned to resonance on 3.65 MHz, was tuned to resonance on 3.55 MHz using a parallel inductor of 2.2 µH. Likewise, the same element was tuned to resonance on 3.75 MHz using a parallel capacitor of 700 pF. The resulting SWR curves show an outstanding bandwidth.

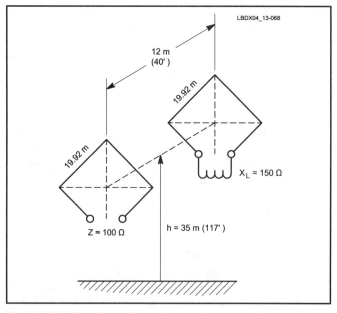

Fig 13-68—Configuration of a 2-element cubical quad antenna designed for 75-meter SSB. Radiation patterns are shown in Fig 13-69. By using a remote tuning system for adjusting the loading of the reflector, the quad can be made to exhibit an F/B of better than 22 dB over the entire operating range. See text for details.

diamond shape does not make any difference. The dimensions remain the same, as well as the results. I will describe a diamond-shaped quad, which has the advantage of making it possible to route the feed line and the loading wires along the fiberglass arms.

I designed this quad with two quad loops of identical length. The total circumference for the quad loop is 1.0033 λ (for a 2-mm-OD conductor or #12 wire). The parasitic element is loaded with a coil or a stub having an inductive reactance of $+j\,150\,\Omega$. The gain is 3.7 dB over a single loop at the same height over the same ground.

In the model I used 3.775 MHz as a central design frequency. This is because the SWR curve rises more sharply on the low side of the design frequency than it does on the high side. You can optimize the quad by changing the reactance of the loading stub as you change the operating frequency.

Figs 13-70 and **13-71** show the gain, F/B and SWR for the 2-element quad with a fixed loading stub or coil ($+150\,\Omega$)

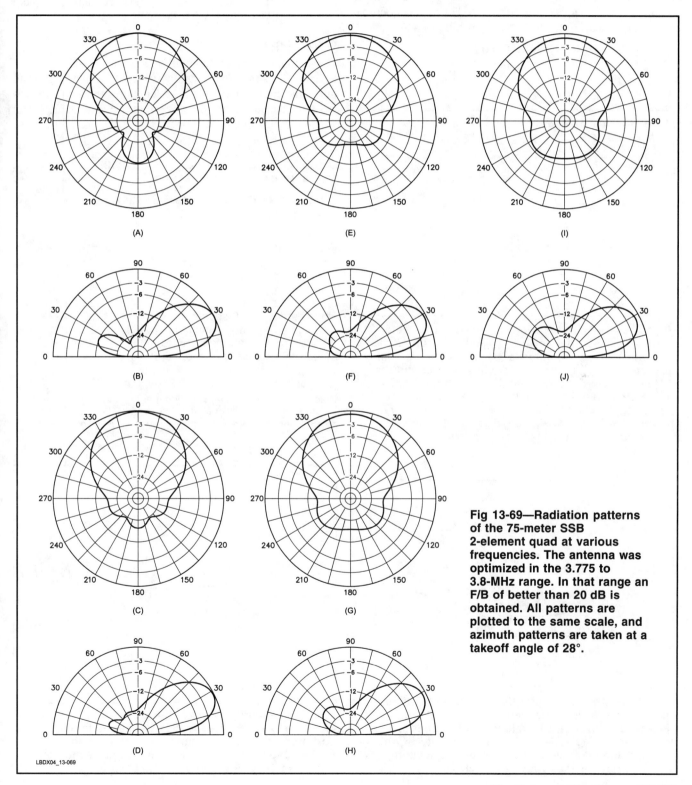

Fig 13-69—Radiation patterns of the 75-meter SSB 2-element quad at various frequencies. The antenna was optimized in the 3.775 to 3.8-MHz range. In that range an F/B of better than 20 dB is obtained. All patterns are plotted to the same scale, and azimuth patterns are taken at a takeoff angle of 28°.

as well as for a design where the loading stub reactance is varied. To make the antenna instantly reversible in direction, you can run two λ/4, 75-Ω lines, one to each element. Using the COAX TRANSFORMER/SMITH CHART module from the NEW LOW BAND SOFTWARE we see that a + j 160-Ω impedance at the end of a λ/4 long 75-Ω transmission line (at 3.775 MHz) looks like a – j 35-Ω impedance. This means that a λ/4, 75-Ω (RG-11) line terminated in a capacitor having a reactance of –35 Ω is all that we need to tune the parasitic element into a reflector. A switch box mounted at the center of the boom houses the necessary relay switching harness and the required variable capacitor to do the job. The required optimal loading impedances can be obtained as follows:

3.750 MHz: X_L = 180 Ω, C = 1322 pF
3.775 MHz: X_L = 160 Ω, C = 1205 pF
3.800 MHz: X_L = 150 Ω, C = 1148 pF
3.825 MHz: X_L = 120 Ω, C = 931 pF
3.850 MHz: X_L = 100 Ω, C = 785 pF

If you tune the reflector for optimum value you will obtain better than 22-dB F/B ratio at all frequencies from 3.75 to 3.85 MHz, and the SWR curve will be much flatter than without the tuned reflector (see Figs 13-70 and 13-71).

The quad can also be made switchable from the SSB to the CW end of 80 meters. There are two methods of loading the elements, inductive loading and capacitive loading (see also the chapter on Large Loops). The capacitive method, which I will describe here, is the most simple.

4.2.1. Capacitive loading

A small single-pole high-voltage relay at the tip of the horizontal fiberglass arms can switch the loading wires in and out of the circuit. The calculated length for the loading wires to switch the quad from the SSB end of the band (3.775 MHz)

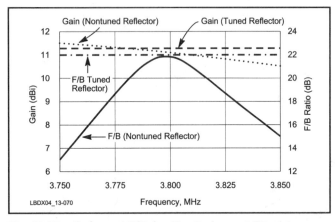

Fig 13-70—Gain and F/B for the 2-element 75-meter quad with fixed reflector tuning, and with adjustable reflector tuning. The antenna is modeled at a height of 35 meters above good ground. With fixed tuning the F/B is 20 dB over 30 kHz, and the gain drops almost 0.5 dB from the low end to the high end of the operating passband (100 kHz). When the reflector tuning is made variable, the gain as well as the F/B remain constant over the operating band.

Table 13-16

Influence of the Number of Sections on the Impedance of a Quad Loop

Taper Arrangement	Calculated Impedance, Ω
Nontapered, 4 × 5 segments	123 – j 20
Nontapered, 4 × 10 segments	130 + j 44
Nontapered, 4 × 20 segments	133 + j 78
Nontapered, 4 × 40 segments	135 + j 95
Nontapered, 4 × 50 segments	135 + j 97
Nontapered, 4 × 60 segments	135 + j 98
32 sections, tapering from 1.0 to 5.0 m	131 + j 85
56 sections, tapering from 0.4 to 2.0 m	134 + j 102
64 sections, tapering from 0.2 to 2.0 m *	135 + j 104
104 sections, tapering from 0.2 to 1.0 m	135 + j 104

Note: See Fig 13-67 regarding the taper procedure.
*Taper arrangement illustrated in Fig 13-67.

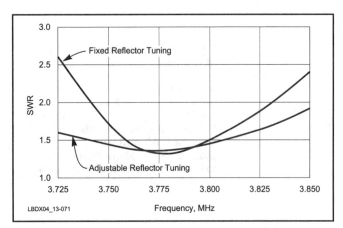

Fig 13-71—SWR curves for the 2-element 75-meter quad. The SWR is plotted versus a nominal input impedance of 100 Ω, which is then matched to a 50-Ω impedance using a λ/4, 75-Ω line. Note that the variable reflector-tuning extends the operating bandwidth considerably for the lower frequencies.

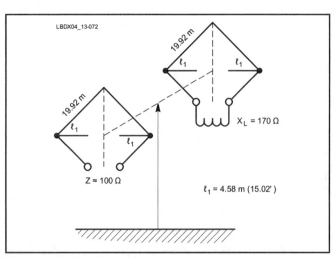

Fig 13-72—Configuration of the 2-element 80-meter quad of Fig 13-68 when loaded to operate in the CW portion of the band. Radiation patterns are shown in Fig 13-73. See text and Fig 13-75 for information on relay switching between SSB and CW.

to the CW end (3.525 MHz) is 4.58 meters. Note that you are switching at a high-voltage point, which means that a high-voltage relay is essential.

As you will have to run a dc feed line to the relay on the tip of the spreader, it is likely that the loading wire will capacitively couple to the feed wire. Use small chokes or ferrite beads on the feed wire to decouple it from the loading wires.

Fig 13-72 shows the configuration and **Fig 13-73** shows the radiation patterns for the 2-element quad tuned for the CW end of the band. The patterns are for a fixed reflector-loading reactance of 170 Ω.

As described above, you can optimize the performance by tuning the loading system as we change frequency. Using the same λ/4, 75-Ω line (cut for 3.775 MHz), you can obtain a constant 22-dB F/B (measured at the peak elevation angle of

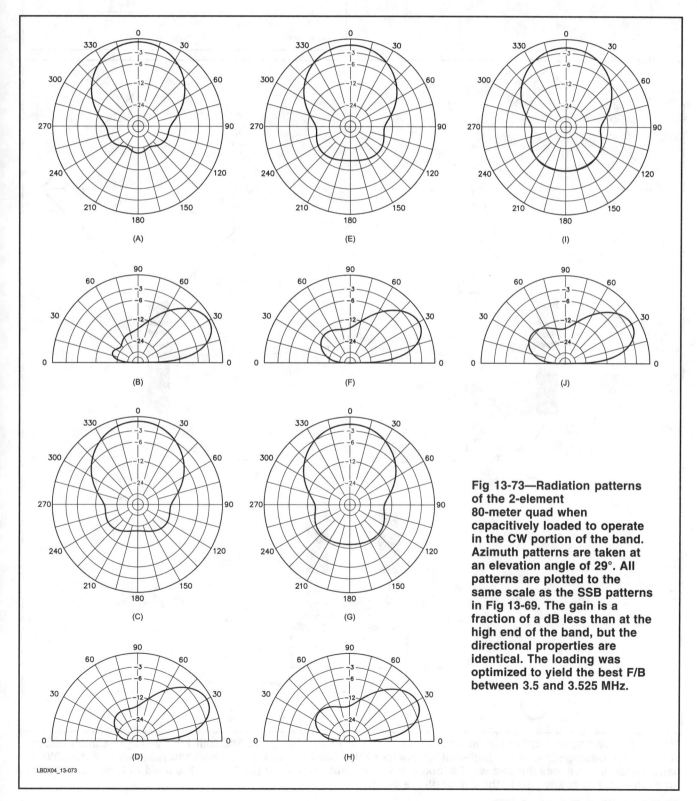

Fig 13-73—Radiation patterns of the 2-element 80-meter quad when capacitively loaded to operate in the CW portion of the band. Azimuth patterns are taken at an elevation angle of 29°. All patterns are plotted to the same scale as the SSB patterns in Fig 13-69. The gain is a fraction of a dB less than at the high end of the band, but the directional properties are identical. The loading was optimized to yield the best F/B between 3.5 and 3.525 MHz.

Fig 13-74—SWR curves for the 2-element 80-meter quad referred to the nominal 100-Ω feed-point impedance. The tuned reflector does not significantly improve the SWR on the high-frequency end. The design was adjusted for the best SWR in the 3.5 to 3.525-MHz region.

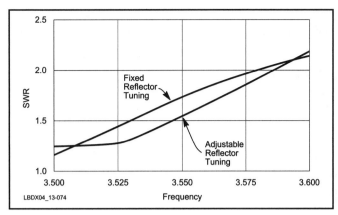

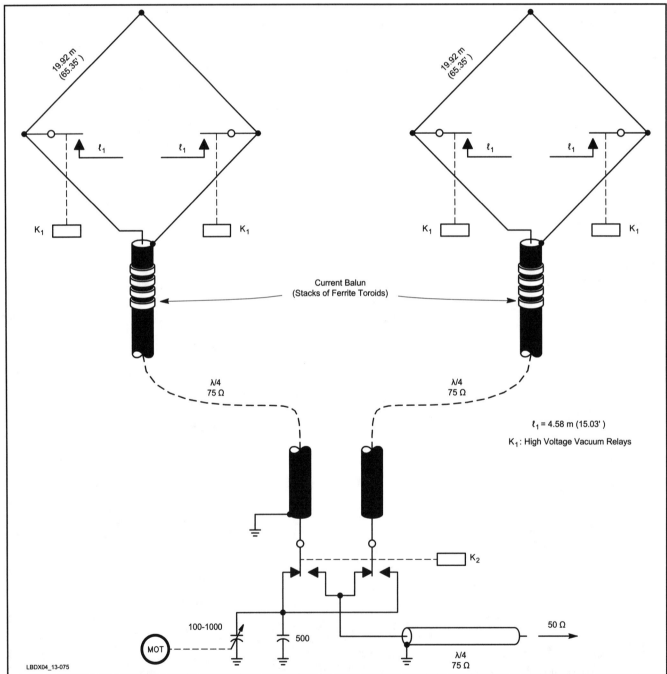

Fig 13-75—Feeding and switching method for the 2-element 80-meter quad. The four high-voltage vacuum relays connect the loading wires to the high-voltage points of the quad, to load the elements to resonance in the CW band. Relay K2 switches directions. The motor-driven variable capacitor (50-1000 pF) is used to tune the reflector for maximum F/B at any part in the CW or phone band.

29°) on all frequencies from 3.5 to 3.6 MHz with the following capacitor values at the end of the 75-Ω line:

3.500 MHz: $X_L = 180\ \Omega$, C = 1083 pF
3.525 MHz: $X_L = 160\ \Omega$, C = 999 pF
3.550 MHz: $X_L = 140\ \Omega$, C = 897 pF
3.575 MHz: $X_L = 120\ \Omega$, C = 795 pF
3.600 MHz: $X_L = 100\ \Omega$, C = 660 pF

The optimized quad has a gain at the low end of 80 meters that is 0.3 dB less than at the high end of the band. The gain is 10.8 dBi at 35 meters over good ground.

Fig 13-74 shows the SWR curve of the quad at the CW end of the band, with both a fixed reflector loading ($X_L = 170\ \Omega$) and a variable setup as explained above. The switching harness for the 2-element quad is shown in **Fig 13-75**. The tuning capacitor at the end of the 75-Ω line going to the reflector can be made of a 500-pF fixed capacitor in parallel with a 100 to 1000-pF variable capacitor. Note that you need a choke balun at both 75-Ω feed lines reaching the loops. This can be in the form of a stack of ferrite beads or as coiled-up coax.

It is also possible to design a 2-element quad array with both elements fed. With the dimensions used in the above design, a phase delay of 135° with identical feed-current magnitudes yields a gain that is very similar to what is obtained with the parasitic reflector. The F/B may be a little better than with the parasitic array. As the array is not fed in quadrature, the feed arrangement is certainly not simpler than for the parasitic array, however. The parasitic array is simpler to tune, since the reflector stub (the capacitor value) can be simply adjusted for best F/B.

4.3. Two-element reduced-size quad

D. Courtier-Dutton, G3FPQ, built a reasonably sized 2-element 40-meter rotatable quad that performs extremely well. The quad side dimensions are 15 meters, and the elements are loaded as shown in **Fig 13-76**. The single loop showed a radiation resistance of 50 Ω. Adding a reflector 12 meters away from the driven element (0.14 λ spacing), dropped the radiation resistance to approximately 30 Ω. The loading wires are spaced 110 cm from the vertical loop wires, and are almost as long as the vertical loop wires. The loading wires are trimmed to adjust the resonant frequency of the element. G3FPQ reports a 90-kHz bandwidth from the 2-element quad with the apex at 40 meters. The middle 7 meters of the spreaders are made of aluminum tubing, and 3.6-meter long tips are made of fiberglass. A front-to-back ratio of up to 30 dB has been reported.

G3FPQ indicates that the length of the reflector element is exactly the same as the length of the driven element for the best F/B ratio. This may seem odd, and is certainly not the case for a full-size quad. The same effect has been found with some 2-element Yagi arrays.

4.4. Three-element 80-meter quad

Fig 13-77 shows the 3-element full-size 80-meter quad at DJ4PT. The boom is 26-meters long, and the boom height is 30 meters. Interlaced on the same boom are five elements for a 40-meter quad.

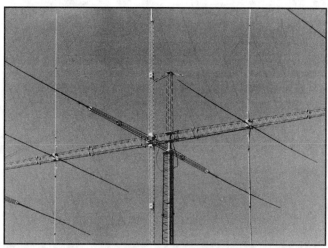

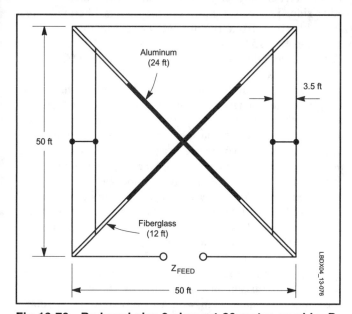

Fig 13-76—Reduced-size 2-element 80-meter quad by D. Courtier-Dutton, G3FPQ. The elements are capacitively loaded, as explained in detail in the chapter on large loop antennas.

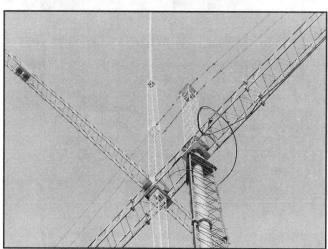

Fig 13-77—This impressive 3-element full-size 80-meter quad, with an interlaced 5-element 40-meter quad, on a 26-meter (87-foot) long boom, sits on top of a self-supporting 30-meter (100-foot) tower at DJ4PT. The antenna was built by DJ6JC (SK).

The greatest challenge in building a quad antenna of such proportions is mechanical in nature. The mechanical design was done by H. Lumpe, DJ6JC, now a Silent Key, who was a well-known professional tower manufacturer in Germany. The center parts of the quad spreaders were made of aluminum lattice sections that are insulated from the boom and broken up at given intervals as well. The tubular sections are made of fiberglass. The driven element is mounted less than 1 meters from the center. This makes it possible to reach the feed point from the tower. To be able to reach the lower tips of the two parasitic elements for tuning, a 26-meter tower was installed exactly 13 meters from the main tower. On top of this smaller tower a special platform was installed from which one can easily tune the parasitic elements.

The weight of the quad is approximately 2000 kg (4400 lb). The monster quad is mounted on top of a 30-meter self-supporting steel tower, also built by DJ6JC. The rotator was placed at the bottom of the tower, and a 20-cm (8-inch) OD rotating pipe with a 10-mm (0.4-inch) thick wall takes care of the rotating job.

4.5. The W6YA 40-meter quad

Jim McCook, W6YA, lives in a fairly typical suburban QTH, and has his neighbors and family accustomed to one crank-up tower (Tri-EX LM-470), on which he must put all of his antennas. Jim has 4-element monoband Yagis for 10, 15 and 20 meters and a WARC triband dipole. McCook set out to make it work on 9 bands. See **Fig 13-78**.

For 40 meters, Jim has extended the 12-meter boom of his 20-meter Yagi to 14.5 meters. At the ends of the boom he mounted fiberglass quad poles, which support two inverted delta loops, separated 6 meters from each other. One loop is tuned as a reflector (3% longer). The driven element is fed through a $\lambda/4$ section of RG-11 75-Ω cable. The inverted delta loops are kept taut by supporting two more abutted quad poles at the bottom. This quad-pole assembly hangs freely, supported only by the loop wires. The assembly pivots around the tower during rotation. The top horizontal sections are allowed to sag slightly (about 2 meters) to minimize interaction with the 20-meter Yagi. This arrangement has been up for 18 years, and has helped Jim to work all but 3 countries on 40 meters! Jim reports a 2:1 SWR bandwidth of 200 kHz and a F/B of 15-20 dB.

The driven element loop has a 14.78-meter "flat top," 14.32-meter sloping length on one side and 4.63 meters on the other side. This offset is to keep the bottom fiberglass-pole assembly free from the tower. The reflector measures 14.78 meters, 15.04 meters and 15.34 meters respectively. The above lengths are for peak performance on 7.020 MHz.

This quad arrangement has low wind load. Jim also uses this arrangement on 80 and even on 160 meters. On 80 and 160, Jim straps the feed point of the driven loop, and feeds the loops with its feed line, at ground level via appropriate matching networks. If you feel tempted to try this combination, I would advise you to use an antenna analyzer to measure the feed-point impedance on both 80 and 160, and design an appropriate network. The feed-point impedance on 80 meters is approximately $90 + j\ 366\ \Omega$, and on 3.8 MHz $120 + j\ 460\ \Omega$. On 1.83 MHz the impedance, including an estimated series-equivalent ground loss resistance of 10 Ω is $25 - j\ 72\ \Omega$. The appropriate matching networks for the different frequencies are shown in **Fig 13-79**.

It goes without saying that a good ground-radial system is essential for this antenna. Jim complements his 9-bands-on-one-tower antenna system with a modified 30-meter rotary dipole, which he center loads for 80 meters. He anticipates adding a second set of loading coils to use the same short loaded dipole on 160 as well.

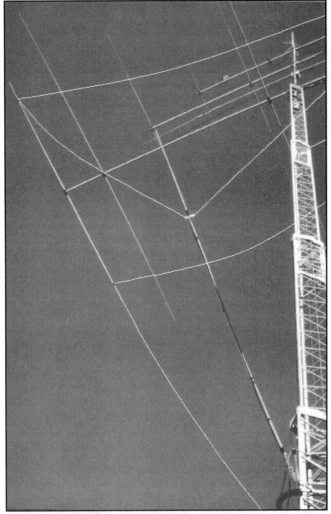

Fig 13-78—Two-element inverted-delta-loop array at W6YA. The top of the loop is about 21 meters high. See text for details.

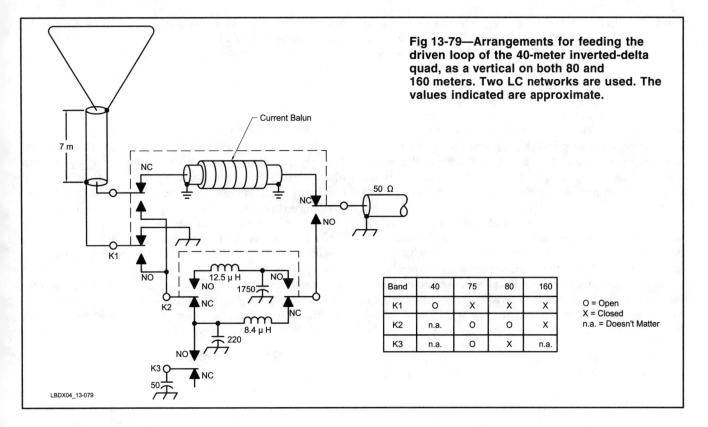

Fig 13-79—Arrangements for feeding the driven loop of the 40-meter inverted-delta quad, as a vertical on both 80 and 160 meters. Two LC networks are used. The values indicated are approximate.

4.6. Quad or Yagi

I must admit I have very little first-hand experience with quad antennas. But I can think of a few disadvantages of quad antennas as compared to Yagi antennas:

- Much better F/B can be obtained with Yagis.
- Quads are three-dimensional; you can't easily assemble a quad on the ground, and then pick it up with a crane and put it on the tower. You must do a lot of assembly work with the boom in the air.
- Wires break.
- It is a non-efficient material user—All the metalwork you put up is not part of the antenna; it is just a support structure.
- So far as electrical performance is concerned, a well-tuned quad antenna should marginally outperform a Yagi with the same boom length, at least as far as gain is concerned. The difference, being in order of a fraction of a dB to maximum of 1 dB, is more of an academic than of a practical nature.

To prevent ice build up on the quad wire, you can feed a current (ac or dc) through the loops. The voltage should be adjusted so as to raise the temperature in the wire just enough to prevent ice loading.

The fact is that the great majority of rotatable arrays on the low bands are Yagis. This seems to indicate that the mechanical issues are probably harder to solve with quads than with Yagis.

5. BUYING A COMMERCIAL LOW-BAND YAGI ANTENNA

5.1. 80-Meter Yagis

5.1.1. The linear-loading approach

To my knowledge, there is no manufacturer who is currently offering a full-size 80-meter rotatable Yagi antenna. The 2 and 3-element shortened Yagis with linear-loaded elements, originally developed by Mike Staal, K6MYC, for KLM more than 20 years ago, have held up over the years. Later M-Square, Mike Staal's new company, and Force-12 sell 80-meter Yagis using linear-loaded Yagis based on his original design. The merits of the linear-loaded design have long been established. The only inherent design compromise seems to be the sacrifice in top-notch directivity, which is caused by some radiation from the slant loading wires (see Section 3.7).

5.1.2. Mechanical issues

When deciding to buy one of these antennas, check the mechanical issues closely. This is what makes an 80-meter Yagi last or not. I must admit I am really scared when I see how some of these antennas are made. I see 80-meter Yagis using booms with a wall thickness that is less than half of the wall on my 40-meter Yagi boom! I see how a simple aluminum plate of a few mm thickness, is connected by a few simple rivets to the boom, and this is supposed to hold the 25-meter long element in a lasting way. I have doubts these antennas can ever stay up in windy areas. In my QTH they would not last one winter!

Mechanical issues are the real issues for a long-lasting 80-meter antenna. So, if you decide to spend a lot of money, take a very close look at the mechanics. An antenna built to withstand high winds and lots of ice loading will inevitably use more aluminum than a flimsy antenna that won't withstand a 90-km/h breeze. And aluminum costs money. There is a price for a good mechanical design. There are no two ways about it.

I know it takes approximately 45 kg of 6061-T6 aluminum to make a full-size 40-meter element that will withstand

160 km/h winds (+30% gusts). I have my doubts that an 80-meter 3-element Yagi, with elements that are 20% longer than for a 40-meter reflector, can be built for a total weight of only 120 kg.

As an example, I modeled the elements of the old KLM 80M-3 Yagi to assess its wind-survival speed. The element mechanical data were taken from the assembly manual of the 80M-3 antenna. The safe wind survival speed turned out to be 90 km/h, without a 30% higher gust factor. An element stress-analysis shows a very unbalanced design: While sections 2 (2-inch OD) and 3 (1.75-inch OD) are loaded to the limit, the three next sections are only loaded to about 60% of their capabilities.

The tip section is only loaded 25%. This does not necessarily mean that the element will disintegrate at 90 km/h, since this assumes that the wind blows at a right angle with respect to the elements. Putting the boom into the wind (perpendicular to the wind direction) will take all the stress off the elements. Provided that the side bracing of the boom is well done, it is likely that the Yagi boom will survive wind speeds above 90 km/h. Using the guidelines explained by Leeson in his book (Ref 964), sections 2, 3 and 4 can be reinforced by doubling the wall thickness to increase the wind survival speed to 123 km/h. In any case, you should add side guying of the central 3-inch section of the elements. Short boom extensions will be required to do this.

5.1.3 Low losses

A very interesting point was brought up by W6ANR. Often the weak point of a design is the lack of long-lasting, low-resistance electrical contacts. Lossy contacts ruin the gain and the pattern of any array. Invest in some good contact grease, Parker screws and heat-shrink tube for assembling the Yagi. Make sure all is done to prevent corrosion in the linear-loading wires. Better still, stay away from these wires and invest in high-Q loading coils.

5.1.4 High-Q coil-loaded Yagis

Creative Design Co, Ltd, manufactures a coil-loaded shortened 80-meter antenna. Their 3-element array has both elements driven in a ZL-Special configuration with 135° phasing. The element spacing is $\lambda/8$ (9 meters). The elements are 24 meters long (or approximately 62% of full size), and the loading is done with high-Q coils and a small capacitance hat about $^2/_3$ out on the elements. The elements are also loaded at the center with hairpin loading coils, which allows precise matching to the phasing line and the coaxial feed line. The array weighs only 80 kg. This is a very popular 80-meter antenna in Japan. Judging from its weight, it is probably not an antenna to put up where I live, though! As described in Section 3.5.1, W6ANR and K7ZV make high-Q loading coils for 80-meter Yagis.

5.1.4 Gain figures

Be very careful when comparing published gain figures. The only thing that really makes sense are free-space dBi gain figures, but the sales and marketing guys like to inflate these low figures and add ground reflection gain, which could be anything up to 6 dB, depending on ground quality and antenna height. Don't let these guys fool you!

I have withheld from publishing a list of commercial low-band antennas as I fear that I might not list all of them. And I worry that doing so might indicate some kind of endorsement on my part. If you plan to buy a commercial low-band antenna, I suggest you get a reference list from the manufacturer, and contact some of the customers—Or better yet, ask around on the Internet.

5.2. 40-Meter Yagis

It is amazing that none of the major antenna manufacturers advertise 2 or 3-element full-size 40-meter Yagis! That fortunately leaves a place for the real hams, the home builders to excel! From the poll I did with over 266 active and successful low-band operators, it appears that an important number use commercially made rotatable Yagi of some kind. Of those that listed the make they're using the breakdown is:

Cushcraft 40-2CD	45%
Force-12	15%
Hy-Gain	10%
Creative Design (Japan)	8%
Mosley	6%
Others	16%

Undoubtedly the most popular commercial low band Yagi remains the 40-meter Cushcraft 40-2CD Yagi. It appears to be the best value (excellent performance/price ratio) on the 40-meter shortened Yagi market.

D. Leeson, W6NL (formerly W6QHS), calculated the wind survival speed of an unmodified 40-2CD as 108 km/h. (Ref 967). The referenced article describes how to increase the wind survival speed to 150 km/h by using internal boom and element reinforcements.

CHAPTER 14

Low Band DXing From a Small Garden

The story to follow is undoubtedly the story of many, and it could be the story of even more people, provided they tried. If you don't have a large garden or a farm, read it. It's the story on 160 meters of a very good friend of mine, George Oliva, K2UO.

"*Having been an avid DXer for many years and having achieved "Number One" Honor Roll status on CW, SSB and MIXED, 5BDXCC, etc, I was in search of a new challenge. Some of the locals had started on 160 meters but I assumed that I didn't have the space for the antennas needed to work Topband on a half-acre lot. My amplifier didn't cover 160 and my tower was a crank-up type so a shunt feed wouldn't work very well. Eventually in 1985, I grew bored of the WARC bands and took on the challenge! I put up what has since become known as my 'stealth' dipole, a full quarter wave on 160, not in a straight line and not very high in the air. I worked 75 countries over the next 36 months with 100 watts and no special receiving antennas. Although most were relatively non-exciting, however, I did manage to snag 3B8CF, D44BC and even VK7BC.*

I next picked up a linear which did cover 160 meters. Now I began to see the need for special receiving antennas, I could now work everything I could hear but knew from the locals and packet clusters that I was not hearing a lot. I asked my 'friendly' neighbor if I could run a wire up the back end of his property line and I was now in business with a 550+ foot single wire, terminated Beverage antenna pointed to about 65 degrees. This antenna is truly amazing. I could now hear stations that I couldn't even imagine hearing on the 'stealth' dipole.

Although I am not the first to get through, I usually make it in the pileups. I have worked Bouvet, Peter I, Heard Island, South Sandwich, and now have 230 countries worked on 160 meters, almost all on CW of course.

When other hams visit my station and look at my Topband antennas, they are amazed at the results I have achieved. The bottom line of all this is that you do not need a super station to work a lot of DX on Topband. What you do need is a little imagination, ingenuity and perseverance to succeed and have a lot of fun."

What better introduction could I have than the above testimony of a dedicated Topband DXer, who's not frustrated living in a (beautiful I must say) but fairly typical suburban house on a $1/2$-acre lot? George did not use his

Fig 14-1— Showing a stealth antenna is easy—you show the sky. Rather than just the sky, here's the view at K2UO's QTH showing his low-profile 10/15/20-meter quad and his beautiful home in a wooded residential area in New Jersey. With his invisible 160-meter stealth dipole, George has worked 200+ countries on Topband.

QTH handicap as an excuse. No, for him it was just another challenge, another hurdle to take.

So don't lament if you don't have a one Million $ QTH. You can work DX on the low bands as well. Maybe you won't be the first in the pileups, but you will get even more satisfaction from succeeding, since you did have to take the extra hurdle!

My good Friend George Oliva, K2UO, holds BSEE and MSEE degrees and is an Associate Director at the US Army's Communications and Electronics Command's Research, Development and Engineering Center at Fort Monmouth, NJ. He is responsible for Research and Development programs involving Information Technology. He got his first amateur license in 1961 and has operated from a few exotic locations such as Lord Howe Island, Guernsey, Turkey and even Belgium. He is a Senior Member of the IEEE and holds several patents

George not only volunteered the above striking testimony, he also volunteered to godfather this section of the book, for which I am very grateful.

1. THE PROBLEM

If you have decided to read on, this is not going to be news for you. But let me nevertheless describe the typical suburban antenna syndrome.

You have this wonderful house, in this wonderful-looking neighborhood, at the right driving distance from your work. A dream, however, may not be a ham's dream. There really isn't enough space for the three towers and the Four-Square you would like to put up, and the neighbors would rather see trees growing than antennas. And your spouse won't really tell it to your face, but thinks one multiband vertical is more than enough. At the very best, one tower is what you can obtain your spouse's permission for.

If you really want to compete with the big guns on the HF bands, you need Yagis. Not a simple tribander, but monoband Yagis. On the low bands though, you can be relatively competitive with rather simple antennas. This is good news! Read on.

2. SET YOURSELF A GOAL

Maybe you should set yourself a goal that is realistic for your circumstances. You can get satisfaction that way as well. Compete with your equals.

But there are nevertheless "fantastic" stories from average suburban QTHs. Here is another testimony of perseverance (or maybe addiction): *"I was a young engineer working for IBM, just emigrated from Europe and lived until 1986 in a Toronto suburb, on a 46 by120-foot city lot surrounded by houses, TVI, power line noise and nasty neighbors. First I had a home-brewed 65-foot TV tiltover tower with used TH6 and 402BA and inverted Vs ($350). Later I thought I struck gold when I found a second-hand Telrex Big Bertha monopole with the antennas for $1200. I designed and built my own antennas (about $200 in material from junkyards). The rig was a used Drake B-line + R4C (about $500). All the rest of the station, the amplifier and the gadgets were home-brewed. I realized that I had a hard time beating the M/M stations in the contests, so I specialized in single-band operation. This netted me about 16 world records and all Canadian monoband records from 160 through 10 meters in CQWW and WPX contests..."*

All that from a 14 by 36-meter city lot! Wow! This was Yuri Blanarovich, VE3BMV, ex-OK3BU, now K3BU. But you are not that addicted? Keep on reading.

This book has explained propagation and focused on various types of antenna configurations for both receiving and transmitting. Factors such as gain, polarization, radiation angle, incoming signal direction and angle, soil conductivity and the many other factors affecting receive and transmit performance. It is up to you as an individual to assess your own situation, set your own goals and use the information in this book in conjunction with basic engineering judgment to experiment in the true amateur spirit.

Every QTH has its own limitations, and you must apply your own skills to optimize your station based on your individual goals. Let's have a look at some simple but very effective antennas that might help overcome some of the limitations.

3. THE FLAGPOLE VERTICAL ANTENNA

A λ/4 vertical for 7 MHz measures 10 meters, about the size of a really good patriot's flagpole. There you have a

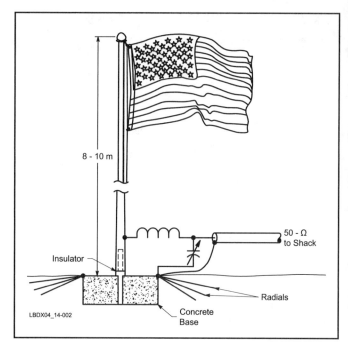

Fig 14-2—Forty-meter flagpole antenna. Any metal flagpole between 8 and 10 meters will do. Use an L-network to match to the 50-Ω feed line.

wonderful full-size 40-meter vertical. If the pole is a metal pole, make sure there is a good electrical contact between the different sections. If you are using a wooden flagpole, you will have to run a wire along the pole. It is best to use small stand-off insulators, so that the wire does not make contact with the wood. If your neighbor is curious about the wire, tell him it's part of a lightning protection system. Being a vertical antenna, the flagpole requires radials, but you can hide these in the ground, so nobody should object. You should of course insulate the flagpole from the ground. If the flagpole is exactly resonant on 40 meters, you can probably feed it directly with a 50-Ω feed line. Chances are the flagpole may be a little shorter, so you can load it at the bottom with a coil.

An L network, as shown in **Fig 14-2** will load and match the antenna at the same time. For a flagpole measuring 8 meters, typical component values (assuming a 5-Ω equivalent ground loss resistance) are: C = 500 pF and L = 2.8 μH. With a 10-meter long flagpole, no matching network will be required on 40 meters.

How about 80 meters? You can transform the 8 to 10-meter tall 40-meter vertical into an efficient inverted L at night, if it has to be a super stealth antenna. See **Fig 14-3**. Connect the top loading wire to the top of the metal flagpole. When you operate 40 meters, or during daytime, hang the top wire along the flagpole (coil up the bottom end so that it does not touch the ground). When you want to operate 80, raise the wire with an invisible nylon fishing line and stretch it toward the house or a tree. The top loading wire can be any thin wire, as it hardly carries any current (all the current is at the base of the flagpole). Now you're all set on 80 meters. For this 40/80-meter flagpole antenna (using an 8-meter long flagpole) the typical L-network component value for 80 meters is: L1 = 1.1 μH and C1 = 1100 pF.

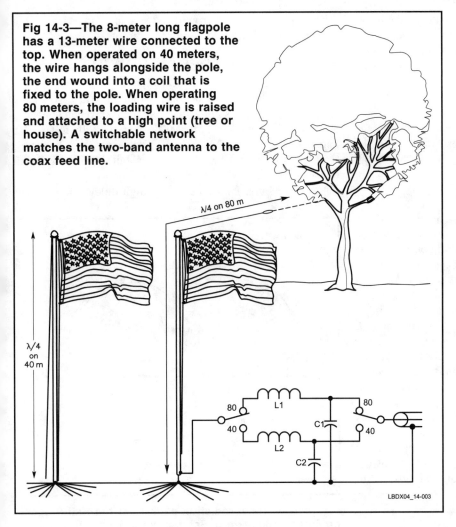

Fig 14-3—The 8-meter long flagpole has a 13-meter wire connected to the top. When operated on 40 meters, the wire hangs alongside the pole, the end wound into a coil that is fixed to the pole. When operating 80 meters, the loading wire is raised and attached to a high point (tree or house). A switchable network matches the two-band antenna to the coax feed line.

Fig 14-4—With a 40-meter trap installed at the top of the flagpole you can get the 80-meter top-loading wire permanently connected. It can be directed to a tree, the house or any other available support.

If your spouse or the neighbors won't object to a permanent tiny wire running from the top of the flagpole (maybe they haven't even seen it), install an 40-meter trap at the top of the flagpole. Disguise it using your imagination. See **Fig 14-4**. Appropriate traps are described in Chapter 9.

For an efficient 160-meter vertical antenna you need at least 15 meters of vertical conductor. Have you looked at the trees in that corner of your lot? They should do as supports. Maybe you need to exercise a bit with a bow and arrow, but if you can shoot a nylon wire over the trees, you're probably set for a good 160-meter antenna. If you use an inverted L or T antenna, you can use the tree-supported vertical on 40, 80 and 160 meters. And your neighbors will hardly see it! Don't forget that this antenna requires a good radial system. But you can put those down during the night.

Don't forget that the open ends of an antenna are always at very high voltage. If you run the outer ends of these wire antennas through the foliage toward a tree, it's a good idea to use Teflon-insulated wire. This will help prevent setting your tree on fire. And, by the way, all these wires don't have to be perfectly horizontal or perfectly vertical. Slopes of up to 20° will not noticeably upset the antenna performance.

You could of course also buy a commercial antenna, and spend lots of money for lots of loss. Use your imagination instead, and put your brains to work instead of your wallet!

4. LOADING YOUR EXISTING TOWER WITH THE HF ANTENNAS ON 80 OR 160 METERS

If you have a tower with one or more HF or VHF Yagis, you can probably turn it into an efficient vertical on 80 or 160 meters. A tower of about 15 meters with a simple tribander will give you the right amount of loading to turn it into an excellent 80-meter vertical.

For 160 meters you will need a little higher tower, but starting about 18 meters with a reasonably sized tribander antenna will get you about 70° electrical length on 160. See Chapter 9 for details on how to shunt-feed these antennas.

If the tower is guyed, make sure the guy wires are broken up in short sections. Short means about λ/4. Better still, use dielectric guy rope, such as Phillystran (Kevlar) or glass-epoxy rods (Fiberglass Reinforced Plastic or FRP).

If you use a crankup tower, you will do better running a solid copper cable along the sections (an old coaxial cable will be fine), as the electrical contact between sections may not be all that good. In case of doubt, climb your tower and measure the resistance.

It is imperative that you run the cables inside the tower all the way down to ground level, and run them underground to the house; otherwise it will be extremely difficult to decouple these cables. It is a good idea to coil all the cables at ground level, to provide enough inductance to form a good common-mode RF choke.

Don't forget that shunt-fed towers do require a good ground system. Run as many radials as you can in as many directions. Don't overly worry if the tower is next to the house—you may lose a couple of dBs in that direction but that's all.

5. HALF SLOPERS

Half slopers are covered in Chapter 10, Section 6. These antennas are popular with those who have a tower with a rotary antenna, and who want to get it working on 80 meters. A minimum height of about 13 meters (depending on the loading on top of the tower) is required to make a good vertical radiator on 80 meters. For a 160-meter sloper to work well, you would need a tower about twice that high. Don't forget that it is not the sloping wire that does most of the radiating, it is the vertical tower. The sloping wire merely serves as a kind of resonating counterpoise for the feed line to push against. As with all vertical antennas the efficiency of a half sloper will depend primarily on the radial system used.

Don't feel tempted to use sloping wires in various (switchable) directions. As the sloping wire only radiates a small part of the total field, this effort would be in vain. As with shunt-fed towers, all cables that run to the top of the tower should run inside the tower, and run underground to the shack to maximize RF decoupling.

6. HALF LOOPS

Half loops are covered in Chapter 10, Section 5. Fed at the bottom of the sloping wire, this antenna is attractive where space is limited. A 15-meter high tree could support the vertical wire, and from the top a slant wire can run to the shack or any other convenient place. If you use a 26-meter long sloping wire, the antenna will be resonant around 3.5 MHz, and have a feed-point impedance of 60 to 75 Ω, good for a direct feed to the transmitter. To make it work on 3.8 MHz, shorten the total length of the antenna by approx 3 meters, or simply feed it through an antenna tuner or L network. This antenna will also work quite well on 160. Its feed impedance will be very high, however. The best feed system is to use a parallel-tuned circuit as shown in Chapter 12, Fig 12-20. Needless to say, the feed point is at very high RF voltage, and the necessary precautions should be taken to prevent accidental touching of the antenna at this high voltage point. **Fig 14-5** shows the radiation patterns for this half-loop for 80 and 160 meters. On 80 meters the antenna shows some directivity, about 4 dB in favor of the direction of the sloping wire. Again, a good ground system is required for this antenna, at both ground connection points.

7. VERY LOW TOPBAND DIPOLES

The saying is that very low dipoles (10 meters up) are only good as receiving antennas. Is that so? **Fig 14-6** shows the layout of K2UO's *Zig-Zag dipole* for 160. When you walk around his lot, you can hardly see the wire. It really is a stealth antenna, but it has given George 230+ countries on Topband. And that's not only "heard" countries, but those worked and confirmed!

In this book I have described high dipoles as efficient

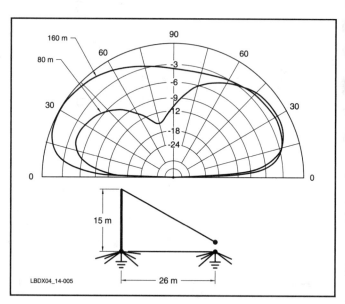

Fig 14-5—Vertical radiation pattern for the Half-Sloper on 80 and 160 meters. See text for details.

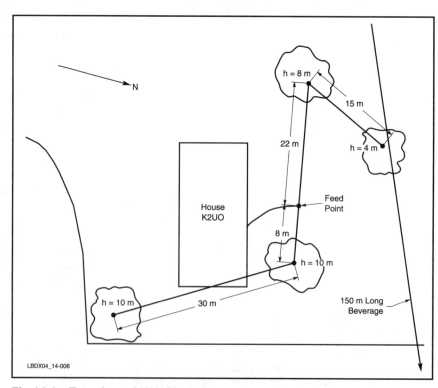

Fig 14-6—Top view of K2UO's 160-meter stealth dipole, which is supported by trees and which is at no point higher than 10 meters!

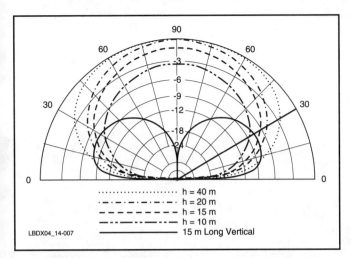

Fig 14-7—Vertical radiation pattern of dipoles at various heights, compared to a short 15-meter long vertical with 5 Ω equivalent ground-loss resistance.

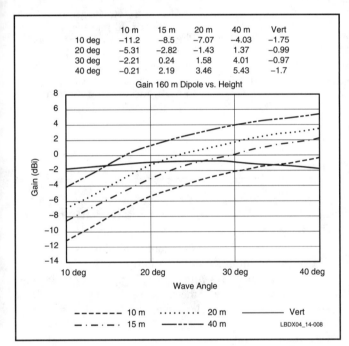

Fig 14-8—Gain of low dipoles compared to a reference 15-meter long vertical.

low-angle radiators. In order to be competitive with vertical antennas at really low angles, a dipole must be at least λ/2 high. I think we will hardly ever find such high antennas on a typical suburban lot though! But low dipoles can still function quite well on the low bands. The antenna at K2UO is a outstanding testimony for such low dipoles.

Fig 14-7 shows the vertical radiation patterns of low λ/2 dipoles, compared to a 15-meter long vertical (R_{rad} = 17 Ω) using a fairly decent radial system (R_{loss} = 5 Ω). A 160-meter dipole between 10 and 15 meters high produces the same signal as our reference vertical (±1 dB) at a wave angle of 30°, which may come as a surprise to some. At very low angles, (10°), the vertical will be 13 dB better than the 10-meter high dipole. **Fig 14-8** shows the gain of the various antennas for wave angles of 10°, 20°, 30° and 40°.

Looking at the patterns in Fig 14-7 we see that the big difference is in the high angles. The low dipole will be much better than the vertical for local coverage, but that means also that the signals from local stations will be much stronger than they would be on a vertical. Although the dipole may have the big advantage of reducing man-made noise (which is generally vertically polarized), it has the disadvantage of producing very strong signals received at high elevation angles.

What may come as an even bigger surprise is that we have learned that not all (though most) of the DX on Topband comes in at very low angles. Especially on 160 meters, however, we know that gray-line enhancement at sunrise or sunset often coincides with an optimum angle of radiation that is rather high, and that definitely gives the advantage to the low dipole. So, you might even beat the big gun with his super low-angle antenna, using a K2UO-style dipole!

As a rule I'd like to stress that it is important that you keep the center of the antenna as clear and as high as possible. The ends are just "capacitance hats"—they don't really radiate a lot, so you can bend and hide them as appropriate without hurting the antenna's performance a lot. If you don't have room for a straight 80-meter long dipole (who has?), rather than loading it with coils, or using a W3DZZ-type dipole, just bend the ends. That's much better, and will introduce less loss than the usual lossy coils. What holds for 160 meters is of course applicable to 80 as well.

K2UO is certainly not the only one who's been successful with low dipoles. Recently I read a similar testimony from Ivo, 5B4ADA (ex-HH2AW): *"My 160-meter antenna is $1/10$-λ high (apex of inverted V is 16.5 meters above ground, wire ends are 1.5 meters above ground). Theoretically, it radiates up most of the RF, but I still have fun working USA, JA, VK, breaking XW3Ø pileup, etc. I had 57-meter long wire in Haiti on a bamboo pole 10 meters above around. Worked many USA and EU stations on 160. Don't be scared with too much theory, get on the air..."*

I would not necessarily agree with the "theory" part of Ivo's statement, since the theory does predict that low dipoles are a viable alternative... to nothing at all.

8. WHY NOT A GAIN ANTENNA FROM YOUR SMALL LOT?

8.1. An Almost Invisible 40-Meter Half-Square Array

I am convinced that on 40 meters you can get up this almost invisible gain antenna. You need to be able to run a horizontal wire about 10 meters up, and 20 meters long. Perhaps from the chimney of the house to a tree in the corner of the lot. **Fig 14-9** shows a 40-meter half-square array that can be squeezed in many small lots. Gain is approx 3.4 dB over a single full-size λ/4 vertical. The ends of the vertical wires are also at very high RF potential, and precautions should be taken to prevent accidental touching. The Half-Square is fed via a parallel-tuned circuit as shown in Chapter 12, Fig 12-20.

You can also feed the Half-Square in one of the top-corners. This may be a good idea if one element is close to the house as shown in Fig 14-9B. When fed in the corner, the feed impedance is about 52 Ω, a perfect match for a 50-Ω feed line. Do not forget to install a current balun on the coaxial feed line. **Fig 14-10** shows the radiation pattern for the 40-meter Half-Square.

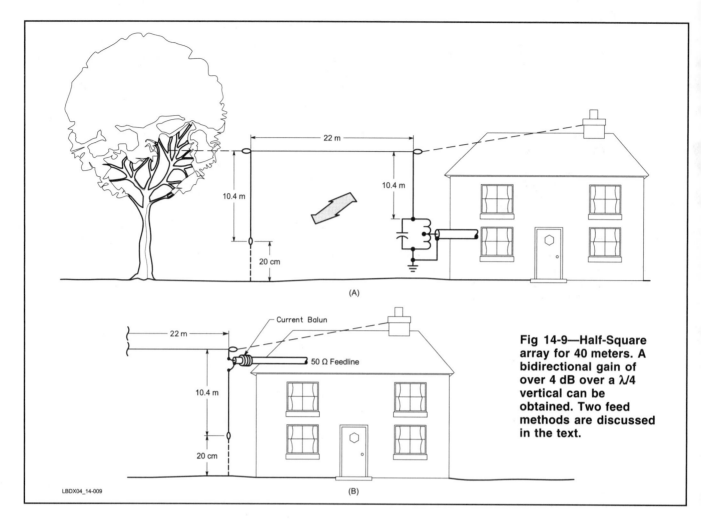

Fig 14-9—Half-Square array for 40 meters. A bidirectional gain of over 4 dB over a λ/4 vertical can be obtained. Two feed methods are discussed in the text.

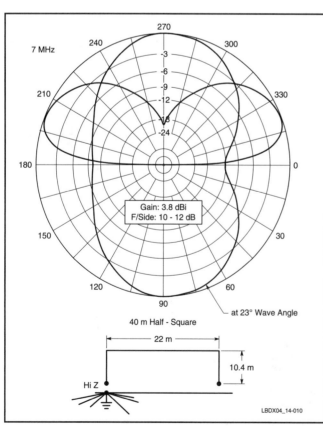

Fig 14-10—Radiation pattern for the 40-meter Half-Square.

8.2. Using the 40-Meter Half-Square on 80 Meters

What about using the 40-meter half-square on 80 meters? A bit of magic turns the antenna into two close-spaced in-phase fed end-fire arrays with top loading. The only thing you need is to short the base of the second element to ground, and feed the array at the other element at ground level (**Fig 14-11**). This 2-element array has a gain of 1.6 dB over a single full-size (20-meter high) vertical and provides excellent low-angle radiation. The antenna has about 4 dB front-to-side ratio. Its feed-point impedance is about 70 Ω excluding ground-loss resistance at each vertical element. This antenna requires a good ground radial system at the base of both elements.

With some ingenuity you could homebrew a switching system that grounds/ungrounds one element, and either feed the other element directly with coax on 80 meters, or feed it via a parallel-tuned network on 40 meters.

8.3. 40-Meter Wire-Type End-Fire Array

Maybe the Half-Square doesn't suit your most wanted direction. You can also turn this into a 2-element parasitic array as shown in **Fig 14-12**. I worked out the example of an array where a maximum height of 8 meters was available as the catenary wire. The elements were top-loaded as shown in Fig 14-12 and **14-13**. The array has a very good F/B and gain, and a feed-point impedance of about 25 Ω. See Fig 14-13. Matching can be done through a λ/4, 35-Ω line, consisting of two parallel 75-Ω coaxial cables (each measuring 7.03 meters

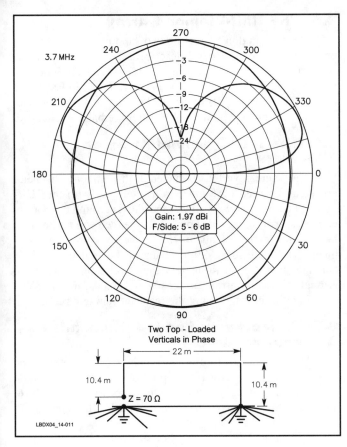

Fig 14-11—The 40-Meter Half-Square can be turned into a 2-element close-spaced top-loaded array for 80 meters, where both elements are also fed in phase. Both vertical and horizontal patterns (at 30° elevation) are shown.

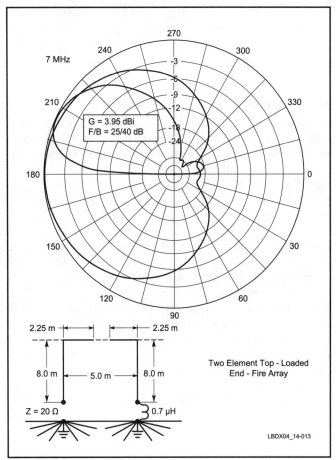

Fig 14-13—Horizontal radiation patterns for the 2-element parasitic array of Fig 14-12. The gain is 3.95 dB, and the F/B is an impressive 25 to 40 dB.

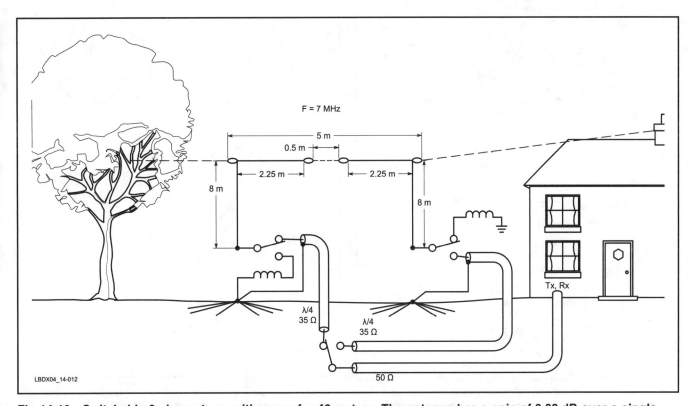

Fig 14-12—Switchable 2-element parasitic array for 40 meters. The antenna has a gain of 3.08 dB over a single vertical.

Low Band DXing From a Small Garden 14-7

for RG-11 or RG-59 (solid PE insulated coax with VF = 0.66).

It is important to install as good a radial system as possible on this array. Where radials from the two elements meet they can be connected to a bus wire, as shown in Chapter 11.

8.4. And an 80-Meter End-Fire Array

Maybe you have two high trees in the back that could help you support a 2-element array for 80 meters. I would recommend a minimum height of the elements of approximately 13 meters; the remainder can be top loaded if necessary. **Fig 14-14** shows two T-loaded 13-meter high verticals, suspended from a single catenary rope, for example between two tall trees.

This array has an excellent F/B ratio and gain, and will certainly put you in the front seat in a pileup if you take care to install a good ground system. When properly adjusted, the array impedance, assuming about 5 Ω equivalent ground loss resistance, is about 20 Ω. The array can be fed via a 37.5-Ω, λ/4 transformer (two parallel 75-Ω cables) as shown in Fig 14-12, or via an unun transformer (20 to 50 Ω).

It is important that the top-loading wires are as shown (points facing one another). If you are forced to try another configuration, I would advise you to model the array exactly as in reality. Needless to say, if you have some really tall trees on your property, this antenna can be scaled up for 160 meters.

8.5. The Half-Diamond Array

Maybe you don't have the two supports required to put up the box-shaped arrays I described above. With just one support, a few good arrays can be created as well. You will require one high support (15 meters); eg, a tree. In **Fig 14-15** I've reshaped the Half-Square to become a Half-Diamond. You lose about 1 dB gain, but the pattern remains unchanged.

8.6. The Half-Diamond Array on 80 Meters

The same inverted-V-shaped Half-Diamond 40-meter array can be used on 80 meters as well. As with the Half-Square (see Section 8.2) all you must do is ground the bottom end of the array at the side opposite to the feed point. The array has an impedance of about 75 Ω. The exact resonant frequency can be tuned by simply changing the total length of the antenna. For use on 40 meters the exact length is not so critical, since the array can easily be tuned for low SWR anywhere in the band using the resonant tuning circuit.

8.7. Capacitively Loaded Diamond Array for 80 Meters

Maybe you can't quite get a height of 15 meters. One solution is to top load the two sloping verticals with a common capacity wire, hanging right down as shown in **Fig 14-16**. Two wires are connected to the apex of the V. Their length and

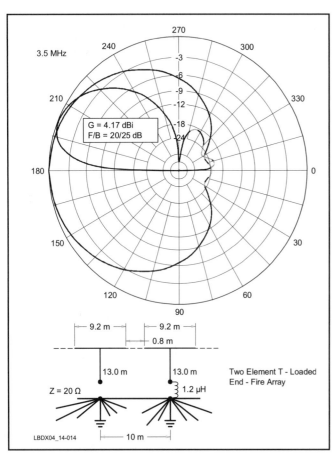

Fig 14-14—Horizontal radiation patterns for the 2-element T-loaded parasitic array for 80 meters. The gain is over 4 dBi and the F/B is 20 to 25 dB.

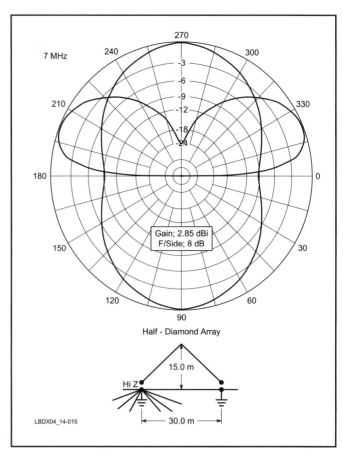

Fig 14-15—Using a single support, you can turn the Half-Square array into a Half-Diamond array, at the sacrifice of about 1 dB of gain.

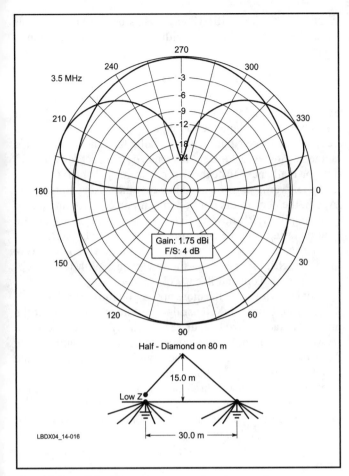

Fig 14-16—This 80-meter array requires only 15 meters of height.

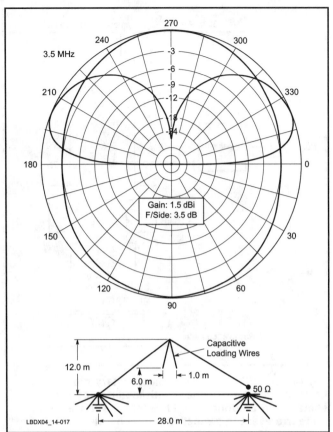

Fig 14-17—This 80-meter capacitively loaded midget Delta-Loop array requires only a 12-meter high support.

the angle between the wires is varied to tune the antenna to the required frequency. The gain of this antenna is still about 1.5 dBi, which is more than 1 dB better than a single full-size (20 meter high) vertical over typical ground. The feed-point impedance, including about 10-Ω loss resistance, is about 50 Ω!

8.8. A Midget Capacitively Loaded Delta Loop for 80 Meters

The Half-Diamond antennas look very much like a Delta Loop with its bottom wire laying on the ground. Let us raise the wire, and turn it into a real Delta Loop. The model shown in **Fig 14-17** has similar dimensions to the antenna in Fig 14-16, and yields the same gain, the same front-to-side ratio, and even the same feed-point impedance. Needless to say, this is once more proof that the Delta Loop is nothing else than two sloping verticals, fed in phase (see Chapter 10, Section 2).

In this example I used capacitive loading in a little different way. This Delta Loop can be tuned anywhere from 3.8 to 3.5 MHz by just changing the length of the bottom capacity wire. L1 is 11.0 meters, and L2 is 6 meters for f = 3.5 MHz. For f = 3.8 MHz the bottom loading wire can be eliminated altogether.

There is nothing magical about these dimensions. Just keep in mind that the capacitive loading wires should be attached at the high-voltage points, and that they carry very high voltage indeed. Where crossing each other, the loading wires should be kept about 20 cm from each other.

Do not fool yourself into thinking that this antenna does not require radials. The radiation is affected just as much by near-field absorption losses under the antenna as in the case of the grounded verticals. In other words, Delta Loops require a ground screen, just as is the case with all antennas that do have radiating elements close to ground!

9. Special Receiving Antennas

Typical suburban QTHs mean rather dense housing, which in turn means a lot of man-made noise. Now that you have used your imagination, and squeezed an efficient vertical—or even a couple—onto your lot, you're faced with a very high noise level. There are basically four ways to tackle this problem:

- Use a horizontally polarized receiving antenna
- Use a directive receiving antenna so that you can null out the main noise source
- Use a noise-reduction system based on phased antennas
- Locate the offending noise sources and kill them (the noise sources, that is!)

Jim McCook, W6YA, swears by his very short rotatable dipole on top of his tower (see **Fig 14-18**). Such a small dipole would fit almost every lot. You can actually rotate it as it has excellent rejection when its ends are turned toward the direc-

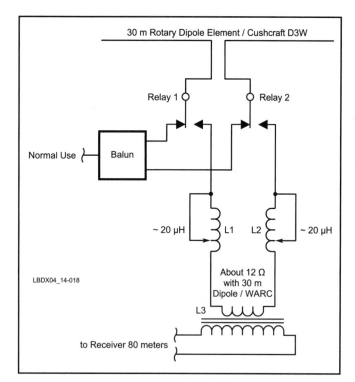

Fig 14-18—W6YA uses a 30-meter rotary dipole, which he resonates on 80 meters with loading coils. The same could be done for 160 meters. As losses are quite irrelevant in a receiving antenna, the Q of the loading coil is not so important.

tions of the noise source. But any low horizontal wire will probably be better than your vertical for receiving.

How about a Beverage? I know your property is not quite like a Texas ranch, but maybe you can run one or even two short ones along the property line. Maybe you can talk your good neighbor into a concession. George, K2UO, has room for only one (150-meter long) Beverage, which partly runs along the property lines of his neighbors. But he says this antenna really was an eye-opener. Even λ/2 long Beverages are better than nothing! If you're after best F/B, tune them for a cone-of-silence length. This means you will need a complex impedance to terminate them (see Chapter 7, Section 2.5). If your neighbor does not care to see the Beverage wire along his property line, maybe you should try a Beverage on Ground (BOG) antenna, as described in Chapter 7, Section 2.11.

How about EWEs, Flags, K9AYs and the like? They are quite popular where space is limited. Don't forget that these antennas should be clear from any transmitting antennas. I would recommend λ/4 as the minimum distance between a special receiving antenna and your transmit antenna. If you cannot achieve that, you can always detune the transmit antenna on receiving (see Chapter 7, Section 3.10).

I have been extremely successful in eliminating man-made noise from a particular source (a chemical plant 10 km away) by using a so-called *noise eliminator*. This is nothing but a circuit in which you combine the inputs from two antennas, the main receiving antenna and a noise-pickup antenna) in such a way that they are added out-of-phase, resulting in complete cancellation of the noise (see Chapter 7,

Section 1.35). The MFJ-1206 unit has been a very valuable asset for me when dealing with noises from one particular source.

10. POWER AND MODE

Power and mode both factor into the results you will receive in terms of working DX as well as "working" your neighbor's TV/telephone/radio, etc! The more power you radiate, the more signal will be available at the DX station's antenna. However, look at the goals you have set for yourself. Do you really need to be the first one to get through the pileup? If the answer is yes, you need power; otherwise, you may not. Again, the question is not how much power is coming out of your amplifier; it's how much power are you radiating and are you radiating it in the right direction at the right angle? You must also remember that it is much easier to work DX on CW than on SSB, especially on the low bands. With few exceptions, DX on Topband is on CW.

It is obvious that the more power you run the easier it will be for you to work DX. That does not mean you will get the most satisfaction from your results though. Also, the more power you run, the more chances there are of creating some kind of TVI, BCI or telephone interference problem in the neighborhood. The first and easiest way to avoid similar problems is not to run high power. But, if you are already handicapped with your pocket-size lot, some power may be one of the few available means to get that evasive DX on the low bands after all. CW as a mode also creates fewer audible interference problems than phone (unless your neighbors copy CW).

11. ACHIEVEMENTS

You have read about K2UO's 200+ countries worked on 160 meters with a Zig-Zag Stealth dipole that's nowhere higher than 10 meters. In Chapter 13 (Section 4.5) I described W6YA's 21-meter tower, on a typical suburban lot, which carries antennas for all nine bands! Don't let space restrictions scare you away from the low bands. Be sure that, if you work the evasive 3B7 on Topband from your small lot QTH, you will get triple the satisfaction of the big gun who maybe got through a couple minutes earlier.

Some time ago I read a very applicable statement: *"...I was always complaining about my shoes, until I saw a man without legs."* Let's have fun with what we have and do our best under the circumstances. Be convinced you're not the only one who's not living in ham's paradise. There are many others in the same boat.

12. THE ULTIMATE ALTERNATIVE

Don't put up any antennas, don't get on the air from your pocket-sized QTH, but save your money and energy, and go on a DXpedition once or even twice a year, and provide us hard-core Low Banders with the new countries we need to be able to prove we're the best!

Of course, there's another solution, if you can afford it. Set up a remote station. I have seen wonders of ingenuity and engineering when visiting some of the top remote-controlled stations on the low band in the US: K9DX, W6RJ and N7JW/K7CA come to mind. (See Chapter 11). While that solution is out of reach for all but a very few, it shows what can be done. The sky is the limit!

CHAPTER 15

From Low-Band DXing to Contesting

When I wrote the First Edition of *Low Band DXing* almost 20 years ago, I was a very active, omnipresent low-band DXer. In Belgium we had just been allocated 160 meters and I really lived and slept on the low bands. Some of you may have wondered why over the past 10 years ON4UN has been almost absent on the bands. Here's why.

My neighbor and friend Peter, ON6TT, (of Peter I and Heard Island fame, and others) stirred up my interest in *contesting*. Peter achieved "mission almost impossible" by helping us in Belgium obtain permission from the PTT to operate high power during international contests. At last we could compete equally with contesters in other countries and at last we did not have to fib about our power levels! While our PTT had been frightened about high power, the experiment with 2-kW licenses for international contests in 2001 led to high-power (1-kW) licenses for all full-license amateurs in Belgium. Over a period of 12 years when 2 kW was allowed only during these contests, our PTT received not a single complaint about interference. Why? Good contest stations are usually built by good engineers and technicians who know what they are doing!

For me, building a competitive contest station has been a unique experience: Getting help from friends and club members and building a team of excellent operators. It not only widened my technical horizons but also my social horizons in terms of working with people, enjoying amateur radio as a group and making new friends from all over the world. One of the highlights of nearly half a century being a ham was being part of three WRTC (World Radio Team Contesting) competitions, either as a competitor, a referee or just a visitor.

Fig 15-1—DL2CC at the ON4UN contest station.

Besides Mark (ON4WW) and Peter (ON6TT), both local neighbors and avid contesters, there are two more friends I met through contesting that I would like mention in particular. They both played an important role in my becoming an avid contester.

I first met Harry Booklan, RA3AUU, at the Clipperton Club convention in Bordeaux, France, in 1993. Harry was then 23 years old and won both the CW as well as the Phone pile-up tests. Harry now has his own telecom company in Russia and has become a very successful businessman. He quickly became my friend as well as one of the fixed assets of our OTxT contest station. As ON9CIB, Harry, and I represented Belgium at WRTC in San Francisco in 1997. Harry ended second in WRTC 2000 in Slovenia, and second again in WRTC 2002 in Helsinki. He is "big" in contesting.

Frank Grossmann, DL2CC (ex DL1SBR), is another highly esteemed operator and friend of the house. Frank is a young computer-programming professional, who runs his own business in Stuttgart. In between jobs he makes the 750-km trip (one-way) several times per year to operate contests from here. Frank was third at WRTC 2002 in Finland.

Both Harry and Frank are superb CW operators. Frank used to be Germany's high-speed champion some years ago. They both turned me into a CW addict, a wonderful addiction that I am proud to admit.

Frank, DL2CC, graciously accepted being my help, counselor and critic for this chapter. Thank you Frank!

1. IN SEARCH OF EXCELLENCE

Amateur Radio is all about satisfaction and self-fulfillment.

My Elmer was ON4GV, who was also my uncle. At his home I saw my very first Amateur Radio station. I was not quite 10 years old. That was around 1950.

Fifty years ago Amateur Radio in my eyes was adventureland and wonderland, all in one. To a young boy, telecommunication was Amateur Radio. You must realize that in the early years after WW II, out in the countryside where I lived with my parents, we still had hand-cranked telephones with manually operated telephone exchanges. These exchanges closed down at 10 pm—we had no chance to telephone anyone during the night. If you wanted to communicate with someone across the ocean you wrote a letter. If you wanted to travel across the water, you took a boat.

But for all I knew, if you wanted to talk to someone anywhere in the world, you needed to be a radio amateur. Being a ham made you an explorer, a discoverer. You could expose yourself to worlds others had hardly ever heard about. *It was this magic, this thrill of radio that lured me into this hobby.*

I will never forget the oh-so-typical smell of Bakelite, wax and tar-filled capacitors and transformers that was very typical for the early-day radios. And the white filament glow of early tubes. Some of the early-day triodes were so "brilliant" that you could literally read a book by them at night. My very first hands-on experiences with electronics (not that it was called that, in those days; we called it simply "radio") were in building small audio amplifiers, using directly heated triodes, such as the E, A416. My father's wooden cigar boxes served as chassis for building three-stage audio amplifiers using heavy 3:1 interstage transformers. This was all about discovering an amazing and intriguing world, the world of radio.

It was technology (a modern word for these early-day sensations) that hooked me to Amateur Radio. For a while my discovery trips were somewhat curtailed and it was not until I was 20 years of age, in high school, that I finally got my license. The challenge then was to prepare for the license, and get it on the first try. The sense of fulfillment, once you got it, was enormous, as was the first antenna you built, the first QSO you made. I was doing something not every one else could do! Now I was part of them...

In the early 1960s I stumbled across some guys working DX on 80 meters. I remember a few calls: GI6TK, GI3CDF, GW3AX and G3FPQ. Only David, G3FPQ, is still there and still a very active low-band DXer. This really seemed like something else—working across the pond and into New Zealand on frequencies where the others would work stations in a 500-km range. What a challenge! This really put you in a separate class among hams.

Working the elusive DX on the low bands was my next challenge, and it became my passion. This time I found out that, in order to be part of those low-band DXers, you needed to have a good signal. That meant you needed to have the know-how to do it. Amateur Radio was no longer a communicating hobby for me, but became an experimenter's hobby: building new, better and bigger antennas, experimenting with and learning about propagation, becoming a better technician and becoming a better operator.

DXing on the low bands is all about overcoming difficult hurdles: everybody can work DX on 10, 15 and 20 meters. There is not much sense of satisfaction involved. In 1987 my last iron curtain was lifted. We finally got 160 meters in Belgium. The last frontier. A vast terrain for chasing difficult game. And yes, Top Band certainly is where the DXer can get the ultimate sense of satisfaction. Technical knowledge and technical achievements are undoubtedly great assets in achieving success in low-band DXing. But, even with a modest station, provided dedication, patience and operating experience, you can be a successful DXer.

If so, what can provide you with the ultimate sense of achievement, of technical excellence, in Amateur Radio? Throughout history, competition has been one of the important leverages for progress in many fields. So is contesting to Amateur Radio. To be a very successful low-band DXer you need to be a very good operator, know propagation, be patient, be persevering and have a "decent" antenna system and station. You can determine yourself whether or not you are successful. You can set your goals as a function of your possibilities, and if you have worked 300 countries on 80 and 200 countries on 160 from an average urban-lot QTH, then you are, by all standards, a very successful DXer.

To be successful in "big game" contesting, you cannot compromise with yourself. You need to have the best antennas, the best station, the best operators, nothing but the best if you want to score high in world ranking. The best multi-operator contest stations are all built, improved, maintained and run by engineers. This is no coincidence.

Undoubtedly international contesting is the ultimate challenge—it provides the truth by excellence. It is truly the Formula 1 competition in Amateur Radio. This is what attracted me to this radio-sport.

Jim Reid, KH7M, was an operator at Stanford University, W6YX, in the years when SSB techniques were worked out there by Art Collins (yes, later from Collins Radio) in the '50s. He was a witness to how Amateur Radio contributed to important advancements in communication technology, now almost half a century ago.

He made an interesting comparison between the world of contesting and the world of car racing.

"Today, about the entire globe, the most sophisticated, and elaborate HF band stations are owned by contesters such as CT1BOH, WB9Z, IR4T, PI4COM, IY4FGM, GW4BLE, JA3ZOH, PY5EG, W3LPL, K3LR, VE3EJ, and so on (ON4UN was also mentioned—thanks Jim!). Each of these stations has elaborate and multiple rig setups, multiple antenna installations, many computers with each operating position networked with logging programs, band mapping programs, propagation monitoring radios, and so on.

These Amateur Radio stations are all Contest/DX stations. They each have the most sophisticated and up to date technology possible. Each has invested into it what would be comparably invested into a stable of Ferrari racing automobiles; and I have seen both, especially in Southern California!!

These guys have pushed and pushed at manufacturers, at antenna designers, at software writers, and continue to do so. The station owners themselves are all first class operators and technicians. They spend virtually all of their time "tinkering" and pushing the state of the HF art, in every way feasible. Every station I listed represents thousands of man-hours of work, every year to maintain and remain in the top ranks of competition in the sport of DX contesting, which of course has no more purpose than the sport of auto racing: fun to have and maintain and win with the BEST.

The owners of these stations have pushed the state of the art of Amateur Radio every bit as directly as the owners of Indianapolis racing machines have pushed the state of the art of tires, lubricants, engines, brakes, frame design, and so on."

In the highly competitive sport of international contesting, it does not suffice to have the best car or engine; you also need the best drivers, the best mechanics, and the best engineers. Contesting is indeed very much like car racing.

In DXing you can get the fulfillment of working all countries on 40 or 80 or even 160 meters. Once it's done, the game is over. Not that the game of working all countries on Top Band will ever be over, I guess. But in contesting, there is a new competition calendar every year. Every year you can measure the station's performance, you can measure your improvements, plan your progress and enjoy the fulfillment of your victories, over and over.

That is why international contesting is the ultimate playground of ever advancing, competitive and self-fulfilling Amateur Radio.

2. WHY CONTESTING?

Now and then I read on the Internet how a proud antenna builder tells us all about the wonderful performance of his new antenna. He is trying to convince the world by telling us all about the rare DX he worked with his new antenna. What does that prove? Very little, really. Working DX is in no way a proof of technical performance of an antenna. Not in the strictly technical sense, in any case.

Working rare DX can be the proof of outstanding operating, dedication and perseverance, when it's done from a modest QTH with small antennas.

If you want to prove the technical capabilities of the station, there are really only two ways. Number one consists of elaborate full-scale field testing and measuring in a precisely controlled environment. This is beyond the reach of almost every ham.

The second possibility consists of testing your weapons, not in a shooting range or in a lab, but on the battlefields. These battlefields are the major international contests. Contests are possibly as close as we can get to a controlled environment, simply because it is extremely unconditioned. In major international contests you are competing against all the best-engineered and equipped stations, under a variety of continuously changing conditions, which really makes it a fair and equal battle and test.

This is why, after having been an "occasional" contester for almost 40 years, I decided to get into some serious contesting, thereby putting emphasis on the low frequency capabilities of my station.

3. WHAT CATEGORY?

In order to convert my station into a successful contesting station the first decision was—"we want to win—but in which category?" In other words, what is the appropriate battleground for the weapons we have?

The biggest and probably best-known stations are the *multi-multi* stations. The most successful of them have two stations per band; that means 12 fully equipped stations. These stations are on each of the six bands, 24 hours per day, and they must catch every single opening. They need to have access to a wide variety of antennas with different wave angles. Therefore, they are generally equipped with various stacked Yagis for the HF bands, even including 40 meters. The second station, whose task it is to look for multipliers, generally has access to a simpler antenna setup. It is located as far away as possible from the running station's antennas, to minimize interference, although eliminating same-band interference is quite impossible.

Interstation interference is the most challenging technical challenge in multi-transmitter station design. With multi-multi contest stations, though, each of the band stations can be completely (galvanically) separated from each other, which certainly helps prevent leakage paths for unwanted coupling between stations. In this respect a multi-multi is simpler to design and make than a so-called small multi-single, where all of the antennas have to be accessible by both stations. This makes eliminating leakage paths much more difficult.

There are a number of *multi-single* stations, which as far as station design is concerned really fall in the category of multi-multi stations. I call them *big multi-single* stations. They have six well-separated stations, one of which "runs" while the five other stations are manned and are checking each of the five other bands simultaneously. In this configuration as well, it is possible to achieve better isolation between bands because there are six completely separate stations. These big multi-single stations normally also have antennas for each band on separate towers. To build a really top-notch multi-multi station you probably require at least 2.5 to 5 acres (1 to 2 hectares or 10,000 to 20,000 meters^2) of land—and that does not include what you need for Beverage antennas.

For a big multi-multi setup, in addition to the financial limitations, there is simply not enough space for putting up additional towers in my backyard. This is why we decided on going for the category of multi-single, or—as I call it—*small multi-single*.

Small multi-single stations can be built much smaller than multi-multi stations. A small multi-single is a station with only two operating positions, one for the "run" station, and one for the "multiplier" station. The operator of the multiplier station has to scan all bands for multipliers. In general both the run as well as the multiplier station will have access to all antennas, which means that a fairly complex antenna switching system is part of the setup. Such switching

Fig 15-2—A 1965 picture showing ON4GV, the author's Elmer, ON4UN and his XYL to be. In the background we recognize a Drake 2B, an SB-1000 and an NCL-2000.

systems increase the potential for unwanted coupling between the two stations. This is what makes designing a small multi-single station technically more difficult than a large multi-single or a multi-multi station. It goes without saying that a station designed for small multi-single is also well suited for single-operator two-radio (SO2R) contesting. In recent years SO2R has become quite popular. In concept and in layout it is very similar to a small multi-single station, where, however, the two operator seats are replaced by a single one.

While it is imperative for a multi-multi station to catch every single band opening, and therefore needs antennas to match all possible elevation angles, this is not necessary for a multi-single station. The run station will run on the bands at the times the takeoff angle of his antenna matches propagation best. In other words, the height of the antennas should be such as to accommodate the average elevation angle, the angle that produces most QSOs for the longest period of time. This means antenna heights between 18 and 30 meters for 10 though 40 meters. The multiplier station may have to call a multiplier with an antenna that is not at the ideal height. He may not get through on the first call, but this is not as important for the multiplier station.

But there is not only multi-operator or SO2R all-band HF contesting. Any station that is successful in DXing could be a candidate for single-operator contesting. And if the station is not equipped for all bands, a single-band effort can be contemplated. Also, if you are not 20 years old any more, single-band contesting is attractive. If you operate the low bands, you have all day to rest. You can still prove the technical excellence of your station on the band of your choice!

4. ANTENNAS

The ON4UN/ OTxT/ ORxT contest station was designed as a multi-single and a single-operator two-radio transmitter

Fig 15-3—The author's contest station is located on a 4000 meter^2 (2 acre) plot about 10 km from the nearby major city of Ghent. On this land are all antennas, except for Beverages, of course. In the foreground you can see part of the 30,000 meters^2 (3 hectares) that are used in winter for putting up 12 Beverages, one per 30°.

station. **Fig 15-3** shows the QTH and some of the antennas. One tower supports the 40-meter Yagi (at 30 meters height) and the 20-meter Yagi (at 25 meters height). As they are both on the same tower, they cannot be rotated independently. A similar combination exists on tower number 2, where a 6-element 15-meter Yagi (at 24 meters) tops a 6-element 10-meter Yagi (at 19 meters). The third tower is quarter-wave 160-meter antenna, which also serves as a support tower for the 80-meter Four-Square.

About 100 meters behind the house is a fourth tower (18 m) with a Force-12 C31XR triband Yagi (formerly a KT34XA).

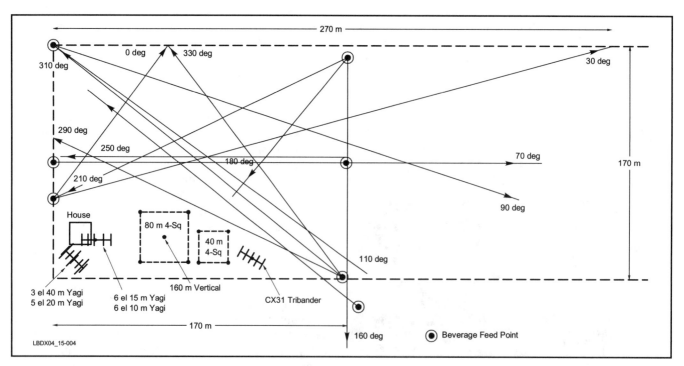

Fig 15-4—Diagram of antenna layout at ON4UN/OTxT. All of the 12 Beverage antennas run over neighboring farm land, which is accessible for this purpose between the end of October and late April. The shortest Beverages are about 165 meters; the longer ones 300 meters.

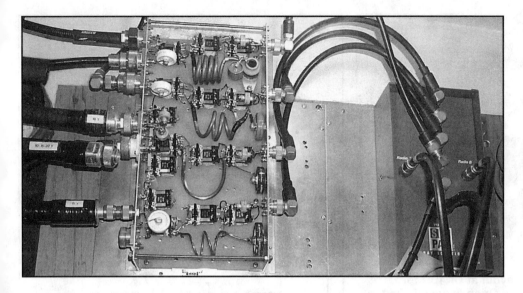

Fig 15-5: The selecting combiner unit for 10 through 40 meters: The feed lines (7/8-inch) from the single-band antennas and the 10-20-meter tribander (C-31XR) arrive on the left. On the right are four band outputs for 10 through 40, which go to the WXØB six-pack. From top to bottom: 40, 20, 10 and 15-meter L-networks. There are 15 relays in the box.

This is what we call the *multiplier antenna*. There are two more multiplier antennas, a 40-meter four-square and an 80-meter low inverted V. See **Fig 15-4.** These multiplier antennas are used whenever the main antenna is not available for the multiplier station; eg, when the run station runs on 20 meters with the big Yagi, the 40-meter Yagi is not available, and the 40-meter four-square must be used to work multipliers on that band.

5. THE OPERATOR'S STATION

For a normal everyday DXing station there is practically no rule on how things have to work in the shack. For a contest station equipped for a team effort it is different. Things have to be simple and ergonomic. A *hired-gun* contest operator is usually someone who doesn't read manuals. He or she wants to sit down and start operating right away. This means that the whole system must be simple and idiot-proof! I remember the days that we had no safety designs and that operators would start transmitting on the wrong antenna or with the amplifier set on the wrong band, resulting in inevitable damage and lots of frustration. Fortunately those days are over now. It is not possible to describe what the ideal contest station should look like. There are too many variables involved. But a really well designed contest station is very different from a run of the mill DXer station.

Building blocks are available from different sources. Array Solutions, WXØB, carries them all, but many suppliers have the individual parts. Check *National Contest Journal* (ARRL publication), where they all advertise.

5.1. Antenna Switching at ON4UN

Since two different antennas are available on the higher bands (40 through 10 meters), I have made provision, for either station to use either one of these antennas (the "run" or the "multiplier" antenna) or split the power 50/50 into those two antennas. When the band is open in two directions, you can thus work in two directions simultaneously. Of course, you must realize that you have more QRM/noise and only half the transmit power in each direction. If you are using one antenna, you can quickly switch directions to work a multiplier. It is imperative that the SWR curves of both antennas are flat and similar, so that you do not have to retune the amp while switching.

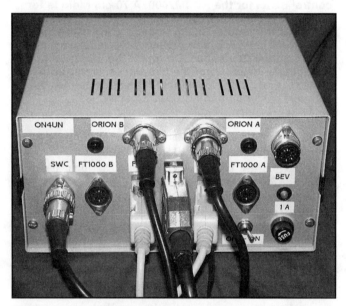

Fig 15-6—The homemade antenna switching logic box. The inputs are the band data outputs from two transceivers and the selection data from the "main/multiplier/both" switch (see Fig 15-7). The outputs are: switching data for bandpass filters and control voltages for switching the 6-pack antenna relay box (WXØB) and for the combiner unit shown in Fig 15-5.

Fig 15-5 shows the switching/combining unit I built.

The unit contains four L-networks (10 through 40 meters) that convert 25 Ω back to 50 Ω. Simple ac-type relays are used for the switching, which results in quite a bit of wiring inductance (about 8 inches of wire). To compensate for the effects of these long wires, I put capacitors at both ends of each relay wire, creating Pi networks. With about 30 pF on both ends, the SWR is less than 1.05:1 even on 10 meters.

The antennas are selected fully automatically, using the band output data from the transceivers, in my case Ten-Tec Orions. I built a control device using the band data output from the two transceivers to generate the logic signals for selecting the antennas. **Fig 15-6** shows the control unit. The switching

Fig 15-7—Early 2004 station layout at ON4UN for contesting. For SO2R a second Orion is added. On top of the Orion is the W2IHY audio equalizer and the control panel for the ACOM2000. A 70-cm radio is for talking to local DXers (ON4WW, ON4MA etc). Behind the Ten-Tec Remote Tuning knob (a great asset for contest-ing) is the Braun preamp/preselector, which is used with the Beverages. On top of the Braun unit is the antenna selector for the two transceivers. Both switches have three positions: (1) Main antenna, (2) Multiplier antenna and (3) Both together.

Fig 15-8—Block diagram of the station and antenna-control circuitry at ON4UN.

logic must prevent the two stations from selecting a stack of the secondary antenna on 10/15/20 while the triband antenna is already in use by the other station. The logic has been developed such that when one station uses the tribander, the other one cannot get it, and will remain on the main antenna. It's first-come, first-served

Relay logic is used throughout so the system is totally immune to RF and very reliable. A total of 11 small relays are used for logic switching. A small switchbox mounting two three-position switches and some LEDs is placed between the two transceivers for the operator to use. The left switch is for Radio A, the right one for Radio B (see **Fig 15-7**). The three positions are:

1. Primary antenna
2. Secondary (multiplier) antenna
3. Both antennas in a 50/50 split.

There are seven LEDs to indicate the status of the switch. When the green LED is on, you are on the main antenna; the orange one stands for the multiplier antenna; and the red one is for both together. Blinking red means that you are trying to select the multiplier antenna, which is not available.

Fig 15-8 shows the block diagram of the system. On 80 and 160 meters, where only one transmit antenna is available, these antennas are fed directly from the six-pack switch. And if for any reason the band data from the switch and the data from the transceiver do not match, the transmitter in the transceiver is inhibited (in the Orion via the TX-EN line).

5.2. Antenna Directions

On the receiving side the visual direction indication of the Beverage antenna selector proved to be very helpful. At the lower left in **Fig 15-9** you can see my Beverage selector

Fig 15-9—Layout of ON4UN station. Note especially the 12-position switch at the lower left. This is used to select receiving Beverages for the low bands.

box, which uses a 12-position rotary switch with LEDs for each azimuth direction.

Point-and-forget rotators: In the heat of the battle you don't want to sit and press that turn-left or turn-right button—just select your direction and press one button. On commercial rotator controls these buttons are too small, however.

5.3. Radio-Computer Interface.

Computers and contesting software is covered in detail in Section 7 of this chapter. Top-notch contesting without top-notch contesting software is a no longer possible. All contesters control a great number of functions of their radio through the computer, starting with sending CW, changing frequencies, bands or modes, etc. You name it. Connection between the computer and the radio is nowadays still done largely by serial ports, although we likely will see them disappear from modern computer in the near future. We will then require converters for changing USB signals to serial-port signals, as long as our radio manufacturers don't go USB. While most radios seem to be giving band data as BCD codes, this requires a decoder in the communicating device. Ten-Tec, with its Orion has chosen simply to provide one line per band, which simplifies switching, since it doesn't require any additional conversion!

5.4. Easy-to-Tune Power Amplifiers

In a typical multi-single or single-op two-transmitter contest setup one transmitter is tuned to the main band, where a pileup is worked. This is usually called the "run station." With the second transceiver the other bands are scanned, and multipliers are picked up between QSOs on the running band. In order to be able to concentrate fully on the operating aspects, band-switching should be automated as much as possible.

Tuning the second linear between bands often has been a problem, because:

- Contest operators often don't know how to properly tune a linear,
- It takes them too long.

The minimum to have is labels stuck to the linear front-panel with all settings for all antennas and modes. Even that sometimes seems to be too difficult for the operators trying to concentrate on moving this very rare and weak KH8 station from one band to another. An amplifier that automatically switches bands, and automatically tunes to preset values of band-switch, load C and tune C, or better yet, performs a fully automatic tune-up, is the answer to that problem.

Since 1998 the ON4UN station has been equipped with an ACOM 2000A linear. The use of this fully auto-tune linear proved to be very helpful for working multipliers by quickly changing bands and antennas.

The ACOM 2000A (see **Fig 15-10**) is an auto-tune nominal 1500-W output amplifier (maximum 2000 W) using two Russian-made 4CX800A (or GU74B) tetrodes (replacement cost in Europe typically $50 to $60). This was the first real auto-tune amateur HF-amplifier I had ever seen. By pressing a button on the remote control panel it automatically tunes itself completely within a half second. The auto-tune function is not limited to recalling preset values—it actually tunes for a match for a load within the 2:1 SWR circle (on some bands up to 3:1)

Fig 15-10—ON4UN's two-radio contesting station, consisting of two Ten-Tec ORIONs. Two computers linked by a network run the logging program, with two separate mini keyboards. When permitted by contest rules, a third computer connects through the Internet to packet clusters around the world.

The amplifier has an absolutely blank front panel, except for an ac on-off switch. This makes it possible to hide the amplifier in any convenient place. All control and monitoring functions are grouped on a remote small control box, which can easily be positioned next to the computer keyboard during operation. The ACOM amplifier can be connected via an RS-232 connector to a PC for either remote control or testing. Its processor keeps track of all the important data (currents, voltages, temperatures).

In case of a fault, you can send the information stored in the INFO BOX for the most-recent 12 faults to the dealer or the factory by means of Baudot code on the telephone—simply put the microphone close to the tiny loudspeaker on the RCU rear, or by means of a personal computer and its inherent communications channels (Internet, modem, etc). Needless to say, the use of this amplifier has greatly increased flexibility and efficiency at the OTxT contest station. Over the past six years, Acom has built a superb quality and service record, and continues to be an excellent choice for serious contesting.

Since Acom introduced fully automatic tuning, Alpha/Power followed. Its well-known Alpha 87A amplifier was previously equipped with a memory-tune system but now it has been improved to included an auto-tune system similar to the one used by Acom. This amplifier has very similar, but not identical, characteristics (also 1500-W nominal output), and is certainly a valid candidate for a top-notch contesting station linear as well. The Alpha 87A uses two 3CX800A7 triodes, which have the disadvantage of being much more expensive than the tetrodes used in the Acom. The Alpha 87A also has an RS-232 interface port, which allows the linear to be controlled remotely. In addition, key parameter measurement values can be monitored remotely. Not only the 87A, but also other models, including the Alpha 89, and the Alpha 91B are very popular with contesters as well as DXers and DXpeditioners.

In many cases, these amplifiers made it possible for us to hear the DXpedition's signals on 160 meters.

6. THE STATION AS SEEN BY THE TECHNICAL PERSON

It is clear that the technical requirements for a top-performing contest station are far superior to what's needed for casual, or even serious, DXing. Think of harmonic suppression. Stations built for multi-transmitter operation must transmit the cleanest signals, and their harmonics must be suppressed far in excess of the standard. Another issue is to keep the contest station "up and running" all year long. It takes good mechanical engineering to keep the antennas up.

6.1. The Operating Table

Even though we were housed in a small shack of 3 × 3.5 meters for years, we nevertheless managed two multi-single first places in Europe. But we were almost sitting on each other's laps! So, one wall was taken out, which made it possible to install a single 7 × 1-meter wide operating table.

The new shack layout was conceived with contesting in mind. To provide the best possible RF and safety ground, the underside of the 7 × 1-meter table was entirely covered with a 1-mm thick aluminum sheath. This sheet provides maximum capacity to the equipment standing on the table, and minimum resistance and especially inductance for good RF grounding. Forty ac outlets are mounted on the aluminum sheath, providing the shortest possible safety ground return for the outlets. Short and wide straps are connected to the sheath and are available on the back side of the table to ground various equipment.

The table is separated from the wall by approx 15 cm, which allows wires to pass and for ventilation as well. The aluminum ground sheath is grounded with a short strap to an excellent RF ground just outside the shack, with a 40-cm heavy-gauge cable going right through the wall.

The table is equipped with three separate mains distribution circuits, each equipped with a professional-grade mains filter. Circuit one powers all the "run" station equipment. Circuit two powers all the "multiplier" station equipment and circuit three powers some other equipment plus the main computer.

6.2. A Monitor Scope at Each Station

In the heat of a phone battle, operators sometimes have the tendency to crank up the microphone gain, resulting in poor and distorted audio, unnecessary splatter and so on. I always have a monitor 'scope connected to the output of each of the stations. I use a second-hand commercial 20-MHz 'scope, and tap off a little RF using a resistive voltage divider across the output of the linear. This way the operator always has the pattern of the transmitted signal right in view (see Fig 15-2).

6.3. The Problem of Interband Interference

In a two or more station setup, interband interference is the number one technical problem. But as the saying goes, every problem is an opportunity. In this field lies the opportunity to excel. Here also lies the opportunity for equipment manufacturers to improve their equipment. Interference can be minimized by using the following techniques:

- Separate the antennas as much as possible
- Use vertical and horizontal polarization to take advantage of the additional attenuation of unlike polarization
- Use band-pass filters between the exciter and the amplifier
- Use amplifiers with Pi-L networks, not simple Pi networks
- Avoid common-mode currents on the feed lines
- Galvanically separate the feed lines of the separate bands
- Use band-reject filters between the amplifier and the antenna
- Push the equipment manufacturers to produce transmitters with much lower in-band noise output.

It is obvious that interference will be heard on the harmonic frequencies. This poses much more of a problem on CW than on phone. The CW band segments are all in the low end of the bands, and the harmonics of 3.503 will be 7.006, 14.012, 21.018 and 28.024 MHz—all right in the CW window. On phone, if you operate on 3.775 kHz, the harmonics will be on 7.550, 15.100, and so on, all outside the band. There is no real problem with the direct harmonic frequencies when operating phone.

Unfortunately most present-day transmitters do not only transmit just the wanted signal; they also transmit a lot of noise around the transmit frequency. This noise can often make it difficult to copy, even many kHz away from the exact harmonic of the transmit frequency, unless effective filtering is applied. And even then, the final improvement will have to come from the designers and manufacturers of our transceivers, putting out equipment producing less in-band noise.

My friend George, W2VJN, covers all of these aspects very thoroughly in his excellent publication *Managing Interstation Interference*, which can be obtained directly from him.

6.3.1. Medium power band-pass filters

There are a few commercial sources for medium-power band-pass filters that are widely used in multi-station contest setups as well as during DXpeditions. I have experience with the ICE, Dunestar and W3NQN units. The ICE units are rated 200 W, and if the SWR is low they will indeed cope with 200 W. The Dunestar filters are rated at 100 W. I have been

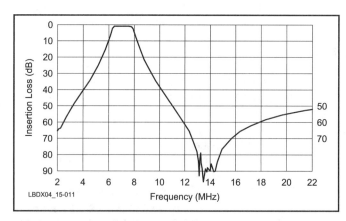

Fig 15-11—ON4UN uses the W3NQN filters switched with the WXØB FM-6 filter box between the transceivers and the amplifiers. This is the response of the 40-meter W3NQN filter. Note the deep null at the second harmonic at 14 MHz.

Table 15-1

	1.8 MHz	3.5 MHz	7 MHz	14 MHz	21 MHz	28 MHz
ICE 160 m	0.4 dB	15 dB	27 dB	40 dB	>45 dB	>45 dB
Dunestar 160 m	1.0 dB	28 dB	>45 dB	>45 dB	>45 dB	>45 dB
W3NQN 160 m	0.2 dB	50 dB	>80 dB	65 dB	50 dB	58 dB
ICE 80 m	25 dB	0.34 dB	17 dB	30 dB	>40 dB	>45 dB
Dunestar 80 m	42 dB	0.64 dB	37 dB	>45 dB	>45 dB	>45 dB
W3NQN 80 m	53 dB	0.4 dB	70 dB	62 dB	55 dB	49 dB
ICE 40 m	>45 dB	38 dB	0.8 dB	32 dB	>45 dB	32 dB
Dunestar 40 m	>45 dB	43 dB	0.6 dB	>50 dB	>45 dB	33 dB
W3NQN 40 m	68 dB	45 dB	0.3 dB	>80 dB	52 dB	48 dB

using all three of them and I did some comprehensive measuring on these units.

The ICE and Dunestar units have insertion losses of between 0.3 and 1.0 dB (that is a lot) depending on band. Due to the circuitry used, the Dunestar filters have significantly steeper shape factors. The W3NQN filters undoubtedly have the best characteristics and show 0.2 to 0.4 dB insertion loss, depending on band. If we look at the performance of the 40-meter filters, we see that the ICE filter will attenuate 20 meters about 32 dB, the Dunestar more than 50 dB and the W3NQN between 80 and 90 dB. **Fig 15-11** shows the response of a 7-MHz W3NQN filter. **Table 15-1** lists some of the major characteristics I measured for 160, 80 and 40 meters.

It is important that the filters be operated at a low SWR. If not, you will likely blow the capacitors. It is important, when driving a linear amplifier through one of these filters, that the linear is switched to the right band. If not, a high input SWR may result. If the exciter is equipped with a built-in tuner, it may try to get the full power into the filter, at a very bad mismatch, which guarantees fried components. Therefore, you shouldn't switch the automatic tuner on when operating with a band-pass filters. Also, it is a good idea to control the selection of the right filter right from the transceiver's band data output, so you do not dump RF of the wrong band into the filter. There must be many contesters and DXpeditioners who have done that. I am sure replacement capacitors for these units must be a hot selling item!

6.3.2. High power filters

If you run power, it really is a must to run filters beyond the amplifier as well, because the amplifier also generates harmonic power. These filters should not only be designed to suppress harmonics, they should attenuate signals on all bands, also on frequencies below the transmit frequency. It is not uncommon for signals from one of the stations of a multi-operator station to mix in the linear with other signals (BC or from another amateur band) and create unwanted mixing products. The ultimate filter is indeed a filter that attenuates all other bands, but gives the highest attenuation to the second harmonic.

The most common way of achieving out-of-band attenuation is by using band-reject filters. These can be made with discrete components or by using coaxial cable.

6.3.2.1. Using discrete components

High-power filters using discrete components can be made much smaller than those using coaxial cable, but the

Fig 15-12—Ten-meter high-power band-reject filter. This simple unit uses one series-tuned trap for each of the five bands to be rejected.

components are hard to come by (high-power, high-voltage, high-current capacitors) and the design requires some expertise and the use of a quality network analyzer.

I have designed a series of such filters, which perform very well. **Fig 15-12** shows a 10-meter band-reject filter that will take 3-kW continuous-duty power. I built it in a box measuring 25 × 6 × 6 cm. The box is made of double-sided glass-epoxy printed board material, which is ideal for the application. With single-pole series-tuned circuits for each band, an attenuation of 38 to 46 dB was obtained on all five bands, with an insertion loss of approx 0.1 dB. **Fig 15-13** shows the response from 1 to 30 MHz for this filter.

The principle for designing such band-reject filters is really quite simple. You design five series-tuned circuits, each one tuned to the frequency you want to suppress and simply connect all these traps in parallel. For the 10-meter filters, all these tuned circuits will exhibit an inductive reactance on 10 meters. You can easily calculate this value: Calculate the impedance of all the coils and all the capacitors (five of each) used in this filter. Since they are connected in series (for each band), you can simply add the values, taking the sign (+ for a coil, – for a capacitor) into account. Then calculate the parallel value of all of these, just as you calculate parallel resistors. Now we can "tune" out this positive reactance by using a parallel capacitor, which resonates the whole thing on 28 MHz. It really is that simple.

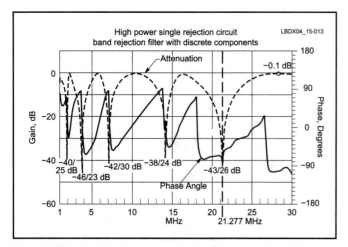

Fig 15-13—Rejection curve for the homemade high-power 10-meter filter in Fig 15-12. The rejection figures quoted are for the band center and band edges. For example, the rejection in the center of the 15-meter band is 43 dB, and 26 dB on the band edges. This measured plot was generated using the Alpha/Power network analyzer.

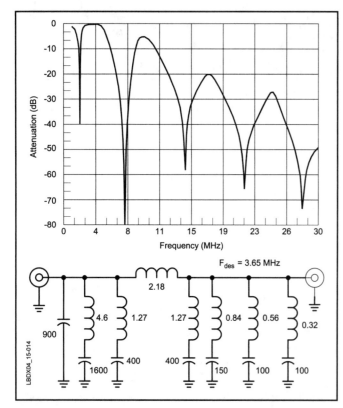

Fig 15-14—High-performance 80-meter filter, using discrete components and two 40-meter traps. This filter has a rejection of 80 dB on 40 meters, and between 60 and 70 dB on 20, 15 and 10 meters. The insertion loss is less than 0.1 dB.

For other bands, series-tuned circuits below the design frequency will show as inductors on the design frequency, and as capacitors above the design frequency. By judiciously choosing the LC ratio of the series-tuned traps, you can now design filters where the positive reactance of a group of traps will cancel the negative reactance of another group, which means there will be no need for a parallel capacitor or inductor to tune the filter to a 1:1 SWR on the operating frequency.

Fig 15-14 shows a high-performance 80-meter filter using a pair of 40-meter traps for improved rejection. The basic configuration is a low-pass section. The effect of the low-pass section can clearly be seen at the overall shape of the rejection curve. Filters like this can easily be modeled using the *ARRL Radio Designer Software*, an ideal tool for this purpose. In this case I designed a symmetrical low-pass filter, and arranged the traps on both sides of the inductor to obtain the same capacitance value. A capacitor (900 pF) had to be added on one side to tune the low-pass filter. The value of the inductor can easily be calculated using the LINE STRETCHER module of the NEW LOW BAND SOFTWARE.

If you want to design your own filters, the sky is really the limit. The biggest problem in making such filters is to obtain suitable capacitors. Inductors can be wound on powdered-iron toroidal cores (#2 material). Make sure you calculate the estimated power that will be dissipated in the cores. On adjacent bands there may be a substantial amount of heating in the cores, and 2-inch cores may be required in some circumstances.

It is beyond the scope of this book to deal with the concept, design and construction of such filters, but they are necessary to make a multi-transmitter station fully competitive.

6.3.2.2. Using coaxial-cable stubs

Let's work out a situation where we want to operate an 80-meter station and a 40-meter station simultaneously in the CW contest. A single quarter-wave long shorted stub, made of RG-213, cut for 80 meters, will provide about 26 dB attenuation on 7 MHz, 24 dB on 14 MHz, 23 dB on 21 MHz and 22 dB on 80 MHz (see **Fig 15-15**). The insertion loss on 80 meters will be less than 0.1 dB.

A quarter-wave RG-213 stub cut for 20 meters typically shows an attenuation of 37 dB. The same stub with RG-58 shows about 25 dB of attenuation. A 10-meter stub made of RG-213 can achieve 40 dB of attenuation.

We can use two identical stubs to almost double the attenuation, but not by merely connecting them in parallel! Connecting a short across a short can in the best case, when the two shorts are equally "good" or "bad," brings you 6 dB additional attenuation. There is one way, however, to obtain much more attenuation.

Look from the linear amplifier into the feed line. With a well designed and built amplifier, the pi-L filter will provide a good deal of attenuation on 40 meters. But there is some 40-meter power at the linear output. Assume it is 50 dB down from the 80-meter fundamental. At the output of the linear we assume a low Z for the second harmonic, an acceptable assumption at the output terminal of the pi-L filter. If we now put the stub (which is a short on 40) right at the output of the transmitter, we are putting a short across a short, which is not very effective. If we insert a quarter-wave coaxial line be-

15-10   Chapter 15

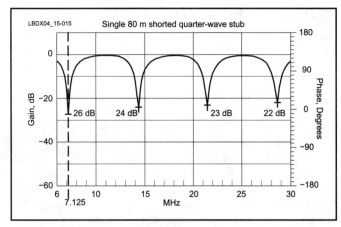

Fig 15-15—Attenuation "dips" are obtained on all of the harmonically related frequencies. (Figs 15-15 to 15-17 generated with the Alpha/Power Analyzer.)

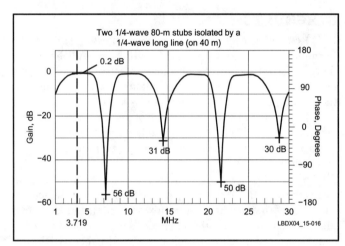

Fig 15-16—Attenuation curve for two stubs separated by a λ/4 feed line (see text).

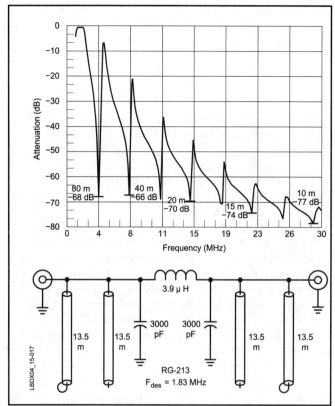

Fig 15-17—Attenuation curves for stubs used in conjunction with discrete components to form a low-pass filter. See text for details.

tween the output of the amplifier and the stub, we have transformed the very low impedance point (on 40 meters) to a high-impedance point. If we now connect the stub at that point, we will have the most effect of the short that the stub represents. All of this holds true only if the output of the amplifier represents a low Z for the second harmonic (40 meters).

In practice it is a good idea to experiment: install the stub right at the output of the amplifier and check the attenuation on 40. Then insert the quarter-wave line between the linear and the stub. If the attenuation is better (which is likely), leave it there. In theory a quarter-wave gives best results (maximum transformation ratio), but anything from λ/8 to 3/8λ should be OK, as long as you stay away from the region of 3λ/8 to 5λ/8.

Fig 15-16 shows the attenuation of a single 80-meter shorted stub (between 6 and 30 MHz).

But you wanted more than 25 dB. You can install another quarter-wave *isolation line* between the first and the second stub. I call it an isolation line because it effectively isolates the two stubs. The reasoning is the same as explained above. Two stubs isolated by a quarter-wave coax (on 40 meters) now exhibit 56 dB of attenuation on 7 MHz, and 50 dB on 21 MHz, but only 31 dB on 20 meters and 30 dB on 10 meters. This is logical, since the isolation line must be an odd number of quarter-waves long on the reject frequency to do its job. This is true on 7 and on 21 MHz only. On 20 and 10 meters we only get the predicted 6-dB improvement. In this case using an isolation line of λ/8 on 40 meters would result in good attenuation on 20 (where the isolation line would be λ/4, on 15 (3λ/8), but not on 10 meters where the isolation line would be λ/2.

Stubs can also be used as elements in a low-pass configuration, in combination with discrete components. The example in **Fig 15-17** is a combination of a simple 160-meter low-pass filter with four stubs. The attenuation pattern is amazingly clean, and gives better than 70 dB on all bands.

The impedance of the coaxial cable used for making stubs is irrelevant. The cable loss is important, however. An 80-meter stub made with RG-58 will yield approx. 15 to 17 dB attenuation, while RG-213 gives 25 dB and 7/8-inch Hardline will give 40 dB!

In a contest station setup, you can also install fixed stubs at the feed points of single-band antennas. At my QTH I have a 160-meter quarter-wave transmit antenna that stands right in the middle of the 80-meter Four-Square. You can hardly imagine how to obtain more coupling between these antennas. Without some kind of stub an ACOM amplifier feeding one of these antennas would switch off when someone would transmit on the other antenna because of the large amount of "alien" RF it saw. I put a shorted λ/4 160-meter stub at the base of the 160-meter antenna and an open-circuit version at the feed point of the 80-meter antenna.

Top Ten devices makes a multiband switched stub system that can be driven from the band-data output of most transceivers. (See **www.qth.com/topten/** or contact **w2vjn@rosenet.net**.)

7. COMPUTERS AND SOFTWARE

While I was a very early user of *CT* (in DOS-days), I was one of the very early users of *Writelog* when 32-bit *Windows* was first introduced. More recently I have switched to N1MM's contesting software. You can't run the latest all bells-and-whistles software on a 66-MHz computer, since performance and speed go hand-in-hand. I use one computer for each of the two radios at ON4UN. They are state-of-the-art computers, running the latest operating software (with a 2-GHz clock and *Windows XP-Professional*).

In addition both are networked by LAN with the main computer. All of this has the advantage of having instantaneous backups on the other machines, just in case something goes wrong. If one of them is a laptop, you are even protected against sudden mains drop out. If we do a Multi-Single operation, a third PC uses *DXConcentrator* program (from ON5OO; see **homeusers.brutele.be/on5oo/Introduction.html**). This computer is connected to the Internet via a broadband router, becoming the "master of ceremonies" computer. The Master of Ceremony operator makes sure that we all get the Packet Cluster information available. He interprets that information and decides when and where to run, and when and where to work multipliers. He is really in charge. We have been using this technique quite satisfactorily for years.

7.1 Connecting Computers

For first Multi-Single as OT3T in 1993, we had linked a variety of PCs ranging from 286s to a 386 66-MHz machine with copper-wire serial cables. Despite pounds of ferrite rods and toroids, we certainly did not achieve a totally RF-free situation. Often a computer would hang, without apparent reason. Several times during the contests logs had to be merged and computers started up again. All of this was certainly far from ideal!

In the second phase we replaced all the copper links with fiber-optic links. This certainly was an expensive improvement, but still not 100% bulletproof. Then we went to an Ethernet solution using software written by David Robbins, K1TTT, and Wayne Wright, W5XD. We found that twisted pair Ethernet cabling worked fine, even in a strong RF environment.

7.2. Connecting to the DX-Cluster

Here too evolution has been staggering. What was advanced five years ago, looks like museum technology today. Having had access to wideband Internet for several years now, I have quit using VHF or UHF radios to connect to the DX-cluster system. If you use ON5OO's excellent *DXConcentrator* software, you have the whole world of DX-information at your fingertips. We typically connect to 16 clusters around the world, and although some of these are interconnected, you can often gain valuable seconds by having access to a cluster that's close to the spotter! Being the first-on-the-spot can be important. I give more details about this excellent program in Chapter 2.

7.3. Computer Noise

In all the years we have been using various computers, I have never had a problem with direct radiation from a computer. Of course, if you use an antenna inside the shack very close to the computer, you will pick up all kinds of hash. It is very important that your antennas are at a sufficient distance from the computers, that the feed lines are well-shielded and well-grounded and that the coax connector makes perfect contact with the receptacle.

Computer screens can be very noisy though. If you buy a new display, make sure you can return it if it radiates too much. You may also want to add extra ferrite cores on the cable between the PC and the monitor. Better still, use an LCD monitor, but I've heard of cases where the monitor's switching power supply needed few more ferrites on the 12-V power cable to silence it. LCD flat-screen monitors are much less fatiguing for the eyes, and this can be important during long contest periods.

8. RESULTS

Remember why we participate in contests. There are the Formula-1 operators who want to win going full bore. But there is also the technical guy, who wants to see the fruit of his labor, his engine, his station, his antennas win in an international competition. This is what it was all about for me.

But at the same time I met a lot of good regular operators and technicians (the always available helpers from the local radio club). This is undoubtedly the important social aspect of contesting.

Since 1993 my station has been tested in 89 international contests (organized by either the ARRL or CQ), which resulted in 54 first places (Europe or worldwide), and 15 second places. This proves that the antennas as well as the station are capable of winning top-notch contests. In the ARRL DX contests the station's 80-meter capabilities were tested in not less than 20 single-band 80-meter operations (10 on CW; 10 on phone). On phone the results were all first place for Europe or worldwide. On CW there were eight first places and two second places.

The results are further confirmed by the 80-meter country and zone totals during CQWW CW contests during the same period ('93-'97). During the 1994 and the 1996 CQWW CW contests we scored 5-band DXCC in one weekend, 10 through 80 in 1994, and 15 through 160 in 1996. The 100 plus countries during a single weekend on 160 meters was a first, I believe.

Twelve entries in the CQWW 160-meter CW contests over the past 16 years have resulted in 12 first places (either Europe or worldwide), while 7 participations in the 160-meter phone contest yielded five first place Europe or worldwide. During most of these contests we scored the highest number of country multipliers worldwide.

The results on Top Band and 80 meters not only speak for the performance of the transmit antennas (Four-Square on 80 and single quarter-wave vertical on 160) but also, and even more importantly, of the receiving capabilities of the station.

9. FURTHER IMPROVEMENTS

To stay competitive in international contesting you must improve the station year after year. Our competitors do the same. This is the driving force that leads to technological and conceptual improvements. Really, it is almost the opposite from DXing. The more successful you are in DXing, the more countries you have worked, the less there are left for you to work, the less pressure there is; the more you can relax. The more successful you are in DXing, the easier your call will be recognized in the pileups (that helps, too). No need to add another couple of dBs for those last two or three countries.

With contesting it is just the opposite. Competition grows and improves steadily, and if you don't match their efforts, you'll be at the losing end. Within the limits of where I live there is not much more I can do antenna wise. Over the years competition got fiercer, especially on the low bands. I hope and believe that some of the work I have put into publishing the techniques and art of Low-Band operations has contributed to the raising of the standards and the increasing of the performance from stations on 160 and 80 meters from all around. That makes me happy.

Another evolution, over the years, is the increase of man-made noise. Whereas 15 years ago, I considered myself living in a semi-rural area where man-made noise was pretty low, I now consider myself living in a semi-residential area! Not that I have any more close neighbors, but there now are two small-industry industrial parks within 2 km of my house.

Every year I spend days chasing noise sources and trying to kill them. In the coming years we will have to continue to fight for our RF spectrum. And BPL (PLC) is a real threat and it should be one of our main concerns in the years to come.

10. THE FUTURE

I have been quite active in contesting over the past 15 years, and I have proven what I wanted to prove, which makes me happy. I built a competitive contest station, so my success in DXing on the low bands (worldwide highest DXCC score on 80 meters, and highest DXCC score on 160 meters outside the USA), are not coincidences.

I have now turned 63, and I don't derive the same pleasure from fighting the same contest battles I did years ago. But that is normal, I guess. And by electing me to the CQ Contest Hall of Fame in 1997, my friends and fellow contesters told me I was a good contester, and that certainly made me happy.

As I wrote before, ham radio is about enjoyment, satisfaction, self fulfillment and maybe a little recognition as well. If we all keep thinking about ham radio in these terms, I am sure we're on the right track!

CHAPTER 16

Literature Review

This literature review lists over 1000 Reference works, catalogued by subject. The Reference numbers are those used in the different chapters of the book. Copies may be obtained directly from the magazines listed at the following addresses:

CQ/Ham Radio: 25 Newbridge Rd, Hicksville, NY 11801.
CQ-DL: DARC, Postfach 1155, D34216 Baunatal 1, Germany.
QST: ARRL HQ, 225 Main St, Newington, CT 06111.
Radio Communication: RSGB HQ, Lambda House, Cranborne Rd, Potters Bar, Herts EN6 3JE, England.
For other magazines, contact the author directly: J. Devoldere, ON4UN, 215 Poelstraat, B9820 Merelbeke, Belgium.

1. PROPAGATION

Ref. 100: K. J. Hortenbach et al, "Propagation of Short Waves Over Long Distances: Predictions and Observations," *Telecommunications Journal*, Jun 1979, p 320.
Ref. 101: D. Reed - W1LC - ed., *The ARRL Handbook for Radio Amateurs*, 81st Ed., ARRL, Newington, CT.
Ref. 102: John Devoldere - ON4UN, "80-Meter DXing," *Communications Technology*, 1978.
Ref. 103: George Jacobs - W3ASK - et al, *The New Shortwave Propagation Handbook*.
Ref. 104: Peter Saveskie - W4LGF, *Radio Propagation Handbook*
Ref. 105: William Orr - W6SAI, *Radio Handbook*, 1991.
Ref. 106: R. D. Straw - N6BV - ed., *The ARRL Antenna Book*, 20th Ed., ARRL - Newington CT, 2003.
Ref. 107: Wayne Overbeck - N6NB - et al, "Computer Programs for Amateur Radio"
Ref. 108: Dale Hoppe - K6UA - et al, "The Grayline Method of DXing," *CQ*, Sep 1975, p 27.
Ref. 109: Rod Linkous - W7OM, "Navigating to 80 Meter DX," *CQ*, Jan 1978, p 16.
Ref. 110: Yuri Blanarovich - VE3BMV, "Electromagnetic Wave Propagation By Conduction," *CQ*, Jun 1980, p 44.
Ref. 111: Guenter Schwarzbeck - DL1BU, "Bedeuting des Vertikalen Abstralwinkels von KW-Antennen," *CQ-DL*, Mar 1985, p 130.
Ref. 112: Guenter Schwarzbeck - DL1BU, "Bedeutung des Vertikalen Abstrahlwinkels von KW-Antennen (Part 2)," *CQ-DL*, Apr 1985, p 184.
Ref. 113: Henry Elwell - N4UH, "Calculator-Aided Propagation Predictions," *Ham Radio*, Apr 1979, p 26.
Ref. 114: Donald C. Mead - K4DE, "How to Determine True North for Antenna Orientation," *Ham Radio*, Oct 1980, p 38.
Ref. 115: Henry Elwell - N4UH, "Antenna Geometry for Optimum Performance," *Ham Radio*, May 1982, p 60.
Ref. 116: Stan Gibilisco - W1GV/4, "Radiation of Radio Signals," *Ham Radio*, Jun 1982, p 26.
Ref. 117: Garth Stonehocker - KØRYW, "Forecasting by Computer," *Ham Radio*, Aug 1982, p 80.
Ref. 118: Bradley Wells - KR7L, "Fundamentals of Grayline Propagation," *Ham Radio*, Aug 1984, p 77.
Ref. 119: Van Brollini - NS6N - et al, "DXing by Computer," *Ham Radio*, Aug 1984, p 81.
Ref. 120: Calvin R. Graf - W5LFM - et al, "High-Frequency Atmospheric Noise - Part 2," *QST*, Feb 1972, p 16.
Ref. 121: Jim Kennedy - K6MIO - et al, "D-Layer Absorption During a Solar Eclipse," *QST*, Jul 1972, p 40.
Ref. 122: Edward P. Tilton - W1HDQ, "The DXer's Crystal Ball," *QST*, Jun 1975, p 23.
Ref. 123: Edward P. Tilton - W1HDQ, "The DXer's Crystal Ball - Part II," *QST*, Aug 1975, p 40.
Ref. 124: Paul Argo - et al, "Radio Propagation and Solar Activity," *QST*, Feb 1977, p 24.
Ref. 125: Kenneth Johnston - W7LIX - et al, "An Eclipse Study on 80 Meters," *QST*, Jul 1979, p 14.
Ref. 126: V. Kanevsky - UL7GW, "Ionospheric Ducting at HF," *QST*, Sep 1979, p 20.
Ref. 127: Robert B. Rose - K6GKU, "MINIMUF: Simplified MUF-Prediction Program for Microcomputers," *QST*, Dec 1982, p 36.
Ref. 128: Tom Frenaye - K1KI, "The KI Edge," *QST*, Jun 1984, p 54.
Ref. 129: Richard Miller - VE3CIE, "Radio Aurora," *QST*, Jan 1985, p 14.
Ref. 130: Pat Hawker - G3VA, "Technical Topics: Trans-equatorial Supermode Theories," *Radio Com-*

munication, Feb 1972, p 94.

Ref. 131: Pat Hawker - G3VA, "Technical Topics: Whispering Galleries," *Radio Communication*, May 1972, p 306.

Ref. 132: Pat Hawker - G3VA, "Technical Topics: Low Angles for Chordal Hops and TEP," *Radio Communication*, Nov 1972, p 746.

Ref. 133: A. P. A. Ashton - G3XAP, "160M DX from Suburban Sites," *Radio Communication*, Dec 1973, p 842.

Ref. 134: Pat Hawker - G3VA, "Technical Topics: Path Deviations - One-Way Propagation - HF Tropo and LDEs," *Radio Communication*, Oct 1974, p 686.

Ref. 135: Pat Hawker - G3VA, "Fading and the Ionosphere," *Radio Communication*, Mar 1978, p 217.

Ref. 136: Pat Hawker - G3VA, "Technical Topics: When Long-Path is Better," *Radio Communication*, Sep 1979, p 831.

Ref. 137: V. Kanevsky - UL7GW, "DX QSOs," *Radio Communication*, Sep 1979, p 835.

Ref. 138: Pat Hawker - G3VA, "Technical Topics: Chordal Hop and Ionospheric Focusing," *Radio Communication*, Apr 1984, p 315.

Ref. 139: Dr. Alexsandr V. Gurevich and Dr. Elena E. Tsedilina : "Long Distance Propagation of HF Radio Waves," ISBN 3-540-15139-7 and ISBN 0-387-15139-7.

Ref. 140: Bob Brown, NM7M, "160 Meter DXing, part 1," *The DX Magazine*, May/Jun 1996, p 22.

Ref. 141: Bob Brown, NM7M, "160 Meter DXing, part 2," *The DX Magazine*, Jul/Aug 1996, p 10.

Ref. 142: Cary Oler and Ted Cohen, N4XX, "The 160 Meter Band: An Enigma Shrouded in Mystery," *CQ Mag*, Part 1—Mar 1998, p 9; Part 2—Apr 1998, p 11.

Ref. 143: Carl Luetzelschwab - K9LA, "Working VK0IR on 160 m"; *The DX Magazine*, Jul/Aug 1997, p 49.

Ref. 144: Carl Luetzelschwab - K9LA,"Popular Great circle paths"; *The Low Band Monitor*, Oct 1997.

Ref. 145: Carl Luetzelschwab - K9LA, "A General Summary of Propagation from VK0IR to the US and VE," *QST*, Oct 1997, p 88.

Ref. 146: Carl Luetzelschwab - K9LA, "Stratwarms and their Effect on HF Propagation," *QST*, Technical Correspondence, Feb 1997, p 76.

Ref. 147: Robert Brown - NM7M, "The Low Band Logs of VK0IR," *Communications Quarterly*, Fall 1997, p 45.

Ref. 148: Bob Brown, NM7M, "Heard & Unheard Islands"; *The DX Magazine*, May/Jun 1997, p 9.

Ref. 149: "Boulder A Index - Propagation Predictor," *The Top Band Monitor*, Aug 1994.

Ref. 150: K0CS: "Boulder A-Index Daily Variations"; *The Top Band Monitor*, Dec 1993.

Ref. 151: NM7M - "An Argument for 160-meter Ducting," *The Low Band Monitor*, Aug 1997.

Ref. 152: Carl Luetzelschwab, K9LA, "160 Meter QSO's After Sunrise," *The Low Band Monitor*, May 1997.

Ref. 153: *Prolab Pro* Manual, High Frequency Radio Propagation Laboratory, by Solar Terrestrial Dispatch (Software manual, available at **http://www.spacew.com/Docs/propman.pdf**).

Ref. 154: Cary Oler and Ted Cohen, N4XX, "The 160 meter Band: An Enigma Shrouded in a Mystery, Part 2 ," *CQ*, Apr 1998.

Ref. 155: Bob Brown, NM7M: "On the Mythology of Solar Flares," *The DX Magazine*, Jan/Feb 1994.

Ref. 156: K9LA: "Do Stratwarms affect 160 m propagation?", *The Low Band Monitor*, Jan 1998.

Ref. 157: Rober Brown, NM7M: *The Little Pistol's Guide to HF Propagation* (Sacramento: Worldradio Books, 1996).

Ref. 158: Cary Oler: "Getting Primed for the Solar Maximum," *CQ*, Nov 1998, p 11.

Ref. 159: Carl Luetzelschwab, K9LA: "Skewed Paths to Europe on the Low Bands," *CQ*, Aug 1999, p 11.

Ref. 160: Karl Thurber, W8FX: "Uncle Sol's Solar Wind and the Earth's Nagnificent Magnetosphere, Part 1," *CQ*, Aug 2000, p 13.

Ref. 161: Karl Thurber, W8FX: "Uncle Sol's Solar Wind and the Earth's Magnificent Magnetosphere, Part 2," *CQ*, Sep 2000, p 30.

Ref. 162: Ken Neubeck, WB2AMU: "Measuring Geomagnetic Weather," *CQ*, Oct 2000, p 28.

Ref. 163: Steve Ireland, VK6VZ: "Go surf the grey and dark lines part 1 (the art of Low and High HF band Dxing)", CQ Feb 2001, p 38

Ref. 164: Steve Ireland, VK6VZ: "Go Surf the Grey and Dark Lines, Part 2, (the Art of Low and High HF Band Dxing)," *CQ*, Mar 2001, p 18.

Ref. 165: Steve Ireland, VK6VZ, Mike Bazley, VK6HD and Bob Brown, NM7N: "Equinoctial and Diurnal Path Witching (A New Perspective on the Long-Path and Short-Path DXing on the Topband, Part 1," *CQ*, Feb 2002, p 22.

Ref. 166: Steve Ireland, VK6VZ, Mike Bazley, VK6HD and Bob Brown, NM7N: "Equinoctial and Diurnal Path Switching (A New Perspective on the Long-Path and Short-Path DXing on the Topband, Part 2," *CQ*, Mar 2002, p 24.

Ref. 167: Steve Nichols, G0KYA: "Mysteries of the Ionosphere," *Radcom*, Jan 1999, p 57.

Ref. 168: "The twilight zone (just what is grey-line propagation)", *Radcom* Jul 2002, p 34

Ref. 169: Luetzelschwab, K9LA: "More on Absorption," *The Low Band Monitor*, Jan 2001.

Ref. 170: Robert R. Brown, NM7M: "What is Wrong with Cycle 23", *The Low Band Monitor*, May 2001

Ref. 171: Robert R Brown, NM7M: "Long Path Propagation Revisited in Year 2000," (book available from NM7M).

Ref. 172: Ian Poole, G3YWX: "Mysteries of the ionosphere", *Radcom*, Jan 1999, p 57.

Ref. 173: Steve Gecewicz, K0CS: "Boulder "A" Index Daily Variations," *The Low Band Monitor*, Dec 1993.

Ref. 174: Margot Tollefson, KB0MPL: "The Boulder "A" Isn't the Way," *The Low Band Monitor*, Apr 1995.

Ref. 175: R Brown, NM7M: "An argument for 160 Meter Ducting," *Top Band Monitor*, Aug 1997.

Ref. 176: Robert R Brown, NM7M: "Bubbles in the Ozone Layer," *QST*, Dec 1999, p 44.

Ref. 177: Robert R Brown, NM7M: "Medium-Frequency Sunrise Enhancements," *QEX*, Jul/Aug 2001, p 3.

Ref. 178: Robert R Brown, NM7M: "On the SSW Path and 160-Meter Propagation," *QEX*, Nov/Dec 2000, p 3.

Ref. 179: Robert R Brown, NM7M: *The Big Gun's Guide to Low-Band Propagation*, (Book available from NM7M.)

Ref. 180: John Devoldere, ON4UN: "The 1999 Solar Eclipse and Amateur Radio," *QST*, Jan 2000, p 35.

Ref. 181: Ken Nuebeck, WB2AMU: "Measuring Geomagnetic 'Weather,'" *CQ*, Oct 2000, p 28.

Ref. 182: Dean Straw, N6BV: "Using HF Propagation Predictions", *The ARRL Antenna Compendium, Vol 6*, p 101

Ref. 183: Dean Straw, N6BV: "Antennas: Here are some verticals on the beach", *The ARRL Antenna Compendium, Vol 6*, page 216.

2. RECEIVERS

Ref. 200: R. D. Straw - N6BV - ed, *The ARRL Handbook for Radio Amateurs*, 2000, ARRL, Newington, CT.

Ref. 201: ON4EG, "Sensibilite des Recepteurs," *CQ-QSO*, Jan 1985.

Ref. 202: ON4EG, "Sensibilite des Recepteurs," *CQ-QSO*, Feb 1985.

Ref. 203: Michael Martin - DJ7JV, "Emfangereingangsteil mit Grossem Dynamik Bereich," *DL-QTC*, Jun 1975.

Ref. 204: R. D. Straw - N6BV - ed, *"The ARRL Handbook for Radio Amateurs,"* 2000, ARRL, Newington, CT.

Ref. 205: John Devoldere - ON4UN, "80-Meter DX-ing," *Communications Technology*, Jun 1978.

Ref. 206: William Orr - W6SAI, *Radio Handbook*, Howard W. Sams & Co. Inc.

Ref. 207: Robert Sternowsky, "Using Preselectors to Improve HF Performance," *Communications International*, May 1980, p 34.

Ref. 208: John Devoldere - ON4UN, "Improved Performance from the Drake R-4B and T4X-B," *CQ*, Mar 1976, p 37.

Ref. 209: Michael Martin - DJ7VY, "Rauscharmer Oszillator fur Empfaenger mit grossem Dynamikbereich," *CQ-DL*, Dec 1976, p 418.

Ref. 210: Wes Hayward - W7ZOI, "Der Dynamische Bereich eines Empfaengers," *CQ-DL*, Mar 1977, p 93.

Ref. 211: Richard Waxweller - DJ7VD, "Hochfrequenz-Zweitongenerator," *CQ-DL*, Sep 1980, p 412.

Ref. 212: Guenter Schwarzbeck - DL1BU, "Testbericht: NF-Filter Datong FL2," *CQ-DL*, Feb 1981, p 56.

Ref. 213: Guenter Schwarzbeck - DL1BU, "Grosssignalverhalten von Kurzwellenempfaengern," *CQ-DL*, Mar 1981, p 117.

Ref. 214: Walter Flor - OE1LO, "KW-Eingangsteile: Eingangsfilter," *CQ-DL*, Aug 1981, p 373.

Ref. 215: Walter Flor - OE1LO, "IM-feste Verstaerker fuer den KW-bereich," *CQ-DL*, Aug 1981, p 473.

Ref. 216: Walter Flor - OE1LO, "KW-Eingangsteile: Extrem IM-feste selektive Vorverstaerker," *CQ-DL*, Aug 1981, p 376.

Ref. 217: Guenter Schwarzbeck - DL1BU, "Geraeteeigenschaften: Besonderheiten zwischen Testbericht und Praxis," *CQ-DL*, Sep 1982, p 424.

Ref. 218: Erich Vogelsang - DJ2IM, "Grundrmauschen und Dynamikbereich bei Kurzwellenempfaengern," *CQ-DL*, Sep 1982, p 432.

Ref. 219: Michael Martin - DJ7YV, "Intermodulationsfster Preselector fur 1.5 - 30 MHz," *CQ-DL*, Jul 1984, p 320.

Ref. 220: Ray Moore, "Designing Communication Receivers for Good Strong-Signal Performance," *Ham Radio*, Feb 1973, p 6.

Ref. 221: Wes Hayward - W7ZOI, "Bandpass Filters for Receiver Preselectors," *Ham Radio*, Feb 1975, p 18.

Ref. 222: Ulrich Rohde - DJ2LR, "High Dynamic Range Receiver Input Stages," *Ham Radio*, Oct 1975, p 26.

Ref. 223: James Fisk - W1DTY, "Receiver Noise Figure Sensitivity and Dynamic Range, What The Numbers Mean," *Ham Radio*, Oct 1975, p 8.

Ref. 224: Marvin Gonsior - W6FR, "Improved Selectivity for Collins S-line Receivers," *Ham Radio*, Jun 1976, p 36.

Ref. 225: Howard Berlin - K3NEZ, "Increased Flexibility for MFJ CW Filters," *Ham Radio*, Dec 1976, p 58.

Ref. 226: Ulrich Rohde - DJ2LR, "I-F Amplifier Design," *Ham Radio*, Mar 1977, p 10.

Ref. 227: Alex Burwasser - WB4ZNV, "Reducing Intermodulation Distortion in High-Frequency Receivers," *Ham Radio*, Mar 1977, p 26.

Ref. 228: Wayne C. Ryder - W6URM, "General Coverage Communications Receiver," *Ham Radio*, Nov 1977, p 10.

Ref. 229: R. Sherwood - WBØJGP - et al, "Receivers: Some Problems and Cures," *Ham Radio*, Dec 1977, p 10.

Ref. 230: R. Sherwood - WBØJGP - et al, "New Product Detector for R-4C," *Ham Radio*, Oct 1978, p 94.

Ref. 231: R. Sherwood - WBØJGP - et al, "Audio Amplifier for the Drake R-4C," *Ham Radio*, Apr 1979, p 48.

Ref. 233: James M. Rohler - NØDE, "Biquad Bandpass Filter," *Ham Radio*, Jun 1979, p 70.

Ref. 234: Sidney Kaiser - WB6CTW, "Measuring Receiver Dynamic Range," *Ham Radio*, Nov 1979, p 56.

Ref. 235: Ulrich Rohde - DJ2LR, "Recent Developments in Circuits and Techniques for High Frequency Communications Receivers," *Ham Radio*, Apr 1980, p 20.

Ref. 236: Edward Wetherhold - W3NQN, "High Performance CW Filter," *Ham Radio*, Apr 1981, p 18.

Ref. 237: D. A. Tong - G4GMQ, "Add-On Selectivity for Communication Receivers," *Ham Radio*, Nov 1981, p 41.

Ref. 238: Ulrich Rohde - DJ2LR, "Communication Receivers for the Year 2000: Part 1," *Ham Radio*, Nov 1981, p 12.

Ref. 239: Jan K. Moller - K6FM, "Understanding Performance Data of High-Frequency Receivers," *Ham Radio*, Nov 1981, p 30.

Ref. 240: Ulrich Rohde - DJ2LR, "Communication Receivers for the Year 2000 - Part 2," *Ham Radio*, Dec 1981, p 6.

Ref. 241: Ulrich Rohde - DJ2LR, "Performance Capability of Active Mixers - Part 1," *Ham Radio,* Mar 1982, p 30.

Ref. 242: Ulrich Rohde - DJ2LR, "Performance Capability of Active Mixers - Part 2," *Ham Radio,* Apr 1982, p 38.

Ref. 243: R. W. Johnson - W6MUR, "Bridged T-Filters for Amateur Use," *Ham Radio,* Oct 1982, p 51.

Ref. 244: Cornell Drentea - WB3JZO, "Designing a Modern Receiver," *Ham Radio,* Nov 1983, p 3.

Ref. 245: Edward Wetherhold - W3NQN, "Elliptic Lowpass Audio Filter Design," *Ham Radio,* Jan 1984, p 20.

Ref. 246: J. A. Dyer - G4OBU, "High Frequency Receiver Performance," *Ham Radio,* Feb 1984, p 33.

Ref. 247: E. A. Andrade - WØDAN, "Recent Trends in Receiver Front-End Design," *QST,* Jun 1962, p 17.

Ref. 248: William K. Squires - W2PUL, "A New Approach to Receiver Front-End Design," *QST,* Sep 1963, p 31.

Ref. 249: Byron Goodman - W1DX, "Some Thoughts on Home Receiver Design," *QST,* May 1965, p 11.

Ref. 250: E. H. Conklin - K6KA, "Front-End Receiving Filters," *QST,* Aug 1967, p 14.

Ref. 251: Doug DeMaw - W1CER, "Rejecting Interference from Broadcast Stations," *QST,* Dec 1967, p 35.

Ref. 252: Rudolf Fisher - DL6WD, "An Engineer's Solid-State Ham-Band Receiver," *QST,* Mar 1970, p 11.

Ref. 253: Douglas A. Blakeslee - W1KLK, "An Experimental Receiver for 75-Meter DX Work,"*QST,* Feb 1972, p 41.

Ref. 254: Wes Hayward - W7ZOI, "Defining and Measuring Receiver Dynamic Range," *QST,* Jul 1975, p 15.

Ref. 255: Doug DeMaw - W1FB, "His Eminence: the Receiver," *QST,* Jun 1976, p 27.

Ref. 256: Wes Hayward - W7ZOI, "CER-Verters," *QST,* Jun 1976, p 31.

Ref. 257: Doug DeMaw - W1FB, "Build this Quickie Preamp," *QST,* Apr 1977, p 43.

Ref. 258: Wes Hayward - W7ZOI, "More Thoughts on Receiver Performance Specification," *QST,* Nov 1979, p 48.

Ref. 259: Ulrich Rohde - DJ2LR, "Increasing Receiver Dynamic Range," *QST,* May 1980, p 16.

Ref. 260: Edward Wetherhold - W3NQN, "Modern Design of a CW Filter Using 88- and 44-mH Surplus Inductors," *QST,* Dec 1980, p 14.

Ref. 261: Doug DeMaw - et al, "Modern Receiver Mixers for High Dynamic Range," *QST,* Jan 1981, p 19.

Ref. 262: Wes Hayward - W7ZOI, "A Progressive Communications Receiver," *QST,* Nov 1981, p 11.

Ref. 263: Robert E. Lee - K2TWK, "Build an Audio Filter with Pizzazz," *QST,* Feb 1982, p 18.

Ref. 264: Harold Mitchell - NØARQ, "88-mH Inductors: A Trap," *QST,* Jan 1983, p 38.

Ref. 265: Gerald B. Hull - AK4L/VE1CER, "Filter Systems for Multi-transmitter Amateur Stations," *QST,* Jul 1983, p 28.

Ref. 266: John K. Webb - W1ETC, "High-Pass Filters for Receiving Applications," *QST,* Oct 1983, p 17.

Ref. 267: Doug DeMaw - W1FB, "Receiver Preamps and How to Use Them," *QST,* Apr 1984, p 19.

Ref. 268: Pat Hawker - G3VA, "Trends in H.F. Receiver Front-ends," *Radio Communication,* Sep 1963, p 161.

Ref. 269: D. A. Tong - G4GMQ, "Audio Filters as an Aid to Reception," *Radio Communication,* Feb 1978, p 114.

Ref. 270: John Bazley - G3HCT, "The Datong Multi-Mode Filter FL-2," *Radio Communication,* Aug 1980, p 783.

Ref. 271: Pat Hawker - G3VA, "Technical Topics: More Thoughts On 'Ideal' HF Receivers," *Radio Communication,* Oct 1982, p 861.

Ref. 272: Edward Wetherhold - W3NQN, "Simplified Elliptic Lowpass Filter Design Using Surplus 88-mH Inductors," *Radio Communication,* Apr 1983, p 318.

Ref. 273: P. E. Chadwick - G3RZP, "Dynamic Range; Intermodulation and Phase Noise," *Radio Communication,* Mar 1984, p 223.

Ref. 274: Pat Hawker - G3VA, "Technical Topics: Comparing Receiver Front-Ends," *Radio Communication,* May 1984, p 400.

Ref. 275: Pat Hawker - G3VA, "Technical Topics: Receivers: Numbers Right or Wrong?," *Radio Communication,* Aug 1984, p 677.

Ref. 276: Pat Hawker - G3VA, "Technical Topics: Receivers of Top Performance," *Radio Communication,* Oct 1984, p 858.

Ref. 277: Edward Wetherhold - W3NQN, "A CW Filter for Radio Amateurs Newcomer," *Radio Communication,* Jan 1985, p 26.

Ref. 278: Pat Hawker - G3VA, "Technical Topics: Whither Experimentation?," *Radio Communication,* Mar 1985, p 189.

Ref. 279: Ian White - G3SEK, "Modern VHF/UHF Front End Design—Part 1," *Radio Communication,* Apr 1985, p 264.

Ref. 280: Ian White - G3SEK, "Modern VHF/UHF Front End Design—Part 2," *Radio Communication,* May 1985, p 367.

Ref. 281: Ian White - G3SEK, "Modern VHF/UHF Front-End Design—Part 3," *Radio Communication,* Jun 1985, p 445.

Ref. 282: Pat Hawker - G3VA, "Technical Topics: Weak Signal Reception," *Radio Communication,* Jul 1985, p 540.

Ref. 283: D. H. G. Fritsch - GØCKZ, "Active Elliptic Audio Filter Design Using Op-Amps (Part 1)," *Radio Communication,* Feb 1986, p 98.

Ref. 284: Ulrich L. Rohde - KA2WEU/DJ2LR, "Designing a State-of-the-Art Receiver," *Ham Radio,* Nov 1987, p 17.

Ref. 285: Robert J. Zavrel - W7SX, "Tomorrow's Receivers," *Ham Radio,* Nov 1987, p 8.

Ref. 286: John Grebenkemper - KI6WX, "Phase Noise and its Effects on Amateur Communications," *QST,* Mar 1988, p 14.

Ref. 287: Zack Lau - KH6CP, "Eliminating AM-Broadcast interference on 160 Meters," *QST,* Apr 1992, p 75.

Ref. 288: Gary Nichols - KD9SV, "Bandpass Filters for 80 and 160 Meters," *QST*, Feb 1989, p 42.
Ref. 289: John Grebenkemper - KI6WX, "Phase Noise and its Effects on Amateur Communications," *QST*, Apr 1988, p 22.
Ref. 290: Dave Hershberger - W9GR, "Low Cost Digital Signal Processing for Radio Amateurs," *QST*, Sep 1992, p 43.
Ref. 291: Bruce C. Hale - KB1MW/7, "An Introduction to Digital Signal Processing," *QST*, Jul 1991, p 35.
Ref. 292: D. DeMaw - W1FB, "Receiver Preamps and How to Use Them," *QST*, Apr 1984, p 19.
Ref. 293: J. Kearman - KR1S, "Audio Filter Roundup," *QST*, Oct 1991, p 36.
Ref. 294: Wes Hayward - W7ZOI, "The Double-tuned Circuit.: An Experimenter's Tutorial," *QST*, Dec 1991, p 29.
Ref. 295: Lew Gordon - K4VX, "Band-Pass filters for HF transceivers," *QST*, Sep 1988, p 17.
Ref. 296: D. DeMaw - W1FB, "A Diode-switched Bandpass Filter," *QST*, Jan 1991, p 24.
Ref. 297: G. E. Myers - K9GZB, "Diode-switched-filter (corrections and amplifications)," *QST*, Aug 1991, p 41.
Ref. 298: KØCS - "A Classic BCI Filter for 160 Meters"; *The Low Band Monitor*, May 1995.

3. TRANSMITTERS

Ref. 300: R. D. Straw - N6BV - ed, *"The ARRL Handbook for Radio Amateurs,"* 2000, ARRL, Newington, CT.
Ref. 301: Bob Heil - K9EID, "Equalise That Microphone," *CQ-DL*, Apr 1985, p 27.
Ref. 302: John Devoldere - ON4UN, "80-Meter DX-ing," *Communications Technology Inc*, Apr 1978.
Ref. 303: William Orr - W6SAI, *Radio Handbook*, Howard W. Sams & Co.- Inc.
Ref. 304: John Devoldere - ON4UN, "Improved Performance from the Drake R-4B and T4X-B," *CQ*, Mar 1976, p 37.
Ref. 305: John Schultz - W4FA, "An Optimum Speech Filter," *CQ*, Oct 1978, p 22.
Ref. 307: L. McCoy - W1ICP, "The Design Electronics QSK-1500," *CQ*, Apr 1985, p 40.
Ref. 308: Guenter Schwarzbeck - DL1BU, "Geraeteeigenschaften: esonderheiten zwischen Testbericht und Praxis," *CQ-DL*, Sep 1982, p 424.
Ref. 309: Leslie Moxon - G6XN, "Performance of RF Speech Clippers," *Ham Radio*, Nov 1972, p 26.
Ref. 310: Charles Bird - K6HTM, "RF Speech Clipper for SSB," *Ham Radio*, Feb 1973, p 18.
Ref. 311: Henry Elwell - W2MB, "RF Speech Processor," *Ham Radio*, Sep 1973, p 18.
Ref. 312: Barry Kirkwood - ZL1BN, "Principles of Speech Processing," *Ham Radio*, Feb 1975, p 28.
Ref. 313: Timothy Carr - W6IVI, "Speech Processor for the Heath SB-102," *Ham Radio*, Jun 1975, p 38.
Ref. 314: Jim Fisk - W1DTY, "New Audio Speech Processing Technique," *Ham Radio*, Jun 1976, p 30.
Ref. 315: Frank C. Getz - K3PDW, "Logarithmic Speech Processor," *Ham Radio*, Aug 1977, p 48.
Ref. 316: Michael James - W1CBY, "Electronic Bias Switching for the Henry 2K4 and 3KA Linear Amplifiers," *Ham Radio*, Aug 1978, p 75.
Ref. 317: Wesley D. Stewart - N7WS, "Split-Band Speech Processor," *Ham Radio*, Sep 1979, p 12.
Ref. 318: J. R. Sheller - KN8Z, "High Power RF Switching With Pin Diodes," *Ham Radio*, Jan 1985, p 82.
Ref. 319: William Sabin - WØIYH, "R.F. Clippers for S.S.B.," *QST*, Jul 1967, p 13.
Ref. 320: J. A. Bryant - W4UX, "Electronic Bias Switching for RF Power Amplifiers," *QST*, May 1974, p 36.
Ref. 321: Robert Myers - W1FBY, "Quasi-Logarithmic Analog Amplitude Limiter," *QST*, Jul 1974, p 22.
Ref. 322: Hal Collins - W6JES, "SSB Speech Processing Revisited," *QST*, Aug 1976, p 38.
Ref. 323: Bob Heil - K9EID, "Equalize Your Microphone and Be Heard!," *QST*, Jul 1982, p 11.
Ref. 324: R. C. V. Macario - G4ADL - et al, "An Assured Speech Processor," *Radio Communication*, Apr 1978, p 310.
Ref. 325: H. Leerning - G3LLL, "Improving the FT-101," *Radio Communication*, Jun 1979, p 516.
Ref. 326: L. McCoy - W1ICP, "The Design Electronics QSK-1500," *CQ*, Apr 1985, p 40.
Ref. 327: Alfred Trossen - DL6YP, "Die AMTOR-II einheit nach G3PLX," *CQ-DL*, Jul 1983, p 316.
Ref. 328: Alfred Trossen - DL6YP, "Die AMTOR-II einheit nach G3PLX," *CQ-DL*, Aug 1983, p 368.
Ref. 329: J. R. Sheller - KN8Z, "High Power RF Switching With Pin Diodes," *Ham Radio*, Jan 1985, p 82.
Ref. 330: Peter Martinez - G3PLX, "AMTOR: an Improved Error-Free RTTY System," *QST*, Jun 1981, p 25.
Ref. 331: Paul Newland - AD7I, "Z-AMTOR: An Advanced AMTOR Code Converter," *QST*, Feb 1984, p 25.
Ref. 332: Peter Martinez - G3PLX, "AMTOR: an Improved Radio Teleprinter System Using a Microprocessor," *Radio Communication*, Aug 1979, p 714.
Ref. 333: Peter Martinez - G3PLX, "AMTOR the Easy Way," *Radio Communication*, Jun 1980, p 610.
Ref. 334: Jon Towle - WB1DNL, "QSK 1500 High-Power RF Switch," *QST*, Sep 1985, p 39.
Ref. 335: Dr. J. R. Sheller - KN8Z, "What Does 'QSK' Really Mean?," *QST*, Jul 1985, p 31.
Ref. 336: Paul Newland - AD7I, "A User's Guide to AMTOR Operation," *QST*, Oct 1985, p 31.
Ref. 337: W. J. Byron - W7DHD, "Designing an amplifier around the 3CX1200A7," *Ham Radio*, Dec 1978, p 33.
Ref. 338: Richard L. Measures - AG6K, "QSK Modification of the Trio Kenwood TL-922 Amplifier," *Ham Radio*, Mar 1989, p 35.
Ref. 339: Paul A. Johnson - W7KBE, "Homebrewing Equipment - From Parts to Metal Work," *Ham Radio*, Mar 1988, p 26.
Ref. 340: Richard L. Measures - AG6K, "Adding 160-meter Coverage to HF Amplifiers," *QST*, Jan 1989, p 23.

Ref. 341: Safford M. North - KG2M, "Putting the Heath SB-200 on 160 Meters," *QST*, Nov 1987, p 33.

Ref. 342: W.J. Byron - W7DHD, "Design Program for the Grounded-grid 3-500Z," *Ham Radio,* Jun 1988, p 8.

4. EQUIPMENT REVIEW

Ref. 400: Guenter Schwarzbeck - DL1BU, "Testbericht: Transceiver TS820," *CQ-DL,* Apr 1977, p 130.

Ref. 401: Guenter Schwarzbeck - DL1BU, "Testbericht und Beschreibung TS-520S," *CQ-DL,* Feb 1978, p 50.

Ref. 402: Guenter Schwarzbeck - DL1BU, "Testbericht un Messdaten FT-901 DM (receiver section)," *CQ-DL,* Feb 1978, p 438.

Ref. 403: Guenter Schwarzbeck - DL1BU, "Testbericht und Messdaten FT-901 DM (transmitter section)," *CQ-DL,* Nov 1978, p 500.

Ref. 404: Guenter Schwarzbeck - DL1BU, "KW-Empfaenger Drake R-4C mit usatzfiltern," *CQ-DL,* Feb 1979, p 56.

Ref. 405: Guenter Schwarzbeck - DL1BU, "KW Transceiver ICOM IC-701," *CQ-DL,* Feb 1979, p 65.

Ref. 406: Guenter Schwarzbeck - DL1BU, "Testbericht: Vorausbericht FT-ONE," *CQ-DL,* Jan 1982, p 11.

Ref. 407: Guenter Schwarzbeck - DL1BU, "Testbericht und Messwerte IC-730," *CQ-DL,* Mar 1982, p 117.

Ref. 408: Guenter Schwarzbeck - DL1BU, "Testbericht und Messwerte FT-102," *CQ-DL,* Aug 1982, p 387.

Ref. 409: Guenter Schwarzbeck - DL1BU, "Testbericht und Messwerte TS930S," *CQ-DL,* Oct 1982, p 484.

Ref. 410: Peter Hart - G3SJX, "The Trio TS-830 HF Transceiver," *Radio Communication,* Jul 1982, p 576.

Ref. 411: Peter Hart - G3SJX, "The Yaesu Musen FT102 HF Transceiver," *Radio Communication,* Jan 1983, p 32.

Ref. 412: Peter Hart - G3SJX, "The ICOM IC740 HF Transceiver," *Radio Communication,* Nov 1983, p 985.

Ref. 413: Peter Hart - G3SJX, "The Yaesu FT77 HF Transceiver," *Radio Communication,* Jun 1984, p 482.

Ref. 414: Peter Hart - G3SJX, "The Yaesu Musen FT980 HF Transceiver," *Radio Communication,* Sep 1984, p 761.

Ref. 415: Peter Hart - G3SJX, "The Ten-Tec Corsair HF Transceiver," *Radio Communication,* Nov 1984, p 957.

Ref. 416: Peter Hart - G3SJX, "The Yaesu Musen FT757GX HF Transceiver," *Radio Communication,* May 1985, p 351.

Ref. 417: Peter Hart - G3SJX, "The Trio TS430S HF Transceiver," *Radio Communication,* Jun 1985, p 441.

Ref. 418: G. Schwarzbeck - DL1BU, "Yaesu FT-1000 Test Review—Part 1," *CQ-DL,* Mar 1991, p 91.

Ref. 419: G. Schwarzbeck - DL1BU, "Yaesu FT-1000 Test Review—Part 2," *CQ-DL,* Apr 1991, p 215.

Ref. 420: G. Schwarzbeck - DL1BU, "Yaesu FT-1000 Test Review—Part 3," *CQ-DL,* May 1991, p 273.

Ref. 421: G. Schwarzbeck - DL1BU, "TS-850S Test Review," *CQ-DL,* Feb 1991, p 79.

Ref. 422: G. Shwarzbeck - DL1BU, "TS-950 SD Test Review," *CQ-DL,* Dec 1989, p 750.

Ref. 423: G. Shwarzbeck - DL1BU, "ICOM IC-761 Test Review," *CQ-DL,* Aug 1988, p 479.

Ref. 424: Peter Hart - G3SJX, "FT-1000 Review," *Radio Communication*, Jun 1991, p 49.

Ref. 425: Peter Hart - G3SJX, "The Yaesu Musen FT767GX HF Transceiver," *Radio Communication*, Jul 1987, p 490.

Ref. 426: Peter Hart - G3SJX, "Kenwood TS-860S Transceiver," *Radio Communication*, Mar 1989, p 47.

Ref. 427: Peter Hart - G3SJX, "Yaesu Musen FT-747GX HF Transceiver," *Radio Communication*, May 1989, p 47.

Ref. 428: Peter Hart - G3SJX, "ICOM IC-725 HF Transceiver," *Radio Communication*, Sep 1989, p 56.

Ref. 429: Peter Hart - G3SJX, "Kenwood TS-950S Transceiver Review," *Radio Communication*, Apr 1990, p 35.

Ref. 430: Peter Hart - G3SJX, "ICOM-781 HF Transceiver," *Radio Communication*, Jul 1980, p 52.

Ref. 431: G. Shwarzbeck - DL1BU, "Ten-Tec Paragon 585 Test Review," *CQ-DL,* May 1988, p 277.

5. OPERATING

Ref. 500: R. D. Straw - N6BV - ed, *The ARRL Handbook for Radio Amateurs*, 2000, ARRL, Newington, CT.

Ref. 501: Erik - SMØAGD, "Split Channel Operation," *CQ-DL,* Apr 1985, p 10.

Ref. 502: John Devoldere - ON4UN, "80-Meter DXing," *Communications Technology,* Apr 1987.

Ref. 503: Wayne Overbeck - N6NB, *Computer Programs for Amateur Radio*,

Ref. 504: Rod Linkous - W7OM, "Navigating to 80 Meter DX," *CQ,* Jan 1978, p 16.

Ref. 505: Larry Brockman, "The DX-list Net. What a Mess," *CQ,* May 1979, p 48.

Ref. 506: Wolfgang Roberts - DL7RT, "Wie werde ich DX-er?," *CQ-DL,* Oct 1981, p 493.

Ref. 507: John Lindholm - W1XX, "Is 160 Your Top Band?" *QST,* Aug 1985, p 45.

Ref. 508: V. Kanevsky - UL7GW, "DX QSOs," *Radio Communication,* Sep 1979, p 835.

Ref. 509: Top Band operators survey, *The Low Band Monitor,* Jun 1996.

Ref. 510: Top band operators survey, *The Low Band Monitor,* Jun 1995.

Ref. 511: Jeff Briggs, K1ZM, *DXing on the Edge, The Thrill of 160 Meters*, ARRL, Newington, CT.

6. ANTENNAS: GENERAL

Ref. 600: R. D. Straw - N6BV - ed, *The ARRL Handbook for Radio Amateurs*, 2000, ARRL, Newington CT.

Ref. 601: J. J. Wiseman, "How Long is a Piece of Wire," *Electronics and Wireless World*, Apr 1985, p 24.

Ref. 602: William Orr - W6SAI - et al, *Antenna Handbook*, Howard W. Sams & Co, Inc,

Ref. 603: Gerald Hall - K1TD - Ed., "The ARRL Antenna Compendium, Volume 1," ARRL, Newington, CT.

Ref. 604: Keith Henney, *Radio Engineering Handbook,* 5th Edition.

Ref. 605: *Ref.erence Data for Radio Engineers*, Howard W. Sams, 5th Edition.

Ref. 606: John Kraus - W8JK, *Antennas*.

Ref. 607: Joseph Boyer - W6UYH, "The Multi-Band Trap Antenna - Part I," *CQ,* Feb 1977, p 26.

Ref. 608: Joseph Boyer - W6UYH, "The Multi-Band Trap Antenna - Part II," *CQ,* Mar 1977, p 51.
Ref. 609: Bill Salerno - W2ONV, "The W2ONV Delta/Slope Antenna," *CQ,* Aug 1978, p 52.
Ref. 610: Cornelio Nouel - KG5B, "Exploring the Vagaries of Traps," *CQ,* Aug 1984, p 32.
Ref. 611: K. H. Kleine - DL3CI, "Der Verkuerzte Dipol," *CQ-DL,* Jun 1977, p 230.
Ref. 612: Hans Wuertz - DL2FA, "Bis zu Einer S-Stufe mehr auf 80 Meter," *CQ-DL,* Dec 1977, p 475.
Ref. 613: Hans Wuertz - DL2FA, "DX-Antennen mit spiegelenden Flaechen," *CQ-DL,* Aug 1979, p 353.
Ref. 614: Hans Wuertz - DL2FA, "DX-Antennen mit spiegelenden Flaechen," *CQ-DL,* Jan 1980, p 18.
Ref. 615: Willi Nitschke - DJ5DW - et al, "Richtungskarakteristik Fusspunktwiederstand etc von Einelementantennen," *CQ-DL,* Nov 1982, p 535.
Ref. 616: Guenter Schwarzbeck - DL1BU, "Bedeuting des Vertikalen Abstralwinkels von KW-Antennen," *CQ-DL,* Mar 1985, p 130.
Ref. 617: Guenter Schwarzbeck - DL1BU, "Bedeutung des Vertikalen Abstrahlwinkels von KW-antennas (part 2)," *CQ-DL,* Apr 1985, p 184.
Ref. 618: E. Vogelsang - DJ2IM, "Vertikaldiagramme typische Kurzwellenantenne," *CQ-DL,* Jun 1985, p 300.
Ref. 619: John Schultz - W2EEY, "Stub-Switched Vertical Antennas," *Ham Radio,* Jul 1969, p 50.
Ref. 620: Malcolm P. Keown - W5RUB, "Simple Antennas for 80 and 40 Meters," *Ham Radio,* Dec 1972, p 16.
Ref. 621: Earl Whyman - W2HB, "Standing-Wave Ratios," *Ham Radio,* Jul 1973, p 26.
Ref. 622: Robert Baird - W7CSD, "Nonresonant Antenna Impedance Measurements," *Ham Radio,* Apr 1974, p 46.
Ref. 623: H. Glenn Bogel - WA9RQY, "Vertical Radiation Patterns," *Ham Radio,* May 1974, p 58.
Ref. 624: Robert Leo - W7LR, "Optimum Height for Horizontal Antennas," *Ham Radio,* Jun 1974, p 40.
Ref. 625: Bob Fitz - K4JC, "High Performance 80-Meter Antenna," *Ham Radio,* May 1977, p 56.
Ref. 626: Everett S. Brown - K4EF, "New Multiband Longwire Antenna Design," *Ham Radio,* May 1977, p 10.
Ref. 627: William A. Wildenhein - W8YFB, "Solution to the Low-Band Antenna Problem," *Ham Radio,* Jan 1978, p 46.
Ref. 628: John Becker - K9MM, "Lightning Protection," *Ham Radio,* Dec 1978, p 18.
Ref. 629: James Lawson - W2PV, "Part V *Yagi Antenna Design*: Ground or Earth Effects," *Ham Radio,* Oct 1980, p 29.
Ref. 630: Henry G. Elwell - N4UH, "Antenna Geometry for Optimum Performance," *Ham Radio,* May 1982, p 60.
Ref. 631: Randy Rhea - N4HI, "Dipole Antenna over Sloping Ground," *Ham Radio,* May 1982, p 18.
Ref. 632: Bradley Wells - KR7L, "Lightning and Electrical Transient Protection," *Ham Radio,* Dec 1983, p 73.
Ref. 633: R. C. Marshall - G3SBA, "An End-Fed Multiband 8JK," *Ham Radio,* May 1984, p 81.
Ref. 634: David Atkins - W6VX, "Capacitively Loaded High-Performance Dipole," *Ham Radio,* May 1984, p 33.
Ref. 635: David Courtier-Dutton - G3FPQ, "Some Notes on a 7-MHz Linear Loaded Quad," *QST,* Feb 1972, p 14.
Ref. 636: John Kaufmann - WA1CQW - et al, "A Convenient Stub-Tuning System for Quad Antennas," *QST,* May 1975, p 18.
Ref. 637: Hardy Lankskov - W7KAR, "Pattern Factors for Elevated Horizontal Antennas Over Real Earth," *QST,* Nov 1975, p 19.
Ref. 638: Robert Dome - W2WAM, "Impedance of Short Horizontal Dipoles," *QST,* Jan 1976, p 32.
Ref. 639: Donald Belcher - WA4JVE - et al, "Loops vs Dipole Analysis and Discussion," *QST,* Aug 1976, p 34.
Ref. 640: Roger Sparks - W7WKB, "Build this C-T Quad Beam for Reduced Size," *QST,* Apr 1977, p 29.
Ref. 641: Ronald K. Gorski - W9KYZ, "Efficient Short Radiators," *QST,* Apr 1977, p 37.
Ref. 642: Byron Goodman - W1DX, "My Feed Line Tunes My Antenna," *QST,* Apr 1977, p 40.
Ref. 643: Doug DeMaw - W1FB, "The Gentlemen's Band: 160 Meters," *QST,* Oct 1977, p 33.
Ref. 644: David S. Hollander - N7RK, "A Big Signal from a Small Lot," *QST,* Apr 1979, p 32.
Ref. 645: Dana Atchley - W1CF, "Putting the Quarter Wave Sloper to Work on 160," *QST,* Jul 1979, p 19.
Ref. 646: Stan Gibilisco - W1GV, "The Imperfect Antenna System and How it Works," *QST,* Jul 1979, p 24.
Ref. 647: John Belrose - VE2CV, "The Half Sloper," *QST,* May 1980, p 31.
Ref. 648: Larry May - KE6H, "Antenna Modeling Program for the TRS-80," *QST,* Feb 1981, p 15.
Ref. 649: Colin Dickman - ZS6U, "The ZS6U Minishack Special," *QST,* Apr 1981, p 32.
Ref. 650: Doug DeMaw - W1FB, "More Thoughts on the 'Confounded' Half Sloper," *QST,* Oct 1981, p 31.
Ref. 651: John S. Belrose - VE2CV, "The Effect of Supporting Structures on Simple Wire Antennas," *QST,* Dec 1982, p 32.
Ref. 652: Gerald Hall - K1TD, "A Simple Approach to Antenna Impedances," *QST,* Mar 1983, p 16.
Ref. 653: Jerry Hall - K1TD, "The Search for a Simple Broadband 80-Meter Dipole," *QST,* Apr 1983, p 22.
Ref. 654: Charles L. Hutchinson - K8CH, "Getting the Most out of Your Antenna," *QST,* Jul 1983, p 34.
Ref. 655: Doug DeMaw - W1FB, "Building and Using 30 Meter Antennas," *QST,* Oct 1983, p 27.
Ref. 656: James Rautio - AJ3K, "The Effects of Real Ground on Antennas - Part 1," *QST,* Feb 1984, p 15.
Ref. 657: James Rautio - AJ3K, "The Effects of Real Ground on Antennas - Part 2," *QST,* Apr 1984, p 34.
Ref. 658: James Rautio - AJ3K, "The Effect of Real Ground on Antennas - Part 3," *QST,* Jun 1984, p 30.
Ref. 659: Doug DeMaw - W1FB, "Trap for Shunt-Fed Towers," *QST,* Jun 1984, p 40.
Ref. 660: James Rautio - AJ3K, "The Effects of Real Ground on Antennas - Part 4," *QST,* Aug 1984, p 31.
Ref. 661: James Rautio - AJ3K, "The Effects of Real Ground on Antennas - Part 5," *QST,* Nov 1984, p 35.

Ref. 662: Robert C. Sommer - N4UU, "Optimizing Coaxial-Cable Traps," *QST*, Dec 1984, p 37.
Ref. 663: Bob Schetgen - KU7G, "Technical Correspondence" *QST*, Apr 1985, p 51.
Ref. 664: Pat Hawker - G3VA, "Technical Topics: Low Angle Operation," *Radio Communication*, Apr 1971, p 262.
Ref. 665: Pat Hawker - G3VA, "Technical Topics: Low-Angle Radiation and Sloping-Ground Sites," *Radio Communication*, May 1972, p 306.
Ref. 666: Pat Hawker - G3VA, "Technical Topics: All-Band Terminated Long-Wire," *Radio Communication*, Nov 1972, p 745.
Ref. 667: A.P.A. Ashton - G3XAP, "160M DX from Suburban Sites," *Radio Communication*, Dec 1973, p 842.
Ref. 668: L. A. Moxon - G6XN, "Gains and Losses in HF Aerials - Part 1," *Radio Communication*, Dec 1973, p 834.
Ref. 669: A. Moxon - G6XN, "Gains and Losses in HF Aerials - Part 2," *Radio Communication*, Jan 1974, p 16.
Ref. 670: Pat Hawker - G3VA, "Technical Topics: Thoughts on Inverted-Vs," *Radio Communication*, Sep 1976, p 676.
Ref. 671: A. P. A. Ashton - G3XAP, "The G3XAP Directional Antenna for the Lower Frequencies," *Radio Communication*, Nov 1977, p 858.
Ref. 672: S. J. M. Whitfield - G3IMW, "3.5 MHz DX Antennas for a Town Garden," *Radio Communication*, Aug 1980, p 772.
Ref. 673: Pat Hawker - G3VA, "Technical Topics: Low Profile 1.8 and 3.5 MHz Antennas," *Radio Communication*, Aug 1980, p 792.
Ref. 674: Pat Hawker - G3VA, "Technical Topics: Half Delta Loop - Sloping One-Mast Yagi," *Radio Communication*, Oct 1983, p 892.
Ref. 675: R. Rosen - K2RR, "Secrets of Successful Low Band Operation - Part 1," *Ham Radio,* May 1986, p 16.
Ref. 676: R. Rosen - K2RR, "Secrets of Successful Low Band Operation - Part 2," *Ham Radio,* Jun 1986.
Ref. 677: J. Dietrich - WAØRDX, "Loops and Dipoles: A Comparative Analysis," *QST*, Sep 1985, p 24.
Ref. 678: Roy Lewallen - W7EL, "*MININEC*: The Other Edge of the Sword," *QST*, Feb 1991, p 18.
Ref. 679: Rich Rosen - K2RR, "Secrets of Successful Low Band Operation," *Ham Radio,* May 1986, p 16
Ref. 680: Yardley Beers - WØJF, "Designing Trap Antennas: a New Approach," *Ham Radio,* Aug 1987, p 60.
Ref. 681: Guenter Schwarzbeck - DL1BU, "Die Antennw und ihre Umgebung," *CQ-DL*, Jan 1988, p 5.
Ref. 682: Maurice C. Hately - GM3HAT, "A No-compromise Multiband Low SWR Dipole," *Ham Radio,* Mar 1987, p 69.
Ref. 683: R. P. Haviland - W4MB, "Design Data for Pipe Masts," *Ham Radio,* Jul 1989, p 38.
Ref. 684: Gary E. O'Neil - N3GO, "Trapping the Mysteries of Trapped Antennas," *Ham Radio,* Oct 1981, p 10.
Ref. 685: J. Belrose VE2CV and P. Bouliane - VE3KLO, "The Off-center-fed Dipole Revisited," *QST*, Aug 1990, p 28.
Ref. 686: John J. Reh - K7KGP, "An Extended Double Zepp Antenna for 12 Meters," *QST*, Dec 1987, p 25.
Ref. 687: James W. Healy - NJ2L, "Feeding Dipole Antennas," *QST*, Jul 1991, p 22.
Ref. 688: Bill Orr, W6SAI, "Antenna Gain," *Ham Radio*, Jan 1990, p 30.
Ref. 689: B. H. Johns - W3JIP, "Coaxial Cable Antenna Traps," *QST*, May 1981, p 15.
Ref. 690: Doug DeMaw, W1FB, "The Paragon Technology NEC-Win Antenna Analysis Software" *CQ* mag, Nov 1996, p 28.
Ref. 691: N. Mullani, KØNM, "The Bent Dipole," *QST*, May 1997, p 56.
Ref. 692: Larry East, W1HUE, "Antenna Trap Design using a Home Computer," *ARRL Antenna Compendium Vol 2*, p 100 (ISBN 0-87259-254-5).
Ref. 693: Moxon, G6XN, *HF Antennas for All Locations*, published by the RSGB.
Ref. 694: **http://www.w8ji.com/loading_inductors.htm.**
Ref. 695: P. Antoniazzi, IW2ACD and M. Arecco, IK2WAQ, "The Art of Making and Measuring LF Coils," *QEX*, Sep 2001, p 26.
Ref. 696: Kuecken, KE2QJ, "Antennas and Transmission Lines," MFJ-3305, 1996.
Ref. 697: *The ARRL Antenna Book*, 20th Ed., ARRL, Newington, CT, 2003.

7. VERTICAL ANTENNAS

Ref. 701: Paul Lee - K6TS, *Vertical Antenna Handbook*, CQ Publishing Inc.
Ref. 702: Wait and Pope, "Input Resistance of LF Unipole Aerials," *Wireless Engineer*, May 1955, p 131.
Ref. 703: J. J. Wiseman, "How Long is a Piece of Wire," *Electronics and Wireless World*, Apr 1985, p 24.
Ref. 704: Carl C. Drumeller - W5JJ, "Using Your Tower as an Antenna," *CQ,* Dec 1977, p 75.
Ref. 705: Karl T. Thurber - W8FX, "HF Verticals - Plain And Simple," *CQ,* Sep 1980, p 22.
Ref. 706: John E. Magnusson - WØAGD, "Improving Antenna Performance," *CQ,* Jun 1981, p 32.
Ref. 707: Larry Strain - N7DF, "A 3.5 to 30 MHz Discage Antenna," *CQ,* Apr 1984, p 18.
Ref. 708: Karl Hille - DL1VU, "Optimierte T-Antenne," *CQ-DL*, Jun 1978, p 246.
Ref. 709: Rolf Schick - DL3AO, "Loop - Dipol und Vertikalantennen - Vergleiche und Erfahrungen," *CQ-DL*, Mar 1979, p 115.
Ref. 710: Guenter Schwarzbeck - DL1BU, "DX Antennen fuer 80 und 160 Meter," *CQ-DL*, Apr 1979, p 150.
Ref. 711: Hans Wurtz - DL2FA, "DX-Antennen mit spiegelenden Flaechen," *CQ-DL*, Aug 1979, p 353.
Ref. 712: Hans Wurtz - DL2FA, "DX-Antennen mit spiegelenden Flaechen," *CQ-DL*, Sep 1979, p 400.
Ref. 713: Hans Wurtz - DL2FA, "DX-Antennen mit spiegelenden Flaechen," *CQ-DL*, Jan 1980, p 18.
Ref. 714: Hans Wurtz - DL2FA, "DX-Antennen mit spiegelenden Flaechen," *CQ-DL*, Jun 1980, p 272.
Ref. 715: Hans Wurtz - DL2FA, "DX-Antennen mit spiegelenden Flaechen," *CQ-DL*, Jul 1980, p 311.
Ref. 716: Hans Wurtz - DL2FA, "DX-Antennen mit spiegelenden Flaechen," *CQ-DL*, Feb 1981, p 61.

Ref. 717: Hans Wurtz - DL2FA, "DX-Antennen mit spiegelenden Flaechen," *CQ-DL*, Jul 1981, p 330.
Ref. 718: Guenter Schwarzbeck - DL1BU, "Groundplane und Vertikalantennenf," *CQ-DL*, Sep 1981, p 420.
Ref. 719: Hans Wurtz - DL2FA, "DX-Antennen mit spiegelenden Flaechen," *CQ-DL*, Apr 1983, p 170.
Ref. 720: Hans Wurtz - DL2FA, "DX-Antennen mit spiegelenden Flaechen," *CQ-DL*, May 1983, p 224.
Ref. 721: Hans Wurtz - DL2FA, "Antennen mit spiegelenden Flaechen," *CQ-DL*, Jun 1983, p 278.
Ref. 722: Hans Adolf Rohrbacher - DJ2NN, "Basic Programm zu Berechnungh von Vertikalen Antennen," *CQ-DL*, Jun 1983, p 275.
Ref. 723: Hans Wurtz - DL2FA, "DX-Antennen mit spiegelenden Flaechen," *CQ-DL*, Jul 1983, p 326.
Ref. 724: John Schultz - W2EEY, "Stub-Switched Vertical Antennas," *Ham Radio*, Jul 1969, p 50.
Ref. 725: John True - W4OQ, "The Vertical Radiator," *Ham Radio*, Apr 1973, p 16.
Ref. 726: John True - W4OQ, "Vertical-Tower Antenna System," *Ham Radio*, May 1973, p 56.
Ref. 727: George Smith - W4AEO, "80- and 40-Meter Log Periodic Antennas," *Ham Radio*, Sep 1973, p 44.
Ref. 728: Robert Leo - W7LR, "Vertical Antenna Characteristics," *Ham Radio*, Mar 1974, p 34.
Ref. 729: Robert Leo - W7LR, "Vertical Antenna Radiation Patterns," *Ham Radio*, Apr 1974, p 50.
Ref. 730: Raymond Griese - K6FD, "Improving Vertical Antennas," *Ham Radio*, Dec 1974, p 54.
Ref. 731: Harry Hyder - W7IV, "Large Vertical Antennas," *Ham Radio*, May 1975, p 8.
Ref. 732: John True - W4OQ, "Shunt-Fed Vertical Antennas," *Ham Radio*, May 1975, p 34.
Ref. 733: H. H. Hunter - W8TYX, "Short Vertical for 7 MHz," *Ham Radio*, Jun 1977, p 60.
Ref. 734: Laidacker M. Seaberg - WØNCU, "Multiband Vertical Antenna System," *Ham Radio*, May 1978, p 28.
Ref. 735: Joseph D. Liga - K2INA, "80-Meter Ground Plane Antennas," *Ham Radio*, May 1978, p 48.
Ref. 736: John M. Haerle - WB5IIR, "Folded Umbrella Antenna," *Ham Radio*, May 1979, p 38.
Ref. 737: Paul A. Scholz - W6PYK, "Vertical Antenna for 40 and 75 Meters," *Ham Radio*, Sep 1979, p 44.
Ref. 738: Ed Marriner - W6XM, "Base-Loaded Vertical for 160 Meters," *Ham Radio*, Aug 1980, p 64.
Ref. 739: John S. Belrose - VE2CV, "The Half-Wave Vertical," *Ham Radio*, Sep 1981, p 36.
Ref. 740: Stan Gibilisco - W1GV/4, "Efficiency of Short Antennas," *Ham Radio*, Sep 1982, p 18.
Ref. 741: John S. Belrose - VE2CV, "Top-Loaded Folded Umbrella Vertical Antenna," *Ham Radio*, Sep 1982, p 12.
Ref. 742: W. J. Byron - W7DHD, "Short Vertical Antennas for the Low Bands - Part 1," *Ham Radio*, May 1983, p 36.
Ref. 743: Forrest Gehrke - K2BT, "Vertical Phased Arrays - Part 1," *Ham Radio*, May 1983, p 18.
Ref. 744: John Belrose - VE2CV, "The Grounded Monopole with Elevated Feed," *Ham Radio*, May 1983, p 87.
Ref. 745: Forrest Gehrke - K2BT, "Vertical Phased Arrays - Part 2," *Ham Radio*, Jun 1983, p 24.
Ref. 746: W. J. Byron - W7DHD, "Short Vertical Antennas for the Low Bands - Part 2," *Ham Radio*, Jun 1983, p 17.
Ref. 747: Forrest Gehrke - K2BT, "Vertical Phased Arrays - Part 3," *Ham Radio*, Jul 1983, p 26.
Ref. 748: Forrest Gehrke - K2BT, "Vertical Phased Arrays - Part 4," *Ham Radio*, Oct 1983, p 34.
Ref. 749: Forrest Gehrke - K2BT, "Vertical Phased Arrays - Part 5," *Ham Radio*, Dec 1983, p 59.
Ref. 750: Marc Bacon - WB9VWA, "Verticals Over REAL Ground," *Ham Radio*, Jan 1984, p 35.
Ref. 751: Robert Leo - W7LR, "Remote Controlled 40 - 80 - and 160 Meter Vertical," *Ham Radio*, May 1984, p 38.
Ref. 752: Harry Hyder - W7IV, "Build a Simple Wire Plow," *Ham Radio*, May 1984, p 107.
Ref. 753: Gene Hubbell - W9ERU, "Feeding Grounded Towers as Radiators," *QST*, Jun 1960, p 33.
Ref. 754: Eugene E. Baldwin - WØRUG, "Some Notes on the Care and Feeding of Grounded Verticals," *QST*, Oct 1963, p 45.
Ref. 755: N. H. Davidson - K5JVF, "Flagpole Without a Flag," *QST*, Nov 1964, p 36.
Ref. 756: Jerry Sevick - W2FMI, "The Ground-Image Vertical Antenna," *QST*, Jul 1971, p 16.
Ref. 757: Jerry Sevick - W2FMI, "The W2FMI 20-Meter Vertical Beam," *QST*, Jun 1972, p 14.
Ref. 758: Jerry Sevick - W2FMI, "The W2FMI Ground Mounted Short Vertical Antenna," *QST*, Mar 1973, p 13.
Ref. 759: Jerry Sevick - W2FMI, "A High Performance 20, 40 and 80 Meter Vertical System," *QST*, Dec 1973, p 30.
Ref. 760: Jerry Sevick - W2FMI, "The Constant Impedance Trap Vertical," *QST*, Mar 1974, p 29.
Ref. 761: Barry A. Boothe - W9UCW, "The Minooka Special," *QST*, Dec 1974, p 15.
Ref. 762: Willi Richartz - HB9ADQ, "A Stacked Multiband Vertical for 80-10 Meters," *QST*, Feb 1975, p 44.
Ref. 763: John S. Belrose - VE2CV, "The HF Discone Antenna," *QST*, Jul 1975, p 11.
Ref. 764: Earl Cunningham - W5RTQ, "Shunt Feeding Towers for Operating on the Low Amateur Frequencies," *QST*, Oct 1975, p 22.
Ref. 765: Dennis Kozakoff - W4AZW, "Designing Small Vertical Antennas," *QST*, Aug 1976, p 24.
Ref. 766: Ronald Gorski - W9KYZ, "Efficient Short Radiators," *QST*, Apr 1977, p 37.
Ref. 767: Richard Lodwig - W2KK, "The Inverted-L Antenna," *QST*, Apr 1977, p 32.
Ref. 768: Asa Collins - K6VV, "A Multiband Vertical Radiator," *QST*, Apr 1977, p 22.
Ref. 769: Walter Schultz - K3OQF, "Slant-Wire Feed for Grounded Towers," *QST*, May 1977, p 23.
Ref. 770: Yardley Beers - WØJF, "Optimizing Vertical Antenna Performance," *QST*, Oct 1977, p 15.
Ref. 771: Walter Schultz - K3OQF, "Designing a Vertical Antenna," *QST*, Sep 1978, p 19.
Ref. 772: John S. Belrose - VE2CV, "A Kite Supported 160M (or 80M) Antenna for Portable Application," *QST*, Mar 1981, p 40.

Ref. 773: Wayne Sandford - K3EQ, "A Modest 45 Foot Tall DX Vertical for 160 - 80 - 40 and 30 Meters," *QST*, Sep 1981, p 27.

Ref. 774: Doug DeMaw - W1FB, "Shunt Fed Towers - Some Practical Aspects," *QST*, Oct 1982, p 21.

Ref. 775: John F. Lindholm - W1XX, "The Inverted L Revisited," *QST*, Jan 1983, p 20.

Ref. 776: Carl Eichenauer - W2QIP, "A Top Fed Vertical Antenna for 1.8 MHz. Plus 3," *QST*, Sep 1983, p 25.

Ref. 777: Robert Snyder - KE2S, "Modified Butternut Vertical for 80-Meter Operation," *QST*, Apr 1985, p 50.

Ref. 778: Doug DeMaw - W1FB, "A Remotely Switched Inverted-L Antenna," *QST*, May 1985, p 37.

Ref. 779: Pat Hawker - G3VA, "Technical Topics: Low Angle Operation," *Radio Communication*, Apr 1971, p 262.

Ref. 780: Pat Hawker - G3VA, "Technical Topics: Improving the T Antenna," *Radio Communication*, Sep 1978, p 770.

Ref. 781: J. Bazley - G3HCT, "A 7 MHz Vertical Antenna," *Radio Communication*, Jan 1979, p 26.

Ref. 782: P. J. Horwood - G3FRB, "Feed Impedance of Loaded 1/4 Vertical Antennas and the Effects of Earth Systems," *Radio Communication*, Oct 1981, p 911.

Ref. 783: Pat Hawker - G3VA, "Technical Topics: The Inverted Groundplane Family," *Radio Communication*, May 1983, p 424.

Ref. 784: Pat Hawker - G3VA, "Technical Topics: More on Groundplanes," *Radio Communication*, Sep 1983, p 798.

Ref. 785: Pat Hawker - G3VA, "Technical Topics: Sloping One-Mast Yagi," *Radio Communication*, Oct 1983, p 892.

Ref. 786: V. C. Lear - G3TKN, "Gamma Matching Towers and Masts at Lower Frequencies," *Radio Communication*, Mar 1986, p 176.

Ref. 787: B. Wermager - KØEOU, "A Truly Broadband Antenna for 80/75 Meters," *QST*, Apr 1986, p 23.

Ref. 788: Pat Hawker - G3VA, "Technical Topics: Elements of Non-Uniform Cross Section," *Radio Communication*, Jan 1986, p 36.

Ref. 789: Andy Bourassa - WA1LJJ, "Build a Top-hat Vertical Antenna for 80/75 Meters," *CQ*, Aug 1990, p 18.

Ref. 790: Carl C. Drumeller - W5JJ, "Using Your Tower as an Antenna," *CQ*, Dec 1977, p 75.

Ref. 791: Carl Huether - KM1H, "Build a High-performance Extended Bandwidth 160 Meter Vertical," *CQ*, Dec 1986, p 38.

Ref. 792: John Belrose - VE2CV, "More on the Half Sloper," *QST*, Feb 1991, p 39.

Ref. 793: Pat Hawker - G3VA, "The Folded Dipole and Monopole," *Radio Communication*, Jul 1987, p 496.

Ref. 794: C. J. Michaels - W7XC, "Evolution of the Short Top-loaded Vertical," *QST*, Mar 1990, p 26.

Ref. 795: Walter J. Schulz - K3OQF, "Calculating the Input Impedance of a Tapered Vertical," *Ham Radio*, Aug 1985, p 24.

Ref. 796: C. J. Michaels - W7XC, "Some Reflections on Vertical Antennas," *QST*, Jul 1987, p 15.

Ref. 797: C.J. Michaels - W7XC, "Loading Coils for 160 Meter Antennas," *QST*, Apr 1990, p 28.

Ref. 798: D. DeMaw - W1FB, "The 160-meter Antenna Dilemma," *QST*, Nov 1990, p 30.

Ref. 799: Al Christman - KB8I, "Elevated Vertical Antenna Systems Q&A," *QST*, May 1989, p 50.

Ref. 7811: Paul Carr, N4PC, "A Short Two Band Vertical for 160 and 80 Meters"; *CQ* mag, Apr 1997, p 20.

Ref. 7812: W1XT "Gladiator Antennas - To Heard Island," *The Low Band Monitor*, Oct 1996.

Ref. 7813; N4KG, "Elevated Radial 80-Meter Reverse Feed," *The Low Band Monitor*, Sep 1996.

Ref. 7814 ; K3ND "The K3ND Low-Band Vertical," *The Low Band Monitor*, Jul 1996.

Ref. 7815: WØCD "The Battle Creek Special," *The Low Band Monitor*, Sep 1993.

Ref. 7816: K1VW "The S92SS 160 M Antenna"; *The Low Band Monitor*, Mar 1997.

Ref. 7817: KØJN, "The UNI-HAT CTSVR Antenna," *The Low Band Monitor*, Jan 1996.

Ref. 7818: Gerd Janzen, "Kurze Antennen," published by Franckh'she Verlaghandlun, Stuttgart, Germany, ISBN 3-440-05469-1 (This book is only available directly from the author, DF6SJ, Hochvogelstrasse 29, D087435 Kempten, Germany).

Ref. 7819: Guy Hamblen, AA7ZQ/2, "A 75/80 Meter Full-size 1/4-1 Vertical," *ARRL Antenna Compendium, Volume 5* (ISBN 0-87259-562-5).

Ref. 7820: Al Christman, KB8I, "Elevated Vertical Antennas for the Low Bands, Varying the Height and Number of Radials," *ARRL Antenna Compendium, Volume 5* (ISBN 0-87259-562-5).

Ref. 7821: John Belrose, VE2CV, "Elevated Radial Wire System for Vertically Polarized Ground-Plane Type Antennas, Part 1" *Communications Quarterly*, Winter 1998, p 29.

Ref. 7822: Dick Weber, K5IU, "Optimal Elevated Radial Vertical Antennas," *Communications Quarterly*, Spring 1997, p 20.

Ref. 7823: Dick Weber, K5IU, "Comments on Belrose's Article in *Comm.Q.* Winter 1998," *Communications Quarterly*, Spring '98, p 5.

Ref. 7824: John Belrose, VE2CV, "Elevated Radial Wire Systems for Vertically Polarized Ground-Plane Type Antennas, Part 2," *Communications Quarterly*, Spring 1998, p 45.

Ref. 7825: Al Christman, KB8I, "Elevated Vertical Antennas for the Low Bands: Varying the Height and Number of Radials," *ARRL Antenna Compendium, Volume 5*, ISBN 0-87259-562-5.

Ref. 7826: Carl. J. Moreschi, N4PY, "A DX Antenna for 160, 80, 40 and 30 Meters," *CQ* mag, Apr 1995, p 36.

Ref. 7827: Bill Orr, W6SAI, "The S92SS Limited Space Antenna for 160 Meters," Radio Fundamentals, *CQ* mag, Aug 1997, p 85.

Ref. 7828: Paul Carr, N4PC, "A Short, Two-Band Vertical for 160 and 80 Meters," *CQ* mag, Apr 1987, p 20.

Ref. 7829: Gale Stewart, K3ND, "The K3ND Low-Band Vertical (160/80)," *The Low Band Monitor*, Jul 1996, p 2.

Ref. 7830: "The S92SS 160 m vertical with 4 short, tuned radials," *The Low Band Monitor*, Mar 1997

Ref. 7831: Dave Bowker, WØRJU, "Low-Band Tri-Bander (Shunt-Fed Tower System)," *The DX Magazine*, Jan/Feb 1994, p 9.

Ref. 7832: Thomas Russell, N4KG, "Simple, Effective, Elevated Ground Plane Antenna," *QST*, Jun 1994, p 45.

Ref. 7833: L. Moxon, G6XN, "Ground Planes, Radial Systems and Asymmetric Dipoles" *ARRL Antenna Compendium Vol 3*, p 19 (ISBN 0-87259-401-7).

Ref. 7834: "Using elevated radials with ground-mounted towers" *IEEE Transactions on Broadcasting*, vol. 37. no. 3, Sep 1991, pp 77-82.

Ref. 7835: J. Devoldere, ON4UN "Radials made Clear," *CQ Contest*, Sep 1996, p 6-9.

Ref. 7991: H. Hille - DL1VU, "Vortschritte in der Entwicklung von Vertikalantennen," *CQ-DL*, Oct 1989, p 631.

Ref. 7992: Doug DeMaw - W1FB, "Trap for Shunt-fed Towers," *QST*, Jun 1984, p 40.

Ref. 7993: W. J. Byron - W7DHD, "Ground Mounted Vertical Antennas," *Ham Radio*, Jun 1990, p 11.

Ref. 7994: Dennis N. Monticelli - AE6C, "A Simple Effective Dual-Band Inverted-L Antenna," *QST*, Jul 1991, p 38.

Ref. 7995: Walter J. Schultz - K3OQF, "The Folded Wire Fed Top Loaded Grounded Vertical," *Ham Radio*, May 1989, p 32.

Ref. 7996: Gary Nichols - KD9SV and Lynn Gerig - WA9FGR, "Low Band Verticals and How to Feed Them," *CQ*, Aug 1990, p 46.

Ref. 7997: "The phase and magnitude of earth currents near transmitting antennas," *Proceedings IRE*, Vol 23, no. 2, Feb 1935, p 168.

Ref. 7998: Bruce Clark, KO1F, "How to Mount a Tower to use as a Low-Band Vertical Antenna"; *CQ* mag, Nov 1994, p 52.

Ref. 7999: Carl Moreschi, N4PY, "A DX Antenna for 160, 80, 40 and 30 Meters"; *CQ* mag, Apr 1995, p 36.

8. ANTENNAS: GROUND SYSTEMS

Ref. 800: Jager, "Effect of Earth's Surface on Antenna Patterns in the Short Wave Range," *Int. Elek. Rundshau*, Aug 1970, p 101.

Ref. 801: G. H. Brown - et al, "Ground Systems as a Factor in Antenna Efficiency," *Proceedings IRE*, Jun 1937, p 753.

Ref. 802: Abbott, "Design of Optimum Buried RF Ground Systems," *Proceedings IRE*, Jul 1952, p 846.

Ref. 803: Monteath, "The Effect of Ground Constants of an Earth System on Vertical Aerials," *Proceedings IRE*, Jan 1958, p 292.

Ref. 804: Caid, "Earth Resistivity and Geological Structure," *Electrical Engineering*, Nov 1935, p 1153.

Ref. 805: "Calculated Pattern of a Vertical Antenna With a Finite Radial-Wire Ground System," *Radio Science*, Jan 1973, p 81.

Ref. 806: John E. Magnusson - WØAGD, "Improving Antenna Performance," *CQ*, Jun 1981, p 32.

Ref. 807: "Improving Vertical Antenna Efficiency: A Study of Radial Wire Ground Systems," *CQ*, Apr 1984, p 24.

Ref. 808: Robert Leo - W7LR, "Vertical Antenna Ground System," *Ham Radio*, May 1974, p 30.

Ref. 809: Robert Sherwood - WBØJGP, "Ground Screen - Alternative to Radials," *Ham Radio*, May 1977, p 22.

Ref. 810: Alan M. Christman - WD8CBJ, "Ground Systems for Vertical Antennas," *Ham Radio*, Aug 1979, p 31.

Ref. 811: H. Vance Mosser - K3ZAP, "Installing Radials for Vertical Antennas," *Ham Radio*, Oct 1980, p 56.

Ref. 813: Bradley Wells - KR7L, "Installing Effective Ground Systems," *Ham Radio*, Sep 1983, p 67.

Ref. 814: Marc Bacon - WB9VWA, "Verticals Over REAL Ground," *Ham Radio*, Jan 1984, p 35.

Ref. 815: Harry Hyder - W7IV, "Build a Simple Wire Plow," *Ham Radio*, May 1984, p 107.

Ref. 816: John Stanley - K4ERO/HC1, "Optimum Ground System for Vertical Antennas," *QST*, Dec 1976, p 13.

Ref. 817: Roger Hoestenbach - W5EGS, "Improving Earth Ground Characteristics," *QST*, Dec 1976, p 16.

Ref. 818: Jerry Sevick - W2FMI, "Short Ground Radial Systems for Short Verticals," *QST*, Apr 1978, p 30.

Ref. 819: Jerry Sevick - W2FMI, "Measuring Soil Conductivity," *QST*, Mar 1981, p 38.

Ref. 820: Archibald C. Doty - K8CFU - et al, "Efficient Ground Systems for Vertical Antennas," *QST*, Feb 1983, p 20.

Ref. 821: Brian Edward - N2MF, "Radial Systems for Ground-Mounted Vertical Antennas," *QST*, Jun 1985, p 28.

Ref. 822: Pat Hawker - G3VA, "Technical Topics: Vertical Polarization and Large Earth Screens," *Radio Communication*, Dec 1978, p 1023.

Ref. 823: P. J. Horwood - G3FRB, "Feed Impedance of Loaded 1/4 Wave Vertical Antenna Systems," *Radio Communication*, Oct 1981, p 911.

Ref. 824: J. A. Frey - W3ESU, "The Minipoise," *CQ*, Aug 1985, p 30.

Ref. 825: Al Christman - KB8I, "Elevated Vertical Antenna Systems," *QST*, Aug 1988, p 35.

Ref. 826: Bill Orr, W6SAI "The S92SS Limited Space Antenna for 160 Meters"; *CQ* mag, Aug 1997, p 85.

Ref. 827: KØCS, "Elevated Radials, Try Two," *Low Band Monitor*, Feb 96.

Ref. 828: KØCS, "Real World Elevated Radials," *Low Band Monitor*, Jul 1994.

9. ANTENNA ARRAYS

Ref. 900: Bill Guimont - W7KW, "Liftoff on 80 Meters," *CQ*, Oct 1979, p 38.

Ref. 901: Guenter Schwarzbeck - DL1BU, "HB9CV Antenna," *CQ-DL*, Jan 1983, p 10.

Ref. 902: Rudolf Fisher - DL6WD, "Das Monster - eine 2 Element Delta-Loop fuer 3.5 MHz," *CQ-DL*, Jul 1983, p 331.

Ref. 903: G. E. Smith - W4AEO, "Log-Periodic Antennas for 40 Meters," *Ham Radio*, May 1973, p 16.

Ref. 904: G. E. Smith - W4AEO, "80- and 40-Meter Log-Periodic Antennas," *Ham Radio*, Sep 1973, p 44.

Ref. 905: G. E. Smith - W4AEO, "Log-Periodic Antenna Design," *Ham Radio*, May 1975, p 14.

Ref. 906: Jerry Swank - W8HXR, "Phased Vertical Array," *Ham Radio*, May 1975, p 24.

Ref. 907: Henry Keen - W5TRS, "Electrically-Steered Phased Array," *Ham Radio,* May 1975, p 52.

Ref. 908: Gary Jordan - WA6TKT, "Understanding the ZL Special Antenna," *Ham Radio,* May 1976, p 38.

Ref. 909: William Tucker - W4FXE, "Fine Tuning the Phased Vertical Array," *Ham Radio,* May 1977, p 46.

Ref. 910: Paul Kiesel - K7CW, "Seven-Element 40-Meter Quad," *Ham Radio,* Aug 1978, p 30.

Ref. 911: Eugene B. Fuller - W2LU, "Sloping 80-meter Array," *Ham Radio,* May 1979, p 70.

Ref. 912: Harold F. Tolles - W7ITB, "Scaling Antenna Elements," *Ham Radio,* Jul 1979, p 58.

Ref. 913: P. A. Scholz - W6PYK - et al, "Log-Periodic Antenna Design," *Ham Radio,* Dec 1979, p 34.

Ref. 914: James Lawson - W2PV, "Yagi Antenna Design: Experiments Confirm Computer Analysis," *Ham Radio,* Feb 1980, p 19.

Ref. 915: James Lawson - W2PV, "Yagi Antenna Design: Multi-Element Simplistic Beams," *Ham Radio,* Jun 1980, p 33.

Ref. 916: Paul Scholz - W6PYK - et al, "Log Periodic Fixed-Wire Beams for 75-Meter DX," *Ham Radio,* Mar 1980, p 40.

Ref. 917: George E. Smith - W4AEO, "Log Periodic Fixed-Wire Beams for 40 Meters," *Ham Radio,* Apr 1980, p 26.

Ref. 918: Ed Marriner - W6XM, "Phased Vertical Antenna for 21 MHz," *Ham Radio,* Jun 1980, p 42.

Ref. 919: William M. Kelsey - N8ET, "Three-Element Switchable Quad for 40 Meters," *Ham Radio,* Oct 1980, p 26.

Ref. 920: Patrick McGuire - WB5HGR, "Pattern Calculation for Phased Vertical Arrays," *Ham Radio,* May 1981, p 40.

Ref. 921: Forrest Gehrke - K2BT, "Vertical Phased Arrays - Part 1," *Ham Radio,* May 1983, p 18.

Ref. 922: Forrest Gehrke - K2BT, "Vertical Phased Arrays - Part 2," *Ham Radio,* Jun 1983, p 24.

Ref. 923: Forrest Gehrke - K2BT, "Vertical Phased Arrays - Part 3," *Ham Radio,* Jul 1983, p 26.

Ref. 924: Forrest Gehrke - K2BT, "Vertical Phased Arrays - Part 4," *Ham Radio,* Oct 1983, p 34.

Ref. 925: Forrest Gehrke - K2BT, "Vertical Phased Arrays - Part 5," *Ham Radio,* Dec 1983, p 59.

Ref. 926: R. C. Marshall - G3SBA, "An End Fed Multiband 8JK," *Ham Radio,* May 1984, p 81.

Ref. 927: Forrest Gehrke - K2BT, "Vertical Phased Arrays - Part 6," *Ham Radio,* May 1984, p 45.

Ref. 928: R. R. Schellenbach - W1JF, "The End Fed 8JK - A Switchable Vertical Array," *Ham Radio,* May 1985, p 53.

Ref. 929: Al Christman - KB8I, "Feeding Phased Arrays - an Alternate Method," *Ham Radio,* May 1985, p 58.

Ref. 930: Dana Atchley - W1HKK, "A Switchable Four Element 80-Meter Phased Array," *QST,* Mar 1965, p 48.

Ref. 931: James Lawson - W2PV, "A 75/80-Meter Vertical Antenna Square Array," *QST,* Mar 1971, p 18.

Ref. 932: James Lawson - W2PV, "Simple Arrays of Vertical Antenna Elements," *QST,* May 1971, p 22.

Ref. 933: Gary Elliott - KH6HCM/W7UXP, "Phased Verticals for 40 Meters," *QST,* Apr 1972, p 18.

Ref. 934: Jerry Sevick - W2FMI, "The W2FMI 20-meter Vertical Beam," *QST,* Jun 1972, p 14.

Ref. 935: Robert Myers - W1FBY - et al, "Phased Verticals in a 40-Meter Beam Switching Array," *QST,* Aug 1972, p 36.

Ref. 936: Robert Jones - KH6AD, "A 7-MHz Parasitic Array," *QST,* Nov 1973, p 39.

Ref. 937: J. G. Botts - K4EQJ, "A Four-Element Vertical Beam for 40/15 Meters," *QST,* Jun 1975, p 30.

Ref. 938: Jarda Dvoracek - OK1ATP, "160-Meter DX with a Two-Element Beam," *QST,* Oct 1975, p 20.

Ref. 939: Dana Atchley - W1CF - et al, "360 Degree - Steerable Vertical Phased Arrays," *QST,* Apr 1976, p 27.

Ref. 940: Richard Fenwick - K5RR - et al, "Broadband Steerable Phased Arrays," *QST,* Apr 1977, p 18.

Ref. 941: Dana Atchley - W1CF, "Updating Phased Array Technology," *QST,* Aug 1978, p 22.

Ref. 942: Bob Hickman - WB6ZZJ, "The Poly Tower Phased Array," *QST,* Jan 1981, p 30.

Ref. 943: W. B. Bachelor - AC3K, "Combined Vertical Directivity," *QST,* Feb 1981, p 19.

Ref. 944: Walter J. Schultz - K3OQF, "Vertical Array Analysis," *QST,* Feb 1981, p 22.

Ref. 945: Edward Peter Swynar - VE3CUI, "40 Meters with a Phased Delta Loop," *QST,* May 1984, p 20.

Ref. 946: Riki Kline - 4X4NJ, "Build a 4X Array for 160 Meters," *QST,* Feb 1985, p 21.

Ref. 947: Trygve Tondering - OZ1TD, "Phased Verticals," *Radio Communication*, May 1972, p 294.

Ref. 948: Pat Hawker - G3VA, "Technical Topics: The Half Square Aerial," *Radio Communication*, Jun 1974, p 380.

Ref. 949: J. L. Lawson - W2PV, "Yagi Antenna Design," *Ham Radio,* Jan 1980, p 22.

Ref. 950: J. L. Lawson - W2PV, "Yagi Antenna Design," *Ham Radio,* Feb 1980, p 19.

Ref. 951: J. L. Lawson - W2PV, "Yagi Antenna Design," *Ham Radio,* May 1980, p 18.

Ref. 952: J. L. Lawson - W2PV, "Yagi Antenna Design," *Ham Radio,* Jun 1980, p 33.

Ref. 953: J. L. Lawson - W2PV, "Yagi Antenna Design," *Ham Radio,* Jul 1980, p 18.

Ref. 953: J. L. Lawson - W2PV, "Yagi Antenna Design," *Ham Radio,* Sep 1980, p 37.

Ref. 954: J. L. Lawson - W2PV, "Yagi Antenna Design," *Ham Radio,* Oct 1980, p 29.

Ref. 955: J. L. Lawson - W2PV, "Yagi Antenna Design," *Ham Radio,* Nov 1980, p 22.

Ref. 956: J. L. Lawson - W2PV, "Yagi Antenna Design," *Ham Radio,* Dec 1980, p 30.

Ref. 957: J. L. Lawson - W2PV, *Yagi Antenna Design*, ARRL, Dec 1986

Ref. 958: Dick Weber - K5IU, "Determination of Yagi Wind Loads Using the Cross-flow Principle," *Communications Quarterly*, Apr 1993.

Ref. 959: Roy Lewallen - W7EL, "The Impact of Current Distribution on Array Patterns," *QST,* Jul 1990, p 39.

Ref. 960: Robert H. Mitchell - N5RM, "The Forty Meter Flame Thrower," *CQ,* Dec 1987, p 36.

Ref. 961: Ralph Fowler - N6YC, "The W8JK Antenna," *Ham Radio,* May 1988, p 9.

Ref. 962: Jurgen A Weigl - OE5CWL, "A Shortened 40-meter Four-element Sloping Dipole Array," *Ham Radio,* May 1988, p 74.

Ref. 963: Al Christman - KB8I, "Phase Driven Arrays for the Low Bands," *QST,* May 1992, p 49.

Ref. 964: D. Leeson - W6QHS, *Physical Design of Yagi Antennas*, ARRL, Newington, CT.

Ref. 965: Dick Weber - K5IU, "Vibration Induced Yagi Fatigue Failures," *Ham Radio,* Aug 1989, p 9.

Ref. 966: Dick Weber - K5IU, "Structural Evaluation of Yagi Elements," *Ham Radio,* Dec 1988, p 29.

Ref. 967: Dave Leeson - W6QHS, "Strengthening the Cushcraft 40-2CD," *QST,* Nov 1991, p 36.

Ref. 968: R. Lahlum - W1MK, "Technical Correspondence," *QST,* Mar 1991, p 39.

Ref. 969: B. Orr, W6SAI, "The HGW Beam for 80 and 160 m," *CQ* mag, Nov 1994, p 120.

Ref. 970: Timothy Hulick, W9QQ, "The Box Antenna"; *CQ* mag, Jan 1997, p 9.

Ref. 971: JH1GNU, "Fishing Pole Low Band Yagis," *Low Band Monitor*, Feb 1996.

Ref. 972: JH1GNU, "Fishing Pole Yagi"; *Low Band Monitor*, Apr 1997.

Ref. 973: KF4HK, "The Comtek Systems ACB-4," *Low Band Monitor,* May 1994.

Ref. 974: KYØA: "Six for the Price of Three, A Novel 160 Meter Array"; *The Top Band Monitor*, Apr 1994.

Ref. 975: D. C. Mitchell, K8UR, "The K8UR Low-band Vertical Array," *CQ* mag, Dec 1989, p 42.

Ref. 976: John Bazley, G3HCT, "Phased Arrays for 7 MHz:, *RadCom*, Mar 1998, p 25.

Ref. 977: Peter Dalton, W6KW: "Designing and Building a 3-element 80-meter Yagi," *CQ* mag, Jul 1998, p 42.

Ref. 978: Dale Hoppe, K6UA, "How to Build a Relatively Small 2-Element 80-Meter Yagi," *CQ* mag, Jun 1998, p 24.

Ref. 979: Nathan A. Miller, NW3Z, and James K. Breakall, WA3FET, "The V-Yagi, A Light Weight Antenna for 40 Meters," *QST,* May 1988, p 38.

Ref. 980: Carl Luetzelschwab, K9LA, "Quad Versus Yagi at Low Heights*,*" *The ARRL Compendium, Vol 4*, p 74 (ISBN 0-87259-491-2).

Ref. 981: Al Christman, KB8I, "The Slant-Wire Special," *The Antenna Compendium Vol. 4*, p 1, (ISBN 0-87259-491-2).

Ref. 982: John Stanley, K4ERO, "The Tuned Guy-Wire, Gain for (Almost) Free," *The ARRL Antenna Compendium Vol. 4*, ARRL, Newington, CT, p 27 (ISBN 0-87259-491-2).

Ref. 983: Al Christman, K3LC (ex KB8I), "Modifying the Slant-Wire Special for More Gain," Technical Correspondence, *QST,* May 1997, p 74.

10. BROADBAND ANTENNAS

Ref. 1000: Larry Strain - N7DF, "3.5 to 30 MHz Discage Antenna," *CQ,* Apr 1984, p 18.

Ref. 1001: F. J. Bauer - W6FPO, "Low SWR Dipole Pairs for 1.8 through 3.5 MHz," *Ham Radio,* Oct 1972, p 42.

Ref. 1002: M. Walter Maxwell - W2DU, "A Revealing Analysis of the Coaxial Dipole Antenna," *Ham Radio,* Jul 1976, p 46.

Ref. 1003: Terry Conboy - N6RY, "Broadband 80-meter Antennas," *Ham Radio,* May 1979, p 44.

Ref. 1004: Mason Logan - K4MT, "Stagger Tuned Dipoles Increase Bandwidth," *Ham Radio,* May 1983, p 22.

Ref. 1005: C. C. Whysall, "The Double Bazooka," *QST,* Jul 1968, p 38.

Ref. 1006: John S. Belrose - VE2CV, "The Discone HF Antenna," *QST,* Jul 1975, p 11.

Ref. 1007: Allen Harbach - WA4DRU, "Broadband 80 Meter Antenna," *QST,* Dec 1980, p 36.

Ref. 1008: Jerry Hall - K1TD, "The Search for a Simple Broadband 80-Meter Dipole," *QST,* Apr 1983, p 22.

Ref. 1009: John Grebenkemper - KA3BLO, "Multiband Trap and Parallel HF Dipoles - a Comparison," *QST,* May 1985, p 26.

Ref. 1010: N. H. Sedgwick - G8WV, "Broadband Cage Aerials," *Radio Communication*, May 1965, p 287.

Ref. 1011: Pat Hawker - G3VA, "Technical Topics: Broadband Bazooka Dipole," *Radio Communication*, Aug 1976, p 601.

Ref. 1012: Frank Witt - AI1H, "The Coaxial Resonator Match and the Broadband Dipole," *QST,* Apr 1989, p 22.

Ref. 1013: Frank Witt - AI1H, "Match Bandwidth of Resonant Antenna Systems," *QST,* Oct 1991, p 21.

Ref. 1014: Rudy Severns, N6LF, "Wideband 80-Meter Dipole," *QST,* Jul 1995, p 27.

11. LOOP ANTENNAS

Ref. 1100: William Orr - W6SAI, *All About Cubical Quads*, Radio Publications Inc.

Ref. 1101: Bill Salerno - W2ONV, "The W2ONV Delta/Sloper Antenna," *CQ,* Aug 1978, p 86.

Ref. 1102: Roy A. Neste - WØWFO, "Dissecting Loop Antennas To Find Out What Makes Them Tick," *CQ,* Aug 1984, p 36.

Ref. 1103: Rolf Schick - DL3AO, "Loop - Dipol und Vertikalantennen - Vergleiche und Erfahrungen," *CQ-DL*, Mar 1979, p 115.

Ref. 1104: Guenter Schwarzbeck - DL1BU, "DX Antennen fuer 80 und 160 Meter," *CQ-DL*, Apr 1979, p 150.

Ref. 1105: Gunter Steppert - DK8NG, "Zweielement Delta Loop mit einem Mast," *CQ-DL*, Aug 1980, p 370.

Ref. 1106: Hans Wurtz - DL2FA, "DX-Antennen mit spiegelenden Flaechen," *CQ-DL*, Apr 1981, p 162.

Ref. 1107: Hans Wurtz - DL2FA, "DX-Antennen mit spiegelenden Flaechen," *CQ-DL*, Dec 1981, p 583.

Ref. 1108: Dieter Pelz - DF3IK et al, "Rahmenantenne — keine Wunderantenne," *CQ-DL*, Sep 1982, p 435.

Ref. 1109: Willi Nitschke - DJ5DW - et al, "Richtkarakteristik - Fusspunktwiderstand etc von Einelementantennen," *CQ-DL*, Dec 1982, p 580.

Ref. 1110: Hans Wurtz - DL2FA, "DX-Antennen mit spiegelenden Flaechen," *CQ-DL*, Feb 1983, p 64.

Ref. 1111: Rudolf Fisher - DL6WD, "Das Monster - eine 2-Element Delta-Loop fuer 3.5 MHz," *CQ-DL*, Jul 1983, p 331.

Ref. 1112: John True - W4OQ, "Low Frequency Loop Antennas," *Ham Radio*, Dec 1976, p 18.

Ref. 1113: Paul Kiesel - K7CW, "7-Element 40-Meter Quad," *Ham Radio*, Jul 1978, p 30.

Ref. 1114: Glenn Williman - N2GW, "Delta Loop Array," *Ham Radio*, Jul 1978, p 16.

Ref. 1115: Frank J. Witt - W1DTV, "Top Loaded Delta Loop Antenna," *Ham Radio*, Dec 1978, p 57.

Ref. 1116: George Badger - W6TC, "Compact Loop Antenna for 40 and 80-Meter DX," *Ham Radio*, Oct 1979, p 24.

Ref. 1117: William M. Kesley - N8ET, "Three Element Switchable Quad for 40 Meters," *Ham Radio*, Oct 1980, p 26.

Ref. 1118: Jerrold Swank - W8HXR, "Two Delta Loops Fed in Phase," *Ham Radio*, Aug 1981, p 50.

Ref. 1119: Hasan Schiers - NØAN, "The Half Square Antenna," *Ham Radio*, Dec 1981, p 48.

Ref. 1120: John S. Belrose - VE2CV, "The Half Delta Loop," *Ham Radio*, May 1982, p 37.

Ref. 1121: V.C. Lear - G3TKN, "Reduced Size, Full Performance Corner Fed Delta Loop," *Ham Radio*, Jan 1985, p 67.

Ref. 1122: Lewis Mc Coy - W1ICP, "The Army Loop in Ham Communication," *QST*, Mar 1968, p 17.

Ref. 1123: J. Wessendorp - HB9AGK, "Loop Measurements," *QST*, Nov 1968, p 46.

Ref. 1124: F.N. Van Zant - W2EGH, "160, 75 and 40 Meter Inverted Dipole Delta Loop," *QST*, Jan 1973, p 37.

Ref. 1125: Ben Vester - K3BC, "The Half Square Antenna," *QST*, Mar 1974, p 11.

Ref. 1126: John Kaufmann - WA1CQW et al, "A Convenient Stub Tuning System for Quad Antennas," *QST*, May 1975, p 18.

Ref. 1127: Robert Edlund - W5DS, "The W5DS Hula-Hoop Loop," *QST*, Oct 1975, p 16.

Ref. 1128: Donald Belcher - WA4JVE et al, "Loops vs Dipole Analysis and Discussion," *QST*, Aug 1976, p 34.

Ref. 1129: Roger Sparks - W7WKB, "Build this C-T Quad Beam for Reduced Size," *QST*, Apr 1977, p 29.

Ref. 1130: John S. Belrose - VE2CV, "The Half-Delta Loop: a Critical Analysis and Practical Deployment," *QST*, Sep 1982, p 28.

Ref. 1131: Richard Gray - W9JJV, "The Two Band Delta Loop Antenna," *QST*, Mar 1983, p 36.

Ref. 1132: Edward Peter Swynar - VE3CUI, "40 Meters with a Phased Delta Loop," *QST*, May 1984, p 20.

Ref. 1133: Doug DeMaw - W1FB et al, "The Full-Wave Delta Loop at Low Height," *QST*, Oct 1984, p 24.

Ref. 1134: Pat Hawker - G3VA, "Technical Topics: Another Look at Transmitting Loops," *Radio Communication*, Jun 1971, p 392.

Ref. 1135: Pat Hawker - G3VA, "Technical Topics: Vertically Polarized Loop Elements," *Radio Communication*, Jun 1973, p 404.

Ref. 1136: Laury Mayhead - G3AQC, "Loop Aerials Close to Ground," *Radio Communication*, May 1974, p 298.

Ref. 1137: Pat Hawker - G3VA, "Technical Topics: The Half Square Aerial," *Radio Communication*, Jun 1974, p 380.

Ref. 1138: Pat Hawker - G3VA, "Miniaturized Quad Elements," *Radio Communication*, Mar 1976, p 206.

Ref. 1139: Pat Hawker - G3VA, "Technical Topics: Radiation Resistance of Medium Loops," *Radio Communication*, Feb 1979, p 131.

Ref. 1140: F. Rasvall - SM5AGM, "The Gain of the Quad," *Radio Communication*, Aug 1980, p 784.

Ref. 1141: Pat Hawker - G3VA, "Technical Topics: Polygonal Loop Antennas," *Radio Communication*, Feb 1981.

Ref. 1142: B. Myers - K1GQ, "The W2PV 80-Meter Quad," *Ham Radio*, May 1986, p 56.

Ref. 1143: Bill Orr - W6SAI, "A Two-Band Loop Antenna," *Ham Radio*, Sep 1987, p 57.

Ref. 1144: R. P. Haviland - W4MB, "The Quad Antenna - Part 1," *Ham Radio*, May 1988.

Ref. 1145: R. P. Haviland - W4MB, "The Quad Antenna - Part 2," *Ham Radio*, Jun 1988, p 54.

Ref. 1146: R. P. Haviland W4MB, "The Quad Antenna - Part 3," *Ham Radio*, Aug 1988, p 34.

Ref. 1147: C. Drayton Cooper - N4LBJ, "The Bi-Square array," *Ham Radio*, May 1990, p 42.

Ref. 1148: Bill Myers - K1GQ, "Analyzing 80 Meter Delta Loop Arrays," *Ham Radio*, Sep 1986, p 10.

Ref. 1149: Dave Donnelly - K2SS, "Yagi vs. Quad - Part 1," *Ham Radio*, May 1988, p 68.

Ref. 1150: D. DeMaw - W1FB, "A Closer Look at Horizontal Loop Antennas," *QST*, May 1990, p 28.

12. RECEIVING ANTENNAS

Ref. 1200: I. Herliz, "Analysis of Action of Wave Antennas," *AIEE Transactions*, Vol 42, May 1942, p 260.

Ref. 1201: "Diversity Receiving System of RCA Communications for Radio Telegraphy," *AIEE Transactions*, Vol 42, May 1923, p 215.

Ref. 1202: Peterson Beverage, "The Wave Antenna: a New Type of Highly Directive Antenna," *Proc IRE*.

Ref. 1203: Dean Baily, "Receiving System for Long-Wave Transatlantic Radio Telephone," *Proc IRE*, Dec 1928.

Ref. 1204: *Antennas for Reception of Standard Broadcast Signals*, FCC Report, Apr 1958.

Ref. 1205: R. D. Straw - N6BV - ed, *The ARRL Handbook for Radio Amateurs*, 2000, ARRL, Newington, CT.

Ref. 1206: Victor Misek - W1WCR, *The Beverage Antenna Handbook,*

Ref. 1207: William Orr - W6SAI, *Radio Handbook*, Howard W. Sams & Co. Inc.

Ref. 1210: Bob Clarke - N1RC, "Six Antennas from Three Wires," 73, Oct 1983, p 10.

Ref. 1211: Davis Harold - W8MTI, "The Wave Antenna," *CQ*, May 1978, p 24.

Ref. 1212: Ulrich Rohde - DJ2LR, "Active Antennas," *CQ*, Dec 1982, p 20.

Ref. 1213: Karl Hille - 9A1VU, "Vom Trafo zum Kurzwellenempgaenger," *CQ-DL*, Mar 1977, p 99.

Ref. 1214: Dieter Pelz - DF3IK et al, "Rahmenantenne - keine Wunderantenne-," *CQ-DL*, Sep 1982, p 435.

Ref. 1215: Hans Wurtz - DL2FA, "Magnetische Antennen," *CQ-DL*, Feb 1983, p 64.

Ref. 1216: Hans Wurtz - DL2FA, "Magnetische Antennen," *CQ-DL*, Apr 1983, p 170.

Ref. 1217: Hans Wurtz - DL2FA, "Elektrisch Magnetische Beam Antennen (EMBA)," *CQ-DL*, Jul 1983, p 326.

Ref. 1218: Guenter Shwarzenbeck - DL1BU, "Rhamen- und Ringantennen," *CQ-DL*, May 1984, p 226.

Ref. 1219: Charles Bird - K6HTM, "160-Meter Loop for Receiving," *Ham Radio,* May 1974, p 46.

Ref. 1220: Ken Cornell - W2IMB, "Loop Antenna Receiving Aid," *Ham Radio,* May 1975, p 66.

Ref. 1221: John True - W4OQ, "Loop Antennas," *Ham Radio,* Dec 1976.

Ref. 1222: Henry Keen - W5TRS, "Selective Receiving Antennas: a Progress Report," *Ham Radio,* May 1978, p 20.

Ref. 1223: Byrd H. Brunemeier - KG6RT, "40-Meter Beverage Antenna," *Ham Radio,* Jul 1979, p 40.

Ref. 1224: David Atkins - W6VX, "Capacitively Loaded Dipole," *Ham Radio,* May 1984, p 33.

Ref. 1225: H. H. Beverage, "A Wave Antenna for 200 Meter Reception," *QST*, Nov 1922, p 7.

Ref. 1226: John Isaacs - W6PZV, "Transmitter Hunting on 75 Meters," *QST*, Jun 1958, p 38.

Ref. 1227: Lewis McCoy - W1ICP, "The Army Loop in Ham Communication," *QST*, Mar 1968, p 17.

Ref. 1228: J. Wessendorp - HB9AGK, "Loop Measurements," *QST*, Nov 1968, p 46.

Ref. 1229: Katashi Nose - KH6IJ, "A 160 Meter Receiving Loop," *QST*, Apr 1975, p 40.

Ref. 1230: Tony Dorbuck - W1YNC, "Radio Direction Finding Techniques," *QST*, Aug 1975, p 30.

Ref. 1231: Robert Edlund - W5DS, "The W5DS Hula-Hoop Loop," *QST*, Oct 1975, p 16.

Ref. 1232: Doug DeMaw - W1FB, "Build this 'Quickie' Preamp," *QST*, Apr 1977, p 43.

Ref. 1233: Larry Boothe - W9UCW, "Weak Signal Reception on 160. Some Antenna Notes," *QST*, Jun 1977, p 35.

Ref. 1234: Doug DeMaw - W1FB, "Low-Noise Receiving Antennas," *QST*, Dec 1977, p 36.

Ref. 1235: Doug DeMaw - W1FB, "Maverick Trackdown," *QST*, Jul 1980, p 22.

Ref. 1236: John Belrose - VE2CE, "Beverage Antennas for Amateur Communications," *QST*, Sep 1981, p 51.

Ref. 1237: H. H. Beverage, "H. H. Beverage on Beverage Antennas," *QST*, Dec 1981, p 55.

Ref. 1238: Doug DeMaw - W1FB et al, "The Classic Beverage Antenna Revisited," *QST*, Jan 1982, p 11.

Ref. 1239: John Webb - W1ETC, "Electrical Null Steering," *QST*, Oct 1982, p 28.

Ref. 1240: John F. Belrose - VE2CV et al, "The Beverage Antenna for Amateur Communications," *QST*, Jan 1983, p 22.

Ref. 1241: Doug DeMaw - W1FB, "Receiver Preamps and How to Use Them," *QST*, Apr 1984, p 19.

Ref. 1242: Pat Hawker - G3VA, "Technical Topics: Beverage Aerials," *Radio Communication*, Oct 1970, p 684.

Ref. 1243: Pat Hawker - G3VA, "Technical Topics: All Band Terminated Longwire," *Radio Communication*, Nov 1972, p 745.

Ref. 1244: Pat Hawker - G3VA, "Technical Topics: A 1.8 MHz Active Frame Aerial," *Radio Communication*, Aug 1976, p 601.

Ref. 1245: Pat Hawker - G3VA, "Technical Topics: Low- Noise 1.8/3.5 MHz Receiving Antennas," *Radio Communication*, Apr 1978, p 325.

Ref. 1246: Pat Hawker - G3VA, "Technical Topics: Radiation Resistance of Medium Loops," *Radio Communication*, Feb 1979, p 131.

Ref. 1247: J.A. Lambert - G3FNZ, "A Directional Active Loop Receiving Antenna System," *Radio Communication*, Nov 1982, p 944.

Ref. 1248: R.C. Fenwick - K5RR, "A Loop Array for 160 Meters," *CQ,* Apr 1986, p 25.

Ref. 1249: Mike Crabtree - AB0X, "Some Thoughts on 160 Meter Receiving Antennas for City Lots," *CQ,* Jan 1988, p 26.

Ref. 1250: J. Wollweber - DF5PY, "Die Magnetisch Antenne - eine Wunderantenne," *CQ-DL,* Mar 1987, p 149.

Ref. 1251: Gary R. Nichols - KD9SV, "Variable gain 160-meter preamp," *Ham Radio,* Oct 1989, p 46.

Ref. 1252: Robert H. Johns - W3JIP, "How to Build an Indoor Transmitting Loop," *CQ,* Dec 1991, p 30.

Ref. 1253: Robert H. Johns - W3JIP, "How to Build an Indoor Transmitting Loop," *CQ,* Jan 1992, p 42.

Ref. 1254: D. DeMaw - W1FB, "On-Ground Low-Noise Receiving Antennas," *QST*, Apr 1988, p 30.

Ref. 1255: Juergen Schaefer - DL7PE, "Die Rahmantenne - eine Befehlsantenne zum Selbstbau," *CQ-DL,* Jan 1990, p 21.

Ref. 1256: D. DeMaw - W1FB, "Preamplifier for 80 and 160 M Loop and Beverage Antennas," *QST*, Aug 1988, p 22.

Ref. 1257: A. G. Lyner, "The Beverage Antenna: a practical way of assessing the radiation pattern using off-air signals," *BBC RD*, 1991/92.

Ref. 1258: B. H. Brunemeier - KG6RT, "Short Beverage for 40 meters," *HR* Jul 1979, p 40.

Ref. 1259 K9FD "The K9FD Receiving Loop," *The Low Band Monitor*, Nov 1995.

Ref. 1260: KY0A and G3SZA, "All Grounds are not equal—optimizing Beverage terminations," *The Low Band Monitor*, Apr 1995.

Ref. 1261: K0CS, "Real World Beverages"; *The Top Band Monitor*, Dec 1993.

Ref. 1262: K0CS "Phased Beverage Primer"; The Top Band Reflector, Sep 1997.

Ref. 1263: Floyd Koontz, WA2WVL "Is this EWE for you?", *QST*, Feb 1995, p 31.

Ref. 1264: Floyd Koontz, WA2WVL "More EWEs for you," *QST*, Jan 1996, p 31.

Ref. 1265: Gary Breed, K9AY "The K9AY Terminated Loop, A compact Directional Receiving Antenna", QST Sept. 1997, pg 43

Ref. 1266: *The ARRL Antenna Compendium, Vol 3,* ARRL 1993.

Ref. 1267: Sergio Cartoceti, IK4AUY "A high level accessory front end for the HF amateur bands" *QEX* Mar/Apr 2003, p 1.

13. TRANSMISSION LINES

Ref. 1300: Wilkinson, "An N-Way Hybrid Power Divider," *IRE Transactions on Microwave*, Jan 1960.

Ref. 1301: R. D. Straw - N6BV, *The ARRL Handbook for Radio Amateurs*, 2000, ARRL, Newington, CT.

Ref. 1302: Pat Hawker - G3VA, *Amateur Radio Techniques*, RSGB.

Ref. 1303: Keith Henney, *Radio Engineering Handbook* - 5th Ed.

Ref. 1304: *Reference Data for Radio Engineers*, 5th Ed, Howard W. Sams & Co. Inc.

Ref. 1305: John Devoldere - ON4UN, "80-Meter DXing," *Communications Technology Inc*, Jan 1979.

Ref. 1306: William Orr - W6SAI, *Radio Handbook*, Howard W. Sams & Co. Inc.

Ref. 1307: R. D. Straw - N6BV ed, *The ARRL Antenna Book*, 20th Ed., ARRL, Newington, CT.

Ref. 1308: Walter Maxwell - W2DU, "Niedriges SWR aus falschem Grund (Teil 1)," *CQ-DL*, Jan 1976, p 3.

Ref. 1309: Walter Maxwell - W2DU, "Niedriges SWR aus falschem Grund (Teil 2)," *CQ-DL*, Feb 1976, p 47.

Ref. 1310: Walter Maxwell - W2DU, "Niedriges SWR aus falschem Grund (Teil 3)," *CQ-DL*, Jun 1976, p 202.

Ref. 1311: Walter Maxwell - W2DU, "Niedriges SWR aus falschem Grund (Teil 5)," *CQ-DL*, Sep 1976, p 272.

Ref. 1312: Jim Fisk - W1HR, "The Smith Chart," *Ham Radio*, Nov 1970, p 16.

Ref. 1313: Richard Taylor - W1DAX, "N-Way Power Dividers and 3-dB Hybrids," *Ham Radio*, Aug 1972, p 30.

Ref. 1314: Joseph Boyer - W6UYH, "Antenna-Transmission Line Analogy," *Ham Radio*, Apr 1977, p 52.

Ref. 1315: Joseph Boyer - W6UYH, "Antenna-Transmission Line Analogy - Part 2," *Ham Radio*, May 1977, p 29.

Ref. 1316: J. Reisert - W1JR, "Simple and Efficient Broadband Balun," *Ham Radio*, Sep 1978, p 12.

Ref. 1317: Jack M. Schulman - W6EBY, "T-Network Impedance Matching to Coaxial Feedlines," *Ham Radio*, Sep 1978, p 22.

Ref. 1318: Charlie J. Carroll - K1XX, "Matching 75-Ohm CATV Hardline to 50-Ohm Systems," *Ham Radio*, Sep 1978, p 31.

Ref. 1319: John Battle - N4OE, "What is Your SWR?," *Ham Radio*, Nov 1979, p 48.

Ref. 1320: Henry Elwell - N4UH, "Long Transmission Lines for Optimum Antenna Location," *Ham Radio*, Oct 1980, p 12.

Ref. 1321: John W. Frank - WB9TQG, "Measuring Coax Cable Loss with an SWR Meter," *Ham Radio*, May 1981, p 34.

Ref. 1322: Stan Gibilisco - W1GV/4, "How Important is Low SWR?" *Ham Radio*, Aug 1981, p 33.

Ref. 1323: Lewis T. Fitch - W4VRV, "Matching 75-Ohm Hardline to 50-Ohm Systems," *Ham Radio*, Oct 1982, p 43.

Ref. 1324: L. A. Cholewski - K6CRT, "Some Amateur Applications of the Smith Chart," *QST*, Jan 1960, p 28.

Ref. 1325: Walter Maxwell - W2DU, "Another Look at Reflections - Part 1," *QST*, Apr 1973, p 35.

Ref. 1326: Walter Maxwell - W2DU, "Another Look at Reflections - Part 2," *QST*, Jun 1973, p 20.

Ref. 1327: Walter Maxwell - W2DU, "Another Look at Reflections - Part 3," *QST*, Aug 1973, p 36.

Ref. 1328: Walter Maxwell - W2DU, "Another Look at Reflections - Part 4," *QST*, Oct 1973, p 22.

Ref. 1329: Walter Maxwell - W2DU, "Another Look at Reflections - Part 5," *QST*, Apr 1974, p 26.

Ref. 1330: Walter Maxwell - W2DU, "Another Look at Reflections - Part 6," *QST*, Dec 1974, p 11.

Ref. 1331: Gerald Hall - K1PLP, "Transmission-Line Losses," *QST*, Dec 1975, p 48.

Ref. 1332: Walter Maxwell - W2DU, "Another Look at Reflections - Part 7," *QST*, Aug 1976, p 16.

Ref. 1333: Charles Brainard - WA1ZRS, "Coaxial Cable: The Neglected Link," *QST*, Apr 1981, p 28.

Ref. 1334: Crawford MacKeand - WA3ZKZ, "The Smith Chart in BASIC," *QST*, Nov 1984, p 28.

Ref. 1336: R. C. Hills - G3HRH, "Some Reflections on Standing Waves," *Radio Communication*, Jan 1964, p 15.

Ref. 1337: Pat Hawker - G3VA, "Technical Topics: Another Look at SWR," *Radio Communication*, Jun 1974, p 377.

Ref. 1338: Garside - G3MYT, "More on the Smith Chart," *Radio Communication*, Dec 1977, p 934.

Ref. 1339: Pat Hawker - G3VA, "Technical Topics: SWR How Important?" *Radio Communication*, Jan 1982, p 41.

Ref. 1340: Kenneth Parker - G3PKR, "Reflected Power Does Not Mean Lost Power," *Radio Communication*, Jul 1982, p 581.

Ref. 1341: Pat Hawker - G3VA, "Technical Topics: Reflected Power is Real Power," *Radio Communication*, Jun 1983, p 517.

Ref. 1342: H. Ashcroft - G4CCM, "Critical Study of the SWR Meter," *Radio Communication*, Mar 1985, p 186.

Ref. 1343: Walter Maxwell - W2DU, "Niedriges SWR aus falschem Grund (Teil 4)," *CQ-DL*, Aug 1976, p 238.

Ref. 1344: Lew McCoy - W1ICP, "Coax as Tuned Feeders: Is It Practical?," *CQ*, Aug 1989, p 13.

Ref. 1345: Chet Smith - K1CCL, "Simple Coaxial Cable Measurements," *QST*, Sep 1990, p 25.

Ref. 1346: J. Althouse - K6NY, "Using a Noise Bridge to Measure Coaxial Cable Impedance," *QST*, May 1991, p 45.

Ref. 1347: A.E. Popodi - OE2APM/AA3K, "Measuring Transmission Line Parameters," *Ham Radio*, Sep 1988, p 22.

Ref. 1348: Byron Goodman - W1DX, "My Feed Line Tunes My Antenna," *QST*, Nov 1991, p 33.

Ref. 1349: Gary E. Myers - K9CZB, "Solving Transmission Line Problems on Your C64," *Ham Radio*, May 1986, p 74.

Ref. 1350: Fred Bonavita - W5QJM, "Working with Balanced Line, part I"; *CQ* mag, Jan 1994, p 56.

Ref. 1351: Fred Bonavita - W5QJM, "Working with Balanced Line, part II"; *CQ* mag, Feb 1994, p 26.

Ref. 1352: Warren Bruene, W5OLY, "Reflected Power Stays in the Coax," *CQ* mag, Jan 1995, p 13.

14. MATCHING

Ref. 1400: R. D. Straw - N6BV - ed, *The ARRL Handbook for Radio Amateurs*, 2000, ARRL, Newington, CT.

Ref. 1401: Harold Tolles - W7ITB, "Gamma-Match Design," *Ham Radio,* May 1973, p 46.

Ref. 1402: Robert Baird - W7CSD, "Antenna Matching Systems," *Ham Radio,* Jul 1973, p 58.

Ref. 1403: Earl Whyman - W2HB, "Standing-Wave Ratios," *Ham Radio,* Jul 1973, p 26.

Ref. 1404: Robert Leo - W7LR, "Designing L-networks," *Ham Radio,* Feb 1974, p 26.

Ref. 1405: G. E. Smith - W4AEO, "Log-Periodic Feed Systems," *Ham Radio,* Oct 1974, p 30.

Ref. 1406: John True - W4OQ, "Shunt-Fed Vertical Antennas," *Ham Radio,* May 1975, p 34.

Ref. 1407: I. L. McNally - K6WX - et al, "Impedance Matching by Graphical Solution," *Ham Radio,* Mar 1978, p 82.

Ref. 1408: Jim Fisk - W1HR, "Transmission-Line Calculations with the Smith Chart," *Ham Radio,* Mar 1978, p 92.

Ref. 1409: Jack M. Schulman - W6EBY, "T-Network Impedance Matching to Coaxial Feed Lines," *Ham Radio,* Sep 1978, p 22.

Ref. 1410: Ernie Franke - WA2EWT, "Appreciating the L Matching Network," *Ham Radio,* Sep 1980, p 27.

Ref. 1411: Robert Leo - W7LR, "Remote-Controlled 40 - 80 - and 160-Meter Vertical," *Ham Radio,* May 1984, p 38.

Ref. 1412: James Sanford - WB4GCS, "Easy Antenna Matching," *Ham Radio,* May 1984, p 67.

Ref. 1413: Chris Bowick - WD4C, "Impedance Matching: A Brief Review," *Ham Radio,* Jun 1984, p 49.

Ref. 1414: Richard Nelson - WBØIKN, "Basic Gamma Matching," *Ham Radio,* Jan 1985, p 29.

Ref. 1415: L. A. Cholewski - K6CRT, "Some Amateur Applications of the Smith Chart," *QST,* Jan 1960, p 28.

Ref. 1416: Gene Hubbell - W9ERU, "Feeding Grounded Towers as Radiators," *QST,* Jun 1960, p 33.

Ref. 1417: J. D. Gooch - W9YRV - et al, "The Hairpin Match," *QST,* Apr 1962, p 11.

Ref. 1418: Eugene E. Baldwin - WØRUG, "Some Notes on the Care and Feeding of Grounded Verticals," *QST,* Oct 1963, p 45.

Ref. 1419: N. H. Davidson - K5JVF, "Flagpole Without a Flag," *QST,* Nov 1964, p 36.

Ref. 1420: Robert Leo - K7KOK, "An Impedance Matching Method," *QST,* Dec 1968, p 24.

Ref. 1421: D. J. Healey - W3PG, "An Examination of the Gamma Match," *QST,* Apr 1969, p 11.

Ref. 1422: Donald Belcher - WA4JVE, "RF Matching Techniques - Design and Example," *QST,* Oct 1972, p 24.

Ref. 1423: James McAlister - WA5EKA, "Simplified Impedance Matching and the Mac Chart," *QST,* Dec 1972, p 33.

Ref. 1424: John Kaufmann - WA1CQW - et al, "A Convenient Stub Tuning System for Quad Antennas," *QST,* May 1975, p 18.

Ref. 1425: Earl Cunningham - W5RTQ, "Shunt Feeding Towers for Operating on the Low Amateur Frequencies," *QST,* Oct 1975, p 22.

Ref. 1426: Doug DeMaw - W1CER, "Another Method of Shunt Feeding Your Tower," *QST,* Oct 1975, p 25.

Ref. 1427: Jerry Sevick - W2FMI, "Simple Broadband Matching Networks," *QST,* Jan 1976, p 20.

Ref. 1428: Walter Schulz - K3OQF, "Slant-Wire Feed for Grounded Towers," *QST,* May 1977, p 23.

Ref. 1429: Bob Pattison - N6RP, "A Graphical Look at the L Network," *QST,* Mar 1979, p 24.

Ref. 1430: Tony Dorbuck - K1FM, "Matching-Network Design," *QST,* Mar 1979, p 26.

Ref. 1431: Herbert Drake Jr. - N6QE, "A Remotely Controlled Antenna-Matching Network," *QST,* Jan 1980, p 32.

Ref. 1432: Doug DeMaw - W1FB, "Ultimate Transmatch Improved," *QST,* Jul 1980, p 39.

Ref. 1433: Colin Dickman - ZS6U, "The ZS6U Minishack Special," *QST,* Apr 1981, p 32.

Ref. 1434: Claude L. Frantz - F5FC/DJØOT, "A New More Versatile Transmatch," *QST,* Jul 1982, p 31.

Ref. 1435: Doug DeMaw - W1FB, "A 'Multipedance' Broadband Transformer," *QST,* Aug 1982, p 39.

Ref. 1436: Don Johnson - W6AAQ, "Mobile Antenna Matching. Automatically!" *QST,* Oct 1982, p 15.

Ref. 1437: Doug DeMaw - W1FB, "Shunt-Fed Towers: Some Practical Aspects," *QST,* Oct 1982, p 21.

Ref. 1438: Crawford MacKaend - WA3ZKZ, "The Smith Chart in BASIC," *QST,* Nov 1984, p 28.

Ref. 1439: J. A. Ewen - G3HGM, "Design of L-Networks for Matching Antennas to Transmitters," *Radio Communication,* Aug 1984, p 663.

Ref. 1440: H. Ashcroft - G4CCM, "Critical Study of the SWR Meter," *Radio Communication,* Mar 1985, p 186.

Ref. 1441: Robert C Cheek - W3VT, "The L-Match," *Ham Radio,* Feb 1989, p 29.

Ref. 1442: W. Orr - W6SAI, "Antenna Matching Systems," *Ham Radio,* Nov 1989, p 54.

Ref. 1443: Will Herzog - K2LB, "Broadband RF Transformers," *Ham Radio,* Jan 1986, p 75.

Ref. 1444: Richard A. Gardner - N1AYW, "Computerizing Smith Chart Network Analysis," *Ham Radio,* Oct 1989, p 10.

Ref. 1445: Joe Carr - K4IPV, "Impedance Matching," *Ham Radio,* Aug 1988, p 27.

Ref. 1446: I. L. McNally - W1CNK, "Graphic Solution of impedance matching problem," *HR,* May 78, p 82.

15. BALUNS

Ref. 1500: J. Schultz - W4FA, "A Selection of Baluns from Palomar Engineers," *CQ,* Apr 1985, p 32.

Ref. 1501: William Orr - W6SAI, "Broadband Antenna Baluns," *Ham Radio,* Jun 1968, p 6.

Ref. 1502: J. Reisert - W1JR, "Simple and Efficient Broadband Balun," *Ham Radio,* Sep 1978, p 12.

Ref. 1503: John J. Nagle - K4KJ, "High Performance Broadband Balun," *Ham Radio,* Feb 1980, p 28.

Ref. 1504: George Badger - W6TC, "A New Class of Coaxial-Line Transformers," *Ham Radio,* Feb 1980, p 12.

Ref. 1505: George Badger - W6TC, "Coaxial-Line Transformers," *Ham Radio,* Mar 1980, p 18.
Ref. 1506: John Nagle - K4KJ, "The Half-Wave Balun: Theory and Application," *Ham Radio,* Sep 1980, p 32.
Ref. 1507: Roy N. Lehner - WA2SON, "A Coreless Balun," *Ham Radio,* May 1981, p 62.
Ref. 1508: John Nagle - K4KJ, "Testing Baluns," *Ham Radio,* Aug 1983, p 30.
Ref. 1509: Lewis McCoy - W1ICP, "Is a Balun Required?" *QST,* Dec 1968, p 28.
Ref. 1510: R. H. Turrin - W2IMU, "Application of Broadband Balun Transformers," *QST,* Apr 1969, p 42.
Ref. 1511: Bruce Eggers - WA9NEW, "An Analysis of the Balun," *QST,* Apr 1980, p 19.
Ref. 1512: William Fanckboner - W9INN, "Using Baluns in Transmatches with High-Impedance Lines," *QST,* Apr 1981, p 51.
Ref. 1513: Doug DeMaw - W1FB, "A 'Multipedance' Broadband Transformer," *QST,* Aug 1982, p 39.
Ref. 1514: R. G. Titterington - G3ORY, "The Ferrite Cored Balun Transformer," *Radio Communication,* Mar 1982, p 216.
Ref. 1515: John S. Belrose - VE2CV, "Transforming the Balun," *QST,* Jun 1991, p 30.
Ref. 1516: D. DeMaw - W1FB, "How to Build and Use Balun Transformers," *QST,* Mar 1987, p 34.
Ref. 1517: J. Sevick - W2FMI, "The 2:1 Unun Matching Transformer," *CQ,* Aug 1992, p 13.
Ref. 1518: J. Sevick - W2FMI, "The 1.5:1 and the 1.33:1 Unun Transformer," *CQ,* Nov 1992, p 26.
Ref. 1519: W. Maxwell - W2DU, "Some Aspects of the Balun Problem," *QST,* Mar 1983, p 38.
Ref. 1520: R. Lewallen - W7EL, "Baluns: What they do and how they do it," *ARRL Antenna Compendium, Vol 1,* p 157.
Ref. 1521: Jerry Sevick, W2FMI, "The Ultimate Multimatch Unun," *CQ* mag, Aug 1993, p 15.
Ref. 1522: Jerry Sevick, "A Multimatch Unun," *CQ* mag, Apr 1993, p 28.
Ref. 1523: Jerry Sevick, "Dual-Ratio Ununs," *CQ* mag, Mar 1993 p 54.
Ref. 1524: Bill Orr, W5SAI "The Coaxial Balun," *CQ* mag, Nov 1993, p 60.
Ref. 1525: Jerry Sevick, "Ununs for Beverage Antenna," *CQ* mag, Dec 1993, p 62.
Ref. 1526: Jerry Sevick, "A Balun Essay," *CQ* mag, Jun 1993, p 50.
Ref. 1527: Jerry Sevick, "More On the 1:1 Balun," *CQ* mag, Apr 1994, p 26.
Ref. 1528: Jerry Sevick: "A Subsequent Look at 4:1 Baluns"; *CQ* mag, Feb 1994, p 28.
Ref. 1829: Jerry Sevick, "The 4:1 UNUN," *CQ* mag, Jan 1993, pp 30.
Ref. 1830: Jerry Sevick, *Transmission Line Transformers*, Noble Publishing, Atlanta, Ga.

16. ANTENNA MEASURING

Ref. 1600: R. D. Straw - N6BV - ed, *The ARRL Handbook for Radio Amateurs*, 2000, ARRL, Newington, CT.
Ref. 1601: John Schultz - W4FA, "The MFJ-202B R.F. Noise Bridge," *CQ,* Aug 1984, p 50.
Ref. 1602: Michael Martin - DJ7YV, "Die HF Stromzange," *CQ-DL,* Dec 1983, p 581.
Ref. 1603: Reginald Brearley - VE2AYU, "Phase-Angle Meter," *Ham Radio,* Apr 1973, p 28.
Ref. 1604: Robert Baird - W7CSD, "Nonresonant Antenna Impedance Measurements," *Ham Radio,* Apr 1974, p 46.
Ref. 1606: John True - W4OQ, "Antenna Measurements," *Ham Radio,* May 1974, p 36.
Ref. 1607: Frank Doting - W6NKU - et al, "Improvements to the RX Noise Bridge," *Ham Radio,* Feb 1977, p 10.
Ref. 1608: Charles Miller - W3WLX, "Gate-Dip Meter," *Ham Radio,* Jun 1977, p 42.
Ref. 1609: Kenneth F. Carr - WØKUS, "Ground Current Measuring on 160 Meters," *Ham Radio,* Jun 1979, p 46.
Ref. 1610: Forrest Gehrke - K2BT, "A Modern Noise Bridge," *Ham Radio,* Mar 1983, p 50.
Ref. 1611: William Vissers - K4KI, "A Sensitive Field Strength Meter," *Ham Radio,* Jan 1985, p 51.
Ref. 1612: Jerry Sevick - W2FMI, "Simple RF Bridges," *QST,* Apr 1975, p 11.
Ref. 1613: David Geiser - WA2ANU, "The Impedance Match Indicator," *QST,* Jul 1980, p 11.
Ref. 1614: Jerry Sevick - W2FMI, "Measuring Soil Conductivity," *QST,* Mar 1981, p 38.
Ref. 1615: Jack Priedigkeit - W6ZGN, "Measuring Impedance with a Reflection-Coefficient Bridge," *QST,* Mar 1983, p 30.
Ref. 1616: Doug DeMaw - W1FB, "Learning to Use Field Strength Meters," *QST,* Mar 1985, p 26.
Ref. 1617: Al Bry - W2MEL, "Beam Antenna Pattern Measurement," *QST,* Mar 1985, p 31.
Ref. 1618: M. R. Irving - G3ZHY, "The 'Peg Antennameter'," *Radio Communication*, May 1972, p 297.
Ref. 1619: Pat Hawker - G3VA, "Technical Topics: Current Probe for Open-Wire Feeders - Wire Antennas - etc," *Radio Communication*, Nov 1984, p 962.
Ref. 1620: D. DeMaw - W1FB, "A Laboratory Style RX Noise Bridge," *QST,* Dec 1987, p 32.
Ref. 1621: John Belrose - VE2VC, "RX Noise Bridges," *QST,* May 1988, p 34.
Ref. 1622: Andrew S. Griffith - W4ULD, "A Dipper Amplifier for Impedance Bridges," *QST,* Sep 1988, p 24.
Ref. 1623: John Grebenkemper - KI6WX, "Improving and Using R-X Noise Bridges," *QST,* Aug 1989, p 27.
Ref. 1624: D. DeMaw - W1FB, "MFJ-931 Artificial RF Ground," *QST,* Apr 1988, p 40.
Ref. 1625: Jim Smith, "How to Build a Snap On RF Current Probe," *CQ* mag, Aug 1996, p 22.
Ref. 1626: Gerd Janzen, DF6SJ/VK2BJZ "HF-Messungen mit einem aktiven Stehwellen-Messgeraet," ISBN 3-88006-170-X (This book is only available directly from the author, DF6SJ, Hochvogelstrasse 29, D087435 Kempten, Germany).

INDEX

[Note: A complete, word-by-word electronic Index is available using the software on the CD-ROM that is located in the back of this book.]

2-Element Broadside	11–37
2-Element End-Fire	11–36
3- and 4-Element Broadside	11–38
3-Element End-Fire	11–39
3B7RF	9–29
3C1BG	2–14
3Y1EE	1–32
3Y0PI	2–13
40-Meter	
End-Fire Array	14–6
Half-Square	14–5
Quad, Reduced Sized	13–49
Yagi, Lightweight	13–17
40-Meter Yagi, Lightweight	
Electrical Performance	13–17
Mechanical Design	13–18
Tuning	13–19
4Nec2	4–3, 7–47, 7–50
4X4NJ	1–34, 7–67
5-Rectangle Array	11–55
5-Square Array	11–57
5B4ADA	1–34
5N0MVE	10–4
6-Circle Array	11–60
8-Circle Array	7–33
80-Meter	
DX Special	8–9
End-Fire Array	14–8
Full-Sized Quad	13–44
Half-Diamond Array	14–8
Three-Element Quad	13–49
80-Meter Yagi Designs	13–20
K6UA 2-Element	13–25
Loaded	13–20
W7CY 2-Element	13–23
9-Circle Array	11–64
9M2AX	1–29

A

A Index (Geomagnetic)	1–19
AA1K	2–16
AA7QZ/2	9–55
AC6LA	4–4, 6–3, 6–4, 6–6
ACOM 2000A	15–6, 15–7
Active Antenna Elements	7–15
Activity	
10 Low-Band Commandments	2–34
Age	2–20
Awards	2–20
Survey	2–20
Time Spent on Each Mode	2–20
Adler, D.	9–19
AEA	11–25
Antenna Analyzer	11–32
CIA-HF	11–33
AG6K	3–16
AH1A	2–13
AI1H	8–9
AKI Special	9–73
Alpha	
87	15–7
91B	15–7
Power Analyzer	15–11
Alvestad, Jan	1–3
AMTOR	3–16
Andrews Corp	6–3
Antenna Model	4–2, 7–7, 7–8, 7–11
Antenna Solver	4–3
antennaeX	4–9
Antennas	15–4
Design Software	4–1
Ground Plane	9–9
Measuring Resonance	11–30
Over Ground	5–7, 8–3
Radiation and Reflection Efficiency	8–3
Reflection Coefficient	8–3
Receiving	7–1
Antennas and Equipment (Survey)	2–26
Antennas, Etc	6–19
Antipodal Focusing	1–17
Anzac	7–99
AO	4–8
Array Solutions	7–90, 7–99, 11–71, 15–5
Arrays	12–1
2-Element Broadside	11–37
2-Element End-Fire	11–36
3- and 4-Element Broadside	11–38
3-Element End-Fire	11–39
3-Element, Parasitic	12–8
5-Rectangle	11–55
5-Square	11–57
6-Circle	11–60
7-Circle	11–63
9-Circle	11–64
Bidirectional End-Fire	11–43
Broadside	7–10
Buried Radials	11–83
Curtain, Bobtail	12–13
Delta Loop	12–11
Designing	11–2
Dipole, 3-Element	12–7
Dipole, ZL Special	12–12
Elements	11–2
Elevated Radials	11–84
End-Fire & Broadside	11–40
Feeding the Elements	7–15
Half Square	12–15
Lazy H	12–13
Loops	7–94
Modeling	11–2
Popular	11–36

Sloping Verticals 11–78
Triangular 11–44
Two-Element 12–3
    Delta Loop 12–5
    Parasitic 12–4
    Unidirectional 12–4
ARRL Antenna Modeling Course 4–2
ARRL Laboratory 3–3
ARRL MicroSmith 6–10
ARRL Radio Designer Software 15–10
Aurora 1–17
    Australis 1–17
    Viewing from Satellites 1–20
Autek 11–27
Awards 2–17
    Standings 2–18
Azimuthal-Equidistant Projection 1–29

B

Balun 5–10, 6–18
    Current 6–19
    Voltage 6–19
Bandwidth 5–11, 8–8
    Continuously Variable 3–7
    CW 3–7
    SSB 3–7
Bandwidth (SWR) 5–11, 8–8
    Dipole 8–15
    Inverted-V 8–20
Barker 9–20
Base Loading 9–34
Battle Creek Special 9–62
BC Interference 11–29
Beverage Antennas 7–42
    Arrays 7–74
    Band Splitter 3–12
    Bi-Directional 7–68
    Bi-Directional Two-Wire 7–71
    Cone-of-Silence 7–47
    Feeding 7–56
    Modeling 7–43
    Slinky 7–69
    Terminating 7–50
    Terrain and Layout 7–63
Beverage, Clarence 9–20
Beverage, Harold 7–42, 7–65
Bi-Square Loop 10–15
Bidirectional End-Fire Array 11–43
Binomial Current Distribution 11–14
Bobtail Curtain 12–13
BOG 7–65
Boom Strength Program 4–7
Braun 3–11, 7–99
Break-In
    Semi, CW 3–18
Brewster, Pseudo 9–2
Broadside Arrays 7–10, 7–20
Broadside/End-Fire Array 11–54
Broadside/End-Fire Arrays 7–20
Brown, Lewis & Epstein 9–14
Brown, Robert 1–1
BS7H 2–2
BV9P 2–2

C

C31XR 15–4
Cable Loss 11–32
Cadweld 9–18
Capacitive Loading 8–14
Capacitively Loaded
    Delta Loop 14–9
    Diamond 14–8
Caron 6–10
Center Loading 9–36
CE0Y 1–32
Ceramic Magnetics 7–57
Chordal Hops 1–16
Christman (K3LC) 11–8
Coax Cable 6–2
    and SWR 6–2
    Attenuation Characteristics 6–3
    Size 6–7
    Stubs 15–10
    Transformer Program 4–5
Cohen, Ted 1–2, 1–15, 1–18
Coil Calculation Program 4–6
Coils
    Making 13–21
Collinear 8–17
Collinear Dipoles 8–17
Collins (W1FC)
    Hybrid Coupler 11–17
Complex Numbers 11–3
Component Ratings
    Capacitors 6–11
    Coils 6–12
Computers
    Connecting Them Together 15–12
    DX-Cluster 15–12
    Interface to Radio 15–7
    Noise 15–12
    Software 15–12
Comtek 11–21, 11–51, 11–76
Conductivity and Dielectric Constants (Earth) ... 5–8
Cone-of-Silence Length 7–47
Conjugate Match 6–2
Contesting 15–1
    What Category? 15–3
    Why Contesting? 15–3, 15–4
Continuously Variable IF Bandwidth 3–7
Coordinates
    Polar & Rectangular 11–2
Corona Discharges 7–3
Coronal Mass Ejections 1–18
Coupled Impedance 11–4
Coupling, Mutual 11–4
Creative Design Co. 13–52
Critical Frequency 1–3
Crooked Paths 1–31
Cross-Fire (W8JI) 11–9
    Feeding 7–17
CT 3–17
CT1BOH 15–2
Current
    Forcing 11–10
    Magnitude & Phase 11–4
Curtain, Bobtail 12–13

CW
- Bandwidth 3–7
- Keying Waveform 3–17
- Operation 3–17
- QSK 3–18
- Switchable Sideband 3–11

D

Declination 1–9
Decoupling Antennas 9–71
- Transmitting 7–92

Delta Loop 10–5
- Bottom-Corner-Fed 10–6
- Compressed 10–6
- Dimensions 10–9
- Equilateral Triangle 10–6
- Feeding 10–9
- Gain and Radiation Angle 10–10
- Horizontal Polarization 10–8
- Horizontal vs Vertical Polarization 10–9
- Radiation Patterns 10–6
- Receiving Loop 7–88
- Supports 10–10
- Two at Right Angles 10–10
- Vertical Polarization 10–6
- Vertical vs Horizontal Polarization 10–9

DF2PY 1–36
DF6SJ/VK2BJZ 8–10
Diamond Receiving Loop 7–89
Dipole 8–1, 8–12
- Bandwidth 8–8, 8–15
- Broadband 8–9
- Calculating Coil Value 8–13
- Center Loading 8–11
- Center-Fed 8–7
- Coil Loading 8–11
- Collinear 8–17
- Efficiency 8–6, 8–15
- Feeding 8–7
- Half-Wave Over Ground 8–1
- Inverted-V 8–18
- Inverted-V, Bandwidth 8–20
- Inverted-V, Height 8–20
- Inverted-V, Length 8–20
- Inverted-V, Radiation Patterns 8–20
- Linear Loading 8–14
- Loading Coils, Away From Center 8–12
- Loading, Capacitive 8–14
- Loading, End 8–14
- Loading, Losses 8–12
- Long 8–15, 8–17
- Long, Feed-point Impedance 8–17
- Losses 8–7, 8–12
- Losses, Dielectric 8–7
- Losses, Ground 8–7
- Matching 8–11
- Modeling 8–25
- Over Real Ground 8–3
- Quarter-Wave Vertical 8–24
- Radiation and Reflection Efficiency 8–3
- Radiation Patterns 8–5, 8–15
- Radiation Resistance 8–6
- Ray Analysis 8–2
- Resonant Frequency vs SWR 8–10
- Resonant vs Nonresonant 8–9
- Shortened 8–10
- Shortened, Efficiency 8–15
- Shortened, Principles 8–10
- Shortened, Radiation Resistance 8–10
- Shortened, Tuned Feeders 8–11
- Shortened, Tuning and Loading 8–10
- Sloping 8–22
- Sloping, Bent-Wire 8–23
- Sloping, Ground Locations 8–3
- Three-Band Antenna 8–18
- Topband, Low 14–4
- Vertical 8–20
- Vertical Radiation Pattern Equations 8–2
- Vertical Radiation Pattern of Horizontal Dipole 8–1
- Vertical, Feeding 8–22
- Vertical, Radiation Resistance 8–21
- Zepp, extended double 8–17

Directivity 5–2, 5–6
- Importance of 5–10
- Merit Figure 5–10

Directivity Merit Figure (DMF) 7–8
DJ2YA 9–1
DJ4AX 1–19, 8–1
DJ4PT 13–49
DJ6JC 13–49
DJ6QT 9–31
DJ7AA 1–19
DJ8WL 1–11, 2–8, 9–32
DK7PE 1–10, 9–61
DL2CC 15–1, 15–7
DL8WPX 2–16
Driving Impedance Program 4–5
DSP 3–8, 3–10
- Audio 3–10
- IF 3–10
- Transmitter 3–19

DU7CC 9–17
DX Atlas 1–25, 1–26, 1–39
DX Clusters 2–15
DX Engineering 7–36, 7–73, 7–76, 7–99, 9–17
DX Windows 2–6
DX-Aid 1–13, 1–25, 1–29, 1–35, 1–37
DX-Edge 1–26
DXing
- 40 Meters 2–7
- 8 Commandments 2–17
- 80 Meters 2–6
- Arranging Skeds 2–15
- Awards 2–17
- Calling CQ 2–12
- Calling the DX Station 2–12
- CW 2–14
- DX Bulletins 2–16
- DX Clusters 2–15
- DXpeditions 2–11
- European 160-Meter Frequencies 2–5
- IARU Bandplan 2–3
- Internet Chat Channels 2–16
- Internet Reflectors, DX Magazines 2–16
- ITU Allocations 2–2
- Myths 2–2
- Nets and Lists 2–14

RIT 2–10
   Split-Frequency Operation 2–7
   Subbands in 160-Meter Contests 2–9
   Zero Beat 2–10
DXpeditions 2–11
   Controlling Pileups 2–12
   DXA 2–13
   Frequencies 2–11
   Pilot Stations 2–13
   Split Frequency 2–12
DXWin 1–26
Dynamic Range 3–5

E

EA3VY 7–87
EAI/TIA-222-E 4–7
Easy-Tune Power Amplifiers 15–7
Efficiency 5–8
   Antenna 5–8, 5–9, 8–6
   Radiation, Monopole 9–8
   Shortened Dipole 8–15
EIA/TIA-222-E 13–6
Eisenstein 9–35
Elements
   Base Loaded 7–19
   Construction 11–71
   Inverted-L 11–72
   Loaded 11–72
   Loading for CW 11–74
   Mechanical Considerations 11–71
   Strength Program 4–7
   T-Loaded 11–72
   Taper Program 4–7
Elevated Radials 9–10, 11–84
Elevation Angles 5–2
   160 Meters 5–4
   40 Meters 5–4
   80 Meters 5–4
ELNEC 4–1, 8–11
End Loading 8–14
End-Fire/Broadside Combination 7–11
ET3PMW 1–10
European 160-Meter Frequencies 2–5
EWE Antenna 7–86
EZNEC 4–2, 4–3, 5–6, 7–9, 7–11, 8–11, 8–26, 10–11, 11–2, 11–77, 11–84, 13–2, 13–17
EZNEC Pro 4–3, 13–25

F

Fair-Rite 7–58, 7–73
Far Field 5–9
Feed Lines 6–1
   75 Ω Cables in 50 Ω Systems 6–17
   Attenuation Characteristics 6–3
   Coaxial Cable 6–2
   Conjugate Match 6–2
   Connectors 6–20
   Loss Mechanism 6–4
   Matching Networks 6–7
   Open Wire 6–4
   Size 6–7
   SWR 6–2
   Tuned Feeders 8–11
   Universal Transmission-Line Program 6–5
Feed System Design 11–7
Feed-Line Analysis Program 4–8
Feeding
   Cross-Fire 7–17
Filters
   Audio 3–10
   Group Delay 3–8
   High-Power 15–9
   IF Position 3–9
   Shape Factor 3–8
FK8CP 1–14, 1–29
Flag Antenna 7–87
Flagpole Vertical 14–2
Flat Lines 11–9
FOØAAA 7–88
FOT 1–5
Four-Square Array 11–45
   "Small" 11–50
   160-Meter 11–83
   8 Directions 11–51
   Direction Switching 11–50
   Feeding 7–23
   K8UR Sloping Dipoles 11–79
   Lahlum/Lewallen Feed 11–47
   Quadrature Fed 11–14
   Receiving Array 7–21
   Sloping Halfwave Dipoles 11–78
   Sloping Quarter-Wave 11–81
   W8JI Cross-Fire Feed 11–47
   WA3FET-Optimized 11–47
Fraunhöfer 5–9
Free Space 5–7
Frequency
   Map 1–6
   Optimum Traffic 1–5
Front-to-Back Ratio 5–9
   Average 5–9
   Gain 5–10
   Geometric 5–9
   Worst-Case 5–9
FT-1000D 3–13
FT-1000MP 3–7, 3–13
FT5ZB 1–17
Further Improvements 15–13

G

G3FPQ 13–49, 15–2
G3PQA 1–17
G3SJX 3–9, 3–13
G4DBN 1–33
Gain 5–9
   Compression 3–5
   vs Directivity 7–2
Gamma Match 9–65, 13–35
Gamma, Omega, Hairpin Program 4–6
Gehrke (K2BT) 11–18
Geochron 1–26
GeoClock 1–26
GM3HAT 8–9
GM3POI 1–14
GNEC 4–3
Gray Line 1–11
Great Circle Map 1–6, 1–29

Ground
    Loss 9–9
    Measurement 9–17
    Modeling vs Measuring 9–19
    Plane Antenna 9–9
    Reflection Coefficient 9–2
    Rods 9–17
GW3YDX 1–14
GW4BLE 15–2
Gyrofrequency, Electron 1–27

H

Hagn 9–20
Hairpin Match 13–28
    Design Guidelines 13–40
    Element Loading 13–40
    Element-to-Boom Capacitance 13–42
Half Diamond 14–8
Half Loops 14–4
Half Slopers 14–4
Half Square 12–15
Half-Wave Vertical
    Radial System 9–50
HB9CV 12–12
Headphones 3–16
Heil 3–17
Henney 9–8
HFTA 5–5, 5–6, 8–3
High-Power Filters 15–9
Highest Possible Frequency (HPF) 1–5
Horizontal Directivity 5–2
HP-8405 7–25
HP-8405A 11–25
Hy-Gain 402BA 9–37
Hybrid Coupler 11–17
    Collins 11–20
    Performance 11–21
    Phase Measurement 11–27

I

I4JMY 7–41
I8UDB 8–22
IARU Bandplan 2–3
Ice Loading 13–13
IK4AUY 7–99
Impedance
    Bridge 11–34
    Coupled 11–4
    Drive 11–5
    Self 11–4
Impedance, Current & Voltage Program 4–5, 6–13
In Search of Excellence 15–1
Induction Field 5–9
Intermodulation Distortion 3–3
International Radio 3–1, 3–8, 3–11, 3–18
Interplanetary Plasma 1–18
Inverted-L Antennas 9–72
Inverted-V 8–18
    Bandwidth 8–20
    Height 8–20
    Length 8–20
    Radiation Patterns 8–20
    Radiation Resistance 8–18

IONCAP 5–2
Ionoprobe 1–2
Ionosondes 1–3
Ionosphere
    Chordal Hops 1–16
    D-Layer 1–7
    E-Layer 1–7
    Path Deviation Mechanism 1–31
    Signal Ducting 1–15
    Sunspot Cycles and Ducting 1–16
Ionospheric Radio 1–15
Isotropic 5–7
ITU Frequency Allocations 2–2
IY4FGM 15–2

J

JA3ZOH 15–2
JA5AQC 2–16
JA5DQH 9–73
JA7AO 2–18
JE3HHT 4–2
JJ1VKL/4S7 1–29
JT1CO 1–29, 2–19

K

K Index 1–19
K1FZ 7–73
K1KI 5–5
K1KP 3–11
K1TD 9–34
K1TTT 15–12
K1UO 1–19, 2–19
K1VR 13–33
K1VR/W1FV Spitfire Array 13–33
K1XX 6–17
K1ZM 1–38, 2–2, 2–19
K2BT 11–4, 11–5, 11–18, 11–32
K2UO 2–2, 14–1
K3BC 12–15
K3BU 1–15, 14–2
K3LC 9–22, 11–4, 11–8, 11–52
K3LR 5–1, 13–1, 13–31, 13–33, 15–2
K3NA 9–15
K3OQF 9–65
K3UL 2–19
K3ZAP 9–18
K4ERO 13–28
K4HKX 10–1
K4JA 7–96, 11–64
K4JA 9-Circle Array 11–66
K4PI 11–32, 11–79
K4TO 1–39
K4VX 3–11, 5–1
K5IU 9–22, 9–25, 13–3, 13–4
    Unequal Radial Currents 9–25
K6LL 6–2
K6SE 1–43, 2–1, 7–84, 7–87, 9–51, 10–10
K6STI 4–4, 4–8, 5–5, 8–3, 13–22
K6UA 1–11, 13–25, 13–32
K7CA 7–36, 7–39, 9–64, 13–33, 14–10
K7EM 9–55
K7NV 13–3
K7ZV 2–16, 13–21, 13–22, 13–52

K8BHZ 7–19, 7–28
   Element 7–19
   Stone-HEX Array 7–28
K8CFU 9–8, 9–19
K8RN 2–14
K8UR 11–79, 11–80
K9AY 7–84, 7–89, 7–96, 14–10
K9DX 1-32, 4-5, 7–53, 7–78
                          7–96, 9–65, 11–1, 11–64, 14–10
K9DX 9-Circle Array 11–66
K9HMB 2–19
K9LA 1–11
K9RJ 2–19, 9–71
KB8I 12–7
KB8NW 2–16
KBØMPL 1–26
KH6DX/W6 2–2
KH7M 15–2
KI6WX 11–32
KK6EK 2–13
KL7RA 1–14
KL7U 1–34
KL7Y 1–34, 1–39
KLM 9–37
KM1H 7–69
KM5KG 4–8, 6–7
   Professional RF Network Designer 4–8
KN4LF 1–31
KN4RID 2–1
KØCKD 2–6
KØHA 1–36, 13–27
KØNM 8–10
KT-34XA 9–65, 15–4
Kurze Antennen 9–33, 9–39
KWM-2 3–13

L

L Network 6–9, 11–10
   Design, Program 6–10
   Program 4–5
Lahlum (W1MK) 11–11
Lahlum (W1MK)/Gehrke (K2BT) 11–18
Lahlum.xls 11–11, 11–14, 11–23, 11–36, 11–48, 11–64, 11–67
Laport 9–8
Lazy H 12–13
Lewallen Phasing Feed System 11–76
Line Stretcher Program 4–6
Linear Amplifiers 3–16
Loading
   Base 9–34
   Capacitance Hat 9–41
   Capacitance Hat Plus Coil 9–44
   Center 9–36
   Dipoles 8–12
   High Radiation Resistance 9–37
   Linear 8–14, 9–37
   Losses, Keeping Low 9–37
   Sloping Wire 9–47
   T-Wire 9–44
   Top 9–35
   Top and Base 9–37
   Tower on 80/160 14–3
Loops
   Arrays 7–94

Bi-Square 10–15
Delta
Delta Array 12–5, 12–11
Delta-Shaped 7–88
Diamond-Shaped Receiving 7–89
Half Loop 10–16
Half Loop, High Angle 10–17
Half Loop, Low-Angle 10–16
Loaded 10–12
Loaded, Capacitive 10–13
Loaded, Inductive 10–12
Modeling 10–11
Quad 10–1
Receiving 7–84
Rectangular (Flag) 7–87
Reduced Size 10–14
Triangular (Pennant) 7–88
Loss
   Cable 11–32
   Ground 9–9
Lossy Elements 7–14
Low-Band Software 4–4

M

Magnetic Disturbances 1–17
Magnetosphere 1–18
Masts 13–15
   Bearings 13–15
Matching 13–43
   Direct Feed 13–43
   Gamma 13–35
   Hairpin 13–38
   Hairpin, Design Guidelines 13–40
   Hairpin, Element Loading 13–40
   Hairpin, Element-to-Boom Capacitance 13–42
   Networks 6–7, 6–12
      Broadband Matching 6–20
      High Impedance 6–14
      L Network 6–9
      Quarter-Wave Sections 6–7
      Replacing Stub 6–12
      Wideband Transformers 6–16
   Omega 13–37
Median Statistical Value 1–5
Merit Figure 5–10
MFJ 11–25
MFJ-1025 .... 7–3, 7–41
MFJ-259B Antenna Analyzer 11–33
MicroDEM 5–6
Microphones 3–16
Minimum Discernable Signal 3–2
MININEC 4–2, 7–43, 8–11, 8–17, 10–2, 11–2, 13–4
MiniProp 1–6
MMANA 4–2
MN8CX 7–57, 7–73
Modeling
   Dipoles 8–25
   Limitations 7–13
   Loops 10–11
   Yagis 13–2
Modified Lewallen Method 11–16
Monopoles
   Radiation Resistance 9–5
Morris, I. L. 4–1

Moxon 9–27
MUF 1–2
MultiNEC 4–4
Mutual Coupling 11–4, 11–34
Mutual Impedance Program 4–5

N

N1DM 11–25
N2FB 4–1
N2MF 9–13
N2MG 2–5
N2PK 11–34
N2PK VNA (Vector Network Analyzer) 11–34
N4KG 1–43, 2–14, 9–27
N4KG Elevated Radials 9–27
N4OE 6–18
N4UH 6–5
N4UK 1–36
N4XX 1–2
N5FG 2–17
N5JA 1–36
N6BV 3–15, 4–1, 5–2, 5–5, 6–4, 8–3, 9–11, 9–15
N6LF 8–9
N6PL 9–9
N6TR 1–22
N7AM 1–33
N7AU 1–19
N7CL 9–12, 9–20
N7JW 7–36, 7–39, 9–64, 13–33, 14–10
N7UA 1–29, 1–33
N8TR 4–1
NA 3–17
NA0Y 1–19
Near Field 5–8
NEC-2 .... 4–2, 4–3, 7–43, 8–11, 8–17, 10–2
NEC-3 11–2
NEC-4 ..... 4–3, 7–43, 9–11, 9–19, 11–2, 11–6, 11–80, 13–25
NEC-GSD 9–19
NEC-Opt 4–4
NEC-Win Plus 4–3
NEC-Win Pro 4–3
NEC4WIN95 4–2
Nets and List Operations 2–14
Network Analyzer 11–32
NI6W 13–3
Nittany Scientific 4–3
NM7M 1–1, 1–16, 1–17, 1–38, 1–39
NOAA 1–21
Noise 7–2
   Atmospheric 1–27
   Blanker 3–13
   Bridge 11–32
   Computer 15–12
   External to Receiver 3–2
   Man-Made 7–4
   Precipitation 1–27
   Propagated 7–4
   Thermal 3–2
   Thunderstorms 7–3
Noise-Canceling Bridge 7–41
NR1DX 2–17, 7–42
NTIA/ITS 1–3
Null Angles 7–6
NW3Z 13–17

O

OH2BH 13–23
Oler, Cary 1–2, 1–15, 1–18, 1–23
Omega Match 9–65, 13–37
ON4GV 15–2
ON4HW 1–43
ON4KST 2–16
ON4UN
   Coax Transformer Program 6–5
   Mini Receiving Four Square 7–27
   Wire Four-Square 11–72
ON4WW 2–13, 15–1
ON5OO 15–12
ON6TT 15–1
ON6WU 4–6, 7–19, 9–61, 11–71, 13–1, 13–3
ON7PC 4–1
Open-Wire Transmission Line 6–4
Operating
   Interband Interference 15–8
   Monitor Scopes 15–8
   Operating Table 15–8
ORION 3–8, 3–9, 3–10, 3–14, 3–17, 3–19, 15–6
OZ8BV 1–33

P

P29XX 2–16
PA3DZN 9–65
PACTOR 3–16
Palomar PCM1 9–27
Paragon Technologies 4–4
Parallel Impedance Program 4–6
Parasitic Receiving Arrays 7–36
Passband Tuning 3–7
Pattern
   Vertical, Ideal Ground 9–1
Pennant Antenna 7–88
Physical Design of Yagi Antennas 13–2
PI4COM 15–2
Pilot Stations 2–13
Polar Coordinates 11–2
Popular Arrays 11–36
Power Amplifiers
   Easy-to-Tune 15–7
Power, Wasted 11–74
Preamplifiers 7–97
Precipitation
   Noise 1–27
   Static 7–3
Professional RF Network Designer 4–8, 4–9
Propagation
   160 Meters 1–45
   40 Meters 1–45
   80 Meters 1–45
   A Index 1–19
   Antipodal Focusing 1–17
   Aurora 1–17
   Aurora from Satellites 1–20
   Calculating Half-Way Local Midnight Peak 1–42
   Chordal Hops 1–16
   Coronal Mass Ejections 1–18
   Critical Frequency 1–3
   Crooked Path 1–31
      Avoiding the Auroral Zones 1–35

  Between Europe and Alaska 1–34
  Examples 1–32
  Midwinter 1–36
  New England to Japan 1–38
 D-Layer 1–7
 Dusk and Dawn 1–11
 E-Layer 1–7
 Frequency Map 1–6
 Frequency of Optimum Traffic (FOT) 1–5
 Gray Line 1–11
 Great Circle Long Path 1–29
  160 Meters 1–29
  40 Meters 1–29
  80 Meters 1–29
 Great Circle Short Path 1–29
 Gyrofrequency, Electron 1–27
 High Wave Angles at Sunrise/Sunset 1–16
 Highest Possible Frequency (HPF) 1–5
 Interpreting Geomagnetic Data 1–22
 Ionoprobe 1–2
 Ionospheric Signal Ducting 1–15
 K Index 1–19
 Location 1–17
 Low band, during high sunspot years 1–8
 Median Statistical Value 1–5
 Midway Midnight Peak 1–11
 MUF 1–2
 Nighttime 1–11
 ON4UN Software 1–44
 Paths 1–29
 Polarization and Power Coupling on 160 Meters 1–27
 Pre-Sunset and Post-Sunrise QSOs 1–42
 Smoothed Sunspot Number 1–2
 Software 1–43
 Solar Flux Index 1–2
 Solar Terrestrial Report 1–3
 Sunrise/Sunset Tables 1–40
 Sunspot Cycle 1–2
 Sunspot Cycles and Ducting 1–16
 Wave Elevation Angle 1–3
 Without Lossy Ground Reflections 1–14
Propagation Software 4–5
Proplab Pro 1–6, 1–15, 1–25
Pseudo-Brewster Angle 9–3
 Equation 9–3
 Radials 9–4
PSK31 2–14
PY1RO 1–29
PY5EG 15–2

Q

Q-Factor 5–11
QRM 7–3
QRN 7–3
QSK 3–18
Quads 13–43
 2-Element, Reduced Size 13–49
 3-Element 80-Meter 13–49
 3-Element, 80-Meter 13–49
 80-Meter 13–44
 Dimensions 10–5
 Feeding 10–5
 Horizontal Polarization 10–2
 Impedance 10–2
 Patterns 10–2
 Quad or Yagi? 13–51
 Rectangular Quad Loop 10–4
 Vertical Polarization 10–2
 W6YA 40-Meter 13–50

R

RA3AUU 15–1
RadCom 3–13
Radials
 Buried 11–83
 Buried, Depth 9–17
 Buried, How Many 9–13
 Buried, What Shape 9–13
 Elevated 9–10, 11–84
 Elevated and On-Ground 9–29
 Elevated, Grounded Tower 9–28
 Elevated, How Many 9–21
 Elevated, Modeling 9–19
 Evaluating a System 9–31
 Plow 9–18
 Single 9–21
Radiation
 Angle Horizontal Antennas, Program 8–3
 Angles 5–2
 Efficiency, Monopoles 9–6
 Horizontal Pattern, Horizontal Dipole 8–6
 Patterns 8–5, 11–2
 Resistance 5–8, 8–6
  in Shortened Dipoles 8–10
  Monopoles 9–5
Radio Engineering Engineering 9–8
Radio Engineering Handbook 9–8
Radio-Computer Interface 15–7
Ray Analysis 8–2
RDF 7–9
Real Ground, Reflection Coefficient 8–3
Receiver
 Areas for Improvement 3–14
 Blocking 3–5
 Blocking Dynamic Range 3–6
 DSP 3–10
 Dynamic Range 3–5
 External Noise and Sensitivity 3–2
 Filter Shape Factor 3–8
 Gain Compression 3–5
 Input Protection 3–13
 Intermodulation 3–3
 Intermodulation Dynamic Range 3–5
 Minimum Discernable Signal 3–2
 Noise 3–2
 Noise Cancelling 3–14
 Passband Tuning 3–7
 Reciprocal Mixing 3–6
 Selectivity 3–7
 Sensitivity 3–1
 Specifications 3–1
 Stability/Frequency Readout 3–11
 Thermal Noise 3–2
Receiver_Levels.xls 3–2
Receiving 5–6
 Antennas 7–1
 Directive Arrays 7–5
 Equipment 3–1

Four Square 7–21
Loops 7–84
Loops, Modeling 7–85
Receiving Directivity Factor (RDF) 7–9
Reciprocal Mixing 3–6
Rectangular
    Coordinates 11–2
    Loop (Flag) 7–87
Redcliff, R. 9–19
Reflection Coefficient 8–3
    Ground 9–2
Resnick, A. 9–19
Rhein-Ruhr DX Association 8–1
Ringing 3–8
RIT 2–10
Roofing Filter 3–5
Rotators 13–15
    Drive Shaft 13–15
    Drive Shaft Calculations 13–16
RS-222-E 13–2

S

S21XX 2–16
S21ZC 1–10
Schultz 9–7
Selectivity 3–7
Selectivity, Static and Dynamic 3–8
Self-Impedance 11–4
Seven-Circle Array 11–63
Short Verticals 9–33
Sideband
    Selectable CW 3–11
Signal Ducting 1–15
Signal Monitoring 3–19
Sky & Telescope 1–23
Sloper
    Half 10–18
Sloping
    Bent-Wire 8–23
    Dipole 8–22
    Radiation Patterns 8–22
    Terrain 5–5, 8–3
SM3CVM 1–33
SM4CAN 1–33
SM4HCM 1–33
SM5MX 2–14
SM6CNS 9–17
Smith Chart Program 4–5
SMØAGD 1–32
Smoothed Sunspot Number 1–2
Snake Antenna 7–65
Software 6–13
    Antenna Design 4–1
    Antenna Optimizing 4–4
    Coax Transformer/Smith Chart 6–5
    Contesting 15–12
    DX Atlas 1–26
    DX-Aid 1–25
    DX-Cluster 15–12
    HFTA 5–6
    Ionoprobe 1–2
    L-Network Design 6–10
    Low Band Software 1–44
    MicroDEM 5–6

    MININEC-Based 4–2
    MiniProp 1–6
    NEC-2 4–2
    NEC-4 4–3
    ON4UN Low-Band 4–4
    Proplab Pro 1–6, 1–25
    Radiation Angle Horizontal Antennas 8–3
    Stub Matching 6–12
    Sunrise/Sunset Times 1–44
    TA 5–5
    The Database 1–44
    TLDetails 6–4
    TLW 6–4
    W6ELProp 1–25
    XLZIZL 6–6
    Yagi Design 13–4, 13–6
    YT 5–5
    YW 13–4
Solar
    27-Day Cycle 1–8
    A Index 1–19
    Coronal Mass Ejections 1–18
    Daily Cycle 1–10
    Flux Index 1–2
    K Index 1–19
    Seasonal Cycle 1–9
    Smoothed Sunspot Number 1–2
    Solar Flux Index 1–2
    Solar Terrestrial Report 1–3
    Sunspot Cycle 1–2
Solar Terrestrial Report 1–3
SP5EWY 10–19
SP7HT 3–6
Special Receiving Antennas 14–9
Speech Processing 3–17
Splatter 3–15
Split-Frequency Operation 2–7
    160 Meters 2–8
    40 Meters 2–9
    80 Meters 2–9
    DXpeditions 2–12
SSB
    Selectivity 3–7
    Splatter 3–15
SSN 1–2
St. Elmo's Fire 7–3
Stability 3–11
Standing-Wave Ratio (SWR) 5–10
    Bandwidth 5–11
    Feed Lines 6–1
    Need for Low SWR 6–18
STD Aurora Monitor 1–23
Stone-HEX Array
    K8BHZ 7–28
Stub Matching 6–12
    Program 4–6, 6–12
    Replacing Stub with Discrete Component .. 6–12
    Series-Connected Discrete Components 6–13
Stubs 11–33
Sunrise/Sunset Tables 1–40
    Using 1–42
Sunspot Cycle 1–2
    27-Day Cycle 1–8
    Seasonal Cycle 1–9

Smoothed Sunspot Number 1–2
Surge Impedance 5–12
SWARM 1–24
SWIM 7–24
Switching at ON4UN 15–5
SWR Bandwidth, Short Verticals 9–39
SWR Program 4–6

T

T Antennas 9–74
TA 5–5, 8–3
Taper Technique 13–44
Tektronix 3–19
Terrain, Sloping 5–5
Test Equipment 11–25
The ARRL Antenna Book 1–4, 5–1, 5–9, 6–3,
                                                     11–3, 11–7, 11–47, 11–71
The ARRL Handbook 3–16
The Operator's Station 15–5
The Radio Works 6–19
The RF Connection 9–18
The Wireman 6–19
Three-Band Antenna (40, 80 160 Meters) 8–18
Three-in-Line End-Fire
    Binomial Current Distribution 11–14
Thunderstorms 7–3
Titanex 11–71
Titanex V160E 9–29, 9–77
TL-922 3–16
TLDetails 6–3, 6–4
TLW 6–4, 13–4
Top Loading 9–35
Top Ten Devices 3–1
Toroid Cores 6–12
Towers 13–14
    Guys Wires as Antenna Elements 13–28
    Raising Antennas 13–17
    Rotating Masts 13–15
    Rotators 13–15
TR 3–17
Transformers, Wideband 6–16
    High Impedance 6–17
Transmission Line Transformers 6–17
Transmitter
    Areas for Improvement 3–20
    Power 3–15
Transmitting 5–1
    Equipment 3–1
Triangular
    Arrays 11–44
    Loop (Pennant) 7–88
TS-850 3–11
TTFD 6–18
Tuned Feeders 8–11

U

UA3AVR 4–3
UA9UCO 1–29
UU4JMG 1–36
UUØJZ 1–42

V

VE1ZZ 1–13, 1–17, 1–36, 2–11, 2–19
VE2CV 9–19, 9–20, 9–27, 10–16, 11–32, 12–12
VE3BMV 1–15
VE3EJ 15–2
VE3NEA 1–2
VE6LB 2–2
VE6WZ 1–36, 13–22
VE7DXR 1–16
VE7VV 1–14, 1–43
Vertical
    Directivity 5–2
    Pattern, Ideal Ground 9–1
Vertical Antenna Handbook 9–9
Vertical Antennas 9–1
    80/160 With Trap 9–57
    Battle-Creek Special 9–62
    Close-Spaced Short 9–75
    Close-to-Earth Elevated Radials 9–10
    Commercial 9–79
    Designing Short 9–40
    Dipole 8–20
        Feeding 8–22
        Radiation Patterns 8–20
        Radiation Resistance 8–21
    Duo-Bander 80/160 9–55
    DXpedition 9–77
    Flagpole 14–2
    Folded Elements 9–38
    Ground & Radial Systems 9–9
    Location 9–76
    Modeling 9–51
    On-Ground Radials 9–10
    ON4UN 160 M 9–60
    Practical 9–53
    Practical, Top Loaded 9–55
    Quarter-Wave 9–1
    Radiation Pattern
        Monopole, Real Ground 9–2
        Sloping Ground 9–5
    Short 9–33
    Short, SWR Bandwidth 9–39
    Tall 9–49
    Using Beam/Tower 9–64
    Yagis 13–27
    3-Element 13–30
VIA-Bravo 11–33
VK1AA 2–18
VK2AVA 1–42
VK5KC 4–2
VK6HD 1–29
VK8XY 2–16
VK9CR 2–16, 9–77
VK9LX 2–14
VKØGC 1–42
VKØHI 2–13
VKØIR 1–12, 1–30, 1–35, 2–13, 2–15, 9–17, 9–38
VOACAP 1–3, 1–26, 5–2
VP6DIA 1–39
VS6DO 1–29
VU2PAI 2–7, 2–17
VY1JA 1–19

W

- W1AW 2–6
- W1BB 2–17
- W1CF 11–44
- W1FB 9–65, 10–18, 11–32
- W1FC 4–6, 11–17
- W1FV 2–19, 7–74, 9–75, 13–33
- W1HGT 2–17
- W1HKK 11–45
- W1JZ 2–19
- W1MK 7–25, 11–1, 11–10, 11–16, 11–25,
  11–27, 11–30, 11–37, 11–69, 11–86
  - Dual-Channel RF Detector 11–25
  - Hybrid Coupler 11–31
- W1NCK 6–10
- W2BML 7–42
- W2DU 6–18, 9–65
- W2FMI 6–17, 6–19, 9–37
- W2IHY 3–16, 3–17
- W2PV 13–2, 13–35
- W2VJN 3–1, 3–18
- W3ESU 9–19
- W3LPL 3–11, 5–1, 6–4, 15–2
- W3NQN 3–12, 7–25, 11–24, 11–30, 15–8
- W3UR 2–16
- W4DR 2–11, 2–18
- W4PA 2–15
- W4PK 9–27
- W4RNL 13–23
- W4ZCY 5–1
- W4ZV 1–2, 1–12, 1–22, 1–34, 2–1, 2–8, 2–18, 10–6
- W5XD 15–12
- W6ANR 13–20, 13–52
- W6BXI 11–32
- *W6ELProp* 1–25
- W6KW 13–20
- W6NKU 11–32
- W6NLZ 1–11
- W6QHS 13–2
- W6RJ 1–35, 13–22, 14–10
- W6SAI 13–36
- W6TC 6–19
- W6YA 13–50, 14–9
- W6YA 40-Meter Quad 13–50
- W7CY 13–23, 13–32
- W7EL 8–26, 11–4, 11–7, 11–10, 13–44
- W7IUV 7–87, 7–98
- W7IV 9–18
- W7LR 6–10, 9–17
- W7OM 1–11
- W7RM 2–1
- W7XC 9–17, 9–62
- W7ZOI 6–10, 7–99
- W8FYR 8–15
- W8JI 1–29, 2–8, 3–6, 5–10, 7–1, 7–33, 7–53,
  9–16, 11–4, 11–48
  - 8-Circle Array 7–33
  - Cross-Fire 11–9
  - Elements (Umbrella Loading) 7–18
  - Parallelogram 7–81
  - Receiving Directivity Factor 7–9
- W8LRL 2–7, 2–17, 2–18, 7–35, 7–79
- W8LT 1–14
- W9KYZ 8–15
- W9UCW 9–63
- WA1ION 7–92
- WA2WVL 7–86, 7–95
- WA3FET 11–16, 11–28, 11–47, 11–53, 13–17
- WA4JVE 10–1
- WA6TKT 12–12
- WA8MOA 9–62
- WA0RDX 10–1
- Wave elevation angle 1–3, 8–1
- WB4ZNH 2–14
- WB9Z 15–2
- Wellbrook Communication 7–90
- Windom 8–7
- Windom Antenna 8–7
- Wire Yagis 13–26
- W0CD 9–62
- W0JF 9–59
- W0UN 11–1, 11–64, 11–79
- W0YG 13–23
- W0ZV 1–17, 1–36, 3–10
- *Writelog* 3–17
- WRTC 15–1
- WX0B 7–90, 7–99, 11–71, 11–76, 13–3, 15–5, 15–8

X

- *XLZIZL* 6–6
- XR0Y/XR0Z 2–13
- XV7SW 2–14
- XY0RR 1–10
- XZ0A 1–17, 9–13

Y

- Yagi 13–2
  - 3-Element 40-Meter 13–3
  - 3-Element 40-Meter, Lightweight 13–17
    - Electrical Performance 13–17
    - Mechanical Design 13–18
    - Tuning 13–19
  - 80-Meter 13–20
    - K6UA 2-Element 13–25
    - Loading Coils 13–20
    - W7CY 2-Element 13–23
  - Alternative Element Designs 13–8
  - Boom Design 13–9
    - Boom Sag 13–10
    - Clamps 13–12
    - Torque Balancing 13–10
    - Weight Balancing 13–10
  - Commercial 13–51
  - Design Program 4–6, 13–4, 13–6
  - Element
    - Sag 13–7
    - Strength 13–6
  - Hairpin
    - Element Loading 13–40
    - Element-to-Boom Capacitance 13–42
  - Horizontal Wire 13–26
  - Ice Loading 13–13
  - K1VR/W1FV Spitfire Array 13–33
  - Loading Coils 13–21
  - Matching 13–14, 13–35
    - Direct Feed 13–43
    - Gamma 13–35

Hairpin	13–38
Omega	13–37
Material Fatigue	13–13
Mechanical	
Design	13–2
Load and Strength	13–4
Modeling	13–2
Rotating Mast Program	4–8
Torque Balancing Program	4–7
Tower Guys as Elements	13–28
Tubing	13–12
Vertical	13–27
Vertical, 3-Element	13–30
Weight Balance Program	4–7
Wind Area Program	4–7
Wind Load Program	4–7

YB1AQS	2–14
YO	4–4, 4–8
YT	5–5
YT6A	5–5
YW	13–4

Z

Zepp	8–17
Zero Beat	2–10
ZL Special	12–12
ZL7DK	1–1, 1–44, 2–13, 2–16, 9–38
ZS4TX	3–17
ZS6EZ	1–39, 2–14

NOTES

NOTES

NOTES

About the Included CD-ROM

On the included CD-ROM you'll find this entire *ON4UN's Low-Band DXing* book, including text, drawings, tables, illustrations and photographs—some in color. Using the industry-standard Adobe *Reader* (included), you can view and print the text of the book, zoom in and out on pages, and copy selected parts of pages to the Clipboard. A powerful search engine—*Acrobat Search*—helps you find topics of interest.

You'll also find a short video *Amateur Radio Today* included on the CD-ROM. This is an entertaining introduction piece that you can show to newcomers to the wonderful world of amateur radio.

Also included on the CD-ROM are a number of very useful software programs written by ON4UN. These are mentioned throughout the book.

INSTALLING YOUR LOW-BAND DXING CD-ROM

Windows

1. Close any open applications and insert the CD-ROM into your CD-ROM drive. (Note: If your system supports "CD-ROM Insert Notification," the CD may automatically run the installation program when inserted.)
2. Select **Run** from the *Windows* **Start** menu.
3. Type **d:\setup** (where d: is the drive letter of your CD-ROM drive; if the CD-ROM is a different drive on your system, type the appropriate letter) and press **Enter**.
4. Follow the instructions that appear on your screen.

Macintosh

No installation procedure is needed for the Macintosh. To install so the CD is not required, drag the "Low-Band DXing" icon onto your hard drive.

USING YOUR LOW-BAND DXING CD

Windows

Note: If your system supports "CD-ROM Insert Notification," the *Low-Band DXing* CD may run automatically when inserted.

1. To view the electronic version of the book after installation, place the CD-ROM in your CD-ROM drive. (Note: This is not necessary if you chose a "complete installation" to your hard disk during setup.)
2. From the **Start** menu, select **Programs**, select **ARRL Software** and then **Low-Band DXing book**.
3. To run the ON4UN Low-Band DXing software, click on the **ON4UN Low-Band DXing** icon. Follow the on-screen instructions.
4. To run the ON4UN Yagi Design software, click on the **ON4UN Yagi** icon.

To view *Amateur Radio Today* or ON4UN's photos, insert the CD-ROM in your CD-ROM drive. A menu will appear with the appropriate options. If "CD-ROM Insert Notification" is disabled on your computer, click the **Start** menu, select **Run**, and type **d:\autorun** (where d: is the drive letter of your CD-ROM drive; if the CD-ROM is a different drive on your system, type the appropriate letter) and press **Enter**.

Macintosh

To view the electronic version of the software, place the CD-ROM in your CD-ROM drive. Adobe *Reader* should launch automatically. Alternatively, open the CD-ROM and double-click **intro.pdf**. Note: The ON4UN software will only run on a PC, not on a Macintosh.

To view *Amateur Radio Today* or ON4UN's photos, insert the CD-ROM in your CD-ROM drive. Open the CD and navigate to either the "Amateur Radio Today" or "Photos" folders and double-click to view.

NOTES ON ON4UN FILES

1. Software Tools for the Book:

1.1. Conversion_calculator.XLS

Converts signal levels typically encountered at receiver inputs between µV, mV, dBµV, µW, etc. You can specify the system impedance (typically 50 or 75 Ω). This is a protected *Excel* XLS file. You must have Microsoft *Excel* on your computer to run it.

1.2. Receiver_Levels.XLS

This spreadsheet tool shows you levels involved with radio signals (from transmit power to received signal). See Chapter 3 of the book. This is a protected *Excel* XLS file. You must have Microsoft *Excel* on your computer to run it.

1.3. RX_noise_figure_and_MDS_calculation.XLS

This spreadsheet tool shows you the relation between receiver bandwidth, temperature and receiver MDS (minimum discernable signal). See Chapter 3 of the book. Again, this is a protected *Excel* XLS file. You must have Microsoft *Excel* on your computer to run it.

1.4. Sunrise_sunset.txt

This is the source code for the sunrise sunset tables mentioned in Section 5.1.7 of Chapter 1 of the book.

1.5. ON4UN Low-Band DXing Software

This directory contains all the programs that make up the **Low Band DXing** software. Start the software by typing MENU, or clicking on the icon labeled **LBDX**. The programs run under DOS and they work in a DOS box, even under windows XP. This software was sold separately since the late 1980s, and I decided to make it freely available to every reader of the new edition of the *Low-Band DXing* book.

1.6. ON4UN-Yagi-Design

This directory contains all the programs that make up the **Yagi Design software**. Start the software by typing YAGMENU, or clicking on the **Yagi** icon. The programs run under DOS and work well in a DOS box, even under windows XP. This software was sold separately since the late 1980s, and I decided to make it freely available to every reader of the new edition of the *Low-Band DXing* book.

2. Low Band Hall of Fame:

This directory contains 235 jpg pictures of well-known low-band DXers. The picture size is approximately 820 by 615 pixels, which is a compromise between quality and file size. Many of the pictures where shot by the author, but a large number were kindly sent by the DXers on the pictures. Many thanks to all.

3. ON4UN DXCC QSLs for 160 Meters:

I scanned in nearly 300 QSL cards, many on two sides. These photos are also 820 pixels in their largest dimension.

4. ON4UN DXCC QSLs for 80 Meters:

The 80-meter DXCC cards, good for 353 confirmed countries, add up to 619 pictures or files.

5. Pictorials at ON4UN:

It took me years to go through old pictures, going back more than 40 years, scanning them in, and putting together a kind of pictorial showing the evolution of my station between the late 1950s and now, the antenna works, and some other projects. It starts with old black and white pictures, then scanned color slides, ending up with digital picture since 1998. Scanning old dusty and scratched pictures and slides is easy; cleaning them up often took several hours for one picture… This is was a major undertaking. Most of it is self-explanatory. Here and there a title is added as well as a year. I trust you will appreciate that I tried only to show "quality pictures" or pictures that might have a specific high interest. I hope you will enjoy looking at some of these pictures. Your comments would of course be very welcome.

The pictures can be used in articles on amateur radio, provided credit is given, and provided I am informed of it.

John Devoldere, ON4UN
john.devoldere@pandora.be

FEEDBACK

Please use this form to give us your comments on this book and what you'd like to see in future editions, or e-mail us at **pubsfdbk@arrl.org** (publications feedback). If you use e-mail, please include your name, call, e-mail address and the book title, edition and printing in the body of your message. Also indicate whether or not you are an ARRL member.

Where did you purchase this book?
☐ From ARRL directly   ☐ From an ARRL dealer

Is there a dealer who carries ARRL publications within:
☐ 5 miles   ☐ 15 miles   ☐ 30 miles   of your location?   ☐ Not sure.

License class:
☐ Novice   ☐ Technician   ☐ Technician Plus   ☐ General   ☐ Advanced   ☐ Amateur Extra

---

Name _____   ARRL member? ☐ Yes ☐ No

Call Sign _____

Daytime Phone (   ) _____   Age _____

Address _____

City, State/Province, ZIP/Postal Code _____   E-mail: _____

If licensed, how long? _____

Other hobbies _____

_____

Occupation _____

For ARRL use only	LBDX
Edition	4 5 6 7 8 9 10 11 12
Printing	2 3 4 5 6 7 8 9 10 11 12

From _____

Please affix postage. Post Office will not deliver without postage.

EDITOR, LOW BAND DXING
ARRL—THE NATIONAL ASSOCIATION FOR AMATEUR RADIO
225 MAIN STREET
NEWINGTON CT 06111-1494

— — — — — — — — — — — — — — — please fold and tape — — — — — — — — — — — — — — —